Atomic Masses of the Elements

Name	Symbol	Atomic Number	Atomic Mass[a]	Name	Symbol	Atomic Number	Atomic Mass[a]
Actinium	Ac	89	(227)	Molybdenum	Mo	42	95.94
Aluminum	Al	13	26.98	Neodymium	Nd	60	144.2
Americium	Am	95	(243)	Neon	Ne	10	20.18
Antimony	Sb	51	121.8	Neptunium	Np	93	(237)
Argon	Ar	18	39.95	Nickel	Ni	28	58.69
Arsenic	As	33	74.92	Niobium	Nb	41	92.91
Astatine	At	85	(210)	Nitrogen	N	7	14.01
Barium	Ba	56	137.3	Nobelium	No	102	(259)
Berkelium	Bk	97	(247)	Osmium	Os	76	190.2
Beryllium	Be	4	9.012	Oxygen	O	8	16.00
Bismuth	Bi	83	209.0	Palladium	Pd	46	106.4
Bohrium	Bh	107	(264)	Phosphorus	P	15	30.97
Boron	B	5	10.81	Platinum	Pt	78	195.1
Bromine	Br	35	79.90	Plutonium	Pu	94	(244)
Cadmium	Cd	48	112.4	Polonium	Po	84	(209)
Calcium	Ca	20	40.08	Potassium	K	19	39.10
Californium	Cf	98	(251)	Praseodymium	Pr	59	140.9
Carbon	C	6	12.01	Promethium	Pm	61	(145)
Cerium	Ce	58	140.1	Protactinium	Pa	91	231.0
Cesium	Cs	55	132.9	Radium	Ra	88	(226)
Chlorine	Cl	17	35.45	Radon	Rn	86	(222)
Chromium	Cr	24	52.00	Rhenium	Re	75	186.2
Cobalt	Co	27	58.93	Rhodium	Rh	45	102.9
Copper	Cu	29	63.55	Roentgenium	Rg	111	(272)
Curium	Cm	96	(247)	Rubidium	Rb	37	85.47
Darmstadtium	Ds	110	(271)	Ruthenium	Ru	44	101.1
Dubnium	Db	105	(262)	Rutherfordium	Rf	104	(261)
Dysprosium	Dy	66	162.5	Samarium	Sm	62	150.4
Einsteinium	Es	99	(252)	Scandium	Sc	21	44.96
Erbium	Er	68	167.3	Seaborgium	Sg	106	(266)
Europium	Eu	63	152.0	Selenium	Se	34	78.96
Fermium	Fm	100	(257)	Silicon	Si	14	28.09
Fluorine	F	9	19.00	Silver	Ag	47	107.9
Francium	Fr	87	(223)	Sodium	Na	11	22.99
Gadolinium	Gd	64	157.3	Strontium	Sr	38	87.62
Gallium	Ga	31	69.72	Sulfur	S	16	32.07
Germanium	Ge	32	72.64	Tantalum	Ta	73	180.9
Gold	Au	79	197.0	Technetium	Tc	43	(98)
Hafnium	Hf	72	178.5	Tellurium	Te	52	127.6
Hassium	Hs	108	(269)	Terbium	Tb	65	158.9
Helium	He	2	4.003	Thallium	Tl	81	204.4
Holmium	Ho	67	164.9	Thorium	Th	90	232.0
Hydrogen	H	1	1.008	Thulium	Tm	69	168.9
Indium	In	49	114.8	Tin	Sn	50	118.7
Iodine	I	53	126.9	Titanium	Ti	22	47.87
Iridium	Ir	77	192.2	Tungsten	W	74	183.8
Iron	Fe	26	55.85	Uranium	U	92	238.0
Krypton	Kr	36	83.80	Vanadium	V	23	50.94
Lanthanum	La	57	138.9	Xenon	Xe	54	131.3
Lawrencium	Lr	103	(260)	Ytterbium	Yb	70	173.0
Lead	Pb	82	207.2	Yttrium	Y	39	88.91
Lithium	Li	3	6.941	Zinc	Zn	30	65.41
Lutetium	Lu	71	175.0	Zirconium	Zr	40	91.22
Magnesium	Mg	12	24.31	—	—	112	(285)
Manganese	Mn	25	54.94	—	—	113	(284)
Meitnerium	Mt	109	(268)	—	—	114	(289)
Mendelevium	Md	101	(258)	—	—	115	(288)
Mercury	Hg	80	200.6				

[a]Values in parentheses are the mass number of the most stable isotope.

A LEARNING PACKAGE BUILT TO ENSURE YOUR SUCCESS, THIS TEXTBOOK IS INTEGRATED WITH THE MOST ADVANCED CHEMISTRY TUTORIAL MEDIA AVAILABLE.

Mastering GOB CHEMISTRY

For instructor-assigned homework, MasteringGOBChemistry™ (www.masteringgob.com) provides the first adaptive-learning online tutorial and assessment system. Based on extensive research of precise concepts students struggle with, the system is able to coach you with feedback specific to your needs, simpler subproblems, and help when you get stuck. The result is targeted tutorial help to optimize your study time and maximize your learning.

REQUIRED for online self-study and instructor-assigned course materials

STUDENT ACCESS KIT

Mastering **GOB CHEMISTRY**

PEARSON Benjamin Cummings

TIMBERLAKE

GENERAL, ORGANIC, AND BIOLOGICAL CHEMISTRY
SECOND EDITION

If your professor requires MasteringGOBChemistry as a component of your course, your purchase of a new copy of Timberlake *General, Organic, and Biological Chemistry, 2e* already includes a free Student Access Kit. You will need this Access Kit to register.

If you did not purchase a new textbook and your professor requires you to enroll in MasteringGOBChemistry, you may purchase online access with a major credit card. Go to www.masteringgob.com and follow the links to purchasing online.

Minimum System Requirements
System requirements are subject to change. See website for the latest requirements.
Windows: 250 MHz CPU; Microsoft® Windows 98, NT, 2000, XP
Macintosh: 233 MHz CPU; Apple® Mac OS® 9.2, 10.2, 10.3
Linux: RedHat® Linux® 8.0, 9.0
All:
- 64 MB RAM
- 1024 x 768 Screen resolution
- Browsers (OS dependent): Firefox® 1.5, Internet Explorer 6.0, Mozilla® 1.7, Netscape® 7.2, Safari® 1.3
- Flash™ Player 7.0, Shockwave, Quicktime

MasteringGOBChemistry is powered by MyCyberTutor by Effective Educational Technologies.

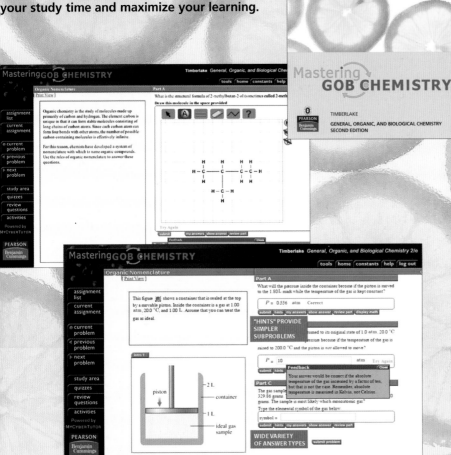

For self-study, MasteringGOBChemistry provides a wealth of targeted self-assessment aids, including quizzes, review questions, flashcards, real-world case studies, interactive activities, and many other resources to enhance your learning.

Second Edition

General, Organic, and Biological
Chemistry

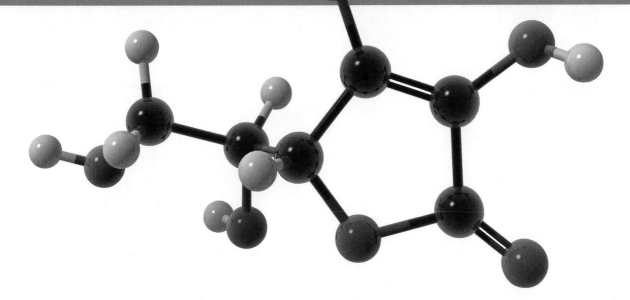

Structures of Life

Second Edition

General, Organic, and Biological
Chemistry

Structures of Life

PEARSON

Prentice
Hall

KAREN C. TIMBERLAKE

Prentice Hall
Upper Saddle River, New Jersey 07458

Editor-in-Chief:	Adam Black
Publisher:	Jim Smith
Project Editor:	Katherine Brayton
Managing Media Producer:	Claire Masson
Editorial Assistants:	Kristin Rose, Grace Joo
Senior Marketing Manager:	Scott Dustan
Managing Editor:	Corinne Benson
Production Supervisor:	Nancy Tabor
Project Management:	Jean Lake
Composition:	Progressive Information Technologies
Art Illustration:	J.B. Woolsey Associates, LLC and Progressive Information Technologies
Photo Researcher:	Laura Murray
Director, Image Resource Center:	Melinda Patelli
Rights and Permissions Coordinator:	Zina Arabia
Text Designer:	Seventeenth Street Studios
Cover Designer:	Jana Anderson
Manufacturing Manager:	Pam Augspurger
Text Printer and Binder:	Courier, Kendallville
Cover Printer:	Phoenix Color Corporation
Cover Photo Credit:	Getty, Laura Ronchi Collection, Bepi Ghiotti–Photographer
Cover Molecular Art:	Imagineering Media Services Inc.

Library of Congress Cataloging-in-Publication Data

Timberlake, Karen.
   General, organic, and biological chemistry : structures of life /
Karen C. Timberlake. -- 2nd ed.
     p. cm.
   Includes index.
   ISBN-13: 978-0-8053-8297-6
   ISBN-10: 0-8053-8297-6
  1. Chemistry. I. Title.
   QD33.2.T56 2007

   540--dc22

2006010692

ISBN 0-8053-8297-6 with MasteringGOBChemistry (Student)

ISBN 0-8053-2185-3 without MasteringGOBChemistry (Student)

ISBN 0-8053-4743-7 with MasteringGOBChemistry (Teacher)

PEARSON
Prentice
Hall

www.prenhall.com/chemistry

3 4 5 6 7 8 9 10-CRK-09 08 07

Brief Contents

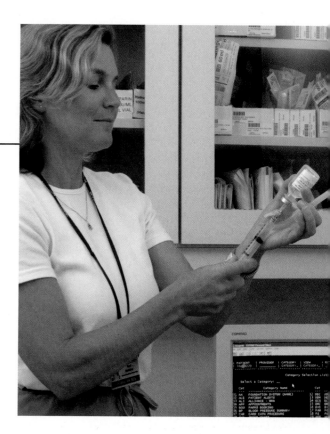

Contents

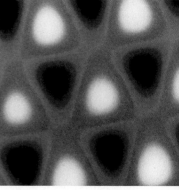

3 Nuclear Chemistry   97

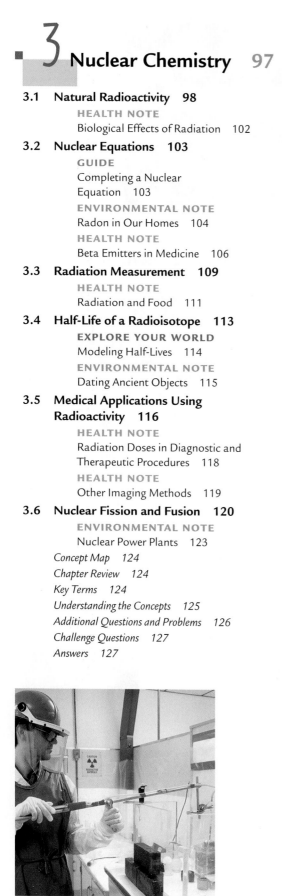

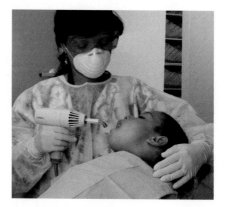

4 Compounds and Their Bonds   130

7 Gases   252

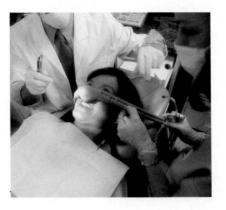

8 Solutions   286

9 Chemical Equilibrium   329

10 Acids and Bases   365

21 Nucleic Acid and Protein Synthesis 744

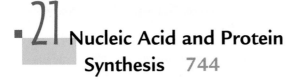

22 Metabolic Pathways for Carbohydrates 789

23 Metabolism and Energy Production 826

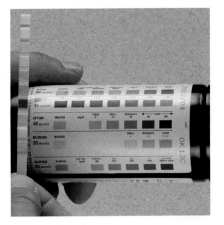

24 Metabolic Pathways for Lipids and Amino Acids 856

Applications and Activities

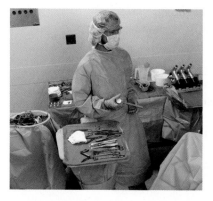

ENVIRONMENTAL NOTE

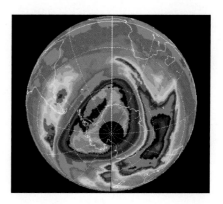

GUIDE TO PROBLEM SOLVING

About the Author

The whole art of teaching is only the art of awakening the natural curiosity of young minds.
—ANATOLE FRANCE

One must learn by doing the thing; though you think you know it, you have no certainty until you try.
—SOPHOCLES

Discovery consists of seeing what everybody has seen and thinking what nobody has thought.
—ALBERT SZENT-GYORGI

Karen Timberlake is Professor Emerita of chemistry at Los Angeles Valley College, where she taught chemistry for allied health and preparatory chemistry for 36 years. She received her bachelor's degree in chemistry from the University of Washington and her master's degree in biochemistry from the University of California at Los Angeles.

Professor Timberlake has been writing chemistry textbooks for 30 years. During that time, her name has become associated with the strategic use of pedagogical tools that promote student success in chemistry and the application of chemistry to real-life situations. More than one million students have learned chemistry using texts, laboratory manuals, and study guides written by Karen Timberlake. In addition to *General, Organic, and Biological Chemistry, Structures of Life*, she is also the author of *Basic Chemistry*, and *Chemistry: An Introduction to General, Organic, and Biological Chemistry, Ninth Edition* with the accompanying *Study Guides with Solutions for Selected Problems, Laboratory Manual*, and *Essentials Laboratory Manual*.

Professor Timberlake belongs to numerous science and educational organizations including the American Chemical Society (ACS) and the National Science Teachers Association (NSTA). In 1987, she was the Western Regional Winner of Excellence in College Chemistry Teaching Award given by Chemical Manufacturers Association. In 2004, she received the McGuffey Award in Physical Sciences from the Textbook Author Association, awarded for textbooks whose excellence has been demonstrated over time. She has participated in education grants for science teaching including the Los Angeles Collaborative for Teaching Excellence (LACTE) and a Title III grant at her college. She often speaks at conferences and educational meetings on using student-centered teaching methods in chemistry to promote learning success of students.

Her husband, Bill, is also a chemistry professor and contributes much to the writing of the textbooks. When the Professors Timberlake are not writing textbooks, they relax by hiking, traveling, trying new restaurants, and playing lots of tennis. The Timberlakes' son, John, and daughter-in-law, Cindy, are involved in writing and publishing materials for English language learners. Their grandson, Daniel, and granddaughter, Emily, don't know what they want to be yet.

TO THE STUDENT

I hope that this textbook helps you discover exciting new ideas and gives you a rewarding experience as you develop an understanding and appreciation of the role of chemistry in your life. If you would like to share your experience with chemistry, or have questions and comments about this text, I would appreciate hearing from you.

Karen Timberlake
Email: khemist@aol.com

Preface

Welcome to the second edition of *General, Organic, and Biological Chemistry: Structures of Life.* Because all too often students find chemistry a series of facts to be memorized, my main objective in writing this text is to relate the structure and behavior of matter to its functions in health and life. This text is designed to be used in a course that includes general, organic, and biological chemistry. It is written for students who are preparing for careers in nursing, dietetics, respiratory therapy, environmental, and agricultural science. This new edition introduces a new problem-solving strategy and new conceptual and challenge problems.

This text contains fundamental topics of general, organic, and biological chemistry. My goal is to provide a learning environment that makes the study of chemistry an engaging and positive experience. It is also my goal to help every student become a critical thinker by instilling the scientific concepts that form the basis for making important decisions about issues concerning health and the environment. Thus, I have utilized materials that

- Help students develop problem-solving skills that lead to success in chemistry;
- Motivate students to learn and enjoy chemistry;
- Relate chemistry to careers that interest students; and
- Provide pedagogy that promotes learning.

The Theme Emphasizes the Relationship Between Structures and Life

Throughout the text, the structures of compounds, organic, and biochemical molecules are related to their function. The discussion of bonding and shapes of molecules in Chapter 4 provides a foundation for understanding the structure of organic and biochemical molecules. The topic of stereochemistry and chiral organic molecules beginning in Chapter 15 is revisited as an important concept in the understanding of structures of carbohydrates, amino acids, and chiral drugs.

The structure of molecules is related to their physical and chemical properties such as solubility in water, density, and boiling point. The structural levels of proteins are related to their function, while chemical processes that denature proteins emphasize the importance of structure to activity. Throughout the text, the atomic structure of matter is highlighted by zoom-in art that relates the atomic level to the macroscopic structures of real-life materials. In this way, the chemical concepts and structures of molecules are continuously related to the behavior and function of biomolecules in the body.

Pedagogical Features

Students are often challenged by the study of chemistry and relevance of chemistry to their career paths. A common view of chemistry is that it is just a lot of facts to be memorized. To change this perception, I have included many features to give students an appreciation of how chemistry is connected to their lives and interests in allied health careers. These pedagogical features include connections to real life, a visual guide to problem solving, and in-chapter problem sets that immediately reinforce the learning of a small group of new concepts.

Interviews with Health Care Professionals Chapter openers begin with an interview with professionals in the health field who discuss the importance of chemistry in their careers.

Learning Goals Directed Learning Learning Goals in every section preview the concepts the student is to learn.

Connections to Real Life Health Notes and Environmental Notes throughout the text relate chemistry to real-life topics in science and medicine that are interesting and motivating to students and support the role of chemistry in the real world.

Guides to Problem Solving As part of a comprehensive learning program, abundant Sample Problems and Study Checks in every chapter model successful problem-solving techniques for the student. Unique Guide to Problem Solving (GPS) strategies illustrate solutions to problems with color blocks that visually guide students through the solution pathway.

Question and Problem Sets Every section includes a comprehensive set of Questions and Problems to encourage students to apply critical thinking and problem solving immediately to a small set of concepts. Students are encouraged to be active learners by working problems and thinking about real-life issues as they progress through each chapter. Within each problem set, the questions are in order of simple to more complex to enable students to build a knowledge base for problem solving. By solving a set of problems after studying each section, students immediately reinforce newly learned concepts rather than waiting until they get to the end of the chapter. Students are encouraged to be active learners by working problems as they progress through each chapter. Answers for all the Study Checks are located at the end of each chapter.

Matched Problem Pairs All the questions and problems in the text are written as pairs of matched problems. Thus, each question or problem is followed with a similar type of question or problem. Answers for the odd-numbered problems are located at the end of each chapter rather than at the end of the text, which gives immediate feedback to problem-solving efforts by students. The even-numbered problems do not have answers, allowing instructors to use them for homework and/or quiz questions.

End-of-Chapter Questions Additional Questions and Problems, also found at the end of the chapter, provide more in-depth questions that integrate the topics from the entire chapter to promote further study and critical thinking. New sets of problems—Understanding the Concepts and Challenge Questions—encourage students to think about the concepts they have learned. These problems can also be utilized in cooperative learning environments for group work. The *Instructor's Manual* contains the answers and solutions to all the problems in the text.

Macro-to-Micro Art Illustrations Throughout the text, a robust and vibrant art program visually connects the real-life world of materials familiar to students with their atomic-level structures. Macro-to-micro illustrations show students that everyday things have an atomic-level organization and structure that determines their behavior and functions. Every figure in the text also contains a question that encourages the student to study that figure and relate the visual representation to the content in the text. The plentiful use of three-dimensional structures takes the descriptions of molecules to a visual level, which stimulates the imagination and aids the understanding of structures by the student.

End-of-Chapter Aids A concept map connects and guides students through all the topics and concepts in the

chapter. Chapter Reviews give a brief overview of the important concepts in each section of the chapter. Key Terms remind students of the new vocabulary presented in each chapter.

New for the Second Edition

In this second edition, the chapters *Chemical Reactions* and *Chemical Quantities* were combined and a new chapter, *Equilibrium*, was written to reinforce the concepts of equilibrium for this text. The chapters *Introduction to Organic Chemistry* and *Alkanes* were combined to make one chapter *Introduction to Organic Chemistry: Alkanes*. New features have been added to every chapter of this second edition that include the following:

- Guides to Problem Solving (GPS) that illustrate problem-solving strategies.
- Concept Maps that visually connect and summarize chapter topics.
- Understanding the Concepts that correlate visual examples to conceptual learning.
- Challenge Questions that integrate concepts from the entire chapter.
- More in-depth problems.
- Answers to odd-numbered problems now included at the end of each chapter.

New Online Tutorial and Homework System

MasteringGOBChemistry™
www.masteringgob.com
MasteringGOBChemistry is the most advanced tutorial, homework, and self-study system ever built. It provides the first library of homework problems pre-tested by students nationally. Sophisticated analysis of student performance (including difficulty, time spent, and most common errors) allows every item to be refined systematically for educational effectiveness and assessment accuracy.

MasteringGOBChemistry allows professors to tutor and assess their classes like never before:

- Provide effective homework and tutorial assignments with automatic grading. Use pre-built weekly assignments or quickly build your own, choosing from an unprecedented variety of ranking and sorting tasks, dynamic visualization activities, interactive tutorials, and end-of-chapter problems.
- Generate ideal exams and tests. Unique tools instantly identify assessment questions of ideal difficulty, duration, and concept coverage for your course.

MasteringGOBChemistry provides students with immediate guidance. Self-tutoring problems provide specific feedback to common wrong answers and give students hints for partial credit. MasteringGOBChemistry gives professors automatic grading and unprecedented diagnostics. Professors can see class statistics at a glance, compare results with the national average, identify the most difficult problem for your class, or zero in on time spent by each student.

Whether professors assign homework activities or not, students still have access to a wealth of self-study media. Interactive activities foster learning through manipulating applets and making predictions. Other targeted self-assessment aids are provided, including quizzes, review questions, flashcards, and real-world case studies.

Throughout *General, Organic, and Biological Chemistry, Second Edition*, icons direct students to MasteringGOBChemistry tutorials, activities, and cases that pertain to the concepts the students are currently exploring.

With MasteringGOBChemistry Tutorials you can turn your guessers into thinkers. These typical end-of-chapter or test bank questions provide self-paced guidance on how to solve typical chemistry problems. The specific wrong answer feedback and numerous hints encourage students to solve chemistry problems on their own with just the right amount of coaching. Pre-tested on thousands of students nationally, the tutorial responsiveness coaches over 90% of students to the correct and final answer.

Written specifically for the GOB chemistry curriculum, its rich interface includes an easy-to-use math palette, orbital diagram tool, molecule building tool, hot-spot clicking on functional groups, dimensional analysis tool, and more!

SELF-STUDY ACTIVITY Students may explore rich interactive tutorials on key concepts in GOB chemistry. These engaging self-paced activities invite students to make predictions, generate graphs, play animations, and more as they confront their common misconceptions and build upon their foundation of chemistry knowledge.

SELF-STUDY CASE Modern case studies are posted online to help students see how chemistry is relevant, fascinating, and vital to the world around them. These popular cases motivate students to see how their GOB studies can be put to good use in the real world. These cases cover topics such as kidney stones, hyperventilation, and food irradiation.

Chapter Organization of the Second Edition

In each textbook I write, I consider it essential to relate every chemical concept to real-life issues of health and environment. In this text, I have added the theme of Structures of Life. All the material in the appendices has been integrated with the appropriate chapter material so that there are no longer any appendices at the end of the book. Because a course of chemistry for allied health may be taught in different time frames, it may be difficult to cover all the chapters in this text. However, each chapter is now a complete package, which allows some chapters to be skipped or the order of presentation to be changed.

Chapters 1–3 Measurements, Atoms and Elements, and Nuclear Chemistry

Chapters 1 through 3 discuss measurement using the metric system, elements and atomic structure, and nuclear radiation and equations. **Chapter 1, Measurements**, begins with a new section on scientific thinking using the scientific method followed by a discussion of measurement and the need to understand numerical structures of the metric system. A new problem feature, Guide to Problem Solving (GPS), now uses color blocks that visually guide students through the solution pathway. Temperature is now included in Chapter 1. In **Chapter 2, Atoms and Elements**, we look at elements, atoms, and subatomic particles. This chapter begins with a new section on classification of matter including pure substances, mixtures, and types of mixtures. A new Environmental Note on *Toxicity of Mercury* was added.

The elements Darmstadtium (Ds) 110 and Roentgenium (Rg) 111 were added along with elements up to 115. The masses of isotopes are now used for atomic mass calculations. Orbital diagrams now illustrate the filling of energy sublevels. A new section, *Periodic Trends*, discusses periodic properties of elements including valence electrons, atomic size, and ionization energy. I placed **Chapter 3, Nuclear Chemistry**, early in the text because it extends the concepts of subatomic particles, atomic number, and mass to radioisotopes. We look at the radioactive particles including the positron. Nuclear equations are written and balanced for both naturally occurring and artificially produced radioactivity. The half-lives of radioisotopes are discussed and the amount of time for a sample to decay is calculated. A new bar graph was added to illustrate half-life and decay. Students are introduced to the application of the atomic world to the field of nuclear medicine.

Chapters 4–6 Compounds and Their Bonds, Chemical Reactions and Quantities, and Energy and Matter

Chapters 4, 5, and 6 take a look at types of bonding, formation of compounds, and energy and matter. **Chapter 4, Compounds and Their Bonds**, describes how the structures of atoms lead to ionic and covalent bonds as well as the shape and polarity of molecules. This introduction to the three-dimensional shape of molecules provides a basis for the shape of organic and biochemical compounds. The discussion of polyatomic ions now follows the formation of ionic compounds. The concept of resonance is now discussed for the electron-dot formulas for compounds with multiple bonds.

Chapter 5, Chemical Reactions and Quantities, includes the quantitative aspects of reactions, such as the mole and molar mass, which are used in calculations of the number of particles in a quantity, mass calculations in reactions, and empirical formulas. Chemical reaction equations are balanced and organized into combination, decomposition, and replacement reactions. Mole and mass relationships among the reactants and products are examined along with calculations of percent yield and limiting reactants. New visual diagrams illustrate *Types of Reactions* and *Limiting Reactants*. The discussion of *Equilibrium* is now a separate chapter. **Chapter 6, Energy and Matter**, discusses the energy in chemical reactions, and calculation of heat in a chemical reaction has been added. The chapter includes the topic of nutritional energy values, states of matter, and the energy involved in changes of state. The attractive forces between particles and their impact on states of matter and changes of state are described.

Chapters 7–10 Gases, Solutions, Equilibrium, and Acids and Bases

Chapters 7, 8, 9 and 10 complete the general chemistry portion of the text with chapters on gases, solutions, chemical equilibrium, and acids and bases. **Chapter 7, Gases**, discusses the properties of a gas and calculates changes in gases using the gas laws. New art was added on gas pressure at different altitudes and figures were resized and moved into the margins. The chapter now includes calculations of the amount of a gas required or produced in a chemical reaction. **Chapter 8, Solutions**, introduces the student to the important health care topics of solutions, electrolytes, saturation and solubility, concentrations, and osmosis and dialysis. Molecular and net ionic equations are written for the formation of insoluble salts accompanied by new macro-to-micro art. This chapter now includes the dilution of solutions and calculations involving the volumes and molarities of solutions in a chemical reaction.

Chapter 9, Equilibrium, is a new chapter on *Chemical Equilibrium* that looks at the rates of reactions and the equilibrium condition when forward and reverse rates for a reaction become equal. Equilibrium expressions for reactions are written and equilibrium constants are calculated. Le Châtelier's principle is used to evaluate the impact on concentrations when a stress is placed on the system. With an understanding of equilibrium, **Chapter 10, Acids and Bases**, discusses acids and bases and their strengths; conjugate acid–base pairs; pH; and buffers. Acid–base titration uses the neutralization reaction between an acid and a base

to calculate quantities of an acid in a sample. The section on *Strengths of Acids and Bases* and *Dissociation Constants* has been rewritten. The section on *Acid–Base Properties of Salt Solution* was rewritten and moved in front of the section on *Buffers*. The section on *Dilutions* was moved to Chapter 8.

Chapters 11–14 Introduction to Organic Chemistry: Alkanes; Unsaturated Hydrocarbons; Alcohols, Phenols, Ethers, and Thiols; and Aldehydes, Ketones, and Chiral Molecules

The second part of the text introduces the student to the world of organic chemistry and the biochemical compounds that form a basis for understanding the chemistry of living systems. **Chapter 11, Introduction to Organic Chemistry: Alkanes**, discusses the structure, nomenclature, and reactions of alkanes. An overview of functional groups and isomers describes the structure of organic chemistry and forms a basis for understanding the biomolecules of living systems. **Chapter 12, Unsaturated Hydrocarbons**, discusses alkenes, alkynes, and aromatic compounds. It includes an *Explore Your World* feature on *Modeling Cis–Trans Isomers*. Polymers and their formation are discussed including examples of synthetic polymers used in everyday items.

Chapter 13, Alcohols, Phenols, Ethers, and Thiols, combines structure and naming of alcohols along with phenols, ethers, and thiols. The section on *Some Important Alcohols and Phenols* was converted to a Health Note. The table on solubility and boiling points was reworked. The *Oxidation of Primary and Secondary Alcohols* is now one section. **Chapter 14, Aldehydes, Ketones, and Chiral Molecules**, discusses the nomenclature and structures of aldehydes and ketones. The section on *Some Important Aldehydes and Ketones* is now a Health Note. The sections *Oxidation and Reduction* and *Addition Reactions* are now in front of the section *Chiral Molecules*. The final section, *Chiral Molecules*, uses simple compounds to introduce chiral molecules and chirality early in the text, which prepares the student for the next chapter, *Carbohydrates*.

Chapters 15–17 Carbohydrates, Carboxylic Acids and Esters, and Lipids

Chapter 15, Carbohydrates, applies the organic chemistry of alcohols, aldehydes, and ketones to biomolecules, which is often the most exciting part for students because it is related to health and medicine. I have combined the sections *Types of Carbohydrates* and *Classification of Monosaccharides*. The section *Cyclic Structures of Monosaccharides* is rewritten. New art is incorporated for lactose and sucrose figures. I added a Health Note on *Blood*

Types and Carbohydrates. **Chapter 16, Carboxylic Acids and Esters**, now includes molecular models in Table of Names and Sources of Carboxylic Acids, and Acidity of Carboxylic Acids, This is followed by **Chapter 17, Lipids**, which contains the same functional groups in larger molecules such as triacylglycerols, glycerophospholipids, and steroids. Structures for fatty acids now include the line-bond formula. Health Notes of interest to students include olestra, trans fatty acids, and lipoproteins. The role of lipids and cholesterol in cell membranes is discussed along with lipids that function as bile salts and steroid hormones.

Chapters 18–21 Amines and Amides, Amino Acids and Proteins, Enzymes and Vitamins, and Nucleic Acid and Protein Synthesis

Chapter 18, Amines and Amides, and **Chapter 19, Amino Acids and Proteins**, continue with the physical and chemical properties of organic compounds followed by related biomolecules. Zwitterions are now part of the early discussion of amino acids. The importance of the structure of proteins from primary to quaternary is related to the shapes and activity of proteins. **Chapter 20, Enzymes and Vitamins**, relates the importance of the three-dimensional shape of proteins to their function as enzymes. The topic of *Enzyme Catalyzed Reactions* now precedes *Lock-and-Key* and *Induced Fit Models* of enzymes. The students learn that the shape of an enzyme is a factor in enzyme regulation and how end products might change the shape to increase or decrease the rate of an enzyme-catalyzed reaction. We also see that proteins change shape and lose function when subjected to pH changes and high temperatures. The important role of water-soluble vitamins as coenzymes is related to enzyme function. New tables for water- and fat-soluble vitamins now include RDA, sources, and deficiency symptoms.

Chapter 21, Nucleic Acids and Protein Synthesis, describes the nucleic acids and their importance as biomolecules that store and direct information for cellular components, growth, and reproduction. The role of complementary base pairing is highlighted in both DNA replication and the formation of mRNA during protein synthesis. I have combined *Components of Nucleic Acids* with *Nucleosides and Nucleotides*, and used new art for the DNA double helix with the proper lengths of G and C, A and T. We look at the genetic code and its relationship to the order of amino acids in a protein and how mutations can occur when the nucleotide sequence is altered. In *Viruses*, we look at how DNA or RNA in viruses utilizes host cells to produce more viruses.

Chapters 22–24 Metabolic Pathways for Carbohydrates, Metabolic Pathways and Energy Production, and Metabolic Pathways for Lipids and Amino Acids

The final three chapters, Chapters 22, 23, and 24, discuss the metabolic pathways of biomolecules from the digestion of food stuffs to the synthesis of ATP. **Chapter 22, Metabolic Pathways for Carbohydrates**, describes the stages of metabolism and the digestion of carbohydrates, our most important fuel. The breakdown of glucose to pyruvate is described using the glycolytic pathway, which is followed under aerobic conditions by the decarboxylation of pyruvate to acetyl CoA. We also look at how the storage molecule glycogen is replenished and how glucose is synthesized from noncarbohydrate sources. In **Chapter 23, Metabolic Pathways and Energy Production**, we look at the entry of acetyl CoA into the citric acid cycle and the production of reduced coenzymes for the electron transport system and oxidative phosphorylation. Details on the structure and function of ATP synthase are included. **Chapter 24, Metabolic Pathways for Lipids and Amino Acids**, discusses the digestion of lipids and proteins and the metabolic pathways that convert fatty acids and amino acids into energy. We also learn how excess carbohydrates are converted to triacylglycerols in adipose tissue and how the intermediates of the citric acid cycle are converted to nonessential amino acids. Finally, we summarize the relationships between the catabolic and anabolic pathways in metabolism.

Instructional Package

General, Organic, and Biological Chemistry: Structures of Life, Second Edition is the nucleus of an integrated teaching and learning package of support material for both students and professors.

For Students

Study Guide with Selected Solutions for *General, Organic, and Biological Chemistry, Second Edition* by Karen Timberlake, is keyed to the learning goals in the text and designed to promote active learning through a variety of exercises with answers as well as mastery exams. The *Study Guide* also contains complete solutions to odd-numbered problems. (ISBN 0-8053-4883-2)

Laboratory Manual for General, Organic, and Biological Chemistry by Karen Timberlake. This best-selling lab manual coordinates 42 experiments with the topics in *General, Organic, and Biological Chemistry, Second Edition*, uses new terms during the lab, and explores chemical concepts. Laboratory investigations develop skills of manipulating equipment, reporting data, solving problems, making calculations, and drawing conclusions. (ISBN 0-8053-4904-9)

MasteringGOBChemistry™
www.masteringgob.com

The first adaptive-learning tutorial system that grades students' homework automatically and provides Socratic tutorials with feedback specific to errors; hints and simpler subproblems upon demand; and motivation with partial credit. Pre-tested on students nationally, the system is uniquely able to respond to students' needs, effectively tutoring and motivating their learning of the concepts and strengthening their problem-solving skills. MasteringGOBChemistry also provides all content formerly found at the ChemistryPlace website including cases, flashcards, quizzes, study goals, a math review, and more.

For Instructors

Instructor Solutions Manual   by Karen Timberlake, includes answers and solutions for all problems in the text. (ISBN 0-8053-0437-1)

Instructor Resource CD (IRCD)   This CD-ROM includes all the art and tables from the book in high resolution (150 dpi) format for use in classroom projection or creating study materials and tests. In addition, the instructor can access the PowerPoint® lecture outlines, featuring over 2000 slides. Also available on the IRCD are downloadable files of the *Instructor Manual and Complete Solutions*, and a set of "clicker questions" suitable for use with classroom-response systems.

Online Instructor Manual for Laboratory Manual contains answers to report pages for the *Laboratory Manual*. **www.prenhall.com**

Printed Test Bank   by Lynn Carlson and Bill Timberlake. Includes 1500 multiple choice, matching, and short-answer questions. (ISBN 0-8053-4763-1)

Computerized Test Bank   Contains over 1500 questions in multiple-choice, matching, and short-answer format. (ISBN 0-8053-4762-3)

Transparency Acetates   A set of 300 full-color transparencies are available and 150 transparency masters with sample problems from the book for use in the classroom. (ISBN 0-8053-764-X)

MasteringGOBChemistry™
www.masteringgob.com

The most advanced online homework and tutorial system available provides thousands of problems and tutorials with automatic grading, immediate wrong-answer specific feedback, and simpler questions upon request. Problems include randomized numerical and algebraic answers and dimensional analysis. Instructors can compare individual student and class results against data collected through pre-testing of students nationwide.

Blackboard, CourseCompass, and WebCT   Course Management Systems provide powerful course management capability.

Also visit the Prentice Hall catalog page for Timberlake's *General, Organic, and Biological Chemistry, Second Edition,* at **www.prenhall.com** to download available instructor supplements.

Acknowledgments

The preparation of a new edition is a continuous effort of many people. As in my work on other textbooks, I am thankful for the support, encouragement, and dedication of many people who put in hours of tireless effort to produce a high quality book that provides an outstanding learning package. Once again the editorial team at Benjamin Cummings has done an exceptional job. I appreciate the work of my publisher, Jim Smith, who supported my vision of this second edition text with a new visual learning strategy and art program. My project editors, Lisa Leung and Katherine Brayton, continually encouraged me on each step during the development of this edition, while skillfully coordinating reviews, art, website materials, supplements, and all the things it takes to make a book come together. Nancy Tabor, production supervisor, and Jean Lake, production editor, brilliantly coordinated all phases of the manuscript to the final pages of a beautiful book. Cathy Cobb, copyeditor, precisely edited the manuscript to make sure the words were correct to help students learn chemistry.

I am especially proud of the art program in this text, which lends beauty and understanding to chemistry. I would like to thank Mark Ong, Art Director, and book designers, Jana Anderson and Randall Goodall, whose creative ideas provided the outstanding design for the cover and pages of the book. Laura Murray, photo researcher, was invaluable in researching and selecting vivid photos for the text so that students can see the beauty of chemistry. The macro-to-micro illustrations designed by J.B. Woolsey Associates give students visual impressions of the atomic and molecular organization of everyday things and are a fantastic learning tool. I want to thank Martha Ghent for the hours of proofreading all the pages. I also appreciate all the hard work in the field put in by the marketing team, Stacy Treco, director of marketing, and Scott Dustan, chemistry marketing manager. Without them, no one would know about this text.

This text also reflects the contributions of many professors who took the time to review and edit the manuscript, provide outstanding comments, help, and suggestions. I am extremely grateful to an incredible group of peers for their careful assessment of all the new ideas for the text, suggested additions, corrections, changes and deletions, and for providing an incredible amount of feedback about the best direction for the text. In addition, I appreciate the time scientists took to let us take photos and discuss their work with them. I admire and appreciate every one of you.

Reviewers

Janet Bjordahl
Lake Area Technical Institute

Dan Black
Snow College

Edwin Geels
Dordt College

Stephen Goldberg
Adelphi University

Carolyn Sweeney Judd
Houston Community College

Eugenii Kozliak
University of North Dakota

Keri Lee
Ithaca College

Lynn Maelia
Mount Saint Mary College

Melvin Merken
Worcester State College

Julia Metzker
Georgia College and State University

Rachel Wang
Spokane Falls Community College

Previous Edition Reviewers

Peter Balanda
Ferris State University

Mark Benvenuto
University of Detroit Mercy

Tom Burkholder
Central Connecticut State University

G. Lynn Carlson
University of Wisconsin, Parkside

Ana Ciereszko
Miami-Dade Community College

Mark Chiu
Seton Hall University

Patricia Draves
University of Central Arkansas

Fabian Fang
California State University, Bakersfield

Don Glover
Bradley University

John Hardee
Henderson State University

Sharon Kapica
County College of Morris

Robert Kolodny
Armstrong-Atlantic State

Patti Landers
University of Oklahoma

Richard Langley
Steven F. Austin State University

Tim Lubben
Northwestern College

Charlene McMahon
Carroll College

Pamela Mork
Concordia College

Barbara Mowery
Thomas Nelson Community College

Sandra Neuendorf
University of Wisconsin, Oshkosh

Elva Mae Nicholson
Eastern Michigan University

Kim Percell
Cape Fear Community College

Shane Phillips
California State University, Stanislaus

Terri Pope
Cuyahoga Community College

Elizabeth Roberts-Kirchhoff
University of Detroit Mercy

Jamie Scheider
Winona State University

Richard Sheardy
Seton Hall University

Kevin Siebenlist
Marquette University

Howard Silverstein
Georgia Perimeter College

Steve Socol
McHenry County College

John Sowa
Seton Hall University

Karen Wiechelman
University of Louisiana at Lafayette

Don Williams
Hope College

Suzanne Williams
Northern Michigan University

David Winters
Tidewater Community College, Virginia Beach

CAREER FOCUS AND REAL WORLD APPLICATIONS

THIS TEXT WAS DESIGNED TO HELP STUDENTS ATTAIN THEIR CAREER GOALS.

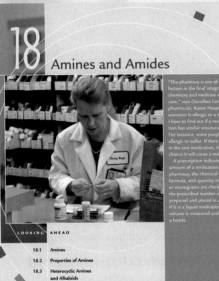

Chapter Opener

Each chapter begins with an interview with a professional in allied health or other relevant fields. These interviews illustrate how health professionals interact with chemistry in their careers.

CAREER FOCUS

■ Physical Therapist

"Physical therapists need to understand how the body, muscles, and joints function in order to recognize when something is not working and which area needs strengthening," says Vincent Leddy, physical therapist. "Chemistry is important to understanding the body's physiology and how chemical changes in the body affect movement. I went into physical therapy because I enjoy teaching movement to children. I am guiding Maggie into her chair but allowing her to move as much as she can by herself. I help her by giving gentle pressure and reassurance that she'll be safe in the transition. I work on getting Maggie to use her body, and the occupational therapist works on Maggie's fine motor skills and how she uses her hand on the switch or picks up objects. By using both physical and occupational therapy, we enhance the child's performance."

Career Focus

Within the chapters are additional examples of professionals using chemistry.

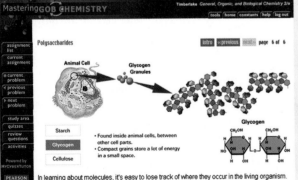

On the Web

The self-study area of MasteringGOBChemistry features in-depth resources for each of the health professions featured in the book and takes students through interactive cases.

Environmental Note

Environmental Notes delve into high interest issues such as global warming, radon, acid rain, pheromones, and ozone depletion.

ENVIRONMENTAL NOTE

■ Global Warming

The amount of carbon dioxide (CO_2) gas in our atmosphere is on the increase as we burn more gasoline, coal, and natural gas. The algae in the oceans and the plants and trees in the forests normally absorb carbon dioxide, but they cannot keep up with the continued increase. The cutting of trees in the rain forests (deforestation) reduces the amount of carbon dioxide removed from the atmosphere. Many of the trees are also burned as land is cleared. It has been estimated that deforestation may account for 15–30% of the carbon dioxide that remains in the atmosphere each year.

The carbon dioxide in the atmosphere acts like the glass in a greenhouse. When sunlight warms the Earth's surface, some of the heat is absorbed by carbon dioxide. It is not yet clear how severe the effects of global warming might be. Some scientists estimate that by around the year 2030, the atmospheric level of carbon dioxide could double and cause the temperature of Earth's atmosphere to rise by 2–5°C. If that should happen, it would have a profound impact on Earth's climate. For example, an increase in the melting of snow and ice could raise the ocean levels by as much as 2 m, which is enough to flood many cities located on the ocean shorelines.

Worldwide efforts are being made to reduce fossil fuel use and to slow or stop deforestation. It will require cooperation throughout the world to avoid the bleak future that some scientists predict should global warming continue unchecked.

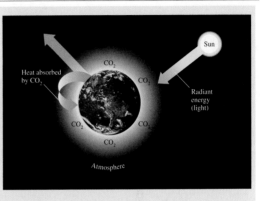

Health Note

A rich array of **Health Notes** in each chapter apply chemical concepts to relevant topics of health and medicine. These topics include weight loss and weight gain, artificial fats, anabolic steroids, alcohol, genetic diseases, viruses, and cancer.

HEALTH NOTE

■ Alpha Hydroxy Acids

Alpha hydroxy acids (AHAs) are naturally occurring carboxylic acids found in fruits, milk, and sugarcane. Cleopatra reportedly bathed in sour milk to smooth her skin. Dermatologists have long been using products with a high concentration of AHAs to remove acne scars and reduce irregular pigmentation and age spots. Now lower concentrations (8–10%) of AHAs have been added to skin care products for the purpose of smoothing fine lines, improving skin texture, and cleansing pores. Several different alpha hydroxy acids may be found in skin care products singly or in combination. Glycolic acid and lactic acid are most frequently used.

Recent studies indicate that products with AHAs increase blisters, rashes, and discoloration of the skin. The FDA does not require product safety reports from cosmetic manufacturers, although they are responsible for marketing safe products. The FDA advises that you test any product containing AHAs on a small area of skin before you use it on a large area.

Alpha Hydroxy Acid (Source)	Structure
Glycolic acid (Sugarcane, sugar beet)	
Lactic acid (Sour milk)	

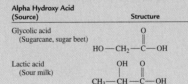

Explore Your World

Explore Your World includes hands-on activities that use everyday materials to encourage students to actively explore selected chemistry topics, either individually or in group-learning environments. Each activity is followed by questions to encourage critical thinking.

EXPLORE YOUR WORLD

■ Solubility of Fats and Oils

Place some water in a small bowl. Add a drop of a vegetable oil. Then add a few more drops of the oil. Now add a few drops of liquid soap and mix. Record your observations.

Place a small amount of fat such as margarine, butter, shortening, or vegetable oil on a dish or plate. Run water over it. Record your observations. Mix some soap with the fat substance and run water over it again. Record your observations.

QUESTIONS

1. Do the drops of oil in the water separate or do they come together? Explain.
2. How does the soap affect the oil layer?
3. Why don't the fats on the dish or plate wash off with water?
4. In general what is the solubility of lipids in water?
5. Why does soap help to wash the fats off the plate?

STUDENT-FRIENDLY APPROACH

KEEPING STUDENTS ENGAGED IS THE ULTIMATE GOAL.

Learning Goals

At the beginning of each section, a **Learning Goal** clearly identifies the key concept of the section, providing a roadmap for studying. All information contained in that section relates back to the Learning Goal.

5.5 The Mole

At the store, you buy eggs by the dozen. In an office, pencils are ordered by the gross and paper by the ream. In a restaurant, soda is ordered by the case. In each of these examples, the terms *dozen, gross, ream,* and *case* count the number of items present. For example, when you buy a dozen eggs, you know you will get 12 eggs in the carton.

24 cans = 1 case

144 pencils = 1 gross

500 sheets = 1 ream

12 eggs = 1 dozen

LEARNING GOAL

Use Avogadro's number to determine the number of particles in a given number of moles.

Avogadro's Number

In chemistry, particles such as atoms, molecules, and ions are counted by the **mole**, a unit that contains 6.02×10^{23} items. This very large number, called **Avogadro's number** after Amedeo Avogadro, an Italian physicist, looks like this when written with 3 significant figures:

Avogadro's Number

$$602\ 000\ 000\ 000\ 000\ 000\ 000\ 000 = 6.02 \times 10^{23}$$

Writing Style

Karen Timberlake is known for her accessible writing style, based on a carefully paced and simple development of chemical ideas, suited to the background of allied health students. She precisely defines terms and sets clear goals for each section of the text. Her clear analogies help students to visualize and understand key chemical concepts.

Concept Maps

Each chapter ends with a **Concept Map** that reviews the key concepts of each chapter and how they fit together.

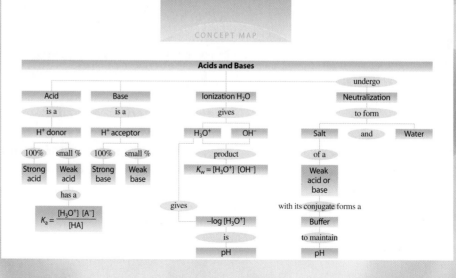

CONCEPT MAP

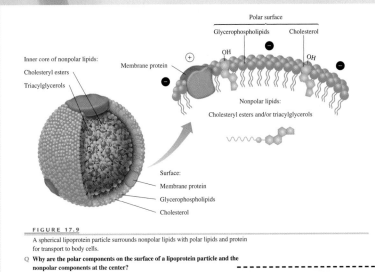

FIGURE 17.9

A spherical lipoprotein particle surrounds nonpolar lipids with polar lipids and protein for transport to body cells.

Q Why are the polar components on the surface of a lipoprotein particle and the nonpolar components at the center?

Art Program

The art program is not only beautifully rendered, but pedagogically effective as well.

Questions paired with figures challenge students to think critically about photos and illustrations.

Macro-to-Micro Art

Photographs and drawings illustrate the atomic structure of recognizable objects, putting chemistry in context and connecting the atomic world to the macroscopic world.

FIGURE 11.1

Propane, C_3H_8, is an organic compound, whereas sodium chloride, NaCl, is an inorganic compound.

Q Why is propane used as a fuel?

PROBLEM SOLVING

MANY TOOLS SHOW STUDENTS HOW TO SOLVE PROBLEMS.

A Visual Guide to Problem Solving

The author understands the learning challenges facing students in this course, so she walks students through the problem-solving process step by step. For each type of problem, she uses a unique, color-coded flow chart that is coordinated with parallel worked examples to visually guide students through each problem-solving strategy.

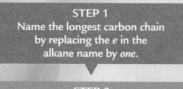

Guide to Naming Ketones

STEP 1
Name the longest carbon chain by replacing the *e* in the alkane name by *one*.

STEP 2
Number the carbon chain starting from the end nearest the carbonyl group and indicate its location.

STEP 3
Name and number any substituents on other carbons in the chain.

STEP 4
For cyclic ketones, add the prefix *cyclo* and number substitutes from the carbonyl carbon as carbon 1.

Sample Problems with Study Checks

Numerous **Sample Problems** appear throughout the text to immediately demonstrate the application of each new concept. The worked-out solutions give step-by-step explanations, provide a problem-solving model, and illustrate required calculations. Each Sample Problem is followed by a **Study Check** question that allows students to test their understanding of the problem-solving strategy.

SAMPLE PROBLEM 8.8

▪ **Percent Concentration (% m/m)**

What is the mass percent of a solution prepared by dissolving 30.0 g NaOH in 120.0 g of H_2O?

SOLUTION

■ STEP 1 **Given**   30.0 g NaOH and 120.0 g H_2O
           Need   mass percent (% m/m) of NaOH

■ STEP 2 **Plan**   The mass percent is calculated by using the mass in grams of the solute and solution in the definition of mass percent.

■ STEP 3 **Equalities/Conversion Factors**

$$\text{mass percent (\% m/m)} = \frac{\text{mass of solute}}{\text{mass of solute } + \text{ mass of solvent}} \times 100$$

$$\text{mass percent (\% m/m)} = \frac{\text{grams of solute}}{\text{grams of solution}} \times 100$$

■ STEP 4 **Set Up Problem**   The mass of the solute and the solution are obtained from the data.

$$\text{mass of solute}\quad =\quad 30.0 \text{ g NaOH}$$
$$\text{mass of solvent}\quad = +120.0 \text{ g } H_2O$$
$$\overline{\text{mass of solution} = 150.0 \text{ g solution}}$$

$$\text{mass percent (\% m/m)} = \frac{30.0 \text{ g NaOH}}{150.0 \text{ g solution}} \times 100$$

$$= 20.0\% \text{ (m/m) NaOH}$$

STUDY CHECK

What is the mass percent of NaCl in a solution made by dissolving 2.0 g of NaCl in 56.0 g of H_2O?

Integrated Questions and Problems

Questions and Problems at the end of each section encourage students to apply concepts and begin problem solving after each section. Answers to odd-numbered problems are given at the end of each chapter.

End-of-Chapter Questions and Problems

Understanding the Concepts questions at the end of the chapter encourage students to think about the concepts they have learned. **Additional Questions and Problems** integrate the topics from the entire chapter to promote further study and critical thinking.

Challenge Questions are designed for group work in cooperative learning environments.

Paired Problems

By pairing each even-numbered problem with a similar odd-numbered problem whose answer is at the end of the chapter, students are able to guide themselves through solving problems.

QUESTIONS AND PROBLEMS

Electrolytes and Nonelectrolytes

8.7 KF is a strong electrolyte, and HF is a weak electrolyte. How are they different?

8.8 NaOH is a strong electrolyte, and CH_3OH is a nonelectrolyte. How are they different?

8.9 The following soluble salts are strong electrolytes. Write a balanced equation for their dissociation in water:
 a. KCl   **b.** $CaCl_2$ **c.** K_3PO_4 **d.** $Fe(NO_3)_3$

8.10 The following salts are strong electrolytes. Write a balanced equation for their dissociation in water:
 a. LiBr   **b.** $NaNO_3$ **c.** $FeCl_3$ **d.** $Mg(NO_3)_2$

8.11 Indicate whether aqueous solutions of the following will contain ions only, molecules only, or molecules and some ions:
 a. acetic acid ($HC_2H_3O_2$), found in vinegar, a weak electrolyte
 b. NaBr, a strong electrolyte
 c. fructose ($C_6H_{12}O_6$), a nonelectrolyte

8.12 Indicate whether aqueous solutions of the following will contain ions only, molecules only, or molecules and some ions:
 a. Na_2SO_4, a strong electrolyte
 b. ethanol, C_2H_5OH, a nonelectrolyte
 c. HCN, hydrocyanic acid, a weak electrolyte

CHALLENGE QUESTIONS

17.87 Identify the lipoprotein in each description as:
 1. chylomicrons   **2.** VLDL   **3.** LDL   **4.** HDL
 a. "good" cholesterol
 b. transports most of the cholesterol to the cells
 c. carries triacylglycerols from the intestine to the fat cells
 d. transports cholesterol to the liver
 e. has the greatest abundance of protein
 f. "bad" cholesterol
 g. carries triacylglycerols synthesized in the liver to the muscles
 h. has the lowest density

17.88 Which of the following has the lower melting point? Explain.
 1. $CH_3-(CH_2)_{16}-COOH$
 2. $CH_3-(CH_2)_7-CH=CH-(CH_2)_7-COOH$

17.89 Draw the structure of glycerophospholipid that is made from stearic acid, palmitic acid, and a phosphate bonded to ethanolamine.

17.90 Olive oil consists of a high percentage of triolean.
 a. Draw the structure for triolean.
 b. How many liters of H_2 gas at STP are needed to completely saturate 100. g of triolean?
 c. How many mL of 0.250 M NaOH are needed to completely saponify 100. g of triolean?

17.91 A sink drain can become clogged with solid fat such as glyceryl tristearate.
 a. How would adding lye (NaOH) to the sink drain remove the blockage?
 b. Write an equation for the reaction that occurs.

UNDERSTANDING THE CONCEPTS

17.73 Palmitic acid is obtained from palm oil as glyceryl tripalmitate. Draw the structure of glyceryl tripalmitate.

17.74 Jojoba wax in candles consists of stearic acid and a 22-carbon saturated alcohol. Write the structure of jojoba wax.

17.75 Sunflower oil can be used to make margarine. A triacylglycerol in sunflower oil consists of 2 linoleic acids and 1 oleic acid.

a. Write two isomers for the triacylglycerol in sunflower oil.
b. Using one of the isomers, write the reaction that would be used when sunflower oil is used to make solid margarine.

17.76 Identify each of the following as saturated, monounsaturated, polyunsaturated, omega-3, or omega-6 fatty acids:

a. $CH_3-(CH_2)_4-CH=CH-CH_2-CH=CH-(CH_2)_7-COOH$
b. linolenic acid
c. $CH_3-(CH_2)_{14}-COOH$
d. $CH_3-(CH_2)_7-CH=CH-(CH_2)_7-COOH$

ADDITIONAL QUESTIONS AND PROBLEMS

17.77 Among the ingredients of a lipstick are beeswax, carnauba wax, hydrogenated vegetable oils, and capric triglyceride. What types of lipids have been used? Draw the structure of capric triglyceride (capric acid is the saturated 10-carbon fatty acid).

17.78 Because peanut oil floats on the top of peanut butter, many brands of peanut butter are hydrogenated. A solid product then forms that is mixed into the peanut butter and does not separate. If a triacylglycerol in peanut oil that contains one palmitic acid, one oleic acid, and one linoleic acid is completely hydrogenated, what is the product?

17.79 Trans fats are produced during the hydrogenation of polyunsaturated oils.

a. What is the typical configuration of a monounsaturated fatty acid?
b. How does a trans fat compare with a cis fat?
c. Draw the structure of *trans*-oleic acid.

17.80 One mole of triolein is completely hydrogenated. What is the product? How many moles of hydrogen are required? How many grams of hydrogen? How many liters of hydrogen are needed if the reaction is run at STP?

17.81 What are the structures of glyceryl stearate and a lecithin containing palmitic acid?

17.82 Some typical meals at fast-food restaurants are listed here. Calculate the number of kilocalories from fat and the

THE WORLD'S MOST POWERFUL ONLINE HOMEWORK AND TUTORIAL SYSTEM.

Mastering GOB CHEMISTRY

This revolutionary online homework and grading system allows instructors to assign problems for a grade. A variety of tutorial and problem types are provided, with each problem type offering a different level of individualized, on-demand help to the student. Students can thus receive help exactly when they need it most—right at the point where they can't get any further. MasteringGOBChemistry™ tracks student performance on multi-step problems so instructors can give partial credit and identify common obstacles that students are having trouble with.

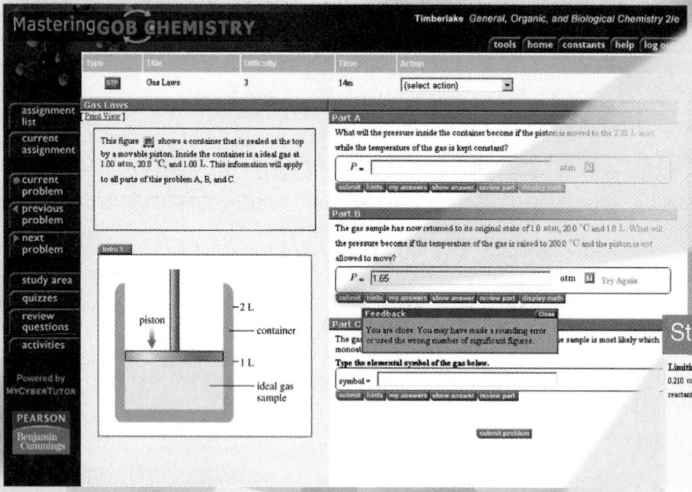

Submit an answer and receive immediate, error-specific feedback

MasteringGOBChemistry is the only system to provide instantaneous feedback specific to the most common wrong answers, accumulated by capturing and researching student errors.

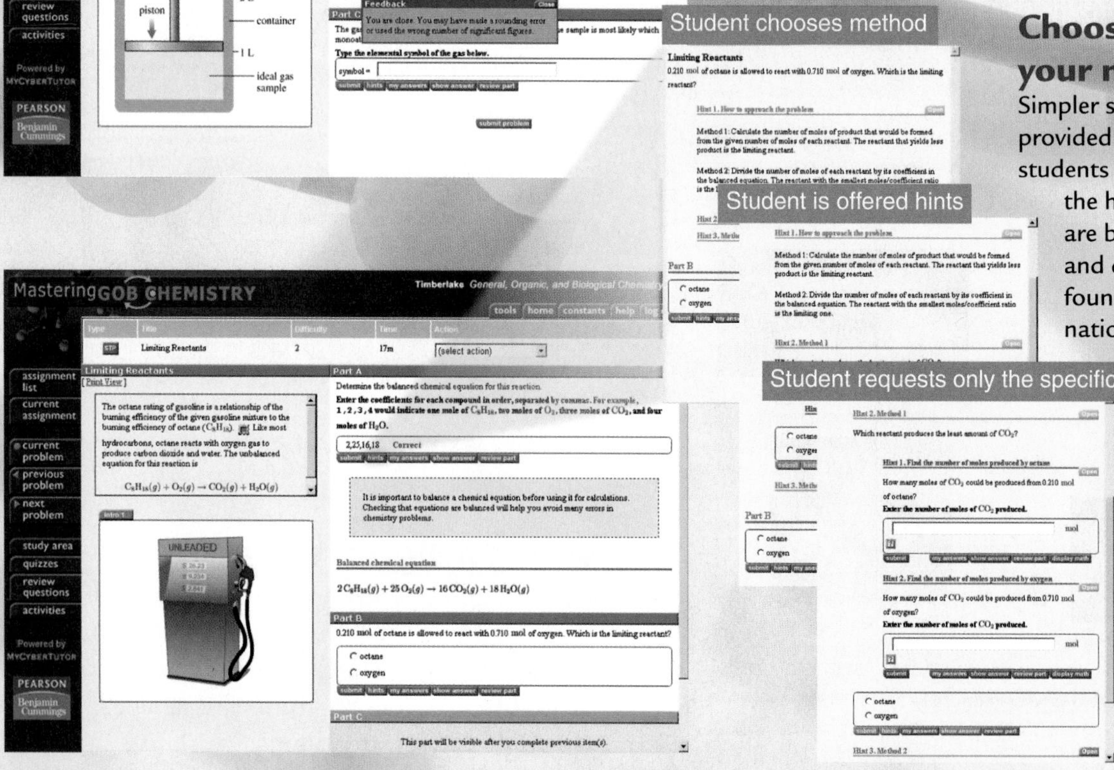

Student chooses method

Student is offered hints

Choose help specific to your needs

Simpler subproblems—"hints"—are provided upon request. From these, students can pick and choose only the help they need. These hints are built on data of key steps and concepts students are found to struggle with nationally.

Student requests only the specific help they need

Partial credit means motivation for method

Mastering GOBChemistry is uniquely able to provide partial credit for the students' methods (based on the simpler subproblems requested and errors made). Credit is at the heart of student motivation and so MasteringGOBChemistry encourage students to focus on their methods as well as their final answer.

A problem library with a difference

Choose from the largest library and widest variety of nationally pre-tested tutorials and problems available. The library provides textbook problems (with randomized numbers and sig-fig feedback) and tutorials that coach students through the full range of general, organic, and biological chemistry topics, incorporating algebraic answers through ranking-tasks and essays.

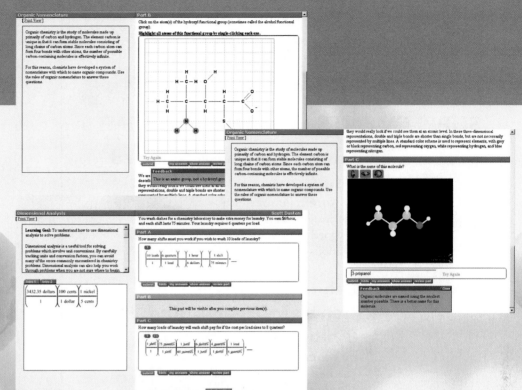

Homework assignments of ideal difficulty and duration

MasteringGOBChemistry is unique in providing instructors with national data on difficulty and duration of every problem and tutorial. This allows instructors to quickly build homework assignments uniquely tailored to the ability of their students and goals of their course. Alternately, instructors can select and customize pre-built assignments to best fit their needs.

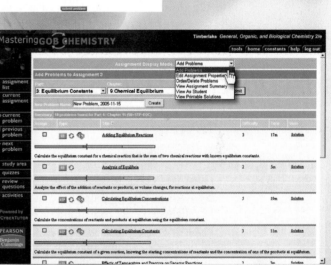

Unprecedented grading and diagnostics

By capturing more detailed work of every student than any other homework system, Mastering-GOBChemistry provides the most powerful gradebook and diagnostics available. Spot students in trouble at a glance with the color-coded gradebook, effortlessly identify the most difficult problem (and step within that problem) in assignments, or critique the detailed work of any student who needs more help. Compare results on any problem and any step with a previous class, or the national average.

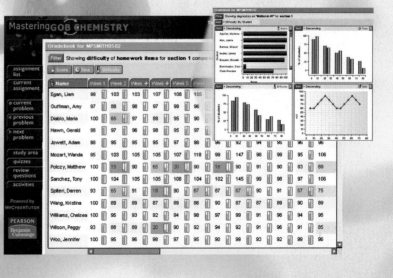

ADDITIONAL SUPPLEMENTS

VALUABLE ANCILLARIES FOR BOTH STUDENTS AND INSTRUCTORS.

SUPPLEMENTS FOR THE STUDENTS

Study Guide with Selected Solutions
by Karen Timberlake
The *Study Guide* is keyed to the learning goals in the text and designed to promote active learning. It includes complete solutions to odd-numbered problems. (0-8053-4883-2)

Laboratory Manual for General, Organic, and Biological Chemistry
by Karen Timberlake
This best-selling *Lab Manual* contains 42 experiments for the standard course sequence of topics. (0-8053-4904-9)

SUPPLEMENTS FOR THE INSTRUCTOR

Instructor Solutions Manual
by Karen Timberlake
Highlights chapter topics and includes suggestions for the laboratory. Contains answers and solutions to all problems in the text. (0-8053-0437-1)

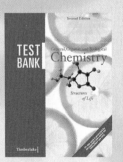

Printed Test Bank
Includes 1500 multiple-choice, matching, and short-answer questions. (0-8053-4763-1)

Computerized Test Bank
Contains 1500 questions in multiple-choice, matching, true-false, and short-answer format on a dual-platform CD-ROM. (0-8053-4762-3)

Instructor Resource CD-ROM
This CD-ROM puts the outstanding art of *General, Organic, and Biological Chemistry* in the palm of your hand. It includes all the art and tables from the book in high resolution (150 dpi) format for use in classroom projection or creating study materials and tests. In addition, customizable PowerPoint lecture outlines, featuring over 2000 slides, are available on the CD. Also available on the IRCD are downloadable files from the *Instructor Solutions Manual* and a set of "clicker questions" suitable for use with classroom-response systems. (0-8053-4882-4)

Transparency Acetates
Includes 300 full-color transparency acetates. (0-8053-4764-X)

Online Instructor Manual to the Laboratory Manual
Contains answers to report pages for the *Laboratory Manual*.

Please visit the Prentice Hall catalog page for Timberlake's *General, Organic, and Biological Chemistry*, 2e, at www.prenhall.com to download available instructor supplements.

1 Measurements

"I use measurement in just about every part of my nursing practice," says registered nurse Vicki Miller. "When I receive a doctor's order for a medication, I have to verify that order. Then I draw a carefully measured volume from an IV or a vial to create that particular dose. Some dosage orders are specific to the size of the patient. I measure the patient's weight and calculate the dosage required for the weight of that patient."

Nurses use measurement each time they determine a patient's temperature, height, weight, or blood pressure. Measurement is used to obtain the correct amounts for injections and medications and to determine the volumes of fluid intake and output. For each measurement, the amounts and units are recorded in the patient's records.

LOOKING AHEAD

Mastering GOB CHEMISTRY

Visit **www.masteringgob.com** for self-study materials and instructor-assigned homework.

hemistry and measurement are an important part of our everyday life. Levels of toxic materials in the air, soil, and water are discussed in our newspaper. We read about radon in our homes, holes in the ozone layer, trans fatty acids, global warming, and DNA analysis. Understanding chemistry and measurement helps us make proper choices about our world.

Think about your day; you probably made some measurements. Perhaps you checked your weight by stepping on a scale. If you did not feel well, you may have taken your temperature. To make some soup, you added 2 cups of water to a package mix. If you stopped at the gas station, you watched the gas pump measure the number of gallons of gasoline you put in the car.

Measurement is an essential part of health careers such as nursing, dental hygiene, respiratory therapy, nutrition, and veterinary technology. The temperature, height, and weight of a patient are measured and recorded. Samples of blood and urine are collected and sent to a laboratory where glucose, pH, urea, and protein are measured by the lab technicians.

By learning about measurement, you will develop skills for solving problems and learn how to work with numbers in chemistry. If you intend to go into a health career, measurement and an understanding of measured values will be an important part of your evaluation of the health of a patient.

LEARNING GOAL

■ Describe the activities that are part of the scientific method.

1.1 Scientific Method: Thinking like a Scientist

When you were very young, you explored the things around you by touching and tasting. When you got a little older, you asked questions about the world you live in. What is lightning? Where does a rainbow come from? Why is water blue? As an adult you may have wondered how antibiotics work. Each day you ask questions and seek answers as you organize and make sense of the world you live in.

When Nobel Laureate Linus Pauling described his student life in Oregon, he recalled that he read many books on chemistry, mineralogy, and physics. "I mulled over the properties of materials: why are some substances colored and others not, why are some minerals or inorganic compounds hard and others soft." He said, "I was building up this tremendous background of empirical knowledge and at the same time asking a great number of questions." Linus Pauling won two Nobel Prizes: the first, in 1954, was in chemistry and for his work on the structure of proteins, and the second, in 1962, was the Peace Prize.

Scientific Method

Although the process of trying to understand nature is unique to each scientist, there is a set of general principles, called the **scientific method**, that describes the thinking of a scientist.

1. **Observations** The first step in the scientific method is to observe, describe, and measure some event in nature. Observations based on measurements are called *data*.

2. **Hypothesis** After sufficient data are collected, a *hypothesis* is proposed that states a possible interpretation of the observations. The hypothesis must be stated in a way that it can be tested by experiments.

3. **Experiments** Experiments are tests that determine the validity of the hypothesis. Often many experiments are performed and a large amount of data collected. If the results of experiments provide different results than predicted by the hypothesis, a new or modified hypothesis is proposed and new experiments are performed.

4. **Theory** When experiments can be repeated by many scientists with consistent results that confirm the hypothesis, the hypothesis becomes a *theory*. Each theory, however, continues to be tested and, based on new data, sometimes needs to be modified or even replaced. Then a new hypothesis is proposed, and the process of experimentation takes place once again.

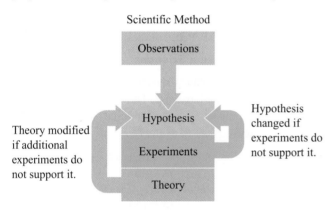

Using the Scientific Method in Everyday Life

 Scientific Method

You may be surprised to realize that you use the scientific method in your everyday life. Let's suppose that you visit a friend in her home. Soon after you arrive, your eyes start to itch and you begin to sneeze. Then you observe that your friend has a new cat. You ask yourself why you are sneezing and form the hypothesis that you are allergic to cats. To test your hypothesis, you leave your friend's home. If the sneezing stops, perhaps your hypothesis is correct. You test your hypothesis further by visiting another friend who also has a cat. If you start to sneeze again, your experimental results indicate that you are allergic to cats. However, if you continue sneezing after you leave your friend's home, your hypothesis is not supported. Now you need to form a new hypothesis, which could be that you have a cold.

SAMPLE PROBLEM 1.1

■ **Scientific Method**

Identify each of the following statements as an observation or a hypothesis.

a. A silver tray turns a dull gray color when left uncovered.
b. Water freezes at 0°C.
c. Ice cubes have a greater volume than the liquid from which they were formed because the water molecules are further apart in the ice structure than in the liquid form.

SOLUTION

a. observation   **b.** observation   **c.** hypothesis

STUDY CHECK

The following statements are found in a student's notebook. Identify each of the following as (**1**) observation, (**2**) hypothesis, or (**3**) experiment.
a. "Today, I planted two tomato seedlings in the garden. Two more tomato seedlings are placed in a closet. I will give all the plants the same amount of water and fertilizer."
b. "After fifty days, the tomato plants in the garden are 3 feet high with green leaves. The plants in the closet are eight inches tall and yellow."
c. "Tomato plants need sunlight to grow."

Answers to all of the Study Checks can be found at the end of this chapter in the Answer section. Checking your answers will let you know if you understand the material in this section.

QUESTIONS AND PROBLEMS

■ Scientific Method: Thinking like a Scientist

In every chapter, each magenta, odd-numbered exercise in Questions and Problems is paired with the next even-numbered exercise. The answers for all the magenta, odd-numbered Questions and Problems are given at the end of this chapter. The complete solutions to the odd-numbered Questions and Problems are in the *Study Guide*.

1.1 Define each of the following terms of the scientific method.
    a. hypothesis   **b.** experiment   **c.** theory   **d.** observation

1.2 Identify each of the following activities in the scientific method as
    (**1**) observation   (**2**) hypothesis   (**3**) experiment   (**4**) theory
    a. Formulate a possible explanation for your experimental results.
    b. Collect data.
    c. Design an experimental plan that will give new information about a problem.
    d. State a generalized summary of your experimental results.

1.3 At a popular restaurant, where Chang is the head chef, the following occur:
    (**1**) Chang determines that sales of the chef's salad have dropped.
    (**2**) Chang decides that the chef's salad needs a new dressing.
    (**3**) In a taste test, four bowls of lettuce are prepared with four new dressings: sesame seed, oil and vinegar, blue cheese, and anchovies.
    (**4**) The tasters rate the dressing with sesame seeds the best.
    (**5**) After two weeks, Chang notes that the orders for the chef's salad with the new sesame dressing have doubled.
    (**6**) Chang decides that the sesame dressing improved the sales of the chef's salad because the sesame dressing improved the taste of the salad.
    Identify each activity as an
    a. observation   **b.** hypothesis   **c.** experiment   **d.** theory

1.4 Lucia wants to develop a process for dyeing shirts so that the color will not fade when the shirt is washed. She proceeds with the following activities:
    (**1**) Lucia notices that the dye in a design on tee shirts fades when the shirt is washed.
    (**2**) Lucia decides that the dye needs something to help it set in the tee shirt fabric.
    (**3**) She places a spot of dye on each of four tee shirts, and then places each one separately in water, salt water, vinegar, and baking soda and water.
    (**4**) After one hour, all the tee shirts are removed and washed with a detergent.
    (**5**) Lucia notices that the dye has faded on the tee shirts in water, salt water, and baking soda, while the dye held up in the tee shirts soaked in vinegar.
    (**6**) Lucia thinks that the vinegar binds with the dye so it does not fade when the shirt is washed.
    Identify each activity as an
    a. observation   **b.** hypothesis   **c.** experiment   **d.** theory

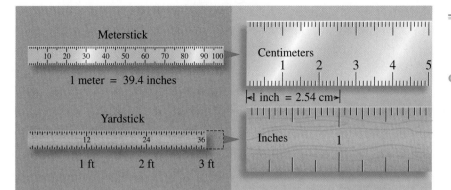

FIGURE 1.1

Length in the metric (SI) system is based on the meter, which is slightly longer than a yard.

Q How many centimeters are in a length of one inch?

1.2 Measurement and Scientific Notation

LEARNING GOAL

Write the names and abbreviations for the units used in measurements of length, volume, and mass; write a number in scientific notation.

The **metric system** is used by scientists and health professionals throughout the world. It is also the common measuring system in all but a few countries. In 1960, a modification of the metric system called the *International System of Units*, Système International (**SI**), was adopted by scientists to provide additional uniformity for units used in the sciences. In this text, we will use metric units and introduce some of the SI units.

Length

The metric and SI unit of length is the **meter (m)**. It is 39.4 inches (in.), which makes a meter slightly longer than a yard (yd). A smaller unit of length, the **centimeter (cm)**, is more commonly used in chemistry and is about as wide as your little finger. For comparison, there is 2.54 cm in 1 in. (See Figure 1.1.)

$$1 \text{ m} = 100 \text{ cm}$$
$$1 \text{ m} = 39.4 \text{ in.}$$
$$2.54 \text{ cm} = 1 \text{ in.}$$

Volume

Volume is the amount of space a substance occupies. A **liter (L)**, which is slightly larger than the quart (qt), is commonly used to measure volume. The **milliliter (mL)** is more convenient for measuring smaller volumes of fluids in hospitals and laboratories. A comparison of metric and U.S. units for volume appears in Figure 1.2.

$$1 \text{ L} = 1000 \text{ mL}$$
$$1 \text{ L} = 1.06 \text{ qt}$$
$$946 \text{ mL} = 1 \text{ qt}$$

Mass

The **mass** of an object is a measure of the quantity of material it contains. You may be more familiar with the term *weight* than with mass. Weight is a measure of the gravitational pull on an object. On Earth, an astronaut with a mass of 75.0 kg has a weight of 165 lb. On the moon where the gravitational pull is one-sixth that of Earth, the astronaut has a weight of 27.5 lb. However, the mass of the astronaut is

FIGURE 1.2

Volume is the space occupied by a substance. In the metric system, volume is based on the liter, which is slightly larger than a quart.

Q **How many milliliters are in 1 quart?**

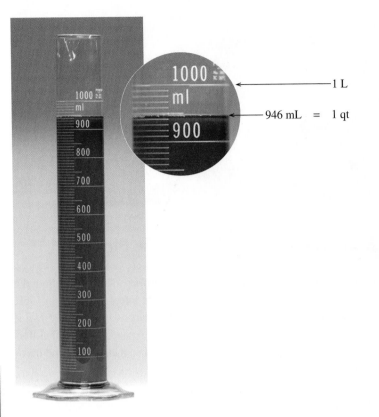

1 L

946 mL = 1 qt

FIGURE 1.3

On an electronic balance, mass is shown in grams as a digital readout.

Q **How many grams are in 1 pound of candy?**

the same as on Earth, 75.0 kg. Scientists measure mass rather than weight because mass does not depend on gravity.

In the metric system, the unit for mass is the **gram (g)**. The SI unit of mass, the **kilogram (kg)**, is used for larger masses such as body weight. It takes 2.20 lb to make 1 kg, and 454 g is needed to equal 1 pound.

$$1 \text{ kg} = 1000 \text{ g}$$
$$1 \text{ kg} = 2.20 \text{ lb}$$
$$454 \text{ g} = 1 \text{ lb}$$

In a chemistry laboratory, a balance is used to measure the mass of a substance as shown in Figure 1.3.

Temperature

You use a thermometer to see how hot something is, or how cold it is outside, or perhaps to determine if you have a fever. (See Figure 1.4.) The **temperature** tells us how hot or cold an object is. A typical laboratory thermometer consists of a glass bulb with a liquid in it that expands as the temperature increases. On a **Celsius (°C) scale**, water freezes at 0°C and boils at 100°C, while on a Fahrenheit (°F) scale, water freezes at 32°C and boils at 212°C. In the SI system, temperature is measured using the **Kelvin (K) scale**, on which the lowest temperature possible is assigned a value of 0 K. Note that the units on the Kelvin scale are called kelvins (K) and do not have a degree sign.

FIGURE 1.4

A thermometer is used to determine the temperature of a substance.

Q **What kinds of temperature readings have you made today?**

TABLE 1.1

Units of Measurement		
Measurement	Metric	SI
Length	Meter (m)	Meter (m)
Volume	Liter (L)	Cubic meter (m³)
Mass	Gram (g)	Kilogram (kg)
Time	Second (s)	Second (s)
Temperature	Celsius (°C)	Kelvin (K)

Time

You probably think of time as years, days, minutes, or seconds. Of these, the SI and metric basic unit of time is the **second (s)**. The standard now used to determine a second is an atomic clock that uses a frequency emitted by cesium atoms. A comparison of metric and SI units for measurement is shown in Table 1.1.

Scientific Notation

In chemistry and science, we use numbers that are very, very small to measure things as tiny as the width of a human hair, which is 0.000 008 m, or things that are even smaller. Or perhaps we want to know the number of hairs on the average scalp, which is about 100 000 hairs. (See Figure 1.5.) In both measurements, it is convenient to write the numbers in scientific notation. (In this section we have added spaces to help make the places easier to count.)

Width of a human hair	0.000 008 m	8×10^{-6} m
Hairs on a human scalp	100 000 hairs	1×10^{5} hairs

Writing a Number in Scientific Notation

When a number is written in **scientific notation**, there are two parts: a coefficient and a power of 10. For example, the number 2400 in scientific notation is 2.4×10^{3}. The coefficient is 2.4 and 10^{3} shows the power of 10. The coefficient is determined by moving the decimal point three places to the left to

8×10^{-6} m

FIGURE 1.5

Humans have an average of 1×10^{5} hairs on their scalps. Each hair is about 8×10^{-6} m wide.

Q Why are large and small numbers written in scientific notation?

TABLE 1.2

Some Powers of Ten

Number	Multiples of Ten	Scientific Notation	
10 000	$10 \times 10 \times 10 \times 10$	1×10^4	
1 000	$10 \times 10 \times 10$	1×10^3	
100	10×10	1×10^2	Some positive powers of ten
10	10	1×10^1	
1	0	1×10^0	
0.1	$\dfrac{1}{10}$	1×10^{-1}	
0.01	$\dfrac{1}{10} \times \dfrac{1}{10} = \dfrac{1}{100}$	1×10^{-2}	Some negative powers of ten
0.001	$\dfrac{1}{10} \times \dfrac{1}{10} \times \dfrac{1}{10} = \dfrac{1}{1000}$	1×10^{-3}	
0.0001	$\dfrac{1}{10} \times \dfrac{1}{10} \times \dfrac{1}{10} \times \dfrac{1}{10} = \dfrac{1}{10\,000}$	1×10^{-4}	

MC GOB Scientific Notation

give a number between 1 and 10. Because we moved the decimal point three places to the left, the power of ten is a positive 3, written as 10^3. For a number greater than one, the power of 10 is positive.

$$2400. = 2.4 \times 1000 = 2.4 \times 10 \times 10 \times 10 = 2.4 \quad \times \quad 10^3$$

$\longleftarrow$ 3 places $\qquad\qquad\qquad\qquad\qquad$ Coefficient $\quad$ Power of ten

When a number less than one is written in scientific notation, the power of ten is negative. For example, to write the number 0.000 86 in scientific notation the decimal point is moved to the right four places to give a coefficient of 8.6, which is between 1 and 10. By moving the decimal point four places to the right, the power of 10 is a negative 4 or 10^{-4}.

$$0.00086 = \frac{8.6}{10\,000} = \frac{8.6}{10 \times 10 \times 10 \times 10} = 8.6 \quad \times \quad 10^{-4}$$

4 places $\longrightarrow$ $\qquad\qquad\qquad\qquad\qquad$ Coefficient $\quad$ Power of ten

Table 1.2 gives some examples of numbers written as positive and negative powers of ten. The powers of tens are a way of keeping track of the decimal point in the decimal number. Table 1.3 gives examples of writing measurements in scientific notation.

Scientific Notation and Calculators

You can enter numbers in scientific notation on many calculators using the EE or EXP key. After you enter the coefficient, push the EXP (or EE) key and enter only

TABLE 1.3

Some Measurements Written in Scientific Notation

Measured Quantity	Measurement	Scientific Notation
Diameter of Earth	12 800 000 m	1.28×10^7 m
Depth of Lake Tanganyika	1 460 m	1.46×10^3 m
Mass of a typical human	68 kg	6.8×10^1 kg
Mass of a hummingbird	0.002 kg	2×10^{-3} kg
Length of a pox virus	0.000 000 3 m	3×10^{-7} m
Mass of a bacterium (mycoplasma)	0.000 000 000 000 000 000 1 kg	1×10^{-19} kg

the power of ten because the EXP function key includes the $\times$ **10** value. To enter a negative power of ten, push the plus/minus (+/−) key or the minus (−) key (depending on your calculator), but **not** the key that performs the subtraction operation "−". Some calculators require entering the sign before the power.

Number to Enter	Method	Display Reads
4×10^6	4 EXP (EE) 6	$4\ 06$ or 4^{06}
2.5×10^{-4}	2.5 EXP (EE) +/− 4	$2.5 - 04$ or 2.5^{-04}

When a calculator answer appears in scientific notation, it is usually shown in the display as a number between 1 and 10 followed by a space and the power of ten. To express this display in scientific notation, write the number, insert $\times$ 10, and use the power of ten as an exponent.

Calculator Display	Expressed in Scientific Notation
7.52^{04}	7.52×10^4
5.8^{-02}	5.8×10^{-2}

On many scientific calculators, a number can be converted into scientific notation using the appropriate keys. For example, the number 0.000 52 can be entered followed by hitting the 2nd or 3rd function key and the SCI key. The scientific notation appears in the calculator display as a coefficient and the power of ten.

0.000 52 [2nd or 3rd function key] [SCI] = 5.2^{-4} or $5.2-04$ = 5.2×10^{-4}
                    Key   Key   Display

Converting Scientific Notation to a Standard Number

When a number in scientific notation has a positive power of 10, the standard number is written by moving the decimal point to the right for the same number of places as the power of 10. Placeholder zeros are used to give additional decimal places.

$$4.3 \times 10^3 = 4.3 \times 1000 = 4.300 = 4300$$

For a number in scientific notation with a negative power of 10, the standard number is written by moving the decimal point to the left for the same number of places. Placeholder zeros are added in front of the coefficient as needed.

$$4.3 \times 10^{-3} = 4.3 \times \frac{1}{1000} = 0004.3 = 0.0043$$

SAMPLE PROBLEM 1.2

■ Scientific Notation

1. Write the following measurements in scientific notation.
   a. 0.000 16 L   **b.** 5 220 000 m
2. Write the following as standard numbers.
   a. 7.2×10^{-3} m   **b.** 2.4×10^5 g

SOLUTION

1. **a.** 1.6×10^{-4} L   **b.** 5.22×10^6 m
2. **a.** 0.0072 m   **b.** 240 000 g

STUDY CHECK

Write the following measurements in scientific notation.

a. 425 000 m   **b.** 0.000 000 8 g

QUESTIONS AND PROBLEMS

■ Measurement and Scientific Notation

In every chapter, each magenta, odd-numbered exercise in Questions and Problems is paired with the next even-numbered exercise. The answers for all the magenta, odd-numbered Questions and Problems are given at the end of this chapter. The complete solutions to the odd-numbered Questions and Problems are in the *Study Guide*.

1.5 State the name of the unit and the type of measurement indicated for each of the following quantities.
   a. 4.8 m   **b.** 325 g   **c.** 1.5 L   **d.** 480 s   **e.** 28°C

1.6 State the name of the unit and the type of measurement indicated for each of the following quantities.
   a. 0.8 L   **b.** 3.6 m   **c.** 14 kg   **d.** 35 g   **e.** 373 K

1.7 Write the following measurements in scientific notation:
   a. 55 000 m   **b.** 480 g   **c.** 0.000 005 cm
   d. 0.000 14 s   **e.** 0.007 85 L   **f.** 670 000 kg

1.8 Write the following measurements in scientific notation:
   a. 180 000 000 g   **b.** 0.000 06 m   **c.** 750 000 g
   d. 0.15 m   **e.** 0.024 s   **f.** 1 500 m^3

1.9 Which number in each pair is larger?
   a. 7.2×10^3 or 8.2×10^2 **b.** 4.5×10^{-4} or 3.2×10^{-2}
   c. 1×10^4 or 1×10^{-4} **d.** 0.000 52 or 6.8×10^{-2}

1.10 Which number in each pair is smaller?
   a. 4.9×10^{-3} or 5.5×10^{-9} **b.** 1250 or 3.4×10^2
   c. 0.000 000 4 or 5×10^{-8} **d.** 4×10^8 or 4×10^{-10}

1.11 Write the following as standard numbers:
   a. 1.2×10^4 **b.** 8.25×10^{-2} **c.** 4×10^6 **d.** 5×10^{-3}

1.12 Write the following as standard numbers:
   a. 3.6×10^{-5} **b.** 8.75×10^4 **c.** 3×10^{-2} **d.** 2.12×10^5

LEARNING GOAL

Determine the number of significant figures in measured numbers.

1.3 Measured Numbers and Significant Figures

Whenever you make a measurement, you use some type of measuring device. For example, you may use a meterstick to measure your height, a scale to check your weight, and a thermometer to take your temperature. **Measured numbers** are the numbers you obtain when you measure a quantity such as your height, weight, or temperature.

Measured Numbers

Suppose you are going to measure the lengths of the objects in Figure 1.6. You would select a ruler with a scale marked on it. By observing the lines on the scale, you determine the measurement for each object. Perhaps the divisions on the scale are marked as 1 cm. Another ruler might be marked in divisions of 0.1 cm. To report the length, you would first read the numerical value of the marked line. Finally, you *estimate* by visually dividing the space between the smallest marked lines. This estimated number is the final digit in a measured number.

For example, in Figure 1.6a, the end of the object falls between the lines marked 4 cm and 5 cm. That means that the length is 4 cm plus an estimated digit. If you estimate that the end is halfway between 4 cm and 5 cm, you would report its length as 4.5 cm. However, someone else might report the length as 4.4 cm. The last digit in a measured number can differ because people do not estimate in the same way. The ruler shown in Figure 1.6b is marked with lines at 0.1 cm. With this ruler, you can estimate the value of the hundredth's place (0.01 cm). Perhaps you would report the length of the object as 4.55 cm, while someone else may report its length as 4.56 cm. Both results are acceptable.

Therefore, there is always *uncertainty* in every measurement. When a measurement ends right on a marked line, a zero is written as the estimated digit. For example, in Figure 1.6c, the measurement for length is written as 3.0 cm, not 3. This means that the uncertainty of the measurement is in the estimated digit.

Significant Figures

In a measured number, the **significant figures** are all the digits including the estimated digit. All nonzero numbers are counted as significant figures. Zeros may or may not be significant depending on their position in a number. Table 1.4 gives the rules and examples of counting significant figures.

When one or more zeros in a large number are significant digits, they are shown by writing the number using scientific notation. For example, if the first zero in the

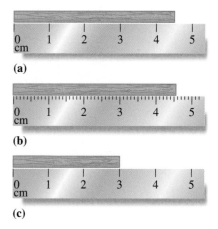

FIGURE 1.6

The lengths of the rectangular objects are measured as (**a**) 4.5 cm and (**b**) 4.55 cm.

Q **What is the length of the object in (c)?**

MC
GOB   SELF-STUDY ACTIVITY
Significant Figures

MC
GOB   Counting Significant Figures

TABLE 1.4

Significant Figures in Measured Numbers

Rule	Measured Number	Number of Significant Figures
1. A number is a *significant figure* if it is		
a. not a zero	4.5 g	2
	122.35 m	5
b. a zero between nonzero digits	205 m	3
	5.082 kg	4
c. a zero at the end of a decimal number	50. L	2
	25.0°C	3
	16.00 g	4
d. any digit in the coefficient of a number written in scientific notation	4.0×10^5 m	2
	5.70×10^{-3} g	3
2. A zero is *not significant* if it is		
a. at the beginning of a decimal number	0.0004 lb	1
	0.075 m	2
b. used as a placeholder in a large number without a decimal point	850 000 m	2
	1 250 000 g	3

measurement 500 m is significant, it can be shown by writing the measurement as 5.0×10^2 m. In this text, we will place a decimal point after a significant zero at the end of a number. For example, a measurement written as 250. g has three significant figures, which includes the zero. It could also be written as 2.50×10^2 g. Unless noted otherwise, we will assume that zeros at the end of large standard numbers are not significant; that is, we would interpret 400 000 g as 4×10^5 g.

Exact Numbers

Exact numbers are numbers obtained by counting items or from a definition that compares two units in the same measuring system. Suppose a friend asks you to tell her the number of coats in your closet or the number of classes you are taking in school. Your answer would be given by counting the items. It was not necessary for you to use any type of measuring tool. Suppose someone asks you to state the number of seconds in one minute. Without using any measuring device, you would give the definition: 60 seconds in one minute. Exact numbers are not measured, do not have a limited number of significant figures, and do not affect the number of significant figures in a calculated answer. For more examples of exact numbers, see Table 1.5.

TABLE 1.5

Examples of Some Exact Numbers

Counted Numbers	Defined Equalities	
	U.S. System	**Metric System**
Eight doughnuts	1 ft = 12 in.	1 L = 1000 mL
Two baseballs	1 qt = 4 cups	1 m = 100 cm
Five capsules	1 lb = 16 ounces	1 kg = 1000 g

SAMPLE PROBLEM 1.3

▪ Significant Figures

Identify each of the following numbers as measured or exact and give the number of significant figures in each measured number.

a. 42.2 g   **b.** 3 eggs   **c.** 0.000 5 cm

d. 450 000 km   **e.** 3.500×10^5 s

SOLUTION

a. measured; three   **b.** exact   **c.** measured; one

d. measured; two   **e.** measured; four

STUDY CHECK

State the number of significant figures in each of the following measured numbers:

a. 0.000 35 g   **b.** 2 000 m   **c.** 2.0045 L

QUESTIONS AND PROBLEMS

■ Measured Numbers and Significant Figures

1.13 Identify the numbers in each of the following statements as measured or exact.
   a. A person weighs 155 lb.
   b. The basket holds 8 apples.
   c. In the metric system, 1 kg is equal to 1000 g.
   d. The distance from Denver, Colorado, to Houston, Texas, is 1720 km.

1.14 Identify the numbers in each of the following statements as measured or exact.
   a. There are 31 students in the laboratory.
   b. The oldest known flower lived 120 000 000 years ago.
   c. The largest gem ever found, an aquamarine, has a mass of 104 kg.
   d. A laboratory test shows a blood cholesterol level of 184 mg/dL.

1.15 In each set of numbers, identify the measured number(s), if any.
   a. 3 hamburgers and 6 oz of meat
   b. 1 table and 4 chairs
   c. 0.75 lb of grapes and 350 g of butter
   d. 60 seconds equals 1 minute

1.16 In each set of numbers, identify the exact number(s), if any.
   a. 5 pizzas and 50.0 g of cheese   **b.** 6 nickels and 16 g of nickel
   c. 3 onions and 3 lb of potatoes   **d.** 5 miles and 5 cars

1.17 For each measurement, indicate if the zeros are significant.
   a. 0.0038 m   **b.** 5.04 cm   **c.** 800. L
   d. 3.0×10^{-3} kg   **e.** 85 000 g

1.18 For each measurement, indicate if the zeros are significant.
   a. 20.05 g   **b.** 5.00 m   **c.** 0.000 02 L
   d. 120 000 years   **e.** 8.05×10^2 g

1.19 How many significant figures are in each of the following measured quantities?
   a. 11.005 g   **b.** 0.000 32 m   **c.** 36 000 000 m
   d. 1.80×10^4 g   **e.** 0.8250 L   **f.** 30.0°C

1.20 How many significant figures are in each of the following measured quantities?
   a. 20.60 L   **b.** 1036.48 g   **c.** 4.00 m
   d. 20.8°C   **e.** 60 800 000 g   **f.** 5.0×10^{-3} L

1.21 In which of the following pairs do both numbers contain the same number of significant figures?
   a. 11.0 m and 11.00 m   **b.** 600.0 K and 60 K
   c. 0.000 75 s and 75 000 s   **d.** 250.0 L and 0.02500 L

1.22 In which of the following pairs do both numbers contain the same number of significant figures?
   a. 0.00575 g and 5.75×10^{-3} g   **b.** 0.0250 m and 0.205 m
   c. 150 000 s and 1.50×10^4 s   **d.** 3.8×10^{-2} L and 3.8×10^5 L

1.23 Write each of the following in scientific notation with two significant figures:
   a. 5 000 L   **b.** 30 000 g   **c.** 100 000 m   **d.** 0.000 25 cm

1.24 Write each of the following in scientific notation with two significant figures:
   a. 5 100 000 g   **b.** 26 000 s   **c.** 40 000 m   **d.** 0.000 820 kg

---

■ 1.4 Significant Figures in Calculations

LEARNING GOAL

■ Adjust calculated answers to the correct number of significant figures.

In the sciences, we measure many things: the length of a bacterium, the volume of a gas sample, the temperature of a reaction mixture, or the mass of iron in a sample. The numbers obtained from these types of measurements are often used in calculations. The number of significant figures in the measured numbers limits the number of significant figures that can be given in the calculated answer.

Using a calculator will usually help do calculations faster. However, calculators cannot think for you. It is up to you to enter the numbers correctly, press the right function keys, and adjust the calculator display to give an answer with the correct number of significant figures.

Rounding Off

To calculate the area of a carpet that measures 5.5 m by 3.5 m, you multiply 5.5 times 3.5 to obtain the number 19.25 as the area in square meters. However, all four digits cannot be given in the answer because they are not all significant figures. Each measurement of length and width has only two significant figures. This means that the calculated result must be *rounded off* to give an answer that also has two significant figures, 19 m². When you obtain a calculator result, determine the number of significant figures for the answer and round off using the following rules.

**MC
GOB** Significant Figures in Calculations

Rules for Rounding Off

1. If the first digit to be dropped is *4 or less*, it and all following digits are simply dropped from the number.
2. If the first digit to be dropped is *5 or greater*, the last retained digit of the number is increased by 1.

		Three Significant Figures	Two Significant Figures
Example 1:	8.4234 rounds off to	8.42	8.4
Example 2:	14.780 rounds off to	14.8	15

SAMPLE PROBLEM 1.4

▪ Rounding Off

Round off each of the following numbers to three significant figures:
a. 35.7823 m   **b.** 0.002627 L   **c.** 3826.8 g   **d.** 1.2836 kg

SOLUTION

a. 35.8 m   **b.** 0.00263 L   **c.** 3830 g   **d.** 1.28 kg

STUDY CHECK

Round off each of the numbers in Sample Problem 1.4 to two significant figures.

Multiplication and Division

In multiplication or division, the final answer is written so it has the same number of significant figures as the measurement with the *fewest significant figures* (SFs).

Example 1

Multiply the following measured numbers: 24.65 × 0.67.

24.65	⊠	0.67	⊜	*16.5155*	⟶	17
Four SFs		Two SFs		Calculator display		Final answer, rounded to two SFs

The answer in the calculator display has more digits than the data allows. The measurement 0.67 has the least number of significant figures, two. Therefore, the calculator answer is rounded off to two significant figures.

Example 2

Solve the following:

$$\frac{2.85 \times 67.4}{4.39}$$

To do this problem on a calculator, enter the numbers and then press the operation keys. In this case, we might press the keys in the following order:

2.85	$\times$	67.4	$\div$	4.39	$=$	43.756264	$\longrightarrow$	43.8
Three SFs		Three SFs		Three SFs		Calculator display		Final answer, rounded to three SFs

All of the measurements in this problem have three significant figures. Therefore, the calculator result is rounded off to give an answer, 43.8, that has three significant figures.

Adding Significant Zeros

Sometimes, a calculator displays a small whole number. To give an answer with the correct number of significant figures, significant zeros may need to be written after the calculator result. For example, suppose the calculator display is 4, but you used measurements that have three significant numbers. The answer 4.00 is obtained by placing two significant zeros after the 4.

$$\frac{8.00}{2.00} =$$

	4	$\longrightarrow$	4.00
Three SFs	Calculator display		Final answer, two zeros added to give three SFs

SAMPLE PROBLEM 1.5

▪Significant Figures in Multiplication and Division

Perform the following calculations of measured numbers. Give the answers with the correct number of significant figures.

a. 56.8×0.37 **b.** $\dfrac{71.4}{11}$ **c.** $\dfrac{(2.075)(0.585)}{(8.42)(0.00450)}$ **d.** $\dfrac{25.0}{5.00}$

SOLUTION

a. 21   **b.** 6.5   **c.** 32.0   **d.** 5.00 (Add significant zeros)

STUDY CHECK

Perform the following calculations of measured numbers. Give the answers with the correct number of significant figures.

a. 45.26×0.01088 **b.** $2.6 \div 324$ **c.** $\dfrac{4.0 \times 8.00}{16}$

Addition and Subtraction

In addition or substraction, the answer is written so it has the same number of decimal places as the measurement having the fewest decimal places.

Example 3

Add:

	2.045	Three decimal places
+	34.1	One decimal place
	36.145	Calculator display
	36.1	Answer, rounded to one decimal place

Example 4

Subtract:

	255	Ones place
−	175.65	Two decimal places
	79.35	Calculator display
	79	Answer, rounded to ones place

When numbers are added or subtracted to give answers ending in zero, the zero does not appear after the decimal point in the calculator display. For example, 14.5 g − 2.5 g = 12.0 g. However, if you do the subtraction on your calculator, the display shows 12. To give the correct answer, a significant zero is written after the decimal point.

Example 5

	14.5 g	One decimal place
−	2.5 g	One decimal place
	12.	Calculator display
	12.0 g	Answer, zero written after the decimal point

SAMPLE PROBLEM 1.6

▪ Significant Figures in Addition and Subtraction

Perform the following calculations and give the answers with the correct number of decimal places:

a. 27.8 cm + 0.235 cm
b. 104.45 mL + 0.838 mL + 46 mL
c. 153.247 g − 14.82 g

SOLUTION

a. 28.0 cm **b.** 151 mL **c.** 138.43 g

STUDY CHECK

Perform the following calculations and give the answers with the correct number of decimal places:

a. 82.45 mg + 1.245 mg + 0.00056 mg **b.** 4.259 L − 3.8 L

QUESTIONS AND PROBLEMS

■ Significant Figures in Calculations

1.25 Why do we usually need to round off calculations that use measured numbers?

1.26 Why do we sometimes add a zero to a number in a calculator display?

1.27 Round off each of the following numbers to three significant figures.
   a. 1.854   **b.** 184.2038   **c.** 0.004738265   **d.** 8807   **e.** 1.832149

1.28 Round off each of the numbers in problem 1.27 to two significant figures.

1.29 For the following problems, give answers with the correct number of significant figures:
   a. 45.7 × 0.034
   b. 0.00278 × 5
   c. $\dfrac{34.56}{1.25}$
   d. $\dfrac{(0.2465)(25)}{1.78}$

1.30 For the following problems, give answers with the correct number of significant figures:
   a. 400 × 185
   b. $\dfrac{2.40}{(4)(125)}$
   c. 0.825 × 3.6 × 5.1
   d. $\dfrac{3.5 \times 0.261}{8.24 \times 20.0}$

1.31 For the following problems, give answers with the correct number of decimal places:
   a. 45.48 cm + 8.057 cm   **b.** 23.45 g + 104.1 g + 0.025 g
   c. 145.675 mL − 24.2 mL   **d.** 1.08 L − 0.585 L

1.32 For the following problems, give answers with the correct number of decimal places:
   a. 5.08 g + 25.1 g   **b.** 85.66 cm + 104.10 cm + 0.025 cm
   c. 24.568 mL − 14.25 mL   **d.** 0.2654 L − 0.2585 L

■ 1.5 Prefixes and Equalities

LEARNING GOAL

Use the numerical values of prefixes to write a metric equality.

The special feature of the metric system of units is that a **prefix** can be attached to any unit to increase or decrease its size by some factor of 10. For example, the prefixes *milli* and *micro* are used to make the smaller units, milligram (mg) and microgram (μg). Table 1.6 lists some of the metric prefixes, their symbols, and their decimal values.

 SELF-STUDY ACTIVITY
Metric System

TABLE 1.6

Metric and SI Prefixes

Prefix[a]	Symbol	Meaning	Numerical Value	Scientific Notation
Prefixes That Increase the Size of the Unit				
tera	T	trillion	1 000 000 000 000	10^{12}
giga	G	billion	1 000 000 000	10^{9}
mega	M	million	1 000 000	10^{6}
kilo	k	thousand	1 000	10^{3}
Prefixes That Decrease the Size of the Unit				
deci	d	tenth	0.1	10^{-1}
centi	c	hundredth	0.01	10^{-2}
milli	m	thousandth	0.001	10^{-3}
micro	μ	millionth	0.000 001	10^{-6}
nano	n	billionth	0.000 000 001	10^{-9}
pico	p	trillionth	0.000 000 000 001	10^{-12}
femto	f	quadrillionth	0.000 000 000 000 001	10^{-15}

[a] The prefixes most often used by chemists are in boldface type.

TABLE 1.7

Daily Values for Selected Nutrients

Nutrient	Amount Recommended
Protein	44 g
Vitamin C	60 mg
Vitamin B$_{12}$	6 μg
Calcium	1 g
Iron	18 mg
Iodine	150 μg
Sodium	2400 mg
Zinc	15 mg

The prefix *centi* is like cents in a dollar. One cent would be a centidollar or $\frac{1}{100}$ of a dollar. That also means that one dollar is the same as 100 cents. The prefix *deci* is like dimes in a dollar. One dime would be a decidollar or $\frac{1}{10}$ of a dollar. That also means that one dollar is the same as 10 dimes.

The U.S. Food and Drug Administration has determined the daily values (DV) of nutrients for adults and children age 4 or older. Some of these recommended daily values, which use prefixes, are listed in Table 1.7.

The relationship of a prefix to a unit can be expressed by replacing the prefix with its numerical value. For example, when the prefix *kilo* in *kilometer* is replaced with its value of 1000, we find that a kilometer is equal to 1000 meters. Other examples follow.

1 **kilo**meter (1 km) = **1000** meters (1000 m)

1 **kilo**liter (1 kL) = **1000** liters (1000 L)

1 **kilo**gram (1 kg) = **1000** grams (1000 g)

SAMPLE PROBLEM 1.7

▪ Prefixes

Fill in the blanks with the correct numerical value:

a. kilogram = _____ grams

b. millisecond = _____ second

c. deciliter = _____ liter

SOLUTION

a. The numerical value of *kilo* is 1000; 1000 grams.

b. The numerical value of *milli* is 0.001; 0.001 second.

c. The numerical value of *deci* is 0.1; 0.1 liter.

STUDY CHECK

Write the correct prefix in the blanks:

a. 1 000 000 seconds = _____ second

b. 0.01 meter = _____ meter

Measuring Length

An ophthalmologist may measure the diameter of the retina of an eye in centimeters (cm), whereas a surgeon may need to know the length of a nerve in millimeters (mm). When the prefix *centi* is used with the unit *meter*, it indicates the unit *centimeter*, a length that is one-hundredth of a meter (0.01 m). A *millimeter* measures a length of 0.001 m. There are 1000 mm in a meter.

If we compare the lengths of a millimeter and a centimeter, we find that 1 mm is 0.1 cm; there are 10 mm in 1 cm. These comparisons are examples of **equalities**, which show the relationship between two units that measure the same quantity. For example, in the equality 1 m = 100 cm, each quantity describes the same length but in a different unit. Note that each quantity in the equality expression has both a number and a unit.

First Quantity		Second Quantity	
1 m	=	100	cm
↑ ↑		↑	↑
Number + unit		Number + unit	

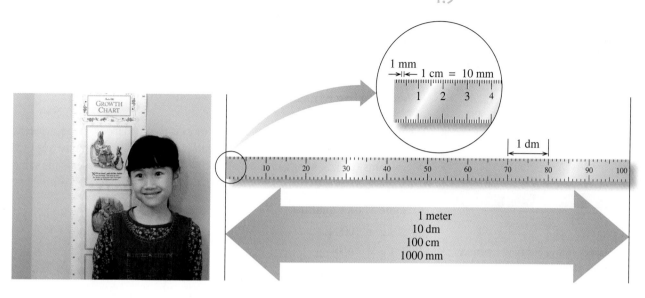

FIGURE 1.7

The metric length of one meter is the same length as 10 dm, 100 cm, and 1000 mm.

Q **How many millimeters (mm) are in 1 centimeter (cm)?**

Some Length Equalities

1 m = 100 cm

1 m = 1000 mm

1 cm = 10 mm

Some metric units for length are compared in Figure 1.7.

Measuring Volume

Volumes of 1 L or smaller are common in the health sciences. When a liter is divided into 10 equal portions, each portion is a deciliter (dL). There are 10 dL in 1 L. Laboratory results for blood work are often reported in mass per deciliter. Table 1.8 lists typical laboratory tests for some substances in the blood.

When a liter is divided into a thousand parts, each of the smaller volumes is a milliliter. In a 1-L container of physiological saline, there are 1000 mL of solution (see Figure 1.8).

Some Volume Equalities

1 L = 10 dL

1 L = 1000 mL

1 dL = 100 mL

The **cubic centimeter** (**cm**^{3} or **cc**) is the volume of a cube whose dimensions are 1 cm on each side. A cubic centimeter has the same volume as a milliliter, and the units are often used interchangeably.

1 cm^3 = 1 cc = 1 mL

TABLE 1.8

Some Typical Laboratory Test Values

Substance in Blood	Typical Range
albumin	3.5–5.0 g/dL
ammonia	20–150 μg/dL
calcium	8.5–10.5 mg/dL
cholesterol	105–250 mg/dL
iron (male)	80–160 μg/dL
protein (total)	6.0–8.0 g/dL

FIGURE 1.8

A plastic intravenous fluid container contains 1000 mL.

Q **How many liters of solution are in the intravenous fluid container?**

When you see *1 cm*, you are reading about length; when you see *1 cc* or *1 cm³* or *1 mL*, you are reading about volume. A comparison of units of volume is illustrated in Figure 1.9.

Measuring Mass

When you get a physical examination, your mass is recorded in kilograms, whereas the results of your laboratory tests are reported in grams, milligrams (mg), or micrograms (μg). A kilogram is equal to 1000 g. One gram represents the same mass as 1000 mg, and one mg equals 1000 μg.

Some Mass Equalities

1 kg = 1000 g

1 g = 1000 mg

1 mg = 1000 μg

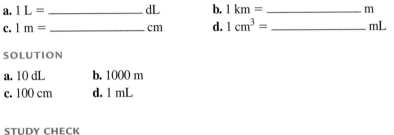

SAMPLE PROBLEM 1.8

■ Writing Metric Relationships

Complete the following list of metric equalities:

a. 1 L = _____ dL   **b.** 1 km = _____ m

c. 1 m = _____ cm   **d.** 1 cm³ = _____ mL

SOLUTION

a. 10 dL   **b.** 1000 m

c. 100 cm   **d.** 1 mL

STUDY CHECK

Complete the following equalities:

a. 1 kg = _____ g   **b.** 1 mL = _____ L

FIGURE 1.9

A cube measuring 10 cm on each side has a volume of 1000 cm³, or 1 L; a cube measuring 1 cm on each side has a volume of 1 cm³ (cc) or 1 mL.

Q **What is the relationship between a milliliter (mL) and a cubic centimeter (cm³)?**

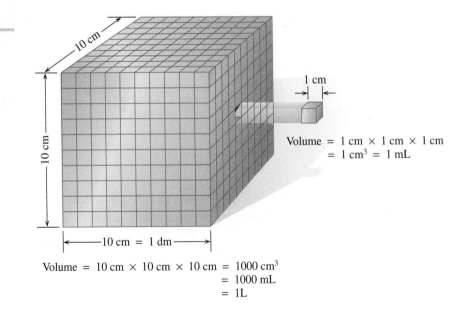

Volume = 1 cm × 1 cm × 1 cm
= 1 cm³ = 1 mL

Volume = 10 cm × 10 cm × 10 cm = 1000 cm³
= 1000 mL
= 1L

QUESTIONS AND PROBLEMS

■ Prefixes and Equalities

1.33 The speedometer in the margin is marked in both km/h and mi/h or mph. What is the meaning of each abbreviation?

1.34 In Canada, a highway sign gives a speed limit as 80 km/h. If you were going the maximum speed allowed in Canada, would you be exceeding the speed limit in the United States of 55 mph?

1.35 How does the prefix *kilo* affect the gram unit in *kilogram*?

1.36 How does the prefix *centi* affect the meter unit in *centimeter*?

1.37 Write the abbreviation for each of the following units:
    a. milligram   **b.** deciliter   **c.** kilometer
    d. kilogram   **e.** microliter

1.38 Write the complete name for each of the following units:
    a. cm   **b.** kg   **c.** dL   **d.** Gm   **e.** μg

1.39 Write the numerical values for each of the following prefixes:
    a. centi   **b.** kilo   **c.** milli   **d.** deci   **e.** mega

1.40 Write the complete name (prefix + unit) for each of the following numerical values:
    a. 0.10 g   **b.** 0.000 001 g   **c.** 1000 g
    d. $\frac{1}{100}$ g   **e.** 0.001 g

1.41 Complete the following metric relationships:
    a. 1 m = _____ cm   **b.** 1 km = _____ m
    c. 1 mm = _____ m   **d.** 1 L = _____ mL

1.42 Complete the following metric relationships:
    a. 1 kg = _____ g   **b.** 1 mL = _____ L
    c. 1 g = _____ kg   **d.** 1 g = _____ mg

1.43 For each of the following pairs, which is the larger unit?
    a. milligram or kilogram   **b.** milliliter or microliter
    c. cm or km   **d.** kL or dL

1.44 For each of the following pairs, which is the smaller unit?
    a. mg or g   **b.** centimeter or millimeter
    c. mm or μm   **d.** mL or dL

1.6 Writing Conversion Factors

LEARNING GOAL

Write a conversion factor for two units that describe the same quantity.

Many problems in chemistry and the health sciences require a change of units. You make changes in units every day. For example, suppose you spent 2.0 hours (hr) on your homework, and someone asked you how many minutes that was. You would answer 120 minutes (min). You knew how to change from hours to minutes because you knew an equality (1 hr = 60 min) that related the two units. To do the problem, the equality is written in the form of a fraction called a **conversion factor**. One of the quantities is the numerator, and the other is the denominator. Be sure to include the units when you write the conversion factors. Two factors are always possible from any equality.

Two Conversion Factors for the Equality 1 hr = 60 min

$$\frac{\text{Numerator} \longrightarrow}{\text{Denominator} \longrightarrow} \quad \frac{60 \text{ min}}{1 \text{ hr}} \quad \text{and} \quad \frac{1 \text{ hr}}{60 \text{ min}}$$

TABLE 1.9

Some Common Equalities

Quantity	U.S.	Metric (SI)	Metric–U.S.
Length	1 ft = 12 in.	1 km = 1000 m	2.54 cm = 1 in.
	1 yard = 3 ft	1 m = 1000 mm	1 m = 39.4 in.
	1 mile = 5280 ft	1 cm = 10 mm	1 km = 0.621 mi
Volume	1 qt = 4 cups	1 L = 1000 mL	946 mL = 1 qt
	1 qt = 2 pints	1 dL = 100 mL	1 L = 1.06 qt
	1 gallon = 4 qts	1 mL = 1 cm^3	
Mass	1 lb = 16 oz	1 kg = 1000 g	1 kg = 2.20 lb
		1 g = 1000 mg	454 g = 1 lb
Time		1 hr = 60 min	
		1 min = 60 s	

These factors are read as "60 minutes per 1 hour," and "1 hour per 60 minutes." The term *per* means "divide." Some common relationships are given in Table 1.9. It is important that the equality you select to construct a conversion factor is a true relationship.

When an equality shows the relationship for two units from the same system (metric or U.S.), it is considered a definition and exact. It is not used to determine significant figures. When an equality shows the relationship of units from two different systems, the number is measured and counts toward the significant figures in a calculation. For example, in the equality 1 lb = 454 g, the measured number 454 has three significant figures. The number one in 1 lb is considered as exact. An exception is the relationship of 1 in. = 2.54 cm: the value 2.54 has been defined as exact.

Metric Conversion Factors

MC GOB Metric Conversions

We can write metric conversion factors for the metric relationships we have studied. For example, from the equality for meters and centimeters, we can write the following factors:

Metric Equality	Conversion Factors	
1 m = 100 cm	$\dfrac{100 \text{ cm}}{1 \text{ m}}$ and	$\dfrac{1 \text{ m}}{100 \text{ cm}}$

Both are proper conversion factors for the relationship; one is just the inverse of the other. The usefulness of conversion factors is enhanced by the fact that we can turn a conversion factor over and use its inverse.

Metric–U.S. System Conversion Factors

Suppose you need to convert from pounds, a unit in the U.S. system, to kilograms in the metric (or SI) system. A relationship you could use is

1 kg = 2.20 lb

The corresponding conversion factors would be

$$\frac{2.20 \text{ lb}}{1 \text{ kg}} \quad \text{and} \quad \frac{1 \text{ kg}}{2.20 \text{ lb}}$$

FIGURE 1.10

In the U.S., the contents of many packaged foods are listed in both U.S. and metric units.

Q **What are some advantages of using the metric system?**

Figure 1.10 illustrates the contents of some packaged foods in both metric and U.S. units.

SAMPLE PROBLEM 1.9

■ Writing Conversion Factors from Equalities

Write conversion factors for the relationship for the following pairs of units:

a. Milligrams and grams   **b.** Quarts and milliliters

SOLUTION

Equality	Conversion Factors	
a. 1 g = 1000 mg	$\dfrac{1 \text{ g}}{1000 \text{ mg}}$ and	$\dfrac{1000 \text{ mg}}{1 \text{ g}}$
b. 1 qt = 946 mL	$\dfrac{1 \text{ qt}}{946 \text{ mL}}$ and	$\dfrac{946 \text{ mL}}{1 \text{ qt}}$

STUDY CHECK

Write the equality and conversion factors for the relationship between inches and centimeters.

Conversion Factors Stated Within a Problem

Many times, an equality is specified within a problem that is true only for that problem. It might be the cost of one kilogram of oranges or the speed of a car in kilometers per hour. Such equalities are easy to miss when you first read a problem. Let's see how conversion factors are written from statements made within a problem.

1. The motorcycle was traveling at a speed of 85 kilometers per hour.

   Equality: 1 hr = 85 km

   Conversion factors: $\dfrac{85 \text{ km}}{1 \text{ hour}}$ and $\dfrac{1 \text{ hour}}{85 \text{ km}}$

2. One tablet contains 500 mg of vitamin C.

   Equality: 1 tablet = 500 mg vitamin C

   Conversion factors: $\dfrac{500 \text{ mg vitamin C}}{1 \text{ tablet}}$ and $\dfrac{1 \text{ tablet}}{500 \text{ mg vitamin C}}$

EXPLORE YOUR WORLD

■ SI and Metric Equalities on Product Labels

Read the labels on some food products on your kitchen shelves and in your refrigerator or use the labels in Figure 1.10. List the amount of product given in different units. Write a relationship for two of the amounts for the same product and container. Look for measurements of grams and pounds or quarts and milliliters.

QUESTIONS

1. Use the stated measurement to derive a metric–U.S. conversion factor.
2. How do your results compare to the conversion factors we have described in this text?

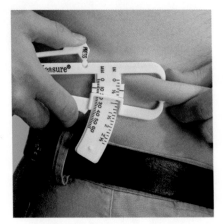

FIGURE 1.11

The thickness of the skin fold at the waist measured in millimeters (mm) is used to determine the amount of body fat.

Q **What is the percent body fat of an athlete with a body mass of 100 kg and 16 kg body fat?**

MC
GOB

Converting Your Medicine Cabinet

Conversion Factors from a Percentage

Sometimes a percentage is given in a problem. The term percent (%) means parts per 100 parts. To write a percentage as a conversion factor, we choose a unit and express the numerical relationship of the parts of this unit to 100 parts of the whole. For example, an athlete might have 18% (percent) body fat by mass. (See Figure 1.11.) The percent quantity can be written as 18 mass units of body fat in every 100 mass units of body mass. Different mass units such as grams, kilograms (kg), or pounds (lb) can be used, but both units in the factor must be the same.

Percent quantity:	18% body fat by mass
Equality:	18 kg body fat = 100 kg body mass
Conversion factors:	$\dfrac{100 \text{ kg body mass}}{18 \text{ kg body fat}}$ and $\dfrac{18 \text{ kg body fat}}{100 \text{ kg body mass}}$

or

Equality:	18 lb body fat = 100 lb body mass
Conversion factors:	$\dfrac{100 \text{ lb body mass}}{18 \text{ lb body fat}}$ and $\dfrac{18 \text{ lb body fat}}{100 \text{ lb body mass}}$

SAMPLE PROBLEM 1.10

■ Conversion Factors Stated in a Problem

Write the conversion factors for each of the following statements:

a. There are 325 mg of aspirin in 1 tablet.
b. One kilogram of bananas costs $1.25 at the grocery store.

SOLUTION

a. $\dfrac{325 \text{ mg aspirin}}{1 \text{ tablet}}$ and $\dfrac{1 \text{ tablet}}{325 \text{ mg aspirin}}$

b. $\dfrac{\$1.25}{1 \text{ kg bananas}}$ and $\dfrac{1 \text{ kg bananas}}{\$1.25}$

STUDY CHECK

What conversion factors can be written for the following statements?

a. A cyclist in the Tour de France bicycle race rides at the average speed of 62.2 km/hr.
b. A 100. g sample of silver has a volume of 9.48 cm^3.

QUESTIONS AND PROBLEMS

MC
GOB

■ Writing Conversion Factors

1.45 Why can two conversion factors be written for an equality such as 1 m = 100 cm?

1.46 How can you check that you have written the correct conversion factors for an equality?

1.47 What equality is expressed by the conversion factor $\dfrac{1000 \text{ g}}{1 \text{ kg}}$?

1.48 What equality is expressed by the conversion factor $\dfrac{1 \text{ m}}{100 \text{ cm}}$?

1.49 Write a numerical relationship and conversion factors for each of the following statements:
a. One yard is 3 feet.
b. One mile is 5280 feet.
c. One minute is 60 seconds.
d. A car goes 27 miles on 1 gallon of gas.
e. Sterling silver is 93% by mass silver.

1.50 Write a numerical relationship and conversion factors for each of the following statements:
a. One gallon is 4 quarts.
b. At the store, oranges are $1.29 per lb.
c. There are 7 days in 1 week.
d. One dollar has four quarters.
e. A ring contains 58% by mass gold.

1.51 Write the numerical relationship and conversion factors for the following pairs of units:
a. centimeters and meters
b. milligrams and grams
c. liters and milliliters
d. deciliters and milliliters

1.52 Write the numerical relationship and corresponding conversion factors for the following pairs of units:
a. centimeters and inches
b. pounds and kilograms
c. pounds and grams
d. quarts and milliliters

1.7 Problem Solving

LEARNING GOAL
Use conversion factors to change from one unit to another.

The process of problem solving in chemistry often requires the conversion of an initial quantity given in one unit to the same quantity but in different units. By using one or more of the conversion factors we discussed in the previous section, the initial unit can be converted to the final unit.

Given quantity × One or more conversion factors = Desired quantity
(Initial unit) ——————————————→ (Final unit)

You may use a sequence similar to the steps in the following guide for problem solving (GPS):

Guide to Problem Solving (GPS) with Conversion Factors

■ STEP 1 **Given/Need** State the initial unit given in the problem and the final unit needed.

■ STEP 2 **Plan** Write out a sequence of units that starts with the initial unit and progresses to the final unit for the answer. Be sure you can supply the equality for each unit conversion.

■ STEP 3 **Equalities/Conversion Factors** For each change of unit in your plan, state the equality and corresponding conversion factors. Recall that equalities are derived from the metric (SI) system, the U.S. system, and statements within a problem.

■ STEP 4 **Set Up Problem** Write the initial quantity and unit and set up conversion factors that connect the units. Be sure to arrange the units in each factor so the unit in the denominator cancels the preceding unit in the numerator. Check that the units cancel properly to give the final unit. Carry out the calculations, count the significant figures in each measured number, and give a final answer with the correct number of significant figures.

Suppose a problem requires the conversion of 164 lb to kilograms. One part of this statement (164 lb) is the given quantity (initial unit), while another part (kilograms) is the final unit needed for the answer. Once you identify these units, you can determine which equalities you need to convert the initial unit to the final unit.

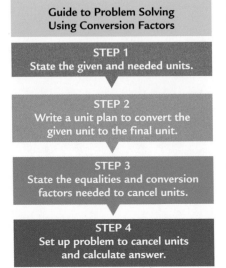

Guide to Problem Solving Using Conversion Factors

STEP 1
State the given and needed units.

STEP 2
Write a unit plan to convert the given unit to the final unit.

STEP 3
State the equalities and conversion factors needed to cancel units.

STEP 4
Set up problem to cancel units and calculate answer.

■ STEP 1 **Given** 164 lb **Need** kg

■ STEP 2 **Plan** It is helpful to decide on a plan of units. When we look at the initial units given and the final units needed, we see that one is a metric unit and the other is a unit in the U.S. system of measurement. Therefore, the connecting conversion factor must be one that includes a metric and a U.S. unit.

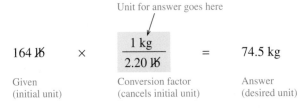

lb   Metric–U.S. factor   kg

■ STEP 3 **Equalities/Conversion Factors** From the discussion on U.S. and metric equalities, we can write the following equality and conversion factors:

$$1 \text{ kg} = 2.20 \text{ lb}$$

$$\frac{2.20 \text{ lb}}{1 \text{ kg}} \quad \text{and} \quad \frac{1 \text{ kg}}{2.20 \text{ lb}}$$

■ STEP 4 **Set Up Problem** Now we can write the setup to solve the problem using the unit plan and a conversion factor. First, write down the initial unit, 164 lb. Then multiply by the conversion factor that has the unit lb in the denominator to cancel out the initial unit. The unit kg in the numerator (top number) gives the final unit for the answer.

Unit for answer goes here

$$164 \text{ lb} \quad \times \quad \frac{1 \text{ kg}}{2.20 \text{ lb}} \quad = \quad 74.5 \text{ kg}$$

Given (initial unit)   Conversion factor (cancels initial unit)   Answer (desired unit)

Take a look at how the units cancel. The unit that you want in the answer is the one that remains after all the other units have cancelled out. This is a helpful way to check that a problem is set up properly.

$$\text{lb} \times \frac{\text{kg}}{\text{lb}} = \text{kg} \qquad \text{Unit needed for answer}$$

The calculation done on a calculator gives the numerical part of the answer. The calculator answer is adjusted to give a final answer with the proper number of significant figures (SFs).

$$164 \times \frac{1}{2.20} = 164 \;\boxed{\div}\; 2.20 = \quad 74.54545455 = 74.5$$

3 SF   3 SF   Calculator answer   3 SF (rounded)

The value of 74.5 combined with the final unit, kg, gives the final answer of 74.5 kg. With few exceptions, answers to numerical problems contain a number and a unit.

SAMPLE PROBLEM 1.11

■ **Problem Solving Using Metric Factors**

The recommended amount of sodium in the diet per day is 2400 mg. How many grams of sodium is that?

SOLUTION

■ **STEP 1**   **Given**   2400 mg   **Need**   g

■ **STEP 2**   **Plan**   When we look at the initial units given and the final units needed, we see that both are metric units. Therefore, the connecting conversion factor must relate two metric units.

> mg   Metric factor   g

■ **STEP 3**   **Equalities/Conversion Factors**   From the discussion on prefixes and metric equalities, we can write the following equality and conversion factors:

$$1 \text{ g} = 1000 \text{ mg}$$

$$\frac{1 \text{ g}}{1000 \text{ mg}} \quad \text{and} \quad \frac{1000 \text{ mg}}{1 \text{ g}}$$

■ **STEP 4**   **Set Up Problem**   We write the setup using the unit plan and a conversion factor starting with the initial unit, 2400 mg. The final answer (g) is obtained by using the conversion factor that cancels the unit mg.

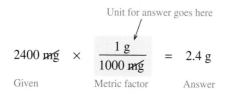

$$2400 \text{ m\cancel{g}} \times \frac{1 \text{ g}}{1000 \text{ m\cancel{g}}} = 2.4 \text{ g}$$

Given   Metric factor   Answer

STUDY CHECK

If 1890 mL of orange juice is prepared from orange juice concentrate, how many liters of orange juice is that?

Using Two or More Conversion Factors

In many problems, two or more conversion factors are needed to complete the change of units. In setting up these problems, one factor follows the other. Each factor is arranged to cancel the preceding unit until the final unit is obtained. Up to this point, we have used the conversion factors one at a time and calculated an answer. You can work all problems in single steps, but if you do, be sure to keep one or two extra digits in the intermediate answers and only round off the final answer to the correct number of significant figures. A more efficient way to do these problems is to use a series of two or more conversion factors set up so that the unit in the denominator of each factor cancels the unit in the preceding numerator. Both these approaches are illustrated in the following sample problem.

<hr>

SAMPLE PROBLEM 1.12

▪ Problem Solving Using Two Factors

A recipe for salsa requires 3 cups of tomato sauce, which can be measured accurately as 3.0 cups. If only metric measures are available, how many liters of tomato sauce are needed? (There are 4 cups in 1 quart.)

SOLUTION

- ▪ **STEP 1 Given**   3.0 cups tomato sauce   **Need**   liters (L)
- ▪ **STEP 2 Plan**   We see that the initial unit, cups, needs to be changed to a final unit of liters, but we do not know an equality for cups and liters. However, we do know the U.S. equality for cups and quarts and the metric–U.S. equality for quarts and liters as shown in the following plan:

$$\text{cups} \quad \boxed{\text{U.S. factor}} \quad \text{quarts} \quad \boxed{\substack{\text{Metric–U.S.} \\ \text{factor}}} \quad \text{liters}$$

- ▪ **STEP 3 Equalities/Conversion Factors**   From the discussion on U.S. and metric equalities, we can write the following equalities and conversion factors:

$$
\begin{array}{c}
1 \text{ qt} = 4 \text{ cups} \\[4pt]
\dfrac{1 \text{ qt}}{4 \text{ cups}} \quad \text{and} \quad \dfrac{4 \text{ cups}}{1 \text{ qt}}
\end{array}
\qquad
\begin{array}{c}
1 \text{ L} = 1.06 \text{ qt} \\[4pt]
\dfrac{1 \text{ L}}{1.06 \text{ qt}} \quad \text{and} \quad \dfrac{1.06 \text{ qt}}{1 \text{ L}}
\end{array}
$$

- ▪ **STEP 4 Set Up Problem**   Working in single steps we can use the U.S. factor to convert from cups to quarts:

$$3.0 \, \cancel{\text{cups}} \times \frac{1 \text{ qt}}{4 \, \cancel{\text{cups}}} = 0.75 \text{ qt}$$

Then we use the metric–U.S. factor to cancel quarts and give liters as the unit.

$$0.75 \, \cancel{\text{qt}} \times \frac{1 \text{ L}}{1.06 \, \cancel{\text{qt}}} = 0.71 \text{ L}$$

When set up as a series, the first factor cancels cups, and the second factor cancels quarts, which gives liters as the final unit for the answer.

$$\cancel{\text{cups}} \times \frac{\cancel{\text{qt}}}{\cancel{\text{cups}}} \times \frac{\text{L}}{\cancel{\text{qt}}} = \text{L}$$

$$\underset{\substack{\text{Given} \\ \text{(initial unit)}}}{3.0 \, \cancel{\text{cups}}} \times \underset{\substack{\text{U.S.} \\ \text{factor}}}{\frac{1 \, \cancel{\text{qt}}}{4 \, \cancel{\text{cups}}}} \times \underset{\substack{\text{Metric-U.S.} \\ \text{factor}}}{\frac{1 \text{ L}}{1.06 \, \cancel{\text{qt}}}} = \underset{\substack{\text{Answer} \\ \text{(desired unit)}}}{0.71 \text{ L}}$$

The calculation is done in a sequence on a calculator to give the numerical part of the answer. The calculator answer is adjusted to give a final answer with the proper number of significant figures (SFs).

$$\underset{\text{2 SFs}}{3.0} \; \boxed{\div} \; \underset{\text{Exact}}{4} \; \boxed{\div} \; \underset{\text{3 SFs}}{1.06} \; = \; \boxed{\underset{\text{Calculator answer}}{0.707547}} \; \underset{\text{2 SFs (rounded)}}{= 0.71}$$

STUDY CHECK

One medium bran muffin contains 4.2 g of fiber. How many ounces (oz) of fiber are obtained by eating three medium bran muffins if 1 lb = 16 oz? (*Hint*: number of muffins ⟶ g of fiber ⟶ lb ⟶ oz.)

Using a sequence of two or more conversion factors is a very efficient way to set up and solve problems, especially if you are using a calculator. Once you have the problem set up, the calculations can be done without writing out the intermediate values. This process is worth practicing until you understand unit cancellation and the mathematical calculations.

Clinical Calculations Using Conversion Factors

Conversion factors are also useful for calculating medications. For example, if an antibiotic is available in 5-mg tablets, the dosage can be written as a conversion factor, 5 mg/1 tablet. In many hospitals, the apothecary unit of grain (gr) is still in use; there are 65 mg in 1 gr. When you do a medication problem, you often start with a doctor's order that contains the quantity to give the patient. The medication dosage is used as a conversion factor.

SAMPLE PROBLEM 1.13

Clinical Factors from a Word Problem

Synthroid is used as a replacement or supplemental therapy for diminished thyroid function. A dosage of 0.200 mg is prescribed with tablets that contain 50 μg Synthroid. How many tablets are required to provide the prescribed medication?

SOLUTION

■ **STEP 1 Given** 0.200 mg Synthroid   **Need** tablets

■ **STEP 2 Plan**   mg → Metric factor → μg → Clinical factor → tablets

■ **STEP 3 Equalities/Conversion Factors**   In the problem, the information for the dosage is given as 50 μg per tablet. We will use this as one of the equalities as well as the metric equality for milligrams and micrograms and write conversion factors for each.

$$1 \text{ mg} = 1000 \ \mu\text{g}$$
$$\frac{1 \text{ mg}}{1000 \ \mu\text{g}} \quad \text{and} \quad \frac{1000 \ \mu\text{g}}{1 \text{ mg}}$$

$$1 \text{ tablet} = 50 \ \mu\text{g}$$
$$\frac{1 \text{ tablet}}{50 \ \mu\text{g}} \quad \text{and} \quad \frac{50 \ \mu\text{g}}{1 \text{ tablet}}$$

■ **STEP 4 Set Up Problem**   The problem can be set up using the metric factor to cancel "milligrams," and then the clinical factor to obtain "tablets" as the final unit.

$$0.200 \ \text{mg} \times \frac{1000 \ \mu\text{g}}{1 \ \text{mg}} \times \frac{1 \text{ tablet}}{50 \ \mu\text{g}} = 4 \text{ tablets}$$

STUDY CHECK

An antibiotic dosage of 500 mg is ordered. If the antibiotic is supplied in liquid form as 250 mg in 5.0 mL, how many mL would be given?

SAMPLE PROBLEM 1.14

■ Using a Percent as a Conversion Factor

Bronze is 80.0% by mass copper and 20.0% by mass tin. A sculptor is preparing to cast a figure that requires 1.75 lb bronze. How many g of copper are needed for the bronze figure?

SOLUTION

■ **STEP 1 Given** 1.75 lb bronze **Need** g copper

■ **STEP 2 Plan**

lb bronze → Metric–U.S. factor → g bronze → Percent factor → g copper

■ **STEP 3 Equalities/Conversion Factors** Now we can write the equalities and conversion factors. One is the U.S.–metric factor for g and lb. The second is the percent factor derived from the information given in the problem.

$$\text{1 lb bronze} = 454 \text{ g}$$
$$\frac{454 \text{ g bronze}}{1 \text{ lb}} \quad \text{and} \quad \frac{1 \text{ lb}}{454 \text{ g bronze}}$$

$$\text{100 g bronze} = 80.0 \text{ g copper}$$
$$\frac{80.0 \text{ g copper}}{100 \text{ g bronze}} \quad \text{and} \quad \frac{100 \text{ g bronze}}{80.0 \text{ g copper}}$$

■ **STEP 4 Set Up Problem** We can set up the problem using conversion factors to cancel each unit, starting with lb bronze, until we obtain the final factor, g copper, in the numerator. After we count the significant figures in the measured quantities, we write the final answer with three significant figures.

$$1.75 \text{ lb bronze} \times \frac{454 \text{ g bronze}}{1 \text{ lb bronze}} \times \frac{80.0 \text{ g copper}}{100 \text{ g bronze}} = 636 \text{ g copper}$$

3 SF   3 SF   3 SF   3 SF

STUDY CHECK

A lean hamburger is 22% fat by weight. How many grams of fat are in 0.25 lb of the hamburger?

QUESTIONS AND PROBLEMS

■ Problem Solving

1.53 When you convert one unit to another, how do you know which unit of the conversion factor to place in the denominator?

1.54 When you convert one unit to another, how do you know which unit of the conversion factor to place in the numerator?

1.55 Use metric conversion factors to solve the following problems.
   a. The height of a student is 175 centimeters. How tall is the student in meters?
   b. A cooler has a volume of 5500 mL. What is the capacity of the cooler in liters?
   c. A hummingbird has a mass of 0.0055 kg. What is the mass of the hummingbird in grams?

1.56 Use metric conversion factors to solve the following problems.
   a. The daily requirement of phosphorus is 800 mg. How many grams of phosphorus are recommended?

(MC GOB) Using Percentage as a Conversion Factor

b. A glass of orange juice contains 0.85 dL of juice. How many milliliters of orange juice is that?

c. A package of chocolate instant pudding contains 2840 mg of sodium. How many grams of sodium is that?

1.57 Solve the following problems using one or more conversion factors.

a. A container holds 0.750 qt of liquid. How many milliliters of lemonade will it hold?

b. In England, a person is weighed in *stones*. If one stone has a weight of 14.0 lb, what is the mass in kilograms of a person who weighs 11.8 stones?

c. The femur, or thighbone, is the longest bone in the body. In a 6-ft-tall person, the femur is 19.5 in. long. What is the length of that femur in millimeters?

d. How many inches thick is an arterial wall that measures 0.50 μm?

1.58 Solve the following problems using one or more conversion factors.

a. You need 4.0 ounces of a steroid ointment. If there are 16 oz in 1 lb, how many grams of ointment does the pharmacist need to prepare?

b. During surgery, a person receives 5.0 pints of plasma. How many milliliters of plasma were given? (1 quart = 2 pints)

c. Solar flares containing hot gases can rise to 120 000 miles above the surface of the sun. What is that distance in kilometers?

d. A filled gas tank contains 18.5 gallons of unleaded fuel. If a car uses 46 L, how many gallons of fuel remain in the tank?

1.59 The singles portion of a tennis court is 27.0 ft wide and 78.0 ft long.

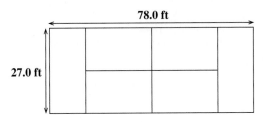

a. What is the length of the court in meters?

b. What is the area of the court in square meters (m^2)?

c. If a serve is measured at 185 km per hour, how many seconds does it take for the tennis ball to travel the length of the court?

1.60 A football field is 300 feet long between goal lines.

a. How many meters does a player run if he catches the ball on his own goal line and scores a touchdown?

b. If a player catches the football and runs 45 yards, how many meters did he gain?

c. If a player runs at a speed of 36 km/hr, how many seconds does it take to run from the 50-yard line to the 20-yard line?

1.61 Using conversion factors, solve the following clinical problems:

a. You have used 250 L of distilled water for a dialysis patient. How many gallons of water is that?

b. A patient needs 0.024 g of a sulfa drug. There are 8-mg tablets in stock. How many tablets should be given?

c. The daily dose of ampicillin for the treatment of an ear infection is 115 mg/kg of body weight. What is the daily dose for a 34-lb child?

1.62 Using conversion factors, solve the following clinical problems:
  a. The physician has ordered 1.0 g of tetracycline to be given every 6 hours to a patient. If your stock on hand is 500-mg tablets, how many will you need for 1 day's treatment?
  b. An intramuscular medication is given at 5.00 mg/kg of body weight. If you give 425 mg of medication to a patient, what is the patient's weight in pounds?
  c. A physician has ordered 325 mg of atropine, intramuscularly. If atropine were available as 0.50 g/mL of solution, how many milliliters would you need to give?

1.63 a. Oxygen makes up 46.7% by mass of the Earth's crust. How many grams of oxygen are present if a sample of the Earth's crust has a mass of 325 g?
  b. Magnesium makes up 2.1% by mass of the Earth's crust. How many grams of magnesium are present if a sample of the Earth's crust has a mass of 1.25 g?
  c. A plant fertilizer contains 15% by mass nitrogen (N). In a container of soluble plant food, there are 10.0 oz of fertilizer. How many grams of nitrogen are in the container?
  d. In a candy factory, the nutty chocolate bars contain 22.0% by mass pecans. If 5.0 kg of pecans were used for candy last Tuesday, how many lb of nutty chocolate bars were made?

1.64 a. Water is 11.2% by mass hydrogen. How many kilograms of water would contain 5.0 g of hydrogen?
  b. Water is 88.8% by mass oxygen. How many grams of water would contain 2.25 kg of oxygen?
  c. Blueberry fiber cakes contain 51% dietary fiber. If a package with a net weight of 12 ounces contains six cakes, how many grams of fiber are in each cake?
  d. A jar of crunchy peanut butter contains 1.43 kg of peanut butter. If you use 8.0% of the peanut butter for a sandwich, how many ounces of peanut butter did you take out of the container?

1.8 Density

Differences in density determine whether an object will sink or float. In Figure 1.12, the density of lead is greater than the density of water, and the lead object sinks. The cork floats because cork is less dense than water.

The mass and volume of any object can be measured. However, the separate measurements do not tell us how tightly packed the substance might be. If we compare the mass of the object to its volume, we obtain a relationship called **density**.

$$\text{Density} = \frac{\text{mass of substance}}{\text{volume of substance}}$$

FIGURE 1.12

Objects that sink in water are more dense than water; objects float if they are less dense.

Q **Why does a cork float and a piece of lead sink?**

Cork (Density = 0.26 g/mL)
Ice (Density = 0.92 g/mL)
H_2O (Density = 1.00 g/mL)
Aluminum (Density = 2.70 g/mL)
Lead (Density = 11.3 g/mL)

Densities of Some Common Substances

Solids (at 25°C)	Density (g/mL)	Liquids (at 25°C)	Density (g/mL)	Gases (at 0°C)	Density (g/L)
Cork	0.26	Gasoline	0.66	Hydrogen	0.090
Wood (maple)	0.75	Ethyl alcohol	0.79	Helium	0.179
Ice	0.92	Olive oil	0.92	Methane	0.714
Sugar	1.59	Water (at 4°C)	1.00	Neon	0.90
Bone	1.80	Plasma (blood)	1.03	Nitrogen	1.25
Aluminum	2.70	Milk	1.04	Air (dry)	1.29
Cement	3.00	Mercury	13.6	Oxygen	1.43
Diamond	3.52			Carbon dioxide	1.96
Silver	10.5				
Lead	11.3				
Gold	19.3				

In the metric system, the densities of solids and liquids are usually expressed as grams per cubic centimeter (g/cm^3) or grams per milliliter (g/mL). The density of gases is usually stated as grams per liter (g/L). Table 1.10 gives the densities of some common substances.

Density of Solids

The density of a solid is calculated from its mass and volume. When a solid is completely submerged, it displaces a volume of water that is *equal to the volume of the solid.* In Figure 1.13, the water level rises from 35.5 mL to 45.0 mL. This means that 9.5 mL of water is displaced and that the volume of the object is 9.5 mL. The density of the zinc is calculated as follows:

$$\text{Density} = \frac{68.60 \text{ g zinc}}{9.5 \text{ mL}} = 7.2 \text{ g/mL}$$

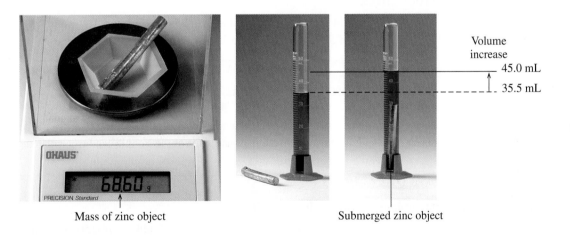

Mass of zinc object　　　　　　Submerged zinc object

FIGURE 1.13

The density of a solid can be determined by volume displacement because a submerged object displaces a volume of water equal to its own volume.

Q **How is the volume of the zinc object determined?**

Volume increase
45.0 mL
35.5 mL

■ Using Volume Displacement to Calculate Density

A lead weight used in the belt of a scuba diver has a mass of 226 g. When the weight is placed in a graduated cylinder containing 200.0 mL of water, the water level rises to 220.0 mL. What is the density of the lead weight (g/mL)?

SOLUTION

■ **STEP 1** **Given** mass = 226 g; water level before object submerged = 200.0 mL; water level after object submerged = 220.0 mL
  Need density (g/mL)

■ **STEP 2** **Plan** To calculate density, substitute the mass (g) and the volume (mL) of the lead weight into the expression for density.

■ **STEP 3** **Equality/Conversion Factor**

$$\text{Density} = \frac{\text{mass of substance}}{\text{volume of substance}}$$

Using Density as a Conversion Factor

■ **STEP 4** **Set Up Problem** The volume of the lead weight is equal to the volume of water displaced, which is calculated as follows:

Water level after object submerged	= 220.0 mL
− Water level before object submerged	= 200.0 mL
Water displaced (volume of lead weight) =	20.0 mL

The density is calculated by dividing the mass (g) by the volume (mL). Be sure to use the volume of water the object displaced and *not* the original volume of water.

$$\text{Density} = \frac{226 \text{ g}}{20.0 \text{ mL}} = \frac{11.3 \text{ g}}{1 \text{ mL}} = 11.3 \text{ g/mL}$$

$\qquad\qquad\qquad$ 3 SF $\qquad\qquad\qquad$ 3 SF

STUDY CHECK

A total of 0.50 lb of glass marbles is added to 425 mL of water. The water level rises to a volume of 528 mL. What is the density (g/mL) of the glass marbles?

Problem Solving Using Density

Density can be used as a conversion factor. For example, if the volume and the density of a sample are known, the mass in grams of the sample can be calculated.

■ Problem Solving Using Density

If the density of milk is 1.04 g/mL, how many grams of milk are in 0.50 qt of milk?

SOLUTION

- **STEP 1 Given** 0.50 qt **Need** g
- **STEP 2 Plan**

$$qt \xrightarrow{\text{U.S.–metric factor}} L \xrightarrow{\text{Metric factor}} mL \xrightarrow{\text{Density factor}} g$$

- **STEP 3 Equalities/Conversion Factors**

$$1 \text{ L} = 1.06 \text{ qt}$$
$$\frac{1 \text{ L}}{1.06 \text{ qt}} \text{ and } \frac{1.06 \text{ qt}}{1 \text{ L}}$$

$$1 \text{ L} = 1000 \text{ mL}$$
$$\frac{1 \text{ L}}{1000 \text{ mL}} \text{ and } \frac{1000 \text{ mL}}{1 \text{ L}}$$

$$1 \text{ mL} = 1.04 \text{ g}$$
$$\frac{1 \text{ mL}}{1.04 \text{ g}} \text{ and } \frac{1.04 \text{ g}}{1 \text{ mL}}$$

- **STEP 4 Set Up Problem**

$$\underset{\text{2 SF}}{0.50 \text{ qt}} \times \underset{\text{3 SF}}{\frac{1 \text{ L}}{1.06 \text{ qt}}} \times \underset{\text{Exact}}{\frac{1000 \text{ mL}}{1 \text{ L}}} \times \underset{\text{3 SF}}{\frac{1.04 \text{ g}}{1 \text{ mL}}} = \underset{\text{2 SF}}{490 \text{ g } (4.9 \times 10^2 \text{ g})}$$

STUDY CHECK

How many mL of mercury are in a thermometer that contains 20.4 g of mercury? (See Table 1.10 for the density of mercury.)

Specific Gravity

Specific gravity (sp gr) is a ratio between the density of a substance and the density of water. Specific gravity is calculated by dividing the density of a sample by the density of water. In this text, we will use the metric value of 1.00 g/mL for the density of water. A substance with a specific gravity of 1.00 has the same density as water. A substance with a specific gravity of 3.00 is three times as dense as water, whereas a substance with a specific gravity of 0.50 is just one-half as dense as water.

$$\text{Specific gravity} = \frac{\text{density of sample}}{\text{density of water}}$$

In the calculations for specific gravity, the units of density must match. Then all units cancel to leave only a number. *Specific gravity is one of the few unitless values you will encounter in chemistry.*

An instrument called a hydrometer is often used to measure the specific gravity of fluids such as battery fluid or a sample of urine. In Figure 1.14, a hydrometer is used to measure the specific gravity of a fluid.

SAMPLE PROBLEM 1.17

- **Problem Solving with Specific Gravity**

John took 2.0 teaspoons (tsp) of cough syrup (sp gr 1.20) for a persistent cough. If there is 5.0 mL in 1 tsp, what was the mass (in grams) of the cough syrup?

EXPLORE YOUR WORLD

- **Sink or Float?**

1. Fill a large container or bucket with water. Place a can of diet and a can of nondiet soft drink in the water. What happens? Using information on the label, how might you account for your observations?
2. Design an experiment to determine the substance that is the most dense in each of the following:
   a. water and vegetable oil
   b. water and ice
   c. rubbing alcohol and ice
   d. vegetable oil, water, and ice

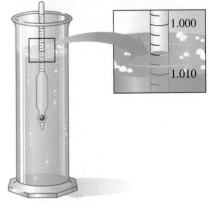

FIGURE 1.14

When the specific gravity of beer measures 1.010 or less with a hydrometer, the fermentation process is complete.

Q **If the hydrometer reading is 1.006, what is the density of the liquid?**

■ **Determination of Percentage of Body Fat**

Body mass is made up of protoplasm, extracellular fluid, bone, and adipose tissue. One way to determine the amount of adipose tissue is to measure the whole-body density. After the on-land mass of the body is determined, the underwater body mass is obtained by submerging the person in water. Because water helps support the body by giving it buoyancy, the underwater body mass is less. A higher percentage of body fat will make a person more buoyant, causing the underwater mass to be even lower. This occurs because fat has a lower density than the rest of the body.

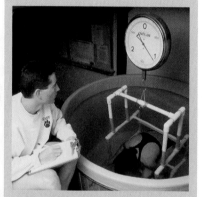

The difference between the on-land mass and underwater mass, known as the buoyant force, is used to determine the body volume. Then the mass and volume of the person are used to calculate body density. For example, suppose a 70.0-kg person has a body volume of 66.7 L. The body density is calculated as

$$\frac{\text{Body mass}}{\text{Body volume}} = \frac{70.0 \text{ kg}}{66.7 \text{ L}}$$
$$= 1.05 \text{ kg/L or } 1.05 \text{ g/mL}$$

When the body density is determined, it is compared with a chart that correlates the percentage of adipose tissue with body density. A person with a body density of 1.05 g/mL has 21% body fat, according to such a chart. This procedure is used by athletes to determine exercise and diet programs.

SOLUTION

■ **STEP 1** **Given**   2.0 tsp   **Need**   grams
■ **STEP 2** **Plan**

tsp → [U.S.–metric factor] → mL → [Density factor] → g

■ **STEP 3** **Equalities/Conversion Factors**   For problem solving, it is convenient to convert the specific gravity value (1.20) to density

$$\text{Density} = (\text{sp gr}) \times 1.00 \text{ g/mL} = 1.20 \text{ g/mL}$$

1 tsp = 5.0 mL		1 mL = 1.20 g	
$\dfrac{5.0 \text{ mL}}{1 \text{ tsp}}$ and $\dfrac{1 \text{ tsp}}{5.0 \text{ mL}}$		$\dfrac{1 \text{ mL}}{1.20 \text{ g}}$ and $\dfrac{1.20 \text{ g}}{1 \text{ mL}}$	

■ **STEP 4** **Set Up Problem**

$$2.0 \text{ tsp} \times \frac{5.0 \text{ mL}}{1 \text{ tsp}} \times \frac{1.20 \text{ g}}{1 \text{ mL}} = 12 \text{ g syrup}$$

STUDY CHECK

An ebony carving has a mass of 275 g. If ebony has a specific gravity of 1.33, what is the volume of the carving?

QUESTIONS AND PROBLEMS

■ **Density**

1.65 In an old trunk, you find a piece of metal that you think may be aluminum, silver, or lead. In lab you find it has a mass of 217 g and a volume of 19.2 cm³. Using Table 1.10, what is the metal you found?

1.66 Suppose you have two 100-mL graduated cylinders. In each cylinder there is 40.0 mL of water. You also have two cubes: one is lead and the other is aluminum. Each cube measures 2.0 cm on each side. After you carefully lower each cube into the water of its own cylinder, what will the new water level be in each of the cylinders?

1.67 Determine the density (g/mL) for each of the following.
  a. A 20.0 mL sample of a salt solution that has a mass of 24.0 g.
  b. A cube of butter that weighs 0.250 lb and has a volume of 130. mL.
  c. A gem has a mass of 45.0 g. When the gem is placed in a graduated cylinder containing 20.0 mL of water, the water level rises to 34.5 mL.
  d. A syrup added to an empty container with a mass of 115.25 g. When 0.100 pint of syrup is added, the total mass of the container and syrup is 182.48 g. (1 qt = 2 pts)

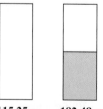

115.25 g   **182.48 g**

1.68 Determine the density (g/mL) for each of the following.
  a. A plastic material that weighs 2.68 lb and has a volume of 3.5 L.
  b. The fluid in a car battery, if it has a volume of 125 mL and a mass of 155 g.
  c. A 5.00-mL urine sample from a patient suffering from diabetes mellitus that has a mass of 5.025 g.
  d. A 10.00 L sample of oxygen gas that has a mass of 0.014 kg.

1.69 Use the density values in Table 1.10 to solve the following problems.
   a. How many liters of ethanol contain 1.5 kg of alcohol?
   b. How many grams of mercury are present in a barometer that holds 6.5 mL of mercury?
   c. A sculptor has prepared a mold for casting a bronze figure. The figure has a volume of 225 mL. If bronze has a density of 7.8 g/mL, how many ounces of bronze are needed in the preparation of the bronze figure?
   d. How many kilograms of gasoline fill a 12.0-gallon gas tank? (1 gallon = 4 qt)

1.70 Use the density values in Table 1.10 to solve the following problems.
   a. A graduated cylinder contains 28.0 mL of water. What is the new water level after 35.6 g of silver metal is submerged in the water?
   b. A fish tank holds 35 gallons of water. How many pounds (lb) of water are in the fish tank?
   c. The mass of an empty container is 88.25 g. The mass of the container and a liquid with a density of 0.758 g/mL is 150.50 g. What is the volume (mL) of the liquid in the container?
   d. A cannon ball made of iron has a volume of 115 cm^3. If iron has a density of 7.86 g/cm^3, what is the mass in kilograms of the cannon ball?

1.71 Solve the following specific gravity problems:
   a. A urine sample has a density of 1.030 g/mL. What is the specific gravity of the sample?
   b. A liquid has a volume of 40.0 mL and a mass of 45.0 g. What is the specific gravity of the liquid?
   c. The specific gravity of a vegetable oil is 0.85. What is its density?

1.72 Solve the following specific gravity problems:
   a. A 5.0% glucose solution has a specific gravity of 1.02. What is the mass of 500. mL of glucose solution?
   b. A bottle containing 325 g of cleaning solution is used for carpets. If the cleaning solution has a specific gravity of 0.850, what volume of solution was used?
   c. Butter has a specific gravity of 0.86. What is the mass, in grams, of 2.15 L of butter?

1.9 Temperature

Temperature is a measure of how hot or cold a substance is compared to another substance. Heat always flows from a substance with a higher temperature to a substance with a lower temperature until the temperatures of both are the same. When you drink hot coffee or touch a hot pan, heat flows to your mouth or hand, which is at a lower temperature. When you touch an ice cube, it feels cold because heat flows from your hand to the colder ice cube.

Celsius and Fahrenheit Temperatures

Temperatures in science, and in most of the world, are measured and reported in *Celsius* (°C) units. In the United States, everyday temperatures are commonly reported in *Fahrenheit* (°F) units. A typical room temperature of 22°C would be the same as 72°F. A normal body temperature of 37.0°C is 98.6°F.

Convert Between Temperature Scales

   On the Celsius and Fahrenheit scales, the temperatures of melting ice and boiling water are used as reference points. On the Celsius scale, the freezing point of water is defined as 0°C, and the boiling point as 100°C. On a Fahrenheit scale, water freezes at 32°F and boils at 212°F. On each scale, the temperature difference between freezing and boiling is divided into smaller units, or degrees. On the

FIGURE 1.15

A comparison of the Fahrenheit, Celsius, and Kelvin temperature scales between the freezing and boiling points of water.

Q What is the difference in the values for freezing on the Celsius and Fahrenheit temperature scales?

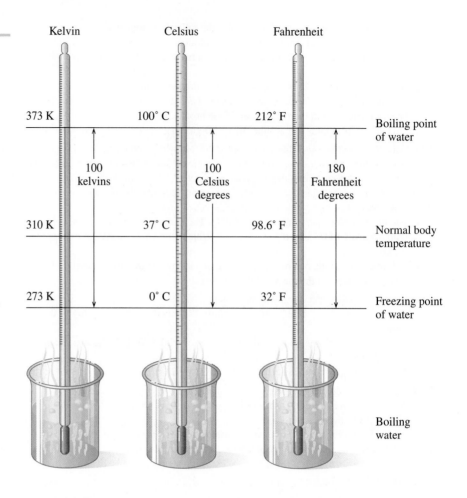

Celsius scale, there are 100 units between the freezing and boiling of water compared with 180 units on the Fahrenheit scale. (See Figure 1.15.)

$$180 \text{ Fahrenheit degrees} = 100 \text{ Celsius degrees}$$

$$\frac{180 \text{ Fahrenheit degrees}}{100 \text{ Celsius degrees}} = \frac{1.8\,^{\circ}\text{F}}{1\,^{\circ}\text{C}}$$

To convert to a Fahrenheit temperature, the Celsius temperature is multiplied by 1.8, and then 32 degrees is added. The 32 degrees adjusts the freezing point of 0°C on the Celsius scale to 32°F on the Fahrenheit scale. Both values, 1.8 and 32, are exact numbers. The equation for this conversion follows:

$$T_F = \frac{1.8\,^{\circ}\text{F}(T_C)}{1\,^{\circ}\text{C}} + 32^{\circ} \quad \text{or} \quad T_F = 1.8(T_C) + 32^{\circ}$$

$\underset{\text{Changes} \atop \text{°C to °F}}{} \qquad \underset{\text{Adjusts} \atop \text{freezing point}}{}$

SAMPLE PROBLEM 1.18

■**Converting Celsius to Fahrenheit**

While traveling in China, you discover that your temperature is 38.2°C. What is your temperature in Fahrenheit degrees?

SOLUTION

- **STEP 1** **Given**   38.2°C   **Need** T_F
- **STEP 2** **Plan**

$$T_C \quad \boxed{\text{Temperature equation}} \quad T_F$$

- **STEP 3** **Equality/Conversion Factor**

$$T_F = 1.8(T_C) + 32°$$

- **STEP 4** **Set Up Problem**   Substitute the Celsius temperature into the equation and solve.

$$T_F = 1.8(38.2) + 32°$$

$$T_F = 68.8° + 32° \qquad \text{1.8 is exact; 32 is exact}$$
$$\quad = 100.8°F \qquad \text{Answer to the first decimal place}$$

In the equation, *the values of 1.8 and 32 are exact numbers*. Therefore, the answer is reported to the same decimal place as the initial temperature.

STUDY CHECK

When making ice cream, rock salt is used to chill the mixture. If the temperature drops to $-11°C$, what is it in °F?

To convert from Fahrenheit to Celsius, the temperature equation is rearranged for T_C.

$$T_C = \frac{T_F - 32°}{1.8}$$

SAMPLE PROBLEM 1.19

■ **Converting Fahrenheit to Celsius**

You are going to cook a pizza at 325°F. What is that temperature in Celsius degrees?

SOLUTION

- **STEP 1** **Given**   325°F   **Need** T_C
- **STEP 2** **Plan**

$$T_F \quad \boxed{\text{Temperature equation}} \quad T_C$$

- **STEP 3** **Equality/Conversion Factor**

$$T_C = \frac{T_F - 32°}{1.8}$$

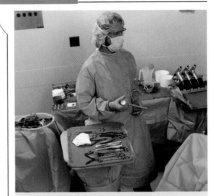

■ **STEP 4 Set Up Problem** To solve for T_C, substitute the Fahrenheit temperature into the equation and solve.

$$T_C = \frac{T_F - 32°}{1.8}$$

$$T_C = \frac{(325° - 32°)}{1.8} \qquad \text{32 is exact; 1.8 is exact}$$

$$= \frac{293°}{1.8} = 163°C \qquad \text{Answer to the one's place}$$

STUDY CHECK

A child has a temperature of 103.6°F. What is this temperature on a Celsius thermometer?

Kelvin Temperature Scale

Scientists tell us that the coldest temperature possible is −273°C (more precisely, −273.15°C). On the Kelvin scale, this temperature, called *absolute zero*, has the value of 0 Kelvin (0 K). Units on the Kelvin scale are called kelvins (K); no degree symbol is used. Because there are no lower temperatures, the Kelvin scale has no negative numbers. Between the freezing and boiling points of water, there are 100 kelvins, which makes a kelvin equal in size to a Celsius unit.

$$1\text{ K} = 1°C$$

To calculate a Kelvin temperature, add 273 to the Celsius temperature:

$$T_K = T_C + 273$$

Table 1.11 gives a comparison of some temperatures on the three scales.

TABLE 1.11

A Comparison of Temperatures

Example	Fahrenheit (°F)	Celsius (°C)	Kelvin (K)
Sun	9937	5503	5776
A hot oven	450	232	505
A desert	120	49	322
A high fever	104	40	313
Room temperature	72	22	295
Water freezes	32	0	273
A northern winter	−76	−60	213
Helium boils	−452	−269	4
Absolute zero	−459	−273	0

■ Variation in Body Temperature

Normal body temperature is considered to be 37.0°C, although it varies throughout the day and from person to person. Oral temperatures of 36.1°C are common in the morning and climb to a high of 37.2°C between 6 P.M. and 10 P.M. Temperatures above 37.2°C for a person at rest are usually an indication of disease. Individuals involved in prolonged exercise may also experience elevated temperatures. Body temperatures of marathon runners can range from 39°C to 41°C as heat production during exercise exceeds the body's ability to lose heat.

Changes of more than 3.5°C from the normal body temperature begin to interfere with bodily functions. Temperatures above 41°C can lead to convulsions, particularly in children, which may cause permanent brain damage. Heatstroke occurs above 41.1°C. Sweat production stops and the skin becomes hot and dry. The pulse rate is elevated, and respiration becomes weak and rapid. The person can become lethargic and lapse into a coma. Damage to internal organs is a major concern, and treatment, which must be immediate, may include immersing the person in an ice-water bath.

In hypothermia, body temperature can drop as low as 28.5°C. The person may appear cold and pale and have an irregular heartbeat. Unconsciousness can occur if the body temperature drops below 26.7°C. Respiration becomes slow and shallow, and oxygenation of the tissues decreases. Treatment involves providing oxygen and increasing blood volume with glucose and saline fluids. Internal temperature may be restored by injecting warm fluids (37.0°C) into the peritoneal cavity.

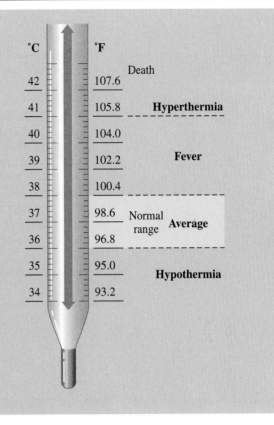

■ Converting from Celsius to Kelvin Temperature

In a cryogenic laboratory, a technician freezes a sample using liquid nitrogen at −196°C. What is the temperature in K?

SOLUTION

To find the Kelvin temperature, we use the equation:

$$T_K = T_C + 273$$
$$T_K = -196°\,C + 273$$
$$= 77\ K$$

STUDY CHECK

On the planet Mercury, the average night temperature is 13 K, and the average day temperature is 683 K. What are these temperatures in Celsius degrees?

QUESTIONS AND PROBLEMS

■ Temperature

1.73 Your friend who is visiting from France just took her temperature. When she reads 99.8, she becomes concerned that she is quite ill. How would you explain the temperature to your friend?

1.74 You have a friend who is using a recipe for flan from a Mexican cookbook. You notice that he set your oven temperature at 175°F. What would you advise him to do?

1.75 Solve the following temperature conversions:

    a. 37.0°C = _____ °F   **b.** 65.3°F = _____ °C

    c. −27°C = _____ K   **d.** 62°C = _____ K

    e. 114°F = _____ °C   **f.** 72°F = _____ K

1.76 Solve the following temperature conversions:

    a. 25°C = _____ °F   **b.** 155°C = _____ °F

    c. −25°F = _____ °C   **d.** 224 K = _____ °C

    e. 545 K = _____ °C   **f.** 875 K = _____ °F

1.77 a. A patient with heat stroke has a temperature of 106°F. What does this read on a Celsius thermometer?

    b. Because high fevers can cause convulsions in children, the doctor wants to be called if the child's temperature goes over 40.0°C. Should the doctor be called if a child has a temperature of 103°F?

1.78 a. Hot compresses are being prepared for the patient in room 32B. The water is heated to 145°F. What is the temperature of the hot water in °C?

    b. During extreme hypothermia, a young woman's temperature dropped to 20.6°C. What was her temperature on the Fahrenheit scale?

CONCEPT MAP

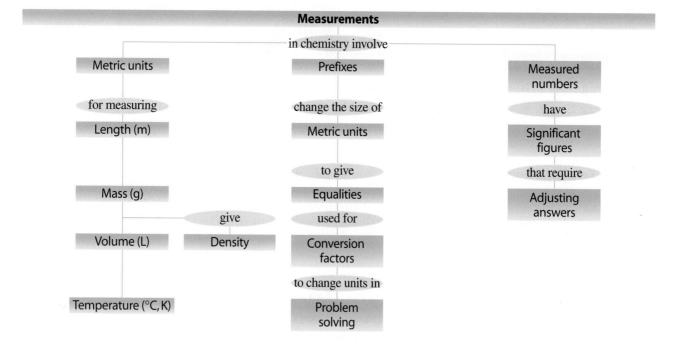

CHAPTER REVIEW

1.1 Scientific Method: Thinking like a Scientist

The scientific method is a process of explaining natural phenomena beginning with observations, hypothesis, and experiments, which may lead to a theory when experimental results support the hypothesis.

1.2 Measurement and Scientific Notation

In science, physical quantities are described in units of the metric or International System (SI). Some important units are meter (m) for length, liter (L) for volume, gram (g) and kilogram (kg) for mass, and Celsius (°C) for temperature. Large and small numbers can be written using scientific notation in which the decimal point is moved to give a coefficient between 1 and 10 and the number of decimal places moved shown as a power of 10. A large number will have a positive power of 10, while a small number will have a negative power of 10.

1.3 Measured Numbers and Significant Figures

A measured number is any number obtained by using a measuring device. An exact number is obtained by counting items or from a definition; no measuring device is used. Significant figures are the numbers reported in a measurement including the estimated digit. Zeros in front of a decimal number or at the end of a nondecimal number are not significant.

1.4 Significant Figures in Calculations

In multiplication or division, the final answer is written so it has the same number of significant figures as the measurement with the fewest significant figures. In addition or subtraction, the final answer is written so it has the same number of decimal places as the measurement with the fewest decimal places.

1.5 Prefixes and Equalities

Prefixes placed in front of a unit change the size of the unit by factors of 10. Prefixes such as *centi*, *milli*, and *micro* provide smaller units; prefixes such as *kilo* provide larger units. An equality relates two metric units that measure the same quantity of length, volume, or mass. Examples of metric equalities are 1 m = 100 cm; 1 L = 1000 mL; 1 kg = 1000 g.

1.6 Writing Conversion Factors

Conversion factors are used to express a relationship in the form of a fraction. Two factors can be written for any relationship in the metric or U.S. system.

1.7 Problem Solving

Conversion factors are useful when changing a quantity expressed in one unit to a quantity expressed in another unit. In the process, a given unit is multiplied by one or more conversion factors that cancel units until the desired answer is obtained.

1.8 Density

The density of a substance is a ratio of its mass to its volume, usually g/mL or g/cm^3. The units of density can be used as a factor to convert between the mass and volume of a substance. Specific gravity (sp gr) compares the density of a substance to the density of water, 1.00 g/mL.

1.9 Temperature

In science, temperature is measured in Celsius units, °C, or kelvins, K. In the United States, the Fahrenheit scale, °F, is still in use.

KEY TERMS

Celsius (°C) temperature scale A temperature scale on which water has a freezing point of 0°C and a boiling point of 100°C.

centimeter (cm) A unit of length in the metric system; there are 2.54 cm in 1 in.

conversion factor A ratio in which the numerator and denominator are quantities from an equality or given relationship. For example, the conversion factors for the relationship 1 kg = 2.20 lb are written as the following:

$$\frac{2.20 \text{ lb}}{1 \text{ kg}} \quad \text{and} \quad \frac{1 \text{ kg}}{2.20 \text{ lb}}$$

cubic centimeter (cm^3, cc) The volume of a cube that has 1-cm sides, equal to 1 mL.

density The relationship of the mass of an object to its volume expressed as grams per cubic centimeter (g/cm^3), grams per milliliter (g/mL), or grams per liter (g/L).

equality A relationship between two units that measure the same quantity.

exact number A number obtained by counting or definition.

experiment A procedure that tests the validity of a hypothesis.

gram (g) The metric unit used in measurements of mass.

hypothesis An unverified explanation of a natural phenomenon.

Kelvin (K) temperature scale A temperature scale on which the lowest possible temperature is 0 K.

kilogram (kg) A metric mass of 1000 g, equal to 2.20 lb. The kilogram is the SI standard unit of mass.

liter (L) The metric unit for volume that is slightly larger than a quart.

mass A measure of the quantity of material in an object.

measured number A number obtained when a quantity is determined by using a measuring device.

meter (m) The metric unit for length that is slightly longer than a yard. The meter is the SI standard unit of length.

metric system A system of measurement used by scientists and in most countries of the world.

milliliter (mL) A metric unit of volume equal to one-thousandth of a L (0.001 L).

observation Information determined by noting and recording a natural phenomenon.

prefix The part of the name of a metric unit that precedes the base unit and specifies the size of the measurement. All prefixes are related on a decimal scale.

scientific method The process of making observations, proposing a hypothesis, testing the hypothesis, and developing a theory that explains a natural event.

scientific notation A form of writing large and small numbers using a coefficient between 1 and 10, followed by a power of 10.

second The unit of time in the SI and metric system.

SI units The International System of units that modifies the metric system.

significant figures The numbers recorded in a measurement.

specific gravity (sp gr) A relationship between the density of a substance and the density of water:

$$\text{sp gr} = \frac{\text{density of sample}}{\text{density of water}}$$

temperature An indicator of the hotness or coldness of an object.

theory An explanation of an observation that has been validated by experiments that support a hypothesis.

volume The amount of space occupied by a substance.

UNDERSTANDING THE CONCEPTS

1.79 According to Sherlock Holmes, "One must follow the rules of scientific inquiry, gathering, observing, and testing data, then formulating, modifying, and rejecting hypotheses, until only one remains." Did Sherlock use the scientific method? Why or why not?

1.80 Read the temperature on each of the Celsius thermometers.

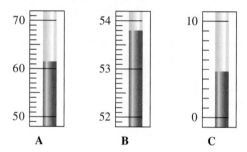

A   B   C

1.81 Indicate if each of the following is answered with an exact number or a measured number.

    a. Number of legs
    b. Height of table
    c. Number of chairs at the table
    d. Area of table top

1.82 Measure the length of each of the objects in figure **(a)**, **(b)**, and **(c)** using the metric rule in the figure. Indicate the number of significant figures for each and the estimated digit for each.

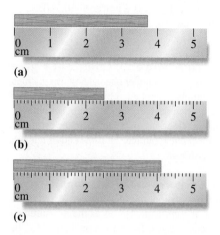

    (a)

    (b)

    (c)

1.83 Measure the length and width of the rectangle using a metric rule.

    a. What is the length and width of this rectangle measured in centimeters?
    b. What is the length and width of this rectangle measured in millimeters?
    c. How many significant figures are in the length measurement?
    d. How many significant figures are in the width measurement?
    e. What is the area of the rectangle in cm^2?
    f. How many significant figures are in the calculated answer for area?

1.84 Each of the following diagrams represents a container of water and a cube. Some cubes float while others sink. Match diagrams A, B, C, or D with one of the following descriptions and explain your choices.

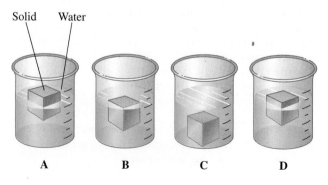

A   B   C   D

a. The cube has a greater density than water.
b. The cube has a density that is 0.60–0.80 g/mL.
c. The cube has a density that is $\frac{1}{2}$ the density of water.
d. The cube has the same density of water.

1.85 What is the density of the solid object that is weighed and submerged in water?

1.86 Consider the following solids.
The solids A, B, and C represent lead, copper, and aluminum. If each has a mass of 10.0 g, what is the identity of each solid?

Density aluminum = 2.70 g/mL
Density gold = 19.3 g/mL
Density silver = 10.5 g/mL

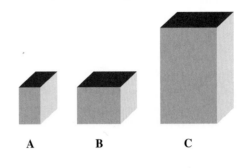

A   B   C

ADDITIONAL QUESTIONS AND PROBLEMS

(MC GOB) For instructor-assigned homework, go to **www.masteringgob.com.**

1.87 Why does the scientific method include a hypothesis?

1.88 Why is experimentation an important part of the scientific method?

1.89 Select the correct phrase(s) to complete the following statement: If experimental results do not support your hypothesis, you should
a. pretend that the experimental results do support your hypothesis.
b. write another hypothesis.
c. do more experiments.

1.90 Select the correct phrase(s) to complete the following statement: A hypothesis becomes a theory when
a. one experiment proves the hypothesis.
b. many experiments by many scientists validate the hypothesis.
c. you decide to call it a theory.

1.91 Classify each of the following statements as an observation, hypothesis, or theory.
a. Aluminum melts at 660°C.
b. Dinosaurs became extinct when a large meteorite struck Earth and caused a huge dust cloud that severely decreased the amount of light reaching Earth.
c. The 100-yard dash was run in 9.8 seconds.

1.92 Classify each of the following statements as an observation, hypothesis, or theory.
a. Analysis of ten ceramic dishes showed that four dishes contained lead levels that exceeded federal safety standards.
b. Marble statues undergo corrosion in acid rain.
c. Statues corrode in acid rain because the acidity is sufficient to dissolve calcium carbonate, the major substance of marble.

1.93 During a workout at the gym, you set the treadmill at a pace of 55.0 meters per minute. How many minutes will you walk if you cover a distance of 7500 feet?

1.94 A fish company delivers 22 kg of salmon, 5.5 kg of crab, and 3.48 kg of oysters to your seafood restaurant.
a. What is the total mass, in kilograms, of the seafood?
b. What is the total number of pounds?

1.95 Bill's recipe for onion soup calls for 4.0 lb of thinly sliced onions. If an onion has an average mass of 115 g, how many onions does Bill need?

1.96 The price of 1 pound (lb) of potatoes is $1.75. If all the potatoes sold today at the store bring in $1420, how many kilograms (kg) of potatoes did grocery shoppers buy?

1.97 The following nutrition information is listed on a box of crackers:
Serving size 0.50 oz (6 crackers)
Fat 4 g per serving **Sodium** 140 mg per serving
a. If the box has a net weight (contents only) of 8.0 oz, about how many crackers are in the box?
b. If you ate 10 crackers, how many ounces of fat are you consuming?
c. How many grams of sodium are used to prepare 50 boxes of crackers?

1.98 An aquarium store unit requires 75 000 mL of water. How many gallons of water are needed?
(1 gal = 4 qt)

1.99 In Mexico, avocados are 48 pesos per kilogram. What is the cost in cents of an avocado that weighs 0.45 lb if the exchange rate is 10.8 pesos to the dollar?

1.100 Celeste's diet restricts her intake of protein to 24 g per day. If she eats an 8.0-oz burger that is 15.0% protein, has she exceeded her protein limit for the day? How many ounces of a burger would be allowed for Celeste?

1.101 A sunscreen preparation contains 2.50% by mass benzyl salicylate. If a tube contains 4.0 ounces of sunscreen, how many kilograms of benzyl salicylate are needed to manufacture 325 tubes of sunscreen?

1.102 An object has a mass of 3.15 oz. When it is submerged in a graduated cylinder initially containing 325.2 mL of water, the water level rises to 442.5 mL. What is the density (g/mL) of the object?

1.103 What is a cholesterol level of 1.85 g/L in units of mg/dL?

1.104 If a recycling center collects 1254 aluminum cans and there are 22 aluminum cans in one pound, what volume in liters of aluminum was collected? (See Table 1.10.)

1.105 The water level in a graduated cylinder initially at 215 mL rises to 285 mL after a piece of lead is submerged. What is the mass in grams of the lead? (See Table 1.10.)

1.106 A graduated cylinder contains 155 mL of water. A 15.0-g piece of iron (density = 7.86 g/cm³) and a 20.0-g piece of lead are added. What is the new water level in the cylinder? (See Table 1.10.)

1.107 How many cubic centimeters (cm³) of olive oil have the same mass as 1.00 L of gasoline? (See Table 1.10.)

1.108 What is the volume, in quarts, of 1.50 kg of ethyl alcohol? (See Table 1.10.)

1.109 a. Some athletes have as little as 3.0% body fat. If such a person has a body mass of 45 kg, how many lb of body fat does that person have?

b. In a process called liposuction, a doctor removes fat deposits from a person's body. If body fat has a density of 0.94 g/mL and 3.0 liters of fat are removed, how many pounds of fat were removed from the patient?

1.110 A mouthwash is 21.6% by mass alcohol. If each bottle contains 0.358 pint of mouthwash with a density of 0.876 g/mL, how many kilograms of alcohol are in 180 bottles of the mouthwash?

1.111 Sterling silver is 92.5% silver by mass with a density of 10.3 g/cm³. If a cube of sterling silver has a volume of 27.0 cm³, how many ounces of pure silver are present?

1.112 A typical adult body contains 55% water. If a person has a mass of 65 kg, how many pounds of water does she have in her body?

1.113 What is −15°F in degrees Celsius and in kelvins?

1.114 The highest recorded body temperature that a person has survived is 46.5°C. Calculate that temperature in °F and in kelvins.

CHALLENGE QUESTIONS

The following groups of questions and problems are related to the topics in this chapter. However, they do not all follow the chapter order and they require you to combine concepts and skills from several sections. These problems will help you increase your critical thinking skills and help you prepare for your next exam.

1.115 Classify each of the following as (1) an observation, (2) an hypothesis, or (3) an experiment.
a. The bicycle tire is flat.
b. If I add air to the bicycle tire, it will expand to the proper size.
c. When I added air to the bicycle tire, it was still flat.
d. The bicycle tire must have a leak in it.

1.116 Classify each of the following as (1) an observation, (2) an hypothesis, or (3) an experiment.
a. A big log in the fire does not burn well.
b. If I chop the log into smaller wood pieces, it will burn better.
c. The smaller pieces of wood burn brighter and make a hotter fire.
d. The small wood pieces are used up faster than burning the big log.

1.117 A balance measures mass to 0.001 g. If you determine the mass of an object that weighs about 30 g, would you record the mass as 30 g, 32.5 g, 31.25 g, 34.075 g, or 3000 g? Explain your choice by writing 2–3 complete sentences that describe your thinking.

1.118 When three students use the same meter stick to measure the length of a paper clip, they obtain results of 5.8 cm, 5.75 cm, and 5.76 cm. If the meter stick has millimeter markings, what are some reasons for the different values?

1.119 A car travels at 55 miles per hour and gets 11 kilometers per liter of gasoline. How many gallons of gasoline are needed for a 3.0-hour trip?

1.120 A 50.0-g silver object and a 50.0-g gold object are both added to 75.5 mL of water contained in a graduated cylinder. What is the new water level in the cylinder?

1.121 In the manufacturing of computer chips, cylinders of silicon are cut into thin wafers that are 3.00 inches in diameter and have a mass of 1.50 g of silicon. How thick (mm) is each wafer if silicon has a density of 2.33 g/cm³? (The volume of a cylinder is $V = \pi r^2 h$.)

ANSWERS

Answers to Study Checks

1.1 a. experiment (3)   **b.** observation (1)   **c.** hypothesis (2)

1.2 a. 4.25×10^5 m   **b.** $8. \times 10^{-7}$ g

1.3 a. two   **b.** one   **c.** five

1.4 a. 36 m   **b.** 0.0026 L
c. 3800 g (3.8×10^3 g)   **d.** 1.3 kg

1.5 a. 0.4924   **b.** 0.0080 or 8.0×10^{-3} **c.** 2.0

1.6 a. 83.70 mg   **b.** 0.5 L

1.7 **a.** mega   **b.** centi

1.8 **a.** 1000   **b.** 0.001 or $\frac{1}{1000}$

1.9 Equality: 1 in. = 2.54 cm

Conversion factors: $\dfrac{1 \text{ in.}}{2.54 \text{ cm}}$ and $\dfrac{2.54 \text{ cm}}{1 \text{ in.}}$

1.10 **a.** $\dfrac{62.2 \text{ km}}{1 \text{ hr}}$; $\dfrac{1 \text{ hr}}{62.2 \text{ km}}$ **b.** $\dfrac{9.48 \text{ cm}^3}{100. \text{ g}}$; $\dfrac{100. \text{ g}}{9.48 \text{ cm}^3}$

1.11 1.89 L

1.12 0.44 oz

1.13 10 mL

1.14 25 g of fat

1.15 2.2 g/mL

1.16 1.50 mL

1.17 207 mL

1.18 12°F

1.19 39.8°C

1.20 night −260.°C; day 410.°C

Answers to Selected Questions and Problems

1.1 **a.** A hypothesis proposes a possible explanation for a natural phenomenon.
  b. An experiment is a procedure that tests the validity of a hypothesis.
  c. A theory is a hypothesis that has been validated many times by many scientists.
  d. An observation is a description or measurement of a natural phenomenon.

1.3 **a.** 1, 4, and 5 are observations.   **b.** 2 is a hypothesis.
  c. 3 is an experiment.   **d.** 6 is a theory.

1.5 **a.** meter; length   **b.** gram; mass
  c. liter; volume   **d.** second; time
  e. Celsius; temperature

1.7 **a.** 5.5×10^4 m   **b.** 4.8×10^2 g
  c. 5×10^{-6} cm   **d.** 1.4×10^{-4} s
  e. 7.85×10^{-3} L   **f.** 6.7×10^5 kg

1.9 **a.** 7.2×10^3 **b.** 3.2×10^{-2}
  c. 1×10^4 **d.** 6.8×10^{-2}

1.11 **a.** 12 000   **b.** 0.0825
  c. 4 000 000   **d.** 0.005

1.13 **a.** measured   **b.** exact
  c. exact   **d.** measured

1.15 **a.** 6 oz of meat   **b.** none
  c. 0.75 lb; 350 g   **d.** none (definitions are exact)

1.17 **a.** not significant   **b.** significant
  c. significant   **d.** significant
  e. not significant

1.19 **a.** 5   **b.** 2   **c.** 2   **d.** 3   **e.** 4   **f.** 3

1.21 Both measurements in part c have two significant figures and both measurements in part d have four significant figures.

1.23 **a.** 5.0×10^3 L   **b.** 3.0×10^4 g
  c. 1.0×10^5 m   **d.** 2.5×10^{-4} cm

1.25 A calculator often gives more digits than the number of significant figures allowed in the answer.

1.27 **a.** 1.85   **b.** 184   **c.** 0.00474
  d. 8810   **e.** 1.83

1.29 **a.** 1.6   **b.** 0.01
  c. 27.6   **d.** 3.5

1.31 **a.** 53.54 cm   **b.** 127.6 g
  c. 121.5 mL   **d.** 0.50 L

1.33 km/h is kilometers per hour; mi/h is miles per hour

1.35 The prefix *kilo* means to multiply by 1000. One kg is the same mass as 1000 g.

1.37 **a.** mg   **b.** dL   **c.** km
  d. kg   **e.** μL

1.39 **a.** 0.01   **b.** 1000   **c.** 0.001
  d. 0.1   **e.** 1 000 000

1.41 **a.** 100 cm   **b.** 1000 m
  c. 0.001 m   **d.** 1000 mL

1.43 **a.** kilogram   **b.** milliliter
  c. km   **d.** kL

1.45 A conversion factor can be inverted to give a second conversion factor.

1.47 1 kg = 1000 g

1.49 **a.** 3 ft = 1 yd; $\dfrac{3 \text{ ft}}{1 \text{ yd}}$ and $\dfrac{1 \text{ yd}}{3 \text{ ft}}$

  b. 1 mile = 5280 feet; $\dfrac{5280 \text{ ft}}{1 \text{ mi}}$ and $\dfrac{1 \text{ mi}}{5280 \text{ ft}}$

  c. 1 min = 60 sec; $\dfrac{60 \text{ sec}}{1 \text{ min}}$ and $\dfrac{1 \text{ min}}{60 \text{ sec}}$

  d. 1 gal = 27 mi; $\dfrac{1 \text{ gal}}{27 \text{ mi}}$ and $\dfrac{27 \text{ mi}}{1 \text{ gal}}$

  e. 93 g silver = 100 g sterling; $\dfrac{93 \text{ g silver}}{100 \text{ g sterling}}$ and $\dfrac{100 \text{ g sterling}}{93 \text{ g silver}}$

1.51 **a.** 100 cm = 1 m; $\dfrac{100 \text{ cm}}{1 \text{ m}}$ and $\dfrac{1 \text{ m}}{100 \text{ cm}}$

  b. 1000 mg = 1 g; $\dfrac{1000 \text{ mg}}{1 \text{ g}}$ and $\dfrac{1 \text{ g}}{1000 \text{ mg}}$

  c. 1 L = 1000 mL; $\dfrac{1000 \text{ mL}}{1 \text{ L}}$ and $\dfrac{1 \text{ L}}{1000 \text{ mL}}$

  d. 1 dL = 100 mL; $\dfrac{100 \text{ mL}}{1 \text{ dL}}$ and $\dfrac{1 \text{ dL}}{100 \text{ mL}}$

1.53 The unit in the denominator must cancel with the preceding unit.

1.55 a. 1.75 m   **b.** 5.5 L   **c.** 5.5 g

1.57 a. 710 mL   **b.** 75.1 kg
    c. 495 mm   **d.** 2.0×10^{-5} in.

1.59 a. 23.8 m   **b.** 196 m^2 **c.** 0.463 s

1.61 a. 66 gal   **b.** 3 tablets   **c.** 1800 mg

1.63 a. 152 g of oxygen   **b.** 0.026 g magnesium
    c. 43 g of N   **d.** 50. lb of chocolate bars

1.65 lead; 11.3 g/mL

1.67 a. 1.20 g/mL   **b.** 0.873 g/mL
    c. 3.10 g/mL   **d.** 1.42 g/mL

1.69 a. 1.9 L   **b.** 88 g
    c. 62 oz   **d.** 30. kg

1.71 a. 1.030   **b.** 1.13   **c.** 0.85 g/mL

1.73 In the U.S. we still use the Fahrenheit temperature scale. In °F, normal body temperature is 98.6. On the Celsius scale, her temperature would be 37.7°C, a mild fever.

1.75 a. 98.6°F   **b.** 18.5°C   **c.** 246 K
    d. 335 K   **e.** 46°C   **f.** 295 K

1.77 a. 41°C
    b. No. The temperature is equivalent to 39°C.

1.79 Yes. Sherlock's investigation includes observations (gathering data), formulating a hypothesis, testing the hypothesis and modifying it until one of the hypotheses is validated.

1.81 a. exact   **b.** measured
    c. exact   **d.** measured

1.83 a. length = 6.96 cm; width = 4.75 cm
    b. length = 69.6 mm; width = 47.5 mm
    c. 3 significant figures
    d. 3 significant figures
    e. 33.1 cm^2
    f. 3 significant figures

1.85 1.8 g/mL

1.87 An hypothesis, which is a possible explanation for an observation, can be tested with experiments.

1.89 b and c

1.91 a. observation
    b. hypothesis or theory
    c. observation

1.93 42 min

1.95 16 onions

1.97 a. 96 crackers   **b.** 0.2 oz of fat
    c. 110 g of sodium

1.99 91 cents

1.101 0.92 kg

1.103 185 mg/dL

1.105 790 g

1.107 720 cm^3

1.109 a. 3.0 lb body fat   **b.** 6.2 lb

1.111 9.07 oz of pure silver

1.113 −26°C, 247 K

1.115 a. (1) Observation   **b. (2)** Hypothesis
    c. (3) Experiment   **d. (2)** Hypothesis

1.117 You should record the mass as 34.075 g. Since your balance will weigh to the nearest 0.001 g, the mass values should be reported to 0.001 g.

1.119 $3.0 \text{ hr} \times \dfrac{55 \text{ mi}}{1 \text{ hr}} \times \dfrac{1 \text{ km}}{0.621 \text{ mi}} \times \dfrac{1 \text{ L}}{11 \text{ km}} \times$
$\dfrac{1.06 \text{ qt}}{1 \text{ L}} \times \dfrac{1 \text{ gal}}{4 \text{ qt}} = 6.4 \text{ gal}$

1.121 Volume: $1.50 \text{ g} \times \dfrac{1 \text{ cm}^3}{2.33 \text{ g}} = 0.644 \text{ cm}^3$

Radius: $3.00 \text{ in.} \times \dfrac{1}{2} \times \dfrac{2.54 \text{ cm}}{1 \text{ in.}} = 3.81 \text{ cm}$

$h = \dfrac{V}{\pi r^2} = \dfrac{0.644 \text{ cm}^3}{3.14 \,(3.81 \text{ cm})^2} = 0.0141 \text{ cm}$

$\times \dfrac{10 \text{ mm}}{1 \text{ cm}} = 0.141 \text{ mm}$

2 Atoms and Elements

"The unique qualities of semiconducting metals make it possible for us to create sophisticated electronic circuits," says Tysen Streib, Global Product Manager, Applied Materials. "Elements from columns 3A, 4A, and 5A of the periodic table often make good semiconductors because they readily form covalently bonded crystals. When small amounts of impurities are added, free-flowing electrons or holes can travel through the crystal with very little interference. Without these covalent bonds and loosely bound electrons, we wouldn't have any of the microchips that we use in computers, cell phones, and thousands of other devices."

Materials scientists study the chemical properties of materials to find new uses for them in products such as cars, bridges, and clothing. They also develop materials that can be used as superconductors or in integrated-circuit chips and fuel cells. Chemistry is important in materials science because it provides information about structure and composition.

Mastering **GOB CHEMISTRY**

Visit **www.masteringgob.com** for self-study materials and instructor-assigned homework.

Every day, we see a variety of materials with many shapes and forms. To a scientist, all of this material is *matter*. There is matter everywhere around you; the orange juice you had for breakfast, the water you put in the coffee maker, the aluminum foil you wrap your sandwich in, your toothbrush and toothpaste, the oxygen you inhale, and the carbon dioxide you exhale are all forms of matter.

All matter is composed of *elements*, of which there are 115 different kinds. Of the 88 elements that occur in nature, one or more is used to make all the substances in our world. Many elements are already familiar to you. Perhaps you use aluminum in the form of foil or drink a soft drink from an aluminum can. You may have a ring or necklace made of gold, or silver, or perhaps platinum. If you play tennis or golf, your racket or clubs may be made from the elements titanium or carbon. In the body, compounds of calcium and phosphorus form the structure of bones and teeth, iron and copper are needed in the formation of red blood cells, and iodine is required for the proper functioning of the thyroid.

Deficiencies in certain of these elements can lead to improper functioning of enzymes and abnormal growth and bone formation. Low levels of iron can lead to anemia, while low levels of iodine can cause hypothyroidism and goiter. Someone working in health care is aware of the physical symptoms that indicate a deficiency of an element. Lab tests assess whether iron, copper, zinc, or iodine are within normal ranges in a patient's blood serum. A dietitian may recommend beef for iron and zinc, whole grains and leafy green vegetables for magnesium, dairy products for calcium, and iodized table salt and seafood for iodine.

LEARNING GOAL

Classify matter as pure substances or mixtures.

 Classification of Matter

2.1 Classification of Matter

Matter is the material that makes up a substance. Thus matter always has mass and occupies space. The materials we use such as water, wood, plates, plastic bags, clothes, and shoes are all made of matter. Because there are so many kinds, we categorize matter by the types of components it contains. A pure substance has a definite composition, while a mixture is made up of two or more substances in varying amounts. (See Figure 2.1.)

Pure Substances

A **pure substance** is matter that has a fixed or definite composition. There are two kinds of pure substances, elements and compounds. **Elements** are the simplest because they are composed of only one type of atom. Perhaps you are aware of the elements silver, iron, and aluminum, which contain matter of one type. **Compounds** are pure substances too, but they consist of a combination of two or more elements always in the same ratio. For example, in all samples of water, H_2O, there is the same proportion of the elements hydrogen and oxygen. In another compound, hydrogen peroxide, H_2O_2, hydrogen and oxygen are also combined, but in a different ratio. Water, H_2O, and hydrogen peroxide, H_2O_2, are different compounds with different properties.

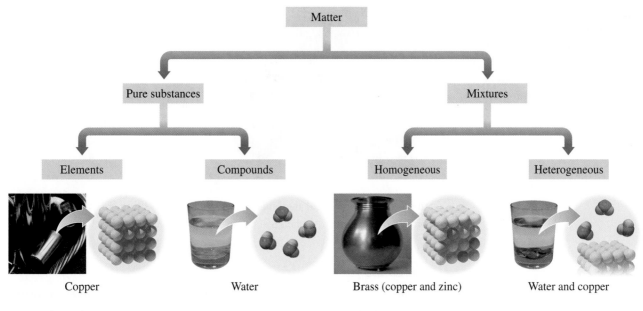

FIGURE 2.1

Matter is organized by its components: elements, compounds, and mixtures. An element or a compound has a fixed composition, whereas a mixture has a variable composition. A homogeneous mixture has a uniform composition, but a heterogeneous mixture does not.

Q **Why are copper and water pure substances, but brass is a mixture?**

An important difference between elements and compounds is that chemical processes can break down compounds into simpler substances such as elements. You may know that ordinary table salt is the compound NaCl, which can be broken down into sodium and chlorine as seen in Figure 2.2. Compounds are not broken

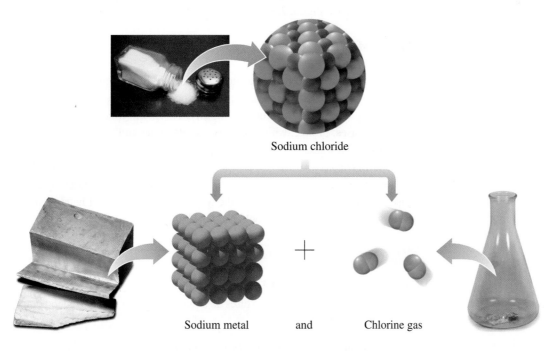

FIGURE 2.2

The decomposition of salt, NaCl, produces the elements sodium and chlorine.

Q **How do elements and compounds differ?**

A mixture of spaghetti and water is separated using a strainer, a physical method of separation.

Q **Why can physical methods be used to separate mixtures but not compounds?**

Physical method of separation

down through physical methods such as boiling or sifting. Elements cannot be broken down further by chemical or physical processes.

Mixtures

Much of the matter in our everyday lives consists of mixtures. In a **mixture**, two or more substances are physically mixed, but not chemically combined. The air we breathe is a mixture of mostly oxygen and nitrogen gases. The steel in buildings and railroad tracks is a mixture of iron, nickel, carbon, and chromium. The brass in knobs and fixtures is a mixture of zinc and copper. Solutions such as tea, coffee, and ocean water are mixtures too. In any mixture, the composition can vary. For example, two sugar–water mixtures may appear the same, but one would taste sweeter because it has a higher ratio of sugar to water. Different types of brass have different properties depending on the ratio of copper to zinc.

Physical processes can be used to separate mixtures because there are no chemical interactions between the components. For example, different coins such as nickels, dimes, and quarters can be separated by size; iron particles mixed with sand can be picked up with a magnet; and water is separated from cooked spaghetti using a strainer. (See Figure 2.3.)

Types of Mixtures

Mixtures can be classified further as homogeneous or heterogeneous. In a *homogeneous mixture*, also called a *solution*, the composition is uniform throughout the sample. Examples of familiar homogeneous mixtures are air, which contains oxygen and nitrogen gases; bronze, which is a mixture of copper and tin; and salt water, a solution of salt and water.

In a *heterogeneous mixture*, the components do not have a uniform composition throughout the sample. For example, a mixture of oil and water is heterogeneous because the oil floats on the surface of the water. Other examples of heterogeneous mixtures include the raisins in a cookie and the bubbles in a soda.

SAMPLE PROBLEM 2.1

Classifying Pure Subtances and Mixtures

Classify each of the following mixtures as a pure substance, heterogeneous mixture, or homogeneous mixture.

a. ice cream float   **b.** chocolate chip cookie

c. copper wire   **d.** vinegar (acetic acid and water)

SOLUTION

a. Heterogeneous; ice cream floats on soda, which is a nonuniform composition.

b. Heterogeneous; the chocolate chips are not uniformly distributed in a cookie.

c. Pure substance; one type of matter.

d. Homogeneous; acetic acid and water in vinegar have a uniform composition.

STUDY CHECK

A salad dressing is prepared with oil, vinegar, and chunks of blue cheese. Is this a homogeneous or heterogeneous mixture?

QUESTIONS AND PROBLEMS

■ **Classification of Matter**

2.1 Classify each of the following as a pure substance or a mixture.
   a. baking soda   **b.** a blueberry muffin
   c. ice   **d.** aluminum foil

2.2 Classify each of the following as a pure substance or a mixture.
   a. a soft drink   **b.** vitamin C
   c. a cheese sandwich   **d.** an iron nail

2.3 Classify each of the following as a compound or element.
   a. a silicon chip   **b.** hydrogen peroxide H_2O_2
   c. oxygen   **d.** vitamin A

2.4 Classify each of the following as a compound or element.
   a. helium gas   **b.** mercury in a thermometer
   c. sugar   **d.** sulfur

2.5 Classify each of the following mixtures as homogeneous or heterogeneous.
   a. vegetable soup   **b.** salt water
   c. tea   **d.** tea with ice and a lemon slice

2.6 Classify each of the following mixtures as homogeneous or heterogeneous.
   a. homogenized milk   **b.** chocolate chip ice cream
   c. gasoline   **d.** bubbly champagne

■ 2.2 Elements and Symbols

LEARNING GOAL

Given the name of an element, write its correct symbol; from the symbol, write the correct name.

Elements are primary substances from which all other things are built. They cannot be broken down into simpler substances. Elements were named for planets, mythological figures, minerals, colors, geographic locations, and famous people. Some sources of names of elements are listed in Table 2.1.

TABLE 2.1

Some Elements and Their Names

Element	Source of Name
uranium	The planet Uranus
titanium	Titans (mythology)
chlorine	*Chloros*, "greenish yellow" (Greek)
iodine	*Ioeides*, "violet" (Greek)
magnesium	Magnesia, a mineral
californium	California
curium	Marie and Pierre Curie

HEALTH NOTE

Latin Names for Elements in Clinical Usage

In medicine, the Latin name natrium may be used for sodium, an important electrolyte in body fluids and cells. An increase in serum sodium, hypernatremia, may occur when water is lost due to profuse sweating, severe diarrhea or vomiting, or when there is inadequate water intake. A decrease in sodium, hyponatremia, may occur when a person takes in a large amount of water or hypotonic fluid replacement solutions. Conditions that occur in cardiac failure, liver failure, and malnutrition can also cause hyponatremia.

The Latin name kalium may be used for potassium, the most common electrolyte inside the cells. Potassium regulates osmotic pressure, acid-base balance, nerve and muscle excitability, and the function of cellular enzymes. Serum potassium measures potassium outside the cells, which amounts to only 2 percent of total body potassium. An increase in serum potassium (hyperkalemia) may occur when cells are severely injured, in renal failure when potassium is not properly excreted, and in Addison's disease. A severe loss of potassium (hypokalemia) may occur during excessive vomiting, diarrhea, renal tubular defects, and glucose/insulin therapy.

 Using and Interpreting Symbols for Elements

Chemical symbols are one- or two-letter abbreviations for the names of the elements. Only the first letter of an element's symbol is capitalized; a second letter is lowercase. That way we know when a different element is indicated. If both letters are capitalized, it represents the symbols of two different elements. For example, the element cobalt has the symbol Co. However, the two capital letters CO specify two elements, carbon (C) and oxygen (O).

One-Letter Symbols	*Two-Letter Symbols*
C carbon	Co cobalt
S sulfur	Si silicon
N nitrogen	Ne neon
I iodine	Ni nickel

Although most of the symbols use letters from the current names, some are derived from their ancient Latin or Greek names. For example, Na, the symbol for sodium, comes from the Latin word *natrium*. The symbol for iron, Fe, is derived from the Latin name *ferrum*. Table 2.2 lists the names and symbols of some common elements. Learning their names and symbols will greatly help your learning of chemistry. A complete list of all the elements and their symbols appears on the inside front cover of this text.

QUESTIONS AND PROBLEMS   MC GOB

Elements and Symbols

2.7 Write the symbols for the following elements:
   a. copper   **b.** silicon   **c.** potassium   **d.** nitrogen
   e. iron   **f.** barium   **g.** lead   **h.** strontium

2.8 Write the symbols for the following elements:
   a. oxygen   **b.** lithium   **c.** sulfur   **d.** aluminum
   e. hydrogen   **f.** neon   **g.** tin   **h.** gold

2.9 Write the name of the element for each symbol.
   a. C   **b.** Cl   **c.** I   **d.** Hg
   e. F   **f.** Ar   **g.** Zn   **h.** Ni

2.10 Write the name of the element for each symbol.
   a. He   **b.** P   **c.** Na   **d.** Mg
   e. Ca   **f.** Br   **g.** Cd   **h.** Si

2.11 What elements are in the following substances?
   a. table salt, $NaCl$
   b. plaster casts, $CaSO_4$
   c. Demerol, $C_{15}H_{22}ClNO_2$
   d. antacid, $CaCO_3$

2.12 What elements are in the following substances?
   a. water, H_2O
   b. baking soda, $NaHCO_3$
   c. lye, $NaOH$
   d. sugar, $C_{12}H_{22}O_{11}$

TABLE 2.2

Names and Symbols of Some Common Elements

Aluminum

Carbon

Copper

Gold

Name[a]	Symbol
Aluminum	Al
Argon	Ar
Arsenic	As
Barium	Ba
Boron	B
Bromine	Br
Cadmium	Cd
Calcium	Ca
Carbon	C
Chlorine	Cl
Chromium	Cr
Cobalt	Co
Copper (*cuprum*)	Cu
Fluorine	F
Gold (*aurum*)	Au
Helium	He
Hydrogen	H
Iodine	I
Iron (*ferrum*)	Fe
Lead (*plumbum*)	Pb
Lithium	Li
Magnesium	Mg
Manganese	Mn
Mercury (*hydrargyrum*)	Hg
Neon	Ne
Nickel	Ni
Nitrogen	N
Oxygen	O
Phosphorus	P
Platinum	Pt
Potassium (*kalium*)	K
Radium	Ra
Silicon	Si
Silver (*argentum*)	Ag
Sodium (*natrium*)	Na
Strontium	Sr
Sulfur	S
Tin (*stannum*)	Sn
Titanium	Ti
Uranium	U
Zinc	Zn

Iron

Silver

Sulfur

[a]Names given in parentheses are based on ancient Latin or Greek words from which the symbols are derived.

ENVIRONMENTAL NOTE

▪ Toxicity of Mercury

Mercury is a silvery, shiny element that is a liquid at room temperature. Mercury can enter the body through inhaled mercury vapor, contact with the skin, or foods or water contaminated with mercury. In the body, mercury destroys proteins and disrupts cell function. Long-term exposure to mercury can damage the brain and kidneys, cause mental retardation, and decrease physical development. Blood, urine, and hair samples are used to test for mercury.

In fresh and salt water, bacteria convert mercury into toxic methylmercury, which primarily attacks the central nervous system (CNS). Because fish absorb methylmercury, we are exposed to mercury by consuming mercury-contaminated fish. As levels of mercury ingested from fish became a concern, the Food and Drug Administration (FDA) set a maximum level of one part mercury per million parts seafood (1 ppm), which is the same as 1 μg mercury in every gram of seafood. Fish higher in the food chain such as swordfish and shark can have such high levels of mercury that the Environmental Protection Agency (EPA) recommends they be consumed no more than once a week.

One of the worst incidents of mercury poisoning occurred in Minamata and Niigata, Japan, in 1950. At that time, the ocean was polluted with high levels of mercury from industrial wastes. Because fish were a major food in the diet, more than 2000 people were affected with mercury poisoning and died or developed neural damage. In the U.S., industry decreased the use of mercury between 1988 and 1997 by 75% by banning mercury in paint and pesticides, reducing mercury in batteries, and regulating mercury in other products.

This mercury fountain, housed in glass, was designed by Calder for the 1937 World's Fair in Paris.

LEARNING GOAL
▪

Use the periodic table to identify the group and the period of an element and decide whether it is a metal, nonmetal, or metalloid.

2.3 The Periodic Table

As more and more elements were discovered, it became necessary to organize them with some type of classification system. By the late 1800s, scientists recognized that certain elements looked alike and behaved in much the same way. In 1872, a Russian chemist, Dmitri Mendeleev, arranged the 60 elements known at that time into groups with similar properties and placed them in order of increasing atomic masses. Today, this arrangement of 115 elements is known as the **periodic table**. (See Figure 2.4.)

Periods and Groups

Each horizontal row in the table is called a **period**. The number of elements in the periods increase going down the periodic table. Each row is counted from the top of the table as Period 1 to Period 7. The first period contains only the elements

Periodic Table of Elements

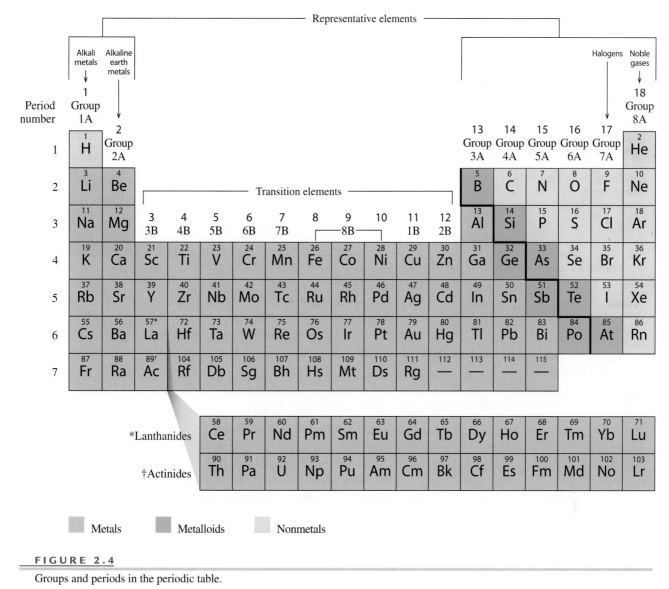

FIGURE 2.4

Groups and periods in the periodic table.

Q **What is the symbol of the alkali metal in Period 3?**

hydrogen (H) and helium (He). The second period contains eight elements: lithium (Li), beryllium (Be), boron (B), carbon (C), nitrogen (N), oxygen (O), fluorine (F), and neon (Ne). The third period also contains eight elements beginning with sodium (Na) and ending with argon (Ar). The fourth period, which begins with potassium (K), and the fifth period, which begins with rubidium (Rb), have 18 elements each. The sixth period, which begins with cesium (Cs), has 32 elements. The seventh period as of today contains the 29 remaining elements although it could go up to 32. (See Figure 2.5.)

Each vertical column on the periodic table contains a **group** (or family) of elements that have similar properties. At the top of each column is a number that is assigned to each group. The elements in the first two columns on the left of the periodic table and the last six columns on the right are called the *representative*

Elements and Symbols in the Periodic Table

▪ Elements Essential to Health

Many elements are essential for the well-being and survival of the human body. The four elements oxygen, carbon, hydrogen, and nitrogen are the most important elements that make up carbohydrates, fats, proteins, and DNA. Most of the hydrogen and oxygen is found in water, which makes up 55–60% of our body mass. Some examples and the amounts present in a 60-kg person are listed in Table 2.3.

TABLE 2.3

Elements Essential to Health

Element	Symbol	Amount in a 60-kg Person
oxygen	O	39 kg
carbon	C	11 kg
hydrogen	H	6 kg
nitrogen	N	1.5 kg
calcium	Ca	1 kg
phosphorus	P	600 g
potassium	K	120 g
sulfur	S	120 g
sodium	Na	86 g
chlorine	Cl	81 g
magnesium	Mg	16 g
iron	Fe	3.6 g
fluorine	F	2.2 g
zinc	Zn	2.0 g
copper	Cu	60 mg
iodine	I	20 mg

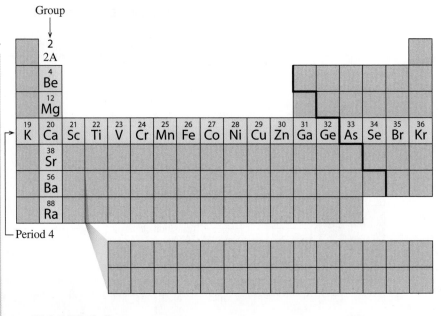

FIGURE 2.5

On the periodic table, each vertical column represents a group of elements and each horizontal row of elements represents a period.

Q **Are the elements Si, P, and S part of a group or a period?**

elements or the *main group elements*. For many years, they have been given group numbers 1A–8A. On some periodic tables, the group numbers may be written with Roman numerals: IA–VIIIA. In the center of the periodic table is a block of elements known as the *transition elements* or *transition metals*, which are designated with the letter "B." A newer numbering system assigns group numbers of 1–18 going across the periodic table. Because both systems of group numbers are currently in use, they are both indicated on the periodic table in this text and are included in our discussions of elements and group numbers.

▪ Classification of Groups

Several groups in the periodic table have special names. (See Figure 2.6.) Group 1A (1) elements, lithium (Li), sodium (Na), potassium (K), rubidium (Rb), cesium (Cs), and francium (Fr), are a family of elements known as the **alkali metals**. (See Figure 2.7.) The elements within this group are soft, shiny metals that are good conductors of heat and electricity and have relatively low melting points. Alkali metals react vigorously with water and form white products when they combine with oxygen.

Although hydrogen (H) is at the top of Group 1A (1), hydrogen is not an alkali metal and has very different properties than the rest of the elements in this group. Thus hydrogen is not included in the classification of alkali metals. In some periodic tables, H is placed at the top of Group 7A (17).

Group 2A (2) elements, beryllium (Be), magnesium (Mg), calcium (Ca), strontium (Sr), barium (Ba), and radium (Ra), are called the **alkaline earth metals**. They are also shiny metals like those in Group 1A (1), but they are not as reactive.

The **halogens** are found on the right side of the periodic table in Group 7A (17). They include the elements fluorine (F), chlorine (Cl), bromine (Br), iodine (I), and

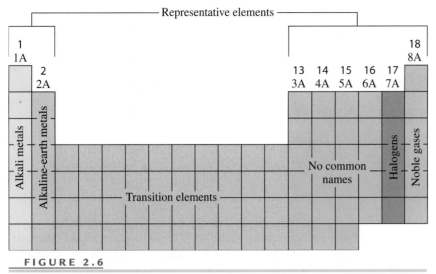

FIGURE 2.6

Certain groups on the periodic table have common names.

Q **What is the common name for the group of elements that includes helium and argon?**

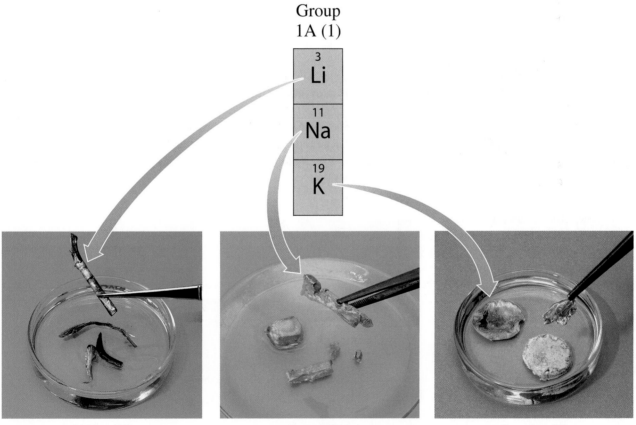

FIGURE 2.7

Lithium (Li), sodium (Na), and potassium (K) are some alkali metals from Group 1A (1).

Q **What physical properties do these alkali metals have in common?**

Group
7A (17)

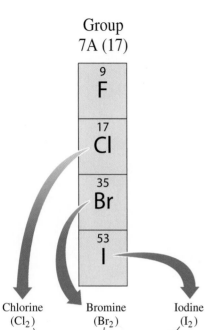

Chlorine
(Cl_2)

Bromine
(Br_2)

Iodine
(I_2)

FIGURE 2.8

Chlorine (Cl_2), bromine (Br_2), and iodine (I_2) are samples of halogens from Group 7A (17).

Q **What elements are in the halogen group?**

astatine (At), as shown in Figure 2.8. The halogens, especially fluorine and chlorine, are strongly reactive and form compounds with most of the elements.

Group 8A (18) contains the **noble gases**, helium (He), neon (Ne), argon (Ar), krypton (Kr), xenon (Xe), and radon (Rn). They are quite unreactive and are seldom found in combination with other elements.

SAMPLE PROBLEM 2.2

■ **Stating Group and Period Number**

Identify the period number and the group number for the following elements:

a. calcium   **b.** tin

SOLUTION

The period is found by counting down the rows of the elements on the periodic table, and the group number is found at the top of the column that contains that element.

a. Calcium (Ca) is in Period 4 and Group 2A (2).
b. Tin (Sn) is in Period 5 and Group 4A (14).

STUDY CHECK

Give the symbols of the elements that are represented by the following period and group numbers:

a. Period 3, Group 5A (15)   **b.** Period 6, Group 8A (18)

Metals, Nonmetals, and Metalloids

Another feature of the periodic table is the heavy zigzag line that separates the elements into the *metals* and the *nonmetals*. The metals are those elements on the left of the line *except for hydrogen*, and the nonmetals are the elements on the right. (See Figure 2.9.)

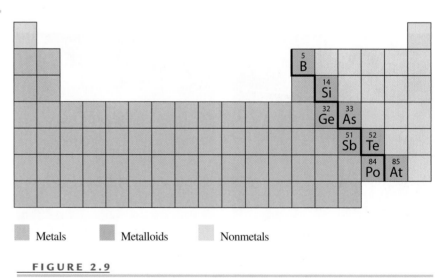

■ Metals   ■ Metalloids   ■ Nonmetals

FIGURE 2.9

Along the heavy zigzag line on the periodic table that separates the metals and nonmetals are metalloids, which exhibit characteristics of both metals and nonmetals.

Q **On which side of the heavy zigzag line are the nonmetals located?**

TABLE 2.4

Some Characteristics of a Metal, a Metalloid, and a Nonmetal

Silver (Ag)	Antimony (Sb)	Sulfur (S)
Metal	Metalloid	Nonmetal
Shiny	Blue-gray, shiny	Dull, yellow
Extremely ductile	Brittle	Brittle
Can be hammered into sheets (malleable)	Shatters when hammered	Shatters when hammered
Good conductor of heat and electricity	Poor conductor of heat and electricity	Poor conductor, good insulator
Used in coins, jewelry, tableware	Used to harden lead, color glass and plastics	Used in gunpowder, rubber, fungicides
Density 10.5 g/mL	Density 6.7 g/mL	Density 2.1 g/mL
Melting point 962°C	Melting point 630°C	Melting point 113°C

In general, most **metals** are shiny solids. They can be shaped into wires (ductile) or hammered into a flat sheet (malleable). Metals are good conductors of heat and electricity. They usually melt at higher temperatures than nonmetals. All of the metals are solids at room temperature, except for mercury (Hg), which is a liquid. Some typical metals are magnesium (Mg), copper (Cu), gold (Au), silver (Ag), iron (Fe), and tin (Sn).

Nonmetals are not very shiny, malleable, or ductile, and they are often poor conductors of heat and electricity. They typically have low melting points and low densities. You may have heard of nonmetals such as hydrogen (H), carbon (C), nitrogen (N), oxygen (O), chlorine (Cl), and sulfur (S).

The elements located along the heavy line are called **metalloids** and include B, Si, Ge, As, Sb, Te, Po, and At. Metalloids are elements that exhibit some properties that are typical of the metals and other properties that are characteristic of the nonmetals. For example, they are better conductors of heat and electricity than the nonmetals, but not as good as the metals. The metalloids are semiconductors because they can act as conductors or insulators. Table 2.4 compares some characteristics of silver, a metal, with those of antimony, a metalloid, and sulfur, a nonmetal.

SAMPLE PROBLEM 2.3

■ Metals, Nonmetals, and Metalloids

Use a periodic table to classify each of the following elements as a metal, nonmetal, or metalloid.

a. Na   **b.** Si   **c.** I   **d.** Sn

SOLUTION

a. Sodium, which lies to the left of the heavy zigzag line, is a metal.
b. Silicon, which lies on the heavy zigzag line, is a metalloid.
c. Iodine, to the right of the zigzag line, is a nonmetal.
d. Tin, which lies on the left of the heavy line, is a metal.

HEALTH NOTE

Some Important Trace Elements in the Body

Some metals and nonmetals known as trace elements are essential to the proper functioning of the body. Although they are required in very small amounts, their absence can disrupt major biological processes and cause illness. The trace elements listed in Table 2.5 are present in the body combined with other elements. The adult DV is the daily recommended amount for an adult.

TABLE 2.5

Some Important Trace Elements in the Body

Element	Adult DV[a]	Biological Function	Deficiency Symptoms	Dietary Sources
Iron (Fe)	10 mg (males) 18 mg (females)	Formation of hemoglobin; enzymes	Dry skin, spoon nails, decreased hemoglobin count, anemia	Liver and other organ meats, oysters, red or dark meat, green leafy vegetables, fortified breads and cereals, egg yolk
Copper (Cu)	2.0–5.0 mg	Necessary in many enzyme systems; growth; aids formation of red blood cells and collagen	Uncommon; anemia; decreased white cell count; bone demineralization	Nuts, organ meats, whole grains, shellfish, eggs, poultry, leafy green vegetables
Zinc (Zn)	15 mg	Amino acid metabolism; enzyme systems; energy production; collagen	Retarded growth and bone formation; skin inflammation; loss of taste and smell; poor healing	Oysters, crab, lamb, beef, organ meats, whole grains
Manganese (Mn)	2.5–5.0 mg	Necessary for some enzyme systems; collagen formation; central nervous system; fat and carbohydrate metabolism; blood clotting	Abnormal skeletal growth; impairment of central nervous system	Whole grains, wheat germ, legumes, pineapple, figs
Iodine (I)	150 μg	Necessary for activity of thyroid gland	Hypothyroidism; goiter; cretinism	Seafood, iodized salt
Fluorine (F)	1.5–4.0 mg	Necessary for solid tooth formation and retention of calcium in bones with aging	Dental cavities	Tea, fish, water in some areas, supplementary drops, toothpaste

[a] DV, daily value.

STUDY CHECK

Identify each of the following elements as a metal or nonmetal.

a. An element with a bright, lustrous reddish color that melts at 2570°C and is malleable, ductile, and an excellent conductor of heat and electricity.

b. A black substance that breaks into smaller chunks when hit with a hammer.

QUESTIONS AND PROBLEMS

The Periodic Table

2.13 Identify the group or period number described by each of the following statements:
    a. contains the elements C, N, and O   **b.** begins with helium
    c. the alkali metals   **d.** ends with neon

2.14 Identify the group or period number described by each of the following statements:
   a. contains Na, K, and Rb   **b.** the row that begins with Li
   c. the noble gases   **d.** contains F, Cl, Br, and I

2.15 Classify the following as an alkali metal, alkaline earth metal, transition element, halogen, or noble gas:
   a. Ca   **b.** Fe   **c.** Xe   **d.** Na   **e.** Cl

2.16 Classify the following as an alkali metal, alkaline earth metal, transition element, halogen, or noble gas:
   a. Ne   **b.** Mg   **c.** Cu   **d.** Br   **e.** Ba

2.17 Give the symbol of the element described by the following:
   a. Group 4A, Period 2   **b.** a noble gas in Period 1
   c. an alkali metal in Period 3   **d.** Group 2, Period 4
   e. Group 13, Period 3

2.18 Give the symbol of the element described by the following:
   a. an alkaline earth metal in Period 2   **b.** Group 15, Period 3
   c. a noble gas in Period 4   **d.** a halogen in Period 5
   e. Group 4A, Period 4

2.19 Is each of the following elements a metal, nonmetal, or metalloid?
   a. calcium   **b.** sulfur
   c. an element that is shiny   **d.** does not conduct heat
   e. located in Group 8A   **f.** phosphorus
   g. boron   **h.** silver

2.20 Is each of the following elements a metal, nonmetal, or metalloid?
   a. located in Group 2A   **b.** a good conductor of electricity   **c.** chlorine
   d. arsenic   **e.** an element that is not shiny   **f.** oxygen
   g. nitrogen   **h.** aluminum

2.4 The Atom

LEARNING GOAL

Describe the electrical charge and location in an atom for a proton, a neutron, and an electron.

All the elements listed on the periodic table are made up of atoms. An **atom** is the smallest particle of an element that retains the characteristics of that element. You have probably seen the element aluminum. Imagine that you are tearing a piece of aluminum foil into smaller and smaller pieces. Now imagine that you have a piece so small that you can no longer break it down further. Then you would have an atom of aluminum, the smallest particle of an element that still retains the characteristics of that element.

The concept of the atom is relatively recent. Although the Greek philosophers in 500 BCE reasoned that everything must contain minute particles they called *atomos*, the idea of atoms did not become a scientific theory until 1808. Then John Dalton (1766–1844) developed an atomic theory that proposed that atoms were responsible for the combinations of elements found in compounds.

SELF-STUDY ACTIVITY
Atoms and Isotopes

The Anatomy of Atoms

Dalton's Atomic Theory

1. All matter is made up of tiny particles called atoms.
2. All atoms of a given element are similar to one another and different from atoms of other elements.
3. Atoms of two or more different elements combine to form compounds. A particular compound is always made up of the same kinds of atoms and always has the same number of each kind of atom.
4. A chemical reaction involves the rearrangement, separation, or combination of atoms. Atoms are never created or destroyed during a chemical reaction.

Atoms are the building blocks of everything we see around us; yet, we cannot see an atom or even a billion atoms with the naked eye. However, when billions and billions of atoms are packed together, the characteristics of each atom are added to those of the next until we can see the characteristics we associate with the element. For example, a small piece of the shiny, copper-colored element we call copper consists of many, many copper atoms. Through a special kind of microscope called a scanning tunneling microscope, we can now see images of individual atoms, such as the atoms of carbon in graphite shown in Figure 2.10.

Parts of an Atom

By the early part of the twentieth century, growing evidence indicated that atoms were not solid spheres, as Dalton had imagined. New experiments showed that atoms were composed of even smaller bits of matter called **subatomic particles**. The chemistry of an element depends upon the subatomic particles that are the building blocks of the atoms.

There are three subatomic particles of interest to us: the proton, neutron, and electron. Two of these carry electrical charges. The **proton** has a positive charge $(+)$, and the **electron** carries a negative charge $(-)$. The **neutron** has no electrical charge; it is neutral.

Like charges repel; they push away from each other. When you brush your hair on a dry day, electrical charges that are alike build up on the brush and in your hair; as a result your hair flies away from the brush. Opposite or unlike charges attract. The crackle of clothes taken from the clothes dryer indicates the presence of electrical charges. The clinginess of the clothing is due to the attraction of opposite, unlike charges, as shown in Figure 2.11.

Mass

All of the subatomic particles are extremely small compared with the things you see around you. One proton has a mass of 1.7×10^{-24}g, and the neutron is about the same. The mass of the electron is even less. Because the mass of a subatomic particle is so minute, chemists use a small unit of mass called an **atomic mass unit (amu)**. An amu is defined as one-twelfth of the mass of the carbon atom with 6 protons and 6 neutrons, a standard with which the mass of every other atom is compared. In biology, the atomic mass unit is called a dalton in honor of John Dalton. On the amu scale, the proton and neutron each have a mass of about 1 amu. Because the electron mass is so small, it is usually ignored in atomic mass calculations. Table 2.6 summarizes some information about the subatomic particles in an atom.

FIGURE 2.10

Graphite, a form of carbon, magnified millions of times by a scanning tunneling microscope. This instrument generates an image of the atomic structure. The round yellow objects are atoms.

Q **Why is a microscope with extremely high magnification needed to see atoms?**

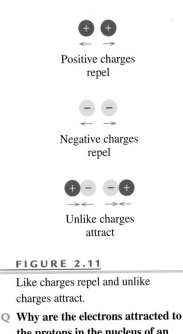

Positive charges repel

Negative charges repel

Unlike charges attract

FIGURE 2.11

Like charges repel and unlike charges attract.

Q **Why are the electrons attracted to the protons in the nucleus of an atom?**

TABLE 2.6

Particles in the Atom

Subatomic Particle	Symbol	Electrical Charge	Approximate Mass (amu)	Location in Atom
Proton	p or p^+	$1+$	1	Nucleus
Neutron	n or n^0	0	1	Nucleus
Electron	e^-	$1-$	$0.0005 \left(\frac{1}{2000} \right)$	Outside nucleus

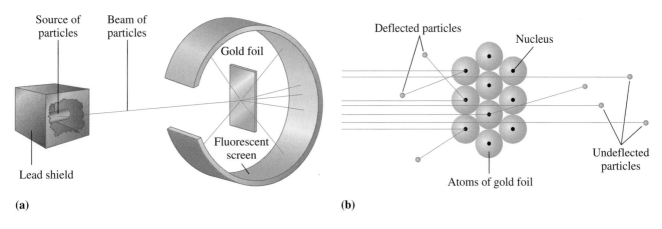

FIGURE 2.12

(a) Positive particles are aimed at a piece of gold foil. (b) Particles that come close to the atomic nuclei are deflected from their straight path.

Q **Why are some particles deflected while most pass through the gold foil undeflected?**

Structure of the Atom

By the early 1900s, scientists knew that atoms contained subatomic particles, but they did not know how those particles were arranged. In 1911, an experiment led Ernest Rutherford to propose that most of the mass of the atom was contained in a small region at the center of the atom. In this experiment, positively charged particles were aimed at a very thin sheet of gold foil as seen in Figure 2.12. He thought that these particles would travel in straight paths and go right on through the gold foil. But much to his surprise, some changed direction as they passed through the gold foil. A few particles were deflected so much that they reversed direction. According to Rutherford, it was as though he had shot a cannon ball at a piece of tissue paper, and it bounced back at him.

From the gold-foil experiment, Rutherford concluded that the particles would change direction only if they came close to some type of dense, positively charged region in the atom. This dense core of an atom called the **nucleus**, where the protons and neutrons are located, has a positive charge and contains most of the mass of an atom. Most of an atom is empty space, which is occupied only by fast-moving electrons. (See Figure 2.13.) If an atom were the size of a football stadium, the nucleus would be about the size of a golf ball placed in the center of the field.

MC GOB SELF-STUDY ACTIVITY
Atoms and Isotopes

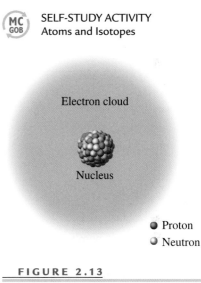

Electron cloud

Nucleus

● Proton
◔ Neutron

FIGURE 2.13

In an atom, the protons and neutrons are found in the nucleus; the electrons are located outside the nucleus.

Q **Why can we say that an atom is mostly empty space?**

▪ Repulsion and Attraction

1. Obtain a tape dispenser with clear tape, a hairbrush, comb, and a piece of paper. Tear off a piece of the tape that is about 20 cm long (the length of your hand). Stick the tape to the edge of a table leaving the end hanging down. Tear off a second piece of tape and slowly bring it close to the first one. What happens? Is there an attraction or repulsion?

   Slide your thumb and finger along the tape you are holding. Bring it close to the piece that is hanging from the table. What happens? Is there an attraction or repulsion? Attach the second tape to the edge of the table.

   Brush your hair and bring the brush close to each piece of tape hanging from the table. What do you observe?

2. Tear a small piece of paper into bits. Brush your hair several times and place the brush just above the bits of paper. Use your knowledge of electrical charges to give an explanation for your observations. Try the same experiment with a comb.

QUESTIONS

1. What happens when objects with like charges are placed close together?
2. What happens when objects with unlike charges are placed close together?

QUESTIONS AND PROBLEMS

■ The Atom

2.21 Is a proton, neutron, or electron described by each of the following?
a. has the smallest mass
b. carries a positive charge
c. is found outside the nucleus
d. is electrically neutral

2.22 Is a proton, neutron, or electron described by each of the following?
a. has a mass about the same as a proton's
b. is found in the nucleus
c. is found in the larger part of the atom
d. carries a negative charge

2.23 What did Rutherford's gold-foil experiment tell us about the organization of the subatomic particles in an atom?

2.24 Why does the nucleus in every atom have a positive charge?

2.25 Which of the following particles have opposite charges?
a. two protons
b. a proton and an electron
c. two electrons
d. a proton and a neutron

2.26 Which of the following pairs of particles have the same charges?
a. two protons
b. a proton and an electron
c. two electrons
d. an electron and a neutron

2.27 On a dry day, your hair flies away when you brush it. How would you explain this?

2.28 Sometimes clothes removed from the dryer cling together. What kinds of charges are on the clothes?

LEARNING GOAL

Given the atomic number and the mass number of an atom, state the number of protons, neutrons, and electrons.

Atomic Number and Mass Number

2.5 Atomic Number and Mass Number

All of the atoms of the same element always have the same number of protons. This feature distinguishes atoms of one element from atoms of all the other elements.

Atomic Number

An **atomic number**, which is equal to the number of protons in the nucleus of an atom, is used to identify each element.

Atomic number = number of protons in an atom

On the inside front cover of this text is a periodic table, which gives all of the elements in order of increasing atomic number. The atomic number is the whole number that appears above the symbol. For example, a hydrogen atom, with atomic number 1, has 1 proton; a lithium atom, with atomic number 3, has 3 protons; an atom of carbon, with atomic number 6, has 6 protons; and gold, with atomic number 79, has 79 protons.

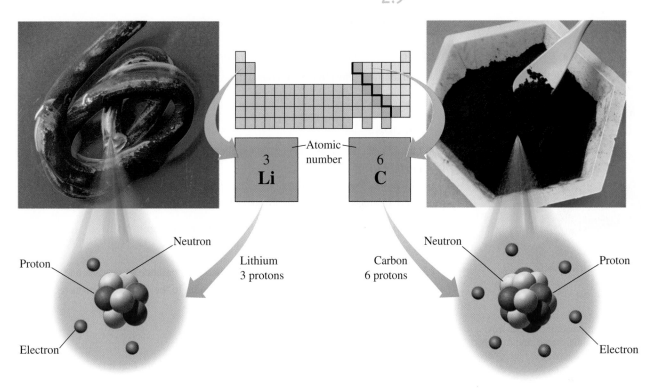

An atom is electrically neutral. That means that the number of protons in an atom is equal to the number of electrons. This electrical balance gives an atom an overall charge of zero. Thus, in every atom, the atomic number also gives the number of electrons.

Mass Number

We now know that the protons and neutrons determine the mass of the nucleus. For any atom, the **mass number** is the sum of the number of protons and neutrons in the nucleus.

Mass number = number of protons + number of neutrons

For example, an atom of oxygen that contains 8 protons and 8 neutrons has a mass number of 16. An atom of iron that contains 26 protons and 30 neutrons has a mass number of 56. Table 2.7 illustrates the relationship between atomic number, mass number, and the number of protons, neutrons, and electrons in some atoms of different elements.

TABLE 2.7

Composition of Some Atoms of Different Elements

Element	Symbol	Atomic Number	Mass Number	Number of Protons	Number of Neutrons	Number of Electrons
hydrogen	H	1	1	1	0	1
nitrogen	N	7	14	7	7	7
chlorine	Cl	17	37	17	20	17
iron	Fe	26	56	26	30	26
gold	Au	79	197	79	118	79

▪ Optician

"When a patient brings in a prescription, I help select the proper lenses, put them into a frame, and fit them properly on the patient's face," says Suranda Lara, optician, Kaiser Hospital. "If a prescription requires a thinner and lighter-weight lens, we formulate that lens. So we have to understand the different materials used to make lenses. Sometimes patients come in with their own glasses that they want to convert to sunglasses. We remove the lenses and put them into a tint bath, which turns them into sunglasses."

Opticians fit and adjust eyewear for patients who have had their eyesight tested by an ophthalmologist or optometrist. Optics and mathematics are used to select materials for frames and lenses that are compatible with patients' facial measurements and lifestyles.

SAMPLE PROBLEM 2.4

▪ Calculating Numbers of Protons and Neutrons

For an atom of phosphorus that has a mass number of 31, determine the following:

a. the number of protons   **b.** the number of neutrons   **c.** the number of electrons

SOLUTION

a. On the periodic table, the atomic number of phosphorus is 15. A phosphorus atom has 15 protons.

b. The number of neutrons in this atom is found by subtracting the atomic number from the mass number. The number of neutrons is 16.

$$\text{Mass number} - \text{atomic number} = \text{number of neutrons}$$
$$31 - 15 = 16$$

c. Because an atom is neutral, there is an electrical balance of protons and electrons. Because the number of electrons is equal to the number of protons, the phosphorus atom has 15 electrons.

STUDY CHECK

How many neutrons are in the nucleus of a bromine atom that has a mass number of 80?

QUESTIONS AND PROBLEMS

▪ Atomic Number and Mass Number

2.29 Would you use atomic number, mass number, or both to obtain the following?
    a. number of protons in an atom   **b.** number of neutrons in an atom
    c. number of particles in the nucleus   **d.** number of electrons in a neutral atom

2.30 What do you know about the subatomic particles from the following?
    a. atomic number   **b.** mass number
    c. mass number − atomic number   **d.** mass number + atomic number

2.31 Write the names and symbols of the elements with the following atomic numbers:
    a. 3   **b.** 9   **c.** 20   **d.** 30
    e. 10   **f.** 14   **g.** 53   **h.** 8

2.32 Write the names and symbols of the elements with the following atomic numbers:
    a. 1   **b.** 11   **c.** 19   **d.** 26
    e. 35   **f.** 47   **g.** 15   **h.** 2

2.33 How many protons and electrons are there in a neutral atom of the following?
    a. magnesium   **b.** zinc   **c.** iodine   **d.** potassium

2.34 How many protons and electrons are there in a neutral atom of the following?
    a. carbon   **b.** fluorine   **c.** calcium   **d.** sulfur

2.35 Complete the following table for neutral atoms.

Name of Element	Symbol	Atomic Number	Mass Number	Number of Protons	Number of Neutrons	Number of Electrons
	Al		27			
		12			12	
Potassium					20	
				16	15	
			56			26

2.36 Complete the following table for neutral atoms.

Name of Element	Symbol	Atomic Number	Mass Number	Number of Protons	Number of Neutrons	Number of Electrons
	N		15			
Calcium			42			
				38	50	
		14			16	
		56	138			

2.6 Isotopes and Atomic Mass

We have seen that all atoms of the same element have the same number of protons and electrons. However, the atoms of any one element are not completely identical because they can have different numbers of neutrons.

Isotopes

Isotopes are atoms of the same element that have different numbers of neutrons. For example, all atoms of the element magnesium (Mg) have 12 protons. However, some magnesium atoms have 12 neutrons, others have 13 neutrons, and still others have 14 neutrons. The differences in numbers of neutrons for these magnesium atoms cause their mass numbers to be different but not their chemical behavior. The three isotopes of magnesium have the same atomic number but different mass numbers.

To distinguish between the different isotopes of an element, we can write an **atomic symbol** that indicates the mass number and the atomic number of the atom.

Atomic Symbol for an Isotope of Magnesium

Mass number $\longrightarrow$

Symbol of element $\longrightarrow$ $\quad ^{24}_{12}\text{Mg}$

Atomic number $\longrightarrow$

An isotope may be referred to by its name or symbol followed by the mass number, such as magnesium-24 or Mg-24. Magnesium has three naturally occurring isotopes, as shown in Table 2.8.

TABLE 2.8

Isotopes of Magnesium

Atomic symbol	$^{24}_{12}\text{Mg}$	$^{25}_{12}\text{Mg}$	$^{26}_{12}\text{Mg}$
Number of protons	12	12	12
Number of electrons	12	12	12
Mass number	**24**	**25**	**26**
Number of neutrons	**12**	**13**	**14**
Mass of isotope (amu)	23.99	24.99	25.98
% abundance	78.70%	10.13%	11.17%

MC GOB SELF-STUDY ACTIVITY
Atoms and Isotopes

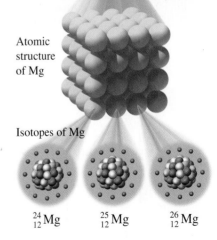

Atomic structure of Mg

Isotopes of Mg

$^{24}_{12}\text{Mg}$ $^{25}_{12}\text{Mg}$ $^{26}_{12}\text{Mg}$

Identifying Protons and Neutrons in Isotopes

State the number of protons and neutrons in the following isotopes of neon (Ne):

a. $^{20}_{10}$Ne   **b.** $^{21}_{10}$Ne   **c.** $^{22}_{10}$Ne

SOLUTION

The atomic number of Ne is 10; each isotope has 10 protons. The number of neutrons in each isotope is found by subtracting the atomic number (10) from each mass number.

a. 10 protons; 10 neutrons (20 − 10)   **b.** 10 protons; 11 neutrons (21 − 10)
c. 10 protons; 12 neutrons (22 − 10)

STUDY CHECK

Write a symbol for the following isotopes:

a. a nitrogen atom with 8 neutrons
b. an atom with 20 protons and 22 neutrons
c. an atom with mass number 27 and 14 neutrons

TABLE 2.9

The Atomic Mass of Some Elements

Element	Isotopes	Atomic Mass (weighted average)
lithium	^{6}Li, ^{7}Li	6.941 amu
carbon	^{12}C, ^{13}C, ^{14}C	12.01 amu
fluorine	^{19}F	19.00 amu
oxygen	^{16}O, ^{17}O, ^{18}O	16.00 amu
sulfur	^{32}S, ^{33}S, ^{34}S, ^{36}S	32.07 amu
copper	^{63}Cu, ^{65}Cu	63.55 amu

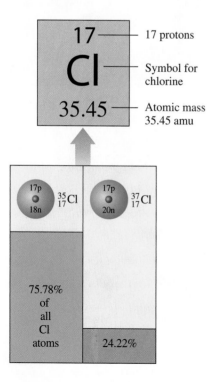

Atomic Mass

In laboratory work, a scientist uses samples that contain many atoms of an element. Among those atoms are all of the various isotopes with their different masses. To obtain a convenient mass to work with, chemists use the mass of an "average atom" of each element. This average atom has an **atomic mass**, which is the weighted average of the mass of all of the naturally occurring isotopes of that element.

To determine an atomic mass of an element, the masses of all the isotopes of that element must be determined experimentally. For example, in a sample of chlorine atoms, there are two isotopes, with ^{35}Cl being the more common isotope. In a large sample of chlorine atoms, there are three atoms of ^{35}Cl for every atom of ^{37}Cl. The atomic mass 35.45 of chlorine indicates a higher percentage of ^{35}Cl.

Most elements consist of several isotopes, and this is one reason atomic masses listed on the periodic table are seldom whole numbers.

Table 2.9 lists the isotopes of some selected elements and their atomic masses. On the periodic table, the atomic mass is given below the symbol of each element.

Calculating Atomic Mass

To calculate an atomic mass, the contribution of each isotope is determined by multiplying the mass of each isotope by its percent abundance and adding the results. For example, in a sample of chlorine (Cl), 75.78% of the Cl atoms have a mass of 34.97 amu and 24.22% of the Cl atoms have a mass of 36.97.

Isotope	Mass (amu)	×	Abundance (%)	=	Contribution to Average Cl Atom
^{35}Cl	34.97	×	$\dfrac{75.78}{100}$	=	26.50 amu
^{37}Cl	36.97	×	$\dfrac{24.22}{100}$	=	8.954 amu
			Atomic mass of Cl	=	35.45 amu

The atomic mass 35.45 amu is the weighted average mass of a sample of Cl atoms, although no individual Cl atom actually has this mass.

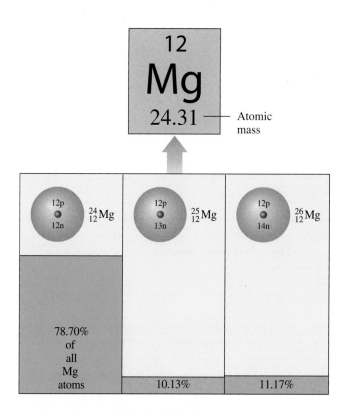

Atomic mass

  Atomic Mass Calculations

SAMPLE PROBLEM 2.6

■ Calculating Atomic Mass

Using Table 2.8, calculate the atomic mass for magnesium.

SOLUTION

$$^{24}\text{Mg} \quad 23.99 \quad \times \quad \frac{78.70}{100} \quad = \quad 18.88 \text{ amu}$$

$$^{25}\text{Mg} \quad 24.99 \quad \times \quad \frac{10.13}{100} \quad = \quad 2.531 \text{ amu}$$

$$^{26}\text{Mg} \quad 25.98 \quad \times \quad \frac{11.17}{100} \quad = \quad 2.902 \text{ amu}$$

$$\text{Atomic mass of Mg} \quad = \quad \overline{24.31 \text{ amu}}$$

STUDY CHECK

There are two naturally occurring isotopes of boron. ^{10}B has a mass of 10.01 amu with an abundance of 19.80% and ^{11}B has a mass of 11.01 amu with an abundance of 80.20%. What is the atomic mass of boron?

QUESTIONS AND PROBLEMS

■ Isotopes and Atomic Mass

2.37 What are the number of protons, neutrons, and electrons in the following isotopes?
   a. $^{27}_{13}\text{Al}$ **b.** $^{52}_{24}\text{Cr}$ **c.** $^{34}_{16}\text{S}$ **d.** $^{56}_{26}\text{Fe}$

2.38 What are the number of protons, neutrons, and electrons in the following isotopes?
   a. $_1^2$H   **b.** $_7^{14}$N   **c.** $_{14}^{26}$Si   **d.** $_{30}^{70}$Zn

2.39 Write the atomic symbols for isotopes with the following:
   a. 15 protons and 16 neutrons
   b. 35 protons and 45 neutrons
   c. 13 electrons and 14 neutrons
   d. a chlorine atom with 18 neutrons

2.40 Write the atomic symbols for isotopes with the following:
   a. an oxygen atom with 10 neutrons
   b. 4 protons and 5 neutrons
   c. 26 electrons and 30 neutrons
   d. a mass number of 24 and 13 neutrons

2.41 There are four isotopes of sulfur with mass numbers 32, 33, 34, and 36.
   a. Write the atomic symbol for each of these atoms.
   b. How are these isotopes alike?
   c. How are they different?
   d. Why is the atomic mass of sulfur listed on the periodic table not a whole number?

2.42 There are four isotopes of strontium with mass numbers 84, 86, 87, 88.
   a. Write the atomic symbol for each of these atoms.
   b. How are these isotopes alike?
   c. How are they different?
   d. Why is the atomic mass of strontium listed on the periodic table not a whole number?

2.43 Copper consists of two isotopes, ^{63}Cu and ^{65}Cu. If the atomic mass for copper on the periodic table is 63.55, are there more atoms of ^{63}Cu or ^{65}Cu in a sample of copper?

2.44 There are four naturally occurring isotopes of iron: ^{54}Fe, ^{56}Fe, ^{57}Fe, and ^{58}Fe. Use the atomic mass of iron listed on the periodic table to identify the most abundant isotope.

2.45 Two isotopes of gallium are naturally occurring with ^{69}Ga at 60.11% (68.93 amu) and ^{71}Ga at 39.89% (70.92 amu). What is the atomic mass of gallium?

2.46 Two isotopes of copper are naturally occurring with ^{63}Cu at 69.09% (62.93 amu) and ^{65}Cu at 30.91% (64.93 amu). What is the atomic mass of copper?

2.7 Electron Energy Levels

When we listen to a radio, use a microwave oven, turn on a light, see the colors of a rainbow, or have an X ray, we are using various forms of *electromagnetic radiation*. Light and other electromagnetic radiation consist of energy particles called photons that move as a wave of energy. In a wave, the distance between the peaks is called the *wavelength*. While all the forms of electromagnetic radiation travel at the speed of light, 3.0×10^8 meters per second, they differ in energy and wavelength. High-energy radiation such as gamma radiation has short wavelengths; low-energy radiation such as radio waves has long wavelengths. (See Figure 2.14.)

When sunlight passes through a prism or crystal, the light separates in a color spectrum: red, orange, yellow, green, blue, indigo, and violet. These are the same colors we see in a rainbow that forms when sunlight passes through raindrops acting as prisms. In this visible spectrum, which is the only part of the electromagnetic spectrum that we can see, red light has the longest wavelength and violet light has the shortest wavelength.

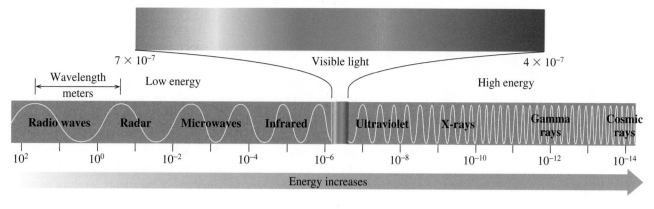

FIGURE 2.14

The electromagnetic spectrum contains radiation with long and short wavelengths in the visible and invisible regions.

Q **How does the energy of ultraviolet light compare to that of a microwave?**

When the light emitted from a heated element is passed through a prism, it does not produce a continuous spectrum. Instead, an **atomic spectrum** is produced that consists of lines of different colors seperated by dark areas. (See Figure 2.15.) This separation of colors indicates that only certain wavelengths of light are produced when an element is heated, which gives each element a unique atomic spectrum.

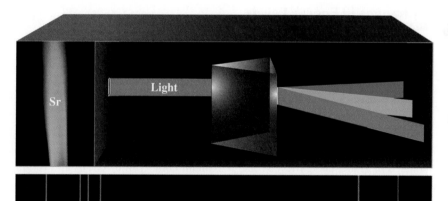

Strontium, Sr

FIGURE 2.15

A spectrum unique to each element is produced as light emitted from the heated element passes through a prism, which separates the light into colored lines.

Q **Why don't the elements form a continuous spectrum as seen with white light?**

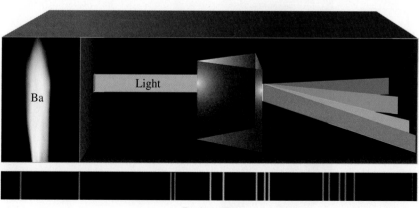

Barium, Ba

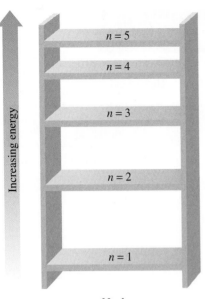

Nucleus

Energy Levels

Electron Energy Levels

Scientists associated the lines in atomic spectra with changes in the energies of the electrons. In an atom, each electron has a fixed or specific energy known as its energy level. The energy of an electron is *quantized*, which means that the energy of an electron can never be between any two specific energy levels.

The energy levels are assigned values called *principal quantum numbers (n)*, which are positive integers ($n = 1$, $n = 2$, . . .). Generally, electrons in the lower energy levels are closer to the nucleus, while electrons in the higher energy levels are farther away. As an analogy, we can think of the energy levels of an atom as similar to the shelves in a bookcase. The lowest energy level is the first shelf; the second energy level would be the second shelf. If we have a stack of books on the floor, it takes less energy to put them on the bottom shelves first, and then the second shelf, and so on. However, we could never get any book to stay in the space between the shelves. Unlike standard bookcases, however, there is a big difference in the energy of the first and second energy levels, and thereafter the energy levels are closer.

Principal Quantum Number (*n*)

$$1 < 2 < 3 < 4 < 5 < 6 < 7$$

Energy of electrons increases $\longrightarrow$

Changes in Energy Levels

By absorbing an amount of energy equal to the difference in energy levels, an electron is raised to a higher energy level. An electron loses energy when it falls to a lower energy level and emits electromagnetic radiation equal to the energy level difference. (See Figure 2.16.) If the electromagnetic radiation emitted has a wavelength in the visible range, we see a color.

There is a limit to the number of electrons in an energy level. Only a few electrons can occupy the lower energy levels, while more electrons can be accommodated in higher energy levels. The maximum number of electrons allowed in any energy level is calculated using the formula $2n^2$ (two times the square of the

FIGURE 2.16

Electrons can absorb a specific amount of energy to move to a higher energy level. When electrons lose energy, photons with specific energies are emitted.

Q **How does the energy of a photon of green light compare to the energy of a photon of red light?**

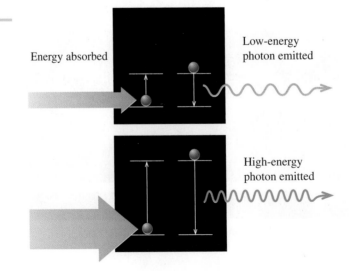

principal quantum number). In Table 2.10, we calculate the maximum number of electrons allowed in the first four energy levels.

TABLE 2.10

Capacity of Energy Levels 1–4

Energy Level (n)	Maximum Number of Electrons ($2n^2$)
1	$2(1)^2 = 2$
2	$2(2)^2 = 8$
3	$2(3)^2 = 18$
4	$2(4)^2 = 32$

Sublevels

Each energy level consists of one or more **sublevels**, which contain electrons with identical energy. The sublevels are identified by the letter s, p, d, and f. The number of sublevels within an energy level is equal to the principal quantum number. The first energy level ($n = 1$) has only one sublevel, $1s$. The second energy level ($n = 2$) has two sublevels, $2s$ and $2p$. The third energy level ($n = 3$) has three sublevels, $3s$, $3p$, and $3d$. The fourth energy level ($n = 4$) has four sublevels, $4s$, $4p$, $4d$, and $4f$. (See Figure 2.17.)

Within each energy level, the s sublevel has the lowest energy. If there are additional sublevels, the p sublevel has the next lowest energy, then the d sublevel, and finally the f sublevel.

Order of Increasing Energy of Sublevels Within an Energy Level

$$s < p < d < f$$

Lowest $\longrightarrow$ Highest
energy   energy

Energy levels from $n = 5$ and higher have as many sublevels as the value of n, but only s, p, d, and f sublevels are needed to hold the electrons of atoms of the elements known today.

Number of Electrons in Sublevels

There is a maximum number of electrons that can occupy each sublevel. An s sublevel holds 1 or 2 electrons. A p sublevel takes up to 6 electrons, a d sublevel can hold up to 10 electrons, and an f sublevel holds a maximum of 14 electrons. The total number of electrons in all the sublevels adds up to give the electrons in each energy level as shown in Table 2.11.

Orbitals

There is no way to know the exact location of an electron in an atom. Instead, scientists describe the location of an electron in terms of probability. A region in an atom where there is the highest probability of finding an electron is called an **orbital**. Suppose you could draw an imaginary circle with a 100-m radius around your chemistry classroom. There is a high probability of finding you within that area when your chemistry class is in session. But once in a while, you may be found outside that circle because you were sick or your car did not start.

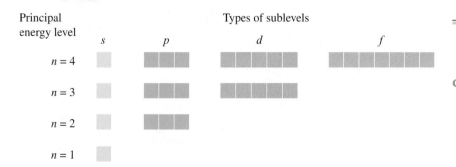

Principal energy level   Types of sublevels

s p d f

$n = 4$

$n = 3$

$n = 2$

$n = 1$

FIGURE 2.17

The number of sublevels in an energy level is the same as the principal quantum number n.

Q **How many sublevels are in energy level $n = 5$?**

TABLE 2.11

Orbitals in Energy Levels 1–4

Energy Level	Orbitals	Maximum Number of Electrons	Energy Level Capacity
1	1s	2	2
2	2s	2	8
	2p 2p 2p	6	
3	3s	2	18
	3p 3p 3p	6	
	3d 3d 3d 3d 3d	10	
4	4s	2	32
	4p 4p 4p	6	
	4d 4d 4d 4d 4d	10	
	4f 4f 4f 4f 4f 4f 4f	14	

The *Pauli exclusion principle* states that an orbital can hold up to 2 electrons, but no more. Within the same orbital, two electrons spin in opposite directions. We represent the spins of the electrons in the same orbital with one arrow pointing up and the other pointing down.

Shapes of Orbitals

Each sublevel within an energy level is composed of the same type of orbitals. There is an *s* orbital for each *s* sublevel; *p* orbitals for each *p* sublevel; *d* orbitals for each *d* sublevel; and *f* orbitals for each *f* sublevel. Each type of orbital has a unique shape. In an *s* orbital, the electrons are most likely found in a region with a spherical shape. Every *s* orbital can hold one or two electrons; there is just one *s* orbital for every *s* sublevel. While the shape of every *s* orbital is spherical, there is an increase in the size of the *s* orbitals in higher energy levels. (See Figure 2.18).

A *p* sublevel consists of three *p* orbitals, each of which has two lobes. The three *p* orbitals are arranged in three different directions (*x*, *y*, and *z* axes) around the nucleus. (See Figure 2.19.) Because each *p* orbital can hold up to two electrons, the

An *s* orbital

FIGURE 2.18

An *s* orbital is a sphere that represents the region of highest probability of finding an *s* electron around the nucleus of an atom.

Q Is the probability high or low of finding an *s* electron outside an *s* orbital?

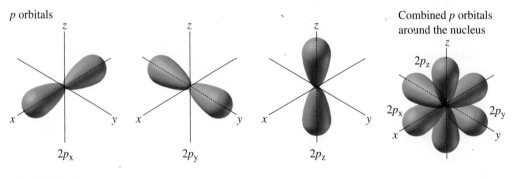

p orbitals

Combined p orbitals around the nucleus

2px   2py   2pz

FIGURE 2.19

Each of the *p* orbitals has a dumbbell shape and is aligned along a different axis. Each *p* orbital holds a maximum of 2 electrons.

Q If there are three *p* orbitals in a *p* sublevel, what is the maximum number of electrons possible in a *p* sublevel?

three *p* orbitals can accommodate six electrons in a *p* sublevel. At higher energy levels, the shape of *p* orbitals is the same, but the volume increases.

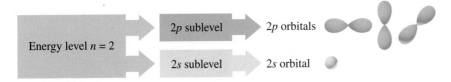

A *d* sublevel consists of five *d* orbitals, which means that a *d* sublevel can hold a maximum of 10 electrons. With a total of seven *f* orbitals, an *f* sublevel can hold up to 14 electrons. The shapes of *d* orbitals and *f* orbitals are more complex and we have not included them in this text.

SAMPLE PROBLEM 2.7

■ Energy Levels, Sublevels, and Orbitals

Indicate the type and number of orbitals in each of the following energy levels or sublevels.

a. 3*p* sublevel   **b.** *n* = 2   **c.** *n* = 3   **d.** 4*d* sublevel

SOLUTION

a. The 3*p* sublevel contains three 3*p* orbitals.
b. The *n* = 2 principal energy level consists of 2*s* (one) and 2*p* (three) orbitals.
c. The *n* = 3 principal energy level consists of 3*s* (one), 3*p* (three), and 3*d* (five) orbitals.
d. The 4*d* sublevel contains five 4*d* orbitals.

STUDY CHECK

What is similar and what is different for 1*s*, 2*s*, and 3*s* orbitals?

QUESTIONS AND PROBLEMS

■ Electron Energy Levels

2.47 Describe the shape of the following orbitals:
   a. 1*s*　　**b.** 2*p*　　**c.** 5*s*

2.48 Describe the shape of the following orbitals:
   a. 3*p*　　**b.** 6*s*　　**c.** 4*p*

2.49 What is similar about the following?
   a. 1*s* and 2*s* orbitals
   b. 3*s* and 3*p* sublevels
   c. 3*p* and 4*p* sublevels
   d. three 3*p* orbitals

2.50 What is similar about the following?
   a. 5*s* and 6*s* orbitals
   b. 3*p* and 4*p* orbitals
   c. 3*s* and 4*s* sublevels
   d. 2*s* and 2*p* orbitals

2.51 Indicate the number of each in the following:
   a. orbitals in the 3*d* sublevel
   b. sublevels in the energy level *n* = 1
   c. orbitals in the 6*s* sublevel
   d. orbitals in the energy level *n* = 3

2.52 Indicate the number of each in the following:
   a. orbitals in the energy level $n = 2$
   b. sublevels in the energy level $n = 4$
   c. orbitals in the $5f$ sublevel
   d. orbitals in the $6p$ sublevel

2.53 Indicate the maximum number of electrons in the following:
   a. $2p$ orbital
   b. $3p$ sublevel
   c. main energy level $n = 4$
   d. $5d$ sublevel

2.54 Indicate the maximum number of electrons in the following:
   a. $3s$ sublevel
   b. $4p$ orbital
   c. main energy level $n = 3$
   d. $4f$ sublevel

LEARNING GOAL

Use the periodic table to write orbital diagrams and electron configurations.

 Electron Configurations

2.8 Electron Configurations

We can now look at how electrons are arranged in the orbitals within an atom. In an **orbital diagram**, boxes (or circles) represent the orbitals containing electrons. We see from the energy diagram (Figure 2.20) that the electrons in the $1s$ orbital have a lower energy level than in the $2s$ orbital.

Period 1   Hydrogen and Helium

We can begin to draw the orbital diagrams and build the electron configurations for the elements H and He in Period 1. The $1s$ orbital (which is also the $1s$ sublevel) is used first because it has the lowest energy. Hydrogen has one electron in the $1s$ sublevel; helium has two. In the orbital diagram, the electrons for helium are shown with opposite spins.

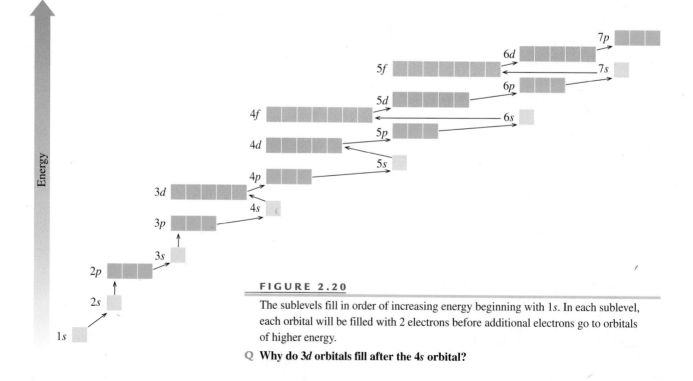

FIGURE 2.20

The sublevels fill in order of increasing energy beginning with $1s$. In each sublevel, each orbital will be filled with 2 electrons before additional electrons go to orbitals of higher energy.

Q Why do $3d$ orbitals fill after the $4s$ orbital?

The **electron configuration** of an atom is "built up" by placing the electrons of an atom in the sublevels in order of increasing energy. The electron configuration for helium is written as

Sublevel   Number of electrons

$1s^2$ Read as "one s two"

Atomic Number	Element	Orbital Diagram	Electron Configuration
		$1s$	
1	H	↑	$1s^1$
2	He	↑↓	$1s^2$

Period 2   Lithium to Neon

Period 2 begins with lithium, which has 3 electrons. The first 2 electrons fill the $1s$ orbital, while the third electron goes into the $2s$ orbital, the sublevel with the next lowest energy. In beryllium, another electron is added to complete the $2s$ orbital. Because $2p$ orbitals have equal energy, electrons from boron to nitrogen add one at a time to give three half-filled $2p$ orbitals. From oxygen to neon, the remaining electrons are paired up using opposite spins until the $2p$ sublevel is complete. In writing the complete electron configurations for the elements in Period 2, begin with the $1s$ sublevel followed by the $2s$ and the $2p$ sublevels.

An electron configuration can also be written in an *abbreviated configuration*. The electron configuration of the preceding noble gas is replaced by writing its symbol inside square brackets. For example, the electron configuration for lithium $1s^2 2s^1$ can be abbreviated as $[He]2s^1$, where $[He]$ replaces $1s^2$.

Atomic Number	Element	Orbital Diagram	Electron Configuration	Abbreviated Configuration
3	Li	$1s$ ↑↓ $2s$ ↑	$1s^2 2s^1$	$[He]\,2s^1$
4	Be	↑↓ ↑↓	$1s^2 2s^2$	$[He]\,2s^2$
5	B	↑↓ ↑↓ $2p$ ↑ □ □	$1s^2 2s^2 2p^1$	$[He]\,2s^2 2p^1$
6	C	↑↓ ↑↓ ↑ ↑ □	$1s^2 2s^2 2p^2$	$[He]\,2s^2 2p^2$
7	N	↑↓ ↑↓ ↑ ↑ ↑	$1s^2 2s^2 2p^3$	$[He]\,2s^2 2p^3$
8	O	↑↓ ↑↓ ↑↓ ↑ ↑	$1s^2 2s^2 2p^4$	$[He]\,2s^2 2p^4$
9	F	↑↓ ↑↓ ↑↓ ↑↓ ↑	$1s^2 2s^2 2p^5$	$[He]\,2s^2 2p^5$
10	Ne	↑↓ ↑↓ ↑↓ ↑↓ ↑↓	$1s^2 2s^2 2p^6$	$[He]\,2s^2 2p^6$

Unpaired electrons

Period 3   Sodium to Argon

In Period 3, electrons enter the orbitals of the 3s and 3p sublevels, but not the 3d sublevel. We notice that the elements sodium to argon, which are directly below the elements lithium to neon in Period 2, have a similar pattern of filling their s and p orbitals. We can write the complete orbital diagram for phosphorus as follows:

For elements in Period 3 and above, we usually write the orbital diagrams for only the electrons in the highest energy levels. In Period 3, the symbol [Ne] replaces the electron configuration of neon $1s^22s^22p^6$. The abbreviated form is convenient to use for electron configurations that contain several sublevel notations.

Atomic Number	Element	Orbital Diagram (3s and 3p orbitals only) 3s 3p	Electron Configuration	Abbreviated Form
11	Na	↑	$1s^22s^22p^63s^1$	[Ne] $3s^1$
12	Mg	↑↓	$1s^22s^22p^63s^2$	[Ne] $3s^2$
13	Al	↑↓ ↑	$1s^22s^22p^63s^23p^1$	[Ne] $3s^23p^1$
14	Si	↑↓ ↑ ↑	$1s^22s^22p^63s^23p^2$	[Ne] $3s^23p^2$
15	P	↑↓ ↑ ↑ ↑	$1s^22s^22p^63s^23p^3$	[Ne] $3s^23p^3$
16	S	↑↓ ↑↓ ↑ ↑	$1s^22s^22p^63s^23p^4$	[Ne] $3s^23p^4$
17	Cl	↑↓ ↑↓ ↑↓ ↑	$1s^22s^22p^63s^23p^5$	[Ne] $3s^23p^5$
18	Ar	↑↓ ↑↓ ↑↓ ↑↓	$1s^22s^22p^63s^23p^6$	[Ne] $3s^23p^6$

SAMPLE PROBLEM 2.8

▪ Orbital Diagrams and Electron Configurations

For each of the following elements, write the stated type of electron notation:

a. orbital diagram for silicon
b. electron configuration for phosphorus
c. abbreviated electron configuration for chlorine

SOLUTION

a. Silicon in Period 3 has atomic number 14, which tells us that it has 14 electrons. To write the orbital diagram, we draw boxes for the orbitals up to 3p.

Add 14 electrons, starting with the $1s$ orbital. Show paired electrons in the same orbital with opposite spins and place the last 2 electrons in different $3p$ orbitals.

b. The electron configuration gives the electrons that fill the sublevel in order of increasing energy. Phosphorus is in Group 5A (15) in Period 3. In Periods 1 and 2, a total of 10 electrons fill sublevels: $1s^2$, $2s^2$, and $2p^6$. In Period 3, 2 electrons go into $3s^2$. The 3 remaining electrons (total of 15) are placed in the $3p$ sublevel.

$$P \quad 1s^2 2s^2 2p^6 3s^2 3p^3$$

c. In chlorine, the previous noble gas is neon. For the abbreviated configuration, write [Ne] for $1s^2 2s^2 2p^6$ and the electrons in the $3s$ and $3p$ sublevels.

$$[Ne]\ 3s^2 3p^5$$

STUDY CHECK

Write the complete and abbreviated electron configuration for sulfur.

Electron Configurations and the Periodic Table

Up until now we have written electron configurations using the energy diagram. As configurations involve more sublevels, this becomes tedious. However, on the periodic table, the atomic numbers are in order of increasing sublevel energy. The electron configurations of the elements are related to their position in the periodic table. Different sections or sublevel blocks within the table correspond to the s, p, d, and f sublevels. (See Figure 2.21.) Therefore, we can "build up" atoms by reading the periodic table from left to right across each period.

MC GOB — SELF-STUDY ACTIVITY
Bohr's Shell Model

1. The **s block elements** include hydrogen and helium and the elements in Groups 1A (1) and Group 2A (2). This means that the final one or two electrons in the elements of the s block are located in s sublevels. The

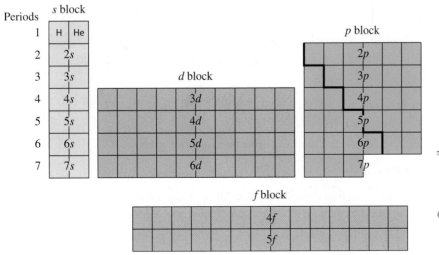

FIGURE 2.21

Electron configuration follows the order of sublevels on the periodic table.

Q If neon is in the Group 8A, Period 2, how many electrons are in the $1s$, $2s$, and $2p$ sublevels of neon?

period number indicates the particular *s* sublevel that is filling: 1*s*, 2*s*, and so on.

2. The *p* **block elements** consists of the elements in Group 3A (13) to Group 8A (18). There are six *p* block elements in each period because each *p* sublevel can hold up to six electrons. The period number indicates the particular *p* sublevel that is filling: 2*p*, 3*p*, and so on.

3. The *d* **block elements** first appear after calcium (atomic number 20) with the ten columns of elements of the transition metals. There are ten elements in the *d* block because each *d* sublevel can hold up to 10 electrons. The particular *d* sublevel is one less $(n - 1)$ than the period number. For example, in Period 4, the first *d* block is the 3*d* sublevel. In Period 5, the second *d* block is the 4*d* sublevel.

4. The *f* **block elements** include all the elements in the two rows at the bottom of the periodic table. There are 14 elements in each *f* block because an *f* sublevel can hold up to 14 electrons. Elements that have atomic numbers higher than 57 (La) have electrons in the 4*f* block. The particular *f* sublevel is two less $(n - 2)$ than the period number. For example, in Period 6, the first *f* block is the 4*f* sublevel. In Period 7, the second *f* block is the 5*f* sublevel.

Writing Electron Configurations Using Sublevel Blocks

Now we can write electron configurations using the sublevel blocks on the periodic table. As before, each configuration begins at H. But now we move across the table writing down each block we come to until we reach the element for which we are writing an electron configuration. For each element, we can use the following steps:

- **STEP 1** Locate the element on the periodic table.
- **STEP 2** Write the filled sublevels in order going across each period (left to right).
- **STEP 3** Count the number of electrons in the sublevel for the given element and complete the configuration.

To write the electron configuration for chlorine (atomic number 17) from the sublevel blocks on the periodic table,

Period		Sublevel Blocks Filled
1	1*s* sublevel (H $\longrightarrow$ He)	$1s^2$
2	2*s* sublevel (Li $\longrightarrow$ Be) then 2*p* sublevel (B $\longrightarrow$ Ne)	$2s^2 \longrightarrow 2p^6$
3	3*s* sublevel (Na $\longrightarrow$ Mg) then 3*p* sublevel (Al $\longrightarrow$ Cl)	$3s^2 \longrightarrow 3p?$ (Cl)

Writing the sublevel blocks in order up to chlorine gives

$$1s^2 2s^2 2p^6 3s^2 3p?$$

Chlorine is the fifth element in the 3*p* block, which means that chlorine has five 3*p* electrons. The complete electron configuration for chlorine is written as

$$1s^2 2s^2 2p^6 3s^2 3p^5 \quad \text{Final sublevel block where Cl appears}$$

Guide to Writing Electron Configurations with Sublevel Blocks

STEP 1
Locate the element on the periodic table.

STEP 2
Write the filled sublevels in order going across each period.

STEP 3
Count the number of electrons in the sublevel for the given element and complete the configuration.

Period 4

Up until now, the filling of sublevels has progressed in order. However, if we look at the sublevel blocks in Period 4, we see that the $4s$ sublevel block fills before the $3d$ orbitals. The $4s$ orbital has a slightly lower energy than the $3d$ orbitals, which means that the $4s$ sublevel has the next lowest energy following the filling of the $3p$ sublevel at the end of Period 3. This order occurs again in Period 5 when the $5s$ orbital fills before the $4d$ orbitals, and again in Period 6 when the $6s$ fills before the $5d$. (See Figure 2.20, page 78.)

At the beginning of Period 4, the 1 and 2 remaining electrons in potassium (19) and calcium (20) go into the $4s$ orbital. In scandium, the remaining electron goes into the $3d$ block, which continues to fill until it has 10 electrons at zinc (30). Once the $3d$ block is complete, the next 6 electrons go into the $4p$ block for elements gallium (31) to krypton (36).

Atomic Number	Element	Electron Configuration	Abbreviated Configuration
$4s$ Block Elements			
19	K	$1s^2 2s^2 2p^6 3s^2 3p^6 4s^1$	$[\text{Ar}]4s^1$
20	Ca	$1s^2 2s^2 2p^6 3s^2 3p^6 4s^2$	$[\text{Ar}]4s^2$
$3d$ Block Elements			
21	Sc	$1s^2 2s^2 2p^6 3s^2 3p^6 4s^2 3d^1$	$[\text{Ar}]4s^2 3d^1$
22	Ti	$1s^2 2s^2 2p^6 3s^2 3p^6 4s^2 3d^2$	$[\text{Ar}]4s^2 3d^2$
23	V	$1s^2 2s^2 2p^6 3s^2 3p^6 4s^2 3d^3$	$[\text{Ar}]4s^2 3d^3$
24	Cr	$1s^2 2s^2 2p^6 3s^2 3p^6 4s^1 3d^5$	$[\text{Ar}]4s^1 3d^5$ (1/2 filled d sublevel is stable)
25	Mn	$1s^2 2s^2 2p^6 3s^2 3p^6 4s^2 3d^5$	$[\text{Ar}]4s^2 3d^5$
26	Fe	$1s^2 2s^2 2p^6 3s^2 3p^6 4s^2 3d^6$	$[\text{Ar}]4s^2 3d^6$
27	Co	$1s^2 2s^2 2p^6 3s^2 3p^6 4s^2 3d^7$	$[\text{Ar}]4s^2 3d^7$
28	Ni	$1s^2 2s^2 2p^6 3s^2 3p^6 4s^2 3d^8$	$[\text{Ar}]4s^2 3d^8$
29	Cu	$1s^2 2s^2 2p^6 3s^2 3p^6 4s^1 3d^{10}$	$[\text{Ar}]4s^1 3d^{10}$ (filled d sublevel is stable)
30	Zn	$1s^2 2s^2 2p^6 3s^2 3p^6 4s^2 3d^{10}$	$[\text{Ar}]4s^2 3d^{10}$
$4p$ Block Elements			
31	Ga	$1s^2 2s^2 2p^6 3s^2 3p^6 4s^2 3d^{10} 4p^1$	$[\text{Ar}]4s^2 3d^{10} 4p^1$
32	Ge	$1s^2 2s^2 2p^6 3s^2 3p^6 4s^2 3d^{10} 4p^2$	$[\text{Ar}]4s^2 3d^{10} 4p^2$
33	As	$1s^2 2s^2 2p^6 3s^2 3p^6 4s^2 3d^{10} 4p^3$	$[\text{Ar}]4s^2 3d^{10} 4p^3$
34	Se	$1s^2 2s^2 2p^6 3s^2 3p^6 4s^2 3d^{10} 4p^4$	$[\text{Ar}]4s^2 3d^{10} 4p^4$
35	Br	$1s^2 2s^2 2p^6 3s^2 3p^6 4s^2 3d^{10} 4p^5$	$[\text{Ar}]4s^2 3d^{10} 4p^5$
36	Kr	$1s^2 2s^2 2p^6 3s^2 3p^6 4s^2 3d^{10} 4p^6$	$[\text{Ar}]4s^2 3d^{10} 4p^6$

Some Exceptions in Sublevel Block Order

Within the filling of the $3d$ sublevel, exceptions occur for chromium and copper. In Cr and Cu, the $3d$ sublevel is close to a half-filled or filled sublevel, which is particularly stable. Thus, chromium has only 1 electron in the $4s$ and 5 electrons in the $3d$ sublevel to give the added stability of a half-filled d sublevel.

A similar exception occurs when copper achieves a completely filled $3d$ sublevel by placing only 1 electron in the $4s$ sublevel and using 10 electrons to complete the $3d$ sublevel. After the $4s$ and $3d$ sublevels are completed, the $4p$ sublevel fills as expected from gallium to krypton, the noble gas that completes Period 4.

SAMPLE PROBLEM 2.9

■ Using Sublevel Blocks to Write Electron Configurations

Use the sublevel blocks on the periodic table to write the electron configuration for bromine.

SOLUTION

a. ■ **STEP 1** Bromine is in the p block and in Period 4.

■ **STEP 2** Beginning with $1s^2$, go across the periodic table writing each filled sublevel block as follows:

Period 1	$1s^2$
Period 2	$2s^2 \longrightarrow 2p^6$
Period 3	$3s^2 \longrightarrow 3p^6$
Period 4	$4s^2 \longrightarrow 3d^{10} \longrightarrow 4p?$

■ **STEP 3** There are 5 electrons in the $4p$ sublevel for Br ($4p^5$), which determine the electron configuration for Br.

$$1s^2 2s^2 2p^6 3s^2 3p^6 4s^2 3d^{10} 4p^5$$

STUDY CHECK

Write the electron configuration for tin.

QUESTIONS AND PROBLEMS

■ Electron Configurations

2.55 Write an orbital diagram for an atom of each of the following:
 a. boron **b.** aluminum **c.** phosphorus **d.** argon

2.56 Write an orbital diagram for an atom of each of the following:
 a. fluorine **b.** sodium **c.** magnesium **d.** sulfur

2.57 Write a complete electron configuration for an atom of each of the following:
 a. nitrogen **b.** sodium **c.** sulfur **d.** arsenic **e.** iron

2.58 Write a complete electron configuration for an atom of each of the following:
 a. carbon **b.** silicon **c.** phosphorus **d.** cobalt **e.** gallium

2.59 Write an abbreviated electron configuration for an atom of each of the following:
 a. magnesium **b.** sulfur **c.** aluminum **d.** titanium **e.** barium

2.60 Write an abbreviated electron configuration for an atom of each of the following:
 a. sodium **b.** oxygen **c.** nickel **d.** tin **e.** silver

2.61 Give the symbol of the element with each of the following electron configurations:
 a. $1s^2 2s^2 2p^6 3s^2 3p^4$ **b.** $1s^2 2s^2 2p^6 3s^2 3p^6 4s^2 3d^7$
 c. $[Ne]3s^2 3p^2$ **d.** $[Ar]4s^2 3d^{10} 4p^5$

2.62 Give the symbol of the element with each of the following electron configurations:
 a. $1s^2 2s^2 2p^4$ **b.** $1s^2 2s^2 2p^6 3s^2 3p^6$
 c. $[Ne]3s^2 3p^1$ **d.** $[Ar]4s^2 3d^4$

2.63 Give the symbol of the element that meets the following conditions:
 a. has 3 electrons in energy level $n = 3$
 b. has two $2p$ electrons
 c. completes the $3p$ sublevel
 d. has 2 electrons in the $4d$ sublevel

2.64 Give the symbol of the element that meets the following conditions:
 a. has 5 electrons in the $3p$ sublevel
 b. has three $2p$ electrons

c. completes the 3*s* sublevel
d. has four 5*p* electrons

2.65 Give the number of electrons in the indicated orbitals for the following:
  a. 3*d* in zinc   **b.** 2*p* in sodium
  c. 4*p* in arsenic   **d.** 5*s* in rubidium

2.66 Give the number of electrons in the indicated orbitals for the following:
  a. 3*d* in manganese   **b.** 5*p* in antimony
  c. 6*p* in lead   **d.** 3*s* in magnesium

2.9 Periodic Trends

LEARNING GOAL

Use the electron configurations of elements to explain periodic trends.

The electron configurations of atoms are an important factor in the physical and chemical behavior of the elements. Going across a period, there is a pattern of regular change in these properties from one group to the next. Known as *periodic properties*, each property increases or decreases across a period and then the trend is repeated again in each successive period. We can use the seasonal changes in temperatures as an analogy for periodic properties. In the winter, temperatures are cold and become warmer in the spring. By summer, the outdoor temperatures are hot, but begin to cool in the fall. By winter, we expect cold temperatures again as the pattern of decreasing and increasing temperatures repeats for another year.

Group Number and Valence Electrons

The chemical properties of representative elements are mostly due to the **valence electrons**, which are the electrons in the outermost energy levels. These valence electrons occupy the *s* and *p* sublevels with the highest quantum number *n*. The group numbers indicate the number of valence (outer) electrons for the elements in each vertical column. For example, the elements in Group 1A (1) such as lithium, sodium, and potassium, all have 1 electron in the outer energy level. Looking at the sublevel block, we can represent the valence electron in the alkali metals of Group 1A (1) as ns^1. All the elements in Group 2A (2), the alkaline earth metals, have two (2) valence electrons, ns^2. The halogens in Group 7A (17) have seven (7) valence electrons, ns^2np^5.

We can see the repetition of the outermost *s* and *p* electrons for the representative elements in Periods 1 to 4 in Table 2.12. Helium is included in Group 8A (18) because it is a noble gas, but it has only 2 electrons in its complete energy level.

Atoms of magnesium

Electron-Dot Symbols

An **electron-dot symbol** is a convenient way to represent the valence electron. Valence electrons are shown as dots placed on the sides, top, or bottom of the symbol for the element. It does not matter on which of the four sides you place the dots. However, 1 to 4 valence electrons are arranged as single dots. When there are more than 4 electrons, the electrons begin to pair up. Any of the following would be an acceptable electron-dot symbol for magnesium, which has 2 valence electrons:

Possible Electron-Dot Symbols for the 2 Valence Electrons in Magnesium

  Mg·   Mg   ·Mg   ·Mg·   Mg·   ·Mg

Electron-dot symbols for selected elements are given in Table 2.13.

Mg·
Electron-dot symbol

$1s^2 2s^2 2p^6 \boxed{3s^2}$

Electron configuration of magnesium

TABLE 2.12

Valence Electrons for Representative Elements in Periods 1–4

1A(1)	2A(2)	3A(13)	4A(14)	5A(15)	6A(16)	7A(17)	8A(18)
1							2
H							He
$1s^1$							$1s^2$
3	4	5	6	7	8	9	10
Li	Be	B	C	N	O	F	Ne
$2s^1$	$2s^2$	$2s^2 2p^1$	$2s^2 2p^2$	$2s^2 2p^3$	$2s^2 2p^4$	$2s^2 2p^5$	$2s^2 2p^6$
11	12	13	14	15	16	17	18
Na	Mg	Al	Si	P	S	Cl	Ar
$3s^1$	$3s^2$	$3s^2 3p^1$	$3s^2 3p^2$	$3s^2 3p^3$	$3s^2 3p^4$	$3s^2 3p^5$	$3s^2 3p^6$
19	20	31	32	33	34	35	36
K	Ca	Ga	Ge	As	Se	Br	Kr
$4s^1$	$4s^2$	$4s^2 4p^1$	$4s^2 4p^2$	$4s^2 4p^3$	$4s^2 4p^4$	$4s^2 4p^5$	$4s^2 4p^6$

TABLE 2.13

Electron-Dot Symbols for Representative Elements in Periods 1–4

	Group Number							
	1A (1)	2A (2)	3A (13)	4A (14)	5A (15)	6A (16)	7A (17)	8A (18)
Valence Electron Configuration	ns^1	ns^2	$ns^2 np^1$	$ns^2 np^2$	$ns^2 np^3$	$ns^2 np^4$	$ns^2 np^5$	$ns^2 np^6$
	H·							He:
	Li·	Ḃe·	·Ḃ·	·Ċ·	·N̈·	·Ö:	·F̈:	:Ḧe:
	Na·	Ṁg·	·Äl·	·S̈i·	·P̈·	·S̈:	·C̈l:	:Är:
	K·	Ċa·	·Ġa·	·Ġe·	·Äs·	·S̈e:	·B̈r:	:K̈r:

SAMPLE PROBLEM 2.10

■**Writing Electron-Dot Symbols**

Write the electron-dot symbol for each of the following elements:

a. bromine
b. aluminum

SOLUTION

a. Because the group number for bromine is 7A (17), bromine has 7 valence electrons.

·B̈r:

b. Aluminum, in Group 3A (3), has 3 valence electrons.

·Äl·

STUDY CHECK

What is the electron-dot symbol for phosphorus?

Atomic Size

Although there are no fixed boundaries to atoms, scientists have a good idea of the typical volume of the electron clouds in atoms. This volume can be described in terms of its *atomic radius*, which is the distance from the nucleus to the valence (outermost) electrons. For representative elements, the atomic radius increases going down each group. We would expect an increase in radius because the outermost electrons in higher energy levels are farther from the nucleus. For example, in the alkali metals, Li has a valence electron in the $2s$ sublevel; Na has a valence electron in the $3s$ sublevel; K has a valence electron in the $4s$ sublevel; and Rb has a valence electron in the $5s$ sublevel. (See Figure 2.22.)

The atomic radii of representative elements typically decrease from left to right on the periodic table. Going across a period, there is an increase in the number of valence electrons with a corresponding increase in protons, which causes a stronger attraction for the outermost electrons. Going across a period, the valence electrons are pulled closer to the nucleus, which makes the atoms smaller.

Ionization Energy

Ionization Energy
Patterns in the Periodic Table

Electrons are held in atoms by their attraction to the nucleus. Therefore, energy is required to remove an electron from an atom. The **ionization energy** is the energy needed to remove the least tightly bound electron from an atom in the gaseous (*g*)

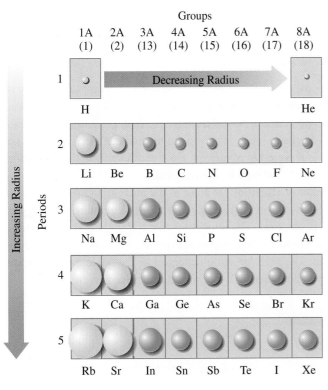

FIGURE 2.22

The atomic radius increases going down a group, but decreases going from left to right across a period.

Q **Why does the atomic radius increase going down a group?**

FIGURE 2.23

Ionization energies for the representative elements tend to decrease going down a group and increase going across a period.

Q Why is the ionization energy for Li less than for O?

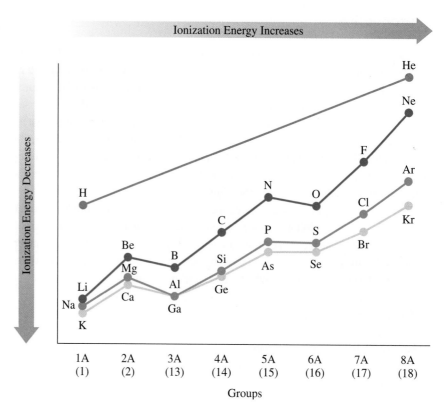

Ionization Energy Increases

Ionization Energy Decreases

| 1A (1) | 2A (2) | 3A (13) | 4A (14) | 5A (15) | 6A (16) | 7A (17) | 8A (18) |

Groups

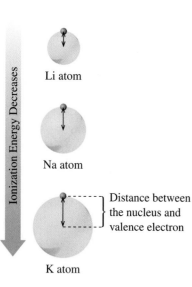

Ionization Energy Decreases

Li atom

Na atom

Distance between the nucleus and valence electron

K atom

state. When an electron is removed from a neutral atom, a particle called a cation, with a 1+ charge, is formed.

$$Na(g) + energy\ (ionization) \longrightarrow Na^+(g) + e^-$$

The ionization energy generally decreases going down a group. Less energy is needed to remove an electron as nuclear attraction decreases for electrons farther from the nucleus. *Going across a period from left to right, the ionization energy generally increases.* As the number of protons in the nucleus increases across a period, more energy is required to remove an electron. In general, the ionization energy is low for the metals and high for the nonmetals.

In Period 1, the valence electrons are close to the nucleus and strongly held. H and He have high ionization energies because a large amount of energy is required to remove an electron. The ionization energy for He is the highest of any element because He has a full, stable, energy level which is disrupted by removing an electron. The high ionization energies of the noble gases indicate that their electron arrangements are especially stable. (See Figure 2.23.)

SAMPLE PROBLEM 2.11

■**Ionization Energy**

Indicate the element in each that has the higher ionization energy and explain your choice.

a. K or Na   **b.** Mg or Cl   **c.** F or N

SOLUTION

a. Na. In Na, the valence electron is closer to the nucleus.
b. Cl. Attraction for the valence electrons increases across a period, going left to right.

c. F. With more protons, the F nucleus has a stronger attraction for valence electrons in the second energy level than N.

STUDY CHECK

Arrange Sn, Sr, and I in order of increasing ionization energy.

QUESTIONS AND PROBLEMS

(MC GOB)

■ Periodic Trends

2.67 Write the group number using both A/B notation and 1–18 numbering of elements that have the following outer electron configuration:
  a. $2s^2$ **b.** $3s^2 3p^3$ **c.** $5s^2 4d^{10} 5p^4$

2.68 Write the group number using both A/B notation and 1–18 numbering of elements that have the following outer electron configuration:
  a. $4s^2 4p^5$ **b.** $4s^1$ **c.** $5s^2 4d^{10} 5p^2$

2.69 Indicate the number of valence (outermost) electrons in each of the following:
  a. aluminum   **b.** Group 5A   **c.** F, Cl, Br, and I

2.70 Indicate the number of valence (outermost) electrons in each of the following:
  a. Li, Na, K, Rb, and Cs
  b. C, Si, Ge, Sn, and Pb
  c. Group 8A

2.71 Write the group number and electron-dot symbol for each element:
  a. sulfur   **b.** nitrogen   **c.** calcium
  d. sodium   **e.** barium

2.72 Write the group number and electron-dot symbol for each element:
  a. carbon   **b.** oxygen   **c.** bromine
  d. lithium   **e.** chlorine

2.73 Using the symbol M for a metal atom, draw the electron-dot symbol for an atom of a metal in the following groups:
  a. Group 1A (1)
  b. Group 2A (2)

2.74 Using the symbol Nm for a nonmetal atom, draw the electron-dot symbol for an atom of a nonmetal in the following groups:
  a. Group 5A (15)
  b. Group 7A (17)

2.75 Place the elements in each set in order of decreasing atomic radius.
  a. Mg, Al, Si   **b.** Cl, Br, I   **c.** I, Sb, Sr

2.76 Place the elements in each set in order of decreasing atomic radius.
  a. Cl, S, P   **b.** Ge, Si, C   **c.** Ba, Ca, Sr

2.77 Select the larger atom in each pair.
  a. Na or Cl   **b.** Na or Rb   **c.** Na or Mg

2.78 Select the larger atom in each pair.
  a. S or Cl   **b.** S or O   **c.** S or Se

2.79 Arrange each set of elements in order of increasing ionization energy.
  a. F, Cl, Br   **b.** Na, Cl, Al   **c.** Na, K, Cs

2.80 Arrange each set of elements in order of increasing ionization energy.
  a. C, N, O   **b.** P, S, Cl   **c.** As, P, N

2.81 Select the element in each pair with the higher ionization energy.
  a. Br or I   **b.** Mg or S   **c.** Si or P

2.82 Select the element in each pair with the higher ionization energy.
  a. O or Ne   **b.** K or Br   **c.** Ca or Ba

CONCEPT MAP

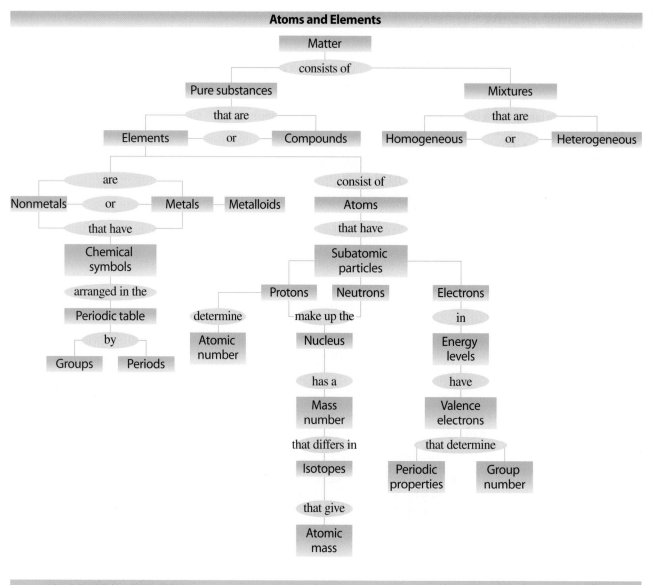

CHAPTER REVIEW

2.1 Classification of Matter

Matter is everything that occupies space and has mass. Matter is classified as pure substances or mixtures. Pure substances, which are elements or compounds, have fixed compositions. Mixtures, which can be homogeneous or heterogeneous, have variable compositions. The substances in mixtures can be separated using physical methods.

2.2 Elements and Symbols

Elements are the primary substances of matter. Chemical symbols are one- or two-letter abbreviations of the names of the elements.

2.3 The Periodic Table

The periodic table is an arrangement of the elements by increasing atomic number. A vertical column on the periodic table containing elements with similar properties is called a group. A horizontal row is called a period. Elements in Group 1A (1) are called the alkali metals; Group 2A (2), alkaline earth metals; Group 7A (17), the halogens; and Group 8A (18), the noble gases. On the periodic table, metals are located on the left of the heavy zigzag line, and nonmetals are to the right of the heavy zigzag line. Elements located on the heavy line are called metalloids.

2.4 The Atom

An atom is the smallest particle that retains the characteristics of an element. Atoms are composed of three subatomic particles. Protons have a positive charge (+), electrons carry a negative charge (−), and neutrons are electrically neutral. The protons and neutrons are found in the tiny, dense nucleus. Electrons are located outside the nucleus.

2.5 Atomic Number and Mass Number

The atomic number gives the number of protons in all the atoms of the same element. In a neutral atom, there is an equal number of protons and electrons. The mass number is the total number of protons and neutrons in an atom.

2.6 Isotopes and Atomic Mass

Atoms that have the same number of protons but different numbers of neutrons are called isotopes. The atomic mass of an element is the average mass of all the atoms in a naturally occurring sample of that element.

2.7 Electron Energy Levels

The atomic spectra of elements are related to the specific energy levels occupied by electrons. When energy is absorbed, an electron moves to a higher energy level; energy is lost when the electron drops to a lower energy level and emits a photon. Each element has it own unique spectrum.

An orbital is a region around the nucleus where an electron with a specific energy is most likely to be found. Each orbital holds a maximum of two electrons, which must have opposite spins. In each principal energy level (n), electrons occupy orbitals within sublevels. An s sublevel contains one s orbital, a p sublevel contains three p orbitals, a d sublevel contains five d orbitals, and an f sublevel contains seven f orbitals. Each type of orbital has a unique shape.

2.8 Electron Configurations

Within a sublevel, electrons enter orbitals in the same energy level one at a time until all the orbitals are half filled. Additional electrons enter with opposite spins until the orbitals in that sublevel are filled with two electrons each. The electrons in an atom can be written in an orbital diagram, which shows the orbitals that are occupied by paired and unpaired electrons. The electron configuration shows the number of electrons in each sublevel. An abbreviated electron configuration places the symbol of a noble gas in brackets to represent the filled sublevels. The periodic table consists of s, p, d, and f sublevel blocks. Beginning with $1s$, an electron configuration is obtained by writing the sublevel blocks in order going across the periodic table until the element is reached.

2.9 Periodic Trends

The properties of elements are related to the valence electrons of the atoms. With only a few minor exceptions, each group of elements has the same arrangement of valence electrons differing only in the energy level. The radius of an atom increases going down a group and decreases going across a period. The energy required to remove a valence electron is the ionization energy, which generally decreases going down a group and generally increases going across a period.

KEY TERMS

alkali metals Elements of Group 1A (1) except hydrogen; these are soft, shiny metals with one outer shell electron.

alkaline earth metals Group 2A (2) elements, which have 2 electrons in their outer shells.

atom The smallest particle of an element that retains the characteristics of the element.

atomic mass The weighted average mass of all the naturally occurring isotopes of an element.

atomic mass unit (amu) A small mass unit used to describe the mass of very small particles such as atoms and subatomic particles; 1 amu is equal to one-twelfth the mass of a ^{12}C atom.

atomic number A number that is equal to the number of protons in an atom.

atomic spectrum A series of lines specific for each element produced by photons emitted by electrons dropping to lower energy levels.

atomic symbol An abbreviation used to indicate the mass number and atomic number of an isotope.

chemical symbol An abbreviation that represents the name of an element.

compound A pure substance consisting of two or more elements, with a definite composition, that can be broken down into simpler substances by chemical methods.

d block elements The block ten-elements wide in Groups 3B (3) to 2B (12) in which electrons fill the five d orbitals in d sublevels.

electron A negatively charged subatomic particle having a very small mass that is usually ignored in mass calculations; its symbol is e^-.

electron configuration A list of the number of electrons in each sublevel within an atom, arranged by increasing energy.

electron-dot symbol The representation of an atom that shows valence electrons as dots around the symbol of the element.

element A pure substance that cannot be separated into any simpler substances by chemical methods.

f block elements The block 14-elements wide in the rows at the bottom of the periodic table in which electrons fill the seven f orbitals in $4f$ and $5f$ sublevels.

group A vertical column in the periodic table that contains elements having similar physical and chemical properties.

halogen Group 7A (17) elements of fluorine, chlorine, bromine, iodine, and astatine.

ionization energy The energy needed to remove the least tightly bound electron from the outermost energy level of an atom.

isotope An atom that differs only in mass number from another atom of the same element. Isotopes have the same atomic number (number of protons) but different numbers of neutrons.

mass number The total number of neutrons and protons in the nucleus of an atom.

matter Anything that has mass and occupies space.

metal An element that is shiny, malleable, ductile, and a good conductor of heat and electricity. The metals are located to the left of the zigzag line in the periodic table.

metalloid Elements with properties of both metals and non-metals located along the heavy zigzag line on the periodic table.

mixture The physical combination of two or more substances that does not change the identities of the mixed substances.

neutron A neutral subatomic particle having a mass of 1 amu and found in the nucleus of an atom; its symbol is n or n^0.

noble gas An element in Group 8A (18) of the periodic table, generally unreactive and seldom found in combination with other elements.

nonmetal An element with little or no luster that is a poor conductor of heat and electricity. The nonmetals are located to the right of the zigzag line in the periodic table.

nucleus The compact, very dense center of an atom, containing the protons and neutrons of the atom.

orbital The region around the nucleus where electrons of a certain energy are more likely to be found. The s orbitals are spherical; the p orbitals have two lobes.

orbital diagram A diagram that shows the distribution of electrons in the orbitals of the energy levels.

***p* block elements** The elements in Groups 3A (13) to 8A (18) in which electrons fill the p orbitals in the p sublevels.

period A horizontal row of elements in the periodic table.

periodic table An arrangement of elements by increasing atomic number such that elements having similar chemical behavior are grouped in vertical columns.

proton A positively charged subatomic particle having a mass of 1 amu and found in the nucleus of an atom; its symbol is p or p^+.

pure substance A type of matter with a fixed composition: elements and compounds.

***s* block elements** The elements in Groups 1A (1) and 2A (2) in which electrons fill the s orbitals.

subatomic particle A particle within an atom; protons, neutrons, and electrons are subatomic particles.

sublevel A group of orbitals of equal energy within principal energy levels. The number of sublevels in each energy level is the same as the principal quantum number (n).

valence electrons Electrons in the outermost energy level of an atom.

UNDERSTANDING THE CONCEPTS

2.83 Identify the following as an element, compound, or mixture:

a.

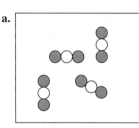

b.

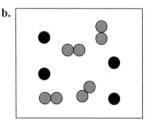

c.

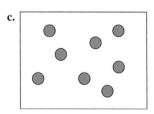

2.84 Consider the mixtures in the following diagrams.

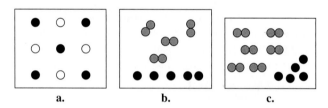

Which diagram(s) illustrate a homogeneous mixture? Explain your choice.

2.85 Which diagram(s) illustrate a heterogeneous mixture? Explain your choice.

2.86 Consider the following atoms in which the chemical symbol of the element is represented by X.

$$^{16}_{8}X \quad ^{16}_{9}X \quad ^{18}_{10}X \quad ^{17}_{8}X \quad ^{18}_{8}X$$

a. What atoms have the same number of protons?
b. Which atoms are isotopes? Of what element?
c. Which atoms have the same mass number?
d. What atoms have the same number of neutrons?

ADDITIONAL QUESTIONS AND PROBLEMS

For instructor-assigned homework, go to **www.masteringgob.com**.

2.87 Classify each of the following as an element, compound, or mixture.
a. carbon in pencils
b. carbon dioxide (CO_2) we exhale
c. orange juice
d. neon gas in lights
e. a salad dressing of oil and vinegar

2.88 Classify each of the following as a homogenous or heterogeneous mixture.
a. hot fudge sundae
b. herbal tea
c. vegetable oil
d. water and sand
e. mustard

2.89 Give the symbol and name of the element found in the following group and period on the periodic table.
  a. Group 2A, Period 3   **b.** Group 7A, Period 4
  c. Group 13, Period 3   **d.** Group 16, Period 2

2.90 The following are trace elements that have been found to be crucial to the biochemical and physiological processes in the body. Indicate whether each is a metal or nonmetal.
  a. zinc   **b.** cobalt   **c.** manganese (Mn)
  d. iodine   **e.** copper   **f.** selenium (Se)

2.91 Indicate if each of the following statements is *true* or *false*.
  a. The proton is a negatively charged particle.
  b. The neutron is 2000 times as heavy as a proton.
  c. The atomic mass unit is based on a carbon atom with 6 protons and 6 neutrons.
  d. The nucleus is the largest part of the atom.
  e. The electrons are located outside the nucleus.

2.92 Indicate if each of the following statements is *true* or *false*.
  a. The neutron is electrically neutral.
  b. Most of the mass of an atom is due to the protons and neutrons.
  c. The charge of an electron is equal, but opposite, to the charge of a neutron.
  d. The proton and the electron have about the same mass.
  e. The mass number is the number of protons.

2.93 For the following atoms, determine the number of protons, neutrons, and electrons.
  a. $^{27}_{13}Al$ **b.** $^{52}_{24}Cr$ **c.** $^{34}_{16}S$
  d. $^{56}_{26}Fe$ **e.** $^{136}_{54}Xe$

2.94 For the following atoms, give the number of protons, neutrons, and electrons.
  a. $^{22}_{10}Ne$ **b.** $^{127}_{53}I$ **c.** $^{75}_{35}Br$
  d. $^{133}_{55}Cs$ **e.** $^{195}_{78}Pt$

2.95 Complete the following table:

Name	Nuclear Symbol	Number of Protons	Number of Neutrons	Number of Electrons
	$^{34}_{16}S$			
		30	40	
magnesium			14	
	$^{220}_{86}Rn$			

2.96

Name	Nuclear Symbol	Number of Protons	Number of Neutrons	Number of Electrons
potassium			22	
	$^{51}_{23}V$			
		48	64	
barium			82	

2.97 **a.** What electron sublevel starts to fill after completion of the 3s sublevel?
  b. What electron sublevel starts to fill after completion of the 4p sublevel?

c. What electron sublevel starts to fill after completion of the 3d sublevel?
  d. What electron sublevel starts to fill after completion of the 3p sublevel?

2.98 **a.** What electron sublevel starts to fill after completion of the 5s sublevel?
  b. What electron sublevel starts to fill after completion of the 4d sublevel?
  c. What electron sublevel starts to fill after completion of the 4f sublevel?
  d. What electron sublevel starts to fill after completion of the 5p sublevel?

2.99 **a.** How many 3d electrons are in Fe?
  b. How many 5p electrons are in Ba?
  c. How many 4d electrons are in I?
  d. How many 6s electrons are in Ba?

2.100 **a.** How many 3d electrons are in Zn?
  b. How many 4p electrons are in Br?
  c. How many 6p electrons are in Bi?
  d. How many 5s electrons are in Cd?

2.101 What do the elements Ca, Sr, and Ba have in common in terms of their electron configuration? Where are they located in the periodic chart?

2.102 What do the elements O, S, and Se have in common in terms of their electron configuration? Where are they located in the periodic chart?

2.103 Name the element that corresponds to each of the following:
  a. $1s^2 2s^2 2p^6 3s^2 3p^3$
  b. alkali metal with the smallest atomic radius
  c. [Kr] $5s^2 4d^{10}$
  d. Group 5A element with highest ionization energy
  e. Period 3 element with largest atomic radius

2.104 Name the element that corresponds to each of the following:
  a. $1s^2 2s^2 2p^6 3s^2 3p^6 4s^1 3d^5$
  b. [Xe] $6s^2 4f^{14} 5d^{10} 6p^5$
  c. halogen with the highest ionization energy
  d. Group 6A element with the smallest ionization energy
  e. Period 4 element with smallest atomic radius

2.105 Why is the ionization energy of Ca higher than K, but lower than Mg?

2.106 Why is the ionization energy of Cl lower than F, but higher than S?

2.107 Of the elements Na, P, Cl, and F, which
  a. is a metal?
  b. has the largest atomic radius?
  c. has the highest ionization energy?
  d. loses an electron most easily?
  e. is found in Group 7A, Period 3?

2.108 Of the elements: Mg, Ca, Br, Kr, which
  a. is a noble gas?
  b. has the smallest atomic radius?
  c. has the lowest ionization energy?
  d. requires the most energy to remove an electron?
  e. is found in Group 2A, Period 4?

CHALLENGE QUESTIONS

2.109 Classify each of the following as a homogeneous mixture or heterogeneous mixture:
 a. lemon-flavored water **b.** stuffed mushrooms
 c. chicken noodle soup **d.** ketchup
 e. hard-boiled egg **f.** eye drops

2.110 For each of the following, write the symbol and name for X and the number of protons and neutrons. Which are isotopes of each other?
 a. $^{37}_{17}X$ **b.** $^{56}_{26}X$ **c.** $^{116}_{50}X$
 d. $^{124}_{50}X$ **e.** $^{116}_{48}X$

2.111 The most abundant isotope of iron is Fe-56.
 a. How many protons, neutrons, and electrons are in this isotope?
 b. What is the symbol of another isotope of iron with 25 neutrons?
 c. What is the symbol of a different atom with the same mass number and 27 neutrons?

2.112 Give the symbol of the element that has the
 a. smallest atomic radius in Group 6A
 b. smallest atomic radius in Period 3
 c. highest ionization energy in Group 15
 d. lowest ionization energy in Period 3

2.113 If the diameter of a sodium atom is 3.14×10^{-8}cm, how many sodium atoms would fit along a line exactly 1 inch long?

2.114 A lead atom has a mass of 3.4×10^{-22}g. How many lead atoms are in a cube of lead that has a volume of 2.00 cm^3 if the density of lead is 11.3 g/cm^3?

2.115 Lead consists of four naturally occurring isotopes. Calculate the atomic mass of lead.

Isotope	Mass	Abundance (%)
^{204}Pb	203.97	1.40
^{206}Pb	205.97	24.10
^{207}Pb	206.98	22.10
^{208}Pb	207.98	52.40

2.116 Indium (In) with an atomic mass of 114.8 consists of two naturally occurring isotopes ^{113}In and ^{115}In. If 4.30% of indium is ^{113}In, which has a mass of 112.90, what is the mass of the ^{115}In?

2.117 Consider three elements with the following noble gas notations:

$X = [Ar]\, 4s^2$ $Y = [Ne]\, 3s^2 3p^4$
$Z = [Ar]\, 4s^2 3d^{10} 4p^4$

 a. Identify each element as a metal, metalloid, or nonmetal.
 b. Which element has the largest atomic radius?
 c. Which elements have similar properties?
 d. Which element has the highest ionization energy?
 e. Which element has the smallest atomic radius?

2.118 Indicate if the following sections of orbital diagrams are or are not possible and explain your reason.

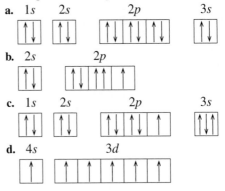

2.119 Consider three elements with the following noble gas notations:

$X = [Ar]\, 4s^2 3d^5$ $Y = [Ar]\, 4s^2 3d^{10} 4p^1$
$Z = [Ar]\, 4s^2 3d^{10} 4p^6$

 a. Identify each element as a metal, metalloid, or nonmetal.
 b. Which element has the smallest atomic radius?
 c. Which elements have similar properties?
 d. Which element has the highest ionization energy?
 e. Which element has a half-filled sublevel?

ANSWERS

Answers to Study Checks

2.1 This is a heterogeneous mixture because it does not have a uniform composition.

2.2 a. P **b.** Rn

2.3 a. metal **b.** nonmetal

2.4 45 neutrons

2.5 a. $^{15}_{7}N$ **b.** $^{42}_{20}Ca$ **c.** $^{27}_{13}Al$

2.6 10.81 amu

2.7 The $1s$, $2s$, and $3s$ orbitals are all spherical but they increase in volume because the electron is most likely to be found farther from the nucleus for higher energy levels.

2.8 $1s^2 2s^2 2p^6 3s^2 3p^4$ Complete electron configuration for sulfur (S)
 $[Ne]3s^2 3p^4$ Abbreviated electron configuration for sulfur (S)

2.9 Tin has the electron configuration:
 $1s^2 2s^2 2p^6 3s^2 3p^6 4s^2 3d^{10} 4p^6 5s^2 4d^{10} 5p^2$

2.10 $\cdot \ddot{P} \cdot$

2.11 Ionization energy increases going across a period: Sr is lowest, Sn is higher, and I is the highest of this set.

Answers to Selected Questions and Problems

2.1 **a.** pure substance   **b.** mixture
    c. pure substance   **d.** pure substance

2.3 **a.** element   **b.** compound
    c. element   **d.** compound

2.5 **a.** heterogeneous   **b.** homogeneous
    c. homogeneous   **d.** heterogeneous

2.7 **a.** Cu   **b.** Si   **c.** K   **d.** N
    e. Fe   **f.** Ba   **g.** Pb   **h.** Sr

2.9 **a.** carbon   **b.** chlorine   **c.** iodine
    d. mercury   **e.** fluorine   **f.** argon
    g. zinc   **h.** nickel

2.11 **a.** sodium, chlorine
    b. calcium, sulfur, oxygen
    c. carbon, hydrogen, chlorine, nitrogen, oxygen
    d. calcium, carbon, oxygen

2.13 **a.** Period 2   **b.** Group 8A (18)
    c. Group 1A (1)   **d.** Period 2

2.15 **a.** alkaline earth metal
    b. transition element
    c. noble gas
    d. alkali metal
    e. halogen

2.17 **a.** C   **b.** He   **c.** Na   **d.** Ca   **e.** Al

2.19 **a.** metal   **b.** nonmetal   **c.** metal
    d. nonmetal   **e.** nonmetal   **f.** nonmetal
    g. metalloid   **h.** metal

2.21 **a.** electron   **b.** proton
    c. electron   **d.** neutron

2.23 Rutherford determined that an atom contains a small, compact nucleus that is positively charged.

2.25 b. a proton and an electron

2.27 In the process of brushing hair, strands of hair become charged with like charges that repel each other.

2.29 **a.** atomic number   **b.** both
    c. mass number   **d.** atomic number

2.31 **a.** lithium, Li   **b.** fluorine, F   **c.** calcium, Ca
    d. zinc, Zn   **e.** neon, Ne   **f.** silicon, Si
    g. iodine, I   **h.** oxygen, O

2.33 **a.** 12, 12   **b.** 30, 30   **c.** 53, 53   **d.** 19, 19

2.35 *See Table 2.14*

2.37 **a.** 13 protons, 14 neutrons, 13 electrons
    b. 24 protons, 28 neutrons, 24 electrons
    c. 16 protons, 18 neutrons, 16 electrons
    d. 26 protons, 30 neutrons, 26 electrons

2.39 **a.** $^{31}_{15}P$ **b.** $^{80}_{35}Br$ **c.** $^{27}_{13}Al$ **d.** $^{35}_{17}Cl$

2.41 **a.** $^{32}_{16}S$ $^{33}_{16}S$ $^{34}_{16}S$ $^{36}_{16}S$
    b. They all have the same number of protons and electrons.
    c. They have different numbers of neutrons, which gives them different mass numbers.
    d. The atomic mass of S listed on the periodic table is the average atomic mass of all the isotopes.

2.43 Since the atomic mass of copper is closer to 63 amu, there are more atoms of ^{63}Cu.

2.45 69.72 amu

2.47 **a.** spherical   **b.** two lobes   **c.** spherical

2.49 **a.** Both are spherical.
    b. Both are part of the third energy level.
    c. Both contain three *p* orbitals.
    d. All have two lobes and belong in the third energy level.

2.51 **a.** There are 5 orbitals in the 3*d* sublevel.
    b. There is one sublevel in the $n = 1$ energy level.
    c. There is one orbital in the 6*s* sublevel.
    d. There are nine orbitals in the $n = 3$ energy level.

2.53 **a.** There is a maximum of 2 electrons in a 2*p* orbital.
    b. There is a maximum of 6 electrons in the 3*p* sublevel.
    c. There is a maximum of 32 electrons in the $n = 4$ energy level.
    d. There is a maximum of 10 electrons in the 5*d* sublevel.

2.55 **a.** 1*s*   2*s*   2*p*

    b. 3*s*   3*p*

    c. 3*s*   3*p*

    d. 3*s*   3*p*

TABLE 2.14

Name of Element	Symbol	Atomic Number	Mass Number	Number of Protons	Number of Neutrons	Number of Electrons
aluminum	Al	13	27	13	14	13
magnesium	Mg	12	24	12	12	12
potassium	K	19	39	19	20	19
sulfur	S	16	31	16	15	16
iron	Fe	26	56	26	30	26

2.57 **a.** N $1s^2\,2s^2\,2p^3$
b. Na $1s^2\,2s^2\,2p^6\,3s^1$
c. S $1s^2 2s^2\,2p^6 3s^2 3p^4$
d. As $1s^2 2s^2 2p^6 3s^2 3p^6 4s^2 3d^{10} 4p^3$
e. Fe $1s^2 2s^2 2p^6 3s^2 3p^6 4s^2 3d^6$

2.59 **a.** Mg $[\text{Ne}]3s^2$
b. S $[\text{Ne}]3s^2 3p^4$
c. Al $[\text{Ne}]3s^2 3p^1$
d. Ti $[\text{Ar}]\,4s^2 3d^2$
e. Ba $[\text{Xe}]\,6s^2$

2.61 **a.** S **b.** Co **c.** Si **d.** Br

2.63 **a.** Al **b.** C **c.** Ar **d.** Zr

2.65 **a.** 10 **b.** 6 **c.** 3 **d.** 1

2.67 **a.** 2A (2) **b.** 5A (15) **c.** 6A (16)

2.69 **a.** 3 **b.** 5 **c.** 7

2.71 **a.** Group 6A (16) $\cdot\overset{\cdot\cdot}{\underset{\cdot\cdot}{\text{S}}}\cdot$
b. Group 5A (15) $\cdot\overset{\cdot}{\underset{\cdot}{\text{N}}}\cdot$
c. Group 2A (2) $\cdot\text{Ca}\cdot$
d. Group 1A (1) Na$\cdot$
e. Group 2A (2) $\cdot\,\text{Ba}\,\cdot$

2.73 **a.** M$\cdot$ **b.** $\cdot$M$\cdot$

2.75 **a.** Mg, Al, Si **b.** I, Br, Cl **c.** Sr, Sb, I

2.77 **a.** Na **b.** Rb **c.** Na

2.79 **a.** Br, Cl, F **b.** Na, Al, Cl **c.** Cs, K, Na

2.81 **a.** Br **b.** S **c.** P

2.83 **a.** compound **b.** mixture **c.** element

2.85 Diagrams b and c illustrate heterogeneous mixtures; they do not have uniform composition throughout the samples.

2.87 **a.** element **b.** compound **c.** mixture
d. element **e.** mixture

2.89 **a.** Mg, magnesium **b.** Br, bromine
c. Al, aluminum **d.** O, oxygen

2.91 **a.** false **b.** false **c.** true
d. false **e.** true

2.93 **a.** 13 protons, 14 neutrons, 13 electrons
b. 24 protons, 28 neutrons, 24 electrons
c. 16 protons, 18 neutrons, 16 electrons
d. 26 protons, 30 neutrons, 26 electrons
e. 54 protons, 82 neutrons, 54 electrons

2.95

Name	Nuclear Symbol	Number of Protons	Number of Neutrons	Number of Electrons
sulfur	$^{34}_{16}\text{S}$	16	18	16
zinc	$^{70}_{30}\text{Zn}$	30	40	30
magnesium	$^{26}_{12}\text{Mg}$	12	14	12
radon	$^{220}_{86}\text{Rn}$	86	134	86

2.97 **a.** $3p$ **b.** $5s$ **c.** $4p$ **d.** $4s$

2.99 **a.** 6 **b.** 6 **c.** 10 **d.** 2

2.101 Ca, Sr, and Ba all have two valence electrons ns^2, which places them in Group 2A (2).

2.103 **a.** phosphorus
b. lithium (H is a nonmetal)
c. cadmium
d. nitrogen
e. sodium

2.105 Calcium has a greater number of protons than K. The least tightly bound electron in Ca is further from the nucleus than in Mg and needs less energy to remove.

2.107 **a.** Na **b.** Na **c.** F **d.** Na **e.** Cl

2.109 **a.** homogeneous **b.** heterogeneous
c. heterogeneous **d.** heterogeneous
e. heterogeneous **f.** homogeneous

2.111 **a.** 26 protons, 30 neutrons, 26 electrons
b. $^{51}_{26}\text{Fe}$
c. $^{51}_{24}\text{Cr}$

2.113 8.10×10^7 sodium atoms

2.115 207.2 amu

2.117 **a.** X is a metal; Y and Z are nonmetals.
b. X has the largest atomic radius.
c. Y and Z have six valence electrons and are in Group 6A (16).
d. Y has the highest ionization energy.
e. Y has the smallest atomic radius.

2.119 **a.** X and Y are metals and Z is a nonmetal.
b. Z **c.** X and Y are both metals.
d. Z **e.** X

3 Nuclear Chemistry

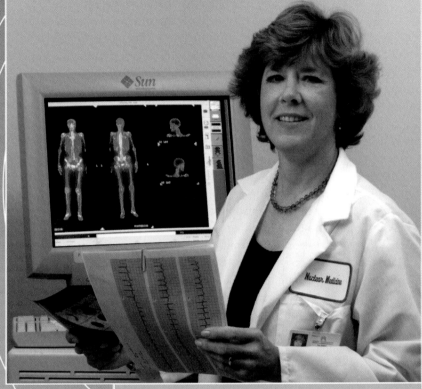

"Everything we do in this department involves radioactive materials," says Julie Goudak, nuclear medicine technologist at Kaiser Hospital. "The radioisotopes are given in several ways. The patient may ingest an isotope, breathe it in, or receive it by an IV injection. We do many diagnostic tests, particularly of the heart function, to determine if a patient needs a cardiac CAT scan."

A nuclear medicine technologist administers isotopes that emit radiation to determine the level of function of an organ such as the thyroid or heart, to detect the presence and size of a tumor, or to treat disease. A radioisotope locates in a specific organ and its radiation is used by a computer to create an image of that organ. From this data, a physician can make a diagnosis and design a treatment program.

LOOKING AHEAD

Mastering GOB CHEMISTRY

Visit **www.masteringgob.com** for self-study materials and instructor-assigned homework.

A female patient, aged 50, complains of nervousness, irritability, increased perspiration, brittle hair, and muscle weakness. Her hands are shaky at times and her heart often beats rapidly. She has been experiencing weight loss. The doctor decides to test for hyperthyroidism. To get a detailed look at the thyroid, a thyroid scan is ordered. The patient is given a small amount of an iodine radioisotope, which will be taken up by the thyroid. The scan shows a higher than normal rate of uptake of the radioactive iodine indicating an overactive thyroid gland, a condition called hyperthyroidism. Treatment for hyperthyroidism includes the use of drugs to lower the level of thyroid hormone, the use of radioactive iodine to destroy thyroid cells, or surgical removal of part or the entire thyroid. In our case, the nuclear physician decides to use radioactive iodine. To begin treatment, the patient drinks a solution containing radioactive iodine. In the following few weeks, the cells that take up the radioactive iodine are destroyed by the radiation. Tests show that the patient's thyroid is smaller, and that the blood level of thyroid hormone is normal once again.

With the discovery of X rays in 1895 and the production of artificial radioactive substances in 1934, the field of nuclear medicine was established. In 1937, the first radioactive isotope was used to treat a patient with leukemia at the University of California at Berkeley. Major strides in the use of radioactivity in medicine occurred in 1946, when a radioactive iodine isotope was successfully used to diagnose thyroid function and to treat hyperthyroidism and thyroid cancer. In the 1970s and 1980s, a variety of radioactive substances were used to produce an image of an organ such as liver, spleen, thyroid gland, kidney, and brain, and to detect heart disease. Today, procedures in nuclear medicine provide information about the function and structure of every organ in the body, which allows the nuclear physician to diagnose and treat diseases early.

LEARNING GOAL

Describe alpha, beta, positron, and gamma radiation.

3.1 Natural Radioactivity

Samples of most elements contain isotopes that differ only in the number of neutrons. While most isotopes contain nuclei that are stable, there are some that contain unstable nuclei. An unstable nucleus is **radioactive** because it spontaneously emits energy to become more stable. The energy emitted from a nucleus, called **radiation**, may be high-energy particles, such as alpha (α) or beta (β) particles, or gamma (γ) rays, which are pure energy. An isotope that emits radiation is called a **radioisotope**. While there are several naturally occurring radioisotopes, many more are produced artificially in nuclear laboratories. You may be familiar with X rays, which are also a type of radiation, although the energy of X rays is lower than that of nuclear radiation. X rays are not emitted from nuclei of unstable atoms but are produced in a different way.

In the symbol for an isotope, the mass number is written in the upper left corner of the symbol, and the atomic number is written in the lower left corner. For

TABLE 3.1

Stable and Radioactive Isotopes of Some Elements

Magnesium	Iodine	Uranium
Stable Isotopes		
$^{24}_{12}\text{Mg}$ magnesium-24	$^{127}_{53}\text{I}$ iodine-127	None
Radioactive Isotopes		
$^{23}_{12}\text{Mg}$ magnesium-23	$^{125}_{53}\text{I}$ iodine-125	$^{235}_{92}\text{U}$ uranium-235
$^{27}_{12}\text{Mg}$ magnesium-27	$^{131}_{53}\text{I}$ iodine-131	$^{238}_{92}\text{U}$ uranium-238

example, a radioactive isotope of iodine used in the diagnosis and treatment of thyroid conditions has a mass number of 131 and an atomic number of 53. We can write its symbol as shown below.

Mass number (protons and neutrons)
Element
Atomic number (protons)

$$^{131}_{53}\text{I}$$

This isotope is also called iodine-131 or I-131. Radioactive isotopes are named by writing the mass number after the element's name or symbol. When necessary, we can obtain the atomic number from the periodic table. Table 3.1 compares some stable, nonradioactive isotopes with some radioactive isotopes.

Types of Radiation

MC GOB — Types of Radiation

Different forms of radiation are emitted from an unstable nucleus when a change takes place among its protons and neutrons. Symbols for different forms of radiation can include the charge and mass number, written in the lower and upper left corner. The release of radiation of any type produces a more stable, lower energy nucleus. One type of radiation consists of alpha particles. An **alpha particle** contains 2 protons and 2 neutrons, which gives it a mass number of 4 and an atomic number of 2. Because it has 2 protons, an alpha particle has a charge of $2+$. That makes it identical to a helium nucleus. In equations, it is written as the Greek letter alpha (α) or as the symbol for helium.

α or ^4_2He
Alpha particle

Another type of radiation occurs when a radioisotope emits **beta particles**. A beta particle, which is a high-energy electron, has a charge of $1-$ and, because its mass is so much less than the mass of a proton, it is given a mass number of 0. It is represented by the Greek letter beta (β) or by the symbol for the electron, with the charge, -1, written on the lower left.

   β or $^{0}_{-1}e$
Beta particle

Beta particles are produced from an unstable nucleus when a neutron is transformed into a proton and an electron. The high-energy electron is emitted from the nucleus as beta radiation. This electron or beta particle did not exist in the nucleus until the following change occurred:

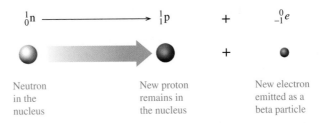

Neutron in the nucleus → New proton remains in the nucleus + New electron emitted as a beta particle

A **positron** is similar to an electron except that a positron has a positive ($+1$) charge. The electron and positron each have a mass number of 0. Instead of the atomic number, the charge is shown when we write the symbols of an electron and a positron as follows.

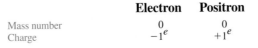

	Electron	**Positron**
Mass number	$^{0}_{-1}e$	$^{0}_{+1}e$
Charge		

A positron is produced by an unstable nucleus when a proton is transformed into a neutron and a positron.

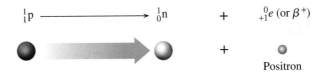

$^{1}_{1}p \longrightarrow {}^{1}_{0}n + {}^{0}_{+1}e \text{ (or } \beta^{+})$

Positron

A positron is an example of *antimatter*, a term physicists use to describe a particle that is the exact opposite of a particle, in this case, an electron. When an electron and a positron collide, their minute masses are completely converted to energy in the form of gamma rays.

$$^{0}_{-1}e + {}^{0}_{+1}e \longrightarrow 2{}^{0}_{0}\gamma$$

When the symbol β is used with no charge, it is taken to mean a beta particle rather than a positron.

Gamma rays are high-energy radiation, released as an unstable nucleus undergoes a rearrangement of its particles to give a more stable, lower energy nucleus. A gamma ray is shown as the Greek letter gamma (γ). Because gamma rays are energy only, there is no mass or charge associated with their symbol.

γ
Gamma ray

Table 3.2 summarizes the types of radiation we will use in nuclear equations.

TABLE 3.2

Some Common Forms of Radiation

Type of Radiation	Symbol		Mass Number	Charge
Alpha particle	α	$^{4}_{2}He$	4	2+
Beta particle	β	$^{0}_{-1}e$	0	1−
Positron	β^{+}	$^{0}_{+1}e$	0	1+
Gamma ray	γ	$^{0}_{0}\gamma$	0	0
Proton	$^{1}_{1}H$	$^{1}_{1}p$	1	1+
Neutron	$^{1}_{0}n$	n	1	0

Radiation Protection

Nuclear radiation is harmful because the particles have tremendous energy. Therefore, it is important that the radiologist, doctor, and nurse working with radioactive isotopes use proper radiation protection. Proper **shielding** is necessary to prevent exposure. Alpha particles are the heaviest of the radiation particles; they travel only a few centimeters in the air before they collide with air molecules, acquire electrons, and become helium atoms. A piece of paper, clothing, and our skin are protection against alpha particles. Lab coats and gloves will also provide sufficient shielding. However, if ingested or inhaled, alpha emitters can bring about serious internal damage because the mass and high charge of alpha particles causes much ionization in a short distance.

Beta particles have a very small mass and move much faster and farther than alpha particles, traveling as much as several meters through air. They can pass through paper and penetrate as far as 4–5 mm into body tissue. External exposure to beta particles can burn the surface of the skin, but they are stopped before they can reach the internal organs. Heavy clothing such as lab coats and gloves are needed to protect the skin from beta particles.

Gamma rays travel great distances through the air and pass through many materials, including body tissues. Only very dense shielding, such as lead or concrete, will stop them. Because they can penetrate so deeply, exposure to gamma rays can be extremely hazardous. Even the syringe used to give an injection of a gamma-emitting radioactive isotope is placed inside a special lead-glass cover. (See Figure 3.1.) Table 3.3 summarizes the shielding materials required for the various types of radiation.

Try to keep the time you must spend in a radioactive area to a minimum. A certain amount of radiation is emitted every minute. Remaining in a radioactive area twice as long exposes a person to twice as much radiation.

Keep your distance! The greater the distance from the radioactive source, the lower the intensity of radiation received. If you double your distance from the radiation source, the intensity of radiation drops to $\left(\frac{1}{2}\right)^2$ or one-fourth of its previous value.

SELF-STUDY ACTIVITY
Radiation and Its Biological Effects

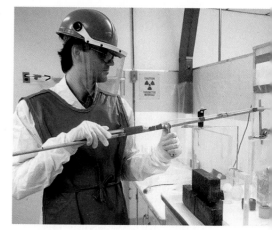

FIGURE 3.1

A person working with radioisotopes wears protective clothing and gloves and stands behind a lead shield.

Q **What types of radiation does the lead shield block?**

TABLE 3.3

Properties of Ionizing Radiation and Shielding Required			
Property	**Alpha**	**Beta**	**Gamma**
Characteristics	Helium nucleus	Electron	High-energy rays
Symbols	α, ^4_2He	β, $^0_{-1}e$	γ, $^0_0\gamma$
Travel distance in air	2–4 cm	200–300 cm	500 m
Tissue depth	0.05 mm	4–5 mm	50 cm or more
Shielding	Paper, clothing	Heavy clothing, lab coats, gloves	Lead, thick concrete
Typical source	radium-226	carbon-14	technetium-99m

QUESTIONS AND PROBLEMS

■ Natural Radioactivity

3.1 **a.** How are an alpha particle and a helium nucleus similar? different?
 b. What symbols are used for alpha particles?

3.2 **a.** How are a beta particle and an electron similar? different?
 b. What symbols are used for beta particles?

■ Biological Effects of Radiation

When high-energy radiation strikes molecules in its path, electrons may be knocked away. The result of this *ionizing radiation* is the formation of unstable ions or radicals. A free radical is a particle that has an unpaired electron. For example, when radiation passes through the human body, it may interact with water molecules, removing electrons and producing H_2O^+ ions or free radicals.

When ionizing radiation strikes the cells of the body, the unstable ions or free radicals that form can cause undesirable chemical reactions. The cells in the body most sensitive to radiation are the ones undergoing rapid division—those of the bone marrow, skin, reproductive organs, and intestinal lining, as well as all cells of growing children. Damaged cells may lose their ability to produce necessary materials. For example, if radiation damages cells of the bone marrow, red blood cells may no longer be produced. If sperm cells or ova or the cells of a fetus are damaged, birth defects may result. In contrast, cells of the nerves, muscles, liver, and adult bones are much less sensitive to radiation because they undergo little or no cellular division.

Cancer cells are another example of rapidly dividing cells. Because cancer cells are highly sensitive to radiation, large doses of radiation are used to destroy them. The surrounding normal tissue, dividing at a slower rate, shows a greater resistance to radiation and suffers less damage. In addition, normal tissue is able to repair itself more readily than cancerous tissue. However, this repair is not always complete. Possible long-term effects of ionizing radiation include a shortened life span, malignant tumors, leukemia, anemia, and genetic mutations.

3.3 Naturally occurring potassium consists of three isotopes, potassium-39, potassium-40, and radioactive potassium-41.
   a. Write the atomic symbols for each isotope.
   b. In what ways are the isotopes similar, and in what ways do they differ?

3.4 Naturally occurring iodine is iodine-127. Medically, radioactive isotopes of iodine-125 and iodine-130 are used.
   a. Write the atomic symbols for each isotope.
   b. In what ways are the isotopes similar, and in what ways do they differ?

3.5 Supply the missing information in the following table:

Medical Use	Atomic Symbol	Mass Number	Number of Protons	Number of Neutrons
Heart imaging	$^{201}_{81}\text{Tl}$			
Radiation therapy		60	27	
Abdominal scan			31	36
Hyperthyroidism	$^{131}_{53}\text{I}$			
Leukemia treatment		32		17

3.6 Supply the missing information in the following table:

Medical Use	Atomic Symbol	Mass Number	Number of Protons	Number of Neutrons
Cancer treatment	$^{60}_{27}\text{Co}$			
Brain scan		99	43	
Blood flow		141	58	
Bone scan		85		47
Lung function	$^{133}_{54}\text{Xe}$			

3.7 Write a symbol for the following:
   a. alpha particle   **b.** neutron
   c. beta particle   **d.** nitrogen-15
   e. iodine-125

3.8 Write a symbol for the following:
   a. proton   **b.** gamma ray
   c. electron   **d.** positron
   e. cobalt-60

3.9 Identify the symbol for X in each of the following:
   a. $^{0}_{-1}\text{X}$ **b.** $^{4}_{2}\text{X}$ **c.** $^{1}_{0}\text{X}$
   d. $^{24}_{11}\text{X}$ **e.** $^{14}_{6}\text{X}$

3.10 Identify the symbol for X in each of the following:
   a. $^{1}_{1}\text{X}$ **b.** $^{32}_{15}\text{X}$ **c.** $^{0}_{0}\text{X}$
   d. $^{59}_{26}\text{X}$ **e.** $^{0}_{+1}\text{X}$

3.11 **a.** Why does beta radiation penetrate further in solid material than alpha radiation?
   b. How does ionizing radiation cause damage to cells of the body?
   c. Why does the x-ray technician leave the room when you receive an X ray?
   d. What is the purpose of wearing gloves when handling radioisotopes?

3.12 **a.** As a nurse in an oncology unit, you sometimes give an injection of a radioisotope. What are three ways you can minimize your exposure to radiation?
   b. Why are cancer cells more sensitive to radiation than nerve cells?
   c. What is the purpose of placing a lead apron on a patient who is receiving routine dental X rays?
   d. Why are the walls in a radiology office built of thick concrete blocks?

3.2 Nuclear Equations

When a nucleus spontaneously breaks down by emitting radiation, the process is called *radioactive decay*. It can be shown as a *nuclear equation* using the symbols for the original radioactive nucleus, the new nucleus, and the radiation emitted.

$$\text{Radioactive nucleus} \longrightarrow \text{new nucleus} + \text{radiation} \ (\alpha, \beta, \gamma, \beta^{+})$$

A nuclear equation is balanced when the sum of the mass numbers and the sum of the atomic numbers of the particles and atoms on one side of the equation are equal to their counterparts on the other side.

The changes in mass number and atomic number of an atom that emits a radioactive particle are shown in Table 3.4.

We will now see how this works in the following examples and sample problems.

LEARNING GOAL

Write an equation showing mass numbers and atomic numbers for radioactive decay.

Alpha Emitters

Alpha emitters are radioisotopes that decay by emitting alpha particles. For example, uranium-238 decays to thorium-234 by emitting an alpha particle.

The alpha particle emitted contains 2 protons, which means that the new nucleus has 2 fewer protons, or 90 protons. That means that the new nucleus has an atomic number of 90 and is therefore thorium (Th). Because the alpha particle has a mass number of 4, the mass number of the thorium isotope is 234, 4 less than that of the original uranium nucleus.

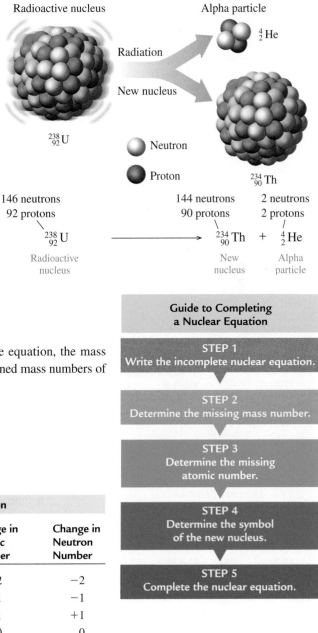

Radioactive nucleus   Alpha particle

Radiation $^{4}_{2}$He

New nucleus

$^{238}_{92}$U

○ Neutron
● Proton $^{234}_{90}$Th

146 neutrons   144 neutrons   2 neutrons
92 protons   90 protons   2 protons

$^{238}_{92}$U $\longrightarrow$ $^{234}_{90}$Th + $^{4}_{2}$He

Radioactive   New   Alpha
nucleus   nucleus   particle

Guide to Completing a Nuclear Equation

In another example of radioactive decay, radium-226 emits an alpha particle to form a new isotope whose mass number, atomic number, and identity we must determine.

■ STEP 1 **Write the incomplete nuclear equation.**

$$^{226}_{88}\text{Ra} \longrightarrow \ ? + ^{4}_{2}\text{He}$$

■ STEP 2 **Determine the missing mass number.**   In the equation, the mass number, 226, of the radium is equal to the combined mass numbers of the alpha particle and the new nucleus.

$$226 \quad = \ ? + 4$$
$$226 - 4 = \ ?$$
$$222 \quad = \ ?(\text{mass number of new nucleus})$$

Guide to Completing a Nuclear Equation

STEP 1
Write the incomplete nuclear equation.

STEP 2
Determine the missing mass number.

STEP 3
Determine the missing atomic number.

STEP 4
Determine the symbol of the new nucleus.

STEP 5
Complete the nuclear equation.

TABLE 3.4

Mass Number and Atomic Number Changes due to Radiation

Decay Process	Radiation Symbol	Change in Mass Number	Change in Atomic Number	Change in Neutron Number
Alpha emission	$^{4}_{2}$He or α	−4	−2	−2
Beta emission	$^{0}_{-1}e$ or β	0	+1	−1
Positron emission	$^{0}_{+1}$ or β^{+}	0	−1	+1
Gamma emission	$^{0}_{0}\gamma$	0	0	0

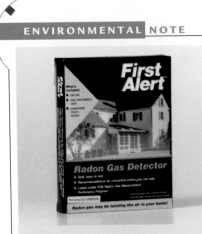

■ Radon in our Homes

The presence of radon has become a much publicized environmental and health issue because of radiation danger. Radioactive isotopes that produce radon, such as radium-226 and uranium-238, are naturally present in many types of rocks and soils. Radium 226 emits an alpha particle and is converted into radon gas, which diffuses out of the rocks and soil.

$$^{226}_{88}\text{Ra} \longrightarrow \, ^{222}_{86}\text{Rn} + \, ^{4}_{2}\text{He}$$

As uranium-238 decays, it also forms radium-226, which in turn produces radon. Uranium-238 has been found in particularly high levels in an area between Pennsylvania and New England.

Outdoors, radon gas poses little danger because it dissipates in the air. However, if the source of radon is under a house or building, the gas can enter the house through cracks in the foundation or other openings, where the radon can be inhaled by those living or working there. Inside the lungs, radon emits alpha particles to form polonium-218, which is known to cause cancer when present in the lungs.

$$^{222}_{86}\text{Rn} \longrightarrow \, ^{218}_{84}\text{Po} + \, ^{4}_{2}\text{He}$$

Some researchers have estimated that 10% of all lung cancer deaths in the United States are due to radon. The Environmental Protection Agency (EPA) recommends that the maximum level of radon not exceed 4 picocuries (pCi) per liter of air in a home. One (1) picocurie (pCi) is equal to 10^{-12} curies (Ci): curies are described in section 3.3. In California, 1% of all the houses surveyed exceeded the EPA's recommended maximum radon level.

■ **STEP 3 Determine the missing atomic number.** The atomic number of radium, 88, must equal the sum of the atomic number of the alpha particle and the new nucleus.

$$88 \quad = ? + 2$$
$$88 - 2 = ?$$
$$86 \quad = ? \text{ (atomic number of new nucleus)}$$

■ **STEP 4 Determine the symbol of the new nucleus.** On the periodic table, the element that has atomic number 86 is radon, Rn. The nucleus of this isotope of Rn is written as

$$^{222}_{86}\text{Rn}$$

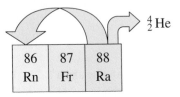

86	87	88
Rn	Fr	Ra

■ **STEP 5 Complete the nuclear equation.**

$$^{226}_{88}\text{Ra} \longrightarrow \, ^{222}_{86}\text{Rn} + \, ^{4}_{2}\text{He}$$

In this nuclear reaction, a radium-226 nucleus decays by releasing an alpha particle and produces a radon-222 nucleus.

SAMPLE PROBLEM 3.1

■ Writing an Equation for Alpha Decay

Smoke detectors, required in homes and apartments, contain an alpha emitter such as americium-241. The alpha particles ionize air molecules, producing a constant stream of electrical current. However, when smoke particles enter the detector, they interfere with the formation of ions in the air, and the electric current is interrupted. This causes the alarm to sound and warns the occupants of the danger of fire. Complete the following nuclear equation for the decay of americium-241:

$$^{241}_{95}\text{Am} \longrightarrow ? + \, ^{4}_{2}\text{He}$$

SOLUTION

■ **STEP 1 Write the incomplete nuclear equation.**

$$^{241}_{95}\text{Am} \longrightarrow ? + \, ^{4}_{2}\text{He}$$

■ **STEP 2 Determine the missing mass number.** In the equation, the mass number, 241, of the americium is equal to the sum of the mass numbers of the alpha particle and the new nucleus. We calculate the mass number of the new nucleus as follows:

$$241 \quad = ? + 4$$
$$241 - 4 = ?$$
$$237 \quad = ? \text{ (mass number of new nucleus)}$$

■ **STEP 3** **Determine the missing atomic number.** The atomic number of americium-95 must equal the sum of the atomic number of the alpha particle and the new nucleus.

$$95 \qquad = ? + 2$$
$$95 - 2 = ?$$
$$93 \qquad = ? \text{ (atomic number of new nucleus)}$$

■ **STEP 4** **Determine the symbol of the new nucleus.** On the periodic table, the element that has atomic number 93 is neptunium, Np. The symbol of this isotope of Np is written as $^{237}_{93}\text{Np}$.

(MC GOB) Alpha, Beta, and Gamma Emitters
Writing Nuclear Equations

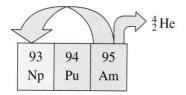

$$^4_2\text{He}$$

93	94	95
Np	Pu	Am

■ **STEP 5** **Complete the nuclear equation.**

$$^{241}_{95}\text{Am} \longrightarrow \, ^{237}_{93}\text{Np} + \, ^4_2\text{He}$$

In this nuclear reaction, an americium-241 nucleus decays by releasing an alpha particle and produces a neptunium-237 nucleus.

STUDY CHECK

Write a balanced nuclear equation for the alpha emitter polonium-214.

Beta Emitters

A beta emitter is a radioactive isotope that decays by emitting beta particles. To form a beta particle, the unstable nucleus converts a neutron into a proton. The newly formed proton adds to the number of protons already in the nucleus and increases the atomic number by 1. However, the mass number of the newly formed nucleus stays the same. For example, carbon-14 decays by emitting a beta particle and forming a nitrogen nucleus.

In the nuclear equation of a beta emitter, the mass number of the radioactive nucleus and the mass number of the new nucleus are the same, and the atomic number of the new nucleus increases by 1, indicating a change of one element into another. At right is the nuclear equation for the beta decay of carbon-14.

Radioactive carbon nucleus

Radiation

Beta particle
$^0_{-1}e$

New nucleus

Stable nitrogen-14 nucleus

$^{14}_6\text{C}$

○ Neutron

● Proton

$^{14}_7\text{N}$

8 neutrons
6 protons
$^{14}_6\text{C}$

Mass number is the same for both nuclei

7 neutrons
7 protons
$^{14}_7\text{N}$

0 neutrons
−1 charge
$^0_{-1}e$

Atomic number of the new nucleus increases by 1

$$^{14}_6\text{C} \longrightarrow \, ^{14}_7\text{N} + \, ^0_{-1}e$$

SAMPLE PROBLEM 3.2

■ **Writing an Equation for Beta Decay**

Cobalt-60 decays by emitting a beta particle. Write the nuclear equation for its decay.

Beta Emitters in Medicine

The radioactive isotopes of several biologically important elements are beta emitters. When a radiologist wants to treat a malignancy within the body, a beta emitter may be used. The short range of penetration into the tissue by beta particles is advantageous for certain conditions. For example, some malignant tumors increase the fluid within the body tissues. A compound containing phosphorus-32, a beta emitter, is injected into the body cavity where the tumor is located. The beta particles travel only a few millimeters through the tissue, so only the malignancy and any tissue within that range are affected. The growth of the tumor is slowed or stopped, and the production of fluid decreases. Phosphorus-32 is also used to treat leukemia, polycythemia vera (excessive production of red blood cells), and lymphomas.

$$^{32}_{15}P \longrightarrow {}^{32}_{16}S + {}^{0}_{-1}e$$

Another beta emitter, iron-59, is used in blood tests to determine the level of iron in the blood and the rate of production of red blood cells by the bone marrow.

$$^{59}_{26}Fe \longrightarrow {}^{59}_{27}Co + {}^{0}_{-1}e$$

SOLUTION

- **STEP 1 Write the incomplete nuclear equation.**

$$^{60}_{27}Co \longrightarrow ? + {}^{0}_{-1}e$$

- **STEP 2 Determine the missing mass number.**   In the equation, the mass number of cobalt-60 is equal to the sum of the mass numbers of the beta particle and the new nucleus.

$$60 \quad = ? + 0$$
$$60 - 0 = ?$$
$$60 \quad = ? \text{ (mass number of new nucleus)}$$

- **STEP 3 Determine the missing atomic number.**   The atomic number of cobalt-60 must equal the sum of the atomic number of the beta particle and the new nucleus.

$$27 \quad = ? - 1$$
$$27 + 1 = ?$$
$$28 \quad = ? \text{ (atomic number of new nucleus)}$$

- **STEP 4 Determine the symbol of the new nucleus.**   On the periodic table, the element that has atomic number 28 is nickel (Ni). This isotope is written as

$$^{60}_{28}Ni$$

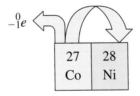

- **STEP 5 Complete the nuclear equation.**

$$^{60}_{27}Co \longrightarrow {}^{60}_{28}Ni + {}^{0}_{-1}e$$

In this nuclear reaction, a cobalt-60 nucleus decays by releasing a beta particle and produces a nickel-60 nucleus.

STUDY CHECK

Write the nuclear equation for iodine-131, a beta emitter.

Gamma Emitters

There are very few pure gamma emitters, although gamma radiation accompanies most alpha and beta radiation. In radiology, one of the most commonly used gamma emitters is technetium (Tc). The excited state called metastable technetium is written as technetium-99m, Tc-99m, or ^{99m}Tc. By emitting energy in the form of gamma rays, the excited nucleus becomes more stable.

$$^{99m}_{43}Tc \longrightarrow {}^{99}_{43}Tc + \gamma$$

Figure 3.2 summarizes the changes in the nucleus for alpha, beta, positron, and gamma radiation.

Producing Radioactive Isotopes

Today, more than 1500 radioisotopes are produced by converting stable, nonradioactive isotopes into radioactive ones. To do this, a stable atom is bombarded by fast-moving alpha particles, protons, or neutrons. When one of these particles is absorbed by a stable nucleus, the nucleus becomes unstable and the atom is now a radioactive isotope, or *radioisotope*. The process of changing one element into another is called *transmutation*.

When boron-10, a nonradioactive isotope, is bombarded by an alpha particle, it is converted to nitrogen-13, a radioisotope. In this bombardment reaction, a neutron is emitted.

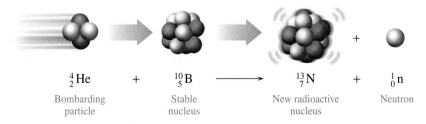

$$\underset{\text{Bombarding}\atop\text{particle}}{^{4}_{2}\text{He}} \quad + \quad \underset{\text{Stable}\atop\text{nucleus}}{^{10}_{5}\text{B}} \quad \longrightarrow \quad \underset{\text{New radioactive}\atop\text{nucleus}}{^{13}_{7}\text{N}} \quad + \quad \underset{\text{Neutron}}{^{1}_{0}\text{n}}$$

All of the known elements that have atomic numbers greater than 92 have been produced by bombardment; none of these elements occurs naturally. Most have been produced in only small amounts and exist for such a short time that it is difficult to study their properties. An example is element 105, dubnium, which is produced when californium-249 is bombarded with nitrogen-15.

$$^{249}_{98}\text{Cf} + {}^{15}_{7}\text{N} \longrightarrow {}^{260}_{105}\text{Db} + 4\,{}^{1}_{0}\text{n}$$

Technetium-99m is a radioisotope used in nuclear medicine for several diagnostic procedures, including the detection of brain tumors and examinations of the liver and spleen. The source of technetium-99m is molybdenum-99, which is produced in a nuclear reactor by neutron bombardment of molybdenum-98.

$$^{98}_{42}\text{Mo} + {}^{1}_{0}\text{n} \longrightarrow {}^{99}_{42}\text{Mo}$$

Many radiology laboratories have a small generator containing the radioactive molybdenum-99, which decays to give the technetium-99m radioisotope.

$$^{99}_{42}\text{Mo} \longrightarrow {}^{99m}_{43}\text{Tc} + {}^{0}_{-1}e$$

The technetium-99m radioisotope decays by emitting gamma rays. Gamma emission is desirable for diagnostic work because the gamma rays pass through the body to the detection equipment.

$$^{99m}_{43}\text{Tc} \longrightarrow {}^{99}_{43}\text{Tc} + \gamma$$

FIGURE 3.2

When the nuclei of alpha, beta, positron, and gamma emitters emit radiation, new, more stable nuclei are produced.

Q **What changes occur in the number of protons and neutrons when an alpha emitter gives off radiation?**

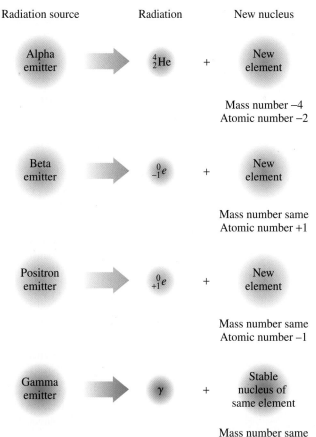

Radiation source	Radiation	New nucleus
Alpha emitter	$^{4}_{2}\text{He}$ +	New element Mass number −4 Atomic number −2
Beta emitter	$^{0}_{-1}e$ +	New element Mass number same Atomic number +1
Positron emitter	$^{0}_{+1}e$ +	New element Mass number same Atomic number −1
Gamma emitter	γ +	Stable nucleus of same element Mass number same Atomic number same

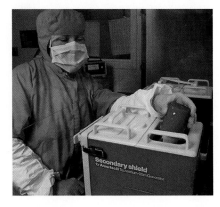

SAMPLE PROBLEM 3.3

■ Writing an Equation for Proton Bombardment

A new isotope is produced when zinc-66 is bombarded by a proton. Write the equation for this nuclear bombardment.

SOLUTION

■ **STEP 1** **Write the incomplete nuclear equation.**

$$^{66}_{30}\text{Zn} + {}^{1}_{1}\text{p} \longrightarrow ?$$

■ **STEP 2** **Determine the missing mass number.** In the equation, the sum of the mass numbers of zinc (66) and the proton (1) must equal the mass number of the new nucleus.

$$66 + 1 = ?$$
$$67 \quad = ? \text{ (mass number of new nucleus)}$$

■ **STEP 3** **Determine the missing atomic number.** The sum of the atomic numbers of zinc (30) and the proton (1) must equal the atomic number of the new nucleus.

$$30 + 1 = ?$$
$$31 \quad = ? \text{ (atomic number of new nucleus)}$$

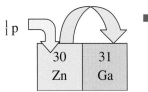

■ **STEP 4** **Determine the symbol of the new nucleus.** On the periodic table, the element that has atomic number 31 is gallium, Ga. The symbol of this isotope of Ga is written as

$$^{67}_{31}\text{Ga}$$

■ **STEP 5** **Complete the nuclear equation.**

$$^{66}_{30}\text{Zn} + {}^{1}_{1}\text{p} \longrightarrow {}^{67}_{31}\text{Ga}$$

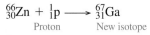

Proton   New isotope

STUDY CHECK

Write the balanced nuclear equation for the bombardment of aluminum-27 by an alpha particle to produce a new isotope and one neutron.

QUESTIONS AND PROBLEMS

■ Nuclear Equations

3.13 Write a balanced nuclear equation for the alpha decay of each of the following radioactive isotopes:

    a. $^{208}_{84}\text{Po}$ **b.** $^{232}_{90}\text{Th}$ **c.** $^{251}_{102}\text{No}$ **d.** $^{220}_{86}\text{Rn}$

3.14 Write a balanced nuclear equation for the alpha decay of each of the following radioactive isotopes:

    a. $^{243}_{96}\text{Cm}$ **b.** $^{252}_{99}\text{Es}$ **c.** $^{251}_{98}\text{Cf}$ **d.** $^{261}_{107}\text{Bh}$

3.15 Write a balanced nuclear equation for the beta decay of each of the following radioactive isotopes:

    a. $^{25}_{11}\text{Na}$ **b.** $^{20}_{8}\text{O}$ **c.** $^{92}_{38}\text{Sr}$ **d.** $^{42}_{19}\text{K}$

3.16 Write a balanced nuclear equation for the beta decay of each of the following radioactive isotopes:

    a. potassium-42   **b.** iron-59   **c.** iron-60   **d.** barium-141

3.17 Write a balanced nuclear equation for the positron decay of each of the following radioactive isotopes:

    a. $^{26}_{14}Si$ **b.** $^{54}_{27}Co$ **c.** $^{77}_{37}Rb$ **d.** $^{93}_{45}Rh$

3.18 Write a balanced nuclear equation for the positron decay of each of the following radioactive isotopes:

    a. $^{8}_{5}B$ **b.** $^{13}_{7}N$ **c.** $^{40}_{19}K$ **d.** $^{118}_{54}Xe$

3.19 Complete each of the following nuclear equations:

    a. $^{28}_{13}Al \longrightarrow ? + ^{0}_{-1}e$ **b.** $? \longrightarrow ^{86}_{36}Kr + ^{1}_{0}n$

    c. $^{66}_{29}Cu \longrightarrow ^{66}_{30}Zn + ?$ **d.** $? \longrightarrow ^{4}_{2}He + ^{234}_{90}Th$

    e. $^{188}_{80}Hg \longrightarrow ? + ^{0}_{+1}e$

3.20 Complete each of the following nuclear equations:

    a. $^{11}_{6}C \longrightarrow ^{7}_{4}Be + ?$ **b.** $^{35}_{16}S \longrightarrow ? + ^{0}_{-1}e$

    c. $? \longrightarrow ^{90}_{39}Y + ^{0}_{-1}e$ **d.** $^{210}_{83}Bi \longrightarrow ? + ^{4}_{2}He$

    e. $? \longrightarrow ^{135}_{59}Pr + ^{0}_{+1}e$

3.21 Complete each of the following bombardment reactions:

    a. $^{9}_{4}Be + ^{1}_{0}n \longrightarrow ?$

    b. $^{32}_{16}S + ? \longrightarrow ^{32}_{15}P$

    c. $? + ^{1}_{0}n \longrightarrow ^{24}_{11}Na + ^{4}_{2}He$

    d. When Al-27 is bombarded by an alpha particle, it forms Si-30. What other particle is produced?

3.22 Complete each of the following bombardment reactions:

    a. $^{40}_{18}Ar + ? \longrightarrow ^{43}_{19}K + ^{1}_{1}p$

    b. $^{238}_{92}U + ^{1}_{0}n \longrightarrow ?$

    c. $? + ^{1}_{0}n \longrightarrow ^{14}_{6}C + ^{1}_{1}p$

    d. When an alpha particle bombards N-14, it forms a radioisotope and a proton. What is the radioisotope that is produced?

3.3 Radiation Measurement

LEARNING GOAL

Describe the measurement of radiation.

One of the most common instruments for detecting beta and gamma radiation is the Geiger counter. It consists of a metal tube filled with a gas such as argon. When radiation enters a window on the end of the tube, it produces charged particles in the gas, which produce an electrical current. Each burst of current is amplified to give a click and a readout on a meter.

$$Ar + radiation \longrightarrow Ar^{+} + e^{-}$$

Radiation is measured in several different ways. We can measure the activity of a radioactive sample or determine the impact of radiation on biological tissue.

Activity

 Measuring Radiation

 SELF-STUDY ACTIVITY
Nuclear Chemistry

When a radiology laboratory obtains a radioisotope, the activity of the sample is measured in terms of the number of nuclear disintegrations per second. The **curie (Ci)**, the original unit of activity, was defined as the number of disintegrations that occur in one second for 1 g of radium, which is equal to 3.7×10^{10} disintegrations per second. The curie was named for Marie Curie, a Polish scientist, who along with her husband Pierre discovered the radioactive elements radium and polonium. A newer unit of radiation activity is the **becquerel (Bq)**, which is one disintegration per second.

Biological Effect

It is often important to know how much radiation the tissues in the body absorb. The **rad (radiation absorbed dose)** is a unit that measures the amount of radiation absorbed by a gram of material such as body tissue. The newer unit for absorbed dose is the **gray (Gy)**, which is equal to 100 rads.

The **rem (radiation equivalent in humans)** measures the biological effects of different kinds of radiation. Alpha particles don't penetrate the skin. But if they should enter the body by some other route, they can cause a lot of damage even though the particles only travel a short distance in tissue. High-energy radiation such as beta particles and high-energy protons and neutrons that penetrate the skin and travel into tissue cause more damage. Gamma rays are damaging because they travel a long way through tissue and create a great deal of ionization.

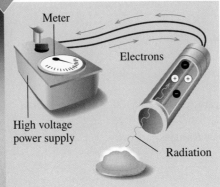

Meter

Electrons

High voltage power supply

Radiation

To determine the **equivalent dose** or rem dose, the absorbed dose (rads) is multiplied by a factor that adjusts for biological damage caused by a particular form of radiation. For beta and gamma radiation the factor is 1, so the biological damage in rems is the same as the absorbed radiation (rads). For high-energy protons and neutrons, the factor is about 10, and for alpha particles it is 20.

$$\text{Biological damage (rem)} = \text{absorbed dose (rad)} \times \text{factor}$$

Often the measurement for an equivalent dose will be in units of millirems (mrem). One rem is equal to 1000 mrem. The newer unit is the **sievert (Sv)**. One sievert is equal to 100 rems. Table 3.5 summarizes the units used to measure radiation.

MC GOB

SELF-STUDY ACTIVITY
Radiation and Its Biological Effects

SAMPLE PROBLEM 3.4

■ **Radiation Measurement**

One treatment of bone pain involves intravenous administration of the radioisotope phosphorous-32, which is primarily incorporated into bone. A typical dose of 7 mCi can produce up to 450 rads in the bone. What is the difference between the units mCi and rads?

SOLUTION

The millicuries (mCi) indicate the activity of the P-32 in terms of nuclei that break down in 1 second. The radiation absorbed dose (rads) is a measure of amount of radiation absorbed by the bone.

TABLE 3.5

Some Units of Radiation Measurement

Measurement	Common Unit	SI Unit	Relationship
Activity	curie (Ci) = 3.7×10^{10} disintegrations/s	becquerel (Bq) = 1 disintegration/s	1 Ci = 3.7×10^{10} Bq
Absorbed dose	rad	gray (Gy)	1 Gy = 100 rad
Biological damage	rem = rad × factor	1 sievert (Sv)	1 Sv = 100 rem

HEALTH NOTE

Radiation and Food

Foodborne illnesses caused by pathogenic bacteria such as *Salmonella, Listeria,* and *Escherichia coli* have become a major health concern in the United States. The Centers for Disease Control and Prevention estimates that each year *E. coli* in contaminated foods infects 20,000 people in the United States, and that 500 people die. *E. coli* has been responsible for outbreaks of illness from contaminated ground beef, fruit juices, lettuce, and alfalfa sprouts.

The Food and Drug Administration (FDA) has approved the use of 0.3 kilogray (0.3 kGy) to 1 kGy of ionizing radiation produced by cobalt-60 or cesium-137 for the treatment of foods. The irradiation technology is much like that used to sterilize medical supplies. Cobalt pellets are placed in stainless steel tubes, which are arranged in racks. When food moves through the series of racks, the gamma rays pass through the food and kill the bacteria.

It is important for consumers to understand that when food is irradiated, it never comes in contact with the radioactive source. The gamma rays pass through the food to kill bacteria, but that does not make the food radioactive. The radiation kills bacteria because it stops their ability to divide and grow. We cook or heat food thoroughly for the same purpose. Radiation has little effect on the food itself because its cells are no longer dividing or growing. Thus irradiated food is not harmed although a small amount of vitamin B_1 and C may be lost.

Currently, tomatoes, blueberries, strawberries, and mushrooms are being irradiated to allow them to be harvested when completely ripe and extend their shelf life. (See Figure 3.3.) The FDA has also approved the irradiation of pork, poultry, and beef in order to decrease potential infections and to extend shelf life. Currently, irradiated vegetable and meat products are available in retail markets in South Africa. Apollo

(a)

(b)

FIGURE 3.3

(a) The FDA requires this symbol to appear on irradiated retail foods. **(b)** After 2 weeks, the irradiated strawberries on the right show no spoilage. Mold is starting to grow on the nonirradiated ones on the left.

Q Why are irradiated foods used on spaceships and in nursing homes?

17 astronauts ate irradiated foods on the moon, and some U.S. hospitals and nursing homes now use irradiated poultry to reduce the possibility of infections among patients. The extended shelf life of irradiated food also makes it useful for campers and military personnel. Soon consumers concerned about food safety will have a choice of irradiated meats, fruits, and vegetables at the market.

STUDY CHECK

If P-32 is a beta emitter, how do the number of rems compare to the rads?

SELF-STUDY CASE
Food Irradiation

Exposure to Radiation

We are all exposed to low levels of radiation every day. Naturally occurring radioactive isotopes are part of the atoms of wood, brick, and concrete in our homes and the buildings where we work and go to school. This radioactivity, called background radiation, is present in the soil, in the food we eat, in the water we drink, and in the air we breathe. For example, one of the naturally occurring isotopes of potassium, potassium-40, is radioactive. It is found in the body because it is always present in any potassium-containing food. Other naturally occurring radioisotopes in air and food are carbon-14, radon-222, strontium-90, and iodine-131. Table 3.6 lists some common sources of radiation.

We are constantly exposed to radiation (cosmic rays) produced in space by the sun. At higher elevations, the amount of radiation from outer space is greater because there are fewer air molecules to absorb the radiation. People living at high

TABLE 3.6

Average Annual Radiation Received by a Person in the United States

Source	Dose (mrem)
Natural	
The ground	15
Air, water, food	30
Cosmic rays	40
Wood, concrete, brick	50
Medical	
Chest X ray	50
Dental X ray	20
Upper gastrointestinal tract X ray	200
Other	
Television	2
Air travel	1
Radon	200[a]
Cigarette smoking	35

[a]Varies widely.

TABLE 3.7

Lethal Doses of Radiation for Some Life-Forms

Life-Form	LD_{50} (rem)
Insect	100 000
Bacterium	50 000
Rat	800
Human	500
Dog	300

altitudes or flying in an airplane receive more radiation from cosmic rays than those who live at sea level. For example, a person living in Denver receives about twice the cosmic radiation as a person living in Los Angeles.

A person living close to a nuclear power plant normally does not receive much additional radiation, perhaps 0.1 millirem (mrem) in 1 year. (One rem equals 1000 mrem.) However, in the accident at the Chernobyl nuclear power plant in 1986, it is estimated that people in a nearby town received as much as 1 rem/hr.

In addition to naturally occurring radiation from construction materials in our homes, we receive radiation from television. In the medical clinic, dental and chest X rays also add to our radiation exposure. The average person in the United States receives about 0.17 rem or 170 mrem of radiation annually.

Radiation Sickness

The larger the dose of radiation received at one time, the greater the effect on the body. Exposure to radiation under 25 rem usually cannot be detected. Whole-body exposure of 100 rem produces a temporary decrease in the number of white blood cells. If the exposure to radiation is 100 rem or higher, the person suffers the symptoms of radiation sickness: nausea, vomiting, fatigue, and a reduction in white-cell count. A whole-body dosage greater than 300 rem can lower the white-cell count to zero. The victim suffers diarrhea, hair loss, and infection. Exposure to radiation of about 500 rem is expected to cause death in 50% of the people receiving that dose. This amount of radiation is called the lethal dose for one-half the population, or the LD_{50}. The LD_{50} varies for different life-forms, as Table 3.7 shows. Radiation dosages of about 600 rem would be fatal to all humans within a few weeks.

QUESTIONS AND PROBLEMS

Radiation Measurement

3.23 **a.** How does a Geiger counter detect radiation?
   b. What are the SI unit and the older unit that describe the activity of a radioactive sample?
   c. What are the SI unit and the older unit that describe the radiation dose absorbed by tissue?
   d. What is meant by the term kilogray?

3.24 **a.** What is background radiation?
   b. What are the SI unit and the older unit that describe the biological effect of radiation?
   c. What is meant by the terms mCi and mrem?
   d. Why is a factor used to determine the dose equivalent?

3.25 The recommended dosage of iodine-131 is 4.20 μCi/kg of body weight. How many microcuries of iodine-131 are needed for a 70.0-kg patient with hyperthyroidism?

3.26 **a.** The dosage of technetium-99m for a lung scan is 20 μCi/kg of body weight. How many millicuries should be given to a 50.0-kg patient? (1 mCi = 1000 μCi)
   b. Suppose a person absorbed 50 mrads of alpha radiation. What would be the dose equivalent in mrems?

3.27 Why would an airline pilot be exposed to more background radiation than the person who works at the ticket counter?

3.28 In radiation therapy, a patient receives high doses of radiation. What symptoms of radiation sickness might the patient exhibit?

3.4 Half-Life of a Radioisotope

LEARNING GOAL

Given the half-life of a radio-isotope, calculate the amount of radioisotope remaining after one or more half-lives.

The **half-life** of a radioisotope is the amount of time it takes for one-half of a sample to decay. For example, ^{131}I has a half-life of 8.0 days. As ^{131}I decays it produces a beta particle and the nonradioactive isotope ^{131}Xe.

$$^{131}_{53}\text{I} \longrightarrow {}^{131}_{54}\text{Xe} + {}^{0}_{-1}\beta$$

Suppose we have a sample that contains 20.0 grams of ^{131}I. We don't know which specific nucleus will emit radiation, but we do know that in 8 days, one-half of all the nuclei in the sample will decay to give ^{131}Xe. That means that after 8.0 days, there are one-half of the number of ^{131}I atoms in the sample, or 10.0 g of ^{131}I, remaining. The decay process has also produced 10.0 g of the product ^{131}Xe. After another half-life or 8.0 days passes, 5.00 g of the 10.0 g of ^{131}I will decay to ^{131}Xe. Now there are 5.00 g of ^{131}I left, while there is a total of 15.0 g of ^{131}Xe.

 SELF-STUDY ACTIVITY
Nuclear Chemistry

$$20 \text{ g } {}^{131}\text{I} \xrightarrow{\text{1 half-life}} 10 \text{ g } {}^{131}\text{I} \xrightarrow{\text{2 half-lives}} 5 \text{ g } {}^{131}\text{I}$$

A third half-life, or another 8.0 days, results in 2.50 g of the ^{131}I decaying to give ^{131}Xe, which leaves 2.50 g of ^{131}I still capable of producing radiation. This information can be summarized in Table 3.8 as follows:

TABLE 3.8

Activity of an ^{131}I Sample with Time				
Time elapsed	0 days	8.0 days	16.0 days	24.0 days
Half-lives	0	1	2	3
^{131}I remaining	20.0 g	10.0 g	5.00 g	2.50 g
^{131}Xe produced	0 g	10.0 g	15.0 g	17.5 g

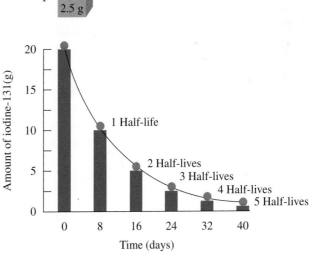

A **decay curve** is a diagram of the decay of a radioactive isotope. Figure 3.4 shows such a curve for the ^{131}I we have discussed.

FIGURE 3.4

The decay curve for iodine-131 shows that one-half of the radioactive sample decays and one-half remains radioactive after each half-life of 8 days.

Q **How many grams of the 20-g sample remain radioactive after 2 half-lives?**

SAMPLE PROBLEM 3.5

■ Using Half-Lives of a Radioisotope

Phosphorus-32, a radioisotope used in the treatment of leukemia, has a half-life of 14 days. If a sample contains 8.0 g of phosphorus-32, how many grams of phosphorus-32 remain after 42 days?

EXPLORE YOUR WORLD

Modeling Half-Lives

Obtain a piece of paper and a licorice stick or celery stalk. Draw a vertical and a horizontal axis on the paper. Label the vertical axis as radioactive atoms and the horizontal axis as minutes. Place the licorice stick or celery against the vertical axis and mark its height for zero minutes. In the next minute, cut the licorice stick or celery in two. (You can eat the half if you are hungry.) Place the shortened licorice stick or celery at 1 minute on the horizontal axis and mark its height. Every minute cut the licorice stick or celery in half again and mark the shorter height at the corresponding time. Keep reducing the length by half until you cannot divide the licorice or celery in half any more. Connect the points you made for each minute. What does the curve look like? How does this curve represent the concept of a half-life for a radioisotope?

MC
GOB Radioactive Half-Lives

SOLUTION

■ **STEP 1 Given** 8.0 g ^{32}P; 42 days; 14 days/half-life

   Need g ^{32}P remaining

■ **STEP 2 Plan**

42 days → | Half-life | → Number of half-lives

8.0 g ^{32}P → | Number of half-lives | → g ^{32}P remaining

■ **STEP 3 Equalities/Conversion Factors**

$$1 \text{ half-life} = 14 \text{ d}$$
$$\frac{14 \text{ d}}{1 \text{ half-life}} \quad \text{and} \quad \frac{1 \text{ half-life}}{14 \text{ d}}$$

■ **STEP 4 Set Up Problem** We can do this problem with two calculations. First, we determine the number of half-lives in the amount of time that has elapsed.

$$\text{Number of half-lives} = 42 \text{ d} \times \frac{1 \text{ half-life}}{14 \text{ d}} = 3 \text{ half-lives}$$

Now, we determine how much of the sample decays in 3 half-lives and how many grams of the phosphorus remain.

$$8.0 \text{ g } ^{32}\text{P} \xrightarrow{14 \text{ d}} 4.0 \text{ g } ^{32}\text{P} \xrightarrow{14 \text{ d}} 2.0 \text{ g } ^{32}\text{P} \xrightarrow{14 \text{ d}} 1.0 \text{ g } ^{32}\text{P}$$

STUDY CHECK

Iron-59 has a half-life of 46 days. If the laboratory received a sample of 8.0 g iron-59, how many grams are still active after 184 days?

Naturally occurring isotopes of the elements usually have long half-lives, as shown in Table 3.9. They disintegrate slowly and produce radiation over a long period of time, even hundreds or millions of years. In contrast, many of the radioisotopes used in nuclear medicine have much shorter half-lives. They disinte-

TABLE 3.9

Half-Lives of Some Radioisotopes

Element	Radioisotope	Half-Life
Naturally Occurring Radioisotopes		
carbon	^{14}C	5730 yr
potassium	^{40}K	1.3×10^9 yr
radium	^{226}Ra	1600 yr
uranium	^{238}U	4.5×10^9 yr
Some Medical Radioisotopes		
chromium	^{51}Cr	28 days
iodine	^{131}I	8 days
iron	^{59}Fe	46 days
technetium	^{99m}Tc	6.0 hr

grate rapidly and produce almost all their radiation in a short period of time. For example, technetium-99m emits half of its radiation in the first 6 hr. This means that a small amount of the radioisotope given to a patient is essentially gone within 2 days. The decay products of technetium 99m are totally eliminated by the body.

SAMPLE PROBLEM 3.6

■ Dating Using Half-Lives

In Los Angeles, the remains of ancient animals have been unearthed at La Brea tar pits. Suppose a bone sample from the tar pits is subjected to the carbon-14 dating method. If the sample shows that two half-lives have passed, when did the animal live?

SOLUTION

We can calculate the age of the bone sample by using the half-life of carbon-14 (5730 years).

$$2 \text{ half-lives} \times \frac{5730 \text{ years}}{1 \text{ half-life}} = 11\,500 \text{ years}$$

We would estimate that the animal lived 11 500 years ago, or about 9500 B.C.E.

ENVIRONMENTAL NOTE

■ Dating Ancient Objects

A technique known as radiological dating is used by geologists, archaeologists, and historians as a way to determine the age of ancient objects. The age of an object derived from plants or animals (such as wood, fiber, natural pigments, bone, and cotton or woolen clothing) is determined by measuring the amount of carbon-14, a naturally occurring radioactive form of carbon. In 1960, Willard Libby received the Nobel Prize for his work developing carbon-14 dating techniques during the 1940s. Carbon-14 is produced in the upper atmosphere by the bombardment of $^{14}_{7}N$ by high-energy neutrons from cosmic rays.

$$^{1}_{0}n \quad + \quad ^{14}_{7}N \quad \longrightarrow \quad ^{14}_{6}C \quad + \quad ^{1}_{1}p$$

Neutron from cosmic rays   Nitrogen in atmosphere   Radioactive carbon-14   Proton

The carbon-14 reacts with oxygen to form radioactive carbon dioxide, $^{14}CO_2$. Because carbon dioxide is continuously absorbed by living plants during the process of photosynthesis, some carbon-14 will be taken into the plant. After the plant dies, no more carbon-14 is taken up, and the amount of carbon-14 contained in the plant steadily decreases as it undergoes radioactive β decay.

$$^{14}_{6}C \longrightarrow ^{14}_{7}N + ^{0}_{-1}e$$

Scientists use the half-life of carbon-14 (5730 years) to calculate the amount of time that has passed since the plant died, a process called **carbon dating**. The smaller the amount of carbon-14 remaining in the sample, the greater the number of half-lives that have passed. Thus, the approximate age of the sample can be determined. For example, a wooden beam found in an ancient Indian dwelling might have one-half of the carbon-14 found in living plants today. Thus, the dwelling was probably con-

structed about 5730 years ago, one half-life of carbon-14. Carbon-14 dating was used to determine that the Dead Sea Scrolls are about 2000 years old.

A radiological dating method used for determining the age of rocks is based on the radioisotope uranium-238, which decays through a series of reactions to lead-206. The uranium-238 isotope has a very long half-life, about 4×10^9 (4 billion) years. Measurements of the amounts of uranium-238 and lead-206 enable geologists to determine the age of rock samples. The older rocks will have a higher percentage of lead-206 because more of the uranium-238 has decayed. The age of rocks brought back from the moon by the *Apollo* missions was determined using uranium-238. They were found to be about 4×10^9 years old, approximately the same age calculated for Earth.

STUDY CHECK

Suppose that a piece of wood found in a tomb had $\frac{1}{8}$ (3 half-lives) of its original C-14 activity. About how many years ago was the wood part of a living tree?

■

QUESTIONS AND PROBLEMS

■ **Half-Life of a Radioisotope**

3.29 What is meant by the term half-life?

3.30 Why are radioisotopes with short half-lives used for diagnosis in nuclear medicine?

3.31 Technetium-99m is an ideal radioisotope for scanning organs because it has a half-life of 6.0 hr and is a pure gamma emitter. Suppose that 80.0 mg were prepared in the technetium generator this morning. How many milligrams would remain after the following intervals?
   a. one half-life   **b.** two half-lives   **c.** 18 hr   **d.** 24 hr

3.32 A sample of sodium-24 with an activity of 12 mCi is used to study the rate of blood flow in the circulatory system. If sodium-24 has a half-life of 15 hr, what is the activity of the sodium after $2\frac{1}{2}$ days?

3.33 Strontium-85, used for bone scans, has a half-life of 65 days. How long will it take for the radiation level of strontium-85 to drop to one-fourth of its original level? To one-eighth?

3.34 Fluorine-18, which has a half-life of 110 min, is used in PET scans (see section 3.5). If 100 mg of fluorine-18 is shipped at 8:00 A.M., how many milligrams of the radioisotope are still active if the sample arrives at the radiology laboratory at 1:30 P.M.?

LEARNING GOAL

■

Describe the use of radioisotopes in medicine.

Medical Applications Using Radioactivity

3.5 Medical Applications Using Radioactivity

When a radiologist wants to determine the condition of an organ in the body, a patient is given a radioisotope that is known to concentrate in that organ. The cells in the body do not differentiate between a nonradioactive atom and a radioactive one. However, radioactive atoms can be detected because they emit radiation and the nonradioactive atoms do not. Some radioisotopes used in nuclear medicine are listed in Table 3.10.

TABLE 3.10

Medical Applications of Radioisotopes

Isotope	Half-Life	Medical Application
Ce-141	32.5 days	Gastrointestinal tract diagnosis; measuring blood flow to the heart
Ga-67	78 hr	Abdominal imaging; tumor detection
Ga-68	68 min	Detect pancreatic cancer
P-32	4.3 days	Treatment of leukemia, excess red blood cells, pancreatic cancer
I-125	60 days	Treatment of brain cancer
I-131	8 days	Imaging thyroid; treatment of Graves' disease, goiter, and hyperthyroidism; treatment of thyroid and prostate cancer
Sr-85	65 days	Detection of bone lesions; brain scans
Tc-99m	6 hr	Imaging of skeleton and heart muscle, brain, liver, heart, lungs, bone, spleen, kidney, and thyroid; *most widely used radioisotope in nuclear medicine*

Scans with Radioisotopes

After a patient receives a radioisotope, the radiologist determines the level and location of radioactivity emitted by the radioisotope. An apparatus called a scanner is used to produce an image of the organ. The scanner moves slowly across the patient's body above the region where the organ containing the radioisotope is located. The gamma rays emitted from the radioisotope in the organ can be used to expose a photographic plate, producing a **scan** of the organ. On a scan, an area of decreased or increased radiation can indicate such conditions as a disease of the organ, a tumor, a blood clot, or edema.

A common method of determining thyroid function is the use of radioactive iodine uptake (RAIU). Taken orally, the radioisotope iodine-131 mixes with the iodine already present in the thyroid. Twenty-four hours later, the amount of iodine taken up by the thyroid is determined. A detection tube held up to the area of the thyroid gland detects the radiation coming from the iodine-131 that has located there. (See Figure 3.5.)

A patient with a hyperactive thyroid will have a higher than normal level of radioactive iodine, whereas a patient with a hypoactive thyroid will record low values. If the patient has hyperthyroidism, treatment is begun to lower the activity of the thyroid. One treatment involves giving the patient a therapeutic dosage of radioactive iodine, which has a higher radiation count than the diagnostic dose. The radioactive iodine goes to the thyroid where its radiation destroys some of the thyroid cells. The thyroid produces less thyroid hormone, bringing the hyperthyroid condition under control.

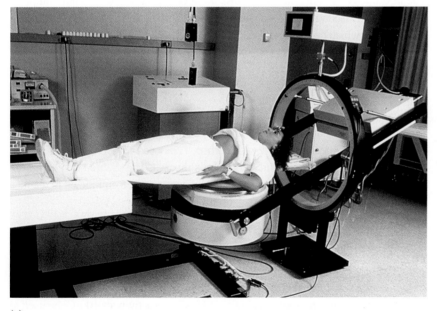

(a)

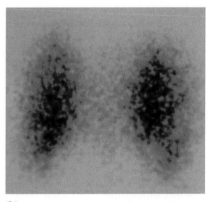

(b)

FIGURE 3.5

(a) A scanner is used to detect radiation from a radioisotope that has accumulated in an organ. **(b)** A scan of the thyroid shows the accumulation of radioactive iodine-131 in the thyroid.

Q **What type of radiation would move through body tissues to create a scan?**

FIGURE 3.6

These PET scans of the brain show a normal brain on the left and a brain affected by Alzheimer's disease on the right.

Q **When positrons collide with electrons, what type of radiation is produced that gives an image of an organ?**

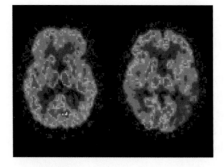

Positron Emission Tomography (PET)

Radioisotopes that emit a positron are used in an imaging method called positron emission tomography (PET). A positron is a particle emitted from the nucleus that has the same mass as an electron but has a positive charge.

$$\beta^+ \quad \text{or} \quad {}^{0}_{+1}e$$
Positron

Carbon-11 is an example of a radioisotope that emits a positron when it decays.

$${}^{11}_{6}C \longrightarrow {}^{0}_{+1}e + {}^{11}_{5}B$$
Positron

A positron is produced when a proton changes to a neutron.

$${}^{1}_{1}p \longrightarrow {}^{0}_{+1}e + {}^{1}_{0}n$$

The new nucleus retains the same mass number but has a lower atomic number. The positron exists for only a moment before it collides with an electron producing gamma rays, which can be detected.

$$\underset{\text{Positron}}{{}^{0}_{+1}e} + \underset{\substack{\text{Electron} \\ \text{in an atom}}}{{}^{0}_{-1}e} \longrightarrow \underset{\substack{\text{Gamma rays} \\ \text{produced}}}{2\gamma}$$

Medically, positron emitters such as carbon-11, oxygen-15, and nitrogen-13 are used to diagnose conditions involving blood flow, metabolism, and particularly brain function. Glucose labeled with carbon-11 is used to detect damage in the brain from epilepsy, stroke, and Parkinson's disease. The gamma rays from the positrons emitted by carbon-11 are detected by computerized equipment to create a three-dimensional image of the organ. (See Figure 3.6.)

SAMPLE PROBLEM 3.7

Medical Application of Radioactivity

In hyperthyroidism, the thyroid produces an overabundance of thyroid hormone. Why would a radioisotope such as iodine-131 with an activity of 10 μCi be used to image the thyroid?

SOLUTION

The thyroid takes up the iodine-131 because the thyroid uses most of the iodine in the body. An activity level of 10 μCi is high enough to determine the level of thyroid function, but low enough to keep cell damage to a minimum.

HEALTH NOTE

Radiation Doses in Diagnostic and Therapeutic Procedures

We can compare the levels of radiation exposure commonly used during diagnostic and therapeutic procedures in nuclear medicine. In diagnostic procedures, the radiologist minimizes radiation damage by using only enough radioisotope to evaluate the condition of an organ or tissue. The doses used in **radiation therapy** are much greater than those used for diagnostic procedures. For example, a therapeutic dose would be used to destroy the cells in a malignant tumor. Although there will be some damage to surrounding tissue, the healthy cells are more resistant to radiation and can repair themselves. (See Table 3.11.)

TABLE 3.11

Radiation Doses Used for Diagnostic and Therapeutic Procedures

Organ/Condition	Dose (rem)
Diagnostic	
Liver	0.3
Thyroid	50.0
Lung	2.0
Therapeutic	
Lymphoma	4500
Skin cancer	5000–6000
Lung cancer	6000
Brain tumor	6000–7000

Other Imaging Methods

COMPUTED TOMOGRAPHY (CT)

Another imaging method used to detect changes within the body is computed tomography (CT). A computer monitors the degree of absorption of 30,000 x-ray beams directed at the brain at successive layers. The differences in absorption based upon the densities of the tissues and fluids in the brain provide a series of images of the brain. This technique is successful in the identification of brain hemorrhages, tumors, and atrophy. (See Figure 3.7.)

MAGNETIC RESONANCE IMAGING (MRI)

Magnetic resonance imaging (MRI) is a powerful imaging technique that does not involve x-ray radiation. It is the least invasive imaging method available. MRI is based on the absorption of energy when the protons in hydrogen atoms are excited by a strong magnetic field. Hydrogen atoms make up 63% of all the atoms in the body. In the nucleus, the spin of the single proton acts like a tiny magnet. With no external field, the spins of the protons have random orientations. However, when placed within a large magnet, the spins of the protons align with the magnetic field. A magnet aligned with the field has a lower energy than one that is aligned against the field. As the MRI scan proceeds, radiofrequency pulses of energy are applied. When a nucleus absorbs certain energy, its proton "flips" and becomes aligned against the field. Because hydrogen atoms in the body are in different chemical environments, frequencies of different energies are absorbed. The energies absorbed are calculated and converted to color images of the body. MRI is particularly useful to image soft tissues because soft tissues contain large amounts of water. (See Figure 3.8.)

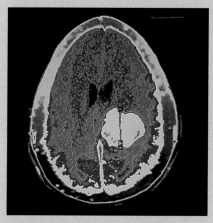

FIGURE 3.7

A CT scan shows a brain tumor (yellow area) in the center of the right side of the brain.

Q **What is the type of radiation used to give a CT scan?**

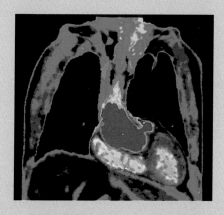

FIGURE 3.8

An MRI scan of the heart and lungs, with the left ventricle shown in red.

Q **What is the source of energy in an MRI?**

STUDY CHECK

The evaluation of the thyroid image determined that the patient had an enlarged thyroid. A few weeks later, that same patient was given 100 mCi of iodine-131 intravenously. Why was the activity of the iodine-131 sample used for treatment so much greater than the activity of the iodine-131 used in the diagnostic procedure?

QUESTIONS AND PROBLEMS

Medical Applications Using Radioactivity

3.35 Bone and bony structures contain calcium and phosphorus.
   a. Why would the radioisotopes of calcium-47 and phosphorus-32 be used in the diagnosis and treatment of bone diseases?
   b. During nuclear tests, scientists were concerned that strontium-85, a radioactive product, would be harmful to the growth of bone in children. Explain.

3.36 a. Technetium-99m emits only gamma radiation. Why would this type of radiation be used in diagnostic imaging rather than an isotope that also emits beta or alpha radiation?

b. A patient with polycythemia vera (excess production of red blood cells) receives radioactive phosphorus-32. Why would this treatment reduce the production of red blood cells in the bone marrow of the patient?

3.37 In a diagnostic test for leukemia, a patient receives 4.0 mL of a solution containing selenium-75. If the activity of the selenium-75 is 45 μCi/mL, what is the dose received by the patient?

3.38 A vial contains radioactive iodine-131 with an activity of 2.0 mCi per milliliter. If the thyroid test requires 3.0 mCi in an "atomic cocktail," how many milliliters are used to prepare the iodine-131 solution?

LEARNING GOAL

Describe the processes of nuclear fission and fusion.

(MC GOB) Fission and Fusion

3.6 Nuclear Fission and Fusion

In the 1930s, scientists bombarding uranium-235 with neutrons discovered that the U-235 nucleus splits into two medium-weight nuclei and produces a great amount of energy. This was the discovery of nuclear **fission**. The energy generated by splitting the atom was called atomic energy. A typical equation for nuclear fission is

$$^{1}_{0}n \; + \; ^{235}_{92}U \; \longrightarrow \; ^{91}_{36}Kr \; + \; ^{142}_{56}Ba \; + \; 3\,^{1}_{0}n \; + \; energy$$

If we could weigh these products with great accuracy, we would find that their total mass is slightly less than the mass of the starting materials. The missing mass has been converted into energy, consistent with the famous equation derived by Albert Einstein:

$$E = mc^2$$

E is the energy released, m is the mass lost, and c is the speed of light, 3×10^8 m/s. Even though the mass loss is very small, when it is multiplied by the speed of light squared the result is a large value for the energy released. The fission of 1 g of uranium-235 produces about as much energy as the burning of 3 tons of coal.

Chain Reaction

Fission begins when a neutron collides with the nucleus of a uranium atom. The resulting nucleus is unstable and splits into smaller nuclei. This fission process also releases several neutrons and large amounts of gamma radiation and energy. The neutrons emitted have high energies and bombard more uranium-235 nuclei. As fission continues, there is a rapid increase in the number of high-energy neutrons capable of splitting more uranium atoms, a process called a **chain reaction**. To sustain a nuclear chain reaction, sufficient quantities of uranium-235 must be brought together to provide a critical mass in which almost all the neutrons immediately collide with more uranium-235 nuclei. So much heat and energy are released that an atomic explosion can occur. (See Figure 3.9.)

Nuclear Fusion

In **fusion**, two small nuclei such as those in hydrogen combine to form a larger nucleus. Mass is lost, and a tremendous amount of energy is released, even more

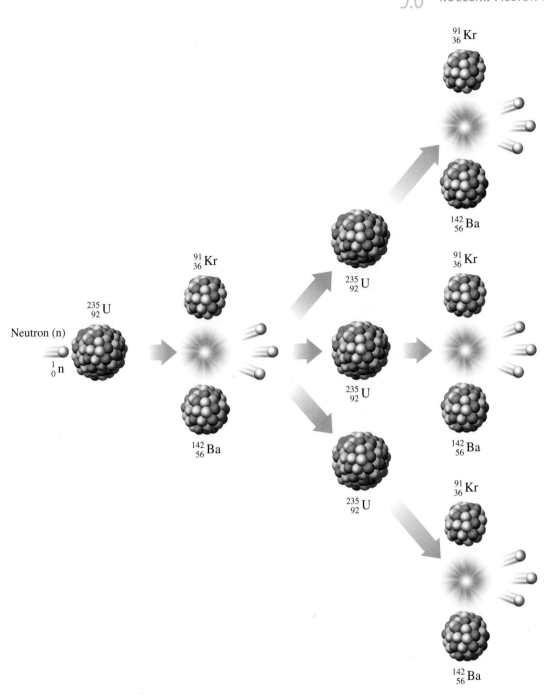

FIGURE 3.9

In a nuclear chain reaction, the fission of each ^{235}U atom produces three neutrons that cause the nuclear fission of more and more ^{235}U atoms.

Q **Why is the fission of ^{235}U called a chain reaction?**

than the energy released from nuclear fission. However, a very high temperature (100 000 000°C) is required to overcome the repulsion of the hydrogen nuclei and cause them to undergo fusion. Fusion reactions occur continuously in the sun and other stars, providing us with heat and light. The huge amounts of energy produced by our sun come from the fusion of 6×10^{11} kilograms of hydrogen every second. The following fusion reaction involves the combination of two isotopes of hydrogen.

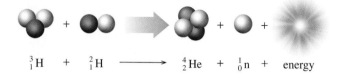

$$^3_1\text{H} \quad + \quad ^2_1\text{H} \quad \longrightarrow \quad ^4_2\text{He} \quad + \quad ^1_0\text{n} \quad + \quad \text{energy}$$

The fusion reaction has tremendous potential as a possible source for future energy needs. One of the advantages of fusion as an energy source is that hydrogen is plentiful in the oceans. Although scientists expect some radioactive waste from fusion reactors, the amount is expected to be much less than that from fission, and the waste products should have shorter half-lives. However, the use of fusion as a source of energy is still in the experimental stage because the extremely high temperatures needed have been difficult to reach and even more difficult to maintain. Research groups around the world are attempting to develop the technology needed to make the harnessing of the fusion reaction for energy a reality in our lifetime.

SAMPLE PROBLEM 3.8

▪ Identifying Fission and Fusion

Classify the following as pertaining to nuclear fission, nuclear fusion, or both:

a. Small nuclei combine to form larger nuclei.
b. Large amounts of energy are released.
c. Very high temperatures are needed for reaction.

SOLUTION

a. fusion   **b.** both fusion and fission   **c.** fusion

STUDY CHECK

Would the following reaction be an example of a fission or fusion reaction?

$$^2_1\text{H} + ^1_1\text{H} \longrightarrow ^3_2\text{He}$$

QUESTIONS AND PROBLEMS

▪ Nuclear Fission and Fusion

3.39 What is nuclear fission?

3.40 How does a chain reaction occur in nuclear fission?

3.41 Complete the following fission reaction:

$$^{235}_{92}\text{U} + ^1_0\text{n} \longrightarrow ^{131}_{50}\text{Sn} + ? + 2\,^1_0\text{n} + \text{energy}$$

3.42 In another fission reaction, U-235 bombarded with a neutron produces Sr-94, another small nucleus, and 3 neutrons. Write the complete equation for the fission reaction.

3.43 Indicate whether each of the following are characteristic of the fission or fusion process or both:
a. Neutrons bombard a nucleus.
b. The nuclear process occurring in the sun.

▪ Nuclear Power Plants

In a nuclear power plant, the quantity of uranium-235 is held below a critical mass, so it cannot sustain a chain reaction. The fission reactions are slowed by placing control rods, which absorb some of the fast-moving neutrons, among the uranium samples. In this way, less fission occurs, and there is a slower, controlled production of energy. The heat from the controlled fission is used to produce steam. The steam drives a generator, which produces electricity. Approximately 10% of the electrical energy produced in the United States is generated in nuclear power plants.

Although nuclear power plants help meet some of our energy needs, there are some problems. One of the most serious problems is the production of radioactive by-products that have very long half-lives. It is essential that these waste products be stored safely for a very long time in a place where they do not contaminate the environment. Early in 1990, the Environmental Protection Agency gave its approval for the storage of radioactive hazardous wastes in chambers 2150 ft underground. In 1998 the Waste Isolation Pilot Plant (WIPP) repository site in New Mexico was ready to receive plutonium waste from former U.S. bomb factories. Although authorities claim the caverns are safe, some people are concerned with the safe transport of the radioactive waste by trucks on the highways.

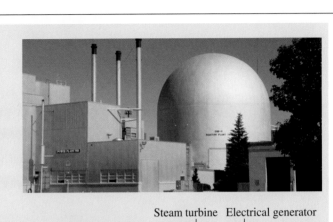

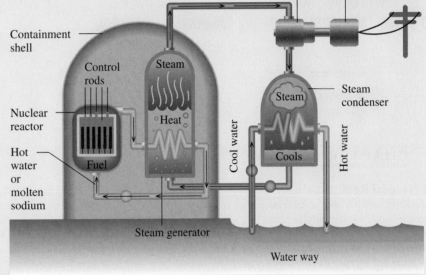

c. A large nucleus splits into smaller nuclei.

d. Small nuclei combine to form larger nuclei.

3.44 Indicate whether each of the following are characteristic of the fission or fusion process or both:

a. Very high temperatures are required to initiate the reaction.

b. Less radioactive waste is produced.

c. Hydrogen nuclei are the reactants.

d. Large amounts of energy are released when the nuclear reaction occurs.

CONCEPT MAP

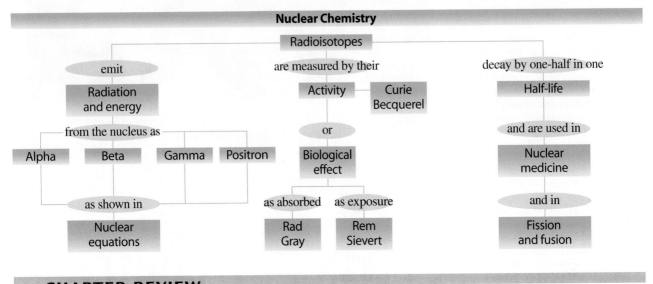

CHAPTER REVIEW

3.1 Natural Radioactivity

Radioactive isotopes have unstable nuclei that break down (decay), spontaneously emitting alpha (α), beta (β), and gamma (γ) radiation. Because radiation can damage the cells in the body, proper protection must be used: shielding, limiting the time of exposure, and distance.

3.2 Nuclear Equations

A balanced equation is used to represent the changes that take place in the nuclei of the reactants and products. The new isotopes and the type of radiation emitted can be determined from the symbols that show the mass numbers and atomic numbers of the isotopes in the nuclear reaction. A radioisotope is produced artificially when a nonradioactive isotope is bombarded by a small particle. Many radioactive isotopes used in nuclear medicine are produced in this way.

3.3 Radiation Measurement

In a Geiger counter, radiation ionizes gas in a metal tube, which produces an electrical current. The curie (Ci) measures the number of nuclear transformations of a radioactive sample. Activity is also measured in becquerel (Bq) units. The amount of radiation absorbed by a substance is measured in rads or the gray (Gy). The

rem and the sievert (Sv) are units used to determine the biological damage from the different types of radiation.

3.4 Half-Life of a Radioisotope

Every radioisotope has its own rate of emitting radiation. The time it takes for one-half of a radioactive sample to decay is called its half-life. For many medical radioisotopes, such as Tc-99m and I-131, half-lives are short. For other isotopes, usually naturally occurring ones such as C-14, Ra-226, and U-238, half-lives are extremely long.

3.5 Medical Applications Using Radioactivity

In nuclear medicine, radioisotopes are given that go to specific sites in the body. By detecting the radiation they emit, an evaluation can be made about the location and extent of an injury, disease, tumor, or the level of function of a particular organ. Higher levels of radiation are used to treat or destroy tumors.

3.6 Nuclear Fission and Fusion

In fission, a large nucleus breaks apart into smaller pieces, releasing one or more types of radiation and a great amount of energy. In fusion, small nuclei combine to form a larger nucleus while great amounts of energy are released.

KEY TERMS

alpha particle A nuclear particle identical to a helium ($_2^4$He, or α) nucleus (2 protons and 2 neutrons).

becquerel (Bq) A unit of activity of a radioactive sample equal to one disintegration per second.

beta particle A particle identical to an electron ($_{-1}^0 e$, or β) that forms in the nucleus when a neutron changes to a proton and an electron.

carbon dating A technique used to date ancient specimens that contain carbon. The age is determined by the amount of active C-14 that remains in the sample.

chain reaction A fission reaction that will continue once it has been initiated by a high-energy neutron bombarding a heavy nucleus such as U-235.

curie (Ci) A unit of radiation equal to 3.7×10^{10} disintegrations/s.

decay curve A diagram of the decay of a radioactive element.

equivalent dose The measure of biological damage from an absorbed dose that has been adjusted for the type of radiation.

fission A process in which large nuclei are split into smaller pieces, releasing large amounts of energy.

fusion A reaction in which large amounts of energy are released when small nuclei combine to form larger nuclei.

gamma ray High-energy radiation (γ) emitted to make a nucleus more stable.

gray (Gy) A unit of absorbed dose equal to 100 rads.

half-life The length of time it takes for one-half of a radioactive sample to decay.

positron A particle with no mass and a positive charge produced when a proton is transformed into a neutron and a positron.

rad (radiation absorbed dose) A measure of an amount of radiation absorbed by the body.

radiation Energy or particles released by radioactive atoms.

radiation therapy The use of high doses of radiation to destroy harmful tissues in the body.

radioactive The process by which an unstable nucleus breaks down with the release of high-energy radiation.

radioisotope A radioactive atom of an element.

rem (radiation equivalent in humans) A measure of the biological damage caused by the various kinds of radiation (rad $\times$ radiation biological factor).

scan The image of a site in the body created by the detection of radiation from radioactive isotopes that have accumulated in that site.

shielding Materials used to provide protection from radioactive sources.

sievert (Sv) A unit of biological damage (equivalent dose) equal to 100 rems.

UNDERSTANDING THE CONCEPTS

3.45 Consider the following nucleus of a radioactive isotope.

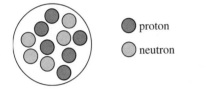

     ● proton
     ○ neutron

    a. What is the nuclear symbol for this isotope?
    b. If this isotope decays by emitting a positron, what does the resulting nucleus look like?

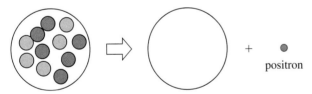

    + ● positron

3.46 Draw in the radioactive nucleus that emits a beta particle to form the following nucleus.

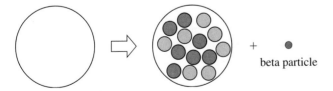

    + ● beta particle

3.47 Draw in the nucleus of the atom to complete the following nuclear reaction.

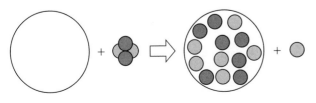

3.48 Complete the following equation by drawing the nucleus of the atom produced.

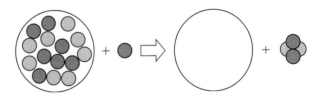

3.49 Carbon dating of small bits of charcoal used in cave paintings has determined that some of the paintings are from 10 000 to 30 000 years old. Carbon-14 has a half-life of 5730 years. In a 1 μg sample of carbon from a live tree, the activity of ^{14}C is 6.4 μCi. If researchers determine that 1 μg charcoal from a prehistoric cave painting in France has an activity of 0.80 μCi, what is the age of the painting?

3.50

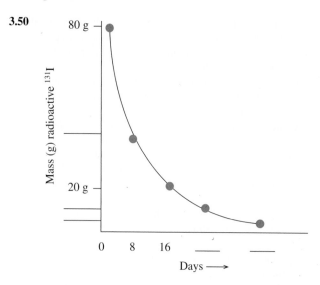

a. Complete the values for the mass of radioactive ^{131}I on the vertical axis.
b. Complete the number of days on the horizontal axis.
c. What is the half-life in days of ^{131}I?

ADDITIONAL QUESTIONS AND PROBLEMS

 For instructor-assigned homework, go to **www.masteringgob.com.**

3.51 Carbon-12 is a nonradioactive isotope of carbon, and carbon-14 is a radioactive isotope. What is similar and different about the two carbon isotopes?

3.52 Give the number of protons, neutrons, and electrons in atoms of the following isotopes:
　a. boron-10　　　　**b.** zinc-72
　c. iron-59　　　　　**d.** gold-198

3.53 Describe alpha, beta, and gamma radiation in terms of the following:
　a. type of radiation
　b. symbols
　c. depth of tissue penetration
　d. type of shielding needed for protection

3.54 When you have dental X rays at the dentist's office, the technician places a heavy lead apron over you, and then leaves the room to take your X rays. What is the purpose of these activities?

3.55 Write the balanced nuclear equations for each of the following emitters:
　a. thorium-225 (α)　　**b.** bismuth-210 (α)
　c. cesium-137 (β)　　　**d.** tin-126 (β)

3.56 Write the balanced nuclear equations for each of the following emitters:
　a. potassium-40 (β)　　**b.** sulfur-35 (β)
　c. platinum-190 (α)　　**d.** radium-210 (α)

3.57 Complete each of the following nuclear equations:
　a. $^{14}_{7}N + ^{4}_{2}He \longrightarrow ? + ^{1}_{1}p$
　b. $^{27}_{13}Al + ^{4}_{2}He \longrightarrow ^{30}_{14}Si + ?$
　c. $^{235}_{92}U + ^{1}_{0}n \longrightarrow ^{90}_{38}Sr + 3^{1}_{0}n + ?$

3.58 Complete each of the following nuclear equations:
　a. $^{59}_{27}Co + ? \longrightarrow ^{56}_{25}Mn + ^{4}_{2}He$
　b. $? \longrightarrow ^{14}_{7}N + ^{0}_{-1}e$
　c. $^{76}_{36}Kr + ^{0}_{-1}e \longrightarrow ?$

3.59 Write the atomic symbols and a complete nuclear equation for the following:
　a. When two oxygen-16 atoms collide, one of the products is an alpha particle. What is the other product?
　b. When californium-249 is bombarded by oxygen-18, a new element, seaborgium-263, and four neutrons are produced.
　c. Radon-222 emits an alpha particle, and the product emits another alpha particle. Write the two nuclear reactions.

3.60 Write the symbols and complete the following nuclear equations:
　a. When polonium-210 emits an alpha particle, the product is lead-206.
　b. Bismuth-211 decays by emitting an alpha particle. The product is another radioisotope that emits a beta particle. Write equations for the two nuclear changes.

3.61 If the amount of radioactive phosphorus-32 in a sample decreases from 1.2 g to 0.30 g in 28 days, what is the half-life of phosphorus-32?

3.62 If the amount of radioactive iodine-123 in a sample decreases from 0.4 g to 0.1 g in 26.2 hours, what is the half-life of iodine-123?

3.63 Iodine-131, a beta emitter, has a half-life of 8.0 days.
　a. Write the nuclear equation for the beta-decay of iodine-131.
　b. How many grams of a 12.0-g sample of iodine-131 would remain after 40 days?
　c. How many days have passed if 48 g of iodine-131 decayed to 3.0 grams of iodine-131?

3.64 Cesium-137, a beta emitter, has a half-life of 30 years.
　a. Write the nuclear equation for the beta-decay of cesium-137.
　b. How many grams of a 16-g sample of cesium-137 would remain after 90 years?
　c. How many years will be needed for 28 grams of cesium-137 to decay to 3.5 g of cesium-137?

3.65 A nurse was accidentally exposed to potassium-42 while doing some brain scans for possible tumors. The error was not discovered until 36 hours later when the activity of the potassium-42 sample was 2.0 μCi. If potassium-42 has a half-life of 12 hr, what was the activity of the sample at the time the nurse was exposed?

3.66 A wooden object from the site of an ancient temple has a carbon-14 activity of 10 counts per minute compared with a reference piece of wood cut today that has an activity of 40 counts per minute. If the half-life for carbon-14 is 5730 years, what is the age of the ancient wood object?

3.67 A 120-mg sample of technetium-99 m is used for a diagnostic test. If technetium-99 m has a half-life of 6.0 hr, how much of the technetium-99 m sample remains 24 hr after the test?

3.68 The half-life of oxygen-15 is 124 seconds. If a sample of oxygen-15 has an activity of 4000 becquerels, how many minutes will elapse before it reaches an activity of 500 becquerels?

3.69 What is the purpose of irradiating meats, fruits, and vegetables?

3.70 The irradiation of foods was approved in the United States in the 1980s.
  a. Why have we not seen many irradiated products in our markets?
  b. Would you buy foods that have been irradiated? Why or why not?

3.71 What is the difference between fission and fusion?

3.72 **a.** What are the products in the fission of uranium-235 that make possible a nuclear chain reaction?
  b. What is the purpose of placing control rods among uranium samples in a nuclear reactor?

3.73 Where does fusion occur naturally?

3.74 Why are scientists continuing to try to build a fusion reactor even though very high temperatures have been difficult to reach and maintain?

CHALLENGE QUESTIONS

3.75 Identify each of the following nuclear reactions as alpha decay, beta decay, positron emission, or gamma radiation.
  a. $^{27m}_{13}Al \longrightarrow ^{27}_{13}Al + ^{0}_{0}\gamma$
  b. $^{8}_{5}B \longrightarrow ^{8}_{4}Be + ^{0}_{+1}e$
  c. $^{90}_{38}Sr \longrightarrow ^{90}_{39}Y + ^{0}_{-1}e$
  d. $^{218}_{85}At \longrightarrow ^{214}_{83}Bi + ^{4}_{2}He$

3.76 Complete and balance each of the following nuclear equations:
  a. $^{23m}_{12}Mg \longrightarrow \underline{\hspace{1cm}} + ^{0}_{0}\gamma$
  b. $^{61}_{30}Zn \longrightarrow ^{61}_{29}Cu + \underline{\hspace{1cm}}$
  c. $^{241}_{95}Am + ^{4}_{2}He \longrightarrow \underline{\hspace{1cm}} + 2^{1}_{0}n$
  d. $^{126}_{50}Sn \longrightarrow \underline{\hspace{1cm}} + ^{0}_{-1}e$

3.77 The half-life for the radioactive decay of calcium-47 is 4.5 days. If a sample has an activity of 4.0 μCi after 18 days, what was the initial activity of the sample?

3.78 A 16 μg sample of sodium-24 decays to 2.0 μg in 45 hr. What is the half-life of ^{24}Na?

3.79 Write a balanced equation for each of the following radioactive emissions:
  a. an alpha particle from Hg-180
  b. a beta particle from Sn-126
  c. a positron from Mn-49

3.80 Write a balanced equation for each of the following radioactive emissions:
  a. an alpha particle from Gd-148
  b. a beta particle from Sr-90
  c. a positron from Al-25

ANSWERS

Answers to Study Checks

3.1 $^{214}_{84}Po \longrightarrow ^{210}_{82}Pb + ^{4}_{2}He$

3.2 $^{131}_{53}I \longrightarrow ^{131}_{54}Xe + ^{0}_{-1}e$

3.3 $^{27}_{13}Al + ^{4}_{2}He \longrightarrow ^{30}_{15}P + ^{1}_{0}n$

3.4 For β, the factor is 1; rads and rems are equal.

3.5 0.50 g

3.6 17,200 years

3.7 An iodine-131 sample with a higher activity is used in radiation therapy when radiation is needed to destroy some of the cells in the thyroid gland.

3.8 fusion

Answers to Selected Questions and Problems

3.1 **a.** Both an alpha particle and a helium nucleus have 2 protons and 2 neutrons. However, an α particle is emitted from a nucleus during radioactive decay.
  b. α, $^{4}_{2}He$

3.3 **a.** $^{39}_{19}K$, $^{40}_{19}K$, $^{41}_{19}K$

b. They all have 19 protons and 19 electrons, but they differ in the number of neutrons.

3.5

Medical Use	Isotope Symbol	Mass Number	Number of Protons	Number of Neutrons
Heart imaging	$^{201}_{81}Tl$	201	81	120
Radiation therapy	$^{60}_{27}Co$	60	27	33
Abdominal scan	$^{67}_{31}Ga$	67	31	36
Hyperthyroidism	$^{131}_{53}I$	131	53	78
Leukemia treatment	$^{32}_{15}P$	32	15	17

3.7 **a.** α, ^4_2He **b.** ^1_0n **c.** β, $^0_{-1}e$

d. $^{15}_7N$ **e.** $^{125}_{53}I$

3.9 **a.** β or e^- **b.** α or He **c.** n

d. Na **e.** C

3.11 **a.** Because β particles move faster than α particles, they can penetrate further into tissue.

b. Ionizing radiation forms reactive species that cause undesirable reactions in the cells.

c. X-ray technicians leave the room to increase the distance between them and the radiation. Also a wall that contains lead shields them.

d. Wearing gloves shields the skin from α and β radiation.

3.13 **a.** $^{208}_{84}Po \longrightarrow ^{204}_{82}Pb + ^4_2He$

b. $^{232}_{90}Th \longrightarrow ^{228}_{88}Ra + ^4_2He$

c. $^{251}_{102}No \longrightarrow ^{247}_{100}Fm + ^4_2He$

d. $^{220}_{86}Rn \longrightarrow ^{216}_{84}Po + ^4_2He$

3.15 **a.** $^{25}_{11}Na \longrightarrow ^{25}_{12}Mg + ^0_{-1}e$

b. $^{20}_8O \longrightarrow ^{20}_9F + ^0_{-1}e$

c. $^{92}_{38}Sr \longrightarrow ^{92}_{39}Y + ^0_{-1}e$

d. $^{42}_{19}K \longrightarrow ^{42}_{20}Ca + ^0_{-1}e$

3.17 **a.** $^{26}_{14}Si \longrightarrow ^{26}_{13}Al + ^0_{+1}e$

b. $^{54}_{27}Co \longrightarrow ^{54}_{26}Fe + ^0_{+1}e$

c. $^{77}_{37}Rb \longrightarrow ^{77}_{36}Kr + ^0_{+1}e$

d. $^{93}_{45}Rh \longrightarrow ^{93}_{44}Ru + ^0_{+1}e$

3.19 **a.** $^{28}_{14}Si$ **b.** $^{87}_{36}Kr$ **c.** $^0_{-1}e$

d. $^{238}_{92}U$ **e.** $^{188}_{79}Au$

3.21 **a.** $^{10}_4Be$ **b.** $^0_{-1}e$

c. $^{27}_{13}Al$ **d.** ^1_1p

3.23 **a.** When radiation enters the Geiger counter, charged particles are produced which create a burst of current that is detected by the instrument.

b. becquerel (Bq), curie (Ci)

c. gray (Gy), rad

d. 1000 Gy

3.25 294 μCi

3.27 When pilots are flying at high altitudes, there is less atmosphere to protect them from cosmic radiation.

3.29 A half-life is the time it takes for one-half of a radioactive sample to decay.

3.31 **a.** 40.0 mg

b. 20.0 mg

c. 10.0 mg

d. 5.00 mg

3.33 130 days, 195 days

3.35 **a.** Since the elements Ca and P are part of bone, their radioactive isotopes will also become part of the bony structures of the body where their radiation can be used to diagnose or treat bone diseases.

b. Strontium (Sr) acts much like calcium (Ca) because both are Group 2A (2) elements. The body will accumulate radioactive strontium in bones in the same way that it incorporates calcium. Radioactive strontium is harmful to children because the radiation it produces causes more damage in cells that are dividing rapidly.

3.37 180 μCi

3.39 Nuclear fission is the splitting of a large atom into smaller fragments with the release of large amounts of energy.

3.41 $^{103}_{42}Mo$

3.43 **a.** fission

b. fusion

c. fission

d. fusion

3.45 **a.** $^{11}_6C$

b.

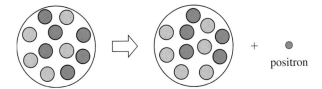

3.47

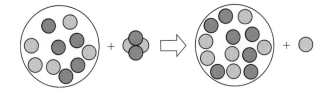

3.49 17 200 years old

3.51 Each has 6 protons and 6 electrons, but carbon-12 has 6 neutrons and carbon-14 has 8 neutrons. Carbon-12 is a stable isotope, but carbon-14 is radioactive and will emit radiation.

3.53 **a.** α radiation consists of a helium nucleus emitted from the nucleus of a radioisotope. β radiation is an electron and γ radiation is high-energy radiation emitted from the nucleus of a radioisotope.

b. α particle: ^4_2He

β particle: $^0_{-1}e$

γ radiation: γ

c. α particles penetrate 0.05 mm into tissue, β particles 4–5 mm, and γ rays 50 cm or more.

d. Paper or clothing will shield you from α particles; heavy clothing, lab coats, and gloves will shield you from β particles; and lead and concrete are needed for shielding from γ radiation.

3.55 **a.** $^{225}_{90}\text{Th} \longrightarrow {}^{221}_{88}\text{Ra} + {}^{4}_{2}\text{He}$

b. $^{210}_{83}\text{Bi} \longrightarrow {}^{206}_{81}\text{Tl} + {}^{4}_{2}\text{He}$

c. $^{137}_{55}\text{Cs} \longrightarrow {}^{137}_{56}\text{Ba} + {}^{0}_{-1}e$

d. $^{126}_{50}\text{Sn} \longrightarrow {}^{126}_{51}\text{Sb} + {}^{0}_{-1}e$

3.57 **a.** $^{17}_{8}\text{O}$ **b.** $^{1}_{1}\text{p}$ **c.** $^{143}_{54}\text{Xe}$

3.59 **a.** $^{16}_{8}\text{O} + {}^{16}_{8}\text{O} \longrightarrow {}^{4}_{2}\text{He} + {}^{28}_{14}\text{Si}$

b. $^{249}_{98}\text{Cf} + {}^{18}_{8}\text{O} \longrightarrow {}^{263}_{106}\text{Sg} + 4\,{}^{1}_{0}\text{n}$

c. $^{222}_{86}\text{Rn} \longrightarrow {}^{218}_{84}\text{Po} + {}^{4}_{2}\text{He}$

d. $^{218}_{84}\text{Po} \longrightarrow {}^{214}_{82}\text{Pb} + {}^{4}_{2}\text{He}$

3.61 14 days

3.63 **a.** $^{131}_{53}\text{I} \longrightarrow {}^{0}_{-1}e + {}^{131}_{54}\text{Xe}$

b. 0.375 g

c. 32 days

3.65 16 μCi

3.67 7.5 mg

3.69 The irradiation of meats, fruits, and vegetables kills bacteria such as *E. coli* that can cause foodborne illnesses. In addition, spoilage is deterred, and shelf life is extended.

3.71 In the fission process, an atom splits into smaller nuclei. In fusion, small nuclei combine (fuse) to form a larger nucleus.

3.73 Fusion occurs naturally in the sun and other stars.

3.75 **a.** gamma radiation
b. positron emission
c. beta decay
d. alpha decay

3.77 $18 \text{ days} \times \dfrac{1 \text{ half-life}}{4.5 \text{ days}} = 4 \text{ half-lives}$

$$\overset{(1)}{} \quad \overset{(2)}{} \quad \overset{(3)}{} \quad \overset{(4)}{}$$
$64\ \mu\text{Ci} \longrightarrow 32\ \mu\text{Ci} \longrightarrow 16\ \mu\text{Ci} \longrightarrow 8.0\ \mu\text{Ci} \longrightarrow 4.0\ \mu\text{Ci}$

3.79 **a.** $^{180}_{80}\text{Hg} \longrightarrow {}^{176}_{78}\text{Pt} + {}^{4}_{2}\text{He}$

b. $^{126}_{50}\text{Sn} \longrightarrow {}^{126}_{51}\text{Sb} + {}^{0}_{-1}e$

c. $^{49}_{25}\text{Mn} \longrightarrow {}^{49}_{24}\text{Cr} + {}^{0}_{+1}e$

4 Compounds and Their Bonds

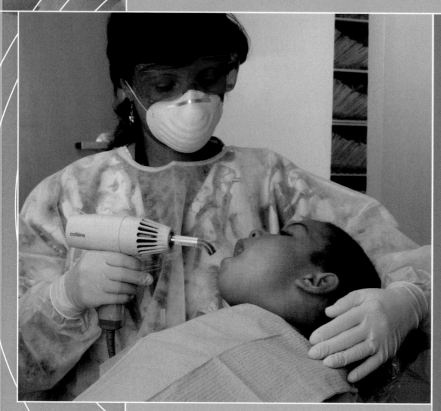

"One way to prevent cavities in children is to apply a thin, plastic coating called a sealant to their teeth," says Dr. Pam Alston, a dentist in private practice. "We look for teeth with deep grooves and pits that trap food. We clean the teeth and apply an etching agent, which helps the sealant bind to the teeth. Then we apply the liquid sealant, which fills in the grooves and pits, and use ultraviolet light to solidify the coating."

The use of fluoride compounds, such as SnF_2 in toothpaste and NaF in water, and mouth rinses have greatly reduced tooth decay. The fluoride ion replaces the hydroxide ion to form $Ca_{10}(PO_4)_6F_2$, which strengthens the enamel and makes it less susceptible to decay. Other compounds used in dentistry are the anesthetic known as laughing gas, N_2O, and Novocaine, $C_{13}H_{20}N_2O_2$.

Mastering GOB CHEMISTRY

Visit **www.masteringgob.com** for self-study materials and instructor-assigned homework.

n nature, atoms of almost all the elements on the periodic table are found in combination with other atoms. Only the atoms of the noble gases — He, Ne, Ar, Kr, Xe, and Rn — are found as individual atoms. As discussed in Chapter 2, a compound is a pure substance, composed of two or more elements, with a definite composition. In a typical ionic compound, one or more electrons are transferred from the atoms of metals to atoms of nonmetals. The attraction that results is called an ionic bond.

We use many ionic compounds every day. When we cook or bake we use ionic compounds such as salt, NaCl, and baking soda, $NaHCO_3$. Epsom salts, $MgSO_4$, may be used to soak sore feet. Milk of magnesia, $Mg(OH)_2$, or calcium carbonate, $CaCO_3$, may be taken to settle an upset stomach. In a mineral supplement, iron is present as iron(II) sulfate, $FeSO_4$. Certain sunscreens contain zinc oxide, ZnO, and the tin(II) fluoride, SnF_2, in toothpaste provides fluoride to help prevent tooth decay.

The structures of ionic crystals result in the beautiful facets seen in gems. Sapphires and rubies are made of aluminum oxide, Al_2O_3. Impurities of chromium make rubies red, and iron and titanium make sapphires blue.

In compounds of nonmetals, covalent bonding occurs by atoms sharing one or more valence electrons. There are many more covalent compounds than there are ionic ones and many simple covalent compounds are present in our everyday lives. For example, water (H_2O), oxygen (O_2), and carbon dioxide (CO_2) are all covalent compounds.

Covalent compounds consist of molecules, which are discrete groups of atoms. A molecule of water (H_2O) consists of two atoms of hydrogen and one atom of oxygen. When you have iced tea, perhaps you add molecules of sugar (sucrose), which is a covalent compound ($C_{12}H_{22}O_{11}$). Other covalent compounds include propane C_3H_8, alcohol C_2H_6O, the antibiotic amoxicillin ($C_{16}H_{19}N_3O_5S$), and the antidepressant Prozac ($C_{17}H_{18}F_3NO$).

4.1 Octet Rule and Ions

LEARNING GOAL

Using the octet rule, write the symbols of the simple ions for the representative elements.

Most of the elements on the periodic table combine to form compounds except for the atoms of the noble gases. One explanation is that noble gases have a particularly stable *octet* of 8 valence electrons except for helium, which is stable with 2 electrons filling its first energy level.

Compounds are the result of the formation of chemical bonds between two or more different elements. Ionic bonds occur when atoms of one element lose valence electrons and the atoms of another element gain valence electrons. Ionic compounds typically occur between metals and nonmetals, which are far apart on the periodic table. For example, atoms of sodium and chlorine form NaCl, an ionic compound. However, atoms of elements that are close together on the periodic table share valence electrons and form covalent compounds. For example, atoms of nitrogen and chlorine form NCl_3, a covalent compound.

Octet Rule and Ions

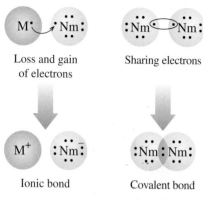

Loss and gain
of electrons

Sharing electrons

Ionic bond

Covalent bond

M is a metal
Nm is a nonmetal

In the formation of either an ionic bond or covalent bond, atoms adjust their valence electrons such that they each have the electron configuration of a noble gas. Known as the **octet rule**, this tendency for atoms to acquire a noble gas arrangement provides a key to our understanding of the ways in which atoms bond and form compounds.

Positive Ions

In ionic bonding, the valence electrons of a metal are transferred to a nonmetal. Because the ionization energies of metals of Groups 1A(1), 2A(2), and 3A(13) are low, these metal atoms readily lose their valence electrons to nonmetals. In doing so, they acquire the electron configuration of a noble gas (usually 8 valence electrons) and form **ions** with positive charges. For example, when a sodium atom loses its single valence electron, the remaining electrons have the noble gas configuration of neon. By losing an electron sodium has 10 electrons instead of 11. Because there are still 11 protons in its nucleus, the atom is no longer neutral. It has become a sodium ion and has an electrical charge, called an **ionic charge**, of 1+. In the symbol for the sodium ion, the ionic charge is written in the upper right-hand corner, as Na^+.

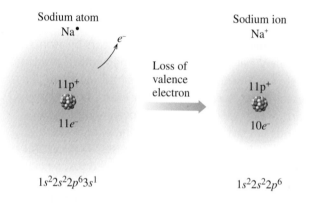

Sodium atom
$Na^\bullet$

Sodium ion
Na^+

e^-

$11p^+$

Loss of
valence
electron

$11p^+$

$11e^-$

$10e^-$

$1s^2 2s^2 2p^6 3s^1$

$1s^2 2s^2 2p^6$

Metals in ionic compounds have lost their valence electrons to form positively charged ions. Positive ions are also called **cations** (pronounced *cat'-ions*). Magnesium, a metal in Group 2A (2), attains a noble gas arrangement like neon by losing 2 valence electrons to form a positive ion with a 2+ ionic charge. A metal ion is named by its element name. Thus, Mg^{2+} is named the magnesium ion.

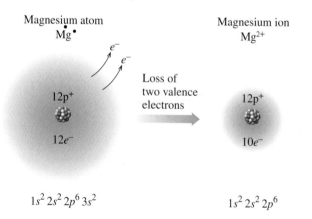

Magnesium atom
$Mg^\bullet$

Magnesium ion
Mg^{2+}

e^-

e^-

$12p^+$

Loss of
two valence
electrons

$12p^+$

$12e^-$

$10e^-$

$1s^2 2s^2 2p^6 3s^2$

$1s^2 2s^2 2p^6$

Negative Ions

Nonmetals form negative ions when they gain one or more electrons to complete their valence energy levels. For example, when an atom of chlorine with 7 valence electrons obtains one more electron, it attains the noble gas arrangement of argon. The resulting particle is a chloride ion having a negative ionic charge (Cl^-). Ions with negative charges are called **anions** (pronounced *an'-ions*). Nonmetal ions are named by using *ide* at the end of their element names.

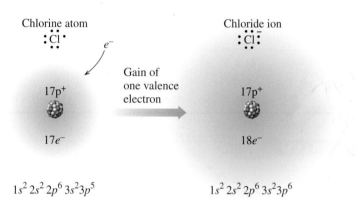

Chlorine atom
:Cl·
e^-

Gain of one valence electron

$17p^+$
$17e^-$

$1s^2 2s^2 2p^6 3s^2 3p^5$

Chloride ion
:Cl:⁻

$17p^+$
$18e^-$

$1s^2 2s^2 2p^6 3s^2 3p^6$

SAMPLE PROBLEM 4.1

Calculating Ionic Charge

Write the symbol and the ionic charge for each of the following ions:

a. a nitride ion that has 7 protons and 10 electrons
b. a calcium ion that has 20 protons and 18 electrons

SOLUTION

a. $7p^+$ and $10e^- = N^{3-}$
b. $20p^+$ and $18e^- = Ca^{2+}$

STUDY CHECK

How many protons and electrons are in a bromide ion, Br^-?

Ionic Charges from Group Numbers

Group numbers can be used to determine the ionic charges for most ions of the representative elements. We have seen that metals lose electrons to form positive ions. The elements in Groups 1A (1), 2A (2), and 3A (13) lose 1, 2, and 3 electrons, respectively. Group 1A (1) metals form ions with 1+ charges, Group 2A(2) metals form ions with 2+ charges, and Group 3A (13) metals form ions with 3+ charges.

The nonmetals from Groups 5A (15), 6A (16), and 7A (17) form negative ions. Group 5A (15) nonmetals usually form ions with 3− charges, Group 6A(16) nonmetals form ions with 2− charges, and Group 7A (17) nonmetals form ions with 1− charges. The nonmetals of Group 4A (14) do not typically form ions. Table 4.1 lists the ionic charges for some common ions of representative elements.

TABLE 4.1

Ionic Charges for Representative Elements

Group Number	Number of Valence Electrons	Electron Change to Give an Octet	Ionic Charge	Examples
Metals				
1A (1)	1	Lose 1	1+	Li^+, Na^+, K^+
2A (2)	2	Lose 2	2+	Mg^{2+}, Ca^{2+}
3A (13)	3	Lose 3	3+	Al^{3+}
Nonmetals				
5A (15)	5	Gain 3	3−	N^{3-}, P^{3-}
6A (16)	6	Gain 2	2−	O^{2-}, S^{2-}
7A (17)	7	Gain 1	1−	F^-, Cl^-, Br^-, I^-

SAMPLE PROBLEM 4.2

■ **Writing Ions**

Consider the elements aluminum and oxygen.

a. Identify each as a metal or a nonmetal.
b. State the number of valence electrons for each.
c. State the number of electrons that must be lost or gained for each to acquire an octet.
d. Write the symbol of each resulting ion, including its ionic charge.

SOLUTION

Aluminum	**Oxygen**
a. metal	nonmetal
b. 3 valence electrons	6 valence electrons
c. loses $3e^-$	gains $2e^-$
d. Al^{3+}	O^{2-}

STUDY CHECK

What are the symbols for the ions formed by potassium and sulfur?

QUESTIONS AND PROBLEMS

MC GOB

■ **Octet Rule and Ions**

4.1 **a.** How does the octet rule explain the formation of a sodium ion?
 b. What noble gas has the same electron arrangement as the sodium ion?
 c. Why are Group 1A (1) and Group 2A (2) elements found in many compounds, but not Group 8A (18) elements?

4.2 **a.** How does the octet rule explain the formation of a chloride ion, Cl^-?
 b. What noble gas has the same electron arrangement as the chloride ion, Cl^-?
 c. Why are Group 7A (17) elements found in many compounds, but not Group 8A (18) elements?

4.3 State the number of electrons that must be lost by atoms of each of the following elements to acquire a noble gas electron arrangement:
 a. Li **b.** Mg **c.** Al **d.** Cs **e.** Ba

■ Some Important Ions in the Body

A number of ions in body fluids have important physiological and metabolic functions. Some of them are listed in Table 4.2.

TABLE 4.2

Ions in the Body

Ion	Occurrence	Function	Source	Result of Too Little	Result of Too Much
Na^+	Principal cation outside the cell	Regulation and control of body fluids	Salt, cheese, pickles	Hyponatremia, anxiety, diarrhea, circulatory failure, decrease in body fluid	Hypernatremia, little urine, thirst, edema
K^+	Principal cation inside the cell	Regulation of body fluids and cellular functions	Bananas, orange juice, milk, prunes, potatoes	Hypokalemia (hypopotassemia), lethargy, muscle weakness, failure of neurological impulses	Hyperkalemia (hyperpotassemia), irritability, nausea, little urine, cardiac arrest
Ca^{2+}	Cation outside the cell; 90% of calcium in the body in bone as $Ca_3(PO_4)_2$ or $CaCO_3$	Major cation of bone; increasing for smooth muscle function	Milk, yogurt, cheese, greens, and spinach	Hypocalcemia, tingling fingertips, muscle cramps, osteoporosis	Hypercalcemia, relaxed muscles, kidney stones, deep bone pain
Mg^{2+}	Cation outside the cell; 70% of magnesium in the body in bone structure	Essential for certain enzymes, muscles, and nerve control	Widely distributed (part of chlorophyll of all green plants), nuts, whole grains	Disorientation, hypertension, tremors, slow pulse	Drowsiness
Cl^-	Principal anion outside the cell	Gastric juice, regulation of body fluids	Salt	Same as for Na^+	Same as for Na^+

4.4 State the number of electrons that must be gained by atoms of each of the following elements to acquire a noble gas electron arrangement:
a. Cl   **b.** O   **c.** N   **d.** I   **e.** P

4.5 Write the symbols of the ions with the following number of protons and electrons:
a. 3 protons, 2 electrons   **b.** 9 protons, 10 electrons
c. 12 protons, 10 electrons   **d.** 26 protons, 23 electrons
e. 30 protons, 28 electrons

4.6 How many protons and electrons are in the following ions?
a. O^{2-} **b.** K^+ **c.** Br^- **d.** S^{2-} **e.** Sr^{2+}

4.7 Write the symbol for the ion of each of the following:
    a. chlorine   **b.** potassium   **c.** oxygen   **d.** aluminum

4.8 Write the symbol for the ion of each of the following:
    a. fluorine   **b.** calcium   **c.** sodium   **d.** lithium

---

LEARNING GOAL

Using charge balance, write the correct formula for an ionic compound.

(MC GOB) Ionic Compounds

4.2 Ionic Compounds

Ionic compounds consist of positive and negative ions. The ions are held together by strong attractions between the oppositely charged ions, called ionic bonds.

Properties of Ionic Compounds

The physical and chemical properties of an ionic compound such as NaCl are very different from those of the original elements. For example, the original elements of NaCl were sodium, a soft, shiny metal, and chlorine, a yellow-green poisonous gas. Yet, as positive and negative ions, they form table salt, NaCl, a white, crystalline substance that is common in our diet. In ionic compounds, the attraction between the ions is very strong, which makes the melting points of ionic compounds high, often more than 300°C. For instance, the melting point of NaCl is 800°C. At room temperature, ionic compounds are solids.

The structure of an ionic solid depends on the arrangement of the ions. In a crystal of NaCl, which has a cubic shape, the larger Cl^- ions (green) are packed close together in a lattice structure as shown in Figure 4.1. The smaller Na^+ ions (silver) occupy the holes between the Cl^- ions.

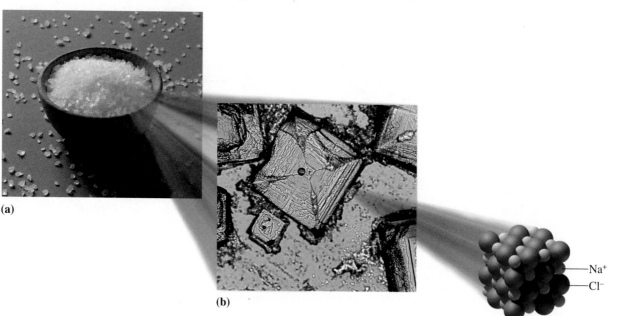

(a)

(b)

(c)

—Na^+
—Cl^-

FIGURE 4.1

(a) The elements sodium and chlorine react to form the ionic compound sodium chloride, the compound that makes up table salt. **(b)** Crystals of NaCl under magnification. **(c)** A diagram of the arrangements of Na^+ and Cl^- packed together in a NaCl crystal.

Q **What is the type of bonding between Na^+ and Cl^- ions in salt?**

Charge Balance in Ionic Compounds

The **formula** of an ionic compound indicates the number and kinds of ions that make up the ionic compound. The sum of the ionic charges in the formula is always zero. That means that the total amount of positive charge is equal to the total amount of negative charge. For example, the NaCl formula indicates that there is one sodium ion, Na^+, for every chloride ion, Cl^-, in the compound. Note that the ionic charges of the ions do not appear in the formula of the ionic compound.

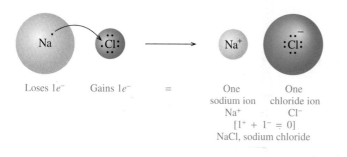

Loses $1e^-$ Gains $1e^-$ =   One sodium ion Na^+ One chloride ion Cl^-
$[1^+ + 1^- = 0]$
NaCl, sodium chloride

Subscripts in Formulas

Consider a compound of magnesium and chlorine. To achieve an octet, a Mg atom loses its 2 valence electrons to form Mg^{2+}. Each Cl atom gains 1 electron to form Cl^-, which has a complete valence energy level. In this example, two Cl^- ions are needed to balance the positive charge of Mg^{2+}. This gives the formula, $MgCl_2$, magnesium chloride, in which a subscript of 2 shows that two Cl^- were needed for charge balance.

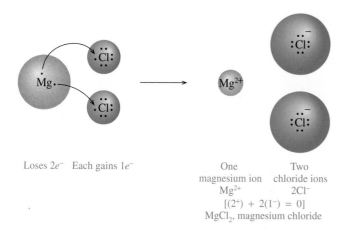

Loses $2e^-$ Each gains $1e^-$ One magnesium ion Mg^{2+} Two chloride ions $2Cl^-$
$[(2^+) + 2(1^-) = 0]$
$MgCl_2$, magnesium chloride

Writing Ionic Formulas from Ionic Charges

The subscripts in the formula of an ionic compound represent the number of positive and negative ions that give an overall charge of zero. Thus, we can now write a formula directly from the ionic charges of the positive and negative ions. Suppose we wish to write the formula of the ionic compound containing Na^+ and S^{2-} ions. To balance the ionic charge of the S^{2-} ion, we will need to place two Na^+ ions in the formula. This gives the formula Na_2S, which has an overall charge of zero. In the formula of an ionic compound, the cation is written first followed by the anion. Appropriate subscripts are used to show the number of ions.

CAREER FOCUS

■ Physical Therapist

"Physical therapists need to understand how the body, muscles, and joints function in order to recognize when something is not working and which area needs strengthening," says Vincent Leddy, physical therapist. "Chemistry is important to understanding the body's physiology and how chemical changes in the body affect movement. I went into physical therapy because I enjoy teaching movement to children. I am guiding Maggie into her chair but allowing her to move as much as she can by herself. I help her by giving gentle pressure and reassurance that she'll be safe in the transition. I work on getting Maggie to use her body, and the occupational therapist works on Maggie's fine motor skills and how she uses her hand on the switch or picks up objects. By using both physical and occupational therapy, we enhance the child's performance."

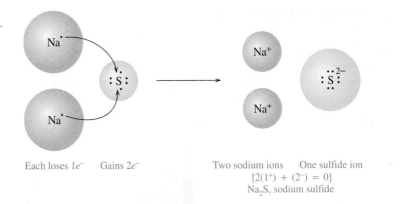

Each loses $1e^-$ Gains $2e^-$ Two sodium ions   One sulfide ion
$$[2(1^+) + (2^-) = 0]$$
Na_2S, sodium sulfide

SAMPLE PROBLEM 4.3

▪ Writing Formulas from Ionic Charges

Use charge balance to write the formula for the ionic compound containing K^+ and N^{3-}.

SOLUTION

Determine the number of each ion needed for charge balance. The charge for nitrogen ($3-$) is balanced by three K^+ ions. Writing the positive ion first gives the formula K_3N.

STUDY CHECK

Use ionic charges to determine the formula of the compound that would form when calcium and oxygen react.   *CaO*

QUESTIONS AND PROBLEMS

▪ Ionic Compounds

4.9 Which of the following pairs of elements are likely to form an ionic compound?
- **a.** lithium and chlorine
- **b.** oxygen and chlorine
- **c.** potassium and oxygen
- **d.** sodium and neon
- **e.** sodium and magnesium

4.10 Which of the following pairs of elements are likely to form an ionic compound?
- **a.** helium and oxygen
- **b.** magnesium and chlorine
- **c.** chlorine and bromine
- **d.** potassium and sulfur
- **e.** sodium and potassium

4.11 Write the correct ionic formula for compounds formed between the following ions:
- **a.** Na^+ and O^{2-}
- **b.** Al^{3+} and Br^-
- **c.** Ba^{2+} and O^{2-}
- **d.** Mg^{2+} and Cl^-
- **e.** Al^{3+} and S^{2-}

4.12 Write the correct ionic formula for compounds formed between the following ions:
- **a.** Al^{3+} and Cl^-
- **b.** Ca^{2+} and S^{2-}
- **c.** Li^+ and S^{2-}
- **d.** K^+ and N^{3-}
- **e.** K^+ and I^-

4.13 Write the correct formula for ionic compounds formed by the following metals and nonmetals:
- **a.** sodium and sulfur
- **b.** potassium and nitrogen
- **c.** aluminum and iodine
- **d.** lithium and oxygen

4.14 Write the correct formula for ionic compounds formed by the following metals and nonmetals:
- **a.** calcium and chlorine
- **b.** barium and bromine
- **c.** sodium and phosphorus
- **d.** magnesium and oxygen

TABLE 4.3

Formulas and Names of Some Common Ions

Group Number	Formula of Ion	Name of Ion	Group Number	Formula of Ion	Name of Ion
	Metals			Nonmetals	
1A (1)	Li^+	lithium	5A (15)	N^{3-}	nitride
	Na^+	sodium		P^{3-}	phosphide
	K^+	potassium	6A (16)	O^{2-}	oxide
2A (2)	Mg^{2+}	magnesium		S^{2-}	sulfide
	Ca^{2+}	calcium	7A (17)	F^-	fluoride
	Ba^{2+}	barium		Cl^-	chloride
3A (13)	Al^{3+}	aluminum		Br^-	bromide
				I^-	iodide

4.3 Naming and Writing Ionic Formulas

As we mentioned in section 4.2, the name of a metal ion is the same as its elemental name. The name of a nonmetal ion is obtained by replacing the end of its elemental name with *ide*. Table 4.3 lists the names of some important metal and nonmetal ions.

Naming Ionic Compounds Containing Two Elements

In the name of an ionic compound made up of two elements, the metal ion is named followed by the name of the nonmetal ion. Subscripts are never mentioned; they are understood as a result of the charge balance of the ions in the compound. (See Table 4.4.)

TABLE 4.4

Names of Some Ionic Compounds

Compound	Metal Ion	Nonmetal Ion	Name
NaF	Na^+	F^-	
	sodium	fluoride	sodium fluoride
$MgBr_2$	Mg^{2+}	Br^-	
	magnesium	bromide	magnesium bromide
Al_2O_3	Al^{3+}	O^{2-}	
	aluminum	oxide	aluminum oxide

SAMPLE PROBLEM 4.4

Naming Ionic Compounds

Write the name of each of the following ionic compounds:

a. Na_2O **b.** Mg_3N_2

LEARNING GOAL

Given the formula of an ionic compound, write the correct name; given the name of an ionic compound, write the correct formula.

MC GOB   Writing Ionic Formulas

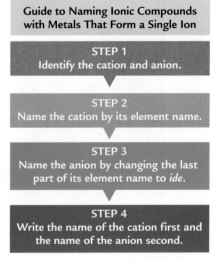

Guide to Naming Ionic Compounds with Metals That Form a Single Ion

STEP 1
Identify the cation and anion.

STEP 2
Name the cation by its element name.

STEP 3
Name the anion by changing the last part of its element name to *ide*.

STEP 4
Write the name of the cation first and the name of the anion second.

SOLUTION

			■ STEP 1		■ STEP 2	■ STEP 3	■ STEP 4
Compound	**Cation**	**Anion**			**Name of Cation**	**Name of Anion**	**Name of Compound**
a. Na_2O	Na^+	O^{2-}			sodium	oxide	sodium oxide
b. Mg_3N_2	Mg^{2+}	N^{3-}			magnesium	nitride	magnesium nitride

STUDY CHECK

Name the compound $CaCl_2$.

Metals with Variable Charge

The transition metals typically form two or more kinds of positive ions because they lose their outer electrons as well as electrons from a lower energy level. For example, in some ionic compounds, iron is in the Fe^{2+} form, but in other compounds, it takes the Fe^{3+} form. Copper also forms two different ions: Cu^+ is present in some compounds and Cu^{2+} in others. When a metal can form two or more ions, it is not possible to predict the ionic charge from the group number. We say that it has a variable valence or variable charge.

When different ions are possible for a metal, a naming system is needed to identify the particular cation in a compound. To do this, a Roman numeral that matches the ionic charge is placed in parentheses after the elemental name of the metal. For the cations of iron, Fe^{2+} is named iron(II), and Fe^{3+} is named iron(III). Table 4.5 lists the ions of some common metals that produce two or more ions.

Figure 4.2 shows some common ions and their location on the periodic table. Typically, the transition metals form more than one positive ion. However, zinc,

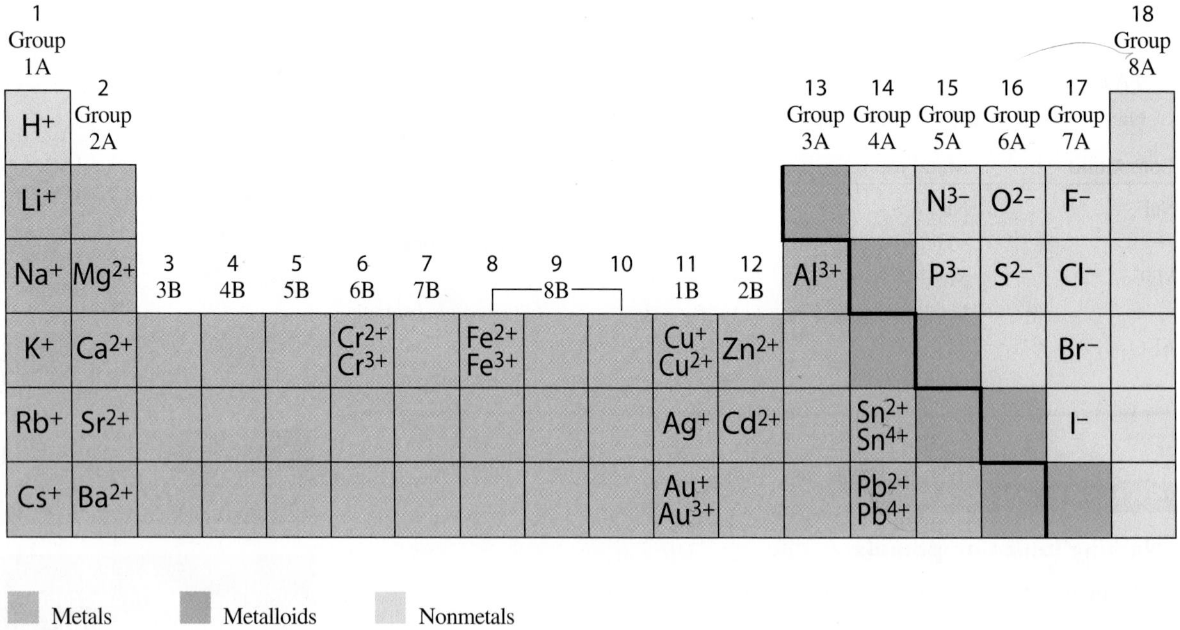

TABLE 4.5

Some Metals That Form More Than One Positive Ion

Element	Possible Ions	Name of Ion
chromium	Cr^{2+}	chromium(II)
	Cr^{3+}	chromium(III)
copper	Cu^+	copper(I)
	Cu^{2+}	copper(II)
gold	Au^+	gold(I)
	Au^{3+}	gold(III)
iron	Fe^{2+}	iron(II)
	Fe^{3+}	iron(III)
lead	Pb^{2+}	lead(II)
	Pb^{4+}	lead(IV)
tin	Sn^{2+}	tin(II)
	Sn^{4+}	tin(IV)

FIGURE 4.2

On the periodic table, positive ions are produced from metals and negative ions are produced from nonmetals.

Q **What are the typical ions produced by calcium, copper, and oxygen?**

cadmium, and silver form only one ion. The ionic charges of silver, cadmium, and zinc are fixed like Group 1A (1), 2A (2), and 3A (13) metals, so their elemental names are sufficient when naming their ionic compounds.

The selection of the correct Roman numeral depends upon the calculation of the ionic charge of the transition metal in the formula. For example, we know that in the formula $CuCl_2$ the positive charge of the copper ion must balance the negative charge of two chloride ions. Because we know that chloride ions each have a $1-$ charge, there must be a total negative charge of $2-$. Balancing the $2-$ by the positive charge gives a charge of $2+$ for Cu (a Cu^{2+} ion):

$CuCl_2$

Cu charge $+$ $2Cl^-$ charge $= 0$

(?) $\quad\quad + 2(1-) \quad\quad = 0$

$(2+) \quad\quad + 2- \quad\quad\quad = 0$

To indicate the copper ion Cu^{2+}, we place (II) after *copper* when naming the compound:

copper(II) chloride

Table 4.6 lists names of some ionic compounds in which the metals form two types of positive ion.

TABLE 4.6

Some Ionic Compounds of Metals That Form Two Kinds of Positive Ions

Compound	Systematic Name
$FeCl_2$	iron(II) chloride
$FeCl_3$	iron(III) chloride
Cu_2O	copper(I) oxide
$CuCl_2$	copper(II) chloride
$SnCl_2$	tin(II) chloride
$PbBr_4$	lead(IV) bromide

SAMPLE PROBLEM 4.5

■ **Naming Ionic Compounds with Variable Charge Metal Ions**

Write the name for Cu_2S.

SOLUTION

■ **STEP 1 Determine the charge of the cation from the anion.** The nonmetal S in Group 6A (16) forms the S^{2-} ion. Because there are two Cu ions to balance the S^{2-}, the charge of each Cu ion is $1+$.

	Metal	Nonmetal
Elements	copper	sulfur
Groups	transition	6A (16)
Ions	Cu?	S^{2-}
Charge balance	$2(+1)$ $\quad+$	$(2-) = 0$
Ions	Cu^+	S^{2-}

■ **STEP 2 Name the cation by its element name and use a Roman numeral in parentheses for the charge.**

copper(I)

■ **STEP 3 Name the anion by changing the last part of its element name to *ide*.**

sulfide

■ **STEP 4 Write the name of the cation first and the name of the anion second.**

copper(I) sulfide

STUDY CHECK

Write the name of the compound whose formula is $AuCl_3$.

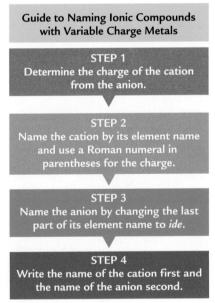

Guide to Naming Ionic Compounds with Variable Charge Metals

STEP 1
Determine the charge of the cation from the anion.

STEP 2
Name the cation by its element name and use a Roman numeral in parentheses for the charge.

STEP 3
Name the anion by changing the last part of its element name to *ide*.

STEP 4
Write the name of the cation first and the name of the anion second.

▪ Writing Formulas of Ionic Compounds

Write the formula for iron(III) chloride.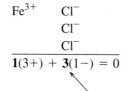

SOLUTION

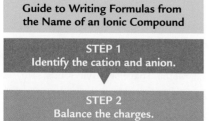

Guide to Writing Formulas from the Name of an Ionic Compound

STEP 1
Identify the cation and anion.

STEP 2
Balance the charges.

STEP 3
Write the formula, cation first, using subscripts from charge balance.

▪ **STEP 1 Identify the cation and anion.** The Roman numeral (III) indicates that the charge of the iron ion is $3+$, Fe^{3+}.

	Metal	**Nonmetal**
Elements	iron(III)	chlorine
Groups	transition	7A (17)
Ions	Fe^{3+}	Cl^-

▪ **STEP 2 Balance the charges.**

$$Fe^{3+} \quad Cl^-$$
$$Cl^-$$
$$Cl^-$$
$$\overline{1(3+) + \mathbf{3}(1-) = 0}$$

Becomes a subscript in the formula

▪ **STEP 3 Write the formula, cation first, using subscripts from the charge balance.**

$$FeCl_3$$

STUDY CHECK

Write the correct formula for chromium(III) oxide. Cr_2O_3

▪ QUESTIONS AND PROBLEMS

▪ Naming and Writing Ionic Formulas

4.15 Write names for the following ionic compounds:
   a. Al_2O_3 **b.** $CaCl_2$ **c.** Na_2O **d.** Mg_3N_2 **e.** KI

4.16 Write names for the following ionic compounds:
   a. $MgCl_2$ **b.** K_3P **c.** Li_2S **d.** LiBr   **e.** MgO

4.17 Why is a Roman numeral placed after the name of most transition metal ions?

4.18 The compound $CaCl_2$ is named calcium chloride; the compound $CuCl_2$ is named copper(II) chloride. Explain why a Roman numeral is used in one name but not the other.

4.19 Write the names of the following Group 4A (14) and transition metal ions (include the Roman numeral when necessary):
   a. Fe^{2+} **b.** Cu^{2+} **c.** Zn^{2+} **d.** Pb^{4+} **e.** Cr^{3+}

4.20 Write the names of the following Group 4A (14) and transition metal ions (include the Roman numeral when necessary):
   a. Ag^+ **b.** Cu^+ **c.** Fe^{3+} **d.** Sn^{2+} **e.** Au^{3+}

4.21 Write names for the following ionic compounds:
   a. $SnCl_2$ **b.** FeO   **c.** Cu_2S **d.** CuS   **e.** $CrBr_3$

4.22 Write names for the following ionic compounds:
   a. Ag_3P **b.** PbS   **c.** SnO_2 **d.** $AuCl_3$ **e.** ZnS

4.23 Indicate the charge on the metal ion in each of the following formulas:
   a. $AuCl_3$ **b.** Fe_2O_3 **c.** PbI_4 **d.** $SnCl_2$

4.24 Indicate the charge on the metal ion in each of the following formulas:
   a. $FeCl_2$ **b.** CuO   **c.** Fe_2S_3 **d.** $CrCl_3$

4.25 Write formulas for the following ionic compounds:
- **a.** magnesium chloride
- **b.** sodium sulfide
- **c.** copper(I) oxide
- **d.** zinc phosphide
- **e.** gold(III) nitride
- **f.** chromium(II) chloride

4.26 Write formulas for the following ionic compounds:
- **a.** iron(III) oxide
- **b.** barium fluoride
- **c.** tin(IV) chloride
- **d.** silver sulfide
- **e.** copper(II) chloride
- **f.** lithium nitride

4.4 Polyatomic Ions

LEARNING GOAL

Write the name and formula of a compound containing a polyatomic ion.

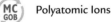 Polyatomic Ions

In a **polyatomic ion**, a group of atoms has acquired an electrical charge. Most polyatomic ions consist of a nonmetal such as phosphorus, sulfur, carbon, or nitrogen bonded to oxygen atoms. These oxygen-containing polyatomic ions have an ionic charge of $1-$, $2-$, or $3-$ because 1, 2, or 3 electrons were added to the atoms in the group to complete their octets. Only one of the common polyatomic ions, NH_4^+, is positively charged.

Naming Polyatomic Ions

The names of the most common polyatomic ions end in *ate*. The *ite* ending is used for the names of related ions that have one less oxygen atom. Recognizing these endings will help you identify polyatomic ions in the name of a compound. The hydroxide ion (OH^-) and cyanide ion (CN^-) are exceptions to this naming pattern. There is no easy way to learn polyatomic ions. You will need to memorize the number of oxygen atoms and the charge associated with each ion, as shown in Table 4.7. By memorizing the formulas and the names of the ions shown in the boxes, you can derive the related ions. For example, the sulfate ion is SO_4^{2-}. We write the formula of the sulfite ion, which has one less oxygen atom, as SO_3^{2-}. The formula of hydrogen carbonate, or *bicarbonate*, can be written by placing a hydrogen cation (H^+) in front of the formula

TABLE 4.7

Names and Formulas of Some Common Polyatomic Ions

Nonmetal	Formula of Ion[a]	Name of Ion
hydrogen	OH^-	hydroxide
nitrogen	NH_4^+	ammonium
	$\boxed{NO_3^-}$	**nitrate**
	NO_2^-	nitrite
chlorine	$\boxed{ClO_3^-}$	**chlorate**
	ClO_2^-	chlorite
carbon	$\boxed{CO_3^{2-}}$	**carbonate**
	HCO_3^-	hydrogen carbonate (or bicarbonate)
	CN^-	cyanide
	$C_2H_3O_2^-$ (CH_3COO^-)	acetate
sulfur	$\boxed{SO_4^{2-}}$	**sulfate**
	HSO_4^-	hydrogen sulfate (or bisulfate)
	SO_3^{2-}	sulfite
	HSO_3^-	hydrogen sulfite (or bisulfite)
phosphorus	$\boxed{PO_4^{3-}}$	**phosphate**
	HPO_4^{2-}	hydrogen phosphate
	$H_2PO_4^-$	dihydrogen phosphate
	PO_3^{3-}	phosphite

[a]Boxed formulas are the most common polyatomic ion for that element.

Fertilizer
$NaNO_3$

Na^+ NO_3^-
          Nitrate ion

Plaster molding
$CaSO_4$

2+   2−

Ca^{2+} SO_4^{2-}
         Sulfate ion

FIGURE 4.3

Many products contain polyatomic ions, which are groups of atoms that carry an ionic charge.

Q **Why does the sulfate ion have a 2− charge?**

for carbonate (CO_3^{2-}) and decreasing the charge from 2− to 1− to give HCO_3^-. Some models of polyatomic ions are shown in Figure 4.3.

Writing Formulas for Compounds Containing Polyatomic Ions

No polyatomic ion exists by itself. Like any ion, a polyatomic ion must be associated with ions of opposite charge. The bonding between polyatomic ions and other ions is one of electrical attraction. For example, the compound sodium sulfate consists of sodium ions (Na^+) and sulfate ions (SO_4^{2-}) held together by ionic bonds.

To write correct formulas for compounds containing polyatomic ions we follow the same rules of charge balance that we used for writing the formulas of simple ionic compounds. The total negative and positive charges must equal zero. For example, consider the formula for a compound containing calcium ions and carbonate ions. The ions are written as

$$Ca^{2+} \qquad CO_3^{2-}$$
calcium ion   carbonate ion

Ionic charge: $(2+) \; + \; (2-) = 0$

Because one ion of each balances the charge, the formula is written as

$$CaCO_3$$
calcium carbonate

When more than one polyatomic ion is needed for charge balance, parentheses are used to enclose the formula of the ion. A subscript is written outside the closing parenthesis to indicate the number of polyatomic ions. Consider the formula for magnesium nitrate. The ions in this compound are the magnesium ion and the nitrate ion, a polyatomic ion.

$$Mg^{2+} \qquad NO_3^-$$
magnesium ion   nitrate ion

To balance the positive charge of 2+, two nitrate ions are needed. The formula, including the parentheses around the nitrate ion, is as follows:

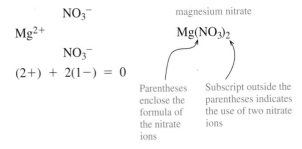

NO_3^-

magnesium nitrate

Mg^{2+} $Mg(NO_3)_2$

NO_3^-

$(2+) \; + \; 2(1-) = 0$

Parentheses enclose the formula of the nitrate ions

Subscript outside the parentheses indicates the use of two nitrate ions

---

SAMPLE PROBLEM 4.7

■ **Writing Formulas for Ionic Compounds with Polyatomic Ions**

Write the formula of aluminum bicarbonate.

SOLUTION

■ **STEP 1 Identify the cation and anion.** The cation is the aluminum ion, Al^{3+}, and the cation is bicarbonate, which is a polyatomic anion, HCO_3^-.

	Cation	Anion
Ions	Al^{3+}	HCO_3^-

Polyatomic Ions in Bones and Teeth

Bone structure consists of two parts: a solid mineral material and a second phase made up primarily of collagen proteins. The mineral substance is a compound called hydroxyapatite, a solid formed from calcium ions, phosphate ions, and hydroxide ions. This material is deposited in the web of collagen to form a very durable bone material.

$$Ca_{10}(PO_4)_6(OH)_2$$
Calcium hydroxyapatite

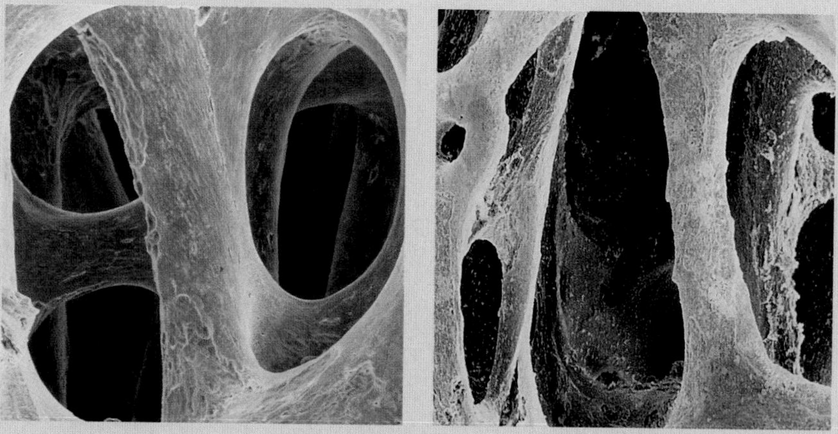

In most individuals, bone material is continuously being absorbed and reformed. After age 40, more bone material may be lost than formed, a condition called osteoporosis. Bone mass reduction occurs at a faster rate in women than in men and at different rates in different parts of the skeleton. The reduction in bone mass can be as much as 50% over a period of 30 to 40 years. Scanning electron micrographs (SEMs) show (a) normal bone and (b) bone in osteoporosis due to calcium loss. It is recommended that persons over 35, especially women, include a daily calcium supplement in their diet.

(a)

(b)

■ STEP 2 **Balance the charges.**

$$Al^{3+}$$

$$HCO_3{}^-$$
$$HCO_3{}^-$$
$$HCO_3{}^-$$

$$\mathbf{1}(3+) \quad + \quad \mathbf{3}(1-) = 0$$

Becomes a subscript in the formula

■ STEP 3 **Write the formula, cation first, using subscripts from the charge balance.** The formula for the compound is written by enclosing the formula of the bicarbonate ion, $HCO_3{}^-$, in parentheses and writing the subscript 3 outside the last parenthesis.

$$Al(HCO_3)_3$$

STUDY CHECK

Write the formula for a compound containing ammonium ion(s) and phosphate ion(s).

AlPO₄ (handwritten)

Naming Compounds Containing Polyatomic Ions

When naming ionic compounds containing polyatomic ions, we write the positive ion, usually a metal, first, and then we write the name of the polyatomic ion. It is important that you learn to recognize the polyatomic ion in the formula and name it correctly. As with other ionic compounds, no prefixes are used.

TABLE 4.8

Some Compounds That Contain Polyatomic Ions

Formula	Name	Use
$BaSO_4$	barium sulfate	Radiopaque medium
$CaCO_3$	calcium carbonate	Antacid, calcium supplement
$Ca_3(PO_4)_2$	calcium phosphate	Calcium replenisher
$CaSO_3$	calcium sulfite	Preservative in cider and fruit juices
$CaSO_4$	calcium sulfate	Plaster casts
$AgNO_3$	silver nitrate	Topical anti-infective
$NaHCO_3$	sodium bicarbonate *or* sodium hydrogen carbonate	Antacid
$Zn_3(PO_4)_2$	zinc phosphate	Dental cements
$FePO_4$	iron(III) phosphate	Food and bread enrichment
K_2CO_3	potassium carbonate	Alkalizer, diuretic
$Al_2(SO_4)_3$	aluminum sulfate	Antiperspirant, anti-infective
$AlPO_4$	aluminum phosphate	Antacid
$MgSO_4$	magnesium sulfate	Cathartic, Epsom salts

Na_2SO_4 $FePO_4$ $Al_2(CO_3)_3$

$Na_2 \boxed{SO_4}$ $Fe \boxed{PO_4}$ $Al_2(\boxed{CO_3})_3$

sodium sulfate   iron(III) phosphate   aluminum carbonate

Table 4.8 lists the formulas and names of some ionic compounds that include polyatomic ions and also gives their uses in medicine and industry.

Guide to Naming Ionic Compounds with Polyatomic Ions

STEP 1
Identify the cation and polyatomic ion (anion).

STEP 2
Name the cation using a Roman numeral, if needed.

STEP 3
Name the polyatomic ion usually ending with *ite* or *ate*.

STEP 4
Write the name of the compound, cation first and the polyatomic ion second.

SAMPLE PROBLEM 4.8

■ Naming Compounds Containing Polyatomic Ions

Name the following ionic compounds:

a. $CaSO_4$ **b.** $Cu(NO_2)_2$

SOLUTION

Formula	■ STEP 1 Cation Anion	■ STEP 2 Name of Cation	■ STEP 3 Name of Anion	■ STEP 4 Name of Compound
a. $CaSO_4$	Ca^{2+} SO_4^{2-}	calcium	sulfate	calcium sulfate
b. $Cu(NO_2)_2$	Cu^{2+} NO_2^{-}	copper(II)	nitrite	copper(II) nitrite

STUDY CHECK

What is the name of $Ca_3(PO_4)_2$?

QUESTIONS AND PROBLEMS

■ Polyatomic Ions

4.27 Write the formulas including the charge for the following polyatomic ions:
   a. hydrogen carbonate   **b.** ammonium
   c. phosphate   **d.** hydrogen sulfate

4.28 Write the formulas including the charge for the following polyatomic ions:
    a. nitrite   **b.** sulfite   **c.** hydroxide   **d.** phosphite

4.29 Name the following polyatomic ions:
    a. SO_4^{2-} **b.** CO_3^{2-} **c.** PO_4^{3-} **d.** NO_3^{-}

4.30 Name the following polyatomic ions:
    a. OH^{-} **b.** HSO_3^{-} **c.** CN^{-} **d.** NO_2^{-}

4.31 Complete the following table with the formula of the compound:

	OH^{-}	NO_2^{-}	CO_3^{2-}	HSO_4^{-}	PO_4^{3-}
Li^{+}					
Cu^{2+}					
Ba^{2+}					

4.32 Complete the following table with the formula of the compound:

	OH^{-}	NO_3^{-}	HCO_3^{-}	SO_3^{2-}	PO_4^{3-}
NH_4^{+}					
Al^{3+}					
Pb^{4+}					

4.33 Write the formula for the polyatomic ion in each of the following and name each compound:
    a. Na_2CO_3 **b.** NH_4Cl **c.** Li_3PO_4 **d.** $Cu(NO_2)_2$ **e.** $FeSO_3$

4.34 Write the formula for the polyatomic ion in each of the following and name each compound:
    a. KOH **b.** $NaNO_3$ **c.** $CuCO_3$ **d.** $NaHCO_3$ **e.** $BaSO_4$

4.35 Write the correct formula for the following compounds:
    a. barium hydroxide   **b.** sodium sulfate   **c.** iron(II) nitrate
    d. zinc phosphate   **e.** iron(III) carbonate

4.36 Write the correct formula for the following compounds:
    a. aluminum chlorate   **b.** ammonium oxide   **c.** magnesium bicarbonate
    d. sodium nitrite   **e.** copper(I) sulfate

4.5 Covalent Compounds

LEARNING GOAL

Given the formula of a covalent compound, write its electron-dot structure.

Earlier, we learned that atoms of metals and nonmetals become more stable by forming ionic compounds. However, atoms of nonmetals have high ionization energies and do not lose electrons easily. Thus, in *covalent compounds*, electrons are not transferred from one atom to another, but are shared between atoms of nonmetals to achieve stability. When atoms share electrons, they form **molecules**.

Formation of a Hydrogen Molecule

SELF-STUDY ACTIVITY
Covalent Bonds

The simplest covalent molecule is hydrogen gas, H_2. When two hydrogen atoms are far apart, they are not attracted to each other. As the atoms move closer, the positive charge of each nucleus attracts the electron of the other atom. This attraction pulls the atoms closer until they share a pair of valence electrons and form a *covalent bond*. In the covalent bond in H_2, the shared electrons give the noble gas configuration of He to each of the H atoms. Thus the atoms bonded in H_2 are more stable than two individual H atoms.

In covalent compounds, the valence electrons and the shared pairs of electrons can be shown using electron-dot formulas. In an electron-dot formula, a shared pair of electrons is written as two dots or a single line between the atomic symbols. This notation is shown in the formation of the covalent bond in the electron-dot formula for the H_2 molecule.

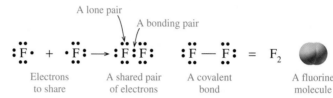

| Electrons to share | A shared pair of electrons | A covalent bond | A hydrogen molecule |

Formation of Octets in Covalent Molecules

In most covalent compounds, atoms share electrons to achieve octets for the valence electrons. For example, a fluorine molecule, F_2, consists of two fluorine atoms. Each fluorine atom has 7 valence electrons. By sharing their unpaired valence electrons, each F atom achieves an octet. In the resulting F_2 molecule, each F atom has the noble gas configuration of neon. In the electron-dot formula, the shared electrons, or *bonding pair*, are written between atoms with the nonbonding pairs of electrons, or *lone pairs*, on the outside. This is shown in the formation of the covalent bond for the F_2 molecule.

A lone pair

A bonding pair

:F• + •F: ⟶ :F:F: :F—F: = F₂

| Electrons to share | A shared pair of electrons | A covalent bond | A fluorine molecule |

Hydrogen (H_2) and fluorine (F_2) are examples of nonmetal elements whose natural state is diatomic; that is, they contain two like atoms. The elements that exist as diatomic molecules are listed in Table 4.9.

Sharing Electrons Between Atoms of Different Elements

In Period 2, the number of electrons that an atom shares and the number of covalent bonds it forms are usually equal to the number of electrons needed to acquire a noble gas arrangement. For example, carbon has 4 valence electrons. Because carbon needs to acquire 4 more electrons for an octet, it forms 4 covalent bonds by sharing its 4 valence electrons.

Methane, a component of natural gas, is a compound made of carbon and hydrogen. To attain an octet, each carbon shares 4 electrons, and each hydrogen shares 1 electron. In this molecule, a carbon atom forms four covalent bonds with four hydrogen atoms. The electron-dot formula for the molecule is written with the carbon atom in the center and the hydrogen atoms on the sides. Table 4.10 gives the formulas of some covalent molecules for Period 2 elements.

While the octet rule is useful, there are exceptions. We have already seen that a hydrogen (H_2) molecule requires just two electrons or a single bond to achieve the stability of the nearest noble gas, helium. In $BeCl_2$, Be only forms 2 covalent bonds. In BF_3, the nonmetal B can share only 3 valence electrons to give a total of 6 valence electrons or 3 bonds. The nonmetals typically form octets. However, atoms such as P, S, Cl, Br, and I can share more of their valence electrons and expand the octets to stable valence shells of 10, 12, or even 14 electrons. In PCl_3, the P atom has an octet, but in PCl_5 the P atom has 10 valence electrons or

Covalent Molecules and the Octet Rule

TABLE 4.9

Elements That Exist as Diatomic, Covalent Molecules

Element	Diatomic Molecule	Name
H	H_2	hydrogen
N	N_2	nitrogen
O	O_2	oxygen
F	F_2	fluorine
Cl	Cl_2	chlorine
Br	Br_2	bromine
I	I_2	iodine

Methane, CH_4

TABLE 4.10

Electron-Dot Formulas for Some Covalent Compounds

CH₄	NH₃	H₂O

Formulas Using Electron Dots Only

$$\begin{array}{c} H \\ H\!:\!\overset{..}{\underset{..}{C}}\!:\!H \\ H \end{array} \qquad \begin{array}{c} H\!:\!\overset{..}{N}\!:\!H \\ H \end{array} \qquad \begin{array}{c} :\overset{..}{\underset{..}{O}}\!:\!H \\ H \end{array}$$

Formulas Using Bonds and Electron Dots

$$\begin{array}{c} H \\ | \\ H-C-H \\ | \\ H \end{array} \qquad \begin{array}{c} H-\overset{..}{N}-H \\ | \\ H \end{array} \qquad \begin{array}{c} :\overset{..}{\underset{..}{O}}-H \\ | \\ H \end{array}$$

Molecular Models

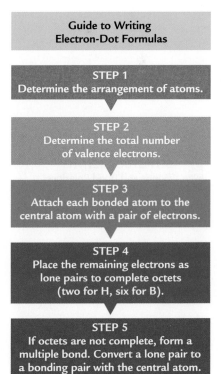

Methane molecule	Ammonia molecule	Water molecule

5 bonds. In H_2S, the S atom has an octet, but in SF_6, there are 12 valence electrons or 6 bonds to the sulfur atom. Table 4.11 gives the bonding patterns for some nonmetals.

 MC GOB Writing Electron-Dot Formulas

TABLE 4.11

Typical Bonding Patterns of Some Nonmetals in Covalent Compounds

1A (1)	3A (13)	4A (14)	5A (15)	6A (16)	7A (17)
ᵃH 1 bond					
	ᵃB 3 bonds	C 4 bonds	N 3 bonds	O 2 bonds	F 1 bond
		Si 4 bonds	P 3 bonds	S 2 bonds	Cl, Br, I 1 bond

ᵃH and B do not form eight-electron octets. H atoms share one electron pair; B atoms share three electron pairs for a set of 6 electrons.

SAMPLE PROBLEM 4.9

■ **Writing Electron-Dot Formulas**

Write the electron-dot formula for PCl_3, phosphorus trichloride.

SOLUTION

■ **STEP 1 Determine the arrangement of atoms.** In PCl_3, the central atom is P.

Cl P Cl

Cl

■ **STEP 2 Determine the total number of valence electrons.** We can use the group numbers to determine the valence electrons for each of the atoms in the molecule.

Guide to Writing Electron-Dot Formulas

STEP 1
Determine the arrangement of atoms.

STEP 2
Determine the total number of valence electrons.

STEP 3
Attach each bonded atom to the central atom with a pair of electrons.

STEP 4
Place the remaining electrons as lone pairs to complete octets (two for H, six for B).

STEP 5
If octets are not complete, form a multiple bond. Convert a lone pair to a bonding pair with the central atom.

Element	Group	Atoms	Valence Electrons	=	Total
P	5A (15)	1 P	$\times 5\,e^-$	=	$5\,e^-$
Cl	7A (17)	3 Cl	$\times 7\,e^-$	=	$21\,e^-$
			Total valence electrons for PCl_3	=	$26\,e^-$

■ **STEP 3 Attach the central atom to each bonded atom by a pair of electrons.**

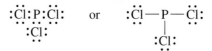

■ **STEP 4 Arrange the remaining electrons as lone pairs to complete octets.**
A total of 6 electrons ($3 \times 2e^-$) are needed to bond the central P atom to three Cl atoms. There are 20 valence electrons left.

$$26 \text{ valence } e^- - 6 \text{ bonding } e^- = 20\ e^- \text{ remaining}$$

The remaining electrons are placed as lone pairs around the outer Cl atoms first, which uses 18 more electrons.

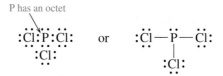

Use the remaining two electrons to complete the octet for the P atom.

P has an octet

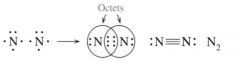

■ **STEP 5 Octets are complete for all the atoms using 26 valence electrons and using single bonds.** No multiple bonds are needed.

STUDY CHECK

Write the electron-dot formula for Cl_2O.

Multiple Covalent Bonds

SELF-STUDY ACTIVITY
Bonds and Bond Polarities

In many covalent compounds, atoms share two or three pairs of electrons to complete their octets. A **double bond** is the sharing of two pairs of electrons, and in a **triple bond**, three pairs of electrons are shared. Atoms of carbon, oxygen, nitrogen, and sulfur are most likely to form multiple bonds. Atoms of hydrogen and the halogens do not form double or triple bonds. Double and triple bonds are formed when single covalent bonds fail to complete the octets of all the atoms in the molecule. For example, in the electron-dot structure for the covalent compound N_2, an octet is achieved when each nitrogen atom shares 3 electrons. Thus, three covalent bonds, or a triple bond, will form.

Octets

·N· ·N· ⟶ :N(:::)N: :N≡N: N_2

Three shared   Triple bond   Nitrogen
pairs   molecule

SAMPLE PROBLEM 4.10

■ Writing Electron-Dot Formulas with Double Bonds

Write the electron-dot formula for CO_2, carbon dioxide.

$$:\ddot{O}:\ddot{C}:\ddot{O}:$$

SOLUTION

We will now use Step 5 in our guide for writing electron-dot formulas from the previous section to complete octets by forming one or more multiple bonds.

■ **STEP 1   Determine the arrangement of atoms.**   In CO_2, the central atom is C.

    O C O

■ **STEP 2   Determine the total number of valence electrons.**

Element	Group	Atoms	Valence Electrons	=	Total
O	6A (16)	2 O	$\times 6\,e^-$	=	$12\,e^-$
C	4A (14)	1 C	$\times 4\,e^-$	=	$4\,e^-$
		Total valence electrons for CO_2		=	$16\,e^-$

■ **STEP 3   Attach the central atom to each bonded atom by a pair of electrons.**

    O:C:O   or   O—C—O

■ **STEP 4   Arrange remaining electrons as lone pairs to complete octets.**
    Four (4) electrons ($2 \times 2\,e^-$) are used to bond the central C atom to the O atoms. The number of valence electrons remaining is

$$16 \text{ valence } e^- - 4 \text{ bonding } e^- = 12 e^- \text{ remaining}$$

    The remaining 12 electrons are placed as 6 lone pairs to complete the octets for the O atoms.

$$:\ddot{O}:C:\ddot{O}:   or   :\ddot{O}—C—\ddot{O}:$$

■ **STEP 5   If octets are not complete, form one or more multiple bonds.**
    Although all the valence electrons are used, the central C atom does not have an octet. To provide an octet for C, it is necessary to convert a lone pair from each O atom to a bonding pair between C and O.

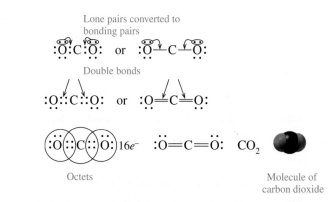

Lone pairs converted to bonding pairs

Double bonds

$:\ddot{O}::C::\ddot{O}:$ or $:\ddot{O}=C=\ddot{O}:$

$\left(:\ddot{O}\left(::\right)C\left(::\right)\ddot{O}:\right) 16e^-$ $:\ddot{O}=C=\ddot{O}:$ CO_2

Octets   Molecule of
                              carbon dioxide

STUDY CHECK

Write the electron dot formula for HCN (atoms arranged as H C N).

Atmosphere

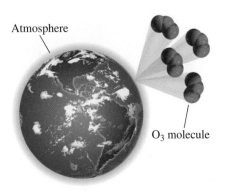

O_3 molecule

MC
GOB   Resonance Structures

Resonance Structures

When a molecule or polyatomic ion contains multiple bonds, it is often possible to write more than one electron-dot formula for the same arrangement of atoms. Suppose we want to write the electron-dot formula for ozone, O_3, a component in the stratosphere that protects us from the ultraviolet rays of the sun. Although all of the valence electrons (18) are used, one of the oxygen atoms does not have an octet. One lone pair must be moved to form a double bond. But which one should be used? One possibility is to form a double bond on the left and the other possibility is to form a double bond on the right. Both electron-dot formulas give complete octets to all the O atoms. However, experiments show that the actual bond length is equivalent to a molecule with a "one and a half" bond between the central O atom and each outside O atom. In this *hybrid*, the electrons are shown spread equally over all the O atoms. When two or more electron-dot formulas can be written, they are called **resonance structures**, which are shown with a double-headed arrow. Although we will write resonance structures of some molecules and ions, the true structure is really a mix or average of the possible structures.

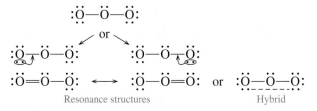

Resonance structures   Hybrid

SAMPLE PROBLEM 4.11

■ Writing Resonance Structures

Write two resonance structures for sulfur dioxide, SO_2.

SOLUTION

- **STEP 1** **Determine the arrangement of atoms.**   In SO_2, the S atom is the central atom.

    O   S   O

- **STEP 2** **Determine the total number of valence electrons.**   We can use the group numbers to determine the valence electrons for each of the atoms in the molecule.

Element	Group	Atoms	Valence Electrons	=	Total
S	6A (16)	1 S	$\times 6\,e^-$	=	$6\,e^-$
O	6A (16)	2 O	$\times 6\,e^-$	=	$12\,e^-$
		Total valence electrons for SO_2		=	$18\,e^-$

- **STEP 3** **Attach the central atom to each bonded atom by a pair of electrons.** We will use a single line to represent a pair of bonding electrons.

    O—S—O

- **STEP 4** **Arrange the remaining electrons to complete octets.**   Four (4) electrons are used to write single bonds between the S atom and the O atoms. The number of valence electrons that remain is

    18 valence e^- − 4 bonding e^- = 14 e^- remaining

The remaining electrons are placed as lone pairs around the O atoms first. One lone pair remains, which is assigned to the S atom.

$$:\ddot{O}—\ddot{S}—\ddot{O}:$$

- **STEP 5 If octets are not complete, form a multiple bond.**   The octet for S is completed using one lone pair from an O atom as a bonding pair and forming a double bond. Because the lone pair can come from either O atom, two resonance structures are possible.

$$:\ddot{O}—\ddot{S}=\ddot{O}: \longleftrightarrow :\ddot{O}=\ddot{S}—\ddot{O}:$$

STUDY CHECK

Write three resonance structures for SO_3.

QUESTIONS AND PROBLEMS

■ Covalent Compounds

4.37 What elements on the periodic table are most likely to form covalent compounds?

4.38 How does the bond that forms between Na and Cl differ from a bond that forms between N and Cl?

4.39 State the number of valence electrons, bonding pairs, and lone pairs in each of the following electron-dot formulas:
  a. $H:H$ **b.** $H:\ddot{Br}:$ **c.** $:\ddot{Br}:\ddot{Br}:$

4.40 State the number of valence electrons, bonding pairs, and lone pairs in each of the following electron-dot formulas:

  a. $H:\ddot{O}:$ **b.** $H:\ddot{N}:H$ with H below   **c.** $:\ddot{Br}:\ddot{O}:\ddot{Br}:$

4.41 Write the electron-dot formula for each of the following molecules:
  a. HF   **b.** SF_2 **c.** NBr_3
  d. CH_3OH (methyl alcohol) H C O H
  e. N_2H_4 (hydrazine) H N N H

4.42 Write the electron-dot formula for each of the following molecules:
  a. H_2O **b.** CCl_4 **c.** SiF_4 **d.** CF_2Cl_2
  e. C_2H_6 (ethane) H C C H

4.43 When is it necessary to write a multiple bond in an electron-dot formula?

4.44 If the available valence electrons for a molecule do not complete all of the octets in an electron-dot formula, what should you do?

4.45 What is resonance?

4.46 When does a covalent compound have resonance?

4.47 Write the electron-dot formulas for each of the following molecules:
  a. CO (carbon monoxide)   **b.** H_2CCH_2 (ethylene)
  c. H_2CO (C is the central atom)

4.48 Write the electron-dot formulas for each of the following molecules
  a. HCCH (acetylene)   **b.** CS_2 (C is the central atom)
  c. $COCl_2$ (C is the central atom)

4.49 Write resonance structures for $ClNO_2$ (N is the central atom).

4.50 Write resonance structures for N_2O (N   N   O).

LEARNING GOAL

Given the formula of a covalent compound, write its correct name; given the name of a covalent compound, write its formula.

TABLE 4.12

Prefixes Used in Naming Covalent Compounds			
1	mono	6	hexa
2	di	7	hepta
3	tri	8	octa
4	tetra	9	nona
5	penta	10	deca

4.6 Naming and Writing Covalent Formulas

When naming a covalent compound, the first nonmetal in the formula is named by its elemental name; the second nonmetal is named by its elemental name with the ending changed to *ide*. Subscripts indicating two or more atoms of an element are expressed as prefixes placed in front of each name. Table 4.12 lists prefixes used in naming covalent compounds. The names of covalent compounds need prefixes because several different compounds can be formed from the same two nonmetals. For example, carbon and oxygen can form two different compounds: carbon monoxide, CO and carbon dioxide, CO_2.

When the vowels *o* and *o* or *a* and *o* appear together, the first vowel is omitted as in carbon monoxide. In the name of a covalent compound, the prefix *mono* is usually omitted, as in NO, nitrogen oxide. Traditionally, however, CO is named carbon monoxide. The prefix *mono* is usually omitted. Table 4.13 lists the formulas, names, and commercial uses of some covalent compounds.

TABLE 4.13

Some Common Covalent Compounds		
Formula	**Name**	**Commercial Uses**
CS_2	carbon disulfide	Manufacture of rayon
CO_2	carbon dioxide	Carbonation of beverages, fire extinguishers, propellant in aerosols, dry ice
NO	nitrogen oxide	Stabilizer
N_2O	dinitrogen oxide	Inhalation anesthetic, "laughing gas"
SiO_2	silicon dioxide	Manufacture of glass
SO_2	sulfur dioxide	Preserving fruits, vegetables; disinfectant in breweries; bleaching textiles
SF_6	sulfur hexafluoride	Electrical circuits

SAMPLE PROBLEM 4.12

Naming Covalent Compounds

Name the covalent compound NCl_3.

SOLUTION

We can use the guide to naming covalent compounds to name these covalent compounds.

- **STEP 1 Name the first nonmetal by its element name.** In NCl_3, the first nonmetal (N) is nitrogen.

- **STEP 2 Name the second nonmetal by changing the last part of its element name to *ide*.** The second nonmetal (Cl) is named chloride.

- **STEP 3 Add prefixes to indicate the number of atoms of each nonmetal.** Because nitrogen is the first nonmetal, the prefix *mono* is understood and not used. The subscript indicating three Cl atoms is shown as the prefix *tri*.

    NCl_3 nitrogen trichloride

Guide to Naming Covalent Compounds with Two Nonmetals

STEP 1
Name the first nonmetal by its element name.

STEP 2
Name the second nonmetal by changing the last part of its element name to *ide*.

STEP 3
Add prefixes to indicate the number of atoms (subscripts).

STUDY CHECK

Write the name of each of the following compounds:

a. $SiBr_4$ **b.** Br_2O

Writing Formulas from the Names of Covalent Compounds

In the name of a covalent compound, the names of two nonmetals are given along with prefixes for the number of atoms of each. To obtain a formula, we write the symbol for each element and a subscript if a prefix indicates two or more atoms.

SAMPLE PROBLEM 4.13

■ **Writing Formulas for Covalent Compounds**

Write the formula for each of the following covalent compounds:

a. sulfur dichloride   **b.** diboron trioxide

SOLUTION

a. sulfur dichloride

- ■ STEP 1 **Write the symbols in order of the elements in the names.**   The first nonmetal is sulfur and the second nonmetal is chlorine.

   S   Cl

- ■ STEP 2 **Write prefixes as subscripts.**   Since there is no prefix for sulfur, we know that there is one atom of sulfur. The prefix *di* in *dichloride* indicates that there are two atoms of chlorine shown as a subscript 2 in the formula.

   SCl_2

b. diboron trioxide

- ■ STEP 1 **Write the symbols in order of the elements in the names.**   The first nonmetal is boron and the second nonmetal is oxygen.

   B   O

- ■ STEP 2 **Write prefixes as subscripts.**   The prefix *di* in *diboron* indicates that there are two atoms of boron shown as a subscript 2 in the formula. The prefix *tri* in *trioxide* indicates that there are three atoms of oxygen shown as a subscript 3 in the formula.

   B_2O_3

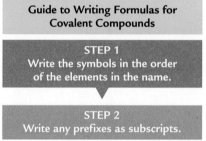

Guide to Writing Formulas for Covalent Compounds

STEP 1
Write the symbols in the order of the elements in the name.

STEP 2
Write any prefixes as subscripts.

STUDY CHECK

What is the formula of iodine pentafluoride?

Summary of Naming Compounds

Throughout this chapter we have examined strategies for naming ionic and covalent compounds. Now we can summarize the rules, as illustrated in Figure 4.4. In general, compounds having two elements are named by stating the first element, followed by the second element with an *ide* ending. If the first element is a metal, the compound

 Naming Covalent Compounds

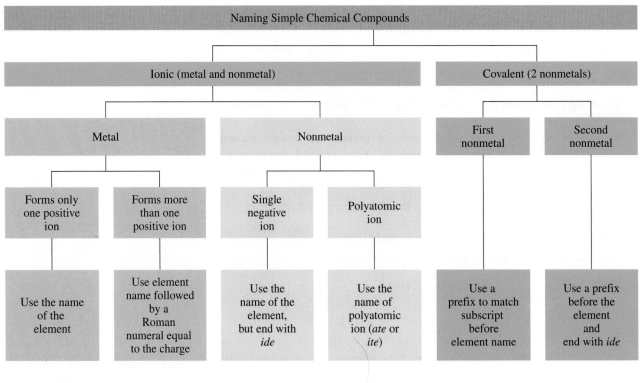

FIGURE 4.4

A flowchart of how ionic and covalent compounds are named.

Q **Why does the name sulfur dichloride have a prefix but magesium chloride does not?**

is ionic; if the first element is a nonmetal, the compound is covalent. For ionic compounds, it is necessary to determine whether the metal can form more than one type of positive ion; if so, a Roman numeral following the name of the metal indicates the particular ionic charge. One exception is the ammonium ion NH_4^+, which is also written first as a positively charged polyatomic ion. In naming covalent compounds having two elements, prefixes are necessary to indicate the number of atoms of each nonmetal as shown in that particular formula. Ionic compounds having three or more elements include some type of polyatomic ion. They are named by ionic rules, but have an *ate* or *ite* ending when the polyatomic ion has a negative charge.

SAMPLE PROBLEM 4.14

■ **Naming Ionic and Covalent Compounds**

Name the following compounds:

a. Na_3P **b.** $CuSO_4$ **c.** SO_3

SOLUTION

a. Na_3P is an ionic compound. Na is a metal that forms a single ion, Na^+, which is named sodium. The single negative ion, P^{3-}, is named phosphide. The compound Na_3P is sodium phosphide.

b. $CuSO_4$ is an ionic compound. Cu is a transition metal that forms more than one ion. Cu^{2+} is the positive ion that balances the charge on SO_4^{2-}. Roman numeral II specifies the charge in the name copper(II). SO_4^{2-} is a polyatomic ion named sulfate. The compound $CuSO_4$ is named copper(II) sulfate.

c. SO_3 is covalent because it consists of two nonmetals. The first element S is named sulfur (no prefix is needed). The subscript three calls for the prefix *tri* and the element oxygen becomes oxide. The compound SO_3 is named sulfur trioxide.

STUDY CHECK

What is the name of $Fe(NO_3)_2$?

QUESTIONS AND PROBLEMS

■ Naming and Writing Covalent Formulas

4.51 Name the following covalent compounds:
   a. PBr_3 **b.** CBr_4 **c.** SiO_2 **d.** N_2O_3 **e.** $SiBr_4$ **f.** PCl_5

4.52 Name the following covalent compounds:
   a. CS_2 **b.** P_2O_5 **c.** Cl_2O **d.** PCl_3 **e.** IBr_3 **f.** SO_3

4.53 Write the formulas of the following covalent compounds:
   a. carbon tetrachloride   **b.** carbon monoxide   **c.** phosphorus trichloride
   d. dinitrogen tetroxide   **e.** boron trifluoride   **f.** sulfur hexafluoride

4.54 Write the formulas of the following covalent compounds:
   a. sulfur dioxide   **b.** silicon tetrachloride   **c.** iodine pentafluoride
   d. dinitrogen oxide   **e.** tetraphosphorus hexoxide   **f.** dinitrogen pentoxide

4.55 Name the compounds that are found in the following sources:
   a. $Al_2(SO_4)_3$ antiperspirant
   b. $CaCO_3$ antacid
   c. N_2O "laughing gas," inhaled anesthetic
   d. Na_3PO_4 cathartic
   e. $(NH_4)_2SO_4$ fertilizer
   f. Fe_2O_3 pigment

4.56 Name the compounds that are found in the following sources:
   a. N_2 Earth's atmosphere
   b. $Mg_3(PO_4)_2$ antacid
   c. $FeSO_4$ iron supplement in vitamins
   d. $MgSO_4$ Epsom salts
   e. Cu_2O fungicide
   f. SnF_2 dental caries prophylactic

4.7 Electronegativity and Bond Polarity

LEARNING GOAL

Use electronegativity to determine the polarity of a bond.

In this chapter, we have seen that atoms form chemical bonds using their valence electrons. In bonds between identical nonmetal atoms, the bonding electrons are shared equally. However, in most compounds, bonds form between atoms of different elements. Then the bonding electrons are attracted to one atom more than the other.

Electronegativity

The ability of an atom to attract bonding electrons is called its **electronegativity**. The electronegativity values assigned to the representative elements are shown in Figure 4.5. Nonmetals have high electronegativity values compared to metals because nonmetals have a greater attraction for electrons than metals. The nonmetals with the highest electronegativity values are fluorine (4.0) at the top of Group 7A (17) and oxygen (3.5) at the top of Group 6A (16). The metal cesium at the bottom of Group 1A (1) has the lowest electronegativity value of 0.7. Smaller atoms tend to have higher electronegativity values because the valence electrons they share are closer to their nuclei. Thus, the values of electronegativity increase going from left to right across each period of the periodic table and increase going up within each group. The values of electronegativity for the transition metals are

SELF-STUDY ACTIVITY
Electronegativity

FIGURE 4.5

The electronegativity of representative elements indicates the ability of atoms to attract shared electrons. Electronegativity values increase across a period and going up a group.

Q **What element on the periodic chart has the strongest attraction for shared electrons?**

Electronegativity increases →

↑ Electronegativity increases

H 2.1						18 Group 8A

1 Group 1A	2 Group 2A		13 Group 3A	14 Group 4A	15 Group 5A	16 Group 6A	17 Group 7A
Li 1.0	Be 1.5		B 2.0	C 2.5	N 3.0	O 3.5	F 4.0
Na 0.9	Mg 1.2		Al 1.5	Si 1.8	P 2.1	S 2.5	Cl 3.0
K 0.8	Ca 1.0		Ga 1.6	Ge 1.8	As 2.0	Se 2.4	Br 2.8
Rb 0.8	Sr 1.0		In 1.7	Sn 1.8	Sb 1.9	Te 2.1	I 2.5
Cs 0.7	Ba 0.9		Tl 1.8	Pb 1.9	Bi 1.9	Po 2.0	At 2.1

also low, but we will not include them in our discussion. Note that there are no electronegativity values for the noble gases because they do not typically form bonds.

Types of Bonding

(MC GOB) Comparing Atom Electronegatives

Earlier we discussed bonding as either *ionic*, in which electrons are transferred, or *covalent*, in which electrons were equally shared. Actually most bonds are somewhere between these types of bonds. The difference in the electronegativity values of two atoms gives an indication of the type of bond that forms. In H—H, the electronegativity difference is zero (2.1 − 2.1 = 0), which means the bonding electrons are shared equally between the two hydrogen atoms. A covalent bond between atoms with identical or very similar electronegativities is a **nonpolar covalent bond**. However, most covalent bonds are between different atoms with different electronegativity values. (For example, in H—Cl, there is an electronegativity difference of 3.0 − 2.1 = 0.9.) (See Figure 4.6.)

FIGURE 4.6

In the nonpolar covalent bond of H_2, electrons are shared equally. In the polar covalent bond of HCl, electrons are shared unequally.

Q **H_2 has a nonpolar covalent bond, but HCl has a polar covalent bond. Explain.**

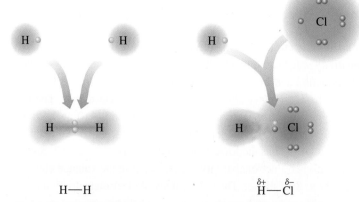

H—H

Equal sharing of electrons in a nonpolar covalent bond

$\overset{\delta+}{H}—\overset{\delta-}{Cl}$

Unequal sharing of electrons in a polar covalent bond (dipole)

When the electrons are shared unequally in a covalent bond, it is called a **polar covalent bond**.

In a polar covalent bond, the shared electrons are attracted to the more electronegative atom, which makes it partially negative. At the other end of the polar bond, the atom with the lower electronegativity becomes partially positive. Because a polar covalent bond has a separation of positive and negative charges or two poles, it is called a **dipole**. The positive and negative ends of a polar covalent bond are indicated by the lowercase Greek letter delta with a positive or negative sign δ^+ and δ^-. Sometimes an arrow pointing from the positive charge to the negative charge ($\longmapsto$) is used to indicate the dipole.

Examples of Dipoles in Polar Covalent Bond

$\overset{\delta+}{C}-\overset{\delta-}{O} \qquad \overset{\delta+}{N}-\overset{\delta-}{O} \qquad \overset{\delta+}{Cl}-\overset{\delta-}{F}$

$\longmapsto \qquad\quad \longmapsto \qquad\quad \longmapsto$

Variations in Bonding

The variations in bonding are continuous; there is no definite point at which one type of bond stops and the next starts. However, for purposes of discussion, we can use some general ranges for predicting the type of bond between atoms. When electronegativity differences are from 0.0 to 0.4, the electrons are shared about equally in a nonpolar covalent bond. For example, H—H ($2.1 - 2.1 = 0$) and C—H ($2.5 - 2.1 = 0.4$) are classified as nonpolar covalent bonds. As electronegativity differences increase, there is also an increase in the polarity of the covalent bond. For example, an O—H bond with a difference of 1.4 is much more polar than an O—F bond with a difference of only 0.5. Differences in electronegativity of about 1.8 or greater generally indicate a bond that is mostly ionic. For example, we would classify K—Cl with an electronegativity difference of $3.0 - 0.8 = 2.2$ as an ionic bond resulting from a transfer of electrons. (See Table 4.14.)

As the electronegativity difference increases, the shared electrons are attracted more strongly to the more electronegative atom. The *polarity*, which depends on the separation of charges, also increases. Eventually, the difference in electronegativity is great enough that the electrons are transferred from one atom to another, which results in an ionic bond. For example, the electronegativity difference for the ionic compound NaCl is $3.0 - 0.9 = 2.1$. Thus for large differences in electronegativity, we would predict an ionic bond. (See Table 4.15.)

TABLE 4.14

Electronegativity Difference and Types of Bonds

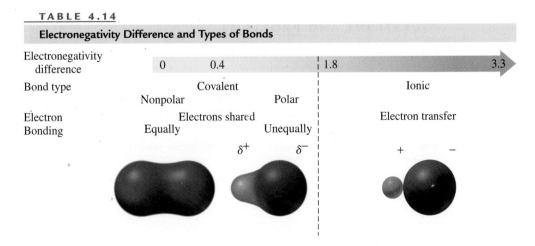

TABLE 4.15

Predicting Bond Type from Electronegativity Differences

Molecule		Type of Electron Sharing	Electronegativity Difference[a]	Bond Type
H_2	H—H	Shared equally	$2.1 - 2.1 = 0$	Nonpolar covalent
Cl_2	Cl—Cl	Shared equally	$3.0 - 3.0 = 0$	Nonpolar covalent
HBr	$\overset{\delta^+}{H}$—$\overset{\delta^-}{Br}$	Shared unequally	$2.8 - 2.1 = 0.7$	Polar covalent
HCl	$\overset{\delta^+}{H}$—$\overset{\delta^-}{Cl}$	Shared unequally	$3.0 - 2.1 = 0.9$	Polar covalent
NaCl	$Na^+ Cl^-$	Electron transfer	$3.0 - 0.9 = 2.1$	Ionic
MgO	$Mg^{2+} O^{2-}$	Electron transfer	$3.5 - 1.2 = 2.3$	Ionic

[a] Values are taken from Figure 4.5.

SAMPLE PROBLEM 4.15

■ Bond Polarity

Using electronegativity values, place the following bonds in order of increasing polarity. Classify each bond as nonpolar covalent, polar covalent, or ionic.

N—N   O—H   Cl—As   O—K

SOLUTION

For each bond, we obtain the electronegativity values and calculate the difference.

Bond	Electronegativity Difference	Type of Bond
N—N	$3.0 - 3.0 = 0.0$	Nonpolar covalent
O—H	$3.5 - 2.1 = 1.4$	Polar covalent
Cl—As	$3.0 - 2.0 = 1.0$	Polar covalent
O—K	$3.5 - 0.8 = 2.7$	Ionic

Least Polar	$\longrightarrow$	Most Polar (ionic)	
N—N	Cl—As	O—H	O—K

STUDY CHECK

For each of the following pairs, identify the more polar bond.

a. Si—S or Si—N   **b.** O—P or N—P

QUESTIONS AND PROBLEMS

(MC GOB)

■ Electronegativity and Bond Polarity

4.57 Describe the trend in electronegativity going across a period.

4.58 Describe the trend in electronegativity going down a group.

4.59 Approximately what electronegativity difference would you expect for a nonpolar covalent bond?

4.60 Approximately what electronegativity difference would you expect for a polar covalent bond?

4.61 Using the periodic table only, arrange the atoms in each of the following sets in order of increasing electronegativity:
    a. Li, Na, K   **b.** Na, P, Cl   **c.** Ca, Br, O

4.62 Using the periodic table only, arrange the atoms in each of the following sets in order of increasing electronegativity:
   a. F, Cl, Br   **b.** B, N, O   **c.** Mg, S, F

4.63 Predict whether each of the following bonds is ionic, polar covalent, or nonpolar covalent:
   a. Si—Br   **b.** Li—F   **c.** Br—F
   d. Br—Br   **e.** N—P   **f.** C—P

4.64 Predict whether each of the following bonds is ionic, polar covalent, or nonpolar covalent:
   a. Si—O   **b.** K—Cl   **c.** S—F
   d. P—Br   **e.** Li—O   **f.** N—P

4.65 For each of the following bonds, indicate the positive end with δ^+ and the negative end with δ^-. Write an arrow to show the dipole for each.
   a. N—F   **b.** Si—Br   **c.** C—O
   d. P—Br   **e.** B—Cl

4.66 For each of the following bonds, indicate the positive end with δ^+ and the negative end with δ^-. Write an arrow to show the dipole for each.
   a. Si—Cl   **b.** Se—F   **c.** Br—F
   d. N—H   **e.** N—P

4.8 Shapes and Polarity of Molecules

Now that we have counted valence electrons and written electron-dot formulas, we can look at the three-dimensional shapes of molecules. The shape is important in our understanding of how molecules interact with enzymes or certain antibiotics or produce our sense of taste and smell.

To predict the three-dimensional shape of a molecule, we use the **valence shell electron-pair repulsion (VSEPR) theory**. In the VSEPR theory, the shape of a molecule depends on minimizing repulsions, which means the electron groups around a central atom are arranged as far apart as possible. Once the electron-dot formulas are written, the specific shape of a molecule can be determined.

Two Electron Groups

In $BeCl_2$, there are two chlorine atoms bonded to a central beryllium atom. Because Be has a strong attraction for its valence electrons, it forms a covalent rather than ionic compound. With only 2 electron groups around the central atom, the electron-dot formula of $BeCl_2$ is an exception to the octet rule. The best arrangement of two electron groups for minimal repulsion is to place them on opposite sides of the Be atom. This gives a **linear** shape and a bond angle of $180°$ to the $BeCl_2$ molecule.

$$:\ddot{Cl}—Be—\ddot{Cl}:$$

Another example of a linear molecule is CO_2. In predicting shape, a *double* or *triple* bond is treated the same as one electron group. Thus, a multiple bond is counted as a single electron group in determining electron repulsion. In CO_2, the two double bonds, which are counted as one electron group each around C, are arranged on opposite sides of the C atom. With two double bonds, the shape of the CO_2 molecule is *linear* and has a bond angle of $180°$.

$$:\ddot{O}=C=\ddot{O}:$$

LEARNING GOAL

Predict the three-dimensional structure of a molecule and classify it as polar or nonpolar.

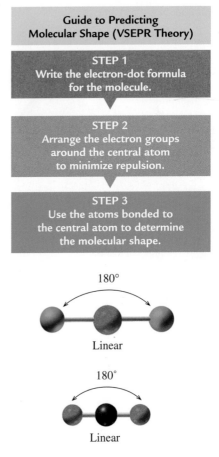

Guide to Predicting Molecular Shape (VSEPR Theory)

STEP 1
Write the electron-dot formula for the molecule.

STEP 2
Arrange the electron groups around the central atom to minimize repulsion.

STEP 3
Use the atoms bonded to the central atom to determine the molecular shape.

$180°$

Linear

$180°$

Linear

Three Electron Groups

In BF_3, the central atom B is attached to fluorine atoms by three electron groups (another exception to the octet rule). The arrangement of three electron groups as far apart as possible is called **trigonal planar** and has bond angles of 120°. In BF_3, each electron group is bonded to an atom, which gives BF_3 a trigonal planar structure. Thus, the BF_3 molecule is flat with all the atoms in the same plane and 120° bond angles.

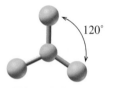

120°

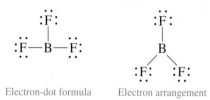

Trigonal planar

Electron-dot formula   Electron arrangement

In the electron-dot formula for SO_2, there is a single bond, a double bond, and a lone pair of electrons surrounding the S atom. Thus there are three electron groups around the S atom that assume a trigonal planar arrangement for minimal repulsion. But only the bonded atoms attached to the central atom determine the shape and structure of a molecule. Therefore, with two O atoms bonded to the central S atom, the structure of the SO_2 molecule is a **bent** shape. In the SO_2 molecule, the bond angle is slightly affected by the lone pair, but is close to 120°.

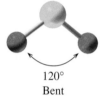

120°
Bent

Electron-dot formula   Electron arrangement

Four Electron Groups

Up to now, the shapes of molecules have been in two dimensions. However, when there are four electron groups around a central atom, the minimum repulsion is obtained by placing four electron groups at the corners of a three-dimensional tetrahedron. A regular tetrahedron consists of four sides that are equilateral triangles. For molecules with four electron groups, the central atom is located in the center of a tetrahedron.

In CH_4, all four electron pairs around the central C atom are bonded to hydrogen atoms. From the electron-dot formula, CH_4 appears planar with 90° bond angles, but this is not the largest angle possible. The best arrangement for minimal repulsion is **tetrahedral**, with the electron groups going toward the corners of a tetrahedron, which creates bond angles of 109.5°. Because CH_4 has four H atoms attached to the central C, the CH_4 molecule has a tetrahedral shape with bond angles of 109.5°.

SELF-STUDY ACTIVITY
Shapes of Molecules

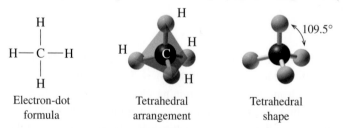

Electron-dot
formula

Tetrahedral
arrangement

Tetrahedral
shape

Now we will look at other molecules with four electron groups, but only two or three attached atoms. For example, ammonia, NH_3, has four electron groups around the central nitrogen atom, which means the electron groups have a tetrahedral arrangement. The three electron groups attached to H atoms and the one lone pair occupy the corners of a tetrahedron. Because one corner has no bonded atom, the NH_3 molecule is said to have a **trigonal pyramidal** shape, which is determined by the attached H atoms. In the NH_3 molecule, the bond angles are somewhat affected by the lone pairs, but are about 109.5°.

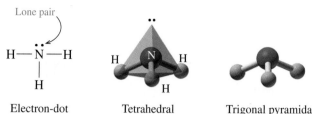

Electron-dot   Tetrahedral   Trigonal pyramidal
formula   arrangement   shape

In water, H_2O, the four electron groups have a tetrahedral arrangement around the central O atom. Two electron groups attached to H atoms and two lone pairs occupy the corners of the tetrahedron. With only two bonded atoms, the H_2O molecule has a *bent* shape. In the H_2O molecule, the bond angle is somewhat affected by the lone pairs, but is close to 109.5°. Table 4.16 gives the molecular shapes for molecules with 2, 3, and 4 electron groups.

MC
GOB   Molecular Shapes

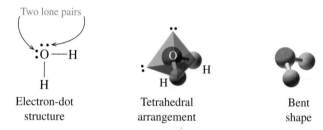

Electron-dot   Tetrahedral   Bent
structure   arrangement   shape

SAMPLE PROBLEM 4.16

■ **Predicting Shapes**

Use VSEPR theory to predict the shape of the following molecules:

a. PH_3 **b.** H_2Se

SOLUTION

We will use the guide to predict the shape of molecules.

a. PH_3

■ **STEP 1 Write the electron-dot formula.**   In the electron-dot formula for PH_3, there are four electron groups.

$$H-\overset{\displaystyle ..}{P}-H$$
$$|$$
$$H$$

■ **STEP 2 Arrange the electron groups around the central atom to minimize repulsion.**   The four electron groups have a tetrahedral arrangement.

■ **STEP 3 Use the atoms bonded to the central atom to determine the molecular shape.**   Three bonded atoms and one lone pair give PH_3 a trigonal pyramidal shape.

b. H_2Se

■ **STEP 1 Write the electron-dot formula.**   In the electron-dot formula for H_2Se, there are four electron groups.

TABLE 4.16

Examples of Shapes of Molecules

Molecule	Electron-Dot Structure	Bonded Atoms	Molecular Shape (angle)	
Two (2) electron groups around the central atom				
$BeCl_2$	:Cl:Be:Cl:	2	linear (180°)	
CO_2	:O::C::O:	2	linear (180°)	
Three (3) electron groups				
BF_3	:F:B:F: with :F: above	3	trigonal planar (120°)	
SO_2	:O:S with :O: below	2	bent (120°)	
Four (4) electron groups around the central atom				
CH_4	H:C:H with H above and H below	4	tetrahedral (109°)	
NH_3	H:N:H with H below	3	pyramidal (109°)	
H_2O	:O:H with H below	2	bent (109°)	

:Se—H
|
H

■ **STEP 2** **Arrange the electron groups around the central atom to minimize repulsion.** The four electron groups around Se would have a tetrahedral arrangement.

■ **STEP 3** **Use the atoms bonded to the central atom to determine the molecular shape.** Two bonded atoms and two lone pairs give H_2Se a bent shape.

STUDY CHECK

Predict the shape of CBr_4.

Polarity of Molecules

We have seen that covalent bonds in molecules can be polar or nonpolar. Molecules can also be polar or nonpolar depending on their shape. Diatomic molecules such as H_2 or Cl_2 are nonpolar because they contain one nonpolar covalent bond.

H—H   Cl—Cl

Nonpolar

Molecules with two or more polar bonds can also be **nonpolar** if the polar bonds have a symmetrical arrangement in the molecule.

In a **polar molecule**, one end of the molecule is more negatively charged than another end. Polarity in a molecule occurs when the polar bonds do not cancel each other. This cancellation depends on the type of atoms, the electron pairs around the central atom, and the shape of the molecule. For example, the HCl molecule is polar because electrons are shared unequally in a polar covalent bond.

H:Cl: H$^{\delta^+}$ $\longleftrightarrow$ Cl$^{\delta^-}$

Positive end   Negative end

In polar molecules with three or more atoms, the shape of the molecule determines whether the dipoles cancel or not. Often there are lone pairs around the central atom. In H_2O, the dipoles do not cancel, which makes the molecule positive at one end and negative at the other end. This gives the molecule a dipole.

In the molecule NH_3, there are three dipoles, but they do not cancel.

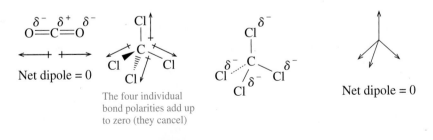

When the polar bonds or dipoles in a molecule cancel each other, the molecule is nonpolar. For example, CO_2 and CCl_4 contain polar bonds. However, the symmetry of the polar bonds cancels the dipoles, which makes CO_2 and CCl_4 molecules nonpolar.

Examples of Nonpolar Molecules with Polar Bonds

O$^{\delta^-}$=C$^{\delta^+}$=O$^{\delta^-}$

$\longleftrightarrow$ + $\longleftrightarrow$

Net dipole = 0

The four individual bond polarities add up to zero (they cancel)

Net dipole = 0

SAMPLE PROBLEM 4.17

■ Polarity of Molecules

Determine whether each of the following molecules is polar or nonpolar:

a. CBr_4 **b.** OF_2

SOLUTION

a. The electron-dot formula for CBr_4 has four bonded atoms.

$$:\ddot{B}r:$$
$$:\ddot{B}r:\ddot{C}:\ddot{B}r:$$
$$:\ddot{B}r:$$

The molecule would have a tetrahedral shape. With four identical atoms bonded to the central and no lone pairs, CBr_4 would be nonpolar.

b. The electron dot formula for OF_2 shows four electron groups with two bonded atoms.

$$:\ddot{O}:\ddot{F}:$$
$$:\ddot{F}:$$

The molecule would have a bent shape. With two lone pairs, the molecule is polar.

STUDY CHECK

Would PCl_3 be a polar or nonpolar molecule?

QUESTIONS AND PROBLEMS

▪ Shapes and Polarity of Molecules

4.67 What is the shape of a molecule with four bonded atoms?

4.68 What is the shape of a molecule with two bonded atoms and two lone pairs?

4.69 In the molecule PCl_3, the four electron groups around the phosphorus atom are arranged in a tetrahedral geometry. However, the shape of the molecule is called pyramidal. Why does the shape of the molecule have a different name from the name of the electron group geometry?

4.70 In the molecule H_2S, the four electron groups around the sulfur atom are arranged in a tetrahedral geometry. However, the shape of the molecule is called bent. Why does the shape of the molecule have a different name from the name of the electron group geometry?

4.71 Compare the electron-dot formulas of PH_3 and NH_3. Why do these molecules have the same shape?

4.72 Compare the electron-dot formulas CH_4 and H_2O. Why do these molecules have approximately the same angles but different shapes?

4.73 Use the VSEPR theory to predict the shape of each molecule:
    a. OF_2 **b.** CCl_4 **c.** HCN **d.** SeO_2

4.74 Use the VSEPR theory to predict the shape of each molecule:
    a. NCl_3 **b.** SCl_2 **c.** CF_4 **d.** CS_2

4.75 The molecule Cl_2 is nonpolar, but HCl is polar. Explain.

4.76 The molecules CH_4 and CH_3Cl both contain four bonds. Why is CH_4 nonpolar whereas CH_3Cl is polar?

4.77 Identify the following molecules as polar or nonpolar:
    a. HBr   **b.** NF_3 **c.** CBr_4 **d.** SO_3

4.78 Identify the following molecules as polar or nonpolar:
    a. H_2S **b.** PBr_3 **c.** $SiCl_4$ **d.** SO_2

CONCEPT MAP

Compounds and Their Bonds

Ionic compounds
contain
Ionic bonds
between

Metals	Nonmetals
that form	that form
Positive ions	Negative ions

use

Charge balance

to write subscripts for the

Chemical formula

Covalent compounds
contain
Covalent bonds
between
Nonmetals
can be

Polar	Nonpolar

use

VSEPR theory	Electronegativity

to determine

Shape	Polarity

CHAPTER REVIEW

4.1 Octet Rule and Ions

The nonreactivity of the noble gases is associated with the stable arrangement of 8 electrons, an octet, in their valence shells; helium needs 2 electrons for stability. Atoms of elements other than the noble gases achieve stability by losing, gaining, or sharing their valence electrons with other atoms in the formation of compounds. Metals of the representative elements form an octet by losing their valence electrons to form positively charged cations: Group 1A (1), 1+, Group 2A (2), 2+, and Group 3A (13), 3+. When they react with metals, nonmetals gain electrons to form octets in their valence shells. As anions, they have negative charges: Group 5A (15), 3−, Group 6A (16), 2−, Group 7A (17), 1−.

4.2 Ionic Compounds

The total positive and negative ionic charge must balance in the formula of an ionic compound. Charge balance in a formula is achieved by using subscripts after each symbol so that the overall charge is zero.

4.3 Naming and Writing Ionic Formulas

In naming ionic compounds, the positive ion is given first, followed by the name of the negative ion. Ionic compounds containing two elements end with *ide*. When the metal can form more than one

positive ion, its ionic charge is determined from the total negative charge in the formula. Typically transition metals form cations with two or more ionic charges. The charge is given as a Roman numeral in the name, such as iron(II) and iron(III) for the cations of iron with 2+ and 3+ ionic charges.

4.4 Polyatomic Ions

A polyatomic ion is a group of nonmetal atoms that carries an electrical charge; for example, the carbonate ion has the formula CO_3^{2-}. Most polyatomic ions have names that end with *ate* or *ite*.

4.5 Covalent Compounds

In a covalent bond, electrons are shared by atoms of nonmetals. By sharing, each of the atoms achieves a noble gas arrangement. In some covalent compounds, double or triple bonds are needed to provide an octet. Resonance structures are possible when two or more electron-dot formulas can be drawn for a molecule with a multiple bond.

4.6 Naming and Writing Covalent Formulas

In the names of covalent compounds, prefixes are used to indicate the subscript in the formula. The ending of the second nonmetal is changed to *ide*.

4.7 Electronegativity and Bond Polarity

Electronegativity is the ability of an atom to attract the electrons it shares with another atom. In general, the electronegativities of metals are low, while nonmetals have high electronegativities. In a nonpolar covalent bond, atoms share the bonding pair of electrons equally. In a polar covalent bond, the bonding electrons are unequally shared because they are attracted to the more electronegative atom forming a dipole. Atoms that form ionic bonds have large differences in electronegativity.

4.8 Shapes and Polarity of Molecules

The shape of a molecule is determined from the electron-dot formula and the number of bonded atoms and lone pairs. The electron arrangement of four electron groups around a central atom is tetrahedral. When all the electron groups are bonded to atoms, the shape has the same name as the electron arrangement. A central atom with two bonded atoms and two lone pairs has a bent shape. A central atom with three bonded atoms and one lone pair has a trigonal pyramidal shape.

Nonpolar molecules contain nonpolar covalent bonds or have an arrangement of bonded atoms that causes the dipoles to cancel out. In polar molecules, the dipoles do not cancel because there are nonidentical bonded atoms or lone pairs.

KEY TERMS

anion A negatively charged ion such as Cl^-, O^{2-}, or SO_4^{2-}.

bent The shape of a molecule with two bonded atoms and two lone pairs.

cation A positively charged ion such as Na^+, Mg^{2+}, Al^{3+}, and NH_4^+.

dipole The separation of positive and negative charge in a polar bond indicated by an arrow that is drawn from the more positive atom to the more negative atom.

double bond A sharing of two pairs of electrons by two atoms.

electronegativity The relative ability of an element to attract electrons.

formula The group of symbols and subscripts that represent the atoms or ions in a compound.

ion An atom or group of atoms having an electrical charge because of a loss or gain of electrons.

ionic charge The difference between the number of protons (positive) and the number of electrons (negative) written in the upper right corner of the symbol for the element or polyatomic ion.

linear The shape of a molecule that has two bonded atoms and no lone pair.

molecule The smallest unit of two or more atoms held together by covalent bonds.

nonpolar covalent bond A covalent bond in which the electrons are shared equally between atoms.

nonpolar molecule A molecule that has only nonpolar bonds or in which the bond dipoles cancel.

octet rule Elements in Groups 1–7A (1, 2, 13–17) react with other elements by forming ionic or covalent bonds to produce a noble gas arrangement, usually 8 electrons in the outer shell.

polar covalent bond A covalent bond in which the electrons are shared unequally between atoms.

polar molecule A molecule containing bond dipoles that do not cancel.

polyatomic ion A group of covalently bonded nonmetal atoms that has an overall electrical charge.

resonance structures Two or more electron-dot formulas that can be written for a molecule by placing a multiple bond between different atoms.

tetrahedral The shape of a molecule with four bonded atoms.

trigonal planar The shape of a molecule with three bonded atoms and no lone pair.

trigonal pyramidal The shape of a molecule that has three bonded atoms and one lone pair.

triple bond A sharing of three pairs of electrons by two atoms.

valence-shell electron-pair repulsion (VSEPR) theory A theory that predicts the shape of a molecule by placing the electron pairs on a central atom as far apart as possible to minimize the mutual repulsion of the electrons.

UNDERSTANDING THE CONCEPTS

4.79 Identify each of the following atoms or ions:

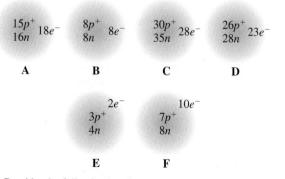

a. Which bonds are polar covalent?
b. Which bonds are nonpolar covalent?
c. Which bonds are ionic?
d. Arrange the covalent bonds in order of decreasing polarity.

4.81 In the following electron-dot formulas, assume X and Y are atoms of nonmetals and all bonds are polar covalent.

$$\text{a. } X—\overset{\displaystyle X}{\underset{\displaystyle ..}{Y}}—X \qquad \text{b. } X—\overset{..}{\underset{..}{Y}}—X \qquad \text{c. } X—\overset{\displaystyle X}{\underset{\displaystyle X}{Y}}—X$$

Match each of the molecules with the correct diagram of its shape and name the shape; indicate if each molecule is polar or nonpolar.

4.80 Consider the following bonds

Ca—O   C—O   K—O   O—O   N—O

4.82 As discussed in the Health Note "Polyatomic Ions in Bone and Teeth," the mineral component of bone and teeth is composed of calcium hydroxyapatite, $Ca_{10}(PO_4)_6(OH)_2$. Name the ions in the formula.

ADDITIONAL QUESTIONS AND PROBLEMS

For instructor-assigned homework, go to
www.masteringgob.com.

4.83 Write the symbol for the ion of each of the following:
a. chloride   **b.** potassium
c. oxide   **d.** aluminum

4.84 Write the symbol for the ion of each of the following:
a. fluoride   **b.** calcium
c. sodium   **d.** lithium

4.85 What is the name of each of the following ions?
a. K^+ **b.** S^{2-} **c.** Ca^{2+} **d.** N^{2-}

4.86 What is the name of each of the following ions?
a. Mg^{2+} **b.** Ba^{2+} **c.** I^- **d.** Cl^-

4.87 Write the formula of the following ionic compounds:
a. gold(III) chloride   **b.** lead(IV) oxide
c. silver chloride   **d.** calcium nitride
e. copper(I) phosphide   **f.** chromium(II) chloride

4.88 Write the formula of the following ionic compounds:
a. tin(IV) oxide   **b.** iron(III) sulfide
c. lead(IV) sulfide   **d.** chromium(III) iodide
e. lithium nitride   **f.** gold(I) oxide

4.89 Write the electron-dot formula for each of the following:
a. Cl_2O **b.** CF_4
c. H_2NOH (N is the central atom)   **d.** H_2CCCl_2

4.90 Write the electron-dot formula for each of the following:
a. H_3COCH_3; the atoms are in the order C O C
b. CS_2; the atoms are in the order S C S
c. NH_3
d. H_2CCHCN the atoms are in the order C C C

4.91 Name each of the following covalent compounds:
a. NCl_3 **b.** SCl_2 **c.** N_2O
d. F_2 **e.** PCl_5 **f.** P_2O_5

4.92 Name each of the following covalent compounds:
a. CBr_4 **b.** SF_6 **c.** Br_2
d. N_2O_4 **e.** SO_2 **f.** CS_2

4.93 Give the formula for each of the following:
a. carbon monoxide   **b.** diphosphorus pentoxide
c. dihydrogen sulfide   **d.** sulfur dichloride

4.94 Give the formula for each of the following:
a. silicon dioxide   **b.** carbon tetrabromide
c. sulfur trioxide   **d.** dinitrogen oxide

4.95 Classify each of the following compounds as ionic or covalent, and give its name.
a. $FeCl_3$ **b.** Na_2SO_4 **c.** N_2O
d. N_2 **e.** PCl_5 **f.** CF_4

4.96 Classify each of the following compounds as ionic or covalent, and give its name.
a. $Al_2(CO_3)_3$ **b.** SF_6 **c.** Br_2
d. Mg_3N_2 **e.** SO_2 **f.** $CrPO_4$

4.97 Write the formulas for the following:
a. tin(II) carbonate   _____
b. lithium phosphide   _____
c. silicon tetrachloride _____
d. iron(III) sulfide   _____
e. carbon dioxide   _____
f. calcium bromide   _____

4.98 Write the formulas for the following:
a. sodium carbonate   _____
b. nitrogen dioxide   _____
c. aluminum nitrate   _____
d. copper(I) nitride   _____
e. potassium phosphate _____
f. lead(IV) oxide   _____

4.99 Select the more polar bond in each of the following pairs:
a. C—N or C—O   **b.** N—F or N—Br
c. Br—Cl or S—Cl   **d.** Br—Cl or Br—I
e. N—F or N—O

4.100 Select the more polar bond in each of the following pairs:
a. C—C or C—O   **b.** P—Cl or P—Br
c. Si—S or Si—Cl   **d.** F—Cl or F—Br
e. P—O or P—S

4.101 Show the dipole arrow for each of the following bonds:
a. Si—Cl   **b.** C—N   **c.** F—Cl
d. C—F   **e.** N—O

4.102 Show the dipole arrow for each of the following bonds:
  a. C—O   **b.** N—F   **c.** O—Cl
  d. S—Cl   **e.** P—F

4.103 Classify each of the following bonds as nonpolar covalent, polar covalent, or ionic.
  a. Si—Cl   **b.** C—C   **c.** Na—Cl
  d. C—H   **e.** F—F

4.104 Classify each of the following bonds as nonpolar covalent, polar covalent, or ionic.
  a. C—N   **b.** Cl—Cl   **c.** K—Br
  d. H—H   **e.** N—F

4.105 Predict the shape and polarity of each of the following molecules: Assume all bonds are polar.
  a. A central atom bonded with identical bonded atoms and no lone pair.
  b. A central atom with two bonded atoms and one lone pair.
  c. A central atom bonded to two identical atoms and no lone pairs.

4.106 Predict the shape and polarity of each of the following molecules: Assume all bonds are polar.
  a. A central atom with four identical bonded atoms and no lone pairs.

b. A central atom with three identical bonded atoms and one lone pair.
c. A central atom with four bonded atoms that are not identical and no lone pair.

4.107 Write the electron-dot formula and determine the shape for each of the following:
  a. NF_3 **b.** $SiBr_4$
  c. $BeCl_2$ **d.** SO_2

4.108 Write the electron-dot formula and determine the shape for each of the following:
  a. SiH_4 **b.** HCCH
  c. $COCl_2$ (C is the central atom)   **d.** BCl_3

4.109 Predict the shape and polarity of each of the following molecules:
  a. H_2S **b.** NF_3 **c.** NCl_3
  d. CH_3Cl **e.** SiF_4

4.110 Predict the shape and polarity of each of the following molecules:
  a. H_2O **b.** CF_4 **c.** GeH_4
  d. PCl_3 **e.** SCl_2

CHALLENGE QUESTIONS

4.111 Consider the following electron-dot formulas for elements X and Y.

$$X\cdot \qquad \cdot \ddot{Y}\cdot$$

  a. What are the group numbers of X and Y?
  b. Will a compound of X and Y be ionic or covalent?
  c. What ions would be formed by X and Y?
  d. What would be the formula of a compound of X and Y?
  e. What would be the formula of a compound of X and chlorine?
  f. What would be the formula of a compound of Y and chlorine?

4.112 Complete the following table for atoms or ions.

Atom or Ion	Number of Protons	Number of Electrons	Electrons Lost/Gained
K^+			
	$12p^+$	$10e^-$	
	$8p^+$		$2e^-$ gained
		$10e^-$	$3e^-$ lost

4.113 One of the ions of tin is tin(IV).
  a. What is the symbol for this ion?
  b. How many protons and electrons are in the ion?
  c. What is the formula of tin(IV) oxide?
  d. What is the formula of tin(IV) phosphate?

4.114 Classify the following compounds as ionic or covalent and name each.
  a. Li_2O **b.** N_2O
  c. CF_4 **d.** Cl_2O
  e. MgF_2 **f.** CO
  g. $CaCl_2$ **h.** K_3PO_4

4.115 Name the following compounds:
  a. $FeCl_2$ **b.** Cl_2O_7
  c. N_2 **d.** $Ca_3(PO_4)_2$
  e. PCl_3 **f.** $Al(NO_3)_2$
  g. $PbCl_4$ **h.** $MgCO_3$
  i. NO_2 **j.** $SnSO_4$
  k. $Ba(NO_3)_2$ **l.** CuS

ANSWERS

Answers to Study Checks

4.1 $35p^+$ and $36e^-$

4.2 K^+ and S^{2-}

4.3 CaO

4.4 calcium chloride

4.5 gold(III) chloride

4.6 Cr_2O_3

4.7 $(NH_4)_3PO_4$

4.8 calcium phosphate

4.9 $:\ddot{Cl}:\ddot{O}:\ddot{Cl}:$ or $:\ddot{Cl}—\ddot{O}—\ddot{Cl}:$

4.10 H:C:::N: or H:C ≡ N: In HCN, there is a triple bond between C and N atoms.

4.11 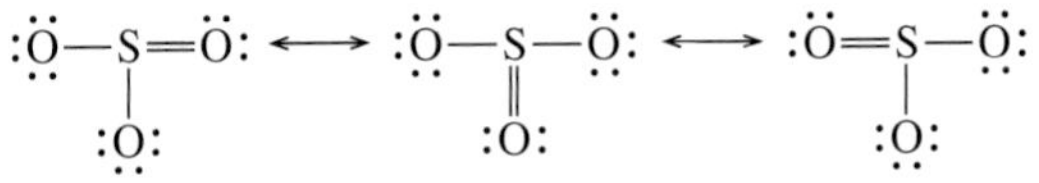

4.12 **a.** silicon tetrabromide
b. dibromine oxide

4.13 IF_5

4.14 iron(II) nitrate

4.15 **a.** Si—N
b. O—P

4.16 tetrahedral

4.17 polar

Answers to Selected Questions and Problems

4.1 **a.** By losing 1 valence electron from the third energy level, sodium achieves an octet in the second energy level.
b. The sodium ion Na^+ has the same electron arrangement as Ne (2, 8).
c. Group 1A (1) and 2A (2) elements become stable by losing electrons to form compounds. Group 8A (18) elements are stable.

4.3 **a.** 1 **b.** 2 **c.** 3 **d.** 1 **e.** 2

4.5 **a.** Li^+ **b.** F^- **c.** Mg^{2+} **d.** Fe^{3+} **e.** Zn^{2+}

4.7 **a.** Cl^- **b.** K^+ **c.** O^{2-} **d.** Al^{3+}

4.9 a, c

4.11 **a.** Na_2O **b.** $AlBr_3$ **c.** BaO **d.** $MgCl_2$ **e.** Al_2S_3

4.13 **a.** Na_2S **b.** K_3N **c.** AlI_3 **d.** Li_2O

4.15 **a.** aluminum oxide **b.** calcium chloride
c. sodium oxide **d.** magnesium nitride
e. potassium iodide

4.17 Most of the transition metals form more than one positive ion. The specific ion is indicated in a name by writing a Roman numeral that is the same as the ionic charge. For example, iron forms Fe^{2+} and Fe^{3+} ions, which are named iron(II) and iron(III).

4.19 **a.** iron(II) **b.** copper(II) **c.** zinc
d. lead(IV) **e.** chromium(III)

4.21 **a.** tin(II) chloride **b.** iron(II) oxide
c. copper(I) sulfide **d.** copper(II) sulfide
e. chromium(III) bromide

4.23 **a.** Au^{3+} **b.** Fe^{3+} **c.** Pb^{4+} **d.** Sn^{2+}

4.25 **a.** $MgCl_2$ **b.** Na_2S **c.** Cu_2O
d. Zn_3P_2 **e.** AuN **f.** $CrCl_2$

4.27 **a.** HCO_3^- **b.** NH_4^+ **c.** PO_4^{3-} **d.** HSO_4^-

4.29 **a.** sulfate **b.** carbonate
c. phosphate **d.** nitrate

4.31

	OH^-	NO_2^-	CO_3^{2-}	HSO_4^-	PO_4^{3-}
Li^+	LiOH	$LiNO_2$	Li_2CO_3	$LiHSO_4$	Li_3PO_4
Cu^{2+}	$Cu(OH)_2$	$Cu(NO_2)_2$	$CuCO_3$	$Cu(HSO_4)_2$	$Cu_3(PO_4)_2$
Ba^{2+}	$Ba(OH)_2$	$Ba(NO_2)_2$	$BaCO_3$	$Ba(HSO_4)_2$	$Ba_3(PO_4)_2$

4.33 **a.** CO_3^{2-}, sodium carbonate
b. NH_4^+, ammonium chloride
c. PO_4^{3-}, lithium phosphate
d. NO_2^-, copper(II) nitrite
e. SO_3^{2-}, iron(II) sulfite

4.35 **a.** $Ba(OH)_2$
b. Na_2SO_4
c. $Fe(NO_3)_2$
d. $Zn_3(PO_4)_2$
e. $Fe_2(CO_3)_3$

4.37 The nonmetallic elements are most likely to form covalent bonds.

4.39 **a.** 2 valence electrons, 1 bonding pair and 0 lone pairs
b. 8 valence electrons, 1 bonding pair and 3 lone pairs
c. 14 valence electrons, 1 bonding pair and 6 lone pairs

4.41 **a.** HF ($8\,e^-$) 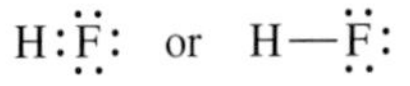
b. SF_2 ($20\,e^-$) 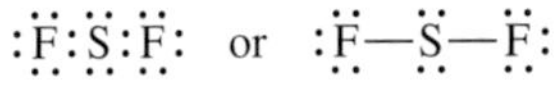
c. NBr_3 ($26\,e^-$) 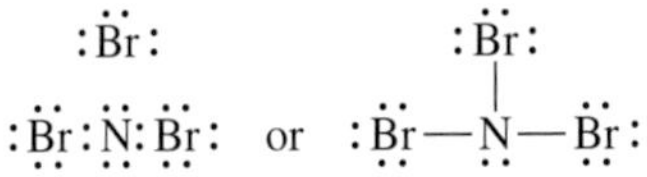
d. CH_3OH ($14\,e^-$) 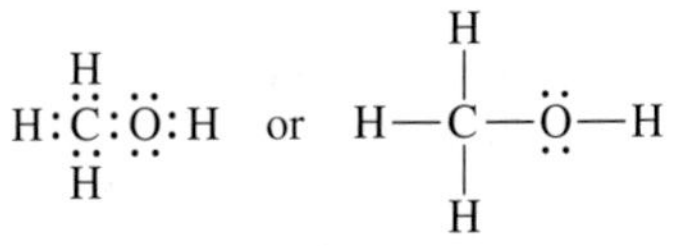
e. N_2H_4 ($14\,e^-$) 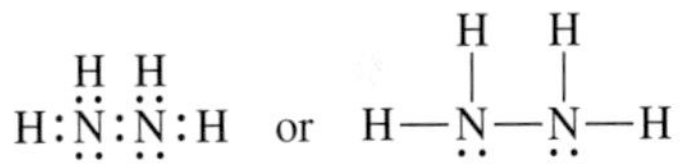

4.43 When using all the valence electrons does not give complete octets, it is necessary to write multiple bonds.

4.45 Resonance occurs when we can write two or more electron-dot formulas for the same molecule or ion that have multiple bonds.

4.47 **a.** CO ($10\,e^-$) :C:::O: or :C≡O:
b. H_2CCH_2 ($12\,e^-$) 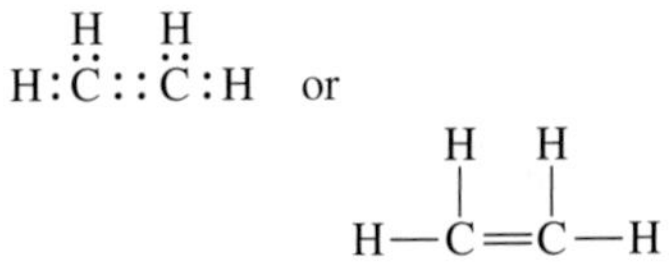
c. H_2CO ($12\,e^-$) 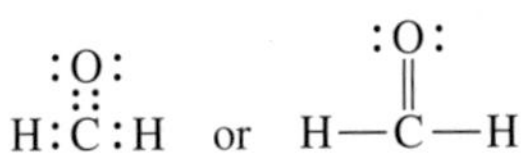

4.49 a. $ClNO_2$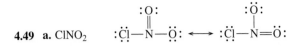

4.51 a. phosphorus tribromide
  b. carbon tetrabromide
  c. silicon dioxide
  d. dinitrogen trioxide
  e. silicon tetrabromide
  f. phosphorus pentachloride

4.53 a. CCl_4 **b.** CO
  c. PCl_3 **d.** N_2O_4
  e. BF_3 **f.** SF_6

4.55 a. aluminum sulfate
  b. calcium carbonate
  c. dinitrogen oxide
  d. sodium phosphate
  e. ammonium sulfate
  f. iron(III) oxide

4.57 The electronegativity increases going across a period.

4.59 A nonpolar covalent bond would have an electronegativity difference of 0.0 to 0.4.

4.61 a. K, Na, Li
  b. Na, P, Cl
  c. Ca, Br, O

4.63 a. polar covalent
  b. ionic
  c. polar covalent
  d. nonpolar covalent
  e. polar covalent
  f. nonpolar covalent

4.65 a. $\overset{\delta+}{N}-\overset{\delta-}{F}$ **b.** $\overset{\delta+}{Si}-\overset{\delta-}{Br}$ **c.** $\overset{\delta+}{C}-\overset{\delta-}{O}$
  d. $\overset{\delta+}{P}-\overset{\delta-}{Br}$ **e.** $\overset{\delta+}{B}-\overset{\delta-}{Cl}$

4.67 tetrahedral

4.69 The four electron groups in PCl_3 have a tetrahedral arrangement, but three bonded atoms around a central atom give a pyramidal shape.

4.71 In both PH_3 and NH_3, there are four electron pairs of which three are bonded to atoms and one is a lone pair. The shapes of both are pyramidal.

4.73 a. bent   **b.** tetrahedral
  c. linear   **d.** bent

4.75 Cl_2 is a nonpolar molecule because there is a nonpolar covalent bond between Cl atoms, which have identical electronegativity values. In HCl, the bond is a polar bond, which makes HCl a polar molecule.

4.77 a. polar   **b.** polar
  c. nonpolar   **d.** nonpolar

4.79 a. P^{3-} ion   **b.** O atom
  c. Zn^{2+} ion   **d.** Fe^{3+} ion
  e. Li^+ **f.** N^{3-}

4.81 a. 2 – pyramidal, polar
  b. 1 – bent, polar
  c. 3 – tetrahedral, nonpolar

4.83 a. Cl^- **b.** K^+ **c.** O^{2-} **d.** Al^{3+}

4.85 a. potassium   **b.** sulfide
  c. calcium   **d.** nitride

4.87 a. $AuCl_3$ **b.** PbO_2 **c.** $AgCl$
  d. Ca_3N_2 **e.** Cu_3P **f.** $CrCl_2$

4.89 a. Cl_2O $(20\ e^-)$

b. CF_4 $(32\ e^-)$

c. H_2NOH $(14\ e^-)$

d. H_2CCCl_2 $(24\ e^-)$

4.91 a. nitrogen trichloride
  b. sulfur dichloride
  c. dinitrogen oxide
  d. fluorine
  e. phosphorus pentachloride
  f. diphosphorus pentoxide

4.93 a. CO **b.** P_2O_5 **c.** H_2S **d.** SCl_2

4.95 a. ionic, iron(III) chloride
  b. ionic, sodium sulfate
  c. covalent, dinitrogen oxide
  d. covalent, nitrogen
  e. covalent, phosphorus pentachloride
  f. covalent, carbon tetrafluoride

4.97 a. $SnCO_3$ **b.** Li_3P **c.** $SiCl_4$
  d. Fe_2S_3 **e.** CO_2 **f.** $CaBr_2$

4.99 a. $C-O$ **b.** $N-F$ **c.** $S-Cl$
  d. $Br-I$ **e.** $N-F$

4.101 a. $Si-Cl$ **b.** $C-N$ **c.** $F-Cl$
  d. $C-F$ **e.** $N-O$

4.103 a. polar covalent
  b. nonpolar covalent
  c. ionic
  d. nonpolar covalent
  e. nonpolar covalent

4.105 a. trigonal planar, nonpolar
  b. bent, polar
  c. linear, nonpolar

4.107 a. NF$_3$ $:\ddot{F}-\overset{..}{\underset{.}{N}}-\ddot{F}:$ trigonal pyramidal

$:\ddot{F}:$

b. SiBr$_4$ $:\ddot{B}r-\overset{:\ddot{B}r:}{\underset{:\ddot{B}r:}{Si}}-\ddot{B}r:$ tetrahedral

c. BeCl$_2$ $:\ddot{C}l-Be-\ddot{C}l:$ linear

d. SO$_2$ $\left[:\ddot{O}=\ddot{S}-\ddot{O}:\right] \longleftrightarrow$

$\left[:\ddot{O}-\ddot{S}=\ddot{O}:\right]$ bent (120°)

4.109 a. bent, polar   **b.** pyramidal, polar
c. pyramidal, polar   **d.** tetrahedral, polar
e. tetrahedral, nonpolar

Answers to Challenge Questions

4.111 a. X is in Group 1A (1); Y is in Group 6A (16)
b. ionic   **c.** X$^+$, Y^{2-} **d.** X$_2$Y
e. XCl   **f.** YCl$_2$

4.113 a. Sn^{4+} **b.** 50 protons, 46 electrons
c. SnO$_2$ **d.** Sn$_3$(PO$_4$)$_4$

4.115 a. iron(II) chloride   **b.** dichlorine heptoxide
c. nitrogen   **d.** calcium phosphate
e. phosphorus trichloride   **f.** aluminum nitrate
g. lead(IV) chloride   **h.** magnesium carbonate
i. nitrogen dioxide   **j.** tin(II) sulfate
k. barium nitrate   **l.** copper(II) sulfide

5 Chemical Reactions and Quantities

"In our food science laboratory I develop a variety of food products, from cake donuts to energy beverages," says Anne Cristofano, senior food technologist at Mattson & Company. "When I started the donut project, I researched the ingredients, then weighed them out in the lab. I added water to make a batter and cooked the donuts in a fryer. The batter and the oil temperature make a big difference. If I don't get the right taste or texture, I adjust the ingredients, such as sugar and flour, or adjust the temperature."

A food technologist studies the physical and chemical properties of food and develops scientific ways to process and preserve it for extended shelf life. The food products are tested for texture, color, and flavor. The results of these tests help improve the quality and safety of food.

Mastering GOB CHEMISTRY

Visit **www.masteringgob.com** for self-study materials and instructor-assigned homework.

Chemical reactions occur everywhere. In the engines of our cars, gasoline burns to provide energy to move the car, play the radio, and run the air conditioner. Chemical reactions in our bodies provide energy to move our muscles. In nature, chemical reactions occur in the leaves of plants to convert carbon dioxide and water into carbohydrates.

Chemists use chemical equations to describe the reactants and products of chemical reactions. Some chemical reactions are simple, while others are quite complex. However, they can all be written with chemical equations. In every chemical reaction, the atoms in the reacting substances, called reactants, are rearranged to give new substances called products. When we know the chemical equation for a reaction, we can determine the amount of product that can be produced. We do much the same thing at home when we use a recipe to make cookies or add the right quantity of water to make soup. At the automotive repair shop, a mechanic adjusts the carburetor or fuel injection system of an engine to allow for the correct amounts of fuel and oxygen so the engine will run properly. In the hospital, a respiratory therapist evaluates the level of CO_2 and O_2 in the blood. A certain amount of oxygen must reach the tissues for efficient metabolic reactions. If the oxygenation of the blood is low, the therapist will oxygenate the patient and recheck the blood levels.

In this chapter, we will describe the quantity of a compound as a mole and calculate the mass related to the formula of a compound. From a balanced equation, we can determine the mass and number of moles of a reactant and calculate the amount of product. Knowing how to determine the quantitative results of a chemical reaction is important to both chemists and medical personnel such as pharmacists and respiratory therapists.

5.1 Chemical Changes

In a **physical change**, the appearance of a substance is altered, but not its composition. When liquid water becomes a gas, or freezes to a solid, it is still water (see Figure 5.1). If we smash a rock or tear a piece of paper, only the size of the material changes. The smaller pieces are still rock or paper because there was no change in the composition of the substances.

In a **chemical change**, the reacting substances change into new substances that have different formulas and different properties. New properties may involve a change in color or the formation of bubbles or a solid. For instance, when silver tarnishes, the bright silver metal (Ag) reacts with sulfur (S) to become the dull, blackish substance we call tarnish (Ag_2S) (see Figure 5.1). Table 5.1 gives examples of some typical physical and chemical changes.

A **chemical reaction** always involves chemical change because atoms of the reacting substances form new combinations with new properties. A chemical reaction takes place when rust forms on a nail or piece of iron: the iron (Fe) reacts with oxygen

LEARNING GOAL

Identify a change in a substance as a chemical or a physical change.

TABLE 5.1

Comparison of Some Chemical and Physical Changes

Chemical Changes	Physical Changes
Rusting nail	Melting ice
Bleaching a stain	Boiling water
Burning a log	Sawing a log in half
Fermenting grapes	Breaking a glass
Souring a milk	Pouring milk

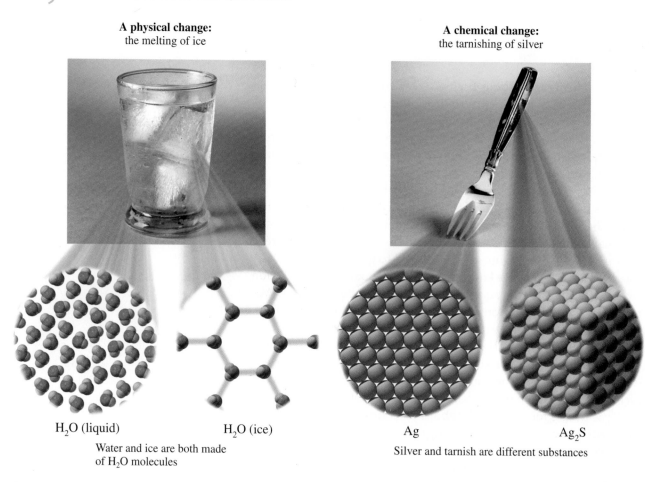

A physical change:
the melting of ice

A chemical change:
the tarnishing of silver

H_2O (liquid) H_2O (ice)   Ag Ag_2S

Water and ice are both made
of H_2O molecules

Silver and tarnish are different substances

FIGURE 5.1

A chemical change produces new substances: a physical change does not.

Q **Why is the formation of tarnish a chemical change?**

 MC GOB Chemical Changes and Physical Changes

(O_2) to produce the new substance, rust (Fe_2O_3). Perhaps you have placed an antacid tablet in a glass of water and noticed it fizzing and bubbling as sodium hydrogen carbonate $(NaHCO_3)$ reacts to form carbon dioxide (CO_2) gas (see Figure 5.2).

SAMPLE PROBLEM 5.1

■ **Classifying Chemical and Physical Change**

Classify each of the following changes as physical or chemical:

a. water freezing into an icicle
b. burning a match
c. digesting a chocolate bar

SOLUTION

a. Physical. Freezing water involves only a change from liquid water to ice. No change has occurred in the substance water.
b. Chemical. Burning a match causes the formation of ash that was not present before striking the match.
c. Chemical. The digestion of the chocolate bar converts its components into new substances.

MC GOB **SELF-STUDY ACTIVITY**
What is Chemistry?

Examples of chemical reactions involve chemical change: iron (Fe) reacts with oxygen (O_2) to form rust (Fe_2O_3) and an antacid ($NaHCO_3$) tablet in water forms bubbles of carbon dioxide (CO_2).

Q **What is the evidence for chemical change in these chemical reactions?**

STUDY CHECK

Classify the following changes as physical or chemical:

a. chopping a carrot   **b.** burning a candle

QUESTIONS AND PROBLEMS

■ **Chemical Changes**

5.1 Classify each of the following changes as chemical or physical:
    a. chewing a gum drop
    b. ignition of fuel in the space shuttle
    c. drying clothes
    d. neutralizing stomach acid with an antacid tablet
    e. formation of snowflakes

5.2 Classify each of the following changes as chemical or physical:
    a. fogging the mirror during a shower
    b. tarnishing of a silver bracelet
    c. breaking a bone
    d. mending a broken bone
    e. burning paper

5.2 Chemical Equations

When you build a model airplane, prepare a new recipe, mix a medication, or clean a patient's teeth, you follow a set of directions. These directions tell you what materials to use and the products you will obtain. In chemistry, a **chemical equation** tells us the materials we need and the products that will form in a chemical reaction.

Writing a Chemical Equation

Suppose you work in a bicycle shop, assembling wheels and bodies into bicycles. You could represent this process by a simple equation.

Equation: Wheels + Body ⟶ Bicycle

When you burn charcoal in a grill, the carbon in the charcoal combines with oxygen to form carbon dioxide. We can represent this reaction by a chemical equation that is much like the one for the bicycle:

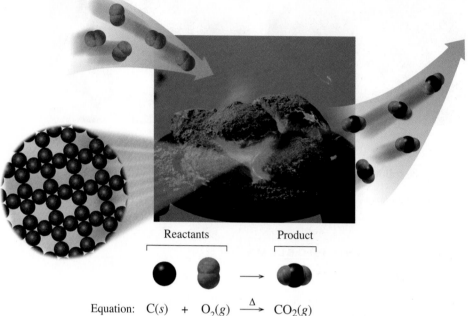

Reactants   Product

Equation: $C(s)$ + $O_2(g)$ $\xrightarrow{\Delta}$ $CO_2(g)$

In an equation, the formulas of the **reactants** are written on the left of the arrow and the formulas of the **products** on the right. When there are two or more formulas on the same side, they are separated by plus ($+$) signs. The delta sign (Δ) indicates that heat was used to start the reaction.

Sometimes, as in the charcoal case, the formulas in an equation may include letters, in parentheses, that give the physical state of the substances as solid (s), liquid (l), or gas (g). If a substance is dissolved in water, it is an aqueous (aq) solution. Table 5.2 summarizes some of the symbols used in equations.

When a reaction takes place, the bonds between the atoms of the reactants are broken and new bonds are formed to give the products. Atoms cannot be gained, lost, or changed into other types of atoms during a chemical reaction. Therefore, a reaction must be written as a **balanced equation**, which shows the same number of atoms for each element on both sides of the arrow. Let's see if the equation we wrote above for burning carbon is balanced:

TABLE 5.2

Some Symbols Used in Writing Equations

Symbol	Meaning
$+$	Separates two or more formulas
$\longrightarrow$	Reacts to form products
Δ	The reactants are heated
(s)	Solid
(l)	Liquid
(g)	Gas
(aq)	Aqueous

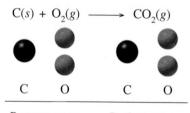

$C(s) + O_2(g) \longrightarrow CO_2(g)$

C   O   C   O

Reactant atoms   =   Product atoms

The answer is yes: this equation is *balanced* because there is one carbon atom and two oxygen atoms on each side of the arrow.

Now consider the reaction in which hydrogen reacts with oxygen to form water. First we write the formulas of the reactants and products:

$H_2(g) + O_2(g) \longrightarrow H_2O(g)$

SELF-STUDY ACTIVITY
Chemical Reactions and Equations

Is the equation balanced? To find out, we add up the atoms of each element on each side of the arrow. No; the equation is *not balanced*. The number of atoms on the left side does not match the number of atoms on the right side. To balance this equation, we place whole numbers called **coefficients** in front of some of the formulas. First we write a coefficient of 2 in front of the H_2O formula to represent the formation of 2 molecules of water. Now the product has four hydrogen atoms. That means we must also write a coefficient of 2 in front of the formula H_2 in the reactants to give 4 H atoms on the reactant side. Coefficients are used to balance an equation, but the subscripts in the formulas are never changed. Now the numbers of hydrogen atoms and oxygen atoms are the same in the reactants as in the products. The equation is *balanced*.

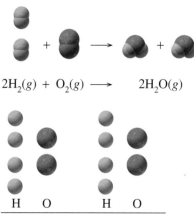

$$2H_2(g) + O_2(g) \longrightarrow 2H_2O(g)$$

H   O   H   O

Reactant atoms = Product atoms

Balancing a Chemical Equation

We have seen that a chemical equation must be balanced. In many cases, we can use a method of trial and error to balance an equation.

To demonstrate the process, let us balance the equation for the reaction of the gas methane, CH_4, with oxygen to produce carbon dioxide and water. This is the principal reaction that occurs in the flame of a laboratory burner or a gas stove.

■ **STEP 1 Write an equation, using the correct formulas.**   As a first step, we write the equation using the correct formulas for the reactants and products.

$$CH_4(g) + O_2(g) \longrightarrow CO_2(g) + H_2O(g)$$

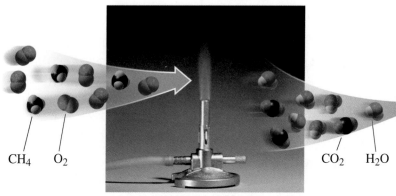

CH_4 O_2 CO_2 H_2O

■ **STEP 2 Determine if the equation is balanced.**   When we compare the atoms on the reactant side and the product side, we see that there are more hydrogen atoms on the left side and more oxygen atoms on the right.

$$CH_4(g) + O_2(g) \longrightarrow CO_2(g) + H_2O(g)$$

1 C	1 C	Balanced
4 H	2 H	Not balanced
2 O	3 O	Not balanced

■ **STEP 3 Balance the equation one element at a time.**   First we balance the hydrogen atoms by placing a 2 in front of the formula for water. Then we balance the oxygen atoms by placing a 2 in front of the formula for oxygen. There are now four oxygen atoms and four hydrogen atoms in both the reactants and products.

$$CH_4(g) + 2O_2(g) \longrightarrow CO_2(g) + 2H_2O(g)$$

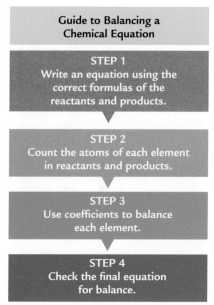

Guide to Balancing a Chemical Equation

STEP 1
Write an equation using the correct formulas of the reactants and products.

STEP 2
Count the atoms of each element in reactants and products.

STEP 3
Use coefficients to balance each element.

STEP 4
Check the final equation for balance.

 Balancing Chemical Equations

■ **STEP 4** **Check to see if the equation is balanced.** Rechecking the balanced equation shows that the numbers of atoms of carbon, hydrogen, and oxygen are the same for both the reactants and the products. The equation is balanced using the lowest possible whole-number coefficients.

$$CH_4(g) + 2O_2(g) \longrightarrow CO_2(g) + 2H_2O(g) \quad \text{Balanced}$$

Reactants		Products
1 C atom	=	1 C atom
4 H atoms	=	4 H atoms
4 O atoms	=	4 O atoms

Suppose you had written the equation as follows:

$$2CH_4(g) + 4O_2(g) \longrightarrow 2CO_2(g) + 4H_2O(g) \quad \text{Incorrect}$$

Although there are equal numbers of atoms on both sides of the equation, this is not written correctly. All the coefficients must be divided by 2 to obtain the lowest possible whole numbers.

<h3>SAMPLE PROBLEM 5.2</h3>

■ **Balancing Equations**

Balance the following equation:

$$Na_3PO_4(aq) + MgCl_2(aq) \longrightarrow Mg_3(PO_4)_2(s) + NaCl(aq)$$

SOLUTION

■ **STEP 1** In the equation, the correct formulas are written.

$$Na_3PO_4(aq) + MgCl_2(aq) \longrightarrow Mg_3(PO_4)_2(s) + NaCl(aq)$$

■ **STEP 2** When we compare the number of ions on the reactant and product sides, we find that the equation is not balanced. In this equation, it is convenient to balance the phosphate ion as a group instead of its individual atoms.

Reactants **Products**
$$Na_3PO_4(aq) + MgCl_2(aq) \longrightarrow Mg_3(PO_4)_2(s) + NaCl(aq)$$

3 Na^+	1 Na^+	Not balanced
1 PO_4^{3-}	2 PO_4^{3-}	Not balanced
1 Mg^{2+}	3 Mg^{2+}	Not balanced
2 Cl^-	1 Cl^-	Not balanced

■ **STEP 3** We begin with the formula of $Mg_3(PO_4)_2$, which is the most complex. A 3 in front of $MgCl_2$ balances magnesium and a 2 in front of Na_3PO_4 balances the phosphate ion. Looking again at each of the ions in the reactants and products, we see that the sodium and chloride ions are not yet equal. A 6 in front of the NaCl balances the equation.

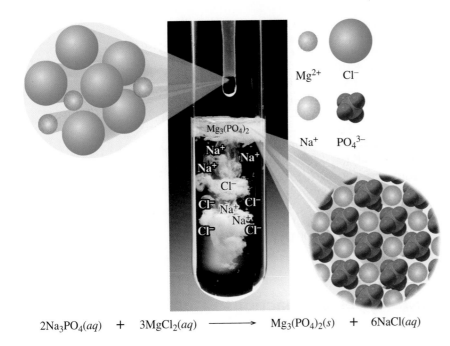

$$2Na_3PO_4(aq) \; + \; 3MgCl_2(aq) \longrightarrow Mg_3(PO_4)_2(s) \; + \; 6NaCl(aq)$$

■ **STEP 4** A check of the atoms indicates the equation is balanced.

Reactants   **Productss**

$2Na_3PO_4(aq) + 3MgCl_2(aq) \longrightarrow Mg_3(PO_4)_2(s) + 6NaCl(aq)$

Balanced

$6\,Na^+ \quad = \quad 6\,Na^+$

$2\,PO_4^{3-} \quad = \quad 2\,PO_4^{3-}$

$3\,Mg^{2+} \quad = \quad 3\,Mg^{2+}$

$6\,Cl^- \quad = \quad 6\,Cl^-$

STUDY CHECK

Balance the following equation:

$$Fe(s) + O_2(g) \longrightarrow Fe_3O_4(s)$$

QUESTIONS AND PROBLEMS

■ **Chemical Equations**

5.3 Determine whether each of the following equations is balanced or not balanced:
   a. $S(s) + O_2(g) \longrightarrow SO_3(g)$ **b.** $2Al(s) + 3Cl_2(g) \longrightarrow 2AlCl_3(s)$
   c. $H_2(g) + O_2(g) \longrightarrow H_2O(g)$ **d.** $C_3H_8(g) + 5O_2(g) \longrightarrow 3CO_2(g) + 4H_2O(g)$

5.4 Determine whether each of the following equations are balanced or not balanced:
   a. $PCl_3(s) + Cl_2(g) \longrightarrow PCl_5(s)$ **b.** $CO(g) + 2H_2(g) \longrightarrow CH_3OH(g)$
   c. $2KClO_3(s) \longrightarrow 2KCl(s) + O_2(g)$ **d.** $Mg(s) + N_2(g) \longrightarrow Mg_3N_2(s)$

5.5 Balance the following equations:
   a. $N_2(g) + O_2(g) \longrightarrow NO(g)$
   b. $HgO(s) \longrightarrow Hg(l) + O_2(g)$
   c. $Fe(s) + O_2(g) \longrightarrow Fe_2O_3(s)$
   d. $Na(s) + Cl_2(g) \longrightarrow NaCl(s)$

5.6 Balance the following equations:

a. $Ca(s) + Br_2(l) \longrightarrow CaBr_2(s)$

b. $P_4(s) + O_2(g) \longrightarrow P_4O_{10}(s)$

c. $Sb_2S_3(s) + HCl(aq) \longrightarrow SbCl_3(s) + H_2S(g)$

d. $Fe_2O_3(s) + C(s) \longrightarrow Fe(s) + CO(g)$

5.7 Balance the following equations:

a. $Mg(s) + AgNO_3(aq) \longrightarrow Mg(NO_3)_2(aq) + Ag(s)$

b. $Al(s) + CuSO_4(aq) \longrightarrow Cu(s) + Al_2(SO_4)_3(aq)$

c. $Pb(NO_3)_2(aq) + NaCl(aq) \longrightarrow PbCl_2(s) + NaNO_3(aq)$

d. $Al(s) + HCl(aq) \longrightarrow AlCl_3(aq) + H_2(g)$

5.8 Balance the following equations:

a. $Zn(s) + H_2SO_4(aq) \longrightarrow ZnSO_4(aq) + H_2(g)$

b. $Al(s) + H_2SO_4(aq) \longrightarrow Al_2(SO_4)_3(aq) + H_2(g)$

c. $K_2SO_4(aq) + BaCl_2(aq) \longrightarrow BaSO_4(s) + KCl(aq)$

d. $CaCO_3(s) \longrightarrow CaO(s) + CO_2(g)$

5.3 Types of Reactions

LEARNING GOAL

Identify a reaction as a combination, decomposition, or replacement.

A great number of reactions occur in nature, in biological systems, and in the laboratory. However, there are some general patterns among all reactions that help us classify reactions. Most reactions fit into four general reaction types.

Combination Reactions

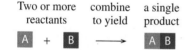

Two or more reactants combine to yield a single product

A + B ⟶ A B

In a **combination reaction**, two or more elements or simple compounds bond together to form one product. For example, sulfur and oxygen combine to form the product sulfur dioxide.

Combination reaction

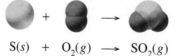

$$S(s) + O_2(g) \longrightarrow SO_2(g)$$

In Figure 5.3, the elements magnesium and oxygen combine to form a single product, magnesium oxide.

$$2Mg(s) + O_2(g) \longrightarrow 2MgO(s)$$

SELF-STUDY ACTIVITY
Chemical Reactions and Equations

Decomposition Reactions

Decomposition

A reactant splits into two or more products

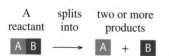

A B ⟶ A + B

In a **decomposition reaction**, a reactant splits into two or more simpler products. For example, calcium carbonate breaks apart into simpler compounds of calcium oxide and carbon dioxide when heated.

$$CaCO_3(s) \xrightarrow{\Delta} CaO(s) + CO_2(g)$$

Replacement Reactions

In a replacement reaction, elements in a compound are replaced by other elements. In a **single replacement reaction**, a reacting element switches place with an element in the other reacting compound.

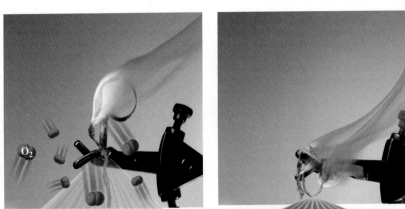

$$2Mg(s) \quad + \quad O_2(g) \quad \longrightarrow \quad 2MgO(s)$$
Magnesium   Oxygen   Magnesium oxide

FIGURE 5.3

In a combination reaction two or more substances combine to form one substance as product.

Q **What happens to the atoms of the reactants in a combination reaction?**

MC GOB   Reaction Types

Single replacement

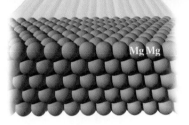

One element   replaces   another element

In the single replacement reaction shown in Figure 5.4, zinc replaces hydrogen in hydrochloric acid, HCl(aq).

$$Zn(s) + 2HCl(aq) \longrightarrow ZnCl_2(aq) + H_2(g)$$

In a **double replacement reaction**, the positive ions in the reacting compounds switch places.

Double replacement

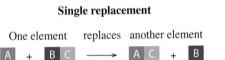

Two elements   replace each other

For example, in the reaction shown in Figure 5.5, barium ions change places with sodium ions in the reactants to form sodium chloride and a white solid precipitate of barium sulfate. The formulas of the products depend on the charges of the ions.

$$BaCl_2(aq) + Na_2SO_4(aq) \longrightarrow BaSO_4(s) + 2NaCl(aq)$$

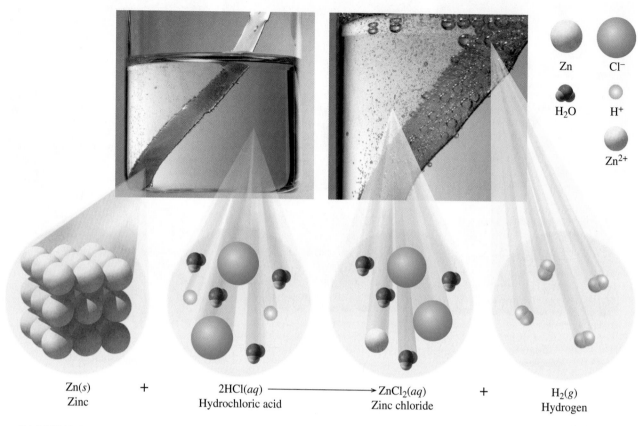

$$Zn(s) \quad + \quad 2HCl(aq) \longrightarrow ZnCl_2(aq) \quad + \quad H_2(g)$$

Zinc   Hydrochloric acid   Zinc chloride   Hydrogen

FIGURE 5.4

In a single replacement reaction, an atom or ion replaces an atom or ion in
a compound.

Q **What changes in the formulas of the reactants identify this equation as a single
replacement?**

SAMPLE PROBLEM 5.3

▪ Identifying Reactions and Predicting Products

Classify the following reactions as combination, decomposition, or single or double
replacement:

a. $2Fe_2O_3(s) + 3C(s) \longrightarrow 3CO_2(g) + 4Fe(s)$

b. $BaCl_2(aq) + K_2SO_4(aq) \longrightarrow BaSO_4(s) + 2KCl(aq)$

SOLUTION

a. In this *single replacement* reaction, a C atom replaces Fe in Fe_2O_3 to form the
compound CO_2 and Fe atoms.

b. There are two reactants and two products, but the positive ions have exchanged
places, which makes this a *double replacement* reaction.

STUDY CHECK

Nitrogen monoxide gas and oxygen gas react to form nitrogen dioxide gas. Write
the balanced equation and identify the reaction type.

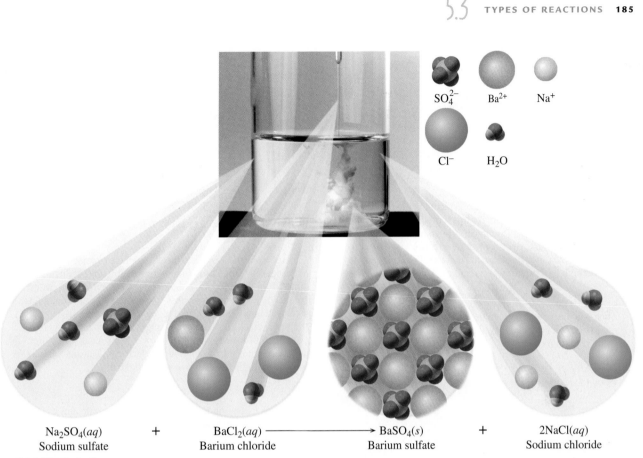

$$\underset{\text{Sodium sulfate}}{\text{Na}_2\text{SO}_4(aq)} \quad + \quad \underset{\text{Barium chloride}}{\text{BaCl}_2(aq)} \longrightarrow \underset{\text{Barium sulfate}}{\text{BaSO}_4(s)} \quad + \quad \underset{\text{Sodium chloride}}{2\text{NaCl}(aq)}$$

FIGURE 5.5

In a double replacement reaction the positive ions in the reactants replace each other.

Q **How do the changes in the formulas of the reactants identify this equation as a double replacement reaction?**

QUESTIONS AND PROBLEMS

▪ Types of Reactions

5.9 **a.** Why is the following reaction called a decomposition reaction?

$$2\text{Al}_2\text{O}_3(s) \xrightarrow{\Delta} 4\text{Al}(s) + 3\text{O}_2(g)$$

b. Why is the following reaction called a single replacement reaction?

$$\text{Br}_2(g) + \text{BaI}_2(s) \longrightarrow \text{BaBr}_2(s) + \text{I}_2(g)$$

5.10 **a.** Why is the following reaction called a combination reaction?

$$\text{H}_2(g) + \text{Br}_2(g) \longrightarrow 2\text{HBr}(g)$$

b. Why is the following reaction called a double replacement reaction?

$$\text{AgNO}_3(aq) + \text{NaCl}(aq) \longrightarrow \text{AgCl}(s) + \text{NaNO}_3(aq)$$

5.11 Classify each of the following reactions as a combination, decomposition, single replacement, or double replacement.
a. $4\text{Fe}(s) + 3\text{O}_2(g) \longrightarrow 2\text{Fe}_2\text{O}_3(s)$
b. $\text{Mg}(s) + 2\text{AgNO}_3(aq) \longrightarrow \text{Mg(NO}_3)_2(aq) + 2\text{Ag}(s)$
c. $\text{CuCO}_3(s) \longrightarrow \text{CuO}(s) + \text{CO}_2(g)$
d. $\text{NaOH}(aq) + \text{HCl}(aq) \longrightarrow \text{NaCl}(aq) + \text{H}_2\text{O}(l)$
e. $\text{Al}_2(\text{SO}_4)_3(aq) + 6\text{KOH}(aq) \longrightarrow 2\text{Al(OH)}_3(s) + 3\text{K}_2\text{SO}_4(aq)$

Smog and Health Concerns

There are two types of smog. One, photochemical smog, requires sunlight to initiate reactions that produce pollutants such as nitrogen oxides and ozone. The other type of smog, industrial or London smog, occurs in areas where coal containing sulfur is burned and the unwanted product sulfur dioxide is emitted.

Photochemical smog is most prevalent in cities where people are dependent on cars for transportation. On a typical day in Los Angeles, for example, nitrogen monoxide (NO) emissions from car exhausts increase as traffic increases on the roads. The nitrogen monoxide is formed when N_2 and O_2 react at high temperatures in car and truck engines.

$$N_2(g) + O_2(g) \xrightarrow{\text{Heat}} 2NO(g)$$

Then NO reacts with oxygen in the air to produce NO_2, a reddish brown gas that is irritating to the eyes and damaging to the respiratory tract.

$$2NO(g) + O_2(g) \longrightarrow 2NO_2(g)$$

When NO_2 is exposed to sunlight, it is converted into NO and oxygen atoms.

$$NO_2(g) + \xrightarrow{\text{Sunlight}} NO(g) + \underset{\text{oxygen atoms}}{O(g)}$$

Oxygen atoms are so reactive that they combine with oxygen molecules in the atmosphere, forming ozone.

$$O(g) + O_2(g) \longrightarrow \underset{\text{ozone}}{O_3(g)}$$

In the upper atmosphere (the stratosphere), ozone is beneficial because it protects us from harmful ultraviolet radiation that comes from the sun. However, in the lower atmosphere, ozone irritates the eyes and respiratory tract, where it causes coughing, decreased lung function, and fatigue. It also causes deterioration of fabrics, cracks rubber, and damages trees and crops.

Industrial smog is prevalent in areas where coal with a high sulfur content is burned to produce electricity. During combustion, the sulfur is converted to sulfur dioxide:

$$S(s) + O_2(g) \longrightarrow SO_2(g)$$

The SO_2 is damaging to plants, suppresses growth, and it is corrosive to metals such as steel. SO_2 is also damaging to humans and can cause lung impairment and respiratory difficulties. The SO_2 in the air reacts with more oxygen to form SO_3. SO_3 combines with water in the air to form sulfuric acid, which makes the rain acidic.

$$2SO_2(g) + O_2(g) \longrightarrow 2SO_3(g)$$

$$SO_3(g) + H_2O(l) \longrightarrow \underset{\text{sulfuric acid}}{H_2SO_4(aq)}$$

The presence of sulfuric acid in rivers and lakes causes an increase in the acidity of the water, reducing the ability of animals and plants to survive.

5.12 Classify each of the following reactions as a combination, decomposition, single replacement, or double replacement.
 a. $CuO(s) + 2HCl(aq) \longrightarrow CuCl_2(aq) + H_2O(l)$
 b. $2Al(s) + 3Br_2(g) \longrightarrow 2AlBr_3(s)$
 c. $Pb(NO_3)_2(aq) + 2NaCl(aq) \longrightarrow PbCl_2(s) + 2NaNO_3(aq)$
 d. $C_6H_{12}O_6(aq) \longrightarrow 2C_2H_6O(aq) + 2CO_2(g)$
 e. $BaCl_2(aq) + K_2CO_3(aq) \longrightarrow BaCO_3(s) + 2KCl(aq)$

LEARNING GOAL

Define the terms *oxidation* and *reduction*.

5.4 Oxidation–Reduction Reactions

Perhaps you have never heard of oxidation and reduction reactions. Yet this type of reaction has many important applications in our everyday lives. When you see a rusty nail, tarnish on a silver spoon, or corrosion on metal, you are observing oxidation.

$$\underset{\text{rust}}{4Fe(s) + 3O_2(g) \longrightarrow 2Fe_2O_3(s)} \qquad \text{Fe is oxidized}$$

When you turn the lights on in your car, an oxidation–reduction reaction within your car battery provides the electricity. On a cold, wintry day, you might build a wood fire. Burning wood is an oxidation–reduction reaction. When you eat foods with starches in them, you digest the starches to give glucose, which is oxidized in your cells to give you energy along with carbon dioxide and water. Every breath you take provides oxygen to carry out oxidation in your cells.

$$C_6H_{12}O_6(aq) + 6O_2(g) \longrightarrow 6CO_2(g) + 6H_2O(l) + \text{energy}$$

Oxidation–Reduction

Oxidation–reduction reactions are defined as reactions with a loss and a gain of electrons among the reactants. In the oxidation part of the reaction, one substance loses electrons. Simultaneously, a reduction also occurs as another substance gains these electrons. You may like to use the shorthand phrase "LEO the lion goes GER" to remind yourself that a loss of electrons is oxidation (LEO), and a gain of electrons is a reduction (GER).

Oxidation–Reduction Involving Ions

When we looked at the formation of ionic compounds, we also saw that metals lost electrons and nonmetals gained electrons. Now we can say that metals are oxidized and nonmetals are reduced.

Consider the elements and ions in the reaction where calcium loses electrons to form calcium ions and where sulfur gains electrons to form sulfide ions. The elemental forms of calcium and sulfur have no ionic charge; they are neutral.

$$Ca + S \longrightarrow Ca^{2+} + S^{2-} = CaS$$

If we write the formation of each ion of the compound as half reactions, we find that one is an oxidation and the other is a reduction. In every oxidation–reduction reaction, the number of electrons exchanged among the reactants must be equal.

$$Ca \longrightarrow Ca^{2+} + 2e^- \qquad \text{Oxidation}$$
$$S + 2e^- \longrightarrow S^{2-} \qquad \text{Reduction}$$

The loss of electrons is an **oxidation**. We say that calcium was oxidized. At the same time, the sulfur atom gained electrons. This gain of electrons is called **reduction**. We say that the sulfur was reduced. The two parts must occur simultaneously and with the same total number of electrons. We can write the two parts as follows.

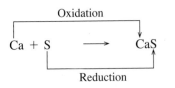

If we look at a single replacement reaction, we also find that it has two parts: an oxidation and a reduction. Consider the reaction of zinc and copper(II) sulfate. (See Figure 5.6.)

$$Zn(s) + CuSO_4(aq) \longrightarrow ZnSO_4(aq) + Cu(s)$$

We can rewrite the equation to show the ions of the compounds.

$$Zn(s) + Cu^{2+}(aq) + SO_4^{2-}(aq) \longrightarrow Zn^{2+}(aq) + SO_4^{2-}(aq) + Cu(s)$$

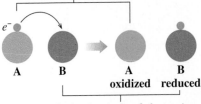

Oxidation (loss of electron)

A   B   A   B
      oxidized reduced

Reduction (gain of electron)

MC GOB Identifying Oxidation-Reduction Reactions

FIGURE 5.6

In this single replacement reaction Zn(s) is oxidized to Zn^{2+} when it provides two electrons to reduce Cu^{2+} to Cu(s):

$$Zn(s) + Cu^{2+}(aq) \longrightarrow Cu(s) + Zn^{2+}(aq)$$

Q **In the oxidation, does Zn(s) lose or gain electrons?**

Now we can show that zinc loses 2 electrons to form Zn^{2+}, and Cu^{2+} gains 2 electrons to form Cu.

$$Zn(s) \longrightarrow Zn^{2+}(aq) + 2e^- \qquad \text{Oxidation}$$
$$Cu^{2+}(aq) + 2e^- \longrightarrow Cu(s) \qquad \text{Reduction}$$

In this single replacement reaction, zinc was oxidized and copper(II) ion was reduced. No change occurred for the sulfate ion.

SAMPLE PROBLEM 5.4

■ Oxidation–Reduction Reactions

In photographic film, the following decomposition reaction occurs in the presence of light. What is oxidized and what is reduced?

$$2AgBr(s) \xrightarrow{\text{Light}} 2Ag(s) + Br_2(g)$$

SOLUTION

In the AgBr compound, silver ion has an ionic charge of 1+ and the bromide ion has an ionic charge of 1−. We can write the reaction as follows.

$$2Ag^+ + 2Br^- \longrightarrow 2Ag(s) + Br_2(g)$$

We can determine the oxidation reaction and the reduction reaction by writing the changes for each ion separately.

$$2Br^- \longrightarrow Br_2(g) + 2e^- \qquad \text{Oxidation}$$
$$2Ag^+ + 2e^- \longrightarrow 2Ag(s) \qquad \text{Reduction}$$

In the reaction, bromide ion is oxidized, and silver ion is reduced.

STUDY CHECK

In the following combination reaction, what is oxidized and what is reduced?

$$2Li(s) + F_2(g) \longrightarrow 2LiF(s)$$

EXPLORE YOUR WORLD

■ Oxidation of Fruits and Vegetables

Freshly cut surfaces of fruits and vegetables discolor when exposed to oxygen in the air. Cut three slices of a fruit or vegetable such as apple, potato, avocado, or banana. Leave one piece on the kitchen counter (uncovered). Wrap one piece in plastic wrap and leave on the kitchen counter. Dip one piece in lemon juice and leave uncovered.

QUESTIONS

1. What changes take place in each sample after 1–2 hours?
2. Why would wrapping fruits and vegetables slow the rate of discoloration?
3. If lemon juice contains vitamin C (an antioxidant), why would dipping a fruit or vegetable in lemon juice affect the oxidation reaction on the surface of the fruit or vegetable?
4. Other kinds of antioxidants are vitamin E, citric acid, and BHT. Look for these antioxidants on the labels of cereals, potato chips, and other foods in your kitchen. Why are antioxidants added to food products that will be stored on our kitchen shelves?

QUESTIONS AND PROBLEMS

■ Oxidation–Reduction Reactions

5.13 In the following reactions, identify what is oxidized, and what is reduced.
    a. $Zn(s) + Cl_2(g) \longrightarrow ZnCl_2(s)$
    b. $Cl_2(g) + 2NaBr(aq) \longrightarrow 2NaCl(aq) + Br_2(g)$
    c. $2PbO(s) \longrightarrow 2Pb(s) + O_2(g)$
    d. $2Fe^{3+}(aq) + Sn^{2+}(aq) \longrightarrow 2Fe^{2+}(aq) + Sn^{4+}(aq)$

5.14 In the following reactions, identify what is oxidized, and what is reduced.
    a. $2Li(s) + F_2(g) \longrightarrow 2LiF(s)$
    b. $Cl_2(g) + 2KI(aq) \longrightarrow 2KCl(aq) + I_2(g)$
    c. $Zn(g) + Cu^{2+}(aq) \longrightarrow Zn^{2+}(aq) + Cu(s)$
    d. $Fe(s) + CuSO_4(aq) \longrightarrow FeSO_4(aq) + Cu(s)$

5.15 In the mitochondria of human cells, energy for the production of ATP is provided by the oxidation and reduction reactions of the iron ions in the cytochromes of the electron transport chain. Identify each of the following reactions as oxidation or reduction.
    a. $Fe^{3+} + e^- \longrightarrow Fe^{2+}$
    b. $Fe^{2+} \longrightarrow Fe^{3+} + e^-$

5.16 Chlorine (Cl_2) is a strong germicide used to disinfect drinking water and to kill microbes in swimming pools. If the product is Cl^-, was the elemental chlorine oxidized or reduced?

5.5 The Mole

At the store, you buy eggs by the dozen. In an office, pencils are ordered by the gross and paper by the ream. In a restaurant, soda is ordered by the case. In each of these examples, the terms *dozen, gross, ream*, and *case* count the number of items present. For example, when you buy a dozen eggs, you know you will get 12 eggs in the carton.

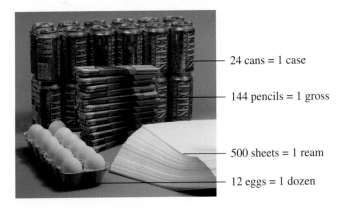

24 cans = 1 case

144 pencils = 1 gross

500 sheets = 1 ream

12 eggs = 1 dozen

■ Avogadro's Number

In chemistry, particles such as atoms, molecules, and ions are counted by the **mole**, a unit that contains 6.02×10^{23} items. This very large number, called **Avogadro's number** after Amedeo Avogadro, an Italian physicist, looks like this when written with 3 significant figures:

 Using Avogadro's Number

 Moles and the Chemical Formula

Avogadro's Number

$602\ 000\ 000\ 000\ 000\ 000\ 000\ 000 = 6.02 \times 10^{23}$

TABLE 5.3

Number of Particles in One-Mole Samples

Substance	Number and Type of Particles
1 mole aluminum	6.02×10^{23} aluminum atoms
1 mole sulfur	6.02×10^{23} sulfur atoms
1 mole water (H_2O)	6.02×10^{23} H_2O molecules
1 mole NaCl	6.02×10^{23} NaCl formula units
1 mole vitamin C ($C_6H_8O_6$)	6.02×10^{23} vitamin C molecules

One mole of any element always contains Avogadro's number of atoms. For example, 1 mole of carbon contains 6.02×10^{23} carbon atoms; 1 mole of aluminum contains 6.02×10^{23} aluminum atoms; 1 mole of sulfur contains 6.02×10^{23} sulfur atoms.

$$1 \text{ mole of an element} = 6.02 \times 10^{23} \text{ atoms of that element}$$

One mole of a compound contains Avogadro's number of molecules or formula units. Molecules are the particles of covalent compounds; **formula units** are the group of ions given by the formula of an ionic compound. For example, 1 mole of CO_2, a covalent compound, contains 6.02×10^{23} molecules of CO_2. One mole of NaCl, an ionic compound, contains 6.02×10^{23} formula units of NaCl (Na^+, Cl^-). Table 5.3 gives examples of the number of particles in some 1-mole quantities.

Avogadro's number tells us that one mole of any element or compound contains 6.02×10^{23} of the particular type of particles that make up a substance. We can use Avogadro's number as a conversion factor to convert between the moles of a substance and number of particles it contains.

$$\frac{6.02 \times 10^{23} \text{ particles}}{1 \text{ mole}} \quad \text{and} \quad \frac{1 \text{ mole}}{6.02 \times 10^{23} \text{ particles}}$$

For example, we use Avogadro's number to convert 4.00 moles iron to atoms of iron.

$$4.00 \text{ moles Fe atoms} \times \underbrace{\frac{6.02 \times 10^{23} \text{ Fe atoms}}{1 \text{ mole Fe atoms}}}_{\text{Avogadro's number as a conversion factor}} = 2.41 \times 10^{24} \text{ Fe atoms}$$

We can also use Avogadro's number to convert 3.01×10^{24} molecules CO_2 to moles of CO_2.

$$3.01 \times 10^{24} \text{ CO}_2 \text{ molecules} \times \underbrace{\frac{1 \text{ mole CO}_2 \text{ molecules}}{6.02 \times 10^{23} \text{ CO}_2 \text{ molecules}}}_{\text{Avogadro's number as a conversion factor}} = 5.00 \text{ moles CO}_2 \text{ molecules}$$

Guide to Calculating the Atoms or Molecules of a Substance

STEP 1
Determine the given number of moles.

STEP 2
Write a plan to convert moles to atoms or molecules.

STEP 3
Use Avogadro's number to write conversion factors.

STEP 4
Set up problem to convert given moles to atoms or molecules.

SAMPLE PROBLEM 5.5

■ **Calculating the Number of Molecules in a Mole**

How many molecules of ammonia, NH_3, are present in 1.75 moles ammonia?

SOLUTION

■ **STEP 1 Given** 1.75 moles NH_3 **Need** molecules of NH_3

■ **STEP 2 Plan** moles of NH_3 Avogadro's number molecules of NH_3

■ **STEP 3 Equalities/Conversion Factors**

$$1 \text{ mole NH}_3 = 6.02 \times 10^{23} \text{ molecules NH}_3$$

$$\frac{6.02 \times 10^{23} \text{ molecules NH}_3}{1 \text{ mole NH}_3} \quad \text{and} \quad \frac{1 \text{ mole NH}_3}{6.02 \times 10^{23} \text{ molecules NH}_3}$$

■ **STEP 4 Set Up Problem** Calculate the number of NH_3 molecules:

$$1.75 \text{ moles NH}_3 \times \frac{6.02 \times 10^{23} \text{ molecules NH}_3}{1 \text{ mole NH}_3} = 1.05 \times 10^{24} \text{ molecules NH}_3$$

STUDY CHECK

How many moles of water, H_2O, contain 2.60×10^{23} molecules of water?

Moles of Elements in a Formula

We have seen that the subscripts in a chemical formula of a compound indicate the number of atoms of each type of element. For example, in a molecule of aspirin, chemical formula $C_9H_8O_4$, there are 9 carbon atoms, 8 hydrogen atoms, and 4 oxygen atoms. The subscripts also state the number of moles of each element in one mole of aspirin: 9 moles of carbon atoms, 8 moles of hydrogen atoms, and 4 moles of oxygen atoms.

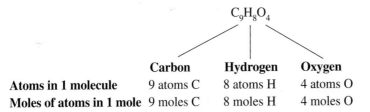

	Carbon	Hydrogen	Oxygen
Atoms in 1 molecule	9 atoms C	8 atoms H	4 atoms O
Moles of atoms in 1 mole	9 moles C	8 moles H	4 moles O

Using the subscripts from the aspirin formula, $C_9H_8O_4$, we can write the following conversion factors for each of the elements in 1 mole of aspirin.

$$\frac{9 \text{ moles C}}{1 \text{ mole C}_9\text{H}_8\text{O}_4} \qquad \frac{8 \text{ moles H}}{1 \text{ mole C}_9\text{H}_8\text{O}_4} \qquad \frac{4 \text{ moles O}}{1 \text{ mole C}_9\text{H}_8\text{O}_4}$$

$$\frac{1 \text{ mole C}_9\text{H}_8\text{O}_4}{9 \text{ moles C}} \qquad \frac{1 \text{ mole C}_9\text{H}_8\text{O}_4}{8 \text{ moles H}} \qquad \frac{1 \text{ mole C}_9\text{H}_8\text{O}_4}{4 \text{ moles O}}$$

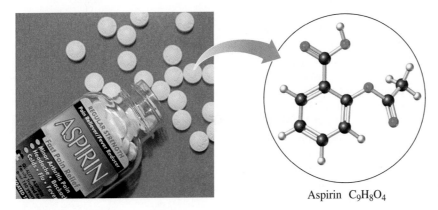

Aspirin $C_9H_8O_4$

Number of atoms in 1 molecule

Carbon (C)   Hydrogen (H)   Oxygen (O)

SAMPLE PROBLEM 5.6

■ Calculating the Moles of an Element in a Compound

How many moles of carbon atoms are present in 1.50 moles aspirin, $C_9H_8O_4$?

SOLUTION

■ **STEP 1** **Given**   1.50 moles $C_9H_8O_4$ **Need**   moles C atoms

■ **STEP 2** **Plan**   Moles $C_9H_8O_4$ $\boxed{\text{Subscript}}$ moles C atoms

■ **STEP 3** **Equalities/Conversion Factors**

$$1 \text{ mole } C_9H_8O_4 = 9 \text{ moles C atoms}$$

$$\frac{9 \text{ moles C}}{1 \text{ mole } C_9H_8O_4} \quad \text{and} \quad \frac{1 \text{ mole } C_9H_8O_4}{9 \text{ moles C}}$$

■ **STEP 4** **Set Up Problem**

$$1.50 \text{ moles } \cancel{C_9H_8O_4} \times \frac{9 \text{ moles C atoms}}{1 \text{ mole } \cancel{C_9H_8O_4}} = 13.5 \text{ moles C atoms}$$

STUDY CHECK

How many moles of aspirin $C_9H_8O_4$ contain 0.480 mole O atoms?

QUESTIONS AND PROBLEMS

■ The Mole

5.17 What is a mole?

5.18 What is Avogadro's number?

5.19 Consider the formula for quinine $C_{20}H_{24}N_2O_2$.
   a. How many moles of hydrogen are in 1.0 mole of quinine?
   b. How many moles of carbon are in 5.0 moles of quinine?
   c. How many moles of nitrogen are in 0.020 mole of quinine?

5.20 Consider the formula for $Al_2(SO_4)_3$, which is used in antiperspirants.
   a. How many moles of sulfur are present in 3.0 moles of $Al_2(SO_4)_3$?
   b. How many moles of aluminum ions are present in 0.40 mole $Al_2(SO_4)_3$?
   c. How many moles of sulfate ions (SO_4^{2-}) are present in 1.5 moles $Al_2(SO_4)_3$?

5.21 Calculate each of the following:
   a. number of C atoms in 0.500 mole C
   b. number of SO_2 molecules in 1.28 moles SO_2
   c. moles of Fe in 5.22×10^{22} atoms Fe
   d. moles of C_2H_5OH in 8.50×10^{24} molecules C_2H_5OH

5.22 Calculate each of the following:
   a. number of Li atoms in 4.5 moles Li
   b. number of CO_2 molecules in 0.0180 mole CO_2
   c. moles of Cu in 7.8×10^{21} atoms Cu
   d. moles of C_2H_6 in 3.754×10^{23} molecules C_2H_6

5.23 Calculate each of the following quantities in 2.00 moles H_3PO_4:
   a. moles of H   **b.** moles of O
   c. atoms of P   **d.** atoms of O

5.24 Calculate each of the following quantities in 0.185 mole $(C_3H_5)_2O$:
   a. moles of C   **b.** moles of O
   c. atoms of H   **d.** atoms of C

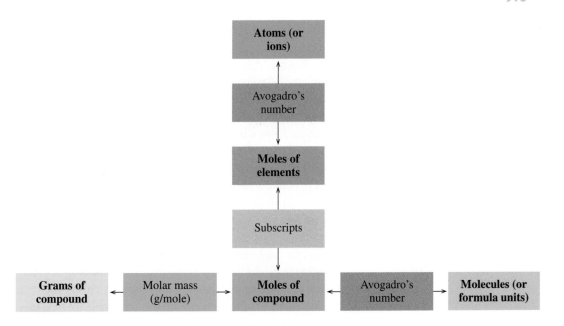

FIGURE 5.8

The moles of a compound are related to its mass in grams by molar mass, to the number of molecules (or formula units) by Avogadro's number, and to the moles of each element by the subscripts in the formula.

Q **What steps are needed to calculate the number of H atoms in 5.00 g CH_4?**

5.29 Calculate the number of grams in each of the following:
   a. 0.500 mole NaCl   **b.** 1.75 moles Na_2O **c.** 0.225 mole H_2O

5.30 Calculate the number of grams in each of the following:
   a. 2.0 moles $MgCl_2$ **b.** 3.5 moles C_3H_8 **c.** 5.00 moles C_2H_6O

5.31 a. The compound $MgSO_4$ is called Epsom salts. How many grams will you need to prepare a bath containing 5.00 moles of Epsom salts?
   b. In a bottle of soda, there is 0.25 mole CO_2. How many grams of CO_2 are in the bottle?

5.32 a. Cyclopropane, C_3H_6, is an anesthetic given by inhalation. How many grams are in 0.25 mole of cyclopropane?
   b. The sedative Demerol hydrochloride has the formula $C_{15}H_{22}ClNO_2$. How many grams are in 0.025 mole of Demerol hydrochloride?

5.33 How many moles are contained in each of the following?
   a. 50.0 g Ag   **b.** 0.200 g C
   c. 15.0 g NH_3 **d.** 75.0 g SO_2

5.34 How many moles are contained in each of the following?
   a. 25.0 g Ca   **b.** 5.00 g S
   c. 40.0 g H_2O **d.** 100.0 g O_2

5.35 A can of Drāno contains 480 g of NaOH. How many moles of NaOH are in the can of Drāno?

5.36 A gold nugget weighs 35.0 g. How many moles of gold are in the nugget?

5.37 How many moles of S are in each of the following quantities?
   a. 25 g S   **b.** 125 g SO_2 **c.** 2.0 moles Al_2S_3

5.38 How many moles of C are in each of the following quantities?
   a. 75 g C   **b.** 0.25 mole C_2H_6 **c.** 88 g CO_2

5.39 How many atoms of N are in each of the following quantities?
   a. 40.0 g N   **b.** 1.5 moles N_2O_4 **c.** 2.0 moles N_2

5.40 How many atoms of Ag are in each of the following quantities?
   a. 5.0 g Ag   **b.** 0.40 mole Ag_2S **c.** 0.75 g AgCl

**MC
GOB** Moles of Reactants and Products

5.7 Mole Relationships in Chemical Equations

In an earlier section, we saw that equations are balanced in terms of the numbers of each type of atom in the reactants and products. However, when experiments are done in the laboratory or medications prepared in the pharmacy, samples contain billions of atoms and molecules so it is impossible to count them individually. What can be measured conveniently is mass, using a balance. Because mass is related to the number of particles through the molar mass, measuring the mass is equivalent to counting the number of particles or moles.

Conservation of Mass

In any chemical reaction, the total amount of matter in the reactants is equal to the total amount of matter in the products. If all of the reactants were weighed, they would have a total mass equal to the total mass of the products. This is known as the *law of conservation of mass*, which says that there is no change in the total mass of the substances reacting in a chemical reaction. Thus, no material is lost or gained as original substances are changed to new substances.

For example, tarnish forms when silver reacts with sulfur to form silver sulfide.

$$2Ag(s) + S(s) \longrightarrow Ag_2S(s)$$

In this reaction, the number of silver atoms that react is two times the number of sulfur atoms. When 200 silver atoms react, 100 sulfur atoms are required. Normally, however, many more atoms would actually be present in this reaction. If we are dealing with molar amounts, then the coefficients in the equation can be interpreted in terms of moles. Thus, 2 moles silver react with each 1 mole sulfur. Since the molar mass of each can be determined, the quantities of silver and sulfur can also be stated in terms of mass in grams of each. Therefore, an equation for a chemical equation can be interpreted several ways, as seen in Table 5.5.

TABLE 5.5

Information Available from a Balanced Equation

		Reactants	Products
Equation	$2Ag(s)$	$+ S(s)$	$\longrightarrow Ag_2S(s)$
Atoms	2 Ag atoms	+ 1 S atom	$\longrightarrow$ 1 Ag$_2$S formula unit
	200 Ag atoms	+ 100 S atoms	$\longrightarrow$ 100 Ag$_2$S formula units
Avogadro's Number of atoms	$2(6.022 \times 10^{23})$ Ag atoms	$+ 1(6.02 \times 10^{23})$ S atoms	$\longrightarrow 1(6.02 \times 10^{23})$ Ag$_2$S formula units
Moles	2 moles Ag	+ 1 mole S	$\longrightarrow$ 1 mole Ag$_2$S
Mass (g)	$2(107.9 \text{ g})$ Ag	$+ 1(32.1 \text{ g})$ S	$\longrightarrow 1(247.9 \text{ g})$ Ag$_2$S
Total mass (g)	247.9 g		$\longrightarrow$ 247.9 g

$$2Ag(s) \quad + \quad S(s) \quad \longrightarrow \quad Ag_2S(s)$$

Mass of reactants $\qquad = \qquad$ Mass of products

Mole–Mole Factors from an Equation

When iron reacts with sulfur, the product is iron(III) sulfide.

$$2Fe(s) + 3S(s) \longrightarrow Fe_2S_3(s)$$

| Iron (Fe) | | Sulfur (S) | | Iron(III) sulfide (Fe$_2$S$_3$) |
| $2Fe(s)$ | $+$ | $3S(s)$ | $\longrightarrow$ | $Fe_2S_3(s)$ |

Because the equation is balanced, we know the proportions of iron and sulfur in the reaction. For this reaction, we see that 2 moles iron reacts with 3 moles sulfur to form 1 mole iron(III) sulfide. From the coefficients, we can write **mole–mole factors** between reactants and between reactants and products.

Fe and S: $\quad \dfrac{2 \text{ moles Fe}}{3 \text{ moles S}} \quad$ and $\quad \dfrac{3 \text{ moles S}}{2 \text{ moles Fe}}$

Fe and Fe$_2$S$_3$: $\quad \dfrac{2 \text{ moles Fe}}{1 \text{ mole Fe}_2\text{S}_3} \quad$ and $\quad \dfrac{1 \text{ mole Fe}_2\text{S}_3}{2 \text{ moles Fe}}$

S and Fe$_2$S$_3$: $\quad \dfrac{3 \text{ moles S}}{1 \text{ mole Fe}_2\text{S}_3} \quad$ and $\quad \dfrac{1 \text{ mole Fe}_2\text{S}_3}{3 \text{ moles S}}$

Using Mole–Mole Factors in Calculations

Whenever you prepare a recipe, adjust an engine for the proper mixture of fuel and air, or prepare medicines in a pharmaceutical laboratory, you need to know the proper amounts of reactants to use and how much of the product will form. Earlier we wrote all the possible conversion factors that can be obtained from the balanced equation: $2Fe(s) + 3S(s) \longrightarrow Fe_2S_3(s)$. Now we will show how mole–mole factors are used in chemical calculations.

SAMPLE PROBLEM 5.10

■ **Using Mole–Mole Factors**

In the reaction of iron and sulfur, how many moles of sulfur are needed to react with 6.0 moles iron?

$$2Fe(s) + 3S(s) \longrightarrow Fe_2S_3(s)$$

SOLUTION

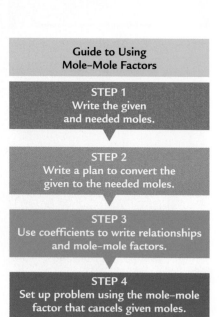

Guide to Using Mole–Mole Factors

STEP 1
Write the given and needed moles.

STEP 2
Write a plan to convert the given to the needed moles.

STEP 3
Use coefficients to write relationships and mole–mole factors.

STEP 4
Set up problem using the mole–mole factor that cancels given moles.

■ **STEP 1** **Write the given and needed number of moles.** In this problem, we need to find the number of moles of S that react with 6.0 moles Fe.

Given moles of Fe   **Need** moles of S

■ **STEP 2** **Write the plan to convert the given to the needed.**

moles of Fe → Mole–mole factor → moles of S

■ **STEP 3** **Use coefficients to write relationships and mole–mole factors.** Use coefficients to write the mole–mole factors for the given and needed substances.

$$2 \text{ moles Fe} = 3 \text{ moles S}$$

$$\frac{2 \text{ moles Fe}}{3 \text{ moles S}} \quad \text{and} \quad \frac{3 \text{ moles S}}{2 \text{ moles Fe}}$$

■ **STEP 4** **Set up problem using the mole–mole factor that cancels given moles.** Use a mole–mole factor to cancel the given moles and provide needed moles.

$$6.0 \text{ moles Fe} \times \frac{3 \text{ moles S}}{2 \text{ moles Fe}} = 9.0 \text{ moles S}$$

The answer is given with 2 significant figures because the given quantity 6.0 moles Fe has 2 SFs. The values in the mole–mole factor are exact.

STUDY CHECK

Using the equation in Sample Problem 5.10, calculate the number of moles of iron(III) sulfide produced from 2.7 moles sulfur.

QUESTIONS AND PROBLEMS

■ **Mole Relationships in Chemical Equations**

5.41 Write all of the mole–mole factors for each of the following equations.
  a. $2SO_2(g) + O_2(g) \longrightarrow 2SO_3(g)$
  b. $4P(s) + 5O_2(s) \longrightarrow 2P_2O_5(s)$

5.42 Write all of the mole–mole factors for each of the following the equations.
  a. $2Al(s) + 3Cl_2(g) \longrightarrow 2AlCl_3(s)$
  b. $4HCl(g) + O_2(g) \longrightarrow 2Cl_2(g) + 2H_2O(g)$

5.43 The reaction of hydrogen with oxygen produces water.

$$2H_2(g) + O_2(g) \longrightarrow 2H_2O(g)$$

a. How many moles of O_2 are required to react with 2.0 moles H_2?
b. If you have 5.0 moles O_2, how many moles of H_2 are needed for the reaction?
c. How many moles of H_2O form when 2.5 moles O_2 reacts?

5.44 Ammonia is produced by the reaction of hydrogen and nitrogen.

$$N_2(g) + 3H_2(g) \longrightarrow 2NH_3(g)$$
ammonia

a. How many moles of H_2 are needed to react with 1.0 mole N_2?
b. How many moles of N_2 reacted if 0.60 mole NH_3 is produced?
c. How many moles of NH_3 are produced when 1.4 moles H_2 reacts?

5.45 Carbon disulfide and carbon monoxide are produced when carbon is heated with sulfur dioxide.

$$5C(s) + 2SO_2(g) \longrightarrow CS_2(l) + 4CO(g)$$

a. How many moles of C are needed to react with 0.500 mole SO_2?
b. How many moles of CO are produced when 1.2 moles C reacts?
c. How many moles of SO_2 are required to produce 0.50 mole CS_2?
d. How many moles of CS_2 are produced when 2.5 moles C reacts?

5.46 In the acetylene torch, acetylene gas (C_2H_2) burns in oxygen to produce carbon dioxide and water.

$$2C_2H_2(g) + 5O_2(g) \longrightarrow 4CO_2(g) + 2H_2O(g)$$

a. How many moles of O_2 are needed to react with 2.00 moles C_2H_2?
b. How many moles of CO_2 are produced when 3.5 moles C_2H_2 reacts?
c. How many moles of C_2H_2 are required to produce 0.50 mole H_2O?
d. How many moles of CO_2 are produced from 0.100 mole O_2?

5.8 Mass Calculations for Reactions

When you perform a chemistry experiment in the laboratory, you use a laboratory balance to obtain a certain mass of reactant. From the mass in grams, you can determine the number of moles of reactant. By using mole–mole factors, you can predict the moles of product that can be produced. Then the molar mass of the product is used to convert the moles back into mass in grams. The following procedure can be used to set up and solve problems that involve calculations of quantities for substances in a chemical reaction.

SAMPLE PROBLEM 5.11

■ **Mass of Products from Moles of Reactant**

In the formation of smog, nitrogen reacts with oxygen to produce nitrogen oxide. Calculate the grams of NO produced when 1.50 moles O_2 react.

$$N_2(g) + O_2(g) \longrightarrow 2NO(g)$$

SOLUTION

■ **STEP 1 Given**   1.50 moles of O_2 **Need**   grams of NO
■ **STEP 2 Plan**

moles of O_2 → Mole–mole factor → moles of NO → Molar mass → grams of NO

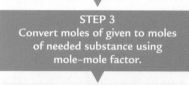

Guide to Calculating the Masses of Reactants and Products in a Chemical Reaction

STEP 1
Use molar mass to convert grams of given to moles (if necessary).

STEP 2
Write a mole–mole factor from the coefficients in the equation.

STEP 3
Convert moles of given to moles of needed substance using mole–mole factor.

STEP 4
Convert moles of needed substance to grams using molar mass.

■ **STEP 3** **Equalities/Conversion Factors** The mole–mole factor that converts moles of O_2 to moles of NO is derived from the coefficients in the balanced equation.

$$1 \text{ mole } O_2 = 2 \text{ moles NO}$$
$$\frac{2 \text{ moles NO}}{1 \text{ mole } O_2} \quad \text{and} \quad \frac{1 \text{ mole } O_2}{2 \text{ moles NO}}$$

$$1 \text{ mole NO} = 30.0 \text{ g}$$
$$\frac{30.0 \text{ g NO}}{1 \text{ mole NO}} \quad \text{and} \quad \frac{1 \text{ mole NO}}{30.0 \text{ g NO}}$$

■ **STEP 4** **Set Up Problem** First, we can change the given, 1.50 moles O_2, to moles of NO.

$$1.50 \text{ moles } O_2 \times \frac{2 \text{ moles NO}}{1 \text{ mole } O_2} = 3.00 \text{ moles NO}$$

Now, the moles of NO can be converted to grams of NO using its molar mass.

$$3.00 \text{ moles NO} \times \frac{30.0 \text{ g NO}}{1 \text{ mole NO}} = 90.0 \text{ g NO}$$

These two steps can also be written as a sequence of conversion factors that lead to the mass in grams of NO.

$$1.50 \text{ moles } O_2 \times \frac{2 \text{ moles NO}}{1 \text{ mole } O_2} \times \frac{30.0 \text{ g NO}}{1 \text{ mole NO}} = 90.0 \text{ g NO}$$

STUDY CHECK

Using the equation in Sample Problem 5.11, calculate the grams of NO that can be produced when 0.734 mole N_2 reacts.

■

SAMPLE PROBLEM 5.12

■ **Mass of Product from Mass of Reactant**

Acetylene (C_2H_2), used in welding, burns with oxygen.

$$2C_2H_2(g) + 5O_2(g) \longrightarrow 4CO_2(g) + 2H_2O(g)$$

How many grams of carbon dioxide are produced when 54.6 g C_2H_2 are burned?

SOLUTION

■ **STEP 1** **Given** grams of C_2H_2 **Need** grams of CO_2
■ **STEP 2** **Plan** Once we convert grams of C_2H_2 to moles of C_2H_2 using its molar mass, we can use a mole–mole factor to find the moles of CO_2. Then the molar mass of CO_2 will give us the grams of CO_2.

grams of C_2H_2 → Molar mass → moles of C_2H_2 → Mole–mole factor → moles of CO_2 → Molar mass → grams of CO_2

■ **STEP 3** **Equalities/Conversion Factors** We need the molar mass of C_2H_2 and CO_2. The mole–mole factor that converts moles of C_2H_2 to moles of CO_2 is derived from the coefficients in the balanced equation.

$$1 \text{ mole } C_2H_2 = 26.0 \text{ g}$$
$$\frac{26.0 \text{ g } C_2H_2}{1 \text{ mole } C_2H_2} \quad \text{and} \quad \frac{1 \text{ mole } C_2H_2}{26.0 \text{ g } C_2H_2}$$

$$2 \text{ moles } C_2H_2 = 4 \text{ moles } CO_2$$
$$\frac{2 \text{ moles } C_2H_2}{4 \text{ moles } CO_2} \quad \text{and} \quad \frac{4 \text{ moles } CO_2}{2 \text{ moles } C_2H_2}$$

$$1 \text{ mole } CO_2 = 44.0 \text{ g}$$
$$\frac{44.0 \text{ g } CO_2}{1 \text{ mole } CO_2} \quad \text{and} \quad \frac{1 \text{ mole } CO_2}{44.0 \text{ g } CO_2}$$

■ **STEP 4 Set Up Problem** Using our plan, we first convert grams of C_2H_2 to moles of C_2H_2.

$$54.6 \text{ g } C_2H_2 \times \frac{1 \text{ mole } C_2H_2}{26.0 \text{ g } C_2H_2} = 2.10 \text{ moles } C_2H_2$$

Then we change moles of C_2H_2 to moles of CO_2 by using the mole–mole factor.

$$2.10 \text{ moles } C_2H_2 \times \frac{4 \text{ moles } CO_2}{2 \text{ moles } C_2H_2} = 4.20 \text{ moles } CO_2$$

Finally, we can convert moles of CO_2 to grams of CO_2.

$$4.20 \text{ moles } CO_2 \times \frac{44.0 \text{ g } CO_2}{1 \text{ mole } CO_2} = 185 \text{ g } CO_2$$

The solution can be obtained using the conversion factors in sequence.

$$54.6 \text{ g } C_2H_2 \times \frac{1 \text{ mole } C_2H_2}{26.0 \text{ g } C_2H_2} \times \frac{4 \text{ moles } CO_2}{2 \text{ moles } C_2H_2} \times \frac{44.0 \text{ g } CO_2}{1 \text{ mole } CO_2} = 185 \text{ g } C$$

STUDY CHECK

Using the equation in Sample Problem 5.12, calculate the grams of CO_2 that can be produced when 25.0 g O_2 reacts.

QUESTIONS AND PROBLEMS

■ **Mass Calculations for Reactions**

5.47 Sodium reacts with oxygen to produce sodium oxide.

$$4Na(s) + O_2(g) \longrightarrow 2Na_2O(s)$$

a. How many grams of Na_2O are produced when 2.50 moles Na react?
b. If you have 18.0 g Na, how many grams of O_2 are required for reaction?
c. How many grams of O_2 are needed in a reaction that produces 75.0 g Na_2O?

5.48 Nitrogen gas reacts with hydrogen gas to produce ammonia by the following equation:

$$N_2(g) + 3H_2(g) \longrightarrow 2NH_3(g)$$

a. If you have 1.80 moles H_2, how many grams of NH_3 can be produced?
b. How many grams of H_2 are needed to react with 2.80 g N_2?
c. How many grams of NH_3 can be produced from 12.0 g H_2?

5.49 Ammonia and oxygen react to form nitrogen and water.

$$4NH_3(g) + 3O_2(g) \longrightarrow 2N_2(g) + 6H_2O(g)$$

Ammonia

a. How many grams of O_2 are needed to react with 8.00 moles NH_3?
b. How many grams of N_2 can be produced when 6.50 g O_2 reacts?
c. How many grams of water are formed from the reaction of 34.0 g NH_3?

5.50 Iron(III) oxide reacts with carbon to give iron and carbon monoxide.

$$Fe_2O_3(s) + 3C(s) \longrightarrow 2Fe(s) + 3CO(g)$$

a. How many grams of C are required to react with 2.50 moles Fe_2O_3?
b. How many grams of CO are produced when 36.0 g C reacts?
c. How many grams of Fe can be produced when 6.00 g Fe_2O_3 reacts?

5.51 Nitrogen dioxide and water react to produce nitric acid HNO_3 and nitrogen monoxide.

$$3NO_2(g) + H_2O(l) \longrightarrow 2HNO_3(aq) + NO(g)$$

a. How many grams of H_2O are required to react with 28.0 g NO_2?
b. How many grams of NO are obtained from 15.8 g NO_2?
c. How many grams of HNO_3 are produced from 8.25 g NO_2?

5.52 Calcium cyanamide reacts with water to form calcium carbonate and ammonia.

$$CaCN_2(s) + 3H_2O(l) \longrightarrow CaCO_3(s) + 2NH_3(g)$$

a. How many grams of water are needed to react with 75.0 g $CaCN_2$?
b. How many grams of NH_3 are produced from 5.24 g $CaCN_2$?
c. How many grams of $CaCO_3$ form if 155 g water react?

5.53 When the ore lead(II) sulfide burns in oxygen, the products are lead(II) oxide and sulfur dioxide.
a. Write the balanced equation for the reaction.
b. How many grams of oxygen are required to react with 0.125 mole lead(II) sulfide?
c. How many grams of sulfur dioxide can be produced when 65.0 g lead(II) sulfide react?
d. How many grams of lead(II) sulfide are used to produce 128 g lead(II) oxide?

5.54 When the gases dihydrogen sulfide and oxygen react, they form the gases sulfur dioxide and water vapor.
a. Write the balanced equation for the reaction.
b. How many grams of oxygen are required to react with 2.50 g dihydrogen sulfide?
c. How many grams of sulfur dioxide can be produced when 38.5 g oxygen react?
d. How many grams of oxygen are required to produce 55.8 g water vapor?

LEARNING GOAL

Given the actual quantity of product, determine the percent yield for a reaction. Identify a limiting reactant when given the quantities of two or more reactants; calculate the amount of product formed from the limiting reactant.

(MC GOB) What Will Run Out First?

5.9 Percent Yield and Limiting Reactants

Up to this point, we have done calculations as though the amount of product were the maximum quantity possible or 100%. In other words, we assumed that all of the reactants were changed completely to product. While this would be an ideal situation, it does not usually happen. As we run a reaction and transfer products from one container to another, some product is lost. There may also be side reactions that use up some of the reactants to give a different product. Thus, in a real experiment, the predicted amount of the desired product is never really obtained.

Suppose we are running a chemical reaction in the laboratory. We first measure out specific quantities of the reactants and place them in a reaction flask. Then we calculate the **theoretical yield** for the reaction, which is the amount of product we could expect if all the reactants were converted to product according to the mole ratios of the equation. The **actual yield** is the amount of product we collect when the reaction ends. Because some product is lost, the actual yield is always less than the theoretical yield. If we know the actual yield and the theoretical yield for a product, we can express the actual yield as a **percent yield**.

$$\text{Percent yield}(\%) = \frac{\text{actual yield}}{\text{theoretical yield}} \times 100\%$$

SAMPLE PROBLEM 5.13

■ Calculating Percent Yield

On a spaceship, LiOH is used to absorb exhaled CO_2 from breathing air.

$$LiOH(s) + CO_2(g) \longrightarrow LiHCO_3(s)$$

What is the percent yield of the reaction if 50.0 g LiOH gives 72.8 g $LiHCO_3$?

SOLUTION

■ **STEP 1** **Given** 50.0 g LiOH and 72.8 g $LiHCO_3$ (actually produced)
Need % yield $LiHCO_3$

■ **STEP 2** **Plan**
Calculation of theoretical yield:

grams of LiOH → Molar mass → moles of LiOH → Mole–mole factor → moles of $LiHCO_3$ → Molar mass → grams of $LiHCO_3$

Calculation percent yield:

$$\frac{\text{Actual yield}}{\text{Theoretical yield}} \times 100\%$$

■ **STEP 3** **Equalities/Conversion Factors**

$$1 \text{ mole LiOH} = 24.0 \text{ g LiOH}$$
$$\frac{1 \text{ mole LiOH}}{24.0 \text{ g LiOH}} \quad \text{and} \quad \frac{24.0 \text{ g LiOH}}{1 \text{ mole LiOH}}$$

$$1 \text{ mole LiHCO}_3 = 1 \text{ mole LiOH}$$
$$\frac{1 \text{ mole LiHCO}_3}{1 \text{ mole LiOH}} \quad \text{and} \quad \frac{1 \text{ mole LiOH}}{1 \text{ mole LiHCO}_3}$$

$$1 \text{ mole LiHCO}_3 = 68.0 \text{ g LiHCO}_3$$
$$\frac{68.0 \text{ g LiHCO}_3}{1 \text{ mole LiHCO}_3} \quad \text{and} \quad \frac{1 \text{ mole LiHCO}_3}{68.0 \text{ g LiHCO}_3}$$

■ **STEP 4** **Set Up Problem**
Calculation of theoretical yield:

$$50.0 \text{ g } \cancel{\text{LiOH}} \times \frac{1 \text{ mole } \cancel{\text{LiOH}}}{24.0 \text{ g } \cancel{\text{LiOH}}} \times \frac{1 \text{ mole } \cancel{\text{LiHCO}_3}}{1 \text{ mole } \cancel{\text{LiOH}}}$$

$$\times \frac{68.0 \text{ g LiHCO}_3}{1 \text{ mole } \cancel{\text{LiHCO}_3}} = 142 \text{ g LiHCO}_3$$

Calculation of percent yield:

$$\frac{\text{Actual yield (given)}}{\text{Theoretical yield (calculated)}} \times 100\% = \frac{72.8 \text{ g LiHCO}_3}{142 \text{ g LiHCO}_3} \times 100\% = 51.3\%$$

A percent yield of 51.3% means that 72.8 g of the theoretical amount of 142 g $LiHCO_3$ was actually produced by the reaction.

STUDY CHECK

For the reaction in Sample Problem 5.13, what is the percent yield if 8.00 g CO_2 produces 10.5 g $LiHCO_3$?

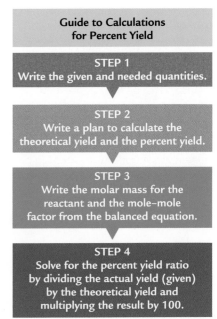

Guide to Calculations for Percent Yield

STEP 1
Write the given and needed quantities.

STEP 2
Write a plan to calculate the theoretical yield and the percent yield.

STEP 3
Write the molar mass for the reactant and the mole–mole factor from the balanced equation.

STEP 4
Solve for the percent yield ratio by dividing the actual yield (given) by the theoretical yield and multiplying the result by 100.

Limiting Reactants

When you make a peanut butter sandwich for lunch, you need 2 slices of bread and 1 tablespoon of peanut butter for each sandwich. As an equation, we could write:

2 slices of bread + 1 tablespoon peanut butter ⟶ 1 peanut butter sandwich

If you have 8 slices of bread, but only one tablespoon of peanut butter, you will run out of peanut butter after you make just one peanut butter sandwich. The amount of peanut butter available has limited the number of sandwiches you can make. You cannot make any more sandwiches once the peanut butter is used up, even though there are several slices of bread left over. On a different day, you might have a full jar of peanut butter, and 8 slices of bread. This time you can only make 4 peanut butter sandwiches before you run out of bread. This time the amount of bread is limiting. You cannot make any more sandwiches after the bread is gone even though there is still peanut butter remaining in the jar.

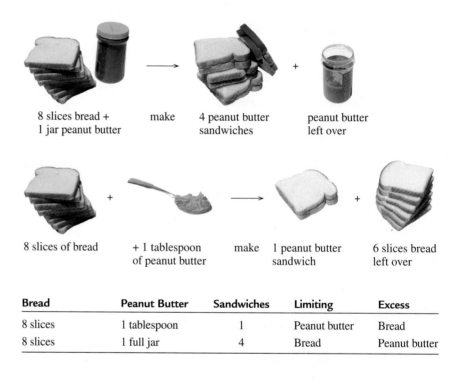

| 8 slices bread + 1 jar peanut butter | make | 4 peanut butter sandwiches | | peanut butter left over |

| 8 slices of bread | + 1 tablespoon of peanut butter | make | 1 peanut butter sandwich | 6 slices bread left over |

Bread	Peanut Butter	Sandwiches	Limiting	Excess
8 slices	1 tablespoon	1	Peanut butter	Bread
8 slices	1 full jar	4	Bread	Peanut butter

Calculating Moles of Product from a Limiting Reactant

In a similar way, the availability of reactants in a chemical reaction can limit the amount of product that forms. In many reactions, the reactants are not combined in quantities that allow each to be used up at exactly the same time. Then one reactant is used up before the other. The reactant that is completely used up is called the **limiting reactant**. The other reactant, called the **excess reactant**, is left over. In the last analogy, the bread was the limiting reactant and the jar of peanut butter was the excess reactant.

Consider the reaction in which hydrogen and chlorine form hydrogen chloride.

$$H_2(g) + Cl_2(g) \longrightarrow 2HCl(g)$$

Suppose the reaction mixture contains 2 moles H_2 and 5 moles Cl_2. From the equation, we see that one mole of hydrogen reacts with one mole of chlorine to produce two moles of hydrogen chloride. Now we need to calculate the amount of product that is possible from each of the reactants. We are looking for the limiting reactant, which is the one that produces the smaller amount of product.

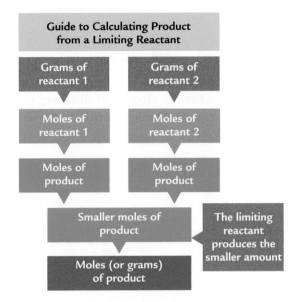

Moles of HCl from H_2

$$2 \text{ moles } H_2 \times \frac{2 \text{ moles HCl}}{1 \text{ mole } H_2} = 4 \text{ moles HCl (smaller amount of product)}$$

Moles of HCl from Cl_2

$$5 \text{ moles } Cl_2 \times \frac{2 \text{ moles HCl}}{1 \text{ mole } Cl_2} = 10 \text{ moles HCl (not possible)}$$

In this reaction mixture, H_2 is the limiting reactant. When 2 moles H_2 are used up, the reaction stops. The excess reactant is the 3 moles Cl_2 left. We can show the changes in each reactant and the product as follows.

	Reactants		**Product**
Equation	H_2 +	Cl_2 $\longrightarrow$	**2HCl**
Initial moles	2 moles	5 moles	0
Moles used/formed	−2 moles	−2 moles	+4 moles
Moles left	0	3 moles	4 moles
Identify as	Limiting reactant	Excess reactant	Product possible

Calculating Mass of Product from a Limiting Reactant

When the quantities of the reactants are given in grams, they must first be converted to moles. Once the limiting reactant is determined, the smaller number of moles of product is converted to grams using molar mass. This calculation is shown in Sample Problem 5.14.

SAMPLE PROBLEM 5.14

■ Mass of Product from a Limiting Reactant

Carbon monoxide and hydrogen gas react to form methanol, CH_3OH.

$$CO(g) + 2H_2(g) \longrightarrow CH_3OH(l)$$

If 48.0 g CO and 10.0 g H_2 react, how many grams of methanol can be produced?

SOLUTION

- ■ **STEP 1** **Given**   48.0 g CO and 10.0 g H_2 **Need**   g CH_3OH

- ■ **STEP 2** **Plan**   Convert the grams of each reactant to moles and calculate the moles of CH_3OH that each reactant can produce. Then convert the number of moles of CH_3OH from the limiting reactant to grams of CH_3OH using molar mass.

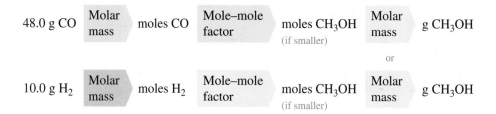

- ■ **STEP 3** **Equalities and Conversion Factors**

1 mole CO = 28.0 g CO	1 mole CO = 1 mole CH_3OH
$\dfrac{1 \text{ mole CO}}{28.0 \text{ g CO}}$ and $\dfrac{28.0 \text{ g CO}}{1 \text{ mole CO}}$	$\dfrac{1 \text{ mole CO}}{1 \text{ mole CH}_3\text{OH}}$ and $\dfrac{1 \text{ mole CH}_3\text{OH}}{1 \text{ mole CO}}$

1 mole H_2 = 2.02 g H_2	2 moles H_2 = 1 mole CH_3OH
$\dfrac{1 \text{ mole H}_2}{2.02 \text{ g H}_2}$ and $\dfrac{2.02 \text{ g H}_2}{1 \text{ mole H}_2}$	$\dfrac{2 \text{ moles H}_2}{1 \text{ mole CH}_3\text{OH}}$ and $\dfrac{1 \text{ mole CH}_3\text{OH}}{2 \text{ moles H}_2}$

1 mole CH_3OH = 32.0 g CH_3OH
$\dfrac{1 \text{ mole CH}_3\text{OH}}{32.0 \text{ g CH}_3\text{OH}}$ and $\dfrac{32.0 \text{ g CH}_3\text{OH}}{1 \text{ mole CH}_3\text{OH}}$

- ■ **STEP 4** **Set Up Problem**   The moles of CH_3OH from each reactant can now be determined in separate calculations.

Grams CH_3OH produced from CO

$$48.0 \text{ g CO} \times \frac{1 \text{ mole CO}}{28.0 \text{ g CO}} \times \frac{1 \text{ mole CH}_3\text{OH}}{1 \text{ mole CO}} = \textbf{1.71 moles CH}_3\textbf{OH}$$
(smaller number of moles)

Grams CH_3OH produced from H_2

$$10.0 \text{ g H}_2 \times \frac{1 \text{ mole H}_2}{2.02 \text{ g H}_2} \times \frac{1 \text{ mole CH}_3\text{OH}}{2 \text{ moles H}_2} = 2.48 \text{ moles CH}_3\text{OH}$$

Because CO produces the smaller number of moles of CH_3OH, CO is the limiting reactant. Now the grams of product CH_3OH from these reactants is calculated by converting the moles of CH_3OH obtained from CO to grams using molar mass.

$$1.71 \text{ moles CH}_3\text{OH} \times \frac{32.0 \text{ g CH}_3\text{OH}}{1 \text{ mole CH}_3\text{OH}} = 54.7 \text{ g CH}_3\text{OH}$$

STUDY CHECK

When silicon dioxide (sand) and carbon are heated, the ceramic material silicon carbide SiC and carbon monoxide are produced. How many grams of SiC are formed from 20.0 g SiO_2 and 50.0 g C?

$$SiO_2(s) + 3C(s) \xrightarrow{\text{heat}} SiC(s) + 2CO(g)$$

QUESTIONS AND PROBLEMS

▪ **Percent Yield and Limiting Reactants**

5.55 Carbon disulfide is produced by the reaction of carbon and sulfur dioxide.

$$5C(s) + 2SO_2(g) \longrightarrow CS_2(g) + 4CO(g)$$

a. What is the percent yield for the reaction if 40.0 g carbon produces 36.0 g carbon disulfide?
b. What is the percent yield for the reaction if 32.0 g sulfur dioxide produces 12.0 g carbon disulfide?

5.56 Iron(III) oxide reacts with carbon monoxide to produce iron and carbon dioxide.

$$Fe_2O_3(s) + 3CO(g) \longrightarrow 2Fe(s) + 3CO_2(g)$$

a. What is the percent yield for the reaction if 65.0 g iron(III) oxide produces 15.0 g iron?
b. What is the percent yield for the reaction if 75.0 g carbon monoxide produces 15.0 g carbon dioxide?

5.57 Aluminum reacts with oxygen to produce aluminum oxide.

$$4Al(s) + 3O_2(g) \longrightarrow 2Al_2O_3(s)$$

The reaction of 50.0 g aluminum and sufficient oxygen has a 75.0% yield. How many grams of aluminum oxide are produced?

5.58 Propane (C_3H_8) burns in oxygen to produce carbon dioxide and water.

$$C_3H_8(g) + 5O_2(g) \longrightarrow 3CO_2(g) + 4H_2O(g)$$

Calculate the mass of CO_2 that can be produced if the reaction of 45.0 g propane and sufficient oxygen has a 60.0% yield.

5.59 When 30.0 g carbon are heated with silicon dioxide, 28.2 g carbon monoxide are produced. What is the percent yield of this reaction?

$$SiO_2(s) + 3C(s) \longrightarrow SiC(s) + 2CO(g)$$

5.60 Calcium and nitrogen react to form calcium nitride.

$$3Ca(s) + N_2(g) \longrightarrow Ca_3N_2(s)$$

If 56.6 g calcium are mixed with nitrogen gas and 32.4 g calcium nitride are produced, what is the % yield of the reaction?

5.61 A taxi company has 10 taxis.
   a. On a certain day, only 8 taxi drivers show up for work. How many taxis can be used to pick up passengers?
   b. On another day, 10 taxi drivers show up for work but 3 taxis are in the repair shop. How many taxis can be driven?

5.62 A clock maker has 15 clock faces. Each clock requires 1 face and 2 hands.
   a. If the clock maker has 42 hands, how many clocks can be produced?
   b. If the clock maker has only 8 hands, how many clocks can be produced?

5.63 Nitrogen and hydrogen react to form ammonia.

$$N_2(g) + 3H_2(g) \longrightarrow 2NH_3(g)$$

Determine the limiting reactant in each of the following mixtures of reactants.
   a. 3.0 moles N_2 and 5.0 moles H_2
   b. 8.0 moles N_2 and 4.0 moles H_2
   c. 3.0 moles N_2 and 12.0 moles H_2

5.64 Iron and oxygen react to form iron(III) oxide.

$$4Fe(s) + 3O_2(g) \longrightarrow 2Fe_2O_3(s)$$

Determine the limiting reactant in each of the following mixtures of reactants.
   a. 2.0 moles Fe and 6.0 moles O_2
   b. 5.0 moles Fe and 4.0 moles O_2
   c. 16.0 moles Fe and 20.0 moles O_2

5.65 For each of the following reactions, calculate the moles of indicated product produced when 2.00 moles of each reactant is used.
   a. $2SO_2(g) + O_2(g) \longrightarrow 2SO_3(g)$ (SO_3)
   b. $3Fe(s) + 4H_2O(l) \longrightarrow Fe_3O_4(s) + 4H_2(g)$ (Fe_3O_4)
   c. $C_7H_{16}(g) + 11O_2(g) \longrightarrow 7CO_2(g) + 8H_2O(g)$ (CO_2)

5.66 For each of the following reactions, calculate the moles of indicated product produced when 3.00 moles of each reactant is used.
   a. $4Li(s) + O_2(g) \longrightarrow 2Li_2O(s)$ (Li_2O)
   b. $Fe_2O_3(s) + 3H_2(g) \longrightarrow 2Fe(s) + 3H_2O(l)$ (Fe)
   c. $Al_2S_3(s) + 6H_2O(l) \longrightarrow 2Al(OH)_3(aq) + 3H_2S(g)$ (H_2S)

5.67 For each of the following reactions, calculate the moles of indicated product produced when 20.0 g of each reactant is used.
   a. $2Al(s) + 3Cl_2(g) \longrightarrow 2AlCl_3(s)$ ($AlCl_3$)
   b. $4NH_3(g) + 5O_2(g) \longrightarrow 4NO(g) + 6H_2O(g)$ (H_2O)
   c. $CS_2(g) + 3O_2(g) \longrightarrow CO_2(g) + 2SO_2(g)$ (SO_2)

5.68 For each of the following reactions, calculate the moles of indicated product produced when 20.0 g of each reactant is used.
   a. $4Al(s) + 3O_2(g) \longrightarrow 2Al_2O_3(s)$ (Al_2O_3)
   b. $3NO_2(g) + H_2O(l) \longrightarrow 2HNO_3(aq) + NO(g)$ (HNO_3)
   c. $4NH_3(g) + 5O_2(g) \longrightarrow 4NO(g) + 6H_2O(g)$ (H_2O)

CONCEPT MAP

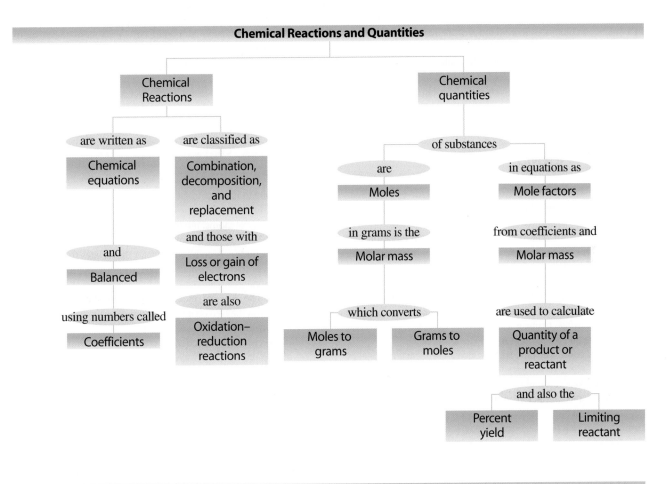

CHAPTER REVIEW

5.1 Chemical Changes

A chemical change occurs when the atoms of the initial substances rearrange to form new substances. When new substances form, a chemical reaction has taken place.

5.2 Chemical Equations

A chemical equation shows the formulas of the substances that react on the left side of a reaction arrow and the products that form on the right side of the reaction arrow. An equation is balanced by writing the smallest whole numbers (coefficients) in front of formulas to equalize the atoms of each element in the reactants and the products.

5.3 Types of Reactions

Many chemical reactions can be organized by reaction type: combination, decomposition, single replacement, or double replacement.

5.4 Oxidation–Reduction Reactions

When electrons are transferred in a reaction, it is an oxidation–reduction reaction. One reactant loses electrons, and another reactant gains electrons. Overall, the number of electrons lost and gained is equal.

5.5 The Mole

One mole of an element contains 6.02×10^{23} atoms; a mole of a compound contains 6.02×10^{23} molecules or formula units.

5.6 Molar Mass

The molar mass (g/mole) of any substance is the mass in grams equal numerically to its atomic mass, or the sum of the atomic masses, which have been multiplied by their subscripts in a formula. It becomes a conversion factor when it is used to change a quantity in grams to moles, or to change a given number of moles to grams.

5.7 Mole Relationships in Chemical Equations

In a balanced equation, the total mass of the reactants is equal to the total mass of the products. The coefficients in an equation describing the relationship between the moles of any two components are used to write mole–mole factors. When the number of moles for one substance is known, a mole–mole factor is used to find the moles of a different substance in the reaction.

5.8 Mass Calculations for Reactions

In calculations using equations, the molar masses of the substances and their mole factors are used to change the number of grams of one substance to the corresponding grams of a different substance.

5.9 Percent Yield and Limiting Reactants

The percent yield of a reaction indicates the percent of product actually produced during a reaction. The percent yield is calculated by dividing the actual yield in grams of a product by the theoretical yield in grams. A limiting reactant is the reactant that produces the smaller amount of product while some excess reactant is left over. When the mass of two or more reactants is given, the mass of a product is calculated from the product produced by the limiting reactant.

KEY TERMS

actual yield The actual amount of product produced by a reaction.

Avogadro's number The number of items in a mole, equal to 6.02×10^{23}.

balanced equation The final form of a chemical equation that shows the same number of atoms of each element in the reactants and products.

chemical change The formation of a new substance with a different composition and properties than the initial substance.

chemical equation A shorthand way to represent a chemical reaction using chemical formulas to indicate the reactants and products and coefficients to show reacting ratios.

chemical reaction The process by which a chemical change takes place.

coefficients Whole numbers placed in front of the formulas to balance the number of atoms or moles of atoms of each element on both sides of an equation.

combination reaction A reaction in which reactants combine to form a single product.

decomposition reaction A reaction in which a single reactant splits into two or more simpler substances.

double replacement reaction A reaction in which parts of two different reactants exchange places.

excess reactant The reactant that remains when the limiting reactant is used up in a reaction.

formula unit The group of ions represented by the formula of an ionic compound.

limiting reactant The reactant used up during a chemical reaction, which limits the amount of product that can form.

molar mass The mass in grams of 1 mole of an element equal numerically to its atomic mass. The molar mass of a compound is equal to the sum of the masses of the elements in the formula.

mole A group of atoms, molecules, or formula units that contains 6.02×10^{23} of these items.

mole–mole factor A conversion factor that relates the number of moles of two compounds derived from the coefficients in an equation.

oxidation The loss of electrons by a substance. Biological oxidation may involve the addition of oxygen, or the loss of hydrogen.

oxidation–reduction reaction A reaction in which the oxidation of one reactant is always accompanied by the reduction of another reactant.

percent yield The ratio of the actual yield of a reaction to the theoretical yield possible for the reaction, multiplied by 100.

physical change A change in which the physical appearance of a substance changes but the chemical composition stays the same.

products The substances formed as a result of a chemical reaction.

reactants The initial substances that undergo change in a chemical reaction.

reduction The gain of electrons by a substance. Biological reduction may involve the loss of oxygen or the gain of hydrogen.

single replacement reaction An element replaces a different element in a compound.

theoretical yield The maximum amount of product that a reaction can produce from a given amount of reactant.

UNDERSTANDING THE CONCEPTS

5.69 Identify some of the physical and chemical changes of a burning wax candle.

5.70 In each of the following, identify the change as physical or chemical.

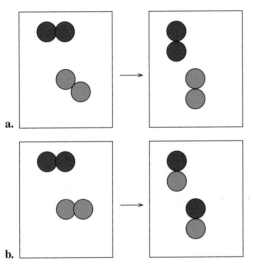

a.

b.

5.71 Balance each of the following by adding coefficients and identify the type of reaction for each.

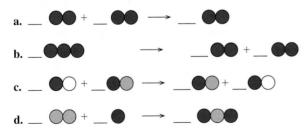

a.

b.

c.

d.

5.72 Allyl sulfide $(C_3H_5)_2S$ is the substance that gives garlic its characteristic odor.
   a. How many moles of sulfur are in 23.2 g $(C_3H_5)_2S$?
   b. How many hydrogen atoms are in 0.75 mole $(C_3H_5)_2S$?
   c. How many grams of carbon are in 4.20×10^{23} molecules $(C_3H_5)_2S$?

d. How many carbon atoms are in 15.0 g $(C_3H_5)_2S$?

5.73 In an experiment, a piece of copper is weighed and allowed to react with oxygen.

$$2Cu(s) + O_2(g) \longrightarrow 2CuO(s)$$

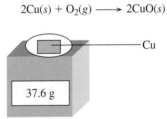

Cu

37.6 g

   a. How many grams of CuO could be produced according to the equation?
   b. How many moles of O_2 are required to completely react the Cu?

5.74 Propane gas, C_3H_8, a fuel for many barbecues, reacts with oxygen according to the following *unbalanced* equation:

$$C_3H_8(g) + O_2(g) \longrightarrow CO_2(g) + H_2O(g)$$

   a. If 5.50 g propane and 11.0 g oxygen are mixed together, how many grams carbon dioxide can be produced?
   b. If the actual mass of carbon dioxide is 7.40 g, what is the percent yield?

ADDITIONAL QUESTIONS AND PROBLEMS

MC GOB For instructor-assigned homework, go to **www.masteringgob.com**.

5.75 Balance each of the following unbalanced equations and identify the type of reaction.
  a. $NH_3(g) + HCl(g) \longrightarrow NH_4Cl(s)$
  b. $Fe_3O_4(s) + H_2(g) \longrightarrow Fe(s) + H_2O(g)$
  c. $Sb(s) + Cl_2(g) \longrightarrow SbCl_3(s)$
  d. $NI_3(s) \longrightarrow N_2(g) + I_2(g)$
  e. $KBr(aq) + Cl_2(aq) \longrightarrow KCl(aq) + Br_2(l)$
  f. $Al_2(SO_4)_3(aq) + NaOH(aq) \longrightarrow$
  $$Na_2SO_4(aq) + Al(OH)_3(s)$$

5.76 Balance each of the following unbalanced equations and identify the type of reaction.
  a. $Li_3N(s) \longrightarrow Li(s) + N_2(g)$
  b. $Mg(s) + N_2(g) \longrightarrow Mg_3N_2(s)$
  c. $Mg(s) + H_3PO_4(aq) \longrightarrow Mg_3(PO_4)_2(s) + H_2(g)$
  d. $Cr_2O_3(s) + H_2(g) \longrightarrow Cr(s) + H_2O(g)$
  e. $Al(s) + Cl_2(g) \longrightarrow AlCl_3(s)$
  f. $MgCl_2(aq) + AgNO_3(aq) \longrightarrow$
  $$Mg(NO_3)_2(aq) + AgCl(s)$$

5.77 Identify each of the following as an oxidation or a reduction reaction.
  a. $Zn^{2+} + 2e^- \longrightarrow Zn$
  b. $Al \longrightarrow Al^{3+} + 3e^-$
  c. $Pb \longrightarrow Pb^{2+} + 2e^-$
  d. $Cl_2 + 2e^- \longrightarrow 2Cl^-$

5.78 Write a balanced chemical equation for each of the following oxidation–reduction reactions.
  a. Sulfur reacts with molecular chlorine to form sulfur dichloride.
  b. Molecular chlorine and sodium bromide react to form molecular bromine and sodium chloride.
  c. Aluminum metal and iron(III) oxide react to produce aluminum oxide and elemental iron.
  d. Copper(II) oxide reacts with elemental C to form elemental copper and carbon dioxide.

5.79 During heavy exercise and workouts, lactic acid, $C_3H_6O_3$, accumulates in the muscles where it can cause pain and soreness.
  a. What is the molar mass of lactic acid?
  b. How many molecules are in 0.500 mole lactic acid?
  c. How many C atoms are in 1.50 moles lactic acid?
  d. How many grams of lactic acid contain 4.5×10^{24} atoms of O?

5.80 Ibuprofen, the anti-inflammatory ingredient in Advil, has the formula $C_{13}H_{18}O_2$.
  a. What is the molar mass of ibuprofen?
  b. How many molecules are in 0.200 mole ibuprofen?
  c. How many H atoms are in 0.100 mole ibuprofen?
  d. How many grams of ibuprofen contain 7.4×10^{25} atoms of C?

5.81 Calculate the molar mass of each of the following:
  a. $FeSO_4$, ferrous sulfate, iron supplement
  b. $Ca(IO_3)_2$, calcium iodate, iodine source in table salt
  c. $C_5H_8NNaO_4$, monosodium glutamate, flavor enhancer

5.82 Calculate the molar mass of each of the following:
  a. $Mg(HCO_3)_2$, magnesium hydrogen carbonate
  b. $Au(OH)_3$, gold(III) hydroxide, used in gold plating
  c. $C_{18}H_{34}O_2$, oleic acid from olive oil

5.83 How many grams are in 0.150 mole of each of the following?
  a. K   **b.** Cl_2 **c.** Na_2CO_3

5.84 How many grams are in 2.25 moles of each of the following?
  a. N_2 **b.** $NaBr$ **c.** C_6H_{14}

5.85 How many moles are in 25.0 g of each of the following compounds?
  a. CO_2 **b.** $Al(OH)_3$ **c.** $MgCl_2$

5.86 How many moles are in 4.00 g of each of the following compounds?
  a. NH_3 **b.** $Ca(NO_3)_2$ **c.** SO_3

5.87 At a winery, glucose ($C_6H_{12}O_6$) in grapes undergoes fermentation to produce ethanol (C_2H_6O) and carbon dioxide.
  $$C_6H_{12}O_6 \longrightarrow 2C_2H_6O + 2CO_2$$
  glucose   ethanol
  a. How many moles of glucose are required to form 124 g of ethanol?
  b. How many grams of ethanol would be formed from the reaction of 0.240 kg of glucose?

5.88 Gasohol is a fuel containing ethanol (C_2H_6O), which burns in oxygen (O_2) to give carbon dioxide and water.
  a. State the reactants and products for this reaction in the form of a balanced equation.
  b. How many moles of O_2 are needed to completely react with 4.0 moles of C_2H_6O?
  c. If a car produces 88 g of CO_2, how many grams of O_2 are used up in the reaction?
  d. If you add 125 g of C_2H_6O to your gas, how many grams of CO_2 and H_2O can be produced from the ethanol?

5.89 Balance the following equation:
  $$NH_3(g) + F_2(g) \longrightarrow N_2F_4(g) + HF(g)$$
  a. How many moles of each reactant are needed to produce 4.00 moles of HF?
  b. How many grams of F_2 are required to react with 1.50 moles of NH_3?
  c. How many grams of N_2F_4 can be produced when 3.40 g of NH_3 reacts?

5.90 When nitrogen dioxide (NO_2) from car exhaust combines with water in the air it forms nitric acid (HNO_3), which causes acid rain, and nitrogen monoxide.
  $$3NO_2(g) + H_2O(l) \longrightarrow 2HNO_3(aq) + NO(g)$$
  a. How many molecules of NO_2 are needed to react with 0.250 mole H_2O?
  b. How many grams of HNO_3 are produced when 60.0 g NO_2 completely reacts?

5.91 Ethane, C_2H_6, reacts with chlorine to form hexachloroethane and hydrogen chloride.

$$C_2H_6(g) + 6Cl_2(g) \longrightarrow C_2Cl_6(g) + 6HCl(g)$$

a. How many grams of chlorine gas must react to produce 1.60 moles hexachloroethane?

b. How many grams of hydrogen chloride are produced from 50.0 g ethane and excess chlorine gas?

5.92 Propane gas, C_3H_8, reacts with oxygen to produce water and carbon dioxide. Propane has a density of 2.02 g/L at room temperature.

$$C_3H_8(g) + 5O_2(g) \longrightarrow 3CO_2(g) + 4H_2O(l)$$
propane

a. How many moles of water form when 5.00 L propane gas (C_3H_8) completely react?

b. How many grams of CO_2 are produced from 18.5 g oxygen gas and excess propane?

c. How many grams of H_2O can be produced from the reaction of 8.50×10^{22} molecules propane gas, C_3H_8?

5.93 Acetylene gas, C_2H_2, burns in oxygen to produce carbon dioxide and water. If 62.0 g CO_2 are produced when 22.5 g C_2H_2 react with sufficient oxygen, what is the percent yield for the reaction?

5.94 When 50.0 g iron(III) oxide reacts with carbon monoxide, 32.8 g iron is produced. What is the percent yield of the reaction?

$$Fe_2O_3(s) + 3CO(g) \longrightarrow 2Fe(s) + 3CO_2(g)$$

5.95 Pentane gas, C_5H_{12}, reacts with oxygen to produce carbon dioxide and water.

$$C_5H_{12}(g) + 8O_2(g) \longrightarrow 5CO_2(g) + 6H_2O(g)$$

a. How many grams of pentane must react to produce 4.0 mol water?

b. How many grams of CO_2 are produced from 32.0 g oxygen and excess pentane?

c. How many grams of CO_2 are formed if 44.5 g C_5H_{12} is mixed with 108 g O_2?

5.96 When nitrogen dioxide (NO_2) from car exhaust combines with water in the air it forms nitric acid (HNO_3), which causes acid rain, and nitrogen monoxide.

$$3NO_2(g) + H_2O(l) \longrightarrow 2HNO_3(aq) + NO(g)$$

a. How many molecules of NO_2 are needed to react with 0.250 mol H_2O?

b. How many grams of HNO_3 are produced when 60.0 g NO_2 completely react?

c. How many grams of HNO_3 can be produced if 225 g NO_2 is mixed with 55.2 g H_2O?

5.97 When a mixture of 12.8 g Na and 10.2 g Cl_2 react, what is the mass of NaCl that is produced?

$$2Na(s) + Cl_2(g) \longrightarrow 2NaCl(s)$$

5.98 If a mixture of 35.8 g CH_4 and 75.5 g S reacts, how many grams of H_2S are produced?

$$CH_4(g) + 4S(g) \longrightarrow CS_2(g) + 2H_2S(g)$$

CHALLENGE QUESTIONS

5.99 Write a balanced equation for each of the following reaction descriptions and identify each type of reaction.

a. An aqueous solution of lead(II) nitrate is mixed with aqueous sodium phosphate to produce solid lead(II) phosphate and aqueous sodium nitrate.

b. Gallium metal heated in oxygen gas forms solid gallium(III) oxide.

c. When solid sodium nitrate is heated, solid sodium nitrite and oxygen gas are produced.

d. Solid bismuth(III) oxide and solid carbon react to form bismuth metal and carbon monoxide gas.

5.100 A toothpaste contains 0.24% by mass sodium fluoride used to prevent dental caries and 0.30% by mass triclosan $C_{12}H_7Cl_3O_2$, a preservative and antigingivitis agent. One tube contains 119 g toothpaste.

a. How many moles of NaF are in the tube of toothpaste?

b. How many fluoride atoms F^- are in the tube of toothpaste?

c. How many grams of sodium ion Na^+ are in 1.50 g toothpaste?

d. How many molecules of triclosan are in the tube of toothpaste?

5.101 A gold bar is 2.31 cm long, 1.48 cm wide, and 0.0758 cm thick.

a. If gold has a density of 19.3 g/mL, what is the mass of the gold bar?

b. How many atoms of gold are in the bar?

c. When the same mass of gold combines with oxygen, the oxide product has a mass of 5.61 g. How many moles of oxygen are combined with the gold?

d. What is the formula of the oxide product?

5.102 The gaseous hydrocarbon acetylene, C_2H_2, used in welders' torches, releases a large amount of heat when it burns according to the following equation:

$$2C_2H_2(g) + 5O_2(g) \longrightarrow 4CO_2(g) + 2H_2O(g)$$

a. How many moles of water are produced from the complete reaction of 64.0 g oxygen?

b. How many moles of oxygen are needed to react completely with 2.25×10^{24} molecules of acetylene?

c. How many grams of carbon dioxide are produced from the complete reaction of 78.0 g acetylene?

d. If the reaction in part c produces 186 g CO_2, what is the percent yield for the reaction?

5.103 Acetylene, C_2H_2, used in welders' torches, burns according to the following equation:

$$2C_2H_2(g) + 5O_2(g) \longrightarrow 4CO_2(g) + 2H_2O(g)$$

a. How many molecules of oxygen are needed to react with 22.0 g acetylene?

b. How many grams of carbon dioxide could be produced from the complete reaction of the acetylene in part a?

c. If the reaction in part a produces 64.0 g CO_2, what is the percent yield for the reaction?

5.104 Consider the *unbalanced* equation for the reaction of sodium and nitrogen to form sodium nitride.

$$Na(s) + N_2(g) \longrightarrow Na_3N(s)$$

a. If 80.0 g sodium is mixed with 20.0 g nitrogen gas, what mass of sodium nitride forms?

b. If the reaction in part a has a percent yield of 75.0%, how much sodium nitride is actually produced?

5.105 Consider the following *unbalanced* equation:

$$Al(s) + O_2(g) \longrightarrow Al_2O_3(s)$$

a. Balance the equation.

b. Identify the type of reaction.

c. How many moles of oxygen must react with 4.50 moles Al?

d. How many grams of aluminum oxide are produced when 50.2 g aluminum react?

e. When 0.500 mole aluminum is reacted in a closed container with 8.00 g oxygen, how many grams of aluminum oxide can form?

f. If 45.0 g aluminum and 62.0 g oxygen undergo a reaction that has a 70.0% yield, what mass of aluminum oxide forms?

ANSWERS

Answers to Study Checks

5.1 **a.** physical
b. chemical

5.2 $3Fe(s) + 2O_2(g) \longrightarrow Fe_3O_4(s)$

5.3 $2NO(g) + O_2(g) \longrightarrow 2NO_2(g)$ Combination reaction

5.4 Lithium is oxidized: $2Li \longrightarrow 2Li^+ + 2e^-$
Fluorine is reduced: $F_2 + 2e^- \longrightarrow 2F^-$

5.5 0.432 mole H_2O

5.6 0.120 mole aspirin

5.7 138.0 g

5.8 24.4 g Au

5.9 0.00621 mole of $CaCO_3$, 0.00550 mole of $MgCO_3$

5.10 0.90 mole Fe_2S_3

5.11 44.0 g NO

5.12 27.5 g CO_2

5.13 84.7%

5.14 13.4 g SiC

Answers to Selected Questions and Problems

5.1 **a.** physical **b.** chemical **c.** physical
d. chemical **e.** physical

5.3 **a.** not balanced **b.** balanced
c. not balanced **d.** balanced

5.5 **a.** $N_2(g) + O_2(g) \longrightarrow 2NO(g)$
b. $2HgO(s) \longrightarrow 2Hg(l) + O_2(g)$
c. $4Fe(s) + 3O_2(g) \longrightarrow 2Fe_2O_3(s)$
d. $2Na(s) + Cl_2(g) \longrightarrow 2NaCl(s)$

5.7 **a.** $Mg(s) + 2AgNO_3(aq) \longrightarrow Mg(NO_3)_2(aq) + 2Ag(s)$
b. $2Al(s) + 3CuSO_4(aq) \longrightarrow 3Cu(s) + Al_2(SO_4)_3(aq)$
c. $Pb(NO_3)_2(aq) + 2NaCl(aq) \longrightarrow$
$\qquad\qquad\qquad\qquad PbCl_2(s) + 2NaNO_3(aq)$
d. $2Al(s) + 6HCl(aq) \longrightarrow 2AlCl_3(aq) + 3H_2(g)$

5.9 **a.** A single reactant splits into two simpler substances (elements).
b. One element in the reacting compound is replaced by the other reactant.

5.11 **a.** combination
b. single replacement
c. decomposition
d. double replacement
e. double replacement

5.13 **a.** Zn is oxidized; Cl_2 is reduced.
b. Br^- in NaBr is oxidized; Cl_2 is reduced.
c. The O^{2-} in PbO is oxidized; the Pb^{2+} is reduced.
d. Sn^{2+} is oxidized; Fe^{3+} is reduced.

5.15 **a.** reduction **b.** oxidation

5.17 1.00 mole contains 6.02×10^{23} atoms of an element, molecules of a covalent substance, or formula units of an ionic substance.

5.19 **a.** 24 moles H **b.** 100 moles C **c.** 0.040 mole N

5.21 **a.** 3.01×10^{23} C atoms
b. 7.71×10^{23} SO_2 molecules
c. 0.0867 mole Fe
d. 14.1 moles C_2H_5OH

5.23 **a.** 6.00 moles H **b.** 8.00 moles O
c. 1.20×10^{24} P atoms **d.** 4.82×10^{24} O atoms

5.25 **a.** 58.5 g **b.** 159.8 g **c.** 73.8 g
d. 342.3 g **e.** 58.3 g **f.** 365.1 g

5.27 **a.** 46.0 g **b.** 112 g **c.** 14.8 g

5.29 **a.** 29.3 g **b.** 109 g **c.** 4.05 g

5.31 **a.** 602 g **b.** 11 g

5.33 **a.** 0.463 mole **b.** 0.0167 mole
c. 0.882 mole **d.** 1.17 moles

5.35 12 moles

5.37 **a.** 0.78 mole S **b.** 1.95 moles S **c.** 6.0 moles S

5.39 **a.** 1.72×10^{24} atoms N
b. 1.8×10^{24} atoms N
c. 2.4×10^{24} atoms N

5.41 **a.** $\dfrac{2 \text{ moles SO}_2}{1 \text{ mole O}_2}$ and $\dfrac{1 \text{ mole O}_2}{2 \text{ moles SO}_2}$

$\dfrac{2 \text{ moles SO}_2}{2 \text{ moles SO}_3}$ and $\dfrac{2 \text{ moles SO}_3}{2 \text{ moles SO}_2}$

$\dfrac{2 \text{ moles SO}_3}{1 \text{ mole O}_2}$ and $\dfrac{1 \text{ mole O}_2}{2 \text{ moles SO}_3}$

b. $\dfrac{4 \text{ moles P}}{5 \text{ moles O}_2}$ and $\dfrac{5 \text{ moles O}_2}{4 \text{ moles P}}$

$\dfrac{4 \text{ moles P}}{2 \text{ moles P}_2\text{O}_5}$ and $\dfrac{2 \text{ moles P}_2\text{O}_5}{4 \text{ moles P}}$

$\dfrac{5 \text{ moles O}_2}{2 \text{ moles P}_2\text{O}_5}$ and $\dfrac{2 \text{ moles P}_2\text{O}_5}{5 \text{ moles O}_2}$

5.43 **a.** 1.0 mole O_2 **b.** 10. moles H_2
c. 5.0 moles H_2O

5.45 **a.** 1.25 moles C   **b.** 0.96 mole CO
c. 1.0 mole SO_2 **d.** 0.50 mole CS_2

5.47 **a.** 77.5 g Na_2O **b.** 6.26 g O_2 **c.** 19.4 g O_2

5.49 **a.** 192 g O_2 **b.** 3.79 g N_2 **c.** 54.0 g H_2O

5.51 **a.** 3.65 g H_2O **b.** 3.43 g NO   **c.** 7.53 g HNO_3

5.53 **a.** $2PbS(s) + 3O_2(g) \longrightarrow 2PbO(s) + 2SO_2(g)$
b. 6.00 g O_2 **c.** 17.4 g SO_2 **d.** 137 g PbS

5.55 **a.** 70.9%   **b.** 63.1%

5.57 70.9 g Al_2O_3

5.59 60.5%

5.61 **a.** 8 taxis can be used to pick up passengers
b. 7 taxis can be driven

5.63 **a.** 5.0 moles H_2 **b.** 4.0 moles H_2 **c.** 3.0 moles N_2

5.65 **a.** 2.00 moles SO_3 **b.** 0.500 mole Fe_3O_4
c. 1.27 moles CO_2

5.67 **a.** 0.188 mole $AlCl_3$ **b.** 0.750 mole H_2O
c. 0.417 mole SO_2

5.69 Physical changes include the following: solid wax melts to form liquid, the candle becomes shorter, the liquid wax changes to solid, the shape of the wax changes, the wick becomes shorter.

Chemical changes include the following: heat and light are emitted, wax and the wick burn in the presence of oxygen.

5.71 **a.** 1, 1, 2 combination reaction
b. 2, 2, 1 decomposition reaction
c. 1, 1, 1, 1 double replacement reaction
d. 1, 4, 2 combination reaction

5.73 **a.** 47.1 g CuO
b. 0.296 mole O_2

5.75 **a.** $NH_3(g) + HCl(g) \longrightarrow NH_4Cl(s)$ combination
b. $Fe_3O_4(s) + 4H_2(g) \longrightarrow 3Fe(s) + 4H_2O(g)$
single replacement
c. $2Sb(s) + 3Cl_2(g) \longrightarrow 2SbCl_3(s)$ combination
d. $2NI_3(s) \longrightarrow N_2(g) + 3I_2(g)$ decomposition
e. $2KBr(aq) + Cl_2(aq) \longrightarrow 2KCl(aq) + Br_2(l)$
single replacement
f. $Al_2(SO_4)_3(aq) + 6NaOH(aq) \longrightarrow$
$3Na_2SO_4(aq) + 2Al(OH)_3(s)$
double replacement

5.77 **a.** reduction   **b.** oxidation
c. oxidation   **d.** reduction

5.79 **a.** 90.1 g   **b.** 3.01×10^{23} molecules
c. 2.71×10^{24} C atoms   **d.** 230 g lactic acid

5.81 **a.** 152.0 g   **b.** 389.9 g   **c.** 169.1 g

5.83 **a.** 5.87 g   **b.** 10.7 g   **c.** 15.9 g

5.85 **a.** 0.568 mole   **b.** 0.321 mole   **c.** 0.262 mole

5.87 **a.** 1.35 moles glucose   **b.** 123 g ethanol

5.89 $2NH_3 + 5F_2 \longrightarrow N_2F_4 + 6HF$
a. 1.33 moles NH_3 and 3.33 moles F_2
b. 143 g F_2
c. 10.4 g N_2F_4

5.91 **a.** 48 g pentane   **b.** 27.5 g CO_2

5.93 81.4%

5.95 **a.** 48 g C_5H_{12} **b.** 27.5 g CO_2 **c.** 92.8 g CO_2

5.97 16.8 g NaCl

5.99 **a.** $3Pb(NO_3)_2(aq) + 2Na_3PO_4(aq) \longrightarrow$
$Pb_3(PO_4)_2(s) + 6NaNO_3(aq)$
double replacement
b. $4Ga(s) + 3O_2(g) \longrightarrow 2Ga_2O_3(s)$ combination
c. $2NaNO_3(s) \longrightarrow 2NaNO_2(s) + O_2(g)$ decomposition
d. $Bi_2O_3(s) + 3C(s) \longrightarrow 2Bi(s) + 3CO(g)$
single replacement

5.101 **a.** 5.00 g gold   **b.** 1.53×10^{22} Au atoms
c. 0.038 mole oxygen   **d.** Au_2O_3

5.103 **a.** 1.27×10^{24} molecules O_2
b. 74.4 g CO_2 **c.** 86.0% yield

5.105 **a.** $4Al(s) + 3O_2(g) \longrightarrow 2Al_2O_3(s)$
b. This is a combination reaction.
c. 3.38 moles oxygen
d. 94.9 g aluminum oxide
e. 17.0 g aluminum oxide
f. 59.6 g aluminum oxide

6 Energy and Matter

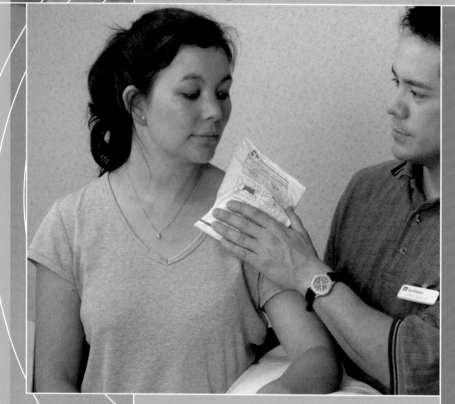

"If you've had first aid for a sports injury," says Cort Kim, physical therapist at the Sunrise Sports Medicine Clinic, "you've likely been treated with a cold pack or hot pack. We use them for several kinds of injury. Here, I'm showing how I can use a cold pack to reduce swelling in my patient's shoulder."

A hot or cold pack is just a packaged chemical reaction. When you hit or open the pack to activate it, your action mixes chemicals together and thus initiates the reaction. In a cold pack, the reaction is one that absorbs heat, chills the pack, and draws heat from the injury. Hot packs use reactions that release energy, thus warming the pack. In both cases, the reaction proceeds at a moderate pace, so that the pack stays active for a long time and doesn't get too cold or hot.

ost of everything we do involves energy. We use energy when we walk, play tennis, study, and breathe. We also use energy when we warm water, cook food, turn on lights, use a washing machine, or drive our cars. Of course, that energy has to come from something. In our bodies, the food we eat provides us with energy. If we don't eat for a while, we run out of energy. In our homes, schools, and automobiles, burning fossil fuels such as oil, propane, or gasoline provides energy.

When we look around us, we see that matter takes the physical form of a solid, liquid, or gas. A familiar example is water, which is a compound that exists in all three states. In an ice cube or an ice rink, water is in the solid state. Water is a liquid when it comes out of a faucet or fills a pool. Water forms a gas when it evaporates from wet clothes or boils in a pan. Substances change states by losing or gaining energy. For example, energy is needed to melt ice cubes and to boil water in a teakettle. In contrast, energy is removed when water vapor (gas) condenses to liquid and liquid water in an ice cube tray freezes.

6.1 Energy

When you are running, walking, dancing, or thinking, you are using energy to do **work**. In fact, **energy** is defined as the ability to do work. Suppose you are climbing a steep hill. Perhaps you become too tired to go on; you do not have sufficient energy to do any more work. Now suppose you sit down and have lunch. In a while you will have obtained energy from the food, and you will be able to do more work and complete the climb. (See Figure 6.1).

Potential and Kinetic Energy

All energy can be classified as potential energy or kinetic energy. **Potential energy** is stored energy, whereas **kinetic energy** is the energy of motion. Any object that is moving has kinetic energy. A boulder resting on top of a mountain has potential energy because of its location. If the boulder rolls down the mountain, the potential energy becomes kinetic energy. Water stored in a reservoir has potential energy. When the water goes over the dam, the potential energy is converted to kinetic energy. Even the food you eat has potential energy. When you digest food, you convert its potential energy to kinetic energy to do work in the body.

Heat and Units of Energy

Heat is the energy associated with the motion of particles in a substance. A frozen pizza feels cold because the particles in the pizza are moving very slowly. As heat is added, the motions of the particles increase, and the pizza becomes warm. Eventually the particles have enough energy to make the pizza hot and ready to eat.

LEARNING GOAL

Identify energy as potential or kinetic and understand the units of energy.

FIGURE 6.1

Work is done as the rock climber moves up the cliff. At the top, the climber has more potential energy than when she started the climb.

Q **What happens to the potential energy of the climber when she descends?**

The SI unit of energy and work is the **joule (J)** pronounced, "jewel." The joule is a small amount of energy, so scientists often use the kilojoule (kJ), 1000 joules. When you heat water for one cup of tea, you use about 75,000 J or 75 kJ of heat.

Energy in joules

Energy in joules	
10^{27}	Energy radiated by sun per second (10^{26})
10^{24}	World reserves of fossil fuel (10^{23})
10^{21}	Energy consumption for one year in US (10^{20})
10^{18}	Solar energy reaching the Earth per second (10^{17})
10^{15}	
10^{12}	Energy use per person in one year in US (10^{11})
10^{9}	Energy from 1 gallon of gasoline (10^{8})
10^{6}	Energy from one serving of pasta, a doughnut, or needed to bicycle one hour (10^{6})
10^{3}	Energy used sleeping one hour (10^{5})
10^{0}	

You may be more familiar with the older unit **calorie (cal)**, from the Latin *caloric*, meaning "heat." The calorie was originally defined as the amount of energy (heat) needed to raise the temperature of one gram of water by 1°C (from 14.5°C to 15.5°C). Now one calorie is defined as exactly 4.184 J. This equality can also be written as a conversion factor.

1 cal = 4.184 J (exact)

$$\frac{4.184 \text{ J}}{1 \text{ cal}} \quad \text{and} \quad \frac{1 \text{ cal}}{4.184 \text{ J}}$$

One **kilocalorie (kcal)** is equal to 1000 calories and a *kilojoule* (kJ) is 1000 joules.

1 kcal = 1000 cal

1 kJ = 1000 J

SAMPLE PROBLEM 6.1

▪ Energy Units

When 10. g octane in gasoline burns in an automobile engine, 8500 J are released. Convert this quantity of energy to each of the following units:

a. calories
b. kilojoules

SOLUTION

a. calories

■ **STEP 1 Given**   8500 J   **Need**   calories (cal)

■ **STEP 2 Plan**   J   ⟩ Energy factor ⟩   cal

■ **STEP 3 Equalities/Conversion Factors**

$$1 \text{ cal} = 4.184 \text{ J}$$
$$\frac{1 \text{ cal}}{4.184 \text{ J}} \quad \text{and} \quad \frac{4.184 \text{ J}}{1 \text{ cal}}$$

■ **STEP 4 Set Up Problem**

$$8500 \text{ J} \times \frac{1 \text{ cal}}{4.184 \text{ J}} = 2000 \text{ or } 2.0 \times 10^3 \text{ cal} \quad \text{(2 SF)}$$

b. kilojoules

■ **STEP 1 Given**   8500 J   **Need**   kilojoules

■ **STEP 2 Plan**   J   ⟩ Energy factor ⟩   kJ

■ **STEP 3 Equalities/Conversion Factors**

$$1 \text{ kJ} = 1000 \text{ J}$$
$$\frac{1000 \text{ J}}{1 \text{ kJ}} \quad \text{and} \quad \frac{1 \text{ kJ}}{1000 \text{ J}}$$

■ **STEP 4 Set Up Problem** $8500 \text{ J} \times \dfrac{1 \text{ kJ}}{1000 \text{ J}} = 8.5 \text{ kJ}$

STUDY CHECK

How many cal are in a jelly doughnut if it provides 26 kcal of energy?

QUESTIONS AND PROBLEMS

■ Energy

6.1 Discuss the changes in the potential and kinetic energy of a roller-coaster ride as the roller-coaster climbs up a ramp and goes down the other side.

6.2 Discuss the changes in the potential and kinetic energy of a ski jumper taking the elevator to the top of the jump and going down the ramp.

6.3 Indicate whether each statement describes potential or kinetic energy:
   a. water at the top of a waterfall   **b.** kicking a ball
   c. the energy in a lump of coal   **d.** a skier at the top of a hill

6.4 Indicate whether each statement describes potential or kinetic energy:
   a. the energy in your food   **b.** a tightly wound spring
   c. an earthquake   **d.** a car speeding down the freeway

6.5 Convert each of the following energy units.
   a. 3500 cal to kcal   **b.** 415 J to cal   **c.** 28 cal to J   **d.** 4.5 kJ to cal

6.6 Convert each of the following energy units.
   a. 8.1 kcal to cal   **b.** 325 J to kJ   **c.** 2550 cal to kJ   **d.** 2.50 kcal to J

LEARNING GOAL

■ Describe the energy changes in exothermic and endothermic reactions.

■ 6.2 Energy in Chemical Reactions

Heat is energy that is lost or gained when a chemical reaction takes place. The *system* is the particular group of reactants and products we are looking at. The *surroundings* are all the things that contain and interact with the system such as the reaction flask, the laboratory room, and the air in the room. In any reaction system, there is a change in energy as reactants break apart and products form. The direction of heat flow depends on whether the products in the reaction have more energy than the reactants or less.

For a chemical reaction to take place, the molecules of the reactants must come in contact with each other. If there are no collisions between the molecules (or atoms), no reaction is possible. When the molecules collide, bonds between atoms are broken and new bonds can form. The energy needed to break apart those bonds is called the **activation energy**. If the energy of a collision is less than the activation energy, the molecules bounce apart without reacting.

The concept of activation energy is analogous to climbing over a hill. To reach a destination on the other side, we must expend energy to climb to the top of the hill. Once we are at the top, we can easily roll down the other side. The energy needed to get us from our starting point to the top of the hill would be our activation energy.

■ Heat of Reaction

Heat of Reaction

The **heat of reaction** is the amount of heat absorbed or released during a reaction that takes place at constant pressure. We determine a heat of reaction, symbol ΔH, as the difference in the energy of the products and the reactants.

$$\Delta H = H_{\text{products}} - H_{\text{reactants}}$$

In the energy diagram for an **endothermic reaction** (*endo* means within), the energy of the products is greater than that of the reactants. In these reactions heat flows out of the surroundings into the system where it is used to convert the reactants to products. For an endothermic reaction, the heat of reaction can be written as one of the reactants. It can also be written as a ΔH value with a positive sign ($+$). Let us look at the equation and ΔH for the endothermic reaction in which 137 kcal of heat is needed to break down 2 moles water into hydrogen and oxygen.

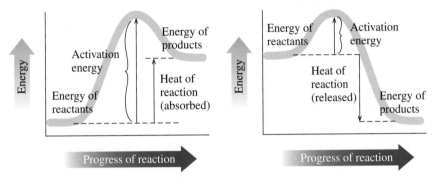

Endothermic reaction   **Exothermic reaction**

Endothermic, Heat Required   **Heat Is a Reactant**

$2H_2O(l) + 137 \text{ kcal} \longrightarrow 2H_2(g) + O_2(g)$

$2H_2O(l) \longrightarrow 2H_2(g) + O_2(g)$ $\Delta H = +137 \text{ kcal}$
                                                          positive sign

In an **exothermic reaction** (*exo* means out), the products have less energy than the reactants. In these reactions heat flows out of the system into the surroundings. For an exothermic reaction, the heat of reaction can be written as one of products. It can also be written as a ΔH value with a negative sign ($-$). The reaction for the formation of ammonia (NH_3) from 3 moles hydrogen (H_2) and 1 mole nitrogen (N_2) is exothermic.

Exothermic, Heat Evolved   **Heat Is a Product**

$3H_2(g) + N_2(g) \longrightarrow 2NH_3(g) + 22.0 \text{ kcal}$

$3H_2(g) + N_2(g) \longrightarrow 2NH_3(g)$ $\Delta H = -22.0 \text{ kcal}$
                                                          negative sign

Reaction	Energy Change	Heat in the Equation	Sign of ΔH
Endothermic	Heat absorbed	Reactant side	Positive sign ($+$)
Exothermic	Heat released	Product side	Negative sign ($-$)

SAMPLE PROBLEM 6.2

■ Exothermic and Endothermic Reactions

In the reaction of 1 mole solid carbon with oxygen gas, the energy of the carbon dioxide produced is 93.9 kcal lower than the energy of the reactants.

a. Is the reaction exothermic or endothermic?
b. Write the equation for the reactions including the heat of the reaction.
c. Write the value of ΔH for this reaction.

SOLUTION

a. When the products have a lower energy than the reactants, the reaction is exothermic.

b. $C(s) + O_2(g) \longrightarrow CO_2(g) + 93.9 \text{ kcal}$

c. $\Delta H = -93.9 \text{ kcal}$

STUDY CHECK

The reaction of hydrogen gas (H_2) and iodine gas (I_2) to form hydrogen iodide (HI) is endothermic and requires 12 kcal of heat.

a. Write a balanced equation including the heat of reaction.

b. Write the value of ΔH for this reaction.

Calculations of Heat in Reactions

The value of ΔH refers to the heat change for the number of moles of each substance in the balanced equation for the reaction. Consider the following decomposition reaction.

$$2H_2O(l) \longrightarrow 2H_2(g) + O_2(g) \qquad \Delta H = 137 \text{ kcal}$$

$$2H_2O(l) + 137 \text{ kcal} \longrightarrow 2H_2(g) + O_2(g)$$

For this reaction, 137 kcal are absorbed by 2 moles H_2O to produce 2 moles H_2 and 1 mole O_2. We can write heat conversion factors for each substance in this reaction.

$$\frac{137 \text{ kcal}}{2 \text{ moles } H_2O} \qquad \frac{137 \text{ kcal}}{2 \text{ moles } H_2} \qquad \frac{137 \text{ kcal}}{1 \text{ mole } O_2}$$

SAMPLE PROBLEM 6.3

Guide to Calculations Using Heat of Reaction (ΔH)

STEP 1
List given and needed data for the equation.

STEP 2
Write a plan using heat of reaction and any molar mass needed.

STEP 3
Write the conversion factors including heat of reaction.

STEP 4
Set up the problem.

■ Calculating Heat in a Reaction

The formation of ammonia from hydrogen and nitrogen has a

$$\Delta H = -22.0 \text{ kcal}$$

$$N_2(g) + 3H_2(g) \longrightarrow 2NH_3(g) \qquad \Delta H = -22.0 \text{ kcal}$$

How much heat in kcal is released when 50.0 g of ammonia forms?

SOLUTION

■ **STEP 1 Given** 50.0 g NH_3 **Need** Heat in kcal to form NH_3

■ **STEP 2 Plan** Use conversion factors that relate the heat released to the moles of NH_3.

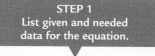 grams of NH_3 → Molar mass → moles of NH_3 → Heat of reaction → kilocalories

■ **STEP 3 Equalities/Conversion Factors**

$$1 \text{ mole } NH_3 = 17.0 \text{ g}$$

$$\frac{17.0 \text{ g}}{1 \text{ mole } NH_3} \quad \text{and} \quad \frac{1 \text{ mole } NH_3}{17.0 \text{ g}}$$

$$2 \text{ moles } NH_3 = 22.0 \text{ kcal}$$

$$\frac{22.0 \text{ kcal}}{2 \text{ moles } NH_3} \quad \text{and} \quad \frac{2 \text{ moles } NH_3}{22.0 \text{ kcal}}$$

■ STEP 4 Set Up Problem

$$50.0 \text{ g NH}_3 \times \frac{1 \text{ mole NH}_3}{17.0 \text{ g NH}_3} \times \frac{22.0 \text{ kcal}}{2 \text{ moles NH}_3} = 32.4 \text{ kcal}$$

STUDY CHECK

Mercury(II) oxide decomposes to mercury and oxygen:

$$2\text{HgO}(s) \longrightarrow 2\text{Hg}(l) + O_2(g) \qquad \Delta H = 182 \text{ kJ}$$

a. Is the reaction exothermic or endothermic?
b. How many kJ are needed to react 25.0 g mercury(II) oxide?

QUESTIONS AND PROBLEMS

■ Energy in Chemical Reactions

6.7 **a.** Why do chemical reactions require an energy of activation?
   b. In an exothermic reaction, is the energy of the products higher or lower than the reactants?
   c. Draw an energy diagram for an exothermic reaction.

6.8 **a.** What is measured by the heat of reaction?
   b. In an endothermic reaction, is the energy of the products higher or lower than the reactants?
   c. Draw an energy diagram for an endothermic reaction.

6.9 Classify the following as exothermic or endothermic reactions:
   a. 55 kcal is released.
   b. The energy level of the products is higher than the reactants.
   c. The metabolism of glucose in the body provides energy.

HEALTH NOTE

■ Hot Packs and Cold Packs

In a hospital, at a first-aid station, or at an athletic event, a *cold pack* may be used to reduce swelling from an injury, remove heat from inflammation, or decrease capillary size to lessen the effect of hemorrhaging. Inside the plastic container of a cold pack, there is a compartment containing solid ammonium nitrate (NH_4NO_3) that is separated from a compartment containing water. The pack is activated when it is hit or squeezed hard enough to break the walls between the compartments and cause the ammonium nitrate to mix with the water (shown as H_2O over the reaction arrow). In an endothermic process, each gram of NH_4NO_3 that dissolves absorbs 79 cal of heat from the water. The temperature drops and the pack becomes cold and ready to use.

Endothermic Reaction in a Cold Pack

$$6.2 \text{ kcal} + NH_4NO_3(s) \xrightarrow{H_2O} NH_4NO_3(aq)$$

Hot packs are used to relax muscles, lessen aches and cramps, and increase circulation by expanding capillary size. Constructed in the same way as cold packs, a hot pack may contain the salt $CaCl_2$. The dissolving of the salt in water is exothermic and releases 160 cal per gram of salt. The temperature rises and the pack becomes hot and ready to use.

Exothermic Reaction in a Hot Pack

$$CaCl_2(s) \xrightarrow{H_2O} CaCl_2(aq) + 18 \text{ kcal}$$

6.10 Classify the following as exothermic or endothermic reactions:
  a. The energy level of the products is lower than the reactants.
  b. In the body, the synthesis of proteins requires energy.
  c. 12 kcal is absorbed.

6.11 Classify the following as exothermic or endothermic reactions and give ΔH for each:
  a. gas burning in a Bunsen burner:

$$CH_4(g) + 2O_2(g) \longrightarrow CO_2(g) + 2H_2O(g) + 210 \text{ kcal}$$

  b. dehydrating limestone:

$$Ca(OH)_2(s) + 65.3 \text{ kJ} \longrightarrow CaO(s) + H_2O(l)$$

  c. formation of aluminum oxide and iron from aluminum and iron(III) oxide:

$$2Al(s) + Fe_2O_3(s) \longrightarrow Al_2O_3(s) + 2Fe(s) + 205 \text{ kcal}$$

6.12 Classify the following as exothermic or endothermic reactions and give ΔH for each:
  a. the combustion of propane:

$$C_3H_8(g) + 5O_2(g) \longrightarrow 3CO_2(g) + 4H_2O(g) + 530 \text{ kcal}$$

  b. the formation of "table" salt:

$$2Na(s) + Cl_2(g) \longrightarrow 2NaCl(s) + 196 \text{ kcal}$$

  c. decomposition of phosphorus pentachloride:

$$PCl_5(g) + 67 \text{ kJ} \longrightarrow PCl_3(g) + Cl_2(g)$$

6.13 The equation for the formation of silicon tetrachloride from silicon and chlorine is

$$Si(s) + 2Cl_2(g) \longrightarrow SiCl_4(g) \qquad \Delta H = -157 \text{ kcal}$$

How many kilocalories are released when 125 g Cl_2 reacts with silicon?

6.14 Methanol (CH_3OH), which is used as a cooking fuel, undergoes combustion to produce carbon dioxide and water.

$$2CH_3OH(l) + 3O_2(g) \longrightarrow 2CO_2(g) + 4H_2O(l) \qquad \Delta H = -726 \text{ kJ}$$

How many kilojoules are released when 75.0 g methanol is burned?

LEARNING GOAL

Use specific heat to calculate heat loss or gain, temperature change, or mass of a sample.

6.3 Specific Heat

Every substance can absorb heat. When you want to bake a potato, you place it in a hot oven. If you are cooking pasta, you add the pasta to boiling water. Every substance has its own characteristic ability to absorb heat. Some substances must absorb more heat than others to reach a certain temperature. These energy requirements for different substances are described in terms of a property called specific heat. **Specific heat** (*SH*) is the amount of heat needed to raise the temperature of exactly 1 g of a substance by exactly 1°C. This temperature change is written as ΔT (*delta T*).

$$\text{Specific heat}\,(SH) = \frac{\text{heat}}{\text{grams} \times \Delta T} = \frac{\text{cal (or J)}}{1 \text{ g} \times 1°C}$$

Now we can write the specific heat for water using our definition of the calorie and joule.

$$\text{Specific heat}\,(SH) \text{ of } H_2O(l) = 1.00\,\frac{\text{cal}}{\text{g}\,°C} = 4.184\,\frac{\text{J}}{\text{g}\,°C}$$

TABLE 6.1

Specific Heats of Some Substances

Substance	(cal/g °C)	Specific Heat (J/g °C)
aluminum, Al	0.214	0.897
copper, Cu	0.0920	0.385
gold, Au	0.0308	0.129
iron, Fe	0.108	0.452
silver, Ag	0.0562	0.235
ammonia, $NH_3(g)$	0.488	2.04
ethanol, $C_2H_5OH(l)$	0.588	2.46
sodium chloride, $NaCl(s)$	0.207	0.864
water, $H_2O(l)$	1.00	4.184

If we look at Table 6.1, we see that 1 g of water requires 1.00 cal to increase its temperature by 1°C. However, adding the same amount of heat will raise the temperature of 1 g of aluminum by about 5°C and 1 g of copper by 10°C. Because of its high specific heat, the water in the body can absorb or release large amounts of heat to maintain an almost constant temperature.

MC GOB Heat
Specific Heat Calculations

SAMPLE PROBLEM 6.4

■ Calculating Specific Heat

What is the specific heat of lead if 13.6 cal are needed to raise the temperature of 35.6 g Pb by 12.5°C?

SOLUTION

■ **STEP 1 Given**

> heat 13.6 cal mass 35.6 g Pb temperature change 12.5°C

> **Need** specific heat (cal/g °C)

■ **STEP 2 Plan** The specific heat (*SH*) is calculated by dividing the heat by the mass (g) and by the temperature difference (ΔT).

$$SH = \frac{\text{heat}}{\text{mass } \Delta T}$$

■ **STEP 3 Write the heat equation.**

■ **STEP 4 Substitute the given values into the equation.**

$$\text{Specific heat } (SH) = \frac{13.6 \text{ cal}}{35.6 \text{ g } \quad 12.5°C} = 0.0306 \frac{\text{cal}}{\text{g °C}}$$

STUDY CHECK

What is the specific heat of sodium if 123 J are needed to raise the temperature of 4.00 g of Na by 25.0°C?

Calculations Using Specific Heat

When we know the specific heat of a substance, we can calculate the heat lost or gained by measuring the mass of the substance and the initial and final temperature. We can substitute these measurements into the specific heat expression that is rearranged to solve for heat, which we call the *heat equation*.

Heat = mass × temperature change × specific heat

Heat = mass × ΔT × SH

cal = grams × °C × $\dfrac{cal}{g\ °C}$

J = grams × °C × $\dfrac{J}{g\ °C}$

SAMPLE PROBLEM 6.5

▪ Calculating Heat with Temperature Increase

How many calories are absorbed by 45.2 g aluminum if its temperature rises from 12.5°C to 76.8°C? (See Table 6.1.)

SOLUTION

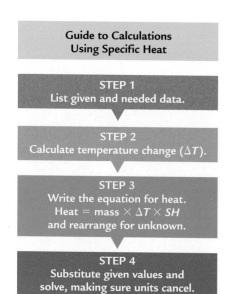

Guide to Calculations Using Specific Heat

STEP 1
List given and needed data.

STEP 2
Calculate temperature change (ΔT).

STEP 3
Write the equation for heat.
Heat = mass × ΔT × SH
and rearrange for unknown.

STEP 4
Substitute given values and solve, making sure units cancel.

▪ **STEP 1**   **Given**   mass = 45.2 g

SH for aluminum = 0.214 cal/g °C

Initial temperature = 12.5°C
Final temperature = 76.8°C

Need   heat   in calories (cal)

▪ **STEP 2** **Calculate the temperature change.**   The temperature change ΔT is the difference between the two temperatures.

$$\Delta T = T_{\text{final}} - T_{\text{initial}} = 76.8°C - 12.5°C = 64.3°C$$

▪ **STEP 3** **Write the heat equation.**

Heat = mass × ΔT × SH

▪ **STEP 4** **Substitute the given values into the equation and solve, making sure units cancel.**

Heat = 45.2 g̶ × 64.3°C̶ × $\dfrac{0.214\ cal}{g̶\ °C̶}$ = 622 cal

STUDY CHECK

Some cooking pans have a layer of copper on the bottom. How many kilocalories are needed to raise the temperature of 125 g copper from 22°C to 325°C? (See Table 6.1)

QUESTIONS AND PROBLEMS

▪ Specific Heat

6.15 If the same amount of heat is supplied to samples of 10.0 g each of aluminum, iron, and copper all at 15°C, which sample would reach the highest temperature? (See Table 6.1)

6.16 Substances A and B are the same mass and at the same initial temperature. When they are heated, the final temperature of A is 55°C higher than the temperature of B. What does this tell you about the specific heats of A and B?

6.17 What is the amount of heat required in each of the following?
a. calories to heat 25 g of water from 15°C to 25°C
b. calories to heat 150 g of water from 0°C to 75°C
c. kilocalories to heat 150 g of water in a kettle from 15°C to 77°C

6.18 What is the amount of heat involved in each of the following?
a. calories given off when 85 g of water cools from 45°C to 25°C
b. calories given off when 25 g of water cools from 86°C to 61°C
c. kilocalories absorbed when 5.0 kg of water warms from 22°C to 28°C

6.19 Calculate the energy in joules and calories
a. required to heat 25.0 g water from 12.5°C to 25.7°C
b. required to heat 38.0 g copper (Cu) from 122°C to 246°C
c. lost when 15.0 g ethanol, C_2H_5OH, cools from 60.5°C to −42.0°C
d. lost when 112 g iron, Fe, cools from 118°C to 55°C

6.20 Calculate the energy in joules and calories
a. required to heat 5.25 g water, H_2O, from 5.5°C to 64.8°C
b. lost when 75.0 g water, H_2O, cools from 86.4°C to 2.1°C
c. required to heat 10.0 g silver (Ag) from 112°C to 275°C
d. lost when 18.0 g gold (Au) cools from 224°C to 118°C

ENVIRONMENTAL NOTE

Global Warming

The amount of carbon dioxide (CO_2) gas in our atmosphere is on the increase as we burn more gasoline, coal, and natural gas. The algae in the oceans and the plants and trees in the forests normally absorb carbon dioxide, but they cannot keep up with the continued increase. The cutting of trees in the rain forests (deforestation) reduces the amount of carbon dioxide removed from the atmosphere. Many of the trees are also burned as land is cleared. It has been estimated that deforestation may account for 15–30% of the carbon dioxide that remains in the atmosphere each year.

The carbon dioxide in the atmosphere acts like the glass in a greenhouse. When sunlight warms the Earth's surface, some of the heat is absorbed by carbon dioxide. It is not yet clear how severe the effects of global warming might be. Some scientists estimate that by around the year 2030, the atmospheric level of carbon dioxide could double and cause the temperature of Earth's atmosphere to rise by 2–5°C. If that should happen, it would have a profound impact on Earth's climate. For example, an increase in the melting of snow and ice could raise the ocean levels by as much as 2 m, which is enough to flood many cities located on the ocean shorelines.

Worldwide efforts are being made to reduce fossil fuel use and to slow or stop deforestation. It will require cooperation throughout the world to avoid the bleak future that some scientists predict should global warming continue unchecked.

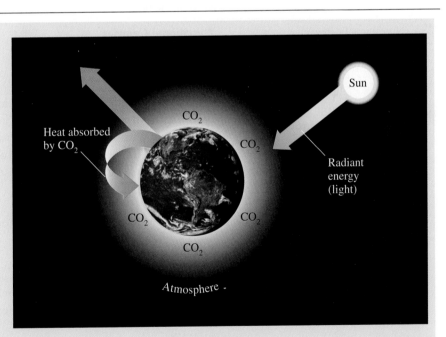

FIGURE 6.2

The caloric value of a food sample temperature change is determined by the combustion of a food sample in a calorimeter.

Q **What happens to the temperature of water in a calorimeter during the combustion of a food sample?**

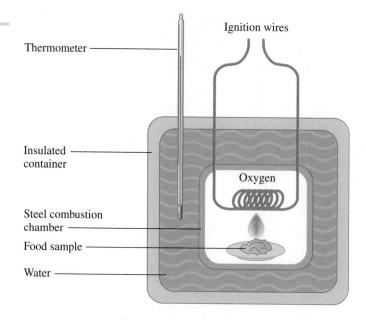

LEARNING GOAL

Use caloric values to calculate the kilocalories (Cal) in a food.

 Nutritional Energy

 SELF-STUDY CASE
Calories from Hidden Sugar

6.4 Energy and Nutrition

When you are watching your food intake, the Calories you are counting are actually kilocalories. In the field of nutrition, it is common to use the **Calorie, Cal** (with an uppercase C) to mean 1000 cal, or 1 kcal. The international unit of nutritional energy is the kilojoule (kJ).

Nutritional Calories

1 Cal = 1 kcal

1 kcal = 1000 cal

The number of Calories in a food is determined by using an apparatus called a calorimeter, shown in Figure 6.2. A sample is placed in a steel container within the calorimeter and water is added to fill a surrounding chamber. When the food burns, the heat released increases the temperature of the surrounding water. By calculating the heat in kilocalories (or Calories) absorbed by the water, we can determine the caloric content of the food.

SAMPLE PROBLEM 6.6

■ Calculating Caloric Values

A 2.3-g sample of butter, a fat, is placed in a calorimeter containing 1900 g of water at an initial temperature of 17°C. After combustion of the butter, the water has a temperature of 28°C. What is the caloric value (kcal/g) of butter?

SOLUTION

The heat absorbed by the water is calculated as follows:

$$1900 \text{ g} \times 11°C \times \frac{1.00 \text{ cal}}{\text{g} °C} = 21\,000 \text{ cal}$$

Mass × ΔT × Specific heat
of water

$$21\,000 \text{ cal} \times \frac{1 \text{ kcal}}{1000 \text{ cal}} = 21 \text{ kcal}$$

Because the fat provided the 21 kcal of heat, the caloric value of butter is calculated as follows:

$$\frac{21 \text{ kcal}}{2.3 \text{ g fat}} = 9.1 \text{ kcal/g fat}$$

STUDY CHECK

A 4.5 g sample of glucose, a carbohydrate, is placed in a calorimeter. The water in the container has a mass of 1500 g and an initial temperature of 15°C. After all the glucose is burned, the water temperature is 27°C. What is the caloric value for glucose?

Caloric Food Values

The **caloric values** are the kilocalories per gram of the three types of food: carbohydrates, fats, and proteins. These values are listed in Table 6.2.

TABLE 6.2

Caloric Values for the Three Food Types

Food Type	Carbohydrate	Fat (lipid)	Protein
Caloric value	$\dfrac{4 \text{ kcal}}{1 \text{ g}}$	$\dfrac{9 \text{ kcal}}{1 \text{ g}}$	$\dfrac{4 \text{ kcal}}{1 \text{ g}}$

If the composition of a food is known in terms of the mass of each food type, the caloric content (the total number of Calories) can be calculated.

$$\text{Kilocalories} = \underset{\substack{\text{Mass of carbohydrate,} \\ \text{fat, or protein}}}{g} \times \underset{\text{Caloric value}}{\frac{\text{kcal}}{g}}$$

The caloric content of a packaged food is listed in the nutritional information on the package, usually in terms of the number of Calories in a serving. The general composition and caloric content of some foods are given in Table 6.3.

TABLE 6.3

General Composition and Caloric Content of Some Foods

Food	Protein (g)	Fat (g)	Carbohydrate (g)	Calories (kcal[a])
Banana, 1 medium	1	trace	26	110
Beans, red kidney, 1 cup	15	1	42	240
Beef, lean, 3 oz	22	5	trace	135
Carrots, raw, 1 cup	1	trace	11	50
Chicken, no skin, 3 oz	20	3	0	110
Egg, 1 large	6	6	trace	80
Milk, 4% fat, 1 cup	9	9	12	165
Milk, nonfat, 1 cup	9	trace	12	85
Oil, olive, 1 tbs	0	14	0	130
Potato, baked	3	trace	23	105
Salmon, 3 oz	17	5	0	115
Steak, 3 oz	20	27	0	325
Yogurt, lowfat, 1 cup	8	4	13	120

[a]Values in kcal are rounded to nearest five.

Snack Crackers

Nutrition Facts
Serving Size 14 crackers (31g)
Servings Per Container About 7

Amount Per Serving	
Calories 120 Calories from Fat 35	

	% Daily Value*
Total Fat 4g	6%
Saturated Fat 0.5g	3%
Polyunsaturated Fat 0.5%	
Monounsaturated Fat 1.5g	
Cholesterol 0mg	0%
Sodium 310mg	13%
Total Carbohydrate 19g	6%
Dietary Fiber Less than 1g	4%
Sugars 2g	
Proteins 2g	

Vitamin A 0%	•	Vitamin C 0%
Calcium 4%	•	Iron 6%

*Percent Daily Values are based on a 2,000 calorie diet. Your daily values may be higher or lower depending on your calorie needs.

		Calories:	2,000	2,500
Total Fat	Less than		65g	80g
Sat Fat	Less than		20g	25g
Cholesterol	Less than		300mg	300mg
Sodium	Less than		2,400mg	2,400mg
Total Carbohydrate			300g	375g
Dietary Fiber			25g	30g

Calories per gram:
Fat 9 • Carbohydrate 4 • Protein 4

SAMPLE PROBLEM 6.7

■ **Caloric Content for a Food**

How many kilocalories are in a piece of chocolate cake that contains 35 g of carbohydrate, 10 g of fat, and 5 g of protein? Round the final answer to the nearest tens place.

SOLUTION

Using the caloric values for carbohydrate, fat, and protein (Table 6.2), we can calculate the total number of kcal:

$$\text{Carbohydrate:} \quad 35 \text{ g} \times \frac{4 \text{ kcal}}{1 \text{ g}} = 140 \text{ kcal}$$

$$\text{Fat:} \quad 10 \text{ g} \times \frac{9 \text{ kcal}}{1 \text{ g}} = 90 \text{ kcal}$$

$$\text{Protein:} \quad 5 \text{ g} \times \frac{4 \text{ kcal}}{1 \text{ g}} = 20 \text{ kcal}$$

Total caloric content: 250 kcal

STUDY CHECK

A 1-oz (28 g) serving of oat-bran hot cereal with half a cup of whole milk contains 22 g of carbohydrate, 7 g of fat, and 10 g of protein. If you eat two servings of the oat bran for breakfast, how many kilocalories will you obtain?

QUESTIONS AND PROBLEMS

■ **Energy and Nutrition**

6.21 Using the following data, determine the kilocalories for each food burned in a calorimeter:
 a. one stalk of celery that heats 505 g water from 25.2°C to 35.7°C
 b. a waffle that heats 4980 g water from 20.6°C to 62.4°C

6.22 Using the following data, determine the kilocalories for each food burned in a calorimeter.
 a. one cup of popcorn that changes the temperature of 1250 g water from 25.5°C to 50.8°C
 b. a sample of olive oil that increases the temperature of 357 g water from 22.7°C to 38.8°C

6.23 Using the caloric values for foods, determine each of the following:
 a. The total Calories for 1 cup of orange juice that has 26 g of carbohydrate, no fat, and 2 g of protein.
 b. The grams of carbohydrate in 1 apple if the apple has no fat and no protein and provides 72 kcal of energy.
 c. The number of Calories in 1 tablespoon of vegetable oil, which contains 14 g of fat and no carbohydrate or protein.
 d. How many Calories are in 1 breakfast roll if it has 30. g of carbohydrate, 15 g of fat, and 5 g of protein?

6.24 Using the caloric values for foods, determine each of the following:
 a. The total Calories for 2 tablespoons of crunchy peanut butter that contains 6 g of carbohydrate, 16 g of fat, and 7 g of protein.
 b. The grams of protein in 1 cup of soup that has 110 Cal with 7 g of fat and 9 g of carbohydrate.

HEALTH NOTE

Losing and Gaining Weight

The number of kilocalories needed in the daily diet of an adult depends on sex and physical activity. Some general levels of energy needs are given in Table 6.4.

When food intake exceeds energy output, a person's body weight increases. Food intake is usually regulated by the hunger center in the hypothalamus, located in the brain. The regulation of food intake is normally proportional to the nutrient stores in the body. If these nutrient stores are low, you feel hungry; if they are high, you do not feel like eating.

Weight reduction occurs when food intake is less than energy output. Many diet products contain cellulose, which has no nutritive value but provides bulk and makes you feel full. Some diet drugs depress the hunger center and must be used with caution because they excite the nervous system and can elevate blood pressure. Because muscular exercise is an important way to expend energy, an increase in daily exercise aids weight loss. Table 6.5 lists some activities and the amount of energy they require.

TABLE 6.4

Typical Energy Requirements for 70-kg Adult		
Sex	Energy (kJ)	Energy (kcal)
Female	10 000	2200
Male	12 500	3000

TABLE 6.5

Energy Expended by a 70-kg Person		
Activity	Energy (kJ/hr)	Energy (kcal/hr)
Sleeping	250	60
Sitting	420	100
Walking	840	200
Swimming	2100	500
Running	3100	750

c. How many grams of sugar (carbohydrate) are in 1 can of cola if there are 140 Cal and no fat and no protein?

d. How many grams of fat are in 1 avocado if there are 405 Calories, 13 g of carbohydrate, and 5 g of protein?

6.25 One cup of clam chowder contains 9 g of protein, 12 g of fat, and 16 g of carbohydrate. How many kilocalories are in the clam chowder?

6.26 A high-protein diet contains 70. g of carbohydrate, 150 g of protein, and 5.0 g of fat. How many kilocalories does this diet provide?

6.5 States of Matter

Matter is anything that occupies space and has mass. This book, the food you eat, the water you drink, your cat or dog, and the air you breathe are just a few examples of matter. On Earth, matter exists in one of three *physical states*: solid, liquid, or gas. All matter is made up of tiny particles. In a **solid**, very strong attractive forces hold the particles close together. They are arranged in such a rigid pattern they can only vibrate slowly in their fixed positions. This gives a solid a definite shape and volume. For many solids, this rigid structure produces a crystal such as seen in quartz and amethyst (see Figure 6.3.)

In a **liquid**, the particles have enough energy to move freely in random directions. They are still close to each other and have enough attractions to maintain a

LEARNING GOAL

Identify the physical state of a substance. Describe the attractive forces between ions, polar molecules, and nonpolar molecules.

MC GOB   Forces Between Molecules

FIGURE 6.3

The solid states of (**a**) quartz and (**b**) amethyst, a purple form of quartz.

Q **Why do these crystals have a definite shape?**

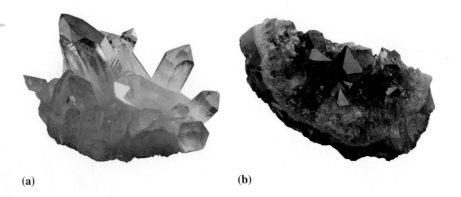

(a)   (b)

FIGURE 6.4

A liquid with a volume of 100 mL takes the shape of its container.

Q **Why does a liquid have a definite volume, but not a definite shape?**

MC GOB **SELF-STUDY ACTIVITY**
Intermolecular Forces

definite volume, but there is no rigid structure. Thus, when oil, water, or vinegar is poured from one container to another, the liquid maintains its own volume but takes the shape of the new container (see Figure 6.4).

The air you breathe is made of gases, mostly nitrogen and oxygen. In a **gas**, the molecules move at high speeds, which creates great distances between molecules. This behavior allows gases to fill their container. Gases have no definite shape or volume of their own; they take the shape and volume of their container, as shown in Figure 6.5. Table 6.6 compares some of the properties of the three states of matter.

Attractive Forces Between Particles

In gases, the interactions between particles are minimal, which allows gas molecules to move far apart from each other. In solids and liquids, there are sufficient interactions between the particles to hold them close together, although some solids have low melting points while others have very high melting points. Such differences in properties are explained by looking at the various kinds of attractive forces between particles.

In ionic compounds, the positive and negative ions are held together by ionic bonds. These are strong attractive forces that require large amounts of energy to pull the ions apart and melt the ionic solid. As a result, ionic solids have high melting points. For example, the ionic solid NaCl melts at 801°C. In compounds held together by covalent bonds, there are other types of attractive forces that exist in the solid. These attractive forces, which are weaker than ionic bonds, include dipole–dipole attractions, hydrogen bonding, and dispersion forces.

In Chapter 5, we learned that some covalent molecules have dipoles. In polar molecules, attractive forces called **dipole–dipole attractions** occur between the positive end of one molecule and the negative end of another. For example, in a sample of HCl, the positive hydrogen atom of one dipole attracts the negative chlorine atom in another molecule.

$$\overset{\delta^+}{H}-\overset{\delta^-}{Cl}\cdots\overset{\delta^+}{H}-\overset{\delta^-}{Cl}$$

A strong dipole occurs when a hydrogen atom is attached to an atom of fluorine, oxygen, or nitrogen, all of which have high electronegativity values. In a special type of dipole–dipole attraction called a **hydrogen bond**, an attractive force occurs between the partially positive hydrogen atom and the strongly electronegative atoms F, O, or N. Hydrogen bonds are a major factor in the formation and structure of biological molecules such as proteins and DNA.

TABLE 6.6
Some Properties of Solids, Liquids, and Gases

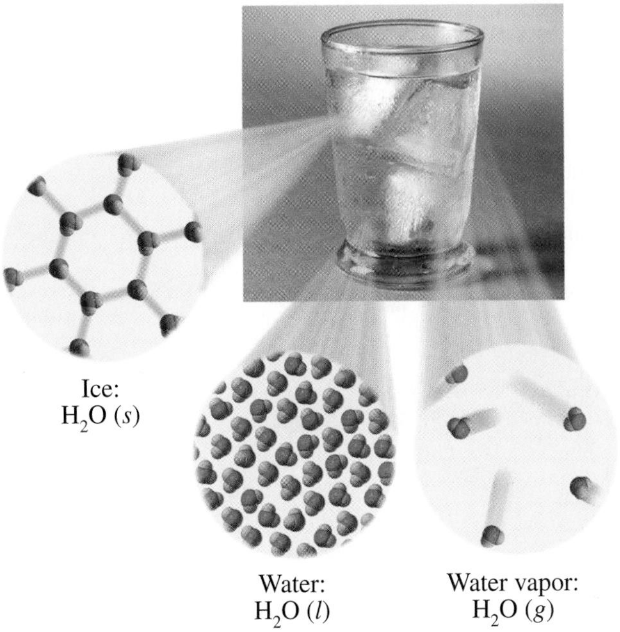

Property	Solid	Liquid	Gas
Shape	Has a definite shape	Takes the shape of the container	Takes the shape of the container
Volume	Has a definite volume	Has a definite volume	Fills the volume of the container
Arrangement of particles	Fixed, very close	Random, close	Random, far apart
Interaction between particles	Very strong	Strong	Essentially none
Movement of particles	Very slow	Moderate	Very fast
Examples	Ice, salt, iron	Water, oil, vinegar	Water vapor, helium, air

Ice: H_2O (s)

Water: H_2O (l)

Water vapor: H_2O (g)

FIGURE 6.5

A gas takes the shape and volume of its container.

Q **Why does a gas fill the volume of a container?**

TABLE 6.7

Melting Points and Attractive Forces of Selected Substances

Substance	Melting Point	Attractive Force
MgF_2	1248°C	Ionic
Na_2S	920°C	
NaCl	801°C	
H_2O	0°C	Hydrogen bonds (H and F, O, or N)
NH_3	−78°C	
CH_3OH	−98°C	
CH_3SH	−123°C	Dipole–dipole (polar molecules)
CH_3F	−142°C	
C_2H_6	−172°C	
CH_4	−183°C	Dispersion (nonpolar molecules)
F_2	−220°C	

Examples of Hydrogen Bonding

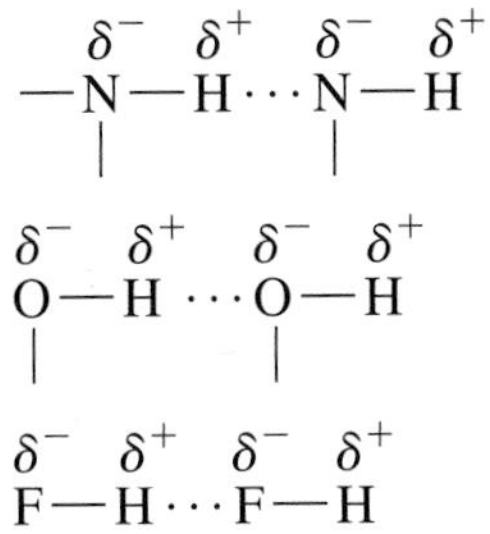

$$\overset{\delta^-}{-N}-\overset{\delta^+}{H}\cdots\overset{\delta^-}{N}-\overset{\delta^+}{H}$$

$$\overset{\delta^-}{O}-\overset{\delta^+}{H}\cdots\overset{\delta^-}{O}-\overset{\delta^+}{H}$$

$$\overset{\delta^-}{F}-\overset{\delta^+}{H}\cdots\overset{\delta^-}{F}-\overset{\delta^+}{H}$$

Hydrogen bonds are the strongest type of attractive forces between polar molecules.

A comparison of the melting points and types of attractive forces of some substances is shown in Table 6.7.

Nonpolar compounds do form solids, but at very low temperatures. Very weak attractions called **dispersion forces** occur when more electrons are momentarily

TABLE 6.8

Comparison of Bonding and Attractive Forces

Type of Force	Particle Arrangement	Energy (kcal/mole)	Example
Between atoms or ions **Ionic bond**		100–1000	$Na^+ \cdots Cl^-$
Covalent bond (X = nonmetal)	 X : X	20–200	Cl—Cl
Between molecules **Hydrogen bond** (X = F, O, or N)	$\delta^+\ \delta^-\quad \delta^+\ \delta^-$ H X -- H X	2–10	$\delta^+\ \delta^-\quad \delta^+\ \delta^-$ H—F $\cdots$ H—F
Dipole–dipole (X and Y = different nonmetals)	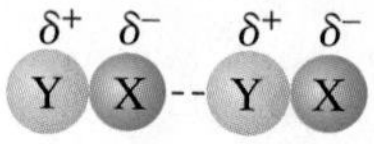 $\delta^+\ \delta^-\quad \delta^+\ \delta^-$ Y X -- Y X	1–5	$\delta^+\ \delta^-\quad \delta^+\ \delta^-$ Br—Cl $\cdots$ Br—Cl
Dispersion (Temporary shift of electrons in nonpolar bonds)	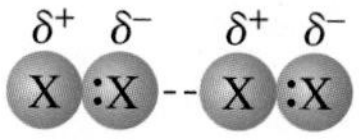 $\delta^+\ \delta^-\quad \delta^+\ \delta^-$ X :X -- X :X	(temporary dipoles) 0.01–2	$\delta^+\ \delta^-\quad \delta^+\ \delta^-$ F—F $\cdots$ F—F

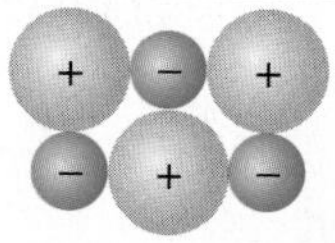

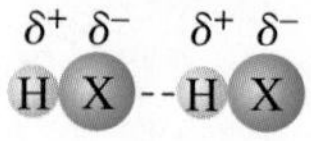

present at one end of the molecule, which forms a temporary dipole. Although these dispersion forces are the weakest of the dipole–dipole interactions, they make it possible for nonpolar molecules to form liquids and solids. The various types of attractions between particles in solids and liquids are summarized in Table 6.8.

SAMPLE PROBLEM 6.8

▪ Attractive Forces Between Particles

Indicate the major type of molecular interaction

a. dipole–dipole   **b.** hydrogen bonding
c. dispersion forces

expected of each of the following:

1. H—F   **2.** F—F   **3.** PCl_3

SOLUTION

1. **(b)** H—F is a polar molecule that interacts with other H—F molecules by hydrogen bonding.
2. **(c)** Because F—F is nonpolar, only dispersion forces provide attractive forces.
3. **(a)** The polarity of the PCl_3 molecules provides dipole–dipole interactions.

STUDY CHECK

Why is the boiling point of H_2S lower than that of H_2O?

QUESTIONS AND PROBLEMS

■ States of Matter

6.27 Indicate whether each of the following describes a gas, a liquid, or a solid:
   a. This substance has no definite volume or shape.
   b. The particles in this substance do not interact strongly with each other.
   c. The particles of this substance are held in a definite structure.

6.28 Indicate whether each of the following describes a gas, a liquid, or a solid:
   a. The substance has a definite volume but takes the shape of the container.
   b. The particles of this substance are very far apart.
   c. This substance occupies the entire volume of the container.

6.29 Identify the major type of interactive force that occurs in the following substances:
   a. BrF   **b.** KCl   **c.** CCl$_4$ **d.** HF   **e.** Cl$_2$

6.30 Identify the major type of interactive force that occurs in the following substances:
   a. HCl   **b.** MgF$_2$ **c.** PBr$_3$ **d.** Br$_2$ **e.** NH$_3$

6.31 Identify the strongest forces between molecules of each of the following:
   a. CH$_3$OH   **b.** Cl$_2$ **c.** HCl   **d.** CCl$_4$ **e.** CH$_3$CH$_3$

6.32 Identify the strongest forces between molecules of each of the following:
   a. O$_2$ **b.** HF   **c.** CH$_3$Cl   **d.** H$_2$O   **e.** NH$_3$

6.33 Identify the substance that would have the higher boiling point in each pair and explain your choice.
   a. HF or HBr   **b.** NaF or HF   **c.** MgBr$_2$ or PBr$_3$ **d.** CH$_4$ or CH$_3$OH

6.34 Identify the substance that would have the higher boiling point in each pair and explain your choice.
   a. MgCl$_2$ or PCl$_3$ **b.** H$_2$O or H$_2$Se
   c. NH$_3$ or PH$_3$ **d.** F$_2$ or HF

6.6 Changes of State

You are probably already familiar with most of these phase changes shown in Figure 6.6. Matter undergoes a **change of state** when it is converted from one state to another state.

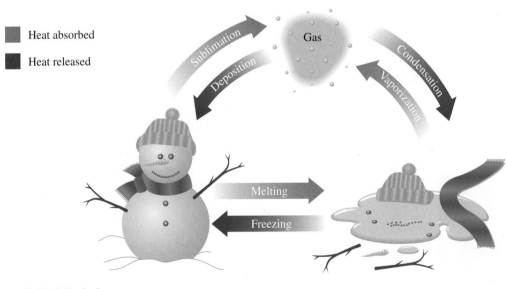

Heat absorbed

Heat released

FIGURE 6.6

A summary of the changes of state.

Q Is heat added or released when liquid water freezes?

When heat is added to a solid, the particles in the rigid structure begin to move faster. At a temperature called the **melting point (mp)**, the particles in the solid gain sufficient energy to overcome the attractive forces that hold them together. The particles in the solid separate and move about in random patterns. The substance is **melting**, changing from a solid to a liquid.

If the temperature of a liquid is lowered, the reverse process takes place. Kinetic energy is lost, the particles slow down, and attractive forces pull the particles close together. The substance is **freezing**. A liquid changes to a solid at the **freezing point (fp)**, which is the same temperature as the melting point. Every substance has its own freezing (melting) point: water freezes (melts) at 0°C; gold freezes (melts) at 1064°C; nitrogen freezes (melts) at −210°C. During any phase change, the temperature of a substance remains constant.

Heat of Fusion

During melting, energy called the **heat of fusion** is needed to separate the particles of a solid. For example, 80. calories of heat are needed to melt exactly 1 g of ice at its melting point.

Heat of Fusion for Water

$$\frac{80. \text{ cal}}{1 \text{ g ice}}$$

The heat of fusion (80. cal/g) is also the heat that must be removed to freeze 1 g of water at its freezing point. Water is sometimes sprayed in fruit orchards during very cold weather. If the air temperature drops to 0°C, the water begins to freeze. Heat is released as the water molecules bond together, which warms the air and protects the fruit.

To determine the heat needed to melt a sample of ice, multiply the mass of the ice by its heat of fusion. There is no temperature change in the calculation because temperature remains constant as long as the ice is melting.

Calculating Heat to Melt (or Freeze) Water

Heat = mass × heat of fusion

cal = g̸ × 80. cal/g̸

SAMPLE PROBLEM 6.9

▪ Heat of Fusion

Ice cubes at 0°C with a mass of 26 g are added to your soft drink.

a. How much heat (cal) must be absorbed to melt all the ice at 0°C?
b. What happens to the temperature of your soft drink. Why?

SOLUTION

a. The heat in calories required to melt the ice is calculated as follows.

▪ **STEP 1 Given** 26 g of $H_2O(s)$ **Need** calories to melt ice

▪ **STEP 2** g ice Heat of fusion cal

Guide to Calculations Using Heat of Fusion/Vaporization

STEP 1
List grams of substance and change of state.

STEP 2
Write the plan to convert grams to heat and desired unit.

STEP 3
Write the heat conversion factor and metric factor if needed.

STEP 4
Set up the problem with factors.

■ Histologist

"While a patient is in surgery for skin cancer, some tissue around the cancer is sent to us," says Mary Ann Pipe, histology technician. "Using the Mohs surgical technique, we place the tissue block on a glass slide, chill it to −30°C in a machine called a cryostat, and freeze it for longer in another machine called a heat extractor. From this frozen block of tissue, we cut extremely thin slices—one-thousandth of an inch—from different depths. We prepare three separate slides from skin at three different depths up to the surface of the skin. We stain the cells pink and blue by placing the slides in hemotoxin, and then in eosin. The slices are a tissue map that the doctor can easily read to determine if the margins around the skin cancer are clear or whether more tissue must be removed."

■ STEP 3 **Equalities/Conversion Factors**

$$1 \text{ g } H_2O(s \longrightarrow l) = 80. \text{ cal}$$

$$\frac{80. \text{ cal}}{1 \text{ g } H_2O} \quad \text{and} \quad \frac{1 \text{ g } H_2O}{80. \text{ cal}}$$

■ STEP 4 **Set Up Problem**

$$26 \text{ g } H_2O \times \frac{80. \text{ cal}}{1 \text{ g } H_2O} = 2100 \text{ cal}$$

b. The soft drink will be colder because heat from the soft drink is providing the energy to melt the ice.

STUDY CHECK

In a freezer, 150 g water at 0°C is placed in an ice cube tray. How much heat in kilocalories must be removed to form ice cubes at 0°C?

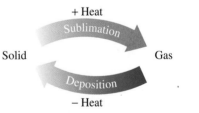

Sublimation

In a process called **sublimation**, the particles on the surface of a solid absorb enough heat to change directly to a gas with no temperature change and without going through the liquid state. For example, dry ice, which is solid carbon dioxide, sublimes at −78°C. It is called "dry" because it does not form a liquid upon warming. In very cold areas, snow does not melt but sublimes directly to vapor. In a frost-free refrigerator, ice on the walls of the freezer sublimes when warm air is circulated through the compartment during the defrost cycle.

Freeze-dried foods prepared by sublimation are convenient for long-term storage and for camping and hiking. A food that has been frozen is placed in a vacuum chamber where it dries as the ice sublimes. The dried food retains all of its nutritional value and needs only water to be edible. A food that is freeze-dried does not need refrigeration because bacteria cannot grow without moisture.

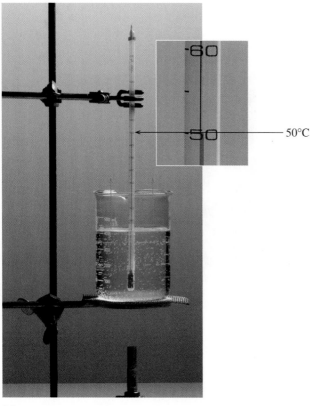
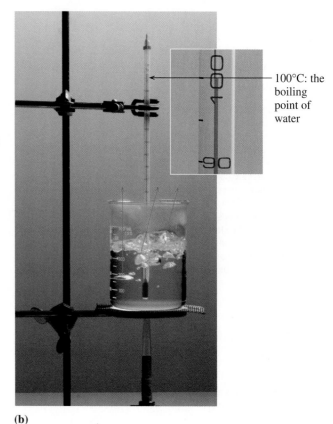

(a)   50°C

100°C: the
boiling
point of
water

(a)   (b)

FIGURE 6.7

(**a**) Evaporation occurs at the surface of a liquid. (**b**) Boiling occurs as bubbles of
gas form throughout the liquid.

Q **Why does water evaporate faster at 80°C than at 20°C?**

Boiling and Condensation

Water in a mud hole disappears, unwrapped food dries out, and clothes hung on a
line dry. **Evaporation** is taking place as water molecules with sufficient energy
escape from the liquid surface and enter the gas phase. (See Figure 6.7a.) The loss
of the "hot" water molecules removes heat, which cools the remaining liquid water.
With additional heat, more and more water molecules evaporate. At the **boiling
point (bp)**, all the molecules of the liquid have the energy needed to change into a
gas. The **boiling** of a liquid occurs as gas bubbles form throughout the liquid, then
rise to the surface and escape. (See Figure 6.7b.)

In a reverse process, water molecules in the gas cool, lose kinetic energy, and
slow down. They undergo the process of **condensation** as water vapor is converted
back to liquid. You may have noticed that condensation occurs when you take a hot
shower and the water vapor forms water droplets on a mirror. Because a substance
loses heat as it condenses, its surroundings become warmer. That is why, when a
rainstorm is approaching, we notice a warming of the air as gaseous water molecules condense to rain.

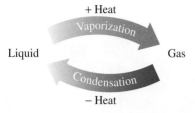

+ Heat
Vaporization
Liquid   Gas
Condensation
− Heat

Heat of Vaporization

The energy needed to vaporize exactly 1 g of liquid to gas at its boiling point is
called the **heat of vaporization**. For water, 540 calories are needed to convert 1 g
of water to vapor at 100°C.

Heat of Vaporization for Water

$$\frac{540 \text{ cal}}{1 \text{ g water}}$$

When 1 g of water condenses, the heat of vaporization, 540 calories, is the amount of energy that must be removed. Therefore, 540 cal/g is also the *heat of condensation* of water. To calculate the amount of heat needed to vaporize (or condense) a sample of water, multiply the mass of the sample by the heat of vaporization. As before, no temperature change occurs during a change of state.

Calculating Heat to Vaporize (or Condense) Water

Heat = mass × heat of vaporization

cal $\;= \cancel{g} \;\; \times$ 540 cal/$\cancel{g}$

HEALTH NOTE

▪ Steam Burns

Hot water at 100°C will cause burns and damage to the skin. However, getting steam on the skin is even more dangerous. Let us consider 25 g hot water at 100°C. If this water falls on a person's skin, the temperature of the water will drop to body temperature, 37°C. The heat released during cooling burns the skin. The amount of heat can be calculated from the temperature change, 63°C.

$$25 \; \cancel{g} \times 63°\cancel{C} \times \frac{1.00 \text{ cal}}{\cancel{g} \; °\cancel{C}} = 1600 \text{ cal heat}$$

For comparison, we can calculate the amount of heat released when 25 g steam at 100°C hits the skin. First, the steam condenses to water (liquid) at 100°C:

$$25 \; \cancel{g} \times \frac{540 \text{ cal}}{1 \; \cancel{g}} = 14\ 000 \text{ cal heat released}$$

Now the temperature of the 25 g water drops from 100°C to 37°C, releasing still more heat as we saw earlier. We can calculate the total amount of heat released from the condensation and cooling of the steam as follows:

Condensation (100°C) = 14 000 cal

Cooling (100°C to 37°C) = 1600 cal

Heat released = 16 000 cal (rounded)

The amount of heat released from steam is 10 times greater than the heat from the same amount of hot water.

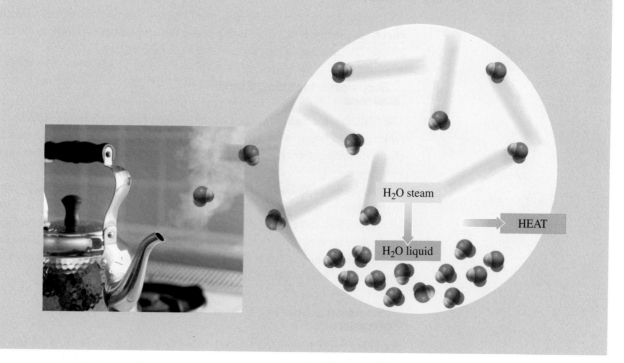

H₂O steam

H₂O liquid

HEAT

SAMPLE PROBLEM 6.10

■ **Using Heat of Vaporization**

In a sauna, 150 g of water is converted to steam at 100°C. How many kilocalories of heat are needed?

SOLUTION

■ **STEP 1** **Given** 150 g $H_2O(l)$ to $H_2O(g)$
Need kilocalories of heat to change state

■ **STEP 2** **Plan** g H_2O | Heat of vaporization | cal | Metric factor | kcal

■ **STEP 3** **Equalities/Conversion Factors**

| 1 g H_2O $(l \rightarrow g)$ = 540 cal | | 1 kcal = 1000 cal | |
| $\dfrac{540 \text{ cal}}{1 \text{ g } H_2O}$ and $\dfrac{1 \text{ g } H_2O}{540 \text{ cal}}$ | | $\dfrac{1000 \text{ cal}}{1 \text{ kcal}}$ and $\dfrac{1 \text{ kcal}}{1000 \text{ cal}}$ | |

■ **STEP 4** **Set up Problem**

$$150 \text{ g } H_2O \times \frac{540 \text{ cal}}{1 \text{ g } H_2O} \times \frac{1 \text{ kcal}}{1000 \text{ cal}} = 81 \text{ kcal}$$

STUDY CHECK

When steam from a pan of boiling water reaches a cool window, it condenses. How much heat in kilocalories (kcal) is released when 25 grams of steam condenses at 100°C?

Heating and Cooling Curves

All the changes of state can be illustrated visually as a diagram called a **heating curve**. (See Figure 6.8a.) In a heating curve, the temperature is shown on the vertical axis. The addition of heat is shown on the horizontal axis.

Steps on a Heating Curve

The first diagonal line indicates a warming of a solid as heat is added. When the melting temperature is reached, solid begins to change to a liquid without any change in temperature. The melting process appears as a flat line, or plateau, on the heating curve.

After all the particles are in the liquid state, the temperature of the liquid begins to rise again as shown by the next diagonal line. As more heat is added, the particles move about more vigorously until the boiling point is reached. Now the liquid begins to boil, indicated as another flat line on the heating curve. Again there is no temperature change as liquid changes into gas. After all the liquid becomes a gas, any additional heat will cause the temperature of the gas to rise again.

Steps on a Cooling Curve

A **cooling curve** is a diagram of the cooling process. (See Figure 6.8b.) When heat is removed from a substance such as water vapor, the gas cools and condenses. On

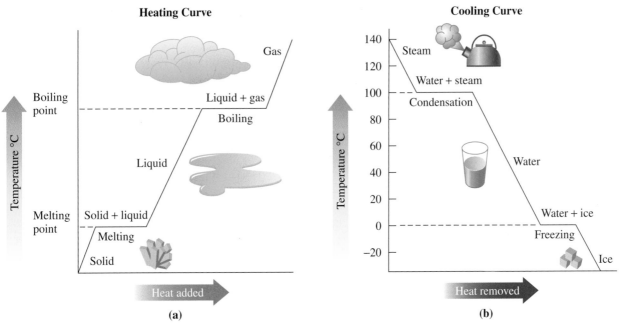

FIGURE 6.8

(a) A heating curve diagrams the temperature increases and changes in state as heat is added. (b) A cooling curve for water.

Q What does the plateau at 100°C represent on the cooling curve for water?

the cooling curve, the temperature is plotted on the vertical axis and the removal of heat on the horizontal axis. As a gas cools, heat is lost and the temperature drops. At the condensation point (same as the boiling point), the gas begins to condense, forming liquid. This process is indicated by a flat line (plateau) on the cooling curve at the condensation point.

After all of the gas has changed into liquid, the particles within the liquid cool as indicated by the downward sloping line that shows the temperature decrease. At the freezing point, a second flat line indicates the change of state from liquid to solid. Once all of the substance is frozen, the loss of more heat lowers the solid's temperature below its freezing point.

Combining Energy Calculations

Up to now, we have calculated one step in a heating or cooling curve. However, many problems require a combination of steps that include a temperature change as well as a change of state. The heat is calculated for each step separately and then added together to find the total energy as seen in Sample Problem 6.11.

MC GOB Heat, Energy, and Changes of State

SAMPLE PROBLEM 6.11

■ Combining Heat Calculations

Calculate the total heat in calories needed to convert 15.0 g of liquid water at 25°C to steam at 100°C.

SOLUTION

■ STEP 1 **Given** 15.0 g water at 25°C

Need Heat (cal) needed to warm water and change to steam.

■ STEP 2 **Plan** When several changes occur, draw a diagram of heating and changes of state.

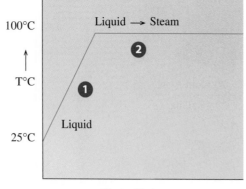

$$\text{Total heat} = \text{calories needed to warm } H_2O \text{ from } 25°C \text{ to } 100°C$$
$$+ \text{ calories to change liquid to steam at } 100°C$$

■ **STEP 3 Equalities/Conversion Factors**

$$SH_{H_2O} = 1 \frac{\text{cal}}{\text{g °C}}$$

$$\frac{1 \text{ cal}}{\text{g °C}} \quad \text{and} \quad \frac{\text{g °C}}{1 \text{ cal}}$$

$$1 \text{ g } H_2O \,(l \rightarrow g) = 540 \text{ cal}$$

$$\frac{540 \text{ cal}}{1 \text{ g } H_2O} \quad \text{and} \quad \frac{1 \text{ g } H_2O}{540 \text{ cal}}$$

■ **STEP 4 Set Up Problem**

$$\Delta T = 100°C - 25°C = 75°C$$

Heat needed to warm H_2O (25°C) to H_2O (100°C):

$$15.0 \text{ g } \times \quad 75°C \quad \times \quad \frac{1.00 \text{ cal}}{\text{g °C}} \quad = \quad 1100 \text{ cal}$$

Heat needed to change H_2O (liquid) to H_2O (gas) at 100°C:

$$15.0 \text{ g } \times \quad \frac{540 \text{ cal}}{1 \text{ g}} \quad = \quad 8100 \text{ cal}$$

Calculate the total heat:

Heating water	1100 cal
Changing liquid to steam (100°C)	8100 cal
Total heat needed	9200 cal

STUDY CHECK

How many kilocalories (kcal) are released when 25.0 g steam at 100°C condenses, cools to 0°C, and freezes? (*Hint:* The solution will require three energy calculations.)

■

QUESTIONS AND PROBLEMS

■ **Changes of State**

6.35 Identify each of the following changes of state as melting, freezing, or sublimation.
    a. The solid structure of a substance breaks down as liquid forms.
    b. Coffee is freeze-dried.
    c. Water on the street turns to ice during a cold wintry night.

6.36 Identify each of the following changes of state as melting, freezing, or sublimation.
   a. Dry ice in an ice-cream cart disappears.
   b. Snow on the ground turns to liquid water.
   c. Heat is removed from 125 g of liquid water at 0°C.

6.37 Calculate the heat needed at 0°C to make each of the following changes of state. Indicate whether heat was absorbed or released.
   a. calories to melt 65 g of ice
   b. calories to melt 17 g of ice
   c. kilocalories to freeze 225 g of water

6.38 Calculate the heat needed at 0°C to make each of the following changes of state. Indicate whether heat was absorbed or released.
   a. calories to freeze 35 g of water
   b. calories to freeze 250 g of water
   c. kilocalories to melt 140 g of ice

6.39 Identify each of the following changes of state as evaporation, boiling, or condensation:
   a. The water vapor in the clouds changes to rain.
   b. Wet clothes dry on a clothesline.
   c. Lava flows into the ocean and steam forms.
   d. After a hot shower, your bathroom mirror is covered with water.

6.40 Identify each of the following changes of state as evaporation, boiling, or condensation:
   a. At 100°C, the water in a pan changes to steam.
   b. On a cool morning, the windows in your car fog up.
   c. A shallow pond dries up in the summer.
   d. Your teakettle whistles when the water is ready for tea.

6.41 **a.** How does perspiration during heavy exercise cool the body?
   b. Why do clothes dry more quickly on a hot summer day than on a cold winter day?

6.42 a. When a sports injury occurs during a game, a spray such as ethyl chloride may be used to numb an area of the skin. Explain how a substance like ethyl chloride that evaporates quickly can numb the skin.
   b. Why does water in a wide, flat, shallow dish evaporate more quickly than the same amount of water in a tall, narrow glass?

6.43 Calculate the heat change at 100°C in each of the following problems. Indicate whether heat was absorbed or released.
   a. calories to vaporize 10.0 g of water
   b. kilocalories to vaporize 50.0 g of water
   c. kilocalories to condense 8.0 kg of steam

6.44 Calculate the heat change at 100°C in each of the following problems. Indicate whether heat was absorbed or released.
   a. calories to condense 10.0 g of steam
   b. kilocalories to condense 75 g of steam
   c. kilocalories to vaporize 44 g of water

6.45 Draw a heating curve for a sample of ice that is heated from −20°C to 140°C. Indicate the segment of the graph that corresponds to each of the following:
   a. solid   **b.** melting point   **c.** liquid   **d.** boiling point   **e.** gas

6.46 Draw a cooling curve for a sample of steam that cools from 110°C to −10°C. Indicate the segment of the graph that corresponds to each of the following:
   a. solid   **b.** freezing point   **c.** liquid
   d. condensation point (boiling point)   **e.** gas

6.47 Using the values for the heat of fusion, specific heat of water, and/or heat of vaporization, calculate the amount of heat energy in each of the following:
   a. calories needed to warm 20.0 g of water at 15°C to 72°C (one step)
   b. calories need to melt 50.0 g of ice at 0°C and to warm the liquid to 65°C (two steps)
   c. kilojoules given off when 15 g of steam condenses at 100°C and the liquid cools to 0°C (two steps)
   d. kilocalories needed to melt 24 g of ice at 0°C, to warm the liquid to 100°C, and to vaporize it at 100°C (three steps)

6.48 Using heat of fusion, specific heat of water, and/or heat of vaporization, calculate the amount of heat energy in each of the following:
   a. calories to condense 125 g of steam at 100°C and to cool the liquid to 15°C (two steps)
   b. joules to melt a 525-g ice cube at 0°C and to warm the liquid to 15°C (two steps)
   c. kilocalories to condense 85 g of steam at 100°C, cool the liquid to 0°C, and freeze it at 0°C (three steps)
   d. calories to warm 55 mL of water (density = 1.0 g/mL) from 10°C to 100°C and vaporize it at 100°C (two steps)

CONCEPT MAP

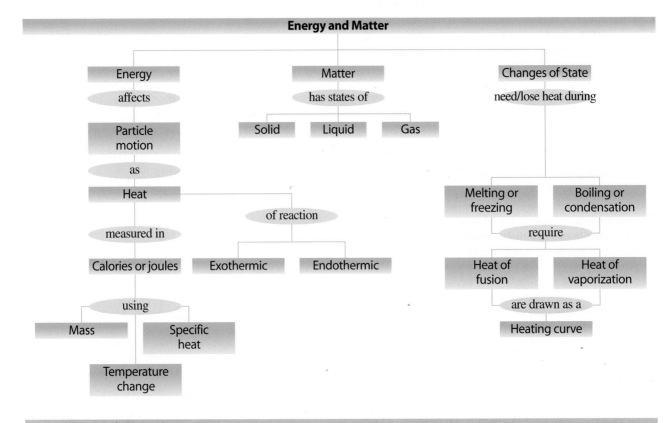

CHAPTER REVIEW

6.1 Energy
Energy is the ability to do work. Potential energy is stored energy; kinetic energy is the energy of motion. Heat, a common form of energy, is measured in calories or joules. One cal is equal to 4.184 J.

6.2 Energy in Chemical Reactions
In chemical reactions, the heat of reaction (ΔH) is the energy difference between the reactants and the products. In an exothermic reaction, the energy of the products is lower than the reactants.

Heat is released and ΔH is negative. In an endothermic reaction, the energy of the products is higher than the reactants. Heat is absorbed and the ΔH is positive.

6.3 Specific Heat
Specific heat is the amount of energy required to raise the temperature of exactly 1 g of a substance by exactly 1°C. The heat lost or gained by a substance is determined by multiplying its mass (g), the temperature change (ΔT), and its specific heat (cal/g °C).

6.4 Energy and Nutrition

The nutritional Calorie is the same amount of energy as 1 kcal or 1000 calories. The caloric content of a food is the sum of kilocalories from carbohydrate, fat, and protein.

6.5 States of Matter

Matter is anything that has mass and occupies space. The three states of matter are solid, liquid, and gas. In ionic solids, oppositely charged ions are held in a rigid structure by ionic bonds. Attractive forces called dipole–dipole attractions and hydrogen bonds hold the solid and liquid states of polar covalent compounds together. Nonpolar compounds form solids and liquids by temporary dipoles called dispersion forces.

6.6 Changes of State

Melting occurs when the particles in a solid absorb enough energy to break apart and form a liquid. The amount of energy required to convert exactly 1 g of solid to liquid is called the heat of fusion. Boiling is the vaporization of a liquid at its boiling point. The heat of vaporization is the amount of heat needed to convert exactly 1 g of liquid to vapor. A heating or cooling curve illustrates the changes in temperature and state as heat is added to or removed from a substance. Plateaus on the graph indicate changes of state with no change in temperature.

KEY TERMS

activation energy The energy needed to break the bonds of reacting molecules.

boiling The formation of bubbles of gas throughout a liquid.

boiling point (bp) The temperature at which a liquid changes to gas (boils) and gas changes to liquid (condenses).

caloric value The kilocalories obtained per gram of the food types: carbohydrate, fat, and protein.

calorie (cal) The amount of heat energy that raises the temperature of exactly 1 g of water exactly 1°C.

Calorie (Cal) The dietary unit of energy, equal to 1000 cal, or 1 kcal.

change of state The transformation of one state of matter to another; for example, solid to liquid, liquid to solid, liquid to gas.

condensation The change of state of a gas to a liquid.

cooling curve A diagram that illustrates temperature changes and changes of states for a substance as heat is removed.

dipole–dipole attractions Attractive forces between oppositely charged ends of polar molecules.

dispersion forces Weak dipole bonding that results from a momentary polarization of nonpolar molecules in a substance.

endothermic reaction A reaction wherein the energy of the products is greater than that of the reactants.

energy The ability to do work.

evaporation The formation of a gas (vapor) by the escape of high-energy molecules from the surface of a liquid.

exothermic reaction A reaction wherein the energy of the reactants is greater than that of the products.

freezing A change of state from liquid to solid.

freezing point (fp) The temperature at which a liquid changes to a solid (freezes), a solid changes to a liquid (melts).

gas A state of matter characterized by no definite shape or volume. Particles in a gas move rapidly.

heat The energy associated with the motion of particles in a substance.

heat of fusion The energy required to melt exactly 1 g of a substance at its melting point. For water, 80. cal are needed to melt 1 g of ice; 80. cal are released when 1 g of water freezes.

heat of reaction The heat (symbol ΔH) absorbed or released when a reaction takes place at constant pressure.

heat of vaporization The energy required to vaporize 1 g of a substance at its boiling point. For water, 540 calories are needed to vaporize exactly 1 g of liquid; 1 g of steam gives off 540 cal when it condenses.

heating curve A diagram that shows the temperature changes and changes of state of a substance as it is heated.

hydrogen bond The attraction between a partially positive H and a strongly electronegative atom of F, O, or N.

joule (J) The SI unit of heat energy; 4.184 J = 1 cal.

kilocalorie (kcal) An amount of heat energy equal to 1000 calories.

kinetic energy The energy of moving particles.

liquid A state of matter that takes the shape of its container but has a definite volume.

matter Anything that has mass and occupies space.

melting The conversion of a solid to a liquid.

melting point (mp) The temperature at which a solid becomes a liquid (melts). It is the same temperature as the freezing point.

potential energy An inactive type of energy that is stored for future use.

solid A state of matter that has its own shape and volume.

specific heat A quantity of heat that changes the temperature of exactly 1 g of a substance by exactly 1°C.

sublimation The change of state in which a solid is transformed directly to a gas without forming a liquid first.

work An activity that requires energy.

UNDERSTANDING THE CONCEPTS

6.49   Determine the energy to heat three cubes (gold, aluminum, and silver) each with a volume of 10.0 cm³ from 15°C to 25°C. Refer to Tables 1.10 and 6.1. What do you notice about the energy needed for each?

6.50 A 70.0-kg person has just eaten a quarter-pound cheeseburger, French fries, and a chocolate shake. According to Table 6.5, determine each of the following:

a. The number of hours of sleep needed to "burn off" the kilocalories in this meal.
b. The number of hours of running needed to "burn off" the kilocalories in this meal.

Item	Protein (g)	Fat (g)	Carbohydrate (g)
Cheeseburger	31	29	34
French fries	3	11	26
Chocolate shake	11	9	60

6.51 The following is a heating curve for chloroform, a solvent for fats, oils, and waxes.

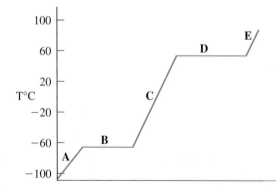

a. What is the melting point of chloroform?
b. What is the boiling point of chloroform?
c. On the heating curve, identify the segments A, B, C, D, and E as solid, liquid, gas, melting, or boiling.
d. At the following temperatures, is chloroform a solid, liquid, or gas?
$-80°C$; $-40°C$; $25°C$; $80°C$

6.52 Associate the following diagrams with a segment on the heating curve for water.

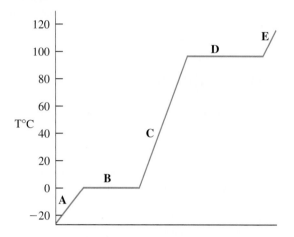

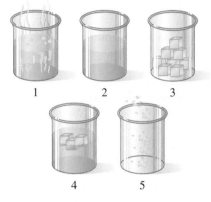

ADDITIONAL QUESTIONS AND PROBLEMS

MC GOB For instructor-assigned homework, go to www.masteringgob.com.

6.53 On a hot day, the beach sand gets hot, but the water stays cool. Compare the specific heat of sand to that of water.

6.54 Why do drops of liquid water form on a glass of iced tea?

6.55 When it rains or snows, the air temperature seems warmer. Explain.

6.56 Water is sprayed on the ground of an orchard when temperatures are near freezing to keep the fruit from freezing. Explain.

6.57 A large bottle of water (883 g) at 4°C is removed from the refrigerator. How many kilojoules (kJ) are absorbed to warm the water to a room temperature of 27°C?

6.58 If you used the 2100 kcal you expend in energy in one day to heat 50 000 g of water at 20°C, what would be its new temperature?

6.59 The formation of nitrogen monoxide, NO, from $N_2(g)$ and $O_2(g)$, requires 21.6 kcal of heat.

$$N_2(g) + O_2(g) \longrightarrow 2NO(g) \qquad \Delta H = 21.6 \text{ kcal}$$

a. How many kcal are required to form 3.00 g NO?
b. What is the complete equation (including heat) for the decomposition of NO?
c. How many kcal are released when 5.00 g NO decomposes to N_2 and O_2?

6.60 The formation of rust (Fe_2O_3) from solid iron and oxygen gas releases 1.7×10^3 kJ.

$$4Fe(s) + 3O_2(g) \longrightarrow 2Fe_2O_3(s)$$
$$\Delta H = -1.7 \times 10^3 \text{ kJ}$$

a. How many kJ are released when 2.00 g Fe react?
b. How many grams of rust form when 150 kcal are released?

6.61 A typical diet in the United States provides 15% of the calories from protein, 45% from carbohydrates, and the remainder from fats. Calculate the grams of protein, carbohydrate, and fat to be included each day in diets having the following caloric requirements:

a. 1200 kcal
b. 1900 kcal
c. 2600 kcal

6.62 Your friend has just eaten a pizza, cola soft drink, and ice cream. What is the total kilocalories your friend obtained from this meal? How many hours will your friend need to swim to "burn off" the kilocalories in this meal if your friend has a mass of 70.0-kg? (See Table 6.5.).

Item	Protein (g)	Fat (g)	Carbohydrate (g)
Pizza	13	10	29
Cola	0	0	51
Ice cream	8	28	44

6.63 If you want to lose 1 pound of "fat," which is 15% water, how many kilocalories do you need to expend?

6.64 Calculate the Cal (kcal) in 1 cup of whole milk: 12 g of carbohydrate, 9 g of fat, and 9 g of protein.

6.65 A hot-water bottle contains 725 g of water at 65°C. If the water cools to body temperature (37°C), how many kilocalories of heat could be transferred to sore muscles?

6.66 Describe the type of compound that could have each of the following types of attractive forces:

a. dipole–dipole
b. hydrogen bonds
c. dispersion forces

6.67 Indicate the major type of attractive force
(1) ionic (2) dipole–dipole
(3) hydrogen bond (4) dispersion forces
that occurs between particles of the following substances:

a. NH_3 **b.** HF **c.** CH_4
d. $CHCl_3$ **e.** H_2O **f.** LiCl

6.68 An ice cube tray holds 325 g water. If the water initially has a temperature of 25°C, how many kilojoules of heat must be removed to cool and freeze the water at 0°C?

6.69 How many kilocalories of heat are released when 85 g steam at 100°C is converted to liquid water at 15°C?

6.70 The melting point of benzene is 5.5°C and its boiling point is 80.1°C. Sketch a heating curve for benzene from 0°C to 100°C.

a. What is the state of benzene at 15°C?
b. What happens on the curve at 5.5°C?
c. What is the state of benzene at 63°C?
d. What is the state of benzene at 98°C?
e. At what temperature will both liquid and gas be present?

CHALLENGE QUESTIONS

6.71 A 25-g sample of an alloy at 98°C is placed in 50.g water at 15°C. If the final temperature reached by the alloy sample and water is 27°C, what is the specific heat (cal/g°C) of the alloy?

6.72 A 0.50 g sample of vegetable oil is placed in a calorimeter. When the sample is burned, 18.9 kJ are given off. What is the caloric value in kcal/g of the oil?

6.73 How many kilocalories of heat are released when 75 g of steam at 100°C is converted to ice at 0°C? (*Hint*: Requires the total of several steps.)

6.74 Identify the most important type of attractions between molecules for each of the following:

a. C_3H_8 **b.** CH_3OH **c.** Br_2
d. HBr **e.** IBr

6.75 A 45.0-g piece of ice at 0.0°C is added to a sample of water at 8.0°C. All of the ice melts and the temperature of the water decreases to 0.0°C. How many kilograms of water were in the sample?

6.76 When peroxide (H_2O_2) is used in rocket fuels, it produces water, oxygen, and heat:

$$2H_2O_2(l) \longrightarrow 2H_2O(l) + O_2(g)$$

a. If 2.00 g H_2O_2 releases 5.76 kJ, what is the heat of reaction?
b. How many kilojoules are released when 275 g peroxide are allowed to react?

6.77 Rearrange the heat equation to solve for each of the following:

a. The mass in grams of water that absorbs 8250 J when its temperature rises from 18.3°C to 92.6°C.
b. The mass in grams of a gold sample that absorbs 225 J when the temperature rises from 15.0°C to 47.0°C.
c. The rise in temperature (ΔT) when a 20.0 g sample of iron absorbs 1580 J.
d. The specific heat of a metal when 8.50 g of the metal absorbs 28 cal and the temperature rises from 12°C to 24°C.

6.78 A 45-g piece of ice at 0°C is added to a sample of water at 8°C. All of the ice melts and the temperature of the water decreases to 0°C. How many grams of water were in the sample?

6.79 A 3.0-kg block of lead is taken from a furnace at 300.°C and placed on a large block of ice at 0°C. The specific heat of lead is 0.13 J/g °C. If all the heat given up by the lead is used to melt ice, how much ice is melted if the temperature of the lead drops to 0°C?

6.80 When 1.0 g gasoline burns, 11 500 calories of energy are given off. If the density of gasoline is 0.74 g/mL, how many kilocalories of energy are obtained from 1.5 gallons of gasoline?

ANSWERS

Answers to Study Checks

6.1 26 000 cal

6.2 **a.** $H_2(g) + I_2(g) + 12$ kcal $\longrightarrow$ 2HI(g)
b. $\Delta H = 12$ kcal

6.3 **a.** endothermic
b. $\Delta H = 10.5$ kJ

6.4 $SH = 1.23$ J/g °C

6.5 3.48 kcal

6.6 4.0 kcal/g

6.7 380 kcal

6.8 H_2O forms hydrogen bonds.

6.9 12 kcal

6.10 14 kcal

6.11 18 kcal

Answers to Selected Questions and Problems

6.1 As the car goes up the ramp, kinetic energy changes to potential energy. As the car descends, potential energy changes to kinetic energy. At the bottom, all the energy is kinetic.

6.3 **a.** potential **b.** kinetic
c. potential **d.** potential

6.5 **a.** 3.5 kcal
b. 99.2 cal
c. 120 J
d. 1100 cal

6.7 **a.** The energy of activation is the energy required to break the bonds of the reacting molecules.
b. In exothermic reactions, the energy of the products is lower than the reactants.
c.

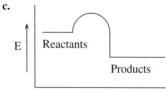

6.9 **a.** exothermic **b.** endothermic **c.** exothermic

6.11 **a.** exothermic $\Delta H = -210$ kcal
b. endothermic $\Delta H = 65.3$ kJ
c. exothermic $\Delta H = -205$ kcal

6.13 138 kcal

6.15 Copper has the lowest specific heat of the samples and will reach the highest temperature.

6.17 **a.** 250 cal **b.** 11 000 cal **c.** 9.3 kcal

6.19 **a.** 1380 J; 330. cal **b.** 1810 J; 434 cal
c. 3780 J; 904 cal **d.** 3200 J; 760 cal

6.21 **a.** 5.30 kcal **b.** 208 kcal

6.23 **a.** 110 Cal **b.** 18 g **c.** 130 Cal **d.** 280 Cal

6.25 210 kcal

6.27 **a.** gas **b.** gas **c.** solid

6.29 **a.** dipole–dipole **b.** ionic **c.** dispersion
d. hydrogen bond **e.** dispersion

6.31 **a.** hydrogen bonding
b. dispersion forces
c. dipole–dipole attraction
d. dispersion forces
e. dispersion forces

6.33 **a.** HF; hydrogen bonds are stronger than dipole–dipole interactions of HBr.
b. NaF; ionic bonds are stronger than the hydrogen bonds in HF.
c. $MgBr_2$; ionic bonds are stronger than the dipole–dipole interactions in PBr_3.
d. CH_3OH; hydrogen bonds are stronger than the dispersion forces in CH_4.

6.35 **a.** melting **b.** sublimation **c.** freezing

6.37 **a.** 5200 cal absorbed
b. 1400 cal absorbed
c. 18 kcal released

6.39 **a.** condensation
b. evaporation
c. boiling
d. condensation

6.41 **a.** The heat from the skin is used to evaporate the water (perspiration). Therefore the skin is cooled.
b. On a hot day, there are more molecules with sufficient energy to become water vapor.

6.43 **a.** 5400 cal absorbed
b. 27 kcal absorbed
c. 4300 kcal released

6.45

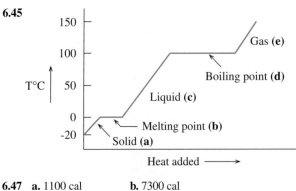

6.47 **a.** 1100 cal   **b.** 7300 cal
    c. 40. kJ   **d.** 17 kcal

6.49 gold 250 J or 59 cal; aluminum 240 J or 58 cal; silver 250 J or 59 cal. The heat needed for all the metal samples is almost the same.

6.51 **a.** −60°C   **b.** 60°C
    c. A represents the solid state. B represents the change from solid to liquid or melting of the substance. C represents the liquid state as temperature increases. D represents the change from liquid to gas or boiling of the liquid. E represents the gas state.
    d. At −80°C, solid; at −40°C, liquid; at 25°C, liquid; at 80°C, gas

6.53 Sand must have a lower specific heat than water. When both substances absorb the same amount of heat, the final temperature of the sand will be higher than that of water.

6.55 When water vapor condenses or liquid water freezes, heat is released, which warms the air.

6.57 85 kJ

6.59 **a.** 1.08 kcal
    b. $2NO(g) \longrightarrow N_2(g) + O_2(g) + 21.6$ kcal
    c. 1.80 kcal

6.61 **a.** 45 g protein, 140 g carbohydrate, 53 g fat
    b. 71 g protein, 210 g carbohydrate, 84 g fat
    c. 98 g protein, 290 g carbohydrate, 120 g fat

6.63 3500 kcal

6.65 20. kcal

6.67 **a.** 3   **b.** 3   **c.** 4
    d. 2   **e.** 3   **f.** 1

6.69 53 kcal

6.71 $\dfrac{0.34 \text{ cal}}{\text{g °C}}$

6.73 54 kcal

6.75 3.0 kg water

6.77 **a.** 26.5 g   **b.** 54.5 g   **c.** 176°C   **d.** $0.27 \dfrac{\text{cal}}{\text{g °C}}$

6.79 350 g ice melt

7 Gases

"When oxygen levels in the blood are low, the cells in the body don't get enough oxygen," says Sunanda Tripathi, registered nurse, Santa Clara Valley Medical Center. "We use a nasal cannula to give supplemental oxygen to a patient. At a flow rate of 2 liters per minute, a patient breathes in a gaseous mixture that is about 28% oxygen compared to 21% in ambient air."

When a patient has a breathing disorder, the flow and volume of oxygen into and out of the lungs are measured. A ventilator may be used if a patient has difficulty breathing. When pressure is increased, the lungs expand. When the pressure of the incoming gas is reduced, the lung volume contracts to expel carbon dioxide. These relationships—known as gas laws—are an important part of ventilation and breathing.

Mastering
GOB CHEMISTRY

Visit **www.masteringgob.com**
for self-study materials
and instructor-assigned
homework.

W e all live at the bottom of a sea of gases called the atmosphere. The most important of these gases is oxygen, which constitutes about 21% of the atmosphere. Without oxygen, life on this planet would be impossible—oxygen is vital to all life processes of plants and animals. Ozone (O_3), formed in the upper atmosphere by the interaction of oxygen with ultraviolet light, absorbs some of the harmful radiation before it can strike Earth's surface. The other gases in the atmosphere include nitrogen (78% of the atmosphere), argon, carbon dioxide (CO_2), and water vapor. Carbon dioxide gas, a product of combustion and metabolism, is used by plants in photosynthesis, a process that produces the oxygen that is essential for humans and animals.

The atmosphere has become a dumping ground for other gases, such as methane, chlorofluorohydrocarbons (CFCs), sulfur dioxide, and nitrogen oxides. The chemical reactions of these gases with sunlight and oxygen in the air are contributing to air pollution, ozone depletion, global warming, and acid rain. Such chemical changes can seriously affect our health and our lifestyle. A knowledge of gases and some of the laws that govern gas behavior can help us understand the nature of matter and allow us to make decisions concerning important environmental and health issues.

7.1 Properties of Gases

LEARNING GOAL

■ Describe the kinetic theory of gases and the properties of gases.

The behavior of gases is quite different from that of liquids and solids. Gas particles are far apart, whereas particles of both liquids and solids are held close together as strong attractive forces become more important at lower temperatures. This means that a gas has no definite shape or volume and will completely fill any container. Because there are great distances between its particles, a gas is less dense than a solid or liquid and can be compressed. A model for the behavior of a gas, called the **kinetic molecular theory of gases**, helps us understand gas behavior.

(MC GOB) SELF-STUDY ACTIVITY
Properties of Gases

Kinetic Molecular Theory of Gases

1. **A gas consists of small particles (atoms or molecules) that move randomly with rapid velocities.** Gas molecules moving in all directions at high speeds cause a gas to fill the entire volume of a container.

2. **The attractive forces between the particles of a gas can be neglected.** Gas particles move far apart and fill a container of any size and shape.

3. **The actual volume occupied by gas molecules is extremely small compared to the volume that the gas occupies.** The volume of the container is considered equal to the volume of the gas. Most of the volume of a gas is empty space, which allows gases to be easily compressed.

4. **The average kinetic energy of gas molecules is proportional to the Kelvin temperature.** Gas particles move faster as the temperature increases. At higher temperatures, gas particles hit the walls of the container with more force, which produces higher pressures.

5. **Gas particles are in constant motion, moving rapidly in straight paths.** When gas particles collide, they rebound and travel in new directions. When they collide with the walls of the container, they exert gas pressure. An

FIGURE 7.1

Gas particles move in straight lines within a container. The gas particles exert pressure when they collide with the walls of the container.

Q Why does heating the container increase the pressure of the gas within it?

increase in the number or force of collisions against the walls of the container causes an increase in the pressure of the gas.

The kinetic theory helps explain some of the characteristics of gases. For example, we can quickly smell perfume from a bottle that is opened on the other side of a room because its particles move rapidly in all directions. They move faster at higher temperatures, and more slowly at lower temperatures. Sometimes tires and gas-filled containers explode when temperatures are too high. From the kinetic theory, we know that gas particles move faster when heated, hit the walls of a container with more force, and cause a buildup of pressure inside a container.

When we talk about a gas, we describe it in terms of four properties: pressure, volume, temperature, and the amount of gas.

Pressure (P)

Gas particles are extremely small. However, when billions and billions of gas particles hit against the walls of a container, they exert a force known as the pressure of the gas. The more molecules that strike the wall, the greater the pressure. (See Figure 7.1.) If we heat the container to make the molecules move faster, they smash into the walls of the container more often and with increased force, which increases the pressure. The gas molecules of oxygen and nitrogen in the air around us are exerting pressure on us all the time. We call the pressure exerted by the air **atmospheric pressure**. (See Figure 7.2.) As you go to higher altitudes the atmospheric

FIGURE 7.2

A column of air extending from the upper atmosphere to the surface of the Earth produces a pressure on each of us of about 1 atmosphere. While there is a lot of pressure on the body, it is balanced by the pressure inside the body.

Q Why is there less pressure at higher altitudes?

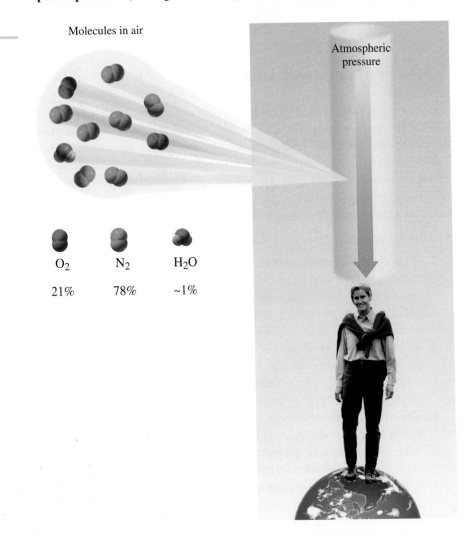

Molecules in air

Atmospheric pressure

O_2 N_2 H_2O

21%   78%   ~1%

pressure is less because there are fewer molecules of oxygen and nitrogen in the air. The most common units used for gas measurement are the atmosphere (atm) and millimeters of mercury (mm Hg). On the TV weather report, you may hear or see the atmospheric pressure given in inches of mercury or, in countries other than the United States, kilopascals. In a chemistry lab the unit torr may be used.

Volume (V)

The volume of gas equals the size of the container in which the gas is placed. When you inflate a tire or a basketball, you are adding more gas particles, which increases the number of particles hitting the walls of the tire or basketball, and its volume increases. Sometimes, on a cool morning, a tire looks flat. The volume of the tire has decreased because a lower temperature decreases the speed of the molecules, which reduces the force of their impacts on the walls of the tire. The most common units for volume measurement are liters (L) and milliliters (mL).

Temperature (T)

The temperature of a gas is related to the kinetic energy of its particles. For example, if we have a gas at 200 K in a rigid container and heat it to a temperature of 400 K, the gas particles will have twice the kinetic energy that they did at 200 K. That also means that the gas at 400 K exerts twice the pressure of the gas at 200 K. Although you measure gas temperature using a Celsius thermometer, all comparisons of gas behavior and all calculations related to temperature must use the Kelvin temperature. No one has created the conditions for absolute zero (K), but we predict that the particles will have zero kinetic energy, and the gas will exert zero pressure at this temperature.

Amount of Gas (n)

When you add air to a bicycle tire, you are increasing the amount of gas, which results in a higher pressure in the tire. Usually we measure the amount of gas by its mass (grams). In gas law calculations, we need to change the grams of gas to moles.

A summary of the four properties of a gas are given in Table 7.1.

TABLE 7.1

Properties That Describe a Gas

Property	Description	Unit(s) of Measurement
Pressure (P)	The force exerted by gas against the walls of the container	atmosphere (atm); mm Hg; torr
Volume (V)	The space occupied by the gas	liter (L); milliliter (mL)
Temperature (T)	Determines the kinetic energy and rate of motion of the gas particles	Celsius (°C); Kelvin (K) *required in calculations*
Amount (n)	The quantity of gas present in a container	grams (g); moles (n) *required in calculations*

SAMPLE PROBLEM 7.1

■ **Properties of Gases**

Identify the property of a gas that is described by each of the following:

a. increases the kinetic energy of gas particles
b. the force of the gas particles hitting the walls of the container
c. the space that is occupied by a gas

■ Forming a Gas

Obtain baking soda and a jar or a plastic bottle. You will also need an elastic glove that fits over the mouth of the jar or a balloon that fits snugly over the top of the plastic bottle. Place a cup of vinegar in the jar or bottle. Sprinkle some baking soda into the fingertips of the glove or into the balloon. Carefully fit the glove or balloon over the top of the jar or bottle. Slowly lift the fingers of the glove or the balloon so that the baking soda falls into the vinegar. Watch what happens. Squeeze the glove or balloon.

QUESTION

1. Describe the properties of gas that you observe as the reaction takes place between vinegar and baking soda.
2. How do you know that a gas was formed?

■ Describe the units of measurement used for pressure and change from one unit to another.

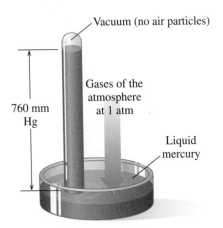

Vacuum (no air particles)

760 mm Hg

Gases of the atmosphere at 1 atm

Liquid mercury

FIGURE 7.3

A barometer: The pressure exerted by the gases in the atmosphere is equal to the downward pressure of a mercury column in a closed glass tube. The height of the mercury column measured in mm Hg is called atmospheric pressure.

Q Why does the height of the mercury column change from day to day?

SOLUTION

a. temperature   **b.** pressure   **c.** volume

STUDY CHECK

When helium is added to a balloon, the number of grams of gas increases. What property of a gas is described?

■

QUESTIONS AND PROBLEMS

■ Properties of Gases

7.1 Use the kinetic molecular theory of gases to explain each of the following:
 a. Gas particles move faster at higher temperatures.
 b. Gases can be compressed much more than liquids or solids.

7.2 Use the kinetic molecular theory of gases to explain each of the following:
 a. A container of non-stick cooking spray explodes when thrown into a fire.
 b. The air in a hot-air balloon is heated to make the balloon rise.

7.3 Identify the property of a gas that is measured in each of the following:
 a. 350 K   **b.** space occupied by a gas
 c. 2.00 g of O_2 **d.** force of gas particles striking the walls of the container

7.4 Identify the property of a gas that is measured in each of the following measurements:
 a. 425 K   **b.** 1.0 atm   **c.** 10.0 L   **d.** 0.50 mole of He

■ 7.2 Gas Pressure

When billions and billions of gas particles hit against the walls of a container, they exert **pressure**, which is defined as a force acting on a certain area.

$$\text{Pressure}(P) = \frac{\text{force}}{\text{area}}$$

The air that covers the surface of Earth, the atmosphere, contains vast numbers of gas particles. Because the air particles have mass, when they collide with the surface of Earth they exert an *atmospheric pressure*.

The atmospheric pressure can be measured using a barometer, as shown in Figure 7.3. At a pressure of exactly 1 atmosphere (atm), the mercury column would be exactly 760 mm high. We say that the atmospheric pressure is 760 mm Hg (millimeters of mercury), which is an **atmosphere (atm)**. One atmosphere of pressure may also be expressed as 760 *torr*, a pressure unit named to honor Evangelista Torricelli, the inventor of the barometer. Because they are equal, units of torr and mm Hg are used interchangeably.

   1 atm = 760 mm Hg = 760 torr

   1 mm Hg = 1 torr

In SI units, pressure is measured in pascals (Pa); 1 atm is equal to 101 325 Pa. Because a pascal is a very small unit, pressures can be reported in kilopascals.

The American equivalent of 1 atm is 14.7 pounds per square inch (lb/in.^{2}). When you use a pressure gauge to check the air pressure in the tires of a car it may read 30–35 psi. Table 7.2 summarizes the various units used in the measurement of pressure.

TABLE 7.2

Units for Measuring Pressure

Unit	Abbreviation	Unit Equivalent to 1 atm
Atmosphere	atm	1 atm (exact)
Millimeters of Hg	mm Hg	760 mm Hg
Torr	torr	760 torr
Inches of Hg	in. Hg	29.9 in. Hg
Pounds per square inch	lb/in.^{2} (psi)	14.7 lb/in.^{2}
Pascal	Pa	101,325 Pa

 Converting Between Units of Pressure

Atmospheric pressure changes with variations in weather and altitude. On a hot, sunny day, the air is more dense, which causes more pressure on the mercury surface. The mercury column rises indicating a higher atmospheric pressure. On a rainy day, the atmosphere exerts less pressure, which causes the mercury column to go lower. In the weather report, this type of weather is called a low-pressure system. Atmosphere pressure is greatest at sea level. Above sea level, the density of the gases in the air decreases, which causes lower atmospheric pressures; the atmospheric pressure is greater in Death Valley because it is below sea level. (See Table 7.3.)

Divers must be concerned about increasing pressures on their ears and lungs when they dive below the surface of the ocean. Because water is denser than air, the pressure on a diver increases rapidly as the diver descends. At a depth of 33 ft below the surface of the ocean, an additional 1 atmosphere of pressure is exerted by the water on a diver, for a total of 2 atm. At 100 ft down, there is a total pressure of 4 atm on a diver. The air tanks a diver carries continuously adjust the pressure of the breathing mixture to match the increase in pressure.

SELF-STUDY CASE
Scuba Diving and Blood Gases

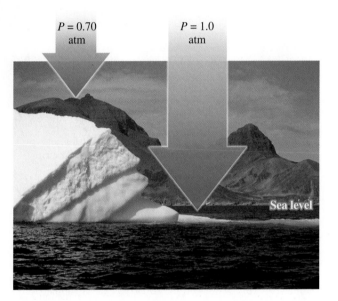

$P = 0.70$ atm

$P = 1.0$ atm

Sea level

SAMPLE PROBLEM 7.2

■ Units of Pressure

A sample of neon gas has a pressure of 0.50 atm. Give the pressure of the neon in millimeters of Hg.

SOLUTION

The equality 1 atm = 760 mm Hg can be written as conversion factors:

$$\frac{760 \text{ mm Hg}}{1 \text{ atm}} \quad \text{or} \quad \frac{1 \text{ atm}}{760 \text{ mm Hg}}$$

Using the appropriate conversion factor, the problem is set up as

$$0.50 \text{ atm} \times \frac{760 \text{ mm Hg}}{1 \text{ atm}} = 380 \text{ mm Hg}$$

STUDY CHECK

What is the pressure in atmospheres for a gas that has a pressure of 655 torr?

TABLE 7.3

Altitude and Atmospheric Pressure

Location	Altitude (km)	Atmospheric Pressure (mm Hg)
Sea level	0	760
Los Angeles	0.09	752
Las Vegas	0.70	700
Denver	1.60	630
Mount Whitney	4.50	440
Mount Everest	8.90	253

Measuring Blood Pressure

The measurement of your blood pressure is one of the important measurements a doctor or nurse makes during a physical examination. Acting like a pump, the heart contracts to create the pressure that pushes blood through the circulatory system. During contraction, the blood pressure is called systolic and is at its highest. When the heart muscles relax, the blood pressure is called diastolic and falls. Normal range for systolic pressure is 100–120 mm Hg, and for diastolic pressure, 60–80 mm Hg, usually expressed as a ratio such as 100/80. These values are somewhat higher in older people. When blood pressures are elevated, such as 140/90, there is a greater risk of stroke, heart attack, or kidney damage. Low blood pressure prevents the brain from receiving adequate oxygen, causing dizziness and fainting.

The blood pressures are measured by a sphygmomanometer, an instrument consisting of a stethoscope and an inflatable cuff connected to a tube of mercury called a manometer. After the cuff is wrapped around the upper arm, it is pumped up with air until it cuts off the flow of blood through the arm. With the stethoscope over the artery, the air is slowly released from the cuff. When the pressure equals the systolic pressure, blood starts to flow again, and the noise it makes is heard through the stethoscope. As air

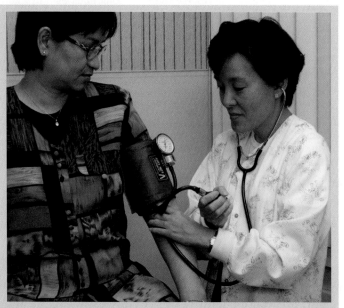

continues to be released, the cuff deflates until no sound in the artery is heard. That second pressure reading is noted as the diastolic pressure, the pressure when the heart is not contracting.

QUESTIONS AND PROBLEMS

Gas Pressure

7.5 What units are used to measure the pressure of a gas?

7.6 Which of the following statement(s) describes the pressure of a gas?
a. the force of the gas particles on the walls of the container
b. the number of gas particles in a container
c. the volume of the container
d. 3.00 atm   **e.** 750 torr

7.7 An oxygen tank contains oxygen (O_2) at a pressure of 2.00 atm. What is the pressure in the tank in terms of the following units?
a. torr   **b.** mm Hg

7.8 On a climb up Mt. Whitney, the atmospheric pressure is 467 mm Hg. What is the pressure in terms of the following units?
a. atm   **b.** torr

7.3 Pressure and Volume (Boyle's Law)

Imagine that you can see air particles hitting the walls inside a bicycle tire pump. What happens to the pressure inside the pump as we push down on the handle? As the air is compressed, the air particles are crowded together. In the smaller volume, more collisions occur, and the air pressure increases.

When a change in one property (in this case, volume) causes a change in the opposite direction of another property (in this case, pressure), the properties have an **inverse relationship**. The relationship between the pressure and volume of a gas is known as **Boyle's law**. The law states that the volume (V) of a sample of gas changes inversely with the pressure (P) of the gas as long as there has been no change in the temperature (T) or amount of gas (n), as illustrated in Figure 7.4.

If we change the volume or pressure of a gas sample without any change in the temperature or in the amount of the gas, the new pressure and volume will give the same PV product as the initial pressure and volume.

Boyle's Law

$$P_1V_1 = P_2V_2 \quad \text{No change in number of moles and temperature}$$

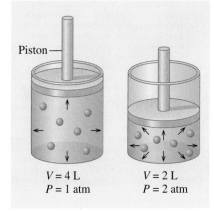

SAMPLE PROBLEM 7.3

■ Calculating Pressure When Volume Changes

A sample of hydrogen gas (H_2) has a volume of 5.0 L and a pressure of 1.0 atm. What is the new pressure if the volume is decreased to 2.0 L at constant temperature?

FIGURE 7.4

Boyle's law: As volume decreases, gas molecules become more crowded, which causes the pressure to increase. Pressure and volume are inversely related.

SOLUTION

■ **STEP 1** **Organize the data in a table.** In this problem, we want to know the final pressure (P_2) for the change in volume. In calculations with gas laws, it is helpful to organize the data in a table.

Q **If the volume of a gas increases, what will happen to its pressure?**

Conditions 1	Conditions 2
$V_1 = 5.0$ L	$V_2 = 2.0$ L
$P_1 = 1.0$ atm	$P_2 = ?$

■ **STEP 2** **Rearrange the gas law for the unknown.** For a PV relationship, we use Boyle's law and solve for P_2 by dividing both sides by V_2.

$$P_1V_1 = P_2V_2$$

$$\frac{P_1V_1}{V_2} = \frac{P_2 V_2}{V_2}$$

$$P_2 = P_1 \times \frac{V_1}{V_2}$$

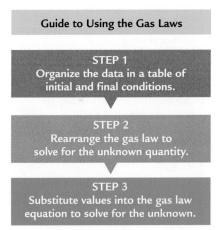

Guide to Using the Gas Laws

STEP 1
Organize the data in a table of initial and final conditions.

STEP 2
Rearrange the gas law to solve for the unknown quantity.

STEP 3
Substitute values into the gas law equation to solve for the unknown.

■ **STEP 3** **Substitute values into the gas law to solve for the unknown.** From the table, we see that the volume has decreased. Because pressure and volume are inversely related, the pressure must increase. When we substitute in the values, we see the ratio of the volumes (volume factor) is greater than 1, which increases the pressure.

$$P_2 = 1.0 \text{ atm} \times \frac{5.0 \text{ L}}{2.0 \text{ L}} = 2.5 \text{ atm}$$

Volume factor increases pressure

Note that the units of volume cancel and the final pressure is in atmospheres.

MC GOB Pressure and Volume

STUDY CHECK

A sample of helium gas has a volume of 150 mL at 750 torr. If the volume expands to 450 mL at constant temperature, what is the new pressure in torr?

SAMPLE PROBLEM 7.4

■ **Calculating Volume When Pressure Changes**

The gauge on a 12-L tank of compressed oxygen reads 4500 mm Hg. How many liters would this same gas occupy at a pressure of 750 mm Hg at constant temperature?

SOLUTION

■ **STEP 1 Organize the data in a table.** Placing our information in a table gives the following:

Conditions 1	Conditions 2
$P_1 = 4500$ mm Hg	$P_2 = 750$ mm Hg
$V_1 = 12$ L	$V_2 = ?$

■ **STEP 2 Rearrange the gas law for the unknown.** Using Boyle's law, we solve for V_2. According to Boyle's law, a decrease in the pressure will cause an increase in the volume.

$$V_2 = V_1 \times \frac{P_1}{P_2}$$

HEALTH NOTE

■ **Pressure–Volume Relationship in Breathing**

The importance of Boyle's law becomes more apparent when you consider the mechanics of breathing. Our lungs are elastic, balloonlike structures contained within an airtight chamber called the thoracic cavity. The diaphragm, a muscle, forms the flexible floor of the cavity.

INSPIRATION

The process of taking a breath of air begins when the diaphragm contracts, and the rib cage expands, causing an increase in the volume of the thoracic cavity. The elasticity of the lungs allows them to expand when the thoracic cavity expands. According to Boyle's law, the pressure inside the lungs will decrease when their volume increases. This causes the pressure inside the lungs to fall below the pressure of the atmosphere. This difference in pressures produces a *pressure gradient* between the lungs and the atmosphere. In a pressure gradient, molecules flow from an area of greater pressure to an area of lower pressure. Thus, we inhale as air flows into the lungs (*inspiration*), until the pressure within the lungs becomes equal to the pressure of the atmosphere.

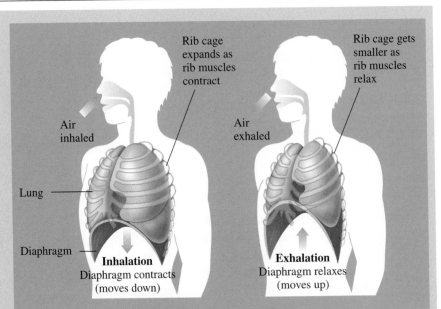

EXPIRATION

Expiration, or the exhalation phase of breathing, occurs when the diaphragm relaxes and moves back up into the thoracic cavity to its resting position. This reduces the volume of the thoracic cavity, which squeezes the lungs and decreases their volume. Now the pressure in the lungs is greater than the pressure of the atmosphere, so air flows out of the lungs. Thus, breathing is a process in which pressure gradients are continuously created between the lungs and the environment as a result of the changes in the volume and pressure.

■ **STEP 3** **Substitute values into the gas law to solve for the unknown.** When we substitute in the values, the ratio of pressures (pressure factor) is greater than 1, which increases the volume.

$$V_2 = 12 \text{ L} \times \frac{4500 \text{ mm Hg}}{750 \text{ mm Hg}} = 72 \text{ L}$$

Pressure factor increases volume

STUDY CHECK

A sample of methane gas (CH_4) has a volume of 125 mL at 0.600 atm pressure. How many milliliters will it occupy at a pressure of 1.50 atm at constant temperature?

QUESTIONS AND PROBLEMS

■ **Pressure and Volume (Boyle's Law)**

7.9 Why do scuba divers need to exhale air when they ascend to the surface of the water?

7.10 Why does a sealed bag of chips expand when you take it to a higher altitude?

7.11 The air in a cylinder with a piston has a volume of 220 mL and a pressure of 650 mm Hg.
  a. To obtain a higher pressure inside the cylinder at constant temperature, should the cylinder change as shown in A or B? Explain your choice.

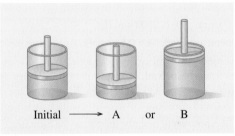

Initial ⟶ A or B

  b. If the pressure inside the cylinder increases to 1.2 atm, what is the final volume of the cylinder? Complete the following data table:

Property	Initial	Final
Pressure (P)		
Volume (V)		

7.12 A balloon is filled with helium gas. When the following changes are made at constant temperature, which of these diagrams (A, B, or C) shows the new volume of the balloon?

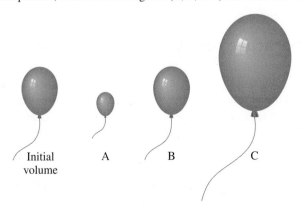

Initial volume A B C

  a. The balloon floats to a higher altitude where the outside pressure is lower.
  b. The balloon is taken inside the house, but the atmospheric pressure remains the same.
  c. The balloon is put in a hyperbaric chamber in which the pressure is increased.

7.13 A gas with a volume of 4.0 L is contained in a closed container. Indicate what changes in pressure must have occurred if the volume undergoes the following changes at constant temperature.
a. The volume is compressed to 2 L.
b. The volume is allowed to expand to 12 L.
c. The volume is compressed to 0.40 L.

7.14 A gas at a pressure of 2.0 atm is contained in a closed container. Indicate the changes in its volume when the pressure undergoes the following changes at constant temperature.
a. The pressure increases to 6.0 atm.
b. The pressure drops to 1.0 atm.
c. The pressure drops to 0.40 atm.

7.15 A 10.0-L balloon contains He gas at a pressure of 655 mm Hg. What is the new pressure of the He gas at each of the following volumes if there is no change in temperature?
a. 20.0 L   **b.** 2.50 L   **c.** 1500 mL

7.16 The air in a 5.00-L tank has a pressure of 1.20 atm. What is the new pressure of the air when the air is placed in tanks that have the following volumes, if there is no change in temperature?
a. 1.00 L   **b.** 2500 mL   **c.** 750 mL

7.17 A sample of nitrogen (N_2) has a volume of 50.0 L at a pressure of 760 mm Hg. What is the volume of the gas at each of the following pressures if there is no change in temperature?
a. 1500 mm Hg   **b.** 2.0 atm   **c.** 0.500 atm

7.18 A sample of methane (CH_4) has a volume of 25 mL at a pressure of 0.80 atm. What is the volume of the gas at each of the following pressures if there is no change in temperature?
a. 0.40 atm   **b.** 2.00 atm   **c.** 2500 mm Hg

7.19 Cyclopropane, C_3H_6, is a general anesthetic. A 5.0-L sample has a pressure of 5.0 atm. What is the volume of the anesthetic given to a patient at a pressure of 1.0 atm?

7.20 The volume of air in a person's lungs is 615 mL at a pressure of 760 mm Hg. Inhalation occurs as the pressure in the lungs drops to 752 mm Hg. To what volume did the lungs expand?

7.21 Use the words *inspiration* and *expiration* to describe the part of the breathing cycle that occurs as a result of each of the following:
a. The diaphragm contracts (flattens out).
b. The volume of the lungs decreases.
c. The pressure within the lungs is less than the atmosphere.

7.22 Use the words *inspiration* and *expiration* to describe the part of the breathing cycle that occurs as a result of each of the following:
a. The diaphragm relaxes, moving up into the thoracic cavity.
b. The volume of the lungs expands.
c. The pressure within the lungs is greater than the atmosphere.

LEARNING GOAL

Use the temperature–volume relationship (Charles' law) to determine the new temperature or volume of a certain amount of gas at a constant pressure.

MC GOB   Temperature and Volume

7.4 Temperature and Volume (Charles' Law)

Suppose that you are going to take a ride in a hot-air balloon. The captain turns on a propane burner to heat the air inside the balloon. As the temperature rises, the air particles move faster and spread out, causing the volume of the balloon to increase. When the air in the balloon becomes less dense than the air outside, the balloon and its passengers rise. In fact, it was in 1787 that Jacques Charles, a balloonist as well as a physicist, proposed that the volume of a gas is related to the temperature. This became **Charles' law**, which states that the volume (V) of a gas is directly related to the temperature (T) when there is no change in the pressure (P) or amount (n) of gas. (See Figure 7.5.) A **direct relationship** is one in which the related properties increase or decrease together. For two conditions, we can write Charles' law as follows.

Charles' Law

$$\frac{V_1}{T_1} = \frac{V_2}{T_2}$$ No change in number of moles and pressure

Remember that all temperatures used in gas law calculations must be converted to their corresponding Kelvin (K) temperatures.

SAMPLE PROBLEM 7.5

■ Calculating Volume When Temperature Changes

A sample of neon gas has a volume of 5.40 L and a temperature of 15°C. Find the new volume of the gas after the temperature has been increased to 42°C at constant pressure.

SOLUTION

■ **STEP 1** **Organize the data in a table.** When the temperatures are given in degrees Celsius, they must be changed to Kelvins.

$$T_1 = 15°C + 273 = 288 \text{ K}$$
$$T_2 = 42°C + 273 = 315 \text{ K}$$

Conditions 1	Conditions 2
$T_1 = 288$ K	$T_2 = 315$ K
$V_1 = 5.40$ L	$V_2 = ?$

■ **STEP 2** **Rearrange the gas law for the unknown.** In this problem, we want to know the final volume (V_2) when the temperature increases. Using Charles' law, we solve for V_2 by multiplying both sides by T_2.

$$\frac{V_1}{T_1} = \frac{V_2}{T_2}$$

$$\frac{V_1}{T_1} \times T_2 = \frac{V_2}{T_2} \times T_2$$

$$V_2 = V_1 \times \frac{T_2}{T_1}$$

■ **STEP 3** **Substitute values into the gas law to solve for the unknown.** From the table, we see that the temperature has increased. Because temperature is directly related to volume, the volume must increase. When we substitute in the values, we see that the ratio of the temperatures (temperature factor) is greater than 1, which increases the volume.

$$V_2 = 5.40 \text{ L} \times \frac{315 \text{ K}}{288 \text{ K}} = 5.91 \text{ L}$$

Temperature factor
increases volume

STUDY CHECK

A mountain climber inhales 486 mL of air at a temperature of −8°C. What volume will the air occupy in the lungs if the climber's body temperature is 37°C?

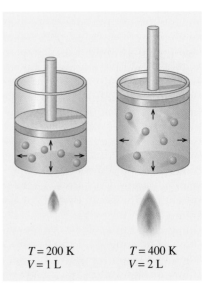

$T = 200$ K $T = 400$ K
$V = 1$ L $V = 2$ L

FIGURE 7.5

Charles' law: The Kelvin temperature of a gas is directly related to the volume of the gas when there is no change in the pressure. When the temperature increases, making the molecules move faster, the volume must increase to maintain constant pressure.

Q If the temperature of a gas decreases at constant pressure, how will the volume change?

QUESTIONS AND PROBLEMS

■ Temperature and Volume (Charles' Law)

7.23 Select the diagram that shows the new volume of a balloon when the following changes are made at constant pressure.

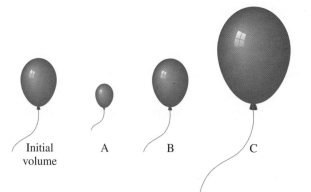

Initial   A   B   C
volume

a. The temperature is changed from 150 K to 300 K.
b. The balloon is placed in a freezer.
c. The balloon is first warmed, and then returned to its starting temperature.

7.24 Indicate whether the final volume of gas in each of the following is the same, larger, or smaller than the initial volume:
a. A volume of 500 mL of air on a cold winter day at 5°C is breathed into the lungs, where body temperature is 37°C.
b. The heater used to heat 1400 L of air in a hot-air balloon is turned off.
c. A balloon filled with helium at the amusement park is left in a car on a hot day.

7.25 What change in volume occurs when gases for hot-air balloons are heated prior to their ascent?

7.26 On a cold, wintry morning, the tires on a car appear flat. How has their volume changed overnight?

7.27 A balloon contains 2500 mL of helium gas at 75°C. What is the new volume of the gas when the temperature changes to the following, if n and P are not changed?
a. 55°C   **b.** 680 K   **c.** −25°C   **d.** 240 K

7.28 A gas has a volume of 4.00 L at 0°C. What final temperature in degrees Celsius is needed to cause the volume of the gas to change to the following, if n and P are not changed?
a. 10.0 L   **b.** 1200 mL   **c.** 2.50 L   **d.** 50.0 mL

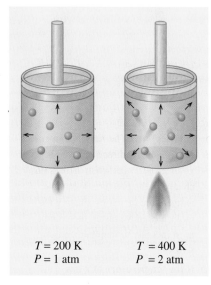

$T = 200$ K $T = 400$ K
$P = 1$ atm $P = 2$ atm

FIGURE 7.6

Gay-Lussac's law: The pressure of a gas is directly related to the temperature of the gas. When the Kelvin temperature of a gas is doubled, the pressure is doubled at constant volume.

Q **How does a decrease in the temperature of a gas affect its pressure at constant volume?**

LEARNING GOAL

Use the temperature–pressure relationship (Gay-Lussac's law) to determine the new temperature or pressure of a certain amount of gas at a constant volume.

7.5 Temperature and Pressure (Gay-Lussac's Law)

If we could watch the molecules of a gas as the temperature rises, we would notice that they move faster and hit the sides of the container more often and with greater force. If we keep the volume of the container the same, we would observe an increase in the pressure. A temperature–pressure relationship, also known as **Gay-Lussac's law**, states that the pressure of a gas is directly related to its Kelvin temperature. This means that an increase in temperature increases the pressure of a gas, and a decrease in temperature decreases the pressure of the gas, provided the volume and number of moles of the gas remain the same. (See Figure 7.6.) The ratio of pressure (P) to temperature (T) is the same under all conditions as long as volume (V) and amount of gas (n) do not change.

Gay-Lussac's Law

$$\frac{P_1}{T_1} = \frac{P_2}{T_2}$$ No change in number of moles and volume

---

SAMPLE PROBLEM 7.6

   The Relationship Between
Temperature and Pressure

■ Calculating Pressure When Temperature Changes

Aerosol containers can be dangerous if they are heated because they can explode. Suppose a container of hair spray with a pressure of 4.0 atm at a room temperature of 25°C is thrown into a fire. If the temperature of the gas inside the aerosol can reaches 402°C, what will be its pressure? The aerosol container may explode if the pressure inside exceeds 8.0 atm. Would you expect it to explode?

SOLUTION

■ **STEP 1 Organize the data in a table.** We must first change the temperatures to kelvins.

$$T_1 = 25°C + 273 = 298 \text{ K}$$
$$T_2 = 402°C + 273 = 675 \text{ K}$$

Conditions 1	Conditions 2
$P_1 = 4.0$ atm	$P_2 = ?$
$T_1 = 298$ K	$T_2 = 675$ K

■ **STEP 2 Rearrange the gas law for the unknown.** Using Gay-Lussac's law, we can solve for P_2.

$$\frac{P_1}{T_1} = \frac{P_2}{T_2}$$

$$\frac{P_1}{T_1} \times T_2 = \frac{P_2}{\cancel{T_2}} \times \cancel{T_2}$$

$$P_2 = P_1 \times \frac{T_2}{T_1}$$

■ **STEP 3 Substitute values into the gas law to solve for the unknown.** From the table, we see that the temperature has increased. Because pressure and temperature are directly related, the pressure must increase. When we substitute in the values, we see the ratio of the temperatures (temperature factor) is greater than 1, which increases pressure.

$$P_2 = 4.0 \text{ atm} \times \frac{675 \text{ K}}{298 \text{ K}} = 9.1 \text{ atm}$$

Temperature factor increases volume

Because the calculated pressure exceeds 8.0 atm, we expect the can to explode.

STUDY CHECK

In a storage area where the temperature has reached 55°C, the pressure of oxygen gas in a 15.0-L steel cylinder is 965 torr. To what temperature (°C) would the gas have to be cooled to reduce the pressure to 850. torr?

TABLE 7.4

Vapor Pressure of Water

Temperature (C°)	Vapor Pressure (mm Hg)
0	5
10	9
20	18
30	32
37	47[a]
40	55
50	93
60	149
70	234
80	355
90	528
100	760

[a]At body temperature.

TABLE 7.5

Pressure and the Boiling Point of Water

Pressure (mm Hg)	Boiling Point (°C)
270	70
467	87
630	93
752	99
760	100
800	100.4
1075	110
1520 (2 atm)	120
2026	130
7600 (10 atm)	180

 Vapor Pressure and Boiling Point

Vapor Pressure and Boiling Point

In Chapter 6, we learned that liquid molecules with sufficient kinetic energy can break away from the surface of the liquid as they become gas particles or vapor. In an open container, all the liquid will eventually evaporate. In a closed container, the vapor accumulates and creates pressure called **vapor pressure**. Each liquid exerts its own vapor pressure at a given temperature. As temperature increases, more vapor forms, and vapor pressure increases. Table 7.4 lists the vapor pressure of water at various temperatures.

A liquid reaches its boiling point when its vapor pressure becomes equal to the external pressure. As boiling occurs, bubbles of the gas form within the liquid and quickly rise to the surface. For example, at an atmospheric pressure of 760 mm Hg, water will boil at 100°C, the temperature at which its vapor pressure reaches 760 mm Hg. (See Table 7.5.)

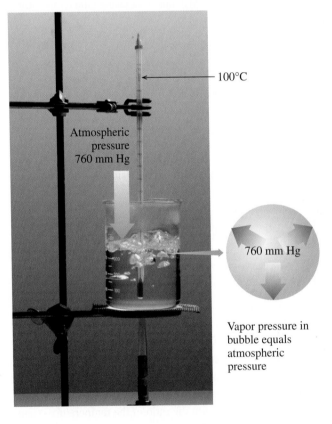

100°C

Atmospheric pressure 760 mm Hg

760 mm Hg

Vapor pressure in bubble equals atmospheric pressure

At higher altitudes, atmospheric pressures are lower and the boiling point of water is lower than 100°C. Earlier, we saw that the typical atmospheric pressure in Denver is 630 mm Hg. This means that water in Denver needs a vapor pressure of 630 mm Hg to boil. Because water has a vapor pressure of 630 mm Hg at 95°C, water boils at 95°C in Denver.

People who live at high altitudes often use pressure cookers to obtain higher temperatures when preparing food. When the external pressure is greater than 1 atm, a temperature higher than 100°C is needed to boil water. Laboratories and hospitals use devices called autoclaves to sterilize laboratory and surgical equipment. An autoclave, like a pressure cooker, is a closed container that increases the total pressure above the liquid so it will boil at higher temperatures.

QUESTIONS AND PROBLEMS

■ **Temperature and Pressure (Gay-Lussac's Law)**

7.29 Why do aerosol cans explode if heated?

7.30 How can the tires on a car have a blowout when the car is driven on hot pavement in the desert?

7.31 In a fire at an oxygen-tank factory, some of the tanks filled with oxygen explode. Explain.

7.32 In Gay-Lussac's law, what properties of gases are considered as directly related and which properties are held constant?

7.33 Solve for the new pressure when each of the following temperature changes occurs, with n and V constant.
   a. A gas sample has a pressure of 1200 torr at 155°C. What is the final pressure of the gas after the temperature has dropped to 0°C?
   b. An aerosol can has a pressure of 1.40 atm at 12°C. What is the final pressure in the aerosol can if it is used in a room where the temperature is 35°C?

7.34 Solve for the new temperature in degrees Celsius when the following pressure changes occur, with constant n and V.
   a. A 10.0-L container of helium gas with a pressure of 250 torr at 0°C is heated to give a pressure of 1500 torr.
   b. A sample of air at 40.°C and 740 mm Hg is cooled to give a pressure of 680 mm Hg.

7.35 Match the terms *vapor pressure, atmospheric pressure*, and *boiling point* to the following descriptions.
   a. the temperature at which bubbles of vapor appear within the liquid
   b. the pressure exerted by a gas above the surface of its liquid
   c. the pressure exerted on the earth by the particles in the air
   d. the temperature at which the vapor pressure of a liquid becomes equal to the external pressure

7.36 In which pair(s) would boiling occur?

Atmospheric Pressure	Vapour Pressure
a. 760 mm Hg	700 mm Hg
b. 480 torr	480 mm Hg
c. 1.2 atm	912 mm Hg
d. 1020 mm Hg	760 mm Hg
e. 740 torr	1.0 atm

7.37 Give an explanation for the following observations:
   a. Water boils at 87°C on the top of Mt. Whitney.
   b. Food cooks more quickly in a pressure cooker than in an open pan.

7.38 Give an explanation for the following observations:
   a. Boiling water at sea level is hotter than boiling water in the mountains.
   b. Water used to sterilize surgical equipment is heated to 120°C at 2.0 atm in an autoclave.

7.6 The Combined Gas Law

All pressure–volume–temperature relationships for gases that we have studied may be combined into a single relationship called the **combined gas law**. This expression is useful for studying the effect of changes in two of these variables on the third as long as the amount of gas (number of moles) remains constant.

LEARNING GOAL

■ Use the combined gas law to find the new pressure, volume, or temperature of a gas when changes in two of these properties are given.

Combined Gas Law

$$\frac{P_1 V_1}{T_1} = \frac{P_2 V_2}{T_2}$$ No change in moles of gas

■ **Using the Combined Gas Law**

A 25.0-mL bubble is released from a diver's air tank at a pressure of 4.00 atm and a temperature of 11°C. What is the volume of the bubble when it reaches the ocean surface, where the pressure is 1.00 atm and the temperature is 18°C?

SOLUTION

■ **STEP 1 Organize the data in a table.** We must first change the temperature to kelvins.

$$T_1 = 11°C + 273 = 284 \text{ K}$$
$$T_2 = 18°C + 273 = 291 \text{ K}$$

Conditions 1	Conditions 2
P_1 = 4.00 atm	P_2 = 1.00 atm
V_1 = 25.0 mL	V_2 = ?
T_1 = 284 K	T_2 = 291 K

■ **STEP 2 Rearrange the gas law for the unknown.** Because the pressure and temperature are both changing, we must use the combined gas law to solve for V_2.

$$\frac{P_1 V_1}{T_1} = \frac{P_2 V_2}{T_2}$$

$$\frac{P_1 V_1}{T_1} \times \frac{T_2}{P_2} = \frac{P_2 V_2 \times T_2}{T_2 \times P_2}$$

$$V_2 = V_1 \times \frac{P_1}{P_2} \times \frac{T_2}{T_1}$$

■ **STEP 3 Substitute the values into the gas law to solve for the unknown.** From the data table, we determine that the pressure decrease and the temperature increase will both increase the volume.

$$V_2 = 25.0 \text{ mL} \times \frac{4.00 \text{ atm}}{1.00 \text{ atm}} \times \frac{291 \text{ K}}{284 \text{ K}} = 102 \text{ mL}$$

Pressure factor increases volume   Temperature factor increases volume

MC GOB   The Combined Gas Law

STUDY CHECK

A weather balloon is filled with 15.0 L of helium at a temperature of 25°C and a pressure of 685 mm Hg. What is the pressure of the helium in the balloon in the upper atmosphere when the temperature is −35°C and the volume becomes 34.0 L?

TABLE 7.6

Summary of Gas Laws

Combined Gas Law	Properties Held Constant	Relationship	
$\dfrac{P_1V_1}{\cancel{T_1}} = \dfrac{P_2V_2}{\cancel{T_2}}$	T, n	$P_1V_1 = P_2V_2$	Boyle's law
$\dfrac{P_1V_1}{T_1} = \dfrac{P_2V_2}{T_2}$	P, n	$\dfrac{V_1}{T_1} = \dfrac{V_2}{T_2}$	Charles' law
$\dfrac{P_1\cancel{V_1}}{T_1} = \dfrac{P_2\cancel{V_2}}{T_2}$	V, n	$\dfrac{P_1}{T_1} = \dfrac{P_2}{T_2}$	Gay-Lussac's law

By remembering the combined gas law, we can derive any of the gas laws by omitting those properties that do not change. (See Table 7.6.)

QUESTIONS AND PROBLEMS

■ The Combined Gas Law

7.39 A sample of helium gas has a volume of 6.50 L at a pressure of 845 mm Hg and a temperature of 25°C. What is the pressure of the gas in atm when the volume and temperature of the gas sample are changed to the following?
  a. 1850 mL and 325 K
  b. 2.25 L and 12°C
  c. 12.8 L and 47°C

7.40 A sample of argon gas has a volume of 735 mL at a pressure of 1.20 atm and a temperature of 112°C. What is the volume of the gas in milliliters when the pressure and temperature of the gas sample are changed to the following?
  a. 658 mm Hg and 281 K
  b. 0.55 atm and 75°C
  c. 15.4 atm and −15°C

7.41 A 100.0-mL bubble of hot gases at 225°C and 1.80 atm escapes from an active volcano. What is the new volume of the bubble outside the volcano where the temperature is −25°C and the pressure is 0.80 atm?

7.42 A scuba diver 40 ft below the ocean surface inhales 50.0 mL of compressed air in a scuba tank at a pressure of 3.00 atm and a temperature of 8°C. What is the pressure of air in the lungs if the gas expands to 150.0 mL at a body temperature of 37°C?

■ 7.7 Volume and Moles (Avogadro's Law)

LEARNING GOAL

Describe the relationship between the amount of a gas and its volume and use this relationship in calculations.

In our study of the gas laws, we have looked at changes in properties for a specified amount (*n*) of gas. Now we will consider how the properties of a gas change when there is a change in number of moles or grams.

■ Avogadro's Law

 Volume and Moles

When you blow up a balloon, its volume increases because you add more air molecules. If a basketball gets a hole in it, and some of the air leaks out, its volume decreases. **Avogadro's law** states that the volume of a gas is directly related to the

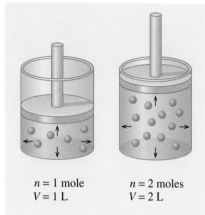

FIGURE 7.7

Avogadro's law: The volume of a gas is directly related to the number of moles of the gas. If the number of moles is doubled, the volume must double at constant temperature and pressure.

Q **If a balloon has a leak, what happens to its volume?**

number of moles of a gas when temperature and pressure are not changed. If the number of moles of a gas are doubled, then the volume will double as long as we do not change the pressure or the temperature. (See Figure 7.7.) Under conditions of constant pressure and temperature we can write Avogadro's law as follows.

Avogadro's Law

$$\frac{V_1}{n_1} = \frac{V_2}{n_2}$$ No change in pressure or temperature

SAMPLE PROBLEM 7.8

■ **Calculating Volume for a Change in Moles**

A balloon with a volume of 220 mL is filled with 2.0 moles helium. To what volume will the balloon expand if 3.0 moles helium are added, to give a total of 5.0 moles helium (the pressure and temperature do not change)?

SOLUTION

■ **STEP 1 Organize the data in a table.** A data table for our given information can be set up as follows:

Conditions 1	Conditions 2
$V_1 = 220$ mL	$V_2 = ?$
$n_1 = 2.0$ moles	$n_2 = 5.0$ moles

■ **STEP 2 Rearrange the gas law for the unknown.** Using Avogadro's law, we can solve for V_2.

$$\frac{V_1}{n_1} = \frac{V_2}{n_2}$$

$$n_2 \times \frac{V_1}{n_1} = \frac{V_2}{n_2} \times n_2$$

$$V_2 = V_1 \times \frac{n_2}{n_1}$$

■ **STEP 3 Substitute the values into the gas law to solve for the unknown.** From the table, we see that the number of moles has increased. Because the number of moles and volume are directly related, the volume must increase at constant pressure and temperature. When we substitute in the values, we see the ratio of the moles (mole factor) is greater than 1, which increases volume.

$$V_2 = 220 \text{ mL} \times \frac{5.0 \text{ moles}}{2.0 \text{ moles}} = 550 \text{ mL}$$

New volume · Initial volume · Mole factor increases volume

STUDY CHECK

At a certain temperature and pressure, 8.00 g oxygen gas has a volume of 5.00 L. What is the volume after 4.00 g oxygen gas is added to the balloon?

STP and Molar Volume

Using Avogadro's law, we can say that any two gases will have equal volumes if they contain the same number of moles of gas at the same temperature and pressure. To help us make comparisons between different gases, arbitrary conditions called standard temperature (273 K) and pressure (1 atm) (abbreviated **STP**) were selected by scientists:

STP Conditions

Standard temperature is 0°C (273 K)
Standard pressure is 1 atm (760 mm Hg)

At STP, it was observed that 1 mole of any gas has a volume of 22.4 L. (See Figure 7.8.) When a gas is at STP, it has a volume of 22.4 L, which is called its **molar volume**.

Molar Volume

The volume of 1 mole of gas at STP = 22.4 L

As long as a gas is at STP conditions (0°C and 1 atm), its molar volume can be used as a conversion factor to convert between the number of moles of gas and its volume.

Molar Volume Conversion Factors

$$\frac{1 \text{ mole gas (STP)}}{22.4 \text{ L}} \quad \text{and} \quad \frac{22.4 \text{ L}}{1 \text{ mole gas (STP)}}$$

Moles of gas	Molar volume 22.4 L/mole	Volume (L) of gas

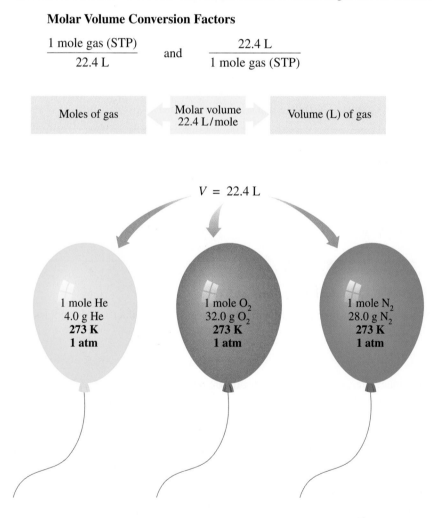

$V = 22.4$ L

1 mole He
4.0 g He
273 K
1 atm

1 mole O_2
32.0 g O_2
273 K
1 atm

1 mole N_2
28.0 g N_2
273 K
1 atm

FIGURE 7.8

Avogadro's law indicates that one mole of any gas at STP has a molar volume of 22.4 L.

Q **What volume of gas is occupied by 16.0 g methane gas, CH_4, at STP?**

■ Using Molar Volume to Find Volume at STP

What is the volume in liters of 64.0 g O_2 gas at STP?

SOLUTION

Once we convert the mass of O_2 to moles O_2, the molar volume of a gas at STP can be used to calculate the volume (L) of O_2.

■ **STEP 1 Given**   64.0 g $O_2(g)$ at STP   **Need**   volume in liters (L)

■ **STEP 2 Write a plan for solution.**

grams of O_2 ⟩ Molar mass ⟩ moles of O_2 ⟩ Molar volume ⟩ liters of O_2

■ **STEP 3 Write conversion factors.**

$$1 \text{ mole } O_2 = 32.0 \text{ g}$$

$$\frac{32.0 \text{ g } O_2}{1 \text{ mole } O_2} \quad \text{and} \quad \frac{1 \text{ mole } O_2}{32.0 \text{ g } O_2}$$

$$1 \text{ mole } O_2 \text{ (STP)} = 22.4 \text{ L}$$

$$\frac{22.4 \text{ L } O_2}{1 \text{ mole } O_2} \quad \text{and} \quad \frac{1 \text{ mole } O_2}{22.4 \text{ L } O_2}$$

■ **STEP 4 Set up problem with factors to cancel units.**

$$64.0 \text{ g } O_2 \times \frac{1 \text{ mole } O_2}{32.0 \text{ g } O_2} \times \frac{22.4 \text{ L } O_2}{1 \text{ mole } O_2} = 44.8 \text{ L } O_2 \text{ (STP)}$$

STUDY CHECK

How many grams of $N_2(g)$ are in 5.6 L $N_2(g)$ at STP?

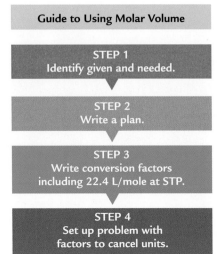

Guide to Using Molar Volume

STEP 1
Identify given and needed.

STEP 2
Write a plan.

STEP 3
Write conversion factors
including 22.4 L/mole at STP.

STEP 4
Set up problem with
factors to cancel units.

Gases in Reactions at STP

We can use the molar volume at STP to determine the moles of a gas in a reaction. Once we know the moles of gas in a reaction, we can use a mole factor to determine the moles of any other substance as we have done before.

■ Gases in Chemical Reactions at STP

When potassium metal reacts with chlorine gas, the product is solid potassium chloride.

$$2K(s) + Cl_2(g) \longrightarrow 2KCl(s)$$

How many grams of potassium chloride are produced when 7.25 L chlorine gas at STP reacts with excess potassium?

SOLUTION

■ **STEP 1 Find moles of gas A using molar volume.**   At STP, we can use molar volume (22.4 L/mole) to determine moles of Cl_2 gas.

$$7.25 \text{ L } Cl_2 \times \frac{1 \text{ mole } Cl_2}{22.4 \text{ L } Cl_2} = 0.324 \text{ mole } Cl_2$$

- **STEP 2** **Determine moles of substance B using mole–mole factor from the balanced equation.**

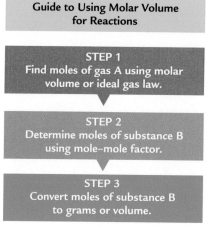

Guide to Using Molar Volume for Reactions

STEP 1
Find moles of gas A using molar volume or ideal gas law.

STEP 2
Determine moles of substance B using mole–mole factor.

STEP 3
Convert moles of substance B to grams or volume.

$$1 \text{ mole Cl}_2 = 2 \text{ moles KCl}$$

$$\frac{2 \text{ moles KCl}}{1 \text{ mole Cl}_2} \quad \text{and} \quad \frac{1 \text{ mole Cl}_2}{2 \text{ moles KCl}}$$

$$0.324 \text{ mole Cl}_2 \times \frac{2 \text{ moles KCl}}{1 \text{ mole Cl}_2} = 0.648 \text{ mole KCl}$$

- **STEP 3** **Convert moles of substance B to grams or volume.** Using the molar mass of KCl, we can determine the grams of KCl.

$$1 \text{ mole KCl} = 74.6 \text{ g KCl}$$

$$\frac{1 \text{ mole KCl}}{74.6 \text{ g KCl}} \quad \text{and} \quad \frac{74.6 \text{ g KCl}}{1 \text{ mole KCl}}$$

$$0.648 \text{ mole KCl} \times \frac{74.6 \text{ g KCl}}{1 \text{ mole KCl}} = 48.3 \text{ g KCl}$$

These steps can also be set up as a continuous solution.

$$7.25 \text{ L Cl}_2 \times \frac{1 \text{ mole Cl}_2}{22.4 \text{ L Cl}_2} \times \frac{2 \text{ moles KCl}}{1 \text{ mole Cl}_2} \times \frac{74.6 \text{ g KCl}}{1 \text{ mole KCl}} = 48.3 \text{ g KCl}$$

STUDY CHECK

Hydrogen gas forms when zinc metal reacts with aqueous HCl.

$$\text{Zn}(s) + 2\text{HCl}(aq) \longrightarrow \text{ZnCl}_2(aq) + \text{H}_2(g)$$

How many liters of hydrogen gas at STP are produced when 15.8 g zinc reacts?

QUESTIONS AND PROBLEMS

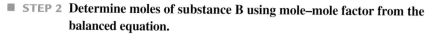

MC GOB

■ Volume and Moles (Avogadro's Law)

7.43 What happens to the volume of a bicycle tire or a basketball when you use an air pump to add air?

7.44 Sometimes when you blow up a balloon and release it, it flies around the room. What is happening to the air that was in the balloon and its volume?

7.45 A sample containing 1.50 moles neon gas has a volume of 8.00 L. What is the new volume of gas in liters when the following changes occur in the quantity of the gas at constant pressure and temperature?
a. A leak allows one-half of the neon atoms to escape.
b. A sample of 25.0 g neon is added to the neon gas already in the container.
c. A sample of 3.50 mole O_2 is added to the neon gas already in the container.

7.46 A sample containing 4.80 g O_2 gas has a volume of 15.0 L. Pressure and temperature remain constant.
a. What is the new volume if 0.500 mole O_2 gas is added?
b. Oxygen is released until the volume is 10.0 L. How many moles of O_2 are removed?
c. What is the volume after 4.00 g He is added to the O_2 gas already in the container?

7.47 Use the molar volume of a gas to solve the following at STP:
  a. the number of moles of O_2 in 44.8 L O_2 gas
  b. the number of moles of CO_2 in 4.00 L CO_2 gas
  c. the volume (L) of 6.40 g O_2
  d. the volume (mL) occupied by 50.0 g neon

7.48 Use molar volume to solve the following problems at STP:
  a. the volume (L) occupied by 2.50 moles N_2
  b. the volume (mL) occupied by 0.420 mole He
  c. the number of grams of neon contained in 11.2 L Ne gas
  d. the number of moles of H_2 in 1620 mL H_2 gas

7.49 Mg metal reacts with HCl to produce hydrogen gas:

$$Mg(s) + 2HCl(aq) \longrightarrow MgCl_2(aq) + H_2(g)$$

What volume of hydrogen at STP is released when 8.25 g Mg reacts?

7.50 Aluminum oxide is formed from its elements.

$$4Al(s) + 3O_2(g) \longrightarrow 2Al_2O_3(s)$$

How many grams of Al will react with 12.0 L O_2 at STP?

7.8 The Ideal Gas Law

LEARNING GOAL

Use the ideal gas law to solve for
P, V, T, or n of a gas when given
three of the four values in the
ideal gas equation.

MC GOB Introduction to the Ideal Gas Law

The four properties used in the measurement of a gas—pressure (P), volume (V),
temperature (T), and amount of a gas (n)—can be combined to give a single expression called the **ideal gas law**, which is written as follows:

Ideal Gas Law

$$PV = nRT$$

Rearranging the ideal gas law shows that the four gas properties equal a constant, R.

$$\frac{PV}{nT} = R$$

To calculate the value of R, we substitute the STP conditions for molar volume
into the expression: 1 mole of any gas occupies 22.4 L at STP (273 K and
1 atm).

$$R = \frac{(1.00 \text{ atm})(22.4 \text{ L})}{(1.00 \text{ mole})(273 \text{ K})} = \frac{0.0821 \text{ L} \cdot \text{atm}}{\text{mole} \cdot \text{K}}$$

The value for R, the **universal gas constant**, is 0.0821 L $\cdot$ atm per moles $\cdot$ K. If we
use 760 mm Hg for the pressure we obtain another useful value for R of 62.4 L $\cdot$ mm
Hg per mole $\cdot$ K.

$$R = \frac{(760 \text{ mm Hg})(22.4 \text{ L})}{(1.00 \text{ mole})(273 \text{ K})} = \frac{62.4 \text{ L} \cdot \text{mm Hg}}{\text{mole} \cdot \text{K}}$$

MC GOB SELF-STUDY ACTIVITY
The Ideal Gas Law

The ideal gas law is a useful expression when you are given the values for any three
of the four properties of a gas.

In working problems using the ideal gas law, the units of each variable must
match the units in the R you select.

Property	Unit
Pressure (P)	atm or mm Hg
Volume (V)	L
Amount (n)	moles
Temperature (T)	K

SAMPLE PROBLEM 7.11

▪ Using the Ideal Gas Law

Dinitrogen oxide, N_2O, which is used in dentistry, is an anesthetic also called "laughing gas." What is the pressure in atmospheres of 0.350 mole N_2O at 22°C in a 5.00 L container?

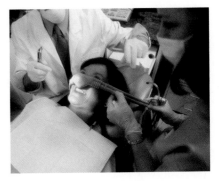

SOLUTION

▪ STEP 1 **Organize the data including R in a table.**   When three of the four quantities (P, V, n, and T) are known, we use the ideal gas law to solve for the unknown quantity. The units must match the units of the gas constant R. The temperature is converted from Celsius to kelvin.

$$P = ?   \quad V = 5.00\text{ L} \quad n = 0.350\text{ mole} \quad R = 0.0821\frac{\text{L}\cdot\text{atm}}{\text{mole}\cdot\text{K}} \quad T = 22°\text{C} + 273 = 295\text{ K}$$

▪ STEP 2 **Rearrange the ideal gas law to solve for the unknown.**   By dividing both sides of the ideal gas law by V, we solve for pressure, P:

$$P\,V = nRT \quad \text{ideal gas law}$$

$$P\,\frac{\cancel{V}}{\cancel{V}} = \frac{nRT}{V}$$

$$P = \frac{nRT}{V}$$

▪ STEP 3 **Substitute values from table to calculate unknown.**

$$P = \frac{0.350\ \cancel{\text{mole}} \times 0.0821\ \dfrac{\cancel{\text{L}}\cdot\text{atm}}{\cancel{\text{mole}}\cdot\cancel{\text{K}}} \times 295\ \cancel{\text{K}}}{5.00\ \cancel{\text{L}}} = 1.70\text{ atm}$$

STUDY CHECK

Chlorine gas, Cl_2, is used to purify water. How many moles of chlorine gas are in a 7.00 L tank if the gas has a pressure of 865 mm Hg and a temperature of 24°C?

SAMPLE PROBLEM 7.12

▪ Calculating Mass Using the Ideal Gas Law

Butane, C_4H_{10}, is used as a fuel for barbecues and as an aerosol propellant. If you have 108 mL butane at 715 mm Hg and 25°C, what is the mass (g) of the butane?

SOLUTION

▪ STEP 1 **Organize the data including R in a table.**   When three of the four quantities (P, V, n, and T) are known, we use the ideal gas law to solve for the unknown quantity. Because the pressure is given in mm Hg, we will use the R in mm Hg. The volume given in milliliters (mL) is converted to a volume in liters (L). The temperature is converted from Celsius to kelvin.

Initial values	Adjusted for units in gas constant R
$P = 715$ mm Hg	715 mm Hg
$V = 108$ mL	$108 \text{ mL} \times \dfrac{1 \text{ L}}{1000 \text{ mL}} = 0.108$ L
$n = ?$ moles of C_4H_{10}	? moles of C_4H_{10}
$R = \dfrac{62.4 \text{ L} \cdot \text{mm Hg}}{\text{mole} \cdot \text{K}}$	$\dfrac{62.4 \text{ L} \cdot \text{mm Hg}}{\text{mole} \cdot \text{K}}$
$T = 25°C$	$25°C + 273 = 298$ K

■ STEP 2 **Rearrange the ideal gas law to solve for the unknown.** By dividing both sides of the ideal gas law by RT, we solve for moles, n:

$$PV = n \; RT \quad \text{ideal gas equation}$$

$$\frac{PV}{RT} = n \; \frac{\cancel{RT}}{\cancel{RT}}$$

$$n = \frac{PV}{RT}$$

■ STEP 3 **Substitute the known values to calculate unknown.**

$$n = \frac{715 \cancel{\text{ mm Hg}} \times 0.108 \cancel{\text{ L}}}{\dfrac{62.4 \cancel{\text{ L}} \cdot \cancel{\text{ mm Hg}}}{\text{mole} \cdot \cancel{\text{K}}} \times 298 \cancel{\text{ K}}} = 0.00415 \text{ mole } (4.15 \times 10^{-3} \text{ mole})$$

Now we convert the moles of butane to grams using its molar mass (58.1 g/mole).

$$0.00415 \cancel{\text{ mole } C_4H_{10}} \times \frac{58.1 \text{ g } C_4H_{10}}{1 \cancel{\text{ mole } C_4H_{10}}} = 0.241 \text{ g } C_4H_{10}$$

STUDY CHECK

What is the volume of 1.20 g carbon monoxide at 8°C if it has a pressure of 724 mm Hg?

QUESTIONS AND PROBLEMS

(MC GOB)

■ The Ideal Gas Law

7.51 Calculate the pressure, in atmospheres, of 2.00 moles helium gas in a 10.0 L container at 27°C.

7.52 What is the volume in liters of 4.0 moles methane gas, CH_4, at 18°C and 1.40 atm?

7.53 An oxygen gas container has a volume of 20.0 L. How many grams of oxygen are in the container if the gas has a pressure of 845 mm Hg at 22°C?

7.54 A 10.0-g sample of krypton has a temperature of 25°C at 575 mm Hg. What is the volume, in milliliters, of the krypton gas?

7.55 A 25.0-g sample of nitrogen, N_2, has a volume of 50.0 L and a pressure of 630. mm Hg. What is the temperature of the gas?

7.56 A 0.226-g sample of carbon dioxide, CO_2, has a volume of 525 mL and a pressure of 455 mm Hg. What is the temperature of the gas?

7.9 Partial Pressures (Dalton's Law)

LEARNING GOAL

Use partial pressures to calculate the total pressure of a mixture of gases.

Many gas samples are a mixture of gases. For example, the air you breathe is a mixture of mostly oxygen and nitrogen gases. In gas mixtures, scientists observed that all gas particles behave in the same way. Therefore, the total pressure of the gases in a mixture is a result of the collisions of the gas particles regardless of what type of gas they are.

Dalton's Law

In a gas mixture, each gas exerts its **partial pressure**, which is the pressure it would exert if it were the only gas in the container. **Dalton's law** states that the total pressure of a gas mixture is the sum of the partial pressures of the gases in the mixture.

MC GOB   Mixture of Gases

Dalton's Law

$$P_{total} = P_1 + P_2 + P_3 + \cdots$$

Total pressure = Sum of the partial pressures
of a gas mixture   of the gases in the mixture

Suppose we have two separate tanks, one filled with helium at 2.0 atm and the other filled with argon at 4.0 atm. When the gases are combined in a single tank with the same volume and temperature, the number of gas molecules, not the type of gas, determines the pressure in a container. There the pressure of the gas mixture would be 6.0 atm, which is the sum of their individual or partial pressures.

$$
\begin{aligned}
P_{total} &= P_{He} + P_{Ar} \\
&= 2.0 \text{ atm} + 4.0 \text{ atm} \\
&= 6.0 \text{ atm}
\end{aligned}
$$

$P_{He} = 2.0$ atm $P_{Ar} = 4.0$ atm

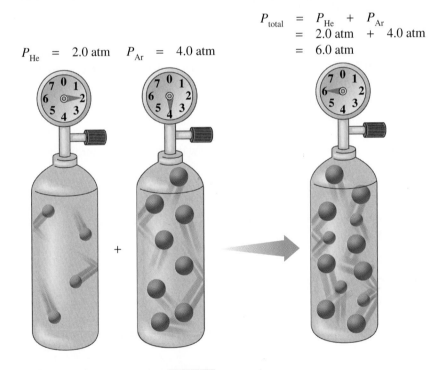

SAMPLE PROBLEM 7.13

■ Calculating the Total Pressure of a Gas Mixture

A 10-L gas tank contains propane (C_3H_8) gas at a pressure of 300. torr. Another 10-L gas tank contains methane (CH_4) gas at a pressure of 500. torr. In preparing a gas fuel mixture, the gases from both tanks are combined in a 10-L container at the same temperature. What is the pressure of the gas mixture?

SOLUTION

Using Dalton's law of partial pressure, we find that the total pressure of the gas mixture is the sum of the partial pressures of the gases in the mixture.

$$P_{total} = P_{propane} + P_{methane}$$
$$= 300. \text{ torr} + 500. \text{ torr}$$
$$= 800. \text{ torr}$$

Therefore, when both propane and methane are placed in the same container, the total pressure of the mixture is 800. torr.

STUDY CHECK

A gas mixture consists of helium with a partial pressure of 315 mm Hg, nitrogen with a partial pressure of 204 mm Hg, and argon with a partial pressure of 422 mm Hg. What is the total pressure in atmospheres?

TABLE 7.7

Typical Composition of Air

Gas	Partial Pressure (mm Hg)	Percentage (%)
Nitrogen, N_2	594.0	78
Oxygen, O_2	160.0	21
Carbon dioxide, CO_2	0.3 ⎤	
Water vapor, H_2O	5.7 ⎦	1
Total air	760.0	100

Air Is a Gas Mixture

The air you breathe is a mixture of gases. What we call the atmospheric pressure is actually the sum of the partial pressures of the gases in the air. Table 7.7 lists partial pressures for the gases in air on a typical day.

SAMPLE PROBLEM 7.14

■ Partial Pressure of a Gas in a Mixture

A mixture of oxygen and helium is prepared for a scuba diver who is going to descend 200 ft below the ocean surface. At that depth, the diver breathes a gas mixture that has a total pressure of 7.0 atm. If the partial pressure of the oxygen in the tank at that depth is 1.5 atm, what is the partial pressure of the helium?

SOLUTION

From Dalton's law of partial pressures, we know that the total pressure is equal to the sum of the partial pressures:

$$P_{total} = P_{O_2} + P_{He}$$

To solve for the partial pressure of helium (P_{He}), we rearrange the expression to give the following:

$$P_{He} = P_{total} - P_{O_2}$$
$$P_{He} = 7.0 \text{ atm} - 1.5 \text{ atm}$$
$$= 5.5 \text{ atm}$$

Thus, in the gas mixture that the diver breathes, the partial pressure of the helium is 5.5 atm.

STUDY CHECK

An anesthetic consists of a mixture of cyclopropane gas, C_3H_6, and oxygen gas, O_2. If the mixture has a total pressure of 825 torr, and the partial pressure of the cyclopropane is 73 torr, what is the partial pressure of the oxygen in the anesthetic?

HEALTH NOTE

Blood Gases

Our cells continuously use oxygen and produce carbon dioxide. Both gases move in and out of the lungs through the membranes of the alveoli, the tiny air sacs at the ends of the airways in the lungs. An exchange of gases occurs in which oxygen from the air diffuses into the lungs and into the blood, while carbon dioxide produced in the cells is carried to the lungs to be exhaled. In Table 7.8, partial pressures are given for the gases in air that we inhale (inspired air), air in the alveoli, and the air that we exhale (expired air).

At sea level, oxygen normally has a partial pressure of 100 mm Hg in the alveoli of the lungs. Because the partial pressure of oxygen in venous blood is 40 mm Hg, oxygen diffuses from the alveoli into the bloodstream. The oxygen combines with hemoglobin, which carries it to the tissues of the body where the partial pressure of oxygen can be very low, less than 30 mm Hg. Oxygen diffuses from the blood where the partial pressure of O_2 is high into the tissues where O_2 pressure is low.

As oxygen is used in the cells of the body during metabolic processes, carbon dioxide is produced, so the partial pressure of CO_2 may be as high as 50 mm Hg or more. Carbon dioxide diffuses from the tissues into the bloodstream and is carried to the lungs. There it diffuses out of the blood, where CO_2 has a partial pressure of 46 mm Hg, into the alveoli, where the CO_2 is at 40 mm Hg and is exhaled. Table 7.9 gives the partial pressures of blood gases in the tissues and in oxygenated and deoxygenated blood.

TABLE 7.8

Partial Pressures of Gases During Breathing

Gas	Partial Pressure (mm Hg)		
	Inspired Air	Alveolar Air	Expired Air
Nitrogen, N_2	594.0	573	569
Oxygen, O_2	160.0	100	116
Carbon dioxide, CO_2	0.3	40	28
Water vapor, H_2O	5.7	47	47
Total	760.0	760	760

TABLE 7.9

Partial Pressures of Oxygen and Carbon Dioxide in Blood and Tissues

Gas	Partial Pressure (mm Hg)		
	Oxygenated Blood	Deoxygenated Blood	Tissues
O_2	100	40	30 or less
CO_2	40	46	50 or greater

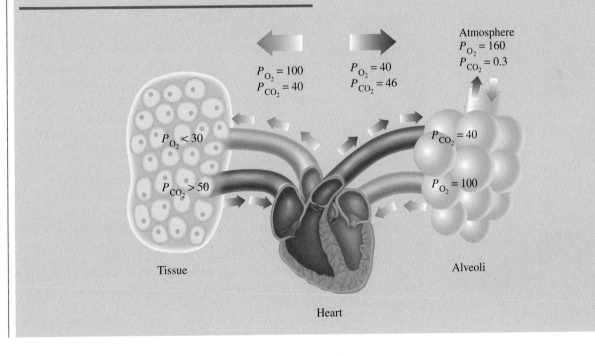

$P_{O_2} = 100$
$P_{CO_2} = 40$

$P_{O_2} = 40$
$P_{CO_2} = 46$

Atmosphere
$P_{O_2} = 160$
$P_{CO_2} = 0.3$

$P_{O_2} < 30$

$P_{CO_2} > 50$

$P_{CO_2} = 40$

$P_{O_2} = 100$

Tissue

Alveoli

Heart

QUESTIONS AND PROBLEMS

Partial Pressures (Dalton's Law)

7.57 A typical air sample in the lungs contains oxygen at 100 mm Hg, nitrogen at 573 mm Hg, carbon dioxide at 40 mm Hg, and water vapor at 47 mm Hg. Why are these pressures called partial pressures?

Hyperbaric Chambers

A burn patient may undergo treatment for burns and infections in a hyperbaric chamber, a device in which pressures can be obtained that are two to three times greater than atmospheric pressure. A greater oxygen pressure increases the level of dissolved oxygen in the blood and tissues, where it fights bacterial infections. High levels of oxygen are toxic to many strains of bacteria. The hyperbaric chamber may also be used during surgery to help counteract carbon monoxide (CO) poisoning, and to treat some cancers.

The blood is normally capable of dissolving up to 95% of the oxygen. Thus, if the partial pressure of the oxygen is 2280 mm Hg (3 atm), 95% of that or 2160 mm Hg of oxygen can dissolve in the blood where it saturates the tissues. In the case of carbon monoxide poisoning, this oxygen can replace the carbon monoxide that has attached to the hemoglobin.

A patient undergoing treatment in a hyperbaric chamber must also undergo decompression (reduction of pressure) at a rate that slowly reduces the concentration of dissolved oxygen in the blood. If decompression is too rapid, the oxygen dissolved in the blood may form gas bubbles in the circulatory system.

If divers do not decompress slowly, they suffer a similar condition called the bends. While below the surface of the ocean, divers breathe air at higher pressures. At such high pressures, nitrogen gas will dissolve in their blood. If they ascend to the surface too quickly, the dissolved nitrogen forms bubbles in the blood that can produce life-threatening blood clots. The gas bubbles can also appear in the joints and tissues of the body and be quite painful. A diver suffering from the bends is placed immediately in a decompression chamber where pressure is first increased and then slowly decreased. The dissolved nitrogen can then diffuse through the lungs until atmospheric pressure is reached.

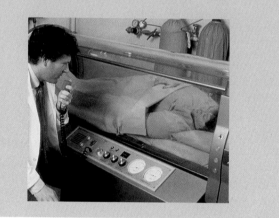

7.58 Suppose a mixture contains helium and oxygen gases. If the partial pressure of helium is the same as the partial pressure of oxygen, what do you know about the number of helium atoms compared to the number of oxygen molecules? Explain.

7.59 In a gas mixture, the partial pressures are nitrogen 425 torr, oxygen 115 torr, and helium 225 torr. What is the total pressure (torr) exerted by the gas mixture?

7.60 In a gas mixture, the partial pressures are argon 415 mm Hg, neon 75 mm Hg, and nitrogen 125 mm Hg. What is the total pressure (atm) by the gas mixture?

7.61 A gas mixture containing oxygen, nitrogen, and helium exerts a total pressure of 925 torr. If the partial pressures are oxygen 425 torr and helium 75 torr, what is the partial pressure (torr) of the nitrogen in the mixture?

7.62 A gas mixture containing oxygen, nitrogen, and neon exerts a total pressure of 1.20 atm. If helium added to the mixture increases the pressure to 1.50 atm, what is the partial pressure (atm) of the helium?

7.63 In certain lung ailments such as emphysema, there is a decrease in the ability of oxygen to diffuse into the blood.
a. How would the partial pressure of oxygen in the blood change?
b. Why does a person with severe emphysema sometimes use a portable oxygen tank?

7.64 An accident to the head can affect the ability of a person to ventilate (breathe in and out), and so can certain drugs.
a. What would happen to the partial pressures of oxygen and carbon dioxide in the blood if a person cannot properly ventilate?
b. When a person with hypoventilation is placed on a ventilator, an air mixture is delivered at pressures that are alternately above the air pressure in the person's lung, and then below. How will this move oxygen gas into the lungs, and carbon dioxide out?

CONCEPT MAP

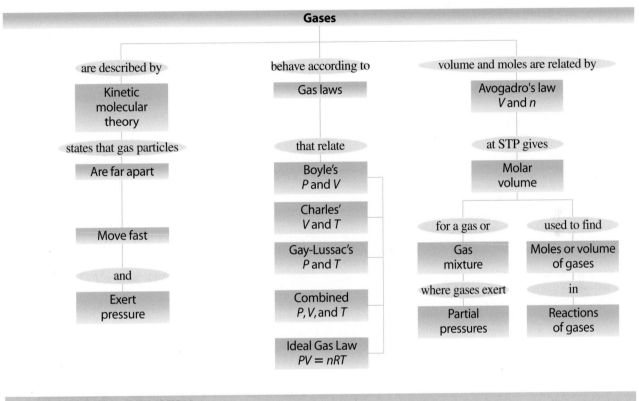

Gases

are described by

Kinetic molecular theory

states that gas particles

Are far apart

Move fast

and

Exert pressure

behave according to

Gas laws

that relate

Boyle's
P and V

Charles'
V and T

Gay-Lussac's
P and T

Combined
P, V, and T

Ideal Gas Law
$PV = nRT$

volume and moles are related by

Avogadro's law
V and n

at STP gives

Molar volume

for a gas or

Gas mixture

where gases exert

Partial pressures

used to find

Moles or volume of gases

in

Reactions of gases

CHAPTER REVIEW

7.1 Properties of Gases
In a gas, particles are so far apart and moving so fast that their attractions are unimportant. A gas is described by the physical properties of pressure (P), volume (V), temperature (T), and amount in moles (n).

7.2 Gas Pressure
A gas exerts pressure, the force of the gas particles striking the surface of a container. Gas pressure is measured in units of torr, mm Hg, and atm.

7.3 Pressure and Volume (Boyle's Law)
The volume (V) of a gas changes inversely with the pressure (P) of the gas if there is no change in the amount and temperature: $P_1V_1 = P_2V_2$. This means that the pressure increases if volume decreases; pressure decreases if volume increases.

7.4 Temperature and Volume (Charles' Law)
The volume (V) of a gas is directly related to its Kelvin temperature (T) when there is no change in the amount and pressure of the gas:

$$\frac{V_1}{T_1} = \frac{V_2}{T_2}$$

Therefore, if temperature increases, the volume of the gas increases; if temperature decreases, volume decreases.

7.5 Temperature and Pressure (Gay-Lussac's Law)
The pressure (P) of a gas is directly related to its Kelvin temperature (T).

$$\frac{P_1}{T_1} = \frac{P_2}{T_2}$$

This means that an increase in temperature increases the pressure of a gas, or a decrease in temperature decreases the pressure, as long as the amount and volume stay constant.

7.6 The Combined Gas Law
Gas laws combine into a relationship of pressure (P), volume (V), and temperature (T).

$$\frac{P_1 V_1}{T_1} = \frac{P_2 V_2}{T_2}$$

This expression is used to determine the effect of changes in two of the variables on the third.

7.7 Volume and Moles (Avogadro's Law)

The volume (V) of a gas is directly related to the number of moles (n) of the gas when the pressure and temperature of the gas do not change.

$$\frac{V_1}{n_1} = \frac{V_2}{n_2}$$

If the moles of gas are increased, the volume must increase; or if the moles of gas are decreased, the volume decreases. At standard temperature (273 K) and pressure (1 atm) abbreviated STP, 1 mole of any gas has a volume of 22.4 L.

7.8 The Ideal Gas Law

The ideal gas law gives the relationship of all the quantities P, V, n, and T that describe and measure a gas. $PV = nRT$. Any of the four variables can be calculated if the other three are known.

7.9 Partial Pressures (Dalton's Law)

In a mixture of two or more gases, the total pressure is the sum of the partial pressures of the individual gases.

$$P_{total} = P_1 + P_2 + P_3 + \cdots$$

The partial pressure of a gas in a mixture is the pressure it would exert if it were the only gas in the container.

KEY TERMS

atmosphere (atm) The pressure exerted by a column of mercury 760 mm high.

atmospheric pressure The pressure exerted by the atmosphere.

Avogadro's law A gas law that states that the volume of gas is directly related to the number of moles of gas in the sample when pressure and temperature do not change.

Boyle's law A gas law stating that the pressure of a gas is inversely related to the volume when temperature and moles of the gas do not change.

Charles' law A gas law stating that the volume of a gas changes directly with a change in Kelvin temperature when pressure and moles of the gas do not change.

combined gas law A relationship that combines several gas laws relating pressure, volume, and temperature.

$$\frac{P_1 V_1}{T_1} = \frac{P_2 V_2}{T_2}$$

Dalton's law A gas law stating that the total pressure exerted by a mixture of gases in a container is the sum of the partial pressures that each gas would exert alone.

direct relationship A relationship in which two properties increase or decrease together.

Gay-Lussac's law A gas law stating that the pressure of a gas changes directly with a change in temperature when the number of moles of a gas and its volume are held constant.

ideal gas law A law that combines the four measured properties of a gas in the equation $PV = nRT$.

inverse relationship A relationship in which two properties change in opposite directions.

kinetic molecular theory of gases A model used to explain the behavior of gases.

molar volume A volume of 22.4 L occupied by 1 mole of a gas at STP conditions of 0°C (273 K) and 1 atm.

partial pressure The pressure exerted by a single gas in a gas mixture.

pressure The force exerted by gas particles that hit the walls of a container.

STP Standard conditions of 0°C (273 K) temperature and 1 atm pressure used for the comparison of gases.

universal gas constant R A numerical value that relates the quantities P, V, n, and T in the ideal gas law, $PV = nRT$.

vapor pressure The pressure exerted by the particles of vapor above a liquid.

UNDERSTANDING THE CONCEPTS

7.65 At 100°C, which of the following gases exerts
  a. the lowest pressure?
  b. the highest pressure?

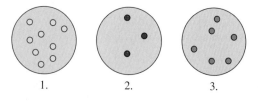

7.66 Indicate which diagram represents the volume of the gas sample in a flexible container when each of the following changes take place.

a. Temperature increases at constant pressure
b. Temperature decreases at constant pressure
c. Pressure increases at constant temperature
d. Pressure decreases at constant temperature
e. Doubling the pressure and doubling the Kelvin temperature

7.67 A balloon is filled with helium gas with a pressure of 1.00 atm and neon gas with a pressure of 0.50 atm. For each of the following changes of the initial balloon, select the diagram (A, B, or C) that shows the final (new) volume of the balloon.

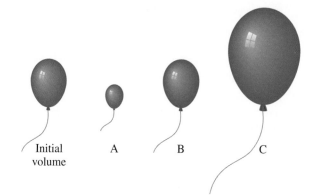

Initial volume   A   B   C

a. The balloon is put in a cold storage unit (P and n constant).
b. The balloon floats to a higher altitude where the pressure is less (n, T constant).
c. All of the neon gas is removed (T and P constant).
d. The Kelvin temperature doubles and 1/2 of the gas atoms leak out (P constant).
e. 2.0 mol O_2 gas is added at constant T and P.

7.68 Indicate if pressure increases, decreases, or stays the same in each of the following:

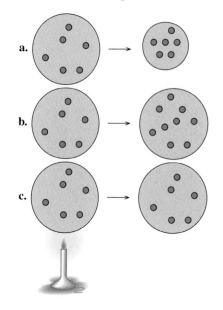

a.

b.

c.

ADDITIONAL QUESTIONS AND PROBLEMS

MC GOB

For instructor-assigned homework, go to
www.masteringgob.com.

7.69 At a restaurant, a customer chokes on a piece of food. You put your arms around the person's waist and use your fists to push up on the person's abdomen, an action called the Heimlich maneuver.
a. How would this action change the volume of the chest and lungs?
b. Why does it cause the person to expel the food item from the airway?

7.70 An airplane is pressurized to 650 mm Hg, which is the atmospheric pressure at a ski resort at 13,000 ft altitude.
a. If air is 21% oxygen, what is the partial pressure of oxygen on the plane?
b. If the partial pressure of oxygen drops below 100 mm Hg, passengers become drowsy. If this happens, oxygen masks are released. What is the total cabin pressure at which oxygen masks are dropped?

7.71 A fire extinguisher has a pressure of 10. atm at 25°C. What is the pressure in atmospheres if the fire extinguisher is used at a temperature of 75°C?

7.72 A weather balloon has a volume of 750 L when filled with helium at 8°C at a pressure of 380 torr. What is the new volume of the balloon, where the pressure is 0.20 atm and the temperature is −45°C?

7.73 A sample of hydrogen (H_2) gas at 127°C has a pressure of 2.00 atm. At what temperature (°C) will the pressure of the H_2 decrease to 0.25 atm?

7.74 A sample of nitrogen (N_2) has a volume of 250 mL at 30°C and a pressure of 745 mm Hg. What is the volume of the nitrogen at STP?

7.75 A 2.00 L container is filled with methane gas CH_4 at a pressure of 2500 mm Hg and a temperature of 18°C. How many grams of methane are in the container?

7.76 A steel cylinder with a volume of 15.0 L is filled with 50.0 g nitrogen gas at 25°C. What is the pressure of the N_2 gas in the cylinder?

7.77 How many molecules of CO_2 are in 35.0 L $CO_2(g)$ at 1.2 atm and 5°C?

7.78 A container is filled with 4.0×10^{22} O_2 molecules at 5°C and 845 mm Hg. What is the volume in mL of the container?

7.79 When heated, calcium carbonate decomposes to give calcium oxide and carbon dioxide gas.

$$CaCO_3(s) \longrightarrow CaO(s) + CO_2(g)$$

If 2.00 moles of $CaCO_3$ react, how many liters of CO_2 gas are produced at STP?

7.80 Magnesium reacts with oxygen to form magnesium oxide. How many liters of oxygen gas at STP are needed to react completely with 8.0 g of magnesium?

$$2Mg(s) + O_2(g) \longrightarrow 2MgO(s)$$

7.81 Your space ship has docked at a space station above Mars. The temperature inside the space station is a carefully controlled 24°C at a pressure of 745 mm Hg. A balloon with a volume of 425 mL drifts into the airlock where the temperature is −95°C and the pressure is 0.115 atm. What is the new volume of the balloon? Assume that the balloon is very elastic.

7.82 How many liters of H_2 gas can be produced at STP from 25.0 g of Zn?

$$Zn(s) + 2HCl(aq) \longrightarrow ZnCl_2(aq) + H_2(g)$$

7.83 Aluminum oxide can be formed from its elements.

$$4Al(s) + 3O_2(g) \longrightarrow 2Al_2O_3(s)$$

What volume of oxygen is needed at STP to completely react 5.4 g of aluminum?

7.84 Glucose $C_6H_{12}O_6$ is metabolized in living systems according to the reaction

$$C_6H_{12}O_6(s) + 6O_2(g) \longrightarrow 6CO_2(g) + 6H_2O(l)$$

How many grams of water can be produced when 12.5 L O_2 reacts at STP?

7.85 A gas mixture with a total pressure of 2400 torr is used by a scuba diver. If the mixture contains 2.0 moles of helium and 6.0 moles of oxygen, what is the partial pressure of each gas in the sample?

7.86 What is the total pressure in mm Hg of a gas mixture containing argon gas at 0.25 atm, helium gas at 350 mm Hg, and nitrogen gas at 360 torr?

7.87 A gas mixture contains oxygen and argon at partial pressures of 0.60 atm and 425 mm Hg. If nitrogen gas added to the sample increases the total pressure to 1250 torr, what is the partial pressure in torr of the nitrogen added?

7.88 A gas mixture contains helium and oxygen at partial pressures of 255 torr and 0.450 atm. What is the total pressure in mm Hg of the mixture after it is placed in a container one-half the volume of the original container?

CHALLENGE QUESTIONS

7.89 A gas sample has a volume of 4250 mL at 15°C and 745 mm Hg. What is the new temperature (°C) after the sample is transferred to a new container with a volume of 2.50 L and a pressure of 1.20 atm?

7.90 In the fermentation of glucose (wine making), a volume of 780 mL CO_2 gas was produced at 37°C and 1.00 atm. What is the volume (L) of the gas when measured at 22°C and 675 mm Hg?

7.91 When sensors in a car detect a collision, they cause the reaction of sodium azide, NaN_3, which generates nitrogen gas to fill the air bags within 0.03 second.

$$2NaN_3(s) \longrightarrow 2Na(s) + 3N_2(g)$$

How many liters of N_2 are produced at STP if the air bag contains 132 g NaN_3?

7.92 Nitrogen dioxide reacts with water to produce oxygen and ammonia.

$$4NO_2(g) + 6H_2O(g) \longrightarrow 7O_2(g) + 4NH_3(g)$$

How many liters of O_2 at STP are produced when 2.5×10^{23} molecules NO_2 react?

7.93 A 1.00-g sample of dry ice (CO_2) is placed in a container that has a volume of 4.60 L and a temperature of 24.0°C. Calculate the pressure in mm Hg inside the container after all the dry ice changes to a gas.

$$CO_2(s) \longrightarrow CO_2(g)$$

7.94 A 250 mL sample of nitrogen (N_2) has a pressure of 745 mm Hg at 30°C. What is the mass of nitrogen?

7.95 Hydrogen gas can be produced in the laboratory through the reaction of magnesium metal with hydrochloric acid.

$$Mg(s) + 2HCl(aq) \longrightarrow MgCl_2(aq) + H_2(g)$$

What is the volume, in liters, of H_2 gas produced at 24°C and 835 mm Hg, from the reaction of 12.0 g Mg?

ANSWERS

Answers to Study Checks

7.1 The mass in grams gives the amount of gas.

7.2 0.862 atm

7.3 250 torr

7.4 50.0 mL

7.5 569 mL

7.6 16°C

7.7 241 mm Hg

7.8 7.50 L

7.9 7.0 g N_2

7.10 5.41 L H_2

7.11 0.327 mole Cl_2

7.12 1.04 L CO

7.13 1.24 atm

7.14 752 torr

Answers to Selected Questions and Problems

7.1 **a.** At a higher temperature, gas particles have greater kinetic energy, which makes them move faster.
b. Because there are great distances between the particles of a gas, they can be pushed closer together and still remain a gas.

7.3 **a.** temperature **b.** volume
c. amount **d.** pressure

7.5   atmospheres (atm), mm Hg, torr, lb/in.^{2}, kPa

7.7   **a.** 1520 torr
b. 1520 mm Hg

7.9   As a diver ascends to the surface, external pressure decreases. If the air in the lungs were not exhaled, its volume would expand and severely damage the lungs. The pressure in the lungs must adjust to changes in the external pressure.

7.11   **a.** The pressure is greater in cylinder A. According to Boyle's law, a decrease in volume pushes the gas particles closer together, which will cause an increase in the pressure.
b.

Property	Initial	Final
Pressure (P)	650 mm Hg	1.2 atm
Volume (V)	220 mL	160 mL

7.13   **a.** The pressure doubles.
b. The pressure falls to one-third the initial pressure.
c. The pressure increases to ten times the original pressure.

7.15   **a.** 328 mm Hg   **b.** 2620 mm Hg
c. 4400 mm Hg

7.17   **a.** 25 L   **b.** 25 L   **c.** 100. L

7.19   25 L

7.21   **a.** inspiration   **b.** expiration   **c.** inspiration

7.23   **a.** C   **b.** A   **c.** B

7.25   When a gas is heated at constant pressure, the volume of the gas increases to fill the hot-air balloon.

7.27   **a.** 2400 mL   **b.** 4900 mL
c. 1800 mL   **d.** 1700 mL

7.29   An increase in temperature increases the pressure inside the can. When the pressure exceeds the pressure limit of the can, it explodes.

7.31   When the temperature increases, the pressure of the oxygen inside the filled tanks increases causing the tanks to explode.

7.33   **a.** 770 torr   **b.** 1.51 atm

7.35   **a.** boiling point   **b.** vapor pressure
c. atmospheric pressure   **d.** boiling point

7.37   **a.** On top of a mountain, water boils below 100°C because the atmospheric (external) pressure is less than 1 atm.
b. Because the pressure inside a pressure cooker is greater than 1 atm, water boils above 100°C. At a higher temperature, food cooks faster.

7.39   **a.** 4.25 atm   **b.** 3.07 atm   **c.** 0.605 atm

7.41   110 mL

7.43   The volume increases because the amount of gas particles is increased.

7.45   **a.** 4.00 L   **b.** 14.6 L   **c.** 26.7 L

7.47   **a.** 2.00 moles O_2 **b.** 0.179 mole CO_2
c. 4.48 L   **d.** 55 400 mL

7.49   7.60 L H_2

7.51   4.93 atm

7.53   29.4 g O_2

7.55   565 K (292°C)

7.57   In a gas mixture, the pressure that each gas exerts as part of the total pressure is called the partial pressure of that gas. Because the air sample is a mixture of gases, the total pressure is the sum of the partial pressures of each gas in the sample.

7.59   765 torr

7.61   425 torr

7.63   **a.** The partial pressure of oxygen will be lower than normal.
b. Breathing a higher concentration of oxygen will help to increase the supply of oxygen in the lungs and blood and raise the partial pressure of oxygen in the blood.

7.65   **a.** 2   **b.** 1

7.67   **a.** A   **b.** C   **c.** A
d. B   **e.** C

7.69   **a.** The volume of the chest and lungs is decreased.
b. The decrease in volume increases the pressure, which can dislodge the food in the trachea.

7.71   12 atm

7.73   −223°C

7.75   4.4 g

7.77 1.1×10^{24} molecules of CO_2

7.79   44.8 L CO_2

7.81   2170 mL

7.83   3.4 L of $O_2(g)$

7.85   He 600 torr, O_2 1800 torr

7.87   370 torr

7.89   −66°C

7.91   68.2 L

7.93   91.5 mm Hg

7.95   11.0 L

8 Solutions

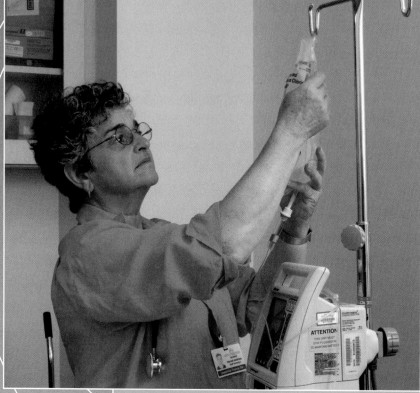

"There is a lot of chemistry going on in the body, including drug interactions," says Josephine Firenze, registered nurse, Kaiser Hospital.

Normally, the body maintains a homeostasis of fluids and electrolytes. Conditions that alter the composition of body fluids can lead to convulsions, coma, or death. To halt the disease process and to establish homeostasis, a patient may be given intravenous fluid therapy. Solutions that are compatible with body fluids such as a 5% glucose or a 0.9% saline are used. An infusion pump delivers the desired number of milliliters per hour to the patient. During IV therapy, a patient is checked for fluid overload as indicated by edema, which is swelling, or a greater fluid input than output.

Mastering GOB CHEMISTRY

Visit **www.masteringgob.com** for self-study materials and instructor-assigned homework.

Solutions are everywhere around us. Most consist of one substance dissolved in another. The air we breathe is a solution of oxygen and nitrogen gases. Carbon dioxide gas dissolved in water makes carbonated drinks. When we make solutions of coffee or tea, we use hot water to dissolve substances from coffee beans or tea leaves. The ocean is also a solution, consisting of many salts such as sodium chloride dissolved in water. In a hospital, the antiseptic tincture of iodine is a solution of iodine dissolved in alcohol.

Our body fluids contain water and dissolved substances such as glucose and urea and electrolytes such as K^+, Na^+, Cl^-, Mg^{2+}, HCO_3^-, and HPO_4^{2-}. Proper amounts of each of these dissolved substances and water must be maintained in the body fluids. Small changes in electrolyte levels can seriously disrupt cellular process and endanger our health. Therefore, the measurement of their concentrations is a valuable diagnostic tool.

In the processes of osmosis and dialysis, water, essential nutrients, and waste products enter and leave the cells of the body. In osmosis, water flows in and out of the cells of the body. In dialysis, small particles in solution as well as water diffuse through semipermeable membranes. The kidneys utilize osmosis and dialysis to regulate the amount of water and electrolytes that are excreted.

8.1 Solutions

LEARNING GOAL

Identify the solute and solvent in a solution. Describe the formation of a solution.

A **solution** is a mixture in which one substance called the **solute** is uniformly dispersed in another substance called the **solvent**. Because the solute and the solvent do not react with each other, they can be mixed in varying proportions. A little salt dissolved in water tastes slightly salty. When more salt dissolves, the water tastes very salty. Usually, the solute (in this case, salt) is the substance present in the smaller amount, whereas the solvent (in this case, water) is present in the larger amount. In a solution, the particles of the solute are evenly dispersed among the molecules of the solvent. (See Figure 8.1.)

Types of Solutes and Solvents

Solutes and solvents may be solids, liquids, or gases. When sugar is dissolved in a glass of water, a liquid sugar solution forms. Sugar is the solute, and water is the solvent. Soda water and soft drinks are prepared by dissolving CO_2 gas in water. The CO_2 gas is the solute, and water is the solvent. Table 8.1 lists some solutes and solvents and their solutions.

Water as a Solvent

Water is one of the most common solvents in nature. In the H_2O molecule, an oxygen atom shares electrons with two hydrogen atoms. Because the oxygen atom is much more electronegative, the O—H bonds are polar. In each of the polar

Solute: The substance present in lesser amount

Salt

Water

Solvent: The substance present in greater amount

FIGURE 8.1

A solution of copper(II) sulfate (CuSO₄) forms as particles of solute dissolve, move away from the crystals, and become evenly dispersed among the solvent (water) molecules.

Q **What does the uniform blue color indicate about the CuSO₄ solution?**

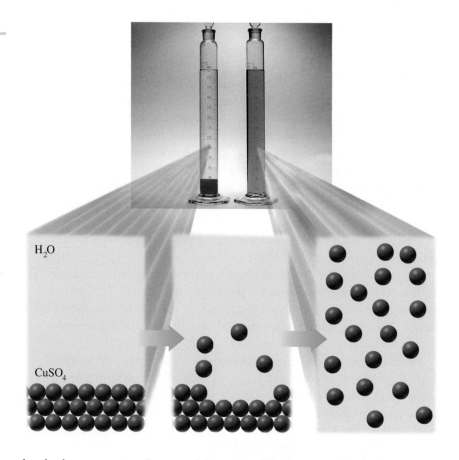

H₂O

CuSO₄

SELF-STUDY ACTIVITY
Hydrogen Bonding

bonds, the oxygen atom has a partial negative (δ^-) charge, and the hydrogen atom has a partial positive (δ^+) charge. Because of the arrangement of the polar bonds, water is a *polar substance*.

Hydrogen bonds occur between molecules where a partially positive hydrogen is attracted to the strongly electronegative atoms of O, N, or F in other molecules. In water, hydrogen bonds are formed by the attraction between the oxygen atom of one water molecule and a hydrogen atom in another water molecule. In the diagram, hydrogen bonds are shown as dots between the water molecules.

TABLE 8.1

Some Examples of Solutions

Type	Example	Solute	Solvent
Gas Solutions			
Gas in a gas	Air	Oxygen (gas)	Nitrogen (gas)
Liquid Solutions			
Gas in a liquid	Soda water	Carbon dioxide (gas)	Water (liquid)
	Household ammonia	Ammonia (gas)	Water (liquid)
Liquid in a liquid	Vinegar	Acetic acid (liquid)	Water (liquid)
Solid in a liquid	Seawater	Sodium chloride (solid)	Water (liquid)
	Tincture of iodine	Iodine (solid)	Alcohol (liquid)
Solid Solutions			
Liquid in a solid	Dental amalgam	Mercury (liquid)	Silver (solid)
Solid in a solid	Brass	Zinc (solid)	Copper (solid)
	Steel	Carbon (solid)	Iron (solid)

Although hydrogen bonds are much weaker than covalent or ionic bonds, there are many of them linking molecules together. As a result hydrogen bonding plays an important role in the properties of water and biological compounds such as proteins, carbohydrates, and DNA.

Partial negative charge

Partial positive charge

Hydrogen bonds

HEALTH NOTE

Water in the Body

The average adult contains about 60% water by weight, and the average infant about 75%. About 60% of the body's water is contained within the cells as intracellular fluids; the other 40% makes up extracellular fluids, which include the interstitial fluid in tissue and the plasma in the blood. These external fluids carry nutrients and waste materials between the cells and the circulatory system.

Every day you lose between 1500 and 3000 mL of water from the kidneys as urine, from the skin as perspiration, from the lungs as you exhale, and from the gastrointestinal tract. Serious dehydration can occur in an adult if there is a 10% net loss in total body fluid, and a 20% loss of fluid can be fatal. An infant suffers severe dehydration with a 5–10% loss in body fluid.

Water loss is continually replaced by the liquids and foods in the diet and from metabolic processes that produce water in the cells of the body. Table 8.2 lists the % by mass of water contained in some foods.

24 Hours

Water gain		Water loss	
		Urine	1500 mL
Liquid	1000 mL	Perspiration	300 mL
Food	1200 mL	Breath	600 mL
Metabolism	300 mL	Feces	100 mL
Total	2500 mL	Total	2500 mL

TABLE 8.2

Percentage of Water in Some Foods

Food	Water (% by mass)	Food	Water (% by mass)
Vegetables		**Meats/Fish**	
Carrot	88	Chicken, cooked	71
Celery	94	Hamburger, broiled	60
Cucumber	96	Salmon	71
Tomato	94	**Grains**	
Fruits		Cake	34
Apple	85	French bread	31
Banana	76	Noodles, cooked	70
Cantaloupe	91	**Milk Products**	
Orange	86	Cottage cheese	78
Strawberry	90	Milk, whole	87
Watermelon	93	Yogurt	88

Formation of Solutions

An ionic compound such as sodium chloride, NaCl, is held together by ionic bonds between positive Na^+ ions and negative Cl^- ions. It dissolves in water because water is a polar solvent. When solid NaCl crystals are placed in water, the process of dissolution begins as the ions on the surface of the crystal come in contact with water molecules. (See Figure 8.2.) The negatively charged oxygen atom at one end of a water molecule attracts the positive Na^+ ions. The positively charged hydrogen atoms at the other end of a water molecule attract the negative Cl^- ions. The attractive forces of several water molecules provide the energy to break the ionic bonds between the Na^+ and Cl^- ions in the NaCl crystal. As the water molecules pull the ions into solution, a new surface of the NaCl crystal is exposed to the solvent. During a process called **hydration**, the dissolved Na^+ and Cl^- ions are surrounded by water molecules, which diminishes their attraction to other ions and helps keep them in solution.

Like Dissolves Like

Gases form solutions easily because their particles are moving so rapidly that they are far apart and attractions to the other gas particles are not important. When solids or liquids form solutions, there must be an attraction between the solute particles and the solvent particles. Then the particles of the solute and solvent will mix together. If there is no attraction between a solute and a solvent, their particles do not mix and no solution forms.

A salt such as NaCl will form a solution with water because the Na^+ and Cl^- ions in the salt are attracted to the partially positive and negative regions of the individual water molecules. You can dissolve table sugar, $C_{12}H_{22}O_{11}$, in water because the molecule has many polar —OH groups that attract water molecules.

However, compounds containing nonpolar molecules such as iodine (I_2), oil, or grease do not dissolve in water because water is polar. Nonpolar solutes require nonpolar solvents for a solution to form. The expression "like dissolves like" is a way of saying that the polarities of a solute and a solvent must be similar in order to form a solution. Figure 8.3 illustrates the formation of some polar and nonpolar solutions.

SAMPLE PROBLEM 8.1

MC
GOB
Polar vs. Nonpolar

■ Polar and Nonpolar Solutes

Indicate whether each of the following substances will dissolve in water. Explain.

a. KCl
b. octane, C_8H_{18}, a compound in gasoline
c. ethanol, C_2H_5OH, a substance in mouthwash

SOLUTION

a. Yes. KCl is an ionic compound.
b. No. C_8H_{18} is a nonpolar substance.
c. Yes. C_2H_5OH is a polar substance.

STUDY CHECK

Will oil, a nonpolar substance, dissolve in hexane, a nonpolar solvent?

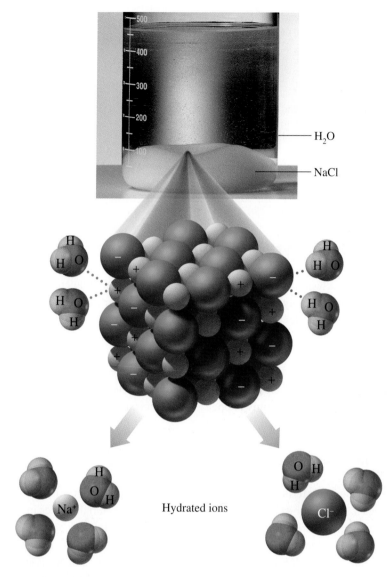

FIGURE 8.2

Ions on the surface of a crystal of NaCl dissolve in water as they are attracted to the polar water molecules that pull the ions into solution and surround them.

Q What helps keep the Na^+ and Cl^- ions in solution?

QUESTIONS AND PROBLEMS

■ Solutions

8.1 Identify the solute and the solvent in each solution composed of the following:
a. 10.0 g of NaCl and 100.0 g of H_2O
b. 50.0 mL of ethanol, C_2H_5OH, and 10.0 mL of H_2O
c. 0.20 L of O_2 and 0.80 L of N_2

8.2 Identify the solute and the solvent in each solution composed of the following:
a. 50.0 g of silver and 4.0 g of mercury
b. 100.0 mL of water and 5.0 g of sugar
c. 1.0 g of I_2 and 50.0 mL of alcohol

8.3 Describe the formation of an aqueous KI solution.

8.4 Describe the formation of an aqueous LiBr solution.

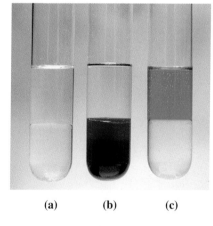

(a)   (b)   (c)

FIGURE 8.3

Like dissolves like. **(a)** The test tubes contain an upper layer of water (polar) and a lower layer of CH_2Cl_2 (nonpolar). **(b)** The nonpolar solute I_2 dissolves in the nonpolar layer. **(c)** The ionic solute $Ni(NO_3)_2$ dissolves in the water.

Q Which layer would dissolve polar molecules of sugar?

EXPLORE YOUR WORLD

■ **Like Dissolves Like**

Mix together small amounts of the following substances:
a. oil and water
b. water and vinegar
c. salt and water
d. sugar and water
e. salt and oil

QUESTIONS

1. Which of the mixtures formed a solution? Which did not?
2. Why do some mixtures form solutions, but others do not?

LEARNING GOAL

Identify solutes as electrolytes or nonelectrolytes.

Electrolytes and Nonelectrolytes

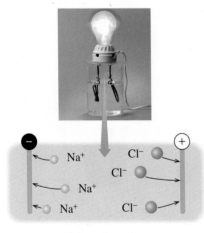

Strong electrolyte

8.5 Water is a polar solvent; CCl_4 is a nonpolar solvent. In which solvent is each of the following more likely to be soluble?
a. KCl, ionic
b. I_2, nonpolar
c. sugar, polar
d. gasoline, nonpolar

8.6 Water is a polar solvent; hexane is a nonpolar solvent. In which solvent is each of the following more likely to be soluble?
a. vegetable oil, nonpolar
b. benzene, nonpolar
c. $LiNO_3$, ionic
d. Na_2SO_4, ionic

■ 8.2 Electrolytes and Nonelectrolytes

Solutes can be classified by their ability to conduct an electrical current. When solutes called **electrolytes** dissolve in water, they separate into ions, which are able to conduct electricity. When solutes called **nonelectrolytes** dissolve in water, they do not separate into ions and their solutions do not conduct electricity.

To test solutions for ions, we can use an apparatus that consists of a battery and a pair of electrodes connected by wires to a light bulb. The light bulb glows when electricity can flow, which only happens when electrolytes provide ions to complete the circuit.

■ Strong Electrolytes

A **strong electrolyte** is a solute that dissociates completely into ions when it dissolves in water. For example, when sodium chloride (NaCl) dissolves in water, the sodium and chloride ions are attracted to water molecules. In a process called *dissociation*, the ions separate from the solid. As ions form, they are hydrated by surrounding water molecules. In the equation for the dissociation of NaCl in water, the H_2O over the arrow indicates that water is needed for the dissociation process, but is not a reactant.

$$NaCl(s) \xrightarrow{H_2O} Na^+(aq) + Cl^-(aq)$$

When we write the equation for the dissociation, the electrical charges must balance. For example, magnesium nitrate dissolves in water to give one magnesium ion for every two nitrate ions. However, only the ionic bonds between Mg^{2+} and NO_3^- are broken, not the covalent bonds within the polyatomic ion. The dissociation for $Mg(NO_3)_2$ is written as follows:

$$Mg(NO_3)_2(s) \xrightarrow{H_2O} Mg^{2+}(aq) + 2NO_3^-(aq)$$

■ Weak Electrolytes

A **weak electrolyte** is a solute that dissolves in water mostly as whole molecules. Only a few of the dissolved molecules separate, which produces a small number of ions in solution. Thus solutions of weak electrolytes do not conduct electrical current as well as solutions of strong electrolytes. For example, an aqueous solution of HF, which is a weak electrolyte, consists of mostly HF molecules, and a few H^+ and F^- ions. Within the solution, a few HF molecules dissociate into ions. As more H^+ and F^- ions form, some recombine to give HF molecules, which is indicated by a backwards arrow. Eventually the rate of formation of ions is equal to the rate at which they recombine. The use of two arrows indicates that the forward and reverse reactions are taking place at the same time.

$$HF(aq) \underset{\text{Recombination}}{\overset{\text{Dissociation}}{\rightleftharpoons}} H^+(aq) + F^-(aq)$$

Nonelectrolytes

Solutes that are nonelectrolytes dissolve in water as molecules and do not separate into ions; thus, solutions of nonelectrolytes do not conduct electricity. For example, sucrose (sugar) is a nonelectrolyte that dissolves in water as whole molecules only.

$$C_{12}H_{22}O_{11}(s) \xrightarrow{H_2O} C_{12}H_{22}O_{11}(aq)$$

Sucrose   Solution of sucrose molecules

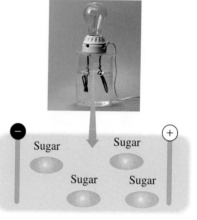

Weak electrolyte

Nonelectrolyte

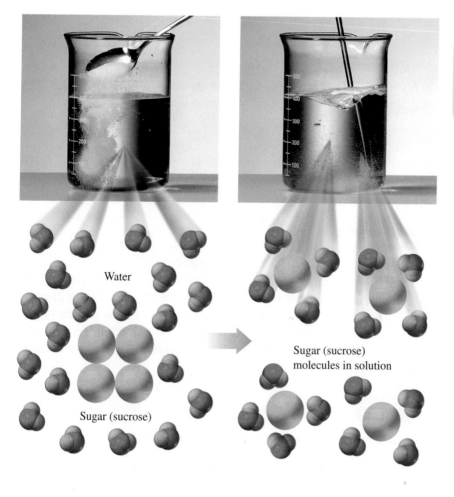

Water

Sugar (sucrose)

Sugar (sucrose)
molecules in solution

Classification of Solutes in Aqueous Solutions

Types of Solute	Dissociation	Contained in Solution	Conducts Electricity	Examples
Strong electrolyte	Completely	Ions only	Yes	Ionic compounds such as $NaCl$, KBr, $MgCl_2$, $NaNO_3$; bases such as $NaOH$, KOH; acids such as HCl, HBr, HNO_3, $HClO_4$
Weak electrolyte	Partially	Mostly molecules and a few ions	Yes, but poorly	HF, H_2O, NH_3, CH_3COOH (acetic acid)
Nonelectrolyte	None	Molecules only	No	Carbon compounds such as CH_3OH, C_2H_5OH, $C_{12}H_{22}O_{11}$

Table 8.3 summarizes the classification of solutes in aqueous solutions.

SAMPLE PROBLEM 8.2

■ **Solutions of Electrolytes and Nonelectrolytes**

Indicate whether solutions of each of the following contain ions, molecules, or both:

a. Na_2SO_4, a strong electrolyte
b. CH_3OH, a nonelectrolyte

SOLUTION

a. A solution of Na_2SO_4 contains the ions of the salt, Na^+ and SO_4^{2-}.
b. A nonelectrolyte such as CH_3OH dissolves in water as molecules.

STUDY CHECK

Boric acid, H_3BO_3, is a weak electrolyte. Would you expect a boric acid solution to contain ions, molecules, or both?

Equivalents

Body fluids typically contain a mixture of several electrolytes, such as Na^+, Cl^-, K^+, and Ca^{2+}. We measure each individual ion in terms of an **equivalent (Eq)**, which is the amount of that ion equal to 1 mole of positive or negative electrical charge. For example, 1 mole of Na^+ ions and 1 mole of Cl^- ions are each 1 equivalent because they each contain 1 mole of charge. For an ion with a charge of $2+$ or $2-$, there are 2 equivalents for each mole. Some examples of ions and equivalents are shown in Table 8.4.

Equivalents of Electrolytes

Ion	Electrical Charge	Number of Equivalents in 1 Mole
Na^+	$1+$	1 Eq
Ca^{2+}	$2+$	2 Eq
Fe^{3+}	$3+$	3 Eq
Cl^-	$1-$	1 Eq
SO_4^{2-}	$2-$	2 Eq

SAMPLE PROBLEM 8.3

▪ Electrolyte Concentration

In body fluids, concentrations of electrolytes are often expressed as milliequivalent (mEq) per liter. A typical concentration for Na^+ in the blood is 138 mEq/L. How many moles of sodium ion are in 1.00 L of blood?

SOLUTION

Using the volume and the electrolyte concentration (in mEq/L) we can find the number of equivalents in 1.00 L of blood.

$$1.00 \, \cancel{L} \times \frac{138 \, \cancel{mEq}}{1 \, \cancel{L}} \times \frac{1 \, Eq}{1000 \, \cancel{mEq}} = 0.138 \, Eq \, Na^+$$

We can then convert equivalents to moles (for Na^+ there is 1 Eq per mole).

$$0.138 \, \cancel{Eq \, Na^+} \times \frac{1 \, mole \, Na^+}{1 \, \cancel{Eq \, Na^+}} = 0.138 \, mole$$

STUDY CHECK

A Ringer's solution for intravenous fluid replacement contains 155 mEq Cl^- per liter of solution. If a patient receives 1250 mL of Ringer's solution, how many moles of chloride were given?

QUESTIONS AND PROBLEMS

MC
GOB

▪ Electrolytes and Nonelectrolytes

8.7 KF is a strong electrolyte, and HF is a weak electrolyte. How are they different?

8.8 NaOH is a strong electrolyte, and CH_3OH is a nonelectrolyte. How are they different?

8.9 The following soluble salts are strong electrolytes. Write a balanced equation for their dissociation in water:
  a. KCl **b.** $CaCl_2$ **c.** K_3PO_4 **d.** $Fe(NO_3)_3$

8.10 The following salts are strong electrolytes. Write a balanced equation for their dissociation in water:
  a. LiBr **b.** $NaNO_3$ **c.** $FeCl_3$ **d.** $Mg(NO_3)_2$

8.11 Indicate whether aqueous solutions of the following will contain ions only, molecules only, or molecules and some ions:
  a. acetic acid ($HC_2H_3O_2$), found in vinegar, a weak electrolyte
  b. NaBr, a strong electrolyte
  c. fructose ($C_6H_{12}O_6$), a nonelectrolyte

8.12 Indicate whether aqueous solutions of the following will contain ions only, molecules only, or molecules and some ions:
  a. Na_2SO_4, a strong electrolyte
  b. ethanol, C_2H_5OH, a nonelectrolyte
  c. HCN, hydrocyanic acid, a weak electrolyte

8.13 Indicate the type of electrolyte represented in the following equations:

  a. $K_2SO_4(s) \xrightarrow{H_2O} 2K^+(aq) + SO_4{}^{2-}(aq)$

  b. $NH_4OH(aq) \underset{}{\overset{H_2O}{\rightleftharpoons}} NH_4{}^+(aq) + OH^-(aq)$

  c. $C_6H_{12}O_6(s) \xrightarrow{H_2O} C_6H_{12}O_6(aq)$

Electrolytes in Body Fluids

The concentrations of electrolytes present in body fluids and in intravenous fluids given to a patient are often expressed in milliequivalents per liter (mEq/L) of solution.

$$1 \text{ Eq} = 1000 \text{ mEq}$$

Table 8.5 gives the concentrations of some typical electrolytes in blood plasma. There is a charge balance because the total number of positive charges is equal to the total number of negative charges. The use of a specific intravenous solution depends on the nutritional, electrolyte, and fluid needs of the individual patient. Examples of various types of solutions are given in Table 8.6.

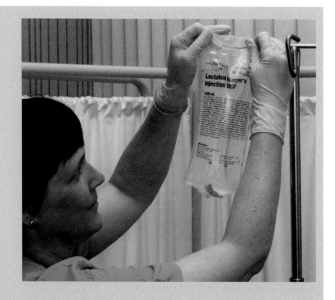

TABLE 8.5

Some Typical Concentrations of Electrolytes in Blood Plasma

Electrolyte	Concentration (mEq/L)
Cations	
Na^+	138
K^+	5
Mg^{2+}	3
Ca^{2+}	4
Total	150
Anions	
Cl^-	110
HCO_3^-	30
HPO_4^{2-}	4
Proteins	6
Total	150

TABLE 8.6

Electrolyte Concentrations in Intravenous Replacement Solutions

Solution	Electrolytes (mEq/L)	Use
Sodium chloride (0.9%)	Na^+ 154, Cl^- 154	Replacement of fluid loss
Potassium chloride with 5% dextrose	K^+ 40, Cl^- 40	Treatment of malnutrition (low potassium levels)
Ringer's solution	Na^+ 147, K^+ 4, Ca^{2+} 4, Cl^- 155	Replacement of fluids and electrolytes lost through dehydration
Maintenance solution with 5% dextrose	Na^+ 40, K^+ 35, Cl^- 40, lactate$^-$ 20, HPO_4^{2-} 15	Maintenance of fluid and electrolyte levels
Replacement solution (extracellular)	Na^+ 140, K^+ 10, Ca^{2+} 5, Mg^{2+} 3, Cl^- 103, acetate$^-$ 47, citrate^{3-} 8	Replacement of electrolytes in extracellular fluids

8.14 Indicate the type of electrolyte represented in the following equations:

a. $CH_3OH(l) \xrightarrow{H_2O} CH_3OH(aq)$

b. $MgCl_2(s) \xrightarrow{H_2O} Mg^{2+}(aq) + 2Cl^-(aq)$

c. $HClO(aq) \underset{H_2O}{\rightleftharpoons} H^+(aq) + ClO^-(aq)$

8.15 Indicate the number of equivalents in each of the following:
a. 1 mole K^+ 　　　　　　　**b.** 2 moles OH^-
c. 1 mole Ca^{2+} 　　　　　　**d.** 3 moles CO_3^{2-}

8.16 Indicate the number of equivalents in each of the following:
a. 1 mole Mg^{2+} 　　　　　　**b.** 0.5 mole H^+
c. 4 moles Cl^- 　　　　　　　**d.** 2 moles Fe^{3+}

8.17 A physiological saline solution contains 154 mEq/L each of Na^+ and Cl^-. How many moles each of Na^+ and Cl^- are in 1.00 L of the saline solution?

8.18 A solution to replace potassium loss contains 40. mEq/L each of K^+ and Cl^-. How many moles each of K^+ and Cl^- are in 1.5 L of the solution?

8.19 A solution contains 40. mEq/L of Cl^- and 15 mEq/L of $HPO_4{}^{2-}$. If Na^+ is the only cation in the solution, what is the Na^+ concentration in milliequivalents per liter?

8.20 A sample of Ringer's solution contains the following concentrations (mEq/L) of cations: Na^+ 147, K^+ 4, and Ca^{2+} 4. If Cl^- is the only anion in the solution, what is the Cl^- concentration in milliequivalents per liter?

8.3 Solubility

The term **solubility** is used to describe the amount of a solute that can dissolve in a given amount of solvent. Many factors such as the type of solute, the type of solvent, and temperature affect solubility. Solubility, usually expressed in grams of solute in 100 grams of solvent, is the maximum amount of solute that can be dissolved at a certain temperature. If a solute readily dissolves when added to the solvent, the solution does not contain the maximum amount of solute. We call the solution an **unsaturated solution**. When a solution contains all the solute that can dissolve, it is a **saturated solution**. If we try to add more solute, undissolved solute will remain on the bottom of the container.

LEARNING GOAL

Define *solubility*; distinguish between an unsaturated and a saturated solution.

 Solubility

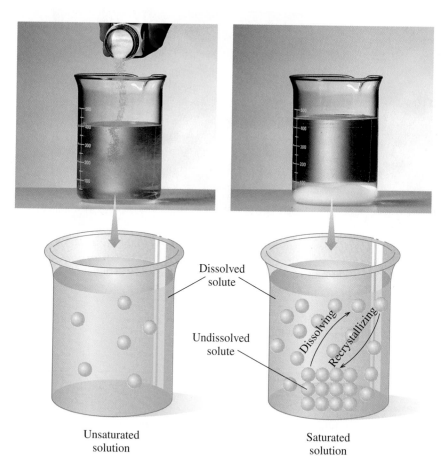

Dissolved solute

Undissolved solute

Dissolving

Recrystallizing

Unsaturated solution

Saturated solution

A solution becomes saturated when the rates of the forward reaction that dissolves the solute and the reverse reaction of recrystallization become equal. Then there is no further change in the amount of solid solute. The rate at which

Gout and Kidney Stones: A Problem of Saturation in Body Fluids

The conditions of gout and kidney stones involve compounds in the body that exceed their solubility levels and form solid products. Gout affects adults, primarily men, over the age of 40. Attacks of gout may occur when the concentration of uric acid in blood plasma exceeds its solubility, which is 7 mg/100 mL of plasma at 37°C. Insoluble deposits of needle-like crystals of uric acid can form in the cartilage, tendons, and soft tissues where they cause painful gout attacks. They may also form in the tissues of the kidneys, where they can cause renal damage. High levels of uric acid in the body can be caused by an increase in uric acid production, failure of the kidneys to remove uric acid, or by a diet with an overabundance of foods containing purines, which are metabolized to uric acid in the body. Foods in the diet that contribute to high levels of uric acid include certain meats, sardines, mushrooms, asparagus, and beans. Drinking alcoholic beverages may also significantly increase uric acid levels and bring about gout attacks.

Treatment for gout involves diet changes and drugs. Depending on the levels of uric acid, a medication, such as probenecid, can be used to help the kidneys eliminate uric acid, or allopurinol, which blocks the production of uric acid by the body.

Kidney stones are solid materials that form in the urinary tract. Most kidney stones are composed of calcium phosphate and calcium oxalate, although they can be solid uric acid. The excessive ingestion of minerals and insufficient water intake can cause the concentration of mineral salts to exceed the solubility of the mineral salts and lead to the formation of kidney stones. When a kidney stone passes through the urinary tract, it causes considerable pain and discomfort, necessitating the use of painkillers and surgery. Sometimes ultrasound is used to break up kidney stones. Persons prone to kidney stones are advised to drink six to eight glasses of water every day to prevent saturation levels of minerals in the urine.

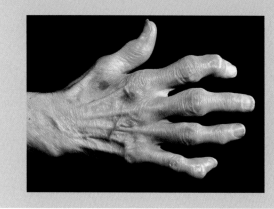

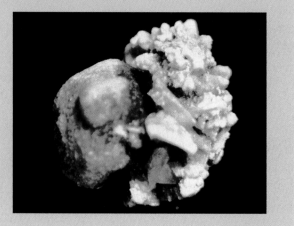

solute particles are removed from the surface of the solid is equal to the rate at which particles in solution crystallize out as solid.

MC
GOB

SELF-STUDY CASE
Kidney Stones and Saturated
Solutions

$$\text{Solid solute} \underset{\text{Crystallizes}}{\overset{\text{Dissolves}}{\rightleftharpoons}} \text{saturated solution}$$

SAMPLE PROBLEM 8.4

Saturated Solutions

At 20°C, the solubility of KCl is 34 g/100 g of water. In the laboratory, a student mixes 45 g of KCl with 100 g of water at a temperature of 20°C. Is the solution saturated or unsaturated?

SOLUTION

Only 34 g of the KCl will dissolve because that is its solubility at 20°C. The solution is saturated.

■ **Preparing Solutions**

In your kitchen, place $\frac{1}{4}$ or $\frac{1}{2}$ cup of cold water in a glass. Begin adding 1 tablespoon of sugar at a time and stir thoroughly. Take a sip of the liquid in the glass as you proceed. As the sugar solution becomes more concentrated, you may need to stir for a few minutes until all the sugar dissolves. Each time observe the solution after several minutes to determine when it is saturated.

Repeat the above activity with $\frac{1}{4}$ or $\frac{1}{2}$ cup of warm water. Count the number of tablespoons of sugar you need to form a saturated solution.

QUESTIONS

1. What do you notice about how sweet each sugar solution tastes?
2. How did you know when you obtained a saturated solution?
3. How much sugar dissolved in the warm water compared to the cold water?

STUDY CHECK

At 50°C, the solubility of $NaNO_3$ is 114 g/100 g of water. How many grams of $NaNO_3$ are needed to make a saturated $NaNO_3$ solution with 50. g of water at 50°C?

Effect of Temperature on Solubility

For most solids, solubilities increase as temperature increases. A few substances show little change in solubility at higher temperatures, and a few are less soluble. (See Figure 8.4.) For example, when you add sugar to iced tea, a layer of undissolved sugar may form on the bottom of the glass. Hot tea can dissolve more sugar than cold tea, which is why sugar is added before the tea is cooled.

The solubility of a gas in water decreases as the temperature increases. At higher temperatures, more gas molecules have the energy to escape from the solution. Perhaps you have observed the bubbles escaping from a cold carbonated soft drink as it warms. At high temperatures, bottles containing carbonated solutions may burst as more gas molecules leave the solution and increase the gas pressure inside the bottle. Biologists have found that increased temperatures in rivers and lakes cause the amount of dissolved oxygen to decrease until the warm water can no longer support a biological community. Electricity-generating plants are required to have their own ponds to use with their cooling towers to lessen the threat of thermal pollution.

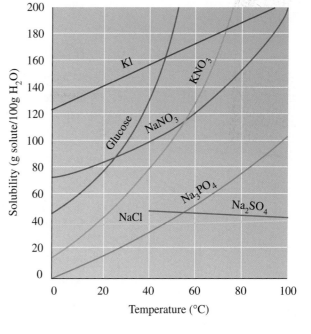

Henry's Law

Henry's law states that the solubility of gas in a liquid is directly related to the pressure of that gas above the liquid. At higher pressures, there are more gas molecules available to enter and dissolve in the liquid. A can of soda is carbonated by using CO_2 gas at high pressure to increase the solubility of the CO_2 in the beverage. When you open the can at atmospheric pressure, the pressure on the CO_2 drops, which decreases the solubility of CO_2. As a result, bubbles of CO_2 rapidly escape from the solution. The burst of bubbles is even more noticeable when you open a warm can of soda.

FIGURE 8.4

In water, most common solids are more soluble as the temperature increases.

Q Compare the solubility of $NaNO_3$ at 20°C and 60°C.

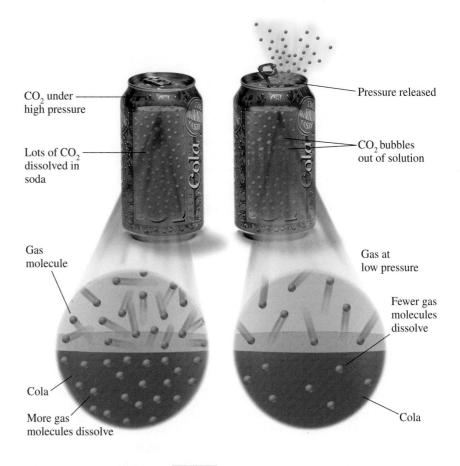

CO$_2$ under high pressure

Lots of CO$_2$ dissolved in soda

Gas molecule

Cola

More gas molecules dissolve

Pressure released

CO$_2$ bubbles out of solution

Gas at low pressure

Fewer gas molecules dissolve

Cola

SAMPLE PROBLEM 8.5

▪ Factors Affecting Solubility

Indicate whether the solubility of the solute will increase or decrease in each of the following situations:

a. using 80°C water instead of 25°C water to dissolve sugar
b. a lake, which contains dissolved O$_2$, warms

SOLUTION

a. An increase in the temperature increases the solubility of the sugar.
b. An increase in the temperature decreases the solubility of O$_2$ gas.

STUDY CHECK

At 10°C, the solubility of KNO$_3$ is 20 g/100 g H$_2$O. Would the value of 5 g/100 g H$_2$O or 65 g/100 g H$_2$O be the more likely solubility at 40°C? Explain.

Soluble and Insoluble Salts

SELF-STUDY ACTIVITY
Solubility

Up to now, we have considered ionic compounds that dissolve in water; they are **soluble salts**. However, some ionic compounds do not separate into ions in water. They are **insoluble salts** that remain as solids even in contact with water.

Salts that are soluble in water typically contain at least one of the following ions: Li$^+$, Na$^+$, K$^+$, NH$_4^+$, NO$_3^-$, or C$_2$H$_3$O$_2^-$. Most salts containing Cl$^-$ are soluble,

TABLE 8.7

Solubility Rules for Ionic Solids in Water

Soluble if salt contains		Insoluble if salt contains
NH_4^+, Li^+, Na^+, K^+ Nitrates NO_3^-, acetates $C_2H_3O_2^-$	but are soluble with	Carbonates CO_3^{2-}, Sulfides S^{2-} Phosphates PO_4^{3-}, Hydroxides, OH^-
Chlorides Cl^-, Bromides Br^-, Iodides I^-	but are not soluble with	Ag^+, Pb^{2+}, or Hg_2^{2+}
Sulfates SO_4^{2-}	but are not soluble with	Ba^{2+}, Pb^{2+}, Ca^{2+}, Sr^{2+}

but $AgCl$, $PbCl_2$, or Hg_2Cl_2 are not; they are insoluble chloride salts. Similarly, most salts containing SO_4^{2-} are soluble, but a few are insoluble as shown in Table 8.7. Most other salts are insoluble and do not dissolve in water. (See Figure 8.5.) In an insoluble salt, attractions between its positive and negative ions are too strong for the polar water molecules to break. We can use the solubility rules to predict whether a salt (a solid ionic compound) would be expected to dissolve in water. Table 8.8 illustrates the use of these rules.

In medicine, the insoluble salt $BaSO_4$ is used as an opaque substance to enhance X rays of the gastrointestinal tract. $BaSO_4$ is so insoluble that it does not dissolve in gastric fluids. (See Figure 8.6.) Other barium salts cannot be used, because they would dissolve in water releasing Ba^{2+}, which is poisonous.

CdS

FeS

PbCrO$_4$

Ni(OH)$_2$

FIGURE 8.5

Mixing certain aqueous solutions produces insoluble salts.

Q **What makes each of these salts insoluble in water?**

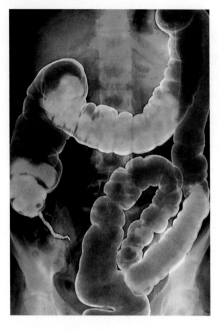

FIGURE 8.6

A barium sulfate enhanced X ray of the abdomen shows the large intestine.

Q **Is BaSO₄ a soluble or an insoluble substance?**

TABLE 8.8

Using Solubility Rules

Ionic Compound	Solubility in Water	Reasoning
K_2S	Soluble	Contains K^+
$Ca(NO_3)_2$	Soluble	Contains NO_3^-
$PbCl_2$	Insoluble	Is an insoluble chloride
$NaOH$	Soluble	Contains Na^+
$AlPO_4$	Insoluble	Contains no soluble ions

SAMPLE PROBLEM 8.6

■ **Soluble and Insoluble Salts**

Predict whether each of the following salts is soluble in water.

a. Na_3PO_4 **b.** $CaCO_3$ **c.** K_2SO_4

SOLUTION

a. Soluble. Salts containing Na^+ are soluble.
b. The salt $CaCO_3$ is not soluble. Most salts containing CO_3^{2-} are not soluble.
c. Soluble. Salts containing K^+ are soluble.

STUDY CHECK

Would you expect the following salts to form solutions in water? Why?

a. $PbCl_2$ **b.** K_3PO_4 **c.** $FeCO_3$

Formation of a Solid

We can use solubility rules to predict whether a solid called a *precipitate* forms when two solutions of ionic compounds are mixed. A solid forms when two ions of an insoluble salt come in contact with one another. For example, when a solution of $AgNO_3$ (Ag^+ and NO_3^-) is mixed with a solution of NaCl (Na^+ and Cl^-), the white insoluble salt AgCl is produced. We can write the reaction as a double replacement equation. However, the molecular equation does not show the individual ions to help us decide which, if any, insoluble salt would form. To help us determine any insoluble salt, we can first write the reactants to show all the ions present when the two solutions are mixed.

$$Ag^+(aq) + NO_3^-(aq) + Na^+(aq) + Cl^-(aq) \longrightarrow$$

Then we look at possible new combinations of cations and anions to see if any would form an insoluble salt. The new combination of AgCl would form an insoluble salt.

Guide to Writing Net Ionic Equations for Formation of an Insoluble Salt

STEP 1
Write the ions of the reactants.

STEP 2
Write the new combinations of ions and determine if any are insoluble.

STEP 3
Write the ionic equation including any solid.

STEP 4
Write the net ionic equation by removing spectator ions.

■ **STEP 1**

Reactants
(initial combinations)

$Ag^+(aq) + NO_3^-(aq)$

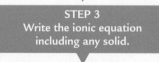

$Na^+(aq) + Cl^-(aq)$

■ **STEP 2**

Mixture
(new combinations)   **Product**

$Ag^+(aq) + Cl^-(aq) \longrightarrow AgCl(s)$

$Na^+(aq) + NO_3^-(aq)$

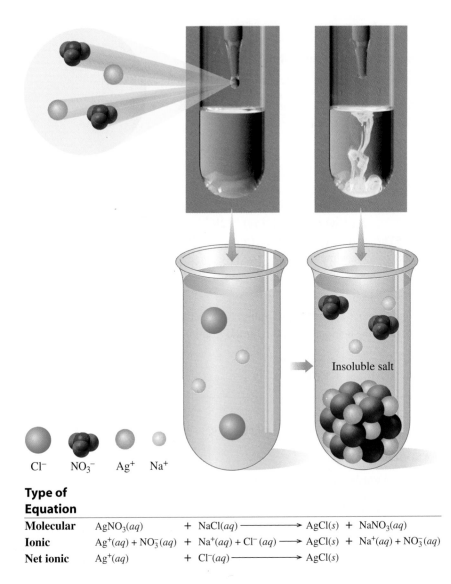

Cl⁻   NO₃⁻   Ag⁺   Na⁺

Insoluble salt

Type of Equation

Molecular	$AgNO_3(aq)$	$+ NaCl(aq) \longrightarrow$	$AgCl(s)$	$+ NaNO_3(aq)$
Ionic	$Ag^+(aq) + NO_3^-(aq)$	$+ Na^+(aq) + Cl^-(aq) \longrightarrow$	$AgCl(s)$	$+ Na^+(aq) + NO_3^-(aq)$
Net ionic	$Ag^+(aq)$	$+ Cl^-(aq) \longrightarrow$	$AgCl(s)$	

■ **STEP 3** Now we can write an **ionic equation** to show that a precipitate of AgCl forms while the ions Na^+ and NO_3^- remain in solution.

$$Ag^+(aq) + NO_3^-(aq) + Na^+(aq) + Cl^-(aq) \longrightarrow$$
$$AgCl(s) + Na^+(aq) + NO_3^-(aq)$$

■ **STEP 4** Now we can remove the Na^+ and NO_3^- ions known as *spectator ions* because they are unchanged during the reaction.

$$Ag^+(aq) + \cancel{NO_3^-(aq)} + \cancel{Na^+(aq)} + Cl^-(aq) \longrightarrow$$
$$AgCl(s) + \cancel{Na^+(aq)} + \cancel{NO_3^-(aq)}$$

Finally, a **net ionic equation** can be written that gives the chemical reaction that occurred.

The Na^+ and NO_3^- ions, the spectator ions, are removed from the ionic equation we wrote above.

$$Ag^+(aq) + Cl^-(aq) \longrightarrow AgCl(s)$$

SAMPLE PROBLEM 8.7

Formation of an Insoluble Salt

Solutions of $BaCl_2$ and K_2SO_4 are mixed and a white solid forms.

a. Write the net ionic equation.
b. What is the white solid that forms?

SOLUTION

a. ■ **STEP 1** $Ba^{2+}(aq) + Cl^-(aq) + K^+(aq) + SO_4^{2-}(aq)$
■ **STEP 2** $BaSO_4(s)$ is insoluble.
■ **STEP 3** $Ba^{2+}(aq) + 2Cl^-(aq) + 2K^+(aq) + SO_4^{2-}(aq) \longrightarrow$
$BaSO_4(s) + 2Cl^-(aq) + 2K^+(aq)$
■ **STEP 4** $Ba^{2+}(aq) + SO_4^{2-}(aq) \longrightarrow BaSO_4(s)$
b. $BaSO_4$ is the white solid.

STUDY CHECK

Predict whether a solid might form in each of the following mixtures of solutions. If so, write the net ionic equation for the reaction.

a. $NH_4Cl(aq) + Ca(NO_3)_2(aq)$
b. $Pb(NO_3)_2(aq) + KCl(aq)$

QUESTIONS AND PROBLEMS

Solubility

8.21 State whether each of the following refers to a saturated or unsaturated solution:
a. A crystal added to a solution does not change in size.
b. A sugar cube completely dissolves when added to a cup of coffee.

8.22 State whether each of the following refers to a saturated or unsaturated solution:
a. A spoonful of salt added to boiling water dissolves.
b. A layer of sugar forms on the bottom of a glass of tea as ice is added.

Use this table for problems 8.23–8.26.

Solubility (g/100 g H₂O)		
Substance	20°C	50°
KCl	34.0	42.6
NaNO₃	88.0	114.0
C₁₂H₂₂O₁₁ (sugar)	203.9	260.4

8.23 Using the above table, determine whether each of the following solutions will be saturated or unsaturated at 20°C:
a. Adding 25.0 g KCl to 100 g H_2O
b. Adding 11.0 g $NaNO_3$ to 25 g H_2O
c. Adding 400.0 g sugar to 125 g H_2O

8.24 Using the above table, determine whether each of the following solutions will be saturated or unsaturated at 50°C:
a. Adding 25.0 g KCl to 50 g H_2O
b. Adding 150.0 g $NaNO_3$ to 75 g H_2O
c. Adding 80.0 g sugar to 25 g H_2O

8.25 A solution containing 80.0 g KCl in 200 g H_2O at 50°C is cooled to 20°C.
a. How many grams of KCl remain in solution at 20°C?
b. How many grams of solid KCl came out of solution after cooling?

8.26 A solution containing 80.0 g of $NaNO_3$ in 75.0 g H_2O at 50°C is cooled to 20°C.
a. How many grams of $NaNO_3$ remain in solution at 20°C?
b. How many grams of solid $NaNO_3$ came out of solution after cooling?

8.27 Explain the following observations:
   a. More sugar dissolves in hot tea than in ice tea.
   b. Champagne in a warm room goes flat.
   c. A warm can of soda has more spray when opened than a cold one.

8.28 Explain the following observations:
   a. An open can of soda loses its "fizz" quicker at room temperature than in the refrigerator.
   b. Chlorine gas in tap water escapes as the sample warms to room temperature.
   c. Less sugar dissolves in iced coffee than in hot coffee.

8.29 Predict whether each of the following ionic compounds is soluble in water:
   a. LiCl   **b.** AgCl   **c.** $BaCO_3$ **d.** K_2O **e.** $Fe(NO_3)_3$

8.30 Predict whether each of the following ionic compounds is soluble in water:
   a. PbS   **b.** NaI   **c.** Na_2S **d.** Ag_2O **e.** $CaSO_4$

8.31 Determine whether a solid forms when solutions containing the following salts are mixed. If so, write the molecular equation (double replacement) and the net ionic equation for the reaction.
   a. KCl and Na_2S **b.** $AgNO_3$ and K_2S **c.** $CaCl_2$ and Na_2SO_4

8.32 Determine whether a solid forms when solutions containing the following salts are mixed. If so, write the molecular equation (double replacement) and the net ionic equation for the reaction.
   a. Na_3PO_4 and $AgNO_3$ **b.** K_2SO_4 and Na_2CO_3 **c.** $Pb(NO_3)_2$ and Na_2CO_3

8.4 Percent Concentration

LEARNING GOAL

Calculate the percent concentration of a solute in a solution; use percent concentration to calculate the amount of solute or solution.

The amount of solute dissolved in a certain amount of solution is called the **concentration** of the solution. Although there are many ways to express a concentration, they all specify a certain amount of solute in a given amount of solution.

$$\text{Concentration of a solution} = \frac{\text{amount of solute}}{\text{amount of solution}}$$

Mass Percent

The **mass percent** (% m/m) concentration of a solution describes the mass of the solute in every 100 grams of solution. The mass in grams of the solution is the sum of the mass of the solute and the mass of the solvent.

$$\text{mass percent (\% m/m)} = \frac{\text{mass of solute (g)}}{\text{mass of solute (g)} + \text{mass of solvent (g)}} \times 100$$

$$= \frac{\text{mass of solute (g)}}{\text{mass of solution (g)}} \times 100$$

Suppose we prepared a solution by mixing 8.00 g of KCl (solute) with 42.00 g of water (solvent). Together the mass of the solute and mass of solvent give the mass of the solution (8.00 g + 42.00 g = 50.00 g). The mass % is calculated by substituting in the values into the mass percent expression.

$$\frac{8.00 \text{ g KCl}}{50.00 \text{ g solution}} \times 100\% = 16.0\% \text{ (m/m)}$$

$\overbrace{8.00 \text{ g KCl} + 42.00 \text{ g H}_2\text{O}}$
   (solute   +   solvent)

Add 8.00 g of KCl

Add water until the solution weighs 50.00 g

SAMPLE PROBLEM 8.8

■ Percent Concentration (% m/m)

What is the mass percent of a solution prepared by dissolving 30.0 g NaOH in 120.0 g of H_2O?

SOLUTION

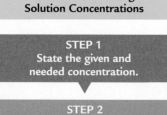

Guide to Calculating Solution Concentrations

STEP 1
State the given and needed concentration.

STEP 2
Write a plan to calculate needed concentration.

STEP 3
Write equalities and conversion factors.

STEP 4
Set up problem to calculate answer.

■ **STEP 1 Given** 30.0 g NaOH and 120.0 g H_2O
 Need mass percent (% m/m) of NaOH

■ **STEP 2 Plan** The mass percent is calculated by using the mass in grams of the solute and solution in the definition of mass percent.

■ **STEP 3 Equalities/Conversion Factors**

$$\text{mass percent (\% m/m)} = \frac{\text{mass of solute}}{\text{mass of solute} + \text{mass of solvent}} \times 100$$

$$\text{mass percent (\% m/m)} = \frac{\text{grams of solute}}{\text{grams of solution}} \times 100$$

■ **STEP 4 Set Up Problem** The mass of the solute and the solution are obtained from the data.

$$\begin{aligned} \text{mass of solute} &= \quad 30.0 \text{ g NaOH} \\ \text{mass of solvent} &= +120.0 \text{ g } H_2O \\ \hline \text{mass of solution} &= 150.0 \text{ g solution} \end{aligned}$$

$$\text{mass percent (\% m/m)} = \frac{30.0 \text{ g NaOH}}{150.0 \text{ g solution}} \times 100$$

$$= 20.0\% \text{ (m/m) NaOH}$$

STUDY CHECK

What is the mass percent of NaCl in a solution made by dissolving 2.0 g of NaCl in 56.0 g of H_2O?

MC GOB

Calculating Percent Concentration
Percent Concentration as a
Conversion Factor

■ Volume Percent

Because the volumes of liquids or gases are easily measured, the concentrations of their solutions are often expressed as **volume percent** (% v/v). The units of volume used in the ratio must be the same, for example, both in milliliters or both in liters.

$$\text{volume percent (\% v/v)} = \frac{\text{volume of solute}}{\text{volume of solution}} \times 100\%$$

We interpret a volume/volume percent as the volume of solute in 100 mL of solution. In the wine industry, a label that reads 12% (v/v) means 12 mL of alcohol in 100 mL of wine.

■ Mass/Volume Percent

A **mass/volume percent** (% m/v), or weight/volume percent (% w/v), is calculated by dividing the grams of the solute by the volume (mL) of solution and multiplying by 100. Widely used in hospitals and pharmacies, the preparation of intravenous solutions and medicines involves the mass/volume percent.

$$\text{mass/volume \%} = \frac{\text{grams of solute}}{\text{milliliters of solution}} \times 100\%$$

For example, a solution prepared by dissolving 5.0 g KI in enough water to give a final volume of 250 mL is a 2.0% (m/v) KI solution.

$$\text{mass/volume \%} = \frac{\overset{\text{Mass of solute}}{5.0 \text{ g KI}}}{\underset{\text{Volume of solution}}{250 \text{ mL solution}}} \times 100\% = 2.0\% \text{ (m/v) KI}$$

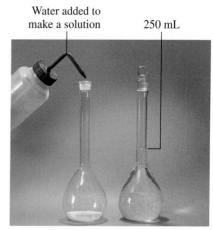

Water added to make a solution   250 mL

5.0 g KI   2.0% (m/v) KI solution

Percent Concentrations as Conversion Factors

In the preparation of solutions, we often need to calculate the amount of solute or solution. Then the percent concentration is useful as a conversion factor. Some examples of percent concentrations, their meanings, and possible conversion factors are given in Table 8.9.

TABLE 8.9

Conversion Factors from Percent Concentrations

Percent Concentration	Meaning	Conversion Factors		
10% (m/m) KCl	There are 10 g of KCl in 100 g of solution.	$\dfrac{10 \text{ g KCl}}{100 \text{ g solution}}$	and	$\dfrac{100 \text{ g solution}}{10 \text{ g KCl}}$
5% (m/v) glucose	There are 5 g of glucose in 100 mL of solution.	$\dfrac{5 \text{ g glucose}}{100 \text{ mL solution}}$	and	$\dfrac{100 \text{ mL solution}}{5 \text{ g glucose}}$
12% (v/v) ethanol	There are 12 mL of ethanol in 100 mL of solution.	$\dfrac{12 \text{ mL ethanol}}{100 \text{ mL solution}}$	and	$\dfrac{100 \text{ mL solution}}{12 \text{ mL ethanol}}$

SAMPLE PROBLEM 8.9

■ **Using Mass/Volume Percent to Find Mass of Solute**

A topical antibiotic is 1.0% (m/v) Clindamycin. How many grams of Clindamycin are in 60. mL of the 1.0% (m/v) solution?

SOLUTION

■ **STEP 1** **Given** 1.0% (m/v) Clindamycin
       Need grams of Clindamycin

■ **STEP 2** **Plan** $\boxed{\begin{array}{c}\text{milliliters}\\\text{of solution}\end{array}} \xrightarrow{\text{\% (m/v) factor}} \begin{array}{c}\text{grams of}\\\text{Clindamycin}\end{array}$

■ **STEP 3** **Equalities/Conversion Factors** The percent (m/v) indicates the grams of a solute in every 100 mL of a solution. The 1.0% (m/v) can be written as two conversion factors.

100 mL solution = 1.0 g Clindamycin

$$\frac{1.0 \text{ g Clindamycin}}{100 \text{ mL solution}} \quad \text{and} \quad \frac{100 \text{ mL solution}}{1.0 \text{ g Clindamycin}}$$

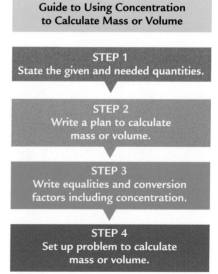

Guide to Using Concentration to Calculate Mass or Volume

STEP 1
State the given and needed quantities.

STEP 2
Write a plan to calculate mass or volume.

STEP 3
Write equalities and conversion factors including concentration.

STEP 4
Set up problem to calculate mass or volume.

■ **STEP 4 Set Up Problem** The volume of the solution is converted to mass of solute using the conversion factor.

$$60. \; \text{mL solution} \times \frac{1.0 \text{ g Clindamycin}}{100 \text{ mL solution}} = 0.60 \text{ g Clindamycin}$$

STUDY CHECK

Calculate the grams of KCl and grams of water in 225 g of an 8.00% (m/m) KCl solution.

QUESTIONS AND PROBLEMS

(MC GOB)

■ **Percent Concentration**

8.33 What is the difference between a 5% (m/m) glucose solution and a 5% (m/v) glucose solution?

8.34 What is the difference between a 10% (v/v) methyl alcohol (CH_3OH) solution and a 10% (m/m) methyl alcohol solution?

8.35 Calculate the mass percent, % (m/m), for the solute in each of the following solutions:
   a. 25 g of KCl and 125 g of H_2O
   b. 12 g of sugar in 225 g of tea with sugar (solution)

8.36 Calculate the mass percent, % (m/m), for the solute in each of the following solutions:
   a. 75 g of NaOH in 325 g of NaOH solution
   b. 2.0 g of KOH in 20.0 g of H_2O

8.37 Calculate the mass/volume percent, % (m/v) for the solute in each of the following solutions:
   a. 75 g of Na_2SO_4 in 250 mL of Na_2SO_4 solution
   b. 39 g of sucrose in 355 mL of a carbonated drink

8.38 Calculate the mass/volume percent, % (m/v), for the solute in each of the following solutions:
   a. 2.50 g KCl in 50.0 mL of solution
   b. 7.5 g of casein in 120 mL of low-fat milk

8.39 Calculate the amount of solute needed to prepare the following solutions:
   a. 50.0 mL of a 5.0% (m/v) KCl solution
   b. 1250 mL of a 4.0% (m/v) NH_4Cl solution

8.40 Calculate the amount of solute needed to prepare the following solutions:
   a. 150 mL of a 40.0% (m/v) $LiNO_3$ solution
   b. 450 mL of a 2.0% (m/v) KCl solution

8.41 A mouthwash contains 22.5% alcohol by volume. If the bottle of mouthwash contains 355 mL, what is the volume in milliliters of the alcohol?

8.42 A bottle of champagne is 11% alcohol by volume. If there are 750 mL of champagne in the bottle, how many milliliters of alcohol are present?

8.43 A patient receives 100. mL of 20.% (m/v) mannitol solution every hour.
   a. How many grams of mannitol are given in 1 hour?
   b. How many grams of mannitol does the patient receive in 15 hours?

8.44 A patient receives 250 mL of a 4.0% (m/v) amino acid solution twice a day.
   a. How many grams of amino acids are in 250 mL of solution?
   b. How many grams of amino acids does the patient receive in 1 day?

8.45 A patient needs 100. g of glucose in the next 12 hours. How many liters of a 5% (m/v) glucose solution must be given?

8.46 A patient received 2.0 g of NaCl in 8 hours. How many milliliters of a 0.90% (m/v) NaCl (saline) solution were delivered?

8.5 Molarity and Dilution

When the solutes of solutions take part in reactions, chemists are interested in the number of reacting particles. For this purpose, chemists use **molarity (M)**, a concentration that states the number of moles of solute in exactly 1 liter of solution. The molarity of a solution can be calculated knowing the moles of solute and the volume of solution.

$$\text{Molarity (M)} = \frac{\text{moles of solute}}{\text{liters of solution}}$$

For example, if 1.0 mole of NaCl were dissolved in enough water to prepare 1.0 L of solution, the resulting NaCl solution has a molarity of 1.0 M. The abbreviation M indicates the units of moles per liter (moles/L).

$$M = \frac{\text{moles of solute}}{\text{liters of solution}} = \frac{1.0 \text{ mole NaCl}}{1 \text{ L}} = 1.0 \text{ M NaCl}$$

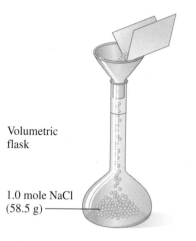

Volumetric flask

1.0 mole NaCl (58.5 g)

SAMPLE PROBLEM 8.10

■ Calculating Molarity

What is the molarity (M) of 60.0 g NaOH in 0.250 L of solution?

SOLUTION

■ **STEP 1 Given** 60.0 g NaOH in 0.250 L of solution
Need molarity (moles/L)

■ **STEP 2 Plan** The calculation of molarity requires the moles of NaOH and the volume of the solution in liters.

$$\text{Molarity(M)} = \frac{\text{moles of solute}}{\text{liters of solution}}$$

g NaOH → Molar mass → $\frac{\text{moles NaOH}}{\text{volume (L)}}$ = M NaOH solution

■ **STEP 3 Equalities/Conversion Factors**

$$1 \text{ mole NaOH} = 40.0 \text{ g NaOH}$$
$$\frac{1 \text{ mole NaOH}}{40.0 \text{ g NaOH}} \quad \text{and} \quad \frac{40.0 \text{ g NaOH}}{1 \text{ mole NaOH}}$$

■ **STEP 4 Set Up Problem**

$$\text{moles NaOH} = 60.0 \text{ g NaOH} \times \frac{1 \text{ mole NaOH}}{40.0 \text{ g NaOH}} = 1.50 \text{ moles NaOH}$$

The molarity is calculated by dividing the moles of NaOH by the volume in liters.

$$\frac{1.50 \text{ moles NaOH}}{0.250 \text{ L}} = \frac{6.00 \text{ moles NaOH}}{1 \text{ L}} = 6.00 \text{ M NaOH}$$

Add water until 1.0 liter mark is reached.

Mix

A 1.0 molar NaCl solution

STUDY CHECK

What is the molarity of a solution that contains 75.0 g KNO_3 dissolved in 0.350 L solution?

TABLE 8.10

Some Examples of Molar Solutions		
Molarity	Meaning	Conversion Factors
6.0 M HCl	6.0 moles HCl in 1 liter of solution	$\dfrac{6.0 \text{ moles HCl}}{1 \text{ L}}$ and $\dfrac{1 \text{ L}}{6.0 \text{ moles HCl}}$
0.20 M NaOH	0.20 mole NaOH in 1 liter of solution	$\dfrac{0.20 \text{ mole NaOH}}{1 \text{ L}}$ and $\dfrac{1 \text{ L}}{0.20 \text{ mole NaOH}}$

Molarity as a Conversion Factor

When we need to calculate the moles of solute or the volume of solution, the molarity is used as a conversion factor. Examples of conversion factors from molarity are given in Table 8.10.

Using the molarity of the solution with the molar mass of the solute, we can calculate the volume of solution needed as illustrated in Sample Problem 8.11.

SAMPLE PROBLEM 8.11

■ **Using Molarity to Find Volume**

How many liters of a 2.00 M NaCl solution are needed to provide 67.3 g NaCl?

SOLUTION

■ **STEP 1** **Given** 67.3 g NaCl from a 2.00 M NaCl solution
Need liters NaCl

■ **STEP 2** **Plan** The volume of NaCl is calculated using the moles of NaCl and molarity of the NaCl solution.

grams NaCl Molar mass moles NaCl Molarity liters NaCl

■ **STEP 3** **Equalities/Conversion Factors**

$$1 \text{ mole NaCl} = 58.5 \text{ g NaCl}$$
$$\frac{1 \text{ mole NaCl}}{58.5 \text{ g NaCl}} \quad \text{and} \quad \frac{58.5 \text{ g NaCl}}{1 \text{ mole NaCl}}$$

The molarity of any solution can be written as two conversion factors.

$$1 \text{ L NaCl} = 2.00 \text{ moles NaCl}$$
$$\frac{1 \text{ L NaCl}}{2.00 \text{ moles NaCl}} \quad \text{and} \quad \frac{2.00 \text{ moles NaCl}}{1 \text{ L NaCl}}$$

■ **STEP 4** **Set Up Problem**

$$\text{liters of NaCl} = 67.3 \text{ g NaCl} \times \frac{1 \text{ mole NaCl}}{58.5 \text{ g NaCl}} \times \frac{1 \text{ L NaCl}}{2.00 \text{ moles NaCl}}$$
$$= 0.575 \text{ L NaCl}$$

STUDY CHECK

How many moles of HCl are present in 750 mL of a 6.0 M HCl solution?

Dilution

In chemistry, we often need to prepare a dilute solution from a more concentrated solution. In a process called **dilution**, we add water to a solution to make a larger volume. For example, you might prepare some orange juice by adding three cans of water to the original can of concentrated orange juice.

Dilution

| 1 can orange juice concentrate | + | 3 cans water | = | 4 cans of orange juice |

When more water is added, the solution volume increases, causing a decrease in the concentration. However, *the amount of solute does not change*.

Amount of solute in the = Amount of solute in the
concentrated solution   diluted solution

Therefore the initial amount of solute before dilution is equal to the final amount of solute in the diluted solution. (See Figure 8.7.) By diluting the solution, we decrease the concentration. The amount of the solute depends on the concentration, C, and the volume, V.

$$C_1V_1 = C_2V_2$$

FIGURE 8.7

When water is added to a concentrated solution, there is no change in the number of particles, but the solute particles can spread out as the volume of the diluted solution increases.

Q **What is the concentration of the diluted solution after an equal volume of water is added to a sample of 6 M HCl?**

We have seen that the concentration of a solution can be expressed in % or molarity, M.

$$\%_1 V_1 = \%_2 V_2$$

or

$$M_1 V_1 = M_2 V_2$$

SAMPLE PROBLEM 8.12

■ Molarity of a Diluted Solution

What is the molarity of a solution prepared when 75.0 mL of a 4.00 M KCl solution is diluted to a volume of 0.500 L?

SOLUTION

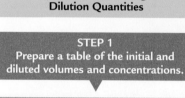

Guide to Calculating Dilution Quantities

STEP 1
Prepare a table of the initial and diluted volumes and concentrations.

STEP 2
Write a plan that solves the dilution expression for the unknown quantity.

STEP 3
Set up problem by placing known quantities in dilution expression.

■ **STEP 1 Give Data in a Table** We make a table of the molar concentrations and volumes of the initial and diluted solutions.

Initial: M_1 = 4.00 M KCl V_1 = 75.0 mL = 0.0750 L

Diluted: M_2 = ? M KCl V_2 = 0.500 L

■ **STEP 2 Plan** The unknown molarity can be calculated by solving the dilution expression for M_2.

$$M_1 V_1 = M_2 V_2$$

$$\frac{M_1 V_1}{V_2} = M_2 \frac{\cancel{V_2}}{\cancel{V_2}}$$

$$M_2 = M_1 \times \frac{V_1}{V_2}$$

■ **STEP 3 Set Up Problem** The diluted concentration is calculated by placing the values from the table into the dilution expression.

$$= 4.00\ M \times \frac{0.075\ \cancel{L}}{0.500\ \cancel{L}} = 0.600\ M\ KCl\ (\text{diluted solution})$$

STUDY CHECK

You need to prepare 600. mL of 2.00 M NaOH solution from a 10.0 M NaOH solution. What volume of the 10.0 M NaOH solution do you use?

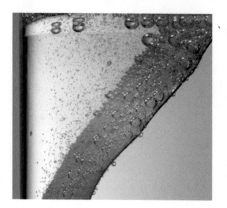

Chemical Reactions in Solutions

Earlier in this section, we learned that the moles of a solute can be determined if we know the molarity and volume of a solution. We can also determine the volume of a solution if we know the number of moles of a solute and its molarity. These are the types of calculations we need to do for the substances involved in a chemical reaction that takes place in aqueous solution. If we have a balanced equation, we can use the molarity and volume of a solution to determine the moles of a substance required or produced in a chemical reaction. We can also use molarity and the number of moles of a solute to determine the volume of a solution as seen in Sample Problem 8.13.

SAMPLE PROBLEM 8.13

Volume of a Solution in a Reaction

Zinc reacts with HCl to produce $ZnCl_2$ and hydrogen gas H_2.

$$Zn(s) + 2HCl(aq) \longrightarrow ZnCl_2(aq) + H_2(g)$$

How many liters of a 1.50 M HCl solution completely react with 5.32 g of zinc?

SOLUTION

■ **STEP 1 Given** 5.32 g Zn and a 1.50 M HCl solution

        Need L HCl solution

■ **STEP 2 Plan** We can use the molar mass of Zn to find the moles of Zn and the mole–mole factor from the balanced equation to convert moles of Zn to moles of HCl. Since the concentration of the HCl solution is given, the molarity can be used to convert moles to volume in liters.

g Zn | Molar mass | moles Zn | Mole–mole factor | moles HCl | Molarity | L HCl

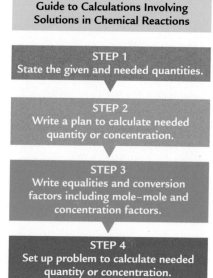

Guide to Calculations Involving Solutions in Chemical Reactions

STEP 1
State the given and needed quantities.

STEP 2
Write a plan to calculate needed quantity or concentration.

STEP 3
Write equalities and conversion factors including mole–mole and concentration factors.

STEP 4
Set up problem to calculate needed quantity or concentration.

■ **STEP 3 Equalities/Conversion Factors**

Molar mass of Zn

$$1 \text{ mole Zn} = 65.4 \text{ g Zn}$$

$$\frac{1 \text{ mole Zn}}{65.4 \text{ g Zn}} \quad \text{and} \quad \frac{65.4 \text{ g Zn}}{1 \text{ mole Zn}}$$

Mole–mole factor

$$1 \text{ mole Zn} = 2 \text{ moles HCl}$$

$$\frac{1 \text{ mole Zn}}{2 \text{ moles HCl}} \quad \text{and} \quad \frac{2 \text{ moles HCl}}{1 \text{ mole Zn}}$$

Molarity of HCl solution

$$1 \text{ L HCl} = 1.50 \text{ moles HCl}$$

$$\frac{1 \text{ L HCl}}{1.50 \text{ moles HCl}} \quad \text{and} \quad \frac{1.50 \text{ moles HCl}}{1 \text{ L HCl}}$$

■ **STEP 4 Set Up Problem** We can write the problem setup as seen in our plan.

$$5.32 \text{ g Zn} \times \frac{1 \text{ mole Zn}}{65.4 \text{ g Zn}} \times \frac{2 \text{ moles HCl}}{1 \text{ mole Zn}} \times \frac{1 \text{ L HCl}}{1.50 \text{ moles HCl}} = 0.108 \text{ L HCl}$$

STUDY CHECK

Using the reaction in Sample Problem 8.13, how many grams of zinc can react with 225 mL of 0.200 M HCl?

QUESTIONS AND PROBLEMS

(MC GOB)

Molarity and Dilution

8.47 Calculate the molarity (M) of the following solutions:
    a. 2.0 moles of glucose in 4.0 L of solution
    b. 4.0 g of KOH in 2.0 L of solution

8.48 Calculate the molarity (M) of the following solutions:
    a. 0.50 mole glucose in 0.200 L of solution
    b. 36.5 g of HCl in 1.0 L of solution

8.49 Calculate the moles of solute needed to prepare each of the following:
    a. 1.0 L of a 3.0 M NaCl solution
    b. 0.40 L of a 1.0 M KBr solution

8.50 Calculate the moles of solute needed to prepare each of the following:
  a. 5.0 L of a 2.0 M $CaCl_2$ solution
  b. 4.0 L of a 0.10 M NaOH solution

8.51 Calculate the grams of solute needed to prepare each of the following solutions:
  a. 2.0 L of a 1.5 M NaOH solution
  b. 4.0 L of a 0.20 M KCl solution

8.52 Calculate the grams of solute needed to prepare each of the following solutions:
  a. 2.0 L of a 6.0 M NaOH solution
  b. 5.0 L of a 0.10 M $CaCl_2$ solution

8.53 What volume in liters provides the following amounts of solute?
  a. 3.0 moles of NaOH from a 2.0 M NaOH solution
  b. 15 moles of NaCl from a 1.5 M NaCl solution

8.54 What volume in liters provides the following amounts of solute?
  a. 0.100 mole of KCl from a 4.0 M KCl solution
  b. 5.0 moles of HCl from a 6.0 M HCl solution

8.55 Calculate the final concentration of each of the following diluted solutions:
  a. 2.0 L of a 6.0 M HCl solution is added to water so that the final volume is 6.0 L.
  b. Water is added to 0.50 L of a 12 M NaOH solution to make 3.0 L of a diluted NaOH solution.
  c. A 10.0-mL sample of 25% (m/v) KOH solution is diluted with water so that the final volume is 100.0 mL.
  d. A 50.0-mL sample of 15% (m/v) H_2SO_4 solution is added to water to give a final volume of 250 mL.

8.56 Calculate the final concentration of each of the following diluted solutions:
  a. 1.0 L of a 4.0 M HNO_3 solution is added to water so that the final volume is 8.0 L.
  b. Water is added to 0.25 L of a 6.0 M KOH solution to make 2.0 L of a diluted KOH solution.
  c. A 50.0-mL sample of 8.0% (m/v) NaOH is diluted with water so that the final volume is 200.0 mL.
  d. A 5.0-mL sample of 50.0% (m/v) acetic acid ($HC_2H_3O_2$) solution is added to water to give a final volume of 25 mL.

8.57 What is the final volume in liters of each of the following diluted solutions?
  a. a 0.20 M HCl solution prepared from 20.0 mL of a 6.0 M HCl solution
  b. a 2.0% (m/v) NaOH solution prepared from 50.0 mL of a 10.0% (m/v) NaOH solution
  c. a 0.50 M H_3PO_4 solution prepared from 0.500 L of a 6.0 M H_3PO_4 solution

8.58 What is the final volume (L) of each of the following diluted solutions?
  a. a 1.0% (m/v) KOH solution prepared from 10.0 mL of a 20.0% KOH solution
  b. a 0.10 M HCl solution prepared from 25 mL of a 6.0 M HCl solution
  c. a 1.0 M NaOH solution prepared from 50.0 mL of a 12 M NaOH solution

8.59 Given the reaction

$$Pb(NO_3)_2(aq) + 2KCl(aq) \longrightarrow PbCl_2(s) + 2KNO_3(aq)$$

  a. How many grams of $PbCl_2$ will be formed from 50.0 mL of 1.50 M KCl and excess $Pb(NO_3)_2$?
  b. How many milliliters of 2.00 M $Pb(NO_3)_2$ will react with 50.0 mL of 1.50 M KCl?

8.60 In the reaction

$$NiCl_2(aq) + 2NaOH(aq) \longrightarrow Ni(OH)_2(s) + 2NaCl(aq)$$

  a. How many milliliters of 0.200 M NaOH are needed to react with 18.0 mL of 0.500 M $NiCl_2$?
  b. How many grams of $Ni(OH)_2$ are produced from the reaction of 35.0 mL of 1.75 M NaOH?

8.61 In the reaction

$$Mg(s) + 2HCl(aq) \longrightarrow MgCl_2(aq) + H_2(g)$$

  a. How many milliliters of a 6.00 M HCl solution are required to react with 15.0 g magnesium?
  b. How many moles of hydrogen gas form when 0.500 L of 2.00 M HCl reacts?

8.62 The calcium carbonate in limestone reacts with HCl to produce a calcium chloride solution and carbon dioxide gas.

$$CaCO_3(s) + 2HCl(aq) \longrightarrow CaCl_2(aq) + H_2O(l) + CO_2(g)$$

a. How many milliliters of 0.200 M HCl can react with 8.25 g $CaCO_3$?
b. How many moles of CO_2 form when 15.5 mL of 3.00 M HCl react with excess $CaCO_3$?

8.6 Properties of Solutions

LEARNING GOAL

Identify a mixture as a solution, a colloid, or a suspension. Describe osmosis and dialysis.

The solute particles in a solution play an important role in determining the properties of that solution. In most of the solutions discussed so far, the solute is dissolved as small particles that are uniformly dispersed throughout the solvent to give a homogeneous solution. When you observe a solution, such as salt water, you cannot visually distinguish the solute from the solvent. The solution appears transparent. The particles are so small that they go through filters and through semipermeable membranes. A **semipermeable membrane** allows solvent molecules such as water and very small solute particles to pass through, but not large solute molecules.

Colloids

The particles in colloidal dispersions, or **colloids**, are much larger than solute particles in a solution. Colloidal particles are large molecules, such as proteins, or groups of molecules or ions. Colloids are homogeneous mixtures that do not separate or settle out. Colloidal particles are small enough to pass through filters but too large to pass through semipermeable membranes. Table 8.11 lists several examples of colloids.

Suspensions

Suspensions are heterogeneous, nonuniform mixtures that are very different from solutions or colloids. The particles of a suspension are so large that they can often be seen with the naked eye. They are trapped by filters and semipermeable membranes.

The weight of the suspended solute particles causes them to settle out soon after mixing. If you stir muddy water, it mixes but then quickly separates as the

HEALTH NOTE

Colloids and Solutions in the Body

Colloids in the body are isolated by semipermeable membranes. For example, the intestinal lining allows solution particles to pass into the blood and lymph circulatory systems. However, the colloids from foods are too large to pass through the membrane, and they remain in the intestinal tract. Digestion breaks down large colloidal particles, such as starch and protein, into smaller particles, such as glucose and amino acids, that can pass through the intestinal membrane and enter the circulatory system. Certain foods, such as bran, a fiber, cannot be broken down by human digestive processes, and they move through the intestine intact.

Because large proteins, such as enzymes, are colloids, they remain inside cells. However, many of the substances that must be obtained by cells, such as oxygen, amino acids, electrolytes, glucose, and minerals, can pass through cellular membranes. Waste products, such as urea and carbon dioxide, pass out of the cell to be excreted.

TABLE 8.11

Examples of Colloids

	Substance Dispersed	Dispersing Medium
Fog, clouds, sprays	Liquid	Gas
Dust, smoke	Solid	Gas
Shaving cream, whipped cream, soapsuds	Gas	Liquid
Styrofoam, marshmallows	Gas	Solid
Mayonnaise, butter, homogenized milk, hand lotions	Liquid	Liquid
Cheese, butter	Liquid	Solid
Blood plasma, paints (latex), gelatin	Solid	Liquid
Cement, pearls	Solid	Solid

TABLE 8.12

Comparison of Solutions, Colloids, and Suspensions

Type of Mixture	Type of Particle	Settling	Separation
Solution	Small particles such as atoms, ions, or small molecules	Particles do not settle	Particles cannot be separated by filters or semipermeable membranes
Colloid	Larger molecules or groups of molecules or ions	Particles do not settle	Particles can be separated by semipermeable membranes but not by filters
Suspension	Very large particles that may be visible	Particles settle rapidly	Particles can be separated by filters

suspended particles settle to the bottom and leave clear liquid at the top. You can find suspensions among the medications in a hospital or in your medicine cabinet. These include Kaopectate, calamine lotion, antacid mixtures, and liquid penicillin. It is important to "shake well before using" to suspend all the particles before giving a medication that is a suspension.

Water-treatment plants make use of the properties of suspensions to purify water. When coagulants such as aluminum sulfate or ferric sulfate are added to untreated water, they react with impurities to form large suspended particles called floc. In the water-treatment plant, a system of filters traps the suspended particles but clean water passes through.

Table 8.12 compares the different types of mixtures and Figure 8.8 illustrates some properties of solutions, colloids, and suspensions.

FIGURE 8.8

Properties of different types of mixtures: **(a)** suspensions settle out; **(b)** suspensions are separated by a filter; **(c)** solution particles go through a semipermeable membrane, but colloids and suspensions do not.

Q **A filter paper can be used to separate suspension particles from a solution, but a semipermeable membrane is needed to separate colloids from a solution. Explain.**

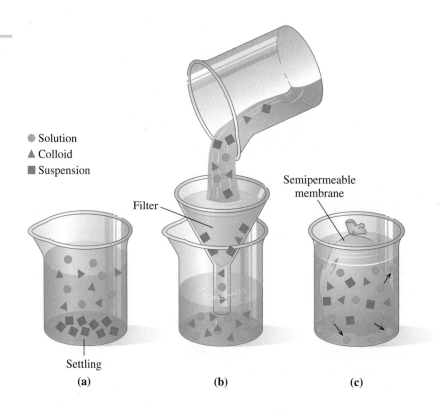

● Solution
▲ Colloid
■ Suspension

Filter

Semipermeable membrane

Settling

(a)   (b)   (c)

SAMPLE PROBLEM 8.14

■ Classifying Types of Mixtures

Classify each of the following as a solution, colloid, or suspension:

a. a mixture that settles rapidly upon standing
b. a mixture whose solute particles pass through both filters and membranes

SOLUTION

a. suspension
b. solution

STUDY CHECK

Enzymes are large protein molecules that catalyze chemical reactions inside the cells of the body. If an aqueous mixture of enzymes cannot pass through the cell membrane, are they solutions or colloids?

Osmosis and Dialysis

The movement of water into and out of the cells of plants as well as our own bodies is an important biological process. In a process called **osmosis**, the solvent water moves through a semipermeable membrane from a solution that has a lower concentration of solute into a solution where the solute concentration is higher. This movement, or osmosis, of water happens in a direction that equalizes (or attempts to equalize) the concentrations on both sides of the membrane. For example, when water is separated from sucrose (sugar) solution by a semipermeable membrane, water molecules, but not the larger sucrose molecules, move through the membrane to dilute the sucrose solution. As more water flows across the membrane, the level of the sugar solution rises as the level of water on the other side decreases.

MC GOB — Osmosis Dialysis

MC GOB — SELF-STUDY ACTIVITY Diffusion Osmosis

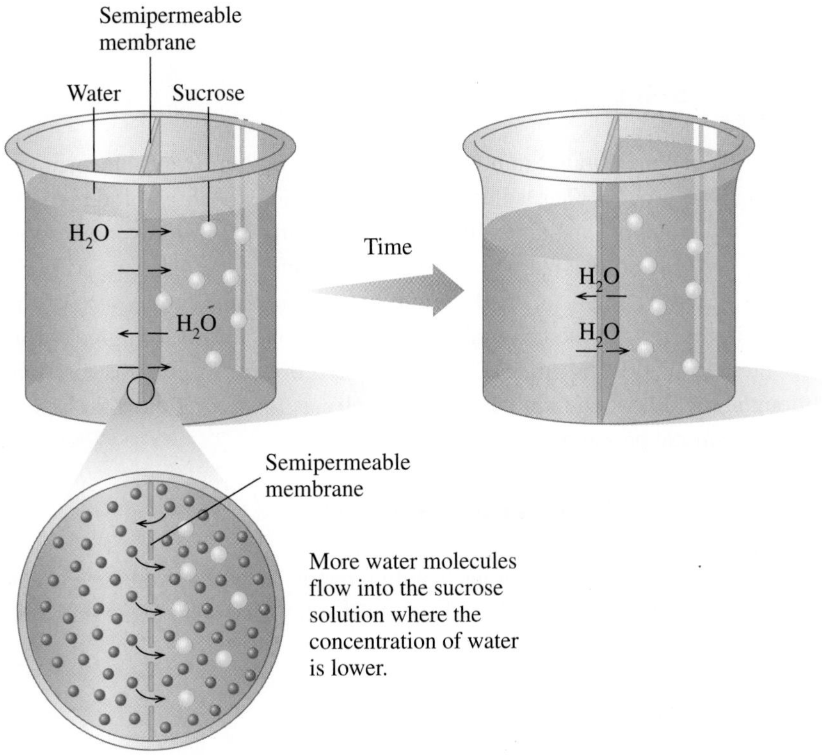

Semipermeable membrane

Water | Sucrose

H_2O

H_2O

Time

H_2O

H_2O

Semipermeable membrane

More water molecules flow into the sucrose solution where the concentration of water is lower.

EXPLORE YOUR WORLD

■ **Everyday Osmosis**

1. Place a few pieces of dry fruit such as raisins, prunes, or banana chips in water. Observe them after 1 hour or more. Look at them again the next day.
2. Place some grapes in a concentrated salt-water solution. Observe them after 1 hour or more. Look at them again the next day.
3. Place one potato slice in water and another slice in a concentrated saltwater solution. After 1–2 hours, observe the shapes and size of the slices. Look at them again the next day.

QUESTIONS

1. How did the shape of the dried fruit change after being in water? Explain.
2. How did the appearance of the grapes change after being in a concentrated salt solution? Explain.
3. How does the appearance of the potato slice that was placed in water compare to the appearance of the potato slice placed in salt water? Explain.
4. At the grocery store, why are sprinklers used to spray water on fresh produce such as lettuce, carrots, and cucumbers?

Osmotic pressure is the pressure that prevents the flow of additional water into the more concentrated solution. The water molecules flow back and forth between the two compartments as the system comes to equilibrium, but the levels remain fixed. The osmotic pressure of a solution depends on the number of solute particles in the solution. Pure water has no osmotic pressure. The greater the number of particles dissolved in a solution, the higher its osmotic pressure.

If a pressure greater than the osmotic pressure is applied to a solution, osmosis is reversed and solvent flows out of the solution and the level in the solvent compartment increases. This process, known as reverse osmosis, is used in desalinization plants in some parts of the world to produce drinking water from sea (salt) water.

SAMPLE PROBLEM 8.15

■ **Osmotic Pressure**

A 2% (m/v) sucrose solution and an 8% (m/v) sucrose solution are separated by a semipermeable membrane.

a. Which sucrose solution exerts the greater osmotic pressure?
b. In what direction does water flow initially?
c. Which solution will have the higher level of liquid at equilibrium?

SOLUTION

a. The 8% (m/v) sucrose solution has the higher solute concentration, more solute particles, and the greater osmotic pressure.
b. Initially, water will flow out of the 2% (m/v) solution into the more concentrated 8% (m/v) solution.
c. The level of the 8% (m/v) solution will be higher.

STUDY CHECK

If a 10% glucose solution is separated from a 5% (m/v) glucose solution by a semipermeable membrane, which solution will decrease in volume?

Isotonic Solutions

Because the cell membranes in biological systems are semipermeable, osmosis is an ongoing process. The solutes in body solutions such as blood, tissue fluids, lymph, and plasma all exert osmotic pressure. Most intravenous solutions are **isotonic solutions**, which exert the same osmotic pressure as body fluids. *Iso* means "equal to," and *tonic* refers to the osmotic pressure of the solution in the cell. In the hospital, isotonic solutions or **physiological solutions** include 0.90% (m/v) NaCl solution and 5.0% (m/v) glucose solution. Although they do not contain the same particles as the body fluids, they exert the same osmotic pressure.

Hypotonic and Hypertonic Solutions

A red blood cell placed in an isotonic solution retains its normal volume because there is an equal flow of water into and out of the cell. (See Figure 8.9a.) However,

Isotonic solution | Hypotonic solution | Hypertonic solution

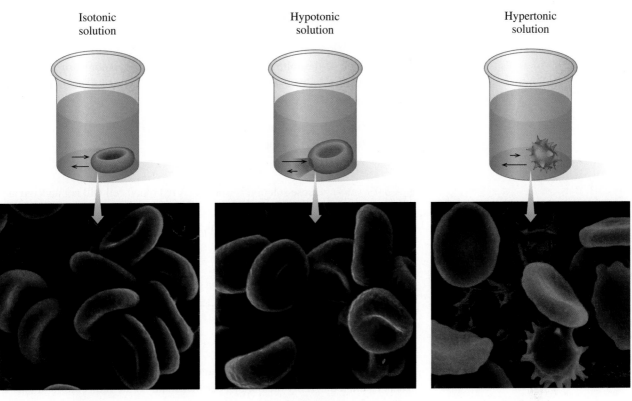

(a) Normal   (b) Hemolysis   (c) Crenation

FIGURE 8.9

(a) In an isotonic solution, a red blood cell retains its normal volume. (b) Hemolysis: In a hypotonic solution, water flows into a red blood cell, causing it to swell and burst. (c) Crenation: In a hypertonic solution, water leaves the red blood cell, causing it to shrink.

Q **What happens to a red blood cell placed in a 4% NaCl solution?**

if a red blood cell is placed in a solution that is not isotonic, the differences in osmotic pressure inside and outside the cell can drastically alter the volume of the cell. When a red blood cell is placed in pure water, a **hypotonic solution** (*hypo* means "lower than"), water flows into the cell by osmosis. (See Figure 8.9b.) The increase in fluid causes the cell to swell, and possibly burst, a process called **hemolysis**. A similar process occurs when you place dehydrated food, such as raisins or dried fruit, in water. The water enters the cells and the food becomes plump and smooth.

If a red blood cell is placed in a **hypertonic solution**, which has a higher solute concentration (*hyper* means "greater than"), water leaves the cell by osmosis. Suppose a red blood cell is placed in a 10% (m/v) NaCl solution. Because the osmotic pressure in the red blood cell is equal to that of a 0.90% (m/v) NaCl solution, the 10% (m/v) NaCl solution has a much greater osmotic pressure. As water is lost, the cell shrinks, a process called **crenation**. (See Figure 8.9c.) A similar process occurs when making pickles, which uses a hypertonic salt solution that causes the cucumbers to shrivel as they lose water.

SAMPLE PROBLEM 8.16

■ Isotonic, Hypotonic, and Hypertonic Solutions

Describe each of the following solutions as isotonic, hypotonic, or hypertonic. Indicate whether a red blood cell placed in each solution will undergo hemolysis, crenation, or no change.

a. a 5.0% (m/v) glucose solution
b. a 0.2% (m/v) NaCl solution

SOLUTION

a. A 5.0% (m/v) glucose solution is isotonic. A red blood cell will not undergo any change.
b. A 0.2% (m/v) NaCl solution is hypotonic. A red blood cell will undergo hemolysis.

STUDY CHECK

What is the effect of a 10% (m/v) glucose solution on a red blood cell?

Dialysis

Dialysis is a process that is similar to osmosis. In dialysis, a semipermeable membrane, called a dialyzing membrane, permits small solute molecules and ions as well as solvent water molecules to pass through, but it retains large particles, such as colloids. Dialysis is a way to separate solution particles from colloids.

Suppose we fill a cellophane bag with a solution containing NaCl, glucose, starch, and protein and place it in pure water. Cellophane is a dialyzing membrane, and the sodium ions, chloride ions, and glucose molecules will pass through it into the surrounding water. However, starch and protein remain inside because they are colloids. Water molecules will flow by osmosis into the cellophane bag. Eventually the concentrations of sodium ions, chloride ions, and glucose molecules inside and outside the dialysis bag become equal. To remove more NaCl or glucose, the cellophane bag must be placed in a fresh sample of pure water.

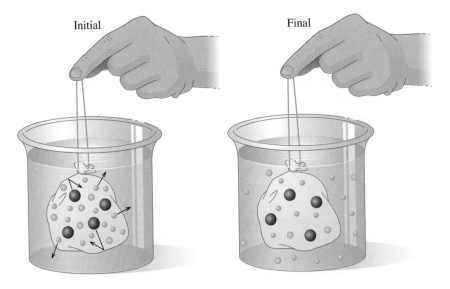

● Solution particles such as Na$^+$, Cl$^-$, glucose
● Colloidal particles such as protein, starch

Dialysis by the Kidneys and the Artificial Kidney

The fluids of the body undergo dialysis by the membranes of the kidneys, which remove waste materials, excess salts, and water. In an adult, each kidney contains about 2 million nephrons. At the top of each nephron, there is a network of arterial capillaries called the glomerulus.

As blood flows into the glomerulus, small particles, such as amino acids, glucose, urea, water, and certain ions, will move through the capillary membranes into the nephron. As this solution moves through the nephron, substances still of value to the body (such as amino acids, glucose, certain ions, and 99% of the water) are reabsorbed. The major waste product, urea, is excreted in the urine.

HEMODIALYSIS

If the kidneys fail to dialyze waste products, increased levels of urea can become life-threatening in a relatively short time. A person with kidney failure must use an artificial kidney, which cleanses the blood by **hemodialysis**.

A typical artificial kidney machine contains a large tank filled with about 100 L of water containing selected electrolytes. In the center of this dialyzing bath (dialysate), there is a dialyzing coil or membrane made of cellulose tubing. As the patient's blood flows through the dialyzing coil, the highly concentrated waste products dialyze out of the blood. No blood is lost because the membrane is not permeable to large particles such as red blood cells.

Dialysis patients do not produce much urine. As a result, they retain large amounts of water between dialysis treatments, which produces a strain on the heart. The intake of fluids for a dialysis patient may be restricted to as little as a few teaspoons of water a day. In the dialysis procedure, the pressure of the blood is increased as it circulates through the dialyzing coil so water can be squeezed out of the blood. For some dialysis patients, 2–10 L of water may be removed during one treatment. Dialysis patients have from two to three treatments a week, each treatment requiring about 5–7 hr. Some of the newer treatments require less time. For many patients, dialysis is done at home with a home dialysis unit.

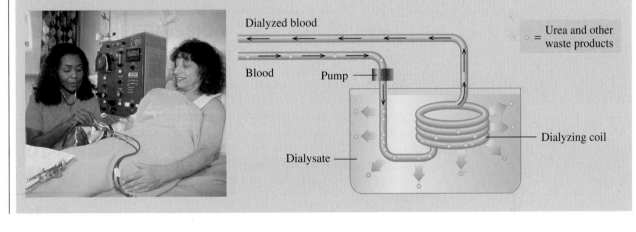

QUESTIONS AND PROBLEMS

Properties of Solutions

8.63 Identify the following as characteristic of a solution, colloid, or suspension:
   a. a mixture that cannot be separated by a semipermeable membrane
   b. a mixture that settles out upon standing

8.64 Identify the following as characteristic of a solution, colloid, or suspension:
   a. Particles of this mixture remain inside a semipermeable membrane, but pass through filters.
   b. The particles of solute in this mixture are very large and visible.

8.65 A 10% (m/v) starch solution is separated from pure water by an osmotic membrane.
   a. Which solution has the higher osmotic pressure?
   b. In which direction will water flow initially?
   c. In which compartment will the volume level rise?

8.66 Two solutions, a 0.1% (m/v) albumin solution and a 2% (m/v) albumin solution, are separated by a semipermeable membrane. (Albumins are colloidal proteins.)
   a. Which compartment has the higher osmotic pressure?
   b. In which direction will water flow initially?
   c. In which compartment will the volume level rise?

8.67 Indicate the compartment (A or B) that will increase in volume for each of the following pairs of solutions separated by semipermeable membranes:

A	B
a. 5.0% (m/v) starch	10% (m/v) starch
b. 4% (m/v) albumin	8% (m/v) albumin
c. 0.1% (m/v) sucrose	10% (m/v) sucrose

8.68 Indicate the compartment (A or B) that will increase in volume for each of the following pairs of solutions separated by semipermeable membranes:

A	B
a. 20% (m/v) starch	10% (m/v) starch
b. 10% (m/v) albumin	2% (m/v) albumin
c. 0.5% (m/v) sucrose	5% (m/v) sucrose

8.69 Are the following solutions isotonic, hypotonic, or hypertonic compared with a red blood cell?
a. distilled H_2O **b.** 1% (m/v) glucose
c. 0.90% (m/v) NaCl   **d.** 5.0% (m/v) glucose

8.70 Will a red blood cell undergo crenation, hemolysis, or no change in each of the following solutions?
a. 1% (m/v) glucose   **b.** 2% (m/v) NaCl
c. 5% (m/v) NaCl   **d.** 0.1% (m/v) NaCl

8.71 Each of the following mixtures is placed in a dialyzing bag and immersed in distilled water. Which substances will be found outside the bag in the distilled water?
a. NaCl solution
b. starch (colloid) and alanine (amino acid) solution
c. NaCl solution and starch (colloid)   **d.** urea solution

8.72 Each of the following mixtures is placed in a dialyzing bag and immersed in distilled water. Which substances will be found outside the bag in the distilled water?
a. KCl solution and glucose solutions
b. an albumin solution (colloid)
c. an albumin solution (colloid), KCl solution, and glucose solution
d. urea solution and NaCl solution

CONCEPT MAP

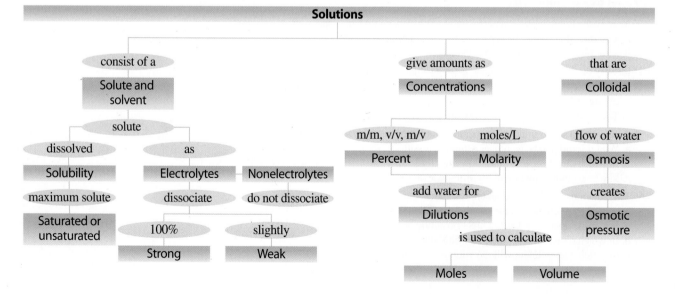

CHAPTER REVIEW

8.1 Solutions
A solution forms when a solute dissolves in a solvent. In a solution, the particles of solute are evenly distributed in the solvent. The solute and solvent may be solid, liquid, or gas. The polar O—H bond leads to hydrogen bonding in water molecules. An ionic solute dissolves in water, a polar solvent, because the polar water molecules attract and pull the ions into solution, where they become hydrated. The expression "like dissolves like" means that a polar or ionic solute dissolves in a polar solvent while a nonpolar solute requires a nonpolar solvent.

8.2 Electrolytes and Nonelectrolytes
Substances that release ions in water are called electrolytes because the solution will conduct an electrical current. Strong electrolytes are completely ionized, whereas weak electrolytes are only partially ionized. Nonelectrolytes are substances that dissolve in water to produce molecules and cannot conduct electrical currents. An equivalent is the amount of a electrolyte that carries one mole of positive or negative charge. One mole of Na^+ is 1 equivalent. One mole of Ca^{2+} has 2 equivalents. In fluid replacement solutions, the concentrations of electrolytes are expressed as mEq/L of solution.

8.3 Solubility
A solution that contains the maximum amount of dissolved solute is a saturated solution. The solubility of a solute is the maximum amount of a solute that can dissolve in 100 g of solvent. A solution containing less than the maximum amount of dissolved solute is unsaturated. An increase in temperature increases the solubility of most solids in water, but decreases the solubility of gases in water. Salts that are soluble in water usually contain Li^+, Na^+, K^+, NH_4^+, NO_3^-, or acetate $C_2H_3O_2^-$. An ionic equation consists of writing all the dissolved substances in an equation for the formation of an insoluble salt as individual ions. A net ionic equation is written by removing all the ions not involved in the chemical change (spectator ions) from the ionic equation.

8.4 Percent Concentration
The concentration of a solution is the amount of solute dissolved in a certain amount of solution. Mass percent expresses the ratio of the mass of solute to the mass of solution multiplied by 100. Percent concentration is also expressed as volume/volume and mass/volume ratios. In calculations of grams or milliliters of solute or solution, the percent concentration is used as a conversion factor.

8.5 Molarity and Dilution
Molarity is the moles of solute per liter of solution. Units of molarity, moles/liter, are used in conversion factors to solve for moles of solute or volume of solution. In dilution, the volume of solvent increases and the solute concentration decreases. If the mass or solution volume and molarity of substances in a reaction are given, the balanced equation can be used to determine the quantities or concentrations of any of the other substances in the reaction.

8.6 Properties of Solutions
Colloids contain particles that do not settle out and pass through filters but not through semipermeable membranes. Suspensions have very large particles that settle out of solution.

In osmosis, solvent (water) passes through a semipermeable membrane from a solution of a lower solute concentration to a solution of a higher concentration. Isotonic solutions have osmotic pressures equal to that of body fluids. A red blood cell maintains its volume in an isotonic solution, but swells and may burst (hemolyze) in a hypotonic solution and shrinks (crenates) in a hypertonic solution. In dialysis, water and small solute particles pass through a dialyzing membrane, while larger particles are retained.

KEY TERMS

colloid A mixture having particles that are moderately large. Colloids pass through filters but cannot pass through semipermeable membranes.

concentration A measure of the amount of solute that is dissolved in a specified amount of solution.

crenation The shriveling of a cell due to water leaving the cell when the cell is placed in a hypertonic solution.

dialysis A process in which water and small solute particles pass through a semipermeable membrane.

dilution A process by which water (solvent) is added to a solution to increase the volume and decrease (dilute) the concentration of the solute.

electrolyte A substance that produces ions when dissolved in water; its solution conducts electricity.

equivalent (Eq) The amount of a positive or negative ion that supplies 1 mole of electrical charge.

hemodialysis A mechanical cleansing of the blood by an artificial kidney using the principle of dialysis.

hemolysis A swelling and bursting of red blood cells in a hypotonic solution due to an increase in fluid volume.

Henry's law The solubility of a gas in a liquid is directly related to the pressure of that gas above the liquid.

hydration The process of surrounding dissolved ions by water molecules.

hypertonic solution A solution that has a higher osmotic pressure than the red blood cells of the body.

hypotonic solution A solution that has a lower osmotic pressure than the red blood cells of the body.

insoluble salt An ionic compound that does not dissolve in water.

ionic equation An equation for a reaction in solution that gives all the individual ions, both reacting ions and spectator ions.

isotonic solution A solution that has the same osmotic pressure as that of the red blood cells of the body.

mass percent The grams of solute in exactly 100 grams of solution.

mass/volume percent The grams of solute in exactly 100 mL of solution.

molarity (M) The number of moles of solute in exactly 1 L of solution.

net ionic equation An equation for a reaction that gives only the reactants undergoing chemical change and products.

nonelectrolyte A substance that dissolves in water as molecules; its solution will not conduct an electrical current.

osmosis The flow of a solvent, usually water, through a semipermeable membrane into a solution of higher solute concentration.

osmotic pressure The pressure that prevents the flow of water into the more concentrated solution.

physiological solution A solution that exerts the same osmotic pressure as normal body fluids.

saturated solution A solution containing the maximum amount of solute that can dissolve at a given temperature. Any additional solute will remain undissolved in the container.

semipermeable membrane A membrane that permits the passage of certain substances while blocking or retaining others.

solubility The maximum amount of solute that can dissolve in exactly 100 g of solvent, usually water, at a given temperature.

soluble salt An ionic compound that dissolves in water.

solute The component in a solution that changes state upon dissolving; if no change in state occurs, it is the component present in the smaller quantity.

solution A homogeneous mixture in which the solute is made up of small particles (ions or molecules) that can pass through filters and semipermeable membranes.

solvent The substance in which the solute dissolves; usually the component present in greatest amount.

strong electrolyte A polar or ionic compound that ionizes completely when it dissolves in water. Its solution is a good conductor of electricity.

suspension A mixture in which the solute particles are large enough and heavy enough to settle out and be retained by both filters and semipermeable membranes.

unsaturated solution A solution that contains less solute than can be dissolved.

volume percent A percent concentration that relates the volume of the solute to the volume of the solution.

weak electrolyte A substance that produces only a few ions along with many molecules when it dissolves in water. Its solution is a weak conductor of electricity.

UNDERSTANDING THE CONCEPTS

8.73 Select the diagram that represents the solution formed by a solute ⬤⬤ that is a
a. nonelectrolyte
b. weak electrolyte
c. strong electrolyte

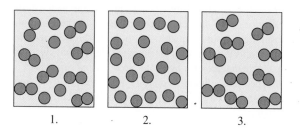

1.   2.   3.

8.74 Match the diagrams with
a. a polar solute and a polar solvent
b. a nonpolar solute and a polar solvent
c. a nonpolar solute and a nonpolar solvent

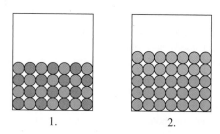

1.   2.

8.75 A pickle is made by soaking a cucumber in a *brine*, a saltwater solution. What makes the smooth cucumber become wrinkled like a prune?

8.76 Do you think solution (1) has undergone heating or cooling to give the solid shown in (2) and (3)?

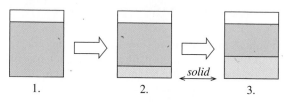

1.   2.   *solid* 3.

8.77 Select the container that represents the dilution of a 4% (m/v) KCl solution to each of the following:
a. 2% (m/v) KCl
b. 1% (m/v) KCl

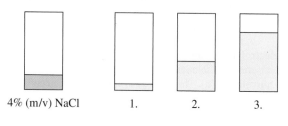

4% (m/v) NaCl   1.   2.   3.

8.78 Why do the lettuce leaves in a salad wilt after a vinagrette dressing containing salt is added?

8.79 A semipermeable membrane separates two compartments, A and B. If the levels of solutions in A and B are equal initially, select the diagram that illustrates the final levels for each of the following:

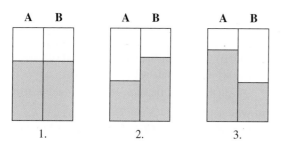

	A	B
a.	2% (m/v) starch	8% (m/v) starch
b.	1% (m/v) starch	1% (m/v) starch
c.	5% (m/v) sucrose	1% (m/v) sucrose
d.	0.1% (m/v) sucrose	1% (m/v) sucrose

8.80 Select the diagram that represents the shape of a red blood cell when placed in each of the following solutions:

   1.   2.   3.

Normal red blood cell

a. 0.90% (m/v) NaCl
b. 10% (m/v) glucose
c. 0.01% (m/v) NaCl
d. 5.0% (m/v) glucose
e. 1% (m/v) glucose

ADDITIONAL QUESTIONS AND PROBLEMS

MC GOB — For instructor-assigned homework, go to www.masteringgob.com.

8.81 Why does iodine dissolve in hexane, but not in water?

8.82 How does temperature and pressure affect the solubility of solids and gases in water?

8.83 If NaCl has a solubility of 36.0 g in 100 g H_2O at 20°C, how many grams of water are needed to prepare a saturated solution containing 80.0 g NaCl?

8.84 If the solid NaCl in a saturated solution of NaCl continues to dissolve, why is there no change in the concentration of the NaCl solution?

8.85 Potassium nitrate has a solubility of 34 g KNO_3 in 100 g H_2O at 20°C. State if each of the following forms an unsaturated or saturated solution at 20°C.
  a. 34 g KNO_3 and 200 g H_2O
  b. 17 g KNO_3 and 50 g H_2O
  c. 68 g KNO_3 and 150 g H_2O

8.86 Potassium fluoride has a solubility of 92 g KF in 100 g H_2O at 18°C. State if each of the following forms an unsaturated or saturated solution at 18°C.
  a. 46 g KF and 100 g H_2O
  b. 46 KF and 50 g H_2O
  c. 184 KF and 150 g H_2O

8.87 Calculate the mass percent of a solution containing 15.5 g of Na_2SO_4 and 75.5 g of H_2O.

8.88 How many grams of K_2CO_3 are in 750 mL of a 3.5% (m/v) K_2CO_3 solution?

8.89 A patient receives all her nutrition from fluids given through the vena cava. Every 12 hours, 750 mL of a solution that is 4% (m/v) amino acids (protein) and 25% (m/v) glucose (carbohydrate) is given along with 500 mL of a 10% (m/v) lipid (fat).
  a. In 1 day, how many grams of amino acids, glucose, and lipid are given to the patient?
  b. How many kilocalories does she obtain in 1 day?

8.90 An 80-proof brandy is 40.0% (v/v) ethyl alcohol. The "proof" is twice the percent concentration of alcohol in the beverage. How many milliliters of alcohol are present in 750 mL of brandy?

8.91 How many milliliters of a 12% (v/v) propyl alcohol solution would you need to obtain 4.5 mL of propyl alcohol?

8.92 How many liters of a 5.0% (m/v) glucose solution would you need to obtain 75 g of glucose?

8.93 If you were in the laboratory, how would you prepare 0.250 L of a 2.00 M KCl solution?

8.94 What is the molarity of a solution containing 15.6 g of KCl in 274 mL of solution?

8.95 A solution is prepared with 70.0 g of HNO_3 and 130.0 g of H_2O. It has a density of 1.21 g/mL.
  a. What is the mass percent of the HNO_3 solution?
  b. What is the total volume of the solution?
  c. What is the mass/volume percent?
  d. What is its molarity (M)?

8.96 What is the molarity of a 15% (m/v) NaOH solution?

8.97 How many grams of solute are in each of the following solutions?
a. 2.5 L of 3.0 M $Al(NO_3)_3$
b. 75 mL of 0.50 M $C_6H_{12}O_6$

8.98 How many grams of solute are in each of the following solutions?
a. 428 mL of a 0.450 M Na_2SO_4 solution
b. 10.5 mL of a 2.50 M $AgNO_3$ solution
c. 28.4 mL of a 6.00 M H_3PO_4 solution

8.99 How many grams of solute are in each of the following solutions?
a. 2.52 L of a 3.00 M KNO_3 solution
b. 75.0 mL of a 0.506 M Na_2SO_4 solution
c. 45.2 mL of a 1.80 M HCl solution

8.100 Indicate whether each of the following is soluble in water.
a. KCl   **b.** $MgSO_4$ **c.** PbS
d. $AgNO_3$ **e.** $Ca(OH)_2$

8.101 Why would a solution made by mixing solutions of $NaNO_3$ and KCl be clear, while a combination of KCl and $Pb(NO_3)_2$ solution produces a solid?

8.102 Write the net ionic equation to show the formation of a precipitate (insoluble salt) when the following solutions are mixed. Write *none* if there is no precipitate.
a. $Ca(NO_3)_2(aq)$ and $Na_2S(aq)$
b. $Na_3PO_4(aq)$ and $Pb(NO_3)_2(aq)$
c. $FeCl_3(aq)$ and $NH_4NO_3(aq)$

8.103 Write the net ionic equation to show the formation of a precipitate (insoluble salt) when the following solutions are mixed. Write *none* if there is no precipitate.
a. $AgNO_3(aq)$ and $NaCl(aq)$
b. $NaCl(aq)$ and $KNO_3(aq)$
c. $Na_2SO_4(aq)$ and $BaCl_2(aq)$

8.104 Calculate the molarity of the solution when water is added to prepare each of the following solutions:
a. 25.0 mL of 18.0 M HCl diluted to 500. mL
b. 50.0 mL of 1.50 M NaCl diluted to 125 mL
c. 4.50 mL of 8.50 M KOH diluted to 75.0 mL

8.105 Calculate the molarity of the solution when water is added to prepare each of the following solutions:
a. 25.0 mL of 0.200 M NaBr diluted to 50.0 mL
b. 15.0 mL of 1.20 M K_2SO_4 diluted to 40.0 mL
c. 75.0 mL of 6.00 M NaOH diluted to 255 mL

8.106 What is the final volume in mL when 5.00 mL of 12.0 M NaOH is diluted to each of the following concentrations?
a. 0.600 M   **b.** 1.00 M   **c.** 2.50 M

8.107 What is the final volume in mL when 25.0 mL of 5.00 M HCl is diluted to each of the following concentrations?
a. 2.50 M HCl   **b.** 1.00 M HCl   **c.** 0.500 M HCl

8.108 Calcium carbonate $CaCO_3$ reacts with stomach acid (HCl, hydrochloric acid) according to the following equation:
$$CaCO_3(s) + 2HCl(aq) \longrightarrow CaCl_2(aq) + H_2O(l) + CO_2(g)$$
Tums, an antacid, contains $CaCO_3$. If Tums is added to 20.0 mL of 0.400 M HCl, how many grams of CO_2 gas are produced?

8.109 The antacid Amphogel contains aluminum hydroxide $Al(OH)_3$. How many milliliters of 6.00 M HCl are required to react with 60.0 mL of 1.00 M $Al(OH)_3$?
$$Al(OH)_3(s) + 3HCl(aq) \longrightarrow AlCl_3(aq) + 3H_2O(aq)$$

8.110 Why would a dialysis unit (artificial kidney) use isotonic concentrations of NaCl, KCl, $NaHCO_3$, and glucose in the dialysate?

8.111 Why would solutions with high salt content be used to prepare dried flowers?

8.112 A patient on dialysis has a high level of urea, a high level of sodium, and a low level of potassium in the blood. Why is the dialyzing solution prepared with a high level of potassium but no sodium or urea?

8.113 Why can't you drink seawater even if you are stranded on a desert island?

CHALLENGE QUESTIONS

8.114 Osmolarity (Osm) is the molar concentration of the particles—molecules and/or ions—in a solution. In physiological solutions, 5% (m/v) glucose ($C_6H_{12}O_6$) and 0.9% (m/v) NaCl, the osmolarity must be the same as for the blood, which is 0.3 Osm. Calculate the osmolarity of these physiological solutions.

8.115 Indicate whether each of the following ionic compounds is soluble (S) or insoluble (I) in water.
a. Na_3PO_4 **b.** $PbBr_2$ **c.** KCl
d. $(NH_4)_2S$ **e.** $MgCO_3$ **f.** $FePO_4$

8.116 Write the net ionic equation to show the formation of a precipitate (insoluble salt) when the following solutions are mixed. Write *none* if no insoluble salt forms.
a. $AgNO_3 + Na_2SO_4$ **b.** $KCl + Pb(NO_3)_2$
c. $CaCl_2 + Mg_3(PO_4)_2$ **d.** $Na_2SO_4 + BaCl_2$

8.117 In a laboratory experiment, a 10.0-mL sample of NaCl solution is poured into an evaporating dish with a mass of 24.10 g. The combined mass of the evaporating dish and NaCl solution is 36.15 g. After heating, the evaporating dish and dry NaCl have a combined mass of 25.50 g.
a. What is the % (m/m) of the NaCl solution?
b. What is the molarity (M) of the NaCl solution?
c. If water is added to 10.0 mL of the initial NaCl solution to give a final volume of 60.0 mL, what is the molarity of the dilute NaCl solution?

8.118 A solution contains 4.56 g KCl in 175 mL of solution. If the density of the KCl solution is 1.12 g/mL, what are the percent (m/m) and molarity, M, for the potassium chloride solution?

8.119 A solution is prepared by dissolving 22.0 g NaOH in 118.0 g water. The NaOH solution has a density of 1.15 g/mL.
   a. What is the % (m/m) concentration of the NaOH solution?
   b. What is the total volume (mL) of the solution?
   c. What is the molarity (M) of the solution?

8.120 How many milliliters of a 1.75 M LiCl solution contain 15.2 g of LiCl?

8.121 How many grams of NaBr are contained in 75.0 mL of a 1.50 M NaBr solution?

8.122 Magnesium reacts with HCl to produce magnesium chloride and hydrogen gas

$$Mg(s) + 2HCl(aq) \longrightarrow MgCl_2(aq) + H_2(g)$$

What is the molarity of the HCl solution if 250. mL of the HCl solution reacts with magnesium to produce 4.20 L of H_2 gas measured at STP?

8.123 How many g NO gas can be produced from 80.0 mL of 4.00 M HNO_3 and excess Cu?

$$3Cu(s) + 8HNO_3(aq) \longrightarrow$$
$$3Cu(NO_3)_2(aq) + 4H_2O(l) + 2NO(g)$$

ANSWERS

Answers to Study Checks

8.1 Yes. Both the solute and solvent are nonpolar substances; "like dissolves like."

8.2 A solution of a weak electrolyte will contain both molecules and ions.

8.3 0.194 mole Cl^-

8.4 57 g $NaNO_3$

8.5 The value of 65 g/100 g H_2O is more likely because the solubility of most solids increases when the temperature increases.

8.6 **a.** No; $PbCl_2$ is an insoluble chloride
   b. Yes; a salt containing K^+ ion is soluble
   c. No; $FeCO_3$ is insoluble

8.7 **a.** no solid forms
   b. $Pb^{2+}(aq) + 2Cl^-(aq) \longrightarrow PbCl_2(s)$

8.8 3.4% (m/m) NaCl solution

8.9 18 g KCl and 207 g H_2O

8.10 2.12 M KNO_3

8.11 4.5 moles HCl

8.12 120. mL

8.13 1.47 g Zn

8.14 colloids

8.15 5% (m/v) glucose

8.16 The red blood cell will shrink (crenate).

Answers to Selected Questions and Problems

8.1 **a.** NaCl, solute; water, solvent
   b. water, solute; ethanol, solvent
   c. oxygen, solute; nitrogen, solvent

8.3 The polar water molecules pull the K^+ and I^- ions away from the solid and into solution, where they are hydrated.

8.5 **a.** water   **b.** CCl_4 **c.** water   **d.** CCl_4

8.7 In a solution of KF, only the ions of K^+ and F^- are present in the solvent. In an HF solution, there are a few ions of H^+ and F^- present but mostly dissolved HF molecules.

8.9 **a.** $KCl(s) \xrightarrow{H_2O} K^+(aq) + Cl^-(aq)$

   b. $CaCl_2(s) \xrightarrow{H_2O} Ca^{2+}(aq) + 2Cl^-(aq)$

   c. $K_3PO_4(s) \xrightarrow{H_2O} 3K^+(aq) + PO_4^{3-}(aq)$

   d. $Fe(NO_3)_3(s) \xrightarrow{H_2O} Fe^{3+}(aq) + 3NO_3^-(aq)$

8.11 **a.** mostly molecules and a few ions   **b.** ions only
   c. molecules only

8.13 **a.** strong electrolyte   **b.** weak electrolyte
   c. nonelectrolyte

8.15 **a.** 1 Eq   **b.** 2 Eq   **c.** 2 Eq   **d.** 6 Eq

8.17 0.154 mole Na^+, 0.154 mole Cl^-

8.19 55 mEq/L

8.21 **a.** saturated   **b.** unsaturated

8.23 **a.** unsaturated   **b.** unsaturated   **c.** saturated

8.25 **a.** 68.0 g KCl   **b.** 12.0 g KCl

8.27 **a.** The solubility of solid solutes typically increases as temperature increases.
   b. The solubility of a gas is less at a higher temperature.
   c. Gas solubility is less at a higher temperature and the CO_2 pressure in the can is increased.

8.29 **a.** soluble   **b.** insoluble   **c.** insoluble
   d. soluble   **e.** soluble

8.31 **a.** no solid forms
   b. $2Ag^+(aq) + S^{2-}(aq) \longrightarrow Ag_2S(s)$
   c. $Ca^{2+}(aq) + SO_4^{2-}(aq) \longrightarrow CaSO_4(s)$

8.33 5% (m/m) is 5 g of glucose in 100 g of solution, whereas 5% (m/v) is 5 g of glucose in 100 mL of solution.

8.35 **a.** 17% (m/m)   **b.** 5.3% (m/m)

8.37 **a.** 30.% (m/v)   **b.** 11% (m/v)

8.39 **a.** 2.5 g KCl   **b.** 50. g NH_4Cl

8.41 79.9 mL of alcohol

8.43 **a.** 20 g mannitol   **b.** 300 g mannitol

8.45 2 L

8.47 **a.** 0.50 M glucose   **b.** 0.036 M KOH

8.49 **a.** 3.0 moles NaCl   **b.** 0.40 mole KBr

8.51 **a.** 120 g NaOH   **b.** 60. g KCl

8.53 **a.** 1.5 L   **b.** 10. L

8.55 **a.** 2.0 M HCl   **b.** 2.0 M NaOH
     c. 2.5% (m/v) KOH   **d.** 3.0 (m/v) H_2SO_4

8.57 **a.** 0.60 L   **b.** 0.25 L   **c.** 6.0 L

8.59 **a.** 10.4 g $PbCl_2$ **b.** 18.8 mL $Pb(NO_3)_2$ solution

8.61 **a.** 206 mL HCl solution   **b.** 0.500 mole H_2 gas

8.63 **a.** solution   **b.** suspension

8.65 **a.** starch   **b.** from pure water into the starch
     c. starch

8.67 **a.** B 10% (m/v) starch   **b.** B 8% (m/v) albumin
     c. B 10% (m/v) sucrose

8.69 **a.** hypotonic   **b.** hypotonic
     c. isotonic   **d.** isotonic

8.71 **a.** NaCl   **b.** alanine
     c. NaCl   **d.** urea

8.73 **a.** 3   **b.** 1   **c.** 2

8.75 The skin of the cucumber acts like a semipermeable membrane and the more dilute solution inside flows into the brine solution.

8.77 **a.** 2   **b.** 3

8.79 **a.** 2   **b.** 1   **c.** 3   **d.** 2

8.81 Because iodine is a nonpolar molecule, it will dissolve in hexane, a nonpolar solvent. Iodine does not dissolve in water because water is a polar solvent.

8.83 222 g water

8.85 **a.** unsaturated solution   **b.** saturated solution
     c. saturated solution

8.87 17.0% (m/m)

8.89 **a.** 60 g of amino acids, 380 g of glucose, and 100 g of lipids
     b. 2700 kcal

8.91 38 mL of solution

8.93 To make a 2.00 M KCl solution, weight out 37.3 g KCl (0.500 mole) and place into a volumetric flask. Add water to dissolve the KCl and give a final volume of 0.250 liter.

8.95 **a.** 35.0% (m/m) HNO_3 **b.** 165 mL
     c. 42.4% (m/v)   **d.** 6.73 M

8.97 **a.** 1600 g $Al(NO_3)_3$ **b.** 6.8 g of $C_6H_{12}O_6$

8.99 **a.** 764 g KNO_3 **b.** 5.39 g Na_2SO_4
     c. 2.97 g HCl

8.101 When solutions of $NaNO_3$ and KCl are mixed, no insoluble products are formed. All the combinations of salts are soluble. When KCl and $Pb(NO_3)_2$ solutions are mixed, the insoluble salt $PbCl_2$ forms.

8.103 **a.** $Ag^+(aq) + Cl^-(aq) \longrightarrow AgCl(s)$
     b. none
     c. $Ba^{2+}(aq) + SO_4^{2-}(aq) \longrightarrow BaSO_4(s)$

8.105 **a.** 0.100 M NaBr   **b.** 0.450 M K_2SO_4
     c. 1.76 M NaOH

8.107 **a.** 50.0 mL HCl   **b.** 125 mL HCl
     c. 250. mL HCl

8.109 30.0 mL HCl solution

8.111 The solution will dehydrate the flowers because water will flow out of the cells of the flowers into the more concentrated salt solution.

8.113 Drinking seawater will cause water to flow out of the body cells and further dehydrate a person.

8.115 **a.** Na^+ salts are soluble.
     b. The halide salts containing Pb^{2+} are insoluble.
     c. K^+ salts are soluble.
     d. Salts containing NH_4^+ ions are soluble.
     e. Salts containing CO_3^{2-} are usually insoluble.
     f. Salts containing PO_4^{3-} and Fe^{3+} are insoluble

8.117 **a.** 11.6% (m/m)   **b.** 2.39 M   **c.** 0.383 M

8.119 **a.** 15.7% (m/m) NaOH   **b.** 122 mL   **c.** 4.51 M

8.121 11.6 g NaBr

8.123 2.40 g NO

9 Chemical Equilibrium

"I use radioactive isotopes to understand the cycling of elements like carbon and phosphorus in the ocean," explains Claudia Benitez-Nelson, a chemical oceanographer and Assistant Professor of Geological Sciences at the University of South Carolina. "For example, I use thorium-234 to trace how and when particles are formed and transported to the bottom of the ocean. I also examine the biological consumption of the nutrient phosphorus by measuring fluctuations over time in the levels of the naturally occurring radioactive isotopes of phosphorus. My knowledge of chemistry is essential for understanding nutrient biogeochemistry and carbon sequestration in the oceans."

Oceanographers study the oceans and the plants and animals that live in the ocean. They study marine life; the chemical compounds in the ocean; the shape and composition of the ocean floor; and the effects of waves and tides.

LOOKING AHEAD

Mastering
GOB CHEMISTRY

Visit **www.masteringgob.com**
for self-study materials
and instructor-assigned
homework.

Earlier we looked at chemical reactions and determined the amounts of substances that react and the products that form. Now we are interested in how fast a reaction goes. If we know how fast a medication acts on the body, we can adjust the time over which the medication is taken. In construction, substances are added to cement to make it dry faster so work can continue. Some reactions such as explosions or the formation of precipitates in a solution are very fast. We know that when we roast a turkey or bake a cake that the reaction is slower. Some reactions such as the tarnishing of silver and the aging of the body are much slower. (See Figure 9.1.) We will see that some reactions need energy to keep running while other reactions produce energy. We burn gasoline in our automobile engines to produce energy to make our cars move. We will also look at the effect of changing the concentrations of reactants or products on the rate of reaction.

Up to now, we have considered a reaction as proceeding in a forward direction from reactants to products. However, in many reactions a reverse reaction also takes place as products collide to reform reactants. When the forward and reverse reactions take place at the same rate, the amounts of reactants and products stay the same. When this balance in forward and reverse rate is reached, we say that the reaction has reached *equilibrium*. At equilibrium both reactants and products are present, though some reaction mixtures contain mostly reactants and form only a few products, while others contain mostly products and few reactants.

FIGURE 9.1

Reaction rates vary greatly for everyday processes. A banana ripens in a few days, silver tarnishes in a few months, while the aging process of humans takes many years.

Q **How would you compare the rates of the reaction that forms sugars in plants by photosynthesis with the reactions that digest sugars in the body?**

Reaction rate increases

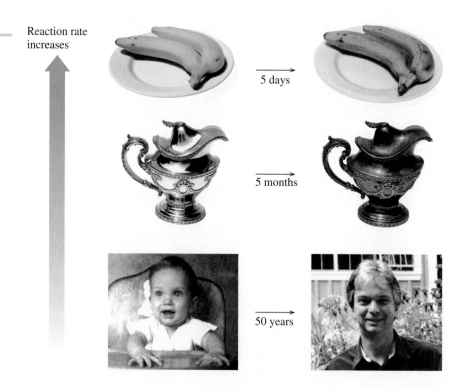

5 days

5 months

50 years

9.1 Rates of Reactions

LEARNING GOAL

Describe how temperature, concentration, and catalysts affect the rate of a reaction.

For a chemical reaction to take place, the molecules of the reactants must come in contact with each other. The **collision theory** indicates that a reaction takes place only when molecules collide with the proper orientation and sufficient energy. Many collisions can occur, but only a few actually lead to the formation of product. For example, consider the reaction of nitrogen and oxygen molecules. (See Figure 9.2.) To form NO product, the collisions between the N_2 and the O_2 molecules must place the atoms in the proper alignment. If the molecules are not aligned properly, no reaction takes place.

Activation Energy

Even when a collision has the proper orientation, there still must be sufficient energy to break the bonds between the atoms of the reactants. The amount of energy required to break the bonds between atoms of reactants is the *activation energy*. In Figure 9.3, this appears as an energy hill. The concept of activation energy is analogous to climbing a hill. To reach a destination on the other side, we must have the energy needed to climb to the top of the hill. Once we are at the top, we can run down the other side. The energy needed to get us from our starting point to the top of the hill would be our activation energy.

In the same way, a collision must provide enough energy to push the reactants to the top of the energy hill. Then the reactants may be converted to products. If the energy provided by the collision is less than the activation energy, the molecules simply bounce apart and no reaction occurs. The features that lead to a successful reaction are summarized next.

Three Conditions Required for a Reaction to Occur

1. **Collision**   The reactants must collide.
2. **Orientation**   The reactants must align properly to break and
                     form bonds.
3. **Energy**   The collision must provide the energy of activation.

Collision that forms products

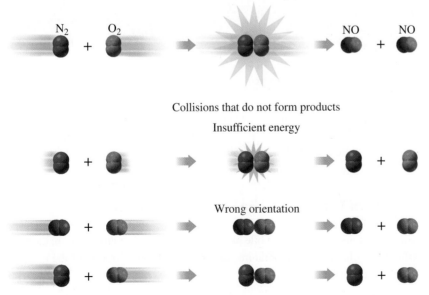

Collisions that do not form products

Insufficient energy

Wrong orientation

FIGURE 9.2

Reacting molecules must collide, have a minimum amount of energy, and the proper orientation to form products.

Q **What happens when reacting molecules collide with the minimum energy but don't have the proper orientation?**

FIGURE 9.3

The activation energy is the energy needed to convert the colliding molecules into product.

Q **What happens in a collision of reacting molecules that have the proper orientation, but not the energy of activation?**

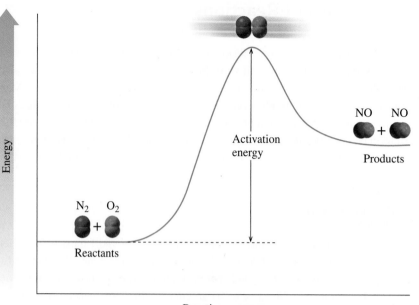

Reaction Rates

The **rate** (or speed) **of reaction** is determined by measuring the amount of a reactant used up, or the amount of a product formed, in a certain period of time.

$$\text{Rate of reaction} = \frac{\text{change in concentration}}{\text{change in time}}$$

Perhaps we can describe the rate of reaction by the analogy of eating a pizza. When we start to eat, we have a whole pizza. As time goes by, there are fewer slices of pizza left. If we know how long it took to eat the pizza, we could determine the rate at which the pizza was consumed. Let's assume 4 slices are eaten every 8 minutes. That gives a rate of $\frac{1}{2}$ slice per minute. After 16 minutes, all 8 slices are gone.

Rate at which pizza slices are eaten

Slices eaten	0	4 slices	6 slices	8 slices
Time (min)	0	8 min	12 min	16 min

$$\text{rate} = \frac{4 \text{ slices}}{8 \text{ min}} = \frac{1 \text{ slice}}{2 \text{ min}} = \frac{\frac{1}{2} \text{ slice}}{1 \text{ min}}$$

Factors That Affect the Rate of a Reaction

Factors that Affect Rate

Some reactions go very fast, while others are very slow. For any reaction, the rate is affected by changes in temperature, changes in the concentration of the reactants, and the addition of catalysts.

Temperature

At higher temperatures, the increase in kinetic energy makes the reacting molecules move faster. As a result, more collisions occur and more colliding molecules have sufficient energy to react and form products. If we want food to cook faster, we use more heat to raise the temperature. When body temperature rises, there is an increase in the pulse rate, rate of breathing, and metabolic rate. On the other hand, we slow down reactions by lowering the temperature. We refrigerate perishable foods to retard spoilage and make them last longer. For some injuries, we apply ice to lessen the bruising process.

Concentrations of Reactants

The rate of a reaction increases when the concentrations of the reactants increases. When there are more reacting molecules, more collisions can occur, and the reaction goes faster. (See Figure 9.4.) For example, a person having difficulty breathing may be given oxygen. The increase in the number of oxygen molecules in the lungs increases the rate at which oxygen combines with hemoglobin and helps the patient breathe more easily.

Catalysts

Another way to speed up a reaction is to lower the energy of activation. We saw that the energy of activation is the energy needed to break apart the bonds of the reacting molecules. If a collision provides less than the activation energy, the bonds do not break apart and the molecules bounce apart. A **catalyst** speeds up a reaction by providing an alternate pathway that has a lower energy of activation. When activation energy is lowered, more collisions provide sufficient energy for reactants to form product. During a reaction, a catalyst is not changed or used up.

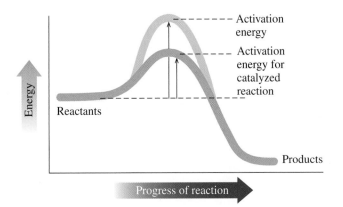

Catalysts have many uses in industry. In the manufacturing of margarine, hydrogen (H_2) is added to vegetable oils. Normally, the reaction is very slow because it has a high activation energy. However, when platinum (Pt) is used as a catalyst, the reaction occurs rapidly. In the body, biocatalysts called enzymes make most metabolic reactions proceed at rates necessary for proper cellular activity. A summary of the factors affecting reaction rates is given in Table 9.1.

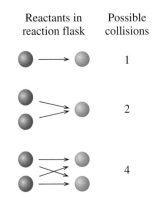

FIGURE 9.4

Increasing the concentration of a reactant increases the number of collisions that are possible.

Q **How many collisions are possible if one more red reactant is added?**

TABLE 9.1

Factors That Increase Reaction Rate

Factor	Reason
More reactants	More collisions
Higher temperature	More collisions, more collisions with energy of activation
Adding a catalyst	Lowers energy of activation

SAMPLE PROBLEM 9.1

■ Factors That Affect the Rate of Reaction

Indicate whether the following changes will increase, decrease, or have no effect upon the rate of reaction:

a. increase in temperature
b. decrease in the number of reactants
c. adding a catalyst

SOLUTION

a. increase
b. decrease
c. increase

STUDY CHECK

How does the lowering of temperature affect the rate of reaction?

QUESTIONS AND PROBLEMS

■ Rates of Reactions

9.1 **a.** What is meant by the rate of a reaction?
  b. Why does bread grow mold more quickly at room temperature than in the refrigerator?

9.2 **a.** How does a catalyst affect the activation energy?
  b. Why is pure oxygen used in cases of respiratory distress?

9.3 In the following reaction, what happens to the number of collisions when more Br_2 molecules are added?

$$H_2(g) + Br_2(g) \longrightarrow 2HBr(g)$$

9.4 In the following reaction, what happens to the number of collisions when the temperature of the reaction is decreased?

$$H_2(g) + Br_2(g) \longrightarrow 2HBr(g)$$

9.5 How would each of the following change the rate of the reaction shown here?

$$2SO_2(g) + O_2(g) \longrightarrow 2SO_3(g)$$

a. adding SO_2
b. raising the temperature
c. adding a catalyst
d. removing some SO_2

9.6 How would each of the following change the rate of the reaction shown here?

$$2NO(g) + 2H_2(g) \longrightarrow N_2(g) + 2H_2O(g)$$

a. adding more NO
b. lowering the temperature
c. removing some H_2
d. adding a catalyst

Catalytic Converters

For many years, manufacturers have been required to include catalytic converters on automobile engines. When gasoline burns, the products found in the exhaust of a car contain high levels of pollutants. These include carbon monoxide (CO) from incomplete combustion, hydrocarbons such as C_7H_{16} from unburned fuel, and nitrogen oxide (NO) from the reaction of N_2 and O_2 at the high temperatures reached within the engine. Carbon monoxide is toxic, and nitrogen oxide is involved in the formation of smog and acid rain.

The purpose of a catalytic converter is to lower the activation energy for reactions that convert each of these pollutants into substances such as CO_2, N_2, O_2, and H_2O, which are already present in the atmosphere.

$$2CO(g) + O_2(g) \longrightarrow 2CO_2(g)$$
$$C_7H_{16}(g) + 11O_2(g) \longrightarrow 7CO_2(g) + 8H_2O(g)$$
$$2NO(g) \longrightarrow N_2(g) + O_2(g)$$

A catalytic converter consists of solid-particle catalysts, such as platinum (Pt) and palladium (Pd), on a ceramic honeycomb that provides a large surface area and facilitates contact with pollutants. As the pollutants pass through the converter,

they react with the catalysts. Today, we all use unleaded gasoline because lead interferes with the ability of the Pt and Pd catalysts in the converter to react with the pollutants.

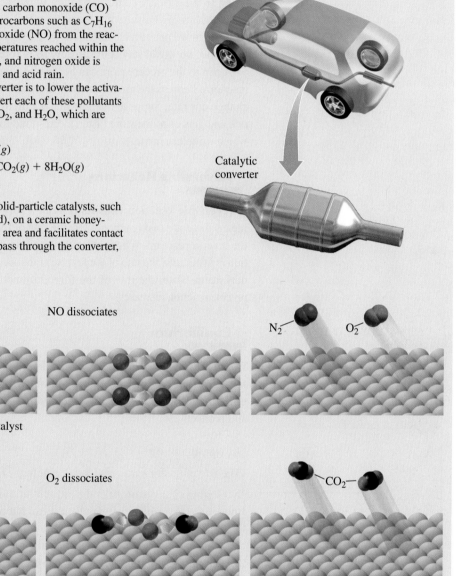

Catalytic converter

$$2NO(g) \longrightarrow N_2(g) + O_2(g)$$

NO absorbed on catalyst   NO dissociates

NO N_2 O_2

Surface of metal (Pt, Pd) catalyst

$$2CO(g) + O_2(g) \longrightarrow 2CO_2(g)$$

CO and O_2 absorbed on catalyst O_2 dissociates CO_2

Surface of metal (Pt, Pd) catalyst

9.2 Chemical Equilibrium

In earlier chapters, we considered the *forward reaction* in an equation and assumed that all of the reactants were converted to products. However, most of the time reactants are not completely converted to products because a *reverse reaction* takes place in which products come together and form the reactants. When a reaction

MC GOB Chemical Equilibrium

consists of both a forward and reverse direction, it is said to be reversible. We have looked at other reversible processes. For example, the melting of solids to form liquids and the freezing of liquids to solids is a reversible physical change. Even in our daily life we have reversible events. We go from home to school and we return from school to home. We go up an escalator and come back down. We put money in our bank account and take money out.

An analogy for a forward and reverse reaction can be found in the phrase "I am going to the grocery store." Although we mention our trip in one direction, we know that we will also return home from the grocery store. Because our trip has both a forward and reverse direction, we can say the trip is reversible. It is not very likely that we would stay at the grocery store forever.

A trip to the grocery store can be used to illustrate another aspect of reversible reactions. Perhaps the grocery store is close and we usually walk. However, we can change our rate. Suppose one day that we drive to the store, which increases our rate and gets us to the store faster. Correspondingly, a car also increases the rate at which we return home.

Reversible Reactions

A **reversible reaction** consists of both a forward and a reverse reaction. That means there are two reaction rates: the rate of the forward reaction and the rate of the reverse reaction. When molecules begin to react, the rate of the forward reaction is faster than the rate of the reverse reaction. As reactants are used up and products accumulate, the rate of the forward reaction decreases and the rate of the reverse reaction increases.

MC GOB SELF-STUDY ACTIVITY
Equilibrium

Equilibrium

Eventually, the rates of the forward and reverse reactions are equal; the reactants form products as often as products form reactants. A reaction reaches **chemical equilibrium** when there is no further change in the concentrations of the reactants and products.

> **At equilibrium:**
>
> The rate of the forward reaction is equal to the rate of the reverse reaction.
>
> No further changes occur in the concentrations of reactants and products, even though the two reactions continue at equal but opposite rates.

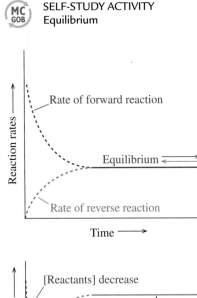

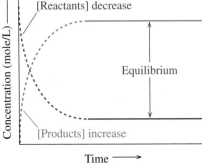

Let us look at the process as the reaction of N_2 and O_2 proceeds to equilibrium. (See Figure 9.5.) Initially, only N_2 and O_2 are present. Soon, a few molecules of NO are produced by the forward reaction. With more time, additional NO molecules are produced. As the concentration of NO increases, more NO molecules collide and react in the reverse direction.

Forward reaction: $N_2(g) + O_2(g) \longrightarrow 2NO(g)$

Reverse reaction: $2NO(g) \longrightarrow N_2(g) + O_2(g)$

As NO product builds up, the rate of the reverse reaction increases while the rate of the forward reaction decreases. Eventually the rates become equal, which means the reaction has reached equilibrium. Even though the concentrations remain constant at equilibrium, the forward and reverse reactions continue to occur. The forward and reverse reactions are usually shown together in a single equation by

$$N_2(g) + O_2(g) \rightleftharpoons 2NO(g)$$

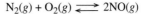

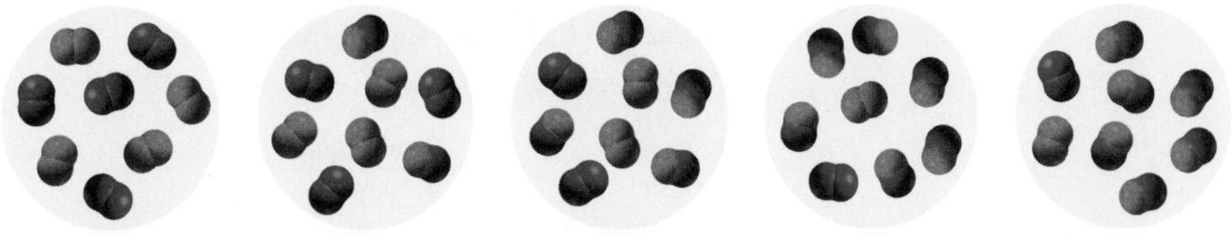

	(a)	(b)	(c)	(d)	(e)
Time (hours)	0	1	2	3	24
Concentration of reactants	8	6	4	2	2
Concentration of products	0	2	4	6	6
Rates of forward and reverse reactions					

FIGURE 9.5

(a) Initially, the reaction flask contains only the reactants N_2 and O_2. (b) The forward reaction between O_2 and N_2 begins to produce 2NO. (c) As the reaction proceeds, there are fewer molecules of O_2 and N_2 and more molecules of NO, which increases the rate of the reverse reaction. (d) At equilibrium, the concentrations of reactants N_2 and O_2 and product NO are constant. (e) The reaction continues with the rate of the forward reaction equal to the rate of the reverse reactions.

Q How do the rates of the forward and reverse reactions compare once a chemical reaction reaches equilibrium?

using a double arrow. A reversible reaction is two opposing reactions that occur at the same time.

$$N_2(g) + O_2(g) \quad \underset{\text{Reverse reaction}}{\overset{\text{Forward reaction}}{\rightleftharpoons}} \quad 2NO(g)$$

SAMPLE PROBLEM 9.2

■ Reversible Reactions

Write the forward and reverse reactions for each of the following:

a. $N_2(g) + 3H_2(g) \rightleftharpoons 2NH_3(g)$ **b.** $2CO(g) + O_2 \rightleftharpoons 2CO_2(g)$

SOLUTION

The equations are separated into forward and reverse reactions.

a. Forward reaction: $N_2(g) + 3H_2(g) \longrightarrow 2NH_3(g)$
     Reverse reaction: $2NH_3(g) \longrightarrow N_2(g) + 3H_2(g)$
b. Forward reaction: $2CO(g) + O_2(g) \longrightarrow 2CO_2(g)$
     Reverse reaction: $2CO_2(g) \longrightarrow O_2(g) + 2CO(g)$

STUDY CHECK

Write the equation for the reaction that contains the following reverse reaction:

$$2HBr(g) \longrightarrow H_2(g) + Br_2(g)$$

SAMPLE PROBLEM 9.3

■ **Reaction Rates and Equilibrium**

Complete each of the following with

1. equal	**2.** not equal
3. faster	**4.** slower
5. do not change	**6.** change

a. Before equilibrium is reached, the concentrations of the reactants and products _____.

b. Initially, reactants placed in a container have a _____ rate of reaction than the rate of reaction of the products.

c. At equilibrium, the rate of the forward reaction is _____ to the rate of the reverse reaction.

d. At equilibrium, the concentrations of the reactants and products _____.

SOLUTION

a. 6   **b.** 3   **c.** 1   **d.** 5

STUDY CHECK

Using the choice of answers in Sample Problem 9.3, complete the following:
As reactants are used up and products accumulate, the rate of the forward reaction becomes _____, while the rate of the reverse reaction becomes _____.

QUESTIONS AND PROBLEMS

■ **Chemical Equilibrium**

9.7   What is meant by the term "reversible reaction"?

9.8   When does a reversible reaction reach equilibrium?

9.9   Which of the following processes are reversible?
   a. breaking a glass
   b. melting snow
   c. heating a pan

9.10 Which of the following processes are at equilibrium?
   a. opposing rates of reaction are equal
   b. concentrations of reactants and products are equal
   c. the concentrations of reactants and products do not change

LEARNING GOAL

■ Calculate the equilibrium constant for a reversible reaction given the concentrations of reactants and products at equilibrium.

9.3 Equilibrium Constants

At equilibrium, reactions occur in opposite directions at the same rate, which means the concentrations of the reactants and products remain constant. We can use a ski lift as an analogy. Early in the morning, skiers at the bottom of the mountain begin to ride the ski lift up to the slopes. As skiers reach the top of the mountain, they ski down. Eventually, the number of people riding up the ski lift becomes equal to the number of people skiing down the mountain. There is no further change in the number of skiers on the slopes; the system is at equilibrium.

Equilibrium Constant Expression

Because the concentrations in a reaction at equilibrium no longer change, they can be used to set up a relationship between the products and the reactants. Suppose we write a general equation for reactants A and B that form products C and D. The small italic letters are the coefficients in the balanced equation.

$$aA + bB \rightleftharpoons cC + dD$$

An **equilibrium constant expression** for the reaction multiplies the concentrations of the products together and divides by the concentrations of the reactants. Each concentration is raised to a power that is its coefficient in the balanced chemical reaction. The square bracket around each substance indicates the concentration is expressed in moles per liter (M). The **equilibrium constant**, K_c, is the numerical value obtained by substituting molar concentrations at equilibrium into the expression. For our general reaction, the equilibrium constant expression is

Equilibrium constant   Equilibrium constant expression

$$K_c = \frac{\text{Products}}{\text{Reactants}} = \frac{[C]^c\,[D]^d}{[A]^a\,[B]^b} \quad \text{Coefficients}$$

We can write the equilibrium constant expression for any reaction if we know the balanced chemical equation. For the reaction of H_2 and I_2

$$H_2(g) + I_2(g) \rightleftharpoons 2HI(g)$$

the equilibrium constant expression is written as

$$K_c = \frac{[HI]^2}{[H_2][I_2]}$$

Guide to Writing the K_c Expression

STEP 1
Write the balanced equilibrium equation.

STEP 2
Write the products in brackets as the numerator and reactants in brackets as the denominator. Do not include pure solids or liquids.

STEP 3
Write the coefficient of each substance in the equation as an exponent.

SAMPLE PROBLEM 9.4

■ Writing Equilibrium Constant Expressions

Write the equilibrium constant expression for each of the following:

a. $2SO_2(g) + O_2(g) \rightleftharpoons 2SO_3(g)$
b. $3H_2(g) + N_2(g) \rightleftharpoons 2NH_3(g)$

SOLUTION

a. In the K_c expression, the products are in the numerator and the reactants are in the denominator. The numerator is the product SO_3 placed in brackets raised to the power of 2, its coefficient in the equation. The reactants SO_2 and O_2 are each placed in brackets in the denominator, with $[SO_2]$ raised to a power of 2.

$$K_c = \frac{[SO_3]^2}{[SO_2]^2\,[O_2]}$$

b. The numerator is the product NH_3 in brackets raised to the power of 2, its coefficient in the equation. The reactants H_2 and N_2 are each placed in brackets in the denominator with a power of 3 for H_2.

$$K_c = \frac{[NH_3]^2}{[H_2]^3\,[N_2]}$$

MC GOB   Equilibrium Constant

Write the balanced chemical equation that would give the following equilibrium constant expression.

$$K_c = \frac{[NO_2]^2}{[NO]^2 [O_2]}$$

Heterogeneous Equilibrium

Up to now, our examples have been reactions that involve only gases. A reaction in which all the reactants and products are in the same state reaches **homogenous equilibrium**. When the reactants and products are in two or more states the equilibrium is termed a **heterogeneous equilibrium**. For example, in the following reaction, the decomposition of calcium carbonate reaches heterogeneous equilibrium with calcium oxide and carbon dioxide. (See Figure 9.6.)

$$CaCO_3(s) \rightleftharpoons CaO(s) + CO_2(g)$$

In contrast to gases, the concentrations of pure solids and pure liquids in a heterogeneous equilibrium are constant; they do not change. Therefore, pure solids and liquids are not included in the equilibrium constant expression. For this heterogeneous equilibrium, the K_c expression does not include the concentration of $CaCO_3(s)$ or $CaO(s)$. It is written as

$$K_c = [CO_2]$$

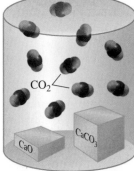
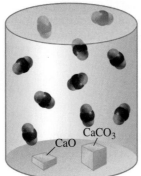

$T = 800°C$ $T = 800°C$

$$CaCO_3(s) \rightleftharpoons CaO(s) + CO_2(g)$$

FIGURE 9.6

At equilibrium at constant temperature, the concentration of CO_2 is the same regardless of the amounts of $CaCO_3(s)$ and $CaO(s)$ in the container.

Q **Why are the concentrations of $CaO(s)$ and $CaCO_3(s)$ not included in K_c for the decomposition of $CaCO_3$?**

SAMPLE PROBLEM 9.5

■ **Heterogeneous Equilibrium Constant Expression**

Write the equilibrium constant expression for the following heterogeneous equilibria:

a. $Si(s) + 2Cl_2(g) \rightleftharpoons SiCl_4(g)$
b. $2Mg(s) + O_2(g) \rightleftharpoons 2MgO(s)$

SOLUTION

In the equilibrium constant expressions for heterogeneous reactions, the concentrations of the pure solids are not included.

a. $K_c = \dfrac{[SiCl_4]}{[Cl_2]^2}$ **b.** $K_c = \dfrac{1}{[O_2]}$

STUDY CHECK

Solid iron(II) oxide and carbon monoxide gas are in equilibrium with solid iron and carbon dioxide gas. Write the equation and the equilibrium constant expression for the reaction.

Calculating Equilibrium Constants

The numerical value of the equilibrium constant is calculated from the equilibrium constant expression by substituting experimentally measured concentrations of the reactants and products at equilibrium into the equilibrium constant

TABLE 9.2

Equilibrium Constant for $H_2(g) + I_2(g) \rightleftharpoons 2HI(g)$ at 427°C				
Experiment	**$[H_2]$**	**$[I_2]$**	**$[HI]$**	$K_c = \dfrac{[HI]^2}{[H_2][I_2]}$
1	0.10 M	0.20 M	1.04 M	54
2	0.20 M	0.20 M	1.47	54
3	0.30 M	0.17 M	1.66	54

expression. For example, the equilibrium constant expression for the reaction of H_2 and I_2 is written

$$H_2(g) + I_2(g) \rightleftharpoons 2HI(g) \qquad K_c = \frac{[HI]^2}{[H_2][I_2]}$$

In an experiment, the molar concentrations for the reactants and products at equilibrium are found to be $[H_2] = 0.10$ M, $[I_2] = 0.20$ M, and $[HI] = 1.04$ M. When we substitute these values into the equilibrium constant expression, we obtain the numerical value of the equilibrium constant.

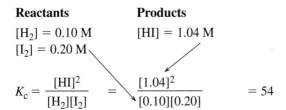

Reactants   **Products**

$[H_2] = 0.10$ M $[HI] = 1.04$ M

$[I_2] = 0.20$ M

$$K_c = \frac{[HI]^2}{[H_2][I_2]} = \frac{[1.04]^2}{[0.10][0.20]} = 54$$

Suppose we look at different equilibrium concentrations of reactants and products for the H_2, I_2, and HI system at 700. K (427°C). When the concentrations of reactants and products are measured in each equilibrium sample and used to calculate the K_c for the reaction, the same value of K_c is obtained for each. (See Table 9.2.) Thus, a reaction at a specific temperature can have only one value for the equilibrium constant.

The units of K_c depend on the specific equation. In the example of $[H_2]$, $[I_2]$, and $[HI]$, K_c has the units of $[M]^2/[M]^2$, which results in a value with no units. However, in the following example, the $[M]$ units would not cancel because $[M]^2/[M]^4 = 1/[M]^2 = [M]^{-2}$. However, we usually do not attach any units to the value of K_c. At 500. K, the value of K_c for the following reaction is 1.7×10^2.

$$N_2(g) + 3H_2(g) \rightleftharpoons 2NH_3(g)$$

$$K_c = \frac{[NH_3]^2}{[N_2][H_2]^3} = 1.7 \times 10^2 \ [M]^{-2} \text{ or usually written as } 1.7 \times 10^2$$

SAMPLE PROBLEM 9.6

■ **Calculating an Equilibrium Constant**

The decomposition of dinitrogen tetroxide forms nitrogen dioxide.

$$N_2O_4(g) \rightleftharpoons 2NO_2(g)$$

What is the value of K_c at 100.°C if a reaction mixture at equilibrium contains $[N_2O_4] = 0.45$ M and $[NO_2] = 0.31$ M?

SOLUTION

Given reactant: $[N_2O_2] = 0.45$ M   product: $[NO_2] = 0.31$ M
Need K_c value

- **STEP 1** **The equilibrium expression is written using any coefficient in the balanced equation as an exponent of the molar concentration.**

$$K_c = \frac{[NO_2]^2}{[N_2O_4]}$$

- **STEP 2** **Substitute the molar concentrations given for the equilibrium mixture into the equilibrium constant expression and calculate the K_c value.**

$$K_c = \frac{[0.31]^2}{[0.45]} = \frac{0.31 \text{ M} \times 0.31 \text{ M}}{0.45 \text{ M}} = 0.21 \text{ M or } 0.21$$

Guide to Calculating the K_c Value

STEP 1
Write the K_c expression for the equilibrium.

STEP 2
Substitute equilibrium (molar) concentrations and calculate K_c.

STUDY CHECK

Ammonia decomposes when heated to give nitrogen and hydrogen.

$$2NH_3(g) \rightleftharpoons 3H_2(g) + N_2(g)$$

Calculate the equilibrium constant if an equilibrium mixture contains $[NH_3] = 0.040$ M, $[N_2] = 0.20$ M, and $[H_2] = 0.60$ M.

QUESTIONS AND PROBLEMS

MC GOB

- **Equilibrium Constants**

9.11 Write the equilibrium constant expression, K_c, for each of the following reactions:
  a. $CH_4(g) + 2H_2S(g) \rightleftharpoons CS_2(g) + 4H_2(g)$
  b. $2NO(g) \rightleftharpoons N_2(g) + O_2(g)$
  c. $2SO_3(g) + CO_2(g) \rightleftharpoons CS_2(g) + 4O_2(g)$

9.12 Write the equilibrium constant expression K_c for each of the following reactions:
  a. $2HBr(g) \rightleftharpoons H_2(g) + Br_2(g)$
  b. $CO(g) + 2H_2(g) \rightleftharpoons CH_3OH(g)$
  c. $CH_4(g) + H_2O(g) \rightleftharpoons CO(g) + 3H_2(g)$

9.13 Identify each of the following as a homogeneous or heterogeneous equilibrium.
  a. $2O_3(g) \rightleftharpoons 3O_2(g)$
  b. $2NaHCO_3(s) \rightleftharpoons Na_2CO_3(s) + CO_2(g) + H_2O(g)$
  c. $CH_4(g) + H_2O(g) \rightleftharpoons 3H_2(g) + CO(g)$
  d. $4HCl(g) + O_2(g) \rightleftharpoons 2H_2O(l) + 2Cl_2(g)$

9.14 Identify each of the following as a homogeneous or heterogeneous equilibrium.
  a. $CO(g) + H_2(g) \rightleftharpoons C(s) + H_2O(g)$
  b. $CO(g) + 2H_2(g) \rightleftharpoons CH_3OH(l)$
  c. $CS_2(g) + 4H_2(g) \rightleftharpoons CH_4(g) + 2H_2S(g)$
  d. $Br_2(g) + Cl_2(g) \rightleftharpoons 2BrCl(g)$

9.15 Write the equilibrium constant expression for each of the reactions in problem 9.13.

9.16 Write the equilibrium constant expression for each of the reactions in problem 9.14.

9.17 What is the K_c for the following equilibrium

$$N_2O_4(g) \rightleftharpoons 2NO_2(g)$$

if $[NO_2] = 0.21$ M and $[N_2O_4] = 0.030$ M?

9.18 What is the K_c for the following equilibrium

$$CO_2(g) + H_2(g) \rightleftharpoons CO(g) + H_2O(g)$$

if [CO] = 0.20 M, [H$_2$O] = 0.30 M, [CO$_2$] = 0.30 M, and [H$_2$] = 0.033 M?

9.19 What is the K_c for the following equilibrium at 1000°C,

$$CO(g) + 3H_2(g) \rightleftharpoons CH_4(g) + H_2O(g)$$

[H$_2$] = 0.30 M, [CO] = 0.51 M, [CH$_4$] = 1.8 M, and [H$_2$O] = 2.0 M?

9.20 What is the K_c for the following equilibrium at 500°C,

$$N_2(g) + 3H_2(g) \rightleftharpoons 2NH_3(g)$$

[H$_2$] = 0.40 M, [N$_2$] = 0.44 M, [NH$_3$] = 2.2 M?

9.4 Using Equilibrium Constants

LEARNING GOAL

Use an equilibrium constant to predict the extent of reaction and to calculate equilibrium concentrations.

We have seen that the values of K_c can be large or small. We can now look at the K_c values to predict how far the reaction proceeds to products at equilibrium. When a K_c is large, the numerator (products) is greater than the concentrations of the reactants in the denominator.

$$\frac{[\text{Products}]}{[\text{Reactants}]} = \text{Large } K_c$$

When a K_c is small, the numerator is smaller than the denominator, which means that the reaction favors the reactants.

$$\frac{[\text{Products}]}{[\text{Reactants}]} = \text{Small } K_c$$

Using a general reaction and its equilibrium constant expression, we can look at the relative concentrations of reactant A and product B.

$$A(g) \rightleftharpoons B(g) \qquad K_c = \frac{[B]}{[A]}$$

For a large K_c, [B] is greater than [A]. For example, if the K_c is 1×10^3, or 1000, [B] would be 1000 times greater than [A] at equilibrium.

$$K_c = \frac{[B]}{[A]} = 1000 \quad \text{or rearranged} \quad [B] = 1000\,[A]$$

For a small K_c, [A] is greater than [B]. For example, if the K_c is 1×10^{-2}, [A] is 100 times greater than [B] at equilibrium.

$$K_c = \frac{[B]}{[A]} = \frac{1}{100} \quad \text{or rearranged} \quad [A] = 100\,[B]$$

Equilibrium mixtures with different values of K_c are illustrated in Figure 9.7.

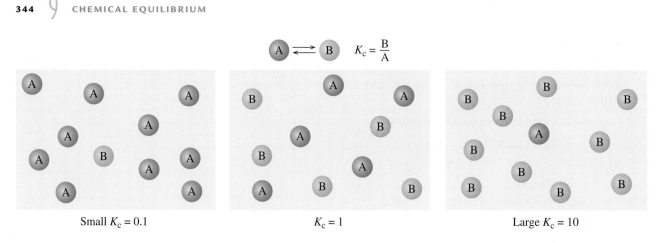

Small $K_c = 0.1$ $K_c = 1$ Large $K_c = 10$

FIGURE 9.7

A reaction with $K_c < 1$ contains a higher concentration of reactants [A] than products [B]. A reaction with K_c of about 1.0 has about the same concentrations of products [B] as reactants [A]. A reaction with $K_c > 1$ has a higher concentration of products [B] than reactants [A].

Q **Does a reaction in which [A] = 100[B] at equilibrium have a K_c greater than, about equal to, or less than 1?**

Equilibrium with a Large K_c

A reaction with a large K_c forms a substantial amount of product by the time equilibrium is established. The greater the value of K_c, the more the equilibrium favors the products. A reaction with a very large K_c essentially goes to completion to give mostly products. Consider the reaction of SO_2 and O_2, which has a large K_c. At equilibrium, the reaction mixture contains mostly product and very little reactant.

$$2SO_2(g) + O_2(g) \rightleftharpoons 2SO_3(g)$$

$$K_c = \frac{[SO_3]^2}{[SO_2]^2\,[O_2]} \quad \frac{\text{Mostly product}}{\text{Little reactant}} = 3.4 \times 10^2 \qquad \text{Reaction favors products}$$

We can start the reaction with only the reactants SO_2 and O_2 or we can start the reaction with just the products SO_3. (See Figure 9.8.) In one reaction, SO_2 and O_2

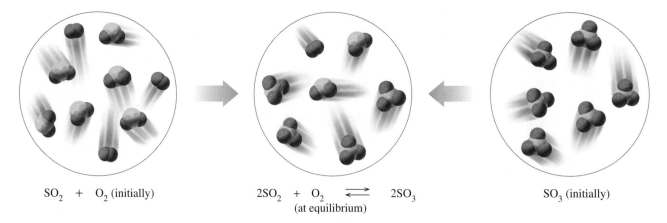

SO_2 + O_2 (initially) $2SO_2$ + O_2 $\rightleftharpoons$ $2SO_3$ SO_3 (initially)
                                      (at equilibrium)

FIGURE 9.8

One sample initially contains SO_2 and O_2, while another sample contains only SO_3. At equilibrium, mostly SO_3 and only small amounts of SO_2 and O_2 are present in the equilibrium mixture.

Q **Why is the same equilibrium mixture obtained from reactants as from products?**

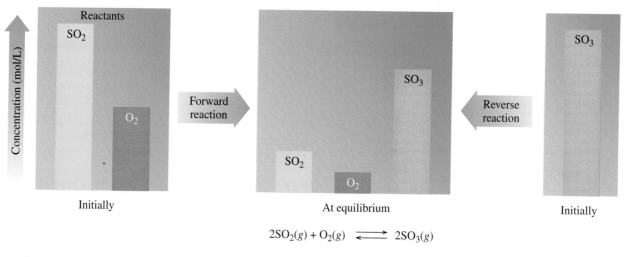

$$2SO_2(g) + O_2(g) \rightleftharpoons 2SO_3(g)$$

FIGURE 9.9

In the reaction of SO_2 and O_2, the equilibrium favors the formation of product SO_3, which results in a large K_c.

Q **Why is an equilibrium mixture obtained after starting with pure SO_3?**

form SO_3 and in the other, SO_3 reacts to form SO_2 and O_2. However, in both equilibrium mixtures, the concentration of SO_3 is much higher than the concentrations of SO_2 and O_2. (See Figure 9.9.) Because there is more product than reactant at equilibrium, the energy of activation for the forward reaction must be lower than the energy of activation for the reverse reaction.

Equilibrium with a Small K_c

For a reaction with a small K_c, the equilibrium mixture contains very small concentrations of products. Consider the reaction for the formation of NO, which has a small K_c. (See Figure 9.10.)

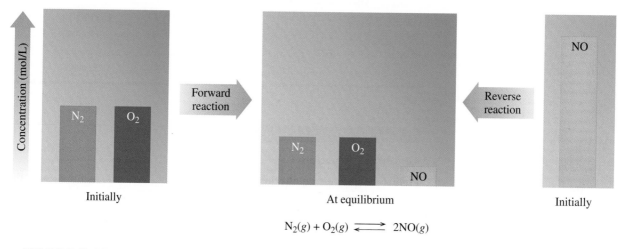

$$N_2(g) + O_2(g) \rightleftharpoons 2NO(g)$$

FIGURE 9.10

At equilibrium, the reaction $N_2 + O_2 \rightleftharpoons 2NO$ favors the reactant and the reaction mixture at equilibrium contains mostly N_2 and O_2, which results in a small K_c.

Q **Starting with only NO in a closed container, how do the forward and reverse reactions change as equilibrium is reached?**

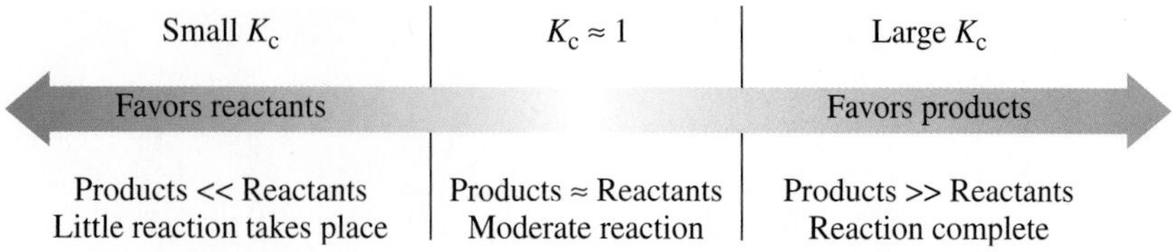

Small K_c	$K_c \approx 1$	Large K_c
← Favors reactants		Favors products →
Products << Reactants Little reaction takes place	Products ≈ Reactants Moderate reaction	Products >> Reactants Reaction complete

FIGURE 9.11

The equilibrium constant K_c indicates how far a reaction goes to products. A reaction with a large K_c contains mostly products; a reaction with a small K_c contains mostly reactants.

Q **Does a reaction with a $K_c = 1.2 \times 10^{15}$ contain mostly reactants or products at equilibrium?**

$$N_2(g) + O_2(g) \rightleftharpoons 2NO(g)$$

$$K_c = \frac{[NO]^2}{[N_2][O_2]} \frac{\text{Little product}}{\text{Mostly reactant}} = 2 \times 10^{-9} \quad \text{Reaction favors reactants}$$

Whether the reaction begins with only the reactants, N_2 and O_2, or with the product NO, the equilibrium mixture contains mostly reactant and very little product. The energy of activation for the forward reaction is much greater than the energy of activation for the reverse reaction. Reactions with very small K_c produce essentially no product.

Reactions with equilibrium constants close to 1 have about the same concentrations of reactants and products. (See Figure 9.11.) Table 9.3 lists some equilibrium constants and the extent of their reaction.

TABLE 9.3

Examples of Reactions with Large and Small K_c Values

Reactants	Products	K_c	Equilibrium Favors
$2CO(g) + O_2(g) \rightleftharpoons$	$2CO_2(g)$	2×10^{11}	Products
$2H_2(g) + S_2(g) \rightleftharpoons$	$2H_2S(g)$	1.1×10^7	Products
$N_2(g) + 3H_2(g) \rightleftharpoons$	$2NH_3(g)$	1.6×10^2	Products
$PCl_5(g) \rightleftharpoons$	$PCl_3(g) + Cl_2(g)$	1.2×10^{-2}	Reactants
$N_2(g) + O_2(g) \rightleftharpoons$	$2NO(g)$	2×10^{-9}	Reactants

SAMPLE PROBLEM 9.7

■ **Extent of Reaction**

Predict whether the equilibrium favors the reactants or products for each of the following reactions:

a. $2H_2(g) + O_2(g) \rightleftharpoons 2H_2O(g)$ $K_c = 2.9 \times 10^{82}$ at 25°C
b. $PCl_5(g) \rightleftharpoons PCl_3(g) + Cl_2(g)$ $K_c = 1.2 \times 10^{-2}$ at 225°C

SOLUTION

a. A large K_c indicates that the equilibrium favors the products.
b. A small K_c indicates that the equilibrium favors the reactants.

STUDY CHECK

The equilibrium constant for the reaction

$$2CO_2(g) \rightleftharpoons 2CO(g) + O_2(g)$$

is 5×10^{-12}. In the reaction mixture, is the concentration of the reactants much greater, much smaller, or about the same as the products?

Calculating Concentrations at Equilibrium

When a reaction goes essentially all to products, we can use the mole factors we studied earlier to calculate the quantity of a product. However, many reactions reach equilibrium without using up all the reactants. If this is the case, then we need to use the equilibrium constant to calculate the amount of product that can be formed in the reaction. For example, if we know the equilibrium constant for a reaction and all the concentrations except one, we can calculate the unknown concentration using the equilibrium constant expression.

(MC GOB) Calculations Using the Equilibrium Constant

SAMPLE PROBLEM 9.8

Calculating Concentration Using an Equilibrium Constant

Phosgene ($COCl_2$) is a toxic substance that is produced by the reaction of carbon monoxide and chlorine; the K_c for the reaction is 5.0.

$$CO(g) + Cl_2(g) \rightleftharpoons COCl_2(g)$$

If the equilibrium concentrations for the reaction are $[Cl_2] = 0.25$ M and $[COCl_2] = 0.80$ M, what is the equilibrium concentration of $CO(g)$?

SOLUTION

Given $[Cl_2] = 0.25$ M, $[COCl_2] = 0.80$ M, $K_c = 5.0$
Need [CO]

■ **STEP 1 Write the K_c expression for the equilibrium.** Using the balanced chemical equation, the equilibrium constant expression is written as

$$K_c = \frac{[COCl_2]}{[CO][Cl_2]}$$

■ **STEP 2 Solve the K_c expression for the unknown concentration.** To rearrange the expression for [CO], we first multiply both sides by $[CO][Cl_2]$.

$$K_c[CO][Cl_2] = \frac{[COCl_2] \times [CO][Cl_2]}{[CO][Cl_2]}$$

$$K_c[CO][Cl_2] = [COCl_2]$$

Dividing both sides by $K_c[Cl_2]$ solves the equilibrium constant expression for [CO].

$$\frac{K_c[CO][Cl_2]}{K_c[Cl_2]} = \frac{[COCl_2]}{K_c[Cl_2]}$$

$$[CO] = \frac{[COCl_2]}{K_c[Cl_2]}$$

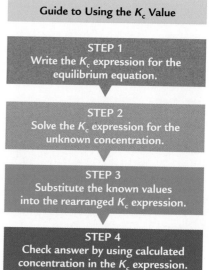

Guide to Using the K_c Value

STEP 1
Write the K_c expression for the equilibrium equation.

STEP 2
Solve the K_c expression for the unknown concentration.

STEP 3
Substitute the known values into the rearranged K_c expression.

STEP 4
Check answer by using calculated concentration in the K_c expression.

■ **STEP 3** **Substitute the known values into the rearranged K_c expression.** Substituting the concentrations for the equilibrium mixture and the K_c value in the equilibrium constant expression solved for [CO] produces a concentration for [CO].

$$[CO] = \frac{[COCl_2]}{K_c[Cl_2]}$$

$$[CO] = \frac{[0.80]}{5.0 \times [0.25]} = 0.64 \text{ M}$$

■ **STEP 4** **Check the answer.** We check the answer by substituting the calculated concentration into the K_c expression.

$$K_c = \frac{[COCl_2]}{[CO][Cl_2]}$$

$$K_c = \frac{[0.80]}{[0.64][0.25]} = 5.0$$
$\nearrow$ Calculated

STUDY CHECK

Ethanol can be produced by reacting ethylene (C_2H_4) with water vapor. At 327°C, the K_c is 9×10^3.

$$C_2H_4(g) + H_2O(g) \rightleftharpoons C_2H_5OH(g)$$

If an equilibrium mixture has concentrations of $[C_2H_4] = 0.020$ M and $[H_2O] = 0.015$ M, what is the equilibrium concentration of C_2H_5OH?

■

QUESTIONS AND PROBLEMS

(MC GOB)

■ Using Equilibrium Constants

9.21 Indicate whether each of the following equilibrium mixtures contain mostly products, mostly reactants, or both reactants and products:
  a. $Cl_2(g) + 2NO(g) \rightleftharpoons 2NOCl(g)$ $K_c = 3.7 \times 10^8$
  b. $H_2O(g) + CH_4(g) \rightleftharpoons CO(g) + 3H_2(g)$ $K_c = 4.7$
  c. $3O_2(g) \rightleftharpoons 2O_3$ $K_c = 1.7 \times 10^{-56}$

9.22 Indicate whether each of the following equilibrium mixtures contain mostly products, mostly reactants, or both reactants and products:
  a. $CO(g) + Cl_2(g) \rightleftharpoons COCl_2(g)$ $K_c = 5.0$
  b. $2HF(g) \rightleftharpoons H_2(g) + F_2(g)$ $K_c = 1.0 \times 10^{-95}$
  c. $2NO(g) + O_2 \rightleftharpoons 2NO_2(g)$ $K_c = 6.0 \times 10^{13}$

9.23 The equilibrium constant, K_c, for the equilibrium

$$H_2(g) + I_2(g) \rightleftharpoons 2HI(g)$$

is 54 at 425°C. If the equilibrium mixture contains 0.030 M HI and 0.015 M I_2, what is the equilibrium concentration of H_2?

9.24 The equilibrium constant, K_c, for the equilibrium

$$N_2O_4(g) \rightleftharpoons 2NO_2(g)$$

is 4.6×10^{-3}. If the equilibrium mixture contains 0.050 M NO_2, what is the concentration of N_2O_4?

9.25 The K_c at 100°C is 2.0 for the reaction

$$2NOBr(g) \rightleftharpoons 2NO(g) + Br_2(g)$$

If the system at equilibrium contains [NO] = 2.0 M and $[Br_2]$ = 1.0 M, what is the [NOBr]?

9.26 An equilibrium mixture at 225°C contains 0.14 M NH_3 and 0.18 M H_2 for the reaction

$$3H_2(g) + N_2(g) \rightleftharpoons 2NH_3(g)$$

If the K_c at this temperature is 1.7×10^2, what is the equilibrium concentration of N_2?

9.5 Changing Equilibrium Conditions: Le Châtelier's Principle

LEARNING GOAL

Use Le Châtelier's principle to describe the changes made in equilibrium concentrations when reaction conditions change.

We have seen that when a reaction reaches equilibrium, the rates of the forward and reverse reactions are equal and the concentrations remain constant. Now we will look at what happens to a system at equilibrium when changes occur in reaction conditions such as changes in temperature, concentration, and pressure.

Le Châtelier's Principle

 Le Châtelier Calculation

In the previous section, we saw that in a system at equilibrium the forward and reverse reactions occur at equal rates. Thus, at equilibrium, the concentrations of the substances do not change. However, any changes in the reaction conditions will disturb the equilibrium. Concentrations can change by adding or removing one of the substances, the volume (pressure) can change, or there may be a change in temperature. When we alter any of the conditions of a system at equilibrium, the rates of the forward and reverse reaction will no longer be equal. We say that a *stress* is placed on the equilibrium. We use Le Châtelier's principle to determine the direction that the equilibrium must shift to relieve that stress and reestablish equilibrium.

Le Châtelier's Principle

When a stress (change in conditions) is placed on a reaction at equilibrium, the equilibrium shifts in the direction that relieves the stress.

Effect of Concentration Changes

We will use the equilibrium for the reaction of PCl_5 to illustrate the stress caused by a change in concentration and how the system reacts to the stress. Consider the following reaction, which has a K_c of 0.042 at 250°C.

$$PCl_5(g) \rightleftharpoons PCl_3(g) + Cl_2(g)$$

For a reaction at a given temperature, there is only one equilibrium constant. Even if there are changes in the concentrations of the components, the K_c value does not change. What will change are the concentrations of the other components in the reaction in order to relieve the stress. For example, we can see that an equilibrium mixture that contains 1.20 M PCl_5, 0.20 M PCl_3, and 0.25 M Cl_2 has a K_c of 0.042.

$$K_c = \frac{[PCl_3][Cl_2]}{[PCl_5]} = \frac{[0.20][0.25]}{[1.20]} = 0.042$$

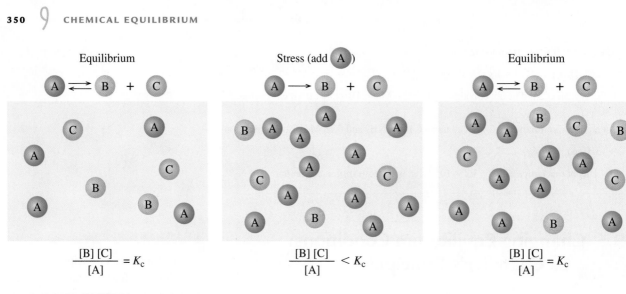

Equilibrium

$$A \rightleftharpoons B + C$$

$$\frac{[B][C]}{[A]} = K_c$$

Stress (add A)

$$A \longrightarrow B + C$$

$$\frac{[B][C]}{[A]} < K_c$$

Equilibrium

$$A \rightleftharpoons B + C$$

$$\frac{[B][C]}{[A]} = K_c$$

FIGURE 9.12

The addition of A places stress on the equilibrium of A $\rightleftharpoons$ B + C. To relieve the stress, the forward reaction converts some A to B + C and the equilibrium is reestablished.

Q **When C is added, does the equilibrium shift to products or reactants? Why?**

Suppose that now we add PCl_5 to the equilibrium mixture to increase $[PCl_5]$ to 2.00 M. If we substitute the concentrations into the equilibrium expression at this point, the ratio of products to reactants is 0.025, which is smaller than the K_c of 0.042.

$$\frac{\text{Products}}{\text{Reactants (added)}} \quad \frac{[PCl_3][Cl_2]}{[PCl_5]} = \frac{[0.20][0.25]}{[2.00]} = 0.025 < K_c$$

Because a K_c cannot change for a reaction at a given temperature, adding more PCl_5 places a stress on the system. (See Figure 9.12.) Forming more of the products can relieve this stress. According to Le Châtelier's principle, adding reactants causes the equilibrium to *shift* toward the products.

Add PCl_5

$$PCl_5(g) \rightleftharpoons PCl_3(g) + Cl_2(g)$$

In our experiment, equilibrium is reestablished with new concentrations of $[PCl_5] = 1.94$ M, $[PCl_3] = 0.26$, and $[Cl_2] = 0.31$ M. The resulting equilibrium mixture now contains more reactants and products, but their new concentrations in the equilibrium expression are once again equal to the K_c.

$$K_c = \frac{[PCl_3][Cl_2]}{[PCl_5]} = \frac{[0.26][0.31]}{[1.94]} = 0.042 = K_c \quad \text{New higher concentrations}$$

Suppose that in another experiment some PCl_5 is removed from the original equilibrium mixture, which lowers the $[PCl_5]$ to 0.76 M. Now the ratio of the products to the reactants is greater than the K_c value of 0.042. The removal of some of the reactant has placed a stress on the equilibrium.

$$\frac{\text{Product}}{\text{Reactant (removed)}} \quad \frac{[PCl_3][Cl_2]}{[PCl_5]} = \frac{[0.20][0.25]}{[0.76]} = 0.066 > K_c$$

In this case, the stress is relieved as the reverse reaction converts some products to reactants. Using Le Châtelier's principle, we see that removing some reactant *shifts* the equilibrium toward the reactants.

> Remove PCl_5
>
> $PCl_5(g) \rightleftharpoons PCl_3(g) + Cl_2(g)$

In this experiment, equilibrium is reestablished with new concentrations of $[PCl_5] = 0.80$ M, $[PCl_3] = 0.16$ M, and $[Cl_2] = 0.21$ M. The resulting equilibrium mixture now contains lower concentrations of reactants and products, but their new concentrations in the equilibrium expression once again are equal to the K_c.

$$K_c = \frac{[PCl_3][Cl_2]}{[PCl_5]} = \frac{[0.16][0.21]}{[0.80]} = 0.042 = K_c \quad \text{New lower concentrations}$$

There can also be changes in the concentrations of other components in this reaction. We could add or remove one of the products in this reaction. Suppose the $[Cl_2]$ is doubled, which makes the product/reactant ratio greater than K_c.

$$\frac{[PCl_3][Cl_2]}{[PCl_5]} = \frac{[0.20][0.50]}{[1.20]} = 0.083 > K_c$$

With an increase in the concentration of Cl_2, the rate of the reverse reaction increases and converts some of the products to reactants. Using Le Châtelier's principle, we see that the addition of a product causes a *shift* toward the reactants.

> Add Cl_2
>
> $PCl_5(g) \rightleftharpoons PCl_3(g) + Cl_2(g)$

On the other hand, we could remove some Cl_2, which would decrease $[Cl_2]$ and *shift* the equilibrium toward the products.

> Remove Cl_2
>
> $PCl_5(g) \rightleftharpoons PCl_3(g) + Cl_2(g)$

In summary, Le Châtelier's principle indicates that a stress caused by adding a substance at equilibrium is relieved by shifting the reaction away from that substance. When a substance is removed, the equilibrium shifts toward that substance. These features of Le Châtelier's principle are summarized in Table 9.4.

TABLE 9.4

Effect of Concentration Changes on Equilibrium $PCl_5(g) \rightleftharpoons PCl_3(g) + Cl_2(g)$

Stress	Shift	Equilibrium Changes		
		$PCl_5(g)$	$PCl_3(g)$	$Cl_2(g)$
Increase PCl_5	Toward products	added	more	more
Decrease PCl_5	Toward reactants	removed	less	less
Increase PCl_3	Toward reactants	more	added	less
Decrease PCl_3	Toward products	less	removed	more
Increase Cl_2	Toward reactants	more	less	added
Decrease Cl_2	Toward products	less	more	removed

Catalysts

Sometimes a catalyst is added to a reaction. Earlier we showed that a catalyst speeds up a reaction by lowering the activation energy. As a result, the rates of the forward and reverse reactions both increase. The time required to reach equilibrium is shorter, but the same ratios of products and reactants are attained. Therefore, a catalyst speeds up the forward and reverse reactions, but it has no effect on the equilibrium constant.

SAMPLE PROBLEM 9.9

Effect of Changes in Concentrations

Consider the effect of each of the following changes on the equilibrium

$$CO(g) + H_2O(g) \rightleftharpoons CO_2(g) + H_2(g)$$

a. Increasing [CO]   **b.** Increasing [H_2]   **c.** Decreasing [H_2O]
d. Decreasing [CO_2]   **e.** Adding a catalyst

SOLUTION

According to Le Châtelier's principle, equilibrium shifts to relieve the stress.

a. Increasing [CO] shifts the equilibrium to products.
b. Increasing [H_2] shifts the equilibrium to reactants.
c. Decreasing [H_2O] shifts the equilibrium to reactants.
d. Decreasing [CO_2] shifts the equilibrium to products.
e. Adding a catalyst does not cause a shift in the equilibrium.

STUDY CHECK

What is the effect of increasing [CO_2] on the equilibrium for the reaction in Sample Problem 9.9?

Effect of Volume (Pressure) Changes on Equilibrium

Reactions that involve gases exert pressure. Although the volume and therefore pressure can change, the value of the equilibrium constant does not change at a given temperature. Using the gas laws, we know that increasing the volume of the container decreases the pressure, while decreasing the volume increases the pressure.

According to Le Châtelier's principle, decreasing the number of moles of gas relieves the stress of increased pressure. This means that the reaction shifts toward the fewer number of moles. Let's look at the effect of decreasing the volume of the equilibrium mixture that originally contained 1.20 M PCl_5, 0.20 M PCl_3, and 0.25 M Cl_2 with a K_c of 0.042.

$$PCl_5(g) \rightleftharpoons PCl_3(g) + Cl_2(g)$$

$$K_c = \frac{[PCl_3][Cl_2]}{[PCl_5]} = \frac{[0.20][0.25]}{[1.20]} = 0.042$$

If we decrease the volume by half, all the molar concentrations are doubled. In the equation there are more moles of products than reactants, so there is an increase in the product/reactant ratio.

$$\frac{[PCl_3][Cl_2]}{[PCl_5]} = \frac{[0.40][0.50]}{[2.40]} = 0.083 > K_c$$

HEALTH NOTE

▪ Oxygen-Hemoglobin Equilibrium and Hypoxia

The transport of oxygen involves an equilibrium between hemoglobin (Hb), oxygen, and oxyhemoglobin.

$$Hb + O_2 \rightleftharpoons HbO_2$$

When the O_2 level is high in the alveoli of the lung, the reaction favors the product HbO_2. In the tissues where O_2 concentration is low, the reverse reaction releases the oxygen from the hemoglobin. The equilibrium expression is written

$$K_c = \frac{[HbO_2]}{[Hb][O_2]}$$

At normal atmospheric pressure, oxygen diffuses into the blood because the partial pressure of oxygen in the alveoli is higher than that in the blood. At altitudes above 8000 ft, the decrease in the amount of oxygen in the air results in a significant reduction of oxygen to the blood and body tissues. At an altitude of 18 000 feet, a person will obtain 29% less oxygen. When oxygen levels are lowered, a person may experience hypoxia, which has symptoms that include increased respiratory rate, headache, decreased mental acuteness, fatigue, decreased physical coordination, nausea, vomiting, and cyanosis. A similar problem occurs in persons with a history of lung

disease that impairs gas diffusion in the alveoli or in persons that have a reduced number of red blood cells, which occurs in smokers.

From the equilibrium expression, we see that a decrease in oxygen will shift the equilibrium to the reactants. Such a shift depletes the concentration of HbO_2 and causes the hypoxia condition.

$$Hb + O_2 \longleftarrow HbO_2$$

Immediate treatment of altitude sickness includes hydration, rest, and if necessary, descending to a lower altitude. The adaptation to lowered oxygen levels requires about 10 days. During this time the bone marrow increases red blood cell production providing hemoglobin. A person living at a high altitude can have 50% more red blood cells than someone at sea level. This increase in hemoglobin causes a shift in the equilibrium back toward HbO_2 product. Eventually the higher concentration of HbO_2 will provide more oxygen to the tissues and the symptoms of hypoxia will lessen.

$$Hb + O_2 \longrightarrow HbO_2$$

For some who climb high mountains, it is important to stop and acclimatize for several days at increasing altitudes. At very high altitudes, it may be necessary to use an oxygen tank.

To relieve the stress, the equilibrium shifts toward the reactants, which will reduce the mole/L of the products and increase the mole/L of the reactant. (See Figure 9.13.)

> Decrease V

$$PCl_5(g) \rightleftharpoons PCl_3(g) + Cl_2(g)$$
1 mole   2 moles

When equilibrium is reestablished, the new concentrations are $[PCl_5] = 2.52$ M, $[PCl_3] = 0.28$ M, and $[Cl_2] = 0.38$ M. The resulting equilibrium mixture contains new concentrations of reactants and products that are now equal to the K_c value.

At volume (1)   **At volume (2)**

$[PCl_3] = 0.20$ M   0.28 M

$[Cl_2]\ = 0.25$ M   0.38 M

$[PCl_5] = 1.20$ M   2.52 M

$$K_c = \frac{[PCl_3][Cl_2]}{[PCl_5]} = \frac{[0.20][0.25]}{[1.20]} = \frac{[0.28][0.38]}{[2.52]} = 0.042$$

On the other hand, when volume increases and pressure decreases, the reaction shifts toward the greater number of moles. Suppose that the volume is doubled. Then the molar concentrations of all the gases decrease by half. Because there

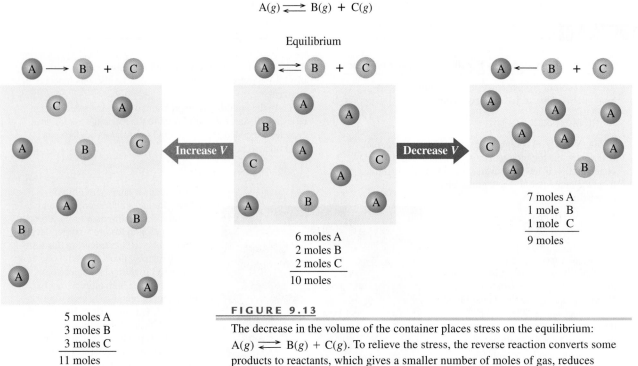

$$A(g) \rightleftharpoons B(g) + C(g)$$

5 moles A
3 moles B
3 moles C
——————
11 moles

6 moles A
2 moles B
2 moles C
——————
10 moles

7 moles A
1 mole B
1 mole C
——————
9 moles

FIGURE 9.13

The decrease in the volume of the container places stress on the equilibrium: $A(g) \rightleftharpoons B(g) + C(g)$. To relieve the stress, the reverse reaction converts some products to reactants, which gives a smaller number of moles of gas, reduces pressure, and reestablishes the equilibrium. When the volume increases, the forward reaction converts reactants to products to increase the moles of gas and relieve the stress.

Q **If you want to increase the products, would you increase or decrease the volume of the reaction container?**

are more moles of products than reactants, there is a decrease in the product/reactant ratio.

$$\frac{[PCl_3][Cl_2]}{[PCl_5]} = \frac{[0.10][0.13]}{[0.60]} = 0.022 < K_c$$

Now the equilibrium has to shift toward the products to relieve the stress; this will increase the concentrations (mole/L) of the products.

Increase V

$$PCl_5(g) \rightleftharpoons PCl_3(g) + Cl_2(g)$$
1 mole $\qquad$ 2 moles

When equilibrium is reestablished, the new concentrations are $[PCl_5] = 0.56$ M, $[PCl_3] = 0.14$ M, and $[Cl_2] = 0.17$ M. The resulting equilibrium mixture contains new concentrations of reactants and products that are now equal to the K_c value.

At volume (1)

$[PCl_3] = 0.20$ M
$[Cl_2]\ = 0.25$ M
$[PCl_5] = 1.20$ M

At volume (2)

0.14 M
0.17 M
0.56 M

$$K_c = \frac{[PCl_3][Cl_2]}{[PCl_5]} = \frac{[0.20][0.25]}{[1.20]} = \frac{[0.14][0.17]}{[0.56]} = 0.042$$

When a reaction has the same number of moles of reactants as products, a volume change does not affect the equilibrium. There is no effect on equilibrium because the molar concentrations of the reactants and products change in the same way. Consider the reaction of H_2 and I_2 to form HI, which has a K_c of 54.

$$H_2(g) + I_2(g) \rightleftharpoons 2HI(g)$$
$$\quad \text{2 moles} \qquad\qquad \text{2 moles}$$

Suppose we start with $[H_2] = 0.060$ M, $[I_2] = 0.015$ M, $[HI] = 0.22$ M.

$$\frac{[HI]^2}{[H_2][I_2]} = \frac{[0.22]^2}{[0.060][0.015]} = 54$$

If the volume is decreased by half, the pressure will double, and all the molar concentrations double. However, the equation has the same number of moles of products as reactants, so there is no effect on the equilibrium. The product/reactant ratio stays the same. We can see this by substituting the increased concentrations into the equilibrium constant expression.

At volume (1)	At volume (2) Concentrations doubled
[H1] = 0.22 M	0.44 M
$[H_2]$ = 0.060 M	0.120 M
$[I_2]$ = 0.015 M	0.030 M

$$K_c = \frac{[HI]^2}{[H_2][I_2]} = \frac{[0.22]^2}{[0.060][0.015]} = \frac{[0.44]^2}{[0.120][0.030]} = 54$$

SAMPLE PROBLEM 9.10

■ Effect of Changes in Volume

Indicate whether the effect of decreasing the volume for each of the following equilibria causes the number of moles of product to increase or decrease.

a. $C_2H_2(g) + 2H_2(g) \rightleftharpoons C_2H_6(g)$
b. $2NO_2(g) \rightleftharpoons 2NO(g) + O_2(g)$
c. $CO(g) + H_2O(g) \rightleftharpoons CO_2(g) + H_2(g)$

SOLUTION

To relieve the stress of decreasing the volume, the equilibrium shifts toward the side with the fewer moles of gaseous components.

a. The equilibrium shifts to C_2H_6 product to reduce the number of moles of gas. The number of moles of product is increased.

$$C_2H_2(g) + 2H_2(g) \longrightarrow C_2H_6(g)$$
$$\qquad \text{3 moles gas} \qquad\qquad \text{1 mole gas}$$

b. The equilibrium shifts to NO_2 reactant to reduce the number of moles of gas. The number of moles of product is decreased.

$$2NO_2(g) \longleftarrow 2NO(g) + O_2(g)$$
$$\text{2 moles gas} \qquad\quad \text{3 moles gas}$$

c. There is no shift in equilibrium because there is no change in the number of moles; the moles of reactant are equal to the moles of product. The number of moles of product does not change.

$$CO(g) + H_2O(g) \rightleftharpoons CO_2(g) + H_2(g)$$
$$\underset{\text{2 moles gas}}{\qquad} \qquad \underset{\text{2 moles gas}}{\qquad}$$

STUDY CHECK

Suppose you want to increase the yield of product in the following reaction. Would you increase or decrease the volume of the reaction container?

$$CO(g) + 2H_2(g) \rightleftharpoons CH_3OH(g)$$

Effect of a Change in Temperature on Equilibrium

In the effects of changes on equilibrium we have seen, shifts in equilibrium occur to reestablish the same value of the equilibrium constant. However, if we change the temperature of a system at equilibrium, we change the value of K_c. When the temperature of an equilibrium system increases, the reaction that is favored is the one that removes the heat. When heat is added to an endothermic reaction, the equilibrium shifts to the products to use up heat. The value of K_c increases because the shift increases the product concentration and decreases the reactant concentration.

Increase T; increase K_c

$$N_2(g) + O_2(g) + heat \rightleftharpoons 2NO(g)$$

If the temperature is lowered, the equilibrium shifts to increase the concentrations of the reactants, and the value of K_c decreases. (See Table 9.5.)

Decrease T; decrease K_c

$$N_2(g) + O_2(g) + heat \rightleftharpoons 2NO(g)$$

For an exothermic reaction, the addition of heat favors the reverse reaction, which uses up heat. The value of K_c for an exothermic reaction decreases when the temperature increases.

TABLE 9.5

Equilibrium Shifts for Temperature Changes in an Endothermic Reaction

K_c	Temperature Change	Equilibrium Shift	Change in K_c Value
$\dfrac{[NO]^2}{[N_2][O_2]}$	Increase	More product $\dfrac{[NO]^2}{[N_2][O_2]}$ Less reactant	Increases
$\dfrac{[NO]^2}{[N_2][O_2]}$	Decrease	Less product $\dfrac{[NO]^2}{[N_2][O_2]}$ More reactant	Decreases

TABLE 9.6

Equilibrium Shifts for Temperature Changes in an Exothermic Reaction

K_c	Temperature Change	Equilibrium Shift	Change in K_c Value
$\dfrac{[SO_3]^2}{[SO_2]^2[O_2]}$	Increase	Less product $$\dfrac{[SO_3]^2}{[SO_2]^2[O_2]}$$ More reactant	⬇ Decreases
$\dfrac{[SO_3]^2}{[SO_2]^2[O_2]}$	Decrease	More product $$\dfrac{[SO_3]^2}{[SO_2]^2[O_2]}$$ Less reactant	⬆ Increases

Increase T; decrease K_c

$$2SO_2(g) + O_2(g) \rightleftharpoons 2SO_3(g) + \text{heat}$$

If heat is removed, the equilibrium of an exothermic reaction favors the products, which provides heat. (See Table 9.6.)

Decrease T; increase K_c

$$2SO_2(g) + O_2(g) \rightleftharpoons 2SO_3(g) + \text{heat}$$

SAMPLE PROBLEM 9.11

■ **Effect of Temperature Change on Equilibrium**

Indicate the change in the concentration of products and the K_c when the temperature of each of the following reactions at equilibrium is increased.

a. $N_2(g) + 3H_2(g) \rightleftharpoons 2NH_3(g) + 92$ kJ
b. $N_2(g) + O_2(g) + 180$ kJ $\rightleftharpoons 2NO(g)$

SOLUTION

a. The addition of heat shifts an exothermic reaction to reactants, which decreases the concentration of the products. The K_c will decrease.
b. The addition of heat shifts an endothermic reaction to products, which increases the concentration of the products. The K_c will increase.

STUDY CHECK

Indicate the change in the concentration of reactants and the K_c when there is a decrease in the temperature of each of the reactions at equilibrium in Sample Problem 9.11.

■

Table 9.7 summarizes the ways we can use Le Châtelier's principle to determine the shift in equilibrium that relieves a stress caused by change in a condition.

TABLE 9.7

Effects of Condition Changes on Equilibrium		
Condition	**Change (Stress)**	**Reaction to Remove Stress**
Concentration	Add reactant	Forward
	Remove reactant	Reverse
	Add product	Reverse
	Remove product	Forward
Volume (container)	Decrease	Toward fewer moles in the gas phase
	Increase	Toward more moles in the gas phase
Temperature	**Endothermic reaction**	
	Raise T	Forward, new value for K_c
	Lower T	Reverse, new value for K_c
	Exothermic reaction	
	Raise T	Reverse, new value for K_c
	Lower T	Forward, new value for K_c
Catalyst	Increases rates equally	No effect

Homeostasis: Regulation of Body Temperature

In a physiological system of equilibrium called homeostasis, changes in our environment are balanced by changes in our bodies. It is crucial to our survival that we balance heat gain with heat loss. If we do not lose enough heat, our body temperature rises. At high temperatures, the body can no longer regulate our metabolic reactions. If we lose too much heat, body temperature drops. At low temperatures, essential functions proceed too slowly.

The skin plays an important role in the maintenance of body temperature. When the outside temperature rises, receptors in the skin send signals to the brain. The temperature-regulating part of the brain stimulates the sweat glands to produce perspiration. As perspiration evaporates from the skin, heat is removed and the body temperature is lowered.

In cold temperatures, epinephrine is released, causing an increase in metabolic rate, which increases the production of heat. Receptors on the skin signal the brain to contract the blood vessels. Less blood flows through the skin, and heat is conserved. The production of perspiration stops to lessen the heat lost by evaporation.

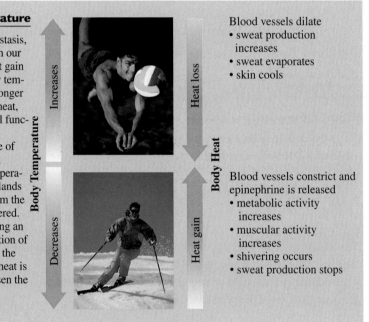

Blood vessels dilate
• sweat production increases
• sweat evaporates
• skin cools

Blood vessels constrict and epinephrine is released
• metabolic activity increases
• muscular activity increases
• shivering occurs
• sweat production stops

QUESTIONS AND PROBLEMS

Changing Equilibrium Conditions: Le Châtelier's Principle

9.27 **a.** Does the addition of reactant to an equilibrium mixture cause the product/reactant ratio to be higher or lower than the K_c?

b. According to Le Châtelier's principle, how is equilibrium in part a established?

9.28 **a.** What is the effect on the K_c when the temperature of an exothermic reaction is lowered?

b. According to Le Châtelier's principle, how is equilibrium in part a established?

9.29 In the lower atmosphere, oxygen is converted to ozone (O_3) by the energy provided from lighting.

$$3O_2(g) + \text{heat} \rightleftharpoons 2O_3(g)$$

For each of the following changes at equilibrium, indicate whether the equilibrium shifts to products, reactants, or does not shift.
a. adding $O_2(g)$
b. adding $O_3(g)$
c. raising the temperature
d. decreasing the volume of the container
e. adding a catalyst

9.30 Ammonia is produced by reacting nitrogen gas and hydrogen gas.

$$N_2(g) + 3H_2(g) \rightleftharpoons 2NH_3(g) + 92 \text{ kJ}$$

For each of the following changes at equilibrium, indicate whether the equilibrium shifts to products, reactions, or does not shift.
a. removing $N_2(g)$
b. lowering the temperature
c. adding $NH_3(g)$
d. adding $H_2(g)$
e. increasing the volume of the container

9.31 Hydrogen chloride can be made by reacting hydrogen gas and chlorine gas.

$$H_2(g) + Cl_2(g) + \text{heat} \rightleftharpoons 2HCl(g)$$

For each of the following changes at equilibrium, indicate whether the equilibrium shifts to products, reactions, or does not shift.
a. adding $H_2(g)$
b. increasing the temperature
c. removing $HCl(g)$
d. adding a catalyst
e. removing $Cl_2(g)$

9.32 When heated, carbon reacts with water to produce carbon monoxide and hydrogen.

$$C(s) + H_2O(g) + \text{heat} \rightleftharpoons CO(g) + H_2(g)$$

For each of the following changes at equilibrium, indicate whether the equilibrium shifts to products, reactions, or does not shift.
a. increasing the temperature
b. adding $C(s)$
c. removing $CO(g)$ as it forms
d. adding $H_2O(g)$
e. decreasing the volume of the container

CONCEPT MAP

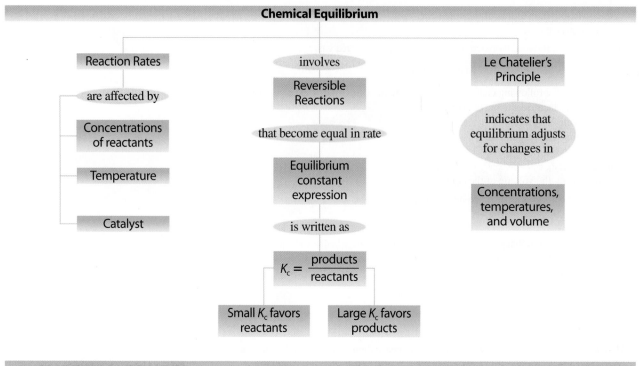

CHAPTER REVIEW

9.1 Rates of Reactions

The rate of a reaction is the speed at which the reactants are converted to products. Increasing the concentrations of reactants, raising the temperature, or adding a catalyst can increase the rate of a reaction.

9.2 Chemical Equilibrium

Chemical equilibrium occurs in a reversible reaction when the rate of the forward reaction becomes equal to the rate of the reverse reaction. At equilibrium, no further change occurs in the concentrations of the reactants and products as the forward and reverse reactions continue.

9.3 Equilibrium Constants

An equilibrium constant, K_c, is the ratio of the concentrations of the products to the concentrations of the reactants with each concentration raised to a power equal to its coefficient in the chemical equation. For heterogeneous reactions, only gases are placed in the equilibrium expression.

9.4 Using Equilibrium Constants

A large value of K_c indicates the equilibrium favors the products and could go nearly to completion whereas a small value of K_c shows that the equilibrium favors the reactants. Equilibrium constants can be used to calculate the concentration of a component in the equilibrium mixture.

9.5 Changing Equilibrium Conditions: Le Châtelier's Principle

The addition of reactants or removal of products favors the forward reaction. Removal of reactants or addition of products favors the reverse reaction. Changing the volume of a reaction container changes the pressure of gases at equilibrium causing a shift to the side with the fewer number of moles. Raising or lowering the temperature for exothermic and endothermic reactions changes the value of K_c and shifts the equilibrium for a reaction.

KEY TERMS

catalyst A substance that increases the rate of reaction by lowering the activation energy.

chemical equilibrium The point at which the forward and reverse reactions take place at the same rate so that there is no further change in concentrations of reactants and products.

collision theory A model for a chemical reaction that states that molecules must collide with sufficient energy in order to form products.

equilibrium constant expression The ratio of the concentrations of products to the concentrations of reactants with each component raised to an exponent equal to the coefficient of that compound in the chemical equation.

equilibrium constant, K_c The numerical value obtained by substituting the equilibrium concentrations of the components into the equilibrium constant expression.

heterogeneous equilibrium An equilibrium system in which the components are in different states.

homogenous equilibrium An equilibrium system in which all components are in the same state.

Le Châtelier's principle When a stress is placed on a system at equilibrium, the equilibrium shifts to relieve that stress.

rate of reaction The speed at which reactants are used to form product(s).

reversible reaction A reaction in which a forward reaction occurs from reactants to products, and a reverse reaction occurs from products back to reactants.

UNDERSTANDING THE CONCEPTS

9.33 Would the reaction shown in the diagrams have a large or small equilibrium constant?

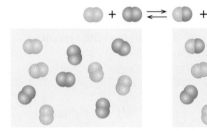

Initial	Equilibrium

9.34 Would the reaction shown in the diagrams have a large or small equilibrium constant?

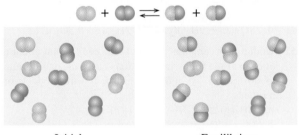

Initial	Equilibrium

9.35 a. Would T_2 be higher or lower than T_1 for the reaction shown in the diagrams?
   b. Would K_c for T_2 be larger or smaller than the K_c for T_1?

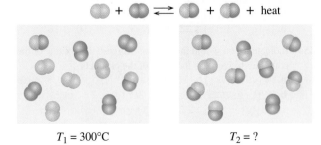

$$T_1 = 300°C \qquad T_2 = ?$$

9.36 a. Would the reaction shown in the diagrams be exothermic or endothermic?
   b. To increase K_c for this reaction, would you raise or lower the temperature?

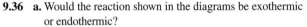

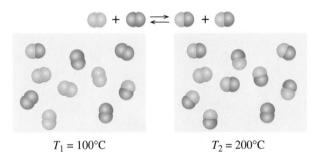

$$T_1 = 100°C \qquad T_2 = 200°C$$

9.37 Consider the following reaction at equilibrium.

$$C_2H_4(g) + Cl_2(g) \rightleftharpoons C_2H_4Cl_2(g) + heat$$

Indicate how each of the following will shift the equilibrium.
   a. raising the temperature of the reaction
   b. decreasing the volume of the reaction container
   c. adding a catalyst
   d. adding Cl_2

9.38 Consider the following reaction at equilibrium.

$$N_2(g) + O_2(g) + heat \rightleftharpoons 2NO(g)$$

Indicate how each of the following will shift the equilibrium.
   a. raising the temperature of the reaction
   b. decreasing the volume of the reaction container
   c. adding a catalyst
   d. adding N_2

ADDITIONAL QUESTIONS AND PROBLEMS

MC GOB For instructor-assigned homework, go to www.masteringgob.com.

9.39 Write the equilibrium constant expression for each of the following reactions.
a. $CH_4(g) + 2O_2(g) \rightleftharpoons CO_2(g) + 2H_2O(g)$
b. $4NH_3(g) + 3O_2(g) \rightleftharpoons 2N_2(g) + 6H_2O(g)$
c. $C(s) + 2H_2(g) \rightleftharpoons CH_4(g)$

9.40 Write the equilibrium constant expression for each of the following reactions.
a. $2C_2H_6(g) + 7O_2(g) \rightleftharpoons 4CO_2(g) + 6H_2O(g)$
b. $2NaHCO_3(s) \rightleftharpoons Na_2CO_3(s) + CO_2(g) + H_2O(g)$
c. $4NH_3(g) + 5O_2(g) \rightleftharpoons 4NO(g) + 6H_2O(g)$

9.41 For each of the following reactions at equilibrium, indicate if the equilibrium mixture contains mostly products, mostly reactants, or both products and reactants.
a. $H_2(g) + Cl_2(g) \rightleftharpoons 2HCl(g)$ $K_c = 1.3 \times 10^{34}$
b. $2NOBr(g) \rightleftharpoons 2NO(g) + Br_2(g)$ $K_c = 2.0$
c. $2NOCl(g) \rightleftharpoons Cl_2(g) + 2NO(g)$ $K_c = 2.7 \times 10^{-9}$

9.42 For each of the following reactions at equilibrium, indicate if the equilibrium mixture contains mostly products, mostly reactants, or both products and reactants.
a. $2H_2O(g) \rightleftharpoons 2H_2(g) + O_2(g)$ $K_c = 4 \times 10^{-48}$
b. $N_2(g) + 3H_2(g) \rightleftharpoons 2NH_3(g)$ $K_c = 0.30$
c. $2SO_2(g) + O_2(g) \rightleftharpoons 2SO_3(g)$ $K_c = 1.2 \times 10^9$

9.43 Write the equation for each of the following equilibrium constant expressions.
a. $K_c = \dfrac{[SO_2][Cl_2]}{[SO_2Cl_2]}$
b. $K_c = \dfrac{[BrCl]^2}{[Br_2][Cl_2]}$
c. $K_c = \dfrac{[CH_4][H_2O]}{[CO][H_2]^3}$
d. $K_c = \dfrac{[N_2O][H_2O]^3}{[O_2]^2[NH_3]^2}$

9.44 Write the equation for each of the following equilibrium constant expressions.
a. $K_c = \dfrac{[CO_2][H_2]}{[CO][H_2O]}$
b. $K_c = \dfrac{[H_2][F_2]}{[HF]^2}$
c. $K_c = \dfrac{[O_2][HCl]^4}{[Cl_2]^2[H_2O]^2}$
d. $K_c = \dfrac{[CS_2][H_2]^4}{[CH_4][H_2S]^2}$

9.45 Consider the reaction
$$2NH_3(g) \rightleftharpoons N_2(g) + 3H_2(g)$$
a. Write the equilibrium constant expression for K_c.
b. What is the K_c for the reaction if at equilibrium the concentrations are $[NH_3] = 0.20$ M; $[H_2] = 0.50$ M; and $[N_2] = 3.0$ M?

9.46 Consider the reaction
$$2SO_2(g) + O_2(g) \rightleftharpoons 2SO_3(g)$$

a. Write the equilibrium constant expression for K_c.
b. What is the K_c for the reaction if at equilibrium the concentrations are $[SO_2] = 0.10$ M; $[O_2] = 0.12$ M; $[SO_3] = 0.60$ M?

9.47 The equilibrium constant for the following reaction is 5.0 at 100°C. If an equilibrium mixture contains $[NO_2] = 0.50$ M, what is the $[N_2O_4]$?
$$2NO_2(g) \rightleftharpoons N_2O_4(g)$$

9.48 The equilibrium constant for the following reaction is 0.20 at 1000°C. If an equilibrium mixture contains solid carbon, $[H_2O] = 0.40$ M, and $[CO] = 0.40$ M, what is the $[H_2]$?
$$C(s) + H_2O(g) \rightleftharpoons CO(g) + H_2(g)$$

9.49 According to Le Châtelier's principle, does the equilibrium shift to products or reactants when O_2 is added to the equilibrium mixture of each of the following reactions?
a. $3O_2(g) \rightleftharpoons 2O_3(g)$
b. $2CO_2(g) \rightleftharpoons 2CO(g) + O_2(g)$
c. $P_4(g) + 5O_2(g) \rightleftharpoons P_4O_{10}(s)$
d. $2SO_2(g) + 2H_2O(g) \rightleftharpoons 2H_2S(g) + 3O_2(g)$

9.50 According to Le Châtelier's principle, what is the effect on the products when N_2 is added to the equilibrium mixture of each of the following reactions?
a. $2NH_3(g) \rightleftharpoons 3H_2(g) + N_2(g)$
b. $N_2(g) + O_2(g) \rightleftharpoons 2NO(g)$
c. $2NO_2(g) \rightleftharpoons N_2(g) + 2O_2(g)$
d. $4NH_3(g) + 3O_2(g) \rightleftharpoons 2N_2(g) + 6H_2O(g)$

9.51 Would decreasing the volume of the equilibrium mixture of each of the following reactions cause the equilibrium to shift, and if so, will the shift be toward products or reactants?
a. $3O_2(g) \rightleftharpoons 2O_3(g)$
b. $2CO_2(g) \rightleftharpoons 2CO(g) + O_2(g)$
c. $P_4(g) + 5O_2(g) \rightleftharpoons P_4O_{10}(s)$
d. $2SO_2(g) + 2H_2O(g) \rightleftharpoons 2H_2S(g) + 3O_2(g)$

9.52 Would increasing the volume of the equilibrium mixture of each of the following reactions cause the equilibrium to shift, and if so, will the shift be toward products or reactants?
a. $2NH_3(g) \rightleftharpoons 3H_2(g) + N_2(g)$
b. $N_2(g) + O_2(g) \rightleftharpoons 2NO(g)$
c. $2NO_2(g) \rightleftharpoons N_2(g) + 2O_2(g)$
d. $4NH_3(g) + 3O_2(g) \rightleftharpoons 2N_2(g) + 6H_2O(g)$

9.53 For each of the following K_c values, indicate whether the equilibrium mixture contains mostly reactants, mostly products, or similar amounts of reactants and products.
a. $N_2(g) + O_2(g) \rightleftharpoons 2NO(g)$ $K_c = 1 \times 10^{-30}$
b. $H_2(g) + Br(g) \rightleftharpoons 2HBr(g)$ $K_c = 2.0 \times 10^{19}$

9.54 Indicate if you would increase or decrease the volume of the container to *increase* the yield of the products
a. $2C(s) + O_2(g) \rightleftharpoons 2CO(g)$
b. $2CH_4(g) \rightleftharpoons C_2H_2(g) + 3H_2(g)$
c. $2H_2(g) + O_2(g) \rightleftharpoons 2H_2O(g)$

CHALLENGE QUESTIONS

9.55 You mix 0.10 mole PCl_5 with 0.050 mole PCl_3 and 0.050 mole Cl_2 in a 1.0 L flask.

$$PCl_5(g) \rightleftarrows PCl_3(g) + Cl_2(g) \quad K_c = 4.2 \times 10^{-2}$$

a. Is the reaction at equilibrium?
b. If not, will the reaction proceed in the forward or reverse direction?

9.56 You mix 0.10 mole each in a 1.0 L container.

$$2NOBr(g) \rightleftarrows 2NO(g) + Br_2(g) \quad K_c = 2.0 \text{ at } 100°C$$

a. Is the reaction at equilibrium?
b. If not, will the reaction proceed in the forward or reverse direction?

9.57 For the reaction

$$PCl_5(g) \rightleftarrows PCl_3(g) + Cl_2(g)$$

a. Write the equilibrium constant expression.
b. Initially 0.60 mole PCl_5 are placed in 1.00 L flask. At equilibrium, there is 0.16 mole PCl_3 in the flask. What are the equilibrium concentrations of the PCl_5 and Cl_2?
c. What is the equilibrium constant for the reaction?
d. If 0.20 mole Cl_2 are added to the equilibrium mixture, will $[PCl_5]$ increase or decrease?

9.58 The K_c at 100°C is 2.0 for the reaction

$$2NOBr(g) \rightleftarrows 2NO(g) + Br_2(g)$$

In an experiment, 1.0 mole of each substance is placed in a 1.0 L container.

a. What is the equilibrium constant expression for the reaction?
b. Is the system at equilibrium?

c. If not, will the rate of the forward or reverse reaction initially speed up?
d. Which concentrations will increase at equilibrium and which will decrease?

9.59 For the reaction

$$C(s) + CO_2(g) \rightleftarrows 2CO(g)$$

the equilibrium mixture contains solid carbon, $[CO] = 0.030$ M, and $[CO_2] = 0.060$ M.

a. What is the value of K_c for the reaction at this temperature?
b. What is the effect of adding more CO_2 to the equilibrium mixture?
c. What is the effect of decreasing the volume of the container?

9.60 $C(s) + H_2O(g) + 31 \text{ kcal} \rightleftarrows CO(g) + H_2(g)$

Indicate how each of the following will affect the equilibrium concentration of CO.

a. add H_2
b. increase the temperature of the reaction
c. increase the volume of the container
d. add $C(s)$
e. decrease the volume of the container
f. add a catalyst
g. decrease the temperature of the reaction
h. remove H_2O

ANSWERS

Answers to Study Checks

9.1 Lowering the temperature will decrease the rate of reaction.

9.2 $H_2(g) + Br_2(g) \rightleftarrows 2HBr(g)$

9.3 (4) slower; (3) faster

9.4 $2NO(g) + O_2(g) \rightleftarrows 2NO_2(g)$

9.5 $FeO(s) + CO(g) \rightleftarrows Fe(s) + CO_2(g)$

$$K_c = \frac{[CO_2]}{[CO]}$$

9.6 $K_c = 27$

9.7 The concentration of the reactants would be much greater than the concentration of the products.

9.8 $[C_2H_5OH] = 2.7$ M

9.9 Increasing $[CO_2]$ will shift the equilibrium toward the reactants.

9.10 Decreasing the volume will increase the yield of product.

9.11 **a.** A decrease in temperature will decrease the concentration of reactants and increase the K_c value.
b. A decrease in temperature will increase the concentration of reactants and decrease the K_c value.

Answer to Selected Questions and Problems

9.1 **a.** The rate of the reaction indicates how fast the products form.
b. Reactions go faster at higher temperatures.

9.3 The number of collisions will increase when the number of Br_2 molecules is increased.

9.5 **a.** increase
b. increase
c. increase
d. decrease

9.7 A reversible reaction is one in which a forward reaction converts reactants to products, while a reverse reaction converts products to reactants.

9.9 **a.** not reversible
b. reversible
c. reversible

9.11 **a.** $K_c = \dfrac{[CS_2][H_2]^4}{[CH_4][H_2S]^2}$ **b.** $K_c = \dfrac{[N_2][O_2]}{[NO]^2}$

c. $K_c = \dfrac{[CS_2][O_2]^4}{[SO_3]^2[CO_2]}$

9.13 **a.** homogeneous equilibrium
    b. heterogeneous equilibrium
    c. homogeneous equilibrium
    d. heterogeneous equilibrium

9.15 **a.** $K_c = \dfrac{[O_2]^3}{[O_3]^2}$ **b.** $K_c = [CO_2][H_2O]$

    c. $K_c = \dfrac{[H_2]^3[CO]}{[CH_4][H_2O]}$ **d.** $K_c = \dfrac{[Cl_2]^2}{[HCl]^4[O_2]}$

9.17 $K_c = 1.5$

9.19 $K_c = 260$

9.21 **a.** mostly products
    b. both reactants and products
    c. mostly reactants

9.23 $[H_2] = 1.1 \times 10^{-3}$ M

9.25 $[NOBr] = 1.4$ M

9.27 **a.** When more reactant is added to an equilibrium mixture, the product/reactant ratio is initially less than K_c.
    b. According to Le Châtelier's principle, equilibrium is reestablished when the forward reaction forms more products to make the product/reactant ratio equal the K_c again.

9.29 **a.** equilibrium shifts to products
    b. equilibrium shifts to reactants
    c. equilibrium shifts to products
    d. equilibrium shifts to products
    e. no shift in equilibrium occurs

9.31 **a.** equilibrium shifts to products
    b. equilibrium shifts to products
    c. equilibrium shifts to products
    d. no shift in equilibrium occurs
    e. equilibrium shifts to reactants

9.33 The reaction would have a small value of the equilibrium constant.

9.35 **a.** T_2 is lower than T_1.
    b. K_c for T_2 is larger than K_c for T_1.

9.37 **a.** Shift toward reactants
    b. Shift toward products
    c. No change
    d. Shift toward products

9.39 **a.** $K_c = \dfrac{[CO_2][H_2O]^2}{[CH_4][O_2]^2}$ **b.** $K_c = \dfrac{[N_2]^2[H_2O]^6}{[NH_3]^4[O_2]^3}$

    c. $K_c = \dfrac{[CH_4]}{[H_2]^2}$

9.41 **a.** mostly products
    b. both products and reactants
    c. mostly reactants

9.43 **a.** $SO_2Cl_2(g) \rightleftharpoons SO_2(g) + Cl_2(g)$
    b. $Br_2(g) + Cl_2(g) \rightleftharpoons 2BrCl(g)$
    c. $CO(g) + 3H_2(g) \rightleftharpoons CH_4(g) + H_2O(g)$
    d. $2O_2(g) + 2NH_3(g) \rightleftharpoons N_2O(g) + 3H_2O(g)$

9.45 **a.** $K_c = \dfrac{[N_2][H_2]^3}{[NH_3]^2}$ **b.** $K_c = 9.4$

9.47 $[N_2O_4] = 1.3$ M

9.49 **a.** equilibrium shifts to products
    b. equilibrium shifts to reactants
    c. equilibrium shifts to products
    d. equilibrium shifts to reactants

9.51 **a.** equilibrium shifts to products
    b. equilibrium shifts to reactants
    c. equilibrium shifts to products
    d. equilibrium shifts to reactants

9.53 **a.** A small K_c indicates that the equilibrium mixture contains mostly reactants.
    b. A large K_c indicates that the equilibrium mixture contains mostly products.

9.55 **a.** The reaction is not at equilibrium.
    b. The reaction will proceed in the reverse direction.

9.57 **a.** $K_c = \dfrac{[PCl_3][Cl_2]}{[PCl_5]}$
    b. At equilibrium, the concentrations are $[Cl_2] = 0.16$ M, $[PCl_5] = 0.84$ M
    c. $K_c = 0.030$
    d. $[PCl_5]$ will increase.

9.59 **a.** $K_c = 0.015$
    b. If more CO_2 is added, the reaction will shift toward the products.
    c. If the container volume is decreased, the equilibrium will shift to the side with the least number of moles, which is the reactant side.

10

Acids and Bases

"In a stat lab, we are sent blood samples of patients in emergency situations," says Audrey Trautwein, clinical laboratory technician, Stat Lab, Santa Clara Valley Medical Center. "We may need to assess the status of a trauma patient in ER or a patient who is in surgery. For example, an acidic blood pH diminishes cardiac function, and affects the actions of certain drugs. In a stat situation, it is critical that we obtain our results fast. This is done using a blood gas analyzer. As I put a blood sample into the analyzer, a small probe draws out a measured volume, which is tested simultaneously for pH, P_{O_2}, and P_{CO_2} as well as electrolytes, glucose, and hemoglobin. In about one minute we have our test results, which are sent to the doctor's computer."

LOOKING AHEAD

Mastering
GOB CHEMISTRY

Visit **www.masteringgob.com**
for self-study materials
and instructor-assigned
homework.

emons, grapefruit, and vinegar taste sour because they contain acids. We have acid in our stomach that helps us digest food; we produce lactic acid in our muscles when we exercise. Acid from bacteria turns milk sour to make cottage cheese or yogurt. Bases are solutions that neutralize acids. Sometimes we take antacids such as milk of magnesia to offset the effects of too much stomach acid.

The pH of a solution describes its acidity. The pH of body fluids, including blood and urine, is regulated primarily by the lungs and the kidneys. Major changes in the pH of the body fluids can severely affect biological activities within the cells. Buffers are present to prevent large fluctuations in pH.

In the environment, the pH of rain, water, and soil can have significant effects. When rain becomes too acidic, it can dissolve marble statues and accelerate the corrosion of metals. In lakes and ponds, the acidity can affect the ability of fish to survive. The acidity of soil around plants affects their growth. If the soil pH is too acidic or too basic, the roots of the plant cannot take up some nutrients. Most plants thrive in soil with a nearly neutral pH, although certain plants such as orchids, camellias, and blueberries require a more acidic soil.

LEARNING GOAL

Describe and name acids and bases.

 SELF-STUDY ACTIVITY
Nature of Acids and Bases

10.1 Acids and Bases

The term *acid* comes from the Latin word *acidus*, which means "sour." We are familiar with the sour tastes of vinegar and lemons and other common acids in foods.

In the nineteenth century, Arrhenius was the first to describe **acids** as substances that produce hydrogen ions (H^+) when they dissolve in water. For example, hydrogen chloride ionizes in water to give hydrogen ions, H^+, and chloride ions, Cl^-. The hydrogen ions, H^+, give acids a sour taste, change blue litmus to red, and corrode some metals.

$$HCl(g) \xrightarrow{H_2O} H^+(aq) + Cl^-(aq)$$

Polar covalent compound   Ionization in water

Naming Acids

When an acid dissolves in water to produce hydrogen ion and a simple nonmetal anion, the prefix *hydro* is used before the name of the nonmetal and its *ide* ending is changed to *ic acid*. For example, hydrogen chloride (HCl) dissolves in water to form HCl(*aq*), which is named hydrochloric acid.

When an acid contains a polyatomic ion, the name of the acid comes from the name of the polyatomic ion. The *ate* in the name is replaced with *ic acid*. If the acid contains a polyatomic ion with an *ite* ending, its name ends with *ous acid*. The names of some common acids and their anions are listed in Table 10.1.

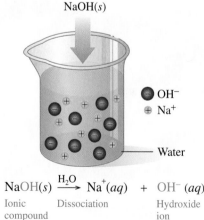

NaOH(*s*)

OH⁻
Na⁺

Water

$$NaOH(s) \xrightarrow{H_2O} Na^+(aq) + OH^-(aq)$$

Ionic compound   Dissociation   Hydroxide ion

TABLE 10.1

Naming Common Acids

Acid	Name of Acid	Anion	Name of Anion
HCl	**hydro**chloric **acid**	Cl^-	chlo**ride**
HBr	**hydro**bromic **acid**	Br^-	brom**ide**
HNO_3	nit**ric acid**	NO_3^-	nit**rate**
HNO_2	nitr**ous acid**	NO_2^-	nit**rite**
H_2SO_4	sulfu**ric acid**	SO_4^{2-}	sulf**ate**
H_2SO_3	sulfur**ous acid**	SO_3^{2-}	sulf**ite**
H_2CO_3	carbon**ic acid**	CO_3^{2-}	carbon**ate**
H_3PO_4	phosphor**ic acid**	PO_4^{3-}	phosph**ate**
$HClO_3$	chlor**ic acid**	ClO_3^-	chlor**ate**
$HClO_2$	chlor**ous acid**	ClO_2^-	chlor**ite**
CH_3COOH	acet**ic acid**	CH_3COO^-	acet**ate**

Bases

You may be familiar with some bases such as antacids, drain openers, and oven cleaners. According to the Arrhenius theory, **bases** are ionic compounds that dissociate into a metal ion and hydroxide ions (OH^-) when they dissolve in water. For example, sodium hydroxide is an Arrhenius base that dissociates in water to give sodium ions, Na^+, and hydroxide ions, OH^-.

Most Arrhenius bases are formed from Groups 1A (1) and 2A (2) metals, such as NaOH, KOH, LiOH, and $Ca(OH)_2$. There are other bases such as $Al(OH)_3$ and $Fe(OH)_3$, but they are fairly insoluble. The hydroxide ions (OH^-) give Arrhenius bases common characteristics such as a bitter taste and soapy feel. Table 10.2 compares some characteristics of acids and bases (Brønsted–Lowry acids are described in the next section).

MC GOB

Acid and Base Formulas
Naming Acids and Bases

Naming Bases

Typical Arrhenius bases are named as hydroxides.

Bases	Name
NaOH	sodium **hydroxide**
KOH	potassium **hydroxide**
$Ca(OH)_2$	calcium **hydroxide**
$Al(OH)_3$	aluminum **hydroxide**

TABLE 10.2

Some Characteristics of Acids and Bases

Characteristic	Acids	Bases
Reaction, Arrhenius	Produce H^+	Produce OH^-
Reaction, Brønsted–Lowry	Donate H^+	Accept H^+
Electrolytes	Yes	Yes
Taste	Sour	Bitter, chalky
Feel	May sting	Soapy, slippery
Litmus paper	Red	Blue
Phenolphthalein	Colorless	Red
Neutralization	Neutralize bases	Neutralize acids

■ **Names of Acids and Bases**

Name the following as acids or bases.

a. H_3PO_4 **b.** NaOH   **c.** HNO_2 **d.** HBr

SOLUTION

a. phosphoric acid   **b.** sodium hydroxide
c. nitrous acid   **d.** hydrobromic acid

STUDY CHECK

Give the name for H_2SO_4 and KOH.

QUESTIONS AND PROBLEMS

■ **Acids and Bases**

10.1 Indicate whether each of the following statements is characteristic of an acid or a base:
   a. has a sour taste   **b.** neutralizes bases
   c. produces H^+ ions in water   **d.** is named potassium hydroxide

10.2 Indicate whether each of the following statements is characteristic of an acid or a base:
   a. neutralizes acids   **b.** produces OH^- in water
   c. has a soapy feel   **d.** turns litmus red

10.3 Name each of the following acid or base:
   a. HCl   **b.** $Ca(OH)_2$ **c.** H_2CO_3 **d.** HNO_3 **e.** H_2SO_3

10.4 Name each of the following acid or base:
   a. $Al(OH)_3$ **b.** HBr   **c.** H_2SO_4 **d.** KOH   **e.** HNO_2

10.5 Write formulas for the following acids and bases:
   a. magnesium hydroxide   **b.** hydrofluoric acid   **c.** phosphoric acid
   d. lithium hydroxide   **e.** copper(II) hydroxide

10.6 Write formulas for the following acids and bases:
   a. barium hydroxide   **b.** hydroiodic acid   **c.** nitric acid
   d. iron(III) hydroxide   **e.** sodium hydroxide

LEARNING GOAL

■ Identify conjugate acid–base pairs for Brønsted–Lowry acids and bases.

10.2 Brønsted–Lowry Acids and Bases

Early in the twentieth century, Brønsted and Lowry expanded the definition of acids and bases. A **Brønsted–Lowry acid** donates a proton (hydrogen ion, H^+) to another substance, and a **Brønsted–Lowry base** accepts a proton.

> A Brønsted–Lowry acid is a proton (H^+) donor.
> A Brønsted–Lowry base is a proton (H^+) acceptor.

A free proton (H^+) does not actually exist in water. Its attraction to polar water molecules is so strong that the proton bonds to the water molecule and forms a **hydronium ion, H_3O^+**.

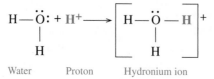

    Water   Proton   Hydronium ion

We can now write the formation of a hydrochloric acid solution as a transfer of a proton from hydrogen chloride to water. By accepting a proton in the reaction, water is acting as a base according to the Brønsted–Lowry concept.

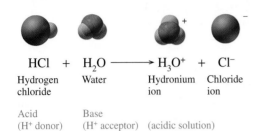

HCl + H₂O ⟶ H₃O⁺ + Cl⁻

| Hydrogen chloride | Water | Hydronium ion | Chloride ion |

Acid   Base
(H⁺ donor)   (H⁺ acceptor)   (acidic solution)

In another reaction ammonia (NH_3) reacts with water. Because the nitrogen atom of NH_3 has a stronger attraction for a proton, water acts as an acid by donating a proton.

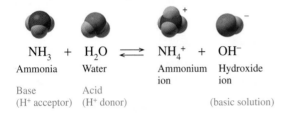

NH_3 + H_2O ⇌ NH_4^+ + OH^-

| Ammonia | Water | Ammonium ion | Hydroxide ion |

Base   Acid
(H⁺ acceptor)   (H⁺ donor)   (basic solution)

SAMPLE PROBLEM 10.2

■ Acids and Bases

In each of the following equations, identify the reactant that is an acid (H⁺ donor) and the reactant that is a base (H⁺ acceptor).

a. $HBr(aq) + H_2O(l) \longrightarrow H_3O^+(aq) + Br^-(aq)$
b. $H_2O(l) + CN^-(aq) \rightleftharpoons HCN(aq) + OH^-(aq)$

SOLUTION

a. HBr, acid; H_2O, base
b. H_2O, acid; CN^-, base

STUDY CHECK

When HNO_3 reacts with water, water acts as a base (H⁺ acceptor). Write the equation for the reaction.

Conjugate Acid–Base Pairs

According to the Brønsted–Lowry theory, the reaction between an acid and base involves proton transfer. When molecules or ions are related by the loss or gain of one H⁺, they make up a **conjugate acid–base pair**. Because protons are transferred in both a forward and reverse reaction, each acid–base reaction contains two conjugate acid–base pairs. In this general reaction, the acid HA has a conjugate base A⁻ and the base B has a conjugate acid BH⁺.

 MC GOB Identifying Conjugate Acid/Base Pairs

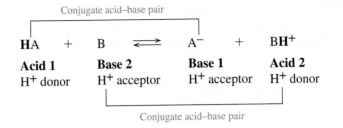

Now we can identify the conjugate acid–base pairs in a reaction such as hydrofluoric acid and water. Because the reaction is reversible, the conjugate acid H_3O^+ can transfer a proton to the conjugate base F^- and re-form the acid HF. Using the relationship of loss and gain of one H^+, we identify the conjugate acid–base pairs as HF and F^- along with H_3O^+ and H_2O.

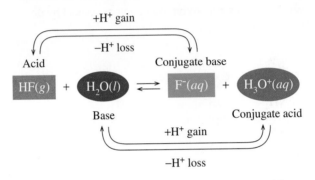

In another proton-transfer reaction, ammonia, NH_3, accepts H^+ from H_2O to form the conjugate acid NH_4^+ and conjugate base OH^-. Each of these conjugate acid–base pairs, NH_3 and NH_4^+ as well as H_2O and OH^-, are related by the loss and gain of one H^+. Table 10.3 gives more examples of conjugate acid–base pairs.

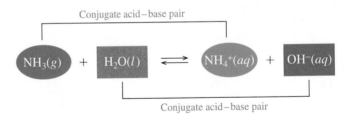

In these two examples, we see that water can act as an acid when it donates an H^+ or a base when it accepts H^+. Substances that can act as both acids and bases are **amphoteric**. For water, the most common amphoteric substance, the acidic or basic behavior depends on whether the other reacting substance is a stronger acid or base. Water donates H^+ when it reacts with a stronger base and accepts H^+ when it reacts with a stronger acid.

SAMPLE PROBLEM 10.3

▪ Conjugate Acid–Base Pairs

Write the formula of the conjugate base of each of the following Brønsted–Lowry acids.

a. $HClO_3$ **b.** H_2CO_3

TABLE 10.3

Some Conjugate Acid–Base Pairs

Acid			Conjugate Base
Strong acids			
perchloric acid	$HClO_4$	ClO_4^-	perchlorate ion
sulfuric acid	H_2SO_4	HSO_4^-	hydrogen sulfate ion
hydroiodic acid	HI	I^-	iodide ion
hydrobromic acid	HBr	Br^-	bromide ion
hydrochloric acid	HCl	Cl^-	chloride ion
nitric acid	HNO_3	NO_3^-	nitrate ion
Weak acids			
hydronium ion	H_3O^+	H_2O	water
hydrogen sulfate ion	HSO_4^-	SO_4^{2-}	sulfate ion
nitrous acid	HNO_2	NO_2^-	nitrite ion
phosphoric acid	H_3PO_4	$H_2PO_4^-$	dihydrogen phosphate ion
acetic acid	CH_3COOH	CH_3COO^-	acetate ion
hydrofluoric acid	HF	F^-	fluoride ion
carbonic acid	H_2CO_3	HCO_3^-	bicarbonate ion
hydrosulfuric acid	H_2S	HS^-	hydrogen sulfide ion
ammonium ion	NH_4^+	NH_3	ammonia
hydrocyanic acid	HCN	CN^-	cyanide ion
bicarbonate ion	HCO_3^-	CO_3^{2-}	carbonate ion
hydrogen sulfide ion	HS^-	S^{2-}	sulfide ion
water	H_2O	OH^-	hydroxide ion

Increasing acid strength ↑ (left)

Increasing base strength ↓ (right)

SOLUTION

The conjugate base forms when the acid donates a proton.

a. ClO_3^- is the conjugate base of $HClO_3$

b. HCO_3^- is the conjugate base of H_2CO_3

STUDY CHECK

Write the conjugate acid of each of the following Brønsted–Lowry bases.

a. HS^- **b.** NO_2^-

SAMPLE PROBLEM 10.4

■ Identifying Conjugate Acid–Base Pairs

Identify the conjugate acid–base pairs in the following reaction:

$$HBr(aq) + NH_3(aq) \longrightarrow Br^-(aq) + NH_4^+(aq)$$

SOLUTION

In the reaction, HBr donates H^+ to NH_3. The resulting Br^- is the conjugate base and NH_4^+ is the conjugate acid. The conjugate acid–base pairs are HBr and Br^- along with NH_3 and NH_4^+.

STUDY CHECK

In the following reaction, identify the conjugate acid–base pairs.

$$HCN(aq) + SO_4^{2-}(aq) \rightleftharpoons CN^-(aq) + H_2SO_4^-(aq)$$

QUESTIONS AND PROBLEMS

Brønsted–Lowry Acids and Bases

10.7 In each of the following equations, identify the acid (proton donor) and base (proton acceptor) for the reactants:
  a. $HI(aq) + H_2O(l) \longrightarrow H_3O^+(aq) + I^-(aq)$
  b. $F^-(aq) + H_2O(l) \rightleftharpoons HF(aq) + OH^-(aq)$

10.8 In each of the following equations, identify the acid (proton donor) and base (proton acceptor) for the reactants:
  a. $CO_3^{2-}(aq) + H_2O(l) \rightleftharpoons HCO_3^-(aq) + OH^-(aq)$
  b. $H_2SO_4(aq) + H_2O(l) \longrightarrow H_3O^+(aq) + HSO_4^-(aq)$

10.9 Write the formula and name of the conjugate base for each of the following acids:
  a. HF **b.** H_2O **c.** H_2CO_3 **d.** HSO_4^-

10.10 Write the formula and name of the conjugate base for each of the following acids:
  a. HCO_3^- **b.** H_3O^+ **c.** HPO_4^{2-} **d.** HNO_2

10.11 Write the formula and name of the conjugate acid for each of the following bases:
  a. CO_3^{2-} **b.** H_2O **c.** $H_2PO_4^-$ **d.** Br^-

10.12 Write the formula and name of the conjugate acid for each of the following bases:
  a. SO_4^{2-} **b.** CN^- **c.** OH^- **d.** ClO_2^-, chlorite ion

10.13 Identify the acid and base on the left side of the following equations and identify their conjugate species on the right side.
  a. $H_2CO_3(aq) + H_2O(l) \rightleftharpoons H_3O^+(aq) + HCO_3^-(aq)$
  b. $NH_4^+(aq) + H_2O(l) \rightleftharpoons H_3O^+(aq) + NH_3(aq)$
  c. $HCN(aq) + NO_2^-(aq) \rightleftharpoons CN^-(aq) + HNO_2(aq)$

10.14 Identify the acid and base on the left side of the following equations and identify their conjugate species on the right side.
  a. $H_3PO_4(aq) + H_2O(l) \rightleftharpoons H_3O^+(aq) + H_2PO_4^-(aq)$
  b. $CO_3^{2-}(aq) + H_2O(l) \rightleftharpoons OH^-(aq) + HCO_3^-(aq)$
  c. $H_3PO_4(aq) + NH_3(aq) \rightleftharpoons NH_4^+(aq) + H_2PO_4^-(aq)$

LEARNING GOAL

Write equations for the dissociation of strong and weak acids; write the equilibrium expression for a weak acid.

10.3 Strengths of Acids and Bases

The *strength* of acids is determined by the moles of H_3O^+ that are produced for each mole of acid that dissolves. The *strength* of bases is determined by the moles of OH^- that are produced for each mole of base that dissolves. In the process called **dissociation**, an acid or base separates into or produces ions in water. Acids and bases vary greatly in their ability to produce H_3O^+ or OH^-. Strong acids and strong bases dissociate completely. In water, weak acids and weak bases dissociate only slightly leaving most of the initial acid or base undissociated.

Strong and Weak Acids

Strong acids give up protons so easily that their dissociation in water is virtually complete. For example, when HCl, one of the strong acids, dissociates in water by transferring a proton to H_2O, the resulting HCl solution contains only the dissolved ions H_3O^+ and Cl^-. We consider the reaction of HCl in water as going 100% to

FIGURE 10.1

A strong acid such as HCl is completely dissociated ($\approx 100\%$) in solution, whereas a solution of a weak acid such as CH_3COOH contains mostly molecules and a few ions.

Q **What is the difference between a strong acid and a weak acid?**

products. Therefore, the equation for a strong acid such as HCl is written with a single arrow to the products.

$$HCl(g) + H_2O(l) \longrightarrow H_3O^+(aq) + Cl^-(aq)$$

Most acids are weak acids. **Weak acids** dissociate slightly in water, which means that only a small percentage of the weak acid transfers a proton to water to produce a small amount of H_3O^+ ions and anions. (See Figure 10.1.) Many of the products you drink or use at home contain weak acids. In carbonated soft drinks, CO_2 dissolves in water to form carbonic acid, H_2CO_3, which remains mostly undissociated in solution. In a weak acid such as H_2CO_3, there is equilibrium between the undissociated H_2CO_3 molecules and the dissociation products, H_3O^+ and HCO_3^-. Therefore, the reaction for a weak acid in water is written with a double arrow. Sometimes a longer reverse arrow is used to indicate that the equilibrium of a weak acid favors the formation of the undissociated reactants.

$$H_2CO_3(aq) + H_2O(l) \rightleftharpoons H_3O^+(aq) + HCO_3^-(aq)$$

Citric acid is a weak acid found in fruits and fruit juices such as lemons, oranges, and grapefruit. Vinegar contains another weak acid known as acetic acid, CH_3COOH. In the vinegar used on salads, acetic acid is present as a 5% acetic acid solution.

$$CH_3COOH(aq) + H_2O(l) \rightleftharpoons H_3O^+(aq) + CH_3COO^-(aq)$$

Strong and Weak Bases

Strong bases are dissociated virtually completely in water. For example, when KOH, a strong base, dissociates in water, the solution consists only of the ions K^+ and OH^-. Essentially no dissolved KOH remains undissociated. Therefore, the equation for a strong base such as KOH is written with a single arrow to products.

$$KOH(s) \xrightarrow{H_2O} K^+(aq) + OH^-(aq)$$

The Arrhenius bases of Groups 1A (1) and 2A (2) such as LiOH, KOH, NaOH, and $Ca(OH)_2$ are strong bases. Sodium hydroxide, NaOH (also known as lye), is used in household products to remove grease in ovens and to clean drains.

Weak bases are poor acceptors of protons and remain mostly undissociated in water. Baking soda ($NaHCO_3$) contains bicarbonate ion, HCO_3^-, which acts as a weak base in water. Detergents, which contain anions of weak acids, act as bases in water. A typical weak base, ammonia, NH_3, is found in window cleaners. In water, a few ammonia molecules accept protons to form NH_4^+ and OH^-.

$$NH_3(g) + H_2O(l) \rightleftharpoons NH_4^+(aq) + OH^-(aq)$$

Direction of Reaction

There is a relationship between the components in each conjugate acid–base pair. The strong acids that donate protons easily have weak conjugate bases that do not readily accept protons. As the strength of the acid decreases, the strength of its conjugate base increases. Weak acids have strong conjugate bases.

In any acid–base reaction, there are two acids and two bases. However, one acid is stronger than the other acid and one base is stronger than the other base. By comparing their relative strengths, we can determine the direction of the reaction. For example, the strong acid H_2SO_4 gives up protons to water. The hydronium ion H_3O^+ produced is a weaker acid than H_2SO_4, and the conjugate base HSO_4^- is a weaker base than water.

$$H_2SO_4(aq) + H_2O(l) \longrightarrow H_3O^+(aq) + HSO_4^-(aq) \text{ Strongly favors products}$$
Stronger acid Stronger base Weaker acid Weaker base

Let's look at another reaction in which water donates a proton to carbonate CO_3^{2-} to form HCO_3^- and OH^-. From Table 10.3 we see that HCO_3^- is a stronger acid than H_2O. We also see that OH^- is a stronger base than CO_3^{2-}. The equilibrium favors the weaker acid and base reactants as shown by the long arrow for the reverse reaction.

$$CO_3^{2-}(aq) + H_2O(l) \xleftarrow{\quad} HCO_3^-(aq) + OH^-(aq) \text{ Strongly favors reactants}$$
Weaker base Weaker acid Stronger acid Stronger base

SAMPLE PROBLEM 10.5

■ Direction of Reaction

Does equilibrium favor the reactants or products in the following reaction?

$$HF(aq) + H_2O(l) \rightleftharpoons H_3O^+(aq) + F^-(aq)$$

SOLUTION

From Table 10.3 we see that HF is a weaker acid than H_3O^+ and H_2O is a weaker base than F^-. Equilibrium favors the reverse direction and therefore the reactants.

$$HF(aq) + H_2O(l) \xleftarrow{\quad} H_3O^+(aq) + F^-(aq)$$
Weaker acid Weaker base Stronger acid Stronger base

STUDY CHECK

Does the reaction of nitric acid and water favor the reactants or the products?

Dissociation Constants

We have seen that reactions of weak acids in water reach equilibrium. If HA is a weak acid, the concentration of H_3O^+ and A^- will be small, which means that the equilibrium will favor the reactants. (See Figure 10.2.)

MC GOB Using Dissociation Constants

$$HA(aq) + H_2O(l) \underset{\longleftarrow}{\overset{\longrightarrow}{}} H_3O^+(aq) + A^-(aq)$$

Dissociation Constants for Weak Acids

As we have seen, acids have different strengths depending on how much they dissociate in water. Because the dissociation of strong acids in water is essentially complete, the reaction is not considered to be an equilibrium situation. However, because weak acids in water dissociate only slightly, the ion products reach an equilibrium with the undissociated weak acid molecules. Thus, an equilibrium expression can be written for weak acids that gives the ratio of the concentrations of products to the weak-acid reactants. As with other equilibrium constants, the molar concentration of the products is divided by the molar concentration of the reactants.

$$\frac{[H_3O^+][A^-]}{[HA][H_2O]}$$

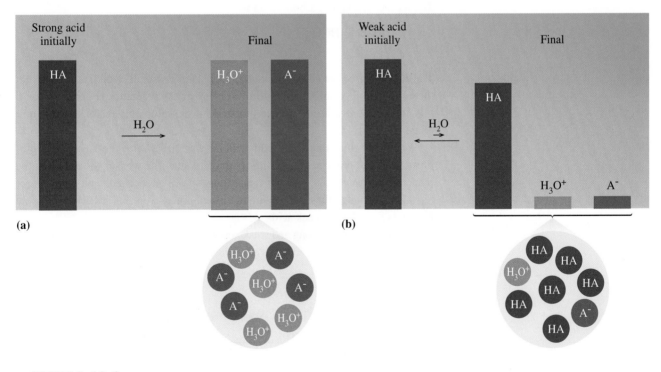

FIGURE 10.2

(a) A strong acid dissociates in water to give H_3O^+ and A^- ions. (b) A weak acid in water dissociates only slightly to form a solution containing only a few H_3O^+ and A^- ions and mostly undissociated HA molecules.

Q **How does the height of the H_3O^+ and A^- in the bar diagram change for a strong acid compared to a weak acid?**

TABLE 10.4

K_a Values for Selected Weak Acids

Acids		
phosphoric acid	H_3PO_4	7.5×10^{-3}
hydrofluoric acid	HF	7.2×10^{-4}
nitrous acid	HNO_2	4.5×10^{-4}
formic acid	HCOOH	1.8×10^{-4}
acetic acid	CH_3COOH	1.8×10^{-5}
carbonic acid	H_2CO_3	4.3×10^{-7}
dihydrogen phosphate	$H_2PO_4^-$	6.2×10^{-8}
hydrocyanic acid	HCN	4.9×10^{-10}
hydrogen phosphate	HPO_4^{2-}	2.2×10^{-13}

Because water is a pure liquid, its concentration, which is constant, is omitted from the equilibrium constant, called the **acid dissociation constant**, K_a (or acid ionization constant). Thus for a weak acid HA, the K_a is written

$$K_a = \frac{[H_3O^+][A^-]}{[HA]} \quad \text{Acid dissociation constant}$$

Let's consider the equilibrium of carbonic acid, which dissociates in water to form bicarbonate ion and hydronium ion.

$$H_2CO_3(aq) + H_2O(l) \rightleftharpoons HCO_3^-(aq) + H_3O^+(aq)$$

The K_a expression for carbonic acid is

$$K_a = \frac{[H_3O^+][HCO_3^-]}{[H_2CO_3]} = 4.3 \times 10^{-7}$$

The K_a measured for carbonic acid at 25°C is quite small, which confirms that the equilibrium of carbonic acid in water favors the reactants. (Recall that usually the concentration units are omitted in the values given for equilibrium constants.)

We can conclude that weak acids will have small K_a values because their equilibria favor the reactants. The smaller the K_a value, the weaker the acid. On the other hand, strong acids, which are essentially 100% dissociated, would have large K_a values, although these values are not usually measured. Table 10.4 gives some K_a values for selected weak acids.

We have described strong and weak acids in several ways. Table 10.5 summarizes the characteristics of acids in terms of strengths and equilibrium position.

TABLE 10.5

Characteristics of Acids

Characteristic	Strong Acids	Weak Acids
Equilibrium position	Toward ionized products	Toward reactants
K_a	Large	Small
$[H_3O^+]$ and $[A^-]$	≈100% of [HA]	Small percent of [HA]
Conjugate bases	Weak	Strong

SAMPLE PROBLEM 10.6

■ Acid Dissociation Constants

Write the expression for the acid dissociation constant for nitrous acid.

SOLUTION

The equation for the dissociation of nitrous acid is written.

$$HNO_2(aq) + H_2O(l) \rightleftharpoons H_3O^+(aq) + NO_2^-(aq)$$

The acid dissociation constant is written as the concentrations of the products divided by the concentration of the undissociated weak acid.

$$K_a = \frac{[H_3O^+][NO_2^-]}{[HNO_2]} = 4.5 \times 10^{-4}$$

STUDY CHECK

Which is the stronger acid, nitrous acid or carbonic acid? Why?

QUESTIONS AND PROBLEMS

■ Strengths of Acids and Bases

10.15 What is meant by the phrase "A strong acid has a weak conjugate base"?

10.16 What is meant by the phrase "A weak acid has a strong conjugate base"?

10.17 Identify the stronger acid in each pair:
    a. HBr or HNO_2 **b.** H_3PO_4 or HSO_4^- **c.** HCN or H_2CO_3

10.18 Identify the stronger acid in each pair:
    a. NH_4^+ or H_3O^+ **b.** H_2SO_4 or HCN   **c.** H_2O or H_2CO_3

10.19 Identify the weaker acid in each pair:
    a. HCl or HSO_4^- **b.** HNO_2 or HF   **c.** HCO_3^- or NH_4^+

10.20 Identify the weaker acid in each pair:
    a. HNO_3 or HCO_3^- **b.** HSO_4^- or H_2O **c.** H_2SO_4 or H_2CO_3

10.21 Predict whether the equilibrium for each of the following reactions favors the reactants or the products.
    a. $H_2CO_3(aq) + H_2O(l) \rightleftharpoons H_3O^+(aq) + HCO_3^-(aq)$
    b. $NH_4^+(aq) + H_2O(l) \rightleftharpoons H_3O^+(aq) + NH_3(aq)$
    c. $HCl(aq) + NH_3(aq) \rightleftharpoons Cl^-(aq) + NH_4^+(aq)$

10.22 Predict whether the equilibrium for each of the following reactions favors the reactants or the products.
    a. $H_3PO_4(aq) + H_2O(l) \rightleftharpoons H_3O^+(aq) + H_2PO_4^-(aq)$
    b. $CO_3^{2-}(aq) + H_2O(l) \rightleftharpoons OH^-(aq) + HCO_3^-(aq)$
    c. $HS^-(aq) + H_2O(l) \rightleftharpoons H_3O^+(aq) + S^{2-}(aq)$

10.23 Write an equation for the acid–base reaction between ammonium ion and sulfate ion. Why does the equilibrium favor the reactants?

10.24 Write an equation for the acid–base reaction between nitrous acid and sulfate ion. Why does the equilibrium favor the reactants?

10.25 Consider the following acids and their dissociation constants:

$$H_2SO_3(aq) + H_2O(l) \rightleftharpoons H_3O^+(aq) + HSO_3^-(aq) \qquad K_a = 1.2 \times 10^{-2}$$
$$HS^-(aq) + H_2O(l) \rightleftharpoons H_3O^+(aq) + S^{2-}(aq) \qquad K_a = 1.3 \times 10^{-19}$$

    a. Which is the stronger acid, H_2SO_3 or HS^-?
    b. What is the conjugate base of H_2SO_3?

c. Which acid has the weaker conjugate base?
d. Which acid has the stronger conjugate base?
e. Which acid produces more ions?

10.26 Consider the following acids and their dissociation constants:

$$HPO_4{}^{2-}(aq) + H_2O(l) \rightleftharpoons H_3O^+(aq) + PO_4{}^{3-}(aq) \qquad K_a = 2.2 \times 10^{-13}$$

$$HCOOH(aq) + H_2O(l) \rightleftharpoons H_3O^+(aq) + HCOO^-(aq) \qquad K_a = 1.8 \times 10^{-4}$$

a. Which is the weaker acid, $HPO_4{}^{2-}$ or $HCOOH$?
b. What is the conjugate base of $HPO_4{}^{2-}$?
c. Which acid has the weaker conjugate base?
d. Which acid has the stronger conjugate base?
e. Which acid produces more ions?

10.27 Phosphoric acid dissociates to form dihydrogen phosphate and hydronium ion. Write the equation and equilibrium constant for the dissociation of the acid.

10.28 Carbonic acid, a weak acid, dissociates to form bicarbonate and hydronium ion. Write the equation and equilibrium constant for the dissociation of the acid.

LEARNING GOAL

■ Use the ion product of water to calculate the $[H_3O^+]$ and $[OH^-]$ in an aqueous solution.

(MC GOB) Ionization of Water

10.4 Ionization of Water

We have seen that in some acid–base reactions water is amphoteric and acts either as an acid or as a base. Does that mean water can be both an acid and a base? Yes, this is exactly what happens with water molecules in pure water. Let's see how this happens. In water, one water molecule donates a proton to another to produce H_3O^+ and OH^-, which means that water can behave as both an acid and a base. Let's take a look at the conjugate acid–base pairs of water.

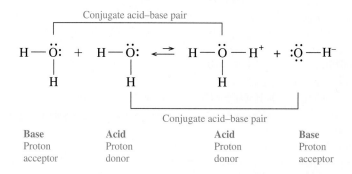

In the ionization of water, there is both a forward and reverse reaction.

$$H_2O(l) + H_2O(l) \rightleftharpoons H_3O^+(aq) + OH^-(aq)$$

In pure water, the transfer of a proton between two water molecules produces equal numbers of H_3O^+ and OH^-. Experiments have determined that, in pure water, the concentrations of H_3O^+ and OH^- at 25°C are each 1.0×10^{-7} M. Square brackets around the symbols indicate their concentrations in moles per liter (M).

Pure water $\qquad [H_3O^+] = [OH^-] = 1.0 \times 10^{-7}$ M

When we multiply these concentrations, it forms the **ion-product constant of water**, K_w, which is 1.0×10^{-14}. The concentration units are omitted in the K_w value.

$$K_w = [H_3O^+] \times [OH^-]$$
$$= (1.0 \times 10^{-7}\text{ M})(1.0 \times 10^{-7}\text{ M}) = 1.0 \times 10^{-14}$$

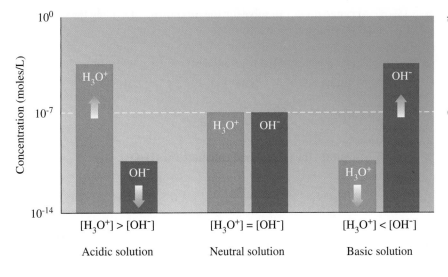

FIGURE 10.3

In a neutral solution, $[H_3O^+]$ and $[OH^-]$ are equal. In acidic solutions, the $[H_3O^+]$ is greater than the $[OH^-]$. In basic solutions, the $[OH^-]$ is greater than the $[H_3O^+]$.

Q **Is a solution that has $[H_3O^+] = 1.0 \times 10^{-3}$ M acidic, basic, or neutral?**

The K_w value of 1.0×10^{-14} is important because it applies to any aqueous solution: all aqueous solutions have H_3O^+ and OH^-.

When the $[H_3O^+]$ and $[OH^-]$ in a solution are equal, the solution is **neutral**. However, most solutions are not neutral and have different concentrations of $[H_3O^+]$ and $[OH^-]$. If acid is added to water, there is an increase in $[H_3O^+]$ and a decrease in $[OH^-]$, which makes an acidic solution. If base is added, $[OH^-]$ increases and $[H_3O^+]$ decreases, which makes a basic solution. (See Figure 10.3.) However, for any aqueous solution, whether it is neutral, acidic, or basic, the product $[H_3O^+] \times [OH^-]$ is equal to K_w (1.0×10^{-14}). Therefore, if the $[H_3O^+]$ is given, K_w can be used to calculate the $[OH^-]$. Or if the $[OH^-]$ is given, K_w can be used to calculate the $[H_3O^+]$. (See Table 10.6.)

$$K_w = [H_3O^+] \times [OH^-]$$

$$[OH^-] = \frac{K_w}{[H_3O^+]} \qquad [H_3O^+] = \frac{K_w}{[OH^-]}$$

TABLE 10.6

Examples of $[H_3O^+]$ and $[OH^-]$ in Neutral, Acidic, and Basic Solutions

Type of Solution	$[H_3O^+]$	$[OH^-]$	K_w
Neutral	1.0×10^{-7} M	1.0×10^{-7} M	1.0×10^{-14}
Acidic	1.0×10^{-2} M	1.0×10^{-12} M	1.0×10^{-14}
Acidic	2.5×10^{-5} M	4.0×10^{-10} M	1.0×10^{-14}
Basic	1.0×10^{-8} M	1.0×10^{-6} M	1.0×10^{-14}
Basic	5.0×10^{-11} M	2.0×10^{-4} M	1.0×10^{-14}

To illustrate these calculations, let us determine the $[H_3O^+]$ for a solution that has an $[OH^-] = 1.0 \times 10^{-6}$ M.

■ **STEP 1** Write the K_w for water.

$$K_w = [H_3O^+][OH^-] = 1.0 \times 10^{-14}$$

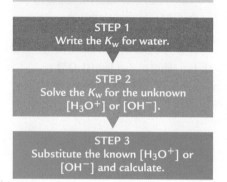

Guide to Calculating [H₃O⁺] and [OH⁻] in Aqueous Solutions

STEP 1
Write the K_w for water.

STEP 2
Solve the K_w for the unknown [H₃O⁺] or [OH⁻].

STEP 3
Substitute the known [H₃O⁺] or [OH⁻] and calculate.

■ **STEP 2** **Arrange the K_w to solve for the unknown.** Dividing through by the [OH⁻] gives

$$\frac{K_w}{[OH^-]} = \frac{[H_3O^+] \times [\cancel{OH^-}]}{[\cancel{OH^-}]} = \frac{1.0 \times 10^{-14}}{[OH^-]}$$

$$[H_3O^+] = \frac{1.0 \times 10^{-14}}{[OH^-]}$$

■ **STEP 3** **Substitute the [OH⁻] and calculate the [H₃O⁺].**

$$[H_3O^+] = \frac{1.0 \times 10^{-14}}{[1.0 \times 10^{-6}]} = 1.0 \times 10^{-8} \text{ M}$$

Because the [OH⁻] of 1.0×10^{-6} M is larger than the [H₃O⁺] of 1.0×10^{-8} M, the solution is basic.

SAMPLE PROBLEM 10.7

■ **Calculating [H₃O⁺] and [OH⁻] in Solution**

A vinegar solution has a [H₃O⁺] = 2.0×10^{-3} M at 25°C. What is the [OH⁻] of the vinegar solution? Is the solution acidic, basic, or neutral?

SOLUTION

■ **STEP 1** **Write the K_w for water.**

$$K_w = [H_3O^+] \times [OH^-] = 1.0 \times 10^{-14}$$

■ **STEP 2** **Arrange the K_w to solve for the unknown.** Rearranging the K_w for OH⁻ gives

$$\frac{K_w}{[H_3O^+]} = \frac{[\cancel{H_3O^+}] \times [OH^-]}{[\cancel{H_3O^+}]} = \frac{1.0 \times 10^{-14}}{[H_3O^+]}$$

$$[OH^-] = \frac{1.0 \times 10^{-14}}{[H_3O^+]}$$

■ **STEP 3** **Substitute the known [H₃O⁺] and calculate.**

$$[OH^-] = \frac{1.0 \times 10^{-14}}{[2.0 \times 10^{-3}]} = 5.0 \times 10^{-12} \text{ M}$$

Because the [H₃O⁺] of 2.0×10^{-3} M is much larger than the [OH⁻] of 5.0×10^{-12} M, the solution is acidic.

STUDY CHECK

What is the [H₃O⁺] of an ammonia cleaning solution with an [OH⁻] = 4.0×10^{-4} M? Is the solution acidic, basic, or neutral?

QUESTIONS AND PROBLEMS

■ **Ionization of Water**

10.29 Why are the concentrations of H₃O⁺ and OH⁻ equal in pure water?

10.30 What is the meaning and value of K_w?

10.31 In an acidic solution, how does the concentration of H_3O^+ compare to the concentration of OH^-?

10.32 If a base is added to pure water, why does the $[H_3O^+]$ decrease?

10.33 Indicate whether the following are acidic, basic, or neutral solutions:
a. $[H_3O^+] = 2.0 \times 10^{-5}$ M
b. $[H_3O^+] = 1.4 \times 10^{-9}$ M
c. $[OH^-] = 8.0 \times 10^{-3}$ M
d. $[OH^-] = 3.5 \times 10^{-10}$ M

10.34 Indicate whether the following are acidic, basic, or neutral solutions:
a. $[H_3O^+] = 6.0 \times 10^{-12}$ M
b. $[H_3O^+] = 1.4 \times 10^{-4}$ M
c. $[OH^-] = 5.0 \times 10^{-12}$ M
d. $[OH^-] = 4.5 \times 10^{-2}$ M

10.35 Calculate the $[OH^-]$ of each aqueous solution with the following $[H_3O^+]$:
a. coffee, 1.0×10^{-5} M
b. soap, 1.0×10^{-8} M
c. cleanser, 5.0×10^{-10} M
d. lemon juice, 2.5×10^{-2} M

10.36 Calculate the $[OH^-]$ of each aqueous solution with the following $[H_3O^+]$:
a. NaOH, 1.0×10^{-12} M
b. aspirin, 6.0×10^{-4} M
c. milk of magnesia, 1.0×10^{-9} M
d. stomach acid, 5.2×10^{-2} M

10.37 Calculate the $[OH^-]$ of each aqueous solution with the following $[H_3O^+]$:
a. vinegar, 1.0×10^{-3} M
b. urine, 5.0×10^{-6} M
c. ammonia, 1.8×10^{-12} M
d. NaOH, 4.0×10^{-13} M

10.38 Calculate the $[OH^-]$ of each aqueous solution with the following $[H_3O^+]$:
a. baking soda, 1.0×10^{-8} M
b. orange juice, 2.0×10^{-4} M
c. milk, 5.0×10^{-7} M
d. bleach, 4.8×10^{-12} M

10.5 The pH Scale

Many kinds of careers such as respiratory therapy, food processing, medicine, agriculture, spa cleaning, and soap manufacturing require personnel to measure the $[H_3O^+]$ and $[OH^-]$ of solutions. The proper levels of acidity are necessary for soil to support plant growth and prevent algae in swimming pool water. Measuring the acidity levels of blood and urine checks the function of the kidneys.

On the pH scale, a number between 0 and 14 represents the H_3O^+ concentration. A pH value less than 7 corresponds to an acidic solution; a pH value greater than 7 indicates a basic solution. (See Figure 10.4.)

Acidic solution	pH < 7	$[H_3O^+] > 1.0 \times 10^{-7}$ M
Neutral solution	pH = 7	$[H_3O^+] = 1.0 \times 10^{-7}$ M
Basic solution	pH > 7	$[H_3O^+] < 1.0 \times 10^{-7}$ M

In the laboratory, a pH meter is commonly used to determine the pH of a solution. There are also indicators and pH papers that turn specific colors when placed in solutions of different pH values. The pH is found by comparing the colors to a color chart. (See Figure 10.5.)

Calculating the pH of Solutions

The pH scale is a log scale that corresponds to the hydrogen-ion concentrations of aqueous solutions. Mathematically, **pH** is the negative logarithm (base 10) of the H_3O^+ concentration.

$$pH = -\log[H_3O^+]$$

Essentially, the negative powers of ten in the concentrations are converted to positive numbers. For example, a lemon juice solution with $[H_3O^+] = 1.0 \times 10^{-2}$ M has a pH of 2.00. This can be calculated using the pH equation. For whole numbers

LEARNING GOAL

Calculate pH from $[H_3O^+]$; given the pH, calculate $[H_3O^+]$ and $[OH^-]$ of a solution.

 SELF-STUDY CASE
Hyperventilation and Blood pH

Logarithms
The pH Scale

SELF-STUDY ACTIVITY
The pH Scale

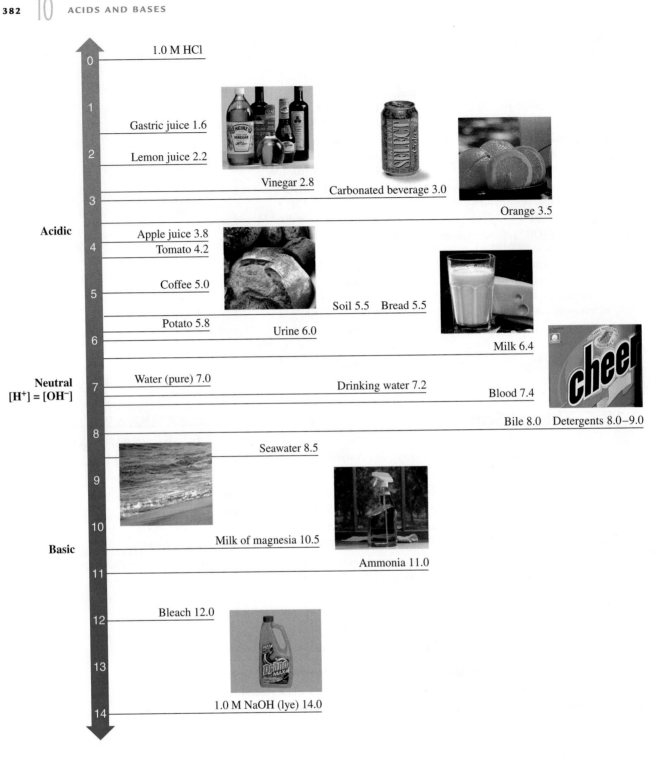

FIGURE 10.4

On the pH scale, values below 7 are acidic, a value of 7 is neutral, and values above 7 are basic.

Q **Is apple juice an acidic, basic, or neutral solution?**

in the $[H_3O^+]$, remember to add the correct number of significant zeros to the resulting pH obtained on your calculator.

$$pH = -\log[1.0 \times 10^{-2}]$$
$$pH = -(-2.00)$$
$$= 2.00$$

(a)

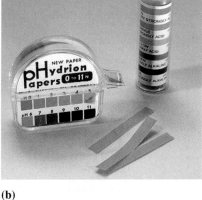
(b)

(c)

FIGURE 10.5

The pH of a solution can be determined using (**a**) a pH meter, (**b**) pH paper, and (**c**) indicators that turn different colors corresponding to different pH values.

Q **If a pH meter reads 4.00, is the solution acidic, basic, or neutral?**

Let's look at how we determine the number of significant figures in the pH. For a logarithm, the number of decimal places in the pH value is equal to the number of significant figures in the $[H_3O^+]$. The number to the left of the decimal point is the power of ten.

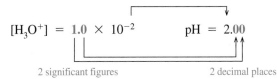

$$[H_3O^+] = 1.0 \times 10^{-2} \qquad pH = 2.00$$

2 significant figures   2 decimal places

Steps for a pH Calculation

The pH of a solution is determined using the *log* key and *changing sign*. For example, to calculate the pH of a vinegar solution with $[H_3O^+] = 2.4 \times 10^{-3}$ M you can use the following steps:

Display Shows

■ STEP 1 **Enter the $[H_3O^+]$ value.** Enter 2.4 and press $\boxed{\text{EE or EXP}}$. 2.4^{03} or $2.4 \ 03$

Enter 3 and press $\boxed{+/-}$ to change the power 2.4^{-03} or $2.4{-}03$
to −3. (For calculators without a change sign
key, consult the instructions for the calculator.)

■ STEP 2 **Press the $\boxed{\text{log}}$ key.** -2.619789

Change the sign. 2.619789

The steps can be combined to give the calculator sequence as follows:

$$pH = -\log[2.4 \times 10^{-3}] = 2.4 \ \boxed{\text{EE or EXP}} \ 3 \ \boxed{+/-} \ \boxed{\text{log}} \ \boxed{+/-}$$
$$= 2.619789$$

Be sure to check the instructions for your calculator. On some calculators, the log key is used first, followed by the concentration.

■ STEP 3 **Adjust significant figures.** In a pH value, the number to the *left* of the decimal point is an *exact* number derived from the power of ten.

HEALTH NOTE

■ Stomach Acid, HCl

When a person sees, smells, thinks about, or tastes food, the gastric glands in the stomach begin to secrete an HCl solution that is strongly acidic. In a single day a person may secrete as much as 2000 mL of gastric juice.

The HCl in the gastric juice activates a digestive enzyme called pepsin that breaks down proteins in food entering the stomach. The secretion of HCl continues until the stomach has a pH of about 2, which is the optimum pH for activating the digestive enzymes without ulcerating the stomach lining. Normally, large quantities of viscous mucus are secreted within the stomach to protect its lining from acid and enzyme damage.

Guide to Calculating pH of an Aqueous Solution

STEP 1
Enter the $[H_3O^+]$ value.

STEP 2
Press the *log* key and *change sign*.

STEP 3
Adjust significant figures to the *right* of the decimal point to equal SFs in the coefficient.

The number of digits to the *right* of the decimal point is equal to the number of significant figures in the coefficient.

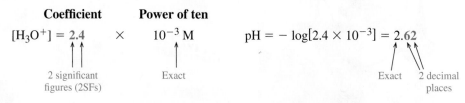

Coefficient   **Power of ten**

$[H_3O^+] = 2.4 \times 10^{-3}\,M$ $pH = -\log[2.4 \times 10^{-3}] = 2.62$

2 significant figures (2SFs)   Exact   Exact   2 decimal places

Because pH is a log scale, a change of one pH unit corresponds to a ten-fold change in $[H_3O^+]$. It is important to note that the pH decreases as the $[H_3O^+]$ increases. For example, a solution with a pH of 2.00 has a $[H_3O^+]$ 10 times higher than a solution with a pH of 3.00 and 100 times higher than a solution with a pH of 4.00.

SAMPLE PROBLEM 10.8

■ **Calculating pH**

Determine the pH for the following solutions:

a. $[H_3O^+] = 1.0 \times 10^{-5}\,M$ **b.** $[H_3O^+] = 5 \times 10^{-8}\,M$

SOLUTION

a. ■ **STEP 1** Enter the $[H_3O^+]$ using the *change sign* key.

                                                     Display

1.0 [EE or EXP] 5 [+/−] 1.0^{-05} or $1.0-05$

■ **STEP 2** Press the *log* key and *change sign* key.

[log] [+/−]   5

■ **STEP 3** Adjust significant figures to the *right* of the decimal point to equal the number of significant figures in the coefficient.

$1.0 \times 10^{-5}\,M$ pH = 5.00

2 SFs ⟶ 2 SFs to the *right* of the decimal point

b. ■ **STEP 1** Enter the concentration using the *change sign* key.

5 [EE or EXP] [+/−] 5^{-08} or $5-08$

■ **STEP 2** Press the *log* key and then the *change sign* key.

[log] [+/−]   7.301029

■ **STEP 3** Adjust significant figures to the *right* of the decimal point to equal the significant figures in the coefficient.

$5 \times 10^{-8}\,M$ pH = 7.3

1 SF ⟶ 1 SF to the *right* of the decimal point

STUDY CHECK

What is the pH of bleach with $[H_3O^+] = 4.2 \times 10^{-12}\,M$?

SAMPLE PROBLEM 10.9

■ **Calculating pH from [OH⁻]**

What is the pH of an ammonia solution with $[OH^-] = 3.7 \times 10^{-3}$ M?

SOLUTION

■ **STEP 1** **Enter the $[H_3O^+]$ using the *change sign* key.** Because $[OH^-]$ is given for the ammonia solution, we have to calculate $[H_3O^+]$ using the ion product of water, K_w. Dividing through by $[OH^-]$ gives $[H_3O^+]$.

$$\frac{K_w}{[OH^-]} = \frac{[H_3O^+]\,[\cancel{OH^-}]}{[\cancel{OH^-}]} = \frac{1.0 \times 10^{-14}}{[OH^-]}$$

$$[H_3O^+] = \frac{1.0 \times 10^{-14}}{[3.7 \times 10^{-3} \text{ M}]} = 2.7 \times 10^{-12} \text{ M}$$

$$2.7 \;\boxed{\text{EE or EXP}}\; 12 \;\boxed{+/-} \;=\; \boxed{2.7^{-12}} \;\text{ or }\; \boxed{2.7 - 12}$$

■ **STEP 2** **Press the *log* key, and then the *change sign* key.**

$\boxed{\text{log}}$ $\boxed{+/-}$ $\boxed{11.56863}$

■ **STEP 3** **Adjust significant figures to the *right* of the decimal point to equal the SFs in the coefficient.**

2.7×10^{-12} M   pH = 11.57

2 SFs ――――――→ 2 SFs *after* the decimal point

STUDY CHECK

Calculate the pH of a sample of acid rain that has $[OH^-] = 2 \times 10^{-10}$ M.

Calculating [H₃O⁺] from pH

In another type of calculation, we may be given the pH of the solution and asked to determine the $[H_3O^+]$. This is a reverse of the pH calculation and may require the use of the 10^x key, which is usually a 2nd function key. On some calculators, this operation is done using the inverse key and the log key.

$$[H_3O^+] = 10^{-pH}$$

SAMPLE PROBLEM 10.10

■ **Calculating [H₃O⁺] from pH**

Calculate $[H_3O^+]$ for each of the following solutions:

a. coffee, pH of 5.0
b. baking soda, pH of 8.25

SOLUTION

a. coffee, pH of 5.0

■ **STEP 1** **Enter the pH value and press the *change sign* key.**

                                                                    Display

5.0 $\boxed{+/-}$ $\boxed{-5.0}$

■ **STEP 2** **Convert −pH to concentration.** Press the *2nd function* key and then the *10^x* key.

[2nd] [10ˣ] 0.00001 or $1.^{-05}$ or $1−05$

Or press the *inverse* key and then the *log* key.

[inv] [log] 0.00001 or $1.^{-05}$ or $1−05$

Write the display in scientific notation with units of concentration. 1×10^{-5} M

■ **STEP 3** **Adjust the significant figures in the coefficient.** The pH value of 5.0 has only one digit to the *right* of the decimal point, which means the $[H_3O^+]$ is written with only one significant figure.

$$[H_3O^+] = 1 \times 10^{-5} \text{ M}$$

b. baking soda, pH of 8.25

■ **STEP 1** **Enter the pH value and press the *change sign* key.**

                                              Display

8.25 [+/−] $−8.25$

■ **STEP 2** **Convert −pH to concentration.** Press the *2nd function* key and then the *10^x* key.

[2nd] [10ˣ] 5.62341^{-09} or $5.62341−09$

Or press the *inverse* key and then the *log* key.

[inv] [log] 5.62341^{-09} or $5.62341−09$

Write the display in scientific notation with units of concentration. 5.62341×10^{-9} M

■ **STEP 3** **Adjust the significant figures in the coefficient.** Because the pH value of 8.25 has two digits to the *right* of the decimal point, the $[H_3O^+]$ is written with two significant figures.

$$[H_3O^+] = 5.6 \times 10^{-9} \text{ M}$$

STUDY CHECK

What is the $[H_3O^+]$ and $[OH^-]$ of a beer that has a pH of 4.50?

A comparison of $[H_3O^+]$, $[OH^-]$, and their corresponding pH values is given in Table 10.7.

TABLE 10.7

A Comparison of $[H_3O^+]$, $[OH^-]$, and Corresponding pH Values

$[H_3O^+]$	pH	$[OH^-]$	
10^0	0	10^{-14}	
10^{-1}	1	10^{-13}	
10^{-2}	2	10^{-12}	
10^{-3}	3	10^{-11}	Acidic
10^{-4}	4	10^{-10}	
10^{-5}	5	10^{-9}	
10^{-6}	6	10^{-8}	
10^{-7}	7	10^{-7}	Neutral
10^{-8}	8	10^{-6}	
10^{-9}	9	10^{-5}	
10^{-10}	10	10^{-4}	
10^{-11}	11	10^{-3}	Basic
10^{-12}	12	10^{-2}	
10^{-13}	13	10^{-1}	
10^{-14}	14	10^0	

QUESTIONS AND PROBLEMS

■ The pH Scale

10.39 Why does a neutral solution have a pH of 7.00?

10.40 If you know the $[OH^-]$, how can you determine the pH of a solution?

10.41 State whether each of the following solutions is acidic, basic, or neutral:
    a. blood, pH 7.38   **b.** vinegar, pH 2.8
    c. drain cleaner, pH 11.2   **d.** coffee, pH 5.5

10.42 State whether each of the following solutions is acidic, basic, or neutral:
    a. soda, pH 3.2   **b.** shampoo, pH 5.7
    c. laundry detergent, pH 9.4   **d.** rain, pH 5.8

10.43 Calculate the pH of each solution given the following $[H_3O^+]$ or $[OH^-]$ values.
    a. $[H_3O^+] = 1.0 \times 10^{-4}$ M   **b.** $[H_3O^+] = 3.0 \times 10^{-9}$ M
    c. $[OH^-] = 1.0 \times 10^{-5}$ M   **d.** $[OH^-] = 2.5 \times 10^{-11}$ M

10.44 Calculate the pH of each solution given the following $[H_3O^+]$ or $[OH^-]$ values.
    a. $[H_3O^+] = 1.0 \times 10^{-8}$ M   **b.** $[H_3O^+] = 5.0 \times 10^{-6}$ M
    c. $[OH^-] = 4.0 \times 10^{-2}$ M   **d.** $[OH^-] = 8.0 \times 10^{-3}$ M

10.45 Complete the following table:

$[H_3O^+]$	$[OH^-]$	pH	Acidic, Basic, or Neutral?
	1.0×10^{-6} M		
		3.00	
2.8×10^{-5} M			

10.46 Complete the following table:

$[H_3O^+]$	$[OH^-]$	pH	Acidic, Basic, or Neutral?
		10.00	
			Neutral
6.4×10^{-12} M			

Using Vegetables and Flowers as pH Indicators

Many flowers and vegetables with strong color, especially reds and purples, contain compounds that change color with changes in pH. Some examples are red cabbage, cranberry juice, and cranberry drinks.

MATERIALS NEEDED

Red cabbage, water, and a saucepan; or cranberry juice or drinks

Several glasses or small glass containers and some tape and a pen or pencil to mark the containers

Several colorless household solutions such as vinegar, lemon juice, other fruit juices, baking soda, antacids, aspirin, window cleaners, soaps, shampoos, and detergents

PROCEDURE

1. Obtain a bottle of cranberry juice or cranberry drink, or use a red cabbage to prepare the red cabbage pH indicator as follows: Tear up several red cabbage leaves and place them in a saucepan and cover with water. Boil for about 5 minutes. Cool and collect the purple solution.

2. Place small amounts of each household solution into separate clear glass containers and mark what each one is. If the sample is a solid or a thick liquid, add a small amount of water. Add some cranberry juice or some red cabbage indicator until you obtain a color.
3. Observe the colors of the various samples. The colors that indicate acidic solutions are the pink and orange colors (pH 1–4) and the pink to lavender colors (5–6). A neutral solution has about the same purple color as the indicator. Bases will give blue to green color (pH 8–11) or a yellow color (pH 12–13).
4. Arrange your samples by color and pH. Classify each of the solutions as acidic (1–6), neutral (7), or basic (8–13).
5. Try to make an indicator using other colorful fruits or flowers.

QUESTIONS

1. Which products that tested acidic listed an acid on their labels?
2. Which products that tested basic listed a base on their labels?
3. How many products were neutral?
4. Which flowers or vegetables behaved as indicators?

LEARNING GOAL

Write balanced equations for reactions of acids and bases.

MC
GOB

Reactions of Acids and Bases

10.6 Reactions of Acids and Bases

Typical reactions of acids and bases include the reactions of acids with metals, bases, and carbonate or bicarbonate ions. For example, when you drop an antacid tablet in water, the bicarbonate ion and citric acid in the tablet react to produce carbon dioxide bubbles, a salt, and water.

Acids and Metals

Acids react with certain metals known as *active metals* to produce hydrogen gas (H_2) and the salt of that metal. Active metals include potassium, sodium, calcium, magnesium, aluminum, zinc, iron, and tin. In these single replacement reactions, the metal loses electrons and the metal ion replaces the hydrogen in the acid.

$$Mg(s) + 2HCl(aq) \longrightarrow MgCl_2(aq) + H_2(g)$$
Metal   Acid   Salt   Hydrogen

$$Zn(s) + 2HCl(aq) \longrightarrow ZnCl_2(aq) + H_2(g)$$
Metal   Acid   Salt   Hydrogen

Acids, Carbonates, and Bicarbonates

When strong acids are added to a carbonate or bicarbonate (also known as hydrogen carbonate), the reaction produces bubbles of carbon dioxide gas, a salt, and water. In the reaction, H^+ is transferred to the carbonate to give carbonic acid, H_2CO_3, which breaks down rapidly to CO_2 and H_2O. The simpler net ionic equation is written by omitting the metal ions and chloride ions that are not reacting.

$$HCl(aq) + NaHCO_3(aq) \longrightarrow CO_2(g) + H_2O(l) + NaCl(aq)$$
$$H^+(aq) + HCO_3^-(aq) \longrightarrow CO_2(g) + H_2O(l)$$

Acid Rain

Rain typically has a pH of 6.2. It is slightly acidic because carbon dioxide in the air combines with water to form carbonic acid. However, in many parts of the world, rain has become considerably more acidic, with pH values as low as 3 being reported. One cause of acid rain is the sulfur dioxide (SO_2) gas produced when coal that contains sulfur is burned.

In the air, the SO_2 gas reacts with oxygen to produce SO_3, which then combines with water to form sulfuric acid, H_2SO_4, a strong acid.

$$S(s) + O_2(g) \longrightarrow SO_2(g)$$
$$2SO_2(g) + O_2(g) \longrightarrow 2SO_3(g)$$
$$SO_3(g) + H_2O(l) \longrightarrow H_2SO_4(aq)$$

In parts of the United States, acid rain has made lakes so acidic they are no longer able to support fish and plant life. Limestone ($CaCO_3$) is sometimes added to these lakes to neutralize the acid. In Eastern Europe, acid rain has brought about an environmental disaster. Nearly 40% of the forests in Poland have been severely damaged, and some parts of the land are so acidic that crops will not grow. Throughout Europe and the United States, monuments made of marble (a form of $CaCO_3$) are deteriorating as acid rain dissolves the marble.

$$2H^+(aq) + CaCO_3(s) \longrightarrow Ca^{2+}(aq) + H_2O(l) + CO_2(g)$$

1935   1994

Marble statue in Washington Square Park

Efforts to slow or stop the damaging effects of acid rain include the reduction of sulfur emissions. This will require installation of expensive equipment in coal-burning plants to absorb more SO_2 before it is emitted. In some outdated plants, this may be impossible, and they will need to be closed. It is a difficult problem for engineers and scientists, but one that must be solved.

$$2HCl(aq) + Na_2CO_3(aq) \longrightarrow CO_2(g) + H_2O(l) + 2NaCl(aq)$$
$$2H^+(aq) + CO_3^{2-}(aq) \longrightarrow CO_2(g) + H_2O(l)$$

Acids and Hydroxides: Neutralization

Neutralization is a reaction between an acid and a base to produce a salt and water. The cation in the salt comes from the base and the anion comes from the acid. In the neutralization reactions of strong acids and strong bases, water is always one of the products. We can write the following equation for the neutralization of HCl and NaOH.

$$HCl(aq) + NaOH(aq) \longrightarrow NaCl(aq) + H_2O(l)$$
Acid   Base   Salt   Water

To see the actual reaction, we write the strong acid and strong base as individual ions.

$$\mathbf{H^+}(aq) + Cl^-(aq) + Na^+(aq) + \mathbf{OH^-}(aq) \longrightarrow$$
$$Na^+(aq) + Cl^-(aq) + \mathbf{H_2O}(l) \quad \text{Ionic equation}$$

In the neutralization reaction, H^+ from the acid reacts with OH^- from the base to form water, leaving the spectator ions from the salt (Na^+ and Cl^-) in solution.

When we omit the spectator ions from the ionic equation, we obtain the *net ionic equation*, which allows us to see that the net reaction for neutralization is the reaction of H^+ and OH^-.

$$H^+(aq) + \cancel{Cl^-(aq)} + \cancel{Na^+(aq)} + OH^-(aq) \longrightarrow$$
$$\cancel{Na^+(aq)} + \cancel{Cl^-(aq)} + H_2O^-(l) \quad \text{Ionic equation}$$

$$H^+(aq) + OH^-(aq) \longrightarrow H_2O(l) \quad \text{Net ionic equation}$$

Balancing Neutralization Equations

In a neutralization reaction, one H^+ always reacts with one OH^-. Therefore, the coefficients in the neutralization equation must be chosen so that the H^+ from the acid is balanced by the OH^- provided by the base. We balance the neutralization of HCl and $Ba(OH)_2$ as follows.

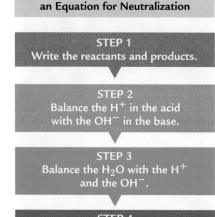

Guide to Balancing an Equation for Neutralization

STEP 1
Write the reactants and products.

STEP 2
Balance the H^+ in the acid with the OH^- in the base.

STEP 3
Balance the H_2O with the H^+ and the OH^-.

STEP 4
Write the salt from the remaining ions.

■ **STEP 1 Write the reactants and products.**

$$HCl(aq) + Ba(OH)_2(s) \longrightarrow H_2O(l) + \text{salt}$$

■ **STEP 2 Balance the H^+ in the acid with the OH^- in the base.** Placing a 2 in front of the HCl provides $2H^+$ for the 2 OH^- in $Ba(OH)_2$.

$$2HCl(aq) + Ba(OH)_2(s) \longrightarrow H_2O(l) + \text{salt}$$

■ **STEP 3 Balance the H_2O with the H^+ and OH^-.**

$$2HCl(aq) + Ba(OH)_2(s) \longrightarrow 2H_2O(l) + \text{salt}$$

■ **STEP 4 Write the salt from the remaining ions in the acid and base.**

$$2HCl(aq) + Ba(OH)_2(s) \longrightarrow 2H_2O(l) + BaCl_2(aq)$$

SAMPLE PROBLEM 10.11

■ **Reactions of Acids**

Write a balanced equation for the reaction of $HCl(aq)$ with each of the following:

a. $Al(s)$ **b.** $K_2CO_3(aq)$ **c.** $Mg(OH)_2(s)$

SOLUTION

a. Al

■ **STEP 1 Write the reactants and products.** When a metal reacts with an acid, the products are H_2 gas and a salt.

$$Al(s) + HCl(aq) \longrightarrow H_2(g) + \text{salt}$$

■ **STEP 2 Determine the formula of the salt.** When $Al(s)$ dissolves, it forms Al^{3+}, which is balanced by 3 Cl^- from HCl.

$$Al(s) + HCl(aq) \longrightarrow H_2(g) + AlCl_3(aq)$$

■ **STEP 3 Balance the equation.**

$$2Al(s) + 6HCl(aq) \longrightarrow 3H_2(g) + 2AlCl_3(aq)$$

b. K_2CO_3

■ **STEP 1 Write the reactants and products.** When a carbonate reacts with an acid, the products are $CO_2(g)$, $H_2O(l)$, and a salt.

$$K_2CO_3(s) + HCl(aq) \longrightarrow CO_2(g) + H_2O(l) + \text{salt}$$

■ STEP 2 **Determine the formula of the salt.**   When $K_2CO_3(s)$ dissolves, it forms K^+, which is balanced by 1 Cl^- from HCl.

$$K_2CO_3(s) + HCl(aq) \longrightarrow CO_2(g) + H_2O(l) + KCl(aq)$$

■ STEP 3 **Balance the equation.**

$$K_2CO_3(aq) + 2HCl(aq) \longrightarrow CO_2(g) + H_2O(l) + 2KCl(aq)$$

c. $Mg(OH)_2$

■ STEP 1 **Write the reactants and products.**   When a base reacts with an acid, the products are $H_2O(l)$ and a salt.

$$Mg(OH)_2(s) + HCl(aq) \longrightarrow H_2O(l) + salt$$

■ STEP 2 **Balance the H^+ in the acid with the OH^- in the base.**   Placing a 2 in front of the HCl provides 2 H^+ for 2 OH^- in $Mg(OH)_2$.

$$Mg(OH)_2(s) + 2HCl(aq) \longrightarrow H_2O(l) + salt$$

■ STEP 3 **Balance the H_2O with the H^+ and OH^-.**

$$Mg(OH)_2(s) + 2HCl(aq) \longrightarrow 2H_2O(l) + salt$$

■ STEP 4 **Write the salt from the remaining ions in the acid and base.**

$$Mg(OH)_2(s) + 2HCl(aq) \longrightarrow 2H_2O(l) + MgCl_2(aq)$$

STUDY CHECK

Write the balanced equation for the reaction between H_2SO_4 and $NaHCO_3$.

Acid–Base Titration

Suppose we need to find the molarity of an HCl solution of unknown concentration. We can do this by a laboratory procedure called **titration** in which we neutralize an acid sample with a known amount of base. In our titration, we first place a measured volume of the acid in a flask and add a few drops of an *indicator* such as phenolphthalein. In an acidic solution, phenolphthalein is colorless. Then we fill a buret with a NaOH solution of known molarity and carefully add NaOH to the acid in the flask, as shown in Figure 10.6.

In the titration, we neutralize the acid by adding a volume of base that contains a matching number of moles of OH^-. We know that neutralization has taken place

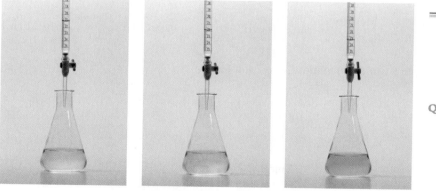

FIGURE 10.6

The titration of an acid. A known volume of an acid is placed in a flask with an indicator and titrated with a measured volume of NaOH to the neutralization point.

Q **What data is needed to determine the molarity of the acid in the flask?**

Antacids

Antacids are substances used to neutralize excess stomach acid (HCl). Some antacids are mixtures of aluminum hydroxide and magnesium hydroxide. These hydroxides are not very soluble in water, so the levels of available OH^- are not damaging to the intestinal tract. However, aluminum hydroxide has the side effects of producing constipation and binding phosphate in the intestinal tract, which may cause weakness and loss of appetite. Magnesium hydroxide has a laxative effect. These side effects are less likely when a combination of the antacids is used.

$$Al(OH)_3(aq) + 3HCl(aq) \longrightarrow AlCl_3(aq) + 3H_2O(l)$$
$$Mg(OH)_2(s) + 2HCl(aq) \longrightarrow MgCl_2(aq) + 2H_2O(l)$$

Some antacids use calcium carbonate to neutralize excess stomach acid. About 10% of the calcium is absorbed into the bloodstream, where it elevates the levels of serum calcium. Calcium carbonate is not recommended for patients who have peptic ulcers or a tendency to form kidney stones.

$$CaCO_3(s) + 2HCl(aq) \longrightarrow$$
$$H_2O(l) + CO_2(g) + CaCl_2(aq)$$

Still other antacids contain sodium bicarbonate. This type of antacid has a tendency to increase blood pH and elevate sodium levels in the body fluids. It also is not recommended in the treatment of peptic ulcers.

$$NaHCO_3(s) + HCl(aq) \longrightarrow NaCl(aq) + CO_2(g) + H_2O(l)$$

The neutralizing substances in some antacid preparations are given in Table 10.8.

TABLE 10.8

Basic Compounds in Some Antacids

Antacid	Base(s)
Amphojel	$Al(OH)_3$
Milk of magnesia	$Mg(OH)_2$
Mylanta, Maalox, Di-Gel, Gelusil, Riopan	$Mg(OH)_2$, $Al(OH)_3$
Bisodol	$CaCO_3$, $Mg(OH)_2$
Titralac, Tums, Pepto-Bismol	$CaCO_3$
Alka-Seltzer	$NaHCO_3$, $KHCO_3$

when the phenolphthalein in the solution changes from colorless to pink. This is called the neutralization *endpoint*. From the volume added and molarity of the NaOH, we can calculate the number of moles of NaOH and then the concentration of the acid.

SAMPLE PROBLEM 10.12

Titration of an Acid

A 25.0 mL sample of an HCl solution is placed in a flask with a few drops of phenolphthalein (indicator). If 32.6 mL of a 0.185 M NaOH is needed to reach the endpoint, what is the concentration (M) of the HCl solution?

$$NaOH(aq) + HCl(aq) \longrightarrow NaCl(aq) + H_2O(l)$$

SOLUTION

■ **STEP 1 Given:** 32.6 mL of 0.185 M NaOH; 25.0 mL HCl = 0.0250 L HCl
    Need: Molarity of HCl

■ **STEP 2 Plan**

32.6 mL → [Metric factor] → L → [Molarity factor] → moles NaOH → [Mole factor] → moles HCl → [Divide by liters.] → M HCl

■ **STEP 3 Equalities/Conversion Factors**

$$1 \text{ L NaOH} = 1000 \text{ mL NaOH}$$
$$\frac{1 \text{ L}}{1000 \text{ mL}} = \frac{1000 \text{ mL}}{1 \text{ L}}$$

$$1 \text{ L NaOH} = 0.185 \text{ mole NaOH}$$
$$\frac{1 \text{ L}}{0.185 \text{ mole NaOH}} = \frac{0.185 \text{ mole NaOH}}{1 \text{ L}}$$

$$1 \text{ mole HCl} = 1 \text{ mole NaOH}$$
$$\frac{1 \text{ mole HCl}}{1 \text{ mole NaOH}} = \frac{1 \text{ mole NaOH}}{1 \text{ mole HCl}}$$

■ **STEP 4 Set up problem.**

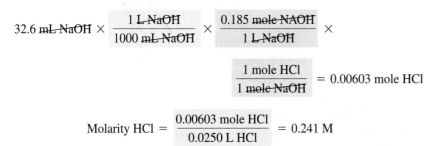

$$32.6 \text{ mL NaOH} \times \frac{1 \text{ L NaOH}}{1000 \text{ mL NaOH}} \times \frac{0.185 \text{ mole NAOH}}{1 \text{ L NaOH}} \times$$

$$\frac{1 \text{ mole HCl}}{1 \text{ mole NaOH}} = 0.00603 \text{ mole HCl}$$

$$\text{Molarity HCl} = \frac{0.00603 \text{ mole HCl}}{0.0250 \text{ L HCl}} = 0.241 \text{ M}$$

STUDY CHECK

What is the molarity of an HCl solution, if 28.6 mL of a 0.175 M NaOH solution is needed to neutralize a 25.0 mL sample of the HCl solution?

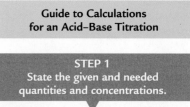

Guide to Calculations for an Acid–Base Titration

STEP 1
State the given and needed quantities and concentrations.

STEP 2
Write a plan to calculate molarity or volume.

STEP 3
State equalities and conversion factors including concentration.

STEP 4
Set up problem to calculate needed quantity.

QUESTIONS AND PROBLEMS

■ **Reactions of Acids and Bases**

10.47 Complete and balance the equation for the following reactions:
  a. $ZnCO_3(s) + HBr(aq) \longrightarrow$
  b. $Zn(s) + HCl(aq) \longrightarrow$
  c. $HCl(aq) + NaHCO_3(s) \longrightarrow$
  d. $H_2SO_4(aq) + Mg(OH)_2(s) \longrightarrow$

10.48 Complete and balance the equations for the following reactions:
  a. $KHCO_3(s) + HCl(aq) \longrightarrow$
  b. $Ca(s) + H_2SO_4(aq) \longrightarrow$
  c. $H_2SO_4(aq) + Al(OH)_3(s) \longrightarrow$
  d. $Na_2CO_3(s) + H_2SO_4(aq) \longrightarrow$

10.49 Balance each of the following neutralization reactions:
  a. $HCl(aq) + Mg(OH)_2(s) \longrightarrow MgCl_2(aq) + H_2O(l)$
  b. $H_3PO_4(aq) + LiOH(aq) \longrightarrow Li_3PO_4(aq) + H_2O(l)$

10.50 Balance each of the following neutralization reactions:
  a. $HNO_3(aq) + Ba(OH)_2(s) \longrightarrow Ba(NO_3)_2(aq) + H_2O(l)$
  b. $H_2SO_4(aq) + Al(OH)_3(aq) \longrightarrow Al_2(SO_4)_3(aq) + H_2O(l)$

10.51 Write a balanced equation for the neutralization of each of the following:
  a. $H_2SO_4(aq)$ and $NaOH(aq)$
  b. $HCl(aq)$ and $Fe(OH)_3(aq)$
  c. $H_2CO_3(aq)$ and $Mg(OH)_2(s)$

10.52 Write a balanced equation for the neutralization of each of the following:
  a. $H_3PO_4(aq)$ and $NaOH(aq)$
  b. $HI(aq)$ and $LiOH(aq)$
  c. $HNO_3(aq)$ and $Ca(OH)_2(s)$

10.53 What is the molarity of an HCl solution if 5.00 mL HCl solution is titrated with 28.6 mL of 0.145 M NaOH solution?

$$HCl(aq) + NaOH(aq) \longrightarrow NaCl(aq) + H_2O(l)$$

10.54 If 29.7 mL of 0.205 M KOH is required to completely neutralize 25.0 mL of a CH_3COOH solution, what is the molarity of the acetic acid solution?

$$CH_3COOH(aq) + KOH(aq) \longrightarrow CH_3COOK(aq) + H_2O(l)$$

10.55 If 38.2 mL of 0.163 M KOH is required to neutralize completely 25.0 mL H_2SO_4 solution, what is the molarity of the acid solution?

$$H_2SO_4(aq) + 2KOH(aq) \longrightarrow K_2SO_4(aq) + 2H_2O(l)$$

10.56 A solution of 0.162 M NaOH is used to neutralize 25.0 mL H_2SO_4 solution. If 32.8 mL NaOH solution is required to reach the endpoint, what is the molarity of the H_2SO_4 solution?

$$H_2SO_4(aq) + 2NaOH(aq) \longrightarrow Na_2SO_4(aq) + 2H_2O(l)$$

10.57 A solution of 0.204 M NaOH is used to neutralize 50.0 mL H_3PO_4 solution. If 16.4 mL of NaOH solution is required to reach the endpoint, what is the molarity of the H_3PO_4 solution?

$$H_3PO_4(aq) + 3NaOH(aq) \longrightarrow Na_3PO_4(aq) + 3H_2O(l)$$

10.58 A solution of 0.312 M KOH is used to neutralize 15.0 mL H_3PO_4 solution. If 28.3 mL KOH solution is required to reach the endpoint, what is the molarity of the H_3PO_4 solution?

$$H_3PO_4(aq) + 3KOH(aq) \longrightarrow K_3PO_4(aq) + 3H_2O(l)$$

10.7 Acid–Base Properties of Salt Solutions

LEARNING GOAL

Predict whether a salt will form an acidic, basic, or neutral solution.

When a salt dissolves in water, it dissociates into cations and anions. Solutions of salts can be acidic, basic, or neutral. Anions and cations from strong acids and bases do not affect pH; however, anions from weak acids and cations from weak bases change the pH of an aqueous solution.

Salts That Form Neutral Solutions

A solution of a salt containing the cation from a strong base and the anion from a strong acid will be neutral. For example, a salt such as $NaNO_3$ forms a neutral solution.

$$NaNO_3(s) \xrightarrow{\text{H}_2\text{O}} Na^+(aq) + NO_3^-(aq) \quad \text{Neutral solution}$$

$$\substack{\text{Does not} \\ \text{change H}^+} \qquad \substack{\text{Does not attract} \\ \text{H}^+ \text{ from water}} \quad \text{(pH = 7.0)}$$

The cation, Na^+, from a strong base does not change H^+, and the anion, NO_3^-, from a strong acid does not attract H^+ from water. Thus there is no effect on the pH of water; the solution is neutral with a pH of 7.0. Other salts such as NaCl, KCl, KNO_3, and KBr also contain cations from strong acids and anions from strong bases and also form neutral solutions.

Some Components of Neutral Salt Solutions

Cations of strong bases: Group 1A (1): Li^+, Na^+, K^+
Group 2A (2): Ca^{2+}, Mg^{2+}, Sr^{2+}, Ba^{2+}

Anions of strong acids: Cl^-, Br^-, I^-, NO_3^-, ClO_4^-

Salts That Form Basic Solutions

A salt solution containing the cation from a strong base and the anion from a weak acid produces a basic solution. Suppose we have a solution of the salt NaF, which contains Na^+ and F^- ions.

$$NaF(s) \xrightarrow{\text{H}_2\text{O}} Na^+(aq) + F^-(aq)$$

$$\substack{\text{Does not} \\ \text{change H}^+} \qquad \substack{\text{Attracts H}^+ \\ \text{from water}}$$

The metal ion Na^+ has no effect on the pH of the solution. However, F^- is the conjugate base of the weak acid HF. Thus, F^- will attract a proton from water and leave OH^- in solution, which makes it basic.

$$F^-(aq) + H_2O(l) \rightleftharpoons HF(aq) + OH^-(aq) \quad \text{Basic solution (pH} > 7.0)$$

Other salts with anions from weak acids such as NaCN, KNO_2, and Na_2SO_4 also produce basic solutions.

Some Components of Basic Salt Solutions

Cations of strong bases: Group 1A (1): Li^+, Na^+, K^+
Group 2A (2): Ca^{2+}, Mg^{2+}, Sr^{2+}, Ba^{2+}

Anions of weak acids: F^-, NO_2^-, CN^-, CO_3^{2-}, SO_4^{2-}, CH_3COO^-, S^{2-}, PO_4^{3-}

Salts That Form Acidic Solutions

A salt solution containing a cation from a weak base and an anion from a strong acid produces an acidic solution. Suppose we have a solution of the salt NH_4Cl, which contains NH_4^+ and Cl^- ions.

$$NH_4Cl(s) \xrightarrow{H_2O} NH_4^+(aq) + Cl^-(aq)$$

Donates H^+ Does not attract
to water H^+ from water

The anion Cl^- has no effect on the pH of the solution. However, as a weak acid, the cation NH_4^+ donates a proton to water, which produces H_3O^+.

$$NH_4^+(aq) + H_2O(l) \rightleftharpoons NH_3(aq) + H_3O^+(aq) \quad \text{Acidic solution (pH} > 7.0)$$

Some Components of Acidic Salt Solutions

Cations of weak bases: NH_4^+ and Be^{2+}, Al^{3+}, Zn^{2+}, Cr^{3+}, Fe^{3+}
(small, highly charged metal ions)

Anions of strong acids: Cl^-, Br^-, I^-, NO_3^-, ClO_4^-

Table 10.9 summarizes the cations and anions of salts that form neutral, basic, and acidic solutions. Table 10.10 summarizes the acid–base properties of some typical salts in water.

TABLE 10.9

Cations and Anions of Salts in Neutral, Basic, and Acidic Salt Solutions			
Type of Solution	**Cations**	**Anions**	**pH**
Neutral	From strong bases Group 1A (1): Li^+, Na^+, K^+ Group 2A (2): Ca^{2+}, Mg^{2+}, Sr^{2+}, Ba^{2+} (but not Be^{2+})	From strong acids Cl^-, Br^-, I^-, NO_3^-, ClO_4^-	7.0
Basic	From strong bases Group 1A (1): Li^+, Na^+, K^+ Group 2A (2): Ca^{2+}, Mg^{2+}, Sr^{2+}, Ba^{2+} (but not Be^{2+})	From weak acids F^-, NO_2^-, CN^-, CO_3^{2-}, SO_4^{2-}, CH_3COO^-, S^{2-}, PO_4^{3-}	>7.0
Acidic	From weak bases NH_4^+, Be^{2+}, Al^{3+}, Zn^{2+}, Cr^{3+}, Fe^{3+} (small, highly charged metal ions)	From strong acids Cl^-, Br^-, I^-, NO_3^-, ClO_4^-	<7.0

TABLE 10.10

Acid–Base Properties of Some Salt Solutions

Typical Salts	Types of Ions	pH	Solution
NaCl, MgBr$_2$, KNO$_3$	cation from a strong base anion from a strong acid	7.0	Neutral
NaF, MgCO$_3$, KNO$_2$	cation from a strong base anion from a weak acid	>7.0	Basic
NH$_4$Cl, FeBr$_3$, Al(NO$_3$)$_3$	cation from a weak base anion from a strong acid	<7.0	Acidic

(MC GOB) Salts of Weak Acids and Bases

Sometimes a salt contains the cation of a weak base and the anion of a weak acid. For example, when NH$_4$F dissociates in water, it produces NH$_4^+$ and F$^-$. We have seen that NH$_4^+$ forms an acidic solution, and F$^-$ forms a basic solution. The ion that reacts to a greater extent with water determines whether the solution is acidic or basic. The salt solution will be neutral only if the ions react with water to the same extent. The determination of these reactions is complex and will not be considered in this text.

SAMPLE PROBLEM 10.13

■ **Predicting the Acid–Base Properties of Salt Solutions**

Predict whether solutions of each of the following salts would be acidic, basic, or neutral:

a. KCN
b. NH$_4$Br
c. NaNO$_3$

SOLUTION

a. KCN

There are only 6 strong acids; all other acids are weak. Bases with cations from Groups 1A(1) and 2A(2) are strong; all other bases are weak. For the salt, KCN, the cation is from a strong base (KOH), but the anion is from a weak acid (HCN).

$$KCN(s) \xrightarrow{H_2O} K^+(aq) + CN^-(aq)$$

The cation K$^+$ has no effect on pH. However, the anion CN$^-$ will attract protons from water to produce a basic solution.

$$CN^-(aq) + H_2O(l) \rightleftharpoons HCN(aq) + OH^-(aq)$$

b. NH$_4$Br

In the salt NH$_4$Br, the cation is from a weak base, but the anion is from a strong acid.

$$NH_4Br(s) \xrightarrow{H_2O} NH_4^+(aq) + Br^-(aq)$$

The anion Br$^-$ has no effect on pH because it is from HBr, a strong acid. However, the cation NH$_4^+$ will donate protons to water to produce an acidic solution.

$$NH_4^+(aq) + H_2O(l) \rightleftharpoons NH_3(aq) + H_3O^+(aq)$$

c. $NaNO_3$

The salt $NaNO_3$ contains a cation from a strong base ($NaOH$) and an anion from a strong acid (HNO_3). Thus, there is no change of pH; the salt solution is neutral.

$$NaNO_3(s) \xrightarrow{H_2O} Na^+(aq) + NO_3^-(aq)$$

STUDY CHECK

Would a solution of Na_3PO_4 be acidic, basic, or neutral?

QUESTIONS AND PROBLEMS

▪ Acid–Base Properties of Salt Solutions

10.59 Why does a salt containing a cation from a strong base and an anion from a weak acid form a basic solution?

10.60 Why does a salt containing a cation from a weak base and an anion from a strong acid form an acidic solution?

10.61 Predict whether each of the following salts will form an acidic, basic, or neutral solution. For acidic and basic solutions, write an equation for the reaction that takes place.
    a. $MgCl_2$ **b.** NH_4NO_3 **c.** Na_2CO_3 **d.** K_2S

10.62 Predict whether each of the following salts will form an acidic, basic, or neutral solution. For acidic and basic solutions, write an equation for the reaction that takes place.
    a. Na_2SO_4 **b.** KBr **c.** $BaCl_2$ **d.** NH_4I

▪ 10.8 Buffers

LEARNING GOAL

Describe the role of buffers in maintaining the pH of a solution.

When a small amount of acid or base is added to pure water, the pH changes drastically. However, if a solution is buffered, there is little change in pH. A **buffer solution** resists a change in pH when small amounts of acid or base are added. For example, blood is buffered to maintain a pH of about 7.4. If the pH of the blood goes even slightly above or below this value, changes in our uptake of oxygen and our metabolic process can be drastic enough to cause death. Even though we are constantly obtaining acids and bases from foods and biological processes, the buffers in the body so effectively absorb those compounds that blood pH remains essentially unchanged. (See Figure 10.7.)

SELF-STUDY ACTIVITY
pH and Buffers

In a buffer, an acid must be present to react with any OH^- that is added, and a base must be available to react with any added H_3O^+. However, that acid and base must not be able to neutralize each other. Therefore, a combination of an acid–base conjugate pair is used in buffers. Most buffer solutions consist of nearly equal concentrations of a weak acid and a salt containing its conjugate base. (See Figure 10.8.) Buffers may also contain a weak base and the salt of the weak base, which contains its conjugate acid.

For example, a typical buffer contains acetic acid (CH_3COOH) and a salt such as sodium acetate ($NaCH_3COO$). As a weak acid, acetic acid dissociates slightly in water to form H_3O^+ and a very small amount of CH_3COO^-. The presence of the salt provides a much larger concentration of acetate ion (CH_3COO^-), which is necessary for its buffering capability.

Preparing Buffer Solutions

$$CH_3COOH(aq) + H_2O(l) \rightleftharpoons H_3O^+(aq) + CH_3COO^-(aq)$$

Large amount   Large amount

FIGURE 10.7

Adding an acid or a base to water changes the pH drastically, but a buffer resists pH change when small amounts of acid or base are added.

Q Why does the pH change several pH units when acid is added to water, but not when acid is added to a buffer?

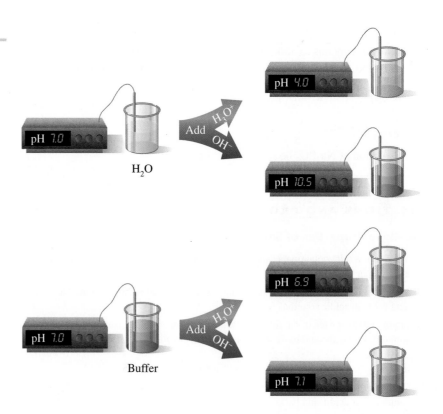

Let's see how this buffer solution maintains the H_3O^+ concentration. When a small amount of acid is added, it will combine with the acetate ion (anion) as the equilibrium shifts to the reactant acetic acid. There will be a small decrease in the $[CH_3COO^-]$ and a small increase in $[CH_3COOH]$, but the $[H_3O^+]$ will not change very much.

$$CH_3COOH(aq) + H_2O(l) \longleftarrow H_3O^+(aq) + CH_3COO^-(aq)$$

If a small amount of base is added to this buffer solution, it is neutralized by the acetic acid and water is produced. The concentration of $[CH_3COOH]$ decreases slightly and the $[CH_3COO^-]$ increases slightly, but again the $[H_3O^+]$ does not change very much.

$$CH_3COOH(aq) + OH^-(aq) \longrightarrow H_2O(aq) + CH_3COO^-(aq)$$

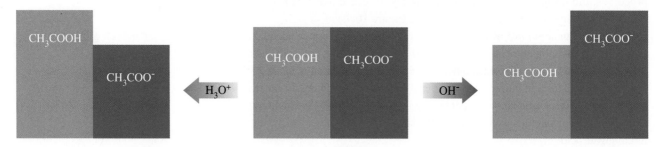

FIGURE 10.8

The buffer described here consists of about equal concentrations of acetic acid (CH_3COOH) and its conjugate base of acetate ion (CH_3COO^-). Adding H_3O^+ to the buffer uses up some CH_3COO^-, whereas adding OH^- neutralizes some CH_3COOH. The pH of the solution is maintained as long as the added amounts of acid or base are small compared to the concentrations of the buffer components.

Q How does this acetic acid/acetate ion buffer maintain pH?

SAMPLE PROBLEM 10.14

▪ **Identifying Buffer Solutions**

Indicate whether each of the following would make a buffer solution:

a. HCl, a strong acid, and NaCl
b. H_3PO_4, a weak acid
c. HF, a weak acid, and NaF

SOLUTION

a. No. A solution of a strong acid and its salt is completely ionized.
b. No. A weak acid is not sufficient for a buffer; the salt of the weak acid is also needed.
c. Yes. This mixture contains a weak acid and its salt.

STUDY CHECK

Will a mixture of NaCl and Na_2CO_3 make a buffer solution? Explain.

Calculating the pH of a Buffer

By rearranging the K_a expression to give $[H_3O^+]$ we can obtain the ratio of the acetic acid/acetate buffer.

MC GOB Calculating the pH of a Buffer

$$K_a = \frac{[H_3O^+][CH_3COO^-]}{[CH_3COOH]}$$

$$[H_3O^+] = K_a \times \frac{[CH_3COOH]}{[CH_3COO^-]}$$

Because K_a is a constant, the $[H_3O^+]$ is determined by the $[CH_3COOH]/[CH_3COO^-]$ ratio. As long as the addition of small amounts of either acid or base changes the ratio of $[CH_3COOH]/[CH_3COO^-]$ only slightly, the changes in $[H_3O^+]$ will be small and the pH will be maintained. It is important to note that the amount of acid or base that is added must be small compared to the supply of the buffer components CH_3COOH and CH_3COO^-. If a large amount of acid or base is added, the buffering capacity of the system may be exceeded.

Buffers are also prepared from other conjugate acid–base pairs such as $H_2PO_4^-/HPO_4^{2-}$, HPO_4^{2-}/PO_4^{3-}, HCO_3^-/CO_3^{2-}, NH_4^+/NH_3. The pH of the buffer solution will depend on the acid–base pair chosen.

SAMPLE PROBLEM 10.15

▪ **pH of a Buffer**

The K_a for acetic acid, CH_3COOH, is 1.8×10^{-5}. What is the pH of a buffer prepared with 1.0 M CH_3COOH and 1.0 M CH_3COO^-?

SOLUTION

$$CH_3COOH(aq) + H_2O(l) \rightleftharpoons H_3O^+(aq) + CH_3COO^-(aq)$$

▪ STEP 1 Write the K_a expression.

$$K_a = \frac{[H_3O^+][CH_3COO^-]}{[CH_3COOH]}$$

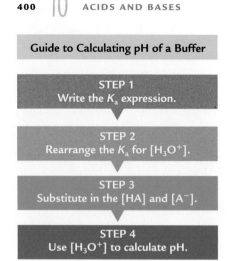

Guide to Calculating pH of a Buffer

STEP 1
Write the K_a expression.

STEP 2
Rearrange the K_a for $[H_3O^+]$.

STEP 3
Substitute in the $[HA]$ and $[A^-]$.

STEP 4
Use $[H_3O^+]$ to calculate pH.

■ **STEP 2 Rearrange K_a for $[H_3O^+]$.**

$$[H_3O^+] = K_a \times \frac{[CH_3COOH]}{[CH_3COO^-]}$$

■ **STEP 3 Substitute $[HA]$ and $[A^-]$.**   Substituting these values in the expression for $[H_3O^+]$ gives

$$[H_3O^+] = 1.8 \times 10^{-5} \times \frac{[1.0\ M]}{[1.0\ M]}$$

$$[H_3O^+] = 1.8 \times 10^{-5}$$

■ **STEP 4 Use $[H_3O^+]$ to calculate pH.**   Using the concentration of $[H_3O^+]$ in the pH expression gives the pH of the buffer.

$$pH = -\log[1.8 \times 10^{-5}] = 4.74$$

STUDY CHECK

One of the conjugate acid–base pairs that buffers the blood is $H_2PO_4^-/HPO_4^{2-}$ with a K_a of 6.2×10^{-8}. What is the pH of a buffer that is 0.50 M $H_2PO_4^-$ and 0.50 M HPO_4^{2-}?

QUESTIONS AND PROBLEMS

■ **Buffers**

10.63 Which of the following represent a buffer system? Explain.
 a. NaOH and NaCl
 b. H_2CO_3 and $NaHCO_3$
 c. HF and KF
 d. KCl and NaCl

10.64 Which of the following represent a buffer system? Explain.
 a. H_3PO_4 **b.** $NaNO_3$ **c.** CH_3COOH and $NaCH_3COO$ **d.** HCl and NaOH

10.65 Consider the buffer system of hydrofluoric acid, HF, and its salt, NaF.

$$HF(aq) + H_2O(l) \rightleftharpoons H_3O^+(aq) + F^-(aq)$$

 a. What is the purpose of the buffer system?
 b. Why is a salt of the acid needed?
 c. How does the buffer react when some H_3O^+ is added?
 d. How does the buffer react when some OH^- is added?

10.66 Consider the buffer system of nitrous acid, HNO_2, and its salt, $NaNO_2$.

$$HNO_2(aq) + H_2O(l) \rightleftharpoons H_3O^+(aq) + NO_2^-(aq)$$

 a. What is the purpose of a buffer system?
 b. What is the purpose of $NaNO_2$ in the buffer?
 c. How does the buffer react when some H_3O^+ is added?
 d. How does the buffer react when some OH^- is added?

10.67 Nitrous acid has a K_a of 4.5×10^{-4}. What is the pH of a buffer solution containing 0.10 M HNO_2 and 0.10 M NO_2^-?

10.68 Acetic acid has a K_a of 1.8×10^{-5}. What is the pH of a buffer solution containing 0.15 M CH_3COOH (acetic acid) and 0.15 M CH_3COO^-?

10.69 Compare the pH of a HF buffer that contains 0.10 M HF and 0.10 M NaF with another HF buffer that contains 0.060 M HF and 0.120 M NaF. See Table 10.4, page 376.

10.70 Compare the pH of a H_2CO_3 buffer that contains 0.10 M H_2CO_3 and 0.10 M $NaHCO_3$ with another H_2CO_3 buffer that contains 0.15 M H_2CO_3 and 0.050 M $NaHCO_3$. See Table 10.4, page 376.

HEALTH NOTE

Buffers in the Blood

The arterial blood has a normal pH of 7.35–7.45. If changes in H_3O^+ lower the pH below 6.8 or raise it above 8.0, cells cannot function properly and death may result. In our cells, CO_2 is continually produced as an end product of cellular metabolism. Some CO_2 is carried to the lungs for elimination, and the rest dissolves in body fluids such as plasma and saliva, forming carbonic acid. As a weak acid, carbonic acid dissociates to give bicarbonate and H_3O^+. More of the anion HCO_3^- is supplied by the kidneys to give an important buffer system in the body fluid: the H_2CO_3/HCO_3^- buffer.

$$CO_2 + H_2O \rightleftharpoons H_2CO_3 \rightleftharpoons H_3O^+ + HCO_3^-$$

Excess H_3O^+ entering the body fluids reacts with the HCO_3^- and excess OH^- reacts with the carbonic acid.

$H_2CO_3 + H_2O \longleftarrow H_3O^+ + HCO_3^-$ Equilibrium shifts left

$H_2CO_3 + OH^- \longrightarrow H_2O + HCO_3^-$ Equilibrium shifts right

For the carbonic acid, we can write the equilibrium expression as

$$K_a = \frac{[H_3O^+][HCO_3^-]}{[H_2CO_3]}$$

To maintain the normal blood pH (7.35–7.45), the ratio of H_2CO_3/HCO_3^- needs to be about 1 to 10, which is obtained by typical concentrations in the blood of 0.0024 M H_2CO_3 and 0.024 M HCO_3^-.

$$[H_3O^+] = K_a \times \frac{[H_2CO_3]}{[HCO_3^-]}$$

$$= 4.3 \times 10^{-7} \times \frac{[0.0024]}{[0.024]} = 4.3 \times 10^{-7} \times 0.10 = 4.3 \times 10^{-8}$$

$$pH = -\log(4.3 \times 10^{-8}) = 7.37$$

In the body, the concentration of carbonic acid is closely associated with the partial pressure of CO_2. Table 10.11 lists the normal values for arterial blood. If the CO_2 level rises, producing more H_2CO_3, the equilibrium produces more H_3O^+, which lowers the pH. This condition is called *acidosis*. Difficulty with ventilation or gas diffusion can lead to respiratory acidosis, which can happen in emphysema or when the medulla of the brain is affected by an accident or depressive drugs.

A lowering of the CO_2 level leads to a high blood pH, a condition called *alkalosis*. Excitement, trauma, or a high temperature may cause a person to hyperventilate, which expels large amounts of CO_2. As the partial pressure of CO_2 in the blood falls below normal, the equilibrium shifts from H_2CO_3 to CO_2 and H_2O. This shift decreases the $[H_3O^+]$ and raises the pH. Table 10.12 lists some of the conditions that lead to changes in the blood pH and some possible treatments. The kidneys also regulate H_3O^+ and HCO_3^- components, but more slowly than the adjustment made by the lungs through ventilation.

TABLE 10.11

Normal Values for Blood Buffer in Arterial Blood

P_{CO_2}	40 mm Hg
H_2CO_3	2.4 mmoles/L of plasma
HCO_3^-	24 mmoles/L of plasma
pH	7.35–7.45

TABLE 10.12

Acidosis and Alkalosis: Symptoms, Causes, and Treatments

Respiratory Acidosis: $CO_2 \uparrow$ pH $\downarrow$

Symptoms:	Failure to ventilate, suppression of breathing, disorientation, weakness, coma
Causes:	Lung disease blocking gas diffusion (e.g., emphysema, pneumonia, bronchitis, and asthma); depression of respiratory center by drugs, cardiopulmonary arrest, stroke, poliomyelitis, or nervous system disorders
Treatment:	Correction of disorder, infusion of bicarbonate

Respiratory Alkalosis: $CO_2 \downarrow$ pH $\uparrow$

Symptoms:	Increased rate and depth of breathing, numbness, light-headedness, tetany
Causes:	Hyperventilation due to anxiety, hysteria, fever, exercise; reaction to drugs such as salicylate, quinine, and antihistamines; conditions causing hypoxia (e.g., pneumonia, pulmonary edema, and heart disease)
Treatment:	Elimination of anxiety-producing state, rebreathing into a paper bag

Metabolic Acidosis: $H^+ \uparrow$ pH $\downarrow$

Symptoms:	Increased ventilation, fatigue, confusion
Causes:	Renal disease, including hepatitis and cirrhosis; increased acid production in diabetes mellitus, hyperthyroidism, alcoholism, and starvation; loss of alkali in diarrhea; acid retention in renal failure
Treatment:	Sodium bicarbonate given orally, dialysis for renal failure, insulin treatment for diabetic ketosis

Metabolic Alkalosis: $H^+ \downarrow$ pH $\uparrow$

Symptoms:	Depressed breathing, apathy, confusion
Causes:	Vomiting, diseases of the adrenal glands, ingestion of excess alkali
Treatment:	Infusion of saline solution, treatment of underlying diseases

CONCEPT MAP

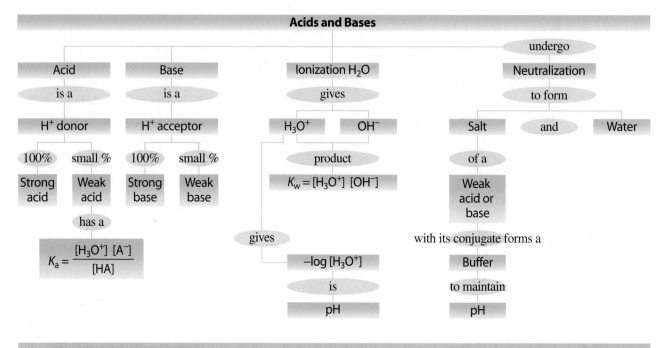

CHAPTER REVIEW

10.1 Acids and Bases
According to the Arrhenius theory, an acid produces H^+, and a base produces OH^- in aqueous solutions.

10.2 Brønsted–Lowry Acids and Bases
According to the Brønsted–Lowry theory, acids are proton (H^+) donors, and bases are proton acceptors. Two conjugate acid–base pairs are present in an acid–base reaction. Each acid–base pair is related by the loss or gain of one H^+. For example, when the aqueous acid HF donates a H^+, the F^- it forms is its conjugate base because F^- is capable of accepting a H^+. The other acid–base pair would be H_2O and H_3O^+.

10.3 Strengths of Acids and Bases
In strong acids, all the H^+ in the acid is donated to H_2O; in a weak acid, only a small percentage of acid molecules produce H_3O^+. Strong bases are hydroxides of Groups 1A (1) and 2A (2) that dissociate completely in water. An important weak base is ammonia, NH_3. In water, weak acids and weak bases produce only a few ions when equilibrium is reached. The reaction for a weak acid can be written as $HA + H_2O \rightleftharpoons H_3O^+ + A^-$. The acid dissociation constant is written as $K_a = \dfrac{[H_3O^+][A^-]}{[HA]}$.

10.4 Ionization of Water
In pure water, a few molecules transfer protons to other water molecules, producing small, but equal, amounts of $[H_3O^+]$ and $[OH^-]$,

such that each has a concentration of 1×10^{-7} mole/L. The ion product, K_w, $[H_3O^+][OH^-] = 1 \times 10^{-14}$, applies to all aqueous solutions. In acidic solutions, the $[H_3O^+]$ is greater than the $[OH^-]$. In basic solutions, the $[OH^-]$ is greater than the $[H_3O^+]$.

10.5 The pH Scale
The pH scale is a range of numbers from 0 to 14 related to the $[H_3O^+]$ of the solution. A neutral solution has a pH of 7. In acidic solutions, the pH is below 7, and in basic solutions the pH is above 7. Mathematically, pH is the negative logarithm of the hydronium ion concentration $(-\log[H_3O^+])$.

10.6 Reactions of Acids and Bases
When an acid reacts with a metal, hydrogen gas and a salt are produced. The reaction of an acid with a carbonate or bicarbonate produces carbon dioxide, a salt, and water. In neutralization, an acid reacts with a base to produce a salt and water. In titration, an acid sample is neutralized with a known amount of a base. From the volume and molarity of the base, the concentration of the acid is calculated.

10.7 Acid–Base Properties of Salt Solutions
A salt of a weak acid contains an anion that removes protons from water and makes the solution basic. A salt of a weak base contains an ion that donates a proton to water producing an acidic solution. Salts of strong acids and bases produce neutral solutions because they contain ions that do not affect the pH.

10.8 Buffers

A buffer solution resists changes in pH when small amounts of acid or base are added. A buffer contains either a weak acid and its salt or a weak base and its salt. The weak acid picks up added OH^-, and the anion of the salt picks up added H^+. Buffers are important in maintaining the pH of the blood.

KEY TERMS

acid A substance that dissolves in water and produces hydrogen ions (H^+), according to the Arrhenius theory. All acids are proton donors, according to the Brønsted–Lowry theory.

acid dissociation constant (K_a) The product of the ions from the dissociation of a weak acid divided by the concentration of the weak acid.

amphoteric Substances that can act as either an acid or a base in water.

base A substance that dissolves in water and produces hydroxide ions (OH^-) according to the Arrhenius theory. All bases are proton acceptors, according to the Brønsted–Lowry theory.

Brønsted–Lowry acids and bases An acid is a proton donor, and a base is a proton acceptor.

buffer solution A mixture of a weak acid or a weak base and its salt that resists changes in pH when small amounts of an acid or base are added.

conjugate acid–base pair An acid and base that differ by one H^+. When an acid donates a proton, the product is its conjugate base, which is capable of accepting a proton in the reverse reaction.

dissociation The separation of an acid or base into ions in water.

hydronium ion, H_3O^+ The ion formed by the attraction of a proton (H^+) to an H_2O molecule.

ion-product constant of water, K_w The product of $[H_3O^+]$ and $[OH^-]$ in solution; $K_w = [H_3O^+][OH^-]$.

neutral The term that describes a solution with equal concentrations of H_3O^+ and OH^-.

neutralization A reaction between an acid and a base to form a salt and water.

pH A measure of the $[H_3O^+]$ in a solution; $pH = -\log[H_3O^+]$.

strong acid An acid that completely ionizes in water.

strong base A base that completely ionizes in water.

titration The addition of base to an acid sample to determine the concentration of the acid.

weak acid An acid that ionizes only slightly in solution.

weak base A base that ionizes only slightly in solution.

UNDERSTANDING THE CONCEPTS

10.71 In each of the following diagrams of acid solutions, determine if each diagram represents a strong acid or a weak acid. The acid has the formula HX.

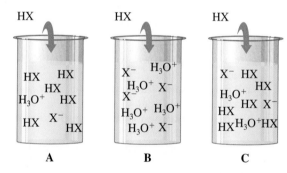

A   **B**   **C**

10.72 Adding a few drops of a strong acid to water will lower the pH appreciably. However, adding the same number of drops to a buffer does not appreciably alter the pH. Why?

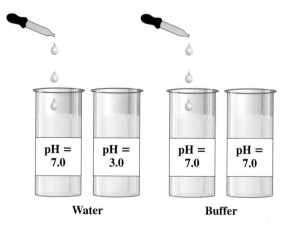

Water   **Buffer**

10.73 Sometimes, during stress or trauma, a person can start to hyperventilate. Then the person might breathe into a paper bag to avoid fainting.

a. What changes occur in the blood pH during hyperventilation?

b. How does breathing into a paper bag help return blood pH to normal?

10.74 In the blood plasma, pH is maintained by the carbonic acid–bicarbonate buffer system.

a. How is pH maintained when acid is added to the buffer system?

b. How is pH maintained when base is added to the buffer system?

ADDITIONAL QUESTIONS AND PROBLEMS

For instructor-assigned homework, go to
www.masteringgob.com.

10.75 Name each of the following:
   a. H_2SO_4 **b.** KOH   **c.** $Ca(OH)_2$
   d. HCl   **e.** HNO_2

10.76 Are the following examples of body fluids acidic, basic, or neutral?
   a. saliva, pH 6.8   **b.** urine, pH 5.9
   c. pancreatic juice, pH 8.0   **d.** bile, pH 8.4
   e. blood, pH 7.45

10.77 What are some similarities and differences between strong and weak acids?

10.78 What are some ingredients found in antacids? What do they do?

10.79 One ingredient in some antacids is $Mg(OH)_2$.
   a. If the base is not very soluble in water, why is it considered a strong base?
   b. What is the neutralization reaction of $Mg(OH)_2$ with stomach acid, HCl?

10.80 Acetic acid, which is the acid in vinegar, is a weak acid. Why?

10.81 Using Table 10.3, determine which is the stronger acid in each of the following pairs:
   a. HF or HCN   **b.** H_3O^+ or NH_4^+
   c. HNO_2 or CH_3COOH **d.** H_2O or HCO_3^-

10.82 Using Table 10.3, determine which is the stronger base in each of the following pairs:
   a. H_2O or Cl^- **b.** OH^- or NH_3
   c. SO_4^{2-} or NO_2^- **d.** CO_3^{2-} or H_2O

10.83 Determine the pH for the following solutions:
   a. $[H_3O^+] = 2.0 \times 10^{-8}$ M
   b. $[H_3O^+] = 5.0 \times 10^{-2}$ M
   c. $[OH^-] = 3.5 \times 10^{-4}$ M
   d. $[OH^-] = 0.0054$ M

10.84 Are the solutions in Problem 10.83 acidic, basic, or neutral?

10.85 What are the $[H_3O^+]$ and $[OH^-]$ for a solution with the following pH values?
   a. 3.00   **b.** 6.48   **c.** 8.85   **d.** 11.00

10.86 Solution A has a pH of 4.5 and solution B has a pH of 6.7.
   a. Which solution is more acidic?
   b. What is the $[H_3O^+]$ in each?
   c. What is the $[OH^-]$ in each?

10.87 What is the $[OH^-]$ in a solution that contains 0.225 g NaOH in 0.250 L of solution?

10.88 What is the $[H_3O^+]$ in a solution that contains 1.54 g HNO_3 in 0.500 L of solution?

10.89 What is the pH of a solution prepared by dissolving 2.5 g HCl in water to make 425 mL of solution?

10.90 What is the pH of a solution prepared by dissolving 1.00 g $Ca(OH)_2$ in water to make 875 mL of solution?

10.91 Calculate the volume (mL) of a 0.150 M NaOH solution that will completely neutralize the following:
   a. 25.0 mL of a 0.288 M HCl solution
   b. 10.0 mL of a 0.560 M H_2SO_4 solution

10.92 How many milliliters of 0.215 M NaOH solution are needed to completely neutralize 2.50 mL of 0.825 M H_2SO_4 solution?

10.93 A solution of 0.205 M NaOH is used to neutralize 20.0 mL H_2SO_4. If 45.6 mL NaOH is required to reach the endpoint, what is the molarity of the H_2SO_4 solution?

$$H_2SO_4(aq) + 2NaOH(aq) \longrightarrow$$
$$Na_2SO_4(aq) + 2H_2O(l)$$

10.94 A 10.0 mL sample of vinegar, which is an aqueous solution of acetic acid, CH_3COOH, requires 16.5 mL of 0.500 M NaOH to reach the endpoint in a titration. What is the molarity of the acetic acid solution?

$$CH_3COOH(aq) + NaOH(aq) \longrightarrow$$
$$CH_3COONa(aq) + H_2O(l)$$

10.95 Will solutions of the following salts be acidic, basic, or neutral?
   a. KF   **b.** NaCN
   c. NH_4NO_3 **d.** NaBr

10.96 Will solutions of the following salts be acidic, basic, or neutral?
   a. K_2SO_4 **b.** KNO_2
   c. $MgCl_2$ **d.** NH_4Cl

10.97 A buffer is made by dissolving H_3PO_4 and NaH_2PO_4 in water.
   a. Write an equation that shows how this buffer neutralizes small amounts of acids.
   b. Write an equation that shows how this buffer neutralizes small amounts of base.
   c. Calculate the pH of this buffer if it is 0.10 M H_3PO_4 and 0.10 M $H_2PO_4^{2-}$; the K_a for H_3PO_4 is 7.5×10^{-3}.

10.98 A buffer is made by dissolving CH_3COOH and CH_3COONa in water.
   a. Write an equation that shows how this buffer neutralizes small amounts of acid.
   b. Write an equation that shows how this buffer neutralizes small amounts of base.
   c. Calculate the pH of this buffer if it is 0.10 M CH_3COOH and 0.10 M CH_3COO^-; the K_a for CH_3COOH is 1.8×10^{-5}.

CHALLENGE QUESTIONS

10.99 Consider the following:
   1. H_2S 2. H_3PO_4 3. HCO_3^-
   a. For each, write the formula of the conjugate base.
   b. For each, write the K_a expression.
   c. Write the formula of the weakest acid.
   d. Write the formula of the strongest acid.

10.100 Identify the conjugate acid–base pairs in each of the following equations and whether the equilibrium mixture contains mostly products or mostly reactants.
 a. $NH_3(aq) + HNO_3(aq) \rightleftharpoons NH_4^+(aq) + NO_3^-(aq)$
 b. $H_2O(l) + HBr(aq) \rightleftharpoons H_3O^+(aq) + Br^-(aq)$
 c. $HNO_2(aq) + HS^-(aq) \rightleftharpoons H_2S(g) + NO_2^-(aq)$
 d. $Cl^-(aq) + H_2O(l) \rightleftharpoons OH^-(aq) + HCl(aq)$

10.101 Complete and balance each of the following:
 a. $ZnCO_3(s) + H_2SO_4(aq) \longrightarrow$
 b. $Al(s) + HCl(aq) \longrightarrow$
 c. $H_3PO_4(aq) + Ca(OH)_2(s) \longrightarrow$
 d. $KHCO_3(s) + HNO_3(aq) \longrightarrow$

10.102 Predict whether a solution of each of the following salts is acidic, basic, or neutral. For salts that form acidic or basic solutions, write a balanced equation for the reaction.

 a. NH_4Br **b.** KNO_2 **c.** $Mg(NO_3)_2$
 d. BaF_2 **e.** K_2S

10.103 Determine each of the following for a 0.050 M KOH solution.
 a. $[H_3O^+]$ **b.** pH
 c. products when reacted with H_3PO_4
 d. milliliters required to neutralize 40.0 mL of 0.035 M H_2SO_4

10.104 Consider the reaction of KOH and HNO_2.
 a. Write the balanced chemical equation.
 b. Calculate the milliliters of 0.122 M KOH required to neutralize 36.0 mL of 0.250 M HNO_2.
 c. Determine whether the final solution would be acidic, basic, or neutral.

ANSWERS

Answers to Study Checks

10.1 Sulfuric acid; potassium hydroxide

10.2 $HNO_3(aq) + H_2O(l) \longrightarrow H_3O^+(aq) + NO_3^-(aq)$

10.3 A base accepts a proton to form its conjugate acid.
 a. H_2S **b.** HNO_2

10.4 The conjugate acid/base pairs are HCN/CN^- and SO_4^{2-}/HSO_4^-.

10.5 $HNO_3 + H_2O \rightleftharpoons H_3O^+ + NO_3^-$
 The products are favored because HNO_3 is a stronger acid than H_3O^+.

10.6 Nitrous acid has a larger K_a than carbonic acid, it dissociates more in H_2O, forms more $[H_3O^+]$, and is a stronger acid.

10.7 $[H_3O^+] = 2.5 \times 10^{-11}$ M; basic

10.8 11.38

10.9 4.3

10.10 $[H_3O^+] = 3.2 \times 10^{-5}$ M; $[OH^-] = 3.1 \times 10^{-10}$ M

10.11 $H_2SO_4(aq) + 2NaHCO_3(s) \longrightarrow$
 $Na_2SO_4(aq) + 2CO_2(g) + 2H_2O(l)$

10.12 0.200 M HCl

10.13 The anion PO_4^{3-} reacts with H_2O forming the weak acid HPO_4^{2-} and OH^-, which makes the solution basic.

10.14 No. Both substances are salts; the mixture has no weak acid present.

10.15 pH = 7.21

Answers to Selected Questions and Problems

10.1 **a.** acid **b.** acid **c.** acid **d.** base

10.3 **a.** hydrochloric acid **b.** calcium hydroxide
 c. carbonic acid **d.** nitric acid
 e. sulfurous acid

10.5 **a.** $Mg(OH)_2$ **b.** HF **c.** H_3PO_4
 d. LiOH **e.** $Cu(OH)_2$

10.7 **a.** HI is the acid (proton donor) and H_2O is the base (proton acceptor).
 b. H_2O is the acid (proton donor) and F^- is the base (proton acceptor)

10.9 **a.** F^-, fluoride ion **b.** OH^-, hydroxide ion
 c. HCO_3^-, bicarbonate ion *or* hydrogen carbonate ion
 d. SO_4^{2-}, sulfate ion

10.11 **a.** HCO_3^-, bicarbonate ion *or* hydrogen carbonate ion
 b. H_3O^+, hydronium ion **c.** H_3PO_4, phosphoric acid
 d. HBr, hydrobromic acid

10.13 **a.** acid H_2CO_3; conjugate base HCO_3^-
 base H_2O; conjugate acid H_3O^+
 b. acid NH_4^+; conjugate base NH_3
 base H_2O; conjugate acid H_3O^+
 c. acid HCN; conjugate base CN^-
 base NO_2^-; conjugate acid HNO_2

10.15 A strong acid is a good proton donor, whereas its conjugate base is a poor proton acceptor.

10.17 **a.** HBr **b.** HSO_4^- **c.** H_2CO_3

10.19 **a.** HSO_4^- **b.** HF **c.** HCO_3^-

10.21 **a.** reactants **b.** reactants **c.** products

10.23 The reactants are favored because NH_4^+ is a weaker acid than HSO_4^-.

$$NH_4^+(aq) + SO_4^{2-}(aq) \rightleftharpoons$$
$$NH_3(aq) + HSO_4^-(aq)$$

10.25 **a.** H_2SO_3 **b.** HSO_3^- **c.** H_2SO_3
 d. HS^- **e.** H_2SO_3

10.27 $H_3PO_4(aq) + H_2O(l) \rightleftharpoons H_3O^+(aq) + H_2PO_4^-(aq)$

$$K_a = \frac{[H_3O^+][H_2PO_4^-]}{[H_3PO_4]}$$

10.29 In pure water, $[H_3O^+] = [OH^-]$ because one of each is produced every time a proton transfers from one water to another.

10.31 In an acidic solution, the $[H_3O^+]$ is greater than the $[OH^-]$.

10.33 **a.** acidic **b.** basic **c.** basic **d.** acidic

10.35 a. 1.0×10^{-9} M   **b.** 1.0×10^{-6} M
c. 2.0×10^{-5} M   **d.** 4.0×10^{-13} M

10.37 a. 1.0×10^{-11} M   **b.** 2.0×10^{-9} M
c. 5.6×10^{-3} M   **d.** 2.5×10^{-2} M

10.39 In a neutral solution, the $[H_3O^+]$ is 1.0×10^{-7} M and the pH is 7.00, which is the negative value of the power of 10.

10.41 a. basic   **b.** acidic   **c.** basic   **d.** acidic

10.43 a. 4.00   **b.** 8.52   **c.** 9.00   **d.** 3.40

10.45

$[H_3O^+]$	$[OH^-]$	pH	Acidic, Basic, or Neutral?
1.0×10^{-8} M	1.0×10^{-6} M	8.00	Basic
1.0×10^{-3} M	1.0×10^{-11} M	3.00	Acidic
2.8×10^{-5} M	3.6×10^{-10} M	4.55	Acidic
1.0×10^{-12} M	1.0×10^{-2} M	12.00	Basic

10.47 a. $ZnCO_3(s) + 2HBr(aq) \longrightarrow$
$\qquad ZnBr_2(aq) + CO_2(g) + H_2O(l)$
b. $Zn(s) + 2HCl(aq) \longrightarrow ZnCl_2(aq) + H_2(g)$
c. $HCl(g) + NaHCO_3(s) \longrightarrow$
$\qquad NaCl(aq) + H_2O(l) + CO_2(g)$
d. $H_2SO_4(aq) + Mg(OH)_2(s) \longrightarrow MgSO_4(aq) + 2H_2O(l)$

10.49 a. $2HCl(aq) + Mg(OH)_2(s) \longrightarrow MgCl_2(aq) + 2H_2O(l)$
b. $H_3PO_4(aq) + 3LiOH(aq) \longrightarrow Li_3PO_4(aq) + 3H_2O(l)$

10.51 a. $H_2SO_4(aq) + 2NaOH(aq) \longrightarrow Na_2SO_4(aq) + 2H_2O(l)$
b. $3HCl(aq) + Fe(OH)_3(aq) \longrightarrow FeCl_3(aq) + 3H_2O(l)$
c. $H_2CO_3(aq) + Mg(OH)_2(s) \longrightarrow MgCO_3(s) + 2H_2O(l)$

10.53 0.829 M HCl

10.55 0.125 M H_2SO_4

10.57 0.0223 M H_3PO_4

10.59 The anion from the weak acid removes a proton from H_2O to make a basic solution.

10.61 a. neutral
b. acidic $NH_4^+(aq) + H_2O(l) \rightleftharpoons$
$\qquad NH_3(aq) + H_3O^+(aq)$
c. basic $CO_3^{2-}(aq) + H_2O(l) \rightleftharpoons HCO_3^-(aq) + OH^-(aq)$
d. basic $S^{2-}(aq) + H_2O(l) \rightleftharpoons HS^-(aq) + OH^-(aq)$

10.63 (b) and (c) are buffer systems. (b) contains the weak acid H_2CO_3 and its salt $NaHCO_3$. (c) contains HF, a weak acid, and its salt KF.

10.65 a. A buffer system keeps the pH constant.
b. To neutralize any H_3O^+ added.
c. The added H_3O^+ reacts with F^- from NaF.
d. The added OH^- is neutralized by the HF.

10.67 $[H_3O^+] = 4.5 \times 10^{-4} \times \dfrac{[0.10 \text{ M}]}{[0.10 \text{ M}]} = 4.5 \times 10^{-4}$

$pH = -\log[4.5 \times 10^{-4}] = 3.35$

10.69 The pH of the 0.10 M HF/0.10 M NaF buffer is 3.14. The pH of the 0.060 M HF/0.120 M NaF buffer is 3.44.

10.71   a. weak acid   **b.** strong acid   **c.** weak acid

10.73 a. Hyperventilation will lower the CO_2 level in the blood, which lowers the H_2CO_3 concentration, which decreases the H_3O^+ and increases the blood pH.
b. Breathing into a bag will increase the CO_2 level, increase the H_2CO_3, increase H_3O^+, and lower the blood pH.

10.75 a. sulfuric acid   **b.** potassium hydroxide
c. calcium hydroxide   **d.** hydrochloric acid
e. nitrous acid

10.77 Both strong and weak acids produce H_3O^+ in water. Weak acids are only slightly ionized, whereas a strong acid exists as ions in solution.

10.79 a. The $Mg(OH)_2$ that dissolves is completely dissociated, making it a strong base.
b. $Mg(OH)_2(aq) + 2HCl(aq) \longrightarrow MgCl_2(aq) + 2H_2O(l)$

10.81 a. HF   **b.** H_3O^+ **c.** HNO_2 **d.** HCO_3^-

10.83 a. pH 7.70   **b.** pH 1.30
c. pH 10.54   **d.** pH = 11.73

10.85 a. $[H_3O^+] = 1.0 \times 10^{-3}$ M; $[OH^-] = 1.0 \times 10^{-11}$ M
b. $[H_3O^+] = 3.3 \times 10^{-7}$ M; $[OH^-] = 3.0 \times 10^{-8}$ M
c. $[H_3O^+] = 1.4 \times 10^{-9}$ M; $[OH^-] = 7.1 \times 10^{-6}$ M
d. $[H_3O^+] = 1.0 \times 10^{-11}$ M; $[OH^-] = 1.0 \times 10^{-3}$ M

10.87 $[OH^-] = 0.0225$ M

10.89 pH = 0.80

10.91 a. 48.0 mL NaOH   **b.** 74.7 mL NaOH

10.93 0.234 M H_2SO_4

10.95 a. basic   **b.** basic   **c.** acidic   **d.** neutral

10.97 a. acid: $H_2PO_4^-(aq) + H_3O^+(aq) \longrightarrow$
$\qquad H_3PO_4(aq) + H_2O(l)$
b. base: $H_3PO_4(aq) + OH^-(aq) \longrightarrow$
$\qquad H_2PO_4^-(aq) + H_2O(l)$
c. pH = 2.12

10.99 a. 1. HS^- 2. $H_2PO_4^-$ 3. CO_3^{2-}

b. 1. $\dfrac{[H_3O^+][HS^-]}{[H_2S]}$ 2. $\dfrac{[H_3O^+][H_2PO_4^-]}{[H_3PO_4]}$

3. $\dfrac{[CO_3^{2-}][H_2O^+]}{[HCO_3^-]}$

c. H_2S **d.** H_3PO_4

10.101 a. $ZnCO_3(s) + H_2SO_4(aq) \longrightarrow$
$\qquad ZnSO_4(aq) + CO_2(g) + H_2O(l)$
b. $2Al(s) + 6HCl(aq) \longrightarrow 2AlCl_3(aq) + 3H_2(g)$
c. $2H_3PO_4(aq) + 3Ca(OH)_2(s) \longrightarrow$
$\qquad Ca_3(PO_4)_2(aq) + 6H_2O(l)$
d. $KHCO_3(s) + HNO_3(aq) \longrightarrow$
$\qquad KNO_3(aq) + CO_2(g) + H_2O(l)$

10.103 $KOH \longrightarrow K^+ + OH^-$
$[OH^-] = 0.050$ M $= 5.0 \times 10^{-2}$ M
a. $[H_3O^+] = \dfrac{1.0 \times 10^{-14}}{[5.0 \times 10^{-2}]} = 2.0 \times 10^{-13}$ M
b. pH $= -\log[2.0 \times 10^{-13}] = 12.70$
c. $3KOH(aq) + H_3PO_4(aq) \longrightarrow K_3PO_4(aq) + 3H_2O(l)$
d. 56 mL KOH solution

11

Introduction to Organic Chemistry: Alkanes

"When we have a hazardous materials spill, the first thing we do is isolate it," says Don Dornell, assistant fire chief, Burlingame Fire Station. "Then our technicians and a county chemist identify the product from its flammability and solubility in water so we can use the proper materials to clean up the spill. We use different methods for alcohol, which mixes with water, than for gasoline, which floats. Because hydrocarbons are volatile, we use foam to cover them and trap the vapors. At oil refineries, we will use foams, but many times we squirt water on the tanks to cool the contents below their boiling points, too. By knowing the boiling point of the product and its density and vapor density, we know if it floats or sinks in water and where its vapors will go."

LOOKING AHEAD

Mastering GOB CHEMISTRY

Visit **www.masteringgob.com**
for self-study materials
and instructor-assigned
homework.

*O*rganic chemistry is the chemistry of carbon compounds that contain, primarily, carbon and hydrogen. The element carbon has a special role in chemistry because it bonds with other carbon atoms to give a vast array of molecules. The variety of molecules is so great that we find organic compounds in many common products we use such as gasoline, medicine, shampoos, plastic bottles, and perfumes. The food we eat is composed of different organic compounds that supply us with fuel for energy and the carbon atoms needed to build and repair the cells of our bodies.

Although many organic compounds occur in nature, chemists have synthesized even more. The cotton, wool, or silk in your clothes contain naturally occurring organic compounds, whereas materials such as polyester, nylon, or plastic have been synthesized through organic reactions. Sometimes it is convenient to synthesize a molecule in the lab even though that molecule is also found in nature. For example, vitamin C synthesized in a laboratory has the same structure as the vitamin C in oranges or lemons. In these chapters, you will learn about the structures and reactions of organic molecules, which will provide a foundation for understanding the more complex molecules of biochemistry.

LEARNING GOAL

Identify properties characteristic of organic or inorganic compounds.

MC GOB SELF-STUDY ACTIVITY
Introduction to Organic Molecules

11.1 Organic Compounds

At the beginning of the nineteenth century, scientists classified chemical compounds as inorganic and organic. An inorganic compound was a substance that was composed of minerals, and an organic compound was a substance that came from an organism, thus the use of the word "organic." It was thought that some type of "vital force," which could only be found in living cells, was required to synthesize an organic compound. This perception was shown to be incorrect in 1828 when the German chemist Friedrick Wöhler synthesized urea, a product of protein metabolism, by heating an inorganic compound, ammonium cyanate.

$$NH_4CNO \xrightarrow{\text{Heat}} H_2N-\overset{\overset{\displaystyle O}{\|}}{C}-NH_2$$

Ammonium cyanate (inorganic) Urea, organic

We now define organic chemistry as the study of carbon compounds. Most **organic compounds** are nonpolar, and as a result, the attractions between molecules are weak, which accounts for the low melting and boiling points of carbon compounds and lack of solubility in water. For example, vegetable oil, which is a mixture of organic compounds, does not dissolve in water, but floats on top.

Many organic compounds undergo combustion and burn vigorously in air. In contrast, many of the inorganic compounds are ionic, which leads to high melting and boiling points. Inorganic compounds that are ionic or polar covalent are usually soluble in water. Most inorganic substances do not burn in air. Table 11.1 contrasts some of the properties associated with organic and inorganic compounds such as propane, C_3H_8, and sodium chloride, NaCl. (See Figure 11.1.)

TABLE 11.1

Some Typical Properties of Organic and Inorganic Compounds

Property	Organic	Example: C_3H_8	Inorganic	Example: NaCl
Bonding	Mostly covalent	Covalent	Many are ionic, some covalent	Ionic
Polarity of bonds	Nonpolar, unless a more electronegative atom is present	Nonpolar	Most are ionic or polar covalent, a few are nonpolar covalent	Ionic
Melting point	Usually low	$-188°C$	Usually high	$801°C$
Boiling point	Usually low	$-42°C$	Usually high	$1413°C$
Flammability	High	Burns in air	Low	Does not burn
Solubility in water	Not soluble, unless a polar group is present	No	Most are soluble, unless nonpolar	Yes

SAMPLE PROBLEM 11.1

■ **Properties of Organic Compounds**

Indicate whether the following properties are characteristic of organic or inorganic compounds:

a. not soluble in water   **b.** high melting point   **c.** burns in air

SOLUTION

a. Many organic compounds are not soluble in water.
b. Inorganic compounds are most likely to have high melting points.
c. Organic compounds are most likely to be flammable.

STUDY CHECK

Octane is not soluble in water. What type of compound is octane?

MC GOB
Distinguishing Organic from Inorganic Compounds

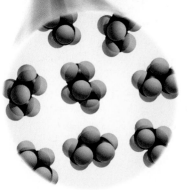

FIGURE 11.1

Propane, C_3H_8, is an organic compound, whereas sodium chloride, NaCl, is an inorganic compound.

Q Why is propane used as a fuel?

Bonding in Organic Compounds

FIGURE 11.2

Three-dimensional representations of methane, CH₄: **(a)** tetrahedron, **(b)** ball-and-stick model, **(c)** space-filling model, **(d)** expanded structural formula.

Q **Why does methane have a tetrahedral shape and not a flat shape?**

The **hydrocarbons** are organic compounds that consist of only carbon and hydrogen. In the simplest hydrocarbon, methane (CH₄), the four valence electrons of carbon are shared with four hydrogen atoms to form an octet. In the electron-dot formula, each shared pair of electrons represents a single bond. In organic molecules, every carbon atom always has four bonds. An **expanded structural formula** is written when we show the bonds between all of the atoms.

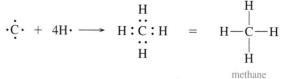

methane

The Tetrahedral Structure of Carbon

The VSEPR theory (Chapter 4) predicts that when four bonds are arranged as far apart as possible, they have a tetrahedral shape. In CH₄ the bonds to hydrogen are directed to the corners of a tetrahedron with bond angles of 109.5°. The three-dimensional structure of methane can be illustrated as a ball-and-stick model or a space-filling model. (See Figure 11.2.)

An organic compound with two atoms of carbon is ethane, C₂H₆. Each carbon atom is bonded to another carbon and three hydrogen atoms. When there are two or more carbon atoms in a molecule, each carbon retains the tetrahedral shape if bonded to four other atoms. In the ball-and-stick model of ethane, C₂H₆, two tetrahedra are attached to each other. (See Figure 11.3.) All the bond angles are close to 109.5°.

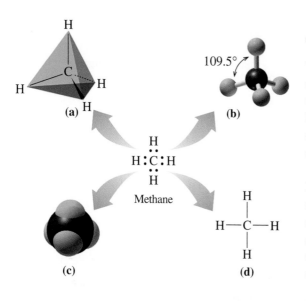

FIGURE 11.3

Three-dimensional representations of ethane, C₂H₆: **(a)** tetrahedral shape of each carbon, **(b)** ball-and-stick model, **(c)** space-filling model, **(d)** expanded structural formula.

Q **How is the tetrahedral shape maintained in a molecule with two carbon atoms?**

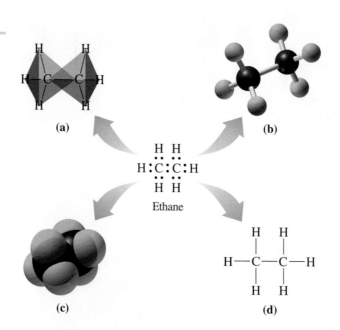

QUESTIONS AND PROBLEMS

▪ Organic Compounds

11.1 Identify the following as formulas of organic or inorganic compounds:
 a. KCl   **b.** C_4H_{10} **c.** CH_3CH_2OH
 d. H_2SO_4 **e.** $CaCl_2$ **f.** CH_3CH_2Cl

11.2 Identify the following as formulas of organic or inorganic compounds:
 a. $C_6H_{12}O_6$ **b.** Na_2SO_4 **c.** I_2
 d. C_4H_9Cl **e.** $C_{10}H_{22}$ **f.** CH_4

11.3 Identify the following properties as most typical of organic or inorganic compounds:
 a. soluble in water   **b.** low boiling point
 c. burns in air   **d.** solid at room temperature

11.4 Identify the following properties as most typical of organic or inorganic compounds:
 a. high melting point   **b.** gas at room temperature
 c. covalent bonds   **d.** produces ions in water

11.5 Match the following physical and chemical properties with the compounds ethane, C_2H_6, or sodium bromide, NaBr.
 a. boils at $-89°C$ **b.** burns vigorously
 c. solid at $250°C$ **d.** dissolves in water

11.6 Match the following physical and chemical properties with the compounds cyclohexane, C_6H_{12}, or calcium nitrate, $Ca(NO_3)_2$.
 a. melts at $500°C$ **b.** insoluble in water
 c. produces ions in water   **d.** is a liquid at room temperature

11.7 Why is the structure of the CH_4 molecule three-dimensional rather than two-dimensional?

11.8 In a propane molecule with three carbon atoms, what is the geometry around each carbon atom?

Propane

11.2 Alkanes

The **alkanes** are a class or family of organic compounds that contain only carbon and hydrogen atoms and only form single bonds. One of the most common uses of alkanes is as fuels. Methane, used in gas heaters and gas cooktops, is an alkane with one carbon atom. The alkanes ethane, propane, and butane contain two, three, and four carbon atoms connected in a row or a *continuous chain*. These names are part of the **IUPAC** (International Union of Pure and Applied Chemistry) **system**, which chemists use to name organic compounds. Alkanes with five or more carbon atoms in a chain are named using Greek prefixes: *pent*(5), *hex*(6), *hept*(7), *oct*(8), *non*(9), and *dec*(10). (See Table 11.2.)

Condensed Structural Formulas

In a **condensed structural formula**, we write each carbon atom and its attached hydrogen atoms as a group. A subscript indicates the number of hydrogen atoms bonded to each carbon atom.

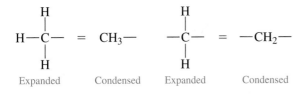

TABLE 11.2

IUPAC Names for the First Ten Continuous-Chain Alkanes

Number of Carbon Atoms	Prefix	Name	Molecular Formula	Condensed Structural Formula
1	meth	methane	CH_4	CH_4
2	eth	ethane	C_2H_6	$CH_3—CH_3$
3	prop	propane	C_3H_8	$CH_3—CH_2—CH_3$
4	but	butane	C_4H_{10}	$CH_3—CH_2—CH_2—CH_3$
5	pent	pentane	C_5H_{12}	$CH_3—CH_2—CH_2—CH_2—CH_3$
6	hex	hexane	C_6H_{14}	$CH_3—CH_2—CH_2—CH_2—CH_2—CH_3$
7	hept	heptane	C_7H_{16}	$CH_3—CH_2—CH_2—CH_2—CH_2—CH_2—CH_3$
8	oct	octane	C_8H_{18}	$CH_3—CH_2—CH_2—CH_2—CH_2—CH_2—CH_2—CH_3$
9	non	nonane	C_9H_{20}	$CH_3—CH_2—CH_2—CH_2—CH_2—CH_2—CH_2—CH_2—CH_3$
10	dec	decane	$C_{10}H_{22}$	$CH_3—CH_2—CH_2—CH_2—CH_2—CH_2—CH_2—CH_2—CH_2—CH_3$

Alkane name   hexane
Molecular formula C_6H_{14}
Ball-and-stick model

Expanded structural formula

Condensed structural formulas

$CH_3—CH_2—CH_2—CH_2—CH_2—CH_3$

Line-bond formula

The molecular formula gives the total number of each kind of atom, but does not indicate the arrangement of the atoms in the molecule.

In **continuous alkanes**, the carbon atoms are connected in a row. However, in chains of three carbon atoms or more, the carbon atoms do not lie in a straight line. The tetrahedral shape of carbon arranges the carbon bonds in a zigzag pattern, which is seen in the ball-and-stick model of hexane. An abbreviated structure called the **line-bond formula** shows only the bonds from carbon to carbon. The ends of the lines and the corners where the lines meet are understood to be carbon atoms attached to the proper number of hydrogen atoms to give four bonds. (See Figure 11.4.)

In an alkane, the groups attached to each carbon are not in fixed positions, but rotate freely about the bond connecting the two carbon atoms. This motion is analogous to the independent rotation of the wheels of a toy car. Thus different arrangements occur during the rotation about a single bond.

Suppose we could look at butane, C_4H_{10}, as it rotates. Sometimes the CH_3 groups line up in front of each other, and at other times they are opposite each other. As the CH_3 groups turn around the single bond, many arrangements are possible. Butane can be depicted by a variety of two-dimensional structural formulas as shown in Table 11.3. It is important to recognize that all of these structural formulas represent the same continuous chain of four carbon atoms.

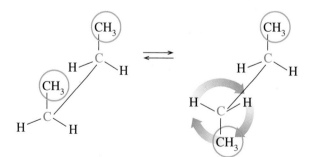

FIGURE 11.4

A ball-and-stick model and some structural formulas of hexane.

Q **Why do the carbon atoms in hexane appear to be arranged in a zigzag chain?**

TABLE 11.3

Some Structural Formulas and Conformations for Butane C$_4$H$_{10}$

Expanded Structural Formula

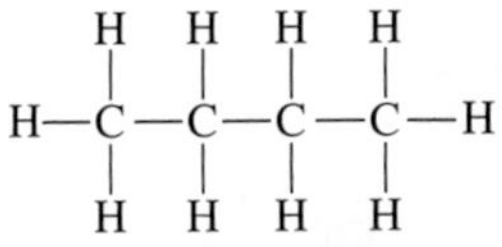

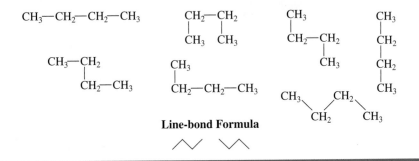

Condensed Structural Formulas

CH$_3$—CH$_2$—CH$_2$—CH$_3$

CH$_2$—CH$_2$
  | |
CH$_3$ CH$_3$

CH$_3$
  |
CH$_2$—CH$_2$
        |
      CH$_3$

CH$_3$
  |
CH$_2$
  |
CH$_2$
  |
CH$_3$

CH$_3$—CH$_2$
         |
       CH$_2$—CH$_3$

CH$_3$
  |
CH$_2$—CH$_2$—CH$_3$

CH$_3$\   /CH$_2$\
       CH$_2$ CH$_3$

Line-bond Formula

SAMPLE PROBLEM 11.2

■ Drawing Structural Formulas for Alkanes

A molecule of butane, C$_4$H$_{10}$, has four carbon atoms in a row. What are its expanded, condensed, and line-bond structural formulas?

SOLUTION

In the expanded structural formula, to give each carbon atom a total of four bonds the four carbon atoms are connected to each other and to hydrogen atoms with single bonds. In the condensed structural formula, each carbon atom and its attached hydrogen atoms are written as CH$_3$—, or —CH$_2$—. The line-bond formula shows only the carbon-to-carbon bonds.

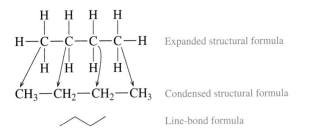

Expanded structural formula

Condensed structural formula

Line-bond formula

STUDY CHECK

Write the expanded, condensed, and line-bond structural formulas of pentane, C$_5$H$_{12}$.

Cycloalkanes

Hydrocarbons can also form cyclic structures called **cycloalkanes**, which have two fewer hydrogen atoms than the corresponding alkanes. Thus, the simplest cycloalkane, cyclopropane, C$_3$H$_6$, has a ring of three carbon atoms bonded to six

TABLE 11.4

Formulas of Some Common Cycloalkanes

Ball and Stick Models

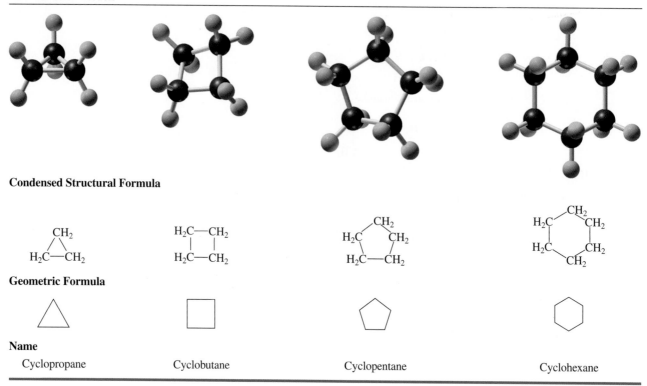

Condensed Structural Formula

CH₂ H₂C—CH₂	H₂C—CH₂ H₂C—CH₂	CH₂ H₂C CH₂ H₂C—CH₂	CH₂ H₂C CH₂ H₂C CH₂ CH₂

Geometric Formula

△   □   ⬠   ⬡

Name

Cyclopropane	Cyclobutane	Cyclopentane	Cyclohexane

hydrogen atoms. A simplified formula, which omits the hydrogen atoms and looks like a geometric figure, is a convenient way to show cyclic structures. Each corner of the triangle represents a carbon atom with four bonds to other carbon and hydrogen atoms.

The ball-and-stick models and their structural formulas for several cycloalkanes are shown in Table 11.4. A cycloalkane is named by adding the prefix *cyclo* to the name of the alkane with the same number of carbon atoms.

SAMPLE PROBLEM 11.3

■ **Naming Alkanes**

Give the IUPAC name for each of the following:

a. CH₃—CH₂—CH₂—CH₂—CH₃   **b.** ⬡   **c.** CH₃—CH₂—CH₃

SOLUTION

a. A chain with five carbon atoms is pentane.
b. The ring of six carbon atoms is named cyclohexane.
c. This alkane is named propane because it has three carbon atoms.

STUDY CHECK

What is the IUPAC name of the following compound?

QUESTIONS AND PROBLEMS

■ Alkanes

11.9 Write the stated type of structural formula for a continuous chain of each of the following:
   a. an expanded structural formula for propane
   b. the condensed structural formula for hexane
   c. a line-bond formula for hexane

11.10 Write the stated type of structural formula for a continuous chain of each of the following:
   a. an expanded structural formula for butane
   b. the condensed structural formula for octane
   c. a line-bond formula for decane

11.11 Give the IUPAC name for each of the following:

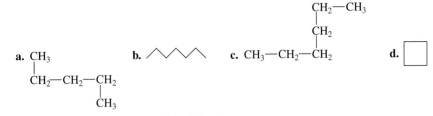

a. CH_3
   CH_2—CH_2—CH_2
                  |
                  CH_3

11.12 Give the IUPAC name for each of the following alkanes.

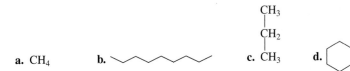

11.13 Write the condensed structural formula or geometric figure for each of the following:
   a. methane   **b.** ethane   **c.** pentane   **d.** cyclopropane

11.14 Write the condensed structural formula or geometric figure for each of the following:
   a. propane   **b.** hexane   **c.** heptane   **d.** cyclopentane

---

■ 11.3 Alkanes with Substituents

LEARNING GOAL

Write the IUPAC names for alkanes with substituents.

When an alkane has four or more carbon atoms, the atoms can be arranged so that a side group called a **branch** or **substituent** is attached to a carbon chain. For example, we can write two different structural formulas for the molecular formula C_4H_{10}. One formula contains the four carbon atoms in a continuous chain. In the other formula, a carbon atom in a *branch* is attached to a chain of three atoms. (See Figure 11.5.) An alkane with at least one branch is called a **branched alkane**. When two compounds have the same molecular formula but different arrangements of atoms, they are called **isomers**.

In another example, we can write the structural formulas of three different isomers with the molecular formula C_5H_{12} as follows:

Isomers of C_5H_{12}

Continuous Chain	Branched Chains	
CH_3—CH_2—CH_2—CH_2—CH_3	$\begin{array}{c} CH_3 \\ \vert \\ CH_3-CH-CH_2-CH_3 \end{array}$	$\begin{array}{c} CH_3 \\ \vert \\ CH_3-C-CH_3 \\ \vert \\ CH_3 \end{array}$

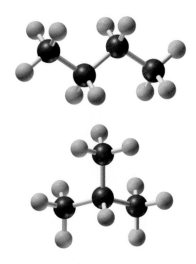

FIGURE 11.5

The isomers of C_4H_{10} have the same number and type of atoms, which are bonded in a different order.

Q **What makes these molecules isomers?**

■ **Isomers**

Identify each pair of structural formulas as isomers or the same molecule.

a. $\underset{\displaystyle CH_2}{\overset{\displaystyle CH_3}{|}}$—$\underset{\displaystyle CH_2}{\overset{\displaystyle CH_3}{|}}$ and CH_2—CH_2—CH_3 with CH_3

b. CH_3—$\overset{\displaystyle CH_3}{\underset{|}{CH}}$—$CH_2$—$CH_2$—$CH_3$ and CH_3—$\overset{\displaystyle CH_3}{\underset{|}{CH}}$—$\overset{\displaystyle CH_3}{\underset{|}{CH}}$—$CH_3$

SOLUTION

a. The structural formulas represent the same molecule because each has four atoms in a continuous chain.

b. These are isomers because the molecular formula C_6H_{14} is identical but the atoms are bonded in a different order. One has one CH_3 side group attached to a five-carbon chain, and the other has two CH_3 side groups attached to a four-carbon chain.

STUDY CHECK

The following compound has the same molecular formula as the compounds in part b of the preceding Sample Problem. Is the following structural formula an isomer or identical to one of the molecules in part b?

CH_3—CH_2—$\overset{\displaystyle CH_3}{\underset{|}{CH}}$—$CH_2$—$CH_3$

TABLE 11.5

Names and Formulas of Some Common Substituents

Substituent	Name
CH_3—	methyl
CH_3—CH_2—	ethyl
CH_3—CH_2—CH_2—	propyl
CH_3—$\overset{\displaystyle}{\underset{|}{CH}}$—$CH_3$	isopropyl
F—, Cl—, Br—, I—	fluoro, chloro, bromo, iodo

MC GOB Naming Alkanes with Substituents

Substituents in Alkanes

In the IUPAC names for alkanes, a carbon branch is named as an **alkyl group**, which is an alkane that is missing one hydrogen atom. The alkyl group is named by replacing the *one* ending of the corresponding alkane name with *yl*. Alkyl groups cannot exist on their own: they must be attached to a carbon chain. When a halogen atom is attached to a carbon chain, it is named as a *halo* group: fluoro (F), chloro (Cl), bromo (Br), or iodo (I). Some of the common groups attached to carbon chains are illustrated in Table 11.5.

Rules for Naming Alkanes with Substituents

In the IUPAC system of naming, the longest carbon chain is numbered to give the location of one or more substituents attached to it. Let's take a look at how we use the IUPAC system to name the following alkane:

CH_3—$\overset{\displaystyle CH_3}{\underset{|}{CH}}$—$CH_2$—$CH_2$—$CH_3$

■ **STEP 1** **Write the alkane name of the longest continuous chain of carbon atoms.** In this alkane, the longest chain has five carbon atoms, which is *pentane*.

$$CH_3-CH-CH_2-CH_2-CH_3$$

with CH_3 branching from the second carbon.

pentane

■ **STEP 2** **Number the carbon atoms starting from the end nearest a substituent.** Once you start numbering, continue in that same direction.

$$\overset{1}{CH_3}-\overset{2}{CH}-\overset{3}{CH_2}-\overset{4}{CH_2}-\overset{5}{CH_3}$$

with CH_3 branching from carbon 2.

pentane

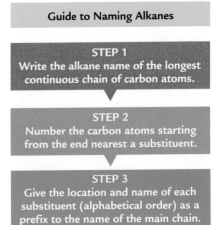

Guide to Naming Alkanes

STEP 1
Write the alkane name of the longest continuous chain of carbon atoms.

STEP 2
Number the carbon atoms starting from the end nearest a substituent.

STEP 3
Give the location and name of each substituent (alphabetical order) as a prefix to the name of the main chain.

■ **STEP 3** **Give the location and name of each substituent as a prefix to the alkane name.** Place a hyphen between the number and the substituent name.

$$\overset{1}{CH_3}-\overset{2}{CH}-\overset{3}{CH_2}-\overset{4}{CH_2}-\overset{5}{CH_3}$$

with CH_3 branching from carbon 2.

2-methyl pentane

List the substituents in alphabetical order.

$$\overset{1}{CH_3}-\overset{2}{CH}-\overset{3}{CH}-\overset{4}{CH_2}-\overset{5}{CH_3}$$

with CH_3 on carbon 2 and Cl on carbon 3.

3-chloro-2-methylpentane

Use a prefix (di-, tri-, tetra-) to indicate a group that appears more than once. Use commas to separate two or more numbers.

$$\overset{1}{CH_3}-\overset{2}{CH}-\overset{3}{CH}-\overset{4}{CH_2}-\overset{5}{CH_3}$$

with CH_3 on carbon 2 and CH_3 on carbon 3.

2,3-dimethylpentane

When there are two or more substituents, the main chain is numbered in the direction that gives the lowest set of numbers.

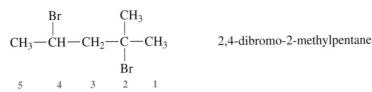

$$\overset{5}{CH_3}-\overset{4}{CH}-\overset{3}{CH_2}-\overset{2}{C}-\overset{1}{CH_3}$$

with Br on carbon 4, CH_3 on carbon 2, and Br below carbon 2.

2,4-dibromo-2-methylpentane

Naming Cycloalkanes

When one substituent is attached to a carbon atom in a ring, the name of the substituent is placed in front of the cycloalkane name. No number is needed for a single alkyl group or halogen atom because the carbon atoms in the cycloalkane are equivalent. However, if two or more groups are attached, the ring is numbered to show the location of each group. The numbering starts by assigning carbon 1 to the

(MC GOB) Naming Cycloalkanes

substituent that gives the lowest numbers to the other substituents. Therefore, we may count clockwise or counterclockwise around a cycloalkane to give the lowest combination of numbers to the substituents.

CH_3

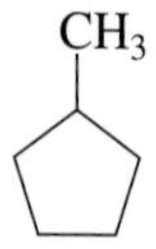

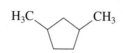

methylcyclopentane

H_3C
CH_3

1,3-dimethylcyclopentane
not 2,4-dimethylcyclohexane

CH_3
Cl

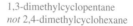

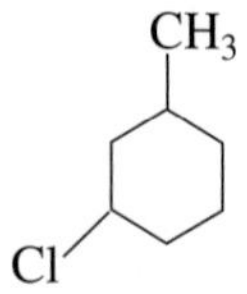

1-chloro-3-methylcyclohexane

SAMPLE PROBLEM 11.5

Writing IUPAC Names

Give the IUPAC name for the following alkane:

CH_3
Br
$CH_3—CH—CH_2—C—CH_2—CH_3$
CH_3

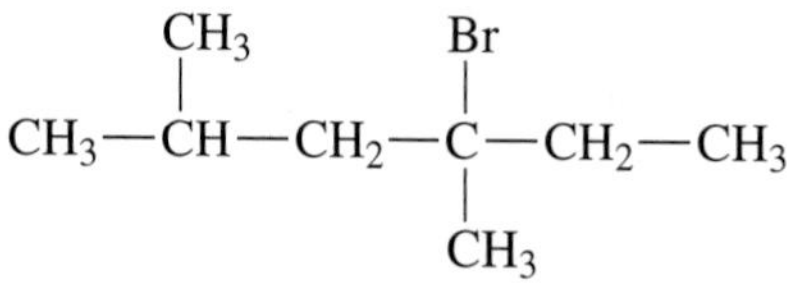

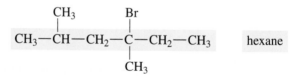

SOLUTION

STEP 1 **Write the alkane name of the longest continuous chain of carbon atoms.** In this alkane, the longest chain has six carbon atoms, which is *hexane*.

CH_3
Br
$CH_3—CH—CH_2—C—CH_2—CH_3$
CH_3

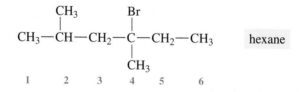

hexane

STEP 2 **Number the carbon atoms starting from the end nearest a substituent.**

CH_3
Br
$CH_3—CH—CH_2—C—CH_2—CH_3$
CH_3
1 2 3 4 5 6

hexane

STEP 3 **Give the location and name of each substituent in front of the name of the longest chain. List the names of different substituents in alphabetical order.** Place a hyphen between the number and the substituent names and commas to separate two or more numbers. A prefix (di-, tri-, tetra-) indicates a group that appears more than once.

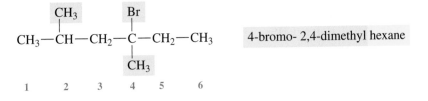

4-bromo- 2,4-dimethyl hexane

STUDY CHECK

Give the IUPAC name for the following compound:

$$CH_3-CH_2-\underset{\underset{CH_3}{|}}{CH}-CH_2-\underset{\underset{CH_3}{|}}{CH}-CH_2-Cl$$

Drawing Structural Formulas for Alkanes

The IUPAC name gives all the information needed to draw the condensed structural formula of an alkane. Suppose you are asked to draw the condensed structural formula of 2,3-dimethylbutane. The alkane name gives the number of carbon atoms in the longest chain. The names in the beginning indicate the substituents and where they are attached. We can break down the name in the following way.

2,3-dimethylbutane

2,3-	di	methyl	but	ane
Substituents on carbons 2 and 3	Two identical groups	CH_3— alkyl groups	4 Carbon atoms in the main chain	Single C—C bonds

SAMPLE PROBLEM 11.6

■ Drawing Structures from IUPAC Names

Write the condensed structural formula for 2,3-dimethylbutane.

SOLUTION

We can use the following guide to draw the condensed structural formula.

- **STEP 1 Draw the main chain of carbon atoms.** For butane, we draw a chain of four carbon atoms.

    C—C—C—C

- **STEP 2 Number the chain and place the substituents on the carbons indicated by the numbers.** The first part of the name indicates two methyl groups (CH_3—), one on carbon 2 and one on carbon 3.

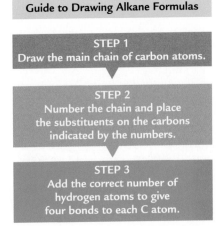

Guide to Drawing Alkane Formulas

STEP 1
Draw the main chain of carbon atoms.

STEP 2
Number the chain and place the substituents on the carbons indicated by the numbers.

STEP 3
Add the correct number of hydrogen atoms to give four bonds to each C atom.

■ **STEP 3** **Add the correct number of hydrogen atoms to give four bonds to each C atom.**

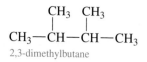

2,3-dimethylbutane

STUDY CHECK

What is the structural formula for 2,4-dimethylpentane?

Haloalkanes

Drawing Haloalkanes and Branched Alkanes

In a **haloalkane**, halogen atoms replace hydrogen atoms in an alkane. The halo substituents are numbered and arranged alphabetically just as we did with the alkyl groups. Many times chemists use the common, traditional name for these compounds rather than the systematic IUPAC name. Simple haloalkanes are commonly named as alkyl halides; the carbon group is named as an alkyl group followed by the halide name.

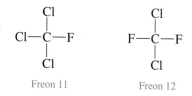

	CH₃—Cl	CH₃—CH₂—Br	CH₃—CH—CH₃ with F
IUPAC:	chloromethane	bromoethane	2-fluoropropane
Common:	methyl chloride	ethyl bromide	isopropyl fluoride

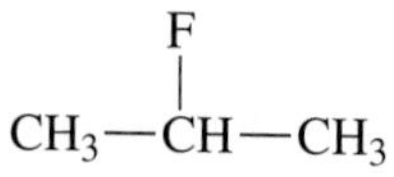

SAMPLE PROBLEM 11.7

■**Naming Haloalkanes**

Freon 11 and Freon 12 are compounds, known as chlorofluorocarbons (CFCs), that were used as refrigerants and aerosol propellants. What are their IUPAC names?

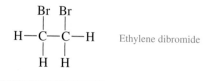

Freon 11   Freon 12

SOLUTION

Freon 11, trichlorofluoromethane; Freon 12, dichlorodifluoromethane.

STUDY CHECK

Ethylene dibromide is the common name of a haloalkane used as a fumigant. What is its IUPAC name?

Br Br
H—C—C—H   Ethylene dibromide
     H H

▪ Common Uses of Haloalkanes

Some common uses of haloalkanes include solvents and anesthetics. For many years, carbon tetrachloride was widely used in dry cleaners and in home spot removers to take oils and grease out of clothes. However, this use was discontinued when carbon tetrachloride was found to be toxic to the liver, where it can cause cancer. Today, dry cleaners use other halogenated compounds such as dichloromethane, 1,1,1-trichloroethane, and 1,1,2-trichloro-1,2,2-trifluoroethane.

CH_2Cl_2 $Cl_3C—CH_3$ $FCl_2C—CClF_2$

dichloro-   1,1,1-trichloroethane   1,1,2-trichloro-
methane   1,2,2-trifluoroethane

General anesthetics are compounds that are inhaled or injected to cause a loss of sensation so that surgery or other procedures can be done without causing pain to the patient. As nonpolar compounds, they are soluble in the nonpolar nerve membranes, where they decrease the ability of the nerve cells to conduct the sensation of pain. Chloroform, $CHCl_3$, was once used as an anesthetic, but it is toxic and may be carcinogenic. One of the most widely used general anesthetics is halothane, also called Fluothane. It has a pleasant odor, is nonexplosive, has few side effects, undergoes few reactions within the body, and is eliminated quickly.

halothane (Fluothane)

For minor surgeries, a local anesthetic such as chloroethane (ethyl chloride) $CH_3—CH_2—Cl$ is applied to an area of the skin. Chloroethane evaporates quickly, which cools the skin and causes a loss of sensation.

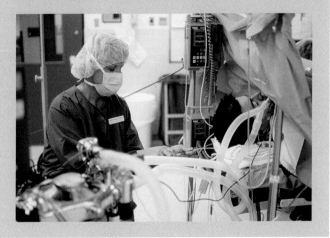

QUESTIONS AND PROBLEMS

▪ Alkanes with Substituents

11.15 Indicate whether each of the following pairs of structural formulas represent isomers or the same molecule.

a.
$$CH_3—\overset{\displaystyle CH_3}{\underset{}{CH}}—CH_3 \quad \text{and} \quad \overset{\displaystyle CH_3}{\underset{\displaystyle CH_3}{CH}}—CH_3$$

b.
$$CH_3—\overset{\displaystyle CH_3}{\underset{}{CH}}—CH_2—CH_3 \quad \text{and} \quad \overset{\displaystyle CH_3}{\underset{}{CH_2}}—CH_2—\overset{\displaystyle CH_3}{\underset{}{CH_2}}$$

c.
$$\overset{\displaystyle CH_3}{\underset{}{CH_2}}—\overset{\displaystyle CH_3}{\underset{}{CH}}—CH_2—CH_3 \quad \text{and} \quad CH_3—\overset{\displaystyle CH_3}{\underset{}{CH}}—\overset{\displaystyle CH_3}{\underset{}{CH}}—CH_3$$

11.16 Indicate whether each of the following pairs of structural formulas represent isomers or the same molecule.

a.
$$CH_3—\overset{\displaystyle CH_3}{\underset{\displaystyle CH_3}{C}}—CH_3 \quad \text{and} \quad \overset{\displaystyle CH_3}{\underset{\displaystyle CH_3}{CH}}—CH_2—CH_3$$

b.
$$CH_3—\overset{\displaystyle CH_3}{\underset{}{CH}}—\overset{\displaystyle CH_3}{\underset{}{CH}}—\overset{\displaystyle CH_3}{\underset{}{CH_2}} \quad \text{and} \quad CH_3—\overset{\displaystyle CH_3}{\underset{}{CH}}—CH_2—\overset{\displaystyle CH_3}{\underset{}{CH}}—CH_3$$

c.
$$CH_3—\overset{\displaystyle CH_3}{\underset{}{CH}}—CH_2—CH_3 \quad \text{and} \quad CH_3—CH_2—\overset{\displaystyle CH_3}{\underset{}{CH}}—CH_3$$

11.17 Give the IUPAC name for each of the following alkanes:

a.
$$CH_3$$
$$|$$
$$CH_3—CH—CH_2—CH_3$$

b.
$$CH_3$$
$$|$$
$$CH_3—C—CH_3$$
$$|$$
$$CH_3$$

c.
$$CH_3\ \ CH_3$$
$$|\ \ \ \ \ |$$
$$CH_3—CH_2—CH—CH—CH_3$$

d.
$$CH_3\ \ \ \ \ \ \ CH_2—CH_3$$
$$|\ \ \ \ \ \ \ \ \ \ \ \ |$$
$$CH_3—C—CH_2—CH—CH_2—CH_3$$
$$|$$
$$CH_3$$

11.18 Give the IUPAC name for each of the following alkanes:

a.
$$CH_3$$
$$|$$
$$CH_3—CH—CH_2—CH_2—CH_3$$

b.
$$CH_3\ \ CH_3$$
$$|\ \ \ \ \ |$$
$$CH_3—CH—CH—CH_3$$

c.
$$CH_3\ \ \ \ \ \ \ \ \ \ CH_3$$
$$|\ \ \ \ \ \ \ \ \ \ \ \ \ |$$
$$CH_3—CH_2—CH—CH_2—CH—CH_3$$

d.
$$CH_2—CH_3$$
$$|$$
$$CH_3—CH_2—CH—CH—CH_2—CH_3$$
$$|$$
$$CH_2—CH_3$$

11.19 Give the IUPAC name for each of the following cycloalkanes:

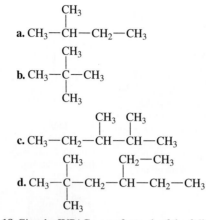

11.20 Give the IUPAC name for each of the following cycloalkanes.

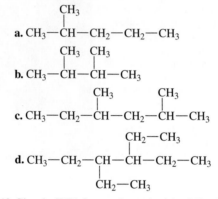

11.21 Draw a condensed structural formula for each of the following alkanes:
a. 2-methylbutane
b. 3,3-dimethylpentane
c. 2,3,5-trimethylhexane
d. 3-ethyl-2,5-dimethyloctane

11.22 Draw a condensed structural formula for each of the following alkanes:
a. 3-ethylpentane
b. 3-ethyl-2-methylpentane
c. 2,2,3,5-tetramethylhexane
d. 4-ethyl-2,2-dimethyloctane

11.23 Draw the structural formulas for each of the following cycloalkanes.
   a. methylcyclopentane
   b. chlorocyclohexane
   c. 1,3-dimethylcyclobutane
   d. 1-bromo-2,3-dimethylcyclopentane

11.24 Draw the structural formula for each of the following cycloalkanes.
   a. bromocyclopropane
   b. ethylcyclohexane
   c. 1,2-dichlorocyclobutane
   d. 1,3-dibromocyclopentane

11.25 Give the IUPAC name for each of the following compounds:
   a. CH_3-CH_2-Br
   b. $CH_3-CH_2-CH_2-F$

   c.
$$CH_3-\underset{\underset{\displaystyle CH_3}{|}}{CH}-Cl$$

   d. $CHCl_3$

11.26 Give the IUPAC name for each of the following compounds:

   a.
$$CH_3-CH_2-\underset{\overset{\displaystyle Cl}{|}}{CH}-CH_3$$
   b. CCl_4

   c.
$$CH_3-\underset{\underset{\displaystyle CH_3}{|}}{\overset{\overset{\displaystyle CH_3}{|}}{C}}-I$$
   d. CH_3F

11.27 Write the condensed structural formula for each of the following compounds:
   a. 2-chloropropane
   b. 2-bromo-3-chlorobutane
   c. methyl bromide
   d. tetrabromomethane

11.28 Write the condensed structural formula for each of the following compounds:
   a. 1,1,2,2-tetrabromopropane
   b. 2-bromopropane
   c. 2,3-dichloro-2-methylbutane
   d. dibromodichloromethane

11.4 Properties of Alkanes

LEARNING GOAL

Identify the properties of alkanes and write a balanced equation for combustion.

Useful alkanes include the gasoline and diesel fuels that power our cars and the heating oils that heat our homes. You may have used a mixture of hydrocarbons such as mineral oil or petrolatum as a laxative or to soften your skin. The different uses of many of the alkanes result from their physical properties including solubility, density, and boiling point.

Solubility and Density

Alkanes are nonpolar, which makes them insoluble in water. However, they are soluble in nonpolar solvents such as other alkanes. Alkanes have densities from 0.65 g/mL to about 0.70 g/mL, which is less dense than water (1.0 g/mL). If there is an oil spill in the ocean, the alkanes in the crude oil remain on the surface and spread over a large area. In the *Exxon Valdez* oil spill in 1989, 40 million liters of

 SELF-STUDY CASE
Hazardous Materials

FIGURE 11.6

In oil spills, large quantities of oil spread over the water.

Q **What physical properties cause oil to remain on the surface of water?**

FIGURE 11.7

The solid alkanes that make up waxy coatings on fruits and vegetables help retain moisture, inhibit mold, and enhance appearance.

Q **Why does the waxy coating help the fruits and vegetables retain moisture?**

oil covered over 25 000 square kilometers of water in Prince William Sound, Alaska. (See Figure 11.6.) If the crude oil reaches the beaches and inlets, there can be considerable damage to beaches, shellfish, fish, birds, and wildlife habitats. Even today there is oil on the surface, or just beneath the surface, in some areas of Prince William Sound. Cleanup includes both mechanical and chemical methods. In one method, a nonpolar compound that is "oil-attracting" is used to pick up oil, which is then scraped off into recovery tanks.

Some Uses of Alkanes

The first four alkanes—methane, ethane, propane, and butane—are gases at room temperature and are widely used as heating fuels. Tanks of liquid propane and butane are used to provide fuels for heating homes and cooking on barbecues. As chlorofluorocarbons (CFCs) are phased out, butane has replaced Freon as a propellant in aerosol containers.

Alkanes having 5–8 carbon atoms (pentane, hexane, heptane, and octane) are liquids at room temperature. They are highly volatile, which makes them useful in fuels such as gasoline. Liquid alkanes with 9–17 carbon atoms have higher boiling points and are found in kerosene, diesel, and jet fuels. Motor oil is a mixture of high-molecular-weight liquid hydrocarbons and is used to lubricate the internal components of engines. Mineral oil is a mixture of liquid hydrocarbons and is used as a laxative and a lubricant. Alkanes with 18 or more carbon atoms are waxy solids at room temperature. The high-molecular-weight alkanes, known as paraffins, are used in waxy coatings of fruits and vegetables to retain moisture, inhibit mold growth, and enhance appearance. (See Figure 11.7.) Petrolatum, or Vaseline, is a mixture of liquid hydrocarbons, which have low boiling points, that are encapsulated in solid hydrocarbons. It is used in ointments and cosmetics and as a lubricant.

Melting and Boiling Points

Alkanes have the lowest melting and boiling points of all the organic compounds. The attractions between nonpolar alkanes in the solid and liquid states result from dispersion forces. In longer carbon chains, the greater number of electrons produces more attractions between molecules, which results in higher melting and boiling points.

CH_4 $CH_3—CH_3$ $CH_3—CH_2—CH_3$
methane, bp $-164°C$ ethane, bp $-89°C$ propane, bp $-42°C$

The boiling points of branched alkanes are generally lower than continuous alkanes with the same number of carbon atoms. The branched chain alkanes tend to be more compact, which reduces the amount of contact between the molecules. Cycloalkanes have higher boiling points than the continuous-chain alkanes. Because rotation of carbon bonds is restricted, cycloalkanes maintain a rigid structure. Those rigid structures are like a set of dishes that can be stacked closely together with many points of contact and therefore attractions to each other. We can compare the boiling points of alkanes and cycloalkanes with five carbon atoms as shown in Table 11.6.

TABLE 11.6

Comparison of Boiling Points of Alkanes and Cycloalkanes with Five Carbons

Formula	Name	Boiling Point (°C)
Cycloalkanes		
	cyclopentane	49
CH$_3$	methylcyclobutane	36.3
Continuous alkane		
CH$_3$—CH$_2$—CH$_2$—CH$_2$—CH$_3$	pentane	36
Branched alkanes		
CH$_3$—CH—CH$_2$—CH$_3$ with CH$_3$	2-methylbutane	28
CH$_3$—C—CH$_3$ with CH$_3$ above and CH$_3$ below	dimethylpropane	10

Combustion

Carbon–carbon single bonds are difficult to break, which makes alkanes the least reactive family of organic compounds. However, alkanes burn readily in oxygen.

The growth of civilization was strongly influenced by the discovery of fire. Heat from burning wood was used to make pottery and glass, extract metals, make weapons, and forge tools. An alkane undergoes **combustion** when it reacts with oxygen to produce carbon dioxide, water, and energy.

$$\text{Alkane} + O_2 \longrightarrow CO_2 + H_2O + \text{energy}$$

Methane is the gas we use to cook our foods and heat our homes. Propane is the gas used in portable heaters and gas barbecues. (See Figure 11.8.) Gasoline, a mixture of liquid hydrocarbons, is the fuel that powers our cars, lawn mowers, and snow blowers. As alkanes, they all undergo combustion. The equations for the combustion of methane (CH_4) and propane (C_3H_8) follow:

$$CH_4 + 2O_2 \longrightarrow CO_2 + 2H_2O + \text{energy}$$
$$C_3H_8 + 5O_2 \longrightarrow 3CO_2 + 4H_2O + \text{energy}$$

In the cells of our bodies, energy is produced by the combustion of glucose. Although a series of reactions is involved, we can write the overall combustion of glucose in our cells as follows:

$$C_6H_{12}O_6 + 6O_2 \xrightarrow{\text{Enzymes}} 6CO_2 + 6H_2O + \text{energy}$$

MC GOB Writing Balanced Equations for Combustion of Alkanes

EXPLORE YOUR WORLD

Combustion

In this exploration, we will look at the behavior of the products of combustion. You will need one or two candles, a Pyrex glass such as a measuring cup, and some matches or wooden splints.

Hold a Pyrex cup upside down, and place a burning match inside it. The match should continue to burn as long as oxygen is available. Light a candle and hold the inverted Pyrex cup above it for 15–20 seconds. Remove the cup from the candle and immediately place a burning match into it. The CO_2 accumulated from the combustion of the candle should extinguish the match.

Add some water and a lot of ice to the same Pyrex cup. It should become cold to the touch. Wipe the bottom of the cup and carefully hold the Pyrex cup over a burning candle. Look for condensation as water is formed from the combustion reaction. This may be more noticeable if two candles are used or the Pyrex cup is held for a few seconds over a gas flame.

QUESTIONS

1. What are the products of combustion?
2. What was the evidence for the production of CO_2?
3. What observations gave evidence of the production of H_2O during combustion?

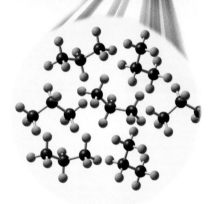

SAMPLE PROBLEM 11.8

■ Combustion

Write a balanced equation for the complete combustion of butane.

SOLUTION

The balanced equation for the complete combustion of butane can be written

$$2C_4H_{10} + 13O_2 \longrightarrow 8CO_2 + 10H_2O$$

STUDY CHECK

Write a balanced equation for the complete combustion of the following.

$$CH_3-\overset{\displaystyle CH_3}{\underset{\displaystyle |}{CH}}-CH_2-CH_3$$

Classifying Carbon Atoms in Hydrocarbons

In an alkane, each carbon atom is classified according to the number of carbon atoms connected to it. A *primary* (1°) *carbon* is bonded to one other carbon atom. A *secondary* (2°) *carbon* has two carbon atoms attached to it. A *tertiary* (3°) *carbon* is bonded to three other carbon atoms.

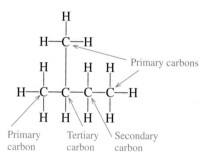

FIGURE 11.8

The propane fuel in the tank undergoes combustion, which provides energy.

Q **What is the balanced equation for the combustion of propane?**

MC GOB SELF-STUDY ACTIVITY
Introduction to Organic Molecules

■ Incomplete Combustion

You may already know that it is dangerous to burn natural gas, oil, or wood in a closed room where ventilation and fresh air are not adequate. A gas heater, fireplace, or wood stove must have proper ventilation. If the supply of oxygen is limited, incomplete combustion produces carbon monoxide. The incomplete combustion of methane in natural gas is written as:

$$2CH_4(g) + 3O_2(g) \longrightarrow 2CO(g) + 4H_2O(g) + heat$$

Limited oxygen supply — carbon monoxide

Carbon monoxide (CO) is a colorless, odorless, poisonous gas. When inhaled, CO passes into the bloodstream, where it attaches to hemoglobin. When CO binds to the hemoglobin, it reduces the amount of oxygen (O_2) reaching the organs and cells. As a result, a healthy person can experience a reduction in exercise capability, visual perception, and manual dexterity.

When the amount of hemoglobin bound to CO (COHb) is 10% or less, a person may experience shortness of breath, mild headache, and drowsiness, which are symptoms that may be mistaken for the flu. Heavy smokers can have as high as 9% COHb in their blood. When as much as 30% of the hemoglobin is bound to CO, a person may experience more severe symptoms including dizziness, mental confusion, severe headache, and nausea. If 50% or more of the hemoglobin is bound to CO, a person could become unconscious and die if not treated immediately with oxygen.

Crude Oil

Crude oil or petroleum contains a wide variety of hydrocarbons. At an oil refinery, the components in crude oil are separated by fractional distillation, a process that removes groups or fractions of hydrocarbons by continually heating the mixture to higher temperatures. (See Table 11.7.) Fractions containing alkanes with longer carbon chains require higher temperatures before they reach their boiling temperature and form gases. The gases are removed and passed through a distillation column where they cool and condense back to liquids. The major use of crude oil is to obtain gasoline, but a barrel of crude oil is only about 35% gasoline. To increase the production of gasoline, heating oils are broken down to give the lower weight alkanes.

TABLE 11.7

Typical Alkane Mixtures Obtained by Distillation of Crude Oil

Distillation Temperatures (°C)	Number of Carbon Atoms	Product
Below 30	1–4	Natural gas
30–200	5–12	Gasoline
200–250	12–16	Kerosene, jet fuel
250–350	15–18	Diesel fuel, heating oil
350–450	18–25	Lubricating oil
Nonvolatile residue	Over 25	Asphalt, tar

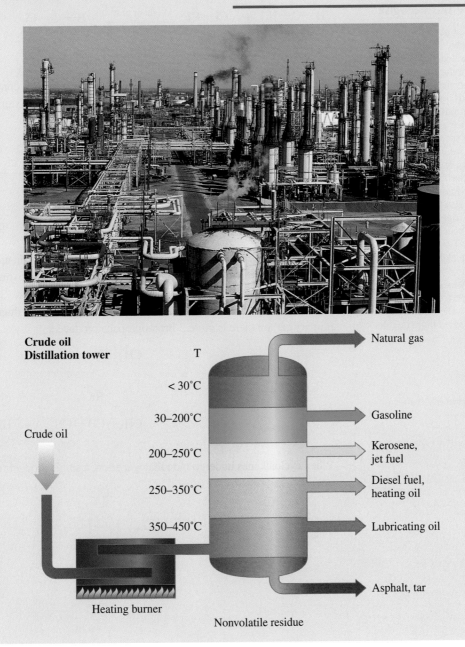

Crude oil Distillation tower

T

< 30°C → Natural gas

30–200°C → Gasoline

Crude oil

200–250°C → Kerosene, jet fuel

250–350°C → Diesel fuel, heating oil

350–450°C → Lubricating oil

→ Asphalt, tar

Heating burner

Nonvolatile residue

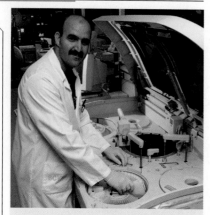
MC
GOB   Halogenation of Alkanes

Halogenation of Alkanes (Substitution)

Alkanes react with halogen to produce a mixture of halogenated compounds. In a **halogenation reaction**, atoms of a halogen bond to a carbon atom. This type of reaction is also called a *substitution* reaction because halogen atoms replace one or more hydrogen atoms in an alkane. When halogenation uses chlorine Cl_2 ($Cl—Cl$), it is also called chlorination, and with bromine Br_2 ($Br—Br$), it is called bromination. When a mixture of an alkane such as methane and chlorine are heated or exposed to light, halogenation can take place.

$$\text{Alkane + halogen} \xrightarrow{\text{Light or heat}} \text{haloalkane and hydrogen halide}$$

The first halogenation of methane occurs when one chlorine atom replaces a hydrogen atom as follows:

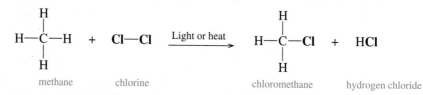

In the presence of heat or light and chlorine, the reaction continues. Chlorine replaces hydrogen atoms in chloromethane, which gives a mixture of halogenated products.

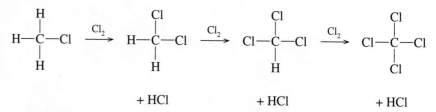

When halogenation occurs with larger alkanes, the halogen atoms may substitute for any of the hydrogen atoms. For example, the first substitution of propane (monohalogenation) produces both 1-bromopropane and 2-bromopropane. Because hydrogens on a secondary carbon are more reactive than hydrogens on a primary carbon there is more 2-bromopropane produced.

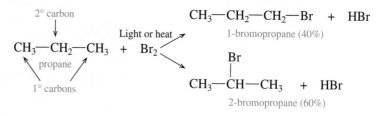

When cycloalkanes undergo monohalogenation, a single product results.

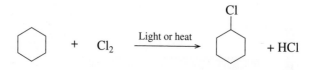

ENVIRONMENTAL NOTE

▪ CFCs and Ozone Depletion

The compounds called chlorofluorocarbons (CFCs) were used as propellants for hairs sprays, paints, and as refrigerants in home and car air conditioners. Two widely used CFCs, Freon 11 (CCl_3F) and Freon 12 (CCl_2F_2), were developed in the 1920s as nontoxic refrigerants, which were safer than the sulfur dioxide and ammonia used at the time.

$$\underset{\text{Freon 11}}{Cl-\overset{\displaystyle Cl}{\underset{\displaystyle Cl}{C}}-F} \qquad \underset{\text{Freon 12}}{F-\overset{\displaystyle Cl}{\underset{\displaystyle Cl}{C}}-F}$$

In the stratosphere, a layer of ozone (O_3) absorbs the ultraviolet (UV) radiation of the sun and acts as a protective shield for plants and animals on earth. Ozone is produced in the stratosphere when oxygen reacts with ultraviolet light and breaks into oxygen atoms that quickly combine with oxygen molecules to form ozone.

$$O_2 \xrightarrow{\text{UV}} O + O$$
$$O_2 + O \longrightarrow O_3$$

In the 1970s scientists became concerned that CFCs entering the atmosphere were accelerating the depletion of ozone and threatening the stability of the ozone layer. CFCs decompose in the upper atmosphere in the presence of UV light to produce highly reactive chlorine atoms.

$$CCl_3F \xrightarrow{\text{UV light}} CCl_2F + Cl$$

The reactive chlorine atoms catalyze the breakdown of ozone molecules.

$$Cl + O_3 \longrightarrow ClO + O_2$$
$$ClO + O_3 \longrightarrow Cl + 2O_2$$

It has been estimated that one chlorine atom can destroy as many as 100 000 ozone molecules. Normally, there is a balance between the ozone and oxygen in the atmosphere but the rapid destruction of ozone has upset that equilibrium.

Reports of polar ozone depletion over Antarctica in March 1985 prompted scientists to call for a freeze on the production of CFCs. In some areas as much as 50% of the

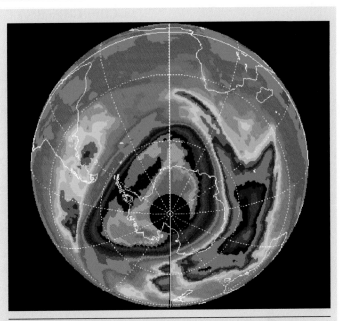

In the color image, the pink areas have the lowest levels of ozone.

ozone had been depleted, and at certain times of the year an ozone hole appears.

There is evidence of thinning in the ozone layer over the Arctic as well but to a somewhat lesser degree due to warmer temperatures. It is interesting that in the lower atmosphere, ozone is an automobile and industrial pollutant, but in the stratosphere ozone is a life-protecting compound.

Today the use of CFCs is being phased out. However, it is expected that ozone levels will remain low for several decades due to the stability of CFCs. Chemical companies are developing substitutes to CFCs that are not as damaging to the ozone. Replacement compounds such as hydrochlorofluorocarbons (HCFCs) contain chlorine atoms, but these compounds break down in the lower atmosphere reducing the amount of chlorine that reaches the stratosphere. Hydrofluorocarbons (HFCs), which contain no chlorine, are being considered as another replacement for CFCs. However, the potential effects of fluorine compounds on ozone destruction must be determined.

SAMPLE PROBLEM 11.9

▪ Halogenation of Alkanes

Write the structural formula of the monohalogenated product formed in the following reactions:

a. $CH_3-CH_3 + Br_2 \xrightarrow{\text{Light or heat}}$ **b.** ⬡ $+ Cl_2 \xrightarrow{\text{Light or heat}}$

SOLUTION

a. CH₃—CH₂—Br

b.

STUDY CHECK

Write the structural formulas of the organic products that form when two bromine atoms replace two hydrogen atoms in the halogenation of ethane.

QUESTIONS AND PROBLEMS

■ Properties of Alkanes

11.29 Heptane, C_7H_{16}, has a density of 0.68 g/mL and boils at 98°C.
    a. What is the condensed structural formula of heptane?
    b. Is it a solid, liquid, or gas at room temperature?
    c. Is it soluble in water?
    d. Will it float or sink in water?

11.30 Nonane, C_9H_{20}, has a density of 0.79 g/mL and boils at 151°C.
    a. What is the condensed structural formula of nonane?
    b. Is it a solid, liquid, or gas at room temperature?
    c. Is it soluble in water?
    d. Will it float or sink in water?

11.31 In each of the following pairs of hydrocarbons, which one would you expect to have the higher boiling point?
    a. pentane or heptane
    b. propane or cyclopropane
    c. hexane or 2-methylpentane

11.32 In each of the following pairs of hydrocarbons, which one would you expect to have the higher boiling point?
    a. propane or butane
    b. hexane or cyclohexane
    c. 2,2-dimethylpentane or heptane

11.33 Write a balanced equation for the complete combustion of each of the following compounds:
    a. ethane, C_2H_6 **b.** cyclopropane, C_3H_6 **c.** octane   **d.** cyclohexane

11.34 Write a balanced equation for the complete combustion of each of the following compounds:
    a. hexane, C_6H_{14} **b.** cyclopentane, C_5H_{10} **c.** nonane   **d.** 2-methylbutane

11.35 Write the condensed structural formulas of the monohalogenated products from the chlorination of each of the following:
    a. ethane   **b.** cyclopentane   **c.** 2-methylpropane

11.36 Write the condensed structural formulas of the monohalogenated products from the bromination of each of the following:
    a. butane   **b.** pentane   **c.** cyclobutane

■ 11.5 Functional Groups

LEARNING GOAL

Classify organic molecules according to their functional groups.

In organic compounds, carbon atoms are most likely to bond with hydrogen, oxygen, nitrogen, sulfur, and halogens such as chlorine. Hydrogen, with one valence electron, forms a single covalent bond. An octet is achieved by nitrogen forming three covalent bonds and oxygen and sulfur each forming two covalent bonds.

TABLE 11.8

Covalent Bonds for Elements in Organic Compounds

Element	Group	Covalent Bonds	Structure of Atoms
H	1A(1)	1	H—
C	4A(4)	4	$-\overset{\textstyle |}{\underset{\textstyle |}{C}}-$
N	5A(15)	3	$-\overset{\textstyle |}{\underset{\textstyle \cdot\cdot}{N}}-$
O, S	6A(16)	2	$-\overset{\cdot\cdot}{\underset{\cdot\cdot}{O}}-$ $-\overset{\cdot\cdot}{\underset{\cdot\cdot}{S}}-$
F, Cl, Br, I	7A(17)	1	$-\overset{\cdot\cdot}{\underset{\cdot\cdot}{X}}:$ (X = F, Cl, Br, I)

The halogens with seven valence electrons form one covalent bond. Table 11.8 lists the number of covalent bonds most often formed by elements found in organic compounds.

Organic compounds number in the millions and more are synthesized every day. It might seem that the task of learning organic chemistry would be overwhelming. However, within this vast number of compounds, there are characteristic structural features called **functional groups**, which are a certain group of atoms that react in a predictable way. Compounds with the same functional group undergo similar chemical reactions. The identification of functional groups allows us to classify organic compounds according to their structure and to name compounds within each family. For now we will focus on recognizing the patterns of atoms that make up the functional groups.

MC
GOB

Identifying Functional Groups
Drawing Organic Compounds
with Functional Groups

Alkenes, Alkynes, and Aromatic Compounds

Earlier we learned that alkanes contain only carbon–carbon single bonds. The **alkenes** contain a functional group that is a double bond between two adjacent carbon atoms; **alkynes** contain a triple bond. **Aromatic compounds** contain benzene, a molecule that has a ring of six carbon atoms with one hydrogen atom attached to each carbon. The benzene structure is represented as a hexagon with a circle in the center.

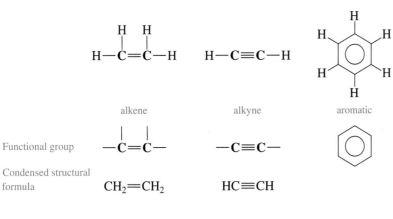

	alkene	alkyne	aromatic
Functional group	$-\overset{\textstyle |}{C}=\overset{\textstyle |}{C}-$	$-C\equiv C-$	
Condensed structural formula	$CH_2{=}CH_2$	$HC{\equiv}CH$	

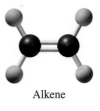

Alkene

Alkyne

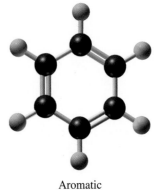

Aromatic

Alcohol

Ether

Aldehyde

Ketone

Alcohols and Ethers

The characteristic functional group in **alcohols** is the **hydroxyl (—OH) group** bonded to a carbon atom. In **ethers**, the characteristic structural feature is an oxygen atom bonded to two carbon atoms. The oxygen atom also has two unshared pairs of electrons, but they are not shown in the structural formulas.

$$CH_3—CH_2—\textbf{OH} \qquad CH_3—\textbf{O}—CH_3$$

alcohol   ketone

Functional group   —O—H   —O—

Aldehydes and Ketones

The aldehydes and ketones contain a **carbonyl group** (C=O), which is a carbon with a double bond to oxygen. In an **aldehyde**, the carbon atom of the carbonyl group is bonded to another carbon and one hydrogen atom. Only the simplest aldehyde, CH_2O, has a carbonyl group attached to two hydrogen atoms. In a **ketone**, the carbonyl group is bonded to two other carbon atoms.

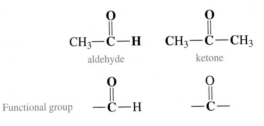

aldehyde   ketone

Functional group   —C—H   —C—

SAMPLE PROBLEM 11.10

▪ **Classifying Organic Compounds**

Classify the following organic compounds according to their functional groups:

a. $CH_3—CH_2—CH_2—OH$ **b.** $CH_3—CH=CH—CH_3$

c. $CH_3—CH_2—\overset{\displaystyle O}{\overset{\displaystyle \|}{C}}—CH_2—CH_3$

SOLUTION

a. alcohol   **b.** alkene   **c.** ketone

STUDY CHECK

Why is $CH_3—CH_2—O—CH_3$ an ether, but $CH_3—\overset{\displaystyle OH}{\underset{\displaystyle |}{CH}}—CH_3$ is an alcohol?

Carboxylic Acids and Esters

In **carboxylic acids**, the functional group is the *carboxyl group*, which is a combination of the *carbo*nyl and hydro*xyl* groups.

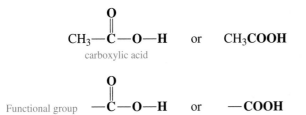

$$CH_3-\overset{\overset{\displaystyle O}{\|}}{C}-O-H \quad \text{or} \quad CH_3COOH$$

carboxylic acid

Functional group
$$-\overset{\overset{\displaystyle O}{\|}}{C}-O-H \quad \text{or} \quad -COOH$$

An **ester** is similar to a carboxylic acid, except the oxygen of the carboxyl group is attached to a carbon and not to hydrogen.

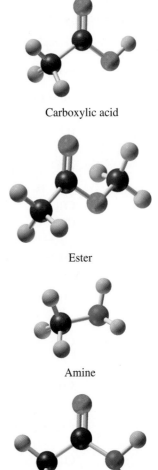

Carboxylic acid

$$CH_3-\overset{\overset{\displaystyle O}{\|}}{C}-O-CH_3 \quad \text{or} \quad CH_3COOCH_3$$

ester

Functional group
$$-\overset{\overset{\displaystyle O}{\|}}{C}-O- \quad \text{or} \quad -COO-$$

Ester

Amines and Amides

In **amines**, the central atom is a nitrogen atom. Amines are derivatives of ammonia, NH_3, in which carbon atoms replace one, two, or three of the hydrogen atoms.

$$NH_3 \qquad CH_3-NH_2 \qquad CH_3-\underset{\underset{\displaystyle CH_3}{|}}{NH} \qquad CH_3-\underset{\underset{\displaystyle CH_3}{|}}{N}-CH_3$$

ammonia   Examples of amines

Amine

In an **amide**, the hydroxyl group of a carboxylic acid is replaced by a nitrogen group.

$$CH_3-\overset{\overset{\displaystyle O}{\|}}{C}-NH_2$$

amide

Amide

A list of the common functional groups in organic compounds is shown in Table 11.9.

SAMPLE PROBLEM 11.11

■ Identifying Functional Groups

Classify the following organic compounds according to their functional groups:

a. $CH_3-CH_2-NH-CH_3$

b. $CH_3-\overset{\overset{\displaystyle O}{\|}}{C}-O-CH_2-CH_3$

c. $CH_3-CH_2-\overset{\overset{\displaystyle O}{\|}}{C}-OH$

SOLUTION

a. amine   **b.** ester   **c.** carboxylic acid

STUDY CHECK

How does a carboxylic acid differ from an ester?

TABLE 11.9

Classification of Organic Compounds

Class	Example	Functional Group	Characteristic
Alkene	$H_2C{=}CH_2$	$\backslash C{=}C /$	Carbon–carbon double bond
Alkyne	$HC{\equiv}CH$	$-C{\equiv}C-$	Carbon–carbon triple bond
Aromatic	(benzene ring with H atoms)	(benzene ring)	Six-atom carbon ring with six hydrogen atoms
Haloalkane	CH_3-Cl	$-F, -Cl, -Br, -I$	One or more halogen atoms
Alcohol	CH_3-CH_2-OH	$-OH$	Hydroxyl group ($-OH$)
Ether	CH_3-O-CH_3	$-O-$	Oxygen atom bonded to two carbons
Thiol	CH_3-SH	$-SH$	A $-SH$ group bonded to carbon
Aldehyde	$CH_3-\overset{\overset{\displaystyle O}{\|\|}}{C}-H$	$-\overset{\overset{\displaystyle O}{\|\|}}{C}-H$	Carbonyl group (carbon–oxygen double bond) with $-H$
Ketone	$CH_3-\overset{\overset{\displaystyle O}{\|\|}}{C}-CH_3$	$-\overset{\overset{\displaystyle O}{\|\|}}{C}-$	Carbonyl group (carbon–oxygen double bond) between carbon atoms
Carboxylic acid	$CH_3-\overset{\overset{\displaystyle O}{\|\|}}{C}-O-H$	$-\overset{\overset{\displaystyle O}{\|\|}}{C}-O-H$	Carboxyl group (carbon–oxygen double bond and $-OH$)
Ester	$CH_3-\overset{\overset{\displaystyle O}{\|\|}}{C}-O-CH_3$	$-\overset{\overset{\displaystyle O}{\|\|}}{C}-O-$	Carboxyl group with $-H$ replaced by a carbon
Amine	CH_3-NH_2	$-N-$	Nitrogen atom with one or more carbon groups
Amide	$CH_3-\overset{\overset{\displaystyle O}{\|\|}}{C}-NH_2$	$-\overset{\overset{\displaystyle O}{\|\|}}{C}-N-$	Carbonyl group bonded to nitrogen

QUESTIONS AND PROBLEMS

■ Functional Groups

11.37 Identify the class of compounds that contains each of the following functional groups.
   a. hydroxyl group attached to a carbon chain
   b. carbon–carbon double bond
   c. carbonyl group attached to a hydrogen atom
   d. carboxyl group attached to two carbon atoms

11.38 Identify the class of compounds that contains each of the following functional groups:
   a. a nitrogen atom attached to one or more carbon atoms
   b. carboxyl group
   c. oxygen atom bonded to two carbon atoms
   d. a carbonyl group between two carbon atoms

11.39 Classify the following molecules according to their functional groups. The possibilities are alcohol, ether, ketone, carboxylic acid, or amine:

ENVIRONMENTAL NOTE

■ Functional Groups in Familiar Compounds

The flavors and odors of foods and many household products can be attributed to the functional groups of organic compounds. As we discuss these familiar products, look for the functional groups we have described.

Ethyl alcohol is the alcohol found in alcoholic beverages. Isopropyl alcohol is another alcohol commonly used to disinfect skin before giving injections and to treat cuts.

$$CH_3-CH_2-OH \qquad CH_3-\overset{\overset{\displaystyle OH}{|}}{CH}-CH_3$$
ethyl alcohol isopropyl alcohol

Ketones and aldehydes are in many items we use or eat each day. Acetone or dimethyl ketone is produced in great amounts commercially. Acetone is used as an organic solvent because it dissolves a wide variety of organic substances. You may be familiar with acetone as fingernail polish remover. Ketones and aldehydes used in the food industry are found in flavorings such as vanilla, cinnamon, and spearmint. When we buy a small bottle of liquid flavorings, the aldehyde or ketone is dissolved in alcohol because the compounds are not very soluble in water. The aldehyde butyraldehyde adds a "buttery" taste to foods and margarine.

$$CH_3-CH_2-CH_2-\overset{\overset{\displaystyle O}{||}}{C}-H$$
butyraldehyde "butter" flavoring

The sour tastes of vinegar and fruit juices and the pain from ant stings are all due to carboxylic acids. Acetic acid is the carboxylic acid that makes up vinegar. Aspirin also contains a carboxylic acid group. Esters found in fruits produce the pleasant aromas and tastes of bananas, oranges, pears, and pineapples. Esters are also used as solvents in many household cleaners, polishes, and glues.

One of the characteristics of fish is their odor, which is due to amines. Amines produced when proteins decay have a particularly pungent and offensive odor.

$$CH_3-\overset{\overset{\displaystyle O}{||}}{C}-OH$$
acetic acid (in vinegar)

$$CH_3-\overset{\overset{\displaystyle O}{||}}{C}-O-CH_2-CH_2-CH_3$$
propyl acetate (pears)

$$CH_3-NH_2$$
methyl amine

$$CH_3-\overset{\overset{\displaystyle O}{||}}{C}-O-CH_2-CH_2-CH_2-CH_2-CH_3$$
pentyl acetate (bananas)

$$H_2N-CH_2-CH_2-CH_2-CH_2-NH_2$$
putrescine

$$H_2N-CH_2-CH_2-CH_2-CH_2-CH_2-NH_2$$
cadaverine

Alkaloids are biologically active amines synthesized by plants to ward off insects and animals. Some typical alkaloids include caffeine, nicotine, histamine, and the decongestant epinephrine. Many are painkillers and hallucinogens such as morphine, LSD, marijuana, and cocaine. Certain parts of our neurons have receptor sites that respond to the various alkaloids. By modifying the structures of certain alkaloids to eliminate side effects, chemists have synthesized painkillers and drugs such as Novocain, codeine, and Valium.

a. $CH_3-CH_2-O-CH_2-CH_3$

b. $CH_3-\overset{\overset{\displaystyle OH}{|}}{CH}-CH_3$ **c.** $CH_3-\overset{\overset{\displaystyle O}{||}}{C}-CH_2-CH_3$

d. $CH_3-CH_2-CH_2-COOH$ **e.** $CH_3-CH_2-NH_2$

11.40 Classify the following molecules according to their functional groups. The possibilities are alkene, aldehyde, carboxylic acid, ester, or amide:

a. $CH_3-CH_2-\overset{\overset{\displaystyle O}{||}}{C}-O-CH_2-CH_3$

b. $CH_3-\overset{\overset{\displaystyle O}{||}}{C}-NH_2$ **c.** $CH_3-CH_2-CH_2-\overset{\overset{\displaystyle O}{||}}{C}-H$

d. $CH_3-CH_2-CH_2-COOH$ **e.** $CH_3-CH=CH-CH_3$

CONCEPT MAP

Introduction to Organic Chemistry: Alkanes

Organic compounds — contain — Carbon atoms — Organic compounds

Organic compounds → tend to be → Nonpolar → with → Low melting and boiling points → and are → Insoluble in water → and are usually → Flammable

Carbon atoms → form → Four covalent bonds

Carbon atoms → and have a → Tetrahedral shape

are written as → Expanded and condensed formulas → and named by the → IUPAC System → as → Alkanes → undergo reactions such as → Halogenation | Combustion

Organic compounds → with groups of atoms called → Functional groups → show similar behavior →

Haloalkanes
Alkenes
Alkynes
Alcohols
Thiols
Ethers
Esters
Aldehydes
Ketones
Carboxylic Acids
Amines
Amides
Aromatic

CHAPTER REVIEW

11.1 Organic Compounds

Organic compounds have covalent bonds and most form nonpolar molecules. Often they have low melting points and low boiling points, are not very soluble in water, produce molecules in solutions, and burn vigorously in air. In contrast, many inorganic compounds are ionic or contain polar covalent bonds and form polar molecules. Many have high melting and boiling points, are usually soluble in water, produce ions in water, and do not burn in air. Carbon atoms share four valence electrons to form four covalent bonds. In the simplest organic molecule, methane, CH_4, the four bonds that bond hydrogen to the carbon atom are directed out to the corners of a tetrahedron with bond angles of 109.5°.

11.2 Alkanes

Alkanes are hydrocarbons that have only C—C single bonds. In the expanded structural formula, a separate line is drawn for every bonded atom. A condensed structural formula depicts groups composed of each carbon atom and its attached hydrogen atoms. In a line-bond formula, only the carbon–carbon bonds are drawn. In

cycloalkanes, the carbon atoms form a ring or cyclic structure. The name is written by placing the prefix *cyclo* before the alkane name with the same number of carbon atoms. The IUPAC system is used to name organic compounds in a systematic manner. The IUPAC name indicates the number of carbon atoms.

11.3 Alkanes with Substituents

For a continuous alkane, the carbon atoms are connected in a chain and bonded to hydrogen atoms. Substituents such as alkyl groups can replace hydrogen atoms on an alkane. A haloalkane contains one or more F, Cl, Br, or I atoms. In the IUPAC system, halogen atoms are named as fluoro, chloro, bromo, or iodo substituents attached to the main chain. In the common name, alkyl halides, the name of the alkyl group precedes the halide, for example methyl chloride.

11.4 Properties of Alkanes

As nonpolar molecules, alkanes are not soluble in water. They are less dense than water. With only weak attractions, they have low melting and boiling points. Although the C—C bonds in alkanes resist most reactions, alkanes undergo combustion. In combustion, or burning, alkanes react with oxygen to produce carbon dioxide and water. In the halogenation of alkanes, halogen atoms replace one or more hydrogen atoms. A source of energy such as UV radiation or heat is required for this substitution reaction. For alkanes of similar mass, cycloalkanes have higher boiling points and branched alkanes have lower boiling points than the continuous, nonbranched chain alkanes.

11.5 Functional Groups

An organic molecule contains a characteristic group of atoms called a functional group that determines the molecule's family name and chemical reactivity. Functional groups are used to classify organic compounds, act as reactive sites in the molecule, and provide a system of naming for organic compounds. Some common functional groups include the hydroxyl group (—OH) in alcohols, the carbonyl group (C=O) in aldehydes and ketones, and a nitrogen atom

—N— in amines, and a nitrogen and carbonyl group in amides.

SUMMARY OF NAMING

Type	Example	Characteristic	Structure
Alkane	Propane	single C—C, C—H bonds	CH_3—CH_2—CH_3
	Methylpropane		CH_3—$\overset{\overset{\displaystyle CH_3}{\vert}}{CH}$—$CH_3$
Haloalkane	1-Chloropropane	halogen atom	CH_3—CH_2—CH_2—Cl
Cycloalkane	Cyclobutane	carbon ring	□

SUMMARY OF REACTIONS

COMBUSTION

Alkane + O_2 ⟶ CO_2 + H_2O + energy

HALOGENATION (SUBSTITUTION BY HALOGEN)

Alkane + halogen $\xrightarrow{\text{Light or heat}}$ haloalkane + hydrogen halide

CH_4 + Cl_2 $\xrightarrow{\text{Light or heat}}$ CH_3Cl + HCl

⬠ + Cl_2 $\xrightarrow{\text{Light or heat}}$ (cyclopentane with Cl) + HCl

KEY TERMS

alcohols A class of organic compounds that contains the hydroxyl (—OH) group bonded to a carbon atom.

aldehydes A class of organic compounds that contains a carbonyl group (C=O) bonded to at least one hydrogen atom.

alkanes Hydrocarbons containing only single bonds between carbon atoms.

alkenes Hydrocarbons that contain carbon–carbon double bonds (C=C).

alkyl group An alkane minus one hydrogen atom. Alkyl groups are named like the alkanes except a *yl* ending replaces *ane*.

alkynes Hydrocarbons that contain carbon–carbon triple bonds (C≡C).

amide A class of organic compounds in which the hydroxyl group of a carboxylic acid is replaced by a nitrogen group.

amines A class of organic compounds that contains a nitrogen atom bonded to one or more carbon atoms.

aromatic A compound that contains benzene. Benzene has a six-carbon ring with only one hydrogen attached to each carbon.

branch A carbon group bonded to the main carbon chain.

branched alkane A single-bonded hydrocarbon containing a substituent bonded to the main chain.

carbonyl group A functional group that contains a double bond between a carbon atom and an oxygen atom (C=O).

carboxylic acids A class of organic compounds that contains the carboxyl functional group.

combustion A chemical reaction in which an alkane reacts with oxygen to produce CO_2, H_2O, and energy.

condensed structural formula A structural formula that shows the arrangement of the carbon atoms in a molecule but groups each carbon atom with its bonded hydrogen atoms (CH_3, CH_2, or CH).

continuous alkane An alkane in which the carbon atoms are connected in a row, one after the other.

cycloalkane An alkane that is a ring or cyclic structure.

esters A class of organic compounds that contains a —COO— group with an oxygen atom bonded to carbon.

ethers A class of organic compounds that contains an oxygen atom bonded to two carbon atoms.

expanded structural formula A type of structural formula that shows the arrangement of the atoms by showing each bond in the hydrocarbon as C—H, C—C, C=C, or C≡C.

functional group A group of atoms that determine the physical and chemical properties and naming of a class of organic compounds.

haloalkane A type of alkane that contains one or more halogen atoms.

halogenation reaction A substitution reaction that occurs in the presence of light or heat in which halogen atoms replace hydrogen atoms in an alkane.

hydrocarbons Organic compounds consisting of only carbon and hydrogen.

hydroxyl group The group of atoms (—OH) characteristic of alcohols.

isomers Organic compounds in which identical molecular formulas have different arrangements of atoms.

IUPAC system An organization known as International Union of Pure and Applied Chemistry that determines the system for naming organic compounds.

ketones A class of organic compounds in which a carbonyl group is bonded to two carbon atoms.

line-bond formula A type of structural formula that shows only the bonds from carbon to carbon.

organic compounds Compounds made of carbon that typically have covalent bonds, nonpolar molecules, low melting and boiling points, are insoluble in water, and flammable.

substituent Groups of atoms such as an alkyl group or a halogen bonded to the main chain or ring of carbon atoms.

UNDERSTANDING THE CONCEPTS

11.41 Suncreens contain compounds that absorb UV light such as oxybenzone and 2-ethylhexyl p-methoxycinnamate.

Identify the functional groups in each of the following UV-absorbing compounds used in suncreens:

a. oxybenzone

b. 2-ethylhexyl-p-methoxycinnamate

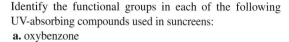
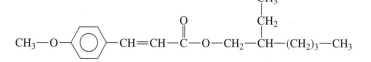

11.42 Oxymetazoline is a vasoconstrictor used in nasal decongestant sprays such as Afrin.

What functional groups are in oxymetazoline?

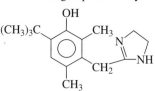

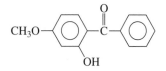

11.43 Decimemide is used as an anticonvulsant.

What functional groups are in decimemide?

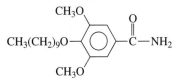

11.44 The odor and taste of pineapples is from ethyl butyrate.

What functional group is in ethyl butyrate?

$$CH_3-CH_2-CH_2-\overset{\overset{\displaystyle O}{\|}}{C}-O-CH_2-CH_3$$

ADDITIONAL QUESTIONS AND PROBLEMS

For instructor-assigned homework, go to
www.masteringgob.com.

11.45 Compare organic and inorganic compounds in terms of:
a. types of bonds **b.** solubility in water
c. melting points **d.** flammability

11.46 Identify each of the following compounds as organic or inorganic:
a. Na_2SO_4 **b.** $CH_2{=}CH_2$
c. Cr_2O_3 **d.** $C_{12}H_{22}O_{11}$

11.47 Match the following physical and chemical properties with the compounds butane, C_4H_{10}, or potassium chloride, KCl.
a. melts at $-138°C$
b. burns vigorously in air
c. melts at $770°C$
d. produces ions in water
e. is a gas at room temperature

11.48 Match the following physical and chemical properties with the compounds cyclohexane, C_6H_{12}, or calcium nitrate, $Ca(NO_3)_2$.
a. contains only covalent bonds
b. melts above $500°C$
c. insoluble in water
d. liquid at room temperature
e. produces ions in water

11.49 Complete the following structures by adding the correct number of hydrogen atoms:
a. C—C **b.** C—C—O
c. $C-C-\overset{\overset{\displaystyle O}{\|}}{C}$ **d.** C—O—C—C

11.50 Identify the functional groups in each.
a. $CH_3{-}NH_2$
b. $CH_3-\overset{\overset{\displaystyle O}{\|}}{C}-CH_3$

c. $CH_3-\overset{\overset{\displaystyle O}{\|}}{C}-CH_3$
d. $CH_3-C{\equiv}CH$
e. $CH_3-O-CH_2-CH_3$
f. ⬡—CH_3

11.51 Write the name of each of the following alkyl groups:
a. CH_3-
b. $CH_3-CH_2-CH_2-$
c. $CH_3-\overset{\overset{\displaystyle CH_3}{|}}{CH}-$
d. CH_3-CH_2-

11.52 Give the IUPAC names for each of the following molecules:
a. $CH_3-CH_2-\overset{\overset{\displaystyle CH_3}{|}}{\underset{\underset{\displaystyle CH_3}{|}}{C}}-CH_3$
b. CH_3-CH_2-Cl
c. $CH_3-CH_2-\overset{\overset{\displaystyle CH_3-CH_2}{|}}{CH}-CH_2-\overset{\overset{\displaystyle Br}{|}}{CH}-CH_3$
d. ⬡

11.53 Write the condensed structural formula for each of the following molecules:
a. 3-ethylhexane
b. 1,3-dimethylcyclopentane
c. 1,3-dichloro-3-methylheptane
d. bromocyclobutane

11.54 Write the condensed structural formula for each of the following molecules:
  a. ethylcyclopropane
  b. 2-methylhexane
  c. isopropylcyclopentane
  d. 1,1-dichloropentane

11.55 Draw the line-bond structural formula for each of the following molecules:
  a. pentane
  b. 2,3-dimethylhexane
  c. 2-bromo-4-methylheptane

11.56 Draw the line-bond structural formula for each of the following molecules:
  a. butane
  b. 2,3,3-trimethylpentane
  c. 3,4,5-trimethyloctane

11.57 In an automobile engine, "knocking" occurs when the combustion of gasoline occurs too rapidly. The octane number of gasoline represents the ability of a gasoline mixture to reduce knocking. A sample of gasoline is compared with heptane, rated 0 because it reacts with severe knocking, and 2,2,4-trimethylpentane, which has a rating of 100 because of its low knocking. Write the condensed structural formula, molecular formula, and equation for the complete combustion of 2,2,4-trimethylpentane.

11.58 Draw the structures of the following halogenated compounds, which are used as refrigerants.
  a. Freon 14, tetrafluoromethane
  b. Freon 114, 1,2-dichloro-1,1,2,2-tetrafluoroethane
  c. Freon C318, octafluorocyclobutane

11.59 Identify the compound in each pair that has the higher boiling point.
  a. pentane or heptane
  b. pentane or cyclopentane
  c. hexane or 2-methylpentane

11.60 Identify the compound in each pair that has the higher boiling point.
  a. butane or octane
  b. cyclohexane or hexane
  c. propane or pentane

11.61 Write a balanced equation for the complete combustion of each of the following:
  a. propane
  b. C_5H_{12}
  c. cyclobutane
  d. octane

11.62 Write a balanced equation for the complete combustion of each of the following:
  a. hexane
  b. methylcyclohexane
  c. cyclopentane
  d. 2-methylpropane

11.63 Draw the structural formulas of the monohalogenated organic products that result from the chlorination of each of the following:
  a. ethane **b.** propane **c.** cyclopentane

11.64 Draw the structural formulas of the monohalogenated organic products that result from the bromination of each of the following:
  a. pentane **b.** cyclohexane **c.** 2-methylbutane

11.65 For each of the definitions, find a corresponding term from the following list: alkane, alkene, alkyne, alcohol, ether, aldehyde, ketone, carboxylic acid, ester, amine, functional group, tetrahedral, isomers.
  a. An organic compound that contains a hydroxyl group bonded to a carbon.
  b. A hydrocarbon that contains one or more carbon–carbon double bonds.
  c. An organic compound in which the carbon of a carbonyl group is bonded to a hydrogen.
  d. A hydrocarbon that contains only carbon–carbon single bonds.
  e. An organic compound in which the carbon of a carbonyl group is bonded to a hydroxyl group.
  f. An organic compound that contains a nitrogen atom bonded to one or more carbon atoms.
  g. The three-dimensional shape of a carbon bonded to four hydrogen atoms.

11.66 For each of the definitions, find a corresponding term from the following list: alkane, alkene, alkyne, alcohol, ether, aldehyde, ketone, carboxylic acid, ester, amine, functional group, tetrahedral, isomers
  a. Organic compounds with identical molecular formulas that differ in the order the atoms are connected.
  b. An organic compound in which the hydrogen atom of a carboxyl group is replaced by a carbon atom.
  c. An organic compound that contains an oxygen atom bonded to two carbon atoms.
  d. A hydrocarbon that contains a carbon–carbon triple bond.
  e. A charactistic group of atoms that make compounds behave and react in a particular way.
  f. An organic compound in which the carbonyl group is bonded to two carbon atoms.

CHALLENGE QUESTIONS

11.67 The density of pentane, a component of gasoline, is 0.63 g/mL. The heat of combustion for pentane is 845 kcal per mole.
  a. Write an equation for the complete combustion of pentane.
  b. What is the molar mass?
  c. How much heat is produced when 1 gallon of pentane is burned (1 gallon = 3.78 liters)?
  d. How many liters of CO_2 at STP are produced from the complete combustion of 1 gallon of pentane?

11.68 Write structural formulas of two esters and a carboxylic acid that each have molecular formula $C_3H_6O_2$.

11.69 Draw all the possible structures of an organic compound with 6 carbon atoms that has a 4-carbon chain.

11.70 Draw all the possible structures of an organic compound with 4 carbon atoms that has a 3-carbon ring and a hydroxyl group.

11.71 Consider the compound propane.
  a. Draw the chemical structure.
  b. Write the equation for the complete combustion of propane.
  c. How many grams O_2 are needed to react with 12.0 L propane gas at STP?
  d. How many grams of CO_2 would be produced from the reaction in part c.?

11.72 Consider the compound ethylcyclopentane.
  a. Draw the chemical structure.
  b. Write the equation for the complete combustion of ethyl-cyclopentane.
  c. Calculate the grams O_2 required for the reaction of 25.0 g ethylcyclopentane.
  d. How many liters of CO_2 would be produced at STP from the reaction in part c.?

ANSWERS

Answers to Study Checks

11.1 Octane is not soluble in water, it is an organic compound.

11.2

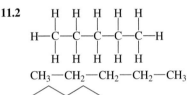

11.3 cyclobutane

11.4 This is another isomer. There is a five-carbon chain with a carbon group bonded to the middle (third) carbon.

11.5 1-chloro-2,4-dimethylhexane

11.6 CH_3—CH—CH_2—CH—CH_3
                     | |
                    CH_3 CH_3

11.7 1,2-dibromoethane

11.8
                      CH_3
                       |
CH_3—CH—CH_2—CH_3 = C_5H_{12}

$C_5H_{12} + 8O_2 \longrightarrow 5CO_2 + 6H_2O$

11.9
            Br
            |
CH_3—CH—Br   Br—CH_2—CH_2—Br

11.10 CH_3—CH_2—O—CH_3 contains the functional group C—O—C: it is an ether.

              OH
              |
CH_3—CH—CH_3

contains the —OH functional group; it is an alcohol.

11.11 A carboxylic acid has a carboxyl group COOH. In an ester, the oxygen atom of the carboxyl group is attached to a carbon atom, not hydrogen.

Answers to Selected Questions and Problems

11.1 **a.** inorganic   **b.** organic   **c.** organic
          d. inorganic   **e.** inorganic   **f.** organic

11.3 **a.** inorganic   **b.** organic
          c. organic   **d.** inorganic

11.5 **a.** ethane   **b.** ethane
          c. NaBr   **d.** NaBr

11.7 VSEPR theory predicts that the four bonds in CH_4 will be as far apart as possible, which means that the hydrogen atoms are at the corners of a tetrahedron.

11.9 **a.**
                    H   H   H
                    | | |
                H—C—C—C—H
                    | | |
                    H   H   H
          b. CH_3—CH_2—CH_2—CH_2—CH_2—CH_3
          c.

11.11 **a.** pentane   **b.** heptane
           c. hexane   **d.** cyclobutane

11.13 **a.** CH_4 **b.** CH_3—CH_3
           c. CH_3—CH_2—CH_2—CH_2—CH_3
           d.

11.15 **a.** same molecule
           b. isomers of C_5H_{12}
           c. isomers of C_6H_{14}

11.17 **a.** 2-methylbutane   **b.** 2,2-dimethylpropane
           c. 2,3-dimethylpentane   **d.** 4-ethyl-2,2-dimethylhexane

11.19 **a.** chlorocyclopentane
           b. methylcyclohexane
           c. 1-bromo-3-methylcyclobutane
           d. 1-bromo-2-chlorocyclopentane

11.21 **a.**
                          CH_3
                          |
           CH_3—CH—CH_2—CH_3

           b.
                            CH_3
                            |
           CH_3—CH_2—C—CH_2—CH_3
                            |
                            CH_3

           c.
                   CH_3 CH_3 CH_3
                    | | |
           CH_3—CH—CH—CH_2—CH—CH_3

           d.
                   CH_3 CH_2—CH_3 CH_3
                    | | |
           CH_3—CH—CH—CH_2—CH—CH_2—CH_2—CH_3

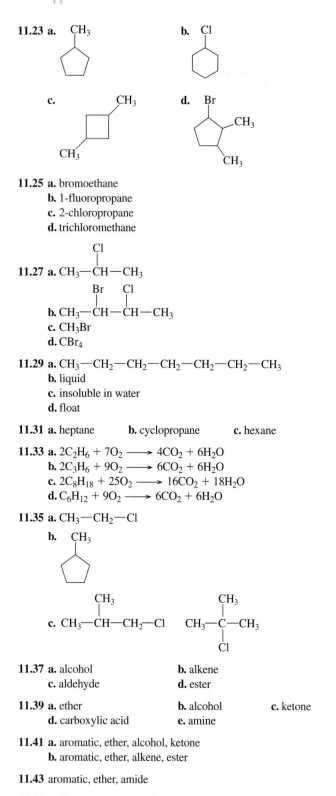

11.23 a. (cyclopentane with CH$_3$) **b.** (cyclohexane with Cl)

c. (cyclobutane with two CH$_3$) **d.** (cyclopentane with Br and two CH$_3$)

11.25 a. bromoethane
  b. 1-fluoropropane
  c. 2-chloropropane
  d. trichloromethane

11.27 a. CH$_3$—CH—CH$_3$ (with Cl on middle carbon)
  b. CH$_3$—CH—CH—CH$_3$ (with Br and Cl)
  c. CH$_3$Br
  d. CBr$_4$

11.29 a. CH$_3$—CH$_2$—CH$_2$—CH$_2$—CH$_2$—CH$_2$—CH$_3$
  b. liquid
  c. insoluble in water
  d. float

11.31 a. heptane **b.** cyclopropane **c.** hexane

11.33 a. 2C$_2$H$_6$ + 7O$_2$ ⟶ 4CO$_2$ + 6H$_2$O
  b. 2C$_3$H$_6$ + 9O$_2$ ⟶ 6CO$_2$ + 6H$_2$O
  c. 2C$_8$H$_{18}$ + 25O$_2$ ⟶ 16CO$_2$ + 18H$_2$O
  d. C$_6$H$_{12}$ + 9O$_2$ ⟶ 6CO$_2$ + 6H$_2$O

11.35 a. CH$_3$—CH$_2$—Cl

  b. (cyclopentane with CH$_3$)

  c. CH$_3$—CH—CH$_2$—Cl (with CH$_3$) CH$_3$—C—CH$_3$ (with CH$_3$ and Cl)

11.37 a. alcohol **b.** alkene
  c. aldehyde **d.** ester

11.39 a. ether **b.** alcohol **c.** ketone
  d. carboxylic acid **e.** amine

11.41 a. aromatic, ether, alcohol, ketone
  b. aromatic, ether, alkene, ester

11.43 aromatic, ether, amide

11.45 a. Organic compounds have covalent bonds; inorganic compounds have ionic as well as polar covalent and a few have nonpolar covalent bonds.
  b. Most organic compounds are insoluble in water, many inorganic compounds are soluble in water.

c. Most organic compounds have low melting points; inorganic compounds have high melting points.
d. Most organic compounds are flammable; inorganic compounds are not usually flammable.

11.47 a. butane **b.** butane
  c. potassium chloride **d.** potassium chloride
  e. butane

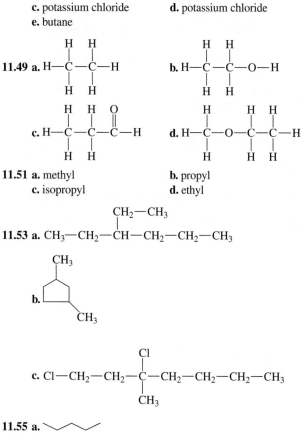

11.49 a. H—C—C—H (ethane structure) **b.** H—C—C—O—H
  c. H—C—C—C—H (with O) **d.** H—C—O—C—C—H

11.51 a. methyl **b.** propyl
  c. isopropyl **d.** ethyl

11.53 a. CH$_3$—CH$_2$—CH—CH$_2$—CH$_2$—CH$_3$ (with CH$_2$—CH$_3$)

  b. (cyclopentane with two CH$_3$)

  c. Cl—CH$_2$—CH$_2$—C—CH$_2$—CH$_2$—CH$_2$—CH$_3$ (with Cl and CH$_3$)

11.55 a. (branched chain)
  b. (branched chain)
  c. (chain with Br)

11.57 Condensed structural formula

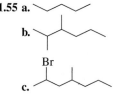

CH$_3$—C—CH$_2$—CH—CH$_3$ (with CH$_3$, CH$_3$, and CH$_3$ substituents)

molecular formula of C$_8$H$_{18}$

The combustion reaction: 2C$_8$H$_{18}$ + 25O$_2$ ⟶ 16CO$_2$ + 18H$_2$O

11.59 a. heptane
  b. cyclopentane
  c. hexane

11.61 a. C$_3$H$_8$ + 5O$_2$ ⟶ 3CO$_2$ + 4H$_2$O
  b. C$_5$H$_{12}$ + 8O$_2$ ⟶ 5CO$_2$ + 6H$_2$O
  c. C$_4$H$_8$ + 6O$_2$ ⟶ 4CO$_2$ + 4H$_2$O
  d. 2C$_8$H$_{18}$ + 25O$_2$ ⟶ 16CO$_2$ + 18H$_2$O

11.63 a. CH_3-CH_2-Cl

b. $CH_3-CH_2-CH_2-Cl$ and $CH_3-\overset{\displaystyle Cl}{\underset{|}{CH}}-CH_3$

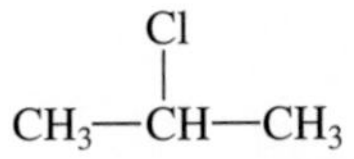

c.

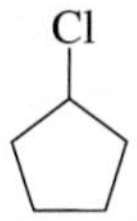

11.65 a. alcohol **b.** alkene **c.** aldehyde
d. alkane **e.** carboxylic acid **f.** amine
g. tetrahedral

11.67 a. $C_5H_{12} + 8O_2 \longrightarrow 5CO_2 + 6H_2O$
b. 72.0 g/mole
c. 2.8×10^4 kcal
d. 3700 L

11.69 $CH_3-\overset{\displaystyle CH_3}{\underset{|}{CH}}-\overset{\displaystyle CH_3}{\underset{|}{CH}}-CH_3$ $CH_3-\overset{\displaystyle CH_3}{\underset{\displaystyle \underset{|}{CH_3}}{\overset{|}{\underset{|}{C}}}}-CH_2-CH_3$

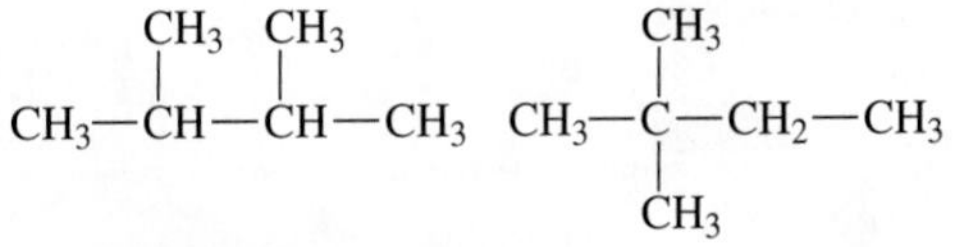

11.71 a. $CH_3-CH_2-CH_3$
b. $C_3H_8 + 5O_2 \longrightarrow 3CO_2 + 4H_2O$
c. 85.7 g O_2
d. 70.7 g CO_2

12

Unsaturated Hydrocarbons

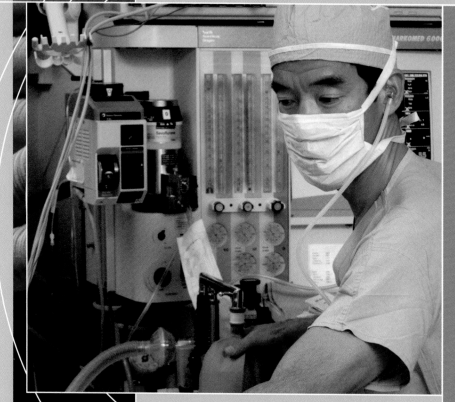

"During surgery, I work with the surgeon to provide a safe level of anesthetics that renders the patient free from pain," says Mark Noguchi, nurse anesthetist (CRNA), Kaiser Hospital. "We do spinal and epidural blocks as well as general anesthetics, which means the patient is totally asleep. We use a variety of pharmaceutical agents including halothane ($C_2HBrClF_3$), and bupivacain ($C_{18}H_{28}N_2O$), as well as muscle relaxants such as midazolam ($C_{18}H_{13}ClFN_3$) to achieve the results we want for the surgical situation. We also assess the patient's overall hemodynamic status. If blood is lost, we replace components such as plasma, platelets, and coagulation factors. We also monitor the heart rate and run EKGs to determine cardiac function."

LOOKING AHEAD

Mastering
GOB CHEMISTRY

Visit **www.masteringgob.com**
for self-study materials
and instructor-assigned
homework.

n Chapter 11, we looked primarily at *saturated hydrocarbons*, which contain all single bonds between their carbon atoms. Now we will investigate unsaturated hydrocarbons, carbon compounds containing one or more carbon–carbon double bonds or triple bonds. When we cook with vegetable oils such as corn oil, safflower oil, or olive oil, we are using unsaturated lipids that have double bonds in their long carbon chains. Saturated fats from animal sources also have long chains of carbon atoms, but they are connected only by the single bonds of alkanes. If we compare the two types of fats, we find considerable differences in their physical and chemical properties. Vegetable oils are liquid at room temperature whereas animal fats are solid. Because double bonds are reactive, unsaturated fats can be oxidized by oxygen in the air, especially at warm temperatures, forming products that have rancid, unpleasant odors. The saturated fats are more resistant to reactions.

12.1 Alkenes and Alkynes

The **unsaturated hydrocarbons** have fewer hydrogen atoms than alkanes. **Alkenes** contain at least one double bond between carbons. Alkenes are useful compounds in industry and have important functions in plants and animals. Ethylene is an important industrial compound used to make a plastic called polyethylene. The unsaturated sites are very reactive, which distinguishes unsaturated compounds from the alkanes. **Alkynes** contain at least one triple bond between carbons. (See Figure 12.1.) Only a small number of alkynes are found in nature. However, ethyne, commonly called acetylene, is used in welding where it burns at a very high temperature.

$$H_2C{=}CH_2$$

ethene (ethylene)
C_2H_4

$$\underset{\underset{H}{|}}{CH_3(CH_2)_7}\underset{\underset{H}{|}}{(CH_2)_{12}CH_3}\\ C{=}C$$

cis-9-tricosene, muscalure,
housefly sex attractant
$C_{23}H_{46}$

$$HC{\equiv}CH$$

ethyne (acetylene)
C_2H_2

Alkene Structure

The simplest alkene C_2H_4 is called ethene, but it is more likely to be called by its common name ethylene. There are two CH_2 groups connected by a double bond, which represents two sets of electrons. We already know that carbon has four electrons and needs four bonds. In a double bond, each carbon atom is attached to three other atoms (one carbon and two hydrogens). According to VSEPR theory (Chapter 4), three groups bonded to each carbon in the double bond are planar and arranged at angles of 120°.

LEARNING GOAL

Write the IUPAC names for alkenes and alkynes; give common names for simple structures.

Ethene

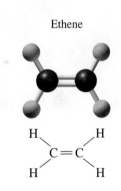

$$\underset{\underset{H}{\diagup}}{H}\overset{\overset{H}{\diagdown}}{C}{=}\underset{\underset{H}{\diagdown}}{\overset{\overset{H}{\diagup}}{C}}$$

Ethyne

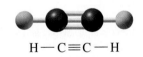

$$H{-}C{\equiv}C{-}H$$

FIGURE 12.1

Ball-and-stick models of ethene and ethyne show the functional groups of double or triple bonds.

Q **Why are these compounds called unsaturated hydrocarbons?**

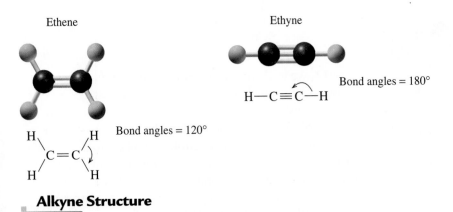

Ethene

Ethyne

Bond angles = 180°

$$H-C\equiv C-H$$

Bond angles = 120°

Alkyne Structure

The simplest alkyne is called ethyne, but it is commonly known as acetylene. In ethyne there are two CH groups connected by a triple bond. We already know that carbon has four electrons and needs four bonds. In a triple bond, each carbon atom is attached to two other atoms (one carbon and one hydrogen). According to VSEPR theory, two groups bonded to each carbon in the triple bond are linear and arranged at angles of 180°.

MC GOB

Drawing Alkenes and Alkynes
Naming Alkenes and Alkynes

Naming Alkenes and Alkynes

The IUPAC names for alkenes and alkynes are similar to those of alkanes. The simplest alkene, ethene (ethylene) is an important plant hormone involved in promoting the ripening of fruit. Commercially grown fruit, such as avocados, bananas, and tomatoes, are often picked before they are ripe. Before the fruit is brought to market, it is exposed to ethylene to accelerate the ripening process. Ethylene also accelerates the breakdown of cellulose in plants, which causes flowers to wilt and leaves to fall from trees.

The IUPAC name of the simplest alkyne is ethyne, although acetylene, its common name, is often used. See Table 12.1 for a comparison of the naming for alkanes, alkenes, and alkynes. For alkenes and alkynes, the longest carbon chain must contain the double or triple bond.

TABLE 12.1

Comparison of Names for Alkanes, Alkenes, and Alkynes

Alkane	Alkene	Alkyne
H_3C-CH_3	$H_2C=CH_2$	$HC\equiv CH$
ethane	ethene (ethylene)	ethyne (acetylene)
$CH_3-CH_2-CH_3$	$CH_3-CH=CH_2$	$CH_3-C\equiv CH$
propane	propene	propyne

EXPLORE YOUR WORLD

Ripening Fruit

Obtain two unripe green bananas. Place one in a plastic bag and seal it. Leave them on the counter. Check the bananas twice a day to observe any difference in the ripening process.

QUESTIONS

1. What compound helps ripen the bananas?
2. What are some possible reasons for any difference in the ripening rate?
3. If you wish to ripen an avocado, what procedure might you use?

■ STEP 1 **Name the longest carbon chain that contains the double or triple bond.** Replace the corresponding alkane ending with *ene* for an alkene and *yne* for an alkyne.

■ STEP 2 **Number the longest chain from the end nearest the double or triple bond.** Indicate the position of the double or triple bond with the number of the first unsaturated carbon. Unsaturated compounds with two or three carbons do not need numbers.

$$\underset{4\quad\ 3\ \quad\ 2\quad\ 1}{CH_3-CH_2-CH=CH_2}\qquad\underset{1\quad\ 2\quad\ 3\quad\ 4}{CH_3-CH=CH-CH_3}\qquad\underset{1\quad\ 2\quad\ 3\quad\ 4}{CH_3-C\equiv C-CH_3}$$

1-butene   2-butene   2-butyne

■ **STEP 3 Give the location and name of each substituent (alphabetical order) as a prefix to the alkene or alkyne name.**

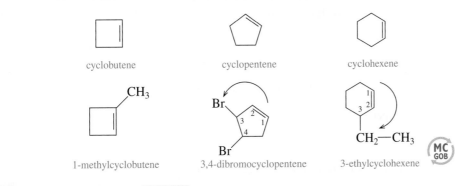

$$CH_2\!=\!CH\!-\!CH_2\!-\!\underset{\displaystyle CH_3}{\overset{\displaystyle |}{CH}}\!-\!CH_3$$
1   2   3   4   5
4-methyl-1-pentene

$$CH_3\!-\!\underset{\displaystyle CH_3}{\overset{\displaystyle |}{C}}\!=\!\underset{\displaystyle CH_3}{\overset{\displaystyle |}{C}}\!-\!CH_3$$
1   2   3   4
2,3-dimethyl-2-butene

$$CH_3\!-\!\underset{\displaystyle Cl}{\overset{\displaystyle |}{CH}}\!-\!C\!\equiv\!CH$$
4   3   2   1
3-chloro-1-butyne

In **cycloalkenes** with a substituent, the double bond is understood to be on carbons 1 and 2 and the ring is numbered to give the lowest number to a substituent.

cyclobutene

cyclopentene

cyclohexene

1-methylcyclobutene

3,4-dibromocyclopentene

3-ethylcyclohexene

MC GOB

SELF-STUDY ACTIVITY
Organic Molecules and Isomers
Introduction to Organic Molecules

SAMPLE PROBLEM 12.1

■ **Naming Alkenes and Alkynes**

Write the IUPAC name for each of the following:

a. $$CH_3\!-\!\underset{\displaystyle CH_3}{\overset{\displaystyle |}{CH}}\!-\!CH\!=\!CH\!-\!CH_3$$

b. $CH_3\!-\!CH_2\!-\!C\!\equiv\!C\!-\!CH_2\!-\!CH_3$

SOLUTION

a. ■ **STEP 1 Name the longest carbon chain that contains the double or triple bond.**   There are five carbon atoms in the longest carbon chain containing the double bond. Replacing the corresponding alkane ending with *ene* gives pentene.

■ **STEP 2 Number the longest chain from the end nearest the double or triple bond.**   The number of the first carbon in the double bond is used to give the location of the double bond.

$$CH_3\!-\!\underset{\displaystyle CH_3}{\overset{\displaystyle |}{CH}}\!-\!CH\!=\!CH\!-\!CH_3$$
5   4   3   2   1

2-pentene

■ **STEP 3 Give the location and name of each substituent (alphabetical order) as a prefix to the alkene or alkyne name.**   The methyl group is located on carbon 4.

$$CH_3\!-\!\underset{\displaystyle CH_3}{\overset{\displaystyle |}{CH}}\!-\!CH\!=\!CH\!-\!CH_3$$
5   4   3   2   1

4-methyl-2-pentene

b. ■ **STEP 1 Name the longest carbon chain that contains the double or triple bond.**   There are six carbon atoms in the longest chain containing

Guide to Naming Alkenes and Alkynes

STEP 1
Name the longest carbon chain with a double or triple bond.

STEP 2
Number the carbon chain starting from the end nearest a double or triple bond.

STEP 3
Give the location and name of each substituent (alphabetical order) as a prefix to the name, if needed.

the triple bond. Replacing the corresponding alkane ending with *yne* gives hexyne.

■ **STEP 2** **Number the main chain from the end nearest the double or triple bond.** The number of the first carbon in the triple bond is used to give the location of the double bond.

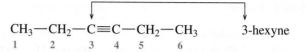

$$CH_3—CH_2—C≡C—CH_2—CH_3 \qquad \text{3-hexyne}$$
$$\;\;1 \qquad 2 \qquad 3 \quad 4 \quad 5 \qquad 6$$

■ **STEP 3** **Give the location and name of each substituent (alphabetical order) as a prefix to the alkene or alkyne name.** There are no substituents in this formula.

STUDY CHECK

Draw the structural formulas for each of the following:

a. 2-pentyne   **b.** 3-methylcyclopentene

QUESTIONS AND PROBLEMS

■ Alkenes and Alkynes

12.1 Identify the following as alkenes, cycloalkenes, or alkynes:

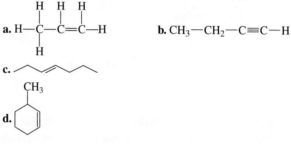

a.
$$\begin{array}{ccc} H & H & H \\ | & | & | \\ H—C—C{=}C—H \\ | \\ H \end{array}$$

b. $CH_3—CH_2—C≡C—H$

c.

d.

12.2 Identify the following as alkenes, cycloalkenes, or alkynes:

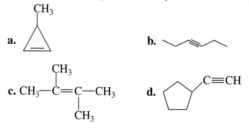

a.

b.

c. $CH_3—\overset{\overset{\displaystyle CH_3}{|}}{C}{=}\overset{\overset{}{}}{C}—CH_3$ with CH_3 below

d.

12.3 Give the IUPAC name for each of the following:

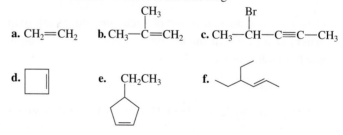

a. $CH_2{=}CH_2$ **b.** $CH_3—\overset{\overset{\displaystyle CH_3}{|}}{C}{=}CH_2$ **c.** $CH_3—\overset{\overset{\displaystyle Br}{|}}{CH}—C≡C—CH_3$

d.

e. CH_2CH_3

f.

▪ Fragrant Alkenes

The odors you associate with lemons, oranges, roses, and lavender are due to volatile compounds that are synthesized by the plants. Often it is unsaturated compounds that are responsible for the pleasant flavors and fragrances of many fruits and flowers. They were some of the first kinds of compounds to be extracted from natural plant material. In ancient times, they were highly valued in their pure forms. Limonene and myrcene give the characteristic odors and flavors to lemons and oranges and bay leaves, respectively. Geraniol and citronellal give roses and lemon grass their distinct aromas. In the food and perfume industries, these compounds are extracted or synthesized and used as perfumes and flavorings.

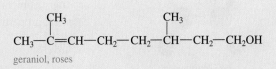

geraniol, roses

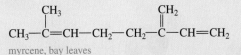

myrcene, bay leaves

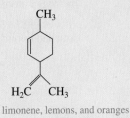

citronellal, lemon grass

limonene, lemons, and oranges

12.4 Give the IUPAC name for each of the following:

a. $CH_2{=}CH{-}CH_2{-}CH_2{-}CH_2{-}CH_3$ **b.** $CH_3{-}C{\equiv}C{-}CH_2{-}CH_2{-}\overset{\overset{\displaystyle CH_3}{|}}{CH}{-}CH_3$ **c.**

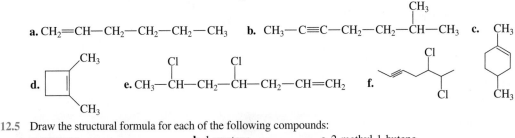

d.   **e.** $CH_3{-}\overset{\overset{\displaystyle CH_3}{|}}{CH}{-}CH_2{-}\overset{\overset{\displaystyle Cl}{|}}{CH}{-}CH_2{-}CH{=}CH_2$ **f.**

12.5 Draw the structural formula for each of the following compounds:
a. propene   **b.** 1-pentene   **c.** 2-methyl-1-butene
d. 3-methylcyclohexene   **e.** 2-chloro-3-hexyne

12.6 Draw the structural formula for each of the following compounds:
a. 1-methylcyclopentene   **b.** 3-methyl-1-butyne
c. 3,4-dimethyl-1-pentene   **d.** 4-ethyl-1-methylcyclohexene
e. 1,2-dichlorocyclopentene

▪ 12.2 Cis–Trans Isomers

In alkenes, there is no rotation around the carbons in the rigid double bond. As a result, any groups connected to the double bond remain fixed on one side or the other. In a **cis isomer**, two groups are on the same side of the double bond. In the **trans isomer**, the groups are on opposite sides of the double bond. For example,

LEARNING GOAL

▪ Write the structural formulas and names for the cis–trans isomers of alkenes.

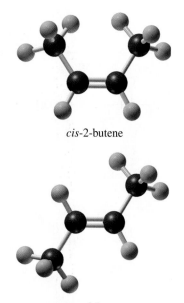

cis-2-butene

trans-2-butene

FIGURE 12.2

Ball-and-stick models of the cis and trans isomers of 2-butene.

Q **What feature in 2-butene accounts for the cis and trans isomers?**

cis–trans isomers can be written for 2-butene. (See Figure 12.2.) In general, trans isomers are more stable than their cis counterparts because the large groups attached to the double bond are further apart. As with any pair of isomers, the cis–trans isomers of 2-butene are different compounds with different physical and chemical properties as shown in the following:

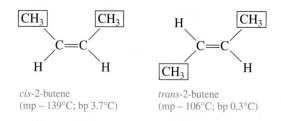

cis-2-butene
(mp − 139°C; bp 3.7°C)

trans-2-butene
(mp − 106°C; bp 0.3°C)

Not every alkene shows cis–trans isomerism. If one of the carbons in the double bond is attached to identical groups, the molecule does not have cis–trans isomers. This is the case of 1-butene and 2-methyl-1-propene, isomers of 2-butene. If the hydrogen atoms are interchanged on carbon 1, the same structure results. Alkynes do not have cis–trans isomers because the carbons in the triple bond are each attached to only one group.

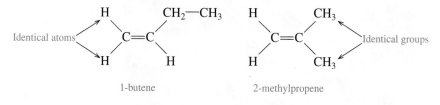

1-butene

2-methylpropene

Pheromones in Insect Communication

Insects and many other organisms emit minute quantities of chemicals called pheromones. Insects use pheromones to send messages to individuals of the same species. Some pheromones warn of danger, others call for defense, mark a trail, or attract the opposite sex. In the last 40 years, the structures of many pheromones have been chemically determined. One of the most studied is bombykol, the sex pheromone produced by the female of the silkworm moth species. The bombykol molecule is a 16 carbon chain with one cis double bond, one trans double bond, and an alcohol group. A few molecules of synthetic bombykol will attract male silkworm moths from distances of over one kilometer. The effectiveness of many of these pheromones depends on the cis or trans configuration of the double bonds in the molecules. A certain species will respond to one isomer but not the other.

Scientists are interested in synthesizing pheromones for use as nontoxic alternatives to pesticides. When used in a trap, bombykol can be used to isolate male silkworm moths. When a synthetic pheromone is released in several areas of a field or crop, the males cannot locate the females, which disrupts the reproductive cycle. This technique has been successful with controlling the oriental fruit moth, the grapevine moth, and the pink bollworm.

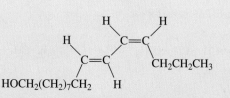

Bombykol, sex attractant for the silkworm moth

■ **Cis–Trans Isomers for Night Vision**

The retinas of the eyes consist of two types of cells: rods and cones. The rods on the edge of the retina allow us to see in dim light, and the cones, in the center, produce our vision in bright light. In the rods, there is a substance called rhodopsin that absorbs light. Rhodopsin is composed of *cis*-11-retinal, an unsaturated compound, attached to a protein. When rhodopsin absorbs light, the *cis*-11-retinal isomer is converted to its trans isomer, which changes its shape. The trans form no longer fits the protein and it separates from the protein. The change from the cis to trans isomer and the separation from the protein generates an electrical signal that the brain converts into an image.

An enzyme (isomerase) converts the trans isomer back to the *cis*-11-retinal isomer and the rhodopsin re-forms. If there is a deficiency of rhodopsin in the rods of the retina, night blindness may occur. One common cause is a lack of vitamin A in the diet. In our diet, we obtain vitamin A from plant pigments containing β-carotene, which is found in foods such as carrots, squash, and spinach. In the small intestine, the β-carotene is converted to vitamin A, which can be converted to *cis*-11-retinal or stored in the liver for future use. Without a sufficient quantity of retinal, not enough rhodopsin is produced to enable us to see adequately in dim light.

■ **Cis–Trans Isomers of Retinal**

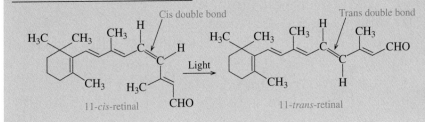

11-*cis*-retinal   →Light→   11-*trans*-retinal

As long as the groups attached to the double bond are different, an alkene will show cis–trans isomers. Another example of cis–trans isomers is the following:

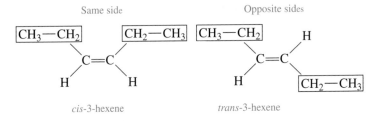

cis-3-hexene   trans-3-hexene

SELF-STUDY ACTIVITY
Geometric Isomers

SAMPLE PROBLEM 12.2

■ **Naming Cis–Trans Isomers**

Name each of the following as cis or trans isomers.

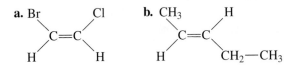

Cis-Trans Isomers

SOLUTION

a. This isomer is a cis isomer because the Br and Cl atoms are on the same side of the double bond: *cis*-1-bromo-2-chloroethene.

b. This isomer is a trans isomer because the CH_3— and —CH_2—CH_3 groups are on the opposite sides of the double bond: *trans*-2-pentene.

▪ Modeling Cis–Trans Isomers

Because cis–trans isomerism is not easy to imagine, here are some things you can do to understand the difference in rotation around a single bond compared to a double bond and how it affects groups that are attached to the carbon atoms in the double bond.

Put the fingertips of your index fingers together. This is a model of a single bond. Consider the index fingers as a pair of carbon atoms and think of your thumbs and other fingers as other parts of a carbon chain. While your index fingers are touching, twist your hands and change the position of the thumbs relative to each other. Notice how the relationship of your other fingers changes.

Now place the tips of your index fingers and middle fingers together in a model of a double bond. As you did before, twist your hands to move the thumbs apart. What happens? Can you change the location of your thumbs relative to each other without breaking the double bond? The difficulty of moving your hands with two fingers touching represents the lack of rotation about a double bond. You have made a model of a cis isomer when both thumbs are on the same side. If you turn one hand over so one thumb points down and one points up, you have made a model of a trans isomer.

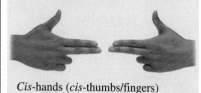

Cis-hands (*cis*-thumbs/fingers)

Trans-hands (*trans*-thumbs/fingers)

LEARNING GOAL

Write the structural formulas and names for the organic products of addition reactions of alkenes and alkynes.

STUDY CHECK

Is the following compound *cis*-3-hexene or *trans*-3-hexene?

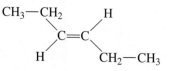

QUESTIONS AND PROBLEMS

▪ **Cis–Trans Isomers**

12.7 Which of the following can be written as cis–trans isomers?
a. CH_2=CH—CH_3 **b.** CH_3—CH_2—CH=CH—CH_3
c.

12.8 Which of the following do not have cis–trans isomers?
a.   **b.** CH_3—CH_2—CH_2—CH=CH_2
c.

12.9 Write the IUPAC name of each of the following using cis or trans prefixes:
a.   **b.**
c.

12.10 Write the IUPAC name of each of the following using cis or trans prefixes:
a.   **b.**
c.

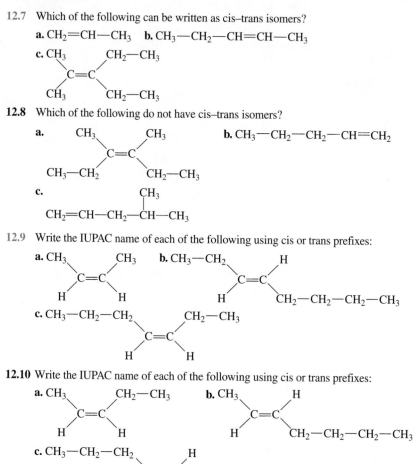

12.11 Draw the structural formula for each of the following:
  a. *trans*-2-butene **b.** *cis*-2-pentene **c.** *trans*-3-heptene
12.12 Draw the structural formula for each of the following:
  a. *cis*-3-hexene **b.** *trans*-2-pentene **c.** *cis*-4-octene

12.3 Addition Reactions

For alkenes and alkynes, the most characteristic reaction is the **addition** of atoms or groups of atoms to the carbons of the double or triple bond. Addition occurs because double and triple bonds are easily broken, which provides electrons for new single bonds. The general equation for the addition of a reactant A—B to an alkene can be written as follows:

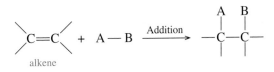

alkene

The addition reactions have different names that depend on the type of reactant we add to the alkene, as shown in Table 12.2.

TABLE 12.2

Reactants and Addition Reactions

Reactant Added	Name of Addition Reaction
H_2	Hydrogenation
Cl_2, Br_2	Halogenation
HCl, HBr, HI	Hydrohalogenation
H_2O	Hydration

Hydrogenation

In a reaction called **hydrogenation**, atoms of hydrogen add to the carbons in a double or triple bond to form alkanes. A catalyst such as platinum (Pt), nickel (Ni), or palladium (Pd) is added to speed up the reaction. The general equation for hydrogenation can be written as follows:

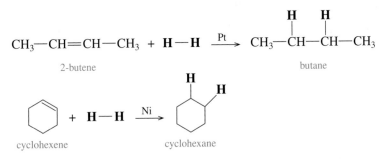

Double bond (unsaturated)   Single bond (saturated)

(MC GOB) Addition Reactions
Types of Addition Reactions

Some examples of the hydrogenation of alkenes and alkynes follow:

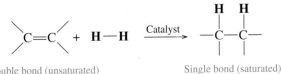

2-butene   butane

cyclohexene   cyclohexane

The hydrogenation of alkynes requires two molecules of hydrogen to form the alkane product.

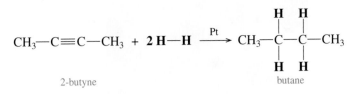

2-butyne   butane

EXPLORE YOUR WORLD

■ Unsaturation in Fats and Oils

Read the labels on some containers of vegetable oils, margarine, peanut butter, and shortenings.

QUESTIONS

1. What terms on the label tell you that the compounds contain double bonds?
2. A label on a bottle of canola oil lists saturated, polyunsaturated, and monounsaturated fats. What do these terms tell you about the type of bonding in the fats?
3. A peanut butter label states that it contains partially hydrogenated vegetable oils or completely hydrogenated vegetable oils. What does this tell you about the type of reaction that took place in preparing the peanut butter?

SAMPLE PROBLEM 12.3

■ Writing Equations for Hydrogenation

Write the structural formula for the product of the following hydrogenation reactions:

a. $CH_3-CH=CH_2 + H_2 \xrightarrow{Pt}$?   **b.** ⬠∥ + $H_2 \xrightarrow{Pt}$?   **c.** $HC≡CH + 2H_2 \xrightarrow{Ni}$?

SOLUTION

In an addition reaction, hydrogen adds to the double or triple bond to give an alkane.

a. $CH_3-CH_2-CH_3$ **b.** ⬠   **c.** H_3C-CH_3

STUDY CHECK

Draw the structural formula of the product of the hydrogenation of 2-methyl-1-butene using a platinum catalyst.

Halogenation

In the **halogenation** reactions of alkenes or alkynes, halogen atoms such as chlorine or bromine are added to the double or triple bonds. The reaction occurs readily, without the use of any catalyst, and adds halogen atoms to yield a di- or tetra-haloalkane product. In the general equation for halogenation, the symbol X—X or X_2 is used for Cl_2 or Br_2.

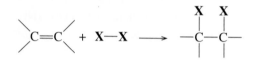

Here are some examples of adding Cl_2 or Br_2 to alkenes:

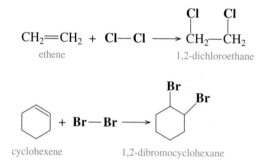

CH_2=CH_2 + Cl—Cl ⟶
ethene

1,2-dichloroethane

cyclohexene

1,2-dibromocyclohexane

Hydrogenation of Unsaturated Fats

Vegetable oils such as corn oil or safflower oil are unsaturated fats composed of fatty acids that contain double bonds. The process of hydrogenation is used commercially to convert the double bonds in the unsaturated fats in vegetable oils to saturated fats such as margarine, which are more solid. Adjusting the amount of added hydrogen produces partially hydrogenated fats such as soft margarine, solid margarine in sticks, and shortenings, which are used in cooking. For example, oleic acid is a typical unsaturated fatty acid in olive oil and has a cis double bond at carbon 9. When oleic acid is hydrogenated, it is converted to stearic acid, a saturated fatty acid.

$CH_3(CH_2)_7$ $(CH_2)_7\overset{O}{\overset{\|}{C}}OH$

C=C

H   H

oleic acid (the cis isomer is found in olive oil and other unsaturated fats)

+ H_2 $\xrightarrow{\text{Pt}}$ $CH_3(CH_2)_7$—CH_2—CH_2—$(CH_2)_7\overset{O}{\overset{\|}{C}}OH$

stearic acid (found in saturated fats)

$$CH_3—C≡CH + 2Cl—Cl \longrightarrow \begin{matrix} & Cl & Cl \\ & | & | \\ CH_3—C—CH \\ & | & | \\ & Cl & Cl \end{matrix}$$

propyne   1,1,2,2-tetrachloropropane

The addition reaction of bromine is sometimes used to test for the presence of double and triple bonds, as shown in Figure 12.3.

SAMPLE PROBLEM 12.4

▪Writing Products of Halogenation

Write the condensed structural formula of the product of the following reaction:

$$\begin{matrix} & CH_3 \\ & | \\ CH_3—C=CH_2 + Br_2 \longrightarrow \end{matrix}$$

SOLUTION

The addition of bromine to an alkene places a bromine atom on each of the carbon atoms of the double bond.

$$\begin{matrix} & CH_3 \\ & | \\ CH_3—C—CH_2 \\ & | & | \\ & Br & Br \end{matrix}$$

STUDY CHECK

What is the name of the product formed when chlorine is added to 1-butene?

▪ Hydrohalogenation

In the reaction called **hydrohalogenation**, a hydrogen halide (HCl, HBr, or HI) adds to an alkene to yield a haloalkane. The hydrogen atom bonds to one carbon of the double bond, and the halogen atom adds to the other carbon. The general reaction, in which HX represents HCl, HBr, or HI, can be written as follows:

$$\begin{matrix} \diagdown & & & & & H & X \\ & C=C & + & H—X \longrightarrow & & | & | \\ \diagup & & \diagup & & & —C—C— \\ & & & & & | & | \end{matrix}$$

alkene   haloalkane (alkyl halide)

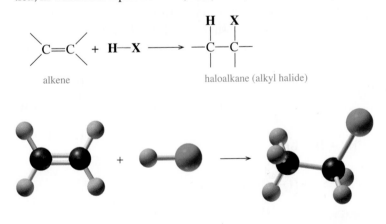

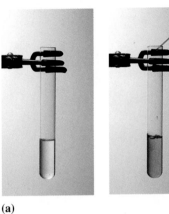

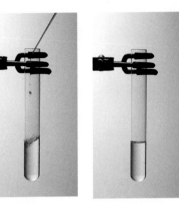

(a)

(b)

FIGURE 12.3

(a) When bromine is added to an alkane in the first test tube, the red color of bromine remains because the alkane does not react or reacts slowly. (b) When bromine is added to an alkene in the second test tube, the red color immediately disappears as bromine atoms add to the double bond.

Q Will the red color disappear when bromine is added to cyclohexane or cyclohexene?

Two examples of hydrohalogenation follow:

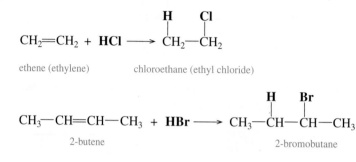

ethene (ethylene)   chloroethane (ethyl chloride)

2-butene   2-bromobutane

Steps in Addition Reactions of H—X to Alkenes

We have seen that in the addition reaction of an alkene, two groups add to the carbons in the double bond to give a saturated compound. To understand how the addition of H—X or H—OH takes place, we can consider the two steps involved when HBr adds to ethene. Initially, a proton from the HBr reacts with a pair of electrons in the double bond. This makes the other carbon into a **carbocation** (a carbon cation), which has only three bonds and a positive charge. The carbocation reacts quickly with the bromide ion Br⁻.

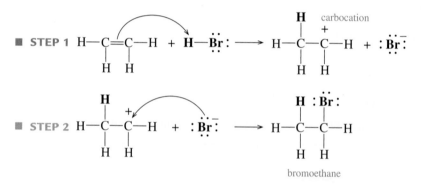

bromoethane

Markovnikov's Rule

When HBr adds to a symmetric alkene, a single product results. However, HBr can also add to a double bond in unsymmetric alkenes. The most stable carbocation that forms is the one attached to the most alkyl groups. Therefore, in the initial step, the proton adds to the carbon that is less substituted, which is the carbon in the double bond that has the greater number of hydrogens.

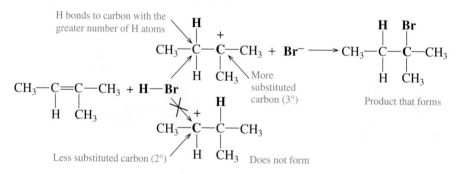

In 1870, Markovnikov, a Russian chemist, observed that the hydrohalogenation addition products of alkenes were limited to the more substituted halide product.

His observation now called **Markovnikov's rule** states that when HX adds to a double bond, the proton is bonded to the carbon atom that already has the most protons. Today, we know that this occurs because the proton adds to the position that produces the most stable carbocation, which also leads to the more substituted halide product.

SAMPLE PROBLEM 12.5

■ **Addition to Alkenes**

Predict the organic product for each of the following reactions:

a. CH_3—CH=CH—CH_3 + HBr $\longrightarrow$

b. 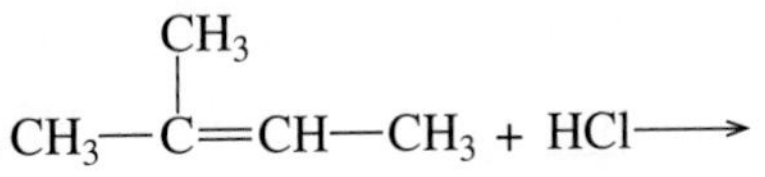
$$CH_3—\overset{\overset{\displaystyle CH_3}{|}}{C}=CH—CH_3 + HCl \longrightarrow$$

SOLUTION

a. This is a symmetric alkene. Only one product forms when the H^+ and Br^- add to the carbons of the double bond

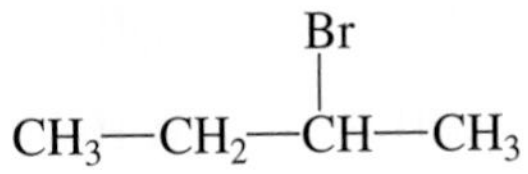

$$CH_3—CH_2—\overset{\overset{\displaystyle Br}{|}}{CH}—CH_3$$

b. In the double bond of this unsymmetric alkene, carbon 3 has the greater number of hydrogen atoms. Using Markovnikov's rule, the H from HCl adds to carbon 3 and the Cl adds to carbon 2. The product is the most substituted halide.

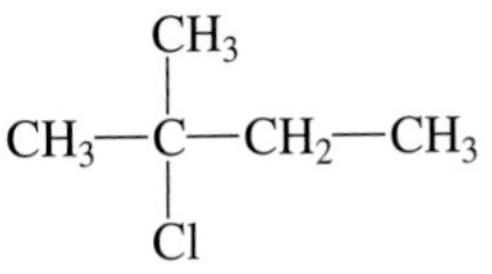

$$CH_3—\overset{\overset{\displaystyle CH_3}{|}}{\underset{\underset{\displaystyle Cl}{|}}{C}}—CH_2—CH_3$$

STUDY CHECK

Draw the structural formula of the organic product obtained when HBr adds to 1-methylcyclopentene.

Hydration

Alkenes react with water (HOH) when the reaction is catalyzed by a strong acid such as H_2SO_4. In this reaction called hydration, H— attaches to one of the carbon atoms in the double bond, and —OH to the other carbon. Hydration is used to prepare alcohols, which have the functional group —OH. In the general equation the acid is represented by H^+.

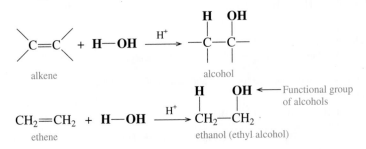

The addition of water to a double bond in unsymmetric alkene follows Markovnikov's rule.

$$CH_3-CH=CH_2 + H-OH \xrightarrow{H^+} CH_3-\overset{\displaystyle \overset{OH}{|}}{CH}-\overset{\displaystyle \overset{H}{|}}{CH_2} \; not \; CH_3-CH_2-CH_2-OH$$

propene   2-propanol

SAMPLE PROBLEM 12.6

▪ Writing Products of Hydration

Write the structural formulas for the products that form in the following hydration reactions:

a. $CH_3-CH_2-CH_2-CH=CH_2 + HOH \xrightarrow{H^+}$ **b.**  [cyclobutene] $+ HOH \xrightarrow{H^+}$

SOLUTION

a. Water adds H— and —OH to the double bond. We use Markovnikov's rule to add the H— to the CH_2 in the double bond, and the —OH to the CH.

$$CH_3-CH_2-CH_2-\overset{\displaystyle \overset{OH}{\downarrow}}{CH}=\overset{\displaystyle \overset{H}{\downarrow}}{CH_2} \xrightarrow{H^+} CH_3-CH_2-CH_2-\overset{\displaystyle \overset{OH}{|}}{CH}-CH_3$$

b. In cyclobutene, the H— adds to one side of the double bond, and the —OH adds to the other side.

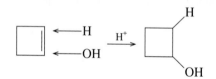

STUDY CHECK

Draw the structural formula for the alcohol obtained by the hydration of 2-methyl-2-butene.

QUESTIONS AND PROBLEMS

▪ Addition Reactions

12.13 Give the condensed structural formulas and names of the products in each of the following reactions:

  a. $CH_3-CH_2-CH_2-CH=CH_2 + H_2 \xrightarrow{Pt}$

  b. $CH_2=\overset{\displaystyle \overset{CH_3}{|}}{C}-CH_2-CH_3 + Cl_2 \longrightarrow$ **c.**  [cyclobutane] $+ Br_2 \longrightarrow$

  d. cyclopentene $+ H_2 \xrightarrow{Pt}$

  e. 2-methyl-2-butene $+ Cl_2 \longrightarrow$ **f.** 2-pentyne $+ 2H_2 \xrightarrow{Pd}$

12.14 Give the condensed structural formulas and names of the products in each of the following reactions:

  a. $CH_3-CH_2-CH=CH_2 + Br_2 \longrightarrow$

  b. cyclohexene $+ H_2 \xrightarrow{Pt}$ **c.** *cis*-2-butene $+ H_2 \xrightarrow{Pt}$

$$\textbf{d. } CH_3-\overset{\displaystyle CH_3}{\underset{\displaystyle |}{C}}=CH-CH_2-CH_3 \ + \ Cl_2 \longrightarrow$$

$$\textbf{e. } \underset{\displaystyle }{\overset{\displaystyle CH_3}{\bigcirc}} \ + \ Br_2 \longrightarrow$$

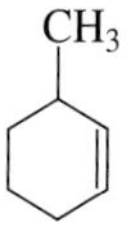

$$\textbf{f. } CH_3-\overset{\displaystyle CH_3}{\underset{\displaystyle |}{CH}}-C\equiv CH \ + \ 2Cl_2 \longrightarrow$$

12.15 Give the condensed structural formulas of the products in each of the following reactions using Markovnikov's rule when necessary:

a. $CH_3-CH=CH-CH_3 \ + \ HBr \longrightarrow$

b. cyclopentene $+ \ HOH \overset{H^+}{\longrightarrow}$

c. $CH_2=CH-CH_2-CH_3 \ + \ HCl \longrightarrow$

d. $CH_3-\overset{\displaystyle CH_3}{\underset{\displaystyle \underset{\displaystyle CH_3}{|}}{\overset{\displaystyle |}{C}}}=C-CH_3 \ + \ HI \longrightarrow$

e. $CH_3-CH_2-\overset{\displaystyle CH_3}{\underset{\displaystyle |}{C}}=CH-CH_3 \ + \ HBr$

f. 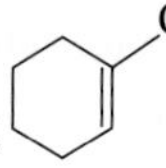 $\underset{\displaystyle }{\overset{\displaystyle CH_3}{\bigcirc}} \ + \ HOH \overset{H^+}{\longrightarrow}$

12.16 Give the condensed structural formulas of the products in each of the following reactions using Markovnikov's rule when necessary:

a. $CH_3-\overset{\displaystyle CH_3}{\underset{\displaystyle |}{C}}=CH-CH_3 \ + \ HCl \longrightarrow$

b. $CH_3-CH_2-CH=CH-CH_2-CH_3 \ + \ HOH \overset{H^+}{\longrightarrow}$

c. $CH_3-\overset{\displaystyle CH_3}{\underset{\displaystyle |}{C}}=CH_2 \ + \ HBr \longrightarrow$

d. 4-methylcyclopentene $+ \ HOH \overset{H^+}{\longrightarrow}$

e.  $\bigcirc \ + \ HBr \longrightarrow$

f. $CH_3-C\equiv C-CH_3 \ + \ 2HCl \longrightarrow$

12.17 Write an equation including any catalysts for the following reactions:
  a. hydrogenation of 2-methylpropene
  b. addition of hydrogen chloride to cyclopentene
  c. addition of bromine to 2-pentene
  d. hydration of propene
  e. addition of chlorine to 2-butyne

12.18 Write an equation including any catalysts for the following reactions:
  a. hydration of 1-methylcyclobutene
  b. hydrogenation of 3-hexene
  c. addition of hydrogen bromide to 2-methyl-2-butene
  d. addition of chlorine to 2,3-dimethyl-2-pentene
  e. addition of HCl to 1-methylcyclopentene

12.4 Polymers of Alkenes

LEARNING GOAL

Draw structural formulas of monomers that form a polymer or a three-monomer section of a polymer.

Polymers are large molecules that consist of small repeating units called **monomers**. In the past hundred years, the plastics industry has made synthetic polymers that are in many of the materials we use every day, such as carpeting, plastic wrap, nonstick pans, plastic cups, and rain gear. In medicine, synthetic polymers are used to replace diseased or damaged body parts such as hip joints, teeth, heart valves, and blood vessels. (See Figure 12.4.)

Many of the synthetic polymers are made by addition reactions of monomers that are small alkenes. Many polymerization reactions required high temperature

SELF-STUDY ACTIVITY
Polymers

and a high pressure (over 1000 atm). In an addition reaction, the polymer grows as monomers are added to the end of the chain. Polyethylene, a polymer made from ethylene $CH_2\!=\!CH_2$, is used in plastic bottles, film, and plastic dinnerware. In the polymerization, a series of addition reactions joins one monomer to the next until a long carbon chain forms that contains as many as 1000 monomers.

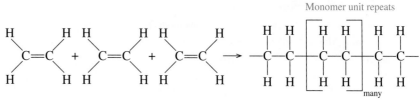

ethene (ethylene) monomers   polyethylene section

Table 12.3 lists several alkene monomers that are used to produce common synthetic polymers and Figure 12.5 shows examples of each. The alkane-like nature of these plastic synthetic polymers makes them unreactive. Thus, they do not decompose easily (they are nonbiodegradable) and have become contributors to pollution. Efforts are being made to make them more degradable.

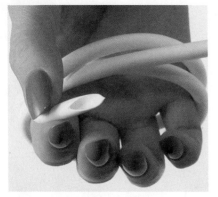

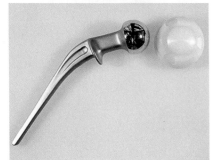

FIGURE 12.4

Synthetic polymers are used to replace diseased veins and arteries. A broken hip is repaired using a metal piece that fits into an artificial plastic cup socket.

Q **Why are the substances in these plastic devices called polymers?**

(MC GOB) Polymers

TABLE 12.3

Some Alkenes and Their Polymers

Monomer	Polymer Section	Common Uses
$CH_2\!=\!CH_2$ ethene (ethylene)	polyethylene	Plastic bottles, film, insulation materials
$CH_2\!=\!CH$—Cl chloroethene (vinyl chloride)	polyvinyl chloride (PVC)	Plastic pipes and tubing, garden hoses, garbage bags
$CH_2\!=\!CH$—CH_3 propene (propylene)	polypropylene	Ski and hiking clothing, carpets, artificial joints
F—$C\!=\!C$—F tetrafluoroethene	polytetrafluoroethylene (Teflon)	Nonstick coatings
$CH_2\!=\!C$—Cl, Cl 1,1-dichloroethene	polydichloroethylene (Saran)	Plastic film and wrap
$H_2C\!=\!CH$ phenylethene (styrene)	polystyrene	Plastic coffee cups and cartons, insulation

Polyethylene

Polyvinyl chloride

Polypropylene

Polytetrafluoroethylene (Teflon)

Polydichloroethylene (Saran)

Polystyrene

FIGURE 12.5

Synthetic polymers provide a wide variety of items that we use every day.

Q **What are some alkenes used to make the polymers in these plastic items?**

It is important to recycle plastic material, rather than add to our growing landfills. You can identify the type of polymer used to manufacture a plastic item by looking for the recycle symbol (arrows in a triangle) found on the label or on the bottom of the plastic container. For example, either the number 5 or the letters PP inside the triangle is a code for a polypropylene plastic.

1	2	3	4	5	6
PETE	HDPE	PV	LDPE	PP	PS
polyethylene terephthalate	high-density polyethylene	polyvinyl chloride	low-density polyethylene	polypropylene	polystyrene

SAMPLE PROBLEM 12.7

■ **Polymers**

What are the starting monomers for the following polymers?

a. polypropylene

b.

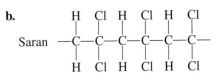

■ **Polymers and Recycling Plastics**

1. Make a list of the items you use or have in your room or home that are made of polymers.
2. Recycling information on the bottom or side of a plastic bottle includes a triangle with a code number that identifies the type of polymer used to make the plastic. Make a collection of several different kinds of plastic bottles. Try to find plastic items with each type of polymer.

QUESTIONS

1. What are the most common types of plastics among the plastic containers in your collection?
2. What are the monomer units of some of the plastics you looked at?

SOLUTION

a. propene (propylene) $CH_2{=}CH{-}CH_3$ b. 1,1-dichloroethene, $CH_2{=}\underset{\underset{Cl}{|}}{\overset{\overset{Cl}{|}}{C}}$

STUDY CHECK

What is the monomer for PVC?

QUESTIONS AND PROBLEMS

■ **Polymers of Alkenes**

12.19 What is a polymer?

12.20 What is a monomer?

12.21 Write an equation that represents the formation of a part of the Teflon polymer from three of the monomer units.

12.22 Write an equation that represents the formation of a part of the polystyrene polymer from three of the monomer units.

12.23 A plastic called polyvinylidene difluoride, PVDF, is made from monomers of 1,1-difluoroethene. Write the structure of the polymer formed from the addition of three monomers of 1,1-difluoroethene.

12.24 An alkene called acrylonitrile is the monomer used to form the polymer used in the fabric material called Orlon. Write an equation that represents the formation of a part of the polyacrylonitrile polymer from three of the monomer units.

Acrylonitrile $CH_2{=}\overset{\overset{CN}{|}}{CH}$

Describe the bonding in benzene; name aromatic compounds, and write their structural formulas.

12.5 Aromatic Compounds

In 1825, Michael Faraday isolated a hydrocarbon called benzene, which had the molecular formula C_6H_6. Because many compounds containing benzene had fragrant odors, the family of benzene compounds became known as **aromatic compounds**. A molecule of **benzene** consists of a ring of six carbon atoms with one hydrogen atom attached to each carbon. Each carbon atom uses 3 valence electrons to bond to the hydrogen atom and two adjacent carbons. That leaves 1 valence electron to share in a double bond with an adjacent carbon. In 1865, August Kekulé proposed that the carbon atoms were arranged in a flat ring with alternating single and double bonds between the carbon atoms. This idea led to two ways of writing the benzene structure, as follows:

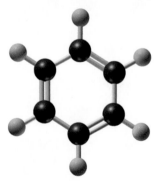

  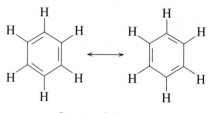

Structures for benzene

However, there is only one structure of benzene. Today we know that all the bonds in benzene are identical. In the benzene ring, scientists determined that the electrons are shared equally, a unique feature that makes aromatic compounds especially stable. To show this, the benzene structure is also represented as a hexagon with a circle in the center.

Naming Aromatic Compounds

Aromatic compounds that contain a benzene ring with a single substituent are usually named as benzene derivatives. However, many of these compounds have been important for many years and still use their common names. Names such as toluene, aniline, and phenol are allowed by IUPAC rules.

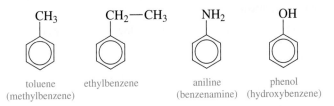

toluene
(methylbenzene)

ethylbenzene

aniline
(benzenamine)

phenol
(hydroxybenzene)

When a benzene ring is a substituent, C_6H_5—, it is named as a phenyl group.

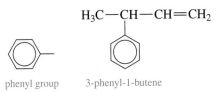

phenyl group

3-phenyl-1-butene

When there are two substituents on benzene, the ring is numbered to give the lowest numbers to the substituents. However, common names use the prefixes **ortho**, **meta**, and **para** to show the substituent arrangement. The prefix *ortho* (*o*) indicates a 1,2 arrangement, *meta* (*m*) is a 1,3 arrangement, and *para* (*p*) is used for 1,4 arrangements.

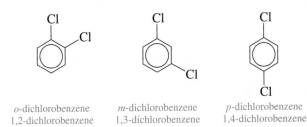

o-dichlorobenzene
1,2-dichlorobenzene

m-dichlorobenzene
1,3-dichlorobenzene

p-dichlorobenzene
1,4-dichlorobenzene

Common names are used for many disubstituted benzenes such as the isomers of dimethylbenzene.

o-xylene
1,2-dimethylbenzene

m-xylene
1,3-dimethylbenzene

p-xylene
1,4-dimethylbenzene

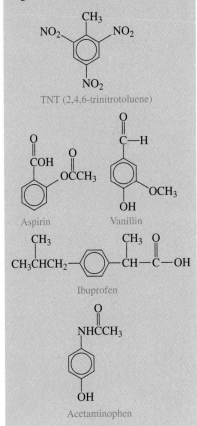
MC GOB   Naming Aromatic Compounds

When there are three or more substituents on the benzene ring, numbers are used to show their arrangement. The substituents are numbered to give the lowest numbers and named alphabetically.

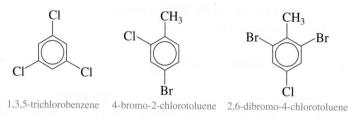

1,3,5-trichlorobenzene   4-bromo-2-chlorotoluene   2,6-dibromo-4-chlorotoluene

---

SAMPLE PROBLEM 12.8

■ **Naming Aromatic Compounds**

Give IUPAC and any common names for each of the following aromatic compounds:

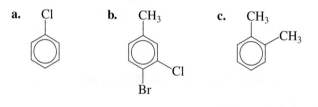

a.   **b.**   **c.**

SOLUTION

a. chlorobenzene

b. 4-bromo-3-chlorotoluene

c. 1,2-dimethylbenzene; *o*-xylene

STUDY CHECK

Name the following compound.

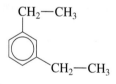

QUESTIONS AND PROBLEMS

■ **Aromatic Compounds**

12.25 Cyclohexane and benzene each have six carbon atoms. How are they different?

12.26 In the Health Note "Some Common Aromatic Compounds," what part of each molecule is the aromatic portion?

12.27 Give the IUPAC and any common names for each of the following:

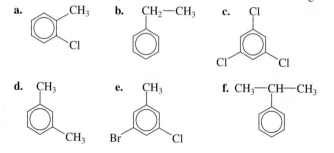

a.   **b.**   **c.**

d.   **e.**   **f.** CH₃—CH—CH₃

Polycyclic Aromatic Hydrocarbons (PAHs)

Large aromatic compounds known as polycyclic aromatic hydrocarbons are formed by fusing together two or more benzene rings edge-to-edge. In a fused-ring compound, neighboring benzene rings share two or more carbon atoms. Naphthalene with two benzene rings is well known for its use in mothballs. Anthracene with three rings is used in the manufacture of dyes.

napthalene   anthracene   phenanthrene

When a polycyclic compound contains phenanthrene, it may act as a carcinogen, a substance known to cause cancer. For example, some aromatic compounds in cigarette smoke cause cancer, as seen in the lung tissue of a heavy smoker. Benz[a]pyrene, a product of combustion, has been identified in coal tar, tobacco smoke, barbecued meats, and automobile exhaust.

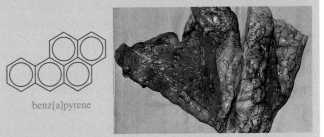

benz[a]pyrene

Compounds containing five or more fused benzene rings such as benz[a]pyrene are potent carcinogens. The molecules interact with the DNA in the cells, causing abnormal cell growth and cancer. Increased exposure to carcinogens increases the chance of DNA alternations in the cells.

12.28 Give the IUPAC and any common names for each of the following:

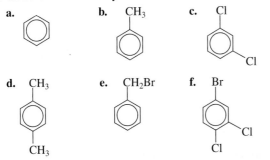

12.29 Draw the structural formulas for each of the following compounds:
   a. methylbenzene
   b. *m*-dichlorobenzene
   c. 1-ethyl-4-methylbenzene
   d. *p*-chlorotoluene

12.30 Draw the structural formulas for each of the following compounds:
   a. benzene
   b. *o*-chloromethylbenzene
   c. propylbenzene
   d. 1,2,4-trichlorobenzene

12.6 Properties of Aromatic Compounds

The symmetrical structure of benzene allows the cyclic structures to stack close together, which contributes to higher melting points and boiling points of benzene and its derivatives. For example, hexane melts at $-95°C$, while benzene melts at $6°C$. Among the disubstituted benzene compounds, the para isomers are more

LEARNING GOAL

Describe the physical and chemical properties of aromatic compounds; draw structural formulas produced by substitution of benzene.

symmetric and have higher melting points than the ortho and meta isomers: *o*-xylene melts at $-26°C$ and *m*-xylene melts at $-48°C$, while *p*-xylene melts at $13°C$.

Aromatic compounds are less dense than water, although they are somewhat denser than other hydrocarbons. Halogenated benzene compounds are denser than water. Aromatic hydrocarbons are insoluble in water and are used as solvents for other organic compounds. Only those containing strongly polar functional groups such as —OH or —COOH will be more soluble. Benzene and other aromatic compounds are resistant to reactions that break up the aromatic system, although they are flammable, as are other hydrocarbon compounds.

Chemical Properties

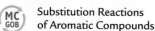

Substitution Reactions of Aromatic Compounds

The most important type of reaction for benzene and aromatic compounds is **substitution**, in which an atom or group of atoms replaces a hydrogen atom on a benzene ring. A substitution reaction, rather than addition, retains the stability of the aromatic bonding system. Substitution reactions of benzene include halogenation, nitration, and sulfonation.

Halogenation

In the chlorination or bromination of benzene, a chlorine or bromine atom replaces a hydrogen atom on the benzene ring. A catalyst such as $FeCl_3$ is required for chlorination; $FeBr_3$ is a catalyst in bromination.

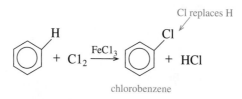

chlorobenzene

When toluene (methylbenzene) undergoes halogenation, a mixture of isomers is obtained as products. However, the presence of a methyl group in toluene has the effect of producing mostly ortho and para isomers. In most substitution reactions of toluene, the meta isomer is produced in very small amounts.

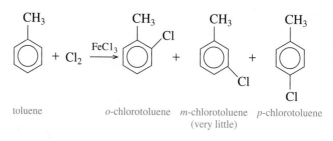

toluene   *o*-chlorotoluene   *m*-chlorotoluene   *p*-chlorotoluene
(very little)

Nitration

When benzene is heated with nitric acid, nitrobenzene is produced. Sulfuric acid (H_2SO_4) is required as a catalyst for the nitration.

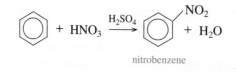

nitrobenzene

Sulfonation

When benzene reacts with a mixture of SO_3 + H_2SO_4, known as "fuming sulfuric acid," the product is benzenesulfonic acid.

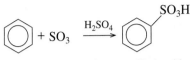

benzenesulfonic acid

The sulfonation of aromatic compounds is one way to produce sulfa drugs.

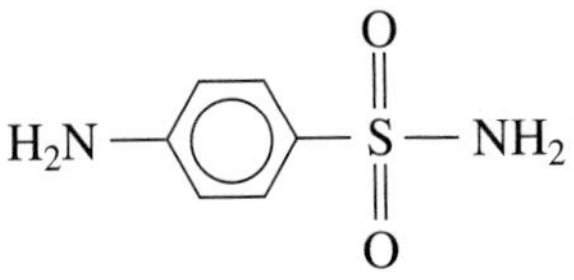

$$H_2N \overset{}{\underset{}{\bigcirc}} \overset{O}{\underset{O}{\overset{\|}{\underset{\|}{S}}}} - NH_2$$

sulfanilamide, a sulfa drug

SAMPLE PROBLEM 12.9

■ **Reactions of Benzene**

Write the structure of the organic product when benzene reacts with the following:

a. Br_2 and $FeBr_3$ **b.** HNO_3 and H_2SO_4

SOLUTION

a. 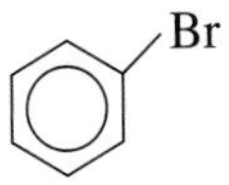   **b.**

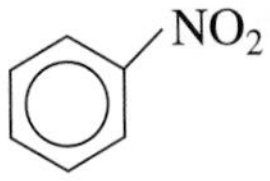

STUDY CHECK

A chemist needs to synthesize chlorobenzene. If benzene is available in the lab, how could she prepare this compound?

QUESTIONS AND PROBLEMS

MC
GOB

■ **Properties of Aromatic Compounds**

12.31 Alkenes undergo addition reactions, but benzene does not. How does benzene react and why?

12.32 When toluene is reacted with chlorine in light, the product is $CH_2—Cl$
How would you explain this result?

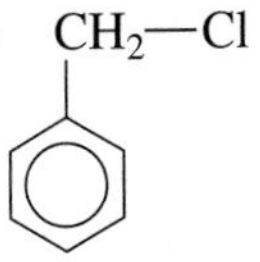

12.33 Draw the structures of the organic product(s), if any, for the following reactants:

   a. benzene + Cl_2 $\xrightarrow{FeCl_3}$

   b. benzene + HNO_3 $\xrightarrow{H_2SO_4}$

12.34 Draw the structures of the organic product(s), if any, for the following reactants:

   a. toluene + Br_2 $\xrightarrow{FeBr_3}$

   b. benzene + SO_3 $\xrightarrow{H_2SO_4}$

CONCEPT MAP

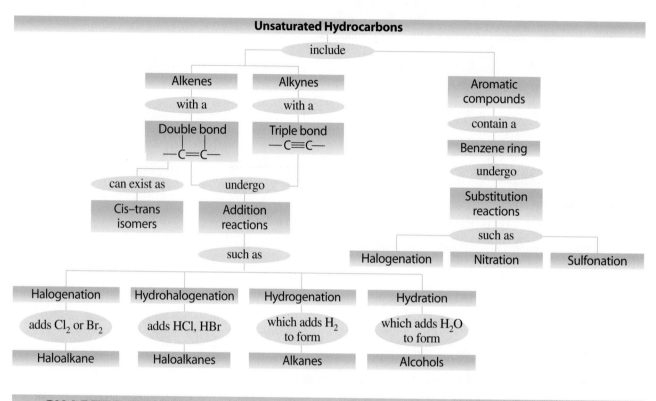

CHAPTER REVIEW

12.1 Alkenes and Alkynes

Alkenes are unsaturated hydrocarbons that contain carbon–carbon double bonds (C=C). Alkynes contain a triple bond (C≡C). The IUPAC names of alkenes end with *ene*, while alkyne names end with *yne*. The main chain is numbered from the end nearest the double or triple bond. In a cycloalkene, the double bond is carbon 1 and 2, and the ring is numbered to give the lowest numbers to any substituents, which are named alphabetically.

12.2 Cis–Trans Isomers

Isomers of alkenes occur when the carbon atoms in the double bond are connected to different atoms or groups. In the cis isomer the attached groups are on the same side of the double bond, whereas in the trans isomer they are connected on the opposite sides of the double bond.

12.3 Addition Reactions

The addition of small molecules to the double bond is a characteristic reaction of alkenes. Hydrogenation adds hydrogen atoms to the double bond of an alkene to yield an alkane. Halogenation adds bromine or chlorine atoms to produce dihaloalkanes. Hydrogen halides and water can also add to a double bond. When there are a different number of groups attached to the carbons in the double bond, the H from the reactant (HX or H—OH) adds to the carbon with the greater number of hydrogen atoms.

12.4 Polymers of Alkenes

Polymers are long-chain molecules that consist of many repeating units of smaller carbon molecules called monomers. In nature, cellulose and starch are polymers of glucose and proteins are polymers of amino acids. Many materials that we use every day are synthetic polymers, including carpeting, plastic wrap, nonstick pans, and nylon. These synthetic materials are often made by addition reactions in which a catalyst links the carbon atoms from various kinds of alkene molecules.

12.5 Aromatic Compounds

Most aromatic compounds contain benzene, a cyclic structure containing six CH units. The structure of benzene is represented as a hexagon with a circle in the center. Many aromatic compounds use the parent name benzene, although common names such as toluene are retained. The benzene ring is numbered and the substituents are listed in alphabetical order. For two substituents, the positions are often shown by the prefixes *ortho* (1,2-), *meta* (1,3-), and *para* (1,4-).

12.6 Properties of Aromatic Compounds

Aromatic compounds undergo substitution reactions such as halogenation, nitration, and sulfonation. They do not undergo addition reactions, which would disrupt their stable aromatic bonding system.

SUMMARY OF NAMING

Type	Example	Characteristic	Structure
Alkene	Propene (propylene)	double bond	$CH_3-CH=CH_2$
Cycloalkene	Cyclopropene	double bond in a carbon ring	 △
Alkyne	Propyne	triple bond	$CH_3-C\equiv CH$
Aromatic	Benzene	Aromatic ring of six carbons	
	Methylbenzene or toluene		
	1,4-dichlorobenzene or *para*-dichlorobenzene		

SUMMARY OF REACTIONS

HYDROGENATION

$$CH_2=CH-CH_3 + H_2 \xrightarrow{Pt} CH_3-CH_2-CH_3$$
propene → propane

$$CH_3-C\equiv CH + 2H_2 \xrightarrow{Pt} CH_3-CH_2-CH_3$$
propyne → propane

HALOGENATION

$$CH_2=CH-CH_3 + Cl_2 \longrightarrow CH_2-CH-CH_3$$
propene → 1,2-dichloropropane (with Cl, Cl substituents)

SUBSTITUTION REACTIONS OF BENZENE

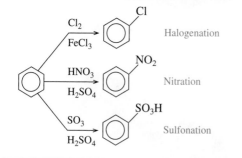

Halogenation (Cl₂ / FeCl₃ → Cl)

Nitration (HNO₃ / H₂SO₄ → NO₂)

Sulfonation (SO₃ / H₂SO₄ → SO₃H)

HYDROHALOGENATION

Markovnikov's rule

$$CH_2=CH-CH_3 + HCl \longrightarrow CH_3-CH-CH_3$$
propene → 2-chloropropane (with Cl substituent)

HYDRATION OF ALKENES

Markovnikov's rule

$$CH_2=CH-CH_3 + H-OH \xrightarrow{H^+} CH_3-CH-CH_3$$
propene → 2-propanol (with OH substituent)

KEY TERMS

addition A reaction in which atoms or groups of atoms bond to a double bond. Addition reactions include the addition of hydrogen (hydrogenation), halogens (halogenation), hydrogen halides (hydrohalogenation), or water (hydration).

alkene An unsaturated hydrocarbon containing a carbon–carbon double bond.

alkyne An unsaturated hydrocarbon containing a carbon–carbon triple bond.

aromatic compounds Compounds that contain the ring structure of benzene.

benzene A ring of six carbon atoms each of which is attached to a hydrogen atom, C_6H_6.

carbocation A carbon cation that has only three bonds and a positive charge and is formed during the addition reactions of hydration and hydrohalogenation.

cis isomer A geometric isomer in which similar groups are connected on the same side of the double bond.

cycloalkene A cyclic hydrocarbon that contains a double bond in the ring.

halogenation The addition of Cl_2 or Br_2 to an alkene or benzene to form halogen-containing compounds.

hydration An addition reaction in which the components of water, H— and —OH, bond to the carbon–carbon double bond to form an alcohol.

hydrogenation The addition of hydrogen (H_2) to the double bond of alkenes to yield alkanes.

hydrohalogenation The addition of a hydrogen halide such as HCl or HBr to a double bond.

Markovnikov's rule When adding HX or HOH to alkenes with different numbers of groups attached to the double bonds, the H adds to the carbon that has the greater number of hydrogen atoms.

meta A method of naming that indicates two substituents at carbons 1 and 3 of benzene.

monomer The small organic molecule that is repeated many times in a polymer.

ortho A method of naming that indicates two substituents at carbons 1 and 2 of a benzene ring.

para A method of naming that indicates two substituents at carbons 1 and 4 of a benzene ring.

polymer A very large molecule that is composed of many small, repeating structural units that are identical.

substitution The reactions of benzene and other aromatic compounds in which an atom or group of atoms replaces a hydrogen on a benzene ring.

trans isomer A geometric isomer in which similar groups are connected to opposite sides of the double bond in an alkene.

unsaturated hydrocarbons A compound of carbon and hydrogen in which the carbon chain contains at least one double (alkene) or triple carbon–carbon bond (alkyne). An unsaturated compound is capable of an addition reaction with hydrogen, which converts the double or triple bonds to single carbon–carbon bonds.

UNDERSTANDING THE CONCEPTS

12.35 Draw a part of the polymer (use four monomers) of Teflon made from 1,1,2,2-tetrafluoroethene.

12.36 A garden hose is made of polyvinylchloride (PVC) from chloroethene (vinyl chloride). Draw a part of the polymer (use four monomers) for PVC.

12.37 Explosives used in mining contain TNT or 2,4,6-trinitrotoluene.

a. If the functional group *nitro* is —NO$_2$, what is the structural formula of TNT?

b. TNT is actually a mixture of isomers of trinitrotoluene. Draw two possible isomers.

12.38 Margarine is produced from the hydrogenation of vegetable oils, which contain unsaturated fatty acids. How many grams of hydrogen are required to completely saturate 75.0 g oleic acid, C$_{18}$H$_{34}$O$_2$, which has one double bond?

ADDITIONAL QUESTIONS AND PROBLEMS

For instructor-assigned homework, go to **www.masteringgob.com.**

12.39 Compare the formulas and bonding in propane, cyclopropane, propene, and propyne.

12.40 Compare the formulas and bonding in butane, cyclobutane, cyclobutene, and 2-butyne.

12.41 Give the IUPAC name for each of the following compounds:

a.

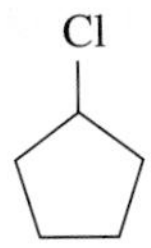

b.

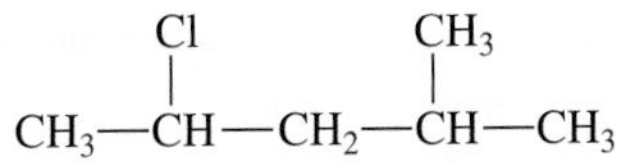

$$CH_3-\underset{\underset{Cl}{|}}{CH}-CH_2-\underset{\underset{CH_3}{|}}{CH}-CH_3$$

c.

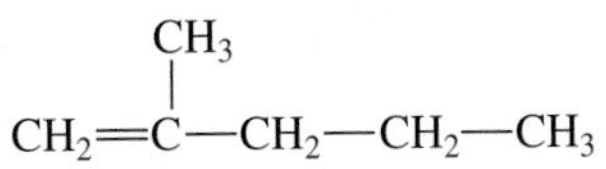

$$CH_2{=}\underset{\underset{CH_3}{|}}{C}-CH_2-CH_2-CH_3$$

d. $CH_3-CH_2-C{\equiv}C-CH_3$ **e.**

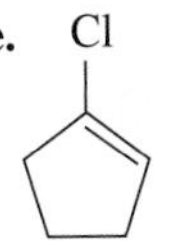

f.

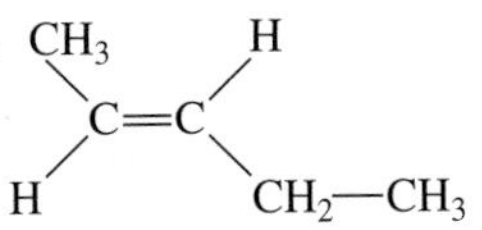

g.

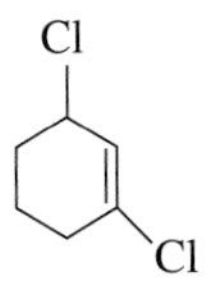

12.42 Write the condensed structures of each of the following compounds:

a. 1,2-dibromocyclopentane
b. 2-pentyne
c. *cis*-2-heptene
d. 3,3-dichloro-2-methylpentane
e. *trans*-3-hexene
f. 2-bromo-3-chlorocyclohexene
g. 2,3-dichloro-1-butene

12.43 Indicate if the following pairs of structures represent isomers, cis–trans isomers, or identical compounds.

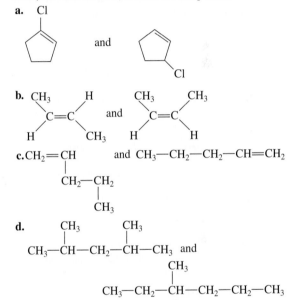

a.

Cl   and   Cl

b.

$$\underset{H}{\overset{CH_3}{C}}{=}\underset{CH_3}{\overset{H}{C}} \quad and \quad \underset{H}{\overset{CH_3}{C}}{=}\underset{H}{\overset{CH_3}{C}}$$

c. $CH_2{=}CH$ and $CH_3-CH_2-CH_2-CH{=}CH_2$
    CH_2-CH_2
    CH_3

d.

$$CH_3-\underset{\underset{CH_3}{|}}{CH}-CH_2-\underset{\underset{CH_3}{|}}{CH}-CH_3 \quad and$$

$$CH_3-CH_2-\underset{\underset{CH_3}{|}}{CH}-CH_2-CH_2-CH_3$$

12.44 Draw the condensed structures and give the names for all the isomers of C$_4$H$_8$ including cyclic and cis–trans isomers.

12.45 Methylcyclopentane is formed by four different alkenes that react with hydrogen (H$_2$) in the presence of a Ni catalyst. Draw the condensed structures of each of these alkenes.

12.46 What role do cis–trans isomers play in night vision?

12.47 Write the cis and trans isomers for each of the following:
    a. 2-pentene   **b.** 3-hexene   **c.** 2-butene   **d.** 2-hexene

12.48 Write the structures of the products, if any, for the following.

a. $CH_3-CH{=}CH-CH_3 \; + \; H_2 \xrightarrow{Ni}$

b.   + H$_2$ $\xrightarrow{Ni}$

c. $CH_3—CH=CH—CH_3$ + HBr ⟶

d.  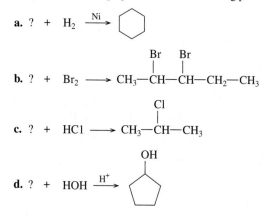 + HBr ⟶

e. $CH_3—CH=CH—CH_3$ + Cl_2 ⟶

f. $CH_3—CH_2—CH_3$ + H_2 $\xrightarrow{Ni}$

g. $CH_3—CH=CH—CH_3$ + HOH $\xrightarrow{H^+}$

h.  + HOH $\xrightarrow{H^+}$

i. $CH_3—\overset{\overset{\displaystyle CH_3}{|}}{C}=\underset{\underset{\displaystyle CH_3}{|}}{C}—CH_3$ + HCl ⟶

12.49 What is the condensed structural formula of the organic compound needed to prepare each of the following products?

a. ? + H_2 $\xrightarrow{Ni}$

b. ? + Br_2 ⟶ $CH_3—\overset{\overset{\displaystyle Br}{|}}{CH}—\overset{\overset{\displaystyle Br}{|}}{CH}—CH_2—CH_3$

c. ? + HCl ⟶ $CH_3—\overset{\overset{\displaystyle Cl}{|}}{CH}—CH_3$

d. ? + HOH $\xrightarrow{H^+}$

12.50 Lucite or Plexiglas is a polymer of methylmethacrylate. Write the part of the polymer that is made from the addition of three of these monomers.

$CH_2=\overset{\overset{\displaystyle CH_3}{|}}{C}—\overset{\overset{\displaystyle O}{\|}}{C}—O—CH_3$

methylmethacrylate

12.51 Copolymers contain more than one type of monomer. One copolymer used in medicine is made of alternating units of styrene and acrylonitrile. Write a section of the copolymer that would have these alternating units. (For structure of the styrene, see Table 12.3.)

$H_2C=\overset{\overset{\displaystyle CN}{|}}{CH}$

acrylonitrile

12.52 Write the structural formulas for each of the following:
a. ethylbenzene
b. *m*-dichlorobenzene
c. 1,2,4-trimethylbenzene
d. 1,4-dimethylbenzene

12.53 Name the organic product(s) produced, if any, in each of the following reactions.
a. benzene and Cl_2 $\xrightarrow{FeCl_3}$
b. toluene and Br_2 $\xrightarrow{FeBr_3}$
c. benzene and SO_3 $\xrightarrow{H_2SO_4}$
d. benzene and Br_2 $\xrightarrow{light}$

12.54 What reactants and catalysts are needed to synthesize the following products?
a. nitrobenzene
b. benzenesulfonic acid
c. bromobenzene

12.55 Name each of the following aromatic compounds:

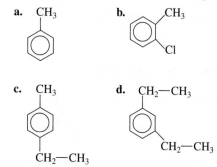

CHALLENGE QUESTIONS

12.56 How many grams of hydrogen are needed to hydrogenate 30.0 g 2-butene?

12.57 Using each of the following carbon chains for C_5H_{10}, write and name all the possible alkenes, including those with cis and trans isomers.

C—C—C—C—C

$C—\overset{\overset{\displaystyle C}{|}}{C}—C—C$

12.58 Acetylene gas reacts with oxygen and burns at high temperature in an acetylene torch.
a. Write the balanced equation for the complete combustion of acetylene.
b. How many grams of oxygen are needed to react with 8.5 L acetylene at STP?
c. How many liters of CO_2 (at STP) are produced when 30.0 g acetylene undergoes combustion?

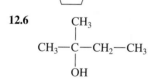

12.59 If a female silkworm moth secretes 50 ng Bombykol, a sex attractant, how many molecules did she secrete? (See Environmental Note "Pheromones in Insect Communication," on page 450.)

ANSWERS

Answers to Study Checks

12.1 a. $CH_3—C≡C—CH_2—CH_3$

b. (structure with CH_3)

12.2 trans-3-hexene

12.3 $CH_3—\overset{\overset{\displaystyle CH_3}{|}}{CH}—CH_2—CH_3$

12.4 1,2-dichlorobutane

12.5 CH_3 Br 1-bromo-1-methylcyclopentane
(cyclopentane structure)

12.6 $CH_3—\overset{\overset{\displaystyle CH_3}{|}}{\underset{\underset{\displaystyle OH}{|}}{C}}—CH_2—CH_3$

12.7 The monomer of PVC, polyvinyl chloride, is chloroethene:

$$\underset{\underset{\displaystyle H}{|}}{\overset{\overset{\displaystyle H}{|}}{C}}=\underset{\underset{\displaystyle H}{|}}{\overset{\overset{\displaystyle Cl}{|}}{C}}$$

12.8 1, 3-diethylbenzene; *m*-diethylbenzene

12.9 Chlorobenzene can be prepared from benzene and chlorine, using $FeCl_3$ as a catalyst.

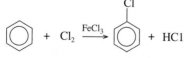

Answers to Selected Questions and Problems

12.1 a. An alkene has a double bond.
    b. An alkyne has a triple bond.

c. An alkene has a double bond.
d. A cycloalkene has a double bond in a ring.

12.3 a. ethene   **b.** 2-methyl-1-propene
    c. 4-bromo-2-pentyne   **d.** cyclobutene
    e. 4-ethylcyclopentene   **f.** 4-ethyl-2-hexene

12.5 a. $CH_3CH=CH_2$

    b. $CH_2=CH—CH_2—CH_2—CH_3$

    c. $CH_2=\overset{\overset{\displaystyle CH_3}{|}}{C}—CH_2—CH_3$ **d.** (cyclohexene structure with CH_3)

    e. $CH_3—\overset{\overset{\displaystyle Cl}{|}}{CH}—C≡C—CH_2—CH_3$

12.7 a. There are no cis–trans isomers.
    b. This alkene has cis–trans isomers.
    c. There are no cis–trans isomers.

12.9 a. *cis*-2-butene   **b.** *trans*-3-octene   **c.** *cis*-3-heptene

12.11 a.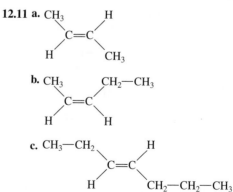

    b. (structure)

    c. (structure)

12.13 a. $CH_3—CH_2—CH_2—CH_2—CH_3$ Pentane

    b. $Cl—CH_2—\overset{\overset{\displaystyle CH_3}{|}}{\underset{\underset{\displaystyle Cl}{|}}{C}}—CH_2—CH_3$ 1,2-dichloro-2-methylbutane

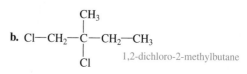

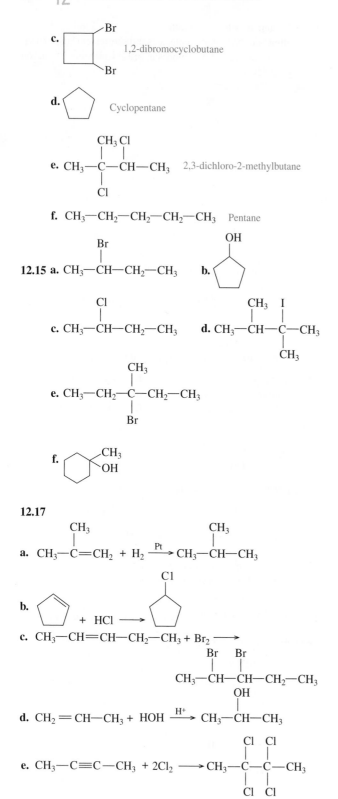

c. 1,2-dibromocyclobutane

d. Cyclopentane

e. CH₃—C(CH₃)(Cl)—CH(Cl)—CH₃ 2,3-dichloro-2-methylbutane

f. CH₃—CH₂—CH₂—CH₂—CH₃ Pentane

12.15 a. CH₃—CH(Br)—CH₂—CH₃

b. (cyclopentanol with OH)

c. CH₃—CH(Cl)—CH₂—CH₃

d. CH₃—CH(CH₃)—C(I)(CH₃)—CH₃

e. CH₃—CH₂—C(CH₃)(Br)—CH₂—CH₃

f. (cyclohexane with CH₃ and OH)

12.17

a. CH₃—C(CH₃)=CH₂ + H₂ —Pt→ CH₃—CH(CH₃)—CH₃

b. (cyclopentene) + HCl ⟶ (chlorocyclopentane)

c. CH₃—CH=CH—CH₂—CH₃ + Br₂ ⟶ CH₃—CH(Br)—CH(Br)—CH₂—CH₃

d. CH₂=CH—CH₃ + HOH —H⁺→ CH₃—CH(OH)—CH₃

e. CH₃—C≡C—CH₃ + 2Cl₂ ⟶ CH₃—C(Cl)₂—C(Cl)₂—CH₃

12.19 A polymer is a very large molecule composed of small units that are repeated many times.

12.21

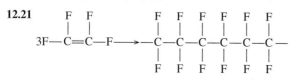

12.23

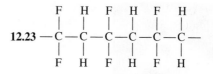

12.25 Cyclohexane, C₆H₁₂, is a cycloalkane in which six carbon atoms in a ring are linked by single bonds. In benzene, C₆H₆, electrons that would be alternating single and double bonds are shared equally by the six carbon atoms.

12.27 a. 2-chlorotoluene; *o*-chlorotoluene
b. ethylbenzene
c. 1,3,5-trichlorobenzene
d. 3-methyltoluene; *m*-xylene
e. 3-bromo-5-chlorotoluene
f. isopropyl benzene

12.29 a. (toluene) **b.** (1,3-dichlorobenzene)
c. (4-ethyltoluene) **d.** (4-chlorotoluene)

12.31 Benzene undergoes substitution reactions because a substitution reaction allows benzene to retain the stability of the aromatic system.

12.33 a. (chlorobenzene) **b.** (nitrobenzene)

12.35

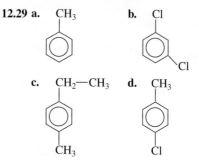

12.37 a.

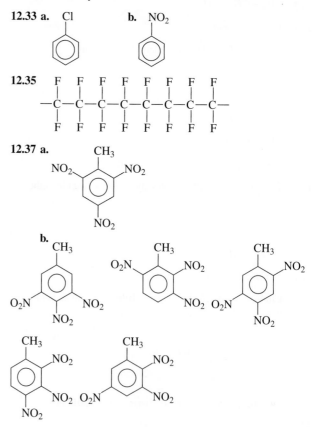

12.39 All the compounds have three carbon atoms. Propane is a saturated alkane, and cyclopropane is a saturated cyclic hydrocarbon. Both propene and propyne are unsaturated hydrocarbons, but propene has a double bond and propyne has a triple bond.

12.41 **a.** chlorocyclopentane
b. 2-chloro-4-methylpentane
c. 2-methyl-1-pentene
d. 2-pentyne
e. 1-chlorocyclopentene
f. *trans*-2-pentene
g. 1,3-dichlorocyclohexene

12.43 **a.** isomers   **b.** cis–trans isomers
c. identical   **d.** isomers

12.45

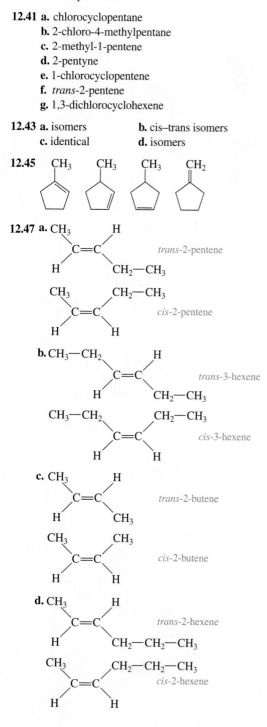

12.47 **a.**

trans-2-pentene

cis-2-pentene

b.

trans-3-hexene

cis-3-hexene

c.

trans-2-butene

cis-2-butene

d.

trans-2-hexene

cis-2-hexene

12.49

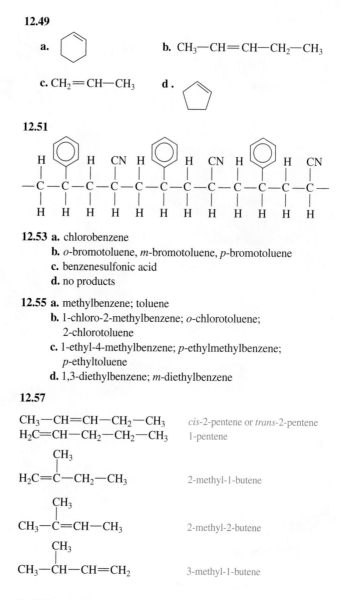

a.   **b.** $CH_3-CH=CH-CH_2-CH_3$

c. $CH_2=CH-CH_3$ **d.**

12.51

12.53 **a.** chlorobenzene
b. *o*-bromotoluene, *m*-bromotoluene, *p*-bromotoluene
c. benzenesulfonic acid
d. no products

12.55 **a.** methylbenzene; toluene
b. 1-chloro-2-methylbenzene; *o*-chlorotoluene; 2-chlorotoluene
c. 1-ethyl-4-methylbenzene; *p*-ethylmethylbenzene; *p*-ethyltoluene
d. 1,3-diethylbenzene; *m*-diethylbenzene

12.57

$CH_3-CH=CH-CH_2-CH_3$ *cis*-2-pentene or *trans*-2-pentene
$H_2C=CH-CH_2-CH_2-CH_3$ 1-pentene

$H_2C=\overset{\overset{\displaystyle CH_3}{|}}{C}-CH_2-CH_3$ 2-methyl-1-butene

$CH_3-\overset{\overset{\displaystyle CH_3}{|}}{C}=CH-CH_3$ 2-methyl-2-butene

$CH_3-\overset{\overset{\displaystyle CH_3}{|}}{CH}-CH=CH_2$ 3-methyl-1-butene

12.59 Bombykol $C_{16}H_{30}O$ = 238.3 g/mole

$$50 \text{ ng} \times \frac{1 \text{ g}}{10^9 \text{ ng}} \times \frac{1 \text{ mole}}{238. \text{ g}} \times \frac{6.02 \times 10^{23} \text{ molecules}}{1 \text{ mole}}$$

$$= 1 \times 10^{14} \text{ molecules}$$

13 Alcohols, Phenols, Thiols, and Ethers

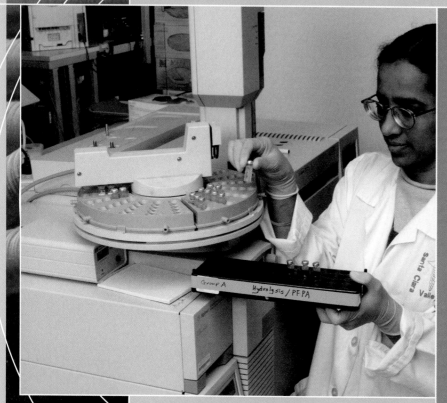

"We use mass spectrometry to analyze and confirm the presence of drugs," says Valli Vairavan, clinical lab technologist—Mass Spectrometry, Santa Clara Valley Medical Center. "A mass spectrometer separates and identifies compounds including drugs by mass. When we screen a urine sample, we look for metabolites, which are the products of drugs that have metabolized in the body. If the presence of one or more drugs such as heroin and cocaine is indicated, we confirm it by using mass spectrometry."

Drugs or their metabolites are detected in urine 24–48 hours after use. Cocaine metabolizes to benzoylecgonine and hydroxycocaine, morphine to morphine-3-glucuronide, and heroin to acetylmorphine. Amphetamines and methamphetamines are detected unchanged.

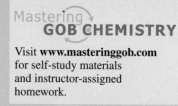

T
n this chapter, we will look at organic compounds that contain single bonds to oxygen atoms or sulfur atoms. Alcohols, which contain the hydroxyl group (—OH) are commonly found in nature and used in industry and at home. For centuries, grains, vegetables, and fruits have been fermented to produce the ethanol present in alcoholic beverages. The hydroxyl group is important in biomolecules such as sugars and starches as well as in steroids such as cholesterol and estradiol. Menthol is a cyclic alcohol with a minty odor and flavor that is used in cough drops, shaving creams, and ointments. The ethers are compounds that contain an oxygen atom connected to two carbon atoms (—O—). Ethers are important solvents in chemistry and medical laboratories. Beginning in 1842, diethyl ether was used for about 100 years as a general anesthesia. Today less flammable and more easily tolerated anesthetics are used. Thiols, which contain an —SH group, give the strong odors we associate with garlic and onions.

13.1 Alcohols, Phenols, and Thiols

LEARNING GOAL

Give IUPAC and common names for alcohols, phenols, and thiols; draw their condensed structural formulas. Classify alcohols as primary, secondary, or tertiary.

In an **alcohol**, a **hydroxyl group** (—OH) replaces a hydrogen atom in an alkane. In a **phenol**, the hydroxyl group is attached to an aromatic ring. Both types of compounds have bent structures similar to water. One hydrogen atom is replaced by an alkyl or aromatic group.

Thiols are a family of sulfur-containing organic compounds that have a *sulfhydryl* (—SH) *group*. They have structures similar to alcohols except that a —SH group takes the place of an —OH group.

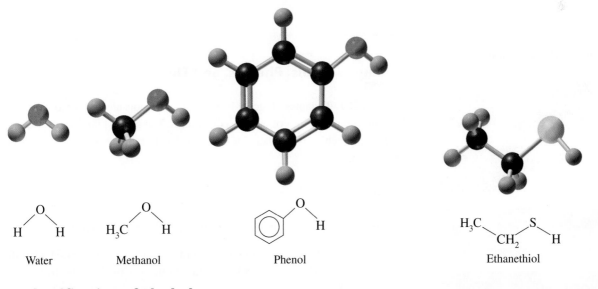

Water   Methanol   Phenol   Ethanethiol

Classification of Alcohols

Alcohols are classified by the number of carbon groups attached to the carbon atom bonded to the hydroxyl (—OH) group. A **primary (1°) alcohol** has one alkyl

MC GOB SELF-STUDY ACTIVITY
Alcohols and Thiols

group attached to the carbon atom bonded to the —OH, a **secondary (2°) alcohol** has two alkyl groups, and a **tertiary (3°) alcohol** has three alkyl groups.

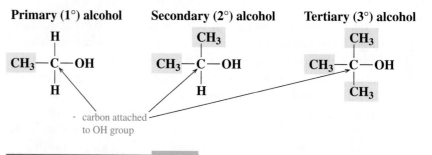

Primary (1°) alcohol   Secondary (2°) alcohol   Tertiary (3°) alcohol

carbon attached to OH group

SAMPLE PROBLEM 13.1

■ Classifying Alcohols

Classify each of the following alcohols as primary, secondary, or tertiary.

a. $CH_3—CH_2—CH_2—OH$

b.
$$CH_3—CH_2—\overset{\displaystyle OH}{\underset{\displaystyle CH_3}{C}}—CH_3$$

SOLUTION

a. One alkyl group attached to the carbon atom bonded to the —OH makes this a primary alcohol.

b. Three alkyl groups attached to the carbon atom bonded to the —OH makes this a tertiary alcohol.

STUDY CHECK

Classify the following as primary, secondary, or tertiary:

H_3C OH

Naming Alcohols, Phenols, and Thiols

The rules for IUPAC names of alcohols, phenols, and thiols are similar to those we used to name other families of organic compounds. The alcohol family is indicated by an *ol* ending, which is numbered to show the location of the hydroxyl group on the main chain.

Naming Alcohols

In the IUPAC system, the alcohol family is indicated by the *ol* ending.

■ **STEP 1 Name the longest carbon chain containing the —OH group.** Replace the *e* in the alkane name with *ol*. Consider the following alcohol.

$CH_3—CH_2—CH_2—OH$ propanol

Guide to Naming Alcohols

STEP 1
Name the longest carbon chain with the —OH group

STEP 2
Number the longest chain starting at the end closest to the —OH

STEP 3
Name substituents counting from the —OH group.

STEP 4
Name a cyclic alcohol as a *cycloalkanol*.

■ **STEP 2 Number the longest chain starting at the end closest to the —OH group.** For simple alcohols, the common name (shown in parentheses) gives the name of the carbon chain as an alkyl group followed by *alcohol.*

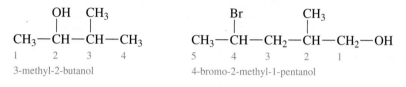

$$CH_3—CH_2—CH_2—OH$$
$$\quad 3 \qquad 2 \qquad 1$$

1-propanol
(propyl alcohol)

■ **STEP 3 Name and number other substituents relative to the —OH group.**

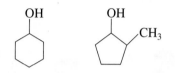

3-methyl-2-butanol

4-bromo-2-methyl-1-pentanol

■ **STEP 4 Name a cyclic alcohol as a *cycloalkanol.*** For other substituents, the ring is numbered with the —OH group on carbon 1.

cyclohexanol   2-methylcyclopentanol

Naming Phenols

The term *phenol* is the IUPAC name for a benzene ring bonded to a hydroxyl group (—OH) and is used in the name of the family of organic compounds derived from phenol. When there is a second substituent, the benzene ring is numbered starting from the carbon 1, which is bonded to the —OH group. The terms *ortho, meta,* and *para* are used for the common names of simple phenols.

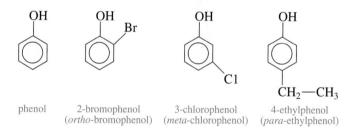

phenol

2-bromophenol
(*ortho*-bromophenol)

3-chlorophenol
(*meta*-chlorophenol)

4-ethylphenol
(*para*-ethylphenol)

Certain disubstituted phenols have common names based on historical uses. The methylphenols are commonly named as *cresols,* while benzenediols have a variety of common names.

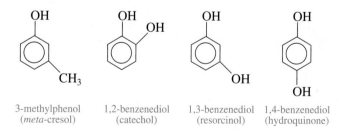

3-methylphenol
(*meta*-cresol)

1,2-benzenediol
(catechol)

1,3-benzenediol
(resorcinol)

1,4-benzenediol
(hydroquinone)

EXPLORE YOUR WORLD

■ **Alcohols in Household Products**

Read the labels on household products such as mouthwashes, cold remedies, alcoholic beverages, rubbing alcohol, and flavoring extracts. Look for names of alcohols such as ethyl alcohol, isopropyl alcohol, thymol, and menthol.

QUESTIONS

1. What part of the name tells you that it is an alcohol?
2. What alcohol is usually meant by the term "alcohol"?
3. What is the percent of alcohol in the products?
4. Write out the structures of the alcohols you find listed on the labels. You may need to use a reference book for some structures.

HEALTH NOTE

Some Important Alcohols and Phenols

Methanol (*methyl alcohol*), the simplest alcohol, is found in many solvents and paint removers. If ingested, methanol is oxidized to formaldehyde, which can cause headaches, blindness, and death. Methanol is used to make plastics, medicines, and fuels. In car racing, it is used as a fuel because it is less flammable and has a higher octane rating than gasoline.

Ethanol (*ethyl alcohol*) has been known since prehistoric times as an intoxicating product formed by the fermentation of grains and starches.

$$C_6H_{12}O_6 \xrightarrow{\text{Fermentation}} 2CH_3{-}CH_2{-}OH + 2CO_2$$

Today, ethanol for commercial uses is produced by allowing ethene and water to react at high temperatures and pressures. It is used as a solvent for perfumes, varnishes, and some medicines, such as tincture of iodine. "Gasohol" is a mixture of ethanol and gasoline used as a fuel.

$$H_2C{=}CH_2 + H_2O \xrightarrow{\text{300°C, 200 atm, catalyst}} CH_3{-}CH_2{-}OH$$

Several of the essential oils of plants, which produce the odor or flavor of the plant, are derivatives of phenol. Eugenol is found in cloves, vanillin in vanilla bean, isoeugenol in nutmeg, and thymol in thyme and mint. Thymol has a pleasant, minty taste and is used in mouthwashes and by dentists to disinfect a cavity before adding a filling compound.

1,2,3-Propanetriol (*glycerol or glycerin*), a trihydroxy alcohol, is a viscous liquid obtained from oils and fats during the production of soaps. The presence of several polar —OH groups makes it strongly attracted to water, a feature that makes glycerin useful as a skin softener in products such as skin lotions, cosmetics, shaving creams, and liquid soaps.

$$HO{-}CH_2{-}\overset{\displaystyle OH}{CH}{-}CH_2{-}OH$$
1,2,3-propanetriol (glycerol)

1,2-Ethanediol (*ethylene glycol*) is used as antifreeze in heating and cooling systems. It is also a solvent for paints, inks, and plastics, and is used in the production of synthetic fibers such as Dacron. If ingested, it is extremely toxic. In the body, it is oxidized to oxalic acid, which forms insoluble salts in the kidneys that cause renal damage, convulsions, and death. Because its sweet taste is attractive to pets and children, ethylene glycol solutions must be carefully stored.

$$HO{-}CH_2{-}CH_2{-}OH \xrightarrow{[O]} HO{-}\overset{\displaystyle O}{\overset{\|}{C}}{-}\overset{\displaystyle O}{\overset{\|}{C}}{-}OH$$
1,2-ethanediol (ethylene glycol)   oxalic acid

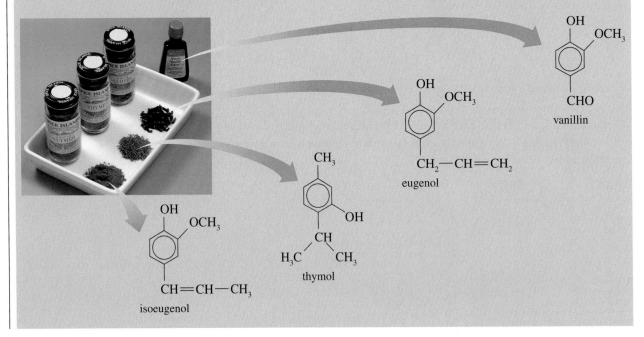

vanillin

eugenol

thymol

isoeugenol

SAMPLE PROBLEM 13.2

■**Naming Alcohols and Phenols**

Give the IUPAC name for each of the following:

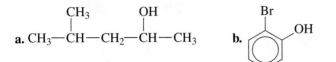

a. $CH_3-CH-CH_2-CH-CH_3$ with CH_3 and OH substituents

b.

 Naming Alcohols, Phenols, and Thiols

 Drawing Alcohols, Phenols, and Thiols

SOLUTION

a. The parent chain is pentane; the alcohol is named pentanol. The carbon chain is numbered to give the position of the —OH group on carbon 2 and the methyl group on carbon 4. The compound is named 4-methyl-2-pentanol.

b. The compound is a *phenol* because the —OH is attached to a benzene ring. The ring is numbered with carbon 1 attached to the —OH in the direction that gives the bromine the lower number. The compound is named 2-bromophenol or *ortho*-bromophenol.

STUDY CHECK

Give the IUPAC name for the following:

$$CH_3-CH-CH_2-CH_2-OH$$

with Cl substituent on the second carbon

Naming Thiols

In the IUPAC system, thiols are named by adding *thiol* to the alkane name of the longest carbon chain bonded to the —SH group. The location of the —SH group is indicated by numbering the main chain from the closest end.

CH_3-OH CH_3-SH $CH_3-CH-CH_2-CH_3$ with SH on second carbon
methanol   methanethiol   2-butanethiol

An important property of thiols is a strong, sometimes disagreeable, odor. Methanethiol is the characteristic odor of oysters and cheddar cheese. (See Figure 13.1.) To help us detect natural gas (methane) leaks, a small amount of ethanethiol is added to the gas supply. There are thiols in the spray emitted when a skunk senses danger. The odor of onions is due to 1-propanethiol, which is also a lachrymator, a substance that makes eyes tear. Garlic contains thiols such as 2-propene-1-thiol. We can break this name down as follows.

2-	prop	ene	-1-	thiol
carbon 2 has C=C	3 carbons in chain	alkene	on carbon 1	—SH group

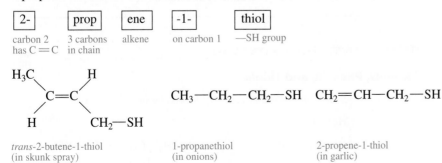

trans-2-butene-1-thiol   1-propanethiol   2-propene-1-thiol
(in skunk spray)   (in onions)   (in garlic)

CH_3—SH
Methanethiol
Oysters and cheese

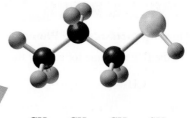

CH_3—CH_2—CH_2—SH
1-Propanethiol
Onions

CH_2=CH—CH_2—SH
2-Propene-1-thiol
Garlic

FIGURE 13.1

Thiols are sulfur-containing compounds with an —SH group.

Q **Why do thiols have structures similar to alcohols?**

SAMPLE PROBLEM 13.3

■ **Thiols**

Draw the condensed structural formula of the following:

a. 1-butanethiol   **b.** cyclohexanethiol

SOLUTION

a. This compound has a —SH group on the first carbon of a butane chain.

   CH_3—CH_2—CH_2—CH_2—SH

b. This compound has a —SH group on cyclohexane.

STUDY CHECK

What is the condensed structural formula of ethanethiol?

QUESTIONS AND PROBLEMS

■ **Alcohols, Phenols, and Thiols**

13.1 Classify each of the following as a primary, secondary, or tertiary alcohol:

$$CH_3$$
a. CH_3—$\overset{|}{CH}$—CH_2—CH_2—OH   **b.** CH_3—CH_2—CH_2—CH_2—OH

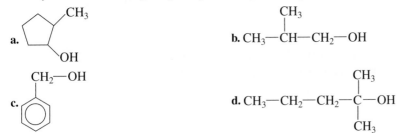

13.2 Classify each of the following as a primary, secondary, or tertiary alcohol:

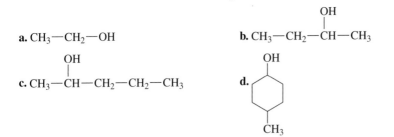

13.3 Give the IUPAC name for each of the following:

a. CH₃—CH₂—OH

b. CH₃—CH₂—CH—CH₃ with OH on the CH

c. CH₃—CH—CH₂—CH₂—CH₃ with OH on the second carbon

d.

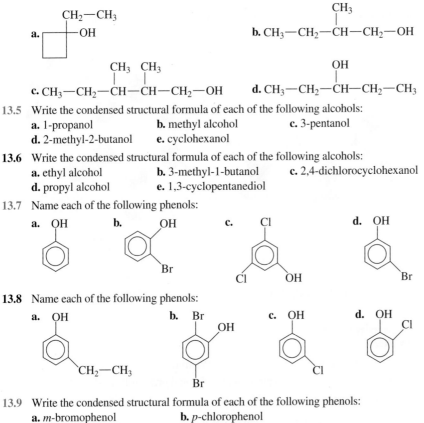

13.4 Give the IUPAC name for each of the following alcohols:

a. cyclobutane with CH₂—CH₃ and —OH

b. CH₃—CH₂—CH—CH₂—OH with CH₃ branch

c. CH₃—CH₂—CH—CH—CH₂—OH with two CH₃ branches

d. CH₃—CH₂—CH—CH₂—CH₃ with OH on third carbon

13.5 Write the condensed structural formula of each of the following alcohols:
  a. 1-propanol   **b.** methyl alcohol   **c.** 3-pentanol
  d. 2-methyl-2-butanol   **e.** cyclohexanol

13.6 Write the condensed structural formula of each of the following alcohols:
  a. ethyl alcohol   **b.** 3-methyl-1-butanol   **c.** 2,4-dichlorocyclohexanol
  d. propyl alcohol   **e.** 1,3-cyclopentanediol

13.7 Name each of the following phenols:

13.8 Name each of the following phenols:

13.9 Write the condensed structural formula of each of the following phenols:
  a. *m*-bromophenol   **b.** *p*-chlorophenol
  c. 2,5-dichlorophenol   **d.** *o*-phenylphenol

13.10 Write the condensed structural formula of each of the following phenols:
a. *o*-ethylphenol **b.** 2,4-dichlorophenol
c. 2,4-dimethylphenol **d.** 2-ethyl-5-methylphenol

13.11 Give the IUPAC name for each of the following thiols:

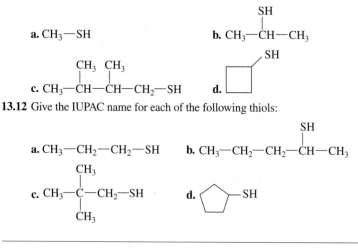

a. CH₃—SH

b. CH₃—CH—CH₃ (with SH above central carbon)

c. CH₃—CH—CH—CH₂—SH (with CH₃ and CH₃ above)

d. (cyclobutane ring with SH)

13.12 Give the IUPAC name for each of the following thiols:

a. CH₃—CH₂—CH₂—SH **b.** CH₃—CH₂—CH₂—CH—CH₃ (with SH above fourth carbon)

c. CH₃—C—CH₂—SH (with CH₃ above and CH₃ below central carbon)

d. (cyclopentane ring)—SH

13.2 Ethers

An **ether** contains an oxygen atom that is attached by single bonds to two carbon groups that are alkyls or aromatic rings. Ethers have a bent structure like water and alcohols except both hydrogen atoms are replaced by alkyl groups.

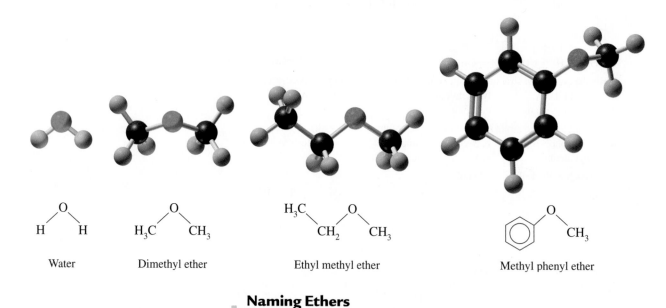

Water   Dimethyl ether   Ethyl methyl ether   Methyl phenyl ether

Naming Ethers

MC GOB Naming Ethers

MC GOB Drawing Ethers

Simple ethers use their common names. Write the name of each alkyl or aromatic group attached to the oxygen atom in alphabetical order followed by the word *ether*.
Complex ethers use the IUPAC names.

■ **STEP 1** Write the alkane name of the larger alkyl group as the main chain.

■ **STEP 2** Name the oxygen and smaller alkyl group as a substituent called an **alkoxy group**.

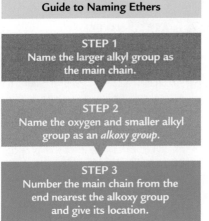

■ **STEP 3** Number the main chain beginning at the end nearest the alkoxy group and give the location of the alkoxy group on the main chain.

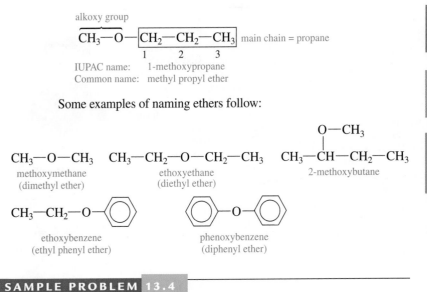

alkoxy group

$$\overbrace{CH_3 - O - }\boxed{CH_2 - CH_2 - CH_3} \text{ main chain = propane}$$
$$\qquad\qquad\qquad 1 \quad\ 2 \quad\ \ 3$$

IUPAC name: 1-methoxypropane
Common name: methyl propyl ether

Some examples of naming ethers follow:

$CH_3 - O - CH_3$ $CH_3 - CH_2 - O - CH_2 - CH_3$ $CH_3 - \underset{\underset{\displaystyle O - CH_3}{|}}{CH} - CH_2 - CH_3$

methoxymethane   ethoxyethane   2-methoxybutane
(dimethyl ether)   (diethyl ether)

$CH_3 - CH_2 - O -$⬡   ⬡$- O -$⬡

ethoxybenzene   phenoxybenzene
(ethyl phenyl ether)   (diphenyl ether)

SAMPLE PROBLEM 13.4

■ **Ethers**

Give a common name and IUPAC name to the following ethers:

a. $CH_3 - CH_2 - O - CH_2 - CH_2 - CH_3$

b. ▢
$\qquad \diagdown O - CH_3$

SOLUTION

a. The groups attached to the oxygen are an ethyl group and a propyl group. The common name is *ethyl propyl ether*. Naming the shorter alkyl group and the oxygen as ethoxy gives the IUPAC name of *1-ethoxypropane*.

b. The groups attached to the oxygen are a methyl group and a cyclobutyl group. The common name is *cyclobutyl methyl ether*. In the IUPAC name, the CH_3O- is named methoxy and the cyclobutane is the longer carbon chain, which gives the name *methoxycyclobutane*.

STUDY CHECK

What is the common name of ethoxybenzene?

Isomers of Alcohols and Ethers

Alcohols and ethers can have the same molecular formula. For example, we can write structural formulas for the isomers with the molecular formula C_2H_6O as follows:

$CH_3 - CH_2 - OH$ $CH_3 - O - CH_3$

ethyl alcohol   dimethyl ether

Guide to Naming Ethers

STEP 1
Name the larger alkyl group as the main chain.

STEP 2
Name the oxygen and smaller alkyl group as an *alkoxy group*.

STEP 3
Number the main chain from the end nearest the alkoxy group and give its location.

Ethers as Anesthetics

Anesthesia is the loss of all sensation and consciousness. A general anesthetic is a substance that blocks signals to the awareness centers in the brain so the person has a loss of memory, a loss of feeling pain, and an artificial sleep. The term *ether* has been associated with anesthesia because diethyl ether was the most widely used anesthetic for more than a hundred years. Although it is easy to administer, ether is very volatile and highly flammable. A small spark in the operating room could cause an explosion. Since the 1950s, anesthetics such as Forane (isoflurane), Ethrane (enflurane),

and Penthrane (methoxyflurane) have been developed that are not as flammable and do not cause nausea. Most of these anesthetics retain the ether group, but the addition of many halogen atoms reduces the volatility and flammability of the ethers. More recently, they have been replaced by halothane (1-bromo-1-chloro-2,2,2-trifluoroethane), discussed in Chapter 11, because of the side effects of the ether-type inhalation anesthetics.

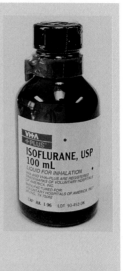

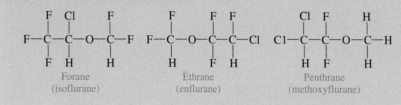

Forane
(isoflurane)

Ethrane
(enflurane)

Penthrane
(methoxyflurane)

SAMPLE PROBLEM 13.5

Isomers

Draw the structural formulas and give the common names of two alcohols and one ether with a molecular formula C_3H_8O.

SOLUTION

To draw the structural formulas for alcohols, the hydroxyl group is bonded to two different atoms in a chain of three carbon atoms. For the ether, two alkyl groups are bonded to an oxygen atom.

$$CH_3-CH_2-CH_2-OH \qquad \overset{\displaystyle OH}{\underset{\displaystyle \text{}}{CH_3-CH-CH_3}} \qquad CH_3-CH_2-O-CH_3$$

propyl alcohol   isopropyl alcohol   ethyl methyl ether

STUDY CHECK

Write the IUPAC names of the unbranched isomers of $C_4H_{10}O$.

Cyclic Ethers

Cyclic ethers contain an oxygen atom in a carbon ring. They are *heterocyclic compounds* because there is a ring with one or more atoms that are not carbon. The cyclic ethers are usually given common names. The five-atom rings with an oxygen atom use common names derived from the aromatic ring *furan*. The four-atom cyclic ethers are not common. The rings are numbered from the oxygen atom as 1.

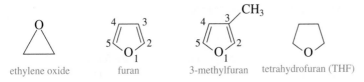

ethylene oxide   furan   3-methylfuran   tetrahydrofuran (THF)

An unsaturated ether ring of six atoms is named *pyran*.

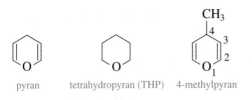

pyran   tetrahydropyran (THP)   4-methylpyran

Cyclic ethers containing two oxygen atoms in a ring of six atoms are called *dioxanes*. The oxygen atoms are numbered because they can take different positions in the ring.

1,4-dioxane   1,3-dioxane

Dioxin is a term used for a group of highly toxic compounds composed of dioxanes bonded to aromatic rings. One of the most toxic is 2,3,7,8-tetrachlorodibenzo-*p*-dioxin (TCDD), now considered carcinogenic (cancer causing) because its structure interferes with DNA. Dioxin is formed during forest fires and as a by-product of many industrial processes involving chlorine such as chemical and pesticide manufacturing and pulp and paper bleaching. The herbicide Agent Orange used in Vietnam was contaminated by highly toxic dioxin, which formed during the synthesis of Agent Orange.

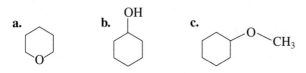

2,4,5-trichlorophenoxyacetic acid
(2,4,5-T; Agent Orange)

2,3,7,8-tetrachlorodibenzo-*p*-dioxin
(TCDD, "dioxin")

SAMPLE PROBLEM 13.6

■ **Cyclic Ethers**

Identify the following as a cyclic alcohol, ether, or cyclic ether.

SOLUTION

a. A cyclic ether has an oxygen atom in the ring.
b. A cyclic alcohol has a hydroxyl group bonded to a cyclic alkane.
c. An ether has an oxygen atom with single bonds to two alkyl or aryl groups.

STUDY CHECK

What is the difference between furan and pyran?

QUESTIONS AND PROBLEMS

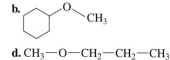

■ Ethers

13.13 Give the IUPAC name and a common name for each of the following ethers:

a. CH₃—O—CH₂—CH₃

b.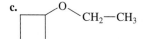

c.

d. CH₃—O—CH₂—CH₂—CH₃

13.14 Give the IUPAC name and a common name for each of the following ethers:

a. CH₃—CH₂—O—CH₂—CH₂—CH₃

b.

c.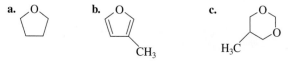

d. CH₃—O—CH₃

13.15 Write the condensed structural formula for each of the following ethers:
a. ethyl propyl ether
b. cyclopropyl ethyl ether
c. methoxycyclopentane
d. 1-ethoxy-2-methylbutane
e. 2,3-dimethoxypentane

13.16 Write the condensed structural formula for each of the following ethers:
a. diethyl ether
b. diphenyl ether
c. ethoxycyclohexane
d. 2-methoxy-2,3-dimethylbutane
e. 1,2-dimethoxybenzene

13.17 Indicate whether each of the following pairs represent isomers, the same compound, or different compounds:
a. 2-pentanol and 2-methoxybutane
b. 2-butanol and cyclobutanol
c. ethyl propyl ether and 2-methyl-1-butanol

13.18 Indicate whether each of the following pairs represent isomers, the same compound, or different compounds:
a. 2-methoxybutane and 3-methyl-2-butanol
b. 1-hexanol and dipropyl ether
c. 2-methyl-2-propanol and diethyl ether

13.19 Give the name for each of the following cyclic ethers:

a.   b.   c.

13.20 Give the name for each of the following cyclic ethers:

a.   b.   c.

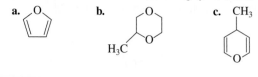

LEARNING GOAL

Describe some physical properties of alcohols, phenols, and ethers.

■ 13.3 Physical Properties of Alcohols, Phenols, and Ethers

In Chapters 11 and 12, we learned that hydrocarbons, which are composed of only carbon and hydrogen, are nonpolar. In this chapter, we looked at compounds containing the element oxygen, which is strongly electronegative. As a result the oxygen and hydrogen atoms in the hydroxyl group —OH form hydrogen bonds. Although ethers also contain an oxygen atom, there is no hydrogen atom attached. Thus, ethers do not hydrogen bond with each other, but they do hydrogen bond with water.

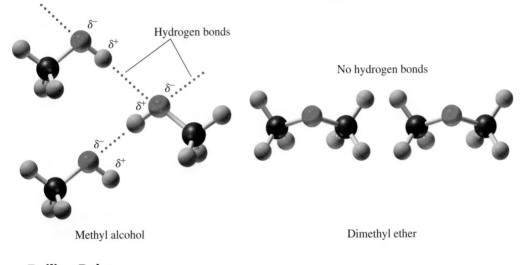

Methyl alcohol

Dimethyl ether

Boiling Points

Higher temperatures are required to provide the energy needed to break the hydrogen bonds between alcohol molecules. Thus, alcohols have higher boiling points than alkanes and ethers of similar mass. The boiling points of ethers are close to those of alkanes of similar mass because hydrogen bonding does not occur.

Physical Properties of Alcohols and Ethers

Solubility in Water

The oxygen atom in alcohols and ethers influences their solubility in water. In alcohols, the polar —OH group can hydrogen bond with water, which makes alcohols with one to four carbon atoms very soluble in water.

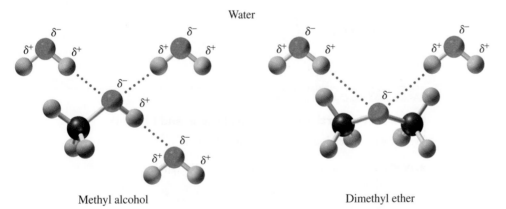

Water

Methyl alcohol

Dimethyl ether

When there are more carbon atoms in the alkyl portion of the alcohol, the effect of the —OH group is diminished. The alkane portion does not participate in hydrogen bonding with water. Thus alcohols with five or more carbon atoms are not very soluble in water.

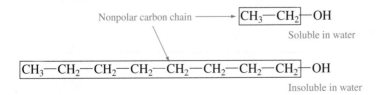

Nonpolar carbon chain ⟶ CH_3—CH_2—OH

Soluble in water

CH_3—CH_2—CH_2—CH_2—CH_2—CH_2—CH_2—CH_2—OH

Insoluble in water

TABLE 13.1

Solubility and Boiling Points of Some Typical Alkanes, Alcohols, and Ethers of Similar Molar Mass

Compound	Structural Formula	Molar Mass (g/mole)	Boiling Point (°C)	Soluble in Water
propane	CH_3—CH_2—CH_3	44	−42	No
dimethyl ether	CH_3—O—CH_3	46	−23	Yes
ethanol	CH_3—CH_2—OH	46	78	Yes
butane	CH_3—CH_2—CH_2—CH_3	58	0	No
ethyl methyl ether	CH_3—O—CH_2—CH_3	60	8	Yes
1-propanol	CH_3—CH_2—CH_2—OH	60	97	Yes

Ethers are more soluble in water than alkanes of similar mass because the oxygen can hydrogen bond with water, but ethers are not as soluble as alcohols. Ethers such as diethyl ether are very useful as solvents for hydrocarbons. However, ether vapors are highly flammable and react with oxygen to form explosive compounds. The utmost care must be taken when working with ethers.

Table 13.1 also compares the boiling points and solubility of some alkanes, alcohols, and ethers of similar mass.

Solubility of Phenols

Phenol is soluble in water because the hydroxyl group ionizes slightly as a weak acid. In fact, an early name for phenol was *carbolic acid*. A concentrated solution of phenol is very corrosive and highly irritating to the skin; it can cause severe burns and ingestion can be fatal. Dilute solutions of phenol were previously used in hospitals as antiseptics, but they have generally been replaced.

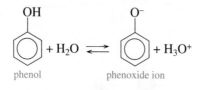

phenol   phenoxide ion

SAMPLE PROBLEM 13.7

▪ Physical Properties of Alcohols, Ethers, and Phenols

Predict which compounds in each pair will be more soluble in water.

a. propane or ethanol   **b.** 1-propanol or 1-hexanol

SOLUTION

a. Ethanol is more soluble because it can form hydrogen bonds with water.
b. The 1-propanol is more soluble because it has a shorter carbon chain.

STUDY CHECK

Dimethyl ether and ethanol both have molar masses of 46. However, ethanol has a much higher boiling point than dimethyl ether. How would you explain this difference in boiling points?

QUESTIONS AND PROBLEMS

■ Physical Properties of Alcohols, Phenols, and Ethers

13.21 Predict the compound with the higher boiling point in the following pairs:
  a. ethane or methanol   **b.** diethyl ether or 1-butanol
  c. 1-butanol or pentane

13.22 Glycerol (1,2,3-propanetriol) has a boiling point of 290°C. 1-Pentanol, which has about the same molar mass as glycerol, boils at 138°C. Why is the boiling point of glycerol so much higher?

13.23 Are each of the following soluble in water? Explain.

  a. CH_3-CH_2-OH
  b. CH_3-O-CH_3
  c. $CH_3-CH_2-CH_2-CH_2-CH_2-CH_2-OH$
  d. OH

13.24 Give an explanation for the following observations:
  a. Ethanol is soluble in water, but propane is not.
  b. Dimethyl ether is soluble in water, but pentane is not.
  c. 1-Propanol is soluble in water, but 1-hexanol is not.

13.4 Reactions of Alcohols and Thiols

In Chapter 11, we learned that hydrocarbons undergo combustion in the presence of oxygen. Alcohols burn with oxygen too. For example, in a restaurant, a dessert may be prepared by pouring a liquor on fruit or ice cream and lighting it. (See Figure 13.2.) The combustion of the ethanol in the liquor proceeds as follows:

$$CH_3-CH_2-OH + 3O_2 \longrightarrow 2CO_2 + 3H_2O + energy$$

■ Dehydration of Alcohols to Form Alkenes

Earlier we saw that alkenes add water to yield alcohols. In a reverse reaction alcohols lose a water molecule when they are heated with an acid catalyst such as H_2SO_4. During the **dehydration** of an alcohol, a H— and —OH are removed from *adjacent carbon atoms of the same alcohol* to produce a water molecule. A double bond forms between the same two carbon atoms to produce an alkene product.

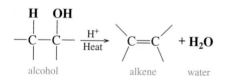

Examples

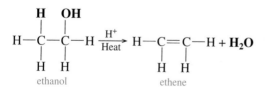

LEARNING GOAL

■ Write equations for the combustion, dehydration, and oxidation of alcohols and thiols.

FIGURE 13.2

A flaming dessert is prepared using a liquor that undergoes combustion.

Q **What is the equation for the combustion of the ethanol in the liquor?**

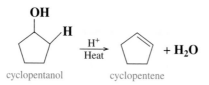

cyclopentanol   cyclopentene

Dehydration and Oxidation of
Alcohols

The dehydration of a secondary alcohol can result in the formation of either of two products. **Saytzeff's rule** states that the major product is the one that results when the hydrogen (H—) is removed from the carbon atom with the smallest number of hydrogen atoms. This occurs because a hydrogen atom on a secondary carbon atom is easier to remove than a hydrogen from a primary carbon atom.

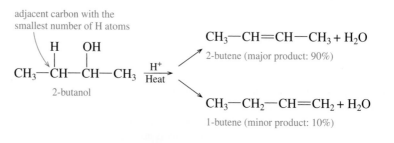

SAMPLE PROBLEM 13.8

▪ Dehydration of Alcohols

Draw the condensed structural formulas for the alkenes produced by the dehydration of the following alcohols:

               OH
               |
a. CH_3—CH_2—CH—CH_2—CH_3 $\xrightarrow[\text{Heat}]{H^+}$

b.

SOLUTION

a. CH_3—CH_2—CH=CH—CH_3 + H_2O

b. The —OH of this alcohol is removed along with an H from an adjacent carbon. Remember that the hydrogens are not drawn in this type of geometric figure.

STUDY CHECK

What is the name of the alkene produced by the dehydration of cyclopentanol?

SAMPLE PROBLEM 13.9

▪ Predicting Reactants

Draw the structural formula of the alcohol that is needed to produce each of the following products.

a.   **b.** CH_3—C=CH—CH_3

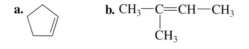

SOLUTION

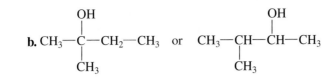

a.

b. $CH_3-\overset{\overset{\displaystyle OH}{|}}{\underset{\underset{\displaystyle CH_3}{|}}{C}}-CH_2-CH_3$ or $CH_3-CH-\overset{\overset{\displaystyle OH}{|}}{CH}-CH_3$

STUDY CHECK

What is the name of an alcohol that forms 2-methylpropene?

Formation of Ethers

Ethers form when the dehydration of alcohols occurs at lower temperatures in the presence of an acid catalyst. Then the components of water are removed from two molecules: an H— from one alcohol and the —OH from another. When the remaining portions of the two alcohols join, an ether is produced.

$$CH_3-OH + HO-CH_3 \xrightarrow[\text{Heat}]{H^+} CH_3-O-CH_3 + H_2O$$

methanol   methanol   dimethyl ether

Oxidation of Alcohols

In organic chemistry, it is convenient to think of **oxidation** as a loss of hydrogen atoms or the addition of oxygen. An oxidation reaction occurs when there is an increase in the number of carbon–oxygen bonds. In a reduction, the product has fewer bonds between carbon and oxygen.

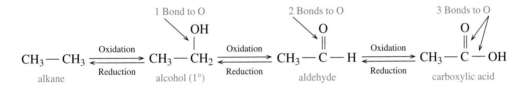

Oxidation of Primary and Secondary Alcohols

The oxidation of a primary alcohol produces an aldehyde, which contains a double bond between carbon and oxygen. The oxidation occurs by removing two hydrogen atoms, one from the —OH group and another from the carbon that is bonded to the —OH. To indicate the presence of an oxidizing agent, such as $KMnO_4$ or $K_2Cr_2O_7$, reactions are often written with the symbol [O].

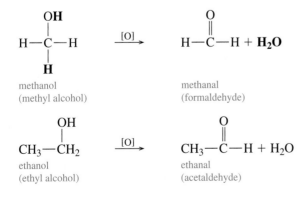

Aldehydes oxidize further, this time by the addition of another oxygen to form a carboxylic acid. This step occurs so readily that it is often difficult to isolate the aldehyde product during oxidation. We will learn more about carboxylic acids in Chapter 16.

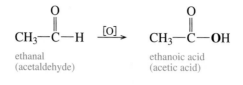

ethanal
(acetaldehyde)

ethanoic acid
(acetic acid)

Oxidation of Methanol

Methanol, "wood alcohol," is a highly toxic alcohol present in products such as windshield washer fluid, Sterno, and paint strippers. Methanol is rapidly absorbed in the gastrointestinal tract. In the liver, it is metabolized to formaldehyde and then formic acid, a substance that causes nausea, severe abdominal pain, and blurred vision. Blindness can occur because the intermediate products destroy the retina of the eye. As little as 4 mL of methanol can produce blindness. The formic acid, which is not readily eliminated from the body, lowers blood pH so severely that just 30 mL of methanol can lead to coma and death.

The treatment for methanol poisoning involves giving sodium bicarbonate to neutralize the formic acid in the blood. In some cases, ethanol is given intravenously to the patient. The enzymes in the liver pick up ethanol molecules to oxidize instead of methanol molecules, which gives more time for the methanol to be eliminated via the lungs without the formation of its dangerous oxidation products.

In the oxidation of secondary alcohols, the products are ketones. One hydrogen is removed from the —OH and another from the carbon bonded to the —OH group. The result is a ketone that has the carbon–oxygen double bond attached to alkyl groups on both sides.

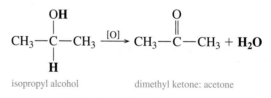

isopropyl alcohol

dimethyl ketone: acetone

Tertiary alcohols do not oxidize readily because there are no hydrogen atoms on the carbon bonded to the —OH group. Because C—C bonds are usually too strong to oxidize, tertiary alcohols resist oxidation.

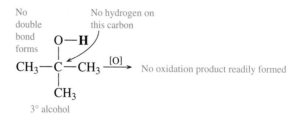

No double bond forms

No hydrogen on this carbon

No oxidation product readily formed

3° alcohol

SAMPLE PROBLEM 13.10

▪ Oxidation of Alcohols

Draw the structural formula of the aldehyde or ketone formed by the oxidation of each of the following:

a.
$$CH_3—CH_2—\overset{\displaystyle OH}{\underset{\displaystyle |}{CH}}—CH_3$$

b. $CH_3—CH_2—CH_2—OH$

SOLUTION

a. The oxidation of a secondary alcohol produces a ketone.

$$CH_3—CH_2—\overset{\displaystyle O}{\overset{\displaystyle ||}{C}}—CH_3$$

b. The oxidation of a primary alcohol produces an aldehyde.

$$CH_3—CH_2—\overset{\displaystyle O}{\overset{\displaystyle ||}{C}}—H$$

STUDY CHECK

Draw the structural formula of the aldehyde or ketone formed by the oxidation of 2-propanol.

 SELF-STUDY ACTIVITY
Alcohol Toxicity

During vigorous exercise, lactic acid accumulates in the muscles and causes fatigue. When the activity level is decreased, oxygen enters the muscles and oxidizes the secondary —OH group in lactic acid to a ketone group in pyruvic acid. Pyruvic acid is metabolized further until it is completely oxidized to CO_2 and H_2O. The muscles in highly trained athletes are capable of taking up greater quantities of oxygen so that vigorous exercise can be maintained for longer periods of time.

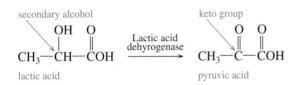

secondary alcohol → $CH_3—\overset{OH}{\underset{|}{CH}}—\overset{O}{\overset{||}{C}}OH$ $\xrightarrow{\text{Lactic acid dehydrogenase}}$ keto group → $CH_3—\overset{O}{\overset{||}{C}}—\overset{O}{\overset{||}{C}}OH$

lactic acid   pyruvic acid

▪ Oxidation of Thiols

Thiols also undergo oxidation by a loss of hydrogen atoms from the —SH groups. The oxidized product is called a **disulfide**.

$$CH_3—S—H + H—S—CH_3 \xrightarrow{[O]} CH_3—S—S—CH_3 + H_2O$$
methanethiol   dimethyl disulfide

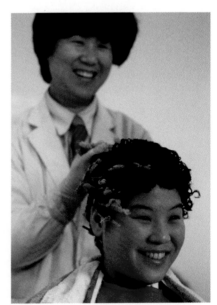

Oxidation of Thiols

Much of the protein in the hair is cross-linked by disulfide bonds, which occur mostly between the thiol groups of the amino acid cysteine.

$$NH_2—\overset{\overset{\text{COOH}}{|}}{CH}—CH_2\textbf{SH} + \textbf{H}SCH_2—\overset{\overset{\text{COOH}}{|}}{CH}—NH_2$$
cysteine

$$\xrightarrow{\text{[O]}} NH_2—\overset{\overset{\text{COOH}}{|}}{CH}—CH_2—S—S—CH_2—\overset{\overset{\text{COOH}}{|}}{CH}—NH_2 + H_2O$$

When a person is given a "perm," a reducing substance is used to break the disulfide bonds. While the hair is still wrapped around the curlers, an oxidizing substance is applied that causes new disulfide bonds to form between different parts of the protein hair strands, which gives the hair a new shape.

HEALTH NOTE

■ Oxidation of Alcohol in the Body

Ethanol is the most commonly abused drug in the United States. When ingested in small amounts, ethanol may produce a feeling of euphoria in the body although it is a depressant. In the liver, enzymes such as alcohol dehydrogenases oxidize ethanol to acetaldehyde, a substance that impairs mental and physical coordination. If the blood alcohol concentration exceeds 0.4%, coma or death may occur. Table 13.2 gives some of the typical behaviors exhibited at various levels of blood alcohol.

$$CH_3CH_2OH \xrightarrow{\text{[O]}} CH_3\overset{\overset{\text{O}}{||}}{CH} \xrightarrow{\text{[O]}} 2CO_2 + H_2O$$
ethanol   ethanal
(ethyl alcohol)   (acetaldehyde)

The acetaldehyde produced from ethanol in the liver is further oxidized to acetic acid, which is converted to car-

bon dioxide and water in the citric acid (Krebs) cycle. Thus, the enzymes in the liver can eventually break down ethanol, but the aldehyde and carboxylic acid intermediates can cause considerable damage while they are present within the cells of the liver.

A person weighing 150 lb requires about one hour to completely metabolize 10 ounces of beer. However, the rate of metabolism of ethanol varies between nondrinkers and drinkers. Typically, nondrinkers and social drinkers can metabolize 12–15 mg of ethanol/dL of blood in one hour, but an alcoholic can metabolize as much as 30 mg of ethanol/dL in one hour. Some effects of alcohol metabolism include an increase in liver lipids (fatty liver), an increase in serum triglycerides, gastritis, pancreatitis, ketoacidosis, alcoholic hepatitis, and psychological disturbances.

When the Breathalyzer test is used for suspected drunk drivers, the driver exhales a volume of breath into a solution containing the orange Cr^{6+} ion. If there is ethyl alcohol present in the exhaled air, the alcohol is oxidized, and the Cr^{6+} is reduced to give a green solution of Cr^{3+}.

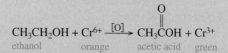

$$CH_3CH_2OH + Cr^{6+} \xrightarrow{\text{[O]}} CH_3\overset{\overset{\text{O}}{||}}{C}OH + Cr^{3+}$$
ethanol   orange   acetic acid   green

Sometimes alcoholics are treated with a drug called Antabuse (disulfiram), which prevents the oxidation of acetaldehyde to acetic acid. As a result, acetaldehyde accumulates in the blood, which causes nausea, profuse sweating, headache, dizziness, vomiting, and respiratory difficulties. Because of these unpleasant side effects, the patient is less likely to use alcohol.

TABLE 13.2

Typical Behaviors Exhibited by a 150-lb Person Consuming Alcohol

Number of Beers (12 oz) or Glasses of Wine (5 oz)	Blood Alcohol Level (w/v %)	Typical Behavior
1	0.025	Slightly dizzy, talkative
2	0.05	Euphoria, loud talking, and laughing
4	0.10	Loss of inhibition, loss of coordination, drowsiness, legally drunk in most states
8	0.20	Intoxicated, quick to anger, exaggerated emotions
12	0.30	Unconscious
16–20	0.40–0.50	Coma and death

QUESTIONS AND PROBLEMS

▪ Reactions of Alcohols and Thiols

13.25 Draw the condensed structural formula of the major-product alkene produced by each of the following dehydration reactions:

a. $CH_3-CH_2-CH_2-CH_2-OH \xrightarrow[\text{Heat}]{H^+}$

b. 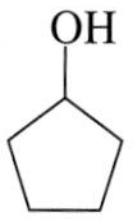 $\xrightarrow[\text{Heat}]{H^+}$

c. 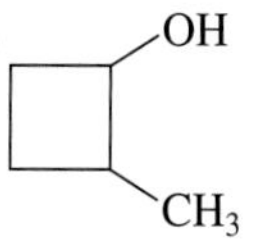 $\xrightarrow[\text{Heat}]{H^+}$

d. $CH_3-CH_2-CH_2-\overset{\overset{\displaystyle OH}{|}}{CH}-CH_3 \xrightarrow[\text{Heat}]{H^+}$

13.26 Draw the condensed structural formula of the major-product alkene produced by each of the following dehydration reactions:

a. $CH_3-\overset{\overset{\displaystyle CH_3}{|}}{CH}-CH_2-OH \xrightarrow[\text{Heat}]{H^+}$

b. $CH_3-\overset{\overset{\displaystyle OH}{|}}{CH}-\overset{\overset{\displaystyle CH_3}{|}}{CH}-CH_2-CH_3 \xrightarrow[\text{Heat}]{H^+}$

c. 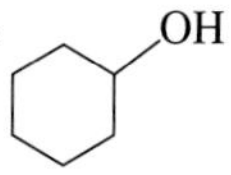 $\xrightarrow[\text{Heat}]{H^+}$

d. 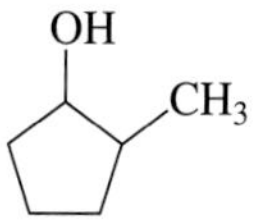 $\xrightarrow[\text{Heat}]{H^+}$

13.27 Write the ether product from the reaction of each of the following:

a. $2CH_3-OH \xrightarrow[\text{Heat}]{H^+}$ **b.** $2CH_3-CH_2-CH_2-OH \xrightarrow[\text{Heat}]{H^+}$

13.28 Write the ether product from the reaction of each of the following:

a. $2CH_3-CH_2-OH \xrightarrow[\text{Heat}]{H^+}$ **b.** $2CH_3-\overset{\overset{\displaystyle CH_3}{|}}{CH}-CH_2-OH \xrightarrow[\text{Heat}]{H^+}$

13.29 What alcohol(s) could be used to produce each of the following compounds?
a. $CH_2=CH_2$
b. $CH_3-O-CH_2-CH_3$
c.

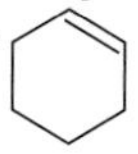

13.30 What alcohol(s) could be used to produce each of the following compounds?
a. $CH_3-CH_2-O-CH_2-CH_3$
b. $CH_3-\overset{\overset{\displaystyle CH_3}{|}}{C}=CH-CH_3$
c.

13.31 Draw the condensed structural formula of the organic product when each of the following alcohols is oxidized [O] (if no reaction, write *none*):

a. CH$_3$—CH$_2$—CH$_2$—CH$_2$—CH$_2$—OH

b.
$$CH_3—CH_2—\overset{\overset{\displaystyle OH}{|}}{CH}—CH_3$$

c.
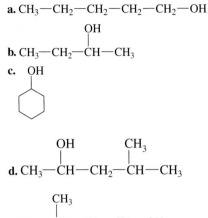

d.
$$CH_3—\overset{\overset{\displaystyle OH}{|}}{CH}—CH_2—\overset{\overset{\displaystyle CH_3}{|}}{CH}—CH_3$$

e. 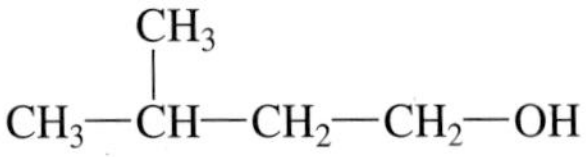
$$CH_3—\overset{\overset{\displaystyle CH_3}{|}}{CH}—CH_2—CH_2—OH$$

13.32 Draw the condensed structural formula of the organic product when each of the following alcohols is oxidized [O] (if no reaction, write *none*):

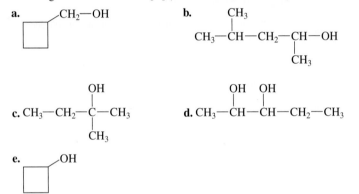

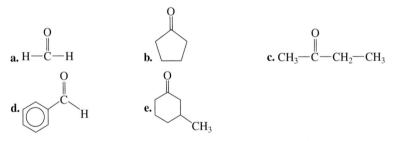

13.33 Draw the condensed structural formula of the alcohol needed to give each of the following oxidation products:

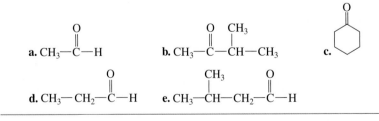

13.34 Draw the condensed structural formula of the alcohol needed to give each of the following oxidation products:

a.
$$CH_3—\overset{\overset{\displaystyle O}{||}}{C}—H$$

b.
$$CH_3—\overset{\overset{\displaystyle O}{||}}{C}—\overset{\overset{\displaystyle CH_3}{|}}{CH}—CH_3$$

c.

d.
$$CH_3—CH_2—\overset{\overset{\displaystyle O}{||}}{C}—H$$

e.
$$CH_3—\overset{\overset{\displaystyle CH_3}{|}}{CH}—CH_2—\overset{\overset{\displaystyle O}{||}}{C}—H$$

CONCEPT MAP

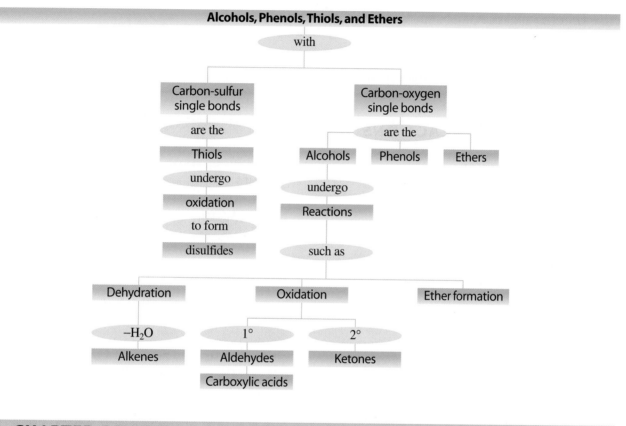

CHAPTER REVIEW

13.1 Alcohols, Phenols, and Thiols

The functional group of an alcohol is the hydroxyl group —OH bonded to a carbon chain. In a phenol, the hydroxyl group is bonded to an aromatic ring. In thiols, the functional group is —SH, which is analogous to the —OH group of alcohols. Alcohols are classified according to the number of alkyl or aromatic groups bonded to the carbon that holds the —OH. In a primary (1°) alcohol one group is attached to the hydroxyl carbon. In a secondary (2°) alcohol two groups are attached, and in a tertiary (3°) alcohol there are three groups bonded to the hydroxyl carbon. In the IUPAC system, the names of alcohols have *ol* endings, and the location of the —OH group is given by numbering the carbon chain. Simple alcohols are generally named by their common names with the alkyl name preceding the term *alcohol*. A cyclic alcohol is named as a cycloalkanol. An aromatic alcohol is named as a phenol.

13.2 Ethers

In an ether, an oxygen atom is connected by single bonds to two alkyl or aromatic groups. In the common names of ethers, the alkyl groups are listed alphabetically followed by the name *ether*. In the IUPAC name, the smaller alkyl group with the oxygen is named as an *alkoxy group* and is attached to the longer alkane chain, which is numbered to give the location of the alkoxy group.

13.3 Physical Properties of Alcohols, Phenols, and Ethers

The —OH group allows alcohols to hydrogen bond, which causes alcohols to have higher boiling points than alkanes and ethers of similar mass. Short-chain alcohols and ethers can hydrogen bond with water, which makes them soluble.

13.4 Reactions of Alcohols and Thiols

At high temperatures, alcohols dehydrate in the presence of an acid to yield alkenes. At lower temperatures, two molecules of alcohol lose H and OH to produce an ether. Primary alcohols are oxidized to aldehydes, which can oxidize further to carboxylic acids. Secondary alcohols are oxidized to ketones. Tertiary alcohols do not oxidize. Thiols undergo oxidation to form disulfides.

SUMMARY OF NAMING

Structure	Family	IUPAC Name	Common Name
CH_3—**OH**	alcohol	methanol	methyl alcohol
⬡—**OH**	phenol	phenol	phenol
CH_3—**SH**	thiol	methanethiol	
CH_3—**O**—CH_3	ether	methoxymethane	dimethyl ether
(furan ring)	cyclic ether	furan	
CH_3—**S**—**S**—CH_3	disulfide	dimethyldisulfide	

SUMMARY OF REACTIONS

COMBUSTION OF ALCOHOLS

$$CH_3—CH_2—OH + 3O_2 \longrightarrow 2CO_2 + 3H_2O$$

ethanol   oxygen   carbon dioxide   water

DEHYDRATION OF ALCOHOLS TO FORM ALKENES

$$CH_3—CH_2—CH_2—OH \xrightarrow[\text{Heat}]{H^+}$$

1-propanol

$$CH_3—CH{=}CH_2 + H_2O$$

propene

FORMATION OF ETHERS

$$CH_3—OH + HO—CH_3 \xrightarrow[\text{Heat}]{H^+}$$

methanol

$$CH_3—O—CH_3 + H_2O$$

dimethyl ether

OXIDATION OF PRIMARY ALCOHOLS TO FORM ALDEHYDES

$$\underset{\text{ethanol}}{\overset{\overset{\displaystyle OH}{|}}{CH_3—CH_2}} \xrightarrow{[O]} \underset{\text{acetaldehyde}}{\overset{\overset{\displaystyle O}{\|}}{CH_3—C—H}} + H_2O$$

OXIDATION OF SECONDARY ALCOHOLS TO FORM KETONES

$$\underset{\text{2-propanol}}{\overset{\overset{\displaystyle OH}{|}}{CH_3—CH—CH_3}} \xrightarrow{[O]} \underset{\text{propanone}}{\overset{\overset{\displaystyle O}{\|}}{CH_3—C—CH_3}} + H_2O$$

OXIDATION OF THIOLS TO FORM DISULFIDES

$$\underset{\text{methanethiol}}{CH_3—S—H + H—S—CH_3} \xrightarrow{[O]} \underset{\text{dimethyl disulfide}}{CH_3—S—S—CH_3} + H_2O$$

OXIDATION OF ALDEHYDES TO CARBOXYLIC ACIDS

$$\underset{\text{acetaldehyde}}{\overset{\overset{\displaystyle O}{\|}}{CH_3—C—H}} \xrightarrow{[O]} \underset{\text{acetic acid}}{\overset{\overset{\displaystyle O}{\|}}{CH_3—C—OH}}$$

KEY TERMS

alcohols Organic compounds that contain the hydroxyl (—OH) functional group, attached to a carbon chain.

alkoxy group A group that contains oxygen bonded to an alkyl group.

cyclic ethers Compounds that contain an oxygen atom in a carbon ring.

dehydration A reaction that removes water from an alcohol in the presence of an acid to form alkenes at high temperature, or ethers at lower temperatures.

disulfides Compounds formed from thiols, disulfides contain the —S—S— functional group.

ether An organic compound in which an oxygen atom is bonded to two alkyl or two aromatic groups, or a mix of the two.

hydroxyl group The —OH functional group.

oxidation The loss of two hydrogen atoms from a reactant to give a more oxidized compound, e.g., primary alcohols oxidize to

aldehydes, secondary alcohols oxidize to ketones. An oxidation can also be the addition of an oxygen atom as in the oxidation of aldehydes to carboxylic acids.

phenol An organic compound that has an —OH group attached to a benzene ring.

primary (1°) alcohol An alcohol that has one alkyl group bonded to the alcohol carbon atom.

Saytzeff's rule In the dehydration of an alcohol, hydrogen is removed from the carbon that already has the smallest number of hydrogen atoms to form an alkene.

secondary (2°) alcohol An alcohol that has two alkyl groups bonded to the carbon atom with the —OH group.

tertiary (3°) alcohol An alcohol that has three alkyl groups bonded to the carbon atom with the —OH.

thiols Organic compounds that contain a thiol group (—SH).

UNDERSTANDING THE CONCEPTS

13.35 A compound called butylated hydroxytoluene or BHT is a common preservative added to cereal. It is named as 1-hydroxy-2,6-dimethylethyl-4-methylbenzene. Draw the structure of BHT.

13.36 The compound 1,3-dihydroxy-2-propanone or dihydroxy-acetone (DHA) used in "sunless" tanning lotions darkens the skin without sun. What is its functional group?

$$HO-CH_2-\overset{\displaystyle O}{\overset{\displaystyle \|}{C}}-CH_2-OH$$

13.37 Identify the functional groups in each of the following:
   a. Urushiol, a substance in poison ivy and poison oak that causes itching and blistering of the skin.

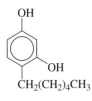

b. Menthol gives a peppermint taste and odor used in candy and throat lozenges.

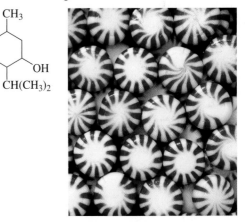

13.38 Identify the functional groups in each of the following:
   a. BHA, an anti-oxidant used as a preservative in foods such as baked goods, butter, meats, and snack foods.

b. Vanillin, a flavoring, obtained from the seeds of the vanilla bean.

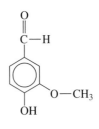

ADDITIONAL QUESTIONS AND PROBLEMS

For instructor-assigned homework, go to
www.masteringgob.com.

13.39 Classify each of the following as primary, secondary, or tertiary alcohols:

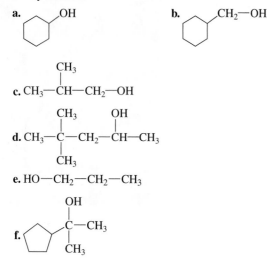

a. [cyclohexane-OH]   **b.** [cyclohexane-CH₂—OH]

c. CH₃—CH—CH₂—OH (with CH₃ branch)

d. CH₃—C(CH₃)(CH₃)—CH₂—CH(OH)—CH₃

e. HO—CH₂—CH₂—CH₃

f. [cyclopentane-C(OH)(CH₃)(CH₃)]

13.40 Classify each of the following as primary, secondary, or tertiary alcohols:

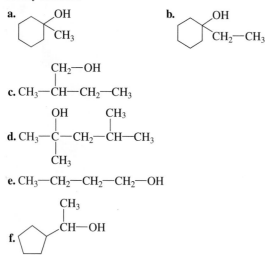

a. [cyclohexane-OH, CH₃]   **b.** [cyclohexane-OH, CH₂—CH₃]

c. CH₃—CH(CH₂—OH)—CH₂—CH₃

d. CH₃—C(CH₃)(OH)—CH₂—CH(CH₃)—CH₃

e. CH₃—CH₂—CH₂—CH₂—OH

f. [cyclopentane-CH(OH)—CH₃]

13.41 Identify each of the following as an alcohol, a phenol, an ether, a cyclic ether, or a thiol:

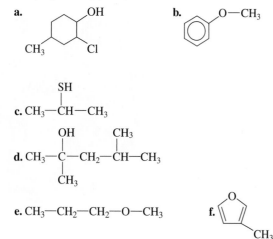

a. [cyclohexane with OH, CH₃, Cl]   **b.** [benzene-O—CH₃]

c. CH₃—CH(SH)—CH₃

d. CH₃—C(OH)(CH₃)—CH₂—CH(CH₃)—CH₃

e. CH₃—CH₂—CH₂—O—CH₃

f. [furan ring with CH₃]

13.42 Identify each of the following as an alcohol, a phenol, an ether, a cyclic ether, or a thiol:

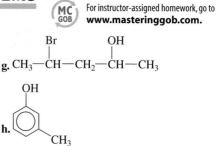

g. CH₃—CH(Br)—CH₂—CH(OH)—CH₃

h. [benzene with OH and CH₃]

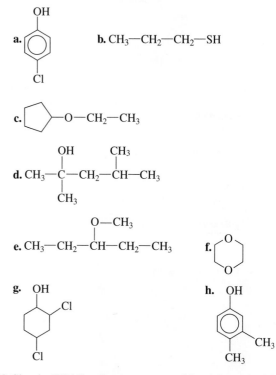

a. [benzene with OH and Cl]   **b.** CH₃—CH₂—CH₂—SH

c. [cyclopentane-O—CH₂—CH₃]

d. CH₃—C(OH)(CH₃)—CH₂—CH(CH₃)—CH₃

e. CH₃—CH₂—CH(O—CH₃)—CH₂—CH₃   **f.** [1,4-dioxane ring]

g. [cyclohexane with OH, Cl, Cl]   **h.** [benzene with OH, CH₃, CH₃]

13.43 Give the IUPAC and common names (if any) for each of the compounds in problem 13.41.

13.44 Give the IUPAC and common names (if any) for each of the compounds in problem 13.42.

13.45 Draw the condensed structural formula of each of the following compounds:
  a. 3-methylcyclopentanol   **b.** *p*-chlorophenol
  c. 2-methyl-3-pentanol   **d.** phenyl ethyl ether
  e. 3-pentanethiol   **f.** *ortho*-cresol
  g. 2,4-dibromophenol

13.46 Draw the condensed structural formula of each of the following compounds:
  a. 3-methoxypentane   **b.** *meta*-chlorophenol
  c. 2,3-pentanediol   **d.** methyl propyl ether
  e. methanethiol   **f.** 3-methyl-2-butanol
  g. 3,4-dichlorocyclohexanol

13.47 Draw the structural formulas of all the alcohols with a molecular formula $C_4H_{10}O$.

13.48 Draw the structural formulas of all the ethers with a molecular formula $C_5H_{12}O$.

13.49 Which compound in each pair would you expect to have the higher boiling point?
  a. butane or 1-propanol
  b. 1-propanol or ethyl methyl ether
  c. ethanol or 1-butanol

13.50 Which compound in each pair would you expect to have the higher boiling point?
  a. propane or ethyl alcohol
  b. 2-propanol or 2-pentanol
  c. diethyl ether or 1-butanol

13.51 Explain why each of the following compounds would be soluble or insoluble in water.
  a. 2-propanol   **b.** dimethyl ether   **c.** 1-hexanol

13.52 Explain why each of the following compounds would be soluble or insoluble in water.
  a. glycerol   **b.** butane   **c.** 1,3-hexanediol

13.53 Draw the condensed structural formula for the major product of each of the following reactions:

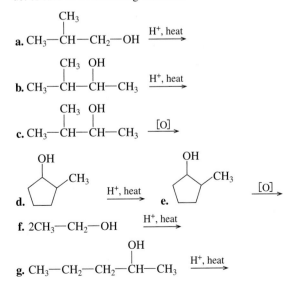

a. CH_3—CH_2—CH_2—OH $\xrightarrow{H^+, \text{ heat}}$

b. CH_3—CH_2—CH_2—OH $\xrightarrow{[O]}$

c. CH_3—CH_2—$\overset{\overset{\displaystyle OH}{|}}{CH}$—$CH_3$ $\xrightarrow{H^+, \text{ heat}}$

d. CH_3—CH_2—$\overset{\overset{\displaystyle OH}{|}}{CH}$—$CH_3$ $\xrightarrow{[O]}$

e. $2CH_3$—CH_2—CH_2—OH $\xrightarrow{H^+}$

f. (cyclohexanol) $\xrightarrow{H^+, \text{ heat}}$

g. (cyclohexanol) $\xrightarrow{[O]}$

13.54 Draw the condensed structural formula for the major product of each of the following reactions:

a. CH_3—$\overset{\overset{\displaystyle CH_3}{|}}{CH}$—$CH_2$—OH $\xrightarrow{H^+, \text{ heat}}$

b. CH_3—$\overset{\overset{\displaystyle CH_3}{|}}{CH}$—$\overset{\overset{\displaystyle OH}{|}}{CH}$—$CH_3$ $\xrightarrow{H^+, \text{ heat}}$

c. CH_3—$\overset{\overset{\displaystyle CH_3}{|}}{CH}$—$\overset{\overset{\displaystyle OH}{|}}{CH}$—$CH_3$ $\xrightarrow{[O]}$

d. (2-methylcyclopentanol) $\xrightarrow{H^+, \text{ heat}}$ **e.** (2-methylcyclopentanol) $\xrightarrow{[O]}$

f. $2CH_3$—CH_2—OH $\xrightarrow{H^+, \text{ heat}}$

g. CH_3—CH_2—CH_2—$\overset{\overset{\displaystyle OH}{|}}{CH}$—$CH_3$ $\xrightarrow{H^+, \text{ heat}}$

13.55 Sometimes several steps are needed to prepare a compound. Using a combination of the reactions we have studied, indicate how you might prepare the following from the starting substance given. For example, 2-propanol could be prepared from 1-propanol by first dehydrating the alcohol to give propene and then hydrating it again to give 2-propanol according to Markovnikov's rule.

CH_3—CH_2—CH_2—OH $\xrightarrow{H^+, \text{ heat}}$ CH_3—CH=CH_2 + H_2O $\xrightarrow{H^+}$

CH_3—$\overset{\overset{\displaystyle OH}{|}}{CH}$—$CH_3$
  2-propanol

  a. Prepare 2-chloropropane from 1-propanol.
  b. Prepare 2-methylpropane from 2-methyl-2-propanol.

  c. Prepare CH_3—$\overset{\overset{\displaystyle O}{||}}{C}$—$CH_3$ from 1-propanol.

13.56 As in problem 13.55, indicate how you might prepare the following from the starting substance given:
  a. Prepare 1-pentene from 1-pentanol.
  b. Prepare chlorocyclohexane from cyclohexanol.
  c. Prepare 1,2-dibromobutane from 1-butanol.

13.57 Identify the functional groups in the following molecule:

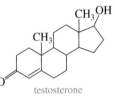

testosterone

13.58 Identify the functional groups in the following molecule:

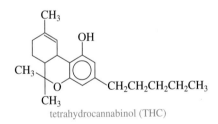

tetrahydrocannabinol (THC)

13.59 Hexylresorcinol, an antiseptic ingredient used in mouthwashes and throat lozenges, has the IUPAC name of 4-hexyl-1,3-benzenediol. Draw its condensed structural formula.

13.60 Menthol, which has a minty flavor, is used in throat sprays and lozenges. Thymol is used as a topical antiseptic to destroy mold. Give each of their IUPAC names. What is similar and what is different about their structures?

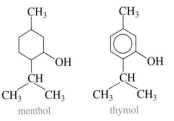

menthol   thymol

CHALLENGE QUESTIONS

13.61 Write the condensed structural formulas for each of the following naturally occurring compounds:
  a. 2,5-dichlorophenol, a defense pheromone of a grasshopper.
  b. Skunk scent, a mixture of 3-methyl-1-butanethiol and *trans*-2-butene-1-thiol.
  c. Pentachlorophenol, which is a wood preservative.

13.62 Dimethyl ether and ethyl alcohol both have the molecular formula C_2H_6O. One has a boiling point of $-24°C$, and the other, $79°C$. Draw the condensed structural formulas of each compound. Decide which boiling point goes with which compound and explain. Check the boiling points in a chemistry handbook.

13.63 A compound with the formula C_4H_8O is synthesized from 2-methyl-1-propanol and oxidizes easily to give a carboxylic acid. What is the structure of the compound?

13.64 Methyl *tert*-butyl ether (MTBE) or methyl 2-methyl-2-propyl ether is a fuel additive for gasoline to boost the octane rating. It increases the oxygen content, which reduces CO emissions to an acceptable level determined by the Clean Air Act.
  a. If fuel mixtures are required to contain 2.7% oxygen by mass, how many grams MTBE must be present in each 100 g gasoline?

  b. How many liters of MTBE would be in a liter of fuel if the density of both gasoline and MTBE is 0.740 g/mL?
  c. Write the equation for the complete combustion of MTBE.
  d. How many liters of air containing 21% (v/v) O_2 are required at STP to completely react (combust) 1.00 L liquid MTBE?

13.65 Draw the structures and give the IUPAC names of all the alcohols that have the formula $C_5H_{12}O$.

ANSWERS

Answers to Study Checks

13.1 tertiary

13.2 3-chloro-1-butanol

13.3 $CH_3—CH_2—SH$

13.4 ethyl phenyl ether

13.5 1-butanol, 2-butanol, methoxy-1-propane, ethoxyethane.

13.6 Both are unsaturated cyclic ethers, but furan has five atoms in the ring and pyran has six atoms.

13.7 Ethanol molecules can hydrogen bond with each other, but ether molecules cannot. Thus, a higher temperature is required to break the hydrogen bonds between ethanol molecules.

13.8 cyclopentene

13.9 2-methyl-1-propanol, 2-methyl-2-propanol

13.10 $CH_3—\overset{\overset{O}{\|}}{C}—CH_3$

Answers to Selected Questions and Problems

13.1 a. 1° **b.** 1° **c.** 3° **d.** 2°

13.3 a. ethanol **b.** 2-butanol
  c. 2-pentanol **d.** 4-methylcyclohexanol

13.5 a. $CH_3—CH_2—CH_2—OH$
  b. $CH_3—OH$
  c. $CH_3—CH_2—\overset{\overset{OH}{|}}{CH}—CH_2—CH_3$
  d. $CH_3—\overset{\overset{OH}{|}}{\underset{\underset{CH_3}{|}}{C}}—CH_2—CH_3$
  e.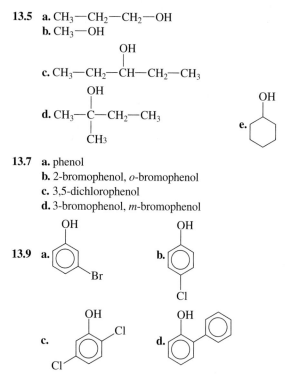

13.7 a. phenol
  b. 2-bromophenol, *o*-bromophenol
  c. 3,5-dichlorophenol
  d. 3-bromophenol, *m*-bromophenol

13.9 a. **b.** **c.** **d.**

13.11 a. methanethiol **b.** 2-propanethiol
  c. 2,3-dimethyl-1-butanethiol **d.** cyclobutanethiol

13.13 a. methoxyethane, ethyl methyl ether
b. methoxycyclohexane, cyclohexyl methyl ether
c. ethoxycyclobutane, cyclobutyl ethyl ether
d. 1-methoxypropane, methyl propyl ether

13.15 a. $CH_3-CH_2-O-CH_2-CH_2-CH_3$

b. $CH_3-CH_2-O-\triangle$

c.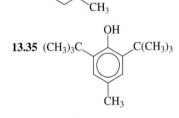

d. $CH_3-CH_2-O-CH_2-\overset{\overset{\displaystyle CH_3}{|}}{CH}-CH_2-CH_3$

e. $CH_3-\overset{\overset{\displaystyle O-CH_3}{|}}{CH}-\underset{\underset{\displaystyle O-CH_3}{|}}{CH}-CH_2-CH_3$

13.17 a. Isomers ($C_5H_{12}O$)
b. Different compounds
c. Isomers ($C_5H_{12}O$)

13.19 a. tetrahydrofuran
b. 3-methylfuran
c. 5-methyl-1,3-dioxane

13.21 a. methanol
b. 1-butanol
c. 1-butanol

13.23 a. yes, hydrogen bonding
b. yes; hydrogen bonding
c. no; long carbon chain diminishes effect of —OH group
d. yes: —OH ionizes

13.25 a. $CH_3-CH_2-CH=CH_2$

b. [cyclopentene structure]

c. [methylcyclobutene structure with CH₃]

d. $CH_3-CH_2-CH=CH-CH_3$

13.27 a. CH_3-O-CH_3
b. $CH_3-CH_2-CH_2-O-CH_2-CH_2-CH_3$

13.29 a. CH_3-CH_2-OH **b.** $CH_3-OH + CH_3-CH_2-OH$

c. [cyclohexanol structure with OH]

13.31 a. $CH_3-CH_2-CH_2-CH_2-\overset{\overset{\displaystyle O}{||}}{CH}$

b. $CH_3-CH_2-\overset{\overset{\displaystyle O}{||}}{C}-CH$

c. [cyclohexanone structure]

d. $CH_3-\overset{\overset{\displaystyle O}{||}}{C}-CH_2-\overset{\overset{\displaystyle CH_3}{|}}{CH}-CH_3$

e. $CH_3-\overset{\overset{\displaystyle CH_3}{|}}{CH}-CH_2-\overset{\overset{\displaystyle O}{||}}{CH}$

13.33 a. CH_3-OH

b. [cyclopentanol structure with OH]

c. $CH_3-CH_2-\overset{\overset{\displaystyle OH}{|}}{CH}-CH_3$

d. [benzene ring with CH₂—OH]

e. [methylcyclohexanol structure with OH and CH₃]

13.35 $(CH_3)_3C$—[benzene ring with OH on top, C(CH₃)₃ on right, CH₃ on bottom]

13.37 a. aromatic, alcohol
b. cycloalkane, alcohol

13.39 a. 2°   **b.** 1°   **c.** 1°
d. 2°   **e.** 1°   **f.** 3°

13.41 a. alcohol   **b.** ether   **c.** thiol
d. alcohol   **e.** ether   **f.** cyclic ether
g. alcohol   **h.** phenol

13.43 a. 2-chloro-4-methylcyclohexanol
b. methoxybenzene; methyl phenyl ether
c. 2-propanethiol
d. 2,4-dimethyl-2-pentanol
e. methoxypropane; methyl propyl ether
f. 3-methyl furan
g. 4-bromo-2-pentanol
h. *meta*-cresol; 3-methylphenol

13.45 a. [methylcyclopentanol structure with OH and CH₃]

b. [benzene ring with OH on top and Cl on bottom]

c. $H_3C-\overset{\overset{\displaystyle CH_3}{|}}{CH}-\overset{\overset{\displaystyle OH}{|}}{CH}-CH_2-CH_3$

d. [benzene ring with O—CH₂—CH₃]

e. $CH_3-CH_2-\overset{\overset{\displaystyle SH}{|}}{CH}-CH_2-CH_3$

f. [benzene ring with CH₃ and OH]

g. [benzene ring with OH on top, Br on right, Br on bottom]

13.47 $CH_3-CH_2-CH_2-CH_2-OH$

$$CH_3-\underset{\underset{\displaystyle CH_3}{|}}{CH}-CH_2-OH$$

$$CH_3-\underset{\underset{\displaystyle OH}{|}}{CH}-CH_2-CH_3$$

$$CH_2-\underset{\overset{\displaystyle OH}{|}}{\underset{\underset{\displaystyle CH_3}{|}}{C}}-CH_3$$

13.49 a. 1-propanol; hydrogen bonding
b. 1-propanol; hydrogen bonding
c. 1-butanol; higher molar mass

13.51 a. soluble; hydrogen bonding
b. soluble; hydrogen bonding
c. insoluble; long carbon chain diminishes effect of polar —OH on hydrogen bonding

13.53 a. $CH_3-CH=CH_2$ **b.** $CH_3-CH_2-\overset{\overset{\displaystyle O}{||}}{C}-H$

c. $CH_3-CH=CH-CH_3$ **d.** $CH_3-CH_2-\overset{\overset{\displaystyle O}{||}}{C}-CH_3$
e. $CH_3-CH_2-CH_2-O-CH_2-CH_2-CH_3$

f.   **g.**

13.55 a. $CH_3-CH_2-CH_2-OH \xrightarrow{H^+,\ heat}$

$CH_3-CH=CH_2 + HCl \longrightarrow CH_3-\overset{\overset{\displaystyle Cl}{|}}{CH}-CH_3$

b. $CH_3-\overset{\overset{\displaystyle OH}{|}}{\underset{\underset{\displaystyle CH_3}{|}}{C}}-CH_3 \xrightarrow{H^+,\ heat} CH_3-\underset{\underset{\displaystyle CH_3}{|}}{C}=CH_2 + H_2 \xrightarrow{Pt}$

$CH_3-\underset{\underset{\displaystyle CH_3}{|}}{CH}-CH_3$

c. $CH_3-CH_2-CH_2-OH \xrightarrow{H^+,\ heat}$

$CH_3-CH=CH_2 + H_2O \xrightarrow{H^+,\ heat} CH_3-\overset{\overset{\displaystyle OH}{|}}{CH}-CH_3$

$\xrightarrow{[O]} CH_3-\overset{\overset{\displaystyle O}{||}}{C}-CH_3$

13.57 a. cycloalkene, ketone, alcohol

13.59

$CH_2-CH_2-CH_2-CH_2-CH_2-CH_3$

13.61 a.

b. $CH_3-\underset{\underset{\displaystyle CH_3}{|}}{CH}-CH_2-CH_2-SH$

c.

13.63 $CH_3-\underset{\underset{\displaystyle CH_3}{|}}{CH}-\overset{\overset{\displaystyle O}{||}}{C}-H$ 2-methylpropanal

13.65

$CH_3-CH_2-CH_2-CH_2-CH_2-OH$ 1-pentanol

$CH_3-\overset{\overset{\displaystyle OH}{|}}{CH}-CH_2-CH_2-CH_3$ 2-pentanol

$CH_3-CH_2-\overset{\overset{\displaystyle OH}{|}}{CH}-CH_2-CH_3$ 3-pentanol

$HO-CH_2-\underset{\underset{\displaystyle CH_3}{|}}{CH}-CH_2-CH_3$ 2-methyl-1-butanol

$HO-CH_2-CH_2-\underset{\underset{\displaystyle CH_3}{|}}{CH}-CH_3$ 3-methyl-1-butanol

$CH_3-\overset{\overset{\displaystyle CH_3}{|}}{\underset{\underset{\displaystyle OH}{|}}{C}}-CH_2-CH_3$ 2-methyl-2-butanol

$CH_3-\overset{\overset{\displaystyle OH}{|}}{CH}-\overset{\overset{\displaystyle CH_3}{|}}{CH}-CH_3$ 3-methyl-2-butanol

$CH_3-\overset{\overset{\displaystyle CH_3}{|}}{\underset{\underset{\displaystyle CH_3}{|}}{C}}-CH_2-OH$ 2,2-dimethyl-1-propanol

14

Aldehydes, Ketones, and Chiral Molecules

"Dentures replace natural teeth that are extracted due to cavities, bad gums, or trauma," says Dr. Irene Hilton, dentist, La Clinica De La Raza. "I make an impression of teeth using alginate, which is a polysaccharide extracted from seaweed. I mix the compound with water and place the gel-like material in the patient's mouth, where it becomes a hard, cementlike substance. I fill this mold with gypsum ($CaSO_4$) and water, which form a solid to which I add teeth made of plastic or porcelain. When I get a good match to the patient's own teeth, I prepare a preliminary wax denture. This is placed in the patient's mouth to check the bite and adjust the position of the replacement teeth. Then a permanent denture is made using a hard plastic polymer (methyl methacrylate)."

LOOKING AHEAD

Mastering GOB CHEMISTRY

Visit **www.masteringgob.com** for self-study materials and instructor-assigned homework.

any of the odors that you associate with solvents, flavorings, paint removers, and perfumes are from organic molecules called aldehydes and ketones. In biology, you may have seen specimens preserved in a solution of formaldehyde. You may have noticed the odor of acetone if you used paint or polish remover. Foods and perfumes with the odors and flavors of vanilla, almond, and cinnamon contain aldehydes that occur in nature.

The feature responsible for this wide variety of flavors and odors is a carbon–oxygen double bond called a *carbonyl group* (C═O). In this chapter we will study two families with carbonyl groups: aldehydes and ketones. They are prevalent in nature and play an important role in biochemical pathways. In later chapters, we will see how the carbonyl group influences the structures of carbohydrates, proteins, and nucleic acids. Aldehydes and ketones are also important compounds in industry, providing the solvents and reactants that make up many common materials we use in our lives.

LEARNING GOAL

Identify compounds with the carbonyl group as aldehydes and ketones. Give the IUPAC and common names for aldehydes and ketones; draw their condensed structural formulas.

 SELF-STUDY ACTIVITY
Aldehydes and Ketones

FIGURE 14.1

The carbonyl group in aldehydes and ketones.

Q **If aldehydes and ketones both contain a carbonyl group, how can you differentiate between compounds from each family?**

14.1 Aldehydes and Ketones

In an **aldehyde**, the carbon of the carbonyl group is bonded to at least one hydrogen atom. That carbon may also be bonded to another hydrogen, a carbon of an alkyl group, or an aromatic ring. (See Figure 14.1.) In a **ketone**, the carbonyl group is bonded to two alkyl groups or aromatic rings.

Structure of the Carbonyl Group

The carbonyl group consists of a carbon–oxygen double bond with bonds at angles of 120° to two other atoms. The double bond in the carbonyl group is similar to that

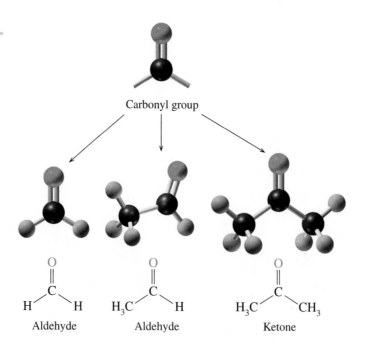

Carbonyl group

of alkenes, except the carbonyl group has a dipole. The oxygen atom with two lone pairs of electrons is much more electronegative than the carbon atom. Therefore, the carbonyl group has a strong dipole with a partial negative charge (δ^-) on the oxygen, and a partial positive charge (δ^+) on the carbon. The polarity of the carbonyl group strongly influences the physical and chemical properties of aldehydes and ketones.

MC GOB Aldehyde or Ketone? Naming Aldehydes and Ketones

There are several ways to write the structural formulas of aldehydes and ketones. In the condensed structural formula, the aldehyde group may be drawn as separate atoms or it may be written as —CHO, with the double bond understood. An aldehyde would not be written as —COH, which looks like a hydroxy group. The keto group (C=O) is sometimes written as CO. For convenience, the line-bond formulas are also used with the functional group atoms written separately. Isomers, which have the same molecular formula, can be written for aldehydes and ketones as follows:

Formulas for Isomers of C_3H_6O

Aldehyde

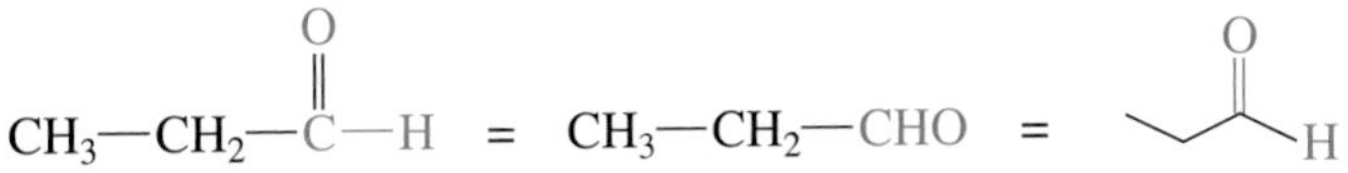

Ketone

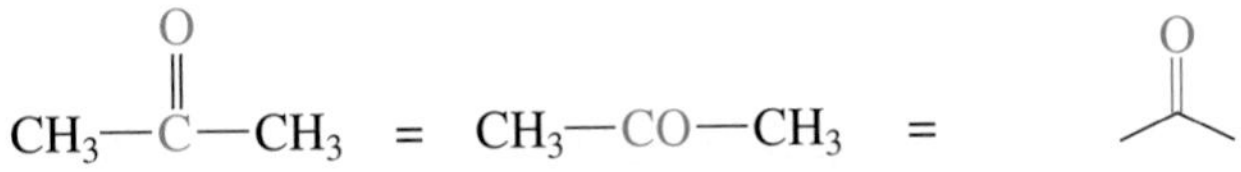

SAMPLE PROBLEM 14.1

Identifying Aldehydes and Ketones

Identify each of the following compounds as an aldehyde or ketone:

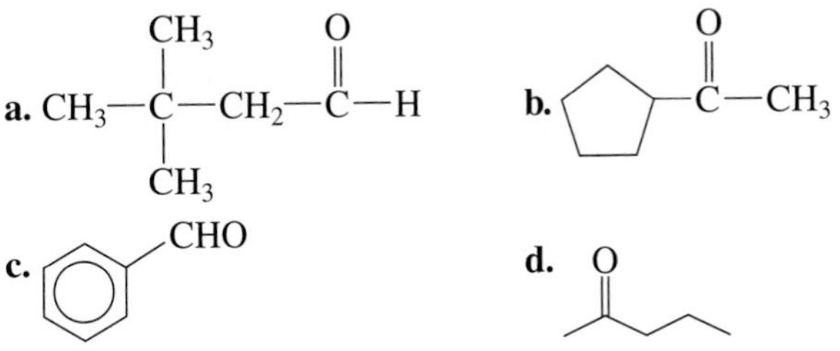

SOLUTION

a. aldehyde **b.** ketone **c.** aldehyde **d.** ketone

STUDY CHECK

Draw the condensed structural formula of a ketone that has a carbonyl group bonded to two ethyl groups.

Naming Aldehydes

In the IUPAC names of aldehydes, the *e* of the alkane name is replaced with *al*.

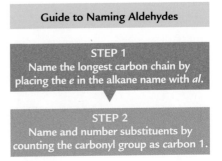

Guide to Naming Aldehydes

STEP 1
Name the longest carbon chain by placing the *e* in the alkane name with *al*.

STEP 2
Name and number substituents by counting the carbonyl group as carbon 1.

STEP 1 Name the longest carbon chain containing the carbonyl group by replacing the *e* in the corresponding alkane name with *al*. No

The carbonyl carbon is at the end of the chain

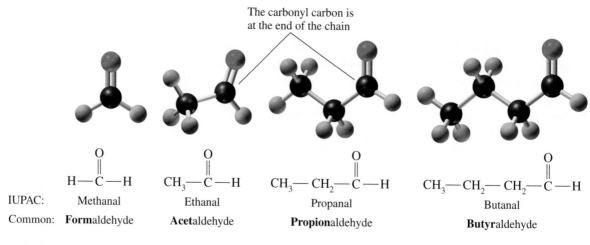

IUPAC: Methanal   Ethanal   Propanal   Butanal

Common: **Form**aldehyde   **Acet**aldehyde   **Propion**aldehyde   **Butyr**aldehyde

FIGURE 14.2

In the structures of aldehydes, the carbonyl group is always the end carbon.

Q **Why is the carbon in the carbonyl group in aldehydes always at the end of the chain?**

number is needed for the aldehyde group because it always appears at the end of the chain.

The IUPAC system names the aldehyde of benzene as benzaldehyde.

benzaldehyde

The first four unbranched aldehydes are often referred to by their common names, which end in *aldehyde*. (See Figure 14.2.) The roots of these common names are derived from Latin or Greek words that indicate the source of the corresponding carboxylic acid. We will look at carboxylic acids in an upcoming chapter.

■ STEP 2 **Name and number any substituents on the carbon chain by counting the carbonyl carbon as carbon 1.**

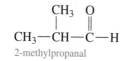

2-methylpropanal

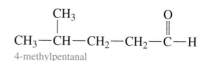

4-methylpentanal

SAMPLE PROBLEM 14.2

■**Naming Aldehydes**

Give the IUPAC names for the following aldehydes:

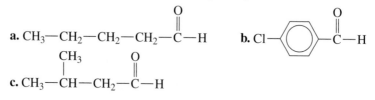

SOLUTION

a. pentanal

b. 4-chlorobenzaldehyde

c. The longest unbranched chain has four atoms with a methyl group on the third carbon. The IUPAC name is 3-methylbutanal.

STUDY CHECK

What are the IUPAC and common names of the aldehyde with three carbon atoms?

Naming Ketones

Aldehydes and ketones are some of the most important classes of organic compounds. Because they have played a major role in organic chemistry for more than a century, the common names for unbranched ketones are still in use. In the common

---

ENVIRONMENTAL NOTE

▪ Vanilla

Vanilla has been used as a flavoring for over a thousand years. After drinking a beverage made from powdered vanilla and cocoa beans with Emperor Montezuma in Mexico, Cortez took vanilla back to Europe where it became popular for flavoring and for scenting perfumes and tobacco. Thomas Jefferson introduced vanilla to the United States in the late 1700s. Today much of the vanilla we use in the world is grown in Mexico, Madagascar, Réunion, Seychelles, Tahiti, Ceylon, Java, the Philippines, and Africa.

The vanilla plant is a member of the orchid family. There are many species of *Vanilla*, but *Vanilla planifolia* (or *V. fragrans*) is considered to produce the best flavor. The vanilla plant grows like a vine and can grow to 100 feet in length. Its flowers are hand-pollinated to produce a green fruit that is picked in 8 or 9 months. It is sun-dried to form a long, dark brown pod, which is called "vanilla bean" because it looks like a string bean. The flavor and fragrance of the vanilla bean comes from the black seeds found inside the dried bean.

The seeds and pod are used to flavor desserts such as custards and ice cream. The extract of vanilla is made by chopping up vanilla beans and mixing them with a 35% alcohol–water mixture. The liquid, which contains the aldehyde vanillin, is drained from the bean residue and used for flavoring.

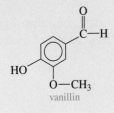
vanillin

names the alkyl groups bonded to the carbonyl group are named as substituents and listed alphabetically followed by *ketone*. Acetone, which is another name for propanone, has been retained by the IUPAC system.

In the IUPAC system, the name of a ketone is obtained by replacing the *e* in the corresponding alkane name with *one*.

■ **STEP 1** **Name the longest carbon chain containing the carbonyl group by replacing the *e* in the corresponding alkane name by *one*.**

■ **STEP 2** **Number the main chain starting from the end nearest the carbonyl group.** Place the number of the carbonyl carbon in front of the ketone name. (Propanone and butanone do not require numbers.)

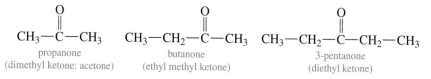

propanone
(dimethyl ketone: acetone)

butanone
(ethyl methyl ketone)

3-pentanone
(diethyl ketone)

■ **STEP 3** **Name and number any substituents on the carbon chain.**

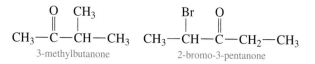

3-methylbutanone

2-bromo-3-pentanone

■ **STEP 4** For cyclic ketones, the prefix *cyclo* is used in front of the ketone name. Any substituents are located by numbering the ring starting with the carbonyl carbon as carbon 1. The ring is numbered so that the substituents have the lowest possible number.

cyclopentanone

3-methylcyclohexanone

2,3-dichlorocyclopentanone

SAMPLE PROBLEM 14.3

■ **Names of Ketones**

Give the IUPAC name for the following ketone:

$$CH_3-CH-CH_2-\overset{\overset{\displaystyle O}{\|}}{C}-CH_3$$
$$\overset{\displaystyle CH_3}{|}$$

SOLUTION

The longest chain is five carbon atoms. Counting from the right, the carbonyl group is on carbon 2 and a methyl group is on carbon 4. The IUPAC name is 4-methyl-2-pentanone.

STUDY CHECK

What is the common name of 3-hexanone?

Guide to Naming Ketones

STEP 1
Name the longest carbon chain by replacing the *e* in the alkane name by *one*.

STEP 2
Number the carbon chain starting from the end nearest the carbonyl group and indicate its location.

STEP 3
Name and number any substituents on other carbons in the chain.

STEP 4
For cyclic ketones, add the prefix *cyclo* and number substitutes from the carbonyl carbon as carbon 1.

HEALTH NOTE

Some Important Aldehydes and Ketones

Formaldehyde, the simplest aldehyde, is a colorless gas with a pungent odor. Industrially, it is a reactant in the synthesis of polymers used to make fabrics, insulation materials, carpeting, pressed wood products such as plywood, and plastics for kitchen counters. An aqueous solution called formalin, which contains 40% formaldehyde, is used as a germicide and to preserve biological specimens. Exposure to formaldehyde fumes can irritate eyes, nose, and upper respiratory tract and cause skin rashes, headaches, dizziness, and general fatigue.

Acetone, or propanone (dimethyl ketone), which is the simplest ketone, is a colorless liquid with a mild odor that has wide use as a solvent in cleaning fluids, paint and nail-polish removers, and rubber cement. It is extremely flammable and care must be taken when using acetone. In the body, acetone may be produced in uncontrolled diabetes, fasting, and high-protein diets when large amounts of fats are metabolized for energy.

Several naturally occurring aromatic aldehydes are used to flavor food and as fragrances in perfumes. Benzaldehyde is found in almonds, vanillin in vanilla beans, and cinnamaldehyde in cinnamon.

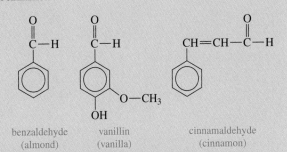

benzaldehyde (almond) vanillin (vanilla) cinnamaldehyde (cinnamon)

The flavor of butter or margarine is from butanedione, muscone is used to make musk perfumes, and oil of spearmint contains carvone.

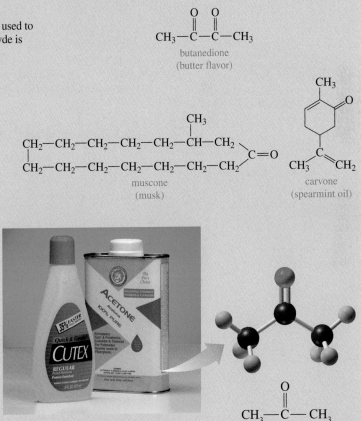

butanedione (butter flavor)

muscone (musk) carvone (spearmint oil)

$$CH_3-\overset{\displaystyle O}{\underset{\displaystyle \|}{C}}-\overset{\displaystyle O}{\underset{\displaystyle \|}{C}}-CH_3$$

butanedione

$$CH_3-\overset{\displaystyle O}{\underset{\displaystyle \|}{C}}-CH_3$$

propanone

QUESTIONS AND PROBLEMS

■ **Aldehydes and Ketones**

14.1 Identify the following compounds as aldehydes or ketones:

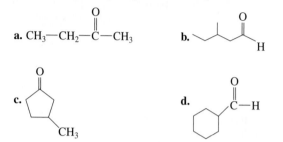

14.2 Identify the following compounds as aldehydes or ketones:

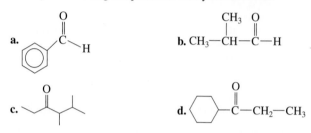

14.3 Indicate if each of the following pairs of formulas represent (1) isomers, (2) the same compound, or (3) different compounds.

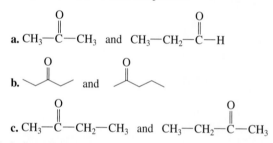

14.4 Indicate if each of the following pairs of formulas represent (1) isomers, (2) the same compound, or (3) different compounds.

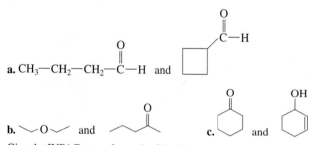

14.5 Give the IUPAC name for each of the following compounds:

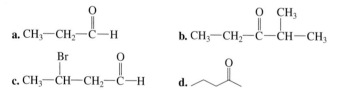

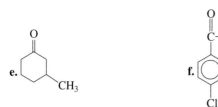

14.6 Give the IUPAC name for each of the following compounds:

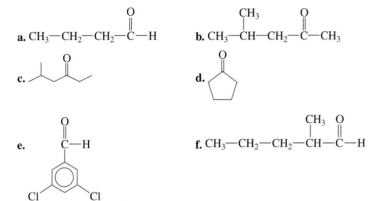

a. $CH_3-CH_2-CH_2-\overset{\overset{\displaystyle O}{\|}}{C}-H$

b. $CH_3-\overset{\overset{\displaystyle CH_3}{|}}{CH}-CH_2-\overset{\overset{\displaystyle O}{\|}}{C}-CH_3$

c.

d.

e.

f. $CH_3-CH_2-CH_2-\overset{\overset{\displaystyle CH_3}{|}}{CH}-\overset{\overset{\displaystyle O}{\|}}{C}-H$

14.7 Give a common name for each of the following compounds:

a. $CH_3-\overset{\overset{\displaystyle O}{\|}}{C}-H$

b. $CH_3-\overset{\overset{\displaystyle O}{\|}}{C}-CH_2-CH_2-CH_3$

c. $H-\overset{\overset{\displaystyle O}{\|}}{C}-H$

14.8 Give the common name for each of the following compounds:

a. $CH_3-\overset{\overset{\displaystyle O}{\|}}{C}-CH_2-CH_3$

b. $CH_3-CH_2-\overset{\overset{\displaystyle O}{\|}}{C}-CH_2-CH_3$

c. $CH_3-CH_2-\overset{\overset{\displaystyle O}{\|}}{C}-H$

14.9 Write the condensed structural formula for each of the following compounds:
  a. acetaldehyde   **b.** 4-methyl-2-pentanone   **c.** 2,3-dibromobutanal
  d. methyl butyl ketone   **e.** 3-methylpentanal

14.10 Write the condensed structural formula for each of the following compounds:
  a. propionaldehyde   **b.** butanal   **c.** 3,4-dichlorohexanal
  d. 4-bromobutanone   **e.** acetone

14.11 The IUPAC name of anisaldehyde used in perfumes is 4-methoxybenzaldehyde. What is the structural formula of anisaldehyde?

14.12 The IUPAC name of vanillin, a naturally occurring compound in vanilla beans, is 4-hydroxy-3-methoxybenzaldehyde. What is the structural formula of vanillin?

14.2 Physical Properties of Aldehydes and Ketones

LEARNING GOAL

Compare the boiling points and solubility of aldehydes and ketones to those of alkanes and alcohols.

At room temperature, formaldehyde (bp $-21°C$) and acetaldehyde (bp $21°C$) are gases. Aldehydes containing from 3 to 10 carbon atoms are liquids. The polar carbonyl group with a partially negative oxygen atom and a partially positive carbon atom has an influence on the boiling points and the solubility of aldehydes and ketones in water.

Properties of Aldehydes
and Ketones

Boiling Points

The polar carbonyl group gives aldehydes and ketones higher boiling points than alkanes and ethers of similar mass. The increase in boiling points is due to dipole–dipole interactions.

Dipole-dipole interaction

$$\underset{}{>}C^{\delta^+}{=}O^{\delta^-}\ \text{IIIIIIII}\ \underset{}{>}C^{\delta^+}{=}O^{\delta^-}\ \text{IIIIIIII}\ \underset{}{>}C^{\delta^+}{=}O^{\delta^-}$$

However, because there is no hydrogen on the oxygen atom, aldehydes and ketones cannot form hydrogen bonds with each other. Thus they have boiling points that are lower than alcohols.

	$CH_3{-}CH_2{-}CH_2{-}CH_3$	$CH_3{-}CH_2{-}O{-}CH_3$	$CH_3{-}CH_2{-}\overset{\overset{O}{\|}}{C}{-}H$	$CH_3{-}\overset{\overset{O}{\|}}{C}{-}CH_3$	$CH_3{-}CH_2{-}CH_2{-}OH$
Name	butane	ethyl methyl ether	propanal	propanone	1-propanol
Molar Mass	58	60	58	58	60
Family	alkane	ether	aldehyde	ketone	alcohol
bp	0°C	8°C	49°C	56°C	97°C

Increasing boiling point →

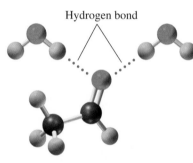

Acetaldehyde

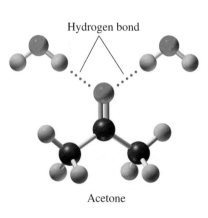

Acetone

FIGURE 14.3

Hydrogen bonding of acetaldehyde and acetone with water.

Q **Would you expect propanal to be soluble in water?**

Solubility of Aldehydes and Ketones in Water

Although aldehydes and ketones do not hydrogen bond with each other, the electronegative oxygen atom does hydrogen bond with water molecules. Carbonyl compounds with one to four carbons are very soluble in water. However, those with five carbon atoms or more are not very soluble because the alkyl portions diminish the effect of the polar carbonyl group. (See Figure 14.3.)

Table 14.1 compares the boiling points of some carbonyl compounds, as well as their solubilities in water.

TABLE 14.1

Comparison of Physical Properties of Some Selected Compounds

Compound	Boiling Point (°C)	Solubility in Water
methanal (formaldehyde)	−21	very soluble
ethanal (acetaldehyde)	21	very soluble
propanal (propionaldehyde)	49	soluble
propanone (acetone)	56	soluble
butanal (butyraldehyde)	75	soluble
butanone	80	soluble
pentanal	103	slightly soluble
2-pentanone	102	slightly soluble
3-pentanone	102	slightly soluble
hexanal	129	not soluble
2-hexanone	127	not soluble
3-hexanone	124	not soluble
acetophenone	202	not soluble

SAMPLE PROBLEM 14.4

■ Boiling Point and Solubility

Would you expect ethanol CH_3CH_2OH to have a higher or lower boiling point than ethanal CH_3CHO? Explain.

SOLUTION

Ethanol would have a higher boiling point because its molecules can hydrogen bond with each other, while molecules of ethanal cannot.

STUDY CHECK

If acetone molecules cannot hydrogen bond with each other, why is acetone soluble in water?

QUESTIONS AND PROBLEMS

■ Physical Properties of Aldehydes and Ketones

14.13 Which compound in each of the following pairs would have the higher boiling point? Explain.

a. CH_3—CH_2—CH_3 or CH_3—$\overset{\overset{\displaystyle O}{\|}}{C}$—H

b. propanal or pentanal   **c.** butanal or 1-butanol

14.14 Which compound in each of the following pairs would have the higher boiling point? Explain.

a.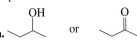

b. pentane or butanone   **c.** propanone or pentanone

14.15 Which compound in each of the following pairs would be more soluble in water? Explain.

a. CH_3—$\overset{\overset{\displaystyle O}{\|}}{C}$—$CH_2$—$CH_3$ or CH_3—$\overset{\overset{\displaystyle O}{\|}}{C}$—$\overset{\overset{\displaystyle O}{\|}}{C}$—$CH_3$

b. propane or acetaldehyde   **c.** acetone or 2-pentanone

14.16 Which compound in each of the following pairs would be more soluble in water? Explain.
a. CH_3—CH_2—CH_3 or CH_3—CH_2—CHO
b. propanone or 3-hexanone
c. propane or propanone

14.17 Would you expect an aldehyde with a formula of $C_8H_{16}O$ to be soluble in water? Explain.

14.18 Would you expect an aldehyde with a formula of C_3H_6O to be soluble in water? Explain.

■ 14.3 Oxidation and Reduction of Aldehydes and Ketones

LEARNING GOAL

Draw the structural formulas of reactants and products for the oxidation or reduction of aldehydes and ketones.

In Chapter 13, we saw that aldehydes produced by the oxidation of primary alcohols oxidize readily to carboxylic acids. In fact they oxidize so easily that even the aldehydes exposed to the air in the laboratory quickly form carboxylic acids. In

Oxidation-Reduction Reactions
of Aldehydes and Ketones

contrast, ketones produced by the oxidation of secondary alcohols do not undergo further oxidation. Let's review examples of the oxidation reactions of primary and secondary alcohols that form aldehydes and ketones.

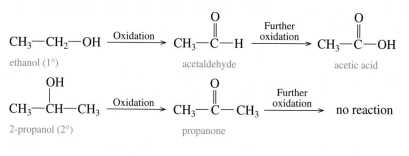

Tollens' Test

The ease of oxidation of aldehydes allows certain mild oxidizing agents to oxidize the aldehyde functional group without oxidizing other functional groups such as alcohols or ethers. In the laboratory, **Tollens' test** may be used to distinguish between an aldehyde and ketone. Tollens' reagent, a solution of Ag^+ ($AgNO_3$) and ammonia, oxidizes aldehydes, but not ketones. The silver ion is reduced to metallic silver, which forms a layer called a "silver mirror" on the inside of the container. Commercially, a similar process is used to make mirrors by applying a mixture of $AgNO_3$ and ammonia on glass with a spray gun. (See Figure 14.4.)

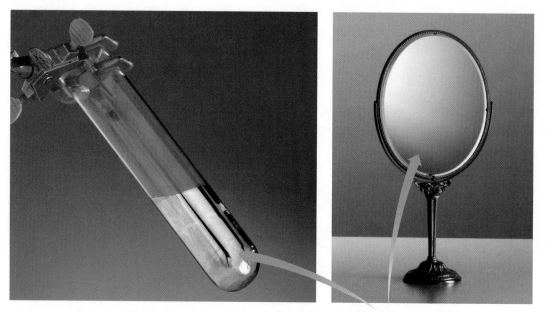

$$Ag^+ + 1e^- \longrightarrow Ag(s)$$

FIGURE 14.4

In Tollens' test, a silver mirror forms when the oxidation of an aldehyde reduces silver ion to metallic silver. The silvery surface of a mirror is formed in a similar way.

Q **What is the product of the oxidation of an aldehyde?**

Another test, called **Benedict's test**, gives a positive test with compounds that have an aldehyde functional group and an adjacent hydroxyl group. When Benedict's reagent containing Cu^{2+} ($CuSO_4$) ions is added to this type of aldehyde, a brick-red solid of Cu_2O forms. (See Figure 14.5.) The test is negative with simple aldehydes and ketones.

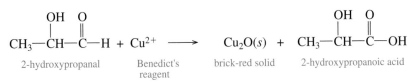

CH₃—CH—C—H + Cu²⁺ ⟶ Cu₂O(s) + CH₃—CH—C—OH

2-hydroxypropanal   Benedict's reagent   brick-red solid   2-hydroxypropanoic acid

Because many sugars such as glucose contain this type of aldehyde grouping, Benedict's reagent can be used to determine the presence of glucose in blood or urine.

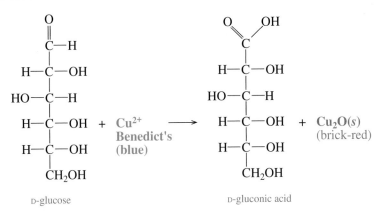

D-glucose   D-gluconic acid

Cu²⁺   Cu₂O(s)

FIGURE 14.5

The blue Cu^{2+} in Benedict's solution forms a brick-red solid of Cu_2O in a positive test for many sugars and aldehydes with adjacent hydroxyl groups.

Q **Which test tube contains an aldehyde with an adjacent hydroxyl group?**

SAMPLE PROBLEM 14.5

■ Alcohol Oxidation

Draw the condensed structural formula of the alcohol needed to give each of the following oxidation products:

a. CH₃—C—CH₂—CH₃   **b.** CH₃—CH—C—H   **c.** CH₃—C—OH

SOLUTION

a. A secondary alcohol oxidizes to a ketone.

CH₃—CH—CH₂—CH₃ —[O]→ CH₃—C—CH₂—CH₃

b. A primary alcohol oxidizes to an aldehyde with a mild oxidizing agent.

CH₃—CH—CH₂—OH —[O]→ CH₃—CH—C—H

c. A primary alcohol oxidizes to an aldehyde, which oxidizes further to a carboxylic acid.

CH₃—CH₂—OH —[O]→ CH₃—C—H —[O]→ CH₃—C—OH

STUDY CHECK

What is the IUPAC name of the alcohol that oxidized to cyclohexanone?

SAMPLE PROBLEM 14.6

■ Tollens' Test

Draw the condensed structural formula of the product of oxidation, if any, when Tollens' reagent is added to each of the following compounds:

a. propanal   **b.** propanone   **c.** 2-methylbutanal

SOLUTION

Tollens' reagent will oxidize aldehydes, but not ketones.

$$
\textbf{a. } CH_3-CH_2-\overset{\overset{\displaystyle O}{\|}}{C}-OH
\qquad \textbf{b. } \text{no reaction}
\qquad \textbf{c. } CH_3-CH_2-\overset{\overset{\displaystyle CH_3}{|}}{CH}-\overset{\overset{\displaystyle O}{\|}}{C}-OH
$$

STUDY CHECK

Why does a silver mirror form when Tollens' reagent is added to a test tube containing benzaldehyde?

Reduction of Aldehydes and Ketones

Aldehydes and ketones are reduced by sodium borohydride ($NaBH_4$) or hydrogen (H_2). **Reduction** decreases the number of carbon–oxygen bonds by the addition of hydrogen or the loss of oxygen. Aldehydes are reduced to primary alcohols, and ketones to secondary alcohols. A catalyst such as nickel, platinum, or palladium is used with hydrogenation.

Aldehydes Reduce to Primary Alcohols

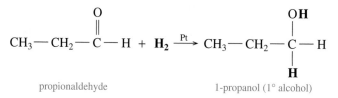

propionaldehyde   1-propanol (1° alcohol)

Ketones Reduce to Secondary Alcohols

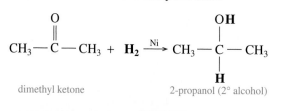

dimethyl ketone   2-propanol (2° alcohol)

SAMPLE PROBLEM 14.7

■ Reduction of Carbonyl Groups

Write an equation for the reduction of cyclopentanone in the presence of a nickel catalyst.

SOLUTION

The reacting molecule is a cyclic ketone that has five carbon atoms. Hydrogen atoms will add to the carbon and oxygen in the carbonyl group to form the corresponding secondary alcohol.

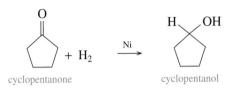

cyclopentanone + H$_2$ $\xrightarrow{\text{Ni}}$ cyclopentanol

STUDY CHECK

What is the name of the product obtained from the hydrogenation of propionaldehyde?

QUESTIONS AND PROBLEMS

MC GOB

▪ Oxidation and Reduction of Aldehydes and Ketones

14.19 Draw the condensed structural formula of the alcohol needed to give each of the following oxidation products:
- **a.** formaldehyde
- **b.** cyclopentanone
- **c.** 2-butanone
- **d.** benzaldehyde
- **e.** 3-methylcyclohexanone

14.20 Draw the condensed structural formula of the alcohol needed to give each of the following oxidation products:
- **a.** acetaldehyde
- **b.** 3-methylbutanone
- **c.** cyclohexanone
- **d.** propionaldehyde
- **e.** 3-methylbutanal

14.21 Draw the condensed structural formula of the organic product when each of the following alcohols is oxidized [O] (if no reaction, write *none*):

a. CH$_3$—CH$_2$—CH$_2$—CH$_2$—CH$_2$—OH

b. 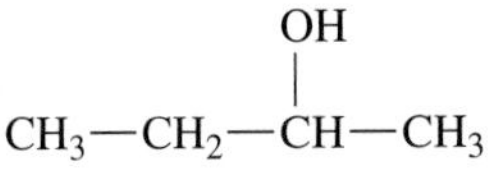 CH$_3$—CH$_2$—CH(OH)—CH$_3$

c. 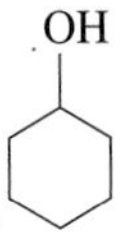 (cyclohexane with OH)

d. 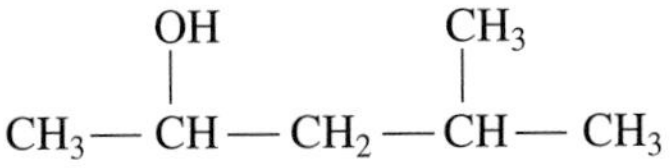 CH$_3$—CH(OH)—CH$_2$—CH(CH$_3$)—CH$_3$

e. 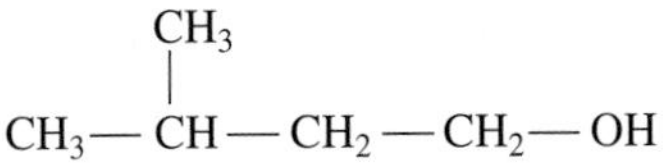 CH$_3$—CH(CH$_3$)—CH$_2$—CH$_2$—OH

14.22 Draw the condensed structural formula of the organic product when each of the following alcohols is oxidized [O] (if no reaction, write *none*):

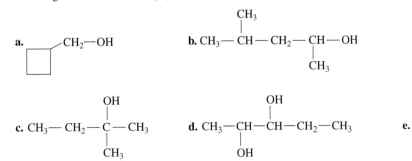

a. (cyclobutane)—CH$_2$—OH

b. CH$_3$—CH(CH$_3$)—CH$_2$—CH(OH)—CH$_3$

c. CH$_3$—CH$_2$—C(OH)(CH$_3$)—CH$_3$

d. CH$_3$—CH(OH)—CH(OH)—CH$_2$—CH$_3$

e. 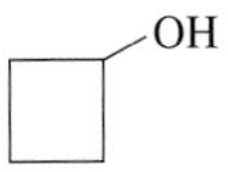 (cyclobutane)—OH

14.23 Give the condensed structural formula of the organic product formed when each of the following is reduced by hydrogen in the presence of a nickel catalyst:
a. butyraldehyde   **b.** acetone   **c.** 3-bromohexanal   **d.** 2-methyl-3-pentanone

14.24 Give the condensed structural formula of the organic product formed when each of the following is reduced by hydrogen in the presence of a nickel catalyst:
a. ethyl propyl ketone   **b.** formaldehyde
c. 3-chlorocyclopentanone   **d.** 2-pentanone

LEARNING GOAL

Write the products of the addition of alcohols to aldehydes and ketones.

Addition of Polar Molecules to a Carbonyl Group

14.4 Addition Reactions of Aldehydes and Ketones

One of the most common reactions of aldehydes and ketones is the addition of polar molecules to the carbonyl group. The carbonyl group is reactive because of the polarity of the C=O double bond. In addition reactions, the partially negative part of the adding molecule bonds with the partially positively charged carbonyl carbon. The partially positive part, usually a proton, combines with the partially negatively charged carbonyl oxygen. This type of addition to the carbonyl group can be illustrated as follows:

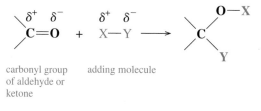

carbonyl group   adding molecule
of aldehyde or
ketone

In general, aldehydes are more reactive than ketones because the carbonyl carbon is more positive in aldehydes. Also, the presence of two alkyl groups makes it more difficult for a molecule to form bonds with the carbon in the carbonyl group.

Addition of Water

The components of water add to aldehydes and ketones to give carbonyl hydrates in the presence of acid or base. The negative —OH group bonds with the carbonyl carbon, while the —H bonds to the negative oxygen. In water, the simplest aldehyde, formaldehyde, forms its hydrate called formalin, which is used to preserve tissues. Other aldehydes form hydrates in water as well, but not with as high a percentage as formaldehyde. The carbonyl group in ketones also reacts with water, but their hydrates are not very stable.

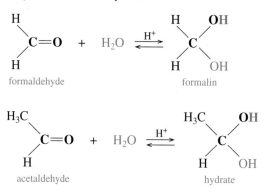

Chloral, which is an aldehyde with chlorine atoms, forms a hydrate known as chloral hydrate, the substance in "knock out" drops.

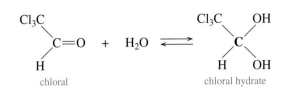

chloral   chloral hydrate

Acetal Formation

Similar to the addition of water to form hydrates, aldehydes and ketones react with alcohols in the presence of an acid catalyst to form **acetals**. (Ketal is an older term previously used for acetals from ketones.) In the acetal product, the two —OR groups are added to the carbonyl carbon and a molecule of water is eliminated.

MC GOB   Formation of Acetals

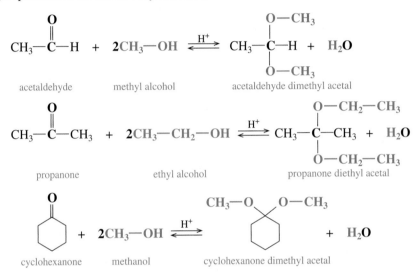

Hemiacetal Intermediate

In the process of forming acetals, an intermediate called a **hemiacetal** forms when one of the two alcohol molecules adds to the carbonyl carbon. The term *hemi* indicates that the hemiacetal is halfway to an acetal. Most of the hemiacetal intermediates are unstable and difficult to isolate from the reaction mixture. In the next step, the second alcohol is added to produce the more stable acetal. Acetals are stable and can be isolated from the reaction mixture.

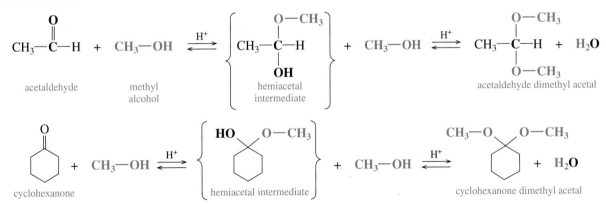

Both the step to the hemiacetal and the step to the acetal are reversible. The forward reaction to form the acetal is favored by removing water from the reaction mixture. The reverse reaction, which is the hydrolysis of an acetal, is favored by adding water to drive the equilibrium back to the ketone or aldehyde.

SAMPLE PROBLEM 14.8

■ **Acetals**

Write the structural formula of the hemiacetal and acetal products when methanol adds to propionaldehyde.

SOLUTION

To form the hemiacetal, the hydrogen from the alcohol adds to the oxygen of the carbonyl group to form a new hydroxyl group and the remaining part of the alcohol adds to the carbon atom in the carbonyl group. The acetal forms when a second molecule of methanol is added to the carbonyl carbon atom.

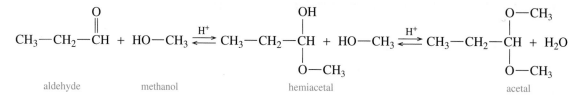

aldehyde   methanol   hemiacetal   acetal

STUDY CHECK

What is the structural formula of the acetal produced when methanol adds to propanone?

Cyclic Hemiacetals

One very important type of hemiacetal that can be isolated is a cyclic hemiacetal that forms when the carbonyl group and the —OH group are in the *same* molecule.

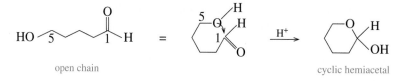

open chain   cyclic hemiacetal

The five- and six-atom cyclic hemiacetals and acetals are more stable than their open-chain structures. For example, glucose, a simple sugar, forms a hemiacetal when the hydroxyl group on carbon 5 bonds with the carbonyl group. The hemiacetal of glucose is so stable that almost all the glucose (99%) exists as the hemiacetal in aqueous solution. We will discuss carbohydrates and their structures in Chapter 15.

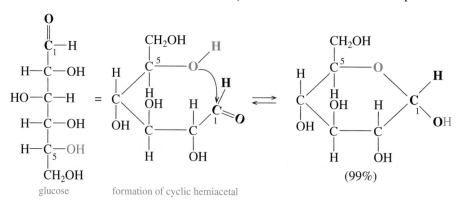

glucose   formation of cyclic hemiacetal

An alcohol can add to the cyclic hemiacetal to form a cyclic acetal. This reaction is also very important in carbohydrate chemistry. It is the linkage used by glucose molecules to bond to other glucose molecules to form long chains.

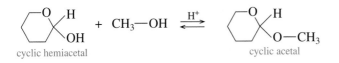

cyclic hemiacetal   cyclic acetal

QUESTIONS AND PROBLEMS

■ Addition Reactions of Aldehydes and Ketones

14.25 Write the structural formula of the organic product formed by the addition of water to each of the following:
    a. acetaldehyde   **b.** formaldehyde

14.26 Write the structural formula of the organic product formed by the addition of water to each of the following:
    a. propanal   **b.** propanone

14.27 Indicate whether each of the following structural formulas is a hemiacetal, acetal, or neither.

    a. $CH_3-CH_2-O-CH_2-OH$ **b.** $CH_3-CH_2-CH_2-\underset{\underset{OH}{|}}{\overset{\overset{O-CH_3}{|}}{CH}}$

    c. $CH_3-\underset{\underset{O-CH_2-CH_3}{|}}{\overset{\overset{O-CH_2-CH_3}{|}}{C}}-CH_2-CH_3$ **d.** (cyclohexane ring with OH and $O-CH_2-CH_3$)

    e. (cyclopentane ring with CH_3-O and $O-CH_3$)

14.28 Indicate whether each of the following structural formulas is a hemiacetal, acetal, or neither.
    a. $CH_3-CH_2-O-CH_2-CH_3$
    b. $HO-CH_2-CH_2-O-CH_2-CH_2-O-CH_3$

    c. $CH_3-\underset{\underset{OH}{|}}{\overset{\overset{O-CH_2-CH_3}{|}}{C}}-CH_3$ **d.** (cyclohexane ring with $O-CH_3$ and $O-CH_3$)   **e.** (cyclopentane ring with $O-CH_3$ and $O-CH_3$)

14.29 Draw the structural formula of the hemiacetal formed by adding one methanol to each of the following compounds.
    a. ethanal   **b.** propanone
    c. cyclopentanone   **d.** butanal

14.30 Draw the structural formula of the hemiacetal formed by adding one ethanol to each of the following compounds.
    a. propanal   **b.** 2-butanone
    c. cyclohexanone   **d.** formaldehyde

14.31 Draw the structural formulas of the acetal formed by adding a second methanol to the compounds in problem 14.29.

14.32 Draw the structural formulas of the acetal formed by adding a second ethanol to the compounds in problem 14.30.

LEARNING GOAL

Identify chiral and achiral carbon atoms in an organic molecule.

14.5 Chiral Molecules

In the preceding chapters, we have looked at some types of isomers. Let's review those now. Molecules are structural isomers when they have the same molecular formula, but different bonding arrangements.

Isomers

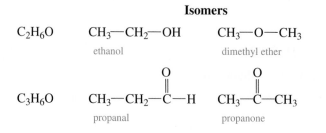

Another group of isomers called **stereoisomers** have identical molecular formulas, too, but they are not structural isomers. In stereoisomers the atoms are bonded in the same sequence, but differ in the way they are arranged in space.

Chirality

When stereoisomers have mirror images that are different, they are said to have "handedness." If you look at the palms of your hands, your thumbs are on opposite sides. If you turn your palms toward each other, you have mirror images. (See Figure 14.6.) The left hand is the mirror image of the right hand. Your hands are not superimposable. Objects such as hands that have nonsuperimposable mirror images are **chiral**. Left and right shoes are chiral; left- and right-handed golf clubs are chiral. When one mirror image can be superimposed on the other, the object is **achiral**. The object has no "handedness."

Molecules in nature also have mirror images, and often the "left-handed" stereoisomer has a different biological effect than does the "right-handed" one. For some compounds, one isomer has a certain odor, and the mirror image has a completely different odor. For example, one enantiomer of limonene smells like oranges, while its mirror image has the odor of lemons. When we think of how difficult it is to put a left-hand glove on our right hand, or a right shoe on

Limonene (oranges)

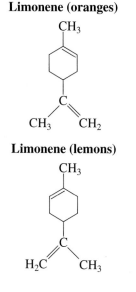

Limonene (lemons)

FIGURE 14.6

A pair of hands are chiral because they have mirror images that cannot be superimposed on each other.

Q Why are your shoes chiral objects?

Left hand

Right hand

Mirror image of right hand

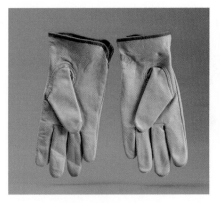

Chiral

Golf club, chiral

Achiral

Achiral

Chiral

Right-handed scissors, chiral

FIGURE 14.7

Everyday objects can be chiral or achiral.

Q **Why are some of the above objects chiral and others are achiral?**

our left foot, or use left-handed scissors if we are right handed, we begin to realize that certain properties of mirror images can be very different. (See Figure 14.7.)

SAMPLE PROBLEM 14.9

■ **Chiral Objects**

Classify each of the following objects as chiral or achiral:

a. left ear   **b.** clear drinking glass   **c.** a glove

SOLUTION

a. Chiral; the left ear cannot be superimposed on the right ear.
b. Achiral; mirror images of a clear drinking glass can be superimposed on each other.
c. Chiral; there is a right glove for the right hand and a left glove for the left hand.

STUDY CHECK

Would a bowling pin be chiral or achiral?

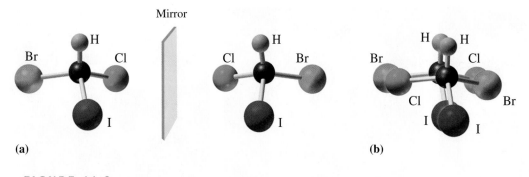

(a) (b)

FIGURE 14.8

(a) The enantiomers of a chiral molecule are mirror images. (b) The enantiomers of a chiral molecule cannot be superimposed on each other.

Q **Why is the carbon atom in this compound a chiral carbon?**

Chiral Carbon Atoms

Chiral Carbon Atoms

A carbon compound is chiral if it has at least one carbon atom bonded to four different atoms or groups. This type of carbon atom is called a **chiral carbon** because there are two different ways that it can bond to four atoms or groups of atoms. The resulting structures are mirror images of each other. Let's look at the mirror images of a carbon bonded to four different atoms. (See Figure 14.8.) If we line up the hydrogen and iodine atoms in the mirror images, the bromine and chlorine atoms appear on opposite sides. No matter how we turn the models, we cannot align all four atoms at the same time. When stereoisomers cannot be superimposed, they are called **enantiomers**. If two or more atoms are the same, the atoms can be aligned (superimposed) and the mirror images represent the same structure. (See Figure 14.9.)

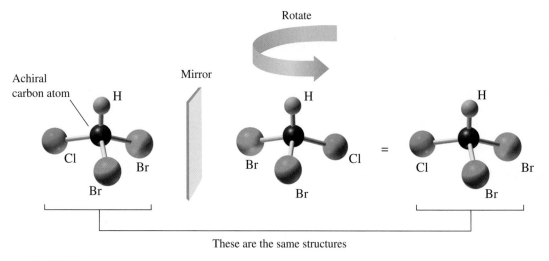

These are the same structures

FIGURE 14.9

The mirror images of an achiral compound can be superimposed on each other.

Q **Why can the mirror images of the compound be superimposed?**

SAMPLE PROBLEM 14.10

■ **Chiral Carbons**

Indicate whether the carbon in red is chiral or not chiral.

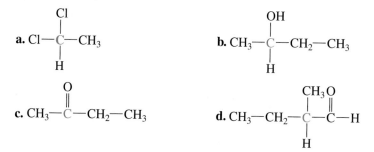

a. Cl—C—CH$_3$ (with Cl above and H below the central carbon)

b. CH$_3$—C—CH$_2$—CH$_3$ (with OH above and H below the central carbon)

c. CH$_3$—C—CH$_2$—CH$_3$ (with O double-bonded above the second carbon)

d. CH$_3$—CH$_2$—C—C—H (with CH$_3$ above and H below the third carbon, and O double-bonded above the fourth carbon)

SOLUTION

a. Not chiral. Two of the substituents on the carbon are the same (Cl). A chiral carbon must be bonded to four different groups or atoms.

b. Chiral. Carbon 2 is bonded to four different groups: one OH, one CH$_3$, one CH$_2$—CH$_3$, and one H.

c. Not chiral. Carbon 2 is bonded to only three groups, not four.

d. Chiral. Carbon 2 is bonded to four different groups: one H, one CH$_3$, one CH$_2$—CH$_3$, and one CHO.

STUDY CHECK

Circle the two chiral carbons in the structural formula of the carbohydrate erythrose.

HO—CH$_2$—CH—CH—C—H (with OH above first CH, OH above second CH, and O double-bonded above the C)

Erythrose

Drawing Fischer Projections

Emil Fischer devised a simplified system for drawing stereoisomers that shows the arrangements of the atoms. Fischer received the Nobel Prize in 1902 for his contributions to carbohydrate and protein chemistry. Using his method called a **Fischer projection**, the bonds to a chiral atom are drawn as intersecting lines with the chiral carbon being at the center where the lines cross. The horizontal lines represent the bonds that come forward in the three-dimensional structure, and the vertical lines represent the bonds that point away.

By convention the carbon chain in the Fischer projection is written vertically with the most highly oxidized carbon at the top. For glyceraldehyde, the carbonyl group, which is the most highly oxidized group in the molecule, is written at the top. The letter L is assigned to the left-handed stereoisomer, which has the —OH group on the left of the chiral carbon. The letter D is assigned to the right-handed structure where the —OH is on the right of the chiral carbon. Let's look at how glyceraldehyde, the simplest sugar, is converted from a three-dimensional view to a Fischer projection. (See Figure 14.10.)

MC GOB Drawing Fischer Projections

FIGURE 14.10

In a Fischer projection, the chiral carbon atom is at the center with horizontal lines for bonds that extend toward the viewer and vertical lines for bonds that point away.

Q Why does glyceraldehyde have only one chiral carbon atom?

Dash-wedge structures of glyceraldehyde

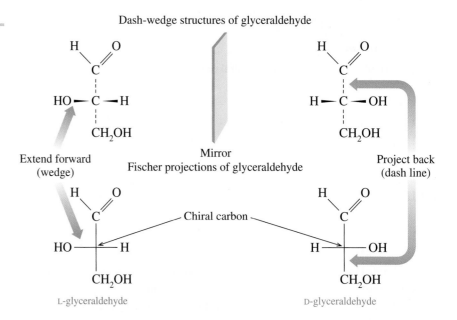

Fischer projections can also be written for larger compounds that have two or more chiral carbons. For example, the mirror images below are different. Each chiral atom is bonded to four different groups. To draw the mirror image of a Fischer projection, the positions of the substituents on the horizontal lines are reversed, while the groups on the vertical line are left unchanged.

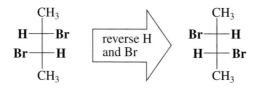

We can also draw the mirror image of the carbohydrate erythrose, which has two chiral carbons.

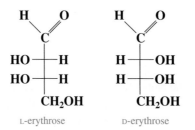

SAMPLE PROBLEM 14.11

■ Fischer Projections

Determine if each Fischer projection is a chiral compound. If so, identify it as the D or L isomer and draw the mirror image.

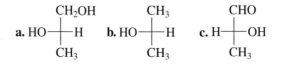

SOLUTION

a. This chiral compound with four different substituents is the L isomer. The mirror image is written by reversing the H and OH on the horizontal lines.

$$
\begin{array}{c}
\text{CH}_2\text{OH} \\
\text{H} \!-\!\!\!\!\begin{array}{c}|\\\text{ }\end{array}\!\!\!\!-\text{OH} \\
\text{CH}_3
\end{array}
$$

b. The compound is achiral because it has two identical groups (CH_3).

c. This chiral compound with four different substituents is the D isomer. The mirror image is written by reversing the H and OH on the horizontal lines.

$$
\begin{array}{c}
\text{CHO} \\
\text{HO} \!-\!\!\!\!\begin{array}{c}|\\\text{ }\end{array}\!\!\!\!-\text{H} \\
\text{CH}_3
\end{array}
$$

STUDY CHECK

Draw the Fischer projections for the D and L stereoisomers of 2-hydroxypropanal.

QUESTIONS AND PROBLEMS

MC
GOB

■ Chiral Molecules

14.33 Identify each of the following structures as chiral or achiral. If chiral, indicate the chiral carbon.

$$
\textbf{a.}\ \text{CH}_3\!-\!\overset{\displaystyle \text{OH}}{\underset{\displaystyle |}{\text{CH}}}\!-\!\text{CH}_3 \qquad \textbf{b.}\ \text{CH}_3\!-\!\overset{\displaystyle \text{Br}}{\underset{\displaystyle |}{\text{CH}}}\!-\!\text{CH}_2\!-\!\text{CH}_3
$$

$$
\textbf{c.}\ \text{CH}_3\!-\!\overset{\displaystyle \text{OH}}{\underset{\displaystyle |}{\text{CH}}}\!-\!\text{CH}_3 \qquad \textbf{d.}\ \text{CH}_3\!-\!\text{CH}_2\!-\!\overset{\displaystyle \text{O}}{\overset{\displaystyle \|}{\text{C}}}\!-\!\text{CH}_3
$$

14.34 Identify each of the following structures as chiral or achiral. If chiral, indicate the chiral carbon.

$$
\textbf{a.}\ \text{CH}_3\!-\!\overset{\displaystyle \text{Cl}}{\underset{\displaystyle \underset{\displaystyle \text{CH}_3}{|}}{\overset{\displaystyle |}{\text{C}}}}\!-\!\text{CH}_2\!-\!\overset{\displaystyle \text{Cl}}{\underset{\displaystyle |}{\text{CH}}}\!-\!\text{CH}_3 \qquad \textbf{b.}\ \text{CH}_3\!-\!\overset{\displaystyle \text{Br}}{\underset{\displaystyle |}{\text{C}}}\!\!=\!\!\text{CH}\!-\!\text{CH}_3
$$

$$
\textbf{c.}\ \text{CH}_3\!-\!\overset{\displaystyle \text{OH}}{\underset{\displaystyle \underset{\displaystyle \text{CH}_3}{|}}{\overset{\displaystyle |}{\text{C}}}}\!-\!\overset{\displaystyle \text{OH}}{\underset{\displaystyle |}{\text{CH}}}\!-\!\text{CH}_3 \qquad \textbf{d.}\ \text{Br}\!-\!\text{CH}_2\!-\!\overset{\displaystyle \text{Cl}}{\underset{\displaystyle |}{\text{CH}}}\!-\!\text{CH}_3
$$

14.35 Identify the chiral carbon in each of the following naturally occurring compounds.
a. citronellol; one enantiomer has the geranium odor.

$$
\text{CH}_3\!-\!\overset{\displaystyle \text{CH}_3}{\underset{\displaystyle |}{\text{C}}}\!\!=\!\!\text{CH}\!-\!\text{CH}_2\!-\!\text{CH}_2\!-\!\overset{\displaystyle \text{CH}_3}{\underset{\displaystyle |}{\text{CH}}}\!-\!\text{CH}_2\!-\!\text{CH}_2\!-\!\text{OH}
$$

b. alanine, amino acid

$$
\text{H}_2\text{N}\!-\!\overset{\displaystyle \text{CH}_3}{\underset{\displaystyle |}{\text{CH}}}\!-\!\overset{\displaystyle \text{O}}{\overset{\displaystyle \|}{\text{C}}}\!-\!\text{OH}
$$

▪ Enantiomers in Biological Systems

Most stereoisomers that are active in biological systems consist of only one enantiomer. Rarely are both enantiomers of biological molecules active. This happens because the enzymes and cell surface receptors on which metabolic reactions take place also have "handedness." Thus, only one of the enantiomers of a reactant or a drug interacts with its enzymes or receptors; the other is inactive. At the target site, the chiral receptor fits the arrangement of the substituents in only one enantiomer. Its mirror image, which is inactive, does not fit properly. This is similar to the idea that your right hand will only fit into a right-handed glove. (See Figure 14.11.)

A substance called carvone exists as two enantiomers. One enantiomer gives the odor of spearmint oil, while the other enantiomer produces the flavor of caraway seeds. Such differences in odor are due to the chiral receptor sites in the nose that fit the shape of one enantiomer, but not the other. Thus, our senses of smell and also taste are sensitive to the chirality of molecules.

ENANTIOMERS OF CARVONE

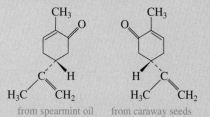

from spearmint oil   from caraway seeds

In the brain, one stereoisomer of LSD causes hallucinations because it affects the production of serotonin, a chemical that is important in sensory perception. However, its enantiomer produces little effect in the brain. The behavior of nicotine and epinephrine (adrenaline) also depends

upon one of their enantiomers. One enantiomer of nicotine is more toxic than the other. In epinephrine (adrenaline), one enantiomer is responsible for the constriction of blood vessels.

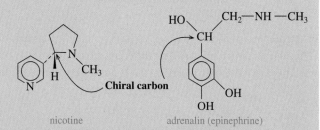

nicotine   adrenalin (epinephrine)

A substance used to treat Parkinson's disease is L-dopa, which is converted to dopamine in the brain, where it raises the serotonin level. However, the D-dopa enantiomer has no biological effect.

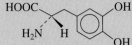 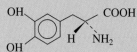

L-dopa, anti-Parkinsonian drug   D-dopa has no biological effect

For many drugs, only one of the enantiomers is biologically active. However, for many years, drugs have been produced that were mixtures of their enantiomers. Today, drug researchers are using *chiral technology* to produce the active enantiomers of chiral drugs. Chiral catalysts are being designed that direct the formation of just one enantiomer rather than both. The benefits of producing only the active enantiomer include using a lower dose, enhancing activity, reducing interactions with other drugs, and eliminating possible harmful side effects from the nonactive enantiomer. Several active enantiomers are now being produced such as L-dopa and the active enantiomer of the popular analgesic ibuprofen used in Advil, Motrin, and Nuprin.

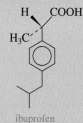

ibuprofen

FIGURE 14.11

(a) The substituents on the biologically active enantiomer bind to all the sites on a chiral receptor; (b) its enantiomer does not bind properly and is not active biologically.

Q Why don't all the substituents of the mirror image of the active enantiomer fit into a chiral receptor site?

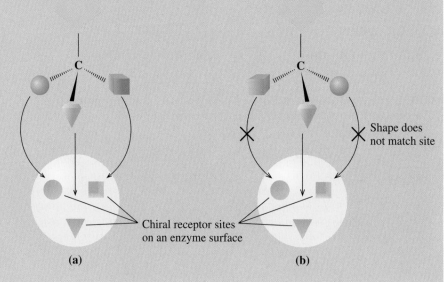

Chiral receptor sites on an enzyme surface

Shape does not match site

(a)   (b)

14.36 Identify the chiral carbon in each of the following naturally occurring compounds.

   a. amphetamine (Benzedrine), stimulant, treatment of hyperactivity

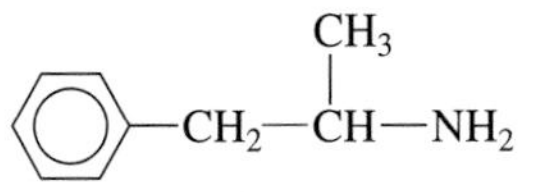

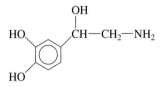

   b. Norepinephrine, increases blood pressure and nerve transmission

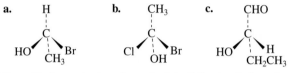

14.37 Draw Fischer projections for each of the following dash-wedge structures.

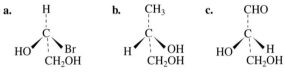

14.38 Draw Fischer projections for each of the following dash-wedge structures.

a.

$$\begin{array}{c} H \\ | \\ C \\ HO \diagup \quad \diagdown Br \\ CH_2OH \end{array}$$

b.

$$\begin{array}{c} CH_3 \\ | \\ C \\ H \diagup \quad \diagdown OH \\ CH_2OH \end{array}$$

c.

$$\begin{array}{c} CHO \\ | \\ C \\ HO \diagup \quad \diagdown H \\ CH_2OH \end{array}$$

14.39 Indicate whether each pair of Fischer projections represent enantiomers or identical structures.

a. Br—⊢—Cl and Cl—⊢—Br
with CH₃ top and CH₃ bottom for both

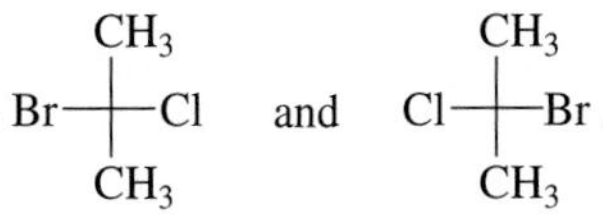

b. HO—⊢—H and H—⊢—OH
with CHO top and CH₃ bottom for both

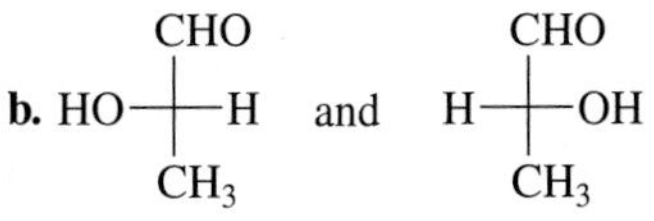

c. Br—⊢—Cl and Br—⊢—Cl
with H top and CH₃ bottom for first; CH₃ top and H bottom for second

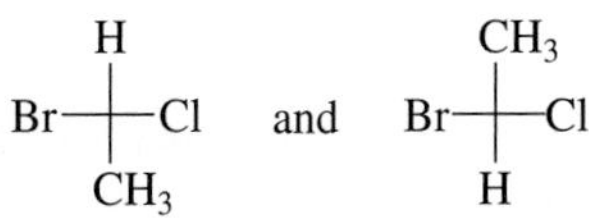

d. H—⊢—OH and HO—⊢—H
with COOH top and CH₃ bottom for both

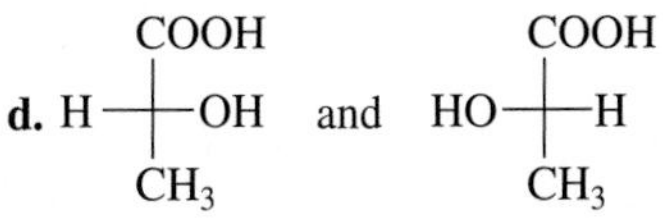

14.40 Indicate whether each pair of Fischer projections represent enantiomers or identical structures.

a. Br—⊢—Cl and Cl—⊢—Br
with CH₂OH top and CH₃ bottom for both

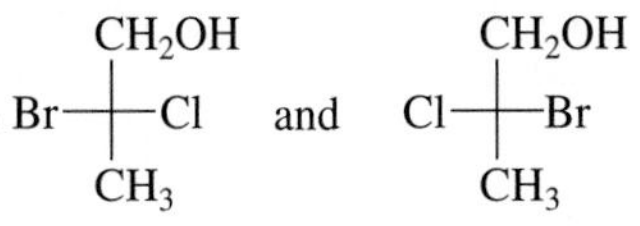

b. H—⊢—H and H—⊢—H
with CHO top and CH₃ bottom for both

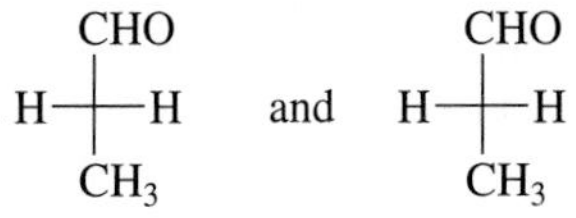

c. H—⊢—OH and HO—⊢—H
with CH₃ top and CH₂CH₃ bottom for both

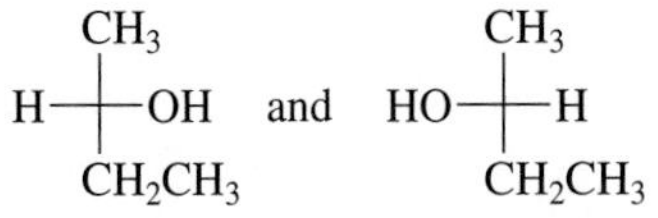

d. H—⊢—NH₂ and H₂N—⊢—H
with COOH top and CH₃ bottom for both

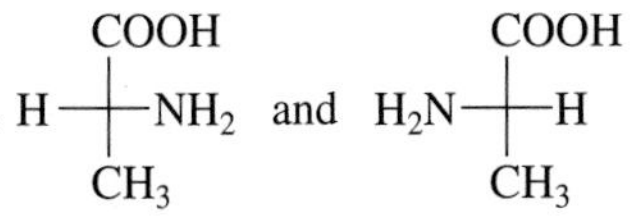

CONCEPT MAP

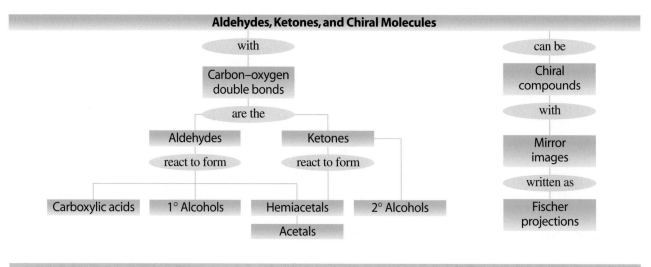

CHAPTER REVIEW

14.1 Aldehydes and Ketones

Aldehydes and ketones contain a carbonyl group ($C=O$), which consists of a double bond between a carbon and an oxygen atom. However, in contrast to the $C=C$ double bond, the $C=O$ is strongly polar. In aldehydes, the carbonyl group appears at the end of carbon chains. In ketones, the carbonyl group occurs between two alkyl groups. In the IUPAC system, the *e* in the corresponding alkane is replaced with *al* for aldehydes, and *one* for ketones. For ketones with more than four carbon atoms in the main chain, the carbonyl group is numbered to show its location. Many of the simple aldehydes and ketones use common names. Many aldehydes and ketones are found in biological systems, flavorings, and drugs.

14.2 Physical Properties of Aldehydes and Ketones

Because they contain a polar carbonyl group, aldehydes and ketones have higher boiling points than alkanes and ethers. However, their boiling points are lower than alcohols because aldehydes and ketones cannot hydrogen bond with each other. Aldehydes and ketones can hydrogen bond with water molecules, which makes carbonyl compounds with one to four carbon atoms soluble in water.

14.3 Oxidation and Reduction of Aldehydes and Ketones

Aldehydes are easily oxidized to carboxylic acids, but ketones do not oxidize further. Aldehydes, but not ketones, react with Tollens' reagent to give silver mirrors. In Benedict's test, aldehydes with adjacent hydroxyl groups reduce blue Cu^{2+} to give a brick-red Cu_2O solid. The reduction of aldehydes with hydrogen produces primary alcohols, while ketones are reduced to secondary alcohols.

14.4 Addition Reactions of Aldehydes and Ketones

Water and alcohols can add to the carbonyl group of aldehydes and ketones. The addition of one alcohol forms a hemiacetal, while the addition of two alcohols forms an acetal. Hemiacetals are not usually stable, except for cyclic hemiacetals, which are the most common form of simple sugars such as glucose.

14.5 Chiral Molecules

Chiral molecules are molecules with mirror images that cannot be superimposed on each other. These types of stereoisomers are called enantiomers. A chiral molecule must have at least one chiral carbon, which is a carbon bonded to four different atoms or groups of atoms. The Fischer projection is a simplified way to draw the arrangements of atoms by placing the chiral carbons at the center of crossed lines. The names of the mirror images are labeled D or L to differentiate between the enantiomers.

SUMMARY OF NAMING

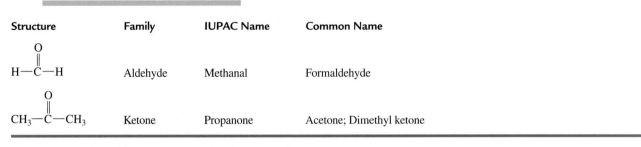

Structure	Family	IUPAC Name	Common Name
H—C—H (aldehyde)	Aldehyde	Methanal	Formaldehyde
CH₃—C—CH₃ (ketone)	Ketone	Propanone	Acetone; Dimethyl ketone

SUMMARY OF REACTIONS

OXIDATION OF ALDEHYDES TO CARBOXYLIC ACIDS

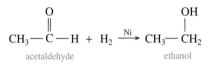

acetaldehyde → acetic acid

REDUCTION OF ALDEHYDES TO PRIMARY ALCOHOLS

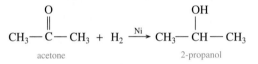

acetaldehyde → ethanol

REDUCTION OF KETONES TO SECONDARY ALCOHOLS

acetone → 2-propanol

ADDITION OF WATER TO ALDEHYDES

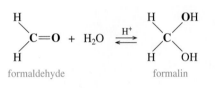

formaldehyde → formalin

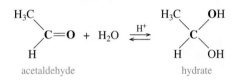

acetaldehyde → hydrate

ADDITION OF ALCOHOLS TO FORM HEMIACETALS AND ACETALS

FROM ALDEHYDES

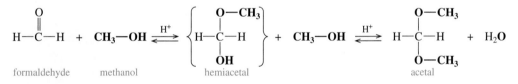

formaldehyde   methanol   hemiacetal   acetal

FROM KETONES

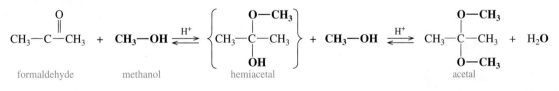

formaldehyde   methanol   hemiacetal   acetal

KEY TERMS

acetal The product of the addition of two alcohols to an aldehyde or ketone.

achiral Molecules with mirror images that are superimposable.

aldehyde An organic compound with a carbonyl functional group and at least one hydrogen.

Benedict's test A test for aldehydes with adjacent hydroxyl groups in which Cu²⁺ (CuSO₄) ions in Benedict's reagent are reduced to a brick-red solid of Cu₂O.

chiral Objects or molecules that have mirror images that cannot be superimposed on each other.

chiral carbon A carbon atom that is bonded to four different atoms or groups of atoms.

enantiomers Stereoisomers that are mirror images that cannot be superimposed on each other.

Fischer projection A system for drawing stereoisomers that shows horizontal lines for bonds coming forward, and vertical lines for bonds going back with the chiral atom at the center.

hemiacetal The product of the addition of one alcohol to the double bond of the carbonyl group in aldehydes and ketones.

ketone An organic compound in which the carbonyl functional group is bonded to two alkyl groups.

reduction A decrease in the number of carbon–oxygen bonds by the addition of hydrogen to a carbonyl bond. Aldehydes are reduced to primary alcohols; ketones to secondary alcohols.

stereoisomers Isomers that have atoms bonded in the same order, but with different arrangements in space.

Tollens' test A test for aldehydes in which Ag^+ in Tollens' reagent is reduced to metallic silver, which forms a "silver mirror" on the walls of the container.

UNDERSTANDING THE CONCEPTS

14.41 Which of the following will give a positive Tollen's test?

a. $CH_3-CH_2-\overset{\overset{\displaystyle O}{\|}}{C}-H$ **b.** $CH_3-\overset{\overset{\displaystyle O}{\|}}{C}-H$

c. $CH_3-O-CH_2-CH_3$ **d.** $CH_3-CH_2-CH_2-OH$

e. $CH_3-\overset{\overset{\displaystyle OH}{|}}{CH}-CH_3$ **f.** $\overset{\overset{\displaystyle O}{\|}}{C}-H$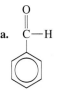

14.42 Citronellal, a constituent of oil of citronella as well as lemon and lemon grass, is used in perfumes and as an insect repellent.

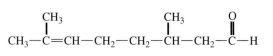

a. Complete the IUPAC name

_____ , _____-di_____ - _____-octenal

b. What does the *en* in octenal signify?

c. What does the *al* in octenal signify?

d. Write the balanced equation for the combustion of citronellal when burned in a candle to repel insects.

14.43 Identify the functional groups in each of the following:

a.

almonds

b.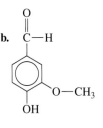

mint

c.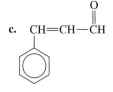

cinnamon

d.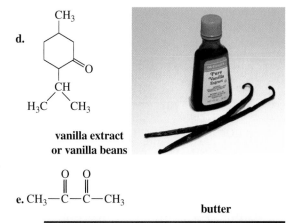

**vanilla extract
or vanilla beans**

e.

butter

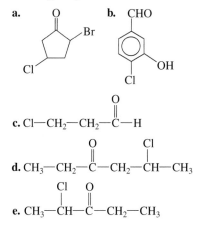

Match each of the formulas with the correct name:

1. 2,3-butadione
2. benzaldehyde
3. 2-isopropyl-5-methylcyclohexanone
4. cinnamaldehyde
5. 4-hydroxy-3-methoxybenzaldehyde

14.44 Draw the condensed structural formulas and line-bond formulas for each of the following:
a. 2-heptanone, an alarm pheromone of bees

b. 2,6-dimethyl-3-heptanone, communication pheromone of bees
c. *trans*-2-hexenal, an alarm pheromone of ants
d. 2,6-dimethyl-5-heptenal, communication pheromone of ants

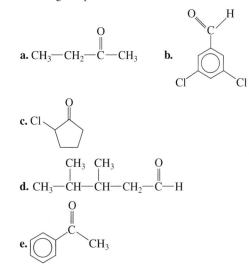

ADDITIONAL QUESTIONS AND PROBLEMS

MC GOB — For instructor-assigned homework, go to **www.masteringgob.com.**

14.45 Write the isomers for the carbonyl compounds of C_4H_8O.

14.46 Why does the C=O double bond have a dipole, while the C=C does not?

14.47 Give the IUPAC and common names (if any) for each of the following compounds:

a.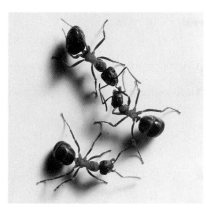

b. CHO ... OH, Cl

c. Cl—CH₂—CH₂—C—H (with O on C)

d. CH₃—CH₂—C—CH₂—CH—CH₃ (with O double bond and Cl)

e. CH₃—CH—C—CH₂—CH₃ (with Cl and O)

14.48 Give the IUPAC and common names (if any) for each of the following compounds:

a. CH₃—CH₂—C—CH₃ (with O double bond)

b. (benzene ring with CHO and two Cl)

c. Cl (cyclopentanone with Cl)

d. CH₃—CH—CH—CH₂—C—H (with two CH₃ and O)

e. (benzene ring with C=O and CH₃)

14.49 Draw the condensed structural formulas of each of the following:
    a. 3-methylcyclopentanone
    b. 4-chlorobenzaldehyde
    c. 3-chloropropionaldehyde
    d. ethyl methyl ketone
    e. 3-methylhexanal

14.50 Draw the condensed structural formulas of each of the following:
    a. propionaldehyde
    b. 2-chlorobutanal
    c. 2-methylcyclohexanone
    d. 3,5-dimethylhexanal
    e. 3-bromocyclopentanone

14.51 Which of the following compounds are soluble in water?
    a. $CH_3-CH_2-CH_2-CH_3$

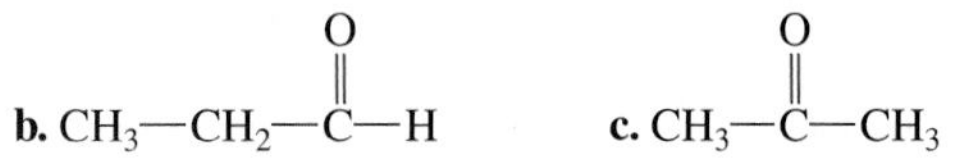

    b. $CH_3-CH_2-\overset{\overset{\displaystyle O}{\|}}{C}-H$
    c. $CH_3-\overset{\overset{\displaystyle O}{\|}}{C}-CH_3$

    d. $CH_3-CH_2-CH_2-OH$

    e. $CH_3-CH_2-\overset{\overset{\displaystyle O}{\|}}{C}-CH_2-CH_2-CH_3$

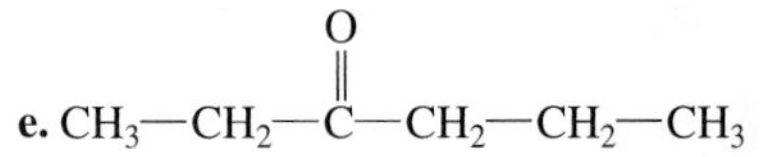

14.52 Which of the following compounds are soluble in water?
    a. $CH_3-CH_2-\overset{\overset{\displaystyle O}{\|}}{C}-CH_3$
    b. $H-\overset{\overset{\displaystyle O}{\|}}{C}-H$

    c. $CH_3-\overset{\overset{\displaystyle O}{\|}}{C}-H$
    d. $CH_3-CH_2-CH_3$

    e. $CH_3-CH_2-\overset{\overset{\displaystyle CH_3}{|}}{CH}-CH_2-CH_2-\overset{\overset{\displaystyle O}{\|}}{C}-H$

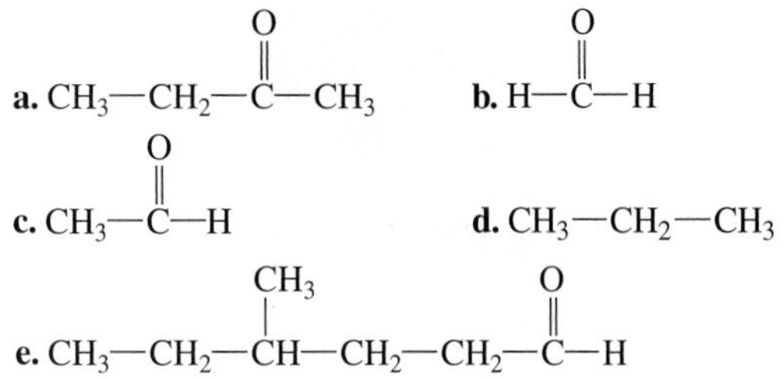

14.53 In each of the following pairs of compounds, select the compound with the higher boiling point:

    a. CH_3-CH_2-OH or $CH_3-\overset{\overset{\displaystyle O}{\|}}{C}-H$

    b. $CH_3-CH_2-CH_2-CH_3$ or $CH_3-CH_2-\overset{\overset{\displaystyle O}{\|}}{C}-H$

    c. $CH_3-CH_2-CH_2-OH$ or $CH_3-\overset{\overset{\displaystyle O}{\|}}{C}-CH_3$

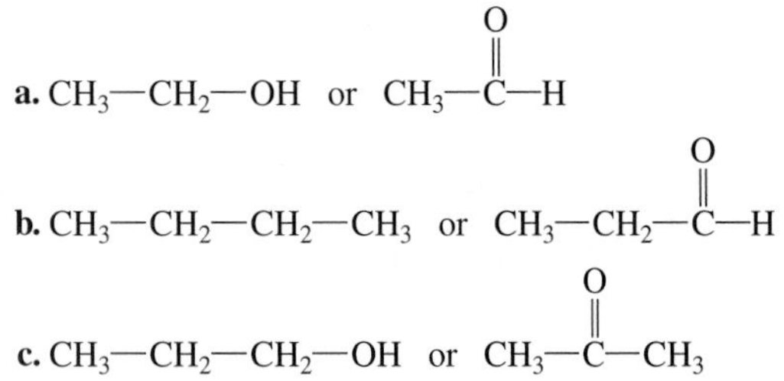

14.54 In each of the following pairs of compounds, select the compound with the higher boiling point:

    a. $CH_3-\overset{\overset{\displaystyle O}{\|}}{C}-H$ or $CH_3-CH_2-CH_2-CH_2-\overset{\overset{\displaystyle O}{\|}}{C}-H$

    b. $CH_3-CH_2-CH_2-CH_3$ or $CH_3-\overset{\overset{\displaystyle O}{\|}}{C}-CH_3$

    c. $CH_3-CH_2-\overset{\overset{\displaystyle O}{\|}}{C}-H$ or $CH_3-\overset{\overset{\displaystyle OH}{|}}{CH}-CH_3$

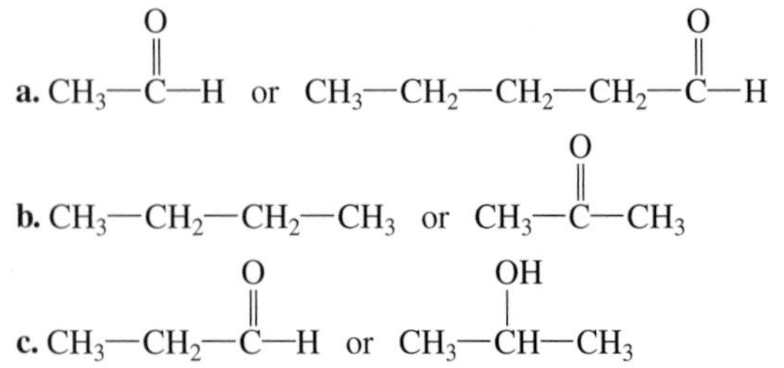

14.55 Identify the chiral carbons, if any, in each of the following compounds.

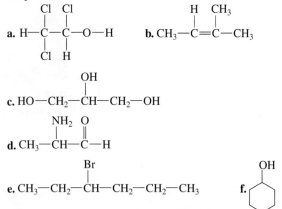

a. $H-\overset{\overset{\displaystyle Cl}{|}}{\underset{\underset{\displaystyle Cl}{|}}{C}}-\overset{\overset{\displaystyle Cl}{|}}{\underset{\underset{\displaystyle H}{|}}{C}}-O-H$
    b. $CH_3-\overset{\overset{\displaystyle H}{|}}{C}=\overset{\overset{\displaystyle CH_3}{|}}{C}-CH_3$

c. $HO-CH_2-\overset{\overset{\displaystyle OH}{|}}{CH}-CH_2-OH$

d. $CH_3-\overset{\overset{\displaystyle NH_2}{|}}{CH}-\overset{\overset{\displaystyle O}{\|}}{C}-H$

e. $CH_3-CH_2-\overset{\overset{\displaystyle Br}{|}}{CH}-CH_2-CH_2-CH_3$
    f.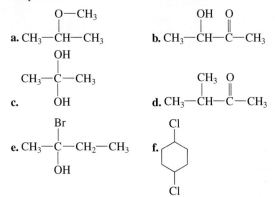

14.56 Identify the chiral carbons, if any, in each of the following compounds.

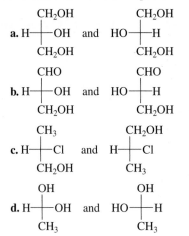

a. $CH_3-\overset{\overset{\displaystyle O-CH_3}{|}}{CH}-CH_3$
    b. $CH_3-\overset{\overset{\displaystyle OH}{|}}{CH}-\overset{\overset{\displaystyle O}{\|}}{C}-CH_3$

c. $CH_3-\overset{\overset{\displaystyle OH}{|}}{\underset{\underset{\displaystyle OH}{|}}{C}}-CH_3$
    d. $CH_3-\overset{\overset{\displaystyle CH_3}{|}}{CH}-\overset{\overset{\displaystyle O}{\|}}{C}-CH_3$

e. $CH_3-\overset{\overset{\displaystyle Br}{|}}{\underset{\underset{\displaystyle OH}{|}}{C}}-CH_2-CH_3$
    f.

14.57 Identify each of the following pairs of Fischer projections as enantiomers or identical compounds:

a.
```
   CH2OH              CH2OH
H──┼──OH    and   HO──┼──H
   CH2OH              CH2OH
```

b.
```
   CHO                CHO
H──┼──OH    and   HO──┼──H
   CH2OH              CH2OH
```

c.
```
   CH3                CH2OH
H──┼──Cl    and   H──┼──Cl
   CH2OH              CH3
```

d.
```
   OH                 OH
H──┼──OH    and   HO──┼──H
   CH3                CH3
```

14.58 Identify each of the following pairs of Fischer projections as enantiomers or identical compounds:

a.
```
   CH2─CH3               CH2─CH3
H──┼──Cl    and   Cl──┼──H
   CH2─OH               CH2─OH
```

b.
```
   CH2─OH               CH2─OH
H──┼──OH    and   HO──┼──H
   CH2─OH               CH2─OH
```

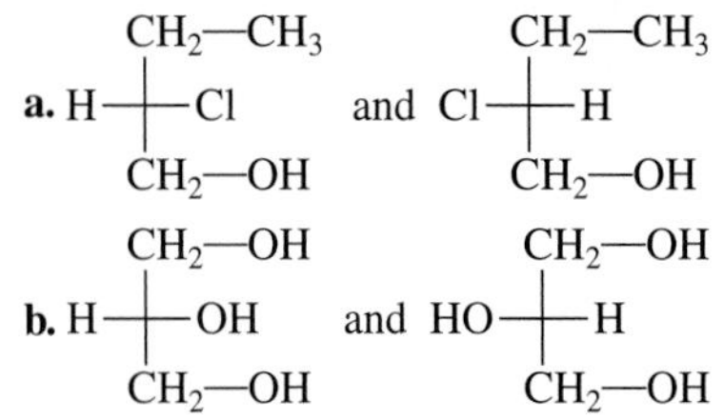

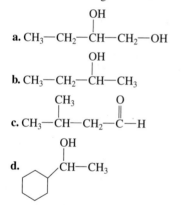

c. H—|—Cl   and H—|—Cl

(CH₃ above, CH₂—OH below each)

d. H—|—OH   and HO—|—H

(CHO above, CH₂—OH below each)

14.59 Draw the structural formula of the organic product when each of the following is oxidized:

a. CH₃—CH₂—CH₂—OH

b. CH₃—CH—CH₂—CH₂—CH₃
(OH above CH)

c. CH₃—CH₂—CH₂—C—H
(O double bond above C)

d. [cyclohexane with OH]

14.60 Draw the structural formula of the organic product when each of the following is oxidized:

a. CH₃—CH₂—CH—CH₂—OH
(OH above CH)

b. CH₃—CH₂—CH—CH₃
(OH above CH)

c. CH₃—CH—CH₂—C—H
(CH₃ above first CH, O double bond above C)

d. [cyclohexane]—CH—CH₃
(OH above CH)

14.61 Draw the structural formula of the organic product when hydrogen and a nickel catalyst reduce each of the following:

a. CH₃—C—CH₃
(O double bond above C)

b. [benzene ring]—CH₂—C—H
(O double bond above C)

c. CH₃—CH—CH₂—C—CH₃
(CH₃ above first CH, O double bond above C)

14.62 Draw the structural formula of the organic product when hydrogen and a nickel catalyst reduce each of the following:

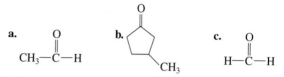

a. CH₃—C—H
(O double bond above C)

b. [cyclopentanone ring with CH₃]

c. H—C—H
(O double bond above C)

14.63 Using reactions such as dehydration, hydrogenation, oxidation, reduction, and hydration, indicate how you might prepare the following from the starting substance given:

a. propene to propanone
b. butanal to 1,2-dibromobutane
c. butanal to butanone

14.64 Using reactions such as dehydration, hydrogenation, oxidation, reduction, and hydration, indicate how you might prepare the following from the starting substance given:

a. pentanal to 1-pentene
b. 1-butanol to butanone
c. cyclohexene to cyclohexanone

14.65 Identify the following as hemiacetals or acetals. Give the names of the carbonyl compounds and alcohols used in their synthesis.

a. CH₃—CH₂—CH
(O—CH₃ above, O—CH₃ below)

b. CH₃—CH₂—C—CH₃
(O—CH₂—CH₃ above, OH below)

c. CH₃—CH₂—O   O—CH₂—CH₃
(cyclohexane ring below)

14.66 Identify the following as hemiacetals or acetals. Give the names of the carbonyl compounds and alcohols used in their synthesis.

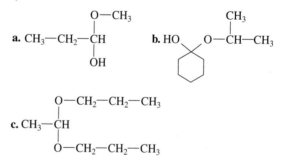

a. CH₃—CH₂—CH
(O—CH₃ above, OH below)

b. HO   O—CH—CH₃
(CH₃ above, cyclohexane ring below)

c. CH₃—CH
(O—CH₂—CH₂—CH₃ above, O—CH₂—CH₂—CH₃ below)

CHALLENGE QUESTIONS

14.67 Use the following structures to answer the true-false questions below:

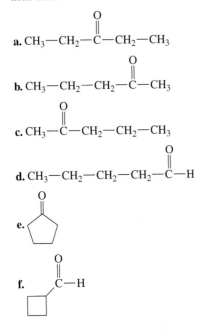

a. CH₃—CH₂—C—CH₂—CH₃

b. CH₃—CH₂—CH₂—C—CH₃

c. CH₃—C—CH₂—CH₂—CH₃

d. CH₃—CH₂—CH₂—CH₂—C—H

e.

f.

True or False:

1. a and b are isomers
2. a and c are the same compound
3. b and c are the same compound
4. e and f are isomers
5. a is chiral
6. d and f are aldehydes
7. b and e are ketones

14.68 a. Write the balanced equation for the reduction of butanone using a Pt catalyst.
  b. How many milliliters of H_2 gas at STP are needed to reduce 1.56 g butanone?

14.69 Compound A has the formula C_3H_8O. When A is heated with acid, compound B (C_3H_6) forms. When A is oxidized, compound C (C_3H_6O) forms that oxidizes further to a carboxylic acid. Another compound D has the formula C_3H_8O. When D is heated with acid, compound B (C_3H_6) forms. When D is oxidized, compound E (C_3H_6O) forms that cannot be oxidized further. What are the structures and names of compounds A, B, C, D, and E?

14.70 Draw the structures and give the IUPAC names of all the aldehydes and ketones that have the formula $C_5H_{10}O$.

ANSWERS

Answers to Study Checks

14.1 CH₃—CH₂—C—CH₂—CH₃
(with O double-bonded to C)

14.2 propanal (IUPAC), propionaldehyde (common)

14.3 ethyl propyl ketone

14.4 The oxygen atom in the carbonyl group of acetone hydrogen bonds with water molecules.

14.5 cyclohexanol

14.6 The oxidation of benzaldehyde reduces Ag^+ to metallic silver, which forms a silvery coating on the walls of the test tube.

14.7 1-propanol

14.8
CH₃—C—CH₃
with O—CH₃ above and O—CH₃ below

14.9 achiral. A bowling pin has a symmetrical shape.

14.10
HO—CH₂—CH—CH—C—H
with OH, OH, O

14.11

CHO   CHO
H—OH   HO—H
CH₃   CH₃

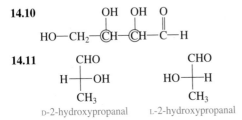

D-2-hydroxypropanal   L-2-hydroxypropanal

Answers to Selected Problems

14.1 a. ketone   **b.** aldehyde
      c. ketone   **d.** aldehyde

14.3 a. 1   **b.** 1   **c.** 2

14.5 a. propanal   **b.** 2-methyl-3-pentanone
      c. 3-bromobutanal   **d.** 2-pentanone
      e. 3-methylcyclohexanone   **f.** 4-chlorobenzaldehyde

14.7 a. acetaldehyde
      b. methyl propyl ketone
      c. formaldehyde

14.9 a. CH₃—C—H (with O)

      b. CH₃—C—CH₂—CH—CH₃ (with O and CH₃)

      c. CH₃—CH—CH—C—H (with Br, Br, O)

      d. CH₃—C—CH₂—CH₂—CH₂—CH₃ (with O)

      e. CH₃—CH₂—CH—CH₂—C—H (with CH₃ and O)

14.11

14.13 a. CH₃—CH₂—C(=O)—H has a polar carbonyl group.
    b. pentanal has a higher molar mass and thus a higher boiling point
    c. 1-butanol because it can hydrogen bond with other 1-butanol molecules

14.15 a. CH₃—C(=O)—C(=O)—CH₃ ; more hydrogen bonding
    b. acetaldehyde; more hydrogen bonding
    c. acetone; lower number of carbon atoms

14.17 No. The long carbon chain diminishes the effect of the carbonyl group.

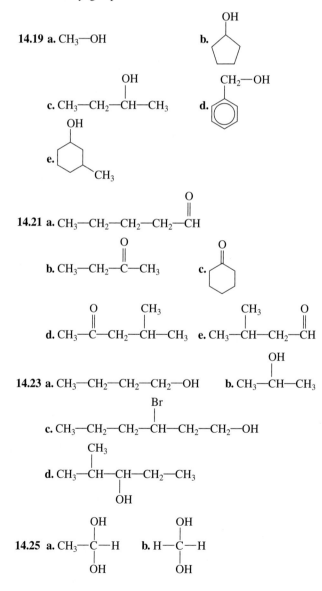

14.27 a. hemiacetal   **b.** hemiacetal   **c.** acetal
    d. hemiacetal   **e.** acetal

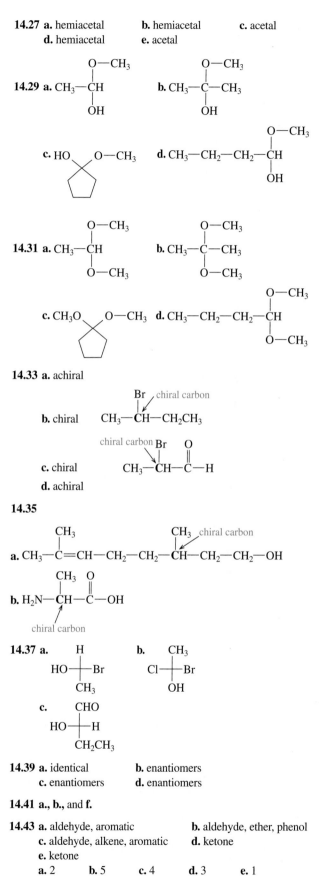

14.43 a. aldehyde, aromatic   **b.** aldehyde, ether, phenol
    c. aldehyde, alkene, aromatic   **d.** ketone
    e. ketone
    a. 2   **b.** 5   **c.** 4   **d.** 3   **e.** 1

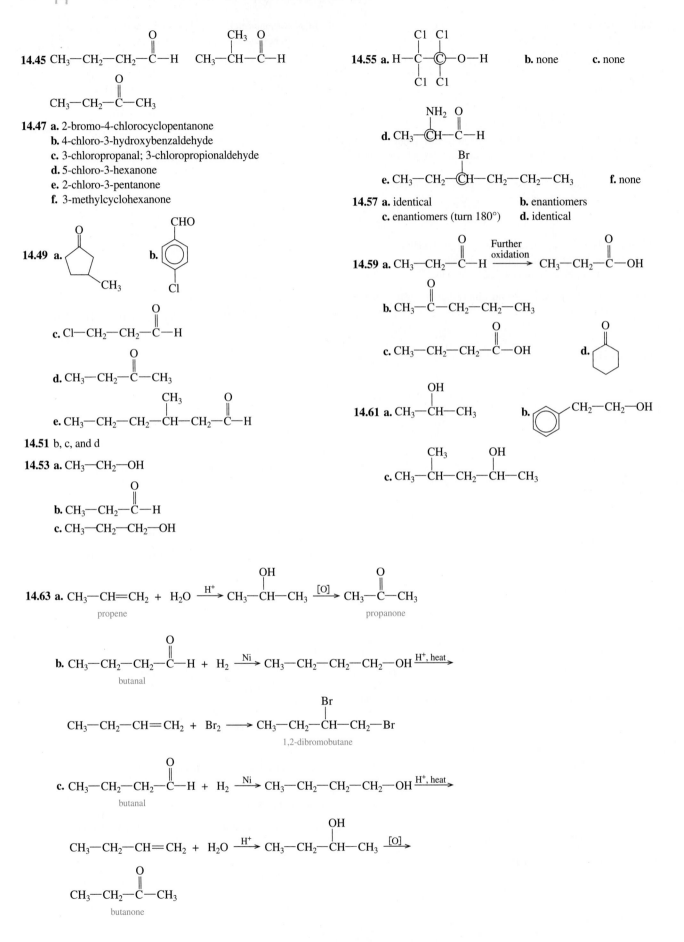

14.45 CH₃—CH₂—CH₂—C(=O)—H   CH₃—CH(CH₃)—C(=O)—H

CH₃—CH₂—C(=O)—CH₃

14.47 a. 2-bromo-4-chlorocyclopentanone
b. 4-chloro-3-hydroxybenzaldehyde
c. 3-chloropropanal; 3-chloropropionaldehyde
d. 5-chloro-3-hexanone
e. 2-chloro-3-pentanone
f. 3-methylcyclohexanone

14.49 a. [cyclopentanone with CH₃]   **b.** [benzaldehyde with CHO and Cl]

c. Cl—CH₂—CH₂—C(=O)—H

d. CH₃—CH₂—C(=O)—CH₃

e. CH₃—CH₂—CH₂—CH(CH₃)—CH₂—C(=O)—H

14.51 b, c, and d

14.53 a. CH₃—CH₂—OH

b. CH₃—CH₂—C(=O)—H

c. CH₃—CH₂—CH₂—OH

14.55 a. H—C(Cl)(Cl)—C(Cl)(Cl)—O—H   **b.** none   **c.** none

d. CH₃—CH(NH₂)—C(=O)—H

e. CH₃—CH₂—CH(Br)—CH₂—CH₂—CH₃   **f.** none

14.57 a. identical   **b.** enantiomers
c. enantiomers (turn 180°)   **d.** identical

14.59 a. CH₃—CH₂—C(=O)—H —Further oxidation→ CH₃—CH₂—C(=O)—OH

b. CH₃—C(=O)—CH₂—CH₂—CH₃

c. CH₃—CH₂—CH₂—C(=O)—OH   **d.** [cyclohexanone]

14.61 a. CH₃—CH(OH)—CH₃   **b.** [benzene ring]—CH₂—CH₂—OH

c. CH₃—CH(CH₃)—CH₂—CH(OH)—CH₃

14.63 a. CH₃—CH=CH₂ + H₂O —H⁺→ CH₃—CH(OH)—CH₃ —[O]→ CH₃—C(=O)—CH₃
propene   propanone

b. CH₃—CH₂—CH₂—C(=O)—H + H₂ —Ni→ CH₃—CH₂—CH₂—CH₂—OH —H⁺, heat→
butanal

CH₃—CH₂—CH=CH₂ + Br₂ → CH₃—CH₂—CH(Br)—CH₂—Br
1,2-dibromobutane

c. CH₃—CH₂—CH₂—C(=O)—H + H₂ —Ni→ CH₃—CH₂—CH₂—CH₂—OH —H⁺, heat→
butanal

CH₃—CH₂—CH=CH₂ + H₂O —H⁺→ CH₃—CH₂—CH(OH)—CH₃ —[O]→

CH₃—CH₂—C(=O)—CH₃
butanone

14.65 a. acetal; propanal and methanol
b. hemiacetal; butanone and ethanol
c. acetal; cyclohexanone and ethanol

14.67 1. true
2. false
3. true
4. true
5. false
6. true
7. true

14.69 $CH_3-CH_2-CH_2-OH$ A. 1-propanol

$CH_3-CH=CH_2$ B. propene

$$\underset{\displaystyle CH_3-CH_2-\overset{\textstyle O}{\overset{\|}{C}}-H}{}$$ C. propanal

$$\underset{\displaystyle CH_3-\underset{\textstyle |}{\overset{\textstyle OH}{CH}}-CH_3}{}$$ D. 2-propanol

$$\underset{\displaystyle CH_3-\overset{\textstyle O}{\overset{\|}{C}}-CH_3}{}$$ E. propanone

15 Carbohydrates

"We use a refractometer to measure sugar content in a small sample of juices from the grapes in different areas of the vineyard," says Leslie Bucher, laboratory director at Bouchaine Winery. "We also measure the alcohol content during fermentation and run tests for sulfur, pH, and total acid."

As grapes ripen, there is an increase in the sugars, which are the monosaccharides fructose and glucose. The sugar content is affected by soil conditions and the amount of sun and water. When the grapes are ripe and sugar content is at a desirable level, they are harvested. During fermentation, enzymes from yeast convert about half the sugar to ethanol, and half to carbon dioxide. Grapes harvested with 22.5% sugar will ferment to give a wine with 12.5–13.5% alcohol content.

LOOKING AHEAD

Mastering GOB CHEMISTRY

Visit **www.masteringgob.com** for self-study materials and instructor-assigned homework.

SAMPLE PROBLEM 15.1

▪ Monosaccharides

Classify each of the following monosaccharides to indicate their carbonyl group and number of carbon atoms:

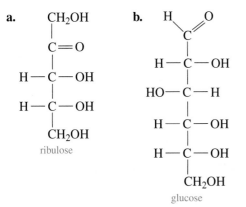

a.

$$CH_2OH$$
$$|$$
$$C=O$$
$$|$$
$$H—C—OH$$
$$|$$
$$H—C—OH$$
$$|$$
$$CH_2OH$$

ribulose

b.

$$\underset{H}{}\overset{}{C}=O$$
$$|$$
$$H—C—OH$$
$$|$$
$$HO—C—H$$
$$|$$
$$H—C—OH$$
$$|$$
$$H—C—OH$$
$$|$$
$$CH_2OH$$

glucose

SOLUTION

a. The structural formula has a ketone group; ribulose is a ketose. Because there are five carbon atoms, it is a pentose. Combining these classifications makes it a ketopentose.

b. The structural formula has an aldehyde group; glucose is an aldose. Because there are six carbon atoms, it is an aldohexose.

STUDY CHECK

The simplest ketose is a triose named dihydroxyacetone. Draw its structural formula.

QUESTIONS AND PROBLEMS

MC
GOB

▪ Carbohydrates

15.1 What reactants are needed for photosynthesis and respiration?

15.2 What is the relationship between photosynthesis and respiration?

15.3 What is a monosaccharide? A disaccharide?

15.4 What is a polysaccharide?

15.5 What functional groups are found in all monosaccharides?

15.6 What is the difference between an aldose and a ketose?

15.7 What are the functional groups and number of carbons in a ketopentose?

15.8 What are the functional groups and number of carbons in an aldohexose?

15.9 Classify each of the following monosaccharides as an aldose or ketose.

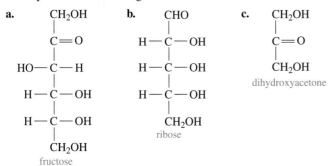

a.

$$CH_2OH$$
$$|$$
$$C=O$$
$$|$$
$$HO—C—H$$
$$|$$
$$H—C—OH$$
$$|$$
$$H—C—OH$$
$$|$$
$$CH_2OH$$

fructose

b.

$$CHO$$
$$|$$
$$H—C—OH$$
$$|$$
$$H—C—OH$$
$$|$$
$$H—C—OH$$
$$|$$
$$CH_2OH$$

ribose

c.

$$CH_2OH$$
$$|$$
$$C=O$$
$$|$$
$$CH_2OH$$

dihydroxyacetone

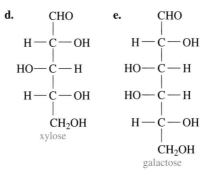

d.
$$
\begin{array}{c}
\text{CHO} \\
\text{H—C—OH} \\
\text{HO—C—H} \\
\text{H—C—OH} \\
\text{CH}_2\text{OH}
\end{array}
$$
xylose

e.
$$
\begin{array}{c}
\text{CHO} \\
\text{H—C—OH} \\
\text{HO—C—H} \\
\text{HO—C—H} \\
\text{H—C—OH} \\
\text{CH}_2\text{OH}
\end{array}
$$
galactose

15.10 Classify each of the monosaccharides in problem 15.9 according to the number of carbon atoms in the chain.

LEARNING GOAL

■ Draw the D or L configurations of glucose, galactose, and fructose.

15.2 Structures of Monosaccharides

In Chapter 14 we learned that chiral compounds exist as mirror images that cannot be superimposed. The monosaccharides, which contain chiral carbons, exist as mirror images.

Fischer Projections

Let's take a look again at the Fischer projection for the simplest aldose, glyceraldehyde. By convention the carbon chain is written vertically with the aldehyde group (most oxidized carbon) at the top. The letter L is assigned to the stereoisomer if the —OH group is on the left of the chiral carbon. In D-glyceraldehyde, the —OH is on the right. The very bottom carbon atom in the Fischer projection of a carbohydrate, —CH₂OH, is not chiral because it does not have four different groups bonded to it.

(MC GOB) SELF-STUDY ACTIVITY
Forms of Carbohydrates

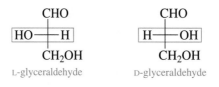

$$
\begin{array}{c}
\text{CHO} \\
\boxed{\text{HO——H}} \\
\text{CH}_2\text{OH}
\end{array}
\qquad
\begin{array}{c}
\text{CHO} \\
\boxed{\text{H——OH}} \\
\text{CH}_2\text{OH}
\end{array}
$$
L-glyceraldehyde   D-glyceraldehyde

(MC GOB) Identifying Chiral Carbons
in Monosaccharides
Drawing Fisher Projections
of Monosaccharides
Identifying D and L Sugars

Most of the carbohydrates we will study have carbon chains with five or six carbon atoms, which means that they have several chiral carbons. Then we use the chiral carbon furthest from the carbonyl group to determine the D or L isomer. The following are the isomers of ribose, which is a five-carbon monosaccharide, and glucose, a six-carbon monosaccharide.

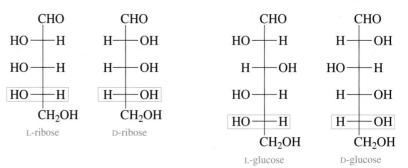

$$
\begin{array}{c}
\text{CHO} \\
\text{HO——H} \\
\text{HO——H} \\
\boxed{\text{HO——H}} \\
\text{CH}_2\text{OH}
\end{array}
\quad
\begin{array}{c}
\text{CHO} \\
\text{H——OH} \\
\text{H——OH} \\
\boxed{\text{H——OH}} \\
\text{CH}_2\text{OH}
\end{array}
\qquad
\begin{array}{c}
\text{CHO} \\
\text{HO——H} \\
\text{H——OH} \\
\text{HO——H} \\
\boxed{\text{HO——H}} \\
\text{CH}_2\text{OH}
\end{array}
\quad
\begin{array}{c}
\text{CHO} \\
\text{H——OH} \\
\text{HO——H} \\
\text{H——OH} \\
\boxed{\text{H——OH}} \\
\text{CH}_2\text{OH}
\end{array}
$$
L-ribose   D-ribose   L-glucose   D-glucose

SAMPLE PROBLEM 15.2

■ **Identifying D and L Isomers of Sugars**

Is the following structure the D or L enantiomer of ribose?

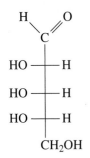

SOLUTION

In ribose, carbon 4 is the chiral atom furthest from the carbonyl group. Because the hydroxyl group on carbon 4 is on the left, this enantiomer is L-ribose.

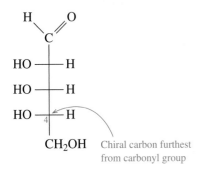

Chiral carbon furthest from carbonyl group

STUDY CHECK

Draw the Fischer projection for D-ribose.

Some Important Monosaccharides

The hexoses glucose, galactose, and fructose are important monosaccharides. Although structural formulas for both the D and L isomers can be written for these hexoses, the D isomers are the more common form of carbohydrates found in nature. We can write their open-chain structures as follows:

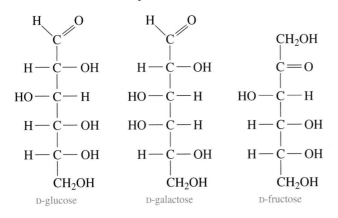

D-glucose   D-galactose   D-fructose

The most common hexose, D-**glucose**, $C_6H_{12}O_6$, also known as dextrose and blood sugar, is found in fruits, vegetables, corn syrup, and honey. It is a building block of the disaccharides sucrose, lactose, and maltose and polysaccharides such as starch, cellulose, and glycogen.

In the body, glucose normally occurs at a concentration of 70–90 mg/dL (1 dL = 100 mL) of blood. Excess glucose is converted to glycogen and stored in the liver and muscle. When the amount of glucose exceeds what is needed for energy or glycogen, the excess glucose is converted to fat, which can be stored in unlimited amounts.

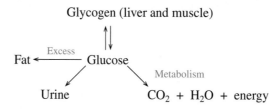

Galactose is an aldohexose that does not occur in the free form in nature. It is obtained as a hydrolysis product of the disaccharide lactose, a sugar found in milk and milk products. Galactose is important in the cellular membranes of the brain and nervous system. The only difference in the structures of D-glucose and D-galactose is the arrangement of the —OH group on carbon 4.

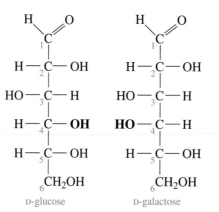

In a condition called *galactosemia*, an enzyme needed to convert galactose to glucose is missing. The accumulation of galactose in the blood and tissues can lead to cataracts, mental retardation, and cirrhosis. The treatment for galactosemia is the removal of all galactose-containing foods, mainly milk and milk products, from the diet. If this is done for an infant immediately after birth, the damaging effects of galactose accumulation can be avoided.

In contrast to glucose and galactose, **fructose** is a ketohexose. The structure of fructose differs from glucose at carbons 1 and 2 by the location of the carbonyl group.

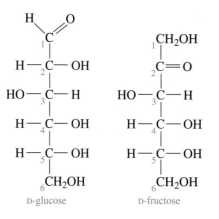

HEALTH NOTE

Hyperglycemia and Hypoglycemia

In the body, glucose normally occurs at a concentration of 70–90 mg/dL (1 dL = 100 mL) of blood. However, the amount of glucose depends on the time that has passed since eating. In the first hour after a meal, the level of glucose rises to about 130 mg/dL of blood, and then decreases over the next 2–3 hours as it is used in the tissues.

A doctor may order a glucose tolerance test to evaluate the body's ability to return to normal glucose concentration in response to the ingestion of a specified amount of glucose. The patient fasts for 12 hours and then drinks a solution containing glucose. A blood sample is taken immediately, followed by more blood samples each half-hour for 2 hours, and then every hour for a total of 5 hours. If the blood glucose exceeds 140 mg/dL in plasma and remains high, hyperglycemia may be indicated. The term *glyc* or *gluco* refers to "sugar." The prefix *hyper* means above or over, and *hypo* is below or under. Thus the blood sugar level in *hyperglycemia* is above normal and below normal in *hypoglycemia*.

An example of a disease that can cause hyperglycemia is diabetes mellitus, which occurs when the pancreas is unable to produce sufficient quantities of insulin. As a result, glucose levels in the body fluids can rise as high as

350 mg/dL plasma. Symptoms of diabetes in people under the age of 40 include thirst, excessive urination, increased appetite, and weight loss. In older persons, diabetes is sometimes a consequence of excessive weight gain.

When a person is hypoglycemic, the blood glucose level rises and then decreases rapidly to levels as low as 40 mg/dL plasma. In some cases, hypoglycemia is caused by overproduction of insulin by the pancreas. Low blood glucose can cause dizziness, general weakness, and muscle tremors. A diet may be prescribed that consists of several small meals high in protein and low in carbohydrate. Some hypoglycemic patients are finding success with diets that include more complex carbohydrates rather than simple sugars.

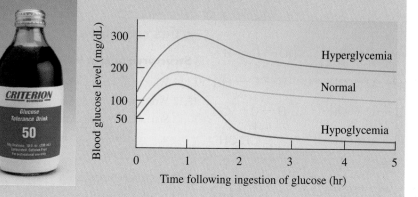

Fructose is the sweetest of the carbohydrates, twice as sweet as sucrose (table sugar). This makes fructose popular with dieters because less fructose, and therefore fewer calories, are needed to provide a pleasant taste. After fructose enters the bloodstream, it is converted to its isomer, glucose. (See Figure 15.2.) Fructose is found in fruit juices and honey; it is also called levulose and fruit sugar. Fructose is also obtained as one of the hydrolysis products of sucrose, the disaccharide known as table sugar.

SAMPLE PROBLEM 15.3

Monosaccharides

Ribulose has the following structure.

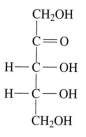

a. Identify the above compound D- or L-ribulose.
b. Write the structure of its mirror image.

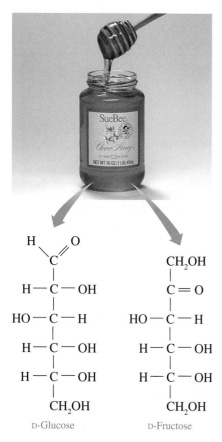

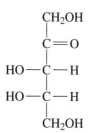

D-Glucose   D-Fructose

FIGURE 15.2

The sweet taste of honey is due to the monosaccharides of D-glucose and fructose.

Q **What are some differences in the structures of D-glucose and D-fructose?**

SOLUTION

a. D-ribulose

b. The mirror image, L-ribulose, has the following structural formula:

$$
\begin{array}{c}
CH_2OH \\
| \\
C{=}O \\
| \\
HO{-}C{-}H \\
| \\
HO{-}C{-}H \\
| \\
CH_2OH
\end{array}
$$

STUDY CHECK

What type of carbohydrate is ribulose?

QUESTIONS AND PROBLEMS

■ Structures of Monosaccharides

15.11 What is a Fischer projection?

15.12 Write the Fischer projection formula for D-glyceraldehyde and L-glyceraldehyde.

15.13 State whether each of the following sugars is the D or L isomer:

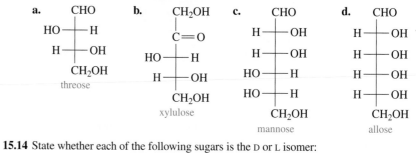

15.14 State whether each of the following sugars is the D or L isomer:

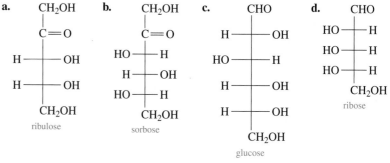

15.15 Write the mirror images for a–d in problem 15.13.

15.16 Write the mirror images for a–d in problem 15.14.

15.17 Draw the structure of D-glucose and L-glucose.

15.18 Draw the structure of D-fructose and L-fructose.

15.19 How does the structure of D-galactose differ from D-glucose?

15.20 How does the structure of D-fructose differ from D-glucose?

15.21 Identify a monosaccharide that fits each of the following descriptions:
   a. also called blood sugar
   b. not metabolized in galactosemia
   c. also called fruit sugar

15.22 Identify a monosaccharide that fits each of the following descriptions:
    a. high blood levels in diabetes
    b. obtained as a hydrolysis product of lactose
    c. the sweetest of the monosaccharides

15.3 Cyclic Structures of Monosaccharides

LEARNING GOAL

Draw and identify the cyclic structures of monosaccharides.

In Chapter 14, we saw that an aldehyde group reacts with one alcohol molecule to form a hemiacetal. For example, acetaldehyde reacts with methanol to form the following hemiacetal. In the product, the carbonyl carbon is bonded by an ether link to the alkyl group and to a new —OH group.

SELF-STUDY ACTIVITY
Forms of Carbohydrates

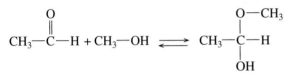

This same reaction occurs when a carbonyl group and —OH group are in the *same* molecule. The product, called a *cyclic hemiacetal*, forms a ring structure that is the most stable form of aldopentoses and aldohexoses. In the following general diagram, the hydroxyl group on carbon 5 bonds with the carbonyl carbon 1 to produce a heterocyclic six-atom ring containing an oxygen atom and a new —OH group on carbon 1.

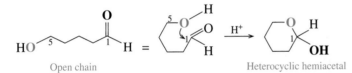

Open chain   Heterocyclic hemiacetal

Drawing Haworth Structures for Cyclic Forms

While the carbonyl group in an aldohexose could react with several of the —OH groups, the equilibrium for aldohexoses favors the formation of six-atom rings. Let's look at how we draw the cyclic hemiacetal for D-glucose starting with the Fischer projection.

Drawing Cyclic Sugars

■ **STEP 1** Think of turning the open chain of glucose clockwise to the right. Then the —OH groups written on the right, other than the one on carbon 5, are drawn down, and the —OH group on the left is up.

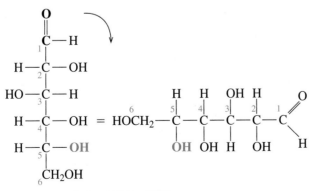

D-glucose (open chain)

Guide to Drawing Haworth Structures

STEP 1
Turn the open chain clockwise 90°.

STEP 2
Fold the chain into a hexagon and bond the O on carbon 5 to carbon 1 of the carbonyl group.

STEP 3
Write the new —OH group on carbon 1 down to give the α anomer or up to give the β anomer.

■ **STEP 2** Fold the chain into a hexagon placing the —CH₂OH up, and the —OH group on carbon 5 close to the carbonyl carbon. Form the cyclic hemiacetal by bonding the oxygen in the —OH group to the carbonyl carbon. This structure is known as a **Haworth structure**.

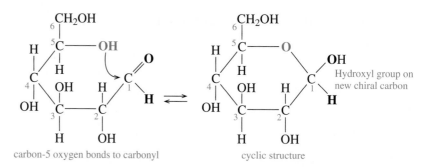

carbon-5 oxygen bonds to carbonyl   cyclic structure

■ **STEP 3** In the cyclic hemiacetal, carbon 1 is bonded to a new —OH group. There are two ways to place the —OH, either up or down, which gives two isomers called **anomers**. The —OH group on carbon 1 is down in the α (alpha) anomer and up in the β (beta) anomer.

MC GOB   Identifying α and β Anomers

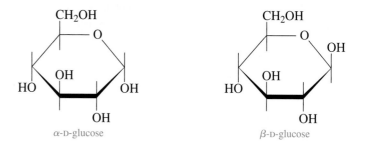

α-D-glucose   β-D-glucose

Such differences in structural forms may seem trivial. However, we can digest starch products such as pasta to obtain glucose because the polysaccharide contains the α isomers of glucose. We cannot digest paper or wood because cellulose consists of only β-D-glucose units. Humans have an α-amylase, an enzyme needed for the digestion of starches, but not a β-amylase for the digestion of cellulose.

In solution, α-D-glucose is in equilibrium with β-D-glucose. In a process called **mutarotation**, each isomer converts from the closed ring to the open chain and back again. As the ring opens and closes, the bond between carbons 1 and 2 can rotate, which allows the hydroxyl (—OH) group on carbon 1 to shift between the α and the β position. Although the open chain is an essential part of mutarotation, only a small amount of open chain is present at any given time.

Haworth Structures for α- and β-D-Glucose

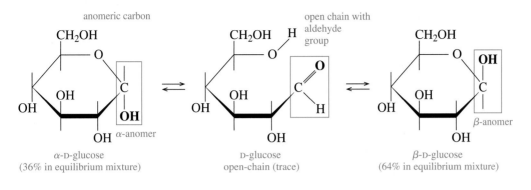

α-D-glucose
(36% in equilibrium mixture)

D-glucose
open-chain (trace)

β-D-glucose
(64% in equilibrium mixture)

Cyclic Structures of Galactose

Galactose is an aldohexose that differs from glucose only in the arrangement of the
—OH group on carbon 4. Thus, its cyclic structure is also similar to glucose,
except that in galactose the —OH on carbon 4 is up. With the formation of a new
hydroxyl group on carbon 1, galactose also exists as α and β anomers and undergoes mutarotation via the open-chain form in solution.

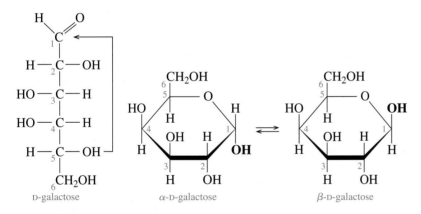

D-galactose

α-D-galactose

β-D-galactose

Cyclic Structures of Fructose

In contrast to glucose and galactose, fructose is a ketohexose. It forms a hemiacetal
when a hydroxyl group on carbon 5 reacts with the ketone group. The cyclic structure for fructose is a five-atom ring with carbon 2 at the right corner. In fructose,
the new hydroxyl group is on carbon 2. There are α and β anomers of fructose that
undergo mutarotation in solution.

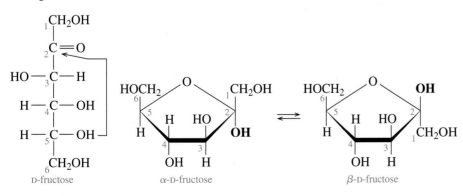

D-fructose

α-D-fructose

β-D-fructose

SAMPLE PROBLEM 15.4

▪ Drawing Cyclic Structures for Sugars

D-Mannose, a carbohydrate found in immunoglobulins, has the following open-chain structure. Draw the cyclic structure for β-D-mannose anomer.

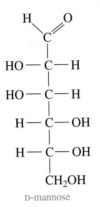

D-mannose

SOLUTION

▪ **STEP 1** Turn the chain on its side.

▪ **STEP 2** Bend it into a hexagon so that the —OH group on carbon 5 is close to the carbon 1 carbonyl group. Draw the —OH groups on the left of the open chain above the ring, and the —OH groups on the right below.

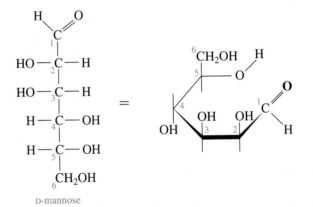

D-mannose

Form the cyclic hemiacetal by bonding the —OH on carbon 5 to the carbonyl group.

▪ **STEP 3** Write the new hydroxyl group up to make the β-D-mannose anomer.

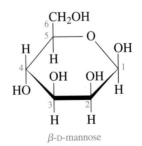

β-D-mannose

STUDY CHECK

Draw the cyclic structure for α-D-glucose.

QUESTIONS AND PROBLEMS

■ Cyclic Structures of Monosaccharides

15.23 What are the kind and number of atoms in the ring portion of the cyclic structure of glucose?

15.24 What are the kind and number of atoms in the ring portion of the cyclic structure of fructose?

15.25 Draw the cyclic structures for the α and β anomers of D-glucose.

15.26 Draw the cyclic structures for the α and β anomers of D-fructose.

15.27 Identify each of the following cyclic structures as the α or β anomer:

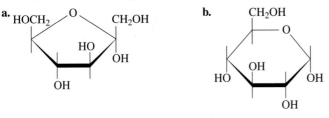

15.28 Identify each of the following cyclic structures as the α or β anomer.

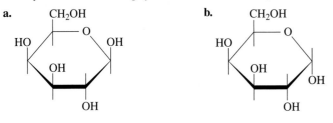

MC GOB · SELF-STUDY ACTIVITY
Forms of Carbohydrates

15.4 Chemical Properties of Monosaccharides

Monosaccharides contain functional groups that can undergo chemical reactions. In an aldose, the aldehyde group can be oxidized to a carboxylic acid. The carbonyl group in both an aldose and ketose can be reduced to give a hydroxyl group. The hydroxyl groups can react with other compounds to form a variety of derivatives that are important in biological structures.

Oxidation of Monosaccharides

Although monosaccharides exist mostly in their cyclic forms, the aldehyde group of the open-chain structure oxidizes easily. When the monosaccharide oxidizes, a carboxylic group is produced. At the same time the Cu^{2+} in Benedict's reagent is reduced to Cu^+, forming a brick-red precipitate of Cu_2O. Monosaccharides that reduce another substance such as Benedict's reagent are called **reducing sugars**.

MC GOB · Oxidation and Reduction of Monosaccharides

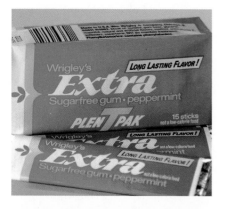

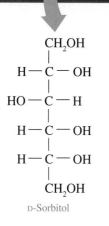

D-Sorbitol

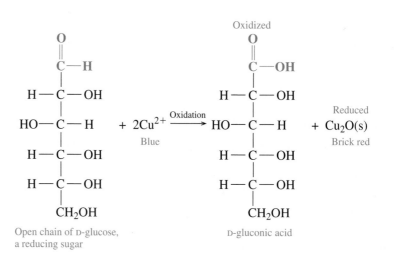

Open chain of D-glucose,
a reducing sugar

D-gluconic acid

Fructose is also a reducing sugar. In the open-chain form, a rearrangement between the hydroxyl group on carbon 1 and the ketone group provides an aldehyde group that can be oxidized.

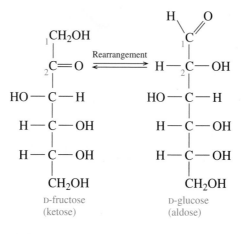

D-fructose
(ketose)

D-glucose
(aldose)

Reduction of Monosaccharides

The reduction of the carbonyl group in monosaccharides produces sugar alcohols, which are also called *alditols*. D-Glucose is reduced to D-glucitol, better known as sorbitol. D-Mannose is reduced to give D-mannitol.

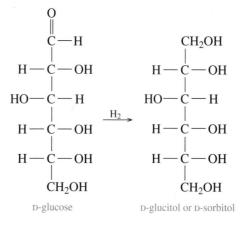

D-glucose

D-glucitol or D-sorbitol

Testing for Glucose in Urine

Normally, blood glucose flows through the kidneys and is reabsorbed into the bloodstream. However, if the blood level exceeds about 160 mg of glucose/dL of blood, the kidneys cannot reabsorb it all, and glucose spills over into the urine, a condition known as glucosuria. A symptom of diabetes mellitus is a high level of glucose in the urine.

Benedict's test can be used to determine the presence of glucose in urine. The amount of cuprous oxide (Cu_2O) formed is proportional to the amount of reducing sugar

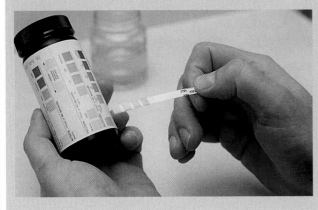

present in the urine. Low to moderate levels of reducing sugar turn the solution green; solutions with high glucose levels turn Benedict's yellow or brick-red. Table 15.1 lists some colors associated with the concentration of glucose in the urine.

In another clinical test that is more specific for glucose, the enzyme glucose oxidase is used. The oxidase enzyme converts glucose to gluconic acid and hydrogen peroxide, H_2O_2. The peroxide produced reacts with a dye in the test strip of the different colors. The level of glucose present in the urine is found by matching the color produced to a color chart on the container.

TABLE 15.1

Glucose Test Results

Color	Glucose Present in Urine	
	%	mg/dL
Blue	0	0
Blue-green	0.25	250
Green	0.50	500
Yellow	1.00	1000
Brick-red	2.00	2000

Sugar alcohols such as sorbitol, xylitol from xylose, and mannitol from mannose are used as sweeteners in many sugar-free products such as diet drinks and sugarless gum as well as products for people with diabetes. However, there are some side effects of these sugar substitutes. Some people experience some discomfort such as gas and diarrhea from the ingestion of sugar alcohols. The development of cataracts in diabetics is attributed to the accumulation of sorbitol in the lens of the eye.

MC GOB SELF-STUDY CASE
Diabetes and Blood Glucose

SAMPLE PROBLEM 15.5

Reducing Sugars

Why is D-glucose called a *reducing sugar*?

SOLUTION

D-Glucose is easily oxidized by Benedict's reagent. A carbohydrate that reduces Cu^{2+} to Cu^+ is called a reducing sugar.

STUDY CHECK

A test using Benedict's reagent turns brick-red with a urine sample. According to Table 15.1, what might this result indicate?

QUESTIONS AND PROBLEMS

▪ Chemical Properties of Monosaccharides

15.29 Draw the product xylitol produced from the reduction of D-xylose.

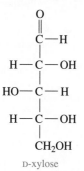

D-xylose

15.30 Draw the product mannitol produced from the reduction of D-mannose.

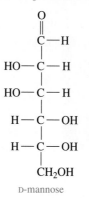

D-mannose

15.31 Write the oxidation and reduction products of D-arabinose. What is the name of the sugar alcohol produced by reduction?

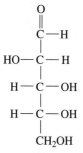

D-arabinose

15.32 Write the oxidation and reduction products of D-ribose. What is the name of the sugar alcohol produced by reduction?

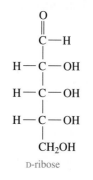

D-ribose

15.5 Disaccharides

A disaccharide is composed of two monosaccharides linked together. The most common disaccharides are maltose, lactose, and sucrose. Their hydrolysis, by an acid or an enzyme, gives the following monosaccharides.

$$\text{Maltose} + H_2O \xrightarrow{H^+} \text{glucose} + \text{glucose}$$

$$\text{Lactose} + H_2O \xrightarrow{H^+} \text{glucose} + \text{galactose}$$

$$\text{Sucrose} + H_2O \xrightarrow{H^+} \text{glucose} + \text{fructose}$$

Maltose, or malt sugar, is a disaccharide obtained from starch. When maltose in barley and other grains is hydrolyzed by yeast enzymes, glucose is obtained, which can undergo fermentation to give ethanol. Maltose is used in cereals, candies, and the brewing of beverages.

When a hydroxyl group in one monosaccharide reacts with a hydroxyl group in another monosaccharide, a **glycosidic bond** forms and the product is a disaccharide. In maltose, the glycosidic bond that joins the two glucose molecules is designated an α-1,4 linkage to show that —OH on carbon 1 of α-D-glucose is joined to carbon 4 of the second glucose. Because the second glucose molecule has a free —OH on carbon 1, there are α and β anomers of maltose. Because the anomeric carbon can open to give a free aldehyde group, maltose is a reducing sugar.

MC GOB Disaccharide Linkages

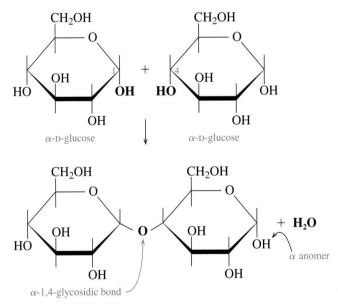

α-D-glucose α-D-glucose

α-1,4-glycosidic bond

α-maltose, a disaccharide

Lactose, milk sugar, is a disaccharide found in milk and milk products. (See Figure 15.3.) It makes up 6–8% of human milk and about 4–5% of cow's milk and is used in products that attempt to duplicate mother's milk. Some people do not produce sufficient quantities of the enzyme needed to hydrolyze lactose, and the sugar remains undigested, causing abdominal cramps and diarrhea. In some commercial milk products, an enzyme called lactase is added to break down lactose. The bond in lactose is a β-1,4-glycosidic bond because the β anomer of galactose forms a bond with a hydroxyl group on carbon 4 of glucose. The

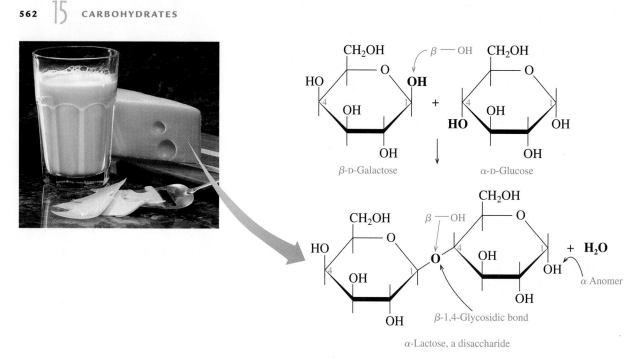

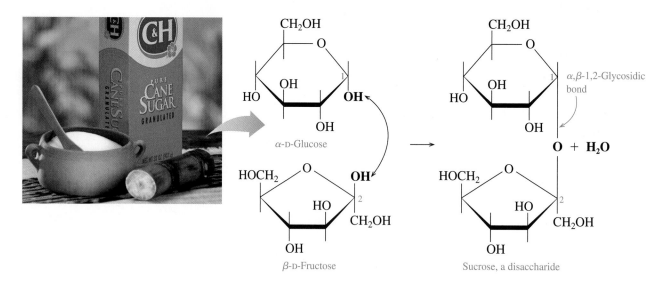

FIGURE 15.3

α-Lactose, a disaccharide found in milk and milk products, contains β-D-galactose
and α-D-glucose.

Q **What type of glycosidic bond links β-D-galactose and α-D-glucose in α-lactose?**

hydroxyl group on carbon 1 of glucose gives both α- and β-lactose. Because the
open chain has an aldehyde group that can be oxidized, lactose is a reducing
sugar.

Sucrose consists of an α-D-glucose and β-D-fructose molecule joined by an
α,β-1,2-glycosidic bond. (See Figure 15.4.) The structure of sucrose

FIGURE 15.4

Sucrose, a disaccharide obtained from sugar beets and sugar cane, contains
α-D-glucose and β-D-fructose.

Q **Why is sucrose a nonreducing sugar?**

differs from the other disaccharides because the glycosidic bond between carbon 1 of glucose and carbon 2 of fructose cannot open to give an aldehyde. Sucrose cannot react with Benedict's reagent; sucrose is not a reducing sugar.

You already know that ordinary table sugar is sucrose, a disaccharide that is the most abundant carbohydrate in the world. Most of the sucrose for table sugar comes from sugar cane (20% by mass) or sugar beets (15% by mass). Both the raw and refined forms of sugar are sucrose. Some estimates indicate that each person in the United States consumes an average of 45 kg (100 lb) of sucrose every year either by itself or in a variety of food products.

SAMPLE PROBLEM 15.6

▪ Glycosidic Bonds in Disaccharides

Melebiose is a disaccharide that has a sweetness of about 30 compared with sucrose (= 100).

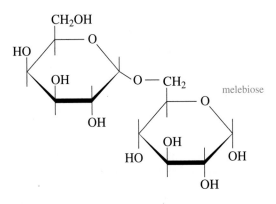

melebiose

a. What are the monosaccharide units in melebiose?
b. What type of glycosidic bond links the monosaccharides?
c. Is the compound drawn as α- or β-melebiose?

SOLUTION

a. The monosaccharide on the left side is α-D-galactose; on the right is α-D-glucose.
b. The monosaccharide units are linked by an α-1,6-glycosidic bond.
c. The downward position of the hydroxyl group on carbon 1 makes it α-melebiose.

STUDY CHECK

Cellobiose is a disaccharide composed of two β-D-glucose molecules linked by a β-1,4-glycosidic linkage. Draw a structural formula for β-cellobiose.

EXPLORE YOUR WORLD

▪ Sugar and Sweeteners

Add a tablespoon of sugar to a glass of water and stir. Taste. Add more tablespoons of sugar, stir, and taste. If you have other carbohydrates such as fructose, honey, cornstarch, arrowroot, or flour, add some of each to separate glasses of water and stir. If you have some artificial sweeteners, add a few drops of the sweetener or a package, if solid, to a glass of water. Taste each.

QUESTIONS

1. Which substance is the most soluble in water?
2. Place the substances in order from the one that tastes least sweet to the sweetest.
3. How does your list compare to Table 15.2?
4. How does the sweetness of sucrose compare with the artificial sweeteners?
5. Check the labels of food products in your kitchen. Look for sugars such as sucrose, fructose, or artificial sweeteners such as aspartame or sucralose on the label. How many grams of sugar are in a serving of the food?

HEALTH NOTE

▪ How Sweet Is My Sweetener?

Although many of the monosaccharides and disaccharides taste sweet, they differ considerably in their degree of sweetness. Dietetic foods contain sweeteners that are noncarbohydrate or carbohydrates that are sweeter. Some examples of sweeteners compared with sucrose are shown in Table 15.2.

Sucralose is made from sucrose by replacing some of the hydroxyl groups with chlorine atoms.

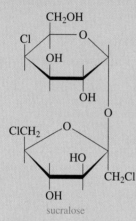

sucralose

Aspartame, which is marketed as NutraSweet, is used in a large number of sugar-free products. It is a noncarbohydrate sweetener made of aspartic acid and a methyl ester of phenylalanine. It does have some caloric value, but it is so sweet that a very small quantity is needed. However, one of the breakdown products, phenylalanine, poses a danger to anyone who cannot metabolize it properly, a condition called phenylketonuria (PKU).

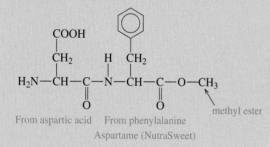

From aspartic acid   From phenylalanine   methyl ester

Aspartame (NutraSweet)

TABLE 15.2

Relative Sweetness of Sugars and Artificial Sweeteners

	Sweetness Relative to Sucrose (= 100)
Monosaccharides	
galactose	30
sorbitol	36
glucose	75
fructose	175
Disaccharides	
lactose	16
maltose	33
sucrose	100 ⟶ reference standard
Artificial Sweeteners (Noncarbohydrate)	
sucralose	60 000
aspartame	18 000
saccharin	45 000

Saccharin has been used as a noncarbohydrate artificial sweetener for the past 25 years. The use of saccharin has been banned in Canada because studies indicate that it may cause bladder tumors. However, it is still approved for use by the FDA in the United States.

saccharin

QUESTIONS AND PROBLEMS

■ Disaccharides

15.33 For each of the following disaccharides, give the monosaccharide units produced by hydrolysis, the type of glycosidic bond, and the identity of the disaccharide including the α or β anomer:

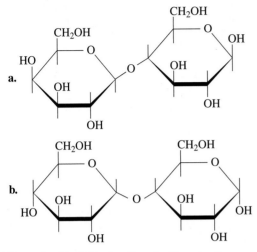

15.34 For each of the following disaccharides, give the monosaccharide units produced by hydrolysis, the type of glycosidic bond, and the identity of the disaccharide including the α or β anomer:

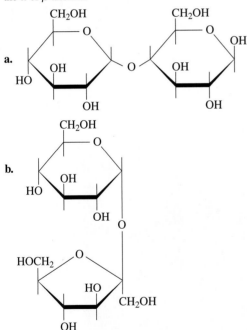

15.35 Indicate whether the sugars in problem 15.33 will undergo oxidation.

15.36 Indicate whether the sugars in problem 15.34 will undergo oxidation.

15.37 Identify disaccharides that fit each of the following descriptions:
    a. ordinary table sugar   **b.** found in milk and milk products
    c. also called malt sugar   **d.** hydrolysis gives galactose and glucose

15.38 Identify disaccharides that fit each of the following descriptions:
    a. not a reducing sugar   **b.** composed of two glucose units
    c. also called milk sugar   **d.** hydrolysis gives glucose and fructose

■ Phlebotomist

As part of the medical team, phlebotomists collect and process blood for laboratory tests. They work directly with patients, calming them if necessary prior to the collection of blood. Phlebotomists are trained to collect blood in a safe manner and provide patient care if fainting occurs. Blood is drawn through venipuncture methods such as syringe, vacutainer, and fingerstick. They also prepare patients for procedures such as glucose tolerance tests. In the preparation of specimens for analysis, a phlebotomist determines media, inoculation method, and reagents for culture setup.

■ Blood Types and Carbohydrates

Every individual's blood can be typed as one of four blood groups, A, B, AB, and O. Although there is some variation among ethnic groups in the United States, the incidence of blood types in the general population is about 43% O, 40% A, 12% B, and 5% AB.

The blood types are determined by three or four monosaccharides that are attached to the red blood cells. All the blood types include *N*-acetylglucosamine, galactose, and fucose. In type A blood, a fourth monosaccharide, *N*-acetylgalactosamine, is bonded to the galactose. The structures of these monosaccharides are as follows:

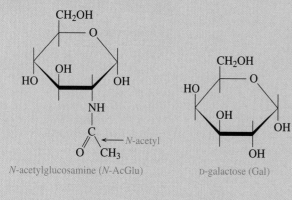

N-acetylglucosamine (*N*-AcGlu)   D-galactose (Gal)

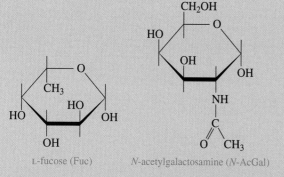

L-fucose (Fuc)   *N*-acetylgalactosamine (*N*-AcGal)

In type B blood, there is a second molecule of galactose. Type AB blood contains both A and B blood types. A person with type A blood produces antibodies against type B, whereas a person with type B blood produces antibodies against A. Type AB blood produces no antibodies, whereas type O produces both. Thus, if a person with type A blood receives a transfusion of type B blood, factors in the recipient's blood will agglutinate the donor's red blood cells. If you become a blood donor, your blood is screened to make sure that an exact match is made with the blood type of the recipient. Table 15.3 summarizes the compatibility of blood groups for transfusion.

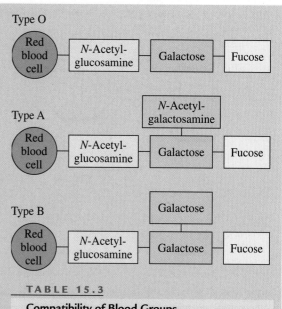

Type O
Red blood cell — *N*-Acetyl-glucosamine — Galactose — Fucose

Type A
Red blood cell — *N*-Acetyl-glucosamine — Galactose — Fucose / *N*-Acetyl-galactosamine

Type B
Red blood cell — *N*-Acetyl-glucosamine — Galactose — Fucose / Galactose

TABLE 15.3

Compatibility of Blood Groups

Blood Group	Can Receive Blood Types	Cannot Receive Blood Types
A	A, O	B, AB
B	B, O	A, AB
AB[a]	A, B, AB, O	Can receive all blood types
O[b]	O	A, B, AB

[a]AB universal recipient
[b]O universal donor

■ 15.6 Polysaccharides

A polysaccharide is a polymer of many monosaccharides joined together. Four biologically important polysaccharides—amylose, amylopectin, cellulose, and glycogen—are all polymers of D-glucose that differ only in the type of glycosidic bonds and the amount of branching in the molecule.

Starch, a storage form of glucose in plants, is found as insoluble granules in rice, wheat, potatoes, beans, and cereals. Starch is composed of two kinds of polysaccharides, amylose and amylopectin. **Amylose**, which makes up about 20% of starch, consists of 250 to 4000 α-D-glucose molecules connected by α-1,4-glycosidic bonds in a continuous chain. Sometimes called a straight-chain polymer, polymers of amylose are actually coiled in helical fashion.

Amylopectin, which makes up as much as 80% of plant starch, is a branched-chain polysaccharide. Like amylose, the glucose molecules are connected by α-1,4-glycosidic bonds. However, at about every 25 glucose units, there is a branch of glucose molecules attached by an α-1,6-glycosidic bond between carbon 1 of the branch and carbon 6 in the main chain. (See Figure 15.5.)

Starches hydrolyze easily in water and acid to give dextrins, which then hydrolyze to maltose and finally glucose. In our bodies, these complex carbohydrates are digested by the enzymes amylase (in saliva) and maltase. The glucose obtained provides about 50% of our nutritional calories.

$$\text{Amylose, amylopectin} \xrightarrow{\text{H}^+ \text{ or amylase}} \text{dextrins} \xrightarrow{\text{H}^+ \text{ or amylase}} \text{maltose} \xrightarrow{\text{H}^+ \text{ or maltase}} \text{many D-glucose units}$$

Glycogen, or animal starch, is a polymer of glucose that is stored in the liver and muscle of animals. It is hydrolyzed in our cells at a rate that maintains the blood level of glucose and provides energy between meals. The structure of glycogen is very similar to that of amylopectin found in plants except that glycogen is more highly branched. In glycogen, the glucose units are joined by α-1,4-glycosidic bonds, and branches occurring about every 10–15 glucose units are attached by α-1,6-glycosidic bonds.

Cellulose is the major structural material of wood and plants. Cotton is almost pure cellulose. In **cellulose**, glucose molecules form a long unbranched chain similar to that of amylose. However, the glucose units in cellulose are linked by β-1,4-glycosidic bonds. The β isomers do not form coils like the α isomers but are aligned in parallel rows that are held in place by hydrogen bonds between hydroxyl groups in adjacent chains, which makes cellulose insoluble in water. This gives a rigid structure to the cell walls in wood and fiber that is more resistant to hydrolysis than the starches. (See Figure 15.6)

Enzymes in our saliva and pancreatic juices hydrolyze the α-1,4-glycosidic bonds of the starches. However, there are no enzymes in humans that are able to hydrolyze the β-1,4-glycosidic bonds of cellulose; we cannot digest cellulose. Some animals such as goats and cows and insects like termites are able to obtain glucose from cellulose. Their digestive systems contain bacteria and protozoa with enzymes such as cellulase that can hydrolyze β-1,4-glycosidic bonds.

Iodine Test

In the **iodine test**, iodine (I_2) is used to test for the presence of starch. The unbranched helical shape of the polysaccharide amylose in starch reacts strongly with iodine to form a deep blue-black complex. Amylopectin, cellulose, and glycogen produce reddish-purple and brown colors. Such colors do not develop when iodine is added to samples of mono- or disaccharides.

SAMPLE PROBLEM 15.7

Structures of Polysaccharides

Identify the polysaccharide described by each of the following:

a. A polysaccharide that is stored in the liver and muscle tissues

---

MC GOB SELF-STUDY ACTIVITY
Polymers

MC GOB Polysaccharides

EXPLORE YOUR WORLD

Polysaccharides

Read the nutrition label on a box of crackers, cereal, bread, chips, or pasta. The major ingredient in crackers is flour, a starch. Chew on a single cracker for 4–5 minutes. An enzyme (amylase) in your saliva breaks apart the bonds in starch.

QUESTIONS

1. How are carbohydrates listed?
2. What other carbohydrates are listed?
3. How did the taste of the cracker change during the time that you chewed it?
4. What happened to the starches in the cracker as the amylase enzyme in your saliva reacted with the amylose and amylopectin?

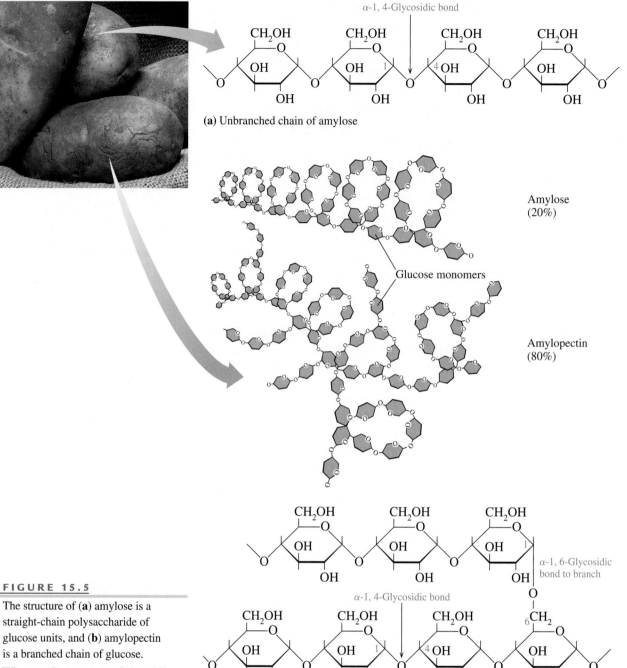

α-1, 4-Glycosidic bond

(a) Unbranched chain of amylose

Amylose (20%)

Glucose monomers

Amylopectin (80%)

α-1, 6-Glycosidic bond to branch

α-1, 4-Glycosidic bond

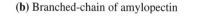

(b) Branched-chain of amylopectin

FIGURE 15.5

The structure of **(a)** amylose is a straight-chain polysaccharide of glucose units, and **(b)** amylopectin is a branched chain of glucose.

Q **What are the two types of glycosidic bonds that link glucose molecules in amylopectin?**

b. An unbranched polysaccharide containing β-1,4-glycosidic bonds

c. A starch containing α-1,4- and α-1,6-glycosidic bonds

SOLUTION

a. glycogen   **b.** cellulose   **c.** amylopectin, glycogen

STUDY CHECK

Cellulose and amylose are both unbranched glucose polymers. How do they differ?

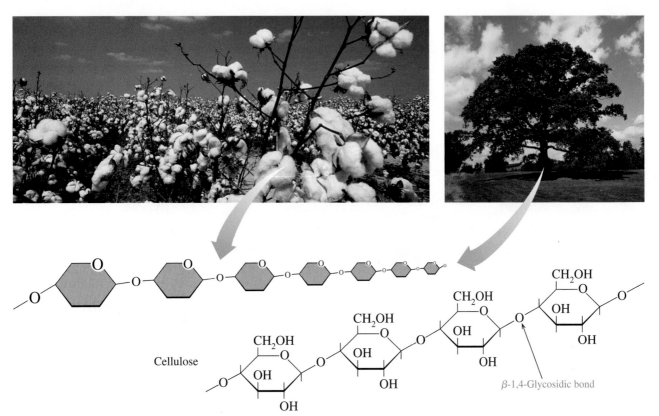

FIGURE 15.6

The polysaccharide cellulose is composed of β-1,4-glycosidic bonds.

Q Why are humans unable to digest cellulose?

QUESTIONS AND PROBLEMS

■ Polysaccharides

15.39 Describe the similarities and differences in the following polysaccharides:
a. amylose and amylopectin   **b.** amylopectin and glycogen

15.40 Describe the similarities and differences in the following polysaccharides:
a. amylose and cellulose   **b.** cellulose and glycogen

15.41 Give the name of one or more polysaccharides that matches each of the following descriptions:
a. not digestible by humans
b. the storage form of carbohydrates in plants
c. contains only α-1,4-glycosidic bonds
d. the most highly branched polysaccharide

15.42 Give the name of one or more polysaccharides that matches each of the following descriptions:
a. the storage form of carbohydrates in animals
b. contains only β-1,4-glycosidic bonds
c. contains both α-1,4- and α-1,6-glycosidic bonds
d. produces maltose during digestion

CONCEPT MAP

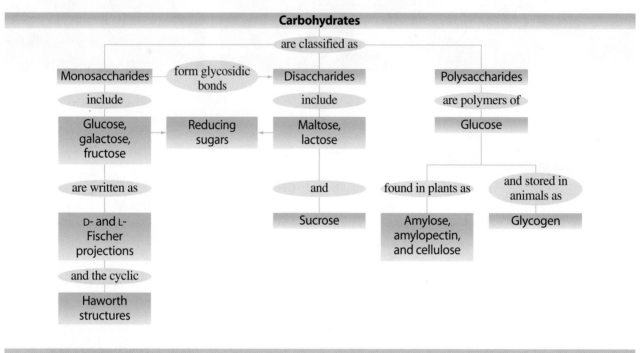

CHAPTER REVIEW

15.1 Carbohydrates

Carbohydrates are classified as monosaccharides (simple sugars), disaccharides (two monosaccharide units), and polysaccharides (many monosaccharide units). Monosaccharides are polyhydroxy aldehydes (aldoses) or ketones (ketoses). Monosaccharides are also classified by their number of carbon atoms: *triose, tetrose, pentose,* or *hexose.*

15.2 Structures of Monosaccharides

Chiral molecules can exist in two different forms, which are mirror images of each other. In a Fischer projection (straight chain), the prefixes D- and L- are used to distinguish between the mirror images. In D-monosaccharides, the —OH is on the right of the chiral carbon farthest from the carbonyl carbon; it is on the left in L-monosaccharides. Important monosaccharides are the aldohexoses glucose and galactose and the ketohexose fructose.

15.3 Cyclic Structures of Monosaccharides

The predominant form of monosaccharides is the cyclic arrangement of five or six atoms. The cyclic structure forms by a reaction between an OH (usually the one on carbon 5 in hexoses) with the carbonyl group of the same molecule. The formation of a new hydroxyl group on carbon 1 (or 2 in fructose) gives α and β anomers of the cyclic monosaccharide.

15.4 Chemical Properties of Monosaccharides

The aldehyde group in an aldose can be oxidized to a carboxylic acid, while the carbonyl group in an aldose or a ketose can be reduced to give a hydroxyl group. Monosaccharides are reducing sugars because the open-chain aldehyde group (also available in ketoses) can be oxidized by a metal ion such as Cu^{2+}.

15.5 Disaccharides

Disaccharides are two monosaccharide units joined together by a glycosidic bond. In the most common disaccharides, maltose, lactose, and sucrose, there is at least one glucose unit.

15.6 Polysaccharides

Polysaccharides are polymers of monosaccharide units. Starch consists of amylose, an unbranched chain of glucose, and amylopectin, a branched polymer of glucose. Glycogen, the storage form of glucose in animals, is similar to amylopectin with more branching. Cellulose is also a polymer of glucose, but in cellulose the glycosidic bonds are β bonds rather than α bonds as in the starches. Humans can digest starches, but not cellulose, to obtain energy. However, cellulose is important as a source of fiber in our diets.

SUMMARY OF CARBOHYDRATES

Carbohydrate	Food Sources	Monosaccharides
Monosaccharides		
glucose	Fruit juices, honey, corn syrup	
galactose	Lactose hydrolysis	
fructose	Fruit juices, honey, sucrose hydrolysis	
Disaccharides		**Monosaccharides**
maltose	Germinating grains, starch hydrolysis	glucose + glucose
lactose	Milk, yogurt, ice cream	glucose + galactose
sucrose	Sugar cane, sugar beets	glucose + fructose
Polysaccharides		
amylose	Rice, wheat, grains, cereals	Unbranched polymer of glucose joined by α-1,4-glycosidic bonds
amylopectin	Rice, wheat, grains, cereals	Branched polymer of glucose joined by α-1,4- and α-1,6-glycosidic bonds
glycogen	Liver, muscles	Highly branched polymer of glucose joined by α-1,4- and α-1,6-glycosidic bonds
cellulose	Plant fiber, bran, beans, celery	Unbranched polymer of glucose joined by β-1,4-glycosidic bonds

SUMMARY OF REACTIONS

FORMATION OF DISACCHARIDES

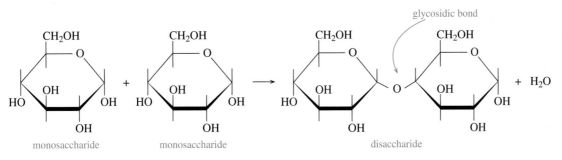

monosaccharide   +   monosaccharide   →   disaccharide   + H_2O

glycosidic bond

OXIDATION AND REDUCTION OF MONOSACCHARIDES

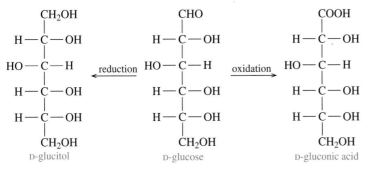

D-glucitol ← reduction ← D-glucose → oxidation → D-gluconic acid

HYDROLYSIS OF DISACCHARIDES

sucrose + H_2O ⟶ glucose + fructose

lactose + H_2O ⟶ glucose + galactose

maltose + H_2O ⟶ glucose + glucose

KEY TERMS

aldose A monosaccharide that contains an aldehyde group.

amylopectin A branched-chain polymer of starch composed of glucose units joined by α-1,4- and α-1,6-glycosidic bonds.

amylose An unbranched polymer of starch composed of glucose units joined by α-1,4-glycosidic bonds.

anomers The isomers of cyclic hemiacetals of monosaccharides that have a hydroxyl group on carbon 1 (or carbon 2). In the α anomer, the OH is drawn downward; in the β isomer the OH is up.

carbohydrate A simple or complex sugar composed of carbon, hydrogen, and oxygen.

cellulose An unbranched polysaccharide composed of glucose units linked by β-1,4-glycosidic bonds that cannot be hydrolyzed by the human digestive system.

disaccharides Carbohydrates composed of two monosaccharides joined by a glycosidic bond.

fructose A monosaccharide found in honey and fruit juices; it is combined with glucose in sucrose. Also called levulose and fruit sugar.

galactose A monosaccharide that occurs combined with glucose in lactose.

glucose The most prevalent monosaccharide in the diet. An aldohexose that is found in fruits, vegetables, corn syrup, and honey. Also known as blood sugar and dextrose. Combines in glycosidic bonds to form most of the polysaccharides.

glycogen A polysaccharide formed in the liver and muscles for the storage of glucose as an energy reserve. It is composed of glucose in a highly branched polymer joined by α-1,4- and α-1,6-glycosidic bonds.

glycosidic bond The bond that forms when the hydroxyl group of one monosaccharide reacts with the hydroxyl group of another monosaccharide. It is the type of bond that links monosaccharide units in di- or polysaccharides.

Haworth structure The cyclic structure that represents the closed chain of a monosaccharide.

iodine test A test for amylose that forms a blue-black color after iodine is added to the sample.

ketose A monosaccharide that contains a ketone group.

lactose A disaccharide consisting of glucose and galactose found in milk and milk products.

maltose A disaccharide consisting of two glucose units; it is obtained from the hydrolysis of starch and in germinating grains.

monosaccharide A polyhydroxy compound that contains an aldehyde or ketone group.

mutarotation The conversion between α and β anomers.

polysaccharides Polymers of many monosaccharide units, usually glucose. Polysaccharides differ in the types of glycosidic bonds and the amount of branching in the polymer.

reducing sugar A carbohydrate with a free aldehyde group capable of reducing the Cu^{2+} in Benedict's reagent.

sucrose A disaccharide composed of glucose and fructose; a nonreducing sugar, commonly called table sugar or "sugar."

UNDERSTANDING THE CONCEPTS

15.43 Isomaltose, obtained from the breakdown of starch, has the following structure.

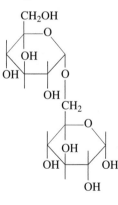

a. Is isomaltose a mono-, di-, or polysaccharide?
b. What are the monosaccharides in isomaltose?
c. What is the glycosidic link in isomaltose?
d. Is this the α or β form of isomaltose?
e. Would isomaltose be a reducing sugar?

15.44 Sophorose, a saccharide found in certain types of beans, has the following structure:

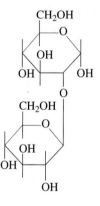

a. Is sophorose a mono-, di-, or polysaccharide?
b. What are the monosaccharides in sophorose?
c. What is the glycosidic link in sophorose?
d. Is this the α or β form of sophorose?
e. Would sophorose be a reducing sugar?

15.45 Melezitose is a saccharide with the following structure:

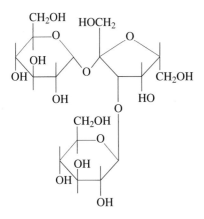

a. Is melezitose a mono-, di-, tri-, or polysaccharide?
b. What are the monosaccharides in melezitose?

15.46 What are the disaccharides and polysaccharides present in each of the following?

(a)   (b)

(c)   (d)

ADDITIONAL QUESTIONS AND PROBLEMS

 For instructor-assigned homework, go to
www.masteringgob.com.

15.47 What are the structural differences in D-glucose and D-galactose?

15.48 What are the structural differences in D-glucose and D-fructose?

15.49 How do D-galactose and L-galactose differ?

15.50 How do α-D-glucose and β-D-glucose differ?

15.51 Consider the sugar D-gulose.

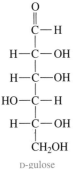

D-gulose

a. What is the Fisher projection for L-gulose?
b. Draw the Haworth structure for α- and β-D-gulose.

15.52 Consider the structures for D-gulose in question 15.51.
a. What is the structure and name of the product formed by the reduction of D-gulose?
b. Write the structure and name of the product formed by the oxidation of D-gulose.

15.53 D-Sorbitol, a sweetener found in seaweed and berries, contains only hydroxyl functional groups. When D-sorbitol is oxidized, it forms D-glucose. What is the structural formula of D-sorbitol?

15.54 Raffinose is a trisaccharide found in Australian manna and in cottonseed meal. It is composed of three different monosaccharides. Identify the monosaccharides in raffinose.

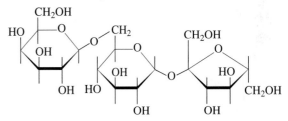

15.55 If α-galactose is dissolved in water, β-galactose is eventually present. Explain how this occurs.

15.56 Why are lactose and maltose reducing sugars, but sucrose is not?

15.57 β-Cellobiose is a disaccharide obtained from the hydrolysis of cellulose. It is quite similar to maltose except it has a β-1,4-glycosidic bond. What is the structure of β-cellobiose?

15.58 The disaccharide trehalose found in mushrooms is composed of two α-D-glucose molecules joined by an α-1,1 glycosidic bond. Draw the structure of trehalose.

CHALLENGE QUESTIONS

15.59 Gentiobiose is found in saffron.
a. Gentiobiose contains two glucose molecules linked by a β-1,6-glycoside bond. Draw the structure of α-gentiobiose.
b. Would gentiobiose be a reducing sugar? Why or why not?

15.60 From the compounds shown that follow, select those that meet the following statements:
a. is the L-enantiomer of mannose   **b.** a ketopentose
c. an aldopentose   **d.** a ketohexose

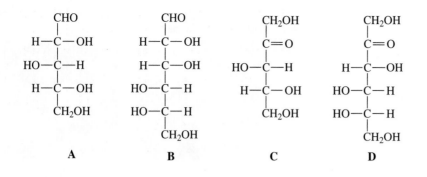

A   B   C   D

![divider bar]

ANSWERS

Answers to Study Checks

15.1
CH₂OH
|
C=O
|
CH₂OH

15.2
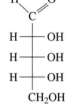

15.3 Ribulose is a ketopentose.

15.4
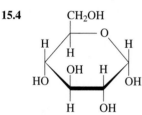

15.5 This indicates a high level of reducing sugar (probably glucose) in the urine. One common cause of this condition is diabetes mellitus.

15.6
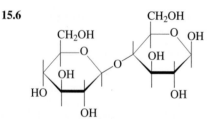

15.7 Cellulose contains glucose units connected by β-1,4-glycosidic bonds, whereas the glucose units in amylose are connected by α-1,4-glycosidic bonds.

Answers to Selected Questions and Problems

15.1 Photosynthesis requires CO₂, H₂O, and the energy from the sun. Respiration requires O₂ from the air and glucose from our foods.

15.3 A monosaccharide cannot be split into smaller carbohydrates. A disaccharide is composed of two monosaccharide units.

15.5 Hydroxyl groups are found in all monosaccharides along with a carbonyl on the first or second carbon.

15.7 A ketopentose contains hydroxyl and ketone functional groups and has five carbon atoms.

15.9 **a.** ketose   **b.** aldose   **c.** ketose
       d. aldose   **e.** aldose

15.11 A Fischer projection is a two-dimensional representation of the three-dimensional structure of a molecule.

15.13 a. D   **b.** D   **c.** L   **d.** D

15.15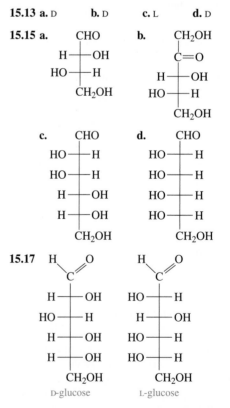

15.17

D-glucose   L-glucose

15.19 In D-galactose the hydroxyl on carbon four extends to the left. In D-glucose this hydroxyl goes to the right.

15.21 a. glucose   **b.** galactose   **c.** fructose

15.23 In the cyclic structure of glucose, there are five carbon atoms and an oxygen.

15.25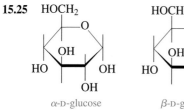

α-D-glucose   β-D-glucose

15.27 a. α-anomer   **b.** α-anomer

15.29

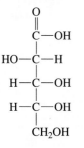

xylitol

15.31 Oxidation product:

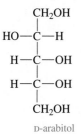

Reduction product (sugar alcohol):

D-arabitol

15.33 a. galactose and glucose; β-1,4 bond; β-lactose
b. glucose and glucose; α-1,4 bond; α-maltose

15.35 a. Can be oxidized   **b.** Can be oxidized

15.37 a. sucrose   **b.** lactose
c. maltose   **d.** lactose

15.39 a. Amylose is an unbranched polymer of glucose units joined by α-1,4 bonds; amylopectin is a branched polymer of glucose joined by α-1,4 and α-1,6 bonds.
b. Amylopectin, which is produced in plants, is a branched polymer of glucose, joined by α-1,4 and α-1,6 bonds. Glycogen, which is produced in animals, is a highly branched polymer of glucose, joined by α-1,4 and α-1,6 bonds.

15.41 a. cellulose   **b.** amylose, amylopectin
c. amylose   **d.** glycogen

15.43 a. disaccharide   **b.** α-glucose   **c.** α-1,6
d. α   **e.** Yes.

15.45 a. trisaccharide   **b.** glucose and fructose

15.47 They differ only at carbon 4 where the —OH in D-glucose is on the right side and in D-galactose it is on the left side.

15.49 D-galactose is the mirror image of L-galactose. In D-galactose, the —OH group on carbon 5 is on the right side whereas in L-galactose, the —OH group on carbon 5 is on the left side.

15.51 a.

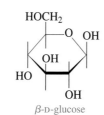

L-gulose

b.

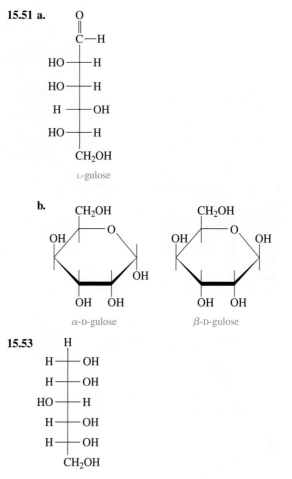

α-D-gulose   β-D-gulose

15.53

15.55 The α-galactose forms an open-chain structure, and when the chain closes, it can form both α- and β-galactose.

15.57

15.59 a.

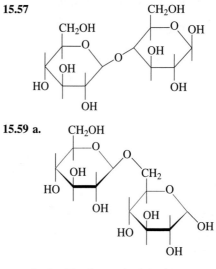

b. Yes. The ring on the right side can open up to form an aldehyde.

16

Carboxylic Acids and Esters

"There are many carboxylic acids, including the alpha hydroxy acids, that are found today in skin products," says Dr. Ken Peterson, pharmacist and cosmetic chemist, Oakland. "When you take a carboxylic acid called a fatty acid and react it with a strong base, you get a salt called soap. Soap has a high pH because the weak fatty acid and the strong base won't have a neutral pH of 7. If you take soap and drop its pH down to 7, you will convert the soap to the fatty acid. When I create fragrances, I use my nose and my chemistry background to identify and break down the reactions that produce good scents. Many fragrances are esters, which form when an alcohol reacts with a carboxylic acid. For example, the ester that smells like pineapple is made from ethanol and butyric acid."

LOOKING AHEAD

Carboxylic acids are similar to the weak acids we studied in Chapter 10. They have a sour or tart taste, produce hydronium ions in water, and neutralize bases. You encounter carboxylic acids when you use a vinegar salad dressing, which is a solution of acetic acid and water, or experience the sour taste of citric acid in a grapefruit or lemon. When a carboxylic acid combines with an alcohol, an ester is produced. Aspirin is an ester as well as a carboxylic acid. Fats known as triglycerides are esters of glycerol and fatty acids, which are long-chain carboxylic acids. Esters produce the pleasant aromas and flavors of many fruits, such as bananas, strawberries, and oranges.

16.1 Carboxylic Acids

LEARNING GOAL

Give the common names, IUPAC names, and condensed structural formulas of carboxylic acids.

In Chapter 14, we described the carbonyl group (C=O) as the functional group in aldehydes and ketones. In a **carboxylic acid**, a hydroxyl group is attached to the carbonyl group, forming a **carboxyl group**. The carboxyl functional group may be attached to an alkyl group or an aromatic group.

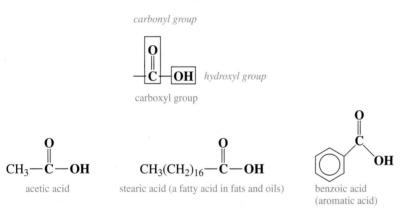

The carboxyl group can be written in several different ways. For example, the condensed structural formula and line-bond formula for propanoic acid can be written as follows:

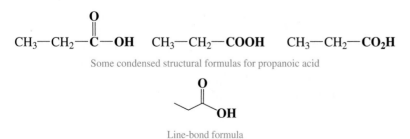

Naming Carboxylic Acids

The IUPAC names of carboxylic acids use the alkane names of the corresponding carbon chains.

- **STEP 1** **For nonaromatics, identify the longest carbon chain containing the carboxyl group and replace the *e* of the alkane name by *oic acid*.**

- **STEP 2** **Number the carbon chain beginning with the carboxyl carbon as carbon 1.**

- **STEP 3** **Give the location and names of substituents on the main chain.**

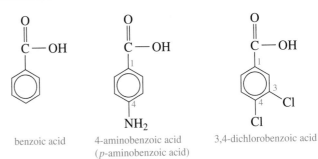

methanoic acid   2-methylpropanoic acid   3-hydroxybutanoic acid

- **STEP 4** **For the aromatic benzoic acid, number the ring from the carboxyl group as carbon 1.** With the carboxyl carbon bonded to carbon 1, the ring is numbered in the direction that gives substituents the smallest possible numbers. As with other aromatic compounds, the prefixes *ortho*, *meta*, and *para* may be used to show the position of one other substituent.

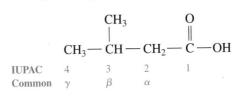

benzoic acid   4-aminobenzoic acid   3,4-dichlorobenzoic acid
                      (*p*-aminobenzoic acid)

Guide to Naming Carboxylic Acids

STEP 1
For nonaromatics, identify the longest carbon chain containing carboxyl group and replace the *e* in the alkane name by *oic acid*.

STEP 2
Number the carbon chain beginning with the carboxyl carbon as carbon 1.

STEP 3
Give the location and names of substituents on the main chain.

STEP 4
For the aromatic benzoic acid, number the ring from the carboxyl group as carbon 1.

Many carboxylic acids are still named by their common names, which are derived from their natural sources. In Chapter 14, we named aldehydes using the prefixes that represent the typical sources of carboxylic acids.

When using the common names, the Greek letters alpha (α), beta (β), and gamma (γ) are assigned to the carbons adjacent to the carboxyl carbon.

	CH₃		O
	CH₃—CH—CH₂—C—OH		
IUPAC	4	3	2	1
Common	γ	β	α	

Formic acid is injected under the skin from bee or red ant stings and other insect bites. Acetic acid is the oxidation product of the ethanol in wines and apple cider. The resulting solution of acetic acid and water is known as vinegar. Butyric acid gives the foul odor to rancid butter. (See Table 16.1.)

TABLE 16.1

Names and Natural Sources of Carboxylic Acids

Condensed Structural Formulas	IUPAC Name	Common Name	
$\overset{\displaystyle O}{\underset{\displaystyle \parallel}{}}$ H—C—OH	methanoic acid	formic acid	
$\overset{\displaystyle O}{\underset{\displaystyle \parallel}{}}$ CH₃—C—OH	ethanoic acid	acetic acid	
$\overset{\displaystyle O}{\underset{\displaystyle \parallel}{}}$ CH₃—CH₂—C—OH	propanoic acid	propionic acid	
$\overset{\displaystyle O}{\underset{\displaystyle \parallel}{}}$ CH₃—CH₂—CH₂—C—OH	butanoic acid	butyric acid	

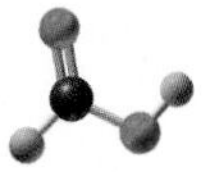

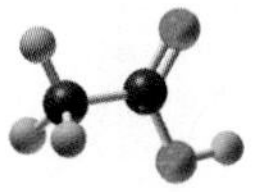

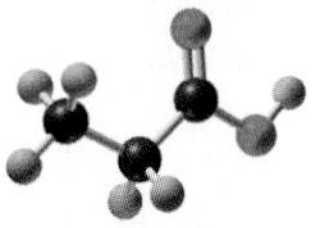

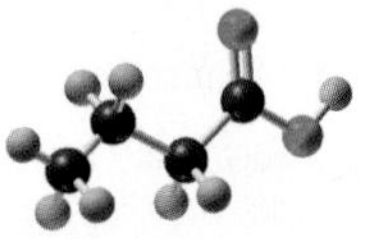

SAMPLE PROBLEM 16.1

MC GOB Naming and Drawing Carboxylic Acids

■ **Naming Carboxylic Acids**

Give the IUPAC and common name, if any, for each of the following carboxylic acids:

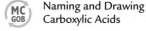

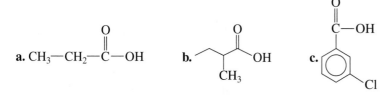

a. CH₃—CH₂—C—OH   **b.**   **c.**

SOLUTION

a. This carboxylic acid has 3 carbon atoms. In the IUPAC system, the *e* in propane is replaced by *oic acid*, to give the name, *propanoic acid*. Its common name is *propionic acid*.

b. This carboxylic acid has a methyl group on the second carbon. It has the IUPAC name *2-methylbutanoic acid*. In the common name, the Greek letter *α* specifies the carbon atom next to the carboxyl carbon, *α-methylbutyric acid*.

c. An aromatic carboxylic acid is named as benzoic acid. Counting from the carboxyl carbon places the Cl on carbon 3, or the *meta* carbon. The name is 3-chlorobenzoic acid or *meta*-chlorobenzoic acid.

STUDY CHECK

Write the condensed structural formula of 3-phenylpropanoic acid.

■

Preparation of Carboxylic Acids

Carboxylic acids can be prepared from primary alcohols or aldehydes. As we saw in Chapter 13, there is an increase in carbon–oxygen bonds as a primary alcohol is oxidized to an aldehyde. Oxidation continues easily as another oxygen is added to

HEALTH NOTE

Alpha Hydroxy Acids

Alpha hydroxy acids (AHAs) are naturally occurring carboxylic acids found in fruits, milk, and sugarcane. Cleopatra reportedly bathed in sour milk to smooth her skin. Dermatologists have long been using products with a high concentration of AHAs to remove acne scars and reduce irregular pigmentation and age spots. Now lower concentrations (8–10%) of AHAs have been added to skin care products for the purpose of smoothing fine lines, improving skin texture, and cleansing pores. Several different alpha hydroxy acids may be found in skin care products singly or in combination. Glycolic acid and lactic acid are most frequently used.

Recent studies indicate that products with AHAs increase sensitivity of the skin to sun and UV radiation. It is recommended that a sunscreen with a sun protection factor (SPF) of at least 15 be used when treating the skin with products that include AHAs. Products containing AHAs at concentrations under 10% and pH values greater than 3.5 are generally considered safe. However, the Food and Drug Administration has reports of AHAs causing skin irritation including

blisters, rashes, and discoloration of the skin. The FDA does not require product safety reports from cosmetic manufacturers, although they are responsible for marketing safe products. The FDA advises that you test any product containing AHAs on a small area of skin before you use it on a large area.

Alpha Hydroxy Acid (Source)	Structure

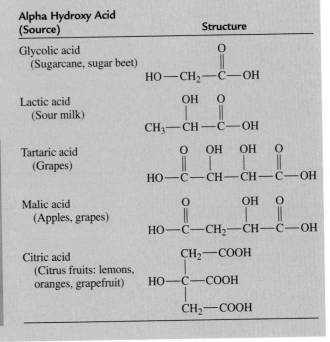

yield a carboxylic acid. For example, when ethyl alcohol in wine comes in contact with the oxygen in the air, vinegar is produced. The oxidation process converts the ethyl alcohol (primary alcohol) to acetaldehyde, and then to acetic acid, the carboxylic acid in vinegar. (See Figure 16.1.)

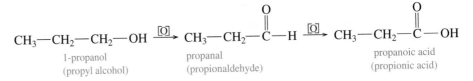

SAMPLE PROBLEM 16.2

Preparation of Carboxylic Acids

Write an equation for the oxidation of 1-propanol and name each product.

SOLUTION

A primary alcohol will oxidize to an aldehyde, which can oxidize further to a carboxylic acid.

FIGURE 16.1

Vinegar is a 5% solution of acetic acid and water.

Q What is the IUPAC name for acetic acid?

$$CH_3-CH_2-CH_2-OH \xrightarrow{[O]} CH_3-CH_2-\overset{\displaystyle O}{\overset{\|}{C}}-H \xrightarrow{[O]} CH_3-CH_2-\overset{\displaystyle O}{\overset{\|}{C}}-OH$$

1-propanol   propanal   propanoic acid
(propyl alcohol)   (propionaldehyde)   (propionic acid)

STUDY CHECK

Write the condensed structural formula of the carboxylic acid produced by the oxidation of 1-butanol.

QUESTIONS AND PROBLEMS

■ **Carboxylic Acids**

16.1 What carboxylic acid is responsible for the pain of an ant sting?

16.2 What carboxylic acid is found in a solution of vinegar?

16.3 Explain the differences in the condensed structural formulas of propanal and propanoic acid.

16.4 Explain the differences in the condensed structural formulas of benzaldehyde and benzoic acid.

16.5 Give the IUPAC and common names (if any) for the following carboxylic acids:

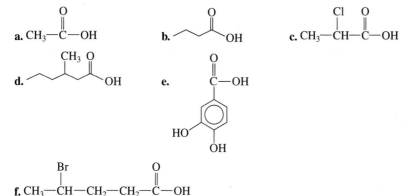

16.6 Give the IUPAC and common names (if any) for the following carboxylic acids:

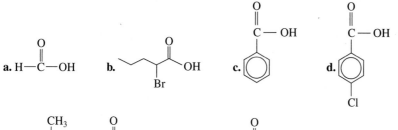

16.7 Draw the condensed structural formulas of each of the following carboxylic acids:
   a. propionic acid   **b.** benzoic acid
   c. 2-chloroethanoic acid   **d.** 3-hydroxypropanoic acid
   e. α-methylbutyric acid   **f.** 3,5-dibromoheptanoic acid

16.8 Draw the condensed structural formulas of each of the following carboxylic acids:
   a. butyric acid   **b.** 3-ethylbenzoic acid
   c. α-hydroxyacetic acid   **d.** 2,4-dibromobutanoic acid
   e. m-methylbenzoic acid   **f.** 4,4-dibromohexanoic acid

16.9 Draw the condensed structural formulas of the carboxylic acids formed by the oxidation of each of the following:

a. CH_3—OH

b.
$$CH_3-\overset{\overset{\displaystyle O}{\|}}{C}-H$$

c. $CH_3-\overset{\overset{\displaystyle CH_3}{|}}{CH}-CH_2-CH_2-OH$

d. 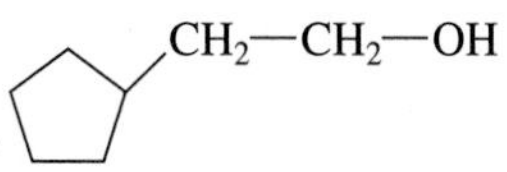 (cyclopentane)$-CH_2-CH_2-OH$

16.10 Draw the condensed structural formulas of the carboxylic acids formed by the oxidation of each of the following:

a. $CH_3-CH_2-CH_2-CH_2-CH_2-CH_2-OH$

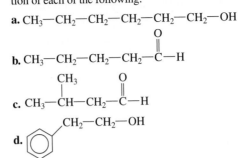

b. $CH_3-CH_2-CH_2-CH_2-\overset{\displaystyle O}{\overset{\|}{C}}-H$

c. $CH_3-\overset{\displaystyle CH_3}{\overset{|}{CH}}-CH_2-\overset{\displaystyle O}{\overset{\|}{C}}-H$

d. [benzene ring]$-CH_2-CH_2-OH$

Describe the boiling points, solubility, and ionization of carboxylic acids in water.

(MC GOB) Properties of Carboxylic Acids

16.2 Properties of Carboxylic Acids

Carboxylic acids are among the most polar organic compounds because the functional group consists of two polar groups: a hydroxyl (—OH) group and a carbonyl (C=O) group. This C=O double bond is similar to that of the aldehydes and ketones.

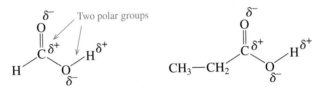

Therefore, carboxylic acids form hydrogen bonds with other carboxylic acid molecules and water. This ability to form hydrogen bonds has a major influence on both their boiling points and solubility in water.

Boiling Points

(MC GOB) SELF-STUDY ACTIVITY
Carboxylic Acids

Carboxylic acids have higher boiling points than alcohols, ketones, and aldehydes of similar mass.

	$CH_3-CH_2-\overset{\displaystyle O}{\overset{\|}{C}}-H$	$CH_3-CH_2-CH_2-OH$	$CH_3-\overset{\displaystyle O}{\overset{\|}{C}}-OH$
Compound	propanal	1-propanol	acetic acid
Molar mass	58	60	60
bp	49°C	97°C	118°C

One reason for the high boiling points of carboxylic acids is the formation of hydrogen bonds between two carboxylic acids to give a dimer. Because the dimers are stable as gases, the mass of the carboxylic acid is effectively doubled, which means that higher temperatures are required to reach the boiling point and form gases.

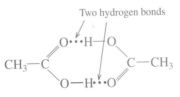

A dimer of two acetic acid molecules

Solubility in Water

Carboxylic acids with one to four carbons are very soluble in water because the carboxyl group forms hydrogen bonds with several water molecules. (See Figure 16.2.) However, as the length of the carbon chain increases, the nonpolar portion reduces solubility. Carboxylic acids having five or more carbons are not very soluble in water. Table 16.2 lists boiling points and solubilities for some selected carboxylic acids.

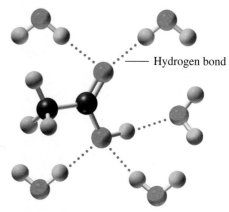

—— Hydrogen bond

SAMPLE PROBLEM 16.3

■ Properties of Carboxylic Acids

Put the following organic compounds in order of increasing boiling points: butanoic acid, pentane, and 2-butanol.

SOLUTION

The boiling point increases when the molecules of a compound can form hydrogen bonds or dipole–dipole interactions. The alkane has the lowest boiling point because alkanes cannot hydrogen bond. Alcohols and carboxylic acids have higher boiling points because they hydrogen bond via the —OH. The higher boiling points of carboxylic acids are due to the formation of stable dimers, which increase the effective mass and therefore the boiling point.

Pentane < 2-butanol < butanoic acid

STUDY CHECK

At its boiling point, acetic acid forms gas molecules that have a mass of 120 rather than the molar mass of 60 g/mole for CH_3COOH. Explain.

Acidity of Carboxylic Acids

One of the most important properties of carboxylic acids is their ionization in water, which makes them weak acids (Chapter 10). In the ionization, a carboxylic acid donates a proton to a water molecule to produce an anion called a **carboxylate ion** and a hydronium ion.

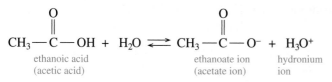

$$CH_3-\overset{\overset{\displaystyle O}{\|}}{C}-OH + H_2O \rightleftharpoons CH_3-\overset{\overset{\displaystyle O}{\|}}{C}-O^- + H_3O^+$$

ethanoic acid   ethanoate ion   hydronium
(acetic acid)   (acetate ion)   ion

Carboxylic acids are more acidic than other organic compounds including phenols, but only a small percentage (~1%) of the carboxylic acid molecules in a dilute solution are ionized, which means that most of the acid is not ionized. The acid dissociation constants of carboxylic acid are between 10^{-4} to 10^{-5} as seen in Table 16.3.

FIGURE 16.2

Acetic acid forms hydrogen bonds with water molecules.

Q Why do the atoms in the carboxyl group hydrogen bond with water molecules?

TABLE 16.2

Properties of Selected Carboxylic Acids

IUPAC name	bp (°C)	Soluble in water?
methanoic acid	101	Yes
ethanoic acid	118	Yes
propanoic acid	141	Yes
butanoic acid	164	Yes
pentanoic acid	187	Slightly
hexanoic acid	205	Slightly
benzoic acid	250	Slightly

TABLE 16.3

Acid Dissociation Constants K_a for Selected Carboxylic Acids

Name	K_a (25°C)
methanoic acid	1.8×10^{-4}
ethanoic acid	1.8×10^{-5}
propanoic acid	1.3×10^{-5}
butanoic acid	1.5×10^{-5}
pentanoic acid	1.5×10^{-5}
benzoic acid	6.5×10^{-5}

SAMPLE PROBLEM 16.4

■**Ionization of Carboxylic Acids in Water**

Write the equation for the ionization of propionic acid in water.

SOLUTION

The ionization of propionic acid produces a carboxylate ion and a hydronium ion.

$$CH_3-CH_2-\overset{\displaystyle O}{\overset{\displaystyle \|}{C}}-OH \;+\; H_2O \;\rightleftharpoons\; CH_3-CH_2-\overset{\displaystyle O}{\overset{\displaystyle \|}{C}}-O^- \;+\; H_3O^+$$

STUDY CHECK

Write an equation for the ionization of formic acid in water.

FIGURE 16.3

Preservatives and flavor enhancers in soups and seasonings are often carboxylic acids or their salts.

Q **What is the carboxylic acid salt produced by the neutralization of butanoic acid and lithium hydroxide?**

Neutralizaton of Carboxylic Acids

Although carboxylic acids are weak acids, they are completely neutralized by strong bases such as NaOH and KOH. The products are water and a **carboxylic acid salt**, which is a carboxylate ion and the metal ion from the base. The carboxylate ion is named by replacing the *oic acid* ending of the acid name with *ate*.

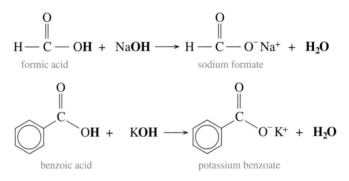

Sodium propionate, a preservative, is added to bread, cheeses, and bakery items to inhibit the spoilage of the food by microorganisms. Sodium benzoate, an inhibitor of mold and bacteria, is added to juices, margarine, relishes, salads, and jams. Monosodium glutamate (MSG) is added to meats, fish, vegetables, and bakery items to enhance flavor, although it causes headaches in some people. (See Figure 16.3.)

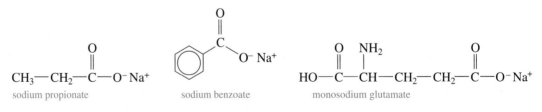

The carboxylic acid salts are solids at room temperature and have high melting points. Because they are ionic compounds, carboxylic acid salts of the alkali metals (Li$^+$, Na$^+$, and K$^+$) and NH$_4^+$ are usually soluble in water.

HEALTH NOTE

Carboxylic Acids in Metabolism

There are several carboxylic acids that are part of the metabolic processes within our cells. For example, during glycolysis, a molecule of glucose is broken down into two molecules of pyruvic acid or actually its carboxylate ion pyruvate. During strenuous exercise when oxygen levels are low (anaerobic), pyruvic acid is reduced to give lactic acid or the lactate ion. The buildup of lactate ion in the muscle leads to fatigue and pain.

$$CH_3-\overset{O}{\underset{}{C}}-\overset{O}{\underset{}{C}}-OH \;+2H \;\xrightarrow{\text{Reduction}}$$

pyruvic acid

$$CH_3-\overset{OH}{\underset{}{C}}H-\overset{O}{\underset{}{C}}-OH$$

lactic acid

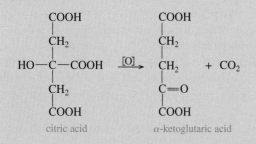

In the citric acid cycle or Krebs cycle, several dicarboxylic acids are oxidized and decarboxylated (loss of CO_2) in order to produce energy for the cell. These carboxylic acids are normally referred to by their common names. At the start of the citric acid cycle, citric acid with six carbons is converted to five-carbon α-ketoglutaric acid. Citric acid is also the acid that gives the sour tastes to citrus fruits such as lemons and grapefruits.

citric acid   α-ketoglutaric acid

The citric acid cycle continues as α-ketoglutaric acid loses CO_2 to give a four-carbon succinic acid. Then a series of reactions converts succinic acid to oxaloacetic acid. We see that some of the functional groups we have studied along with reactions such as hydration and oxidation are part of the metabolic processes that take place in our cells.

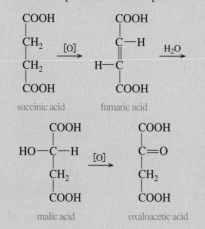

succinic acid   fumaric acid

malic acid   oxaloacetic acid

At the pH of the aqueous environment in the cells, the carboxylic acids are ionized, which means it is actually the carboxylate ions that take part in the reactions of citric acid cycle. For example, in water, succinic acid is in equilibrium with its carboxylate ion succinate.

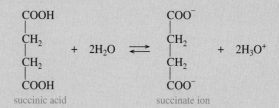

succinic acid   succinate ion

SAMPLE PROBLEM 16.5

■ **Neutralization of a Carboxylic Acid**

Write the equation for the neutralization of propionic acid with sodium hydroxide.

SOLUTION

The neutralization of an acid with a base produces the salt of the acid and water.

propionic acid   sodium propionate

STUDY CHECK

What carboxylic acid will give potassium butyrate when it is neutralized by KOH?

QUESTIONS AND PROBLEMS

■ **Properties of Carboxylic Acids**

16.11 Identify the compound in each pair that has the higher boiling point.
  a. acetic acid or butanoic acid
  b. 1-propanol or propanoic acid
  c. butanone or butanoic acid

16.12 Identify the compound in each pair that has the higher boiling point.
  a. acetone or propanoic acid
  b. propanoic acid or hexanoic acid
  c. ethanol or acetic acid

16.13 Select the compound in each group that is the most soluble in water:
  a. propanoic acid, hexanoic acid, benzoic acid
  b. pentane, 1-butanol, propanoic acid

16.14 Select the compound in each group that is the most soluble in water:
  a. butanone, butanoic acid, butane
  b. acetic acid, pentanoic acid, octanoic acid

16.15 Write equations for the ionization of each of the following carboxylic acids in water:

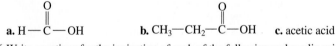

a. H—C—OH   **b.** CH₃—CH₂—C—OH   **c.** acetic acid

16.16 Write equations for the ionization of each of the following carboxylic acids in water.

  a. CH₃—CH—C—OH   **b.** α-hydroxyacetic acid   **c.** butanoic acid

16.17 Write equations for the reaction of each of the following carboxylic acids with NaOH:
  a. formic acid   **b.** propanoic acid   **c.** benzoic acid

16.18 Write equations for the reaction of each of the following carboxylic acids with KOH:
  a. acetic acid   **b.** 2-methylbutanoic acid   **c.** *p*-chlorobenzoic acid

16.19 Give the IUPAC and common names, if any, of the carboxylic acid salts in problem 16.17.

16.20 Give the IUPAC and common names, if any, of the carboxylic acid salts in problem 16.18.

16.3 Esters

LEARNING GOAL

Name an ester; write equations for the formation and hydrolysis of an ester.

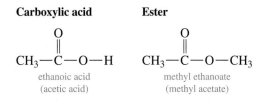

A carboxylic acid reacts with an alcohol to form an **ester**. In an ester, the —H of the carboxylic acid is replaced by an alkyl group. Aspirin is an ester as well as a carboxylic acid. Fats and oils in our diets contain esters of glycerol and fatty acids, which are long-chain carboxylic acids. The aromas and flavors of many fruits including bananas, oranges, and strawberries are due to esters.

Carboxylic acid | **Ester**

$$CH_3{-}\overset{\displaystyle O}{\overset{\|}{C}}{-}O{-}H \qquad CH_3{-}\overset{\displaystyle O}{\overset{\|}{C}}{-}O{-}CH_3$$

ethanoic acid (acetic acid) | methyl ethanoate (methyl acetate)

MC GOB: Formation of Esters from Carboxylic Acids; Writing Esterification Equations

Esterification

In a reaction called **esterification**, a carboxylic acid reacts with an alcohol when heated in the presence of an acid catalyst (usually H_2SO_4). In the reaction, water is produced from the —OH removed from the carboxylic acid and an —H lost by the alcohol.

$$\underset{\text{acetic acid}}{CH_3{-}\overset{\displaystyle O}{\overset{\|}{C}}{-}O{-}H} + \underset{\text{methyl alcohol}}{H{-}O{-}CH_3} \xrightleftharpoons{H^+,\ \text{heat}} \underset{\text{methyl acetate}}{CH_3{-}\overset{\displaystyle O}{\overset{\|}{C}}{-}O{-}CH_3} + H{-}O{-}H$$

If we use acetic acid and 1-propanol, we can write an equation for the formation of the ester that is responsible for the flavor and odor of pears.

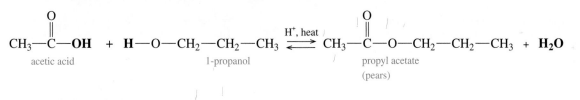

$$\underset{\text{acetic acid}}{CH_3{-}\overset{\displaystyle O}{\overset{\|}{C}}{-}\mathbf{OH}} + \underset{\text{1-propanol}}{\mathbf{H}{-}O{-}CH_2{-}CH_2{-}CH_3} \xrightleftharpoons{H^+,\ \text{heat}} \underset{\text{propyl acetate (pears)}}{CH_3{-}\overset{\displaystyle O}{\overset{\|}{C}}{-}O{-}CH_2{-}CH_2{-}CH_3} + \mathbf{H_2O}$$

SAMPLE PROBLEM 16.6

Writing Esterification Equations

The ester that gives the flavor and odor of apples can be synthesized from butyric acid and methyl alcohol. What is the equation for the formation of the ester in apples?

SOLUTION

$$\underset{\text{butyric acid}}{CH_3{-}CH_2{-}CH_2{-}\overset{\displaystyle O}{\overset{\|}{C}}{-}\mathbf{OH}} + \underset{\text{methyl alcohol}}{\mathbf{H}{-}O{-}CH_3} \xrightleftharpoons{H^+,\ \text{heat}}$$

$$\underset{\text{methyl butyrate}}{CH_3{-}CH_2{-}CH_2{-}\overset{\displaystyle O}{\overset{\|}{C}}{-}O{-}CH_3} + \mathbf{H_2O}$$

Salicylic Acid and Aspirin

Chewing on a piece of willow bark was used as a way of relieving pain for many centuries. By the 1800s, chemists discovered that salicylic acid was the agent in the bark responsible for the relief of pain. However, salicylic acid, which has both a carboxylic group and a hydroxyl group, irritates the stomach lining. A less irritating ester of salicylic acid and acetic acid, called acetylsalicyclic acid or "aspirin," was prepared in 1899 by the Bayer chemical company in Germany. In some aspirin preparations, a buffer is added to neutralize the carboxylic acid group and lessen its irritation of the stomach. Aspirin is used as an analgesic (pain reliever), antipyretic (fever reducer), and anti-inflammatory agent.

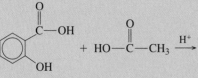

Oil of wintergreen, or methyl salicylate, has a spearmint odor and flavor. Because it can pass through the skin, methyl salicylate is used in skin ointments where it acts as a counterirritant, producing heat to soothe sore muscles.

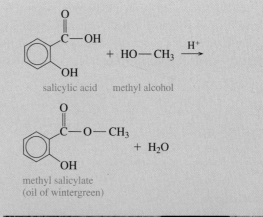

STUDY CHECK

What carboxylic acid and alcohol are needed to form the following ester, which gives the flavor and odor to apricots?

$$CH_3-CH_2-\overset{\overset{\displaystyle O}{\|}}{C}-O-CH_2-CH_2-CH_2-CH_2-CH_3$$

QUESTIONS AND PROBLEMS

MC
GOB

Esters

16.21 Identify each of the following as an aldehyde, a ketone, a carboxylic acid, or an ester:

a. $CH_3-\overset{\overset{\displaystyle O}{\|}}{C}-H$

b. $CH_3-\overset{\overset{\displaystyle O}{\|}}{C}-O-CH_3$

c. $CH_3-CH_2-\overset{\overset{\displaystyle O}{\|}}{C}-CH_3$

d. $CH_3-CH_2-\overset{\overset{\displaystyle O}{\|}}{C}-O-H$

Plastics

Terephthalic acid (an acid with two carboxyl groups) is produced in large quantities for the manufacture of polyesters such as Dacron and plastics. When terephthalic acid reacts with ethylene glycol, ester bonds can form on both ends of the molecules, allowing many molecules to combine until they have formed a long polymer known as a *polyester*.

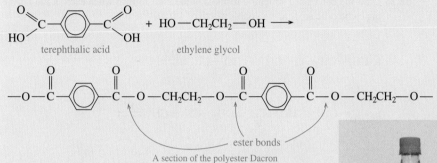

A section of the polyester Dacron

Dacron polyester is used to make permanent press fabrics, carpets, and clothes. In medicine, artificial blood vessels and valves are made of Dacron, which is biologically inert and does not clot the blood. The polyester can also be made as a film called Mylar and as a plastic known as PETE (**p**oly**e**thylene**te**rphthalate). PETE is used for plastic soft drink bottles as well as for containers of salad dressings, shampoos, and dishwashing liquids.

Today PETE is the most widely recycled of all the plastics. In 1992, 365 million pounds (166 million kilograms) of PETE were recycled. After it is separated from other plastics, PETE can be changed into other useful items including polyester fabric for T-shirts and coats, fill for sleeping bags, door mats, and tennis ball containers.

16.22 Identify each of the following as an aldehyde, a ketone, a carboxylic acid, or an ester:

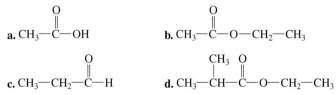

16.23 Write the condensed structural formula of the ester formed when each of the following react with methyl alcohol:

a. acetic acid

b. butyric acid

c. benzoic acid

16.24 Write the condensed structural formula of the ester formed when each of the following react with methyl alcohol:

a. formic acid

b. propionic acid

c. 2-methylpentanoic acid

16.25 Draw the condensed structural formulas of the ester formed when each of the following carboxylic acids and alcohols react:

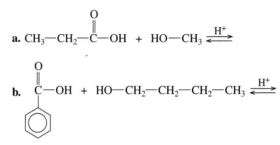

a. $CH_3-CH_2-\overset{\overset{\displaystyle O}{\|}}{C}-OH$ + $HO-CH_2-CH_2-CH_3$ $\overset{H^+}{\rightleftharpoons}$

b. $CH_3-CH_2-CH_2-CH_2-\overset{\overset{\displaystyle O}{\|}}{C}-OH$ + $HO-\overset{\overset{\displaystyle CH_3}{|}}{CH}-CH_3$ $\overset{H^+}{\rightleftharpoons}$

16.26 Draw the condensed structural formula of the ester formed when each of the following carboxylic acids and alcohols react:

a. $CH_3-CH_2-\overset{\overset{\displaystyle O}{\|}}{C}-OH$ + $HO-CH_3$ $\overset{H^+}{\rightleftharpoons}$

b. $\overset{\overset{\displaystyle O}{\|}}{C}-OH$ + $HO-CH_2-CH_2-CH_2-CH_3$ $\overset{H^+}{\rightleftharpoons}$

LEARNING GOAL

Write the IUPAC and common names for esters; draw condensed structural formulas.

Guide to Naming Esters

STEP 1
Write the name of the carbon chain from the alcohol as an *alkyl* group.

STEP 2
Write the name of the carboxylic acid as *carboxylate*.

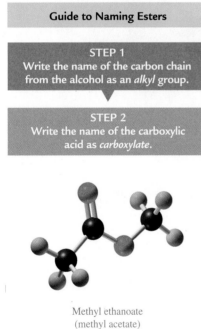

Methyl ethanoate
(methyl acetate)

■ 16.4 Naming Esters

The name of an ester consists of two words taken from the names of the alcohol and the acid. The first word indicates the *alkyl* part of the alcohol. The second word is the *carboxylate* name of the carboxylic acid. The IUPAC names of esters use the IUPAC names for the acid, while the common names of esters use the common names for the acid. Let's take a look at the following ester and break it into two parts, one from the alcohol and one from the acid. By writing and naming the alcohol and carboxylic acid that produced the ester, we can determine the name of the ester.

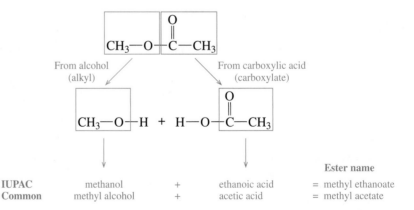

				Ester name
IUPAC	methanol	+	ethanoic acid	= methyl ethanoate
Common	methyl alcohol	+	acetic acid	= methyl acetate

The following examples of some typical esters show the IUPAC as well as the common names of esters.

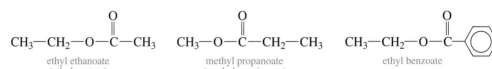
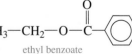

$CH_3-CH_2-O-\overset{\overset{\displaystyle O}{\|}}{C}-CH_3$ $CH_3-O-\overset{\overset{\displaystyle O}{\|}}{C}-CH_2-CH_3$ $CH_3-CH_2-O-\overset{\overset{\displaystyle O}{\|}}{C}-$⬡

ethyl ethanoate
(ethyl acetate)

methyl propanoate
(methyl propionate)

ethyl benzoate

TABLE 16.4

Some Esters in Fruits and Flavorings

Condensed Structural Formula and Name	Flavor/Odor
$CH_3-\overset{\displaystyle O}{\overset{\|}{C}}-O-CH_2-CH_2-CH_3$ propyl ethanoate (propyl acetate)	Pears
$CH_3-\overset{\displaystyle O}{\overset{\|}{C}}-O-CH_2-CH_2-CH_2-CH_2-CH_3$ pentyl ethanoate (pentyl acetate)	Bananas
$CH_3-\overset{\displaystyle O}{\overset{\|}{C}}-O-CH_2-CH_2-CH_2-CH_2-CH_2-CH_2-CH_2-CH_3$ octyl ethanoate (octyl acetate)	Oranges
$CH_3-CH_2-CH_2-\overset{\displaystyle O}{\overset{\|}{C}}-O-CH_2-CH_3$ ethyl butanoate (ethyl butyrate)	Pineapples
$CH_3-CH_2-CH_2-\overset{\displaystyle O}{\overset{\|}{C}}-O-CH_2-CH_2-CH_2-CH_2-CH_3$ pentyl butanoate (pentyl butyrate)	Apricots

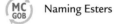

Many of the fragrances of perfumes and flowers and the flavors of fruits are due to esters. Small esters are volatile so we can smell them and soluble in water so we can taste them. Several of these are listed in Table 16.4.

SAMPLE PROBLEM 16.7

■ Naming Esters

Write the IUPAC and common names of the following ester:

$$CH_3-CH_2-\overset{\displaystyle O}{\overset{\|}{C}}-O-CH_2-CH_2-CH_3$$

MC GOB   Naming Esters

SOLUTION

The alcohol part of the ester is propyl, and the carboxylic acid part is propanoic (propionic) acid.

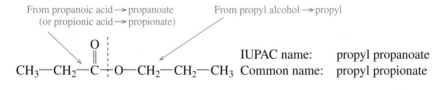

From propanoic acid → propanoate
(or propionic acid → propionate)   From propyl alcohol → propyl

$$CH_3-CH_2-\overset{\displaystyle O}{\overset{\|}{C}}\!\!\dashv\!O-CH_2-CH_2-CH_3$$

IUPAC name: propyl propanoate
Common name: propyl propionate

STUDY CHECK

Draw the condensed structural formula of pentyl acetate.

QUESTIONS AND PROBLEMS

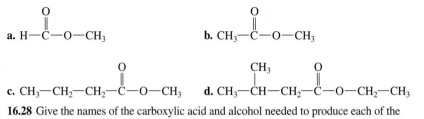

▪ Naming Esters

16.27 Give the names of the carboxylic acid and alcohol needed to produce each of the following esters:

a. $H-\overset{\displaystyle O}{\overset{\|}{C}}-O-CH_3$ b. $CH_3-\overset{\displaystyle O}{\overset{\|}{C}}-O-CH_3$

c. $CH_3-CH_2-CH_2-\overset{\displaystyle O}{\overset{\|}{C}}-O-CH_3$ d. $CH_3-\overset{\overset{\displaystyle CH_3}{\displaystyle |}}{CH}-CH_2-\overset{\displaystyle O}{\overset{\|}{C}}-O-CH_2-CH_3$

16.28 Give the names of the carboxylic acid and alcohol needed to produce each of the following esters:

a. $CH_3-CH_2-\overset{\displaystyle O}{\overset{\|}{C}}-O-CH_2-CH_3$ b. $CH_3-CH_2-CH_2-CH_2-CH_2-\overset{\displaystyle O}{\overset{\|}{C}}-O-CH_3$

c. $CH_3-CH_2-\underset{\underset{\displaystyle CH_3}{\displaystyle |}}{CH}-\overset{\displaystyle O}{\overset{\|}{C}}-O-CH_3$ d. $CH_3-CH_2-\overset{\displaystyle O}{\overset{\|}{C}}-O-CH_2-CH_2-CH_2-CH_3$

16.29 Name each of the following esters:

a. $CH_3-O-\overset{\displaystyle O}{\overset{\|}{C}}-H$ b. $CH_3-O-\overset{\displaystyle O}{\overset{\|}{C}}-CH_3$

c. $CH_3-O-\overset{\displaystyle O}{\overset{\|}{C}}-CH_2-CH_2-CH_3$ d. $CH_3-\overset{\overset{\displaystyle CH_3}{\displaystyle |}}{CH}-CH_2-\overset{\displaystyle O}{\overset{\|}{C}}-O-CH_2-CH_3$

16.30 Name each of the following esters:

a. $CH_3-CH_2-O-\overset{\displaystyle O}{\overset{\|}{C}}-CH_2-CH_2-CH_3$

b. $CH_3-O-\overset{\displaystyle O}{\overset{\|}{C}}-CH_2-CH_2-CH_2-CH_2-CH_3$

c. $CH_3-O-\overset{\displaystyle O}{\overset{\|}{C}}-CH_2-\overset{\overset{\displaystyle CH_3}{\displaystyle |}}{CH}-CH_3$

d. $CH_3-CH_2-\overset{\displaystyle O}{\overset{\|}{C}}-O-CH_2-CH_2-CH_2-CH_3$

16.31 Draw the condensed structural formulas of each of the following esters:
a. methyl acetate   b. butyl formate
c. ethyl pentanoate   d. 2-bromopropyl propanoate

16.32 Draw the condensed structural formulas of each of the following esters:
a. hexyl acetate   b. propyl propionate
c. ethyl-2-hydroxybutanoate   d. methyl benzoate

16.33 What is the ester responsible for the flavor and odor of the following fruit?
a. banana   b. orange   c. apricot

16.34 What flavor would you notice if you smelled or tasted the following?
a. ethyl butanoate   b. propyl acetate   c. octyl acetate

16.5 Properties of Esters

Esters have higher boiling points than alkanes, but lower than alcohols and carboxylic acids of similar mass. Because ester molecules do not have hydroxyl groups, they cannot hydrogen bond to each other.

LEARNING GOAL

Describe the boiling points and solubility of esters. Draw the condensed structural formulas of the hydrolysis products.

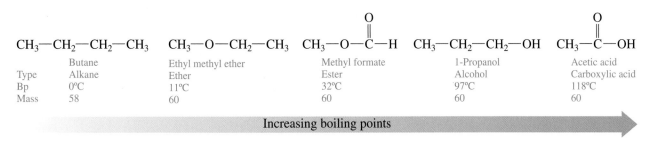

	Butane	Ethyl methyl ether	Methyl formate	1-Propanol	Acetic acid
Type	Alkane	Ether	Ester	Alcohol	Carboxylic acid
Bp	0°C	11°C	32°C	97°C	118°C
Mass	58	60	60	60	60

Increasing boiling points ⟶

Solubility in Water

Esters having only a few carbon atoms are soluble in water. The partially negative oxygen of the carbonyl group forms hydrogen bonds with the partially positive hydrogen atoms of water molecules. The solubility of esters decreases as the number of carbon atoms increases.

MC GOB Solubility of Esters
Hydrolysis of Esters

Acid Hydrolysis of Esters

In **hydrolysis**, water splits apart esters when heated in the presence of a strong acid, usually H_2SO_4 or HCl. The products of acid hydrolysis are the carboxylic acid and alcohol. Therefore, hydrolysis is the reverse of the esterification reaction. When hydrolysis of biological compounds occurs in the cells, an enzyme replaces the acid as the catalyst. In the hydrolysis reaction, the —OH from a water molecule bonds to the carbonyl group of the ester to form the carboxylic acid.

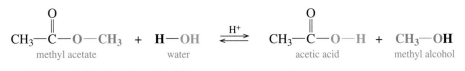

$$CH_3-\overset{\overset{\displaystyle O}{\|}}{C}-O-CH_3 \ + \ H-OH \ \underset{}{\overset{H^+}{\rightleftarrows}} \ CH_3-\overset{\overset{\displaystyle O}{\|}}{C}-O-H \ + \ CH_3-OH$$

methyl acetate   water   acetic acid   methyl alcohol

SAMPLE PROBLEM 16.8

■ Acid Hydrolysis of Esters

Aspirin that has been stored for a long time may undergo hydrolysis in the presence of water and heat. What are the hydrolysis products of aspirin? Why does a bottle of old aspirin smell like vinegar?

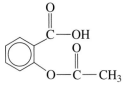

aspirin, acetylsalicylic acid

SOLUTION

To write the hydrolysis products, separate the compound at the ester bond. Complete the formula of the carboxylic acid by adding —OH (from water) to the carbonyl group and an —H to complete the alcohol. The acetic acid in the products gives the vinegar odor to a sample of aspirin that has hydrolyzed.

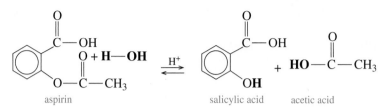

aspirin   salicylic acid   acetic acid

STUDY CHECK

What are the names of the products from the acid hydrolysis of ethyl propionate?

Base Hydrolysis of Esters (Saponification)

When an ester undergoes hydrolysis with a strong base such as NaOH or KOH, the products are the carboxylic acid salt and the corresponding alcohol. The base hydrolysis reaction is also called **saponification**, which refers to the reaction of a long-chain fatty acid with NaOH to make soap. The carboxylic acid, which is produced in acid hydrolysis, is converted in an irreversible reaction to its carboxylate ion by the strong base.

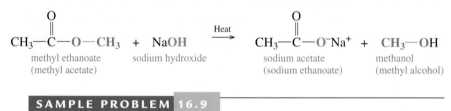

methyl ethanoate   sodium hydroxide   sodium acetate   methanol
(methyl acetate)   (sodium ethanoate)   (methyl alcohol)

SAMPLE PROBLEM 16.9

■ Base Hydrolysis of Esters

Ethyl acetate is a solvent widely used for fingernail polish, plastics, and lacquers. Write the equation of the hydrolysis of ethyl acetate by NaOH.

SOLUTION

The hydrolysis of ethyl acetate by NaOH gives the salt of acetic acid and ethyl alcohol.

$$CH_3-\overset{\overset{\displaystyle O}{\|}}{C}-O-CH_2-CH_3 \ + \ NaOH \ \xrightarrow{\text{Heat}} \ CH_3-\overset{\overset{\displaystyle O}{\|}}{C}-O^-Na^+ \ + \ HO-CH_2-CH_3$$

ethyl acetate   sodium acetate   ethyl alcohol

STUDY CHECK

Write the condensed structural formulas of the products from the hydrolysis of methyl benzoate by KOH.

ENVIRONMENTAL NOTE

Cleaning Action of Soaps

For many centuries, soaps were made by heating a mixture of animal fats (tallow) with lye, a basic solution obtained from wood ashes. In the soap-making process, fatty acids, which are long-chain carboxylic acids, undergo saponification with the strong base in lye.

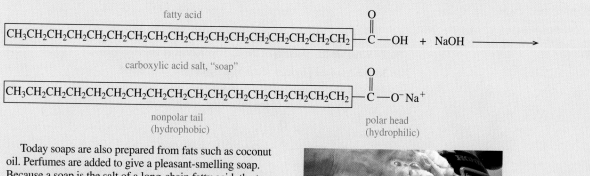

fatty acid

$$CH_3CH_2CH_2CH_2CH_2CH_2CH_2CH_2CH_2CH_2CH_2CH_2CH_2CH_2CH_2CH_2 \overset{\overset{\displaystyle O}{\|}}{C}-OH \ + \ NaOH \longrightarrow$$

carboxylic acid salt, "soap"

$$CH_3CH_2CH_2CH_2CH_2CH_2CH_2CH_2CH_2CH_2CH_2CH_2CH_2CH_2CH_2CH_2 \overset{\overset{\displaystyle O}{\|}}{C}-O^-Na^+$$

nonpolar tail
(hydrophobic)

polar head
(hydrophilic)

Today soaps are also prepared from fats such as coconut oil. Perfumes are added to give a pleasant-smelling soap. Because a soap is the salt of a long-chain fatty acid, the two ends of a soap molecule have different polarities. The long carbon chain end is nonpolar and *hydrophobic* (water-fearing). It is soluble in nonpolar substances such as oil or grease; but it is not soluble in water. The carboxylate salt end is ionic and *hydrophilic* (water-loving). It is very soluble in water but not in oils or grease.

When a soap is used to clean grease or oil, the nonpolar ends of the soap molecules dissolve in the nonpolar fats and oils that accompany dirt. The water-loving salt ends of the soap molecules extend outside where they can dissolve in water. The soap molecules coat the oil or grease, forming clusters called *micelles*. The ionic ends of the soap molecules provide polarity to the micelles, which makes them soluble in water. As a result, small globules of oil and fat coated with soap molecules are pulled into the water and rinsed away.

One of the problems of using soaps is that the carboxylate end reacts with ions in water such as Ca^{2+} and Mg^{2+} and forms insoluble substances.

$$2CH_3(CH_2)_{16}COO^- + Mg^{2+} \longrightarrow [CH_3(CH_2)_{16}COO^-]_2Mg^{2+}$$

stearate ion   magnesium   magnesium stearate
                      ion   (insoluble)

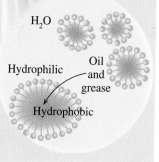

Soap molecule

Fatty acid
chain

H_2O

Hydrophilic

Oil
and
grease

Hydrophobic

QUESTIONS AND PROBLEMS

Properties of Esters

16.35 For each of the following pairs of compounds, select the compound that has the higher boiling point:

a. $CH_3-\overset{\overset{\displaystyle O}{\|}}{C}-O-CH_3$ or $CH_3-\overset{\overset{\displaystyle O}{\|}}{C}-OH$

b. $CH_3-\overset{\overset{\displaystyle O}{\|}}{C}-O-CH_3$ or $CH_3-CH_2-CH_2-CH_2-OH$

c. $CH_3-CH_2-CH_2-CH_3$ or $CH_3-O-\overset{\overset{\displaystyle O}{\|}}{C}-CH_3$

16.36 For each of the following pairs of compounds, select the compound that has the higher boiling point:

a.
$$\underset{\text{H}}{\overset{\text{O}}{\underset{\|}{\text{C}}}}\text{—O—CH}_3 \quad \text{or} \quad \text{CH}_3\text{—CH}_2\text{—CH}_2\text{—OH}$$

b.
$$\text{CH}_3\text{—}\overset{\text{O}}{\overset{\|}{\text{C}}}\text{—O—CH}_3 \quad \text{or} \quad \text{CH}_3\text{—CH}_2\text{—}\overset{\text{O}}{\overset{\|}{\text{C}}}\text{—OH}$$

c.
$$\text{CH}_3\text{—O—CH}_2\text{—CH}_3 \quad \text{or} \quad \text{CH}_3\text{—O—}\overset{\text{O}}{\overset{\|}{\text{C}}}\text{—H}$$

16.37 What are the products of the acid hydrolysis of an ester?

16.38 What are the products of the base hydrolysis of an ester?

16.39 Draw the structural formulas of the products from the acid- or base-catalyzed hydrolysis of each of the following compounds:

a. $\text{CH}_3\text{—CH}_2\text{—}\overset{\text{O}}{\overset{\|}{\text{C}}}\text{—O—CH}_3 + \text{NaOH} \longrightarrow$

b. $\text{CH}_3\text{—}\overset{\text{O}}{\overset{\|}{\text{C}}}\text{—O—CH}_2\text{—CH}_2\text{—CH}_3 + \text{H}_2\text{O} \xrightarrow{\text{H}^+}$

c. $\text{CH}_3\text{—CH}_2\text{—CH}_2\text{—}\overset{\text{O}}{\overset{\|}{\text{C}}}\text{—O—CH}_2\text{—CH}_3 + \text{H}_2\text{O} \xrightarrow{\text{H}^+}$

d. $\bigcirc\text{—}\overset{\text{O}}{\overset{\|}{\text{C}}}\text{—O—CH}_2\text{—CH}_3 + \text{H}_2\text{O} \xrightarrow{\text{H}^+}$

e. $\bigcirc\text{—}\overset{\text{O}}{\overset{\|}{\text{C}}}\text{—O—CH}_2\text{—CH}_3 + \text{NaOH} \longrightarrow$

16.40 Draw the structural formulas of the products from the acid- or base-catalyzed hydrolysis of each of the following compounds:

a. $\text{CH}_3\text{—CH}_2\text{—}\overset{\text{O}}{\overset{\|}{\text{C}}}\text{—O—CH}_2\text{—CH}_2\text{—CH}_2\text{—CH}_3 + \text{H}_2\text{O} \xrightarrow{\text{H}^+}$

b. $\text{H—}\overset{\text{O}}{\overset{\|}{\text{C}}}\text{—O—CH}_2\text{—CH}_3 + \text{NaOH} \longrightarrow$

c. $\text{CH}_3\text{—CH}_2\text{—}\overset{\text{O}}{\overset{\|}{\text{C}}}\text{—O—CH}_3 + \text{H}_2\text{O} \xrightarrow{\text{H}^+}$

d. $\text{CH}_3\text{—CH}_2\text{—}\overset{\text{O}}{\overset{\|}{\text{C}}}\text{—O—}\bigcirc + \text{H}_2\text{O} \xrightarrow{\text{H}^+}$

e. $\bigcirc\text{—CH}_2\text{—}\overset{\text{O}}{\overset{\|}{\text{C}}}\text{—O—CH}_2\text{—CH}_3 + \text{NaOH} \longrightarrow$

CONCEPT MAP

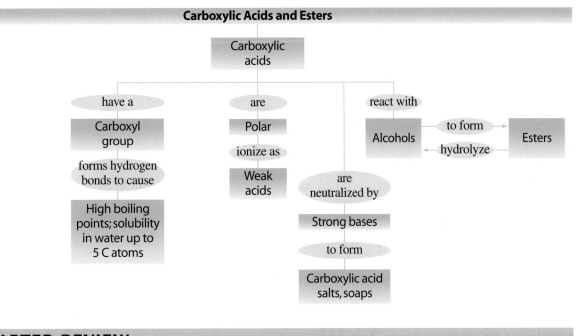

Carboxylic Acids and Esters

Carboxylic acids

have a → Carboxyl group → forms hydrogen bonds to cause → High boiling points; solubility in water up to 5 C atoms

are → Polar → ionize as → Weak acids

are neutralized by → Strong bases → to form → Carboxylic acid salts, soaps

react with → Alcohols → to form → Esters / hydrolyze

CHAPTER REVIEW

16.1 Carboxylic Acids

A carboxylic acid contains the carboxyl functional group, which is a hydroxyl group connected to the carbonyl group.

16.2 Properties of Carboxylic Acids

The carboxyl group contains polar bonds of O—H and C=O, which makes a carboxylic acid with one to four carbon atoms very soluble in water. As weak acids, carboxylic acids ionize slightly by donating a proton to water to form carboxylate and hydronium ions. Carboxylic acids are neutralized by base, producing the carboxylate salt and water.

16.3 Esters

In an ester, an alkyl or aromatic group has replaced the H of the hydroxyl group of a carboxylic acid. In the presence of a strong acid, a carboxylic acid reacts with an alcohol to produce an ester. A molecule of water is removed: —OH from the carboxylic acid and —H from the alcohol molecule.

16.4 Naming Esters

The names of esters consist of two words, one from the alcohol and the other from the carboxylic acid with the *ic* ending replaced by *ate*.

16.5 Properties of Esters

Esters undergo acid hydrolysis by adding water to yield the carboxylic acid and alcohol (or phenol). Base hydrolysis or saponification of an ester produces the carboxylate salt and an alcohol.

SUMMARY OF NAMING

Family	Condensed Structural Formula	IUPAC Name	Common Name
carboxylic acid	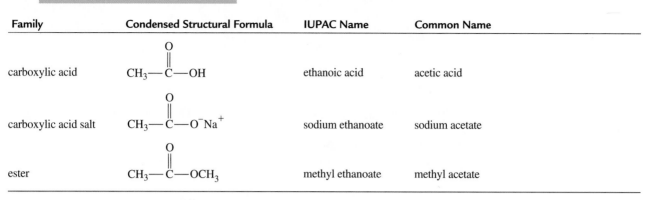	ethanoic acid	acetic acid
carboxylic acid salt		sodium ethanoate	sodium acetate
ester		methyl ethanoate	methyl acetate

SUMMARY OF REACTIONS

IONIZATION OF A CARBOXYLIC ACID IN WATER

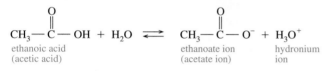

NEUTRALIZATION OF A CARBOXYLIC ACID

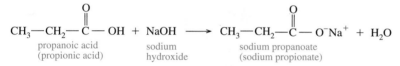

ESTERIFICATION: CARBOXYLIC ACID AND AN ALCOHOL

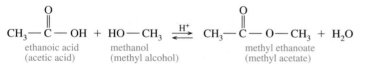

ACID HYDROLYSIS OF AN ESTER

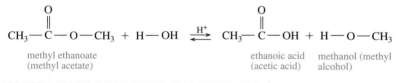

BASE HYDROLYSIS OF AN ESTER (SAPONIFICATION)

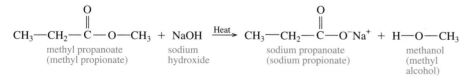

KEY TERMS

carboxyl group A functional group found in carboxylic acids composed of carbonyl and hydroxyl groups.

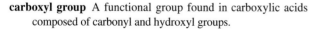

carboxylate ion The anion produced when a carboxylic acid donates a proton to water.

carboxylic acids A family of organic compounds containing the carboxyl group.

carboxylic acid salt The product of neutralization of a carboxylic acid; a carboxylate ion and the metal ion from the base.

esterification The formation of an ester from a carboxylic acid and an alcohol with the elimination of a molecule of water in the presence of an acid catalyst.

esters A family of organic compounds in which an alkyl group replaces the hydrogen atom in a carboxylic acid.

hydrolysis The splitting of a molecule by the addition of water. Esters hydrolyze to produce a carboxylic acid and an alcohol.

saponification The hydrolysis of an ester with a strong base to produce a salt of the carboxylic acid and an alcohol.

UNDERSTANDING THE CONCEPTS

16.41 Propyl acetate is the ester that gives the odor and smell of pears.

a. What is the structure of propyl acetate?
b. Write an equation for the formation of propyl acetate.
c. Write an equation for the acid hydrolysis of propyl acetate.
d. Write an equation for the base hydrolysis of propyl acetate with NaOH.
e. How many mL of 0.208 M NaOH are needed to complete hydrolyze (saponify) 1.58 g propyl acetate?

16.42 Ethyl octanoate is a flavor component of mangos.

a. What is the structure of ethyl octanoate?
b. Write an equation for the formation of ethyl octanoate.
c. Write an equation for the acid hydrolysis of ethyl octanoate.
d. Write an equation for the base hydrolysis of ethyl octanoate with NaOH.
e. How many mL of 0.315 M NaOH are needed to complete hydrolyze (saponify) 2.840 g ethyl octanoate?

ADDITIONAL QUESTIONS AND PROBLEMS

For instructor-assigned homework, go to **www.masteringgob.com.**

16.43 Give the IUPAC and common names (if any) for each of the following compounds:

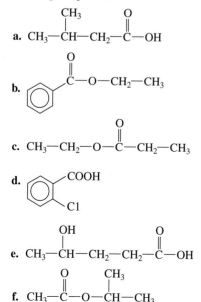

16.44 Give the IUPAC and common names (if any) for each of the following compounds:

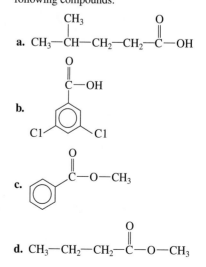

e.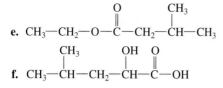

$$CH_3-CH-CH_2-CH-C-OH$$

with CH_3 and OH O groups

f.

16.45 Draw the structural formulas of at least three carboxylic acids with the molecular formula $C_5H_{10}O_2$.

16.46 Draw the structural formulas of at least three esters with the formula $C_4H_8O_2$.

16.47 Draw the structural formulas of each of the following:
　a. methyl acetate
　b. p-chlorobenzoic acid
　c. β-chloropropionic acid
　d. ethyl butanoate
　e. 3-methylpentanoic acid
　f. ethyl benzoate

16.48 Draw the structural formulas of each of the following:
　a. α-bromobutyric acid
　b. ethyl butyrate
　c. 2-methyloctanoic acid
　d. 3,5-dimethylhexanoic acid
　e. propyl acetate
　f. 3,4-dibromobenzoic acid

16.49 In each of the following pairs of compounds, select the compound that would have the higher boiling point:

　a. $CH_3-CH_2-CH_2-OH$ or $CH_3-\overset{\overset{\displaystyle O}{\|}}{C}-OH$

　b. $CH_3-CH_2-CH_2-CH_3$ or $CH_3-CH_2-\overset{\overset{\displaystyle O}{\|}}{C}-OH$

　c. $CH_3-\overset{\overset{\displaystyle O}{\|}}{C}-OH$ or $CH_3-CH_2-CH_2-\overset{\overset{\displaystyle O}{\|}}{C}-OH$

16.50 In each of the following pairs of compounds, select the compound that would have the higher boiling point:

　a. $CH_3-CH_2-CH_2-OH$ or $CH_3-\overset{\overset{\displaystyle O}{\|}}{C}-O-CH_3$

　b. $CH_3-O-\overset{\overset{\displaystyle O}{\|}}{C}-CH_3$ or $CH_3-CH_2-\overset{\overset{\displaystyle O}{\|}}{C}-OH$

　c. $CH_3-\overset{\overset{\displaystyle O}{\|}}{C}-O-CH_3$ or $CH_3-CH_2-CH_2-CH_3$

16.51 Why does acetic acid have a higher boiling point than either 1-propanol or methyl formate when they all have the same molar mass?

16.52 Propionic acid, 1-butanol, and butanal all have the same molar mass. The possible boiling points are 76°C, 118°C, and 141°C. Match the compounds with the boiling points and explain your choice.

16.53 Which of the following compounds are soluble in water?
　a. $CH_3-CH_2-CH_2-CH_2-CH_3$

　b. $CH_3-CH_2-\overset{\overset{\displaystyle O}{\|}}{C}-O^-Na^+$

　c. $CH_3-\overset{\overset{\displaystyle O}{\|}}{C}-O-CH_3$

　d. $CH_3-CH_2-CH_2-OH$

　e. $CH_3-CH_2-\overset{\overset{\displaystyle O}{\|}}{C}-OH$

16.54 Which of the following compounds are soluble in water?
　a. $CH_3-CH_2-CH_2-\overset{\overset{\displaystyle O}{\|}}{C}-OH$

　b. $CH_3-CH_2-\overset{\overset{\displaystyle O}{\|}}{C}-O-CH_2-CH_2-CH_3$

　c. $CH_3-CH_2-CH_2-CH_3$

　d. $CH_3-(CH_2)_8-CH_2-OH$

　e. $CH_3-CH_2-CH_2-O-CH_2-CH_2-CH_3$

16.55 Write the products of the following reactions:

　a. $CH_3-CH_2-\overset{\overset{\displaystyle O}{\|}}{C}-OH + H_2O \rightleftharpoons$

　b. $CH_3-CH_2-\overset{\overset{\displaystyle O}{\|}}{C}-OH + KOH \longrightarrow$

　c. $CH_3-CH_2-\overset{\overset{\displaystyle O}{\|}}{C}-OH + CH_3-OH \overset{H^+}{\rightleftharpoons}$

　d. benzene ring$-\overset{\overset{\displaystyle O}{\|}}{C}-OH$ $+ CH_3-CH_2-OH \overset{H^+}{\rightleftharpoons}$

16.56 Write the products of the following reactions:

　a. $CH_3-\overset{\overset{\displaystyle O}{\|}}{C}-OH + NaOH \longrightarrow$

　b. $CH_3-\overset{\overset{\displaystyle O}{\|}}{C}-OH + H_2O \rightleftharpoons$

　c. $CH_3-\overset{CH_3}{\underset{}{CH}}-\overset{\overset{\displaystyle O}{\|}}{C}-OH + KOH \longrightarrow$

　d. $CH_3-\overset{CH_3}{\underset{}{CH}}-\overset{\overset{\displaystyle O}{\|}}{C}-OH + CH_3-OH \overset{H^+}{\rightleftharpoons}$

16.57 Give the IUPAC names of the carboxylic acid and alcohol needed to prepare each of the following esters:

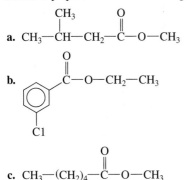

a. $CH_3-CH-CH_2-C-O-CH_3$ (with CH_3 and O groups)

b. (benzene ring with $C-O-CH_2-CH_3$ and Cl)

c. $CH_3-(CH_2)_4-C-O-CH_3$ (with O)

16.58 Give the IUPAC names of the carboxylic acid and alcohol needed to prepare each of the following esters:

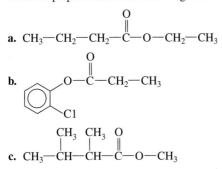

a. $CH_3-CH_2-CH_2-C-O-CH_2-CH_3$ (with O)

b. (benzene ring with $O-C-CH_2-CH_3$ and Cl)

c. $CH_3-CH-CH-C-O-CH_3$ (with CH_3, CH_3, O)

16.59 Write the products of the following reactions:

a. $CH_3-CH_2-C-O-CH-CH_3 + H_2O \xrightleftharpoons{H^+}$ (with O and CH_3 groups)

b. $CH_3-CH-C-O-CH_2-CH_2-CH_3 + NaOH \longrightarrow$ (with CH_3 and O)

16.60 Write the products of the following reactions:

a. $CH_3-CH_2-C-O-CH-CH_3 + NaOH \longrightarrow$ (with O and CH_3 groups)

b. $CH_3-CH-C-O-CH_2-CH_2-CH_3 + H_2O \xrightleftharpoons{H^+}$ (with CH_3 and O)

CHALLENGE QUESTIONS

16.61 Using the reactions we have studied, indicate how you might prepare the following from the starting substance given:
   a. acetic acid from ethene
   b. butyric acid from 1-butanol

16.62 Using the reactions we have studied, indicate how you might prepare the following from the starting substance given:
   a. pentanoic acid from 1-pentanol
   b. ethyl acetate from 2 molecules of ethanol

16.63 Methyl benzoate is not soluble in water; but when it is heated with KOH, it dissolves. Write an equation for the reaction and explain what happens. When HCl is added to the solution, a white solid forms. What is the solid?

16.64 Hexanoic acid is soluble in NaOH solution, but hexanal is not. Explain.

16.65 Salicylic acid could be named *o*-hydroxybenzoic acid.
   a. What two reactive functional groups are present?
   b. Draw the structure of the ester product that forms when the hydroxyl group of salicylic acid reacts with acetic acid.
   c. Draw the structure of methyl salicylate, oil of wintergreen, formed when salicylic acid forms an ester with methyl alcohol.

16.66 What volume of 0.100 M NaOH is needed to neutralize 3.00 g of benzoic acid?

ANSWERS

Answers to Study Checks

16.1

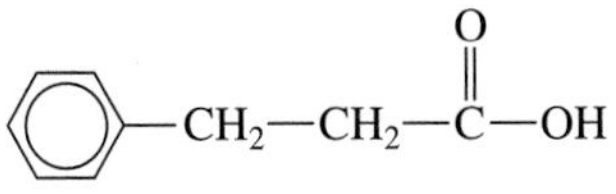

(benzene ring)$-CH_2-CH_2-C-OH$ (with O)

16.2 $CH_3-CH_2-CH_2-C-OH$ (with O)

16.3 Two carboxylic acid molecules hydrogen bond, forming a dimer, which has a mass twice that of the single acid molecule.

16.4 $H-C-OH + H_2O \rightleftharpoons H-C-O^- + H_3O^+$ (with O groups)

16.5 butanoic acid, butyric acid

16.6 propanoic (propionic) acid and 1-pentanol

16.7 $CH_3-C-O-CH_2-CH_2-CH_2-CH_2-CH_3$ (with O)

16.8 propanoic (propionic) acid and ethanol

16.9

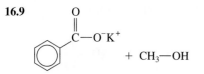

+ CH_3—OH

Answers to Selected Problems

16.1 methanoic acid (formic acid)

16.3 Each compound contains three carbon atoms. They differ because propanal, an aldehyde, contains a carbonyl group bonded to a hydrogen. In propanoic acid, the carbonyl group connects to a hydroxyl group forming a carboxyl group.

16.5 **a.** ethanoic acid (acetic acid)
b. butanoic acid (butyric acid)
c. 2-chloropropanoic acid (α-chloropropionic acid)
d. 3-methylhexanoic acid
e. 3,4-dihydroxybenzoic acid
f. 4-bromopentanoic acid

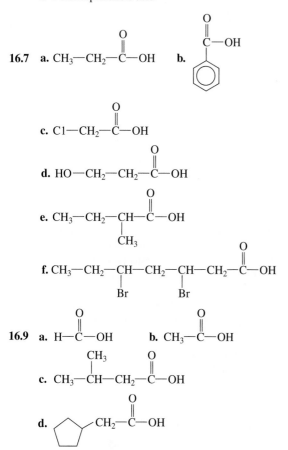

16.11 **a.** Butanoic acid has a higher molar mass and would have a higher boiling point.
b. Propanoic acid can form more hydrogen bonds and would have a higher boiling point.
c. Butanoic acid can form hydrogen bonds and would have a higher boiling point.

16.13 **a.** propanoic acid
b. propanoic acid

16.15 **a.** H—$\overset{\overset{\displaystyle O}{\|}}{C}$—OH + H_2O $\rightleftharpoons$ H—$\overset{\overset{\displaystyle O}{\|}}{C}$—$O^-$ + H_3O^+

b. CH_3—CH_2—$\overset{\overset{\displaystyle O}{\|}}{C}$—OH + H_2O $\rightleftharpoons$

CH_3—CH_2—$\overset{\overset{\displaystyle O}{\|}}{C}$—$O^-$ + H_3O^+

c. CH_3—$\overset{\overset{\displaystyle O}{\|}}{C}$—OH + H_2O $\rightleftharpoons$

CH_3—$\overset{\overset{\displaystyle O}{\|}}{C}$—$O^-$ + H_3O^+

16.17 **a.** H—$\overset{\overset{\displaystyle O}{\|}}{C}$—OH + NaOH $\longrightarrow$

H—$\overset{\overset{\displaystyle O}{\|}}{C}$—$O^-Na^+$ + H_2O

b. CH_3—CH_2—$\overset{\overset{\displaystyle O}{\|}}{C}$—OH + NaOH $\longrightarrow$

CH_3—CH_2—$\overset{\overset{\displaystyle O}{\|}}{C}$—$O^-Na^+$ + H_2O

c. (benzoic acid) + NaOH $\longrightarrow$ (sodium benzoate) + H_2O

16.19 **a.** sodium methanoate, sodium formate
b. sodium propanoate, sodium propionate
c. sodium benzoate

16.21 **a.** aldehyde   **b.** ester
c. ketone   **d.** carboxylic acid

16.23 **a.** CH_3—$\overset{\overset{\displaystyle O}{\|}}{C}$—O—$CH_3$

b. CH_3—CH_2—CH_2—$\overset{\overset{\displaystyle O}{\|}}{C}$—O—$CH_3$

c. $\overset{\overset{\displaystyle O}{\|}}{C}$—O—$CH_3$ (benzene ring)

16.25 **a.** CH_3—CH_2—$\overset{\overset{\displaystyle O}{\|}}{C}$—O—$CH_2$—$CH_2$—$CH_3$

b. CH_3—CH_2—CH_2—CH_2—$\overset{\overset{\displaystyle O}{\|}}{C}$—O—$\overset{\overset{\displaystyle CH_3}{|}}{CH}$—$CH_3$

16.27 a. formic acid (methanoic acid) and methyl alcohol
(methanol)
  b. acetic acid (ethanoic acid) and methyl alcohol
(methanol)
  c. butyric acid (butanoic acid) and methyl alcohol
(methanol)
  d. β-methylbutyric acid (3-methylbutanoic acid) and ethyl
alcohol (ethanol)

16.29 a. methyl formate (methyl methanoate)
  b. methyl acetate (methyl ethanoate)
  c. methyl butyrate (methyl butanoate)
  d. ethyl β-methyl butyrate (ethyl 3-methyl butanoate)

16.31 a. $CH_3-\overset{\overset{\displaystyle O}{\|}}{C}-O-CH_3$

  b. $H-\overset{\overset{\displaystyle O}{\|}}{C}-O-CH_2-CH_2-CH_2-CH_3$

  c. $CH_3-CH_2-CH_2-CH_2-\overset{\overset{\displaystyle O}{\|}}{C}-O-CH_2-CH_3$

  d. $CH_3-CH_2-\overset{\overset{\displaystyle O}{\|}}{C}-O-CH_2-\overset{\overset{\displaystyle Br}{|}}{C}H-CH_3$

16.33 a. pentyl ethanoate (pentyl acetate)
  b. octyl ethanoate (octyl acetate)
  c. pentyl butanoate (pentyl butyrate)

16.41 a. $CH_3-CH_2-CH_2-O-\overset{\overset{\displaystyle O}{\|}}{C}-CH_3$

  b. $CH_3-CH_2-CH_2-OH + HO-\overset{\overset{\displaystyle O}{\|}}{C}-CH_3 \xrightarrow{\underset{\longleftarrow}{H^+, \text{ heat}}} CH_3-CH_2-CH_2-O-\overset{\overset{\displaystyle O}{\|}}{C}-CH_3 + H_2O$

  c. $CH_3-CH_2-CH_2-O-\overset{\overset{\displaystyle O}{\|}}{C}-CH_3 + H_2O \xrightarrow{\underset{\longleftarrow}{H^+}} CH_3-CH_2-CH_2-OH + HO-\overset{\overset{\displaystyle O}{\|}}{C}-CH_3$

  d. $CH_3-CH_2-CH_2-O-\overset{\overset{\displaystyle O}{\|}}{C}-CH_3 + NaOH \xrightarrow{\text{Heat}} CH_3-CH_2-CH_2-OH + Na^+ \ {}^-O-\overset{\overset{\displaystyle O}{\|}}{C}-CH_3$
  e. 74.4 mL of 0.208 M NaOH

16.43 a. 3-methylbutanoic acid; β-methylbutyric acid
  b. ethyl benzoate
  c. ethyl propanoate; ethyl propionate
  d. 2-chlorobenzoic acid; *ortho*-chlorobenzoic acid
  e. 4-hydroxypentanoic acid
  f. 2-propyl ethanoate; isopropyl acetate

16.45

$CH_3-CH_2-CH_2-CH_2-\overset{\overset{\displaystyle O}{\|}}{C}-OH$ $CH_3-CH_2-\overset{\overset{\displaystyle CH_3}{|}}{C}H-\overset{\overset{\displaystyle O}{\|}}{C}-OH$

$CH_3-\overset{\overset{\displaystyle CH_3}{|}}{C}H-CH_2-\overset{\overset{\displaystyle O}{\|}}{C}-OH$ $CH_3-\overset{\overset{\displaystyle CH_3}{|}}{\underset{\underset{\displaystyle CH_3}{|}}{C}}-\overset{\overset{\displaystyle O}{\|}}{C}-OH$

16.35 a. $CH_3-\overset{\overset{\displaystyle O}{\|}}{C}-OH$
  b. $CH_3-CH_2-CH_2-CH_2-OH$
  c. $CH_3-O-\overset{\overset{\displaystyle O}{\|}}{C}-CH_3$

16.37 The products of the acid hydrolysis of an ester are an alcohol and a carboxylic acid.

16.39

  a. $CH_3-CH_2-\overset{\overset{\displaystyle O}{\|}}{C}-O^-Na^+$ and CH_3-OH

  b. $CH_3-\overset{\overset{\displaystyle O}{\|}}{C}-OH$ and $CH_3-CH_2-CH_2-OH$

  c. $CH_3-CH_2-CH_2-\overset{\overset{\displaystyle O}{\|}}{C}-OH$ and CH_3-CH_2-OH

  d. $\underset{}{\bigcirc}-\overset{\overset{\displaystyle O}{\|}}{C}-OH$ and CH_3-CH_2-OH

  e. $\underset{}{\bigcirc}-\overset{\overset{\displaystyle O}{\|}}{C}-O^-Na^+$ and CH_3CH_2OH

16.47 a. CH₃—O—C(=O)—CH₃   **b.**
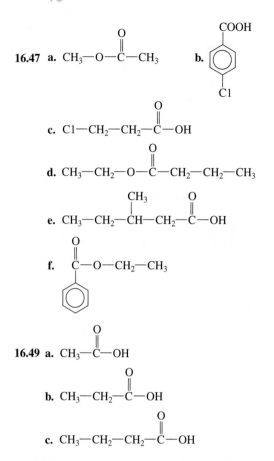

c. Cl—CH₂—CH₂—C(=O)—OH

d. CH₃—CH₂—O—C(=O)—CH₂—CH₂—CH₃

e. CH₃—CH₂—CH(CH₃)—CH₂—C(=O)—OH

f. C(=O)—O—CH₂—CH₃ (phenyl)

16.49 a. CH₃—C(=O)—OH

b. CH₃—CH₂—C(=O)—OH

c. CH₃—CH₂—CH₂—C(=O)—OH

16.51 The presence of two polar groups in the carboxyl group allows hydrogen bonding and the formation of a dimer that doubles the effective molar mass.

16.53 b, c, d, and e are all soluble in water

16.55 a. CH₃—CH₂—C(=O)—O⁻ + H₃O⁺

b. CH₃—CH₂—C(=O)—O⁻K⁺ + H₂O

c. CH₃—CH₂—C(=O)—O—CH₃ + H₂O

d. C(=O)—O—CH₂—CH₃ + H₂O (phenyl)

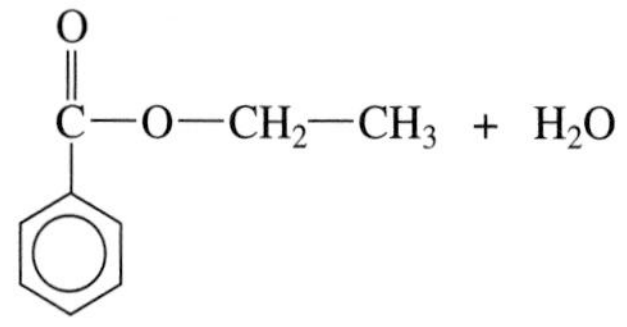

16.57 a. 3-methylbutanoic acid and methanol
b. 3-chlorobenzoic acid and ethanol
c. hexanoic acid and methanol

16.59 a. CH₃—CH₂—C(=O)—OH and HO—CH(CH₃)—CH₃

b. CH₃—CH(CH₃)—C(=O)—O⁻Na⁺ and HO—CH₂—CH₂—CH₃

16.61

a. CH₂=CH₂ + H₂O $\xrightarrow{H^+}$ CH₃—CH₂—OH $\xrightarrow{[O]}$ CH₃—C(=O)—OH

b. CH₃—CH₂—CH₂—CH₂—OH $\xrightarrow{[O]}$ CH₃—CH₂—CH₂—C(=O)—OH

16.63

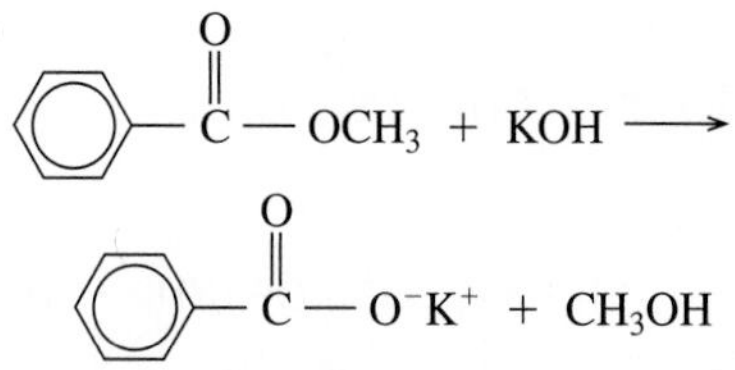

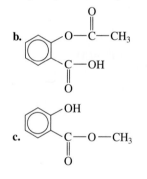

A soluble salt, potassium benzoate, is formed. When acid is added, the salt is converted to the mostly insoluble benzoic acid.

16.65 a. hydroxyl and carboxylic acid

b.
(structure: benzene ring with O—C(=O)—CH₃ and C(=O)—OH)

c.
(structure: benzene ring with OH and C(=O)—O—CH₃)

17

Lipids

"In our toxicology lab, we measure the drugs in samples of urine or blood," says Penny Peng, assistant supervisor of chemistry, Toxicology Lab, Santa Clara Valley Medical Center. "But first we extract the drugs from the fluid and concentrate them so they can be detected in the machine we use. We extract the drugs by using different organic solvents such as methanol, ethyl acetate, or methylene chloride, and by changing the pH. We evaporate most of the organic solvent to concentrate any drugs it may contain. A small sample of the concentrate is placed into a machine called a gas chromatograph. As the gas moves over a column, the drugs in it are separated. From the results, we can identify as many as 10 to 15 different drugs from one urine sample."

LOOKING AHEAD

Mastering
GOB CHEMISTRY

Visit **www.masteringgob.com**
for self-study materials
and instructor-assigned
homework.

hen we talk of fats and oils, waxes, steroids, cholesterol, and fat-soluble vitamins, we are discussing lipids. All the lipids are naturally occurring compounds that vary considerably in structure but share a common feature of being soluble in nonpolar solvents, but not in water. Fats, which are one family of lipids, have many functions in the body: they store energy and protect and insulate internal organs. Other types of lipids are found in nerve fibers and in hormones, which act as chemical messengers. Because they are not soluble in water, a major function of lipids is to build the cell membranes that separate the internal contents of cells from the surrounding aqueous environment.

Many people are concerned about the amounts of saturated fats and cholesterol in our diets. Researchers suggest that saturated fats and cholesterol are associated with diseases such as diabetes, cancers of the breast, pancreas, and colon, and artherosclerosis, a condition in which deposits of lipid materials accumulate in the coronary blood vessels. These plaques restrict the flow of blood to the tissue, causing necrosis (death) of the tissue. In the heart, this could result in a *myocardial infarction* (heart attack).

The American Institute for Cancer Research has recommended that our diet contain more fiber and starch by adding more vegetables, fruits, and whole grains with moderate amounts of foods with low levels of fat and cholesterol such as fish, poultry, lean meats, and low-fat dairy products. They also suggest that we limit our intake of foods high in fat and cholesterol such as eggs, nuts, french fries, fatty or organ meats, cheeses, butter, and coconut and palm oil.

LEARNING GOAL

Describe the classes of lipids.

17.1 Lipids

Lipids are a family of biomolecules that have the common property of being soluble in organic solvents but not in water. The word "lipid" comes from the Greek word *lipos*, meaning "fat" or "lard." Typically, the lipid content of a cell can be extracted using an organic solvent such as ether or chloroform. Lipids are an important feature in cell membranes, fat-soluble vitamins, and steroid hormones.

Types of Lipids

Classes of Lipids

Within the lipid family, there are specific structures that distinguish the different types of lipids. Lipids such as waxes, fats, oils, and glycerophospholipids are esters that can be hydrolyzed to give fatty acids along with other products including an alcohol. Sphingolipids contain an alcohol called sphingosine, and glycosphingolipids contain a carbohydrate. Steroids are characterized by the steroid nucleus of four fused carbon rings. They do not contain fatty acids and cannot be hydrolyzed. Figure 17.1 illustrates the general structure of lipids we will discuss in this chapter.

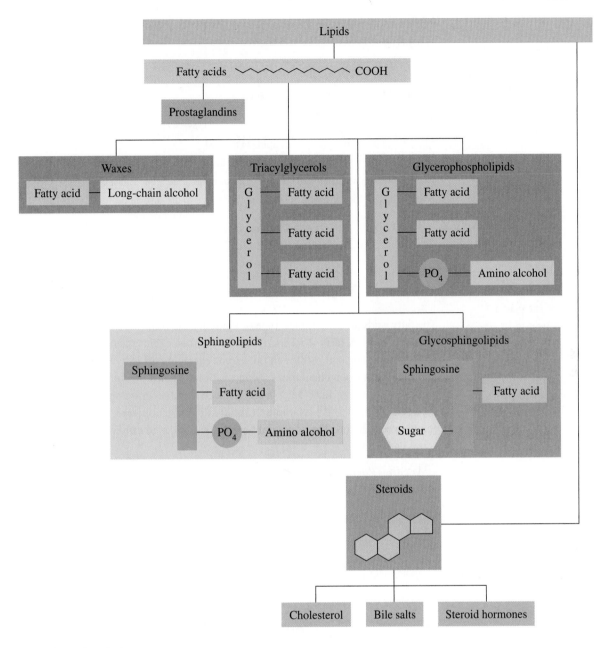

FIGURE 17.1

Structures for some classes of lipids that are naturally occurring compounds in cells and tissues.

Q **What property do waxes, triacylglycerols, and steroids have in common?**

SAMPLE PROBLEM 17.1

▪ **Classes of Lipids**

What type of lipid does not contain fatty acids?

SOLUTION

The steroids are a group of lipids with no fatty acids.

What type of lipid contains a carbohydrate?

QUESTIONS AND PROBLEMS

■ Lipids

17.1 What are some functions of lipids in the body?

17.2 What are some of the different kinds of lipids?

17.3 Lipids are not soluble in water. Are lipids polar or nonpolar molecules?

17.4 Which of the following solvents might be used to dissolve an oil stain?
a. water   **b.** CCl_4 **c.** diethyl ether
d. benzene   **e.** NaCl solution

LEARNING GOAL

Write structures of fatty acids and identify as saturated or unsaturated.

SELF-STUDY ACTIVITY
Fats

17.2 Fatty Acids

The fatty acids are the simplest type of lipids and are found as components in more complex lipids. A **fatty acid** contains a long carbon chain attached to a carboxylic acid group at one end. Although the carboxylic acid part is hydrophilic, the long hydrophobic carbon chain makes long-chain fatty acids insoluble in water. Most naturally occurring fatty acids have an even number of carbon atoms, usually between 10 and 20. An example of a fatty acid is lauric acid, a 12-carbon acid found in coconut oil. The structural formula of lauric acid can be written in several forms as follows.

Writing Formulas for Lauric Acid

$$CH_3-(CH_2)_{10}-\overset{\overset{\textstyle O}{\|}}{C}-OH \qquad CH_3-(CH_2)_{10}-COOH$$

$$CH_3-CH_2-CH_2-CH_2-CH_2-CH_2-CH_2-CH_2-CH_2-CH_2-CH_2-C\overset{\textstyle O}{\underset{\textstyle OH}{\diagdown}}$$
Condensed structural formula

Line-bond structural formula

Fatty Acids

    Saturated fatty acids such as lauric acid contain only single bonds between carbons. **Monounsaturated fatty acids** have one double bond in the carbon chain, and **polyunsaturated fatty acids** have two or more double bonds. Table 17.1 lists some of the typical fatty acids in lipids.

    In Chapter 12 we saw that compounds with double bonds can have cis and trans stereoisomers. This is also true of unsaturated fatty acids. For example, oleic acid, a monounsaturated fatty acid found in olives and corn, has one double bond at carbon 9. We can show its cis and trans structural formulas using the line-bond notation. The cis configuration is most prevalent in naturally occurring unsaturated fatty acids. In the cis isomer, the geometry of the carbon chain is not linear, but has a "kink" at the double bond site. As we will see, the cis bond has a major impact on the properties of unsaturated fatty acids.

TABLE 17.1

Structures and Melting Points of Common Fatty Acids

Name	Carbon Atoms	Source	Melting Point (°C)	Structures
Saturated Fatty Acids				
lauric acid	12	coconut	43	CH_3—$(CH_2)_{10}$—COOH
myristic acid	14	nutmeg	54	CH_3—$(CH_2)_{12}$—COOH
palmitic acid	16	palm	62	CH_3—$(CH_2)_{14}$—COOH
stearic acid	18	animal fat	69	CH_3—$(CH_2)_{16}$—COOH
Monounsaturated Fatty Acids				
palmitoleic acid	16	butter	0	CH_3—$(CH_2)_5$—CH=CH—$(CH_2)_7$—COOH
oleic acid	18	olives, corn	13	CH_3—$(CH_2)_7$—CH=CH—$(CH_2)_7$—COOH
Polyunsaturated Fatty Acids				
linoleic acid	18	soybean, safflower, sunflower	−9	CH_3—$(CH_2)_4$—CH=CH—CH_2—CH=CH—$(CH_2)_7$—COOH
linolenic acid	18	corn	−17	CH_3—CH_2—CH=CH—CH_2—CH=CH—CH_2—CH=CH—$(CH_2)_7$—COOH

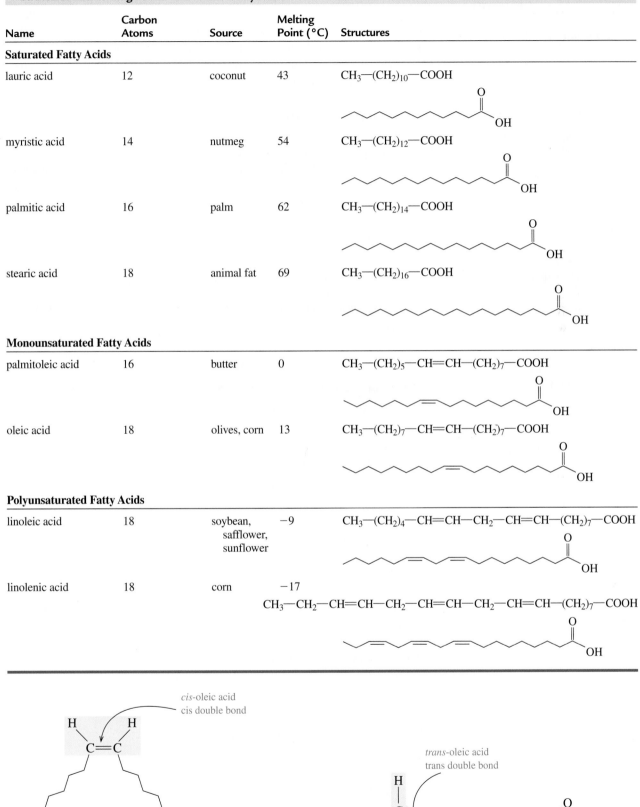

cis-oleic acid
cis double bond

trans-oleic acid
trans double bond

■ Solubility of Fats and Oils

Place some water in a small bowl. Add a drop of a vegetable oil. Then add a few more drops of the oil. Now add a few drops of liquid soap and mix. Record your observations.

Place a small amount of fat such as margarine, butter, shortening, or vegetable oil on a dish or plate. Run water over it. Record your observations. Mix some soap with the fat substance and run water over it again. Record your observations.

QUESTIONS

1. Do the drops of oil in the water separate or do they come together? Explain.
2. How does the soap affect the oil layer?
3. Why don't the fats on the dish or plate wash off with water?
4. In general what is the solubility of lipids in water?
5. Why does soap help to wash the fats off the plate?

MC GOB   Structures and Properties of Fatty Acids

The human body is capable of synthesizing most fatty acids from carbohydrates or other fatty acids. However, humans do not synthesize sufficient amounts of fatty acids that have more than one double bond, such as linoleic acid, linolenic acid, and arachidonic acid. These fatty acids are called *essential* fatty acids because they must be provided by the diet. A deficiency of essential fatty acids can cause skin dermatitis in infants. However, the role of fatty acids in adult nutrition is not well understood. Adults do not usually have a deficiency of essential fatty acids.

Properties of Fatty Acids

The saturated fatty acids fit close together in a regular pattern, which allows strong attractions to occur between the carbon chains. As a result, a significant amount of energy and high temperatures are required to separate the fatty acids and melt the fat. As the length of the carbon chain increases, more interactions occur between the carbon chains, requiring higher melting points. Saturated fatty acids are usually solids at room temperature.

In unsaturated fatty acids, the cis double bonds cause the carbon chain to bend, which gives the molecules an irregular shape. As a result, fewer interactions occur between carbon chains. Consequently, the melting points of unsaturated fats are lower than those of saturated fats. (See Figure 17.2.) Most unsaturated fats are liquid oils at room temperature.

SAMPLE PROBLEM 17.2

■ Structures and Properties of Fatty Acids

Consider the structural formula of oleic acid.

$$CH_3-(CH_2)_7-CH=CH-(CH_2)_7-\overset{\overset{\displaystyle O}{\|}}{C}-OH$$

a. Why is the substance called an acid?
b. How many carbon atoms are in oleic acid?
c. Is it a saturated or unsaturated fatty acid?
d. Is it most likely to be solid or liquid at room temperature?
e. Would it be soluble in water?

SOLUTION

a. Oleic acid contains a carboxylic acid group.
b. It contains 18 carbon atoms.
c. It is an unsaturated fatty acid.
d. It is liquid at room temperature.
e. No, its long hydrocarbon chain makes it insoluble in water.

STUDY CHECK

Palmitoleic acid is a fatty acid with the following formula:

$$CH_3-(CH_2)_5-CH=CH-(CH_2)_7-\overset{\overset{\displaystyle O}{\|}}{C}-OH$$

a. How many carbon atoms are in palmitoleic acid?
b. Is it a saturated or unsaturated fatty acid?
c. Is it most likely to be solid or liquid at room temperature?

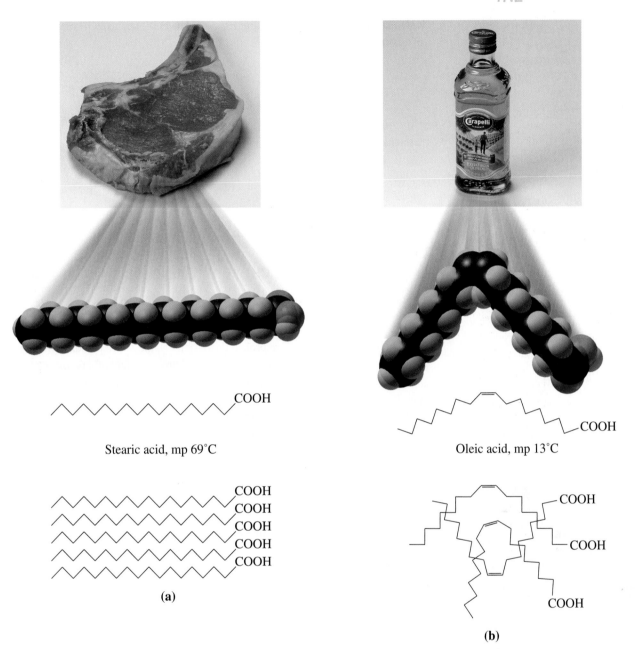

Stearic acid, mp 69°C

Oleic acid, mp 13°C

(a)

(b)

FIGURE 17.2

(**a**) In saturated fatty acids, the molecules fit closely together to give high melting points.
(**b**) In unsaturated fatty acids, molecules cannot pack closely together, resulting in lower melting points.

Q **Why does the cis double bond affect the melting points of unsaturated fatty acids?**

Prostaglandins

Prostaglandins are hormone-like substances produced in low amounts in most cells of the body. The variety of prostaglandins is formed from the unsaturated fatty acid arachidonic acid with 20 carbons. Prostaglandins are sometimes referred to as eicosanoids (*eicos* is the Greek word for 20). Most prostaglandins have a hydroxyl group on carbon 11 and carbon 15 and a trans double bond at carbon 13. Those with a ketone group on carbon 9 are designated as PGE, and as PGF when there is an hydroxyl group on carbon 9.

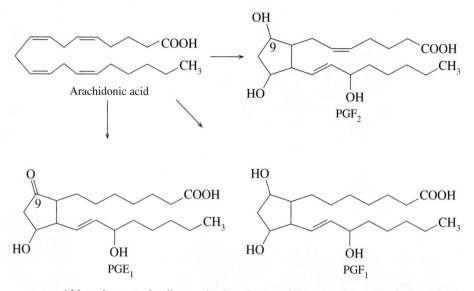

Arachidonic acid

PGF₂

PGE₁

PGF₁

Although prostaglandins are broken down quickly, they have potent physiological effects. Some prostaglandins increase blood pressure, and others lower blood pressure. Other prostaglandins stimulate contraction and relaxation in the smooth muscle of the uterus during the birth process. When tissues are injured, arachidonic acid present in the blood is converted to prostaglandins such as PGE and PGF that produce inflammation and pain in the area.

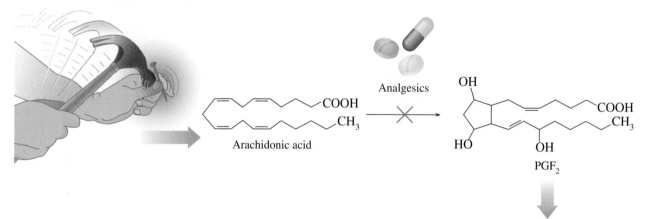

Arachidonic acid

Analgesics

PGF₂

Pain, fever, inflammation

The treatment of pain, fever, and inflammation is based on inhibiting the enzymes that convert arachidonic acid to prostaglandins. Several nonsteroidal anti-inflammatory drugs (NSAIDs), such as aspirin, block the production of prostaglandins, and in doing so decrease pain and inflammation and reduce fever (antipyretics). Ibuprofen has similar anti-inflammatory and analgesic effects. Other NSAIDs include naproxen (Aleve and Naprosyn), ketoprofen (Actron), and nabumetone (Relafen). Long-term use of such products can result in liver, kidney, and gastrointestinal damage. Some forms of PGE are being tested as inhibitors of gastric secretion for use in the treatment of stomach ulcers.

HEALTH NOTE

Omega-3 Fatty Acids in Fish Oils

Over the past several decades, Americans have been changing their diets to include more polyunsaturated fats and fewer saturated fats. This change is a response to research that indicates that atherosclerosis and heart disease are associated with high levels of saturated fats in the diet. However, this association does not seem to be correct for the Inuit people of Alaska, who have a high-fat diet and high levels of blood cholesterol, but a very low occurrence of atherosclerosis and heart attacks. The fats in the Inuit diet are primarily from fish rather than from land animals, as in many other people's diets.

Both fish and vegetable oils have high levels of polyunsaturated fats. The fatty acids in vegetable oils are omega-6 acids, which means that the first double bond occurs at carbon 6 counting from the methyl group. Two common omega-6 acids are linoleic acid and arachidonic acid. However,

the fatty acids in the fish oils are mostly the omega-3 type, in which the first double bond occurs at the third carbon counting from the methyl group. Three common omega-3 fatty acids in fish are linolenic acid, eicosapentaenoic acid (EPA), and docosahexaenoic acid (DHA).

In atherosclerosis and heart disease, cholesterol forms plaque that adhere to the walls of the blood vessels. Blood pressure rises as blood has to squeeze through a smaller opening in the blood vessel. As more plaque forms, there is also a possibility of blood clots blocking the blood vessels and causing a heart attack. Omega-3 fatty acids lower the tendency of blood platelets to stick together, thereby reducing the possibility of blood clots. However, high levels of omega-3 fatty acids can increase bleeding if the ability of the platelets to form blood clots is reduced too much. It does seem that a diet that includes fish such as salmon, tuna, and herring can provide higher amounts of the omega-3 fatty acids, which help lessen the possibility of developing heart disease.

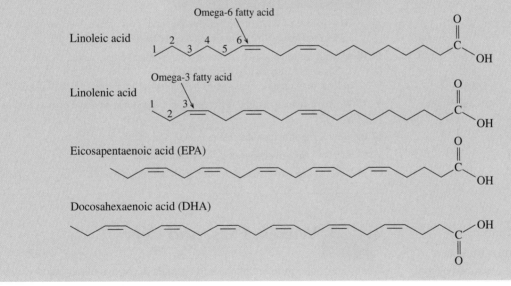

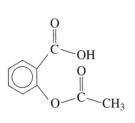

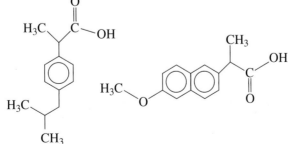

aspirin (acetylsalicyclic acid)   ibuprofen (Advil, Motrin)   naproxen (Aleve, Naprosyn)

■**Fatty Acids**

17.5 Describe some similarities and differences in the structures of a saturated fatty acid and an unsaturated fatty acid.

17.6 Stearic acid and linoleic acid both have 18 carbon atoms. Why does stearic acid melt at 69°C, but linoleic acid melts at −9°C?

17.7 Write the line-bond structure of the following fatty acids:
a. palmitic acid
b. oleic acid

17.8 Write the line-bond structure of the following fatty acids:
a. stearic acid
b. linoleic acid

17.9 Which of the following fatty acids are saturated and which are unsaturated?
a. lauric acid   **b.** linolenic acid
c. palmitoleic acid   **d.** stearic acid

17.10 Which of the following fatty acids are saturated and which are unsaturated?
a. linoleic acid   **b.** palmitic acid
c. myristic acid   **d.** oleic acid

17.11 How does the structure of a fatty acid with a cis double bond differ from the structure of a fatty acid with a trans double bond?

17.12 In each pair, identify the fatty acid with the lower melting point. Explain.
a. myristic acid and stearic acid
b. stearic acid and linoleic acid
c. oleic acid and linolenic acid

17.13 Describe the position of the first double bond in an omega-3 and an omega-6 fatty acid. (See Health Note "Omega-3 Fatty Acids in Fish Oils.")

17.14 a. What are some sources of omega-3 and omega-6 fatty acids? (See Health Note "Omega-3 Fatty Acids in Fish Oils.")
b. How may omega-3 fatty acids help in lowering the risk of heart disease?

17.15 What are some structural differences in arachidonic acid and prostaglandins such as PGE_1?

17.16 What is the structural difference in PGE and PGF?

17.17 What are some functions of prostaglandins in the body?

17.18 How does an anti-inflammatory drug reduce inflammation?

LEARNING GOAL

Write the structural formula of a wax, fat, or oil produced by the reaction of a fatty acid and an alcohol or glycerol.

Ester bond
↓
fatty acid — long-chain alcohol

Wax

17.3 Waxes, Fats, and Oils

Waxes are found in many plants and animals. Coatings of carnauba wax on fruits and the leaves and stems of plants help to prevent loss of water and damage from pests. Waxes on the skin, fur, and feathers of animals and birds provide a waterproof coating. A **wax** is an ester of a saturated fatty acid and a long-chain alcohol, each containing from 14 to 30 carbon atoms.

The formulas of some common waxes are given in Table 17.2. Beeswax obtained from honeycombs and carnauba wax obtained from palm trees are used to give a protective coating to furniture, cars, and floors. Jojoba wax is used in making candles and cosmetics such as lipstick. Lanolin, a mixture of waxes obtained from wool, is used in hand and facial lotions to aid retention of water, which softens the skin.

TABLE 17.2

Some Typical Waxes

Type	Structural Formula	Source	Uses
Beeswax	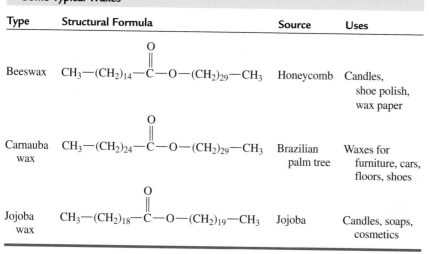	Honeycomb	Candles, shoe polish, wax paper
Carnauba wax		Brazilian palm tree	Waxes for furniture, cars, floors, shoes
Jojoba wax		Jojoba	Candles, soaps, cosmetics

Fats and Oils: Triacylglycerols

In the body, fatty acids are stored as fats and oils known as **triacylglycerols**. These substances, also called *triglycerides*, are triesters of glycerol (a trihydroxy alcohol) and fatty acids. The general formula of a triacylglycerol follows.

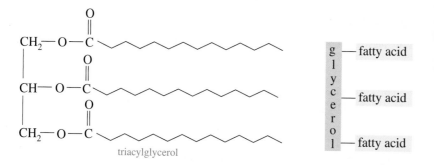

triacylglycerol

A triacyglycerol is produced by **esterification**, a reaction in which the hydroxyl groups of glycerol form ester bonds with the carboxyl groups of fatty acids. For example, glycerol and three molecules of stearic acid form glyceryl tristearate, which is commonly named tristearin. In large structures, bonds between carbon atoms may be omitted.

MC GOB SELF-STUDY ACTIVITY Triacylglycerols

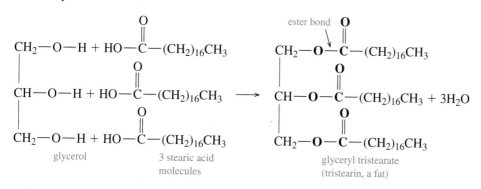

Most fats and oils are mixed triacylglycerols that contain two or three different fatty acids. For example, a mixed triacylglycerol might be made from lauric acid, myristic acid, and palmitic acid. One possible structure for the mixed triacylglyc-erol follows:

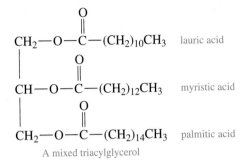

A mixed triacylglycerol

Triacylglycerols are the major form of energy storage for animals. Animals that hibernate eat large quantities of plants, seeds, and nuts that contain high levels of fats and oils. They gain as much as 14 kilograms a week. As the external temperature drops, the animal goes into hibernation. The body temperature drops to nearly freezing, and there is a dramatic reduction in cellular activity, respiration, and heart rate. Animals who live in extremely cold climates will hibernate for 4–7 months. During this time, stored fat is the only source of energy.

MC GOB Structures and Properties of Waxes, Fats and Oils

SAMPLE PROBLEM 17.3

■ Writing Structures for a Triacylglycerol

Draw the structural formula of triolein, a simple triacylglycerol that uses oleic acid.

SOLUTION

Triolein is the triacylglycerol of glycerol and three oleic acid molecules. Each fatty acid is attached by an ester bond to one of the hydroxyl groups in glycerol.

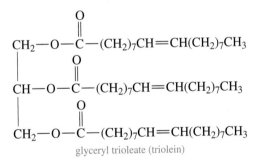

glyceryl trioleate (triolein)

STUDY CHECK

Write the structure of the triacylglycerol containing 3 molecules of myristic acid.

Melting Points of Fats and Oils

A **fat** is a triacylglycerol that is solid at room temperature, such as fats in meat, whole milk, butter, and cheese. Most fats come from animal sources.

An **oil** is a triacylglycerol that is usually liquid at room temperature. The most commonly used oils come from plant sources. Olive oil and peanut oil are monounsaturated because they contain large amounts of oleic acid. Oils from

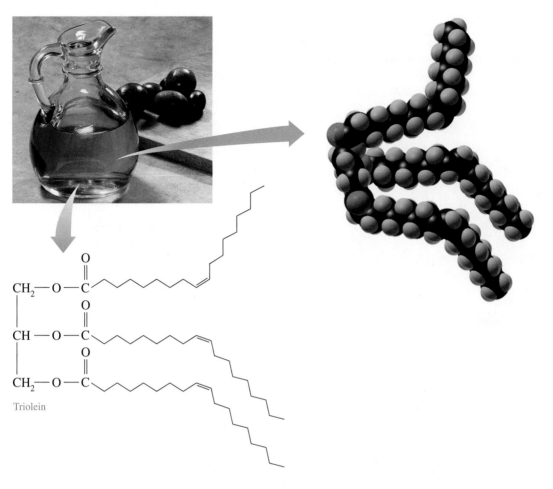

Triolein

FIGURE 17.3

Vegetable oils such as olive oil, corn oil, and safflower oil contain unsaturated fats.

Q **Why is olive oil a liquid at room temperature?**

corn, cottonseed, safflower, and sunflower are polyunsaturated because they contain large amounts of fatty acids with two or more double bonds. (See Figure 17.3.) A few oils such as palm oil and coconut oil are solid at room temperature because they consist mostly of saturated fatty acids.

The amounts of saturated, monounsaturated, and polyunsaturated fatty acids in some typical fats and oils are shown in Figure 17.4. Saturated fatty acids have higher melting points than unsaturated fatty acids because they pack together more tightly. Animal fats usually contain more saturated fatty acids than do vegetable oils. Therefore the melting points of animal fats are higher than those of vegetable oils.

QUESTIONS AND PROBLEMS

■ **Waxes, Fats, and Oils**

17.19 Draw the structure of an ester in beeswax formed from myricyl alcohol, $CH_3(CH_2)_{29}OH$, and palmitic acid.

17.20 Draw the structure of an ester in jojoba wax formed from arachidic acid, a 20-carbon saturated fatty acid, and 1-docosanol, $CH_3(CH_2)_{21}OH$.

17.21 Draw the structure of a triacylglycerol that contains stearic acid and glycerol.

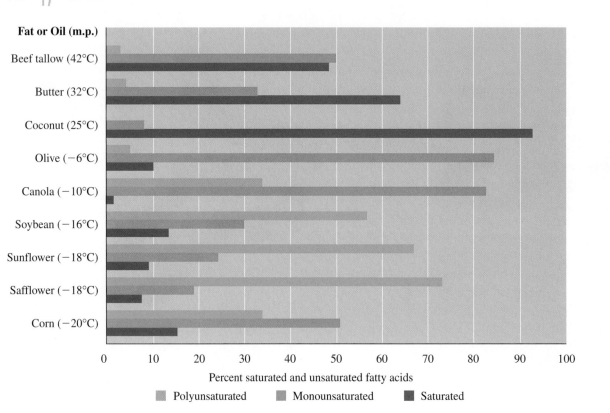

Fat or Oil (m.p.)

Beef tallow (42°C)
Butter (32°C)
Coconut (25°C)
Olive (−6°C)
Canola (−10°C)
Soybean (−16°C)
Sunflower (−18°C)
Safflower (−18°C)
Corn (−20°C)

Percent saturated and unsaturated fatty acids

■ Polyunsaturated   ■ Monounsaturated   ■ Saturated

FIGURE 17.4

Fats and oils have low melting points because they have a higher percentage of unsaturated fatty acids than do animal fats.

Q Why is the melting point of butter higher than olive or canola oil?

17.22 A mixed triacylglycerol contains two palmitic acid molecules to every one oleic acid molecule. Write two possible structures (isomers) for the compound.

17.23 Draw the structure of tripalmitin.

17.24 Draw the structure of triolein.

17.25 Safflower oil is called a polyunsaturated oil, whereas olive oil is a monounsaturated oil. Explain.

17.26 Why does olive oil have a lower melting point than butter fat?

17.27 Why does coconut oil, a vegetable oil, have a melting point similar to fats from animal sources?

17.28 A label on a bottle of 100% sunflower seed oil states that it is lower in saturated fats than all the leading oils.
a. How does the percentage of saturated fats in sunflower seed oil compare to that of safflower, corn, and canola oil? (See Figure 17.4.)
b. Is the claim valid?

LEARNING GOAL

■ Draw the structure of the product when a triacylglycerol is hydrogenated, hydrolyzed, or oxidized.

■ 17.4 Chemical Properties of Triacylglycerols

The chemical reactions of the triacylglycerols (fats and oils) are the same as we discussed for alkenes (Chapter 12), and carboxylic acids and esters (Chapter 16). We will look at the addition of hydrogen to the double bonds and the hydrolysis of the ester bonds of fats and oils.

Olestra: A Fat Substitute

In 1968, food scientists designed an artificial fat called *olestra* as a source of nutrition for premature babies. However, olestra could not be digested and was never used for that purpose. Then scientists realized that olestra had the flavor and texture of a fat without the calories.

Olestra is manufactured by obtaining the fatty acids from the fats in cottonseed or soybean oils and bonding the fatty acids with the hydroxyl groups on sucrose. Chemically, olestra is composed of six to eight long-chain fatty acids attached by ester links to a sugar (sucrose) rather than to a glycerol molecule found in fats. This makes olestra a very large molecule, which cannot be absorbed through the intestinal walls. The enzymes and bacteria in the intestinal tract are unable to break down the olestra molecule and it travels through the intestinal tract undigested.

The large molecule of olestra also combines with fat-soluble vitamins (A, D, E, and K) as well as the carotenoids from the foods we eat before they can be absorbed through the intestinal wall. Carotenoids are plant pigments in fruits and vegetables that protect against cancer, heart disease, and macular degeneration, a form of blindness in the elderly. The FDA now requires manufacturers to add the four vitamins, but not the carotenoids, to olestra products. There have been reports of some adverse reactions, including diarrhea, abdominal cramps, and anal leakage, indicating that olestra may act as a laxative in some people. However, the manufacturers contend there is no direct proof that olestra is the cause of those effects.

Snack foods made with olestra are now in supermarkets nationwide. Since there are already low-fat snacks on the market, it remains to be seen whether olestra will have any significant effect on reducing the problem of obesity.

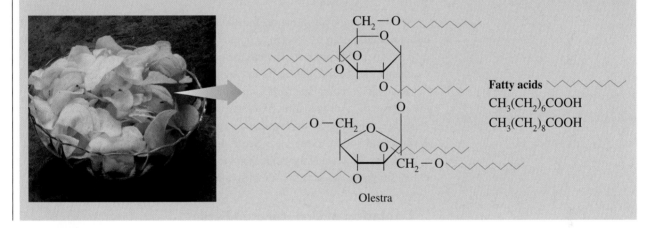

Fatty acids

$CH_3(CH_2)_6COOH$

$CH_3(CH_2)_8COOH$

Olestra

Hydrogenation

The **hydrogenation** of unsaturated fats converts carbon–carbon double bonds to single bonds. The hydrogen gas is bubbled through the heated oil in the presence of a nickel or platinum catalyst.

MC GOB

Hydrogenation and Hydrolysis of Triacylglycerols

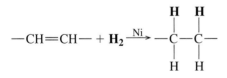

For example, when hydrogen adds to all of the double bonds of triolein the product is the saturated fat tristearin.

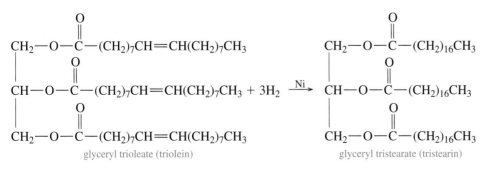

glyceryl trioleate (triolein)   glyceryl tristearate (tristearin)

FIGURE 17.5

Many soft margarines, stick margarines, and solid shortenings are produced by the partial hydrogenation of vegetable oils.

Q **How does hydrogenation change the structure of the fatty acids in the vegetable oils?**

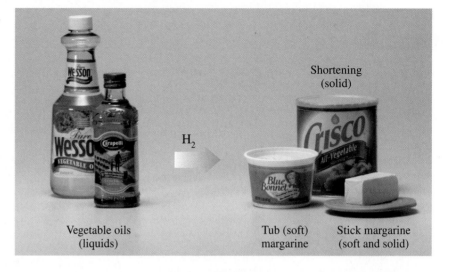

Shortening (solid)

H$_2$

Vegetable oils (liquids)

Tub (soft) margarine

Stick margarine (soft and solid)

In commercial hydrogenation, the addition of hydrogen is stopped before all the double bonds in an oil become completely saturated. Complete hydrogenation gives a very brittle product, whereas the partial hydrogenation of a liquid vegetable oil changes it to a soft, semisolid fat. As the oil becomes more saturated, the melting point increases and the fat becomes more solid at room temperature. Control of the degree of hydrogenation gives the various types of partially hydrogenated vegetable oil products on the market today—soft margarines, solid stick margarines, and solid shortenings. (See Figure 17.5.) Although these products now contain more saturated fatty acids than the original oils, they contain no cholesterol, unlike similar products from animal sources, such as butter and lard.

Oxidation of Unsaturated Fats

A fat or oil becomes rancid when its double bonds are oxidized in the presence of oxygen and microorganisms. The products are short-chain fatty acids and aldehydes that have disagreeable odors.

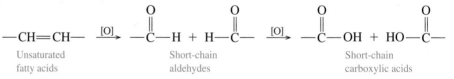

Unsaturated fatty acids

Short-chain aldehydes

Short-chain carboxylic acids

If a vegetable oil does not contain an antioxidant, it will oxidize rather easily. You can detect an oil that has become rancid by its unpleasant odor. If an oil is covered tightly and stored in a refrigerator, the process of oxidation can be slowed and the oil will last longer.

Oxidation also occurs in the oils that accumulate on the surface of the skin during heavy exercise. At body temperature, microorganisms on the skin promote rapid oxidation of the oils as they are exposed to oxygen and water. The resulting short-chain aldehydes and fatty acids account for the body odor associated with a workout and heavy perspiration.

Trans Fatty Acids and Hydrogenation

In the early 1900s, margarine became a popular replacement for the highly saturated fats such as butter and lard. Margarine is produced by partially hydrogenating the unsaturated fats in vegetable oils such as safflower oil, corn oil, canola oil, cottonseed oil, and sunflower oil. Fats that are more saturated are more resistant to oxidation.

In vegetable oils, the unsaturated fats usually contain cis double bonds. As hydrogenation occurs, double bonds are converted to single bonds. However, a small amount of the cis double bonds are converted to trans double bonds, which causes a change in the overall structure of the fatty acids. If the label on a product states that the oils have been "partially hydrogenated," that product will also contain trans fatty acids. In the United States, it is estimated that 2–4% of our total calories come from trans fatty acids.

The concern about trans fatty acids is that their altered structure may make them behave like saturated fatty acids in the body. In the 1980s, research indicated that trans fatty acids have an effect on blood cholesterol similar to that of saturated fats, although study results vary. Several studies reported that trans fatty acids raise the levels of LDL-cholesterol, low-density lipoproteins containing cholesterol that can accumulate in the arteries. (LDLs and HDLs are described in the section on lipoproteins later in the chapter.) Some studies also report that trans fatty acids lower HDL-cholesterol, high-density lipoproteins that carry cholesterol to the liver to be excreted. But other studies did not report any decrease in HDL-cholesterol. In some American and European studies, an increased risk of breast cancer was associated with increased intake of trans fatty acids. However, these studies are not conclusive and not all studies have supported such findings. Current evidence does not yet indicate that the intake of trans fatty acids is a significant risk factor for heart disease. The trans fatty acids controversy will continue to be debated as more research is done.

Foods containing trans fatty acids include milk, bread, fried foods, ground beef, baked goods, stick margarine, butter, soft margarine, cookies, crackers, and vegetable shortening. The American Heart Association recommends that margarine should have no more than 2 grams of saturated fat per tablespoon and a liquid vegetable oil should be the first ingredient. They also recommend the use of soft margarine, which is lower in trans fatty acids because soft margarine is only slightly hydrogenated, and diet margarine because it has less fat and therefore fewer trans fatty acids.

Many health organizations agree that fat should account for less than 30% of daily calories (the current average for Americans is 34%) and saturated fat should be less than 10% of total calories. Lowering the overall fat intake would also decrease the amount of trans fatty acids. The Food and Drug Administration and the U.S. Department of Agriculture are encouraging the use of new food labels to inform consumers of the fat content of food. The best advice may be to reduce total fat in the diet by using fats and oils sparingly, cooking with little or no fat, substituting olive oil or canola oil for other oils, and limiting the use of coconut oil and palm oil, which are high in saturated fatty acids.

There are several products including peanut butter and butterlike spreads on the market that have 0% trans fatty acids. On the labels, they state that their products are nonhydrogenated, which avoids the production of the undesirable trans fatty acid. However, in the list of natural vegetable oils, such as soy and canola oil, there is also palm oil. Because palm oil has a melting point of 30°C, palm oil increases the overall melting point of the spread and gives a product that is solid at room temperature. However, palm oil contains high amounts of saturated fatty acids and has a similar effect in the body as does stearic acid (18 carbons) and fats derived from animal sources. Health experts recommend that we limit the amount of saturated fats, including palm oil, in our diets.

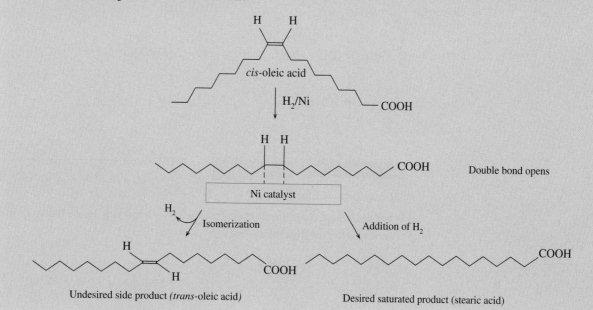

cis-oleic acid

H₂/Ni

Double bond opens

Ni catalyst

H₂ — Isomerization

Addition of H₂

Undesired side product (trans-oleic acid)

Desired saturated product (stearic acid)

Hydrolysis

Triacylglycerols are hydrolyzed (split by water) in the presence of strong acids or digestive enzymes called *lipases*. The products of hydrolysis of the ester bonds are glycerol and three fatty acids. The polar glycerol is soluble in water, but the fatty acids with their long hydrocarbon chains are not.

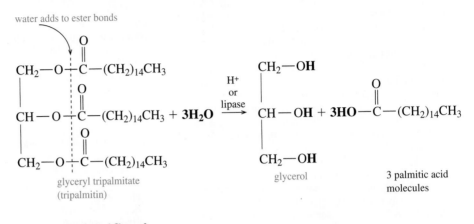

water adds to ester bonds

glyceryl tripalmitate (tripalmitin)

glycerol

3 palmitic acid molecules

Saponification

When a fat is heated with a strong base such as sodium hydroxide, saponification of the fat gives glycerol and the sodium salts of the fatty acids, which are soaps. When NaOH is used, a solid soap is produced that can be molded into a desired shape; KOH produces a softer, liquid soap. Oils that are polyunsaturated produce softer soaps. Names like "coconut" or "avocado shampoo" tell you the sources of the oil used in the reaction.

Fat or oil + strong base ⟶ glycerol + salts of fatty acids (soaps)

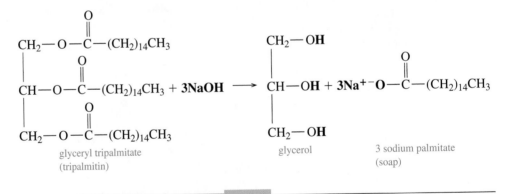

glyceryl tripalmitate (tripalmitin)

glycerol

3 sodium palmitate (soap)

SAMPLE PROBLEM 17.4

■ **Reactions of Lipids**

Write the equation for the reaction catalyzed by the enzyme lipase that hydrolyzes trilaurin during the digestion process.

SOLUTION

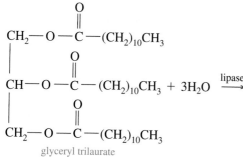

glyceryl trilaurate
(trilaurin)

glycerol   3 lauric acid
molecules

STUDY CHECK

What is the name of the product formed when a triacylglycerol containing oleic acid and linoleic acid is completely hydrogenated?

QUESTIONS AND PROBLEMS

Chemical Properties of Triacylglycerols

17.29 Write an equation for the hydrogenation of glyceryl trioleate, a fat containing glycerol and three oleic acid units.

17.30 Write an equation for the hydrogenation of glyceryl trilinolenate, a fat containing glycerol and three linolenic acid units.

17.31 A label on a container of margarine states that it contains partially hydrogenated corn oil.
a. How has the liquid corn oil been changed?
b. Why is the margarine product solid?

17.32 Why should a bottle of vegetable oil that has no preservatives be tightly covered and refrigerated?

17.33 a. Write an equation for the acid hydrolysis of glyceryl trimyristate (trimyristin).
b. Write an equation for the NaOH saponification of glyceryl trimyristate (trimyristin).

17.34 a. Write an equation for the acid hydrolysis of glyceryl trioleate (triolein).
b. Write an equation for the NaOH saponification of glyceryl trioleate (triolein).

17.35 Compare the structure of a triacylglycerol to the structure of olestra.

17.36 An oil is partially hydrogenated.
a. Are all or just some of the double bonds converted to single bonds?
b. What happens to some of the cis double bonds during hydrogenation?
c. How can you reduce the amount of trans fatty acids in your diet?

17.37 Write the product of the hydrogenation of the following triacylglycerol.

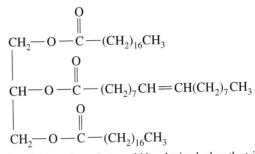

17.38 Write all the products that would be obtained when the triacylglycerol in problem 17.37 undergoes complete hydrolysis.

17.5 Glycerophospholipids

The **glycerophospholipids** are a family of lipids similar to triacylglycerols except that one hydroxyl group of glycerol is replaced by the ester of phosphoric acid and an amino alcohol, bonded through a phosphodiester bond. We can compare the general structures of a triacylglycerol and a glycerophospholipid as follows:

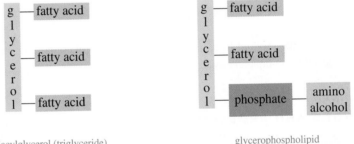

triacylglycerol (triglyceride)   glycerophospholipid

Phosphate Esters

In this group of compounds, ester bonds form between a hydroxyl group of an alcohol and phosphoric acid to give ester products similar to those formed with carboxylic acids.

$$HO-\overset{\overset{O}{\|}}{\underset{\underset{OH}{|}}{P}}-OH \ + \ HO-CH_3 \ \longrightarrow \ HO-\overset{\overset{O}{\|}}{\underset{\underset{OH}{|}}{P}}-O-CH_3 \ + \ H_2O$$

phosphoric acid +   alcohol $\longrightarrow$ phosphate ester   + water

The phosphate ester forms a diester by reaction with another alcohol.

$$HO-\overset{\overset{O}{\|}}{\underset{\underset{OH}{|}}{P}}-O-CH_3 \ + \ HO-CH_3 \ \longrightarrow \ CH_3-O-\overset{\overset{O}{\|}}{\underset{\underset{OH}{|}}{P}}-O-CH_3 \ + \ H_2O$$

phosphate ester   +   alcohol $\longrightarrow$ phosphate diester   + water

Three amino alcohols found in glycerophospholipids are choline, serine, and ethanolamine. In the body, at a pH of 7.4, these amino alcohols are ionized.

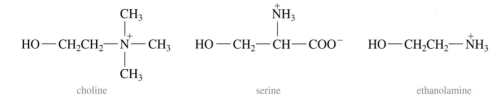

choline   serine   ethanolamine

Lecithins and **cephalins** are two types of glycerophospholipids that are particularly abundant in brain and nerve tissues as well as in egg yolks, wheat germ, and yeast. Lecithins contain choline, and cephalins contain ethanolamine and sometimes serine. In the following structural formulas, palmitic acid is used as an example of a fatty acid.

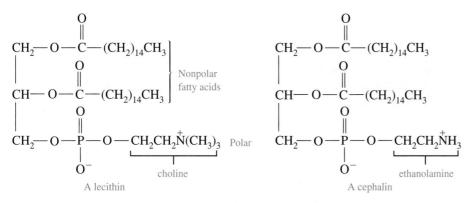

A lecithin   A cephalin

Glycerophospholipids contain both polar and nonpolar regions, which allow them to interact with both polar and nonpolar substances. The ionized alcohol and phosphate portion called "the head" is polar and can hydrogen bond with water. (See Figure 17.6.) The two fatty acids connected to the glycerol molecule represent

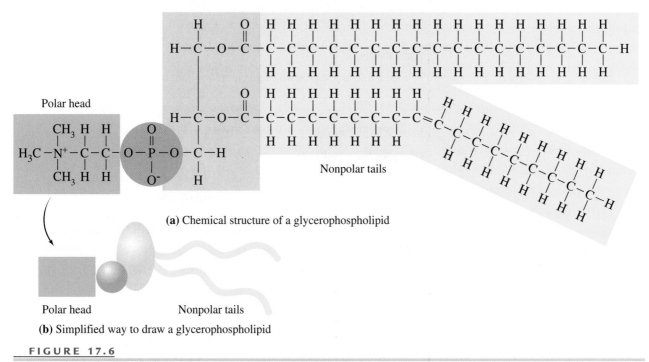

(a) Chemical structure of a glycerophospholipid

(b) Simplified way to draw a glycerophospholipid

FIGURE 17.6

(a) In a glycerophospholipid, a polar "head" contains the ionized amino alcohol and phosphoric acid groups, while the two fatty acids make up the nonpolar "tails."
(b) A simplified drawing indicates the polar region and the nonpolar region.

Q Why are glycerophospholipids polar?

the nonpolar "tails" of the phospholipid. The hydrocarbon chains that make up the "tails" are only soluble in other nonpolar substances, mostly lipids.

Glycerophospholipids are the most abundant lipids in cell membranes, where they play an important role in cellular permeability. They make up much of the myelin sheath that protects nerve cells. In the body fluids, they combine with the less polar triglycerides and cholesterol to make them more soluble as they are transported in the body.

SAMPLE PROBLEM 17.5

■ **Drawing Glycerophospholipid Structures**

Draw the structure of cephalin, using stearic acid for the fatty acids and serine for the amino alcohol. Describe each of the components in the glycerophospholipid.

SOLUTION

In general, glycerophospholipids are composed of a glycerol molecule in which two carbon atoms are attached to fatty acids such as stearic acid. The third carbon atom is attached via an ester bond to phosphate linked to an amino alcohol. In this example, the amino alcohol is serine.

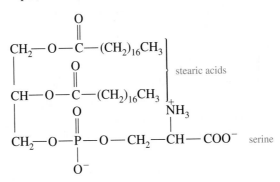

STUDY CHECK

What are the four components of glycerophospholipids?

QUESTIONS AND PROBLEMS

■ **Glycerophospholipids**

17.39 Describe the differences between triacylglycerols and glycerophospholipids.

17.40 Describe the differences between lecithin and cephalin.

17.41 Draw the structure of a glycerophospholipid containing two molecules of palmitic acid and ethanolamine. What is another name for this type of phospholipid?

17.42 Draw the structure of a glycerophospholipid that contains choline and palmitic acids.

17.43 Identify the following glycerophospholipid and list its components:

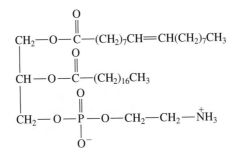

17.44 Identify the following glycerophospholipid and list its components:

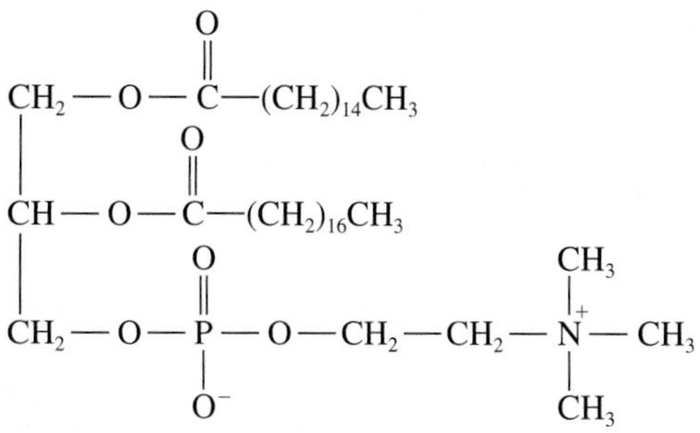

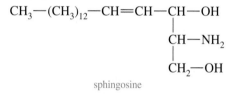

17.6 Sphingolipids

Sphingolipids, like phospholipids, have polar and nonpolar regions with a long-chain amino alcohol called sphingosine attached to a fatty acid, phosphate, and an amino alcohol.

$$CH_3-(CH_3)_{12}-CH=CH-CH-OH$$

$$CH-NH_2$$

$$CH_2-OH$$

sphingosine

When the —NH_2 group in sphingosine bonds to a fatty acid by an *amide* link, a *ceramide* is produced. A sphingolipid forms when the —OH group forms an ester bond with a phosphate attached to an amino alcohol.

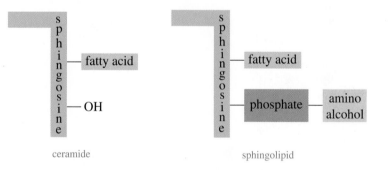

ceramide   sphingolipid

One of the most abundant sphingolipids is sphingomyelin. It is the white matter of the myelin sheath, a coating surrounding the nerve cells that increases the speed of nerve impulses and insulates and protects the nerve cells. In sphingomyelin, the sphingosine is linked to a phosphate ester of choline, an amino alcohol.

LEARNING GOAL

Describe the types of lipids that contain sphingosine.

 SELF-STUDY ACTIVITY
Phospholipids

 Sphingolipids in Disease

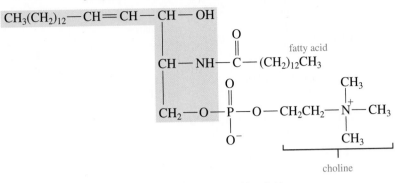

sphingosine

fatty acid

choline

sphingomyelin, a sphingolipid

Glycosphingolipids

Glycosphingolipids are sphingolipids that contain a carbohydrate such as galactose or glucose. They are important components of nerve membranes and muscle. In a glycosphingolipid, one or more monosaccharides, bonded by a glycosidic bond, replace the —OH of the ceramide.

Cerebrosides

Although they are only present in membranes in small amounts, certain types of glycosphingolipids on the cell surface are important to cellular recognition and tissue immunity. In glycosphingolipids such as **cerebrosides**, one monosaccharide (usually galactose) replaces the —OH of the ceramide.

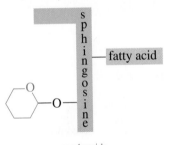

sphingosine

fatty acid

cerebroside

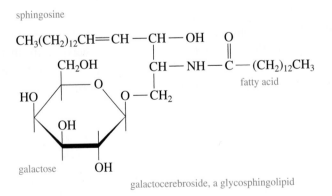

sphingosine

fatty acid

galactose

galactocerebroside, a glycosphingolipid

Gangliosides

Gangliosides are similar to cerebrosides but contain two or more monosaccharides, such as glucose and galactose. Gangliosides are important in the membranes of neurons and act as receptors for hormones, viruses, and certain drugs. In Tay–Sachs disease, the ganglioside known as GM$_2$ accumulates because of a genetic defect in hexosaminidase A, an enzyme needed for the removal of the *N*-acetyl-D-galactosamine.

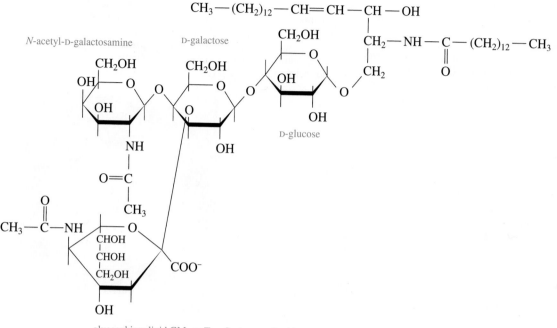

glycosphingolipid GM$_2$ or Tay–Sachs ganglioside

SAMPLE PROBLEM 17.6

▪ Glycosphingolipid

In Fabry's disease, the ganglioside shown here accumulates due to a deficiency of α-galactosidase. Identify the components A–E in this glycosphingolipid.

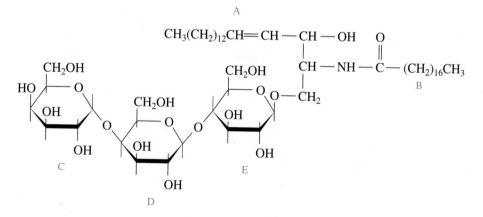

SOLUTION

In this glycosphingolipid, the components are sphingosine (A); stearic acid (B), an 18-carbon fatty acid; two galactose units (C, D); and one glucose (E).

STUDY CHECK

How do we know that this glycosphingolipid is a ganglioside rather than a cerebroside?

HEALTH NOTE

Lipid Diseases

Many lipid diseases (*lipidoses*) involve the excessive accumulation of a sphingolipid or glycolipid because an enzyme needed for its breakdown is deficient or absent. The accumulation of these glycolipids may enlarge the spleen, liver, and bone marrow cells (Gaucher's disease) and cause mental retardation, seizures, blindness, or death in early infancy. Some lipid storage diseases are listed in Table 17.3.

In multiple sclerosis, sphingomyelins are lost from the myelin sheath which is the protective membrane surrounding the neurons in the brain and spinal cord. As the disease progresses, the myelin sheath deteriorates. Scars form on the neurons and impair the transmission of nerve signals. The symptoms of multiple sclerosis include various levels of muscle weakness and loss of coordination and vision depending on the amount of damage. The cause of multiple sclerosis is not yet known, although some researchers suggest that a virus is involved.

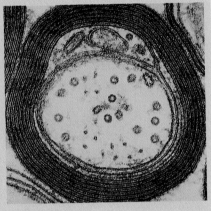

myelin sheath

TABLE 17.3

Lipid Diseases

Names of Disease	Lipid Stored	Type	Enzyme Absent
Fabry's	gal-gal-glucosylceramide	ganglioside	α-galactosidase
Gaucher's	glucosylceramide	cerebroside	β-glucosidase
Niemann–Pick	sphingomyelin	sphingolipid	sphingomyelinase
Tay–Sachs	GM$_2$ ganglioside	ganglioside	hexosaminidase A

QUESTIONS AND PROBLEMS

Sphingolipids

17.45 Describe the difference between glycerophospholipids and sphingolipids.

17.46 Describe the differences between a cerebroside and a ganglioside.

17.47 Draw the structure of a cerebroside containing palmitic acid and galactose.

17.48 What amino alcohol is found in sphingomyelin? Draw the structure of a sphingomyelin containing palmitic acid.

LEARNING GOAL

Describe the structures of steroids.

Steroids: Cholesterol, All Dressed Up

17.7 Steroids: Cholesterol, Bile Salts, and Steroid Hormones

Steroids are compounds containing the steroid nucleus, which consists of three cyclohexane rings and one cyclopentane ring fused together. Although they are large molecules, steroids do not hydrolyze to give fatty acids and alcohols. The four rings in the steroid nucleus are designated A, B, C, and D. The carbon atoms are numbered beginning with the carbons in ring A and ending with the two methyl groups.

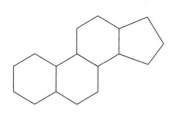

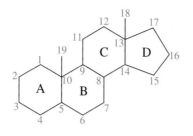

Steroid

Cholesterol

Attaching other atoms and groups of atoms to the steroid structure forms a wide variety of steroid compounds. **Cholesterol**, which is one of the most important and abundant steroids in the body, is a *sterol* because it contains an oxygen atom as a hydroxyl (—OH) group on carbon 3. Like many steroids, cholesterol has a double bond between carbon 5 and carbon 6, methyl groups at carbon 10 and carbon 13, and a carbon chain at carbon 17. In other steroids, the oxygen atom typically at carbon 3 forms a carbonyl (C=O) group.

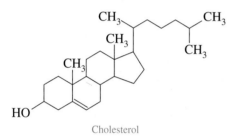

Cholesterol

Cholesterol in the Body

Cholesterol is a component of cellular membranes, myelin sheath, and brain and nerve tissue. It is also found in the liver, bile salts, and skin, where it forms vitamin D. In the adrenal gland, it is used to synthesize steroid hormones. Cholesterol in the body is obtained from eating meats, milk, and eggs, and it is also synthesized by the liver from fats, carbohydrates, and proteins. There is no cholesterol in vegetable and plant products.

If a diet is high in cholesterol, the liver produces less. A typical daily American diet includes 400–500 mg of cholesterol, one of the highest in the world. The American Heart Association has recommended that we consume no more than 300 mg of cholesterol a day. The cholesterol contents of some typical foods are listed in Table 17.4 (p. 632).

When cholesterol exceeds its saturation level in the bile, gallstones may form. Gallstones are composed of almost 100% cholesterol with some calcium salts, fatty acids, and phospholipids. High levels of cholesterol are also associated with the accumulation of lipid deposits (plaque) that line and narrow the coronary arteries. (See Figure 17.7.) Clinically, cholesterol levels are considered elevated if the total plasma cholesterol level exceeds 200 mg/dL.

Saturated fats in the diet may stimulate the production of cholesterol by the liver. A diet that is low in foods containing cholesterol and saturated fats appears to be

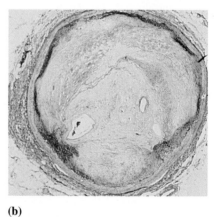

(a)

(b)

FIGURE 17.7

Excess cholesterol forms plaque that can block an artery, resulting in a heart attack. (**a**) A normal, open artery shows no buildup of plaque. (**b**) An artery that is almost completely clogged by atherosclerotic plaque.

Q **What property of cholesterol would cause it to form deposits along the coronary arteries?**

TABLE 17.4
Cholesterol Content of Some Foods

Food	Serving Size	Cholesterol (mg)
Liver (beef)	3 oz	370
Egg	1	250
Lobster	3 oz	175
Fried chicken	3½ oz	130
Hamburger	3 oz	85
Chicken (no skin)	3 oz	75
Fish (salmon)	3 oz	40
Butter	1 tablespoon	30
Whole milk	1 cup	35
Skim milk	1 cup	5
Margarine	1 tablespoon	0

helpful in reducing the serum cholesterol level. Other factors that may also increase the risk of heart disease are family history, lack of exercise, smoking, obesity, diabetes, gender, and age.

SAMPLE PROBLEM 17.7

■ Cholesterol

Observe the structure of cholesterol for the following questions:

a. What part of cholesterol is the steroid nucleus?
b. What features have been added to the steroid nucleus in cholesterol?
c. What classifies cholesterol as a sterol?

SOLUTION

a. The four fused rings form the steroid nucleus.
b. The cholesterol molecule contains an alcohol group (—OH) on the first ring, one double bond in the second ring, and a branched carbon chain on the fourth ring.
c. The alcohol group determines the sterol classification.

STUDY CHECK

Why is cholesterol in the lipid family?

Bile Salts

The *bile* salts are synthesized in the liver from cholesterol and stored in the gallbladder. When bile is secreted into the small intestine, the bile salts mix with the water-insoluble fats and oils in our diets. The bile salts with their nonpolar and polar regions act much like soaps, breaking apart and emulsifying large globules of fat. The emulsions that form have a larger surface area for the lipases, enzymes that digest fat. The bile salts also help in the absorption of cholesterol into the intestinal mucosa.

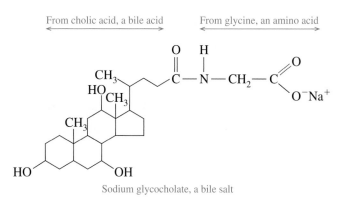

From cholic acid, a bile acid   From glycine, an amino acid

Sodium glycocholate, a bile salt

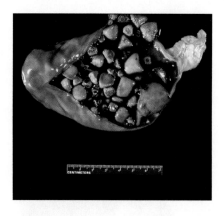

When large amounts of cholesterol accumulate in the gallbladder, cholesterol can precipitate out and form gallstones. (See Figure 17.8.) If a gallstone passes into the bile duct, the pain can be severe. If the gallstone obstructs the duct, bile cannot be excreted. Then bile pigments known as bilirubin enter the blood where they cause jaundice, which gives a yellow color to the skin and eyes.

FIGURE 17.8

Gallstones form in the gallbladder when cholesterol levels are high.

Q **What type of steroid is stored in the gallbladder?**

Lipoproteins: Transporting Lipids

In the body, lipids must be transported through the bloodstream to tissues where they are stored, used for energy, or to make hormones. However, most lipids are nonpolar and insoluble in the aqueous environment of blood. They are made more soluble by combining them with phospholipids and proteins to form water-soluble complexes called **lipoproteins**. In general, lipoproteins are spherical particles with an outer surface of polar proteins and phospholipids that surround hundreds of nonpolar molecules of triacylglycerols and cholesteryl ester. (See Figure 17.9.) Cholesteryl esters are the prevalent form of cholesterol in the blood. They are formed by the esterification of the hydroxyl group in cholesterol with a fatty acid.

MC GOB   The Body's Lipid Limousines

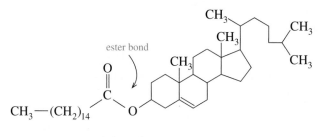

ester bond

$CH_3—(CH_2)_{14}$

cholesteryl ester

There are several types of lipoproteins that differ in density, lipid composition, and function. They include chylomicrons, very-low-density lipoprotein (VLDL), low-density lipoprotein (LDL), and high-density lipoprotein (HDL). (See Table 17.5.)

The chylomicrons formed in the mucosal cells of the small intestine and the VLDLs formed in the liver transport triacylglycerols, phospholipids, and cholesterol to the tissues for storage or to the muscles for energy. (See Figure 17.10.) The LDLs transport cholesterol to tissues to be used for the synthesis of cell membranes, steroid hormones, and bile salts. When the level of LDL exceeds the amount of cholesterol needed by the tissues, the LDLs deposit cholesterol in the arteries, which can restrict blood flow and increase the risk of developing heart disease and/or myocardial infarctions (heart attacks). This is why LDL cholesterol is called "bad" cholesterol.

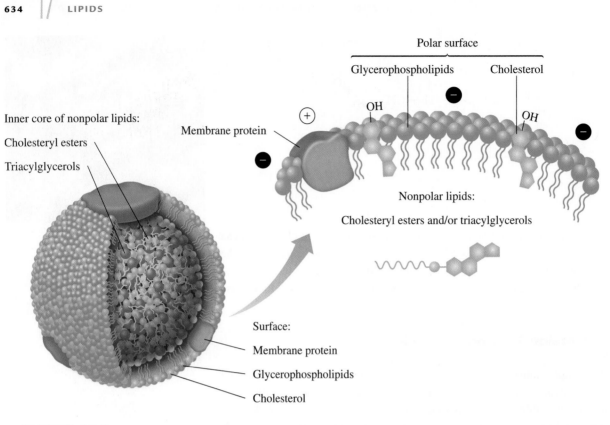

Inner core of nonpolar lipids:

Cholesteryl esters

Triacylglycerols

Surface:

Membrane protein

Glycerophospholipids

Cholesterol

Polar surface

Glycerophospholipids   Cholesterol

Membrane protein

Nonpolar lipids:

Cholesteryl esters and/or triacylglycerols

FIGURE 17.9

A spherical lipoprotein particle surrounds nonpolar lipids with polar lipids and protein for transport to body cells.

Q **Why are the polar components on the surface of a lipoprotein particle and the nonpolar components at the center?**

TABLE 17.5

Composition and Properties of Plasma Lipoproteins

	Chylomicron	VLDL	LDL	HDL
Density (g/mL)	0.940	0.950–1.006	1.006–1.063	1.063–1.210
Composition (% by mass)				
triacylglycerol	86	55	6	4
phospholipids	7	18	22	24
cholesterol	2	7	8	2
cholesteryl esters	3	12	42	15
protein	2	8	22	55

The HDLs remove excess cholesterol from the tissues and carry it to the liver where it is converted to bile salts and eliminated. When HDL levels are high, cholesterol that is not needed by the tissues is carried to the liver for elimination rather than deposited in the arteries, which gives the HDLs the name of "good" cholesterol. Most of the cholesterol in the body is synthesized in the liver, although some comes from the diet. However, a person on a high-fat diet reabsorbs cholesterol from the bile salts causing less cholesterol to be eliminated. In addition, higher levels of saturated fats stimulate the synthesis of cholesterol by the liver.

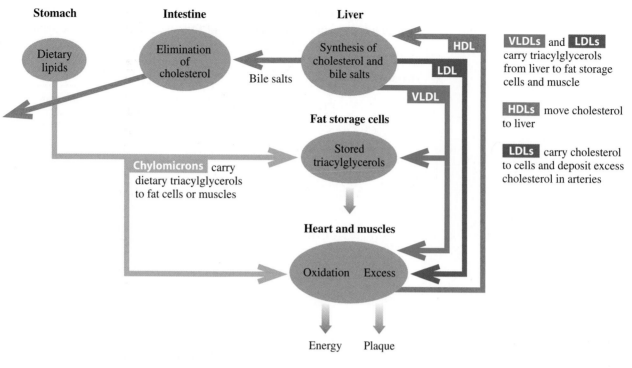

FIGURE 17.10

Lipoproteins such as HDLs and LDLs transport nonpolar lipids and cholesterol to cells and the liver.

Q **What type of lipoprotein transports cholesterol to the liver?**

Because high cholesterol levels are associated with the onset of arteriosclerosis and heart disease, the serum levels of LDL and HDL are generally determined in a medical examination. For adults, recommended levels for total cholesterol are less than 200 mg/dL with LDL less than 130 mg/dL and HDL over 40 mg/dL. A lower level of serum cholesterol decreases the risk of heart disease. Higher HDL levels are found in people who exercise regularly and eat less saturated fat.

Steroid Hormones

The word *hormone* comes from the Greek "to arouse" or "to excite." Hormones are chemical messengers that serve as a communication system from one part of the body to another. The *steroid* hormones, which include the sex hormones and the adrenocortical hormones, are closely related in structure to cholesterol and depend on cholesterol for their synthesis.

Two important male sex hormones, *testosterone* and *androsterone*, promote the growth of muscle and of facial hair and the maturation of the male sex organs and of sperm.

The *estrogens*, a group of female sex hormones, direct the development of female sexual characteristics: the uterus increases in size, fat is deposited in the breasts, and the pelvis broadens. *Progesterone* prepares the uterus for the implantation of a fertilized egg. If an egg is not fertilized, the levels of progesterone and estrogen drop sharply, and menstruation follows. Synthetic forms of the female sex hormones are used in birth-control pills. As with other kinds of steroids, side effects include weight gain and a greater risk of forming blood clots. The structures of some steroid hormones follow:

Hormone	Biological Effects
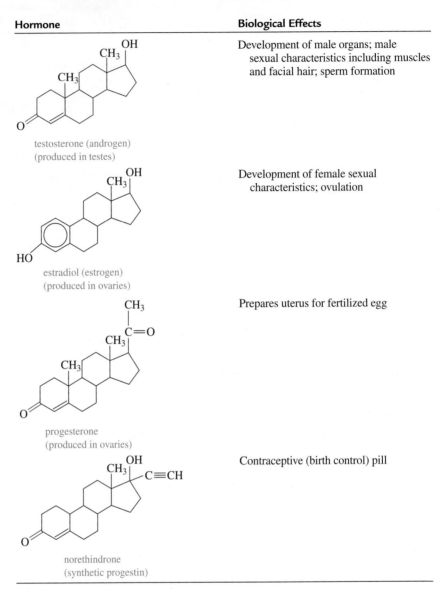 testosterone (androgen) (produced in testes)	Development of male organs; male sexual characteristics including muscles and facial hair; sperm formation
estradiol (estrogen) (produced in ovaries)	Development of female sexual characteristics; ovulation
progesterone (produced in ovaries)	Prepares uterus for fertilized egg
norethindrone (synthetic progestin)	Contraceptive (birth control) pill

Adrenal Corticosteroids

The adrenal glands, located on the top of each kidney, produce the corticosteroids. *Aldosterone*, a mineralocorticoid, is responsible for electrolyte and water balance by the kidneys. *Cortisone*, a glucocorticoid, increases the blood glucose level and stimulates the synthesis of glycogen in the liver from amino acids. Synthetic corticoids such as *prednisone* are derived from cortisone and used medically for reducing inflammation and treating asthma and rheumatoid arthritis, although health problems can result from long-term use.

HEALTH NOTE

Anabolic Steroids

Some of the physiological effects of testosterone are to increase muscle mass and decrease body fat. Derivatives of testosterone called *anabolic steroids* that enhance these effects have been synthesized. Although they have some medical uses, anabolic steroids have been used in rather high dosages by some athletes in an effort to increase muscle mass. Such use is illegal.

Use of anabolic steroids in attempting to improve athletic strength can cause side effects including hypertension, fluid retention, increased hair growth, sleep disturbances, and acne. Over a long period of time, their use can be devastating and may cause irreversible liver damage and decreased sperm production.

Some Anabolic Steroids

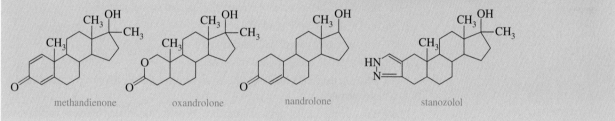

methandienone   oxandrolone   nandrolone   stanozolol

Corticosteroids

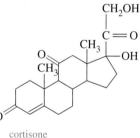

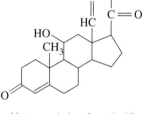

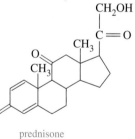

cortisone
(produced in adrenal gland)

aldosterone (mineralocorticoid)
(produced in adrenal gland)

prednisone
(synthetic corticoid)

Biological Effects

| Increases the blood glucose and glycogen levels from fatty acids and amino acids | Increases the reabsorption of Na⁺ in kidneys; retention of water | Reduces inflammation; treatment of asthma and rheumatoid arthritis |

Increases the blood glucose and glycogen levels from fatty acids and amino acids

Increases the reabsorption of Na^+ in kidneys; retention of water

Reduces inflammation; treatment of asthma and rheumatoid arthritis

SAMPLE PROBLEM 17.8

Steroid Hormones

What are the functional groups on the steroid nucleus in the sex hormones estradiol and testosterone?

SOLUTION

Estradiol contains a benzene ring and two alcohol groups. Testosterone contains a ketone group, a double bond, and an alcohol group.

STUDY CHECK

What are the similarities and differences in the structures of testosterone and the anabolic steroid nandrolone?

QUESTIONS AND PROBLEMS

■ Steroids: Cholesterol, Bile Salts, and Steroid Hormones

17.49 Draw the structure for the steroid nucleus.

17.50 Which of the following compounds are derived from cholesterol?
   a. glyceryl tristearate
   b. cortisone
   c. bile salts
   d. testosterone
   e. estradiol

17.51 What is the function of bile salts in digestion?

17.52 Why are gallstones composed of cholesterol?

17.53 What is the general structure of lipoproteins?

17.54 Why are lipoproteins needed to transport lipids in the bloodstream?

17.55 How do chylomicrons differ from very-low-density lipoproteins?

17.56 How do LDLs differ from HDLs?

17.57 Why are LDLs called "bad" cholesterol?

17.58 Why are HDLs called "good" cholesterol?

17.59 What are the similarities and differences between the sex hormones estradiol and testosterone?

17.60 What are the similarities and differences between the adrenal hormone cortisone and the synthetic corticoid prednisone?

17.61 Which of the following are male sex hormones?
   a. cholesterol
   b. aldosterone
   c. estrogen
   d. testosterone
   e. choline

17.62 Which of the following are adrenal steroids?
   a. cholesterol
   b. aldosterone
   c. estrogen
   d. testosterone
   e. choline

LEARNING GOAL

Describe the composition and function of the lipid bilayer in cell membranes.

SELF-STUDY ACTIVITY
Membrane Structure

Diffusion
Osmosis
Active Transport

■ 17.8 Cell Membranes

The membrane of a cell separates the contents of a cell from the external fluids. It is semipermeable so that nutrients can enter the cell and waste products can leave. The main components of a cell membrane are glycerophospholipids and sphingolipids. Earlier in this chapter we saw that phospholipids have a nonpolar region or tail with long-chain fatty acids and a polar region or head from phosphoric acid and amino alcohols that ionize at physiological pH. The lipid composition of the membranes of human red blood cells and bacterial cells is given in Table 17.6.

In a cell membrane, two rows of phospholipids are arranged like a sandwich. Their nonpolar tails, which are hydrophobic ("water-fearing"), move to the center,

TABLE 17.6

Lipid Composition of Cell Membranes

Type of Lipid	Human Red Blood Cells	Bacterial Cells
Glycerophospholipids		
choline	19	0
ethanolamine	18	65
serine	8	0
Triacylglycerol	0	18
Sphingomyelin	18	0
Glycosphingolipids	10	0
Cholesterol	25	0
Others	2	17

Data adapted from Mathews, C. K.; Van Holde, K. K.; Ahem, K. G. *Biochemistry*; Addison Wesley/Longman/ Benjamin Cummings: New York, 2000, p. 322.

while their polar heads, which are hydrophilic ("water-loving"), align on the outer edges of the membrane. This double row arrangement of phospholipids is called a **lipid bilayer**. (See Figure 17.11.) One row of phospholipids forms the outside surface of the membrane, which is in contact with the external fluids, and the other row forms the inside surface of the membrane, which is in contact with the internal contents of the cell.

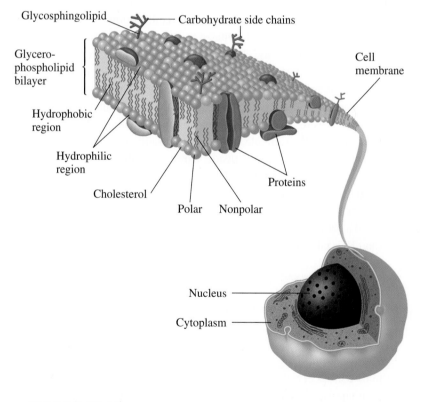

FIGURE 17.11

The fluid mosaic model of a cell membrane. Proteins and cholesterol are embedded in a lipid bilayer of phospholipids. The bilayer forms a membrane-type barrier with polar heads at the membrane surfaces and the nonpolar tails in the center away from the water.

Q **What types of fatty acids are found in the phospholipids of the lipid bilayer?**

MC
GOB
The Fluid Mosaic Model
of Membranes

Most of the phospholipids in the lipid bilayer contain unsaturated fatty acids. Because of the kinks in the carbon chains at the cis double bonds, the phospholipids do not fit closely together. As a result, the lipid bilayer is not a rigid, fixed structure, but one that is dynamic and fluid-like. In this liquid-like bilayer, there are also proteins, carbohydrates, and cholesterol molecules. For this reason, the model of biological membranes is referred to as the **fluid mosaic model** of membranes.

In the fluid mosaic model, peripheral proteins emerge on just one of the surfaces, outer or inner. The integral proteins extend through the entire lipid bilayer and appear on both surfaces of membrane. Some proteins and some lipids on the outer surface of the cell membrane are attached to carbohydrates to form glycoproteins and glyco-sphingolipids. These carbohydrate chains project into the surrounding fluid environment where they are responsible for cell recognition and communication with chemical messengers such as hormones and neurotransmitters. In animals, cholesterol makes up 20–25% of the lipid bilayer. Embedded among the nonpolar tails of the fatty acids, cholesterol molecules add strength and rigidity to the cell membrane.

Transport Through Cell Membranes

Ions and molecules flow in and out of the cell in several ways. In the simplest transport mechanism called diffusion or passive transport, ions and small molecules migrate from a higher concentration to a lower concentration. For example, some ions as well as small molecules such as O_2, urea, and water diffuse through cell membranes. If their concentration is greater outside the cell than inside, they diffuse into the cell. If water has a high concentration in the cell, it diffuses out of the cell.

Another type of transport called facilitated transport increases the rate of diffusion for substances that diffuse too slowly by passive diffusion to meet cell needs. This process utilizes the integral proteins that extend from one edge of the cell membrane to the other. These protein channels allow transport of chloride ion (Cl^-), bicarbonate ion (HCO_3^-), and glucose molecules in and out of the cell.

Certain ions such as K^+, Na^+, and Ca^{2+}, move across a cell membrane against a concentration gradient. For example, the K^+ concentration is greater inside a cell, and the Na^+ concentration is greater outside. However, in the conduction of nerve impulses and contraction of muscles, K^+ moves into the cell, and Na^+ moves out. To move an ion from a lower to a higher concentration requires energy, which is accomplished by a process known as active transport. In active transport, a protein complex called a Na^+/K^+ pump breaks down ATP to ADP, which releases energy to move Na^+ and K^+ against their concentration gradients. (See Figure 17.12.)

SAMPLE PROBLEM 17.9

▪ Lipid Bilayer in the Cell Membranes

Describe the role of phospholipids in the lipid bilayer.

SOLUTION

Phospholipids consist of polar and nonpolar parts. In a cell membrane, an alignment of the nonpolar sections toward the center with the polar sections on the outside produces a barrier that prevents the contents of a cell from mixing with the fluids on the outside of the cell.

STUDY CHECK

Why are protein channels needed in the lipid bilayer?

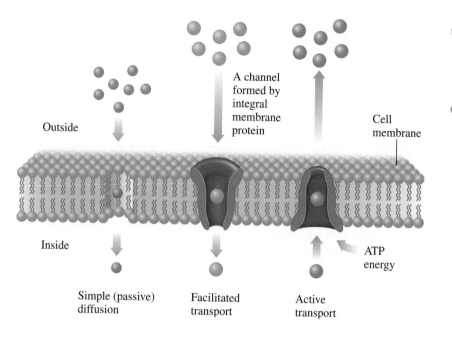

Outside

A channel formed by integral membrane protein

Cell membrane

Inside

ATP energy

Simple (passive) diffusion

Facilitated transport

Active transport

FIGURE 17.12

Substances are transported across a cell membrane by either simple diffusion, facilitated transport, or by active transport.

Q **What is the difference between simple diffusion and facilitated transport?**

QUESTIONS AND PROBLEMS

■ Cell Membranes

17.63 What types of lipids are found in cell membranes?

17.64 Describe the structure of a lipid bilayer.

17.65 What is the function of the lipid bilayer in a cell membrane?

17.66 How do the unsaturated fatty acids in the phospholipids affect the structure of cell membranes?

17.67 Where are proteins located in cell membranes?

17.68 What components are attached to carbohydrates on the outer surface of a cell membrane?

17.69 What is the function of the carbohydrates on a cell membrane?

17.70 Why is a cell membrane semipermeable?

17.71 What are some ways that substances move in and out of cells?

17.72 Identify the type of transport described by each of the following:
   a. A molecule moves through a protein channel.
   b. O_2 moves into the cell from a higher concentration outside the cell.
   c. An ion moves from low to high concentration in the cell.

CONCEPT MAP

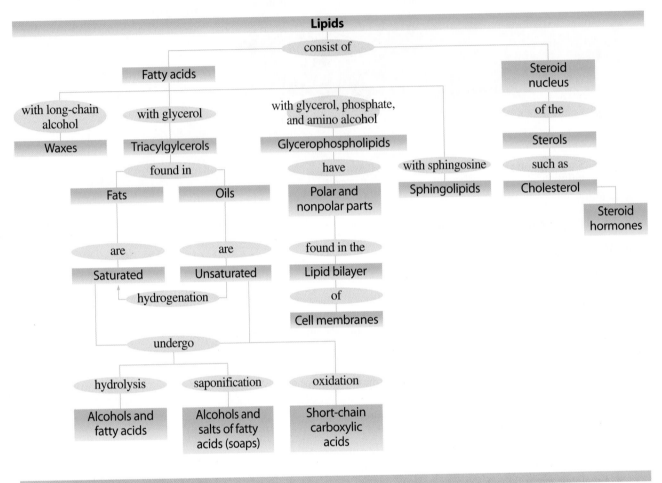

CHAPTER REVIEW

17.1 Lipids
Lipids are nonpolar compounds that are not soluble in water. Classes of lipids include waxes; fats and oils; glycerophospholipids; and steroids.

17.2 Fatty Acids
Fatty acids are unbranched carboxylic acids that typically contain an even number (12–18) of carbon atoms. Fatty acids may be saturated, monounsaturated with one double bond, or polyunsaturated with two or more double bonds. The double bonds in unsaturated fatty acids are almost always cis.

17.3 Waxes, Fats, and Oils
A wax is an ester of a long-chain fatty acid and a long-chain alcohol. The triacylglycerols of fats and oils are esters of glycerol with

three long-chain fatty acids. Fats contain more saturated fatty acids and have higher melting points than most vegetable oils.

17.4 Chemical Properties of Triacylglycerols
The hydrogenation of unsaturated fatty acids converts double bonds to single bonds. The oxidation of unsaturated fatty acids produces short-chain fatty acids with disagreeable odors. The hydrolysis of the ester bonds in fats or oils produces glycerol and fatty acids. In saponification, a fat heated with a strong base produces glycerol and the salts of the fatty acids or soaps.

17.5 Glycerophospholipids
Glycerophospholipids are esters of glycerol with two fatty acids and a phosphate group attached to an amino alcohol.

17.6 Sphingolipids

In sphingolipids, the alcohol sphingosine forms a bond with a fatty acid and a phosphate–amino alcohol group. In glycosphingolipids, sphingosine is bonded to a fatty acid and one or more monosaccharides.

17.7 Steroids: Cholesterol, Bile Salts, and Steroid Hormones

Steroids are lipids containing the steroid nucleus, which is a fused structure of four rings. Steroids include cholesterol, bile salts, and vitamin D. Lipids, which are nonpolar, are transported through the aqueous environment of the blood by forming lipoproteins. Lipoproteins such as chylomicrons and LDL transport triacylglycerols from the intestines and the liver to fat cells for storage and muscles for energy. HDLs transport cholesterol from the tissues to the liver for elimination. The steroid hormones are closely related in structure to cholesterol and depend on cholesterol for their synthesis. The sex hormones, such as estrogen and testosterone, are responsible for sexual characteristics and reproduction. The adrenal corticosteroids such as aldosterone and cortisone regulate water balance and glucose levels in the cells.

17.8 Cell Membranes

All animal cells are surrounded by a semipermeable membrane that separates the cellular contents from the external fluids. The membrane is composed of two rows of glycerophospholipids in a lipid bilayer. Nutrients and waste products move through the cell membrane using simple transport (diffusion), facilitated transport, or active transport.

KEY TERMS

cephalins Phospholipids found in brain and nerve tissues that incorporate the amino alcohol serine or ethanolamine.

cerebroside A glycolipid consisting of sphingosine, a fatty acid, and a monosaccharide (usually galactose).

cholesterol The most prevalent of the steroid compounds found in cellular membranes; needed for the synthesis of vitamin D, hormones, and bile acids.

esterification The reaction of an alcohol such as glycerol with acids to form ester bonds.

fat Another term for solid triacylglycerols.

fatty acids Long-chain carboxylic acids found in fats.

fluid mosaic model The concept that cell membranes are lipid bilayer structures that contain an assortment of polar lipids and proteins in a dynamic, fluid arrangement.

ganglioside A glycolipid consisting of sphingosine, a fatty acid, and two or more monosaccharides.

glycerophospholipids Polar lipids of glycerol attached to two fatty acids and a phosphate group connected to an amino group such as choline, serine, or ethanolamine.

glycosphingolipids The phospholipid that combines sphingosine with a fatty acid bonded to the nitrogen group and one or more monosaccharides bonded by a glycosidic link, which replaces the —OH group of sphingosine.

hydrogenation The addition of hydrogen to unsaturated fats.

lecithins Glycerophospholipids containing choline as the amino alcohol.

lipid bilayer A model of a cell membrane in which phospholipids are arranged in two rows interspersed with proteins arranged at different depths.

lipids A family of compounds that is nonpolar in nature and not soluble in water; includes fats, waxes, phospholipids, and steroids.

lipoprotein A combination of nonpolar lipids with glycerophospholipids and proteins to form a polar complex that can be transported through body fluids.

monounsaturated fatty acids Fatty acids with one double bond.

oil Another term for liquid triacylglycerols.

polyunsaturated fatty acids Fatty acids that contain two or more double bonds.

prostaglandins A number of compounds derived from arachidonic acid that regulate several physiological processes.

saturated fatty acids Fatty acids that have no double bonds; they have higher melting points than unsaturated lipids and are usually solid at room temperatures.

sphingolipids Phospholipids in which sphingosine has replaced glycerol.

steroids Types of lipid composed of a multicyclic ring system.

triacylglycerols A family of lipids composed of three fatty acids bonded through ester bonds to glycerol, a trihydroxy alcohol.

wax The ester of a long-chain alcohol and a long-chain saturated fatty acid.

UNDERSTANDING THE CONCEPTS

17.73 Palmitic acid is obtained from palm oil as glyceryl tripalmitate. Draw the structure of glyceryl tripalmitate.

17.74 Jojoba wax in candles consists of stearic acid and a 22-carbon saturated alcohol. Write the structure of jojoba wax.

17.75 Sunflower oil can be used to make margarine. A triacylglycerol in sunflower oil consists of 2 linoleic acids and 1 oleic acid.

a. Write two isomers for the triacylglycerol in sunflower oil.
b. Using one of the isomers, write the reaction that would be used when sunflower oil is used to make solid margarine.

17.76 Identify each of the following as saturated, monounsaturated, polyunsaturated, omega-3, or omega-6 fatty acids:

a. $CH_3{-}(CH_2)_4{-}CH{=}CH{-}CH_2{-}CH{=}CH{-}(CH_2)_7{-}COOH$
b. linolenic acid
c. $CH_3{-}(CH_2)_{14}{-}COOH$
d. $CH_3{-}(CH_2)_7{-}CH{=}CH{-}(CH_2)_7{-}COOH$

ADDITIONAL QUESTIONS AND PROBLEMS

For instructor-assigned homework, go to **www.masteringgob.com.**

17.77 Among the ingredients of a lipstick are beeswax, carnauba wax, hydrogenated vegetable oils, and capric triglyceride. What types of lipids have been used? Draw the structure of capric triglyceride (capric acid is the saturated 10-carbon fatty acid).

17.78 Because peanut oil floats on the top of peanut butter, many brands of peanut butter are hydrogenated. A solid product then forms that is mixed into the peanut butter and does not separate. If a triacylglycerol in peanut oil that contains one palmitic acid, one oleic acid, and one linoleic acid is completely hydrogenated, what is the product?

17.79 Trans fats are produced during the hydrogenation of polyunsaturated oils.

a. What is the typical configuration of a monounsaturated fatty acid?
b. How does a trans fat compare with a cis fat?
c. Draw the structure of *trans*-oleic acid.

17.80 One mole of triolein is completely hydrogenated. What is the product? How many moles of hydrogen are required? How many grams of hydrogen? How many liters of hydrogen are needed if the reaction is run at STP?

17.81 What are the structures of glyceryl stearate and a lecithin containing palmitic acid?

17.82 Some typical meals at fast-food restaurants are listed here. Calculate the number of kilocalories from fat and the percentage of total kilocalories due to fat (1 gram of fat = 9 kcal). Would you expect the fats to be mostly saturated or unsaturated? Why?

a. a chicken dinner, 830 kcal, 46 g fat

b. a quarter-pound cheeseburger, 518 kcal, 29 g fat

c. pepperoni pizza (three slices), 560 kcal, 18 g fat

d. beef burrito, 470 kcal, 21 g fat

e. deep-fried fish (three pieces), 480 kcal, 28 g fat

17.83 Identify the following as fatty acids, soaps, triacylglycerols, wax, glycerophospholipid, sphingolipid, or steroid:

a. beeswax	**b.** cholesterol
c. lecithin	**d.** glyceryl tripalmitate (tripalmitin)
e. sodium stearate	**f.** safflower oil
g. sphingomyelin	**h.** whale blubber
i. adipose tissue	**j.** progesterone
k. cortisone	**l.** stearic acid

17.84 Why would an animal that lives in a cold climate have more unsaturated triacylglycerols in its body fat than an animal that lives in a warm climate?

17.85 Identify the components in each of the following lipids as:

1. glycerol	**2.** fatty acid
3. phosphate	**4.** amino alcohol
5. steroid nucleus	**6.** sphingosine

a. estrogen

b. cephalin

c. wax

d. triacylglycerol

e. glycerophospholipid

f. sphingomyelin

17.86 Which of the following are found in cell membranes?

a. cholesterol

b. triacylglycerols

c. carbohydrates

d. proteins

e. waxes

f. glycerophospholipids

g. sphingolipids

h. prostaglandins

CHALLENGE QUESTIONS

17.87 Identify the lipoprotein in each description as:

1. chylomicrons   **2.** VLDL   **3.** LDL   **4.** HDL

a. "good" cholesterol

b. transports most of the cholesterol to the cells

c. carries triacylglycerols from the intestine to the fat cells

d. transports cholesterol to the liver

e. has the greatest abundance of protein

f. "bad" cholesterol

g. carries triacylglycerols synthesized in the liver to the muscles

h. has the lowest density

17.88 Which of the following has the lower melting point? Explain.

1. $CH_3-(CH_2)_{16}-COOH$

2. $CH_3-(CH_2)_7-CH=CH-(CH_2)_7-COOH$

17.89 Draw the structure of glycerophospholipid that is made from stearic acid, palmitic acid, and a phosphate bonded to ethanolamine.

17.90 Olive oil consists of a high percentage of triolein.

a. Draw the structure for triolein.

b. How many liters of H_2 gas at STP are needed to completely saturate 100. g of triolein?

c. How many mL of 0.250 M NaOH are needed to completely saponify 100. g of triolein?

17.91 A sink drain can become clogged with solid fat such as glyceryl tristearate.

a. How would adding lye (NaOH) to the sink drain remove the blockage?

b. Write an equation for the reaction that occurs.

ANSWERS

Answers to Study Checks

17.1 a glycosphingolipid

17.2 **a.** 16   **b.** unsaturated   **c.** liquid

17.3

$$CH_2-O-\overset{\overset{\displaystyle O}{\|}}{C}-(CH_2)_{12}-CH_3$$
$$CH-O-\overset{\overset{\displaystyle O}{\|}}{C}-(CH_2)_{12}-CH_3$$
$$CH_2-O-\overset{\overset{\displaystyle O}{\|}}{C}-(CH_2)_{12}-CH_3$$

17.4 tristearin

17.5 Glycerophospholipids contain glycerol, fatty acids, a phosphate, and an amino alcohol.

17.6 Cerebrosides contain only one monosaccharide, and gangliosides contain two or more monosaccharide units.

17.7 Cholesterol is not soluble in water; it is classified with the lipid family.

17.8 Testosterone and nandrolone both contain a steroid nucleus with one double bond and a ketone group in the first ring, and a methyl and alcohol group on the five-carbon ring. Nandrolone does not have the second methyl group at the first and second ring fusion that is seen in the structure of testosterone.

17.9 Protein channels allow ions and polar molecules to flow in and out of the cell through the lipid bilayer.

Answers to Selected Questions and Problems

17.1 Lipids provide energy and protection and insulation for the organs in the body. Lipids are also an important part of cell membranes.

17.3 Because lipids are not soluble in water, a polar solvent, they are nonpolar molecules.

17.5 All fatty acids contain a long chain of carbon atoms with a carboxylic acid group. Saturated fats contain only carbon-to-carbon single bonds; unsaturated fats contain one or more double bonds.

17.7 **a.** palmitic acid

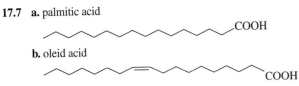

   b. oleid acid

17.9 **a.** saturated
   b. unsaturated
   c. unsaturated
   d. saturated

17.11 In a cis fatty acid, the hydrogen atoms are on the same side of the double bond, which produces a bend in the carbon chain. In a trans fatty acid, the hydrogen atoms are on opposite sides of the double bond, which gives a carbon chain without any bend.

17.13 In an omega-3 fatty acid, there is a double bond on carbon 3 counting from the methyl group, whereas in an omega-6 fatty acid, there is a double bond beginning at carbon 6 counting from the methyl group.

17.15 Arachidonic acid contains four double bonds and no side groups. In PGE$_1$, a part of the chain forms cyclopentane and there are hydroxyl and ketone functional groups.

17.17 Prostaglandins affect blood pressure and stimulate contraction and relaxation of smooth muscle.

17.19

$$CH_3(CH_2)_{14}\overset{\overset{\displaystyle O}{\|}}{C}O(CH_2)_{29}CH_3$$

17.21

$$CH_2O\overset{\overset{\displaystyle O}{\|}}{C}(CH_2)_{16}CH_3$$
$$CHO\overset{\overset{\displaystyle O}{\|}}{C}(CH_2)_{16}CH_3$$
$$CH_2O\overset{\overset{\displaystyle O}{\|}}{C}(CH_2)_{16}CH_3$$

17.23

$$CH_2O\overset{\overset{\displaystyle O}{\|}}{C}(CH_2)_{14}CH_3$$
$$CHO\overset{\overset{\displaystyle O}{\|}}{C}(CH_2)_{14}CH_3$$
$$CH_2O\overset{\overset{\displaystyle O}{\|}}{C}(CH_2)_{14}CH_3$$

17.25 Safflower oil contains fatty acids with two or three double bonds; olive oil contains a large amount of oleic acid, which has only one (monounsaturated) double bond.

17.27 Although coconut oil comes from a vegetable, it has large amounts of saturated fatty acids and small amounts of unsaturated fatty acids.

17.29

$$CH_2O\overset{\overset{\displaystyle O}{\|}}{C}(CH_2)_7CH=CH(CH_2)_7CH_3$$
$$CHO\overset{\overset{\displaystyle O}{\|}}{C}(CH_2)_7CH=CH(CH_2)_7CH_3 + H_2 \xrightarrow{Pt}$$
$$CH_2O\overset{\overset{\displaystyle O}{\|}}{C}(CH_2)_7CH=CH(CH_2)_7CH_3$$

$$CH_2O\overset{\overset{\displaystyle O}{\|}}{C}(CH_2)_{16}CH_3$$
$$CHO\overset{\overset{\displaystyle O}{\|}}{C}(CH_2)_{16}CH_3$$
$$CH_2O\overset{\overset{\displaystyle O}{\|}}{C}(CH_2)_{16}CH_3$$

17.31 a. Some of the double bonds in the unsaturated fatty acids have been converted to single bonds.
b. It is mostly saturated fatty acids.

17.33 a.

$$
\underset{\substack{|\\ \text{CH}_2\text{OC(CH}_2)_{12}\text{CH}_3}}{\overset{\text{O}}{\parallel}}
$$

CH$_2$OC(CH$_2$)$_{12}$CH$_3$
|
CHOC(CH$_2$)$_{12}$CH$_3$ + 3H$_2$O $\xrightarrow{\text{H}^+}$
|
CH$_2$OC(CH$_2$)$_{12}$CH$_3$

CH$_2$OH
|
CHOH + 3CH$_3$(CH$_2$)$_{12}$COH
|
CH$_2$OH

b.

CH$_2$OC(CH$_2$)$_{12}$CH$_3$
|
CHOC(CH$_2$)$_{12}$CH$_3$ + 3 NaOH $\longrightarrow$
|
CH$_2$OC(CH$_2$)$_{12}$CH$_3$

CH$_2$OH
|
CHOH + 3CH$_3$(CH$_2$)$_{12}$CO$^-$Na$^+$
|
CH$_2$OH

17.35 A triacylglycerol is composed of glycerol with three hydroxyl groups that form ester links with three long-chain fatty acids. In olestra, six to eight long-chain fatty acids form ester links with the hydroxyl groups on sucrose, a sugar. The olestra cannot be digested because our enzymes cannot break down the large olestra molecule.

17.37

CH$_2$O—C—(CH$_2$)$_{16}$CH$_3$
|
CHO—C—(CH$_2$)$_{16}$CH$_3$
|
CH$_2$O—C—(CH$_2$)$_{16}$CH$_3$

17.39 A triacylglycerol consists of glycerol and three fatty acids. A glycerophospholipid consists of glycerol, two fatty acids, a phosphate group, and an amino alcohol.

17.41

CH$_2$OC(CH$_2$)$_{14}$CH$_3$
|
CHOC(CH$_2$)$_{14}$CH$_3$
|
CH$_2$OPOCH$_2$CH$_2$NH$_3^+$
|
O$^-$

This is a cephalin

17.43 This phospholipid is a cephalin. It contains glycerol, oleic acid, stearic acid, a phosphate, and ethanolamine.

17.45 A sphingolipid contains the amino alcohol sphingosine (instead of glycerol) and only one fatty acid. A glycerophospholipid consists of glycerol, two fatty acids, a phosphate group, and an amino alcohol.

17.47

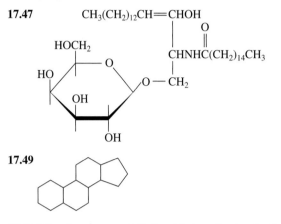

CH$_3$(CH$_2$)$_{12}$CH=CHOH
|
CHNHC(CH$_2$)$_{14}$CH$_3$

17.49

17.51 Bile salts act to emulsify fat globules, allowing the fat to be more easily digested.

17.53 Lipoproteins are large, spherically shaped structures that transport lipids in the bloodstream. They consist of an outside layer of phospholipids and proteins surrounding an inner core of hundreds of nonpolar lipids and cholesterol esters.

17.55 Chylomicrons have a lower density than VLDLs. They pick up triacylglycerols from the intestine, whereas VLDLs transport triacylglycerols synthesized in the liver.

17.57 "Bad" cholesterol is the cholesterol carried by LDLs that can form deposits in the arteries called plaque, which narrow the arteries.

17.59 Both estradiol and testosterone contain the steroid nucleus and a hydroxyl group. Testosterone has a ketone group, a double bond, and two methyl groups. Estradiol has a benzene ring, a hydroxyl group in place of the ketone, and a methyl group.

17.61 d. Testosterone is a male sex hormone.

17.63 Phospholipids with smaller amounts of glycolipids and cholesterol.

17.65 The lipid bilayer in a cell membrane surrounds the cell and separates the contents of the cell from the external fluids.

17.67 The peripheral proteins in the membrane emerge on the inner or outer surface only, whereas the integral proteins extend through the membrane to both surfaces.

17.69 The carbohydrates as glycoproteins and glycosphingolipids on the surface of cells act as receptors for cell recognition and chemical messengers such as neurotransmitters.

17.71 Substances move through cell membrane by simple transport, facilitated transport, and active transport.

17.73

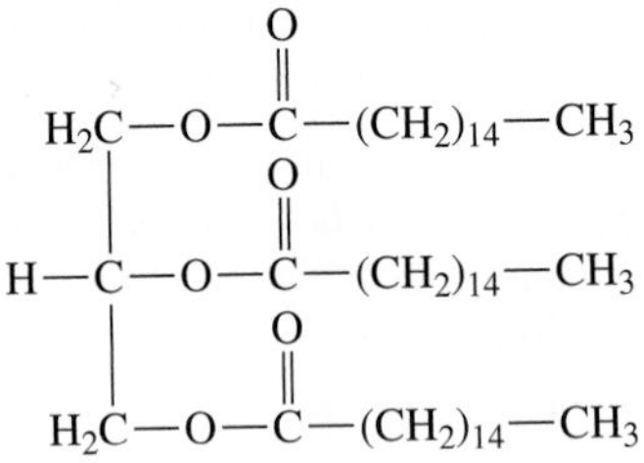

$$H_2C-O-\overset{\overset{\displaystyle O}{\|}}{C}-(CH_2)_{14}-CH_3$$

$$H-C-O-\overset{\overset{\displaystyle O}{\|}}{C}-(CH_2)_{14}-CH_3$$

$$H_2C-O-\overset{\overset{\displaystyle O}{\|}}{C}-(CH_2)_{14}-CH_3$$

17.75 a.

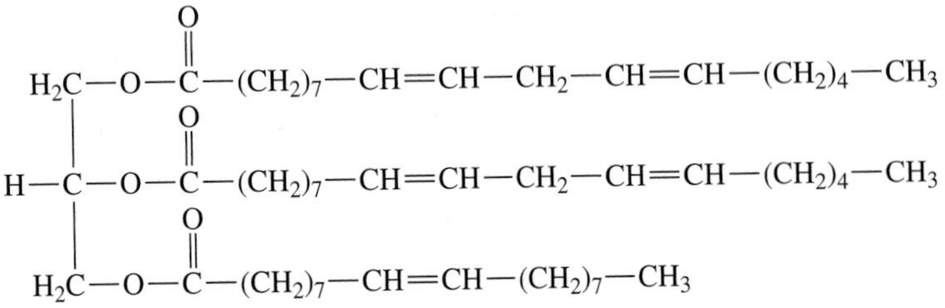

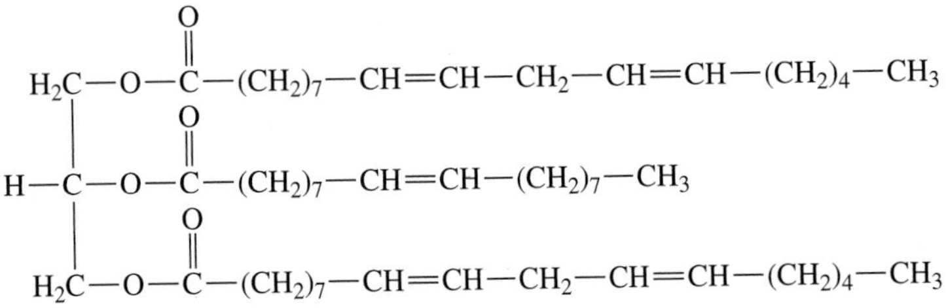

$$H_2C-O-\overset{\overset{\displaystyle O}{\|}}{C}-(CH_2)_7-CH=CH-CH_2-CH=CH-(CH_2)_4-CH_3$$

$$H-C-O-\overset{\overset{\displaystyle O}{\|}}{C}-(CH_2)_7-CH=CH-CH_2-CH=CH-(CH_2)_4-CH_3$$

$$H_2C-O-\overset{\overset{\displaystyle O}{\|}}{C}-(CH_2)_7-CH=CH-(CH_2)_7-CH_3$$

$$H_2C-O-\overset{\overset{\displaystyle O}{\|}}{C}-(CH_2)_7-CH=CH-CH_2-CH=CH-(CH_2)_4-CH_3$$

$$H-C-O-\overset{\overset{\displaystyle O}{\|}}{C}-(CH_2)_7-CH=CH-(CH_2)_7-CH_3$$

$$H_2C-O-\overset{\overset{\displaystyle O}{\|}}{C}-(CH_2)_7-CH=CH-CH_2-CH=CH-(CH_2)_4-CH_3$$

b.

$$H_2C-O-\overset{\overset{\displaystyle O}{\|}}{C}-(CH_2)_7-CH=CH-CH_2-CH=CH-(CH_2)_4-CH_3$$

$$H-C-O-\overset{\overset{\displaystyle O}{\|}}{C}-(CH_2)_7-CH=CH-(CH_2)_7-CH_3 + 5H_2 \xrightarrow{\text{Ni}}$$

$$H_2C-O-\overset{\overset{\displaystyle O}{\|}}{C}-(CH_2)_7-CH=CH-CH_2-CH=CH-(CH_2)_4-CH_3$$

$$H_2C-O-\overset{\overset{\displaystyle O}{\|}}{C}-(CH_2)_{16}-CH_3$$

$$H-C-O-\overset{\overset{\displaystyle O}{\|}}{C}-(CH_2)_{16}-CH_3$$

$$H_2C-O-\overset{\overset{\displaystyle O}{\|}}{C}-(CH_2)_{16}-CH_3$$

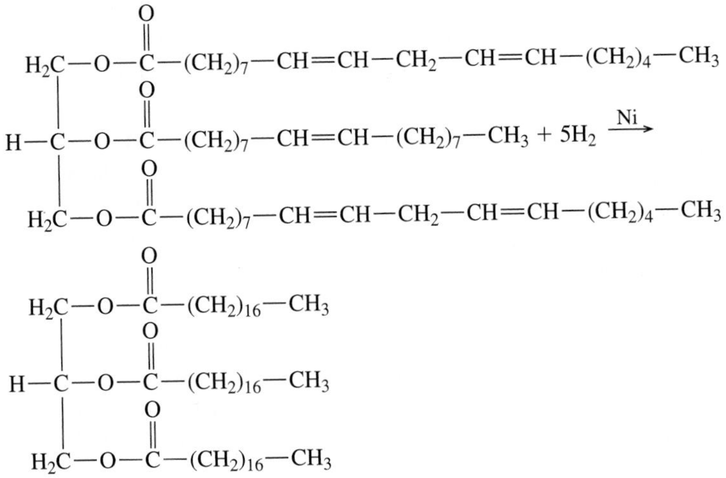

17.77 Beeswax and carnauba are waxes. Vegetable oil and capric triglyceride are triacylglycerols.

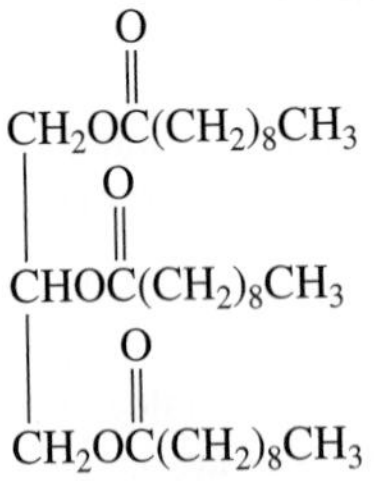

$$CH_2O\overset{\overset{\displaystyle O}{\|}}{C}(CH_2)_8CH_3$$

$$CHO\overset{\overset{\displaystyle O}{\|}}{C}(CH_2)_8CH_3$$

$$CH_2O\overset{\overset{\displaystyle O}{\|}}{C}(CH_2)_8CH_3$$

capric triacylglycerol

17.79 a. A typical fatty acid has a cis double bond.
b. A trans fatty acid has a trans double bond.
c.

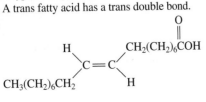

17.81

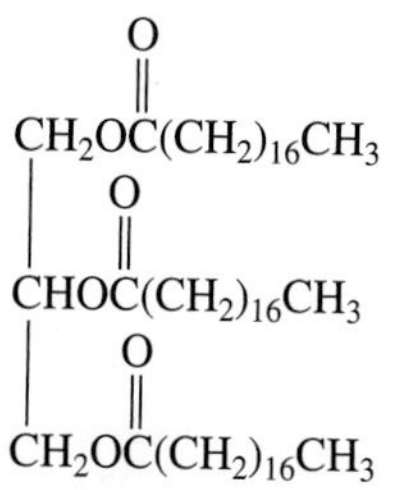

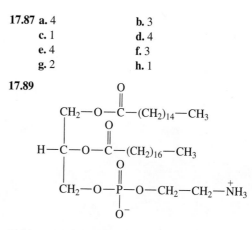

glyceryl stearate

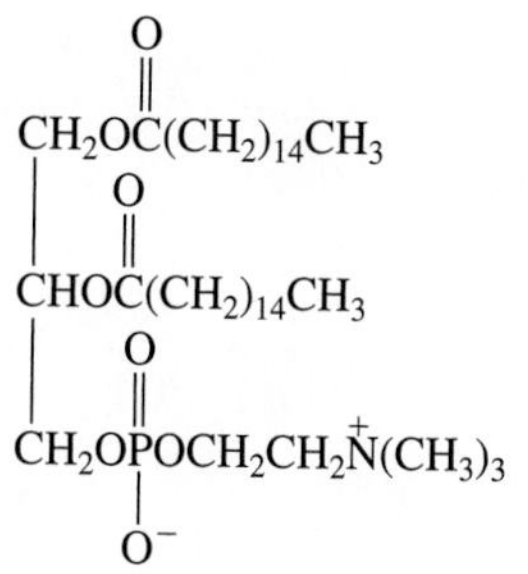

lecithin

17.83 Stearic acid (**l.**) is a fatty acid. Sodium stearate (**e.**) is a soap. Glyceryl tripalmitate (**d.**), safflower oil (**f.**), whale blubber (**h.**), and adipose tissue (**i.**) are triacylglycerols. Beeswax (**a.**) is a wax. Lecithin (**c.**) is a glycerophospholipid. Sphingomyelin (**g.**) is a sphingolipid. Cholesterol (**b.**), progesterone (**j.**), and cortisone (**k.**) are steroids.

17.85 a. 5
   b. 1, 2, 3, 4
   c. 2
   d. 1, 2
   e. 1, 2, 3, 4
   f. 2, 3, 4, 6

17.87 a. 4   **b.** 3
   c. 1   **d.** 4
   e. 4   **f.** 3
   g. 2   **h.** 1

17.89

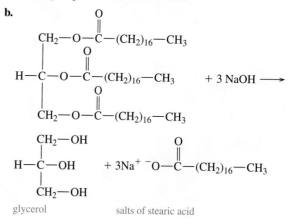

17.91 a. Adding NaOH would hydrolyze the tristearate lipid, breaking it up to wash down the drain.

b.

$$CH_2-O-\overset{\displaystyle O}{\overset{\|}{C}}-(CH_2)_{16}-CH_3$$

$$H-C-O-\overset{\displaystyle O}{\overset{\|}{C}}-(CH_2)_{16}-CH_3 \quad + 3\,NaOH \longrightarrow$$

$$CH_2-O-\overset{\displaystyle O}{\overset{\|}{C}}-(CH_2)_{16}-CH_3$$

$$CH_2-OH$$
$$H-C-OH \quad + 3Na^+ \; {}^-O-\overset{\displaystyle O}{\overset{\|}{C}}-(CH_2)_{16}-CH_3$$
$$CH_2-OH$$

glycerol   salts of stearic acid

18

Amines and Amides

"The pharmacy is one of the many factors in the final integration of chemistry and medicine in patient care," says Dorothea Lorimer, pharmacist, Kaiser Hospital. "If someone is allergic to a medication, I have to find out if a new medication has similar structural features. For instance, some people are allergic to sulfur. If there is sulfur in the new medication, there is a chance it will cause a reaction."

A prescription indicates a specific amount of a medication. At the pharmacy, the chemical name, formula, and quantity in milligrams or micrograms are checked. Then the prescribed number of capsules is prepared and placed in a container. If it is a liquid medication, a specific volume is measured and poured into a bottle.

LOOKING AHEAD

Mastering
GOB CHEMISTRY

Visit **www.masteringgob.com**
for self-study materials
and instructor-assigned
homework.

A mines and amides are organic compounds that contain nitrogen. Many nitrogen-containing compounds are important to life as components of amino acids, proteins, and nucleic acids: DNA and RNA. Many amines that exhibit strong physiological activity are used in medicine as decongestants, anesthetics, and sedatives. Examples include dopamine, histamine, epinephrine, and amphetamine.

Alkaloids such as caffeine, nicotine, cocaine, and digitalis, which demonstrate powerful physiological activity, are naturally occurring amines obtained from plants. Amides are derived from carboxylic acids and amines. In biochemistry, the amide bond is called the peptide bond, which links amino acids in proteins. Some medically important amides include acetaminophen (Tylenol) used to reduce fever; phenobarbital, a sedative and anticonvulsant medication; and penicillin, an antibiotic.

18.1 Amines

Amines are considered as derivatives of ammonia (NH_3) in which one or more hydrogen atoms is replaced with alkyl or aromatic groups. For example, in methylamine, a methyl group replaces one hydrogen atom in ammonia. The bonding of two methyl groups gives dimethylamine, and the three methyl groups in trimethylamine replace all the hydrogen atoms in ammonia.

Classification of Amines

When we classified alcohols in Chapter 13, we looked at the number of carbon atoms bonded to the alcohol carbon. Amines are classified in a similar way by counting the number of carbon atoms directly bonded to a nitrogen atom. In a *primary (1°) amine*, one carbon is bonded to a nitrogen atom. In a *secondary (2°) amine*, two carbons are bonded to the nitrogen atom, and a *tertiary (3°) amine* has three carbon atoms bonded to the nitrogen.

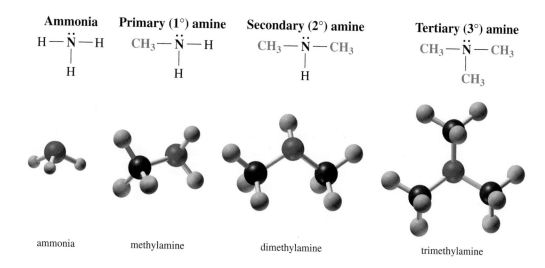

Ammonia	Primary (1°) amine	Secondary (2°) amine	Tertiary (3°) amine
$H-\ddot{N}-H$	$CH_3-\ddot{N}-H$	$CH_3-\ddot{N}-CH_3$	$CH_3-\ddot{N}-CH_3$
H	H	H	CH_3

ammonia   methylamine   dimethylamine   trimethylamine

Line-Bond Formulas for Amines

 MC **GOB** Know What Amine?

MC **GOB** Drawing Amines

We can draw line-bond formulas for amines just as we did for other organic compounds. For example, we can write the following line-bond formulas and classify each of the amines.

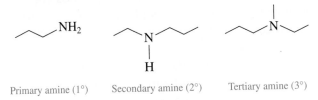

Primary amine (1°)   Secondary amine (2°)   Tertiary amine (3°)

SAMPLE PROBLEM 18.1

▪ Classifying Amines

Classify the following amines as primary, secondary, or tertiary:

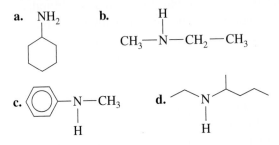

SOLUTION

a. This is a primary amine because there is one alkyl group (cyclohexyl) attached to a nitrogen atom.

b. This is a secondary amine. There are two alkyl groups (methyl and ethyl) attached to the nitrogen atom.

c. This is a secondary amine with two carbon groups, methyl and phenyl, bonded to the nitrogen atom.

d. The nitrogen atom in this line-bond formula is bonded to two carbon atoms, which makes it a secondary amine.

STUDY CHECK

Classify the following amine as primary, secondary, or tertiary:

$$CH_3{-}CH_2{-}N{-}CH_2{-}CH_3$$
$$\mid$$
$$CH_3$$

Naming Amines

MC **GOB** SELF-STUDY ACTIVITY
Amine and Amide Functional Groups

There are several systems in use for naming amines. For simple amines, the common names are often used. In the common name, the alkyl groups bonded to the nitrogen atom are listed in alphabetical order. The prefixes *di* and *tri* are used to indicate two and three identical substituents.

HEALTH NOTE

Amines in Health and Medicine

In response to allergic reactions or injury to cells the body increases the production of histamine, which causes blood vessels to dilate and increases the permeability of the cells. Redness and swelling occur in the area. Administering an antihistamine such as diphenylhydramine helps block the effects of histamine.

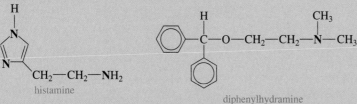

histamine

diphenylhydramine

In the body, hormones called biogenic amines carry messages between the central nervous system and nerve cells. Epinephrine (adrenaline) and norepinephrine (noradrenaline) are released by the adrenal medulla in "fight or flight" situations to raise the blood glucose level and move the blood to the muscles. Used in remedies for colds, hay fever, and asthma, the norepinephrine contracts the capillaries in the mucous membranes of the respiratory passages. The prefix *nor* in a drug name means there is one less CH_3— group on the nitrogen atom. Parkinson's disease is a result of a deficiency in another biogenic amine called dopamine.

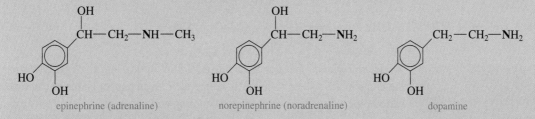

epinephrine (adrenaline)   norepinephrine (noradrenaline)   dopamine

Produced synthetically, amphetamines (known as "uppers") are stimulants of the central nervous system much like epinephrine, but they also increase cardiovascular activity and depress the appetite. They are sometimes used to bring about weight loss, but they can cause chemical dependency. Benzedrine and Neo-Synephrine (phenylephrine) are used in medications to reduce respiratory congestion from colds, hay fever, and asthma. Sometimes, Benzedrine is taken internally to combat the desire to sleep, but it has side effects. Methedrine is used to treat depression and in the illegal form is known as "speed" or "crank." The prefix *meth* means that there is one more methyl group on the nitrogen atom.

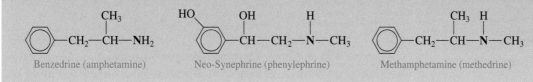

Benzedrine (amphetamine)   Neo-Synephrine (phenylephrine)   Methamphetamine (methedrine)

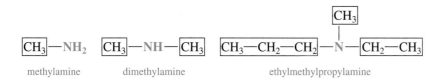

methylamine   dimethylamine   ethylmethylpropylamine

The IUPAC names for amines are similar to the names we used for alcohols, except that the *e* in the parent alkane name is replaced with *amine*.

CH_4 CH_3—OH CH_3—NH_2

methane   methan**ol**   methan**amine**

Guide to IUPAC Naming of Amines

STEP 1
Name the longest carbon chain bonded to the N atom by replacing the *e* with *amine*.

STEP 2
Number the carbon chain and show the position of the amine group and other substituents.

STEP 3
In secondary and tertiary amines, use the prefix *N-* to name smaller alkyl groups attached to the N atom.

■ **STEP 1** Identify the longest carbon chain bonded to the nitrogen atom. Replace the *e* in the corresponding alkane name with *amine*.

■ **STEP 2** Number the carbon chain to show the position of the amine group and any other substituents.

$$CH_3—CH_2—NH_2$$
ethan**amine**

$$CH_3—CH_2—CH_2—NH_2$$
1-propan**amine**

$$CH_3—\overset{\overset{\displaystyle NH_2}{|}}{CH}—CH_3$$
2-propan**amine**

$$CH_3—\overset{\overset{\displaystyle NH_2}{|}}{CH}—CH_2—CH_3$$
2-butan**amine**

$$CH_3—\overset{\overset{\displaystyle CH_3}{|}}{CH}—CH_2—CH_2—NH_2$$
3-methyl-1-butan**amine**

■ **STEP 3** In secondary and tertiary amines, the largest alkyl group attached to the nitrogen is named as the parent amine. The smaller alkyl groups are named with the prefix *N-* followed by the alkyl name and are listed alphabetically.

$$CH_3—CH_2—\overset{\overset{\displaystyle CH_3}{|}}{N}—CH_3$$
N, N-dimethylethanamine

$$CH_3—CH_2—CH_2—\overset{\overset{\displaystyle CH_3}{|}}{N}—CH_3$$
N, N-dimethyl-1-propanamine

$$CH_3—CH_2—CH_2—\overset{\overset{\displaystyle CH_3}{|}}{N}—CH_2—CH_3$$
N-ethyl-*N*-methyl-1-propanamine

Heterocyclic amines contain a nitrogen atom in the ring. For example, some of the pungent aroma and taste we associate with black pepper is due to a compound called piperidine, a six-atom ring containing a nitrogen atom. The fruit from the black pepper plant is dried and ground to give the black pepper we use to season our foods.

Piperidine

An amine with two amine functional groups is named as a *diamine*. For example, the amines 1,4-butanediamine and 1,5-pentanediamine contribute to the odors of decaying flesh.

H_2N ⌇ NH_2
1,4-butanediamine
(putrescine)

H_2N ⌇ NH_2
1,5-pentanediamine
(cadaverine)

In amines where another functional group takes priority, the —NH_2 group is named as a substituent *amino* group and numbered to show its location. For the major functional groups we have studied, the increasing priority follows the increase in oxidation:

$$\text{low priority} \quad —NH_2 < —OH < \overset{\overset{\displaystyle O}{\|}}{—C—} < \overset{\overset{\displaystyle O}{\|}}{—C—H} < \overset{\overset{\displaystyle O}{\|}}{—C—OH} \quad \text{high priority}$$

$$CH_3—\overset{\overset{\displaystyle NH_2}{|}}{CH}—CH_2—OH$$
2-amino-1-propanol

$$CH_3—\overset{\overset{\displaystyle NH_2}{|}}{CH}—CH_2—\overset{\overset{\displaystyle O}{\|}}{C}—CH_3$$
4-amino-2-pentanone

$$CH_3—\overset{\overset{\displaystyle NH_2}{|}}{CH}—CH_2—\overset{\overset{\displaystyle O}{\|}}{C}—OH$$
3-aminobutanoic acid

Aromatic Amines

The aromatic amines use the name *aniline*, which is approved by IUPAC.

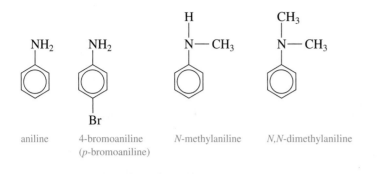

aniline   4-bromoaniline   *N*-methylaniline   *N,N*-dimethylaniline
(*p*-bromoaniline)

SAMPLE PROBLEM 18.2

■ Naming Amines

Give the common name for each of the following amines:

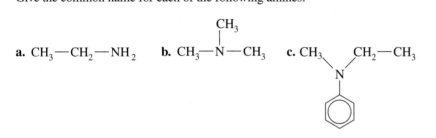

SOLUTION

a. This amine has one ethyl group attached to the nitrogen atom; its name is *ethylamine*.

b. This amine has three methyl groups attached to the nitrogen atom; its name is *trimethylamine*.

c. This amine is a derivative of aniline with a methyl and an ethyl group attached to the nitrogen atom; its name is *N-ethyl-N-methylaniline*.

STUDY CHECK

Draw the structure of *N*-ethyl-1-propanamine.

QUESTIONS AND PROBLEMS

■ Amines

18.1 What is a primary amine?

18.2 What is a tertiary amine?

18.3 Classify each of the following amines as primary, secondary, or tertiary:

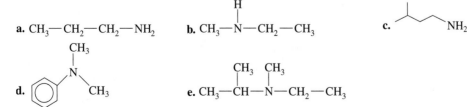

18.4 Classify each of the following amines as primary, secondary, or tertiary:

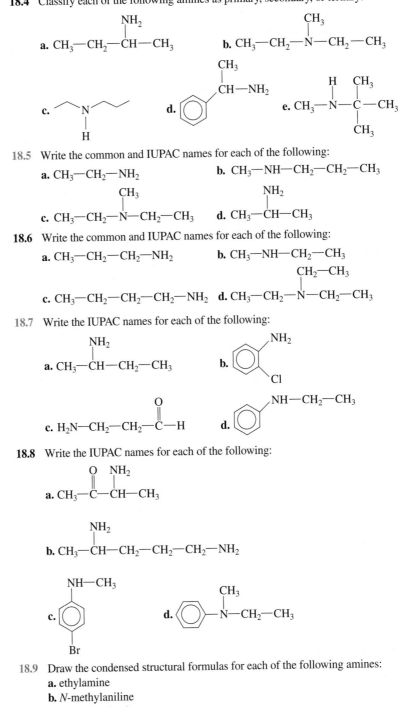

a. $CH_3-CH_2-\underset{\underset{NH_2}{|}}{CH}-CH_3$

b. $CH_3-CH_2-\underset{\underset{CH_3}{|}}{N}-CH_2-CH_3$

c. (structure with N—H, ethyl and propyl groups)

d. (benzene ring with $\underset{\underset{CH_3}{|}}{CH}-NH_2$)

e. $CH_3-\underset{\overset{H}{|}}{N}-\underset{\overset{CH_3}{|}}{\underset{\underset{CH_3}{|}}{C}}-CH_3$

18.5 Write the common and IUPAC names for each of the following:

a. $CH_3-CH_2-NH_2$

b. $CH_3-NH-CH_2-CH_2-CH_3$

c. $CH_3-CH_2-\underset{\underset{CH_3}{|}}{N}-CH_2-CH_3$

d. $CH_3-\underset{\underset{NH_2}{|}}{CH}-CH_3$

18.6 Write the common and IUPAC names for each of the following:

a. $CH_3-CH_2-CH_2-NH_2$

b. $CH_3-NH-CH_2-CH_3$

c. $CH_3-CH_2-CH_2-CH_2-NH_2$

d. $CH_3-CH_2-\underset{\underset{CH_2-CH_3}{|}}{N}-CH_2-CH_3$

18.7 Write the IUPAC names for each of the following:

a. $CH_3-\underset{\underset{NH_2}{|}}{CH}-CH_2-CH_3$

b. (benzene ring with NH_2 and Cl)

c. $H_2N-CH_2-CH_2-\overset{\overset{O}{||}}{C}-H$

d. (benzene ring with $NH-CH_2-CH_3$)

18.8 Write the IUPAC names for each of the following:

a. $CH_3-\overset{\overset{O}{||}}{C}-\underset{\underset{NH_2}{|}}{CH}-CH_3$

b. $CH_3-\underset{\underset{NH_2}{|}}{CH}-CH_2-CH_2-CH_2-NH_2$

c. (benzene ring with $NH-CH_3$ at top and Br at bottom)

d. (benzene ring with $\underset{\underset{CH_3}{|}}{N}-CH_2-CH_3$)

18.9 Draw the condensed structural formulas for each of the following amines:
 a. ethylamine
 b. *N*-methylaniline
 c. butylpropylamine
 d. 2-pentanamine

18.10 Draw the condensed structural formulas for each of the following amines:
 a. dimethylamine
 b. *p*-chloroaniline
 c. *N,N*-diethylaniline
 d. 1-amino-3-pentanone

18.2 Properties of Amines

Amines have higher boiling points than alkanes of similar mass, but lower than the alcohols.

$$CH_3—CH_3 \qquad CH_3—NH_2 \qquad CH_3—OH$$

ethane   methanamine   methanol
−84°C   −7°C   65°C

Because amines contain a polar N—H bond, they form hydrogen bonds. However, nitrogen is not as electronegative as oxygen, which makes the hydrogen bonds in amines weaker. The —NH$_2$ in primary amines can form more hydrogen bonds, which gives them higher boiling points than the secondary amines of the same mass. It is not possible for tertiary amines to hydrogen bond with each other (no N—H bonds), which makes their boiling points much lower and similar to those of alkanes.

$$CH_3—CH_2—CH_2—NH_2 \qquad CH_3—CH_2—NH—CH_3 \qquad CH_3—\overset{\displaystyle CH_3}{\underset{}{N}}—CH_3$$

propylamine (1°)   ethylmethylamine (2°)   trimethylamine (3°)
bp 48°C   bp 36°C   bp 3°C

Solubility in Water

Like alcohols, the smaller amines, including tertiary ones, are soluble in water because they form hydrogen bonds with water. (See Figure 18.1.) However, in amines with more than six carbon atoms, the effect of hydrogen bonding is diminished, and their solubility in water decreases.

SAMPLE PROBLEM 18.3

Boiling Points of Amines

If the compounds trimethylamine and ethylmethylamine have the same molar mass, why is the boiling point of trimethylamine (3°C) lower than that of ethylmethylamine (37°C)?

SOLUTION

With a polar N—H bond, hydrogen bonds form between ethylmethylamine molecules. Thus, a higher temperature is required to break the hydrogen bonds and form a gas. However, trimethylamine, which is a tertiary amine, has no N—H bond and cannot form hydrogen bonds between the amine molecules.

STUDY CHECK

Why is CH$_3$—CH$_2$—NH$_2$ more soluble in water than

$$CH_3—CH_2—CH_2—CH_2—\overset{\displaystyle H}{\underset{}{N}}—CH_2—CH_2—CH_3 \, ?$$

Amines React as Bases

In Chapter 10, we saw that ammonia (NH$_3$) acts as a Brønsted–Lowry base because it accepts a proton (H$^+$) from water to produce an ammonium ion (NH$_4^+$) and a hydroxide ion (OH$^-$).

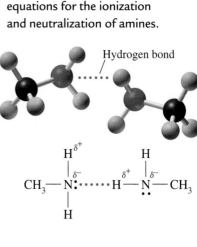

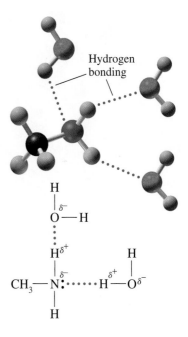

FIGURE 18.1

Hydrogen bonding occurs between amines and water molecules.

Q **Why are tertiary (3°) amines soluble in water?**

 Boiling Points and Solubility of Amines

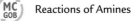 Reactions of Amines

SELF-STUDY ACTIVITY
Amines as Bases

$$\ddot{N}H_3 + H_2O \rightleftharpoons NH_4^+ + OH^-$$

ammonia ammonium ion hydroxide ion

Ionization of an Amine in Water

In water, amines also act as Brønsted–Lowry bases because the lone electron pair on the nitrogen atom accepts a proton from water. The products are an alkyl ammonium ion and hydroxide ion. The name of the alkyl ammonium ion is similar to the common amine name, but *amine* is replaced by *ammonium ion*.

$$CH_3 - \ddot{N}H_2 + H_2O \rightleftharpoons CH_3 - \overset{+}{N}H_3 + OH^-$$

methylamine methylammonium ion hydroxide ion

Secondary amines also accept a proton to form dialkyl ammonium ions.

$$\begin{array}{ccc} CH_3 - \ddot{N}H + H_2O & \rightleftharpoons & CH_3 - \overset{+}{N}H_2 & + & OH^- \\ | & & | \\ CH_3 & & CH_3 \end{array}$$

dimethylamine dimethylammonium ion hydroxide ion

Basicity of Amines

Because amines act as weak bases by accepting protons from water and producing hydroxide ions, their aqueous solutions are basic. We can write the equilibrium constant K for methylamine as follows:

$$K = \frac{[CH_3 - NH_3^+][OH^-]}{[CH_3 - NH_2]} = 4.4 \times 10^{-4}$$

Most of the K values for amines are less than 10^{-3}, which means that the equilibrium favors the undissociated amine molecules. Aqueous solutions of amines have basic pH values and turn red litmus paper blue. We can compare the strengths of some amines by looking at their K values as follows:

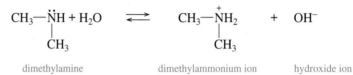

	Ammonia	*1° amine*		*2° amine*	*3° amine*
	NH_3	$CH_3 - NH_2$	$CH_3 - CH_2 - NH_2$	$CH_3 - NH - CH_3$	$CH_3 - \overset{\displaystyle CH_3}{\overset{\displaystyle \mid}{N}} - CH_3$
K	1.8×10^{-5}	4.4×10^{-4}	5.6×10^{-4}	5.1×10^{-4}	5.3×10^{-5}

Amine Salts

When you squeeze lemon juice on fish, the "fishy odor" of the amines is removed by converting them to amine salts. In a neutralization reaction, an amine acts as a base and reacts with an acid to form an **amine salt**. The lone pair of electrons on the nitrogen atom accepts a proton H^+ from an acid to give an amine salt; no water is formed. An amine salt is named by replacing the *amine* part of the amine name with *ammonium* followed by the name of the negative ion.

Neutralization of an Amine

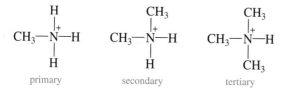

The ammonium ions are classified as primary, secondary, and tertiary.

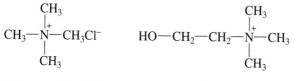

In a **quaternary ammonium ion**, a nitrogen atom bonds to four carbon groups. In the quaternary ion, the nitrogen atom has a positive charge just as it does in other amine salts. Choline, an amino alcohol present in glycerophospholipids is a quaternary ammonium ion.

$$CH_3-\overset{\overset{\displaystyle CH_3}{|}}{\underset{\underset{\displaystyle CH_3}{|}}{N}}-CH_3 Cl^- \qquad HO-CH_2-CH_2-\overset{\overset{\displaystyle CH_3}{|}}{\underset{\underset{\displaystyle CH_3}{|}}{N}}-CH_3$$

tetramethylammonium chloride   choline

The quaternary salts differ from other amine salts because the nitrogen atom is not bonded to an H atom. Thus, quaternary salts do not react with bases.

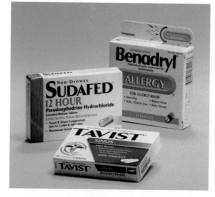

FIGURE 18.2

Decongestants and products that relieve itch and skin irritations can contain ammonium salts.

Q **Why are ammonium salts used in drugs rather than the biologically active amines?**

Properties of Amine Salts

As ionic compounds, amine salts are solids at room temperature, odorless, and soluble in water and body fluids. For this reason, amines used as drugs are converted to their amine salts. The amine salt of ephedrine is used as a bronchodilator and in decongestant products such as Sudafed. The amine salt of diphenhydramine is used in products such as Benadryl for relief of itching and pain from skin irritations and rashes. (See Figure 18.2.) In pharmaceuticals, the naming of the amine salt follows an older method of giving the amine name followed by the name of the acid (see following examples).

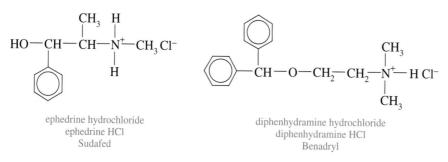

ephedrine hydrochloride
ephedrine HCl
Sudafed

diphenhydramine hydrochloride
diphenhydramine HCl
Benadryl

When an amine salt reacts with a strong base such as NaOH, it is converted back to the amine, which is also called the free amine or free base.

$$CH_3-NH_3^+ Cl^- + NaOH \longrightarrow CH_3-NH_2 + NaCl + H_2O$$

The narcotic cocaine is typically extracted from coca leaves using an acidic solution to give a white, solid amine salt, which is cocaine hydrochloride. This is the form in which cocaine is smuggled and sold illegally on the street to be snorted or injected. "Crack cocaine" is the free amine or free base of the amine obtained by treating the cocaine hydrochloride with NaOH and ether, a process known as "free-basing." The solid product is known as "crack cocaine" because it makes a cracking noise when heated. The free amine is rapidly absorbed when smoked and gives stronger highs than the cocaine hydrochloride. Unfortunately, these effects of crack cocaine have caused a rise in addiction to cocaine.

(MC GOB) **SELF-STUDY CASE**
Death by Chocolate?

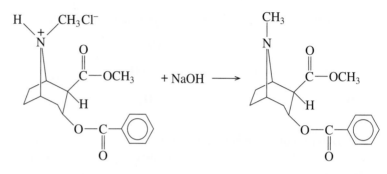

cocaine hydrochloride   cocaine ("free base")

SAMPLE PROBLEM 18.4

■ **Reactions of Amines**

Write an equation that shows ethylamine
a. ionizing as a weak base in water   **b.** neutralized by HCl

SOLUTION

a. In water, ethylamine acts as a weak base by accepting a proton from water to produce ethylammonium hydroxide.

$$CH_3—CH_2—NH_2 + H—OH \rightleftharpoons CH_3—CH_2—NH_3^+ + OH^-$$

b. $CH_3—CH_2—NH_2 + HCl \rightleftharpoons CH_3—CH_2—NH_3^+Cl^-$

STUDY CHECK

What is the condensed structural formula of the salt formed by the reaction of trimethylamine and HCl?

■

QUESTIONS AND PROBLEMS   (MC GOB)

■ **Properties of Amines**

18.11 Indicate the compound in each pair that would have the higher boiling point:
    a. $CH_3—CH_2—NH_2$ or $CH_3—CH_2—OH$
    b. $CH_3—NH_2$ or $CH_3—CH_2—CH_2—NH_2$

            CH_3
            $|$
    c. $CH_3—N—CH_3$ or $CH_3—CH_2—CH_2—NH_2$

18.12 Indicate the compound in each pair that would have the higher boiling point:
    a. $CH_3—CH_2—CH_2—CH_3$ or $CH_3—CH_2—CH_2—NH_2$
    b. $CH_3—NH_2$ or $CH_3—CH_2—NH_2$

                        NH_2
                        $|$
    c. $CH_3—CH_2—CH_2—OH$ or $CH_3—CH—CH_3$

18.13 Propylamine (molar mass 59) has a boiling point of 48°C and ethylmethylamine (molar mass 59) has a boiling point of 37°C. Butane (molar mass 58) has a much lower boiling point −1°C. Explain.

18.14 Assign the boiling point of 3°C, 48°C, or 97°C to the appropriate compound: 1-propanol, propylamine, and trimethylamine.

18.15 Indicate if each of the following is soluble in water:

a. CH_3—CH_2—NH_2 **b.** CH_3—NH—CH_3

c.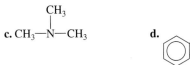

d.
$$\overset{\displaystyle NH_2}{\underset{\displaystyle |}{CH_3}}—CH—CH_2—CH_3$$

18.16 Indicate if each of the following is soluble in water:

a. CH_3—CH_2—CH_2—NH_2 **b.** CH_3—CH_2—CH_2—NH—CH_2—CH_3

c.
$$CH_3—\overset{\displaystyle CH_3}{\underset{\displaystyle |}{N}}—CH_3$$

d.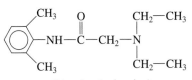

18.17 Write an equation for the ionization of each of the following amines in water:
a. methylamine   **b.** dimethylamine   **c.** aniline

18.18 Write an equation for the ionization of each of the following amines in water:
a. ethylamine   **b.** propylamine   **c.** *N*-methylaniline

18.19 Write the condensed structural formula of the amine salt obtained when each of the amines in problem 18.17 reacts with HCl.

18.20 Write the condensed structural formula of the amine salt obtained when each of the amines in problem 18.18 reacts with HCl.

18.21 Novocain, a local anesthetic, is the hydrochloride salt of procaine.

procaine

a. What is the formula of the amine salt formed when procaine reacts with HCl?
b. Why is Novocain, the amine salt, used rather than the amine procaine?

18.22 Lidocaine (xylocaine) is used as a local anesthetic and cardiac depressant.

lidocaine (xylocaine)

a. What is the formula of the amine salt formed when lidocaine reacts with HCl?
b. Why is the amine salt of lidocaine used rather than the amine?

18.3 Heterocyclic Amines and Alkaloids

LEARNING GOAL

Identify heterocyclic amines; distinguish between the types of heterocyclic amines.

A **heterocyclic amine** is a cyclic organic compound that contains one or more nitrogen atoms in the ring. The heterocyclic amine rings typically consist of 5 or 6 atoms and one or more nitrogen atoms. Of the five-atom rings, the simplest one is pyrrolidine, which is a ring of four carbon atoms and a nitrogen atom, all with

Identifying Types of Heterocyclic
Amines

single bonds. Pyrrole is a five-atom ring with one nitrogen atom and two double
bonds. Imidazole is a five-atom ring that contains two nitrogen atoms.

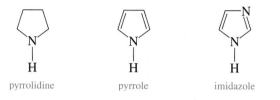

pyrrolidine   pyrrole   imidazole

Many of the six-atom heterocyclic amines are aromatic. Pyridine is similar to
benzene except that it has a nitrogen atom in place of a carbon atom. Pyrimidine,
which is found in nucleic acids, is also similar to benzene except that it has two
nitrogen atoms. In purine, another component of nucleic acids, a pyrimidine ring is
fused with imidazole.

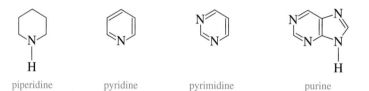

piperidine   pyridine   pyrimidine   purine

SAMPLE PROBLEM 18.5

■ Heterocyclic Amines

Identify each of the following heterocyclic amines:

a. 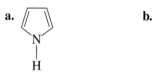   **b.**

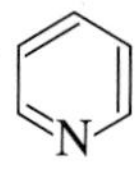

SOLUTION

a. This five-atom ring with one nitrogen atom is pyrrole.
b. This aromatic ring with one nitrogen atom is pyridine.

STUDY CHECK

Identify the following heterocyclic amine.

Alkaloids: Amines in Plants

Alkaloids are physiologically active nitrogen-containing compounds produced by
plants. The term *alkaloid* refers to the "alkali-like" or basic characteristics we have
seen for amines. Certain alkaloids are used in anesthetics, in antidepressants, and
as stimulants, although many are habit forming.

Naturally Addictive Amines

As a stimulant, nicotine increases the level of adrenaline in the blood, which
increases the heart rate and blood pressure. Nicotine is responsible for the addiction of smoking. Nicotine has a simple alkaloid structure that includes a pyrrolidine
ring. Coniine, which is obtained from hemlock, is an extremely toxic alkaloid that
contains a piperidine ring.

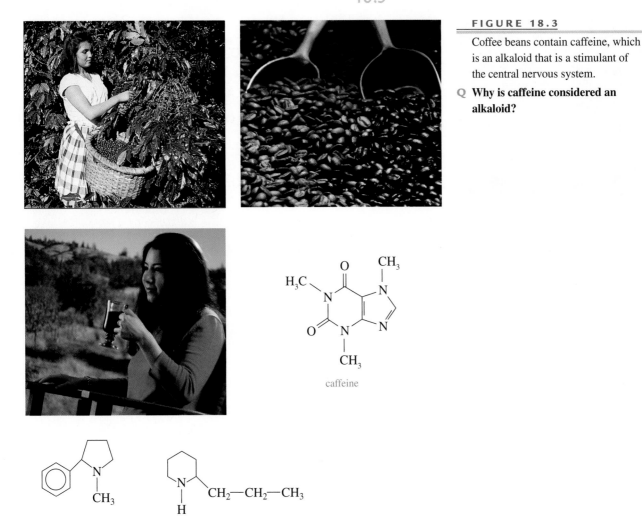

FIGURE 18.3

Coffee beans contain caffeine, which is an alkaloid that is a stimulant of the central nervous system.

Q **Why is caffeine considered an alkaloid?**

caffeine

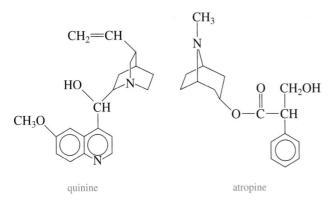

nicotine   coniine

Caffeine contains an imidazole ring and is a central nervous system stimulant. Present in coffee, tea, soft drinks, chocolate, and cocoa, caffeine increases alertness, but may cause nervousness and insomnia. Caffeine is also used in certain pain relievers to counteract the drowsiness caused by an antihistamine. (See Figure 18.3.)

Several alkaloids are used in medicine. Quinine obtained from the bark of the cinchona tree has been used in the treatment of malaria since the 1600s. Atropine from belladonna is used in low concentrations to accelerate slow heart rates and as an anesthetic for eye examinations.

quinine   atropine

For many centuries morphine and codeine, alkaloids found in the oriental poppy plant, have been used as effective painkillers. (See Figure 18.4.) Codeine, which is

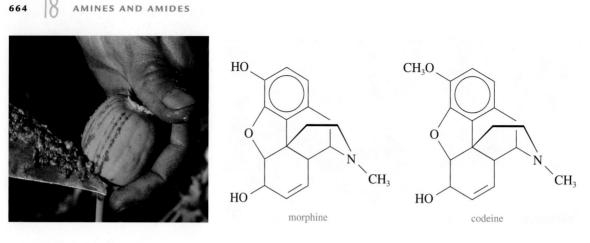

FIGURE 18.4

The green, unripe poppy seed capsule contains a milky sap (opium) that is the source of the alkaloids morphine and codeine.

Q **Where is the piperidine ring in the structures of morphine and codeine?**

structurally similar to morphine, is used in some prescription painkillers and cough syrups. Heroin, obtained by a chemical modification of morphine, is strongly addicting and is not used medically.

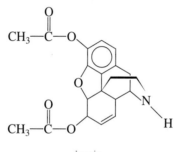

heroin

Synthesizing Drugs

One area of research in pharmacology is the synthesis of compounds that retain the anesthetic characteristic of naturally occurring alkaloids such as cocaine and morphine without the addictive side effects. For example, cocaine is an effective anesthetic, but addictive. Research chemists modified the structure of cocaine, but kept the benzene group and nitrogen atom. The synthetic products procaine and lidocaine retain the anesthetic qualities of the natural alkaloid without the addictive side effects.

The structure of morphine was also modified to make a synthetic alkaloid, meperidine, or Demerol, which acts as an effective painkiller.

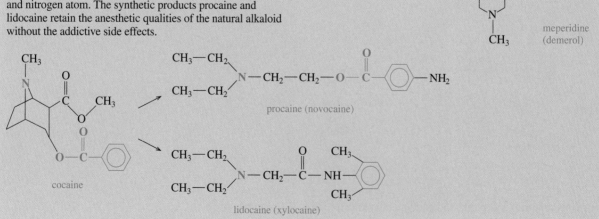

SAMPLE PROBLEM 18.6

■ **Heterocyclic Amines**

Identify the heterocyclic amines in the alkaloids nicotine and caffeine.

SOLUTION

In nicotine, the heterocyclic amine is the five-atom ring of pyrrolidine. Caffeine contains a purine, which is pyrimidine and imidazole fused together.

STUDY CHECK

What is the heterocyclic amine in meperidine (Demerol)?

QUESTIONS AND PROBLEMS

■ **Heterocyclic Amines and Alkaloids**

18.23 Identify the following as amines or heterocyclic amines:

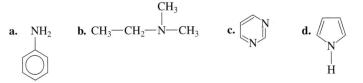

18.24 Identify the following as amines or heterocyclic amines:

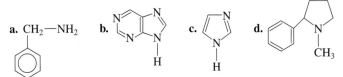

18.25 Identify the type of heterocyclic amines in problem 18.23.

18.26 Identify the type of heterocyclic amines in problem 18.24.

18.27 Low levels of serotonin in the brain appear to be associated with depressed states. What type of heterocyclic amine is serotonin?

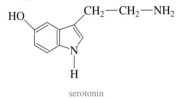

serotonin

18.28 LSD is made from lysergic acid, which is produced by a fungus that grows on rye. What types of heterocyclic amines are in lysergic acid?

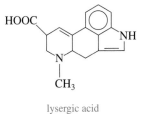

lysergic acid

18.4 Amides

The **amides** are derivatives of carboxylic acids in which an amino group replaces the hydroxyl group.

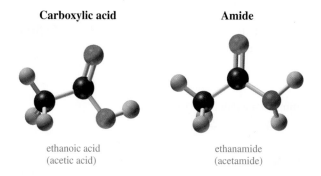

ethanoic acid
(acetic acid)

ethanamide
(acetamide)

Preparation of Amides

 Amidation Reactions

An amide is produced in a reaction called **amidation**, in which a carboxylic acid reacts with ammonia or a primary or secondary amine. A molecule of water is eliminated, and the fragments of the carboxylic acid and amine molecules join to form the amide, much like the formation of ester. Because a hydrogen atom must be lost from the amines, only primary and secondary amines undergo amidation.

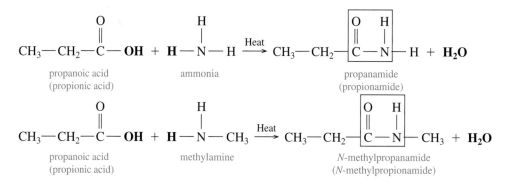

propanoic acid
(propionic acid)

ammonia

propanamide
(propionamide)

propanoic acid
(propionic acid)

methylamine

N-methylpropanamide
(*N*-methylpropionamide)

SAMPLE PROBLEM 18.7

■Amidation

Give the structural formula of the amide product in each of the following reactions:

a.

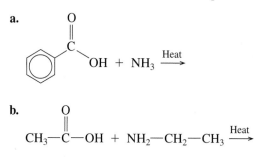

SOLUTION

a. The structural formula of the amide product can be written by attaching the carbonyl group from the acid to the nitrogen atom of the amine. —OH is removed from the acid and —H from the amine to form water.

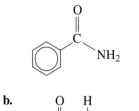

b.

$$CH_3—\overset{\overset{\displaystyle O}{\|}}{C}—\overset{\overset{\displaystyle H}{|}}{N}—CH_2—CH_3$$

STUDY CHECK

What are the condensed structural formulas of the carboxylic acid and amine needed to prepare the following amide?

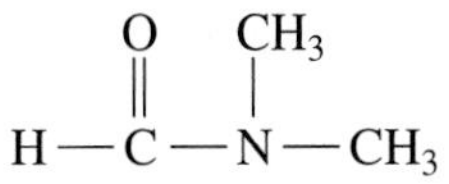

$$H—\overset{\overset{\displaystyle O}{\|}}{C}—\overset{\overset{\displaystyle CH_3}{|}}{N}—CH_3$$

Naming Amides

In both the common and IUPAC names, amides are named by dropping the *ic acid* or *oic acid* from the carboxylic acid names (IUPAC or common) and adding the suffix *amide*.

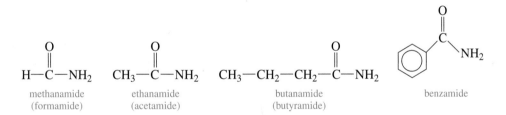

methanamide (formamide)   ethanamide (acetamide)   butanamide (butyramide)   benzamide

When alkyl groups are attached to the nitrogen atom, the name of an amide is preceded by *N-* or *N,N-* depending on whether there are one or two groups.

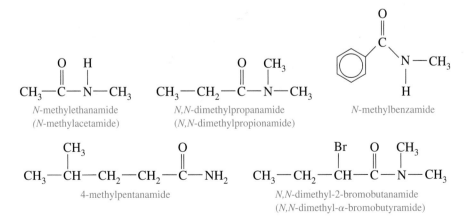

N-methylethanamide (*N*-methylacetamide)   *N,N*-dimethylpropanamide (*N,N*-dimethylpropionamide)   *N*-methylbenzamide

4-methylpentanamide   *N,N*-dimethyl-2-bromobutanamide (*N,N*-dimethyl-α-bromobutyramide)

■ **Naming Amides**

Give the common and IUPAC names for each of the following amides:

a. 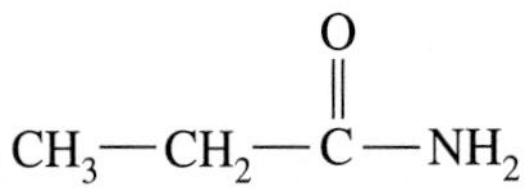 $CH_3—CH_2—\overset{\displaystyle O}{\overset{\|}{C}}—NH_2$ **b.** 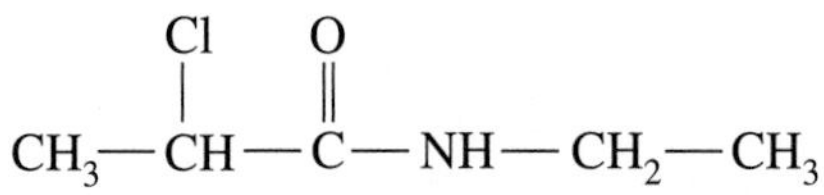 $CH_3—\overset{\displaystyle Cl}{\overset{|}{C}H}—\overset{\displaystyle O}{\overset{\|}{C}}—NH—CH_2—CH_3$

SOLUTION

a. The IUPAC name of the carboxylic acid is propanoic acid; the common name is propionic acid. Replacing the *oic acid* or *ic acid* ending with *amide* gives the IUPAC name of *propanamide* and common name of *propionamide*.
b. The ethyl group attached to the nitrogen atom is named *N-ethyl*. The amide is named *N-ethyl-2-chloropropanamide* (IUPAC), and *N-ethyl-α-chloropropion-amide* (common).

STUDY CHECK

Draw the condensed structural formula of *N,N*-dimethyl-*p*-chlorobenzamide.

Physical Properties of Amides

The amides do not have the properties of bases that we saw for the amines. Only formamide is a liquid at room temperature, while the other amides are solids. For primary amides, the —NH_2 group can form hydrogen bonds, which gives primary amides high melting points. The melting points of the secondary amides are lower because the number of hydrogen bonds decreases. Tertiary amides have even lower melting points because they cannot form hydrogen bonds with other tertiary amides.

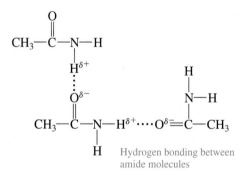

Hydrogen bonding between amide molecules

The amides with one to five carbon atoms are soluble in water because they can hydrogen bond with water molecules.

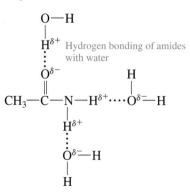

Hydrogen bonding of amides with water

Amides in Health and Medicine

The simplest natural amide is urea, an end product of protein metabolism in the body. The kidneys remove urea from the blood and provide for its excretion in urine. If the kidneys malfunction, urea is not removed and builds to a toxic level, a condition called uremia. Urea is also used as a component of fertilizer, to increase nitrogen in the soil.

$$NH_2-\overset{\overset{\displaystyle O}{\|}}{C}-NH_2 \quad \text{Urea}$$

Synthetic amides are used as substitutes for sugar and aspirin. Saccharin is a very powerful sweetener and is used as a sugar substitute. The sweetener Aspartame is made from two amino acids: aspartic acid and phenylalanine.

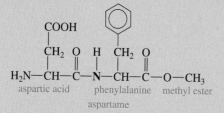

aspartic acid   phenylalanine   methyl ester

aspartame

Many barbiturates are cyclic amides of barbituric acid that act as sedatives in small dosages or sleep inducers in larger dosages. They are often habit forming. Barbiturate drugs include phenobarbital (Luminal), pentobarbital (Nembutal), and secobarbital (Seconal).

Aspirin substitutes contain phenacetin or acetaminophen, which is used in Tylenol. Like aspirin, acetaminophen reduces fever and pain, but it has little anti-inflammatory effect.

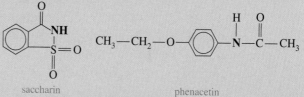

saccharin   phenacetin

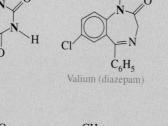

acetaminophen

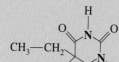

Luminal (phenobarbital)

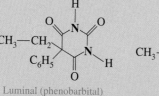

Nembutal (pentobarbital)

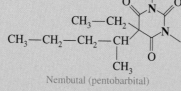

Valium (diazepam)

Seconal (secobarbital)

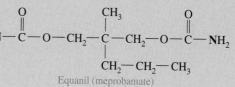

Equanil (meprobamate)

MC GOB   Pharmaceutical Amides

QUESTIONS AND PROBLEMS

⬭ **Amides**

18.29 Draw the condensed structural formulas of the amides formed in each of the following reactions:

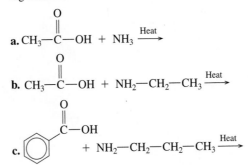

a. CH₃—C(=O)—OH + NH₃ →(Heat)

b. CH₃—C(=O)—OH + NH₂—CH₂—CH₃ →(Heat)

c. (benzene ring)—C(=O)—OH + NH₂—CH₂—CH₂—CH₃ →(Heat)

18.30 Draw the condensed structural formulas of the amides formed in each of the following reactions:

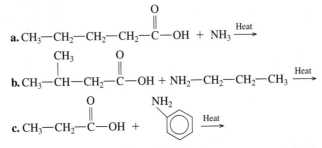

a. CH₃—CH₂—CH₂—CH₂—C(=O)—OH + NH₃ →(Heat)

b. CH₃—CH(CH₃)—CH₂—C(=O)—OH + NH₂—CH₂—CH₂—CH₃ →(Heat)

c. CH₃—CH₂—C(=O)—OH + (benzene ring with NH₂) →(Heat)

18.31 Give the IUPAC and common names (if any) for each of the following amides:

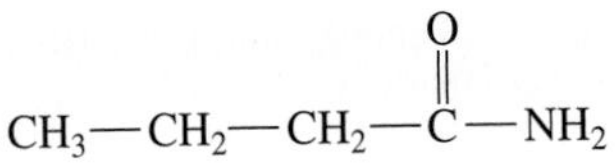

a. 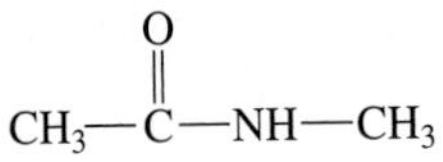 CH₃—C(=O)—NH—CH₃

b. CH₃—CH₂—CH₂—C(=O)—NH₂

c. H—C(=O)—NH₂

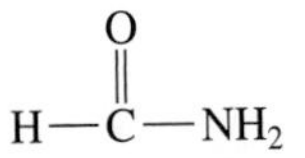

d. 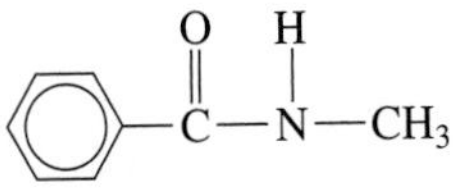 (benzene ring)—C(=O)—N(H)—CH₃

18.32 Give the IUPAC and common names (if any) for each of the following amides:

a. CH₃—CH₂—C(=O)—N(H)—CH₂—CH₃

b. CH₃—CH₂—CH₂—CH₂—CH₂—C(=O)—NH₂

c. CH₃—C(=O)—N(CH₃)—CH₂—CH₂—CH₃

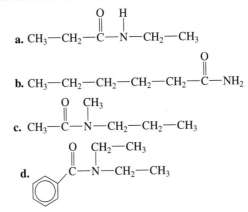

d. (benzene ring)—C(=O)—N(CH₂—CH₃)—CH₂—CH₃

18.33 Draw the condensed structural formulas for each of the following amides:
a. propionamide
b. 2-methylpentanamide
c. methanamide
d. *N*-ethylbenzamide
e. *N*-ethylbutyramide

18.34 Draw the condensed structural formulas for each of the following amides:
a. formamide
b. *N,N*-dimethylbenzamide
c. 3-methylbutyramide
d. 2,2-dichlorohexanamide
e. *N*-propyl-3-chloropentanamide

18.35 In each pair, identify the compound that has the higher melting point:
   a. acetamide or *N*-methylacetamide
   b. butane or propionamide
   c. *N,N*-dimethylpropanamide or *N*-methylpropanamide

18.36 In each pair, identify the compound that has the higher melting point:
   a. propane or acetamide
   b. *N*-methylacetamide or propanamide
   c. *N,N*-dimethylpropanamide or *N*-methylpropanamide

18.5 Hydrolysis of Amides

LEARNING GOAL

Write equations for the hydrolysis of amides.

As we have seen, amide bonds are formed by the elimination of water. The reverse reaction called **hydrolysis** occurs when water is added back to the amide bond to split the molecule. When an acid is used, the hydrolysis products of an amide are the carboxylic acid and the ammonium salt. In base hydrolysis, the amide produces the salt of the carboxylic acid and ammonia or amine.

Hydrolysis of Amides

Acid Hydrolysis of Amides

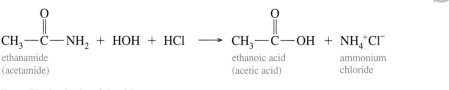

ethanamide   ethanoic acid   ammonium
(acetamide)   (acetic acid)   chloride

Base Hydrolysis of Amides

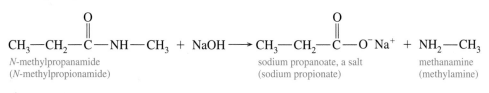

N-methylpropanamide   sodium propanoate, a salt   methanamine
(*N*-methylpropionamide)   (sodium propionate)   (methylamine)

SAMPLE PROBLEM 18.9

■ **Hydrolysis of Amides**

Write the structural formulas for the products for the hydrolysis of *N*-methylpentanamide with NaOH.

SOLUTION

Hydrolysis of the amide with a base produces a carboxylate salt (sodium pentanoate) and the corresponding amine (methylamine).

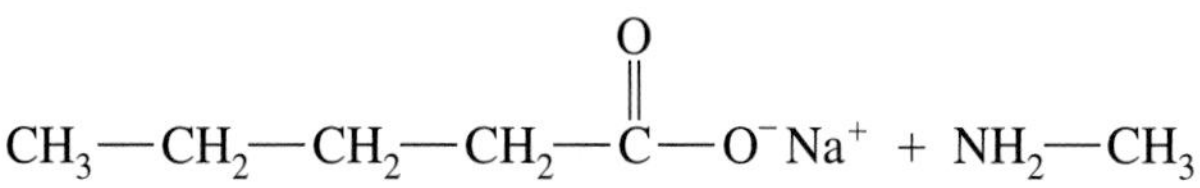

$$CH_3—CH_2—CH_2—CH_2—\overset{\overset{\displaystyle O}{\|}}{C}—O^-Na^+ \ + \ NH_2—CH_3$$

STUDY CHECK

What are the structures of the products from the hydrolysis of *N*-methylbutyramide with HBr?

QUESTIONS AND PROBLEMS

(MC GOB)

■ Hydrolysis of Amides

18.37 Write the condensed structural formulas for the products of the acid hydrolysis of each of the following amides with HCl:

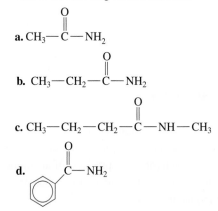

a. $CH_3 - \overset{\overset{\displaystyle O}{\|}}{C} - NH_2$

b. $CH_3 - CH_2 - \overset{\overset{\displaystyle O}{\|}}{C} - NH_2$

c. $CH_3 - CH_2 - CH_2 - \overset{\overset{\displaystyle O}{\|}}{C} - NH - CH_3$

d. $\overset{\overset{\displaystyle O}{\|}}{C} - NH_2$ (benzene ring)

e. *N*-ethylpentanamide

18.38 Write the condensed structural formulas for the products of the base hydrolysis of each of the following amides with NaOH:

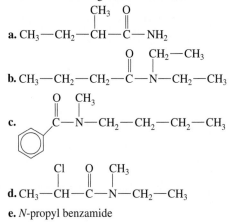

a. $CH_3 - CH_2 - \overset{\overset{\displaystyle CH_3}{|}}{CH} - \overset{\overset{\displaystyle O}{\|}}{C} - NH_2$

b. $CH_3 - CH_2 - CH_2 - \overset{\overset{\displaystyle O}{\|}}{C} - \overset{\overset{\displaystyle CH_2 - CH_3}{|}}{N} - CH_2 - CH_3$

c. $\overset{\overset{\displaystyle O}{\|}}{C} - \overset{\overset{\displaystyle CH_3}{|}}{N} - CH_2 - CH_2 - CH_2 - CH_3$ (benzene ring)

d. $CH_3 - \overset{\overset{\displaystyle Cl}{|}}{CH} - \overset{\overset{\displaystyle O}{\|}}{C} - \overset{\overset{\displaystyle CH_3}{|}}{N} - CH_2 - CH_3$

e. *N*-propyl benzamide

CONCEPT MAP

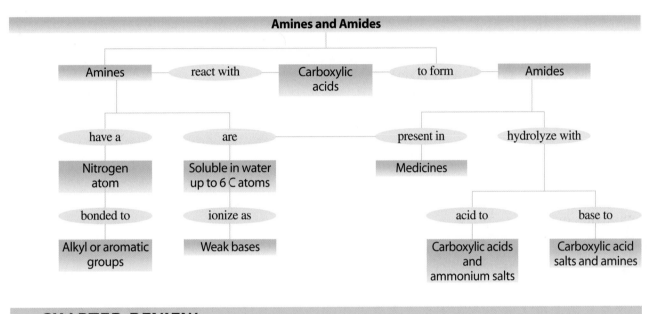

CHAPTER REVIEW

18.1 Amines

A nitrogen atom attached to one, two, or three alkyl or aromatic groups forms a primary, secondary, or tertiary amine. In the IUPAC system, the *amine* suffix is added to the alkane name of the longer carbon chain. Groups attached to the nitrogen atom use a *N*-prefix. When other functional groups are present, the —NH$_2$ is named as an amino group. In the common names of simple amines, the alkyl groups are listed alphabetically followed by the suffix *-amine*.

18.2 Properties of Amines

Primary and secondary amines form hydrogen bonds, which make their boiling points higher than alkanes of similar mass, but lower than alcohols. Amines with up to six carbon atoms are soluble in water. In water, amines act as weak bases because the nitrogen atom accepts protons from water to produce ammonium and hydroxide ions. When amines react with acids, they form amine salts, which are named as ammonium salts. As ionic compounds, amine salts are solids, soluble in water, and odorless compared to the amines. Quaternary ammonium salts contain four carbon groups bonded to the nitrogen atom.

18.3 Heterocyclic Amines and Alkaloids

Heterocyclic amines are cyclic organic compounds that contain one or more nitrogen atoms in the ring. The amine rings typically consist of 5 or 6 atoms and one or more nitrogen atoms. Alkaloids such as caffeine and nicotine are naturally occurring amines derived from plants. Many are known for their physiological activity.

18.4 Amides

Amides are derivatives of carboxylic acids in which the hydroxyl group is replaced by —NH$_2$, a primary, or secondary amine group. Amides are formed when carboxylic acids react with ammonia or primary or secondary amines in the presence of heat. Amides are named by replacing the *ic acid* or *oic acid* with *amide*. Any carbon group attached to the nitrogen atom is named using the *N*-prefix.

18.5 Hydrolysis of Amides

Hydrolysis of an amide by an acid produces an amine salt. Hydrolysis by a base produces the salt of the carboxylic acid.

SUMMARY OF NAMING

Family	Condensed Structural Formula	IUPAC name	Common Name
Amine	$CH_3-CH_2-NH_2$	ethanamine	ethylamine
Amine salt	$CH_3-CH_2-NH_3{}^+Cl^-$	ethylammonium chloride	ethylammonium chloride
Amide	$CH_3-\overset{\overset{\displaystyle O}{\|\|}}{C}-NH_2$	ethanamide	acetamide

SUMMARY OF REACTIONS

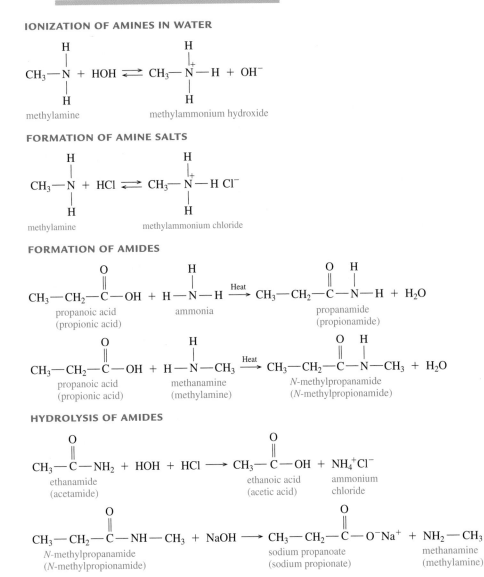

IONIZATION OF AMINES IN WATER

$$CH_3-\overset{\overset{\displaystyle H}{\|}}{\underset{\underset{\displaystyle H}{\|}}{N}} + HOH \rightleftharpoons CH_3-\overset{\overset{\displaystyle H}{\|}}{\underset{\underset{\displaystyle H}{\|}}{\overset{+}{N}}}-H + OH^-$$

methylamine   methylammonium hydroxide

FORMATION OF AMINE SALTS

$$CH_3-\overset{\overset{\displaystyle H}{\|}}{\underset{\underset{\displaystyle H}{\|}}{N}} + HCl \rightleftharpoons CH_3-\overset{\overset{\displaystyle H}{\|}}{\underset{\underset{\displaystyle H}{\|}}{\overset{+}{N}}}-H \ Cl^-$$

methylamine   methylammonium chloride

FORMATION OF AMIDES

$$CH_3-CH_2-\overset{\overset{\displaystyle O}{\|\|}}{C}-OH + H-\overset{\overset{\displaystyle H}{\|}}{N}-H \xrightarrow{Heat} CH_3-CH_2-\overset{\overset{\displaystyle O}{\|\|}}{C}-\overset{\overset{\displaystyle H}{\|}}{N}-H + H_2O$$

propanoic acid   ammonia   propanamide
(propionic acid)   (propionamide)

$$CH_3-CH_2-\overset{\overset{\displaystyle O}{\|\|}}{C}-OH + H-\overset{\overset{\displaystyle H}{\|}}{N}-CH_3 \xrightarrow{Heat} CH_3-CH_2-\overset{\overset{\displaystyle O}{\|\|}}{C}-\overset{\overset{\displaystyle H}{\|}}{N}-CH_3 + H_2O$$

propanoic acid   methanamine   N-methylpropanamide
(propionic acid)   (methylamine)   (N-methylpropionamide)

HYDROLYSIS OF AMIDES

$$CH_3-\overset{\overset{\displaystyle O}{\|\|}}{C}-NH_2 + HOH + HCl \longrightarrow CH_3-\overset{\overset{\displaystyle O}{\|\|}}{C}-OH + NH_4{}^+Cl^-$$

ethanamide   ethanoic acid   ammonium
(acetamide)   (acetic acid)   chloride

$$CH_3-CH_2-\overset{\overset{\displaystyle O}{\|\|}}{C}-NH-CH_3 + NaOH \longrightarrow CH_3-CH_2-\overset{\overset{\displaystyle O}{\|\|}}{C}-O^-Na^+ + NH_2-CH_3$$

N-methylpropanamide   sodium propanoate   methanamine
(N-methylpropionamide)   (sodium propionate)   (methylamine)

KEY TERMS

alkaloids Amines having physiological activity that are produced in plants.

amidation The formation of an amide from a carboxylic acid and ammonia or an amine.

amides Organic compounds containing the carbonyl group attached to an amino group or a substituted nitrogen atom.

amines Organic compounds containing a nitrogen atom attached to one, two, or three hydrocarbon groups.

amine salt An ionic compound produced from an amine and an acid.

heterocyclic amine A cyclic organic compound that contains one or more nitrogen atoms in the ring.

hydrolysis The splitting of a molecule by the addition of water. Amides yield the corresponding carboxylic acid and amine or their salts.

quaternary ammonium ion An amine ion in which the nitrogen atom is bonded to four carbon groups.

UNDERSTANDING THE CONCEPTS

18.39 The sweetener aspartame is made from two amino acids: aspartic acid and phenylalanine. Identify the functional groups in aspartame.

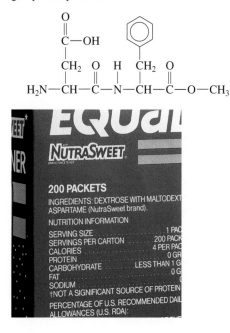

18.40 Some aspirin substitutes contain phenacetin to reduce fever. Identify the functional groups in phenacetin.

ADDITIONAL QUESTIONS AND PROBLEMS

For instructor-assigned homework, go to **www.masteringgob.com.**

18.41 The insect repellant DEET can be made from 3-methyl benzoic acid and *N,N*-diethylamine. What is the structure of DEET?

18.42 Nylon 66 is a polymer used to make shirts and jackets. The structure of one unit of Nylon 66 is shown below. What are the structures of the carboxylic acid and amine that are polymerized to make Nylon 66?

18.43 There are four amine isomers with the molecular formula C_3H_9N. Draw their condensed structural formulas. Name and classify each as a primary, secondary, or tertiary amine.

18.44 Name and classify each of the following compounds as a primary, secondary, or tertiary "amine, or as a quaternary ammonium salt."

a.
$$CH_3—\overset{\overset{\displaystyle CH_2—CH_3}{|}}{N}—CH_2—CH_3$$

b. $CH_3—CH_2—CH_2—CH_2—NH_2$

c. $CH_3—CH_2—CH_2—NH—CH_2—CH_3$

d.

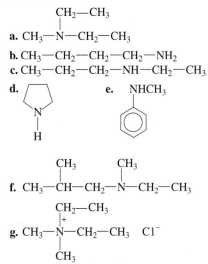

e. NHCH₃

f.
$$CH_3—\overset{\overset{\displaystyle CH_3}{|}}{CH}—CH_2—\overset{\overset{\displaystyle CH_3}{|}}{N}—CH_2—CH_3$$

g.
$$CH_3—\overset{\overset{\displaystyle CH_2—CH_3}{|}}{\underset{\underset{\displaystyle CH_3}{|}}{\overset{+}{N}}}—CH_2—CH_3 \quad Cl^-$$

18.45 Draw the structure of each of the following compounds:
 a. 3-pentanamine
 b. cyclohexylamine
 c. dimethylammonium chloride
 d. triethylamine
 e. 3-amino-2-hexanol
 f. tetramethylammonium bromide
 g. *N,N*-dimethylaniline

18.46 In each pair, indicate the compound that has the higher boiling point:
 a. 1-butanol or butanamine
 b. trimethylamine or propylamine
 c. butylamine or diethylamine
 d. butane or propylamine

18.47 In each pair, indicate the compound that is more soluble in water:
 a. ethylamine or butylamine
 b. trimethylamine or *N*-ethylcyclohexylamine
 c. butylamine or pentane
 d. $NH_2—CH_2—CH_2—CH_2—CH_2—CH_2—NH_2$
   or
   $CH_3—CH_2—CH_2—CH_2—CH_2—NH_2$

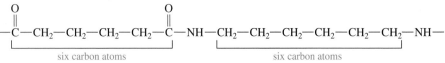

18.48 Give the IUPAC name for each of the following amides:

a. H—C(=O)—NH$_2$

b. CH$_3$—CH$_2$—C(=O)—NH$_2$

c. CH$_3$—C(=O)—N(H)—CH$_3$

d. CH$_3$—CH$_2$—CH$_2$—C(=O)—NH—CH$_2$—CH$_3$

e. CH$_3$—C(=O)—N(CH$_3$)—CH$_2$—CH$_2$—CH$_2$—CH$_3$

18.49 Indicate the name of the alkaloid in each of the following:
   a. malaria treatment
   b. tobacco
   c. coffee and tea
   d. a painkiller in oriental poppy plant

18.50 Identify the heterocyclic amines in each of the following:
   a. caffeine   **b.** Demerol
   c. nicotine   **d.** quinine

18.51 Write the structure of each product of the following reactions:
   a. CH$_3$—CH$_2$—NH$_2$ + H$_2$O $\rightleftharpoons$
   b. CH$_3$—CH$_2$—NH$_2$ + HCl $\rightleftharpoons$

   c. CH$_3$—CH$_2$—NH—CH$_3$ + H$_2$O $\rightleftharpoons$
   d. CH$_3$—CH$_2$—NH—CH$_3$ + HCl $\rightleftharpoons$
   e. CH$_3$—CH$_2$—CH$_2$—NH$_3$$^+Cl^-$ + NaOH $\longrightarrow$

   f. CH$_3$—CH$_2$—NH$_2$$^+Cl^-$ (N—CH$_3$) + NaOH $\longrightarrow$

18.52 Toradol is used in dentistry to relieve pain. Name the functional groups in this molecule.

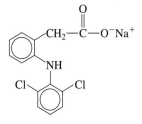

18.53 Voltaren is indicated for acute and chronic treatment of the symptoms of rheumatoid arthritis. Name the functional groups in this molecule.

CHALLENGE QUESTIONS

18.54 Many amine-containing drugs are given to patients in their salt form, such as hydrochloride or sulfate. What might be the reason?

18.55 Using a reference book such as the *Merck Index* or *Physicians' Desk Reference*, look up the structural formula of the following medicinal drugs and list the functional groups in the compounds. You may need to refer to the cross-index of names in the back of the reference book.
   a. Keflex, an antibiotic
   b. Inderal, a β-channel blocker used to treat heart irregularities
   c. ibuprofen, an anti-inflammatory agent
   d. Aldomet (methyldopa)
   e. Percodan, a narcotic pain reliever
   f. triamterene, a diuretic

18.56 Kevlar is a lightweight polymer used in tires and bulletproof vests. Part of the strength of Kevlar is due to hydrogen bonds between polymer chains. The polymer chain is shown below.

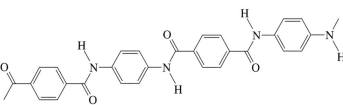

   a. What is the structure of the carboxylic acid and amine that are polymerized to make Kevlar?
   b. What feature of Kevlar will give the hydrogen bonds between the polymer chains?

18.57 Use the K of methylamine to calculate the pH of a 1.0 M solution of methylamine.

ANSWERS

Answers to Study Checks

18.1 tertiary (3°)

18.2 $CH_3—CH_2—\underset{\underset{H}{|}}{\overset{\overset{H}{|}}{N}}—CH_2—CH_2—CH_3$

18.3 Hydrogen bonding makes amines with six or fewer carbon atoms soluble in water.

18.4 $CH_3—\underset{\underset{CH_3}{|}}{\overset{\overset{CH_3}{|}}{\overset{+}{N}}}—H\ Cl^-$

18.5 pyrimidine

18.6 piperidine

18.7 $H—\overset{\overset{O}{||}}{C}—OH$ and $H—\underset{}{\overset{\overset{CH_3}{|}}{N}}—CH_3$

18.8 $Cl—\langle\ \rangle—\overset{\overset{O}{||}}{C}—\underset{}{\overset{\overset{CH_3}{|}}{N}}—CH_3$

18.9 $CH_3—CH_2—CH_2—\overset{\overset{O}{||}}{C}—OH$ and $CH_3—NH_3^+Br^-$

Answers to Selected Problems

18.1 In a primary amine, there is one alkyl group (and two hydrogens) attached to a nitrogen atom.

18.3 **a.** primary **b.** secondary **c.** primary
d. tertiary **e.** tertiary

18.5 **a.** ethylamine; ethanamine
b. methylpropylamine; N-methyl-1-propanamine
c. diethylmethylamine; N-methyl-N-ethylethanamine
d. isopropylamine; 2-propanamine

18.7 **a.** 2-butanamine **b.** 2-chloroaniline
c. 3-aminopropanal **d.** N-ethylaniline

18.9 **a.** $CH_3—CH_2—NH_2$ **b.** (benzene ring with NH—CH₃)

c. $CH_3—CH_2—CH_2—CH_2—\underset{\underset{H}{|}}{\overset{\overset{H}{|}}{N}}—CH_2—CH_2—CH_3$

d. $CH_3—\underset{\underset{NH_2}{|}}{CH}—CH_2—CH_2—CH_3$

18.11 **a.** $CH_3—CH_2—OH$
b. $CH_3—CH_2—CH_2—NH_2$
c. $CH_3—CH_2—CH_2—NH_2$

18.13 As a primary amine, propylamine can form two hydrogen bonds, which gives it the highest boiling point. Ethylmethylamine, a secondary amine, can form one hydrogen bond, and

butane cannot form hydrogen bonds. Thus butane has the lowest boiling point of the three compounds.

18.15 **a.** yes **b.** yes **c.** no **d.** yes

18.17 **a.** $CH_3NH_2 + H_2O \rightleftharpoons CH_3—NH_3^+ + OH^-$
b. $CH_3—NH—CH_3 + H_2O \rightleftharpoons$

$CH_3—\overset{+}{N}H_2—CH_3 + OH^-$

c. (benzene with NH_2) $+ H_2O \rightleftharpoons$ (benzene with NH_3^+) $+ OH^-$

18.19 **a.** $CH_3—NH_3^+\ Cl^-$
b. $CH_3—\overset{+}{N}H_2—CH_3\ Cl^-$
c. (benzene with $NH_3^+\ Cl^-$)

18.21 **a.** $H_2N—\langle\ \rangle—\overset{\overset{O}{||}}{C}—O—CH_2—CH_2—\underset{\underset{CH_2—CH_3}{|}}{\overset{\overset{CH_2—CH_3}{|}}{\overset{+}{N}}}—H\ Cl^-$

b. Amine salts are soluble in body fluids.

18.23 **a.** amine **b.** amine
c. heterocyclic amine **d.** heterocyclic amine

18.25 **a.** pyrimidine **b.** pyrrole

18.27 pyrrole

18.29 **a.** $CH_3—\overset{\overset{O}{||}}{C}—NH_2$

b. $CH_3—\overset{\overset{O}{||}}{C}—\underset{}{\overset{\overset{H}{|}}{N}}—CH_2—CH_3$

c. (benzene)$—\overset{\overset{O}{||}}{C}—\underset{}{\overset{\overset{H}{|}}{N}}—CH_2—CH_2—CH_3$

18.31 **a.** N-methylethanamide (N-methylacetamide)
b. butanamide (butyramide)
c. methanamide (formamide)
d. N-methylbenzamide

18.33 **a.** $CH_3—CH_2—\overset{\overset{O}{||}}{C}—NH_2$

b. $CH_3—CH_2—CH_2—\underset{\underset{}{\overset{\overset{CH_3}{|}}{CH}}}{}—\overset{\overset{O}{||}}{C}—NH_2$

c. $H—\overset{\overset{O}{||}}{C}—NH_2$

d.
$$\underset{\text{(benzene ring)}}{C_6H_5}-\overset{\displaystyle O}{\underset{\displaystyle \|}{C}}-\overset{\displaystyle H}{\underset{\displaystyle |}{N}}-CH_2-CH_3$$

e. $CH_3-CH_2-CH_2-\overset{\displaystyle O}{\underset{\displaystyle \|}{C}}-\overset{\displaystyle H}{\underset{\displaystyle |}{N}}-CH_2-CH_3$

18.35 a. acetamide
b. propionamide
c. *N*-methylpropanamide

18.37 a. $CH_3-\overset{\displaystyle O}{\underset{\displaystyle \|}{C}}-OH + NH_4{}^+Cl^-$

b. $CH_3-CH_2-\overset{\displaystyle O}{\underset{\displaystyle \|}{C}}-OH + NH_4{}^+Cl^-$

c. $CH_3-CH_2-CH_2-\overset{\displaystyle O}{\underset{\displaystyle \|}{C}}-OH + CH_3-NH_3{}^+Cl^-$

d. $\underset{\text{(benzene ring)}}{}\overset{\displaystyle O}{\underset{\displaystyle \|}{C}}-OH + NH_4{}^+Cl^-$

e. $CH_3-CH_2-CH_2-CH_2-\overset{\displaystyle O}{\underset{\displaystyle \|}{C}}-OH +$
$CH_3-CH_2-NH_3{}^+Cl^-$

18.39 amine, carboxylic acid, amide, aromatic ester

18.41
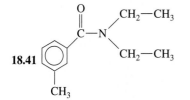

18.43 $CH_3-CH_2-CH_2-NH_2$
propylamine 1°

$CH_3-CH_2-NH-CH_3$
ethylmethylamine 2°

$CH_3-\overset{\displaystyle CH_3}{\underset{\displaystyle |}{N}}-CH_3$
trimethylamine 3°

$CH_3-\overset{\displaystyle CH_3}{\underset{\displaystyle |}{CH}}-NH_2$
isopropylamine 1°

18.45 a. $CH_3-CH_2-\overset{\displaystyle NH_2}{\underset{\displaystyle |}{CH}}-CH_2-CH_3$ **b.** (cyclohexyl)NH_2

c. $CH_3-\overset{\displaystyle CH_3}{\underset{\displaystyle |}{NH_2{}^+}} Cl^-$

d. $CH_3-CH_2-\overset{\displaystyle CH_2-CH_3}{\underset{\displaystyle |}{N}}-CH_2-CH_3$

e. $CH_3-\overset{\displaystyle OH}{\underset{\displaystyle |}{CH}}-\overset{\displaystyle NH_2}{\underset{\displaystyle |}{CH}}-CH_2-CH_2-CH_3$

f. $CH_3-\overset{\displaystyle CH_3}{\underset{\displaystyle \underset{\displaystyle CH_3}{|}}{\overset{\displaystyle |}{N^+}}}-CH_3 \; Br^-$

g. $\overset{\displaystyle CH_3}{\diagdown}\underset{}{N}\overset{\displaystyle CH_3}{\diagup}$ (attached to benzene ring)

18.47 a. ethylamine
b. trimethylamine
c. butylamine
d. $NH_2-CH_2-CH_2-CH_2-CH_2-CH_2-NH_2$

18.49 a. quinine   **b.** nicotine
c. caffeine   **d.** morphine, codeine

18.51 a. $CH_3-CH_2-NH_3{}^+ + OH^-$
b. $CH_3-CH_2-NH_3{}^+Cl^-$
c. $CH_3-CH_2-\overset{+}{N}H_2-CH_3 + OH^-$
d. $CH_3-CH_2-\overset{+}{N}H_2-CH_3 \; Cl^-$
e. $CH_3-CH_2-CH_2-NH_2 + NaCl + H_2O$
f. $CH_3-CH_2-\overset{\displaystyle CH_3}{\underset{\displaystyle |}{N}}H + NaCl + H_2O$

18.53 carboxylate salt, aromatic, amine, haloaromatic

18.55 a. aromatic, amine, amide, carboxylic acid, cycloalkene
b. aromatic, ether, alcohol, amine
c. aromatic, carboxylic acid
d. phenol, amine, carboxylic acid
e. aromatic, ether, alcohol, amine, ketone
f. aromatic, amine

18.57 The pH is 12.32.

19 Amino Acids and Proteins

"This lamb is fed with Lamb Lac, which is a chemically formulated replacement for ewe's milk," says part-time farmer Dennis Samuelson. "Its mother had triplets and didn't have enough milk to feed them all, so they weren't thriving the way the other lambs were. The Lamb Lac includes dried skim milk, dried whey, milk proteins, egg albumin, the amino acids methionine and lysine, vitamins, and minerals."

A veterinary technician diagnoses and treats diseases of animals, takes blood and tissue samples, and administers drugs and vaccines. Agricultural technologists assist in the study of farm crops to increase productivity and ensure a safe food supply. They look for ways to improve crop yields, develop safer methods of weed and pest control, and design methods to conserve soil and water.

LOOKING AHEAD

Mastering GOB CHEMISTRY

Visit **www.masteringgob.com** for self-study materials and instructor-assigned homework.

The word "protein" is derived from the Greek word *proteios*, meaning "first." Made of amino acids, proteins provide structure in membranes, build cartilage and connective tissue, transport oxygen in blood and muscle, direct biological reactions as enzymes, defend the body against infection, and control metabolic processes as hormones. They can even be a source of energy.

Compared with many of the compounds we have studied, protein molecules can be gigantic. Insulin has a molar mass of 5700, and hemoglobin has a molar mass of about 64 000. Some virus proteins are still larger, having molar masses of more than 40 million. Yet all proteins in humans are polymers made up of 20 different amino acids. Each kind of protein is composed of amino acids arranged in a specific order that determines the characteristics of the protein and its biological action.

Proteins perform many functions in the body: making up skin and hair, moving muscles, carrying oxygen, and regulating metabolism. All of these different functions of proteins depend on the structures and chemical behavior of amino acids, the building blocks of proteins. We will see how peptide bonds link amino acids and how the order of the amino acids in these protein polymers directs the formation of unique three-dimensional structures.

19.1 Proteins and Amino Acids

The many kinds of proteins perform different functions in the body. There are proteins that form structural components such as cartilage, muscles, hair, and nails. Wool, silk, feathers, and horns are some other proteins made by animals. Proteins called enzymes regulate biological reactions such as digestion and cellular metabolism. Still other proteins, hemoglobin and myoglobin, carry oxygen in the blood and muscle. (See Figure 19.1). Table 19.1 gives examples of proteins that are classified by their functions in biological systems.

Amino Acids

Proteins are composed of molecular building blocks called amino acids. An **amino acid** contains two functional groups, an amino group ($-NH_2$) and a carboxylic acid group ($-COOH$), and a hydrogen atom bonded to a central carbon atom. At the pH of most body fluids, the carboxyl group and the amino group are usually ionized. The carboxyl group loses a H^+, giving $-COO^-$, and the amino group accepts a H^+ to give an ammonium ion, $-NH_3^+$. Amino acids with this structure are called α (alpha)-amino acids. Although there are many possible amino acids, only 20 are present in human proteins. The unique characteristics of the 20 amino acids are due to a side chain (R), which can be an alkyl, hydroxyl, thiol, amino, sulfide, aromatic, or heterocyclic group.

LEARNING GOAL

Classify proteins by their functions in the cells. Draw the structure for an amino acid.

Protein Building Blocks

Proteins 'R' Us

SELF-STUDY ACTIVITY
Functions of Proteins

FIGURE 19.1

The horns, feathers, and wool of animals are made of proteins.

Q **What class of protein would be in horns?**

TABLE 19.1

Classification of Some Proteins and Their Functions

Class of Protein	Function in the Body	Examples
Structural	Provide structural components	*Collagen* is in tendons and cartilage. *Keratin* is in hair, skin, wool, and nails.
Contractile	Move muscles	*Myosin* and *actin* contract muscle fibers.
Transport	Carry essential substances throughout the body	*Hemoglobin* transports oxygen. *Lipoproteins* transport lipids.
Storage	Store nutrients	*Casein* stores protein in milk. *Ferritin* stores iron in the spleen and liver.
Hormone	Regulate body metabolism and nervous system	*Insulin* regulates blood glucose level. *Growth hormone* regulates body growth.
Enzyme	Catalyze biochemical reactions in the cells	*Sucrase* catalyzes the hydrolysis of sucrose. *Trypsin* catalyzes the hydrolysis of proteins.
Protection	Recognize and destroy foreign substances	*Immunoglobulins* stimulate immune responses.

General Structure of an α-Amino Acid

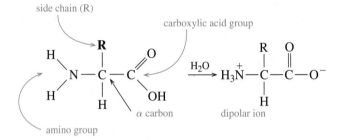

Classification of Amino Acids

Nonpolar amino acids, which have alkyl or aromatic side chains, are *hydrophobic* ("water-fearing"). **Polar amino acids**, which have polar side chains such as hydroxyl (—OH), thiol (—SH), and amide (—CONH$_2$) that form hydrogen bonds with water, are *hydrophilic* ("water attracting"). The **acidic amino acids** have side chains that contain a carboxylic acid group (—COOH) and can ionize as a weak acid. The side chains of the **basic amino acids** contain an amino group that can ionize as a weak base. The structures of the side chains (R), common names, three-letter abbreviations, and isoelectric points (pI) of the 20 amino acids in proteins are listed in Table 19.2.

SAMPLE PROBLEM 19.1

■ **Structural Formulas of Amino Acids**

Write the structural formulas and abbreviations for the following amino acids:

a. alanine (R = —CH$_3$)
b. serine (R = —CH$_2$OH)

TABLE 19.2

The 20 Amino Acids in Proteins

Nonpolar Amino Acids

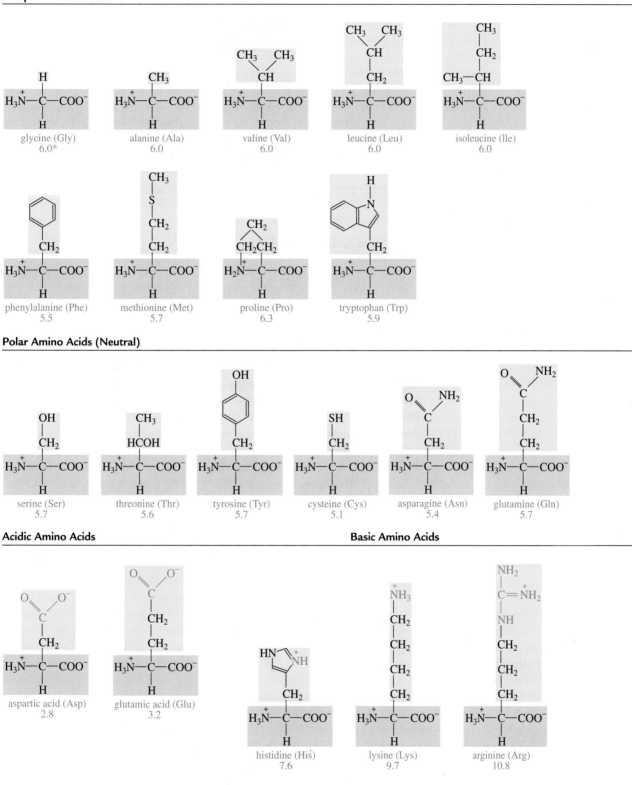

glycine (Gly)
6.0*

alanine (Ala)
6.0

valine (Val)
6.0

leucine (Leu)
6.0

isoleucine (Ile)
6.0

phenylalanine (Phe)
5.5

methionine (Met)
5.7

proline (Pro)
6.3

tryptophan (Trp)
5.9

Polar Amino Acids (Neutral)

serine (Ser)
5.7

threonine (Thr)
5.6

tyrosine (Tyr)
5.7

cysteine (Cys)
5.1

asparagine (Asn)
5.4

glutamine (Gln)
5.7

Acidic Amino Acids   **Basic Amino Acids**

aspartic acid (Asp)
2.8

glutamic acid (Glu)
3.2

histidine (His)
7.6

lysine (Lys)
9.7

arginine (Arg)
10.8

*Isoelectric points (pI)

SOLUTION

a. The structure of the amino acids is written by attaching the side group (R) to the central carbon atom of the general structure of an amino acid.

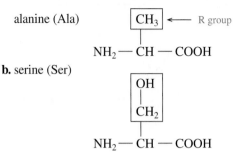

alanine (Ala)

b. serine (Ser)

STUDY CHECK

Classify the amino acids in the sample problem as polar or nonpolar.

Amino Acid Stereoisomers

All of the α-amino acids except for glycine are chiral because the α carbon is attached to four different groups. Thus amino acids can exist as D and L enantiomers. We can write Fischer projections for amino acids as we did in Chapter 14 for aldehydes with the carboxylic acid group (the most highly oxidized carbon) at the top and the R group at the bottom. For the L isomer, the amino group, NH_2, is on the left, and in the D isomer, it is on the right. In biological systems, only L amino acids are incorporated into proteins. There are D amino acids found in nature, but not in proteins. Let's take a look at the enantiomers for glyceraldehyde and how the stereoisomers of alanine and cysteine are similar.

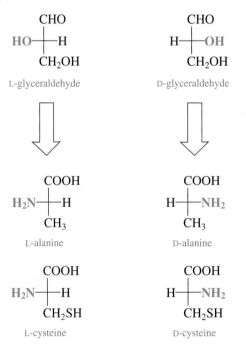

SAMPLE PROBLEM 19.2

■ **Chiral Amino Acids**

Write the Fischer projection for L-serine.

SOLUTION

In L-serine, the —COOH is at the top and the R group —CH$_2$OH is at the bottom. The L-isomer has the —NH$_2$ on the left.

L-serine

STUDY CHECK

How does the Fischer projection for D-serine differ from L-serine?

QUESTIONS AND PROBLEMS

■ **Proteins and Amino Acids**

19.1 Classify each of the following proteins according to its function:
a. hemoglobin, oxygen carrier in the blood
b. collagen, a major component of tendons and cartilage
c. keratin, a protein found in hair
d. amylase, an enzyme that hydrolyzes starch

19.2 Classify each of the following proteins according to its function:
a. insulin, a hormone needed for glucose utilization
b. antibodies, proteins that disable foreign proteins
c. casein, milk protein
d. lipases that hydrolyze lipids

19.3 Describe the functional groups found in all α-amino acids.

19.4 How does the polarity of the side chain in leucine compare to the side chain in serine?

19.5 Draw the structural formula for each of the following amino acids:
a. alanine   **b.** threonine   **c.** glutamic acid   **d.** phenylalanine

19.6 Draw the structural formula for each of the following amino acids:
a. lysine   **b.** aspartic acid   **c.** leucine   **d.** tyrosine

19.7 Classify the amino acids in problem 19.5 as hydrophobic (nonpolar), hydrophilic (polar, neutral), acidic, or basic.

19.8 Classify the amino acids in problem 19.6 as hydrophobic (nonpolar), hydrophilic (polar, neutral), acidic, or basic.

19.9 Give the name of the amino acid represented by each of the following three-letter abbreviations:
a. Ala   **b.** Val   **c.** Lys   **d.** Cys

19.10 Give the name of the amino acid represented by each of the following three-letter abbreviations:
a. Trp   **b.** Met   **c.** Pro   **d.** Gly

19.11 Draw the Fischer projections for the following amino acids.
a. L-valine   **b.** D-cysteine

19.12 Draw the Fischer projections for the following amino acids.
a. L-threonine   **b.** D-valine

(MC GOB) pH, pI, and Amino Acid Ionization

■ 19.2 Amino Acids as Acids and Bases

Although it is convenient to write amino acids with carboxyl (—COOH) group and amine (—NH_2) group, they are usually ionized. The *dipolar* form of an amino acid called a **zwitterion** has a net charge of zero.

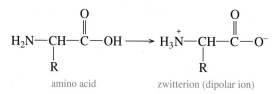

amino acid   zwitterion (dipolar ion)

Solid amino acids have very high melting points because the zwitterion has the properties of a salt. The ionic charges of the amino acids make them more soluble in water, but not in organic solvents.

At a specific pH known as the **isoelectric point (pI)**, the positive and negative charges are equal, which gives the amino acid an overall charge of zero. However, when the pH is different from the pI, the zwitterion accepts or donates H^+. In a solution that is more acidic than the pI, the —COO^- group acts as a base and accepts an H^+, which gives an overall positive charge to the amino acid.

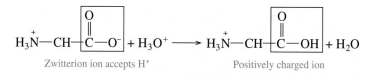

Zwitterion ion accepts H^+ Positively charged ion

In a solution more basic than the pI, the —NH_3^+ group acts as an acid and loses an H^+, which gives the amino acid an overall negative charge.

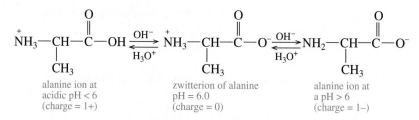

Zwitterion donates H^+ Negatively charged ion

Let's take a look at the changes in all the ionic forms of alanine from its zwitterion (pI = 6.0), to the positive ion in a more acidic solution, and the negative ion in basic solution.

$$NH_3^+—CH—\overset{\overset{\textstyle O}{\|}}{C}—OH \underset{H_3O^+}{\overset{OH^-}{\rightleftharpoons}} NH_3^+—CH—\overset{\overset{\textstyle O}{\|}}{C}—O^- \underset{H_3O^+}{\overset{OH^-}{\rightleftharpoons}} NH_2—CH—\overset{\overset{\textstyle O}{\|}}{C}—O^-$$
$$\quad\ \ CH_3 \qquad\qquad\qquad\qquad\quad\ CH_3 \qquad\qquad\qquad\qquad\quad\ CH_3$$

alanine ion at   zwitterion of alanine   alanine ion at
acidic pH < 6   pH = 6.0   a pH > 6
(charge = 1+)   (charge = 0)   (charge = 1−)

The zwitterions for polar and nonpolar amino acids typically exist at pH values of 5.0 to 6.0. However, the zwitterions of the acidic amino acids form at pH values of about 3 because the carboxyl group in their side chain must pick up H^+. The basic amino acids with amino groups in their side chains form zwitterions at pI values from 7.6 to 10.8. The pI values are included in the list of the amino acids in Table 19.2.

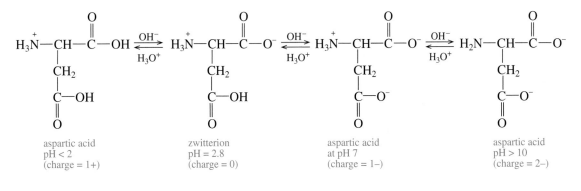

aspartic acid
pH < 2
(charge = 1+)

zwitterion
pH = 2.8
(charge = 0)

aspartic acid
at pH 7
(charge = 1–)

aspartic acid
pH > 10
(charge = 2–)

Electrophoresis

It is possible to separate a mixture of amino acids using a laboratory method called **electrophoresis**. A buffered amino acid mixture is applied to a gel on a thin plate or piece of filter paper, which is connected to two electrodes. A voltage applied to the electrodes causes the positively charged amino acids to move toward the negative electrode and the negatively charged amino acids to move toward the positive electrode. Any amino acid at its isoelectric point with zero net charge would not move. After several hours, the sample is removed. It can be sprayed with a dye such as ninhydrin to make the amino acids visible. They are identified by their direction and rate of migration toward the electrodes. They are recovered separately by cutting up the filter paper or removing the amino acids from the gel.

Suppose we have a mixture of valine (pI 6.0), aspartic acid (pI 2.8), and lysine (pI 9.7) in a buffer of pH 6.0. When the mixture is placed between two electrodes at a high voltage, the aspartic acid, which would have a negative charge at pH 6.0, would move to the positive electrode (anode). (See Figure 19.2.) The lysine, which would be positively charged at a pH of 6.0, would move toward the negative electrode (cathode). Valine, which is neutral at pH 6.0 would not move in the presence of an electric field. Electrophoresis is a method used in medicine to screen for the sickle cell trait in newborn infants.

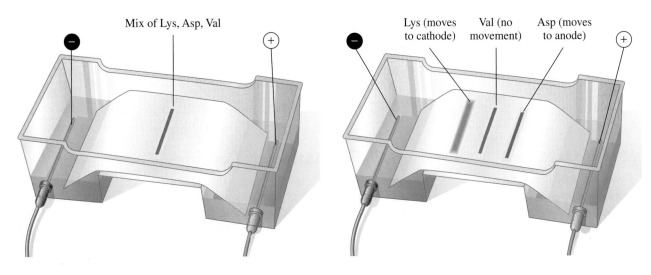

Mix of Lys, Asp, Val

Lys (moves to cathode)   Val (no movement)   Asp (moves to anode)

FIGURE 19.2

A positively charged amino acid (pH < pI) moves toward the negative electrode; a negatively charged amino acid (pH > pI) moves toward the positive electrode; an amino acid with no net charge (pH = pI) does not migrate.

Q **How would the three amino acids migrate if the mixture were buffered to pH 9.7, the pI of lysine?**

SAMPLE PROBLEM 19.3

■ Amino Acids in Acid or Base

Serine exists in its zwitterion form at a pH of 5.7. Draw the structural formula for the zwitterion of serine.

SOLUTION

As a zwitterion, both the carboxylic acid group and the amino group are ionized.

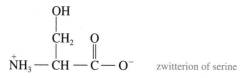

zwitterion of serine

STUDY CHECK

Draw the structure of serine at a pH of 2.0.

QUESTIONS AND PROBLEMS

■ Amino Acids as Acids and Bases

19.13 Write the zwitterion of each of the following amino acids:
  a. glycine   **b.** cysteine   **c.** serine   **d.** alanine

19.14 Write the zwitterion of each of the following amino acids:
  a. phenylalanine   **b.** methionine   **c.** leucine   **d.** valine

19.15 Write the positive ion (acidic ion) of each of the amino acids in problem 19.13 at a pH below 1.0.

19.16 Write the negative ion (basic ion) of each of the amino acids in problem 19.13 at a pH above 12.0.

19.17 Would the following ions of valine exist at a pH above, below, or at pI?

  a. H_2N—CH—COO$^-$ **b.** $H_3\overset{+}{N}$—CH—COOH   **c.** $H_3\overset{+}{N}$—CH—COO$^-$
           | | |
           CH   CH   CH
          ╱ ╲   ╱ ╲   ╱ ╲
       CH_3 CH_3 CH_3 CH_3 CH_3 CH_3

19.18 Would the following ions of serine exist at a pH above, below, or at pI?

  a. $H_3\overset{+}{N}$—CH—COO$^-$ **b.** $H_3\overset{+}{N}$—CH—COOH   **c.** H_2N—CH—COO$^-$
           | | |
          CH_2OH CH_2OH CH_2OH

LEARNING GOAL

Draw the structure of a dipeptide from the zwitterions of two amino acids.

■ 19.3 Formation of Peptides

The linking of two or more amino acids forms a **peptide**. A **peptide bond** is an amide bond that forms when the —COO$^-$ group of one amino acid reacts with the —NH$_3^+$ group of the next amino acid. We can write the amidation reaction for the zwitterion forms of two amino acids.

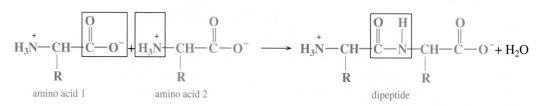

amino acid 1   amino acid 2   dipeptide

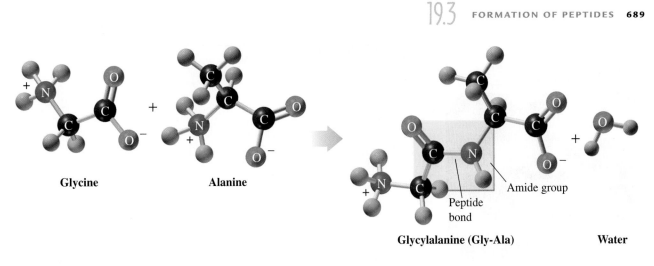

Glycine   **Alanine**   **Glycylalanine (Gly-Ala)**   **Water**

Amide group

Peptide bond

FIGURE 19.3

A peptide bond between glycine and alanine as zwitterions form the dipeptide glycylalanine.

Q **What functional groups in glycine and alanine form the peptide bond?**

Two amino acids linked together by a peptide bond form a *dipeptide*. We can write the formation of the dipeptide glycylalanine between glycine and alanine as follows. (See Figure 19.3.) In a peptide, the amino acid written on the left with the unreacted or free amino group ($-NH_3^+$) is called the **N terminal** amino acid. The **C terminal** amino acid is the last amino acid in the chain with the unreacted or free carboxyl group ($-COO^-$).

MC GOB SELF-STUDY ACTIVITY Structure of Proteins

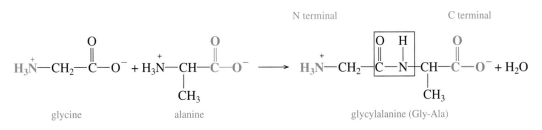

glycine   alanine   glycylalanine (Gly-Ala)

N terminal   C terminal

Naming Peptides

In naming a peptide, each amino acid beginning from the N terminal is named with a *yl* ending followed by the full name of the amino acid at the C terminal. For example, a tripeptide consisting of alanine, glycine, and serine is named as ala**nyl**gly**cyl**serine. For convenience, the order of amino acids in the peptide is often written as the sequence of three-letter abbreviations.

MC GOB Peptide Bonds: Acid Meets Amino

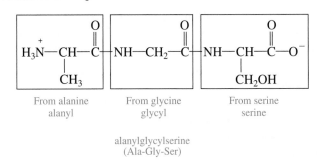

From alanine   From glycine   From serine
alanyl   glycyl   serine

alanylglycylserine
(Ala-Gly-Ser)

▪ Rehabilitation Specialist

"I am interested in the biomechanical part of rehabilitation, which involves strengthening activities that help people return to the activities of daily living," says Minna Robles, rehabilitation specialist. "Here I am fitting a patient with a wrist extension splint that allows her to lift her hand. This exercise will also help the muscles and soft tissues in that wrist area to heal. An understanding of the chemicals of the body, how they interact, and how we can affect the body on a chemical level is important in understanding our work. One technique we use is called myofacial release. We apply pressure to a part of the body, which helps to increase circulation. By increasing circulation, we can move the soft tissues better, which improves movement and range of motion."

LEARNING GOAL

Identify the primary and secondary structures of a protein.

 Proteins are Chains of Amino Acids

SAMPLE PROBLEM 19.4

▪ Identifying a Tripeptide

For the tripeptide:

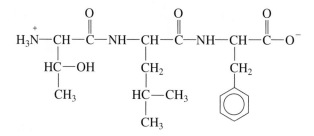

a. What amino acid is the N terminal? What amino acid is the C terminal?
b. What is the three-letter abbreviation name for the tripeptide?

SOLUTION

a. Threonine is the N terminal; phenylalanine is the C terminal.
b. Thr-Leu-Phe

STUDY CHECK

What is the full name of the tripeptide in Sample Problem 19.4?

QUESTIONS AND PROBLEMS

▪ Formation of Peptides

19.19 Draw the structural formula of each of the following peptides and give the abbreviation for their names:
   a. alanylcysteine
   b. serylphenylalanine
   c. glycylalanylvaline
   d. valylisoleucyltryptophan

19.20 Draw the structural formula of each of the following peptides and give the abbreviation for their names:
   a. methionylaspartic acid
   b. alanyltryptophan
   c. methionylglutaminyllysine
   d. histidylglycylglutamylalanine

19.4 Protein Structure: Primary and Secondary Levels

When there are more than 50 amino acids in a chain, the polypeptide is usually called a **protein**. Each protein in our cells has a unique sequence of amino acids that determines its biological function.

Primary Structure

The particular sequence of amino acids in a peptide or protein is referred to as the **primary structure**. For example, a hormone that stimulates the thyroid to release thyroxine consists of a tripeptide Glu-His-Pro.

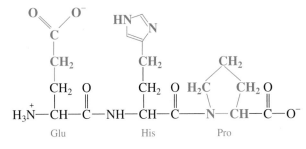

Glu   His   Pro

Although other sequences are possible for these three amino acids, only the tripeptide with the Glu-His-Pro sequence of amino acids has hormonal activity. Sequences such as His-Pro-Glu or Pro-His-Glu do not produce hormonal activity. Thus the biological function of peptides and proteins depends on the order of the amino acids.

The first protein to have its primary structure determined was insulin, which is a hormone that regulates the glucose level in the blood. In the primary structure of human insulin, there are two polypeptide chains. In chain A, there are 21 amino acids, and chain B has 30 amino acids. The polypeptide chains are held together by disulfide bonds formed by the side chains of the cysteine amino acids in each of the

HEALTH NOTE

■ Polypeptides in the Body

Enkephalins and endorphins are natural painkillers produced in the body. They are polypeptides that bind to receptors in the brain to give relief from pain. This effect appears to be responsible for the runner's high, for the temporary loss of pain when severe injury occurs, and for the analgesic effects of acupuncture.

The *enkephalins*, found in the thalamus and the spinal cord, are pentapeptides, the smallest molecules with opiate activity. The amino acid sequence of an enkephalin is found in the longer amino acid sequence of the endorphins.

Four groups of *endorphins* have been identified: α-endorphin contains 16 amino acids, β-endorphin contains 31 amino acids, γ-endorphin has 17 amino acids, and δ-endorphin has 27 amino acids. Endorphins may produce their sedating effects by preventing the release of substance P, a polypeptide with 11 amino acids, which has been found to transmit pain impulses to the brain.

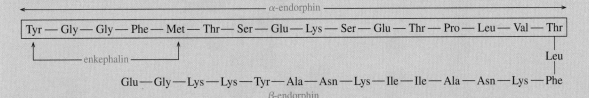

When cells are damaged, a polypeptide called bradykinin is released which stimulates the release of prostaglandins.

Arg—Pro—Pro—Gly—Phe—Ser—Pro—Phe—Arg

bradykinin

Two hormones produced by the pituitary gland are the nonapeptides (nine-amino-acid peptide) oxytocin and vasopressin. Oxytocin stimulates uterine contractions in labor and vasopressin is an antidiuretic hormone that regulates blood pressure by adjusting the amount of water reabsorbed by the kidneys. The structures of these nonapeptides are very similar. Only the amino acids in positions 3 and 8 are different. However, the difference of two amino acids greatly affects how the two hormones function in the body.

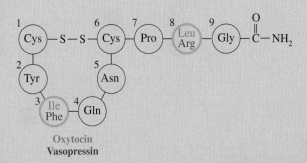

Oxytocin
Vasopressin

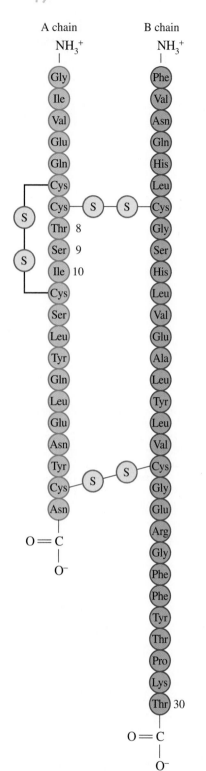

A chain

B chain

NH$_3^+$

NH$_3^+$

FIGURE 19.4

The sequence of amino acids in human insulin is its primary structure.

Q **What kinds of bonds occur in the primary structure of a protein?**

chains. (See Figure 19.4.) This primary structure of insulin in humans is very similar to the primary structure of insulin in cows (bovine). Only the three amino acids at positions 8, 9, and 10 in chain A and position 30 in chain B vary from one species to another. For many years, bovine insulin obtained from the pancreas of cows was used to treat diabetics who lacked insulin. Today human insulin produced through genetic engineering is used in the treatment of diabetes.

Secondary Structure

The **secondary structure** of a protein describes the way the amino acids next to or near to each other along the polypeptide are arranged in space. The three most common types of secondary structure are the *alpha helix*, the *beta-pleated sheet*, and the *triple helix* found in collagen.

The corkscrew shape of an **alpha helix** (α helix) is held in place by hydrogen bonds between each N—H group and the oxygen of a C=O group in the next turn of the helix, four amino acids down the chain. (See Figure 19.5.) Because many

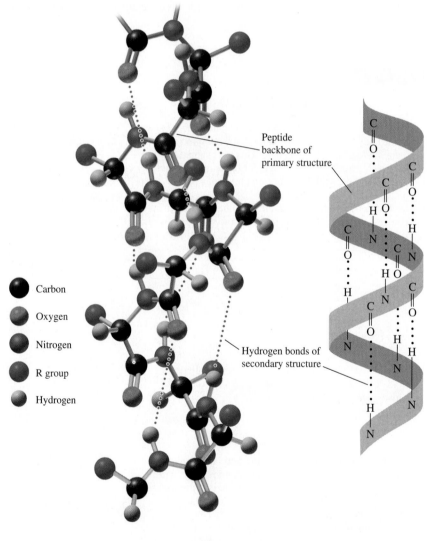

Carbon

Oxygen

Nitrogen

R group

Hydrogen

Peptide backbone of primary structure

Hydrogen bonds of secondary structure

FIGURE 19.5

The α (alpha) helix acquires a coiled shape from hydrogen bonds between the N—H of the peptide bond in one loop and the C=O of the peptide bond in the next loop.

Q **What are partial charges of the H in N—H and the O in C=O that permits hydrogen bonds to form?**

hydrogen bonds form along the peptide backbone, this portion of the protein takes the shape of a strong, tight coil that looks like a telephone cord. All the side chains (R groups) of the amino acids are located on the outside of the helix.

Another type of secondary structure is known as the **beta-pleated sheet** (**β-pleated sheet**). In a β-pleated sheet, polypeptide chains are held together side by side by hydrogen bonds between the peptide chains. In a β-pleated sheet of silk fibroin, the small R groups of the prevalent amino acids, glycine, alanine, and serine, extend above and below the sheet. This results in a series of β-pleated sheets that are stacked close together. The hydrogen bonds holding the β-pleated sheets tightly in place account for the strength and durability of fibrous proteins such as silk. (See Figure 19.6.)

In some proteins, the polypeptide chain consists of mostly he α helix secondary structure, whereas other proteins consist of mostly the β-pleated sheet structure. Another group of proteins have a mixture with some sections of the polypeptide chain in α helixes and other sections in the β-pleated sheet structure. The tendency to form a certain type of secondary structure depends on the amino acids in a particular segment of the polypeptide chain. Amino acids such as valine, proline, serine, and aspartic acid are found in β-pleated sheet regions. The α helix region has large amounts of amino acids like alanine, histidine, leucine, and methionine.

Collagen, the most abundant protein, makes up as much as one-third of all the protein in vertebrates. It is found in connective tissue, blood vessels, skin, tendons, ligaments, the cornea of the eye, and cartilage. The strong structure of collagen is a

 The Shapes of Protein Chains: Helices and Sheets

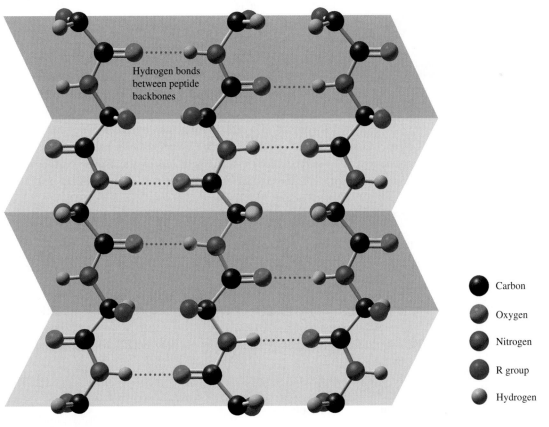

Carbon

Oxygen

Nitrogen

R group

Hydrogen

FIGURE 19.6

In a β (beta)-pleated sheet secondary structure, hydrogen bonds form between the peptide chains.

Q How do the hydrogen bonds differ in a β (beta)-pleated sheet from the helix?

Triple helix   3 α Helix peptide chains

FIGURE 19.7

Hydrogen bonds between polar R groups in three polypeptide chains form the triple helixes that combine to make fibers of collagen.

Q What are some of the amino acids in collagen that form hydrogen bonds between the polypeptide chains?

result of three polypeptides woven together like a braid to form a **triple helix**, as seen in Figure 19.7. Collagen has a high content of glycine (33%), proline (22%), alanine (12%), and smaller amounts of hydroxyproline, and hydroxylysine. The hydroxy forms of proline and lysine contain —OH groups that create hydrogen bonds across the peptide chains and give strength to the collagen triple helix. When several triple helixes wrap together, they form the fibrils that make up connective tissues and tendons.

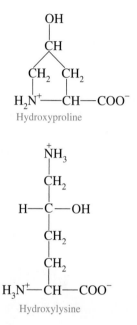

When a diet is deficient in vitamin C, collagen fibrils are weakened because the enzymes needed to form hydroxyproline and hydroxylysine require vitamin C. Without the —OH groups of hydroxyproline and hydroxylysine, there is less hydrogen bonding between collagen fibrils. As a person ages, additional cross links form between the fibrils, which make collagen less elastic. Bones, cartilage, and tendons become more brittle, and wrinkles are seen as the skin loses elasticity.

SAMPLE PROBLEM 19.5

▪ Identifying Secondary Structures

Indicate the secondary structure (α helix, β-pleated sheet, or triple helix) described in each of the following statements:

a. a coiled peptide chain held in place by hydrogen bonding between peptide bonds in the same chain

b. a structure that has hydrogen bonds between polypeptide chains arranged side by side.

SOLUTION

a. α helix   **b.** β-pleated sheet

STUDY CHECK

What is the secondary structure in collagen?

QUESTIONS AND PROBLEMS

■ Protein Structure: Primary and Secondary Levels

19.21 What type of bonding occurs in the primary structure of a protein?

19.22 How can two proteins with exactly the same number and type of amino acids have different primary structures?

19.23 Two peptides each contain one molecule of valine and two molecules of serine. What are their possible primary structures?

19.24 What are three different types of secondary protein structure?

19.25 What happens to the primary structure of a protein when a protein forms a secondary structure?

19.26 In an α helix, how does bonding occur between the amino acids in the polypeptide chain?

19.27 What is the difference in bonding between an α helix and a β-pleated sheet?

19.28 How is the secondary structure of a β-pleated sheet different from that of a triple helix?

HEALTH NOTE

■ Essential Amino Acids

Of the 20 amino acids used to build the proteins in the body, only 10 can be synthesized in the body. The other 10 amino acids, listed in Table 19.3, are **essential amino acids** that cannot be synthesized and must be obtained from the proteins in the diet.

TABLE 19.3

Essential Amino Acids	
arginine (Arg)*	methionine (Met)
histidine (His)*	phenylalanine (Phe)
isoleucine (Ile)	threonine (Thr)
leucine (Leu)	tryptophan (Trp)
lysine (Lys)	valine (Val)

* Required in diets of children, not adults

Complete proteins, which contain all of the essential amino acids are found in most animal products such as eggs, milk, meat, fish, and poultry. However, gelatin and plant proteins such as grains, beans, and nuts are *incomplete proteins* because they are deficient in one or more of the essential amino acids. Diets that rely on plant foods for protein must contain a variety of protein sources to obtain all the essential amino acids. For example, a diet of rice and beans contains all the essential amino acids because they have complementary proteins. Rice contains the methionine and tryptophan deficient in beans, while beans contain the lysine that is lacking in rice. (See Table 19.4.)

TABLE 19.4

Amino Acid Deficiency in Selected Vegetables and Grains	
Food Source	**Amino Acids Missing**
Eggs, milk, meat, fish, poultry	none
Wheat, rice, oats	lysine
Corn	lysine, tryptophan
Beans	methionine, tryptophan
Peas	methionine
Almonds, walnuts	lysine, tryptophan
Soy	low in methionine

LEARNING GOAL

■
Distinguish between the tertiary and quaternary structures of a protein.

(MC GOB) SELF-STUDY ACTIVITY
Tertiary and Quaternary Structure

(MC GOB) Proteins Fold to Function

(MC GOB) Proteins Chains Can Work Together

19.5 Protein Structure: Tertiary and Quaternary Levels

The **tertiary structure** of a protein involves attractions and repulsions between the side chain groups of the amino acids in the polypeptide chain. As interactions occur between different parts of the peptide chain, segments of the chain twist and bend until the protein acquires a specific three-dimensional shape.

Cross-Links in Tertiary Structures

The tertiary structure of a protein is stabilized by interactions between the R groups of the amino acids in one region of the polypeptide chain with R groups of amino acids in other regions of the protein. (See Figure 19.8.) Table 19.5 lists the stabilizing interactions of tertiary structures.

1. **Hydrophobic interactions** are interactions between two nonpolar R groups. Within a protein, the amino acids with nonpolar side chains push as far away from the aqueous environment as possible, which forms a hydrophobic center at the interior of the protein molecule.

2. **Hydrophilic interactions** are attractions between the external aqueous environment and amino acids that have polar or ionized side chains. The polar side chains pull toward the outer surface to hydrogen bond with water.

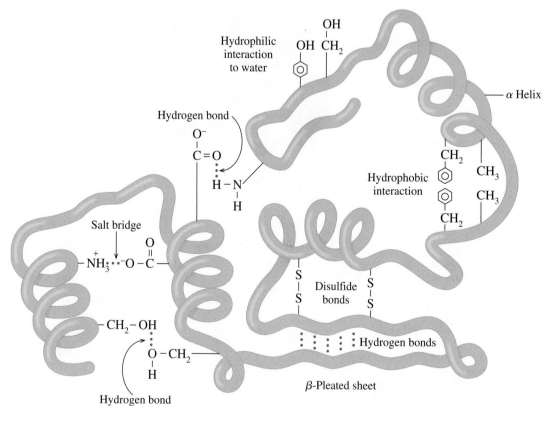

FIGURE 19.8

Interactions between amino acid R groups fold a protein into a specific three-dimensional shape called its tertiary structure.

Q **Why would one section of the protein chain move to the center while another section remains on the surface of the tertiary structure?**

3. **Salt bridges** are ionic bonds between side groups of basic and acidic amino acids, which have positive and negative charges. For example, at a pH of 7.4, the side chain of lysine has a positive charge, and the side chain of glutamic acid has a negative charge. The attraction of the oppositely charged side chains forms a strong bond called a salt bridge. If the pH changes, the basic and acidic side chains lose their ionic charges and cannot form salt bridges, which causes a change in the shape of the protein.

4. **Hydrogen bonds** form between polar amino acids. For example, a hydrogen bond can occur between the —OH of serine and the —NH$_2$ of glutamine.

5. **Disulfide bonds** (—S—S—) are covalent bonds that form between the —SH groups of cysteines in the polypeptide chain. In some proteins, there are several disulfide bonds between the R groups of cysteine in the polypeptide chain.

TABLE 19.5

Some Cross-Links in Tertiary Structures

	Nature of Bonding
Hydrophobic interactions	Attractions between nonpolar groups
Hydrophilic interactions	Attractions between polar or ionized groups and water on the surface of the tertiary structure
Salt bridges	Ionic interactions between ionized acidic and basic amino acids
Hydrogen bonds	Occur between H and O or N
Disulfide bonds	Strong covalent links between sulfur atoms of two cysteine amino acids

---

SAMPLE PROBLEM 19.6

▪ Cross-Links in Tertiary Structures

What type of interaction would you expect between the R groups of the following amino acids?

a. cysteine and cysteine   **b.** glutamic acid and lysine

SOLUTION

a. Because cysteine has an R group containing —SH, a disulfide bond will form.
b. An ionic bond (salt bridge) can form by the interaction of the —COO$^-$ of glutamic acid and the —NH$_3^+$ of lysine.

STUDY CHECK

Would you expect to find valine and leucine in a globular protein on the outside or the inside of the tertiary structure? Why?

Globular and Fibrous Proteins

A group of proteins known as **globular proteins** have compact, spherical shapes because their secondary structures of the polypeptide chain fold over on top of each other. It is the globular proteins that carry out the work of the cells: functions such as synthesis, transport, and metabolism.

Myoglobin is a globular protein that stores oxygen in skeletal muscle. High concentrations of myoglobin have been found in the muscles of sea mammals, such as seals and whales, that stay under the water for long periods. Myoglobin contains 153 amino acids in a single polypeptide chain with about three-fourths of the chain in the α helix secondary structure. The polypeptide chain, including its helical regions, forms a compact tertiary structure by folding upon itself. (See Figure 19.9.) Within the tertiary structure, a pocket of amino acids and a heme group binds and stores oxygen (O$_2$).

The **fibrous proteins** are proteins that consist of long, thin, fiber-like shapes. They are typically involved in the structure of cells and tissues. Two types of fibrous protein are the α- and β-keratins. The **α-keratins** are the proteins that make up hair, wool, skin, and nails. In hair, three α helixes coil together like a braid to form a fibril. Within the fibril, the α helices are held together by disulfide (—S—S—) linkages between the R groups of the many cysteine amino acids in hair. Several

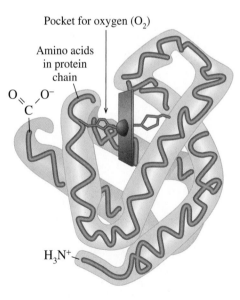

Pocket for oxygen (O$_2$)

Amino acids in protein chain

FIGURE 19.9

Myoglobin is a globular protein with a heme pocket in its tertiary structure that binds oxygen to be carried to the tissues.

Q Would hydrophilic amino acids be found on the outside or inside of the myoglobin structure?

Prions and Mad Cow Disease

Up until recently, researchers thought that only virus or bacteria were responsible for transmitting diseases. Now a group of diseases have been found in which the infectious agents are proteins called *prions*. Bovine spongiform encephalopathy (BSE), or "mad cow disease," is a fatal brain disease of cattle in which the brain fills with cavities resembling a sponge. In the noninfectious form of the prion PrP^c, the N-terminal portion is a random coil. Although the noninfectious form may be ingested from meat products, its structure can change to what is known as PrP^{sc} or *prion-related protein scrapie*. In this infectious form, the end of the peptide chain folds into a β-pleated sheet, which has disastrous effects on the brain and spinal cord. The conditions that cause this structural change are not yet known.

The human variant is called Creutzfeldt–Jakob (CJD) disease. Around 1955, Dr. Carleton Gajdusek was studying a disease known as Kuru, a neurological disease that was killing members of a tribe in Papua New Guinea. Because their diets were low in protein, it was a ritual to eat members of the tribe who died. As a result, the infectious agent Kuru was transmitted from one member to another. After Gajdusek identified the infectious agent in Kuru as similar to the prions that cause BSE, he received the Nobel prize.

BSE was diagnosed in Great Britain in 1986. The protein is present in nerve tissue, but is not found in meat. Control measures that exclude brain and spinal cord from animal feed are now in place to reduce the incidence of BSE.

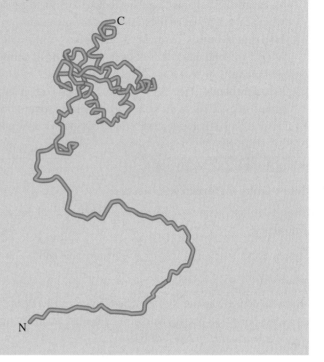

fibrils bind together to form a strand of hair. (Figure 19.10.) The β-keratins are the type of proteins found in the feathers of birds and scales of reptiles. In β-keratins, the proteins consist of large amounts of β-pleated sheet structure.

Quaternary Structure: Hemoglobin

When a biologically active protein consists of two or more polypeptide subunits, the structural level is referred to as a **quaternary structure**. Hemoglobin, a globular protein that transports oxygen in blood, consists of four polypeptide chains or subunits, two α chains and two β chains. The subunits are held together in the quaternary structure by the same interactions that stabilize the tertiary structure, such as hydrogen bonds and salt bridges between side groups, disulfide links, and hydrophobic attractions. (See Figure 19.11.) Each subunit of the hemoglobin contains a heme group that binds oxygen. In the adult hemoglobin molecule, all four subunits ($\alpha_2\beta_2$) must be combined for the hemoglobin to properly function as an oxygen carrier. Therefore, the complete quaternary structure of hemoglobin can bind and transport four molecules of oxygen.

Hemoglobin and myoglobin have similar biological functions. Hemoglobin carries oxygen in the blood, whereas myoglobin carries oxygen in muscle. Myoglobin, a single polypeptide chain with a molar mass of 17 000, has about one-fourth the molar mass of hemoglobin (64 000). The tertiary structure of the single polypeptide myoglobin is almost identical to the tertiary structure of each of the subunits of hemoglobin. Myoglobin stores just one molecule of oxygen, just as each subunit of hemoglobin carries one oxygen molecule. The similarity in

α helix

Alpha keratin

FIGURE 19.10

The fibrous proteins of α-keratin wrap together to form fibrils that make up hair and wool. The proteins called β-keratins are found in the feathers of birds and scales of reptiles.

Q Why does hair have a large amount of cysteine amino acids?

α Chain   β Chain

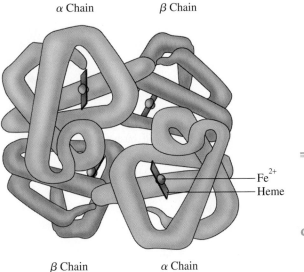

Fe²⁺
Heme

β Chain   α Chain

FIGURE 19.11

The quaternary structure of hemoglobin consists of four polypeptide subunits, each containing a heme group that binds an oxygen molecule.

Q What is the difference between a tertiary structure and a quaternary structure?

tertiary structures allows each protein to bind and release oxygen in a similar manner. Table 19.6 and Figure 19.12 summarize the structural levels of proteins.

SAMPLE PROBLEM 19.7

■ **Identifying Protein Structure**

Indicate whether the following conditions are responsible for primary, secondary, tertiary, or quaternary protein structures:

a. Disulfide bonds form between portions of a protein chain.
b. Peptide bonds form a chain of amino acids.

SOLUTION

a. Disulfide bonds help to stabilize the tertiary structure of a protein.
b. The sequence of amino acids in a polypeptide is a primary structure.

STUDY CHECK

What structural level is represented by the grouping of two subunits in insulin?

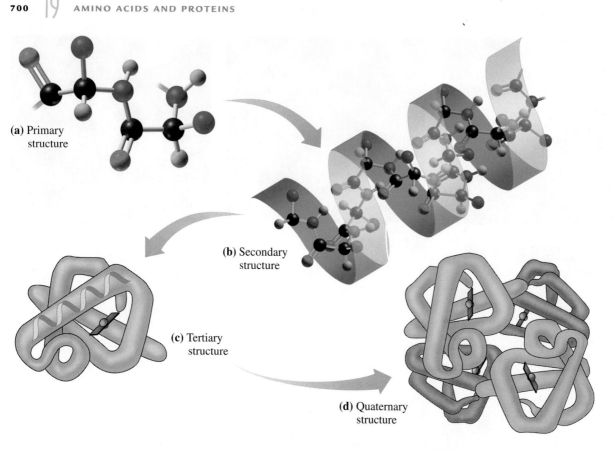

(a) Primary structure

(b) Secondary structure

(c) Tertiary structure

(d) Quaternary structure

FIGURE 19.12

Proteins consist of **(a)** primary, **(b)** secondary, **(c)** tertiary, and sometimes **(d)** quaternary structural levels.

Q **What is the difference between a primary structure and a tertiary structure?**

TABLE 19.6

Summary of Structural Levels in Proteins

Structural Level	Characteristics
Primary	The sequence of amino acids
Secondary	The coiled α helix, β-pleated sheet, or a triple helix formed by hydrogen bonding between peptide bonds along the chain
Tertiary	A folding of the protein into a compact, three-dimensional shape stabilized by interactions between side R groups of amino acids
Quaternary	A combination of two or more protein subunits to form a larger, biologically active protein

---

QUESTIONS AND PROBLEMS

■ Protein Structure: Tertiary and Quaternary Levels

19.29 What type of interaction would you expect from the following R groups in a tertiary structure?

a. two cysteine residues   **b.** glutamic acid and lysine
c. serine and aspartic acid   **d.** two leucine residues

19.30 In myoglobin, about one-half of the 153 amino acids have nonpolar side chains.

a. Where would you expect those amino acids to be located in the tertiary structure?
b. Where would you expect the polar side chains to be?
c. Why is myoglobin more soluble in water than silk or wool?

Sickle-Cell Anemia

Sickle-cell anemia is a disease caused by an abnormality in the shape of one of the subunits of the hemoglobin protein. In the β chain, the sixth amino acid, glutamic acid, which is polar, is replaced by valine, a nonpolar amino acid.

Because valine has a hydrophobic side chain, it draws the hydrophobic pocket that binds to oxygen to the surface of the hemoglobin. The affected red blood cells (RBC) change from a rounded shape to a crescent shape, like a sickle, which interferes with their ability to transport adequate quantities of oxygen. Hydrophobic attractions cause several sickle-cell hemoglobin molecules to stick together, which forms long fibers of sickle-cell hemoglobin. The clumps of insoluble fibers clog capillaries, where they cause inflammation, pain, and organ damage. Critically low oxygen levels may occur in the affected tissues.

In sickle-cell anemia, both genes for the altered hemoglobin must be inherited. However, a few sickled cells are found in persons who carry one gene for sickle-cell hemoglobin, a condition that is also known to provide protection from malaria.

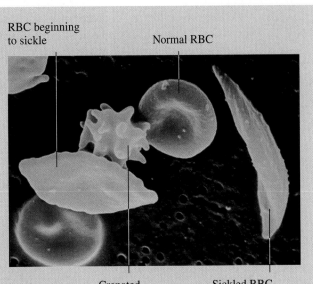

RBC beginning to sickle

Normal RBC

Crenated RBC

Sickled RBC

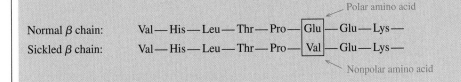

Polar amino acid

Normal β chain: Val — His — Leu — Thr — Pro — Glu — Glu — Lys —

Sickled β chain: Val — His — Leu — Thr — Pro — Val — Glu — Lys —

Nonpolar amino acid

19.31 A portion of a polypeptide chain contains the following sequence of amino acid residues:

-Leu-Val-Cys-Asp-

a. Which R groups can form a disulfide cross-link?
b. Which amino acid residues are likely to be found on the inside of the protein structure? Why?
c. Which amino acid residues would be found on the outside of the protein? Why?
d. How does the primary structure of a protein affect its tertiary structure?

19.32 State whether the following statements apply to primary, secondary, tertiary, or quaternary protein structure:

a. Side groups interact to form disulfide bonds or ionic bonds.
b. Peptide bonds join amino acids in a polypeptide chain.
c. Several polypeptides are held together by hydrogen bonds between adjacent chains.
d. Hydrogen bonding between carbonyl oxygen atoms and nitrogen atoms of amide groups causes a polypeptide to coil.
e. Hydrophobic side chains seeking a nonpolar environment move toward the inside of the folded protein.
f. Protein chains of collagen form a triple helix.
g. An active protein contains four tertiary subunits.

LEARNING GOAL

■ Describe the hydrolysis and denaturation of proteins.

(MC GOB) Protein Demolition

19.6 Protein Hydrolysis and Denaturation

Peptide bonds can be hydrolyzed to give the individual amino acids. This is the process that occurs in the stomach when enzymes such as pepsin or trypsin catalyze the hydrolysis of proteins to give amino acids. This disrupts the primary structure by breaking the covalent amide bonds that link the amino acids. In the digestion of proteins, the amino acids are absorbed through the intestinal walls and carried to the cells where they can be used to synthesize proteins.

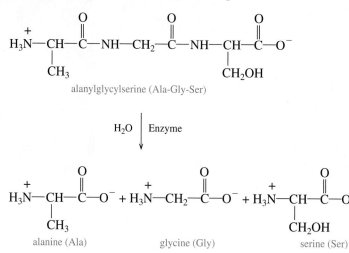

alanylglycylserine (Ala-Gly-Ser)

H_2O | Enzyme

alanine (Ala)　　　　glycine (Gly)　　　　serine (Ser)

Denaturation of Proteins

Denaturation of a protein occurs when there is a disruption of any of the bonds that stabilize the secondary, tertiary, or quaternary structure. However, the covalent amide bonds of the primary structure are not affected.

When the interactions between the R groups are undone or altered, a globular protein unfolds like a loose piece of spaghetti. With the loss of its overall shape, the protein is no longer biologically active. (See Figure 19.13.)

Denaturing agents include heat, acids and bases, organic compounds, heavy metal ions, and mechanical agitation.

Heat

Heat denatures proteins by breaking apart hydrogen bonds and the hydrophobic attraction between nonpolar side groups. Few proteins can remain biologically active above 50°C. Whenever you cook food, you are using heat to denature protein. The nutritional value of the proteins in food is not changed, but they are made more digestible. High temperatures are also used to disinfect surgical instruments and gowns by denaturing the proteins of any bacteria present.

Acids and Bases

Placing a protein in an acid or base affects the hydrogen bonding between polar R groups and disrupts the ionic bonds (salt bridges). In the preparation of yogurt and cheese, a bacteria that produces lactic acid is added to denature the milk protein and produce solid casein. Tannic acid, a weak acid used in burn ointments, coagulates proteins at the site of the burn, forming a protective cover and preventing further loss of fluid from the burn.

EXPLORE YOUR WORLD

■ Denaturation of Milk Protein

Place some milk in five glasses. Add the following to the milk samples in glasses 1–4. The fifth glass of milk is a reference sample.
1. Vinegar, drop by drop. Stir.
2. One-half teaspoon of meat tenderizer. Stir.
3. One teaspoon of fresh pineapple juice. (Canned juice has been heated.)
4. One teaspoon of fresh pineapple juice that you have heated to boiling.

QUESTIONS

1. How did the appearance of the milk change in each of the samples?
2. What enzyme is listed on the package label of the tenderizer?
3. How does the effect of the heated pineapple juice compare with that of the fresh juice? Explain.
4. Why is cooked pineapple used when making gelatin (a protein) desserts?

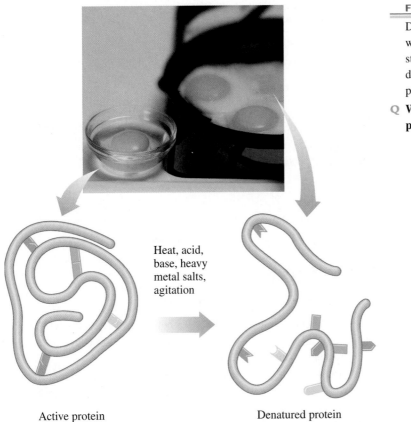

FIGURE 19.13

Denaturation of a protein occurs when the bonds of the tertiary structure are disrupted, which destroys the shape and renders the protein biologically inactive.

Q **What are some ways in which proteins are denatured?**

Heat, acid, base, heavy metal salts, agitation

Active protein

Denatured protein

Organic Compounds

Ethanol and isopropyl alcohol act as disinfectants by forming their own hydrogen bonds with a protein and disrupting the hydrophobic interactions. An alcohol swab is used to clean wounds or to prepare the skin for an injection because the alcohol passes through the cell walls and coagulates the proteins inside the bacteria.

Heavy Metal Ions

When heavy metal ions like Ag^+, Pb^{2+}, and Hg^{2+} form ionic bonds or react with the disulfide (—S—S—) bonds, the denatured protein solidifies. In hospitals, dilute (1%) solutions of $AgNO_3$ are placed in the eyes of newborn babies to destroy the bacteria that causes gonorrhea. These heavy metal ions can be toxic. If ingested, they act as poisons by severely disrupting body proteins, especially enzymes. An antidote is a high-protein food such as milk, eggs, or cheese that will tie up the heavy metal ions until the stomach can be pumped.

Agitation

The whipping of cream and the beating of egg whites are examples of using mechanical agitation to denature protein. The whipping action stretches the polypeptide chains until the stabilizing interactions are disrupted.

SAMPLE PROBLEM 19.8

■ **Effects of Denaturation**

What happens to the tertiary structure of a globular protein when it is placed in an acidic solution?

SOLUTION

An acid causes denaturation by disrupting the hydrogen bonds and the ionic bonds between the R groups. A loss in interactions causes the tertiary structure to lose stability. As the protein unfolds, both the shape and biological function are lost.

STUDY CHECK

Why is a dilute solution of $AgNO_3$ used to disinfect the eyes of newborn infants?

QUESTIONS AND PROBLEMS

■ **Protein Hydrolysis and Denaturation**

19.33 What products would result from the complete hydrolysis of Gly-Ala-Ser?

19.34 Would the hydrolysis products of the tripeptide Ala-Ser-Gly be the same or different from the products in problem 19.33? Explain.

19.35 What dipeptides would be produced from the partial hydrolysis of His-Met-Gly-Val?

19.36 What tripeptides would be produced from the partial hydrolysis of Ser-Leu-Gly-Gly-Ala?

19.37 What structural level of a protein is affected by hydrolysis?

19.38 What structural level of a protein is affected by denaturation?

19.39 Indicate the changes in protein structure for each of the following:
  a. An egg placed in water at 100°C is soft boiled in about 3 min.
  b. Prior to giving an injection, the skin is wiped with an alcohol swab.
  c. Surgical instruments are placed in a 120°C autoclave.
  d. During surgery, a wound is closed by cauterization (heat).

19.40 Indicate the changes in protein structure for each of the following:
  a. Tannic acid is placed on a burn.
  b. Milk is heated to 60°C to make yogurt.
  c. To avoid spoilage, seeds are treated with a solution of $HgCl_2$.
  d. Hamburger is cooked at high temperatures to destroy *E. coli* bacteria that may cause intestinal illness.

CONCEPT MAP

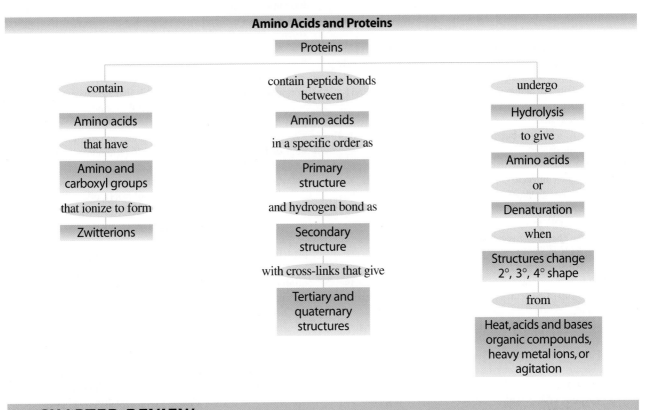

Amino Acids and Proteins

Proteins

contain → **Amino acids** → that have → **Amino and carboxyl groups** → that ionize to form → **Zwitterions**

contain peptide bonds between → **Amino acids** → in a specific order as → **Primary structure** → and hydrogen bond as → **Secondary structure** → with cross-links that give → **Tertiary and quaternary structures**

undergo → **Hydrolysis** → to give → **Amino acids** → or → **Denaturation** → when → **Structures change 2°, 3°, 4° shape** → from → **Heat, acids and bases organic compounds, heavy metal ions, or agitation**

CHAPTER REVIEW

19.1 Proteins and Amino Acids

Some proteins are enzymes or hormones, whereas others are important in structure, transport, protection, storage, and muscle contraction. A group of 20 amino acids provides the molecular building blocks of proteins. Attached to the central (alpha) carbon of each amino acid is an amino group, a carboxyl group, and a unique side group (R). The R group gives an amino acid the property of being nonpolar, polar, acidic, or basic.

19.2 Amino Acids as Acids and Bases

Amino acids exist as dipolar ions called zwitterions, as positive ions at low pH, and as negative ions at high pH levels. At the isoelectric point, zwitterions are neutral.

19.3 Formation of Peptides

Peptides form when an amide bond links the carboxyl group of one amino acid and the amino group of a second amino acid. Long chains of amino acids are called proteins.

19.4 Protein Structure: Primary and Secondary Levels

The primary structure of a protein is its sequence of amino acids. In the secondary structure, hydrogen bonds between peptide

groups produce a characteristic shape such as an α helix, β-pleated sheet, or a triple helix.

19.5 Protein Structure: Tertiary and Quaternary Levels

In globular proteins, the polypeptide chain, including α-helical and β-pleated sheet regions, folds upon itself to form a tertiary structure. A tertiary structure is stabilized by interactions that move hydrophobic R groups to the inside and hydrophilic R groups to the outside surface, and by attractions between R groups that form hydrogen, disulfide, and salt bridges. In a quaternary structure, two or more tertiary subunits must combine for biological activity. They are held together by the same interactions found in tertiary structures.

19.6 Protein Hydrolysis and Denaturation

Denaturation of a protein occurs when heat or other denaturing agents destroy the structure of the protein (but not the primary structure) until biological activity is lost.

KEY TERMS

acidic amino acid An amino acid that has a carboxylic acid side chain (—COOH), which ionizes as a weak acid.

α (alpha) helix A secondary level of protein structure, in which hydrogen bonds connect the NH of one peptide bond with the C=O of a peptide bond later in the chain to form a coiled or corkscrew structure.

α-keratins Fibrous proteins that contain mostly α helixes found in hair, nails, and skin.

amino acid The building block of proteins, consisting of an amino group, a carboxylic acid group, and a unique side group attached to the alpha carbon.

basic amino acid An amino acid that contains an amino (—NH₂) group that can ionize as a weak base.

β (beta)-pleated sheet A secondary level of protein structure that consists of hydrogen bonds between peptide links in parallel polypeptide chains.

C terminal The amino acid that is the last amino acid in a peptide chain with an unreacted or free carboxyl group (—COO⁻).

collagen The most abundant form of protein in the body, which is composed of fibrils of triple helixes with hydrogen bonding between —OH groups of hydroxyproline and hydroxylysine.

denaturation The loss of secondary and tertiary protein structure caused by heat, acids, bases, organic compounds, heavy metals, and/or agitation.

disulfide bonds Covalent —S—S— bonds that form between the —SH group of cysteines in a protein to stabilize the tertiary structure.

electrophoresis The use of electrical current to separate proteins or other charged molecules with different isoelectric points.

essential amino acids Amino acids that must be supplied by the diet because they are not synthesized by the body.

fibrous protein A protein that is insoluble in water; consists of polypeptide chains with α helixes or β-pleated sheets, that make up the fibers of hair, wool, skin, nails, and silk.

globular proteins Proteins that acquire a compact shape from attractions between the R group of the amino acid residues in the protein.

hydrogen bonds Attractions between the polar R groups such as —OH, —NH₂, and —COOH of amino acids in a polypeptide chain.

hydrophilic interactions the attraction between polar R groups on the protein surface and water

hydrophobic interactions The attraction between nonpolar R groups in a tertiary structure of a globular protein.

isoelectric point (pI) The pH at which an amino acid is in the neutral zwitterion form.

N terminal The amino acid in a peptide written on the left with the unreacted or free amino group (—NH₃⁺).

nonpolar amino acids Amino acids that are not soluble in water because they contain a nonpolar R group.

peptide The combination of two or more amino acids joined by peptide bonds; dipeptide, tripeptide, and so on.

peptide bond The amide bond that joins amino acids in polypeptides and proteins.

polar amino acids Amino acids that are soluble in water because their R group is polar; hydroxyl (OH), thiol (SH), carbonyl (C=O), amino (NH₂), or carboxyl (COOH).

primary structure The sequence of the amino acids in a protein.

protein A term used for biologically active polypeptides that have many amino acids linked together by peptide bonds.

quaternary structure A protein structure in which two or more protein subunits form an active protein.

salt bridge The ionic bond formed between side groups of basic and acidic amino acids in a protein.

secondary structure The formation of an α helix, β-pleated sheet, or triple helix.

tertiary structure The folding of the secondary structure of a protein into a compact structure that is stabilized by the interactions of R groups such as ionic and disulfide bonds.

triple helix The protein structure found in collagen consisting of three polypeptide chains woven together like a braid.

zwitterion The dipolar form of an amino acid consisting of two oppositely charged ionic regions, —NH₃⁺ and —COO⁻.

UNDERSTANDING THE CONCEPTS

19.41 Seeds and vegetables are often deficient in one or more essential amino acids. Using the following table, state whether the following combinations would provide all the essential amino acids:

Source	Lysine	Tryptophan	Methionine
Oatmeal	No	Yes	Yes
Rice	No	Yes	Yes
Garbanzo beans	Yes	No	Yes
Lima beans	Yes	No	No
Cornmeal	No	Yes	Yes

a. rice and garbanzo beans
b. lima beans and cornmeal
c. a salad of garbanzo beans and lima beans
d. rice and lima beans
e. rice and oatmeal
f. oatmeal and lima beans

19.42 If cysteine, an amino acid prevalent in hair, has a pI of 5.0, which structure would it have in solutions with the following pH values?

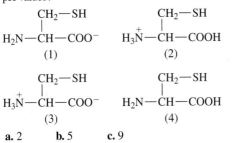

CH₂—SH
|
H₂N—CH—COO⁻
(1)

CH₂—SH
|
H₃N⁺—CH—COOH
(2)

CH₂—SH
|
H₃N⁺—CH—COO⁻
(3)

CH₂—SH
|
H₂N—CH—COOH
(4)

a. 2   **b.** 5   **c.** 9

19.43 For each of the following pairs of side chains, identify the amino acids and the type of cross-link that forms between them.

a. —CH₂—C(=O)—NH₂ and HO—CH₂—

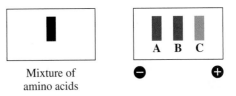

b. —CH₂—C(=O)—O⁻ and H₃N⁺—(CH₂)₄—

c. —CH₂—SH and HS—CH₂—

d. —CH₂—CH(CH₃)—CH₃ and CH₃—

19.44 Consider a mixture of the amino acids lysine, valine, and aspartic acid at pH 6.0 that is subjected to an electric current.

a. Indicate which amino acid would migrate toward the positive electrode (+), the negative electrode (−), or remain stationary.

Mixture of amino acids

A B C

⊖   ⊕

b. If the mixture is the result of hydrolyzing a tripeptide, what are the possible amino acid sequences containing one molecule of each?

c. In an enzyme, which of these amino acids would be found in
  1. hydrophobic regions
  2. hydrophilic regions
  3. hydrogen bonds
  4. salt bridges

ADDITIONAL QUESTIONS AND PROBLEMS

For instructor-assigned homework, go to **www.masteringgob.com.**

19.45 **a.** What are some functions of α-keratins?
  b. What amino acids give strength to the α-keratins?

19.46 **a.** Where is collagen found?
  b. What type of secondary structure is used to form collagen?

19.47 **a.** Draw the structure of Ser-Lys-Asp.
  b. Would you expect to find this segment at the center or at the surface of a globular protein? Why?

19.48 **a.** Draw the structure of Val-Ala-Leu.
  b. Would you expect to find this segment at the center or at the surface of a globular protein? Why?

19.49 Would you expect the following segments in a polypeptide to have an α helix or β-pleated sheet secondary structure?
  a. a segment with a high content of Val, Pro, and Ser
  b. a segment with a high content of His, Met, and Leu.

19.50 What type of interaction would you expect from the following R groups in a tertiary structure?
  a. threonine and asparagine
  b. valine and alanine
  c. arginine and aspartic acid

19.51 If serine were replaced by valine in a protein, how would the tertiary structure be affected?

19.52 If you eat rice, what other vegetable protein source(s) could you eat to ingest all essential amino acids?

19.53 Draw the structure of each of the following amino acids at pH 4.
  a. serine   **b.** alanine   **c.** lysine

19.54 Draw the structure of each of the following amino acids at pH 11.
  a. cysteine   **b.** aspartic acid   **c.** valine

CHALLENGE QUESTIONS

19.55 Indicate the charge of each of the following amino acids at the following pH values as:
  a. 0   **b.** +1   **c.** +2   **d.** −2   **e.** −1
  1. serine at pH 6.0
  2. glutamic acid at pH 10
  3. lysine at pH 2
  4. histidine at pH 7.6
  5. isoleucine at pH 3
  6. leucine at pH 9

19.56 The proteins placed on a gel for electrophoresis have the following isoelectric points: albumin, 4.9, hemoglobin, 6.8, and lysozyme, 11.0. A buffer of pH 6.8 is placed on the gel.
 a. Which protein will migrate toward the positive electrode?
 b. Which protein will migrate toward the negative electrode?
 c. Which protein will remain at the same place it was originally placed?

19.57 What are some differences between the following pairs?
 a. secondary and tertiary protein structures
 b. essential and nonessential amino acids
 c. polar and nonpolar amino acids
 d. di- and tripeptides
 e. an ionic bond (salt bridge) and a disulfide bond
 f. fibrous and globular proteins
 g. α helix and β-pleated sheet
 h. tertiary and quaternary structures of proteins

19.58 How does denaturation of a protein differ from its hydrolysis?

ANSWERS

Answers to Study Checks

19.1 a. nonpolar   **b.** polar

19.2 In the Fischer projection of D-serine, the —NH₂ group is on the right side.

19.3

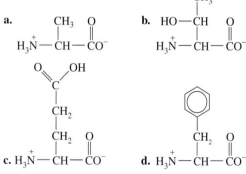

19.4 threonylleucylphenylalanine

19.5 a triple helix

19.6 Both are nonpolar and would be found on the inside of the tertiary structure.

19.7 quaternary

19.8 The heavy metal Ag⁺ denatures the protein in bacteria that cause gonorrhea.

Answers to Selected Questions and Problems

19.1 a. transport   **b.** structural
    c. structural   **d.** enzyme

19.3 All amino acids contain a carboxylic acid group and an amino group on the α carbon.

19.5

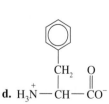

19.7 a. hydrophobic nonpolar   **b.** hydrophilic polar
    c. acidic   **d.** hydrophobic nonpolar

19.9 a. alanine   **b.** valine
    c. lysine   **d.** cysteine

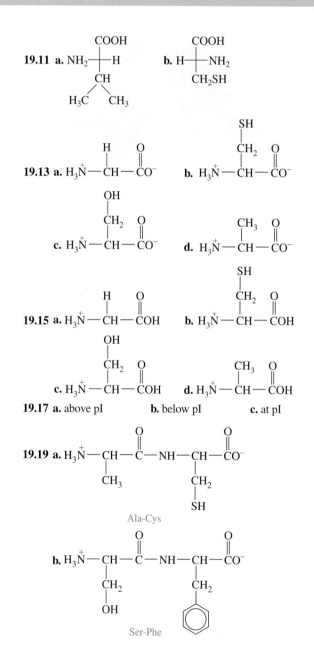

19.11 (figures a, b)

19.13 (figures a, b, c, d)

19.15 (figures a, b, c, d)

19.17 a. above pI   **b.** below pI   **c.** at pI

19.19 a. Ala-Cys
     b. Ser-Phe

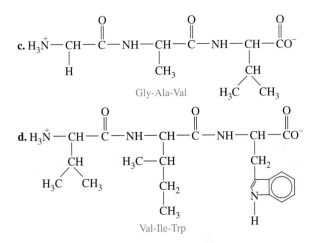

Gly-Ala-Val

Val-Ile-Trp

19.21 Amide bonds form to connect the amino acids that make up the protein.

19.23 Val-Ser-Ser, Ser-Val-Ser, or Ser-Ser-Val

19.25 The primary structure remains unchanged and intact as hydrogen bonds form between carbonyl oxygen atoms and amino hydrogen atoms in the secondary structure.

19.27 In the α helix, hydrogen bonds form between the carbonyl oxygen atom and the amino hydrogen atom of the fourth amino acid in the sequence. In the β-pleated sheet, hydrogen bonds occur between parallel peptides or across sections of a long polypeptide chain.

19.29 a. a disulfide bond **b.** salt bridge
  c. hydrogen bond **d.** hydrophobic interaction

19.31 a. cysteine
  b. Leucine and valine will be found on the inside of the protein because they are hydrophobic.
  c. The cysteine and aspartic acid would be on the outside of the protein because they are polar.
  d. The order of the amino acids (the primary structure) provides the R groups whose interactions determine the tertiary structure of the protein.

19.33 The products would be the amino acids glycine, alanine, and serine.

19.35 His-Met; Met-Gly; Gly-Val

19.37 Hydrolysis splits the amide linkages in the primary structure.

19.39 a. Placing an egg in boiling water coagulates the protein of the egg.
  b. Using an alcohol swab coagulates the protein of any bacteria present.
  c. The heat from an autoclave will coagulate the protein of any bacteria on the surgical instruments.
  d. Heat will coagulate the surrounding protein to close the wound.

19.41 a. yes **b.** yes **c.** no
  d. yes **e.** no **f.** yes

19.43 a. asparagine and serine; hydrogen bond
  b. aspartic acid and lysine; salt bridge
  c. cysteine and cysteine; disulfide bond
  d. leucine and alanine; hydrophobic interaction

19.45 a. α-keratins are fibrous proteins that provide structure to hair, wool, skin, and nails.
  b. α-keratins have a high content of cysteine.

19.47 a.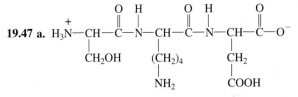

  b. This segment contains polar R groups, which would be found on the surface of a globular protein where they can hydrogen bond with water.

19.49 a. β-pleated sheet **b.** α helix

19.51 Serine is a polar amino acid, whereas valine is nonpolar. Valine would be in the center of the tertiary structure. However, serine would pull that part of the chain to the outside surface of the protein where it forms hydrophilic bonds with water.

19.53

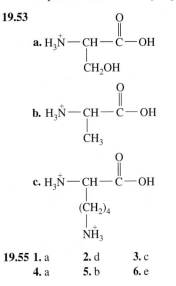

19.55 1. a **2.** d **3.** c
  4. a **5.** b **6.** e

19.57 a. The secondary structure of a protein depends on hydrogen bounds to form a helix or a pleated sheet; the tertiary structure is determined by the interaction of R groups and determines the three-dimensional structure of the protein.
  b. Nonessential amino acids can be synthesized by the body; essential amino acids must be supplied by the diet.
  c. Polar amino acids have hydrophilic side groups, whereas nonpolar amino acids have hydrophobic side groups.
  d. Dipeptides contain two amino acids, whereas tripeptides contain three.
  e. An ionic bond is an interaction between a basic and acidic side group; a disulfide bond links the sulfides of two cysteines.
  f. Fibrous proteins consist of three to seven α helixes coiled like a rope. Globular proteins form a compact spherical shape.
  g. The α helix is the secondary shape like a staircase or corkscrew. The β-pleated sheet is a secondary structure that is formed by many protein chains side by side like a pleated sheet.
  h. The tertiary structure of a protein is its three-dimensional structure. The quaternary structure involves the grouping of two or more peptide units for the protein to be active.

20

Enzymes and Vitamins

"At a time when we have a shortage of health care professionals, I think of myself as a physician extender," says Pushpinder Beasley, orthopedic physician assistant, Kaiser Hospital. "We can put a significant amount of time into our patient care. Just today, I examined a child's knee. One of the most common injuries to children is disruption of either knee ligaments or the soft tissue around the knees. In this child's case, we were checking her anterior ligaments, also known as ACL. I think an important role of the health care professional is to earn the trust of young people."

As part of a health care team, physician assistants examine patients, order laboratory tests, make diagnoses, report patient progress, order therapeutic procedures and, in most states, prescribe medications.

LOOKING AHEAD

Mastering
GOB CHEMISTRY

Visit **www.masteringgob.com**
for self-study materials
and instructor-assigned
homework.

Every second, thousands of chemical reactions occur in the cells of the human body. For example, many reactions occur to digest the food we eat, convert the products to chemical energy, and synthesize proteins and other macromolecules in our cells. In the laboratory, we can carry out reactions that hydrolyze polysaccharides, fats, or proteins, but we must use a strong acid or base, high temperatures, and long reaction times. In the cells of our body, these reactions must take place at rates that meet our physiological and metabolic needs. To make this happen, enzymes catalyze the chemical reactions in our cells, with a different enzyme for every reaction. Digestive enzymes in the mouth, stomach, and small intestine catalyze the hydrolysis of carbohydrate, fats, and proteins. Enzymes in the mitochondria extract energy from biomolecules to give us energy.

Every enzyme responds to what comes into the cells and to what the cells need. Enzymes keep reactions going when our cells need certain products, and turn off reactions when they don't need those products.

Many enzymes require cofactors to function properly. A cofactor can be an inorganic metal ion or an organic compound such as a vitamin. We obtain minerals such as zinc (Zn^{2+}) and iron (Fe^{3+}) and vitamins from our diets. A lack of minerals and vitamins can lead to certain nutritional diseases. For example, rickets is a deficiency of vitamin D and scurvy occurs when a diet is low in vitamin C.

20.1 Enzymes

Biological catalysts known as **enzymes** catalyze nearly all the chemical reactions that take place in the body. As *catalysts*, enzymes increase the rate of a reaction by changing the way a reaction takes place, but is itself not changed at the end of the reaction. An uncatalyzed reaction in a cell may take place eventually, but not at a rate fast enough for survival. For example, the hydrolysis of proteins in our diet would eventually occur without a catalyst, but not fast enough to meet the body's requirements for amino acids. The chemical reactions in our cells must occur at incredibly fast rates under mild conditions of pH 7.4 and a body temperature of 37°C. Enzymes permit cells to use energy and materials efficiently while responding to cellular needs.

As catalysts, enzymes lower the activation energy for the reaction. (See Figure 20.1.) As a result, less energy is required to convert reactant molecules to products, which allows more reacting molecules to form product. However, as a catalyst an enzyme does not affect the equilibrium position, which means that there is an increase in the rates of both the forward and reverse directions. The rates of enzyme-catalyzed reactions are much faster than the rates of the uncatalyzed reactions. Some enzymes can increase the rate of a biological reaction by a factor of a billion or trillion or even a hundred million trillion compared to the rate of the uncatalyzed reaction. For example, an enzyme in the blood called carbonic anhydrase converts carbon dioxide (CO_2) and water (H_2O) to carbonic acid (H_2CO_3). In one minute, one molecule of anhydrase catalyzes the reaction of about one million (10^6) molecules.

$$CO_2 + H_2O \underset{\text{anhydrase}}{\overset{\text{carbonic}}{\rightleftharpoons}} H_2CO_3$$

LEARNING GOAL

Describe how enzymes function as biological catalysts, and name and classify them.

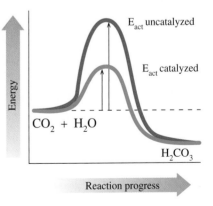

FIGURE 20.1

The activation energy needed for the reaction of CO_2 and H_2O is lowered by the enzyme carbonic anhydrase.

Q **Why are enzymes needed in biological reactions?**

Names and Classification of Enzymes

(MC GOB) Enzymes and Activation Energy

The names of enzymes describe the compound or the reaction that is catalyzed. The actual names of enzymes are derived by replacing the end of the name of the reaction or reacting compound with the suffix *ase*. For example, an *oxidase* catalyzes an oxidation reaction, and a *dehydrogenase* removes hydrogen atoms. The com-

TABLE 20.1

Classification of Enzymes

Class	General Reactions Catalyzed	Typical Subclasses	Function
1. Oxidoreductases	Oxidation–reduction reactions	Oxidases Reductases Dehydrogenases	Oxidation Reduction Remove 2H to form double bonds
2. Transferases	Transfer of functional groups	Transaminases Kinases	Transfer amino groups Transfer phosphate groups
3. Hydrolases	Hydrolysis reactions	Peptidases Lipases Amylases	Hydrolyze peptide bonds Hydrolyze ester bonds in lipids Hydrolyze 1,4-glycosidic bonds in amylose
4. Lyases	Addition of a group to a double bond or removal of a group from a double bond	Decarboxylases Dehydrases Deaminases	Remove CO_2 Remove H_2O Remove NH_3
5. Isomerases	Rearrangement of atoms to form isomers	Isomerases Epimerases	Convert cis and trans Convert D and L isomers
6. Ligases	Bonding of molecules using energy from hydrolysis of ATP (See Section 22.2)	Synthetases Carboxylases	Combine molecules Add CO_2

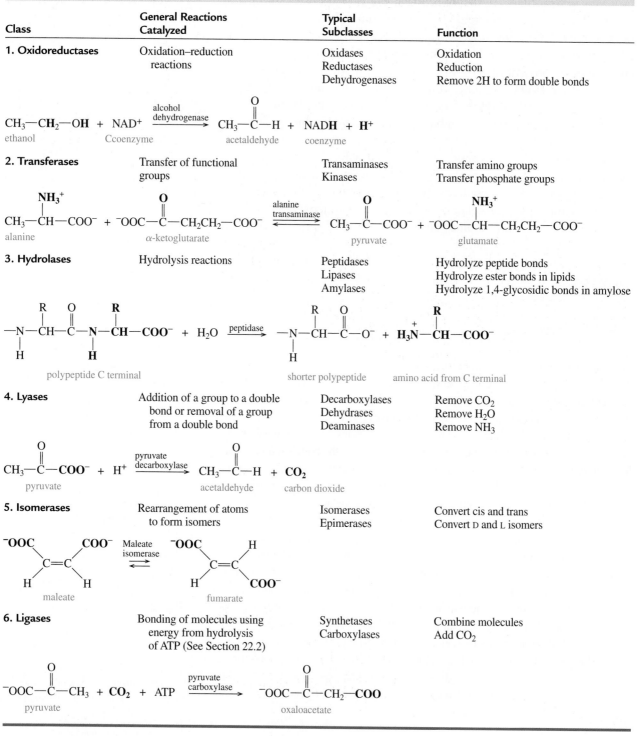

pound sucrose is hydrolyzed by the enzyme *sucrase*, and a lipid is hydrolyzed by a *lipase*. Some enzymes use names that end in the suffix *in*, such as *papain* found in papaya, *rennin* found in milk, and *pepsin* and *trypsin*, enzymes that catalyze the hydrolysis of proteins.

The International Commission on Enzymes has classified enzymes according to the six general types of reactions they catalyze. (See Table 20.1.)

SAMPLE PROBLEM 20.1

■ Naming Enzymes

What chemical reaction do the following enzymes catalyze?

a. amino transferase   **b.** lactate dehydrogenase

SOLUTION

a. catalyzes the transfer of an amino group
b. catalyzes the removal of hydrogen from lactate

STUDY CHECK

What is the class of the enzyme lipase that catalyzes the hydrolysis of ester bonds in triglycerides?

QUESTIONS AND PROBLEMS

■ Enzymes

20.1 Why do cellular reactions require enzymes?

20.2 How do enzymes make cellular reactions proceed at faster rates?

20.3 What types of reaction are catalyzed by each of the following classes of enzymes?
    a. oxidoreductases   **b.** transferases   **c.** hydrolases

20.4 What types of reaction are catalyzed by each of the following classes of enzymes?
    a. lyases   **b.** isomerases   **c.** ligases

20.5 What class of enzyme catalyzes each of the following reactions?
    a. hydrolysis of sucrose   **b.** addition of oxygen
    c. converting glucose ($C_6H_{12}O_6$) to fructose ($C_6H_{12}O_6$)
    d. moving an amino group from one molecule to another

20.6 What class of enzyme catalyzes each of the following reactions?
    a. addition of water to a double bond   **b.** removing hydrogen atoms
    c. splitting peptide bonds in proteins   **d.** removing CO_2 from pyruvate

20.7 Identify the class of enzyme that catalyzes each of the following reactions:

a. $CH_3-\overset{\overset{\textstyle O}{\|}}{C}-COO^- + H^+ \longrightarrow CH_3-\overset{\overset{\textstyle O}{\|}}{C}-H + CO_2$

b. $CH_3-\overset{\overset{\textstyle NH_3^+}{|}}{CH}-COO^- + {}^-OOC-\overset{\overset{\textstyle O}{\|}}{C}-CH_2-CH_3 \longrightarrow$

$CH_3-\overset{\overset{\textstyle O}{\|}}{C}-COO^- + {}^-OOC-\overset{\overset{\textstyle NH_3^+}{|}}{CH}-CH_2-CH_3$

20.8 Identify the class of enzyme that would catalyze each of the following reactions:

a. $CH_3-\overset{\overset{\textstyle O}{\|}}{C}-COO^- + CO_2 + ATP \longrightarrow {}^-OOC-CH_2-\overset{\overset{\textstyle O}{\|}}{C}-COO^- + ADP + P_i$

b. $CH_3-CH_2-OH + NAD^+ \longrightarrow CH_3-\overset{\overset{\textstyle O}{\|}}{C}-H + NADH + H^+$

20.9 Name the enzyme that catalyzes each of the following reactions:
   a. oxidizes succinate
   b. adds water to fumarate
   c. removes 2H from an alcohol

20.10 Name the enzyme that catalyzes each of the following reactions:
   a. hydrolyzes sucrose
   b. transfers an amino group from aspartate
   c. removes a carboxylate group from pyruvate

LEARNING GOAL

Describe the role of an enzyme in an enzyme-catalyzed reaction.

SELF-STUDY ACTIVITY
How Enzymes Work

20.2 Enzyme Action

Nearly all enzymes are globular proteins. Each has a unique three-dimensional shape that recognizes and binds a small group of reacting molecules, which are called **substrates**. The tertiary structure of an enzyme plays an important role in how that enzyme catalyzes reactions.

Active Site

In a catalyzed reaction, an enzyme must first bind to a substrate in a way that favors catalysis. A typical enzyme is much larger than its substrate. However, within its large tertiary structure, there is a region called the **active site** where the enzyme binds a substrate or substrates and catalyzes the reaction. This active site is often a small pocket that closely fits the structure of the substrate. (See Figure 20.2.) Within the active site, the side chains of amino acids bind the substrate with hydrogen bonds, salt bridges, or hydrophobic attractions. The active site of a particular enzyme fits the shape of only a few types of substrates, which makes enzymes very specific about the type of substrate they bind.

Some enzymes show absolute specificity by only catalyzing one reaction of one specific substrate. Other enzymes catalyze a reaction for a group of substrates. Still other enzymes catalyze a reaction for a specific type of bond in a substrate. Types of enzyme specificity are listed in Table 20.2.

Enzyme Catalyzed Reaction

The proper alignment of a substrate within the active site forms an **enzyme–substrate (ES) complex**. This combination of enzyme and substrate provides an alternative pathway for the reaction that has a lower activation energy. Within the active site, amino acid side chains take part in catalyzing the chemical reaction. For example, acidic and basic side chains remove protons from or provide protons for the substrate. As soon as the catalyzed reaction is complete,

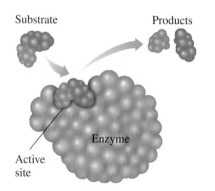

Substrate   Products

Enzyme

Active
site

FIGURE 20.2

On the surface of an enzyme, a small region called an active site binds a substrate and catalyzes a reaction of that substrate.

Q Why does an enzyme catalyze a reaction of only certain substrates?

TABLE 20.2

Types of Enzyme Specificity

Type	Reaction Type	Example
Absolute	Catalyze one type of reaction for a single substrate	Urease catalyzes only the hydrolysis of urea
Group	Catalyze one type of reaction for similar substrates	Hexokinase adds a phosphate group to hexoses
Linkage	Catalyze one type of reaction for a specific type of bond	Chymotrypsin catalyzes the hydrolysis of peptide bonds

the products are released from the enzyme so it can bind to a new substrate molecule. We can write the catalyzed reaction of an enzyme (E) with a substrate (S) to form product (P) as follows:

Step 1   E + S $\rightleftharpoons$ ES

Step 2   ES $\longrightarrow$ E + P

$\overline{\qquad\qquad\qquad\qquad\qquad\qquad\qquad\qquad\qquad}$

              E + S $\rightleftharpoons$ ES $\longrightarrow$ E + P

              Enzyme + substrate   ES complex   Enzyme + product

Let's consider the hydrolysis of sucrose by sucrase. When sucrose binds to the active site of sucrase, the glycosidic bond of sucrose is placed into a geometry favorable for reaction. The amino acid side chains catalyze the cleavage of the sucrose to give the products glucose and fructose.

Sucrase + sucrose $\rightleftharpoons$ sucrase sucrose complex $\longrightarrow$
   E   +   S   ES complex   sucrase + glucose + fructose
                                                        E   +   P$_1$ +   P$_2$

Because the structures of the products are no longer attracted to the active site, they are released and the sucrase binds another sucrose substrate. (See Figure 20.3.)

Lock-and-Key and Induced Fit Models

In an early theory of enzyme action called the **lock-and-key model**, the active site is described as having a rigid, nonflexible shape. Thus only those substrates with shapes that fit exactly into the active site are able to bind with that enzyme. The shape of the active site is analogous to a lock, and the proper substrate is the key that fits into the lock. (See Figure 20.4a.)

While the lock-and-key model explains the binding of substrates for many enzymes, certain enzymes have a broader range of specificity than the lock and key model allows. In the **induced-fit model**, there is an interaction between both the enzyme and substrate. (See Figure 20.4b.) The active site adjusts to fit the shape of

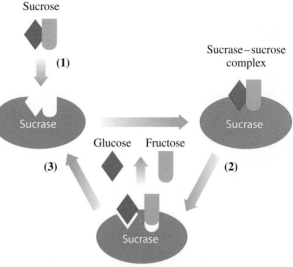

Sucrose

Sucrase–sucrose complex

FIGURE 20.3

(**1**) Sucrose binds to the active site in sucrase. (**2**) Sucrose is hydrolyzed by the enzymatic action of sucrase. (**3**) The products glucose and fructose dissociate from the active site, which allows the enzyme to bind to another sucrose molecule.

Q **Why does the enzyme-catalyzed hydrolysis of sucrose go faster than the hydrolysis of sucrose in the chemistry laboratory?**

MC GOB   Models of Enzyme Action

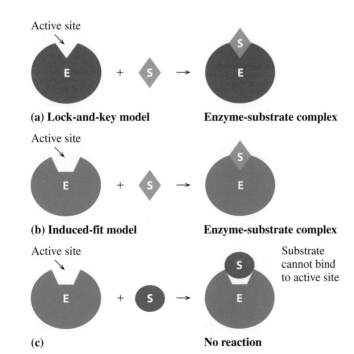

(a) Lock-and-key model   **Enzyme-substrate complex**

(b) Induced-fit model   **Enzyme-substrate complex**

(c)   **No reaction**

FIGURE 20.4

(**a**) In the lock-and-key model, a substrate fits the shape of the active site and forms an enzyme–substrate complex. (**b**) In the induced-fit model, a flexible active site and substrate adapt to provide a close fit to a substrate and proper orientation for reaction. (**c**) A substrate that does not fit or induce a fit in the active site cannot undergo catalysis by the enzyme.

Q **How does the induced-fit model differ from the lock-and-key model?**

■ **Isoenzymes as Diagnostic Tools**

Isoenzymes are different forms of an enzyme that catalyze the same reaction in different cells or tissues of the body. They consist of quaternary structures with slight variations in the amino acids in the polypeptide subunits. For example, there are five isoenzymes of *lactate dehydrogenase (LDH)* that catalyze the conversion between lactate and pyruvate.

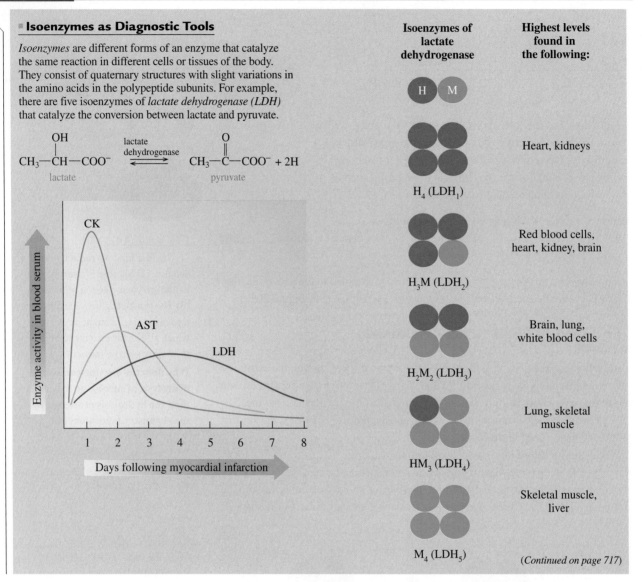

the substrate more closely. At the same time the substrate adjusts its shape to better adapt to the geometry of the active site. As a result, the reacting section of the substrate becomes aligned exactly with the groups in the active site that catalyze the reaction.

In the induced-fit model, substrate and enzyme work together to acquire a geometrical arrangement that lowers the activation energy. A different substrate would not induce these structural changes and no catalysis would occur. (See Figure 20.4c.)

SAMPLE PROBLEM 20.2

■ **The Enzyme Active Site**

What is the function of the active site in an enzyme?

SOLUTION

The active site in an enzyme binds the substrate and contains the amino acid side chains that bind the substrate and catalyze the reaction.

▪ Isoenzymes as Diagnostic Tools (*continued*)

Each LDH isoenzyme contains a mix of polypeptide subunits, M and H. In the liver and muscle, lactate is converted to pyruvate by a LDH_5 isoenzyme with M subunits designated M_4. In the heart, the same reaction is catalyzed by a LDH_1 isoenzyme (H_4) containing four H subunits. Different combinations of the M and H subunits are found in the LDH isoenzymes of the brain, red blood cells, kidney, and white blood cells.

The different forms of an enzyme allow a medical diagnosis of damage or disease to a particular organ or tissue. In healthy tissues, isoenzymes function within the cells. However, when a disease damages a particular organ, cells die, which releases cell contents including the isoenzymes into the blood. Measurements of the elevated levels of specific isoenzymes in the blood serum help to identify the disease and its location in the

body. For example, an elevation in the serum LDH_5, which is the M_4 isoenzyme of lactate dehydrogenase, indicates liver damage or disease. When a myocardial infarction (MI), or heart attack, damages the cells in heart muscle, an increase in the level of LDH_1 (H_4) isoenzyme is detected in the blood serum. (See Table 20.3.)

Another isoenzyme used diagnostically is creatine kinase (CK), which consists of two types of polypeptide subunits. One subunit (B) is prevalent in the brain and the other predominates in skeletal muscle (M). Normally only the CK_3 is present in low amounts in the blood serum. However, in a patient who has suffered an MI, the levels of CK_2 will be elevated soon after the heart attack. Table 20.4 lists some enzymes used to diagnose tissue damage and diseases of certain organs.

TABLE 20.3

Isoenzymes of Lactate Dehydrogenase and Creatinine Kinase

Isoenzyme	Abundant in	Subunits
Lactate Dehydrogenase (LDH)		
LDH_1	Heart, kidneys	H_4
LDH_2	Red blood cells, heart, kidney, brain	MH_3
LDH_3	Brain, lung, white blood cells	M_2H_2
LDH_4	Lung, skeletal muscle	M_3H
LDH_5	Skeletal muscle, liver	M_4
Creatinine Kinase (CK)		
CK_1	Brain, lung	BB
CK_2	Heart	MB
CK_3	Skeletal muscle, red blood cells	MM

TABLE 20.4

Serum Enzymes Used in Diagnosis of Tissue Damage

Condition	Diagnostic Enzymes Elevated
Heart attack, or liver disease (cirrhosis, hepatitis)	Lactate dehydrogenase (LDH) Aspartate transaminase (AST)
Heart attack	Creatine kinase (CK)
Hepatitis	Alanine transaminase (ALT)
Liver (carcinoma) or bone disease (rickets)	Alkaline phosphatase (ALP)
Pancreatic disease	Amylase, cholinesterase, lipase (LPS)
Prostate carcinoma	Acid phosphatase (ACP)

STUDY CHECK

How do the lock-and-key and the induced-fit models differ in their description of the active site in an enzyme?

QUESTIONS AND PROBLEMS

MC
GOB

▪ Enzyme Action

20.11 Match the following three terms, (1) enzyme–substrate complex, (2) enzyme, and (3) substrate, with these phrases:
  a. has a tertiary structure that recognizes the substrate
  b. the combination of an enzyme with the substrate
  c. has a structure that fits the active site of an enzyme

20.12 Match the following three terms, (1) active site, (2) lock-and-key model, and (3) induced-fit model, with these phrases:
  a. the portion of an enzyme where catalytic activity occurs
  b. an active site that adapts to the shape of a substrate
  c. an active site that has a rigid shape

20.13 a. Write an equation that represents an enzyme-catalyzed reaction.
  b. How is the active site different from the whole enzyme structure?

20.14 a. Why does an enzyme speed up the reaction of a substrate?

b. After the products have formed, what happens to the enzyme?

20.15 What are isoenzymes?

20.16 How is the LDH isoenzyme in the heart different from LDH isoenzyme in the liver?

20.17 A patient arrives in emergency complaining of chest pains. What enzymes would you test for in the blood serum?

20.18 A patient who is an alcoholic has elevated levels of LDH and AST. What condition might be indicated?

LEARNING GOAL

Describe the effect of temperature, pH, concentration of enzyme, and concentration of substrate on enzyme activity.

20.3 Factors Affecting Enzyme Activity

The **activity** of an enzyme describes how fast an enzyme catalyzes the reaction that converts a substrate to product. This activity is strongly affected by reaction conditions, which include the temperature, pH, concentration of the substrate, and concentration of the enzyme.

Temperature

MC
GOB   Denaturation and Enzyme Activity

Enzymes are very sensitive to temperature. At low temperatures, most enzymes show little activity because there is not a sufficient amount of energy for the catalyzed reaction to take place. At higher temperatures, enzyme activity increases as reacting molecules move faster to cause more collisions with enzymes. Enzymes are most active at **optimum temperature**, which is 37°C or body temperature for most enzymes. (See Figure 20.5.) At temperatures above 50°C, the tertiary structure and thus the shape of most proteins is destroyed, which causes a loss in enzyme activity. For this reason, equipment in hospitals and laboratories is sterilized in autoclaves where the high temperatures denature the enzymes in harmful bacteria.

pH

Enzymes are most active at their **optimum pH**, the pH that maintains the proper tertiary structure of the protein. (See Figure 20.6.) A pH value above or below the

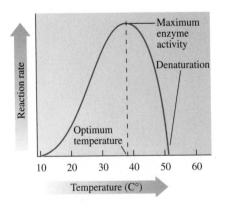

FIGURE 20.5

An enzyme attains maximum activity at its optimum temperature, usually 37°C. Lower temperatures slow the rate of reaction and temperatures above 50°C cause denaturation of the enzyme protein and loss of catalytic activity.

Q **Why is 37°C the optimum temperature for many enzymes?**

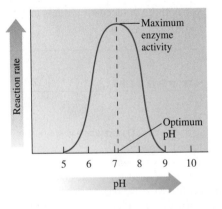

FIGURE 20.6

Enzymes are most active at their optimum pH. At a higher or lower pH, denaturation of the enzyme causes a loss of catalytic activity.

Q **Why does the digestive enzyme pepsin have an optimum pH of 2?**

TABLE 20.5

Optimum pH for Selected Enzymes

Enzyme	Location	Substrate	Optimum pH
pepsin	stomach	peptide bonds	2
urease	liver	urea	5
sucrase	small intestine	sucrose	6.2
pancreatic amylase	pancreas	amylose	7
trypsin	small intestine	peptide bonds	8
arginase	liver	arginine	9.7

optimum pH causes a change in the three-dimensional structure of the enzyme that disrupts the active site. As a result the enzyme cannot bind substrate properly and no reaction occurs.

Enzymes in most cells have optimum pH values at physiological pH values around 7.4. However, enzymes in the stomach have a low optimum pH because they hydrolyze proteins at the acidic pH in the stomach. For example, pepsin, a digestive enzyme in the stomach has an optimum pH of 2. Between meals, the pH in the stomach is 4 or 5 and pepsin shows little or no digestive activity. When food enters the stomach, the secretion of HCl lowers the pH to about 2, which activates pepsin.

If small changes in pH are corrected, an enzyme can regain its structure and activity. However, large variations from optimum pH permanently destroy the structure of the enzyme. Table 20.5 lists the optimum pH values for selected enzymes.

Enzyme and Substrate Concentration

In any catalyzed reaction, the substrate must first bind with the enzyme to form the substrate–enzyme complex. Increasing the enzyme concentration when the substrate concentration remains constant increases the rate of the catalyzed reaction and thus enzyme activity. At higher concentrations more enzyme molecules are available to bind and catalyze the reaction of substrate molecules. As long as the substrate concentration is greater than the enzyme concentration, there is a direct relationship between the enzyme concentration and enzyme activity. (See Figure 20.7a.)

MC GOB Enzyme and Substrate Concentrations

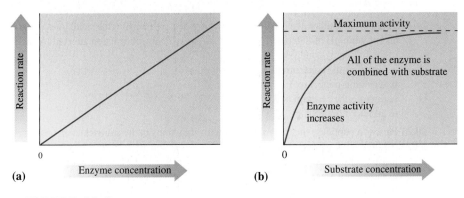

(a) Enzyme concentration

(b) Substrate concentration

Maximum activity

All of the enzyme is combined with substrate

Enzyme activity increases

FIGURE 20.7

(**a**) The rate of reaction increases when the enzyme concentration increases with substrate. (**b**) Increasing the substrate concentration increases the rate of reaction until the enzyme molecules are saturated with substrate.

Q **What happens to the rate of reaction when substrate saturates the enzyme?**

When enzyme concentration is kept constant, increasing the substrate concentration increases the rate of the catalyzed reaction as long as there are more enzyme molecules present than substrate molecules. At some point an increase in substrate concentration saturates the enzyme. With all the available enzyme molecules bonded to substrate, the rate of the catalyzed reaction reaches its maximum. Adding more substrate molecules cannot increase the rate further. (See Figure 20.7b.)

SAMPLE PROBLEM 20.3

■ Factors Affecting Enzymatic Activity

Describe what effect the changes in parts a and b would have on the rate of the reaction catalyzed by urease.

$$H_2N\!-\!\overset{\overset{\displaystyle O}{\|}}{C}\!-\!NH_2 \ + \ H_2O \ \xrightarrow{\text{urease}} \ 2NH_3 \ + \ CO_2$$
urea

a. increasing the urea concentration
b. lowering the temperature to 10°C

SOLUTION

a. An increase in urea concentration will increase the rate of reaction until all of the enzyme molecules are bound to the urea substrate. No further increase in rate occurs.
b. Because 10°C is lower than the optimum temperature of 37°C, the lower temperature will decrease the rate of the reaction.

STUDY CHECK

If urease has an optimum pH of 5, what is the effect of lowering the pH to 3?

QUESTIONS AND PROBLEMS

■ Factors Affecting Enzyme Action

20.19 Trypsin, a peptidase that hydrolyzes polypeptides, functions in the small intestine at an optimum pH of 8. How is the rate of a trypsin-catalyzed reaction affected by each of the following conditions?
a. lowering the concentration of polypeptides
b. changing the pH to 3
c. running the reaction at 75°C
d. adding more trypsin

20.20 Pepsin, a peptidase that hydrolyzes proteins, functions in the stomach at an optimum pH of 2. How is the rate of a pepsin-catalyzed reaction affected by each of the following conditions?
a. increasing the concentration of proteins
b. changing the pH to 5
c. running the reaction at 0°C
d. using less pepsin

20.21 The following graph shows the curves for pepsin, urease, and trypsin. Estimate the optimum pH for each.

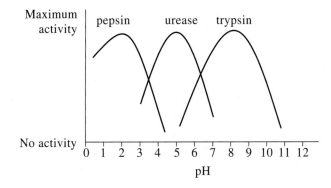

20.22 Refer to the graph in problem 20.21 to determine if the reaction rate in each condition will be at the optimum rate or not.

    a. trypsin, pH 5   **b.** urease, pH 5   **c.** pepsin, pH 4

    d. trypsin, pH 8   **e.** pepsin, pH 2

LEARNING GOAL

Describe reversible and irreversible inhibition.

20.4 Enzyme Inhibition

Molecules called **inhibitors** cause enzymes to lose catalytic activity. Although inhibitors act differently, they all prevent the active site from binding with a substrate. Some inhibitions are *reversible* which means that the enzyme regains activity when the inhibitor dissociates from the enzyme. In an *irreversible inhibition*, an inhibitor cannot be removed, which makes the loss of enzyme activity irreversible.

Reversible Inhibition

Reversible inhibition can be competitive or noncompetitive. In competitive inhibition, an inhibitor competes for the active site, whereas in noncompetitive inhibition, the inhibitor acts on a site that is not the active site.

A **competitive inhibitor** has a structure that is so similar to the substrate it can bond to the enzyme just like the substrate. Thus the competitive inhibitor competes with the substrate for the active site on the enzyme. As long as the inhibitor occupies the active site, the substrate cannot bind to the enzyme and no reaction takes place. (See Figure 20.8.)

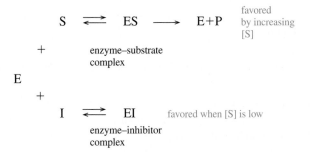

As long as the concentration of the inhibitor is substantial, there is a loss of enzyme activity. However, increasing the substrate concentration displaces more of

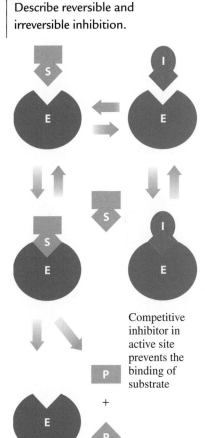

Competitive inhibitor in active site prevents the binding of substrate

FIGURE 20.8

With a structure similar to the substrate for an enzyme, a competitive inhibitor also fits the active site and competes with the substrate when both are present.

Q Why does increasing the substrate concentration reverse the inhibition by a competitive inhibitor?

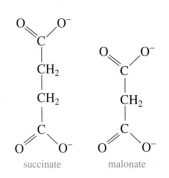

succinate   malonate

the inhibitor molecules. As more enzyme molecules bind to substrate (ES), enzyme activity is regained.

Malonate is a competitive inhibitor of the enzyme succinate dehydrogenase. Because malonate has a structure similar to succinate, the two substances compete for the active site on the dehydrogenase. As long as malonate (inhibitor) occupies the active site, no reaction occurs. When more succinate is added, more active sites will fill with substrate, and there will be less inhibition.

Some bacterial infections are treated with competitive inhibitors called antimetabolites. Sulfanilamide, one of the first sulfa drugs, competes with PABA (*p*-aminobenzoic acid), which is an essential substance (metabolite) in the growth cycle of bacteria.

Substrate Needed for Bacterial Growth   **Inhibitor**

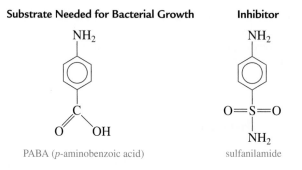

PABA (*p*-aminobenzoic acid)   sulfanilamide

 Enzyme Inhibition

The structure of a **noncompetitive inhibitor** does not resemble the substrate and does not compete for the active site. Instead a noncompetitive inhibitor binds to a site on the enzyme that is not the active site. When the noncompetitive inhibitor is bonded to the enzyme, the shape of the enzyme is distorted. Inhibition occurs because the substrate cannot fit in the active site, or it does not fit properly. Without the proper alignment of substrate with the amino acid side groups, no catalysis can take place. (See Figure 20.9.)

 SELF-STUDY ACTIVITY
Enzyme Inhibition

Because a noncompetitive inhibitor is not competing for the active site, the addition of more substrate does not reverse this type of inhibition. However, enzyme activity can be regained by lowering the concentration of the noncompetitive inhibitor making more enzyme molecules available. Examples of noncompetitive inhibitors are the heavy metal ions Pb^{2+}, Ag^+, and Hg^{2+} that bond with amino acid side groups such as $-COO^-$, or $-OH$. Catalytic activity is restored when chemical reagents remove the inhibitors.

Irreversible Inhibition

In irreversible inhibition, a molecule causes an enzyme to lose all enzymatic activity. Most irreversible inhibitors are toxic substances that destroy enzymes. Usually an irreversible inhibitor forms a covalent bond with an amino acid side group within the active site, which prevents the substrate from entering the active site or prevents catalytic activity.

Insecticides and nerve gases act as irreversible inhibitors of acetylcholinesterase, an enzyme needed for nerve conduction. The compound DFP (diisopropyl fluorophosphate) forms a covalent bond with the side chain $-CH_2OH$ of serine in the active site. When acetylcholinesterase is inhibited, the transmission of nerve impulses is blocked, and paralysis occurs.

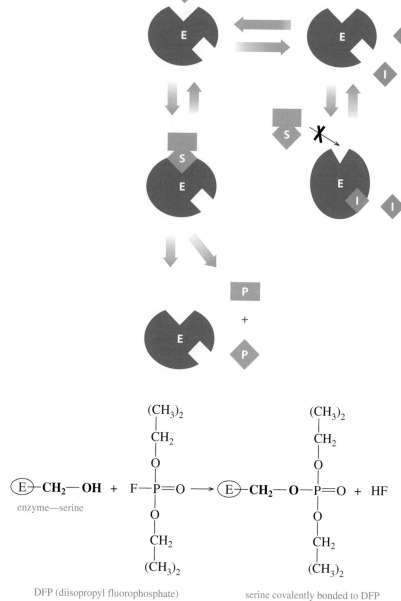

FIGURE 20.9

A noncompetitive inhibitor (I) binds to an enzyme at a site other than the active site, which distorts the enzyme and prevents the proper binding and catalysis of the substrate at the active site.

Q **Why won't an increase in substrate concentration reverse the inhibition by a noncompetitive inhibitor?**

DFP (diisopropyl fluorophosphate)   serine covalently bonded to DFP

Antibiotics produced by bacteria, mold, or yeast are irreversible inhibitors used to inhibit bacterial growth. For example, penicillin inhibits an enzyme needed for the formation of cell walls in bacteria, but not human cell membranes. Without a complete cell wall, bacteria cannot survive, and the infection is stopped. However, some bacteria are resistant to penicillin because they produce penicillinase, an enzyme that breaks down penicillin. Over the years, derivatives of penicillin to which bacteria have not yet become resistant have been produced. Some irreversible enzyme inhibitors are listed in Table 20.6.

TABLE 20.6

Selected Irreversible Enzyme Inhibitors

Name	Structure	Source	Inhibitory Action
cyanide	CN^-	bitter almonds	Bonds to metal ions in enzymes in the electron transport chain
sarin	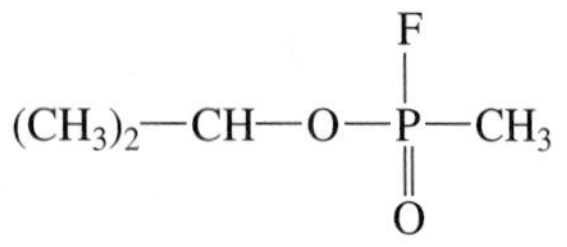 $(CH_3)_2-CH-O-\overset{\overset{\textstyle F}{\vert}}{\underset{\underset{\textstyle O}{\parallel}}{P}}-CH_3$	nerve gas	Similar to DFP
parathion	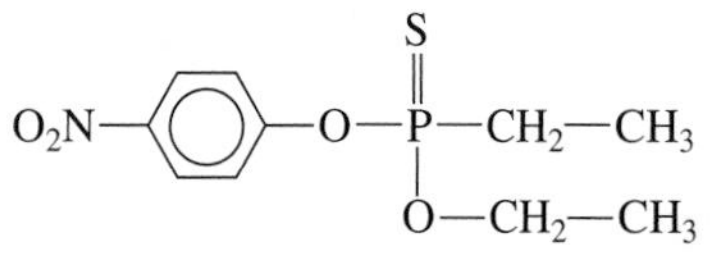 $O_2N-\!\!\!\bigcirc\!\!\!-O-\overset{\overset{\textstyle S}{\parallel}}{\underset{\underset{\textstyle O-CH_2-CH_3}{\vert}}{P}}-CH_2-CH_3$	insecticide	Similar to DFP
penicillin	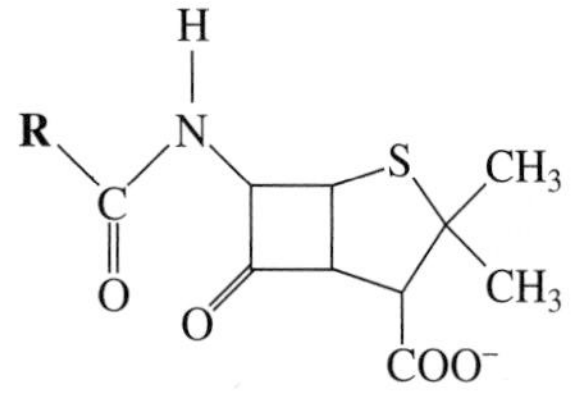	*Penicillium* fungus	Inhibits enzymes that build cell walls in bacteria

R Groups for Penicillin Derivatives

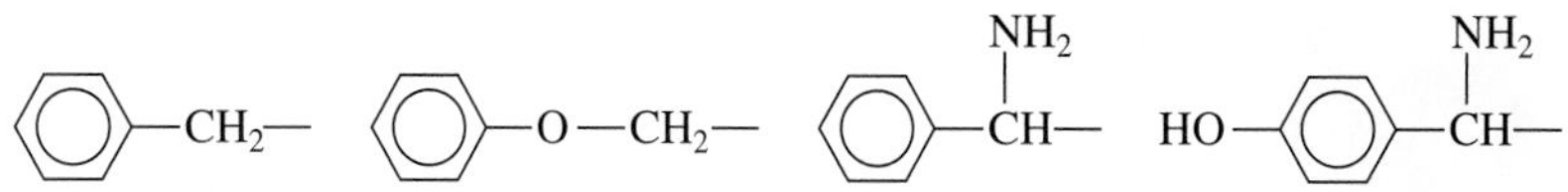

$\bigcirc\!\!\!-CH_2-$ $\bigcirc\!\!\!-O-CH_2-$ $\bigcirc\!\!\!-\overset{\overset{\textstyle NH_2}{\vert}}{CH}-$ $HO-\!\!\!\bigcirc\!\!\!-\overset{\overset{\textstyle NH_2}{\vert}}{CH}-$

SAMPLE PROBLEM 20.4

■ **Enzyme Inhibition**

State the type of reversible inhibition in the following:

a. The inhibitor has a structure that is similar to the substrate.
b. This inhibitor binds to the surface of the enzyme, changing its shape in such a way that it cannot bind to substrate.

SOLUTION

a. competitive inhibition
b. noncompetitive inhibition

STUDY CHECK

Hydrogen cyanide (HCN) forms covalent bonds with catalase, an enzyme that contains iron (Fe^{3+}). Is HCN a competitive or noncompetitive inhibitor?

QUESTIONS AND PROBLEMS

■ **Enzyme Inhibition**

20.23 Indicate whether the following describe a competitive or a noncompetitive enzyme inhibitor:
  a. The inhibitor has a structure similar to the substrate.
  b. The effect of the inhibitor cannot be reversed by adding more substrate.
  c. The inhibitor competes with the substrate for the active site.
  d. The structure of the inhibitor is not similar to the substrate.
  e. The addition of more substrate reverses the inhibition.

20.24 Oxaloacetate is an inhibitor of succinate dehydrogenase:

$$
\begin{array}{cc}
COO^- & COO^- \\
| & | \\
CH_2 & CH_2 \\
| & | \\
CH_2 & C\!\!=\!\!O \\
| & | \\
COO^- & COO^- \\
\text{succinate} & \text{oxaloacetate}
\end{array}
$$

    a. Would you expect oxaloacetate to be a competitive or a noncompetitive inhibitor? Why?

    b. Would oxaloacetate bind to the active site or elsewhere on the enzyme?

    c. How would you reverse the effect of the inhibitor?

20.25 Methanol and ethanol are oxidized by alcohol dehydrogenase. In methanol poisoning, a high concentration of ethanol is given intravenously as an antidote.

    a. Compare the structures of methanol and ethanol.

    b. Would ethanol compete for the active site or bind to a different site?

    c. Would ethanol be a competitive or noncompetitive inhibitor of methanol oxidation?

20.26 In humans, the antibiotic amoxycillin (a type of penicillin) is used to treat certain bacterial infections.

    a. Does the antibiotic inhibit enzymes in humans?

    b. Why does the antibiotic kill bacteria, but not humans?

    c. Are antibiotics reversible or irreversible inhibitors?

20.5 Control of Enzyme Activity

In an enzyme-catalyzed reaction, compounds are produced in the amounts and at the times they are needed. This means that the rate of a catalyzed reaction must be controlled so it can speed up when more molecules of a compound are needed and slow down when that compound is no longer needed.

Zymogens

Many enzymes are active as soon as they are synthesized and acquire their tertiary structure. However, **zymogens** or *proenzymes* are produced as an inactive form and stored in an organ. Zymogens are transported to the part of the body where they are needed and activated by a reaction that removes a peptide section from the zymogen.

Most protein hormones such as insulin as well as digestive enzymes and enzymes needed for blood clotting are initially synthesized as zymogens. (See Table 20.7.) For example, the hormone *insulin* is synthesized in an inactive form

LEARNING GOAL

Describe the role of zymogens, feedback control, and allosteric enzymes in regulating enzyme activity.

SELF-STUDY ACTIVITY
Enzyme Inhibition

TABLE 20.7

Example of Zymogens and Their Active Forms

Zymogen	Produced in	Activated in	Active Form
proinsulin	pancreas	blood	insulin
chymotrypsinogen	pancreas	small intestine	chymotrypsin
pepsinogen	gastric mucosa	stomach	pepsin
trypsinogen	pancreas	small intestine	trypsin
fibrinogen	blood	damaged tissues	fibrin
prothrombin	blood	damaged tissues	thrombin

Proinsulin (Zymogen)

Activation

Insulin (Active)

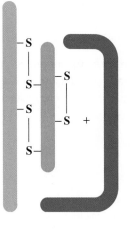

Regulating Enzyme Action

called *proinsulin*. To form insulin (Chapter 19), a polypeptide containing 33 amino acids is removed.

The zymogen of a digestive enzyme called pepsinogen is produced in the gastric mucosal cells that line the stomach. As food enters the stomach, HCl is secreted. The acidic conditions of about pH 2 cause the removal of a peptide containing 42 amino acids from pepsinogen to form pepsin, which is the active enzyme that digests proteins in our foods.

$$\text{pepsinogen} \xrightarrow{\text{H}^+} \text{pepsin} + \text{peptides} + \text{amino acids}$$

Several *digestive enzymes* such as *trypsinogen*, *chymotrypsinogen*, and *procarboxypeptidase* are produced as inactive enzymes and stored in the pancreas. After food is ingested and reaches the small intestine, hormones trigger the release of the zymogens of the digestive enzymes from the pancreas. When the zymogens enter the small intestine, they are converted into active enzymes by proteases that remove peptide sections from their protein chains. The result is the active form of the enzyme. For example, an enzyme called enteropeptidase removes a hexapeptide from trypsinogen to give active trypsin. Trypsin in turn removes peptide sections from the zymogens chymotrypsinogen and procarboxypeptidase to give the active forms chymotrypsin and carboxypeptidase.

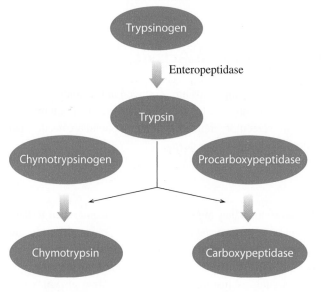

If these digestive enzymes were active in the pancreas, they would digest the proteins that make up the pancreas. This can lead to conditions such as pancreatitis, which is an inflammation of the pancreas.

Feedback Control

Certain enzymes known as **allosteric enzymes** are capable of binding a regulator molecule that is different from the substrate. The binding of a regulator causes a change in the shape of the enzymes and therefore in the active site. There are both positive and negative regulators. A *positive regulator* speeds up a reaction by causing a change in the shape of the active site that permits the substrate to bind more effectively. A *negative regulator* slows down the rate of catalysis by preventing the proper binding of the substrate. In **feedback control**, the end

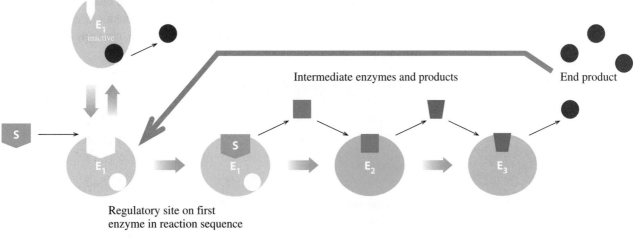

Regulatory site on first
enzyme in reaction sequence

FIGURE 20.10

In feedback control, end product binds to a regulatory site on the first enzyme in the reaction pathway, which prevents the formation of all intermediate compounds needed in the synthesis of the end product.

Q **Why don't the intermediate enzymes in a reaction sequence have regulatory sites?**

product acts as a negative regulator. (See Figure 20.10.) When the end product is present in sufficient amounts for the cell, some end product molecules bind to the first enzyme in the reaction pathway (E_1), which is an allosteric enzyme. The inhibition of the reaction of the initial substrate stops the production of any intermediate compounds in the reaction pathway. The entire enzyme-catalyzed reaction sequence shuts down.

When the end product is needed again, the regulator dissociates from the allosteric enzyme (E_1). The enzyme returns to its active form and catalysis of the initial substrate begins once again. Thus feedback control allows the reaction only when the end product is needed by the cell. This prevents the accumulation of unneeded end product, thereby conserving the materials in the cell.

Let's look at feedback control in a reaction pathway with five enzymes that converts the amino acid threonine to isoleucine, another amino acid. When isoleucine accumulates in the cell, it binds to the first enzyme in the pathway, threonine deaminase, E_1.

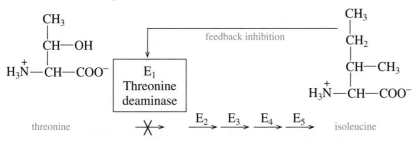

The binding of isoleucine changes the shape of the deaminase, which prevents threonine from binding with the active site. The entire reaction pathway is turned off. None of the intermediate products from the other enzymes in the pathway can inhibit the first enzyme. As isoleucine is utilized in the cell, its concentration decreases, which causes the threonine deaminase to release the end product inhibitor. The tertiary shape of the deaminase returns to its active form and the reaction sequence once again converts threonine to isoleucine.

SAMPLE PROBLEM 20.5

■ **Enzyme Regulation**

How is the rate of a reaction sequence regulated in feedback control?

SOLUTION

When the end product of a reaction sequence is produced at sufficient levels for the cell, some product molecules bind to the first enzyme in the sequence, which shuts down all the reactions that follow and stops the production of end product.

STUDY CHECK

Why is pepsin, a digestive enzyme, produced as a zymogen?

QUESTIONS AND PROBLEMS

■ **Control of Enzyme Activity**

20.27 Why are many of the enzymes that act on proteins synthesized as zymogens?

20.28 The zymogen trypsinogen produced in the pancreas is activated in the small intestine where it catalyzes the digestion and hydrolysis of proteins. Explain how the activation of the zymogen while still in the pancreas can lead to an inflammation of the pancreas called pancreatitis.

20.29 In feedback inhibition, how does the end product of a reaction sequence regulate enzyme activity?

20.30 Why don't the second or third enzymes in a reaction sequence function as regulatory enzymes?

20.31 How does an allosteric enzyme function as a regulatory enzyme?

20.32 What is the difference between a negative regulator and a positive regulator?

20.33 Indicate if the following statements describe (1) a zymogen, (2) a positive regulator, (3) a negative regulator, or (4) allosteric enzyme.
a. It slows down a reaction, but its shape is different from that of the substrate.
b. It is the first enzyme in a sequence of reactions leading to an end product.
c. It is produced as an inactive enzyme.

20.34 Indicate if the following statements describe (1) a zymogen, (2) a positive regulator, (3) a negative regulator, or (4) allosteric enzyme.
a. It is activated when a peptide section is removed from its protein chain.
b. It speeds up a reaction, but it is not the substrate.
c. When it binds to end product, it stops the formation of more end product.

LEARNING GOAL

■ Describe the types of cofactors found in enzymes.

■ 20.6 Enzyme Cofactors and Vitamins

Enzymes are known as **simple enzymes** when their functional forms consist only of proteins. However, many enzymes require small molecules or metal ions called **cofactors** to catalyze reactions properly. When the cofactor is a small organic molecule, it is known as a **coenzyme**. If an enzyme requires a cofactor, neither the protein structure nor the cofactor alone has catalytic activity.

Forms of Active Enzymes

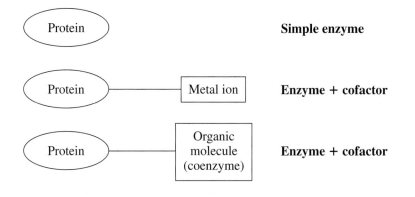

Metal Ions

Many enzymes must contain a metal ion to carry out their catalytic activity. The metal ions are bonded to one or more of the amino-acid side chains. The metal ions from the minerals that we obtain from foods in our diet have various functions in catalysis. Ions such as Fe^{2+} and Cu^{2+} are used by oxidases where they lose or gain electrons in oxidation and reduction reactions. Other metals ions such as Zn^{2+} stabilize the amino acid side chains during hydrolysis reactions. Some metal cofactors required by enzymes are listed in Table 20.8.

Let's look at an example of a metal ion in an enzyme-catalyzed reaction. The enzyme carboxypeptidase A cleaves the C terminal amino acid of a protein when that amino acid has a bulky hydrophobic or aromatic side chain. (See Figure 20.11.) With the substrate in the active site, the Zn^{2+} helps to stabilize the negative charge on the oxygen atom of the carbonyl group and promotes the hydrolysis of the peptide bond.

MC
GOB

Enzyme Cofactors and Vitamins

TABLE 20.8
Enzymes and the Metal Ions Required as Cofactors

Metal Ion Cofactor	Function	Enzyme
Cu^{2+}	oxidation–reduction	cytochrome oxidase
Fe^{2+}/Fe^{3+}	oxidation–reduction	catalase
	oxidation–reduction	cytochrome oxidase
Zn^{2+}	used with NAD^+	alcohol dehydrogenase
		carbonic anhydrase
		carboxypeptidase A
Mg^{2+}	hydrolyzes phosphate esters	glucose-6-phosphatase
Mn^{2+}	removes electrons	arginase
Ni^{2+}	hydrolyzes amides	urease

SAMPLE PROBLEM 20.6

▪ **Cofactors**

Indicate whether each of the following enzymes are active as a simple enzyme or require a cofactor.

a. a polypeptide that needs Mg^{2+} for catalytic activity

b. an active enzyme composed only of a polypeptide chain

c. an enzyme that consists of a quaternary structure attached to vitamin B_6

FIGURE 20.11

A Zn^{2+} cofactor aids in the hydrolysis of the peptide bond of a bulky C terminal amino acid by helping to stabilize the carbonyl oxygen.

Q **When would the Zn^{2+} be utilized as a cofactor by other enzymes?**

Carboxypeptidase A

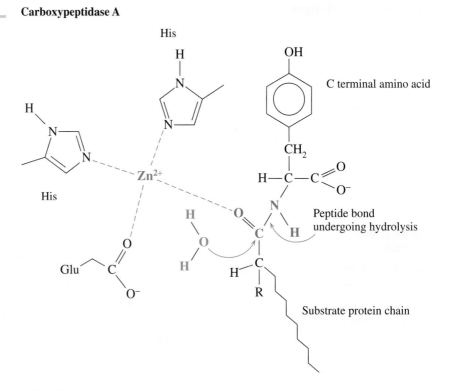

SOLUTION

a. The enzyme requires a cofactor.

b. An active enzyme that consists of only a polypeptide chain is a simple enzyme.

c. The enzyme requires a cofactor.

STUDY CHECK

Which of the nonprotein portions of the enzymes in Sample Problem 20.6 is a coenzyme?

Vitamins and Coenzymes

Vitamins are organic molecules that are essential for normal health and growth. They are required in trace amounts and must be obtained from the diet because they are not synthesized in the body. Before vitamins were discovered, it was known that lime juice prevented the disease scurvy in sailors and that cod liver oil could prevent rickets. In 1912, scientists found that, in addition to carbohydrates, fats, and proteins, certain other factors called vitamins must be obtained from the diet.

Vitamins are classified into two groups by solubility: water-soluble and fat-soluble. **Water-soluble vitamins** have polar groups such as —OH and —COOH, which make them soluble in the aqueous environment of the cells. The **fat-soluble vitamins** are nonpolar compounds, which are soluble in the fat (lipid) components of the body such as fat deposits and cell membranes.

Water Soluble Vitamins

Most water-soluble vitamins are not stored in the body and excess amounts are eliminated in the urine each day. Therefore, the water-soluble vitamins must be in the foods of our daily diets. Because many water-soluble vitamins are easily destroyed by heat, oxygen, and ultraviolet light, care must be taken in food preparation,

TABLE 20.9

Function, Sources, and Deficiency Symptoms of Water-Soluble Vitamins

Vitamin	Function	RDA	Source	Deficiency Symptom
B_1, thiamin	decarboxylation	2 mg	Liver, yeast, whole grain bread, cereals, milk	Beriberi: Fatigue, poor appetite, weight loss, nerve degeneration, heart failure
B_2, riboflavin	electron transfer	1.7 mg	Beef liver, chicken, eggs, green leafy vegetables, dairy foods, peanuts, whole grains	Dermatitis, dry skin, tongue inflammation, cataracts
B_3, niacin	oxidation–reduction	13–18 mg	Brewer's yeast, chicken, beef, fish, liver, brown rice, whole grains	Pellagra: dermatitis, muscle fatigue, loss of appetite, diarrhea, mouth sores, mental disorders
B_5, pantothenic acid	acetyl group transfer	10 mg	Salmon, beef, liver, eggs, brewer's yeast, whole grains, fresh vegetables	Fatigue, retarded growth, muscle cramps, anemia
B_6, pyridoxine	transamination	1 mg	Meat, liver, fish, nuts, whole grains, spinach	Dermatitis, fatigue, anemia, retarded growth
B_{12}, cobalamin	methyl group transfer	3 μg	Liver, beef, kidney, chicken, fish, milk products	Pernicious anemia, malformed red blood cells, nerve damage
C, ascorbic acid	collagen synthesis, healing of wounds	60 mg	Blueberries, oranges, strawberries, cantaloupe, tomatoes, peppers, broccoli, cabbage, spinach	Scurvy: bleeding gums, weakened connective tissues, slow-healing wounds, anemia
biotin	carboxylation	0.3 mg	Liver, yeast, nuts, eggs	Dermatitis, loss of hair, fatigue, anemia, nausea, depression
folic acid	methyl group transfer	0.4 mg	Green leafy vegetables, beans, meat, seafood, yeast, asparagus, whole grains enriched with folic acid	Abnormal red blood cells, anemia, intestinal-tract disturbances, loss of hair, growth impairment, depression, and spina bifida

processing, and storage. Because refining grains such as wheat causes a loss of vitamins, the Committee on Food and Nutrition of the National Research Council began to recommend dietary enrichment of cereal grains, in the 1940s. Thiamine (B_1), riboflavin (B_2), niacin, and iron were in the first group of added nutrients recommended. We now see the Recommended Daily Allowance (RDA) for many vitamins and minerals on food product labels such as cereals and bread.

The water-soluble vitamins are required by many enzymes as cofactors to carry out certain aspects of catalytic action. (See Table 20.9.) The coenzymes do not remain bonded to a particular enzyme, but are used over and over again by different enzymes to facilitate an enzyme-catalyzed reaction. (See Figure 20.12.) Thus, only small amounts of coenzymes are required in the cells.

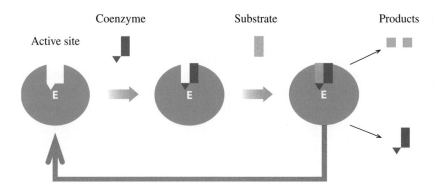

FIGURE 20.12

The active forms of many enzymes require the combination of the protein with a coenzyme.

Q **What is the function of water-soluble vitamins in enzymes?**

Thiamin (vitamin B_1) was the first B vitamin to be identified, thus the abbreviation B_1. The coenzyme thiamine pyrophosphate (TPP) is obtained when a synthetase adds two phosphate groups from ATP to the alcohol group of thiamine. (ATP is discussed in Section 22.2.)

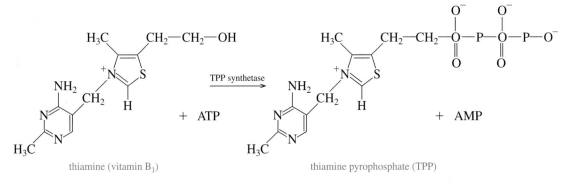

thiamine (vitamin B_1)   thiamine pyrophosphate (TPP)

The TPP coenzyme is involved in the decarboxylation reactions of α-keto carboxylic acids and reactions that cleave bonds to carbonyl carbons of α-hydroxyketones.

Riboflavin (vitamin B_2) is used to make the coenzymes flavin adenine dinucleotide (FAD) and flavin mononucleotide (FMN). The *ribo* part of the name comes from the sugar alcohol ribitol in the *riboflavin* molecule. The coenzymes FAD and FMN are used by enzymes called *flavoenzymes* to catalyze oxidation–reduction reactions of carbohydrates, fats, and proteins.

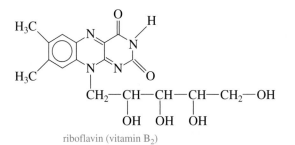

riboflavin (vitamin B_2)

Niacin (vitamin B_3) is a component of coenzymes nicotinamide adenine dinucleotide (NAD^+) and $NADP^+$, the phosphate form of NAD^+. The name niacin was assigned to the vitamin because its actual name, *nicotinic acid*, might be confused with nicotine. These coenzymes participate in oxidation–reduction, energy-production reactions in carbohydrate, fat, and protein metabolism.

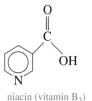

niacin (vitamin B_3)

Pantothenic acid (vitamin B_5) is part of a complex coenzyme known as coenzyme A. Coenzyme A transfers a two-carbon acetyl group from pyruvate to the citric acid cycle for the production of energy. Coenzyme A is also involved in the conversion of amino acids and lipids to glucose as well as the synthesis of cholesterol and steroid hormones.

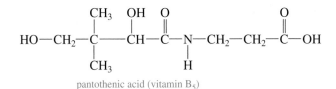

pantothenic acid (vitamin B_5)

Pyridoxine (vitamin B_6) and *pyridoxal* (an aldehyde) are converted to the coenzyme pyridoxal phosphate (PLP). The PLP coenzyme participates in enzyme-catalyzed reactions such as transaminations of amino acids and decarboxylations.

pyridoxine (vitamin B_6)   pyridoxal (vitamin B_6)   pyridoxal phosphate (PLP)

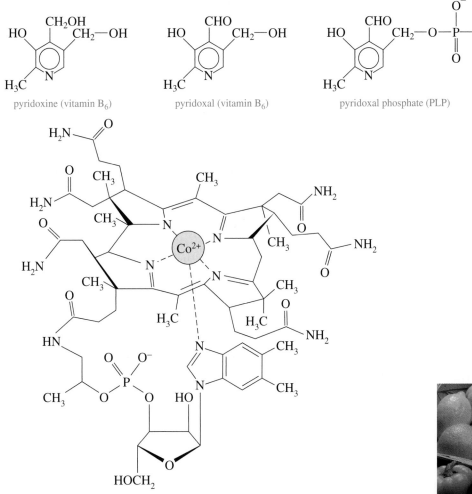

vitamin B_{12} (cobalamin)

Cobalamin (vitamin B_{12}) is a coenzyme consisting of four pyrrole rings with a cobalt ion (Co^{2+}) in the center. In its coenzyme form cobalamin participates in the transfer of methyl groups, molecular rearrangements, the formation of red blood cells, and the synthesis of acetylcholine for nerve cells. Because vitamin B_{12} is not present in plants, strict vegetarians can experience symptoms of pernicious anemia.

Ascorbic acid (vitamin C) has a simple chemical structure compared to most of the other vitamins. Its major function in the cells is its role in the synthesis of hydroxyproline and hydroxylysine, which are needed to form collagen. Collagen is the protein found in tendons, connective tissue, bone structure, and skin. (See Figure 20.13.)

FIGURE 20.13

Oranges, lemons, peppers, and tomatoes contain vitamin C or ascorbic acid.

Q **What happens to excess vitamin C that may be consumed in a day?**

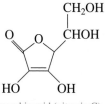

ascorbic acid (vitamin C)

Biotin is a coenzyme for enzymes that transfer a carboxyl group in the reaction of pyruvate to oxaloacetate or acetyl-CoA to malonyl-CoA, which occurs in the synthesis of fatty acids.

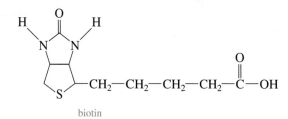

biotin

Folic acid (folate) is composed of a pyrimidine ring, *p*-aminobenzoic acid (PABA), and glutamate. The vitamin was discovered in the 1930s when people with a form of anemia were cured with extracts from liver or yeast. Folic acid is also found in spinach leaves, hence the name *folium*, Latin for leaf. In the cells, an enzyme called dihydrofolate reductase adds hydrogen atoms to the atoms in the heterocyclic ring of folate to yield the coenzyme tetrahydrofolate (THF). This coenzyme is used in reactions that transfer single-carbon groups and synthesize purines and pyrimidine to make DNA and RNA. It also plays a role with cobalamin in the production of red blood cells.

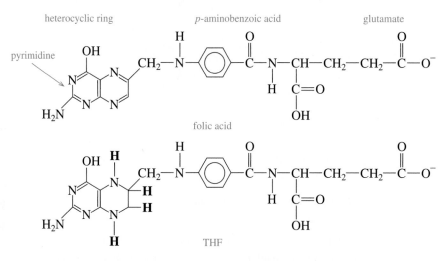

folic acid

THF

Some compounds related to folate bring about remissions in people with leukemia. For example, 4-aminofolate, referred to medically as *methotrexate*, acts as a competitive inhibitor of the dihydrofolate reductase that forms THF. The growth of cells, including tumor cells, depends on THF to build purines and thymine. By inhibiting the reductase enzyme with methotrexate, THF cannot be produced, and the rapid growth of tumor cells is blocked.

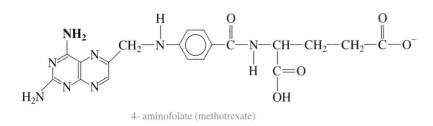

4- aminofolate (methotrexate)

Fat-Soluble Vitamins

The fat-soluble vitamins A, D, E, and K are not involved as coenzymes, but they are important in processes such as vision, formation of bone, protection from oxidation, and proper blood clotting. Because the fat-soluble vitamins are stored in the body and not eliminated, it is possible to take too much, which could be toxic. (See Table 20.10.)

Vitamin A consists of three different forms depending on the oxidation of the functional group: *retinol* (alcohol), *retinal* (aldehyde), and *retinoic acid* (carboxylic acid). Vitamin A is obtained from animal sources in the diet or the β-carotenes of plants, which are converted to vitamin A in the liver. The retinol in the retinas of the eyes accumulates in the rod and cone cells where it plays a role in vision. Vitamin A is also involved in the synthesis of RNA and glycoproteins. (See Figure 20.14.)

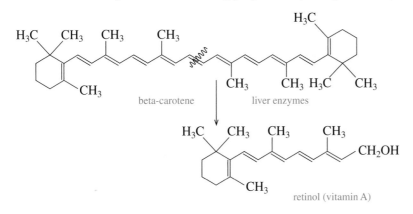

beta-carotene

liver enzymes

retinol (vitamin A)

FIGURE 20.14

Yellow and green fruits and vegetables contain vitamin A.

Q **Why is vitamin A called a fat-soluble vitamin?**

The most prevalent form of vitamin D is vitamin D_3 or *cholecalciferol*. Technically, this is not a vitamin because it is not required in the diet. In skin, vitamin D_3 is synthesized from 7-dehydrocholesterol by the ultraviolet rays from sunlight. In regions of limited sunlight, vitamin D_3 is added to milk products to avoid vitamin D_3 deficiency. Its function in the body is to regulate the absorption of phosphorus and calcium during bone growth.

TABLE 20.10

Function, Sources, and Deficiency Symptoms of Fat-Soluble Vitamins

Vitamin	Function	RDA	Sources	Deficiency Symptoms
A, retinol	Vision, synthesis of RNA	3 mg	Yellow and green fruits and vegetables	Night blindness, immune system repression, slowed growth rickets
D, cholecalciferol	Regulation of absorption of P and Ca	10 μg	Sunlight, cod liver oil, enriched milk, eggs	Rickets, weak bone structure, osteomalacia
E, tocopherol	Antioxidant, cell protection	10 mg	Meats, whole grains, vegetables	Hemolysis, anemia
K_2, menaquinone	Blood clotting	80 μg	Liver, spinach, cauliflower	Prolonged bleeding time, bruising

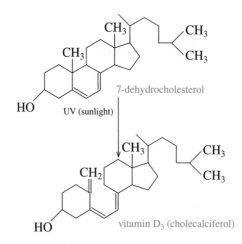

7-dehydrocholesterol

UV (sunlight)

vitamin D_3 (cholecalciferol)

Vitamin E or *tocopherol* has a major role in cells as an antioxidant but not much is known about the mechanism of its activity. It appears to protect the cells in the body by removing damaging chemicals and by preventing the oxidation of unsaturated fatty acids. It has been used to reduce the damage to the retinas that can be caused by the high oxygen levels needed for respiration by premature infants.

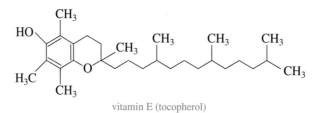

vitamin E (tocopherol)

Vitamin K_1, or *phylloquinone*, a substance found in plants, has a large saturated side chain. Vitamin K_2 or menaquinone, found in animals, has a very long unsaturated side chain. Vitamin K_2 takes part in the synthesis of zymogens needed for blood clotting.

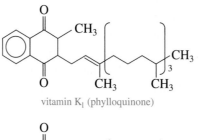

vitamin K_1 (phylloquinone)

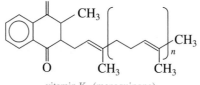

vitamin K_2 (menaquinone)

SAMPLE PROBLEM 20.7

■ Vitamins

Why do you need a certain amount of thiamine and riboflavin in your diet every day, but not vitamin A or D?

SOLUTION

Water-soluble vitamins like thiamine and riboflavin are not stored in the body whereas fat-soluble vitamins such as A and D are stored in the liver. Any excess of thiamine or riboflavin are eliminated in the urine and must be replenished each day from the diet.

STUDY CHECK

Why are fresh fruits rather than cooked fruits recommended as a source of vitamin C?

QUESTIONS AND PROBLEMS

■ Enzyme Cofactors and Vitamins

20.35 Is the enzyme described in each of the following statements a simple enzyme or one that requires a cofactor?
a. requires vitamin B_1 (thiamine)
b. needs Zn^{2+} for catalytic activity
c. its active form consists of two polypeptide chains

20.36 Is the enzyme described in each of the following statements a simple enzyme or one that requires a cofactor?
a. requires vitamin B_2 (riboflavin)
b. its active form is composed of 155 amino acids
c. uses Cu^{2+} during catalysis

20.37 Give the abbreviation for each of the following coenzymes:
a. tetrahydrofolate
b. nicotinamide adenine dinucleotide

20.38 Give the abbreviation for each of the following coenzymes:
a. flavin adenine dinucleotide
b. thiamine pyrophosphate

20.39 Identify a vitamin that is a component of each of the following coenzymes:
a. coenzyme A   **b.** tetrahydrofolate (THF)   **c.** NAD^+

20.40 Identify a vitamin that is a component of each of the following coenzymes:
a. thiamine pyrophosphate
b. FAD
c. pyridoxal phosphate

20.41 What vitamin may be deficient in the following conditions?
a. rickets   **b.** scurvy   **c.** pellagra

20.42 What vitamin may be deficient in the following conditions?
a. poor night vision   **b.** pernicious anemia   **c.** beriberi

20.43 The RDA for pyridoxine (vitamin B_6) is 2 mg daily. Why will it not improve your nutrition to take 100 mg of pyridoxine daily?

20.44 The RDA for vitamin A is 3 mg daily. What would happen if you took 25 mg of vitamin A every day?

20.45 What is the change in the structure of pyridoxine (B_6) that yields the coenzyme PLP?

20.46 What is the change in the structure of folate that yields the coenzyme THF?

CONCEPT MAP

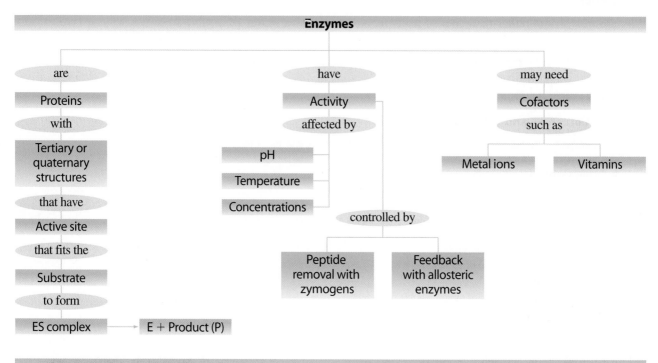

CHAPTER REVIEW

20.1 Enzymes

Enzymes are globular proteins that act as biological catalysts by lowering activation energy and accelerating the rate of cellular reactions. The names of most enzymes ending in *ase* describe the compound or reaction catalyzed by the enzyme. Enzymes are classified by the main type of reaction they catalyze, such as oxidoreductase, transferase, or isomerase.

20.2 Enzyme Action

Within the tertiary structure of an enzyme, a small pocket called the active site binds the substrates. In the lock-and-key model, a substrate precisely fits the shape of the active site. In the induced-fit model, substrates induce the active site to change structure to give an optimal fit by the substrate. In the enzyme–substrate complex, catalysis takes place when amino-acid side chains react with a substrate. The products are released and the enzyme is available to bind another substrate molecule.

20.3 Factors Affecting Enzyme Action

Enzymes are most effective at optimum temperature and pH, usually 37°C and 7.4. The rate of an enzyme-catalyzed reaction decreases considerably at temperature and pH values above or below the optimum. An increase in substrate concentration increases the reaction rate of an enzyme-catalyzed reaction. If an enzyme is saturated, adding more substrate will not increase the rate further.

20.4 Enzyme Inhibition

An inhibitor reduces the activity of an enzyme or makes it inactive. A competitive inhibitor has a structure similar to the substrate and competes for the active site. When the active site is occupied, the enzyme cannot catalyze the reaction of the substrate. A noncompetitive inhibitor attaches elsewhere on the enzyme, changing the shape of both the enzyme and its active site. As long as the noncompetitive inhibitor is attached, the altered active site cannot bind with substrate.

20.5 Control of Enzyme Activity

Insulin and digestive enzymes are produced as inactive forms called zymogens. They are converted to active forms by removing a peptide portion from their protein chains. The rate of an enzyme-catalyzed reaction can be increased or decreased by regulator molecules that bind to a regulator site on an allosteric enzyme. The regulator molecule changes the shape of the enzyme and therefore the shape of the active site. A positive regulator increases the rate whereas a negative regulator decreases the rate. In feedback inhibition, the end product of a reaction sequence binds to a regulator site on the first enzyme, which is an allosteric enzyme, to decrease product formation.

20.6 Enzyme Cofactors and Vitamins

Simple enzymes are biologically active as a protein only, whereas other enzymes require small organic molecules or metals ions called cofactors. A cofactor may be a metal ion such as Cu^{2+} or Fe^{2+}, or an organic molecule called a coenzyme. A vitamin is a small organic molecule needed for health and normal growth. Vitamins are obtained in small amounts through the foods in the diet. The water-soluble vitamins B and C function as coenzymes. The fat-soluble vitamins are A, D, E, and K. Vitamin A is important in vision, vitamin D for proper bone growth, vitamin E is an antioxidant, and vitamin K is required for proper blood clotting.

KEY TERMS

active site A pocket in a part of the tertiary enzyme structure that binds substrate and catalyzes a reaction.

activity The rate at which an enzyme catalyzes the reaction that converts substrate to product.

allosteric enzyme An enzyme that regulates the rate of a reaction when a regulator molecule attaches to a site other than the active site.

antibiotic An irreversible inhibitor produced by bacteria, mold, or yeast that is toxic to bacteria.

coenzyme An organic molecule, usually a vitamin, required as a cofactor in enzyme action.

cofactor A nonprotein metal ion or an organic molecule that is necessary for a biologically functional enzyme.

competitive inhibitor A molecule with a structure similar to the substrate that inhibits enzyme action by competing for the active site.

enzymes Globular proteins that catalyze biological reactions.

enzyme–substrate (ES) complex An intermediate consisting of an enzyme that binds to a substrate in an enzyme-catalyzed reaction.

fat-soluble vitamins Vitamins that are not soluble in water and can be stored in the liver and body fat.

feedback control A type of inhibition in which an end product inhibits the first enzyme in a sequence of enzyme-catalyzed reactions.

induced-fit model A model of enzyme action in which a substrate induces an enzyme to modify its shape to give an optimal fit with the substrate structure.

inhibitors Substances that make an enzyme inactive by interfering with its ability to react with a substrate.

lock-and-key model A model of an enzyme in which the substrate, like a key, exactly fits the shape of the lock, which is the specific shape of the active site.

noncompetitive inhibitor A substance that changes the shape of the enzyme, which prevents the active site from binding substrate properly.

optimum pH The pH at which an enzyme is most active.

optimum temperature The temperature at which an enzyme is most active.

simple enzyme An enzyme that is active as a polypeptide only.

substrate The molecule that reacts in the active site in an enzyme-catalyzed reaction.

vitamins Organic molecules, which are essential for normal health and growth, obtained in small amounts from the diet.

water-soluble vitamins Vitamins that are soluble in water, cannot be stored in the body, are easily destroyed by heat, ultraviolet light, and oxygen, and function as coenzymes.

zymogen An inactive form of an enzyme that is activated by removing a peptide portion from one end of the protein.

UNDERSTANDING THE CONCEPTS

20.47 Ethylene glycol ($HO-CH_2-CH_2-OH$) is a major component of antifreeze. In the body, it is first converted to $HOOC-CHO$ (oxoethanoic acid) and then to $HOOC-COOH$ (oxalic acid), which is toxic.

  a. What class of enzyme catalyzes both of the reactions of ethylene glycol?

  b. The treatment for the ingestion of ethylene glycol is an intravenous solution of ethanol. How might this help prevent toxic levels of oxalic acid in the body?

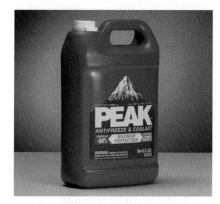

20.48 Adults who are lactose intolerant cannot break down the disaccharide in milk products. To help digest dairy food, a product known as Lactaid can be added to milk and the milk then refrigerated for 24 hours.

  a. What enzyme is present in Lactaid and what is the major class of this enzyme?

  b. What might happen to the enzyme if the digestion product were stored in a warm area?

20.49 Fresh pineapple contains the enzyme bromelain that degrades proteins.
  a. The directions on a Jello package say not to add fresh pineapple. However, canned pineapple where pineapple is heated to high temperatures can be added. Why?
  b. Fresh pineapple is used in a marinade to tenderize tough meat. Why?

20.50 Beano contains an enzyme that breaks down polysaccharides into smaller, more digestible sugars, which diminishes gas formation that can occur after eating foods such as vegetables and beans.

a. The label says "contains alpha-galactosidase." What class of enzyme is present in Beano?
b. What is the substrate for the enzyme?
c. The directions indicate you should not heat or cook with Beano. Why?

ADDITIONAL QUESTIONS AND PROBLEMS

For instructor-assigned homework, go to
www.masteringgob.com.

20.51 Why do the cells in the body have so many enzymes?

20.52 Are all the possible enzymes present at the same time in a cell?

20.53 How are enzymes different from the catalysts used in chemistry laboratories?

20.54 Why do enzymes function only under mild conditions?

20.55 Indicate whether each of the following would be a substrate (S) or an enzyme (E):
  a. lactose   **b.** lactase
  c. urease   **d.** trypsin
  e. pyruvate   **f.** transaminase

20.56 Indicate whether each of the following would be a substrate (S) or an enzyme (E):
  a. glucose   **b.** hydrolase
  c. maleate isomerase **d.** alanine
  e. amylose   **f.** amylase

20.57 Give the substrate of each of the following enzymes:
  a. urease   **b.** lactase
  c. aspartate transaminase **d.** phenylalanine hydroxylase

20.58 Give the substrate of each of the following enzymes:
  a. maltase   **b.** fructose oxidase
  c. phenolase   **d.** sucrase

20.59 Predict the major class of each of the following enzymes:
  a. acyltransferase **b.** oxidase
  c. lipase   **d.** decarboxylase

20.60 Predict the major class for each of the following enzymes:
  a. cis–trans isomerase **b.** reductase
  c. carboxylase   **d.** peptidase

20.61 How would the lock-and-key theory explain that sucrase hydrolyses sucrose, but not lactose?

20.62 How does the induced-fit model of enzyme action allow an enzyme to catalyze a reaction of a group of substrates?

20.63 If a blood test indicates a high level of LDH and CK, what could be the cause?

20.64 If a blood test indicates a high level of ALT, what could be the cause?

20.65 Indicate whether an enzyme is saturated or unsaturated in each of the following conditions:
  a. adding more substrate does not increase the rate of reaction
  b. doubling the substrate concentration doubles the rate of reaction

20.66 Indicate whether each of the following enzymes would be functional:
  a. pepsin, a digestive enzyme, at pH 2
  b. an enzyme at 37°C, if the enzyme is from a type of thermophilic bacteria that have an optimum temperature of 100°C

20.67 How does a reversible inhibition differ from an irreversible inhibition?

20.68 How does a competitive reversible inhibition differ from a noncompetitive reversible inhibition?

20.69 **a.** What type of an inhibitor is the antibiotic amoxicillin?
  b. Why can antibiotics be used to treat bacterial infections?

20.70 **a.** A gardener using Parathion develops a headache, dizziness, nausea, blurred vision, excessive salivation, and muscle twitching. What might be happening to the gardener?
  b. Why must humans be careful when using insecticides?

20.71 The enzyme pepsin is produced in the pancreas as the zymogen pepsinogen.
  a. How and where does pepsinogen become the active form pepsin?
  b. Why are proteases such as pepsin produced in inactive forms?

20.72 Thrombin is an enzyme that helps produce blood clotting when an injury and bleeding occurs.
  a. What would be the name of the zymogen of thrombin?
  b. Why would the active form of thrombin be produced only when an injury occurs to tissue?

20.73 What is an allosteric enzyme?

20.74 Why can some regulator molecules speed up a reaction, while others slow it down?

20.75 In feedback control, what type of regulator modifies the catalytic activity of the reaction pathway?

20.76 Why aren't the intermediate products in a reaction sequence used in feedback control?

20.77 Are each of the following statements describing a simple enzyme or one that requires a cofactor?
  a. contains Mg^{2+} in the active site
  b. has catalytic activity as a tertiary protein structure
  c. requires folic acid for catalytic activity

20.78 Are each of the following statements describing a simple enzyme or one that contains a cofactor?

a. contains riboflavin or vitamin B_2
b. has four subunits of polypeptide chains
c. requires Fe^{3+} in the active site for catalytic activity

20.79 Match the following vitamins with their coenzymes:
  a. pantothenic acid (B_5) NAD^+
  b. niacin (B_3)   TPP
  c. thiamin (B_1)   coenzyme A

20.80 Match the following vitamins with their coenzymes:
  a. folate   pyridoxal phosphate
  b. riboflavin (B_2)   THF
  c. pyridoxine   FAD

20.81 Why are only small amounts of vitamins needed in the cells when there are several enzymes that require coenzymes?

20.82 Why is there a daily requirement for vitamins?

20.83 Match each of the following vitamins with their deficiency symptoms or conditions:
  a. niacin   night blindness
  b. vitamin A   weak bone structure
  c. vitamin D   pellagra

20.84 Match each of the following vitamins with their deficiency symptoms or conditions:
  a. cobalamin   bleeding
  b. vitamin C   anemia
  c. vitamin K   scurvy

CHALLENGE QUESTIONS

20.85 Lactase is an enzyme that hydrolyzes lactose to glucose and galactose.
  a. What are the reactants and products of the reaction?
  b. Draw an energy diagram for the reaction with and without lactase.
  c. How does lactase make the reaction go faster?

20.86 Maltase is an enzyme that hydrolyzes maltose into two glucose molecules.
  a. What are the reactants and products of the reaction?
  b. Draw an energy diagram for the reaction with and without maltase.
  c. How does maltase make the reaction go faster?

20.87 What is the class of the enzyme that would catalyze each of the following reactions?

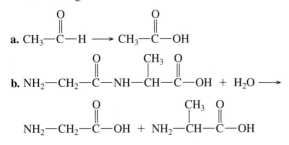

c. $CH_3-CH=CH-CH_3 + H_2O \longrightarrow$

$$\underset{\displaystyle CH_3-CH_2-\overset{\displaystyle \overset{OH}{|}}{CH}-CH_3}{}$$

20.88 What is the class of the enzyme that would catalyze each of the following reactions?

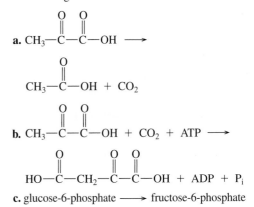

c. glucose-6-phosphate $\longrightarrow$ fructose-6-phosphate

ANSWERS

Answers to Study Checks

20.1 hydrolase

20.2 In the lock-and-key model, the shape of a substrate fits the shape of the active site exactly. In the induced-fit model, the substrate induces the active site to adjust its shape to fit the substrate.

20.3 At a pH lower than the optimum pH, denaturation will decrease the activity of urease.

20.4 HCN is a noncompetitive inhibitor.

20.5 Pepsin hydrolyzes proteins in the foods we ingest. It is synthesized as a zymogen, pepsinogen, to prevent its digestion of the proteins that make up the organs in the body.

20.6 vitamin B_6

20.7 Water-soluble vitamins are easily destroyed by heat.

Answers to Selected Questions and Problems

20.1 The chemical reactions can occur without enzymes, but the rates are too slow. Catalyzed reactions, which are many times faster, provide the amounts of products needed by the cell at a particular time.

20.3 **a.** oxidation–reduction
b. transfer of a group from one substance to another
c. hydrolysis (splitting) of molecules with the addition of water

20.5 **a.** hydrolase **b.** oxidoreductase
c. isomerase **d.** transferase

20.7 **a.** lyase **b.** transferase

20.9 **a.** succinate oxidase
b. fumarate hydrase
c. alcohol dehydrogenase

20.11 **a.** enzyme **b.** enzyme–substrate complex
c. substrate

20.13 **a.** $E + S \rightleftharpoons ES \longrightarrow E + P$
b. The active site is a region or pocket within the tertiary structure of an enzyme that accepts the substrate, aligns the substrate for reaction, and catalyzes the reaction.

20.15 Isoenzymes are slightly different forms of an enzyme that catalyze the same reaction in different organs and tissues of the body.

20.17 A doctor might run tests for the enzymes CK, LDH, and AST to determine if the patient had a heart attack.

20.19 **a.** The rate would decrease.
b. The rate would decrease.
c. The rate would decrease.
d. The rate would increase if [S] < [E].

20.21 pepsin, pH 2; urease, pH 5; trypsin, pH 8

20.23 **a.** competitive **b.** noncompetitive
c. competitive **d.** noncompetitive
e. competitive

20.25 **a.** methanol, CH_3—OH; ethanol, CH_3—CH_2—OH
b. Ethanol has a similar structure to methanol and could compete for the active site.
c. Ethanol is a competitive inhibitor of methanol oxidation.

20.27 Enzymes that act on proteins are proteases and would digest the proteins of the organ where they are produced if they were active immediately upon synthesis.

20.29 In feedback inhibition, the product binds to the first enzyme in a series and changes the shape of the active site. If the active site can no longer bind the substrate effectively, the reaction will stop.

20.31 When a regulator molecule binds to an allosteric site, the shape of the enzyme is altered, which makes the active site more reactive or less reactive and thereby increases or decreases the rate of the reaction.

20.33 **a.** 3; negative regulator
b. 4; allosteric enzyme
c. 1; zymogen

20.35 **a.** an enzyme that requires a cofactor
b. an enzyme that requires a cofactor
c. a simple enzyme

20.37 **a.** THF **b.** NAD^+

20.39 **a.** pantothenic acid (vitamin B_5)
b. folic acid
c. niacin (vitamin B_3)

20.41 **a.** vitamin D or cholecalciferol
b. ascorbic acid or vitamin C
c. niacin or vitamin B_3

20.43 Vitamin B_6 is a water-soluble vitamin, which means that each day any excess of vitamin B_6 is eliminated from the body.

20.45 The side chain —CH_2OH on the ring is oxidized to —CHO, and the other —CH_2OH forms a phosphate ester.

20.47 **a.** oxidoreductase
b. Ethanol would act as a competitive inhibitor of ethylene glycol, saturate the alcohol dehydrogenase enzyme, and allow ethylene glycol to be removed from the body without producing oxalic acid.

20.49 **a.** Fresh pineapple contains an enzyme that breaks down protein, which means that Jello would not turn solid. The high temperatures used to prepare canned pineapple will denature the enzyme so it no longer can break down protein.
b. The enzyme in fresh pineapple juice can be used to tenderize tough meat because the enzyme breaks down proteins.

20.51 The many different reactions that take place in cells require different enzymes because enzymes react with only a certain type of substrate.

20.53 Enzymes are catalysts that are proteins and function only at mild temperature and pH. Catalysts used in chemistry laboratories are usually inorganic materials that can function at high temperatures and in strongly acidic or basic conditions.

20.55 a. S   **b.** E
      c. E   **d.** E
      e. S   **f.** E

20.57 a. urea   **b.** lactose
      c. aspartate   **d.** phenylalanine

20.59 a. transferase   **b.** oxidoreductase
      c. hydrolase   **d.** lyase

20.61 Sucrose fits the shape of the active site in sucrase, but lactose does not.

20.63 A heart attack may be the cause.

20.65 a. saturated   **b.** unsaturated

20.67 In a reversible inhibition, the inhibitor can dissociate from the enzyme, whereas in irreversible inhibition, the inhibitor forms a strong covalent bond with the enzyme and does not dissociate. Irreversible inhibitors act as poisons to enzymes.

20.69 a. Antibiotics such as amoxicillin are irreversible inhibitors.
      b. Antibiotics inhibit enzymes needed to form cell walls in bacteria, not humans.

20.71 a. When pepsinogen enters the stomach, the low pH cleaves a peptide from its protein chain to form pepsin.
      b. An active protease would digest the proteins of the pancreas rather than the proteins in the foods entering the stomach.

20.73 An allosteric enzyme contains sites for regulators that alter the enzyme and speed up or slow down the rate of the catalyzed reaction.

20.75 This would be a negative regulator because the end product of the reaction pathway binds to the enzyme to decrease or stop the first reaction in the reaction pathway.

20.77 a. requires a cofactor
      b. simple enzyme
      c. requires a cofactor (coenzyme)

20.79 a. pantothenic acid (B_5); coenzyme A
      b. niacin (B_3); NAD^+
      c. thiamin (B_1); TPP

20.81 A vitamin combines with an enzyme only when the enzyme and coenzyme are needed to catalyze a reaction. When the enzyme is not needed, the vitamin dissociates for use by other enzymes in the cell.

20.83 a. niacin; pellagra
      b. vitamin A; night blindness
      c. vitamin D; weak bone structure

20.85 a. The reactant is lactose and the products are glucose and galactose.
      b.

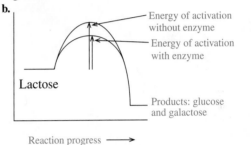

      c. By lowering the energy of activation, the enzyme furnishes a lower energy pathway by which the reaction can take place.

20.87 a. oxidoreductase   **b.** hydrolase   **c.** lyase

21 Nucleic Acids and Protein Synthesis

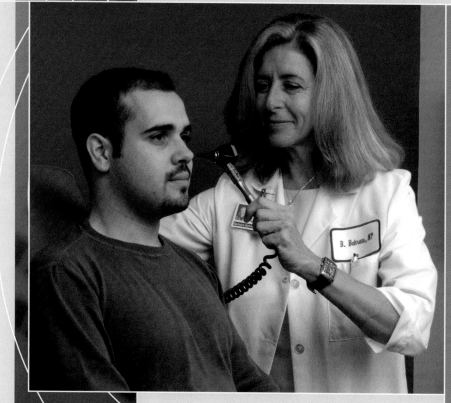

"I run the Hepatitis C Clinic, where patients are often anxious when diagnosed," says Barbara Behrens, nurse practitioner, Hepatitis C Clinic, Kaiser Hospital. "The treatment for hepatitis C can produce significant reactions such as a radical drop in blood count. When this happens, I get help to them within 24 hours. I monitor our patients very closely and many call me whenever they need to."

Hepatitis C is an RNA virus or retrovirus that causes liver inflammation, often resulting in chronic liver disease. Unlike many viruses to which we eventually develop immunity, the hepatitis C virus undergoes mutations so rapidly that scientists have not been able to produce vaccines. People who carry the virus are contagious throughout their lives and are able to pass the virus to other people.

LOOKING AHEAD

Mastering GOB CHEMISTRY

Visit **www.masteringgob.com** for self-study materials and instructor-assigned homework.

ucleic acids are the molecules in our cells that store and direct information for cellular growth and reproduction. Deoxyribonucleic acid (DNA), the genetic material in the nucleus of a cell, contains all the information needed for the development of a complete living system. The way you grow, your hair, your eyes, your physical appearance, the activities of the cells in your body are all determined by a set of directions contained in the DNA of your cells.

The nucleic acids are large molecules found in the nuclei of cells that contain all the information needed to direct the activities of a cell and its reproduction. All of the genetic information in the cell is called the *genome*. Every time a cell divides, the information in the genome is copied and passed on to the new cells. This replication process must duplicate the genetic instructions exactly. Some sections of DNA called *genes* contain the information to make a particular protein.

As a cell requires protein, another type of nucleic acid, RNA, translates the genetic information in DNA and carries that information to the ribosomes where the synthesis of protein takes place. However, mistakes sometimes occur that lead to mutations that affect the synthesis of a certain protein.

21.1 Components of Nucleic Acids

There are two closely related types of nucleic acids: *deoxyribonucleic acid* (**DNA**), and *ribonucleic acid* (**RNA**). Both are unbranched polymers of repeating monomer units known as *nucleotides*. A DNA molecule may contain several million nucleotides; smaller RNA molecules may contain up to several thousand. Each nucleotide has three components: a nitrogenous base, a five-carbon sugar, and a phosphate group. (See Figure 21.1.)

Nitrogen Bases

The **nitrogen-containing bases** in nucleic acids are derivatives of *pyrimidine* or *purine*.

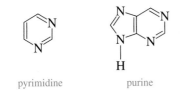

pyrimidine   purine

In DNA, there are two purines, adenine (A) and guanine (G), and two pyrimidines, cytosine (C) and thymine (T). RNA contains the same bases, except thymine (5-methyluracil) is replaced by uracil (U). (See Figure 21.2.)

LEARNING GOAL

Describe the nitrogen bases and ribose sugars that make up the nucleic acids DNA and RNA.

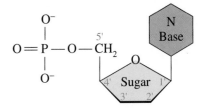

FIGURE 21.1

A diagram of the general structure of a nucleotide found in nucleic acids.

Q **In a nucleotide, what types of groups are bonded to a five-carbon sugar?**

FIGURE 21.2

DNA contains the nitrogen bases A, G, C, and T; RNA contains A, G, C, and U.

Q **Which nitrogen bases are found in DNA?**

Nucleic Acid Building Blocks

Pyrimidines

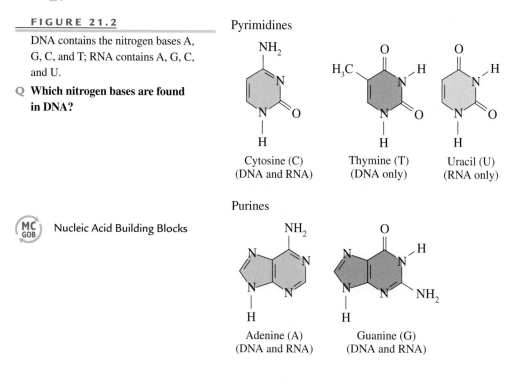

Cytosine (C)
(DNA and RNA)

Thymine (T)
(DNA only)

Uracil (U)
(RNA only)

Purines

Adenine (A)
(DNA and RNA)

Guanine (G)
(DNA and RNA)

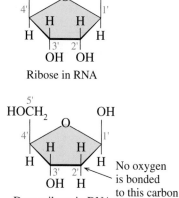

Ribose in RNA

Deoxyribose in DNA

No oxygen is bonded to this carbon

Ribose and Deoxyribose Sugars

The nucleotides of RNA and DNA contain five-carbon pentose sugars. The atoms in the pentose sugars are numbered with primes ($1'$, $2'$, $3'$, $4'$, and $5'$) to differentiate them from the atoms in the nitrogen bases. In RNA, the five-carbon sugar is *ribose*, which gives the letter R in the abbreviation RNA. In DNA, the five-carbon sugar is *deoxyribose*, which is similar to ribose except that there is no hydroxyl group (—OH) on C2$'$. The *deoxy* prefix means "without oxygen" and provides the D in DNA.

SAMPLE PROBLEM 21.1

■ **Components of Nucleic Acids**

Identify each of the following bases as a purine or pyrimidine.

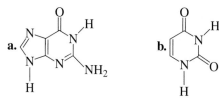

SOLUTION

a. Guanine is a purine.   **b.** Uracil is a pyrimidine.

STUDY CHECK

Indicate if the bases in Sample Problem 21.1 are found in RNA, DNA, or both.

Nucleosides and Nucleotides

A **nucleoside** is produced when a pyrimidine or a purine forms a glycosidic bond to Cl' of a sugar, either ribose or deoxyribose. For example, adenine, a purine, and ribose form a nucleoside called aden**osine**.

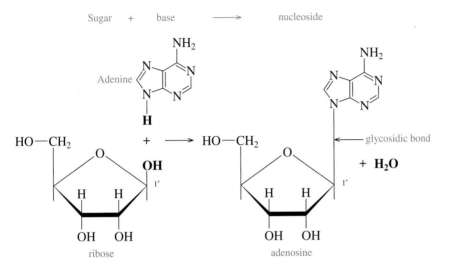

Sugar + base $\longrightarrow$ nucleoside

ribose

adenosine

Nucleotides are formed when the C5'—OH group of ribose or deoxyribose in a nucleoside forms a phosphate ester. Other hydroxyl groups on ribose can form phosphate esters too, but only the 5'-monophosphate nucleotides are found in RNA and DNA. All the nucleotides in RNA and DNA are shown in Figure 21.3.

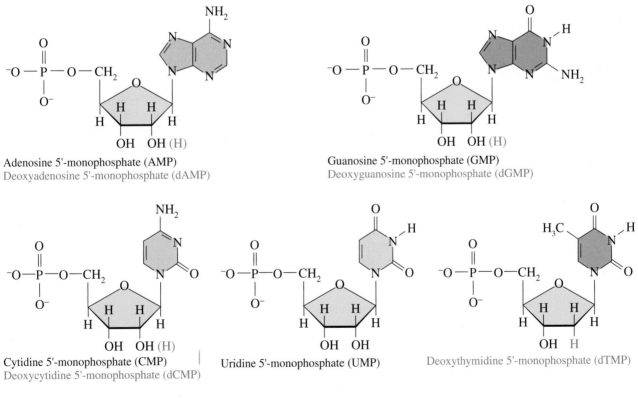

Adenosine 5'-monophosphate (AMP)
Deoxyadenosine 5'-monophosphate (dAMP)

Guanosine 5'-monophosphate (GMP)
Deoxyguanosine 5'-monophosphate (dGMP)

Cytidine 5'-monophosphate (CMP)
Deoxycytidine 5'-monophosphate (dCMP)

Uridine 5'-monophosphate (UMP)

Deoxythymidine 5'-monophosphate (dTMP)

FIGURE 21.3

The nucleotides of RNA are identical to those of DNA except in DNA the sugar is deoxyribose and deoxythymidine replaces uridine.

Q What are two differences in the nucleotides of RNA and DNA?

■ Occupational Therapist

"Occupational therapists teach children and adults the skills they need for the job of living," says occupational therapist Leslie Wakasa. "When working with the pediatric population, we are crucial in educating children with disabilities, their families, caregivers, and school staff in ways to help them be as independent as they can be in all aspects of their daily lives. It's rewarding when you can show children how to feed themselves, which is a huge self-esteem issue for them. The opportunity to help people become more independent is very rewarding."

A combination of technology and occupational therapy helps children who are nonverbal to communicate and interact with their environment. By leaning on a red switch, Alex is learning to use a computer.

TABLE 21.1

Names of Nucleosides and Nucleotides in DNA and RNA

Base	Nucleosides	Nucleotides
RNA		
adenine (A)	adenosine (A)	adenosine 5′-monophosphate (AMP)
guanine (G)	guanosine (G)	guanosine 5′-monophosphate (GMP)
cytosine (C)	cytidine (C)	cytidine 5′-monophosphate (CMP)
uracil (U)	uridine (U)	uridine 5′-monophosphate (UMP)
DNA		
adenine (A)	deoxyadenosine (A)	deoxyadenosine 5′-monophosphate (dAMP)
guanine (G)	deoxyguanosine (G)	deoxyguanosine 5′-monophosphate (dGMP)
cytosine (C)	deoxycytidine (C)	deoxycytidine 5′-monophosphate (dCMP)
thymine (T)	deoxythymidine (T)	deoxythymidine 5′-monophosphate (dTMP)

Naming Nucleotides

The name of a nucleotide is obtained from the name of the nucleoside followed by 5′-monophosphate. Nucleotides of DNA have the prefix *deoxy* added to the beginning of the nucleoside name. Although the letters A, G, C, U, and T represent the bases, they are often used in the abbreviations of the respective nucleotides as well. The names and abbreviations of the bases, nucleosides, and nucleotides in DNA and RNA are listed in Table 21.1.

Formation of Nucleoside Di- and Triphosphates

Any of the nucleoside 5′-monophosphates such as AMP can bond to additional phosphate groups. For example, adding another phosphate to AMP gives ADP (*adenosine 5′-diphosphate*) and ATP (*adenosine 5′-triphosphate*) when there are a total of three phosphates. (See Figure 21.4.) The other nucleotides also form di- and triphosphates. For example, GMP adds phosphate to yield GDP and GTP, dCMP forms dCDP and dCTP, and so on. Of the triphosphates, ATP is of particular interest because it is the major source of energy for most energy-requiring activities in the cell. GTP is an energy source for protein synthesis, and CTP is an intermediate in phospholipid synthesis.

SAMPLE PROBLEM 21.2

■ Nucleotides

Identify which nucleic acid (DNA or RNA) contains each of the following nucleotides; state the components of each nucleotide:

a. deoxyguanosine 5′-monophosphate (dGMP)
b. adenosine 5′-monophosphate (AMP)

SOLUTION

a. This DNA nucleotide consists of deoxyribose, guanine, and phosphate.
b. This RNA nucleotide contains ribose, adenine, and phosphate.

STUDY CHECK

What is the name and abbreviation of the DNA nucleotide of cytosine?

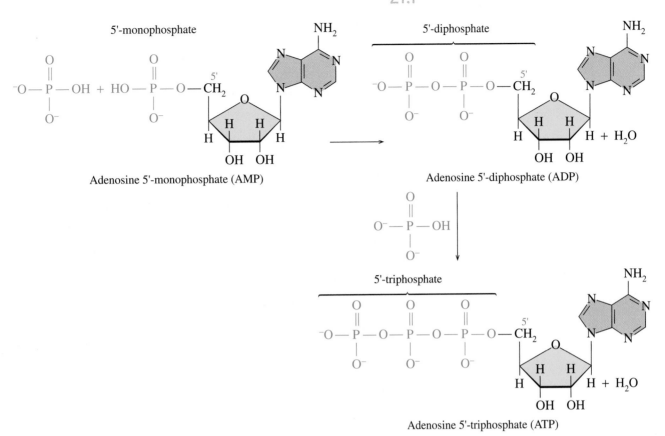

5'-monophosphate

Adenosine 5'-monophosphate (AMP)

5'-diphosphate

Adenosine 5'-diphosphate (ADP)

5'-triphosphate

Adenosine 5'-triphosphate (ATP)

FIGURE 21.4

The addition of more phosphate groups forms adenosine 5′-diphosphate (ADP) and adenosine 5′-triphosphate (ATP).

Q How does the structure of deoxyguanidine triphosphate (dGTP) differ from ATP?

QUESTIONS AND PROBLEMS

■ Components of Nucleic Acids

21.1 Identify each of the following bases as a purine or pyrimidine:
  a. thymine   **b.**

21.2 Identify each of the following bases as a purine or pyrimidine:
  a. guanine   **b.**

21.3 Identify the bases in problem 21.1 as present in RNA, DNA, or both.

21.4 Identify the bases in problem 21.2 as present in RNA, DNA, or both.

21.5 What are the names and abbreviations of the four nucleotides in DNA?

21.6 What are the names and abbreviations of the four nucleotides in RNA?

21.7 Identify each of the following as a nucleoside or nucleotide:
  a. adenosine   **b.** deoxycytidine
  c. uridine   **d.** cytidine 5′-monophosphate

21.8 Identify each of the following as a nucleoside or nucleotide:
 a. deoxythymidine **b.** guanosine
 c. adenosine **d.** uridine 5′-monophosphate

21.9 Draw the structure of deoxyadenosine 5′-monophosphate (dAMP).

21.10 Draw the structure of uridine 5′-monophosphate (UMP).

21.2 Primary Structure of Nucleic Acids

The **nucleic acids** consist of polymers of many nucleotides in which the 3′-OH group of the sugar in one nucleotide bonds to the phosphate group on the 5′-carbon atom in the sugar of the next nucleotide. This phosphate link between the sugars in adjacent nucleotides is referred to as a **phosphodiester bond**. As more nucleotides are added through phosphodiester bonds, a backbone forms that consists of alternating sugar and phosphate groups.

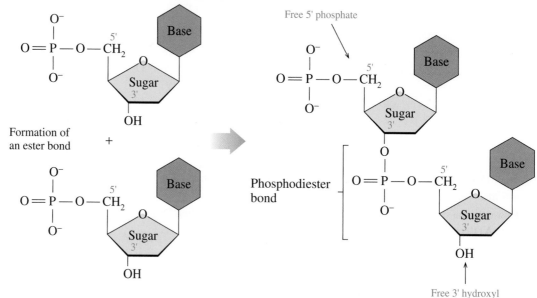

Each nucleic acid has its own unique sequence of bases, which is known as its **primary structure**. It is this sequence of bases that carries the genetic information from one cell to the next. Along a polynucleotide of a DNA or RNA chain, the bases attached to each of the sugars extend out from the nucleic acid backbone. In any nucleic acid, the sugar at the one end has an unreacted or free 5′-phosphate terminal end, and the sugar at the other end has a free 3′-hydroxyl group.

A nucleic acid sequence is read from the sugar with the free 5′-phosphate to the sugar with the free 3′-hydroxyl group. The order of bases is often written using only the letters of the bases. Thus, the nucleotides in a section of RNA shown in Figure 21.5 is read as 5′-ACGU-3′.

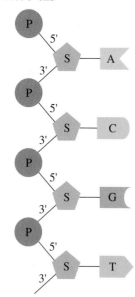

RNA (ribonucleic acid)

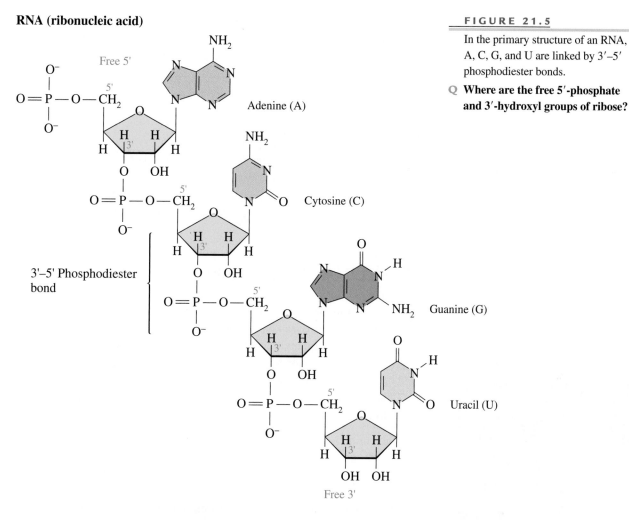

Adenine (A)

Cytosine (C)

Guanine (G)

Uracil (U)

3'–5' Phosphodiester bond

Free 5'

Free 3'

SAMPLE PROBLEM 21.3

■ Bonding of Nucleotides

Draw the structure of an RNA dinucleotide formed by two CMP.

SOLUTION

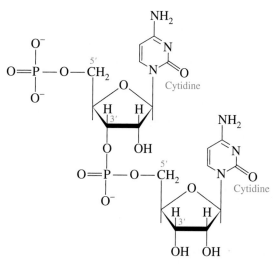

Cytidine

Cytidine

STUDY CHECK

In the dinucleotide of cytidine shown in the solution to Sample Problem 21.3, identify the free 5′-phosphate group and the free 3′-hydroxyl (—OH) group.

 DNA and RNA: Chains of Nucleotides

QUESTIONS AND PROBLEMS

■ Primary Structure of Nucleic Acids

21.11 How are the nucleotides held together in a nucleic acid chain?

21.12 How do the ends of a nucleic acid polymer differ?

21.13 Write the structure of the dinucleotide GC that would be in RNA.

21.14 Write the structure of the dinucleotide AT that would be in DNA.

LEARNING GOAL

■ Describe the double helix of DNA.

21.3 DNA Double Helix

In the 1940s, biologists determined that DNA in a variety of organisms had a specific relationship between bases: the percent of adenine (A) was equal to the percent of thymine (T), and the percent of guanine (G) was equal to cytosine (C). Although different organisms vary in the amounts of bases, adenine is paired only with thymine, and guanine is paired only with cytosine in 1:1 ratios. (See Table 21.2.) This relationship known as *Chargaff's rules* can be summarized as follows:

Amount of purines = Amount of pyrimidines

$$A = T$$

$$G = C$$

In 1953, James Watson and Francis Crick proposed that DNA was a **double helix** that consists of two polynucleotide strands winding about each other like a spiral staircase. (See Figure 21.6.) The sugar-phosphate backbones are analogous to the outside railings with the nitrogen bases arranged like steps along the inside. One strand goes in the 5′-3′ direction, and the other strand goes in the opposite 3′-5′ direction.

Complementary Base Pairs

Each of the bases along a polynucleotide strand forms hydrogen bonds to a specific base on the opposite DNA strand. Adenine bonds only to thymine, and guanine

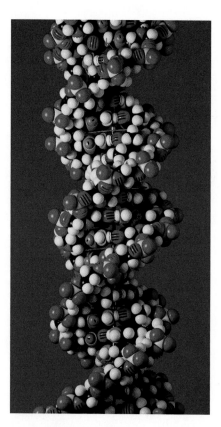

FIGURE 21.6

This space-filling model shows the double helix that is the characteristic shape of DNA molecules.

Q **What is meant by the term *double helix*?**

TABLE 21.2

Percent Bases in the DNAs of Selected Organisms

Organism	%A	%T	%G	%C
Humans	30	30	20	20
Chicken	28	28	22	22
Salmon	28	28	22	22
Corn (maize)	27	27	23	23
Neurospora	23	23	27	27

Source: Handbook of Biochemistry, 2nd ed. Sober, H.E., Ed.; CRC Press: Cleveland, OH, 1970.

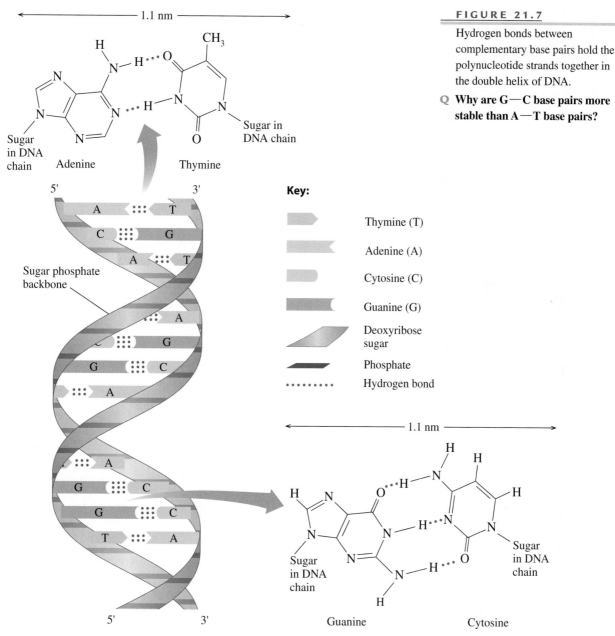

FIGURE 21.7

Hydrogen bonds between complementary base pairs hold the polynucleotide strands together in the double helix of DNA.

Q **Why are G—C base pairs more stable than A—T base pairs?**

bonds only to cytosine. (See Figure 21.7.) The pairs A—T and G—C are called **complementary base pairs**. The specificity of the base pairing is due to the fact that adenine and thymine form two hydrogen bonds, while cytosine and guanine form three hydrogen bonds. This explains why DNA has equal amounts of A and T bases and equal amounts of G and C.

Since each base pair contains a purine and a pyrimidine, the total width of the two pairs of bases A—T and G—C is the same. Thus the two polynucleotide strands of a DNA are the same distance apart all along the DNA polymer. There are no C—T, C—A, G—T, or G—A pairs in DNA because such base-pair combinations cannot form as many hydrogen bonds or maintain a constant width between the two DNA backbones. As we shall see, complementary base pairing plays a crucial role in cell replication and the transfer of hereditary information.

X-ray diffraction patterns of DNA indicate that DNA is a right-handed or alpha (α) helix. In one complete turn, there are about 10 pairs of nucleotides.

MC
GOB   The Double Helix

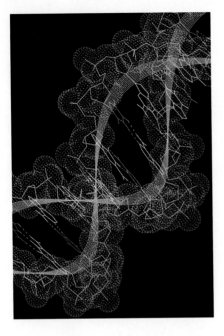

FIGURE 21.8

A computer-generated model of a DNA molecule.

Q **What is the complementary strand of a DNA section of 5′—GGCCTT—3′?**

(See Figure 21.8.) In mitochondria, bacteria, and viruses, DNA molecules are compact, highly coiled molecules. In the chromosomes, DNA strands are wrapped around proteins called *histones*, a structure that provides the most stable and orderly arrangement for the long DNA molecules.

SAMPLE PROBLEM 21.4

■ **Complementary Base Pairs**

Write the base sequence of the complementary segment for the following segment of a strand of DNA.

5′—A—C—G—A—T—C—T—3′

SOLUTION

In the complementary strand of DNA, the base A pairs with T, and G pairs with C.

Given segment of DNA: 5′—A—C—G—A—T—C—T—3′

Complementary segment: 3′—T—G—C—T—A—G—A—3′

STUDY CHECK

What is the sequence of bases that is complementary to a portion of DNA with a base sequence of 5′—G—G—T—T—A—A—C—C—3′?

QUESTIONS AND PROBLEMS

■ **DNA Double Helix**

21.15 How are the two strands of nucleic acid in DNA held together?

21.16 What is meant by complementary base pairing?

21.17 Complete the base sequence in a second DNA strand if a portion of one strand has the following base sequence:
    a. 5′—AAAAAA—3′   **b.** 5′—GGGGGG—3′
    c. 5′—AGTCCAGGT—3′   **d.** 5′—CTGTATACGTTA—3′

21.18 Complete the base sequence in a second DNA strand if a portion of one strand has the following base sequence:
    a. 5′—TTTTTT—3′   **b.** 5′—CCCCCCCCC—3′
    c. 5′—ATGGCA—3′   **d.** 5′—ATATGCGCTAAA—3′

LEARNING GOAL

Describe the process of DNA replication.

 DNA Replication

21.4 DNA Replication

DNA is found in *eukaryotic cells* of animals and plants including algae as well as in *prokaryotic cells* of bacteria. DNA from both types of cells is chemically similar and has the same function, which is to preserve genetic information. However, there are several differences in the cell types. Eukaryotic cells contain DNA in a nucleus with a nuclear membrane and combined with proteins called *histones*, have several organelles within membranes such as mitochondria and lysosomes, and divide by mitosis. Prokaryotic cells contain DNA as a circular chromosome with no membrane and not associated with histones, contain organelles without membranes, and divide by a simpler process called binary fission. As cells divide, copies of DNA must be produced in order to transfer the genetic information to the new cells. This is the process of DNA replication.

Replication and Energy

In DNA **replication**, the strands in the parent DNA separate, which allows each of the original strands to makes copies by synthesizing complementary strands. The replication process begins when an enzyme called *helicase* catalyzes the unwinding of a portion of the double helix by breaking the hydrogen bonds between the complementary bases. These single strands now act as templates for the synthesis of new complementary strands. (See Figure 21.9.) Within the nucleus, nucleoside triphosphates for each base are available so that each exposed base on the template strand can form hydrogen bonds with its complementary base in the nucleoside triphosphate.

SELF-STUDY ACTIVITY
DNA Replication

FIGURE 21.9

In DNA replication, the separate strands of the parent DNA are the templates for the synthesis of complementary strands, which produces two exact copies of DNA.

Q **How many strands of the parent DNA are in each of the new double-stranded copies of DNA?**

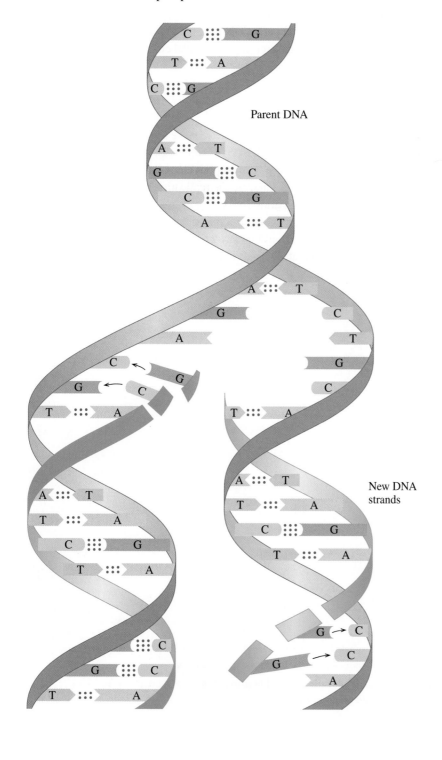

When a nucleoside triphosphate bonds to a sugar at the end of a growing strand, two phosphate groups are cleaved, which provides the energy for the reaction. For example, a T in the template strand forms a hydrogen bond with an A in ATP, and a G on the template strand forms hydrogen bonds with a CTP. After the base pairs are formed, *DNA polymerase* catalyzes the formation of phosphodiester bonds between the nucleotides. The hydrolysis of two phosphate groups releases energy for the new bonds. In this way, energy is provided to join each new nucleotide to the backbone of a growing DNA daughter strand. (See Figure 21.10.)

Eventually the entire double helix of the parent DNA is copied. In each new DNA molecule, one strand of the double helix is from the original DNA and one is a newly synthesized strand. This process, called *semi-conservative replication*, produces two new DNAs called *daughter DNA* that are identical to each other and exact copies of the original parent DNA. In the process of DNA replication, complementary base pairing ensures the correct placements of bases in the new DNA strands.

Direction of Replication

Now that we have seen the overall process, we can take a look at some of the details that are important in understanding DNA replication. The unwinding of DNA by *helicase* occurs simultaneously in several sections along the parent DNA molecule. As a result *DNA polymerase* can catalyze the replication process at each of these open DNA sections called **replication forks**. However, DNA polymerase catalyzes only phosphodiester bonds between the 5′ phosphate of one nucleotide and the 3′ hydroxyl of the next. That means that DNA polymerases have to move in opposite directions along the separated strands of DNA. The new DNA strand that grows

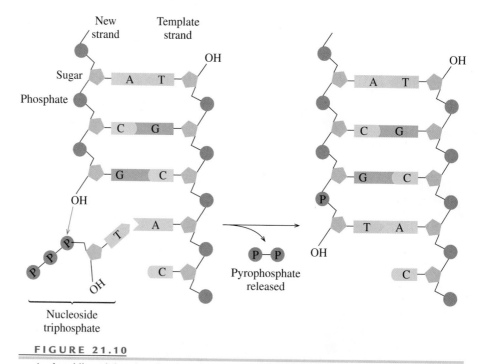

FIGURE 21.10

As thymidine triphosphate bonds to the 3′—OH of the adjacent sugar, two phosphates are removed, which provides energy for the synthesis reaction.

Q **Why are nucleoside triphosphates used to provide the complementary bases instead of nucleoside monophosphates?**

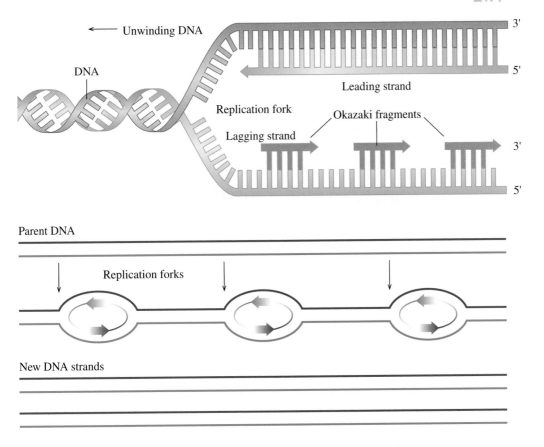

FIGURE 21.11

At each replication fork, DNA polymerase synthesizes a continuous DNA strand in the 5′ to 3′ direction. In the new 3′ to 5′ DNA strand, small Okazaki fragments are produced that are joined by DNA ligase.

Q **Why is only one of the new DNA strands synthesized in a continuous direction?**

in the 5′ to 3′ direction, the *leading strand*, is synthesized continuously. The other new DNA, the *lagging strand*, is synthesized in the opposite direction, which is in the reverse 3′ to 5′ direction. In this lagging strand, short sections called **Okazaki fragments** are synthesized at the same time by several DNA polymerases and connected by DNA *ligase* to give a single 3′ to 5′ DNA strand. (See Figure 21.11.)

SAMPLE PROBLEM 21.5

■**DNA Replication**

In an original DNA strand, there is a portion with the bases 5′—A—G—T—3′. What nucleotides are placed in a growing DNA daughter strand for this sequence?

SOLUTION

Complementary base pairing allows only one possible nucleotide to pair with each base on the original strand. Thymine will pair only with adenine, cytosine with guanine, and adenine with thymine to give the sequence 3′—T—C—A—5′.

STUDY CHECK

Why would the new DNA strand you wrote for problem 21.5 be synthesized as Okazaki fragments that require a DNA ligase?

QUESTIONS AND PROBLEMS

■ **DNA Replication**

21.19 What is the function of the enzyme helicase in DNA replication?

21.20 What is the function of the enzyme DNA polymerase in DNA replication?

21.21 What process ensures that the replication of DNA produces identical copies?

21.22 Why are Okazaki fragments needed in the synthesis of the lagging strand?

LEARNING GOAL

Identify the different RNAs; describe the synthesis of mRNA.

■ 21.5 RNA and Transcription

Ribonucleic acid, RNA, which makes up most of the nucleic acid found in the cell, is involved with transmitting the genetic information needed to operate the cell. Similar to DNA, RNA molecules are unbranched polymers of nucleotides. However, there are several important differences.

1. The sugar in RNA is ribose rather than the deoxyribose found in DNA.
2. The nitrogen base uracil replaces thymine.
3. RNA molecules are single, not double stranded.
4. RNA molecules are much smaller than DNA molecules.

■ Types of RNA

MC GOB Types of RNA

There are three major types of RNA in the cells: *messenger RNA*, *ribosomal RNA*, and *transfer RNA*. Ribosomal RNA (**rRNA**), the most abundant type of RNA, is contained in the ribosomes. Ribosomes, which are the sites for protein synthesis, consist of two subunits, a large subunit and a small subunit. (See Figure 21.12.) Cells that synthesize large numbers of proteins have thousands of ribosomes in their cells.

Messenger RNA (**mRNA**) carries genetic information from the DNA in the nucleus to the ribosomes in the cytoplasm for protein synthesis. Each gene, a segment of DNA, produces a separate mRNA molecule when a certain protein is needed in the cell, but then the mRNA is broken down quickly. The size of an mRNA depends on the number of nucleotides in that particular gene.

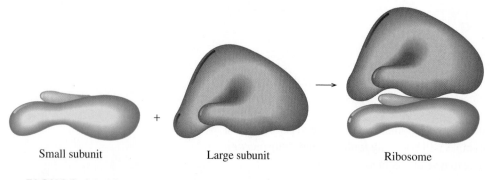

Small subunit   +   Large subunit   Ribosome

FIGURE 21.12

A typical prokaryotic ribosome consists of a small subunit and a large subunit.

Q **Why would there be many thousands of ribosomes in a cell?**

Transfer RNA (**tRNA**), the smallest of the RNA molecules, interprets the genetic information in mRNA and brings specific amino acids to the ribosome for protein synthesis. Only the tRNAs can translate the genetic information into amino acids for proteins. There are one or more different tRNAs for each of the 20 amino acids. The structures of the transfer RNAs are similar, consisting of 70–90 nucleotides. Hydrogen bonds between some of the complementary bases in the chain produce loops that give some double-stranded regions.

Although the structure of tRNA is complex, we draw tRNA as a cloverleaf to illustrate its features. All tRNA molecules have a 3′ end with the nucleotide sequence ACC, which is known as the *acceptor stem*. An enzyme attaches an amino acid by forming an ester bond with the free —OH at the end of the acceptor stem. Each tRNA contains an **anticodon**, which is a series of three bases that complements three bases on an mRNA. (See Figure 21.13.) Table 21.3 summarizes the three types of RNA.

TABLE 21.3

Types of RNA Molecules

Type	Abbreviation	Percentage of Total RNA	Function in the Cell
Ribosomal RNA	rRNA	75–80	Major component of the ribosomes
Messenger RNA	mRNA	5–10	Carries information for protein synthesis from the DNA in the nucleus to the ribosomes
Transfer RNA	tRNA	10–15	Brings amino acids to the ribosomes for protein synthesis

SAMPLE PROBLEM 21.6

■ **Types of RNA**

What is the function of mRNA in a cell?

SOLUTION

mRNA carries the instructions for the synthesis of a protein from the DNA to the ribosomes.

STUDY CHECK

What is the function of tRNA in a cell?

Transcription: Synthesis of mRNA

We now look at the processes involved in transferring genetic information encoded in the DNA to the production of proteins. First, the genetic information is copied from a gene in DNA to make a messenger RNA (mRNA), a process called **transcription**. As a carrier molecule, the mRNA moves out of the nucleus and goes to the ribosomes. During **translation**, tRNA molecules convert the information in the mRNA into amino acids, which are placed in the

SELF-STUDY ACTIVITY
Transcription

FIGURE 21.13

(a) In the L shape of a transfer RNA, some sections of the ribosephosphate backbone form regions of complementary base bonding.
(b) A typical tRNA molecule has an acceptor stem at the 3′ end of the nucleic acid where an amino acid attaches, and an anticodon loop that complements three bases on mRNA.

Q **Why will different tRNAs have different bases in the anticodon loop?**

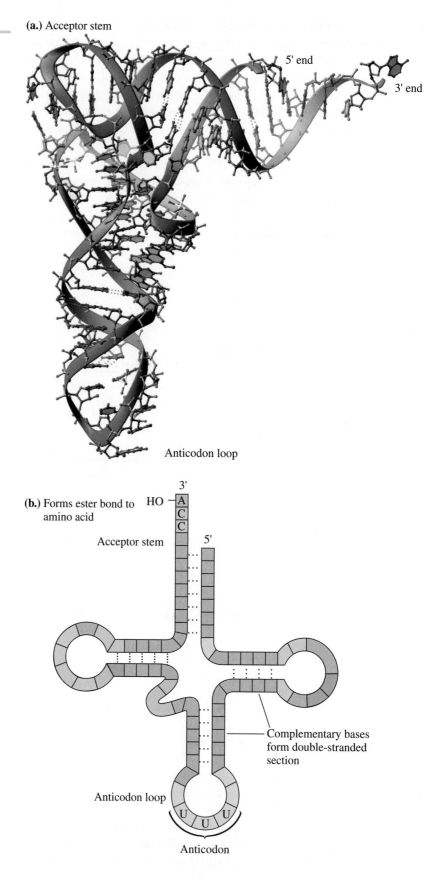

(a.) Acceptor stem

5' end

3' end

Anticodon loop

(b.) Forms ester bond to amino acid

HO

3'

Acceptor stem

5'

Complementary bases form double-stranded section

Anticodon loop

Anticodon

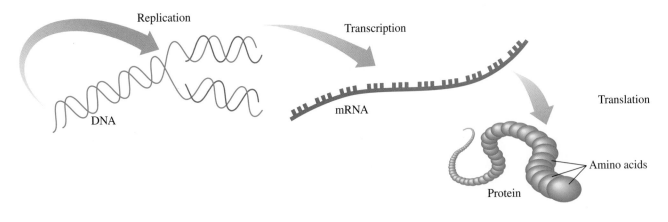

FIGURE 21.14

The genetic information in DNA is replicated in cell division and used to produce messenger RNAs that are converted into amino acids for protein synthesis.

Q **What is the difference between transcription and translation?**

proper sequence to synthesize a protein. (See Figure 21.14.) The language that relates the series of nucleotides in mRNA with the amino acids specified is the *genetic code*.

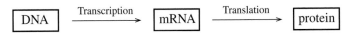

Transcription begins when the section of a DNA that contains the gene to be copied unwinds. Within this unwound DNA, a polymerase enzyme identifies a nucleotide sequence of TATAAA, which is the starting or initiation point that is the signal to begin mRNA synthesis. Just as in DNA synthesis, C is paired with G, T pairs with A, and A pairs with U (not T). The polymerase enzyme moves along the unwound DNA in a 3′ to 5′ direction, forming bonds between the complementary bases in nucleoside triphosphates. Eventually the RNA polymerase reaches the termination point that signals the end of transcription, the new mRNA is released, and the DNA returns to its double helix structure. (See Figure 21.15.)

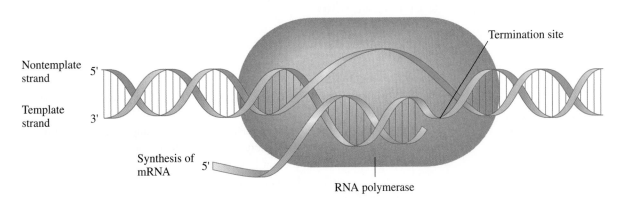

FIGURE 21.15

DNA undergoes transcription when RNA polymerase makes a complementary copy of a gene using the 3′ to 5′ strand as the template.

Q **Why is the mRNA connected in a 5′ to 3′ direction?**

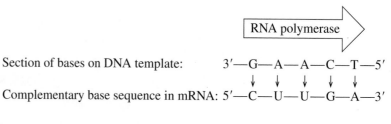

Section of bases on DNA template: 3′—G—A—A—C—T—5′

Complementary base sequence in mRNA: 5′—C—U—U—G—A—3′

■ RNA Synthesis

The sequence of bases in a part of the DNA template for mRNA is 3′—CGATCA—5′. What is the corresponding mRNA produced?

SOLUTION

The nucleotides in DNA pair up with the ribonucleotides as follows: G ⟶ C, C ⟶ G, T ⟶ A, and A ⟶ U.

Portion of DNA template: 3′—C—G—A—T—C—A—5′

Complementary bases in mRNA: 5′—G—C—U—A—G—U—3′

STUDY CHECK

What is the DNA template that codes for the mRNA having the ribonucleotide sequence 5′—GGGUUUAAA—3′?

Processing of mRNA

In eukaryotes, the genes contain sections known as **exons** that code for proteins that are mixed in with sections called **introns** that do not code for protein. A newly formed mRNA is called a pre-mRNA because it is a copy of the entire DNA template including the noncoding introns. Before the newly synthesized pre-mRNA leaves the nucleus, it undergoes processing to remove the intron sections. The splicing of the pre-RNA produces a mature, functional mRNA that leaves the nucleus to deliver the genetic information to the ribosomes for the synthesis of protein. (See Figure 21.16.)

Regulation of Transcription

The synthesis of mRNA does not occur randomly; rather, mRNA is synthesized in response to cellular needs for a particular protein. This regulation takes place at the transcription level where the absence or presence of end products speeds up or slows down the synthesis of mRNAs for specific enzymes. For example, *E. coli* that grow on lactose need β-galactosidase to hydrolyze lactose to glucose and galactose. However, when lactose is low or not present in the cell, there is little or no β-galactosidase present. The transcription of the mRNA for the enzyme is turned off. When lactose enters the cell, β-galactosidase levels rise because lactose induces the synthesis of its enzymes. In this process known as **enzyme induction**, lactose activates the transcription of the genes that produce the mRNAs that code for β-galactosidase.

In prokaryotes, each group of related proteins is regulated by an **operon**, which is a section of DNA containing a control site preceding the **structural genes** that code

SELF-STUDY ACTIVITY
The *lac* Operon in *E. coli*

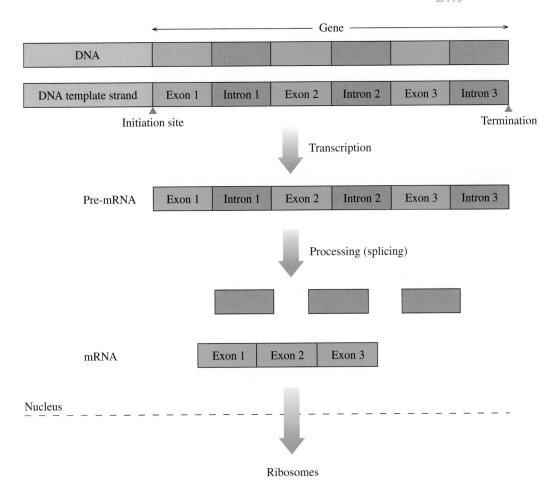

FIGURE 21.16

A pre-RNA, containing the exons' and introns' copies from the gene, is processed to remove the introns and form a mature mRNA that codes for a protein.

Q **What is the difference between exons and introns?**

for proteins. The **control site** consists of a *promoter* and an *operator*. Preceding the operon is a regulatory gene. (See Figure 21.17.) A **regulatory gene** produces an mRNA for the synthesis of a **repressor** protein that binds to the operator and blocks the synthesis of β-galactosidase by RNA polymerase. When lactose enters the cell, lactose inactivates the repressor and it dissociates from the operator. Without a repressor, RNA polymerase transcribes the genes that code for the lactose enzymes.

MC GOB Activating and Inhibiting Genes

SAMPLE PROBLEM 21.8

▪ Transcription

Indicate whether transcription takes place in each of the following conditions:

a. A repressor binds to the operator in the control site.

b. An inducer binds to the repressor protein.

SOLUTION

a. No. A repressor blocks the synthesis of mRNA by RNA polymerase.

b. Yes. The lactose inactivates the repressor, which activates the transcription of the genes.

(a) The lactose operon

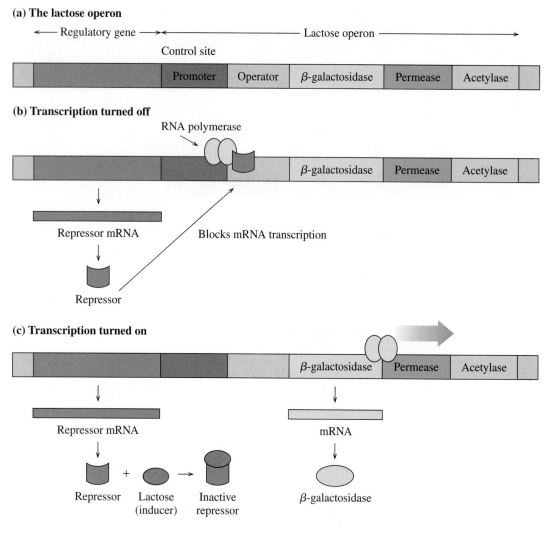

(b) Transcription turned off

(c) Transcription turned on

FIGURE 21.17

(a) The lactose operon consists of a control site and structural genes. (b) Without lactose, a repressor protein blocks the transcription of enzymes for lactose. (c) Lactose, an inducer, inactivates the repressor to allow the transcription of enzymes for lactose hydrolysis.

Q **Why is transcription blocked when no lactose is present in the cell?**

STUDY CHECK

How does a regulatory gene control the use of materials in a cell?

QUESTIONS AND PROBLEMS

■ RNA and Transcription

21.23 What are the three different types of RNA?

21.24 What are the functions of each type of RNA?

21.25 What is the composition of a ribosome?

21.26 What is the smallest RNA?

21.27 What is meant by the term "transcription"?

21.28 What bases in mRNA are used to complement the bases A, T, G, and C in DNA?

21.29 Write the corresponding section of mRNA produced from the following section of DNA template:

3′—CCGAAGGTTCAC—5′

21.30 Write the corresponding section of mRNA produced from the following section of DNA template:

3′—TACGGCAAGCTA—5′

21.31 What are introns and exons?

21.32 What kind of processing do mRNA molecules undergo before they leave the nucleus?

21.33 What is an operon?

21.34 Why does the operon model control protein synthesis at the transcription level?

21.35 How is the lactose operon turned off in *E. coli* that grow on lactose?

21.36 How is the lactose operon activated in *E. coli* that grow on lactose?

21.6 The Genetic Code

LEARNING GOAL

Describe the function of the codons in the genetic code.

The overall function of the RNAs in the cell is to facilitate the task of synthesizing protein. After the genetic information encoded in DNA is transcribed into mRNA molecules, the mRNAs move out of the nucleus to the ribosomes in the cytoplasm. At the ribosomes, the genetic information in the mRNAs is converted into a sequence of amino acids in protein.

$$\text{mRNA} \xrightarrow[\text{ribosomes in the cytoplasm}]{\text{Translation at the}} \text{protein}$$

Codons

Genetic information from DNA is encoded in the mRNA as a sequence of nucleotides. In the **genetic code**, a sequence of three bases (triplet), called a **codon**, specifies each amino acid in the protein. Early work on protein synthesis showed that repeating triplets of uracil (UUU) produced a polypeptide that contained only phenylalanine. Therefore, a sequence of 5′—UUU—UUU—UUU—3′ codes for three phenylalanines.

MC GOB Genetic Code

Codons in mRNA	5′—UUU—UUU—UUU—3′
Translation	↓ ↓ ↓
Amino acid sequence	Phe — Phe — Phe

The codons have now been determined for all 20 amino acids. A total of 64 codons are possible from the triplet combinations of A, G, C, and U. Three of these, UGA, UAA, and UAG, are stop signals that code for the termination of protein synthesis. All the other three-base codons shown in Table 21.4 specify amino acids, which means that one amino acid can have several codons. For example, glycine is the amino acid coded by the codons GGU, GGC, GGA, and GGG. The triplet AUG has two roles in protein synthesis. At the beginning of an mRNA, the codon AUG signals the start of protein synthesis. In the middle of a series of codons, the AUG codon specifies the amino acid methionine.

TABLE 21.4

mRNA Codons: The Genetic Code for Amino Acids

First Letter	Second Letter				Third Letter
	U	**C**	**A**	**G**	
U	UUU⎱ Phe UUC⎰ UUA⎱ Leu UUG⎰	UCU⎱ UCC⎰ Ser UCA⎰ UCG⎰	UAU⎱ Tyr UAC⎰ UAA STOP UAG STOP	UGU⎱ Cys UGC⎰ UGA STOP UGG Trp	U C A G
C	CUU⎱ CUC⎰ Leu CUA⎰ CUG⎰	CCU⎱ CCC⎰ Pro CCA⎰ CCG⎰	CAU⎱ His CAC⎰ CAA⎱ Gln CAG⎰	CGU⎱ CGC⎰ Arg CGA⎰ CGG⎰	U C A G
A	AUU⎱ AUC⎰ Ile AUA⎰ ᵃAUG Met/start	ACU⎱ ACC⎰ Thr ACA⎰ ACG⎰	AAU⎱ Asn AAC⎰ AAA⎱ Lys AAG⎰	AGU⎱ Ser AGC⎰ AGA⎱ Arg AGG⎰	U C A G
G	GUU⎱ GUC⎰ Val GUA⎰ GUG⎰	GCU⎱ GCC⎰ Ala GCA⎰ GCG⎰	GAU⎱ Asp GAC⎰ GAA⎱ Glu GAG⎰	GGU⎱ GGC⎰ Gly GGA⎰ GGG⎰	U C A G

ᵃCodon that signals the start of a peptide chain.

STOP codons signal the end of a peptide chain.

SAMPLE PROBLEM 21.9

■ **Codons**

What is the sequence of amino acids coded by the following codons in mRNA?

5′—GUC—AGC—CCA—3′

SOLUTION

According to Table 21.4, GUC codes for valine, AGC for serine, and CCA for proline. The sequence of amino acids is Val-Ser-Pro.

STUDY CHECK

The codon UGA does not code for an amino acid. What is its function?

QUESTIONS AND PROBLEMS

(MC GOB)

■ **The Genetic Code**

21.37 What is a codon?

21.38 What is the genetic code?

21.39 What amino acid is coded for by each codon?
    a. CUU   **b.** UCA
    c. GGU   **d.** AGG

21.40 What amino acid is coded for by each codon?
    a. AAA   **b.** GUC
    c. CGG   **d.** GCA

21.41 When does the codon AUG signal the start of a protein, and when does it code for the amino acid methionine?

21.42 The codons UAA and UAG do not code for amino acids. What is their role as codons in mRNA?

21.7 Protein Synthesis: Translation

Once the mRNA is synthesized, it migrates out of the nucleus into the cytoplasm to the ribosomes. At the ribosomes, the *translation* process involves tRNA molecules, amino acids, and enzymes, all which convert the codons on mRNA into amino acids to make a protein.

Activation of tRNA

We have seen that the genetic code carried by the mRNA consists of triplets of nucleotides that correspond to one of 20 different amino acids. Now the tRNA are utilized to translate the codons into specific amino acids. This adaptor role of tRNA is possible because it reads both the triplet language of mRNA and picks up the amino acid that corresponds to that codon. The *anticodon* in the loop at the bottom of a tRNA contains a triplet of bases that complement a codon on mRNA. An amino acid is attached to the acceptor stem of the tRNA by enzymes called *aminoacyl-tRNA synthetases*. There is a different synthetase for each amino acid. The synthetase enzymes use energy from ATP to catalyze the formation of ester bonds between the carboxylic acid groups on the amino acids and the hydroxyl groups of the acceptor stem. (See Figure 21.18.) The matching of a tRNA to the correct amino acid is essential. If the wrong amino acid is attached, it would be placed into the protein and make an incorrect protein. However, each synthetase checks the attachment of an amino acid and tRNA and hydrolyzes any incorrect combinations.

LEARNING GOAL

Describe the process of protein synthesis from mRNA.

MC GOB

SELF-STUDY ACTIVITY
Translation

MC GOB

Following the Instructions in DNA

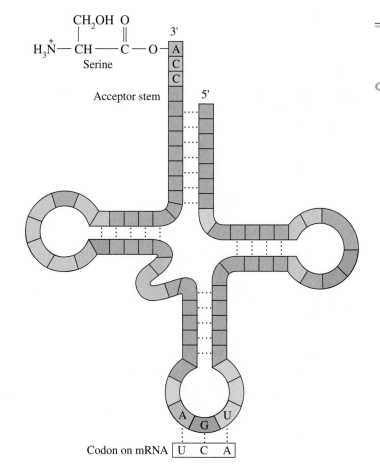

Codon on mRNA [U C A]

FIGURE 21.18

An activated tRNA with anticodon AGU bonds to serine at the acceptor stem.

Q **What is the codon for serine for this tRNA?**

Many Antibiotics Inhibit Protein Synthesis

Several antibiotics stop bacterial infections by interfering with the synthesis of proteins needed by the bacteria. Some antibiotics act only on bacterial cells by binding to the ribosomes in bacteria but do not act on human cells. A description of some of these antibiotics is given in Table 21.5.

TABLE 21.5

Antibiotics That Inhibit Protein Synthesis in Bacterial Cells

Antibiotic	Effect on Ribosomes to Inhibit Protein Synthesis
Chloramphenicol	Inhibits peptide bond formation and prevents the binding of tRNA
Erythromycin	Inhibits peptide chain growth by preventing the translocation of the ribosome along the mRNA
Puromycin	Causes release of an incomplete protein by ending the growth of the polypeptide early
Streptomycin	Prevents the proper attachment of tRNAs
Tetracycline	Prevents the binding of tRNAs

Initiation of Protein Synthesis

Protein synthesis begins when an mRNA combines with the smaller subunit of a ribosome. Because the first codon in an mRNA is a *start* codon, AUG, a tRNA with anticodon UAC and carrying the amino acid methionine forms hydrogen bonds with the AUG codon. The larger ribosome completes the ribosomal unit and translation is ready to begin. (See Figure 21.19.) Protein synthesis takes place on the ribosome at two adjacent sites on the larger subunit. A tRNA that has the anticodon for the second codon with the second amino acid bonds to mRNA. A peptide bond forms between the amino acids, and the first tRNA dissociates from the ribosome and returns to the cytoplasm. After the first tRNA detaches from the ribosome, the ribosome shifts to the next codon on the mRNA, a process called **translocation**. A new tRNA can now attach to the open binding site so that its amino acid forms a peptide bond with the previous amino acid. Again the earlier tRNA detaches and the ribosome shifts down the mRNA to read the next codon. Each time a peptide bond joins the new amino acid to the growing polypeptide chain. Sometimes several ribosomes, called a polysome, translate the same strand of mRNA at the same time to produce several copies of the peptide chain.

Termination of Protein Synthesis

After all the amino acids for a particular protein have been linked together by peptide bonds, the ribosome encounters a stop codon. Because there are no tRNAs to complement the termination codon, protein synthesis ends. An enzyme releases the completed polypeptide chain from the ribosome. The initiating amino acid methionine is often removed from the beginning of the peptide chain. Now the amino acids in the chain form the three-dimensional structure that makes the polypeptide into a biologically active protein.

SAMPLE PROBLEM 21.10

Protein Synthesis: Translation

What order of amino acids would you expect in a peptide for the mRNA sequence of 5′—UCA—AAA—GCC—CUU—3′?

SOLUTION

Each of the codons specifies a particular amino acid. Using Table 21.4, we write a peptide with the following amino acid sequence:

RNA codons: 5′—UCA—AAA—GCC—CUU—3′

                      ↓   ↓   ↓   ↓

Amino acid sequence: Ser — Lys — Ala — Leu

STUDY CHECK

Where would protein synthesis stop in the following series of bases in an mRNA?
5′—GGG—AGC—AGU—UAG—GUU—3′

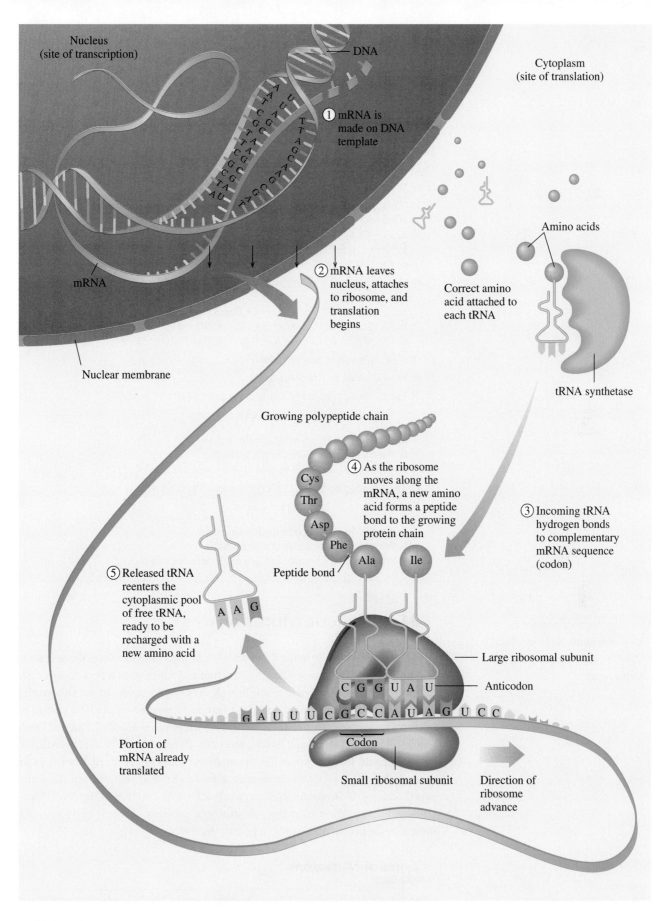

FIGURE 21.19

In the translation process, the mRNA synthesized by transcription attaches to a ribosome and tRNAs pick up their amino acids and place them in a growing peptide chain.

Q How is the correct amino acid placed in the peptide chain?

■ Protein Synthesis: Translation

21.43 What is the difference between a *codon* and an *anticodon*?

21.44 Why are there at least 20 different tRNAs?

21.45 What are the three steps of translation?

21.46 Where does protein synthesis take place?

21.47 What amino acid sequence would you expect from each of the following mRNA segments?
  a. 5′—AAA—AAA—AAA—3′
  b. 5′—UUU—CCC—UUU—CCC—3′
  c. 5′—UAC—GGG—AGA—UGU—3′

21.48 What amino acid sequence would you expect from each of the following mRNA segments?
  a. 5′—AAA—CCC—UUG—GCC—3′
  b. 5′—CCU—CGA—AGC—CCA—UGA—3′
  c. 5′—AUG—CAC—AAA—GAA—GUA—CUU—3′

21.49 How is a peptide chain extended?

21.50 What is meant by "translocation"?

21.51 The following portion of DNA is in the template DNA strand:

  3′—GCT—TTT—CAA—AAA—5′

  a. What is the corresponding mRNA section?
  b. What are the anticodons of the tRNAs?
  c. What amino acids will be placed in the peptide chain?

21.52 The following portion of DNA is in the template DNA strand:

  3′—TGT—GGG—GTT—ATT—5′

  a. What is the corresponding mRNA section?
  b. What are the anticodons of the tRNAs?
  c. What amino acids will be placed in the peptide chain?

LEARNING GOAL

Describe some ways in which DNA is altered to cause mutations.

21.8 Genetic Mutations

A **mutation** is a change in the DNA nucleotide sequence that alters the sequence of amino acids, which may alter the structure and function of a protein in a cell. Some mutations are known to result from X rays, overexposure to sun (ultraviolet or UV light), chemicals called mutagens, and possibly some viruses. If a change in DNA occurs in a somatic cell (a cell other than a reproductive cell) the altered DNA will be limited to that cell and its daughter cells. If there is uncontrolled growth, the mutation could lead to cancer. If the mutation occurs in germ cell DNA (egg or sperm), then all the DNA produced in a new individual will contain the same genetic change. If the genetic change greatly affects the catalysis of metabolic reactions or the formation of important structural proteins, the new cells may not survive or the person may exhibit a genetic disease.

Types of Mutations

Consider a triplet of bases CCG in the coding strand of DNA, which produces the codon GGC in mRNA. At the ribosome, tRNA would place the amino acid glycine in the peptide chain. (See Figure 21.20a.) Now, suppose that T replaces the first C

MC GOB Genetic Mutations

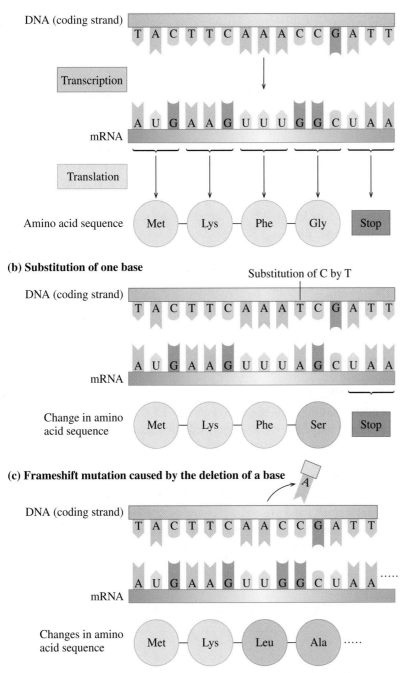

(a) **Normal DNA and protein synthesis**

(b) **Substitution of one base**

(c) **Frameshift mutation caused by the deletion of a base**

FIGURE 21.20

An alteration in the DNA coding strand (template) produces a change in the sequence of amino acids in the protein, which may lead to a mutation. **(a)** A normal DNA leads to the correct amino order in a protein. **(b)** The substitution of a base in DNA leads to a change in the mRNA codon and a change in the amino acid. **(c)** The deletion of a base causes a frameshift mutation, which changes the amino acid order.

Q **When would a substitution mutation cause protein synthesis to stop?**

in the DNA triplet, which gives TCG as the triplet. Then the codon produced in the mRNA is AGC, which brings the amino acid serine to the peptide chain. The replacement of one base in the coding strand of DNA with another is called a **substitution** mutation. The change in the codon can lead to the insertion of a different amino acid at that point in the polypeptide. Substitution is the most common way in which mutations occur. (See Figure 21.20b.)

In a **frameshift mutation**, a base is added to or deleted from the normal order of bases in the coding strand of DNA. Suppose now an A is deleted from the triplet AAA, which gives a new triplet of AAC. The next triplet becomes CGA rather than

▪ A Model for DNA Replication and Mutation

1. Cut out 16 rectangular pieces of paper. For strand 1 of a segment of DNA, write on two pieces these nucleotide symbols with hydrogen bonds: A═, T═, G≡, and C≡. For strand 2, write on two pieces each: ═A, ═T, ≡G, and ≡C. Mix up the pieces for strand 1 and randomly place in vertical order. From the second group, strand 2 nucleotides, select the complementary bases to complete this section of DNA.

2. Using the nucleotides from experiment 1, make a strand 1 of A—T—T—G—C—C. Arrange the corresponding bases to complete the DNA section. In the original column, change the G to an A. What does this do to the complementary strand? How could this change result in a mutation?

CCG and so on. All the triplets shift over by one base, which changes all the codons that follow and leads to a different sequence of amino acids from that point. Figure 21.20c illustrates a frameshift mutation by deletion.

Effect of Mutations

When a mutation causes a change in the amino acid sequence, the structure of the resulting protein can be altered severely and it may lose biological activity. If the protein is an enzyme, it may no longer bind to its substrate or react with the substrate at the active site. When an altered enzyme cannot catalyze a reaction, certain substances may accumulate until they act as poisons in the cell, or substances vital to survival may not be synthesized. If a defective enzyme occurs in a major metabolic pathway or in the building of a cell membrane, the mutation can be lethal. When a protein deficiency is genetic, the condition is called a **genetic disease**.

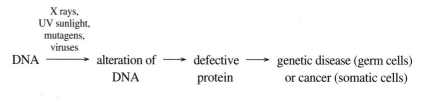

SAMPLE PROBLEM 21.11

▪ Mutations

An mRNA has the sequence of codons 5′—CCC—AGA—GGG—3′. If a base substitution in the DNA changes the mRNA codon of AGA to ACA, how is the amino acid sequence affected in the resulting protein?

SOLUTION

The initial mRNA sequence of CCC—AGA—GGG codes are for the amino acids proline, arginine, and glycine. When the mutation occurs, the new sequence of mRNA codons CCC—ACA—GGG are for proline, threonine, and glycine. The amino acid arginine is replaced by threonine.

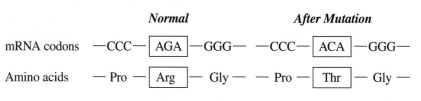

STUDY CHECK

How might the protein made from this mRNA be affected by this mutation?

Genetic Diseases

A genetic disease is the result of a defective enzyme caused by a mutation in its genetic code. For example, phenylketonuria, PKU, results when DNA cannot direct the synthesis of the enzyme phenylalanine hydroxylase, required for the conversion of phenylalanine to tyrosine. In an attempt to break down the phenylalanine, other enzymes in the cells convert it to phenylpyruvate. The accumulation of phenylalanine and phenylpyruvate in the blood can lead to severe brain damage and mental retardation. If PKU is detected in a newborn baby, a diet is prescribed that eliminates all the foods that contain phenylalanine. Preventing the buildup of the phenylpyruvate ensures normal growth and development.

The amino acid tyrosine is needed in the formation of melanin, the pigment that gives the color to our skin and hair. If the enzyme that converts tyrosine to melanin is defective, no melanin is produced, a genetic disease known as albinism. Persons and animals with no melanin have no skin or hair pigment. (See Figure 21.21.) Table 21.6 lists some other common genetic diseases and the type of metabolism or area affected.

FIGURE 21.21

This peacock with albinism does not produce the melanin needed to make bright colors of its feathers.

Q **Why are traits such as albinism related to the gene?**

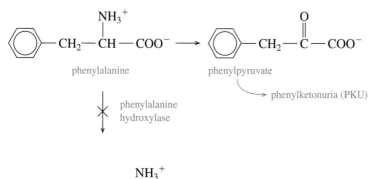

TABLE 21.6

Some Genetic Diseases

Genetic Disease	Result
Galactosemia	The transferase enzyme required for the metabolism of galactose-1-phosphate is absent. Accumulation of Gal-1-P leads to cataracts and mental retardation.
Cystic fibrosis	The most common inherited disease. Thick mucus secretions make breathing difficult and block pancreatic function.
Down syndrome	The leading cause of mental retardation, occurring in about 1 of every 800 live births. Mental and physical problems including heart and eye defects are the result of the formation of three chromosomes, usually chromosome 21, instead of a pair.
Familial hypercholesterolemia	A mutation of a gene on chromosome 19 results in high cholesterol levels that lead to early coronary heart disease in people 30–40 years old.
Muscular dystrophy (Duchenne)	One of 10 forms of MD. A mutation in the X chromosome results in the low or abnormal production of *dystrophin* by the X gene. This muscle-destroying disease appears at about age 5 with death by age 20 and occurs in about 1 of 10,000 males.
Huntington's disease (HD)	Appearing in middle age, HD affects the nervous system, leading to total physical impairment. It is the result of a mutation in a gene on chromosome 4, which can now be mapped to test people in families with HD.
Sickle-cell anemia	A defective hemoglobin from a mutation in a gene on chromosome 11 decreases the oxygen-carrying ability of red blood cells, which take on a sickled shape, causing anemia and plugged capillaries from red blood cell aggregation.
Hemophilia	One or more defective blood-clotting factors lead to poor coagulation, excessive bleeding, and internal hemorrhages.
Tay-Sachs disease	Hexosaminidase A is defective, causing an accumulation of gangliosides resulting in mental retardation, loss of motor control, and early death.

QUESTIONS AND PROBLEMS

■ Genetic Mutations

21.53 What is a substitution mutation?

21.54 How does a substitution mutation in the genetic code for an enzyme affect the order of amino acids in that protein?

21.55 What is the effect of a mutation on the amino acids in the polypeptide?

21.56 How can a mutation decrease the activity of a protein?

21.57 How is protein synthesis affected if the normal base sequence TTT in the DNA template is changed to TTC?

21.58 How is protein synthesis affected if the normal base sequence CCC is changed to ACC?

21.59 Consider the following portion of mRNA produced by the normal order of DNA nucleotides:

$$5'—ACA—UCA—CGG—GUA—3'$$

a. What is the amino acid order produced for normal DNA?
b. What is the amino acid order if a mutation changes UCA to ACA?
c. What is the amino acid order if a mutation changes CGG to GGG?
d. What happens to protein synthesis if a mutation changes UCA to UAA?
e. What happens if the A is removed from the beginning of a chain?

21.60 Consider the following portion of mRNA produced by the normal order of DNA nucleotides:

$$5'—CUU—AAA—CGA—GUU—3'$$

a. What is the amino acid order produced for normal DNA?
b. What is the amino acid order if a mutation changes CUU to CCU?
c. What is the amino acid order if a mutation changes CGA to AGA?
d. What happens to protein synthesis if a mutation changes AAA to UAA?
e. What happens if a G is added to the beginning of a chain?

21.61 a. A base substitution changes a codon for an enzyme from GCC to GCA. Why is there no change in the amino acid order in the protein?
b. In sickle-cell anemia, a base substitution in hemoglobin replaces glutamine (a polar amino acid) with valine. Why does the replacement of one amino acid cause such a drastic change in biological function?

21.62 a. A base substitution for an enzyme replaces leucine (a nonpolar amino acid) with alanine. Why does this change in amino acids have little effect on the biological activity of the enzyme?
b. A base substitution replaces cytosine in the codon UCA with adenine. How would this substitution affect the amino acids in the protein?

LEARNING GOAL

Describe the preparation and uses of recombinant DNA.

SELF-STUDY ACTIVITY
Applications of DNA Technology

MC GOB

Restriction Enzymes

MC GOB Recombinant DNA

21.9 Recombinant DNA

Over the past two decades, new techniques called genetic engineering have permitted scientists to cut, splice, and recombine DNA from different kinds of cells. The new synthetic forms of DNA, which contain DNA fragments from different organisms, are known as **recombinant DNA**. Recombinant DNA technology is now used to produce human insulin for diabetics, the antiviral substance interferon, bloodclotting factor VIII, and human growth hormone.

Preparing Recombinant DNA

Most of the work in recombinant DNA is done with *Escherichia coli (E. coli)* bacteria. The DNA in prokaryotes is present in the bacterial cells as several

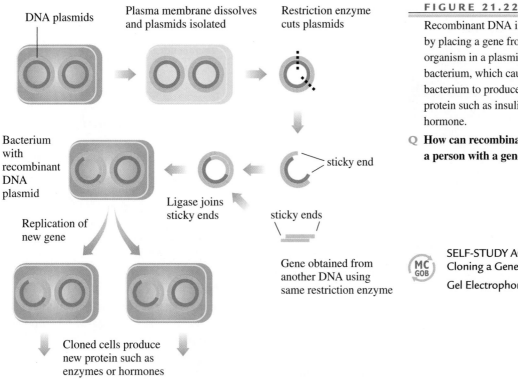

FIGURE 21.22

Recombinant DNA is formed by placing a gene from another organism in a plasmid DNA of the bacterium, which causes the bacterium to produce a nonbacterial protein such as insulin or growth hormone.

Q **How can recombinant DNA help a person with a genetic disease?**

SELF-STUDY ACTIVITY
Cloning a Gene in Bacteria

Gel Electrophoresis of DNA

small circular molecules called *plasmids*. These plasmids, which are easy to isolate, are also capable of replication. Initially, *E. coli* cells are soaked in a detergent solution to dissolve the plasma membrane. The contents of the cells including the plasmids are released and collected. A *restriction enzyme*, which breaks phosphodiester bonds in DNA between specific nucleotides, is used to cut open the circular DNA strands in the plasmids. (See Figure 21.22.) The same enzymes are also used to cut out a piece of DNA from the chromosome of a different organism, such as the gene that produces insulin or growth hormone. The cut-out genes are then mixed with the plasmids that were cut open. The ends of the foreign DNA piece and the ends of the opened plasmids are joined by a DNA ligase. Then the altered plasmids containing the recombined DNA are placed in a fresh culture of *E. coli* bacteria where they can be reabsorbed into the bacterial cells. The new gene that was inserted in the plasmids is copied as the genetically engineered *E. coli* cells start to replicate. In one day, one *E. coli* bacterium is capable of producing a million copies of itself including the foreign DNA, a process known as gene cloning. If the inserted DNA coded for the human insulin protein, the altered plasmids would begin to synthesize human insulin. Eventually, a large number of cells with the new DNA produce the insulin protein. Table 21.7 lists some of the products developed through recombinant DNA technology that are now used therapeutically.

SELF-STUDY ACTIVITY
Analyzing DNA Fragments
Using Gel Electrophoresis

DNA Fingerprinting

In a process referred to as DNA fingerprinting or Southern transfer, restriction enzymes are used to cut fragments from DNA. The resulting DNA fragments called RFLPs (restriction fragment length polymorphisms) are then sorted by size using gel electrophoresis. The gel is treated with a radioactive isotope that adheres

SELF-STUDY ACTIVITY
DNA Fingerprinting

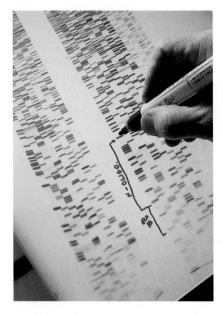

A scientist analyzes a nucleotide sequence in DNA from a human gene. Dark and light bands on the film represent the order of nucleotides. The marked sequences are involved in the growth of melanoma cancer cells.

Q **What causes DNA fragments to appear on x-ray film?**

MC GOB SELF-STUDY ACTIVITY
The Human Genome Project:
Human Chromosome 17

TABLE 21.7

Therapeutic Products of Recombinant DNA

Product	Therapeutic Use
Human insulin	Treat diabetes
Erythropoietin (EPO)	Treat anemia; stimulate production of erythrocytes
Human growth hormone (HGH)	Stimulate growth
Interferon	Treat cancer and viral disease
Tumor necrosis factor (TNF)	Destroy tumor cells
Monoclonal antibodies	Transport drugs needed to treat cancer and transplant rejection
Epidermal growth factor (EGF)	Stimulate healing of wounds and burns
Interleukins	Stimulate immune system; treat cancer
Prourokinase	Destroy blood clots; treat myocardial infarctions

to specific base sequences in the RFLPs. A piece of x-ray film placed over the gel is exposed by the radioactivity emitted from RFLPs, which gives a pattern of dark and light bands known as a DNA fingerprint. (See Figure 21.23.) It has been estimated that the odds of two persons (who are not identical twins) producing the same DNA fingerprint is less than one in a billion.

One application of Southern transfer is genetic screening in which an individual can be screened for genes that are responsible for genetic diseases. It is being used to screen for genes responsible for sickle-cell disease, cystic fibrosis, breast cancer, colon cancer, Huntington's disease, and amyotropic lateral sclerosis, also called Lou Gehrig's disease.

Human Genome Project

In the 1970s, scientists began to use restriction enzymes to map the location of genomes in viruses and plasmids. By 1987, the genome of *E. coli* was determined. Recently, these techniques combined with new computer programs have helped scientists compile the map of the human genome. In some recent reports, the total number of human genes is about 30,000. It now appears that most of the genome is not functional and perhaps carried over for millions of years. Large blocks of genes are copied from one human chromosome to another even though they no longer code for needed proteins. Thus, the coding portions of the genes seems to be even less than previously thought, making up only about 1% of the total genome. The results of the genome project will help us identify defective genes that lead to genetic disease.

Polymerase Chain Reaction

The process of gene cloning using recombinant DNA requires living cells such as *E. coli*. In 1987 a process called **polymerase chain reaction (PCR)** made it possible to produce multiple copies of (amplify) the DNA in a short time. In the PCR technique, a sequence of a DNA is selected to copy, and the DNA is heated to separate the strands. Primers that are complementary to a small group of nucleotides on each side of the sequence to be copied are added to the ends of the templates. The DNA strands with their primers are mixed with DNA polymerase and a mixture of deoxyribonucleotides and complementary strands

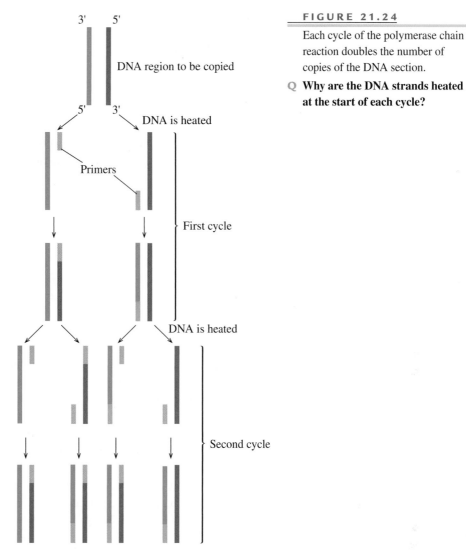

FIGURE 21.24

Each cycle of the polymerase chain reaction doubles the number of copies of the DNA section.

Q **Why are the DNA strands heated at the start of each cycle?**

for the DNA section are produced. Then the process is repeated with the new batch of DNA. After several cycles of the PCR process, millions of copies of the initial DNA section are produced. (See Figure 21.24.)

SAMPLE PROBLEM 21.12

▪ Recombinant DNA

What is the function of restriction enzymes in recombinant DNA?

SOLUTION

Restriction enzymes are used to cut out a particular piece of DNA from a gene and to cut open the circular plasmids in a bacterium where the foreign DNA attaches.

STUDY CHECK

What is gene cloning?

■Recombinant DNA

21.63 Why are *E. coli* bacteria used in recombinant DNA procedures?

21.64 What is a plasmid?

21.65 How are plasmids obtained from *E. coli*?

21.66 Why are restriction enzymes mixed with the plasmids?

21.67 How is a gene for a particular protein inserted into a plasmid?

21.68 Why is DNA polymerase useful in criminal investigations?

21.69 What is a DNA fingerprint?

21.70 What beneficial proteins are produced from recombinant DNA technology?

LEARNING GOAL

Describe the methods by which a virus infects a cell.

TABLE 21.8

Some Diseases Caused by Viral Infection	
Disease	**Virus**
Common cold	Coronavirus (over 100 types)
Influenza	Orthomyxovirus
Warts	Papovavirus
Herpes	Herpesvirus
Leukemia, cancers, AIDS	Retrovirus
Hepatitis	Hepatitis A virus (HAV), hepatitis B virus (HBV), hepatitis C virus (HCV)
Mumps	Paramyxovirus
Epstein–Barr	Epstein–Barr virus (EBV)

 Biological Pirates

21.10 Viruses

Viruses are small particles of 3 to 200 genes that cannot replicate without a host cell. A typical virus contains a nucleic acid, DNA or RNA, but not both, inside a protein coat. It does not have the necessary material such as nucleotides and enzymes to make proteins and grow. The only way a virus can replicate is to invade a host cell and take over the materials necessary for protein synthesis and growth. Some infections caused by viruses invading human cells are listed in Table 21.8. There are also viruses that attack bacteria, plants, and animals.

A viral infection begins when an enzyme in the protein coat makes a hole in the host cell, allowing the nucleic acids to enter and mix with the materials in the host cell. (See Figure 21.25.) If the virus contains DNA, the host cell begins to replicate the viral DNA in the same way it would replicate normal DNA. Viral DNA produces viral RNA, and a protease produces a protein coat to form a viral particle that leaves the cell. (See Figure 21.26.) So many virus particles are synthesized that the cell bursts and releases new viruses to infect more cells.

Vaccines are inactive forms of viruses that boost the immune response by causing the body to produce antibodies to the virus. Several childhood diseases such as polio, mumps, chicken pox, and measles can be prevented through the use of vaccines.

Reverse Transcription

A virus that contains RNA as its genetic material is a **retrovirus**. Once inside the host cell, it must first make viral DNA using a process known as reverse transcription. A retrovirus contains a polymerase enzyme called *reverse transcriptase* that uses the viral RNA template to synthesize complementary strands of DNA. Once produced, the DNA strands form double-stranded DNA using the nucleotides present in the host cell. This newly formed viral DNA, called a *provirus*, joins the DNA of the host cell.

AIDS

In the early 1980s, a disease called AIDS (acquired immune deficiency syndrome) began to claim an alarming number of lives. An HIV-1 virus (human immunodeficiency virus type 1) is now known to be the AIDS-causing agent.

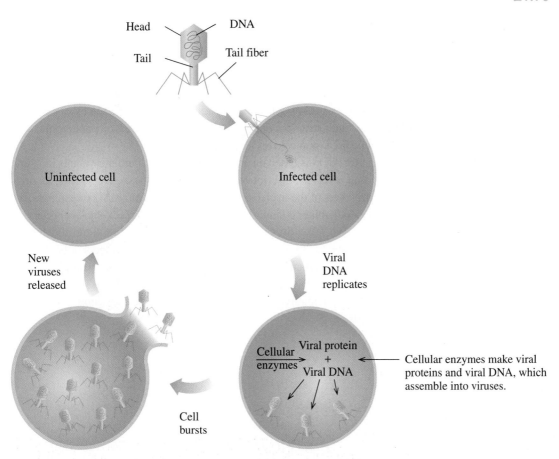

FIGURE 21.25

After a virus attaches to the host cell, it injects its viral DNA and uses the host cell's amino acids to synthesize viral protein and nucleic acids, enzymes, and ribosomes to make viral RNA. When the cell bursts, the new viruses are released to infect other cells.

Q **Why does a virus need a host cell for replication?**

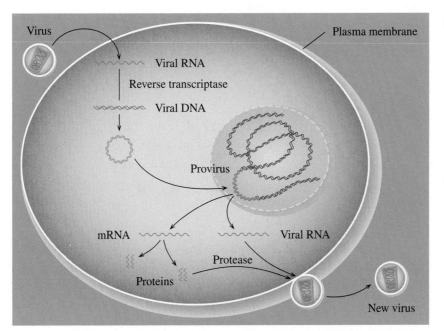

FIGURE 21.26

After a retrovirus injects its viral RNA into a cell, it forms a DNA strand by reverse transcription. The DNA forms a double-stranded DNA called a provirus, which joins the host cell DNA. When the cell replicates, the provirus produces the viral RNA needed to produce more virus particles.

Q **What is reverse transcription?**

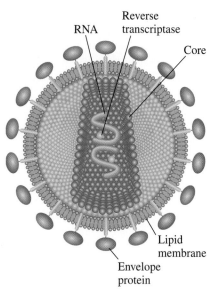

RNA

Reverse transcriptase

Core

Lipid membrane

Envelope protein

FIGURE 21.27

The HIV virus that causes AIDS syndrome destroys the immune system in the body.

Q **Is the HIV virus a DNA virus or an RNA retrovirus?**

SELF-STUDY ACTIVITY
HIV Reproductive Cycle

(See Figure 21.27.) HIV is a retrovirus that infects and destroys T4 lymphocyte cells, which are involved in the immune response. After the HIV-1 virus binds to receptors on the surface of a T4 cell, the virus injects viral RNA into the host cell. As a retrovirus, the genes of the viral RNA direct the formation of viral DNA. The gradual depletion of T4 cells reduces the ability of the immune system to destroy harmful organisms. The AIDS syndrome is characterized by opportunistic infections such as *Pneumocystis carinii*, which causes pneumonia, and *Kaposi's sarcoma*, a skin cancer.

Treatment for AIDS is based on attacking the HIV-1 at different points in its life cycle including reverse transcription and protein synthesis. Nucleoside analogs mimic the structures of the nucleosides used for DNA synthesis. For example, the drug AZT (azidothymine) is similar to thymine, and ddI (dideoxyinosine) is similar to guanosine. Two other drugs include dideoxycytidine (ddC) and didehydro-3′-deoxythymidine (d4T). Such compounds are found in the "cocktails" that are providing extended remission of HIV infections. When a nucleoside analog is incorporated into viral DNA, the lack of a hydroxyl group on the 3′ carbon in the sugar prevents the formation of the sugar–phosphate bonds and stops the replication of the virus.

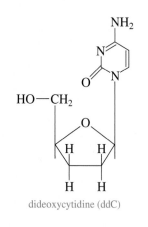

AZT
azidothymine

dideoxyinosine (ddI)

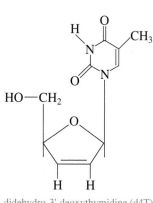

dideoxycytidine (ddC)

didehydro-3'-deoxythymidine (d4T)

The newest and most powerful anti-HIV drugs are the protease inhibitors such as saquinavir (Invirase), indinavir, and ritonavir. The inhibition of protease prevents the synthesis of proteins needed to make more copies of the virus. Researchers are not yet certain how long protease inhibitors will be beneficial for a person with AIDS, but they are encouraged by the current studies.

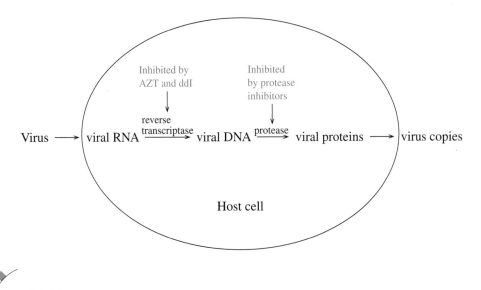

■ Cancer

In an adult body, many cells do not continue to reproduce. When cells in the body begin to grow and multiply without control, they invade neighboring cells and appear as a tumor or growth (neoplasm). When tumors interfere with normal functions of the body, they are cancerous. If they are limited, they are benign. Cancer can be caused by chemical and environmental substances, by radiation, or by oncogenic viruses.

Some reports estimate that 70–80% of all human cancers are initiated by chemical and environmental substances. A carcinogen is any substance that increases the chance of inducing a tumor. Known carcinogens include dyes, cigarette smoke, and asbestos. More than 90% of all persons with lung cancer are smokers. A carcinogen causes cancer by reacting with molecules in a cell, probably DNA, and altering the growth of that cell. Some known carcinogens are listed in Table 21.9.

Radiant energy from sunlight or medical radiation is another type of environmental factor. Skin cancer has become one of the most prevalent forms of cancer. It appears that DNA damage in the exposed areas of the skin causes mutations. The cells lose their ability to control protein synthesis, and uncontrolled cell division leads to cancer. The incidence of malignant melanoma, one of the most serious skin cancers, has been rapidly increasing. Some possible factors for this increase may be the popularity of suntanning as well as the reduction of the ozone layer, which absorbs much of the harmful radiation from sunlight.

Oncogenic viruses cause cancer when cells are infected. Several viruses associated with human cancers are listed in Table 21.10. Some cancers such as retinoblastoma and breast cancer appear to occur more frequently in families. There is some indication that a missing or defective gene may be responsible.

TABLE 21.9

Some Chemical and Environmental Carcinogens

Carcinogen	Tumor Site
Asbestos	Lung, respiratory tract
Arsenic	Skin, lung
Cadmium	Prostate, kidneys
Chromium	Lung
Nickel	Lung, sinuses
Aflatoxin	Liver
Nitrites	Stomach
Aniline dyes	Bladder
Vinyl chloride	Liver

TABLE 21.10

Human Cancers Caused by Oncogenic Viruses

Virus	Disease
RNA viruses	
Human T-cell lymphotropic virus-type I (HTLV-I)	Leukemia
DNA viruses	
Epstein–Barr virus (EBV)	Burkitt's lymphoma (cancer of white blood B cells)
	Nasopharyngeal carcinoma
	Hodgkin's disease
Hepatitis B virus (HBV)	Liver cancer
Herpes simplex virus (type 2)	Cervical and uterine cancer
Papilloma virus	Cervical and colon cancer, genital warts

SAMPLE PROBLEM 21.13

▪ Viruses

Why are viruses unable to replicate on their own?

SOLUTION

Viruses only contain packets of DNA or RNA, but not the necessary replication machinery that includes enzymes and nucleosides.

STUDY CHECK

How do protease inhibitors affect the life cycle of the HIV-1 virus?

QUESTIONS AND PROBLEMS

▪ Viruses

21.71 What type of genetic information is found in a virus?

21.72 Why do viruses need to invade a host cell?

21.73 A virus contains viral RNA.
  a. Why would reverse transcription be used in the life cycle of this type of virus?
  b. What is the name of this type of virus?

21.74 What is the purpose of a vaccine?

21.75 How do nucleoside analogs disrupt the life cycle of the HIV-1 virus?

21.76 How do protease inhibitors disrupt the life cycle of the HIV-1 virus?

CONCEPT MAP

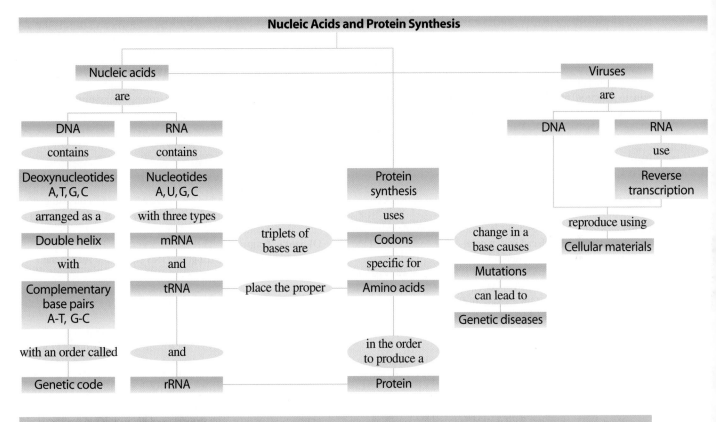

CHAPTER REVIEW

21.1 Components of Nucleic Acids

Nucleic acids, deoxyribonucleic acid (DNA), and ribonucleic acid (RNA) are polymers of nucleotides. A nucleoside is a combination of a pentose sugar and a nitrogen base. A nucleotide is composed of three parts: a nitrogenous base, a sugar, and a phosphate group. In DNA, the sugar is deoxyribose and the nitrogen-containing base can be adenine, thymine, guanine, or cytosine. In RNA, the sugar is ribose and uracil replaces thymine.

21.2 Primary Structure of Nucleic Acids

Each nucleic acid has its own unique sequence of bases known as its primary structure. In a nucleic acid polymer, the 3′OH of each ribose sugar in RNA or deoxyribose sugar in DNA forms a phosphodiester bond to the phosphate group of the 5′-carbon atom group of the sugar in the next nucleotide to give a backbone of alternating sugar and phosphate groups. There is a free 5′-phosphate at one end of the polymer, and a free 3′ OH group at the other end.

21.3 DNA Double Helix

A DNA molecule consists of two strands of nucleotides that are wound around each other like a spiral staircase. The two strands are held together by hydrogen bonds between complementary base pairs, A with T, and G with C.

21.4 DNA Replication

During DNA replication, DNA polymerase makes new DNA strands along each of the original DNA strands that serve as templates. Complementary base pairing ensures the correct pairing of bases to give identical copies of the original DNA.

21.5 RNA and Transcription

The three types of RNA differ by function in the cell: ribosomal RNA makes up most of the structure of the ribosomes, messenger RNA carries genetic information from the DNA to the ribosomes, and transfer RNA places the correct amino acids in the protein. Transcription is the process by which RNA polymerase produces mRNA from one strand of DNA. The bases in the mRNA are complementary to the DNA, except U is paired with A in RNA. The production of mRNA occurs when certain proteins are needed in the cell. In enzyme induction, the appearance of a substrate in a cell removes a repressor, which allows RNA polymerase to produce mRNA at the structural genes.

21.6 The Genetic Code

The genetic code consists of a sequence of three bases (triplet) that specifies the order for the amino acids in a protein. There are 64 codons for the 20 amino acids, which means there are several codons for some amino acids. The codon AUG signals the start of transcription and codons UAG, UGA, and UAA signal it to stop.

21.7 Protein Synthesis: Translation

Proteins are synthesized at the ribosomes in a translation process that includes three steps: initiation, translocation, and termination. During translation, tRNAs bring the appropriate amino acids to the ribosome and peptide bonds form. When the polypeptide is released, it takes on its secondary and tertiary structures and becomes a functional protein in the cell.

21.8 Genetic Mutations

A genetic mutation is a change of one or more bases in the DNA sequence that alters the structure and ability of the resulting protein to function properly. In a substitution, one base may be altered, and a frameshift mutation inserts or deletes a base and changes all the codons after the base change.

21.9 Recombinant DNA

A recombinant DNA is prepared by inserting a DNA segment, a gene, into the DNA present in plasmids of *E. coli* bacteria. As the altered bacterial cells replicate, the protein expressed by the foreign DNA segment is produced. In criminal investigation, large quantities of DNA are obtained from smaller amounts by DNA polymerase chain reactions.

21.10 Viruses

Viruses containing DNA or RNA must invade host cells to use the machinery within the cell for the synthesis of more viruses. For a retrovirus containing RNA, a viral DNA is synthesized by reverse transcription using the nucleotides and enzymes in the host cell. In the treatment of AIDS, nucleoside analogs inhibit the reverse transcriptase of the HIV-1 virus, and protease inhibitors disrupt the catalytic activity of protease needed to produce proteins for the synthesis of more viruses.

KEY TERMS

anticodon The triplet of bases in the center loop of tRNA that is complementary to a codon on mRNA.

codon A sequence of three bases in mRNA that specifies a certain amino acid to be placed in a protein. A few codons signal the start or stop of transcription.

complementary base pairs In DNA, adenine is always paired with thymine (A—T or T—A), and guanine is always paired with cytosine (G—C or C—G). In forming RNA, adenine is paired with uracil (A—U).

control site A section of DNA composed of a promoter and operator that regulates protein synthesis.

DNA Deoxyribonucleic acid; the genetic material of all cells containing nucleotides with deoxyribose sugar, phosphate, and the four nitrogenous bases adenine, thymine, guanine, and cytosine.

double helix The helical shape of the double chain of DNA that is like a spiral staircase with a sugar–phosphate backbone on the outside and base pairs like stair steps on the inside.

enzyme induction A model of cellular regulation in which protein synthesis is induced by a substrate.

exons The sections in a DNA template that code for proteins.

frameshift mutation A mutation that inserts or deletes a base in a DNA sequence.

genetic code The sequence of codons in mRNA that specifies the amino acid order for the synthesis of protein.

genetic disease A physical malformation or metabolic dysfunction caused by a mutation in the base sequence of DNA.

introns The sections in DNA that do not code for protein.

mRNA Messenger RNA; produced in the nucleus by DNA to carry the genetic information to the ribosomes for the construction of a protein.

mutation A change in the DNA base sequence that alters the formation of a protein in the cell.

nitrogen-containing base Nitrogen-containing compounds found in DNA and RNA: adenine (A), thymine (T), cytosine (C), guanine (G), and uracil (U).

nucleic acids Large molecules composed of nucleotides, found as a double helix in DNA, and as the single strands of RNA.

nucleoside The combination of a pentose sugar and a nitrogen-containing base.

nucleotides Building blocks of a nucleic acid consisting of a nitrogen-containing base, a pentose sugar (ribose or deoxyribose), and a phosphate group.

Okazaki fragments The short segments formed by DNA polymerase in the daughter DNA strand that runs in the 3′ to 5′ direction.

operon A group of genes, including a control site and structural genes, whose transcription is controlled by the same regulatory gene.

phosphodiester bond The phosphate link that joins the 3′ hydroxyl group in one nucleotide to the phosphate group on the 5′-carbon atom in the next nucleotide.

polymerase chain reaction (PCR) A strand of DNA is copied many times by mixing it with DNA polymerase and a mixture of deoxyribonucleotides.

primary structure The sequences of nucleotides in nucleic acids.

recombinant DNA DNA spliced from different organisms to form new, synthetic DNA.

regulatory gene A gene in front of the control site that produces a repressor.

replication The process of duplicating DNA by pairing the bases on each parent strand with their complementary base.

replication forks The open sections in unwound DNA strands where DNA polymerase begins the replication process.

repressor A protein that interacts with the operator gene in an operon to prevent the transcription of mRNA.

retrovirus A virus that contains RNA as its genetic material and that synthesizes a complementary DNA strand inside a cell.

RNA Ribonucleic acid, a type of nucleic acid that is a single strand of nucleotides containing adenine, cytosine, guanine, and uracil.

rRNA Ribosomal RNA; the most prevalent type of RNA; a major component of the ribosomes.

structural genes The sections of DNA that code for the synthesis of proteins.

substitution A mutation that replaces one base in a DNA with a different base.

transcription The transfer of genetic information from DNA by the formation of mRNA.

translation The interpretation of the codons in mRNA as amino acids in a peptide.

translocation The shift of a ribosome along mRNA from one codon (three bases) to the next codon during translation.

tRNA Transfer RNA; an RNA that places a specific amino acid into a peptide chain at the ribosome. There is one or more tRNA for each of the 20 different amino acids.

virus Small particles containing DNA or RNA in a protein coat that require a host cell for replication.

UNDERSTANDING THE CONCEPTS

21.77 Answer the following questions for the given section of DNA.
   a. Complete the bases in the parent and template strands.

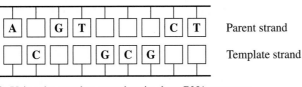
Parent strand

Template strand

   b. Using the template strand, write the mRNA sequence.

   c. Write the 3-letter symbols of the amino acids that would go into the peptide from the mRNA you wrote in part b.

21.78 Suppose a mutation occurs in the DNA section in problem 21.77 and the first base in the parent chain, adenine, is replaced by guanine.
   a. What type of mutation has occurred?
   b. Using the template strand that results from this mutation, write the order of bases in the altered mRNA.

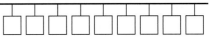

   c. Write the 3-letter symbols of the amino acids that would go into the peptide from the mRNA you wrote in part b.

   d. What effect, if any, might this mutation have on the structure and/or function of the resulting protein?

ADDITIONAL QUESTIONS AND PROBLEMS

For instructor-assigned homework, go to
www.masteringgob.com.

21.79 Identify each of the following nitrogen bases as a pyrimidine or a purine:
   a. cytosine   **b.** adenine   **c.** uracil
   d. thymine   **e.** guanine

21.80 Indicate if each of the nitrogen bases in problem 21.79 are found in DNA only, RNA only, or both DNA and RNA.

21.81 Identify the nitrogen base and sugar in each of the following nucleosides:
   a. deoxythymidine   **b.** adenosine
   c. cytidine   **d.** deoxyguanosine

21.82 Identify the nitrogen base and sugar in each of the following nucleotides:
   a. CMP   **b.** dAMP
   c. dGMP   **d.** UMP

21.83 How do the bases thymine and uracil differ?

21.84 How do the bases cytosine and uracil differ?

21.85 Draw the structure of CMP.

21.86 Draw the structure of dGMP.

21.87 What is similar about the primary structure of RNA and DNA?

21.88 What is different about the primary structure of RNA and DNA?

21.89 If the DNA double helix in salmon contains 28% adenine, what is the percent of thymine, guanine, and cytosine?

21.90 If the DNA double helix in humans contains 20% cytosine, what is the percent of guanine, adenine, and thymine?

21.91 In DNA, how many hydrogen bonds form between each of the following?
   a. adenine and thymine
   b. guanine and cytosine

21.92 Write the complementary base sequence for each of the following DNA segments:
   a. 5′—TTACGGACCGC—3′
   b. 5′—ATAGCCCTTACTGG—3′
   c. 3′—GGCCTACCTTAACGACG—5′

21.93 Write the complementary base sequence for each of the following DNA segments:
   a. 5′—GACTTAGGC—3′
   b. 3′—TGCAAACTAGCT—5′
   c. 5′—ATCGATCGATCG—3′

21.94 How are the Okazaki fragments joined to the growing DNA strand?

21.95 In DNA replication, what is the difference in the synthesis of the leading strand and the lagging strand?

21.96 How can replication occur at several places along a DNA double helix?

21.97 Where are the DNA strands of the original DNA found in the double helix of each of the daughter DNA molecules?

21.98 Match the following statements with rRNA, mRNA, or tRNA.
   a. have two subunits
   b. brings amino acids to the ribosomes for protein synthesis
   c. acts as a template for protein synthesis

21.99 Match the following statements with rRNA, mRNA, or tRNA.
   a. is the smallest type of RNA
   b. makes up the highest percent of RNA in the cell
   c. carries genetic information from the nucleus to the ribosomes

21.100 What are the possible codons for each of the following amino acids?
   a. valine   **b.** proline   **c.** histidine

21.101 What are the possible codons for each of the following amino acids?
   a. threonine   **b.** serine   **c.** cysteine

21.102 What is the amino acid for each of the following codons?
   a. CAA   **b.** GGC   **c.** AAC

21.103 What is the amino acid for each of the following codons?
   a. AAG   **b.** AUU   **c.** CGG

21.104 Endorphins are polypeptides that reduce pain. What is the amino acid order for the following mRNA that codes for a pentapeptide that is an endorphin called methionine enkephalin?
   5′—AUG—UAC—GGU—GGA—UUU—AUG—UAA—3′

21.105 Endorphins are polypeptides that reduce pain. What is the amino acid order for the following mRNA that codes for a pentapeptide that is an endorphin called leucine enkephalin?
   5′—AUG—UAC—GGU—GGA—UUU—CUA—UAA—3′

21.106 What is the anticodon on tRNA for each of the following codons in an mRNA?
   a. GUG   **b.** CCC   **c.** GAA

21.107 What is the anticodon on tRNA for each of the following codons in an mRNA?
   a. AGC   **b.** UAU   **c.** CCA

21.108 How does polymerase chain reaction (PCR) produce many copies of a DNA section?

CHALLENGE QUESTIONS

21.109 Oxytocin is a nonapeptide with nine amino acids. How many nucleotides would be found in the mRNA for this protein?

21.110 Why are there no base pairs in DNA between adenine and guanine, or thymine and cytosine?

21.111 What is the difference between a DNA virus and a retrovirus?

21.112 A protein contains 35 amino acids. How many nucleotides would be found in the mRNA for this protein?

ANSWERS

Answers to Study Checks

21.1   **a.** Guanine is found in both RNA and DNA.
      b. Uracil is found only in RNA.

21.2   deoxycytidine 5′-monophosphate (dCMP)

21.3

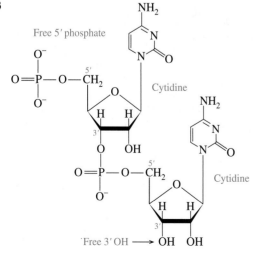

21.4   3′—C—C—A—A—T—T—G—G—5′

21.5   In DNA, the lagging strand is synthesized in the 3′–5′ direction in short sections that are joined by DNA ligase.

21.6   Each type of tRNA matches a specific codon to a specific amino acid and brings the amino acids to the ribosomes for protein synthesis.

21.7   3′—CCCAAATTT—5′

21.8   A regulatory gene produces mRNA for the production of a protein repressor, which binds to the operator and blocks protein synthesis.

21.9   UGA is a stop codon that signals the termination of translation.

21.10   at UAG

21.11   If the substitution of an amino acid in the polypeptide affects an interaction essential to functional structure on the binding of a substrate, the resulting protein could be less effective or nonfunctional.

21.12   Gene cloning is the process by which recombinant DNA technology inserts the DNA of a gene into the plasmid of

E. coli bacteria that multiply rapidly to make many copies of the gene.

21.13 A protease inhibitor modifies the three-dimensional structure of protease, which prevents the enzymes from synthesizing the proteins needed to produce more viruses.

Answers to Selected Questions and Problems

21.1 **a.** pyrimidine   **b.** pyrimidine

21.3 **a.** DNA   **b.** both DNA and RNA

21.5 deoxyadenosine 5′-monophosphate (dAMP), deoxythymidine 5′-monophosphate (dTMP), deoxycytidine 5′-monophosphate (dCMP), and deoxyguanosine 5′-monophosphate (dGMP)

21.7 **a.** nucleoside   **b.** nucleoside
c. nucleoside   **d.** nucleotide

21.9

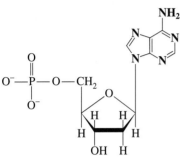

21.11 The nucleotides in nucleic acids are held together by phosphodiester bonds between the 3′-OH of a sugar (ribose or deoxyribose) and a phosphate group on the 5′-carbon of another sugar.

21.13

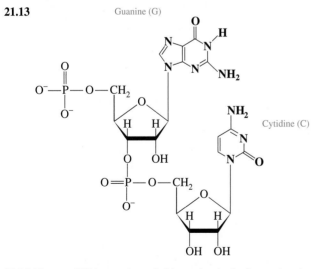

21.15 The two DNA strands are held together by hydrogen bonds between the bases in each strand.

21.17 **a.** 3′—TTTTTT—5′
b. 3′—CCCCCC—5′
c. 3′—TCAGGTCCA—5′
d. 3′—GACATATGCAAT—5′

21.19 The enzyme helicase unwinds the DNA helix to prepare the parent DNA strand for the synthesis of daughter DNA strands.

21.21 The DNA strands separate and the DNA polymerase pairs each of the bases with its complementary base and produces two exact copies of the original DNA.

21.23 Ribosomal RNA, messenger RNA, and transfer RNA.

21.25 A ribosome consists of a small subunit and a large subunit.

21.27 In transcription, the sequence of nucleotides on a DNA template (one strand) is used to produce the base sequences of a messenger RNA.

21.29 5′—GGCUUCCAAGUG—3′

21.31 In eukaryotic cells, genes contain sections called exons that code for protein and sections called introns that do not code for protein.

21.33 An operon is a section of DNA that regulates the synthesis of one or more proteins.

21.35 When the lactose level is low in *E. coli*, a repressor produced by the mRNA from the regulatory gene binds to the operator blocking the synthesis of mRNA from the genes and preventing the synthesis of protein.

21.37 A three-base sequence in mRNA that codes for a specific amino acid in a protein.

21.39 **a.** leucine   **b.** serine
c. glycine   **d.** arginine

21.41 When AUG is the first codon, it signals the start of protein synthesis. Thereafter, AUG codes for methionine.

21.43 A codon is a base triplet in the mRNA. An anticodon is the complementary triplet on a tRNA for a specific amino acid.

21.45 Initiation, translocation, and termination.

21.47 **a.** —Lys—Lys—Lys—
b. —Phe—Pro—Phe—Pro—
c. —Tyr—Gly—Arg—Cys—

21.49 The new amino acid is joined by a peptide bond to the peptide chain. The ribosome moves to the next codon, which attaches to a tRNA carrying the next amino acid.

21.51 **a.** 5′—CGA—AAA—GUU—UUU—3′
b. GCU, UUU, CAA, AAA
c. Using codons in mRNA: Arg—Lys—Val—Phe

21.53 A base in DNA is replaced by a different base.

21.55 If the resulting codon still codes for the same amino acid, there is no effect. If the new codon codes for a different amino acid, there is a change in the order of amino acids in the polypeptide.

21.57 The normal triplet TTT forms a codon AAA, which codes for lysine. The mutation TTC forms a codon AAG, which also codes for lysine. There is no effect on the amino acid sequence.

21.59 a. —Thr—Ser—Arg—Val—
b. —Thr—Thr—Arg—Val—
c. —Thr—Ser—Gly—Val
d. —Thr—STOP   Protein synthesis would terminate early. If this occurs early in the formation of the polypeptide, the resulting protein will probably be nonfunctional.
e. The new protein will contain the sequence —His—His—Gly—.

21.61 a. GCC and GCA both code for alanine.
b. A vital ionic cross-link in the tertiary structure of hemoglobin cannot be formed when the polar glutamine is replaced by valine, which is nonpolar. The resulting hemoglobin is malformed and less capable of carrying oxygen.

21.63 *E. coli* bacterial cells contain several small circular plasmids of DNA that can be isolated easily. After the recombinant DNA is formed, *E. coli* multiply rapidly, producing many copies of the recombinant DNA in a relatively short time.

21.65 *E. coli* are soaked in a detergent solution that dissolves the cell membrane and releases the cell contents including the plasmids, which are collected.

21.67 When a gene has been obtained using restriction enzymes, it is mixed with the plasmids that have been opened by the same enzymes. When mixed together in a fresh *E. coli* culture, the sticky ends of the DNA fragments bond with the sticky ends of the plasmid DNA to form a recombinant DNA.

21.69 In DNA fingerprinting, restriction enzymes cut a sample DNA into fragments, which are sorted by size by gel electrophoresis. After tagging the DNA fragments with a radioactive isotope, a piece of x-ray film placed over the gel is exposed by the radioactivity to give a pattern of dark and light bands known as a DNA fingerprint.

21.71 DNA or RNA, but not both.

21.73 a. A viral RNA is used to synthesize a viral DNA to produce the proteins for the protein coat, which allows the virus to replicate and leave the cell.
b. retrovirus

21.75 Nucleoside analogs such as AZT and ddI are similar to the nucleosides required to make viral DNA in reverse transcription. However, they interfere with the ability of the DNA to form and thereby disrupt the life cycle of the HIV-1 virus.

21.77 a.

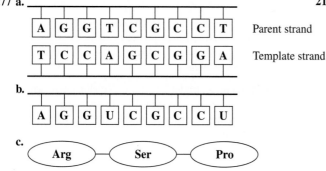

Parent strand

Template strand

b.

c.

21.79 a. pyrimidine   **b.** purine   **c.** pyrimidine
d. pyrimidine   **e.** purine

21.81 a. thymine and deoxyribose   **b.** adenine and ribose
c. cytosine and ribose   **d.** guanine and deoxyribose

21.83 They are both pyrimidines, but thymine has a methyl group.

21.85

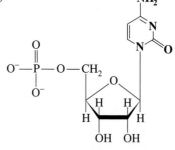

21.87 They are both polymers of nucleotides connected through phosphodiester bonds between alternating sugar and phosphate groups with bases extending out from each sugar.

21.89 28% T, 22% G, and 22% C

21.91 a. two   **b.** three

21.93 a. 3′—CTGAATCCG—5′
b. 5′—ACGTTTGATCGT—3′
c. 3′—TAGCTAGCTAGC—5′

21.95 DNA polymerase synthesizes the leading strand continuously in the 5′ to 3′ direction. The lagging strand is synthesized in small segments called Okazaki fragments because it must grow in the 3′ to 5′ direction.

21.97 One strand of the parent DNA is found in each of the two copies of the daughter DNA molecule.

21.99 a. tRNA   **b.** rRNA   **c.** mRNA

21.101 a. ACU, ACC, ACA, and ACG
b. UCU, UCC, UCA, UCG, AGU, and AGC
c. UGU and UGC

21.103 a. lysine   **b.** isoleucine   **c.** arginine

21.105 start—Tyr—Gly—Gly—Phe—Leu—stop

21.107 a. UCG   **b.** AUA   **c.** GGU

21.109 Three nucleotides are needed for each amino acid plus a start and stop triplet, which makes a minimum total of 33 nucleotides.

21.111 A DNA virus attaches to a cell and injects viral DNA that uses the host cell to produce copies of the DNA to make viral RNA. A retrovirus injects viral RNA from which complementary DNA is produced by reverse transcription.

22 Metabolic Pathways for Carbohydrates

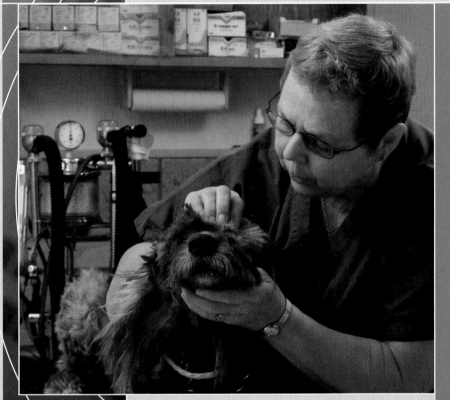

"I was checking this dog's ears for foxtails and her eyes for signs of conjunctivitis," says Joyce Rhodes, veterinary assistant at the Sonoma Animal Hospital. "We always check a dog's teeth for tartar, because dental care is very important to the well-being of the animal. When I do need to give a medication to an animal, I use my chemistry to prepare the proper dose that the pet should take. Dosages may be in milligrams, kilograms, or milliliters."

As a member of the veterinary health care team, a veterinary technician (VT) assists a veterinarian in the care and handling of animals. A VT takes medical histories, collects specimens, performs laboratory procedures, prepares an animal for surgery, assists in surgical procedures, takes X rays, talks with animal owners, and cleans teeth.

LOOKING AHEAD

Mastering **GOB CHEMISTRY**

Visit **www.masteringgob.com** for self-study materials and instructor-assigned homework.

W hen we eat food such as a tuna fish sandwich, the polysaccharides, lipids, and proteins are digested to smaller molecules that are absorbed into the cells of our body. As glucose, fatty acids, and amino acids are broken down further, energy is released. This energy is used in the cells to synthesize high-energy compounds such as adenosine triphosphate (ATP). Our cells utilize ATP energy when they do work such as contracting muscles, synthesizing large molecules, sending nerve impulses, and moving substances across cell membranes.

All the chemical reactions that take place in living cells to break down or build molecules are known as *metabolism*. In a metabolic pathway, reactions are linked together in a series, each catalyzed by a specific enzyme to produce an end product. In this and the following chapters, we will look at these pathways and the way they produce energy and cellular compounds.

LEARNING GOAL

Describe three stages of metabolism.

22.1 Metabolism and Cell Structure

The term **metabolism** refers to all the chemical reactions that provide energy and the substances required for continued cell growth. There are two types of metabolic reactions: catabolic and anabolic. In **catabolic reactions**, complex molecules are broken down to simpler ones with an accompanying release of energy. **Anabolic reactions** utilize energy available in the cell to build large molecules from simple ones.

We can think of the catabolic processes in metabolism as consisting of three stages. (See Figure 22.1.) Let's use that tuna fish sandwich for our example. In stage 1 of metabolism, the processes of digestion break down the large macromolecules into small monomer units. For example, the polysaccharides in bread break down to monosaccharides, the lipids in the mayonnaise break down to glycerol and fatty acids, and the proteins from the tuna yield amino acids. These digestion products diffuse into the bloodstream for transport to cells. In stage 2, they are broken down to two- and three-carbon compounds such as pyruvate and acetyl CoA in the cells. Stage 3 begins with the oxidation of the two-carbon acetyl CoA in the citric acid cycle, which produces several reduced coenzymes. As long as the cells have oxygen, the hydrogen ions and electrons are transferred to the electron transport chain, where most of the energy in the cell is produced.

Cell Structure for Metabolism

MC GOB   Metabolism and Cell Structure

To understand the relationship of metabolic reactions, we need to look at where metabolic reactions take place in the cells. (See Figure 22.2.) In Chapter 21, we described two types of cells: prokaryotic and eukaryotic. *Prokaryotic* (before nucleus) cells have no nucleus and form single-celled organisms such as bacteria. The cells in plants and animals are *eukaryotic* cells, which have a nucleus.

In a eukaryotic cell, a *plasma membrane* is a lipid bilayer that separates the materials inside the cell from the aqueous environment surrounding the cell. In addition, the outer surface of the plasma membrane contains structures that allow cells to communicate with each other. The *nucleus* contains the genes that control DNA replication and protein synthesis of the cell. The **cytoplasm** consists of all the materials between the nucleus and the plasma membrane. The **cytosol**, which is the fluid part of the cytoplasm, is an aqueous solution of electrolytes and enzymes that catalyze many of the cell's chemical reactions.

Stages of Metabolism

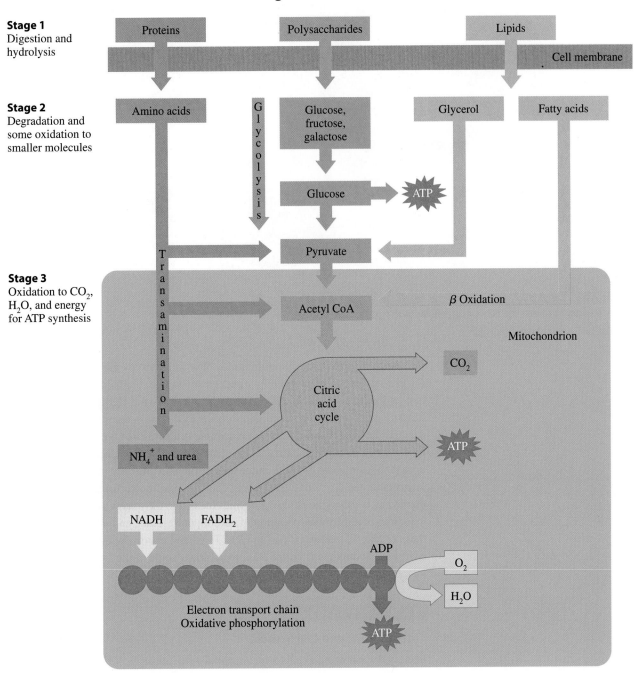

FIGURE 22.1

In the three stages of catabolic metabolism, foods are digested and degraded into smaller molecules, which are oxidized to produce energy.

Q **Where is most of the ATP energy produced in the cells?**

Within the cytoplasm are specialized structures called *organelles* that carry out specific functions in the cell. We have already seen (Chapter 21) that the *ribosomes* are the sites of protein synthesis using mRNA templates. The *endoplasmic reticulum* consists of two forms, a rough endoplasmic reticulum where proteins are processed for secretion and phospholipids are synthesized, and

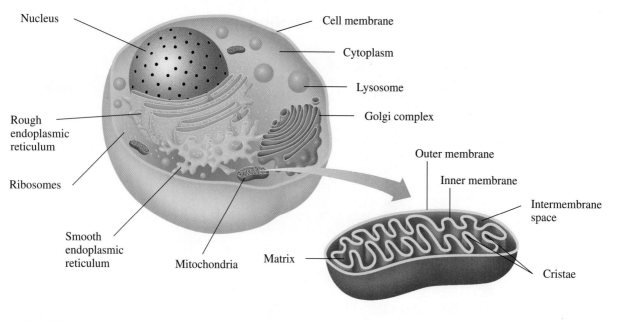

FIGURE 22.2

The diagram illustrates the major components of a typical eukaryotic cell.

Q **What is the cytoplasm in a cell?**

a smooth endoplasmic reticulum where fats and steroids are synthesized. The *Golgi complex* modifies proteins it receives from the rough endoplasmic reticulum and secretes these modified proteins into the extracellular fluid as well as forming glycoproteins and cell membranes. *Lysosomes* contain many hydrolytic enzymes that break down recyclable cellular structures that are no longer needed by the cell. The **mitochondria** are the energy-producing factories of the cells. A mitochondrion consists of an outer membrane and an inner membrane with an intermembrane space between them. The fluid section surrounded by the inner membrane is called the *matrix*. Enzymes located in the matrix and along the inner membrane catalyze the oxidation of carbohydrates, fats, and amino acids. All of these oxidation pathways lead to CO_2, H_2O, and energy, which is used to form energy-rich compounds. Table 22.1 summarizes some of the functions of the cellular components in eukaryotic cells.

SAMPLE PROBLEM 22.1

■ **Metabolism and Cell Structure**

Identify the following as catabolic or anabolic reactions.

a. Digestion of polysaccharides
b. Synthesis of proteins
c. Oxidation of glucose to CO_2 and H_2O

SOLUTION

a. The breakdown of large molecules is a catabolic reaction.
b. The synthesis of large molecules requires energy and involves anabolic reactions.
c. The breakdown of monomers such as glucose involves catabolic reactions.

STUDY CHECK

What is the difference between the cytoplasm and cytosol?

Locations and Functions of Components in Eukaryotic Cells

Component	Description and Function
cell plasma membrane	Separates the contents of a cell from the external environment and contains structures that communicate with other cells
cytoplasm	Consists of all of the cellular contents between the plasma membrane and nucleus
cytosol	Is the fluid part of the cytoplasm that contains enzymes for many of the cell's chemical reactions including glycolysis, glucose, and fatty acid synthesis
endoplasmic reticulum	Rough type processes proteins for secretion and synthesizes phospholipids; smooth type synthesizes fats and steroids
Golgi complex	Modifies and secretes proteins from the endoplasmic reticulum and synthesizes glycoproteins and cell membranes
lysosomes	Contain hydrolytic enzymes that digest and recycle old cell structures
mitochondria	Contain the structures for the synthesis of ATP from energy-producing reactions
nucleus	Contains genetic information for the replication of DNA and the synthesis of protein
ribosomes	Are the sites of protein synthesis using mRNA templates

QUESTIONS AND PROBLEMS

■ Metabolism and Cell Structure

22.1 What stage of metabolism involves the digestion of polysaccharides?

22.2 What stage of metabolism involves the conversion of small molecules to CO_2, H_2O, and energy for the synthesis of ATP?

22.3 What is meant by a catabolic reaction in metabolism?

22.4 What is meant by an anabolic reaction in metabolism?

22.5 Match each of the following organelles with their function:

(1) lysosome   (2) Golgi complex   (3) smooth endoplasmic reticulum
a. synthesis of fats and steroids   **b.** contains hydrolytic enzymes
c. modifies products from rough endoplasmic reticulum

22.6 Match each of the following organelles with their function:
(1) mitochondria   (2) rough endoplasmic reticulum
(3) plasma membrane
a. separates cell contents from external surroundings
b. sites of energy production
c. synthesizes proteins for secretion

22.2 ATP and Energy

In our cells, the energy released from the oxidation of the food we eat is used to form a compound called *adenosine triphosphate* (**ATP**). As we saw in Chapter 21, the ATP molecule is composed of the nitrogen base adenine, a ribose sugar, and three phosphate groups. (See Figure 22.3.)

Hydrolysis of ATP Yields Energy

In the cells, there are a variety of "high-energy" compounds. The most important of these is ATP, which transfers a phosphate group to water when it undergoes

LEARNING GOAL

Describe the role of ATP in catabolic and anabolic reactions.

SELF-STUDY ACTIVITY
ATP

FIGURE 22.3

Adenosine triphosphate hydrolyzes to form ADP and AMP, along with a release of energy.

Q **How much energy is released when a phosphate group is cleaved from one mole of ATP?**

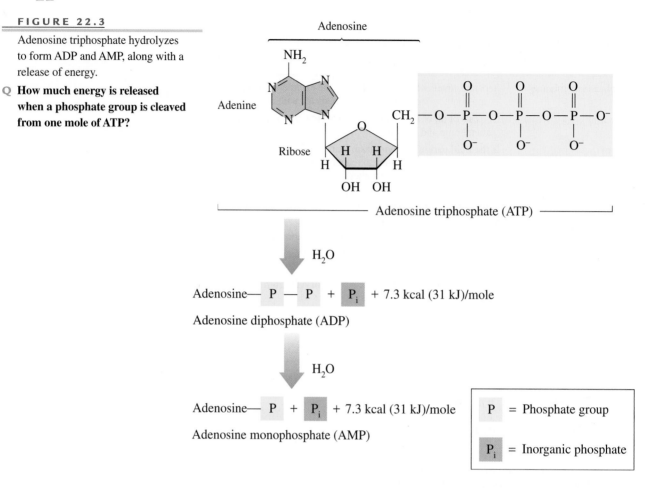

CO$_2$ + H$_2$O + NH$_4^+$

Catabolic reactions (energy producing)

Oxidation of carbohydrates, fats, and proteins

ADP + P$_i$ (energy used)   ATP (energy stored)

Anabolic reactions (energy requiring)

Muscle contraction, transport, and synthesis of cellular compounds

FIGURE 22.4

ATP, the energy-storage molecule, connects the energy-producing reactions with the energy-requiring reactions that do work in the cells.

Q **What type of reaction provides energy for ATP synthesis?**

hydrolysis. The cleavage of one phosphate group releases energy of 7.3 kcal per mole of ATP or 31 kJ per mole of ATP. The equilibrium strongly favors the products, which are adenosine diphosphate (**ADP**) and hydrogen phosphate ion HPO$_4^{2-}$ abbreviated as P$_i$ (inorganic phosphate).

We can write this reaction in its simplified form as

$$ATP + H_2O \longrightarrow ADP + P_i + 7.3 \text{ kcal (31 kJ)/mole}$$

Every time we contract muscles, move substances across cellular membranes, send nerve signals, or synthesize an enzyme, we use energy from ATP hydrolysis. In a cell that is doing work (anabolic processes), 1–2 million ATP molecules may be hydrolyzed in one second. The amount of ATP hydrolyzed in one day can be as much as our body mass, even though only about 1 gram of ATP is present in all our cells at any given time.

When we take in food, the resulting catabolic reactions provide energy to regenerate ATP in our cells. Then 7.3 kcal (31 kJ)/mole is used to make ATP from ADP and P$_i$. (See Figure 22.4.)

$$ADP + P_i + 7.3 \text{ (31 kJ) kcal/mole} \longrightarrow ATP$$

ATP Is Coupled with Reactions

ATP and other energy-rich compounds are coupled with metabolic reactions and processes that require energy. ATP is very useful in the body because it can be used to drive reactions that require energy and do not occur on their own. For example, the glucose obtained from carbohydrates must add a phosphate group to start its breakdown in the cell. However, the cost of adding a phosphate to glucose is 3.3 kcal (14 kJ)/mole, which means that the reaction does not occur spontaneously in the cell. By coupling the reaction with the hydrolysis of an energy-rich compound such as ATP, the reaction takes place because the energy from hydrolysis pushes or "drives" the energy-requiring reaction.

 ATP: Energy Rich

Glucose + P$_i$ + 3.3 kcal (14 kJ)/mole ⟶ glucose-6-phosphate	Requires energy
ATP ⟶ ADP + P$_i$ + 7.3 kcal (31 kJ)/mole	Provides energy
Glucose + ATP ⟶ ADP + glucose-6-phosphate + 4.0 kcal (17 kJ)/mole	

The coupling of a reaction that requires energy with a reaction that supplies energy is a very important concept in biochemical pathways. Many of the reactions essential to a cell for survival cannot proceed by themselves, but can be made to proceed by coupling them with a reaction that releases energy. Similar kinds of coupled reactions are also used to transmit nerve impulses, transport substances across membranes to higher concentrations, and to contract muscles.

HEALTH NOTE

■ ATP Energy and Ca^{2+} Needed to Contract Muscles

Our muscles consist of thousands of parallel fibers. Within these muscle fibers are fibrils composed of two kinds of proteins called filaments. Arranged in alternating rows, the thick filaments of myosin overlap the thin filaments containing actin. During a muscle contraction, the thin filaments slide inward over the thick filaments causing a shortening of the muscle fibers.

Calcium ion, Ca^{2+}, and ATP play an important role in muscle contraction. An increase in the Ca^{2+} concentration in the muscle fibers causes the filaments to slide, while a decrease stops the process. In a relaxed muscle, the Ca^{2+} concentration is low. When a nerve impulse reaches the muscle, calcium channels in the membrane open and Ca^{2+} flows into the fluid surrounding the filaments. The Ca^{2+} combines with troponin, a protein that covers the binding site on actin. When the troponin detaches from actin, myosin can bind to actin, and start to pull the actin filaments inward. The energy used for the con-

traction is provided by splitting ATP that is attached to myosin to ADP + P$_i$ and energy.

Muscle contraction continues as long as both ATP and Ca^{2+} levels are high around the filaments. Relaxation of the muscle occurs as the nerve impulse ends and the Ca^{2+} channels close. The Ca^{2+} concentration is lowered as ATP energy is used to pump the remaining calcium ions out of the area. Troponin reattaches to the actin, which causes the thin filaments to return to their relaxed state. In rigor mortis, Ca^{2+} concentration remains high within the muscle fibers causing a continued state of rigidity for approximately 24 hours. After that, cellular deterioration causes a decrease in calcium ions and muscles relax.

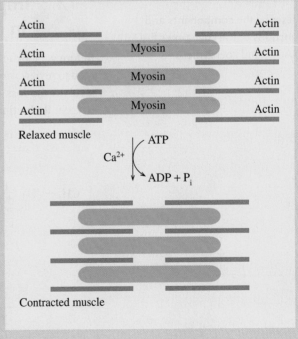

 ATP Powers Reactions

■**Hydrolysis of ATP**

Write an equation for the hydrolysis of ATP.

SOLUTION

The hydrolysis of ATP produces ADP, P_i, and energy.

$$\text{ATP} + H_2O \longrightarrow \text{ADP} + P_i + \text{energy}$$

STUDY CHECK

What are the components of the ATP molecule?

QUESTIONS AND PROBLEMS

■**ATP and Energy**

22.7   Why is ATP considered an energy-rich compound?

22.8   What is meant when we say that the hydrolysis of ATP is used to "drive" a reaction?

22.9   Phosphoenolpyruvate (PEP) is a high-energy phosphate compound that releases 14.8 kcal/mole of energy when it hydrolyzes to pyruvate and P_i. This reaction can be coupled with the synthesis of ATP from ADP and P_i.
   a. Write an equation for the energy-releasing reaction of PEP.
   b. Write an equation for the energy-requiring reaction that forms ATP.
   c. Write the overall equation for the coupled reaction including the net energy change.

22.10 The phosphorylation of glycerol to glycerol-3-phosphate requires 2.2 kcal/mole and is driven by the hydrolysis of ATP.
   a. Write an equation for the energy-releasing reaction of ATP.
   b. Write an equation for the energy-requiring reaction that forms glycerol-3-phosphate.
   c. Write the overall equation for the coupled reaction including the net energy change.

LEARNING GOAL

■ Describe the components and functions of the coenzymes FAD, NAD^+, and coenzyme A.

22.3 Important Coenzymes in Metabolic Pathways

Before we look at the metabolic reactions that extract energy from our food, we need to review some ideas about oxidation and reduction reactions. An **oxidation** involves the loss of hydrogen and electrons by a substance or an increase in oxygen. **Reduction** is the gain of hydrogen and electrons or a decrease in oxygen.

Oxidation: Loss of H, loss of e⁻, or increase of O

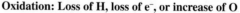

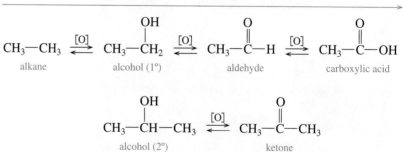

Reduction: Gain of H, gain of e⁻, or decrease of O

When an enzyme catalyzes an oxidation, hydrogen atoms are removed from a substrate as hydrogen ions, $2H^+$, and electrons, $2e^-$.

$$2\text{ H atoms (removed in oxidation)} \longrightarrow 2H^+ + 2e^-$$

Those hydrogen ions and electrons are picked up by a coenzyme for the reaction, which is reduced. As we saw in Chapter 20, the structures of many coenzymes include the water-soluble B vitamins we obtain from the foods in our diets. Now we can look at the structure of several important coenzymes.

MC GOB Central Coenzymes of Metabolism

NAD$^+$

NAD$^+$ (nicotinamide adenine dinucleotide) is an important coenzyme in which the vitamin *niacin* provides the *nicotinamide* group, which is bonded to adenosine diphosphate (ADP). (See Figure 22.5). The NAD$^+$ coenzyme participates in reactions that produce a carbon–oxygen (C=O) double bond such as the oxidation of alcohols to aldehydes and ketones. The NAD$^+$ is reduced when the carbon in the pyridine ring of nicotinamide accepts a hydrogen ion and two electrons leaving one H$^+$. Let's look at the reactions that take place when ethanol is oxidized in the liver to acetaldehyde using NAD$^+$.

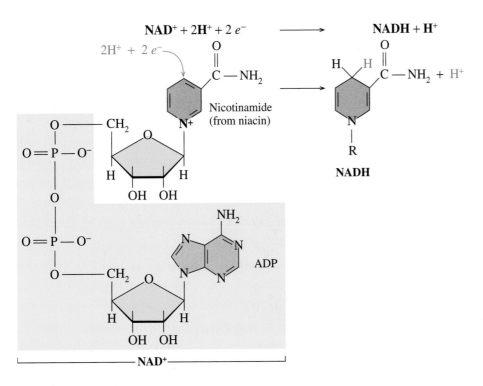

FIGURE 22.5

The coenzyme NAD$^+$ (nicotinamide adenine dinucleotide), which consists of adenine diphosphate, and a nicotinamide portion from the vitamin niacin, ribose, is reduced to NADH + H$^+$.

Q **Why is the conversion of NAD$^+$ to NADH and H$^+$ called a reduction?**

Oxidation

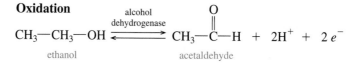

$$CH_3-CH_2-OH \underset{\text{alcohol dehydrogenase}}{\rightleftharpoons} CH_3-\overset{\overset{\displaystyle O}{\|}}{C}-H + 2H^+ + 2\,e^-$$

ethanol   acetaldehyde

Reduction

$$NAD^+ + 2H^+ + 2\,e^- \rightleftharpoons NADH + H^+$$

Overall oxidation–reduction reaction

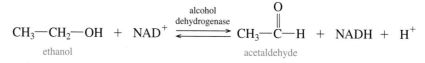

$$CH_3-CH_2-OH + NAD^+ \underset{\text{alcohol dehydrogenase}}{\rightleftharpoons} CH_3-\overset{\overset{\displaystyle O}{\|}}{C}-H + NADH + H^+$$

ethanol   acetaldehyde

FAD

FAD (flavin adenine dinucleotide) is a coenzyme derived from adenosine diphosphate (ADP) and riboflavin. Riboflavin (vitamin B) consists of ribitol, a sugar alcohol, and flavin. As a coenzyme, two nitrogen atoms in the flavin part of the FAD coenzyme accept the hydrogen, which reduces the FAD to FADH$_2$. (See Figure 22.6.)

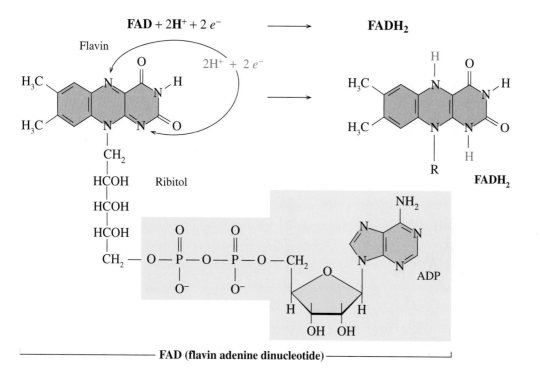

FIGURE 22.6

The coenzyme FAD (flavin adenine dinucleotide) made from riboflavin (vitamin B$_2$) and adenine diphosphate is reduced to FADH$_2$.

Q **What is the type of reaction in which FAD accepts hydrogen?**

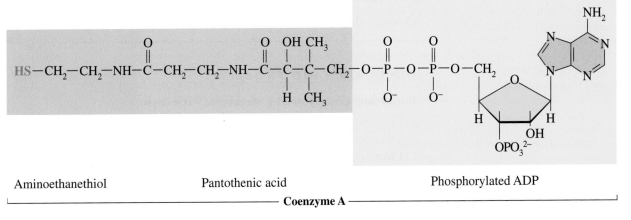

Aminoethanethiol   Pantothenic acid   Phosphorylated ADP

Coenzyme A

FIGURE 22.7

Coenzyme A is derived from a phosphorylated adenine diphosphate (ADP) and pantothenic acid bonded by an amide linked to aminoethanethiol, which contains the —SH reactive part of the molecule.

Q **What part of coenzyme A reacts with a two-carbon acetyl group?**

FAD typically participates in oxidation reactions that produce a carbon–carbon (C=C) double bond.

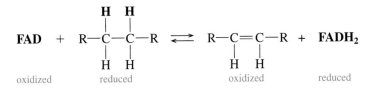

Coenzyme A

Coenzyme A (CoA) is made up of several components: pantothenic acid (vitamin B_3), adenosine diphosphate (ADP), and aminoethanethiol. (See Figure 22.7.) The main function of coenzyme A is to activate acyl groups (indicated by the letter A in CoA), particularly the acetyl group. When the free thiol (—SH) group of CoA bonds to the two-carbon acetyl group, or to longer chain acyl groups such as fatty acids, the products are energy-rich thioesters. Then the acetyl or acyl groups can be transferred to other substrates.

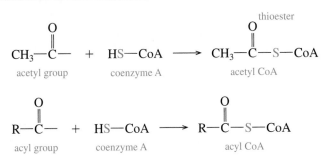

SAMPLE PROBLEM 22.3

■ **Coenzymes in Metabolic Pathways**

What vitamin is part of the coenzyme FAD?

SOLUTION

FAD, flavin adenine dinucleotide, is made from the vitamin riboflavin.

What is the abbreviation of the reduced form of FAD?

QUESTIONS AND PROBLEMS

■ Important Coenzymes in Metabolic Pathways

22.11 Identify one or more coenzymes with each of the following components:
    a. pantothenic acid   **b.** niacin   **c.** ribitol

22.12 Identify one or more coenzymes with each of the following components:
    a. riboflavin   **b.** adenine   **c.** aminoethanethiol

22.13 Give the abbreviation for the following:
    a. the reduced form of NAD^+ **b.** the oxidized form of $FADH_2$

22.14 Give the abbreviation for the following:
    a. the reduced form of FAD   **b.** the oxidized form of NADH

22.15 What coenzyme picks up hydrogen when a carbon–carbon double bond is formed?

22.16 What coenzyme picks up hydrogen when a carbon–oxygen double bond is formed?

Give the sites and products of digestion for carbohydrates.

Breakdown of Carbohydrates

22.4 Digestion of Carbohydrates

In the first stage of catabolism, foods undergo **digestion**, a process that converts large molecules to smaller ones that can be absorbed by the body. We begin the digestion of carbohydrates as soon as we chew food. (See Figure 22.8.) An enzyme produced in the salivary glands called *amylase* hydrolyzes some of the α-glycosidic bonds in amylose and amylopectin, producing smaller polysaccharides called dextrins, which contain three to eight glucose units, maltose, and some glucose. After swallowing, the partially digested starches enter the acid environment of the stomach, where the low pH soon stops further carbohydrate digestion.

Digestion of Disaccharides

In the small intestine, which has a pH of about 8, an α-amylase produced in the pancreas hydrolyzes the remaining polysaccharides to maltose and glucose. A branching enzyme hydrolyzes the glycosidic bonds in amylopectin. Then enzymes produced in the mucosal cells that line the small intestine hydrolyze maltose as well as lactose and sucrose. The hydrolysis reactions for the three common dietary disaccharides are written as follows.

$$\text{lactose} + H_2O \xrightarrow{\text{lactase}} \text{galactose} + \text{glucose}$$

$$\text{sucrose} + H_2O \xrightarrow{\text{sucrase}} \text{fructose} + \text{glucose}$$

$$\text{maltose} + H_2O \xrightarrow{\text{maltase}} \text{glucose} + \text{glucose}$$

The monosaccharides are absorbed through the intestinal wall into the bloodstream, which carries them to the liver where fructose and galactose are converted to glucose.

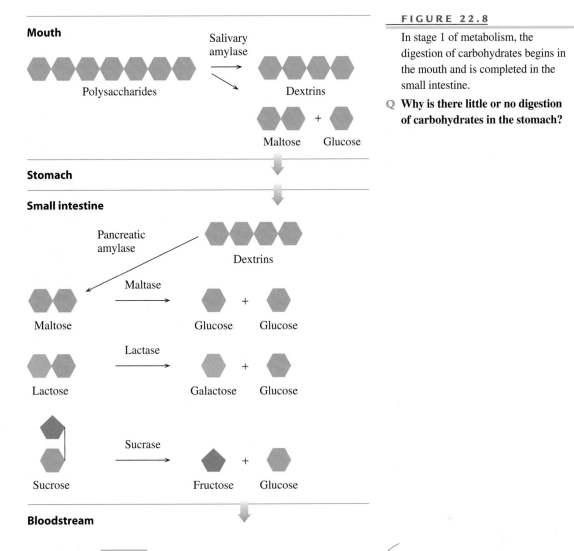

Mouth

Polysaccharides → (Salivary amylase) → Dextrins

Maltose + Glucose

Stomach

Small intestine

Dextrins → (Pancreatic amylase) → Maltose

Maltose → (Maltase) → Glucose + Glucose

Lactose → (Lactase) → Galactose + Glucose

Sucrose → (Sucrase) → Fructose + Glucose

Bloodstream

SAMPLE PROBLEM 22.4

■ Digestion of Carbohydrates

Indicate the carbohydrate that undergoes digestion in each of the following sites:

a. mouth   **b.** stomach   **c.** small intestine

SOLUTION

a. Starches amylose and amylopectin (α-1,4-glycosidic bonds only).
b. Essentially no digestion occurs in the stomach.
c. Dextrins, maltose, sucrose, and lactose.

STUDY CHECK

Describe the digestion of amylose, a polymer of glucose molecules joined by α-glycosidic bonds.

QUESTIONS AND PROBLEMS

■ Digestion of Carbohydrates

22.17 What is the general type of reaction that occurs during the digestion of carbohydrates?

22.18 Why is α-amylase produced in the salivary glands and in the pancreas?

EXPLORE YOUR WORLD

■ Carbohydrate Digestion

1. Obtain a cracker or small piece of bread and chew it for 4–5 minutes. During that time observe any change in the taste.
2. Some milk products contain Lactaid, which is the lactase that digests lactose. Look for the brands of milk and ice cream that contain Lactaid or lactase enzyme.

QUESTIONS

1a. How does the taste of the cracker or bread change after you have chewed it for 4–5 minutes? What could be an explanation?
1b. What part of carbohydrate digestion occurs in the mouth?
2a. Write an equation for the digestion of lactose.
2b. Where does lactose undergo digestion?

Lactose Intolerance

The disaccharide in milk is lactose, which is broken down by *lactase* in the intestinal tract to monosaccharides that are a source of energy. Infants and small children produce lactase to break down the lactose in milk. It is rare for an infant to lack the ability to produce lactase. As they mature, many people experience a decrease in the production of the lactase enzyme. By the time they are adults, many people have little or no lactase production, causing lactose intolerance. This condition may affect 25% of the people in the United States. A deficiency of lactase occurs in adults in many parts of the world, but in the United States it is more prevalent among the African–American, Hispanic, and Asian populations.

When lactose is not broken down into glucose and galactose, it cannot be absorbed through the intestinal wall and remains in the intestinal tract. In the intestines, the lactose undergoes fermentation to products that include lactic acid and gases such as methane (CH_4) and CO_2. Symptoms of lactose intolerance, which appear approximately $\frac{1}{2}$ to 1 hour after ingesting milk or milk products, include nausea, abdominal cramps, and diarrhea. The severity of the symptoms depends on how much lactose is present in the food and how much lactase a person produces.

TREATMENT OF LACTOSE INTOLERANCE

One way to reduce the reaction to lactose is to avoid products that contain lactose. However, it is important to consume foods that provide the body with calcium.

Many people with lactose intolerance seem to tolerate yogurt, which is a good source of calcium. Although there is lactose in yogurt, the bacteria in yogurt may produce some lactase, which helps to digest the lactose. A person who is lactose intolerant should also know that some foods that may not seem to be dairy products contain lactose. For example, baked goods, cereals, breakfast drinks, salad dressings, and even lunchmeat can contain lactose in their ingredients. Labels must be read carefully to see if the ingredients include "milk" or "lactose."

The enzyme lactase is now available in many forms such as tablets that are taken with meals, drops that are added to milk, or as additives in many dairy products such as milk. When lactase is added to milk that is left in the refrigerator for 24 hours, the lactose level is reduced by 70–90%. Lactase pills or chewable tablets are taken when a person begins to eat a meal that contains dairy foods. If taken too far ahead of the meal, the lactase will be degraded by stomach acid. If taken following a meal, the lactose will have entered the lower intestine.

22.19 Complete the following equation by filling in the missing words:

a. _____ + $H_2O \longrightarrow$ galactose + glucose

b. Sucrose + $H_2O \longrightarrow$ _____ + _____

c. Maltose + $H_2O \longrightarrow$ glucose + _____

22.20 Give the site and the enzyme for each of the reactions in problem 22.19.

LEARNING GOAL

Describe the conversion of glucose to pyruvate in glycolysis.

MC
GOB

SELF-STUDY ACTIVITY
Glycolysis

22.5 Glycolysis: Oxidation of Glucose

The major source of energy is the glucose produced when we digest the carbohydrates in our food, or from glycogen, a polysaccharide stored in the liver and skeletal muscle. Glucose in the bloodstream enters our cells for further degradation in a pathway called glycolysis. Early organisms used this pathway to produce energy from simple nutrients long before there was any oxygen in Earth's atmosphere. Glycolysis is an **anaerobic** process; no oxygen is required.

In **glycolysis**, a six-carbon glucose molecule is broken down to yield two three-carbon pyruvate molecules. (See Figure 22.9.) All the reactions in glycolysis take place in the cytoplasm of the cell where the enzymes for glycolysis are located. In the first five reactions (1–5), the energy of 2 ATPs is used to add phosphate groups to form sugar phosphates. (See Figure 22.10.) Then the six-carbon sugar phosphate is cleaved to yield two three-carbon sugar phosphate molecules. In the last five reactions (6–10), the phosphate groups in these energy-rich trioses are hydrolyzed, which generates energy in the form of 4 ATPs. The final products are 2 pyruvate molecules and 2 reduced NADH molecules.

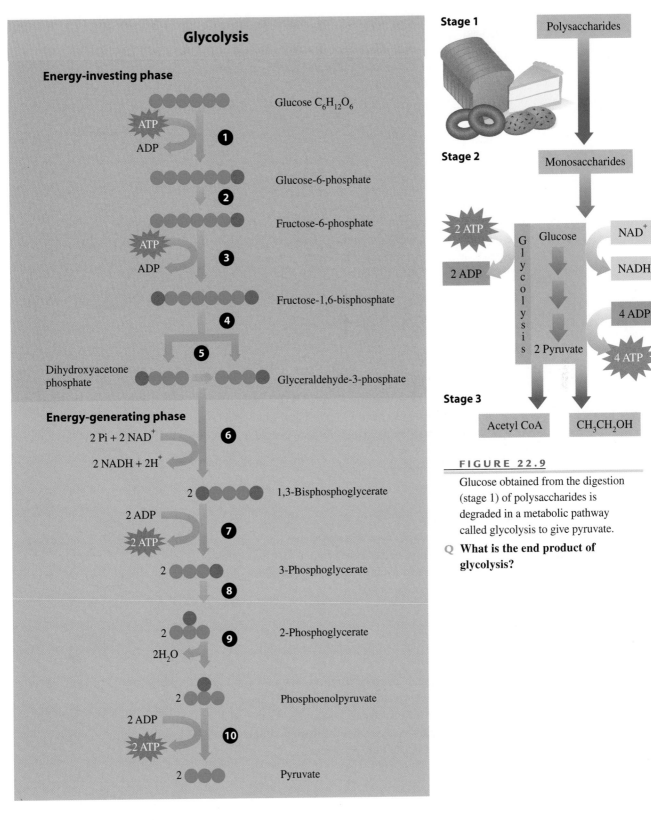

FIGURE 22.10

In glycolysis, the six-carbon glucose molecule is degraded to yield two three-carbon pyruvate molecules. A net of two ATPs are produced along with two NADH.

Q **Where in the glycolysis pathway is glucose cleaved to yield two three-carbon compounds?**

FIGURE 22.9

Glucose obtained from the digestion (stage 1) of polysaccharides is degraded in a metabolic pathway called glycolysis to give pyruvate.

Q **What is the end product of glycolysis?**

 The Glycolysis Pathway

Energy-Investing Reactions: Steps 1–5

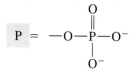

Reaction 1   Phosphorylation: first ATP invested——
Glucose is converted to glucose-6-phosphate by a reaction with ATP catalyzed by *hexokinase*.

$$P = \quad -O-\overset{\overset{\displaystyle O}{\|}}{\underset{\underset{\displaystyle O^-}{|}}{P}}-O^-$$

Reaction 2   Isomerization——————————
The enzyme *phosphoglucoisomerase* converts glucose-6-phosphate, an aldose, to fructose-6-phosphate, a ketose.

Reaction 3   Phosphorylation: second ATP invested——
A second ATP reacts with fructose-6-phosphate to give fructose-1,6-bisphosphate. The word *bisphosphate* is used to show that the phosphates are on different carbons in fructose and not connected to each other.

Reaction 4   Cleavage: two trioses form—————————
Fructose-1,6-bisphosphate is "split" into two triose phosphates—dihydroxyacetone phosphate and glyceraldehyde 3-phosphate—catalyzed by *aldolase*.

Reaction 5   Isomerization of a triose——————
In reaction 5, *triosephosphate isomerase* converts one of the triose products, dihydroxyacetone phosphate, to the other, glyceraldehyde-3-phosphate. Now all 6 carbon atoms from glucose are in two identical triose phosphates.

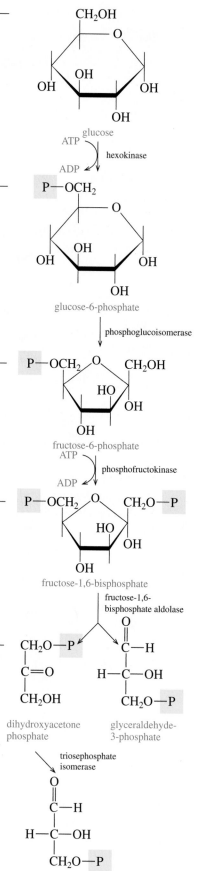

Energy-Generating Reactions: Steps 6–10

Reaction 6   First energy-rich compound

The aldehyde group of glyceraldehyde-3-phosphate is oxidized and phosphorylated by *glyceraldehyde-3-phosphate dehydrogenase*. The coenzyme NAD^+ is reduced to NADH and H^+. The 1,3-bisphosphoglycerate is an extremely high-energy compound that releases 12 kcal per mole when the phosphate bond is broken.

Reaction 7   Formation of first ATP

The energy-rich 1,3-bisphosphoglycerate now drives the formation of ATP when *phosphoglycerate kinase* transfers a phosphate group to ADP. This process is called a substrate-level phosphorylation. At this point, glycolysis pays back the two ATP invested in the early reactions.

Reaction 8   Formation of 2-phosphoglycerate

A *phosphoglycerate mutase* transfers the phosphate group from carbon 3 to carbon 2 to yield 2-phosphoglycerate.

Reaction 9   Second energy-rich compound

An *enolase* catalyzes the removal of water to yield phosphoenolpyruvate, a high-energy compound that releases 15 kcal per mole when the phosphate bond is hydrolyzed.

Reaction 10   Formation of second ATP

This last reaction, catalyzed by *pyruvate kinase*, transfers the phosphate to ADP in a second direct substrate phosphorylation to yield pyruvate and ATP.

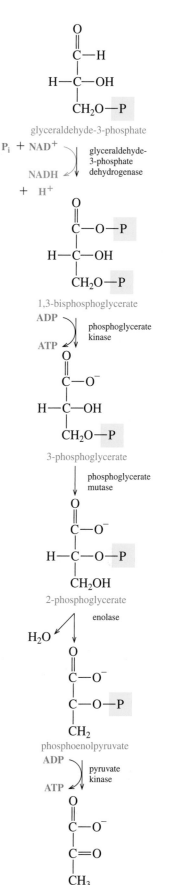

Summary of Glycolysis

In the glycolysis pathway, glucose is converted to two pyruvates. Initially, two ATP molecules were required to form sugar phosphate. Later, four ATP were generated, which gives a net gain of two ATP. Overall, glycolysis yields two ATP and two NADH for each glucose that is converted to two pyruvates.

$$C_6H_{12}O_6 + 2\,NAD^+ \xrightarrow[\text{glucose}]{\substack{2\,ADP + 2\,P_i \quad 2\,ATP}} 2\,CH_3-\overset{\displaystyle O}{\overset{\|}{C}}-COO^- + 2\,NADH + 4H^+$$

It appears right now that glycolysis does a lot of work to produce only 2 ATP, 2 NADH, and pyruvate. However, under aerobic conditions, stage 3 operates to reoxidize NADH to produce more ATP, and pyruvate enters the citric acid cycle where it generates considerably more energy. We will look at the oxidative pathways of stage 3 in Chapter 23.

Other Hexoses Enter Glycolysis

Other monosaccharides enter glycolysis after they are converted to intermediates of the pathway. Digestion of carbohydrates produces galactose from lactose in milk products and fructose from fruits and sucrose. Galactose reacts with ATP to yield galactose-1-phosphate, which is converted to glucose-6-phosphate, an intermediate of glycolysis. In the liver, fructose is converted to fructose-1-phosphate, which is cleaved in a reaction similar to reaction 4 to give dihydroxyacetone phosphate and glyceraldehyde. Dihydroxyacetone phosphate isomerizes to glyceraldehyde-3-phosphate, and glyceraldehyde is phosphorylated to glyceraldehyde-3-phosphate, an intermediate in reaction 6. In muscle and kidney, fructose is phosphorylated to fructose-6-phosphate, which enters glycolysis in reaction 3.

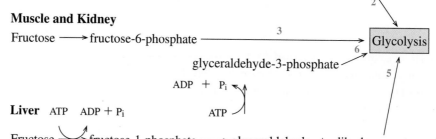

Regulation of Glycolysis

Metabolic pathways such as glycolysis do not run at the same rates all the time. The amount of glucose that is broken down is controlled by the requirements in the cells for pyruvate, ATP, and other intermediates of glycolysis. Within the glycolysis sequence, three enzymes respond to the levels of ATP and other products continually speed up or slow down the flow of glucose into the pathway.

Reaction 1   Hexokinase

The amount of glucose entering the glycolysis pathway decreases when high levels of glucose-6-phosphate are present in the cell. This phosphorylation product inhibits hexokinase, which prevents glucose from reacting with ATP. This is a feedback inhibition, which is a type of enzyme regulation we discussed in Chapter 20.

Reaction 3   Phosphofructokinase

The reaction catalyzed by phosphofructokinase is a very important control point for glycolysis. Once fructose-1,6-bisphosphate is formed, it must continue through the remaining reactions to pyruvate. As an allosteric enzyme, phosphofructokinase is inhibited by high levels of ATP and activated by high levels of ADP and AMP. High levels of ADP and AMP indicate that the cell has used up much of its ATP. As a regulator, phosphofructokinase increases the rate of pyruvate production for ATP synthesis when the cell needs to replenish ATP and slows or stops the reaction when ATP is plentiful.

Reaction 10   Pyruvate Kinase

In the last reaction of glycolysis, high levels of ATP as well as acetyl CoA inhibit pyruvate kinase, which is another allosteric enzyme.

Summary of Regulation

These are examples of how metabolic pathways shut off enzymes to stop the production of molecules that are not needed. Since pyruvate can be used to synthesize ATP, several enzymes in glycolysis respond to ATP levels in the cell. When ATP levels are high, enzymes in glycolysis slow or stop the synthesis of pyruvate. With phosphofructokinase and pyruvate kinase inhibited by ATP, glucose-6-phosphate accumulates and inhibits the first reaction and glucose does not enter the glycolysis pathway. The glycolysis pathway is shut down until ATP is once again needed in the cell. When ATP levels are low or AMP/ADP levels are high, these enzymes are activated and pyruvate production starts again.

SAMPLE PROBLEM 22.5

■ **Glycolysis**

What are the steps in glycolysis that generate ATP?

SOLUTION

ATP is produced when phosphate groups are transferred directly to ADP from 1,3-bisphosphoglycerate (step 7) and from phosphoenolpyruvate (step 10).

STUDY CHECK

If four ATP molecules are produced in glycolysis, why is there a net yield of two ATP?

QUESTIONS AND PROBLEMS

(MC GOB)

■ **Glycolysis: Oxidation of Glucose**

22.21 What is the starting compound of glycolysis?

22.22 What is the end product of glycolysis?

22.23 How is ATP used in the initial steps of glycolysis?

22.24 How many ATP molecules are used in the initial steps of glycolysis?

22.25 What trioses are obtained when fructose-1,6-diphosphate splits?

22.26 Why does one of the triose products undergo isomerization?

22.27 How does direct phosphorylation account for the production of ATP in glycolysis?

22.28 Why are there two ATP molecules formed for one molecule of glucose?

22.29 Indicate the enzyme that catalyzes the following reactions in glycolysis:
a. phosphorylation　　**b.** direct transfer of a phosphate group

22.30 Indicate the enzyme that catalyzes the following reactions in glycolysis:
a. isomerization　　**b.** formation of a ketotriose and an aldotriose

22.31 How many ATP or NADH are produced (or required) in each of the following steps in glycolysis?
a. glucose to glucose-6-phosphate
b. glyceraldehyde-3-phosphate to 1,3-bisphosphoglycerate
c. glucose to pyruvate

22.32 How many ATP or NADH are produced (or required) in each of the following steps in glycolysis?
a. 1,3-bisphosphoglycerate to 3-phosphoglycerate
b. fructose-6-phosphate to fructose-1,6-bisphosphate
c. phosphoenolpyruvate to pyruvate

22.33 Which step in glycolysis involves the following?
a. the first ATP molecule is hydrolyzed
b. direct substrate phosphorylation occurs
c. six-carbon sugar splits into two three-carbon molecules

22.34 Which step in glycolysis involves the following?
a. isomerization takes place
b. NAD$^+$ is reduced
c. a second ATP molecule is synthesized

22.35 How do galactose and fructose, obtained from the digestion of carbohydrates, enter glycolysis?

22.36 What are three enzymes that regulate glycolysis?

22.37 Indicate whether each of the following would activate or inhibit phosphofructokinase.
a. low levels of ATP　　**b.** high levels of ATP

22.38 Indicate whether each of the following would activate or inhibit pyruvate kinase.
a. low levels of ATP　　**b.** high levels of fructose-1,6-bisphosphate

---

LEARNING GOAL

Give the conditions for the conversion of pyruvate to lactate, ethanol, and acetyl coenzyme A.

 Pathways for Pyruvate

22.6 Pathways for Pyruvate

The pyruvate produced from glucose can now enter pathways that continue to extract energy. The available pathway depends on whether there is sufficient oxygen in the cell. During **aerobic** conditions, oxygen is available to convert pyruvate to acetyl coenzyme A (CoA). When oxygen levels are low, pyruvate is reduced to lactate. In yeast cells, which are anaerobic, pyruvate is converted to ethanol.

Aerobic Conditions

In glycolysis, two ATP molecules were generated when glucose was converted to pyruvate. However, much more energy is still available. The greatest amount of the energy is obtained from glucose when oxygen levels are high in the cells. Under aerobic conditions, pyruvate moves from the cytoplasm (where glycolysis took

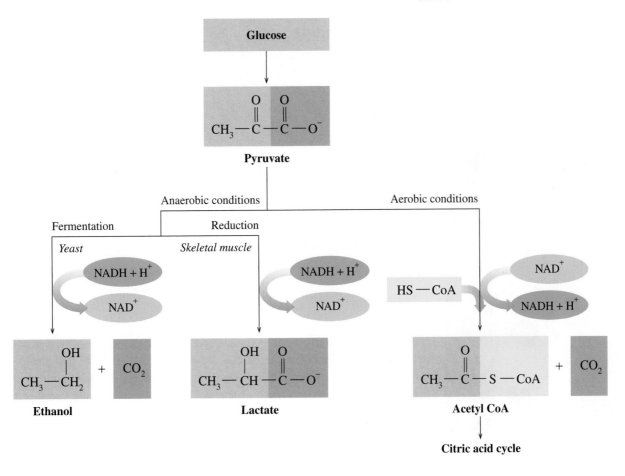

FIGURE 22.11

Pyruvate is converted to acetyl CoA under aerobic conditions and to lactate or ethanol (in certain microorganisms) under anaerobic conditions.

Q **During vigorous exercise, why does lactate accumulate in the muscles?**

place) into the matrix of the mitochondria to be oxidized further. In a complex reaction, pyruvate is oxidized and a carbon atom is removed from pyruvate as CO_2. The coenzyme NAD^+ is required for the oxidation. The resulting two-carbon acetyl compound is attached to CoA, producing **acetyl CoA**, an important intermediate in many metabolic pathways. (See Figure 22.11.)

$$CH_3-\overset{O}{\overset{\|}{C}}-\overset{O}{\overset{\|}{C}}-O^- + HS-CoA + NAD^+ \xrightarrow{\text{pyruvate dehydrogenase}} CH_3-\overset{O}{\overset{\|}{C}}-S-CoA + CO_2 + NADH$$

pyruvate acetyl CoA

Anaerobic Conditions

When we engage in strenuous exercise, the oxygen stored in our muscle cells is quickly depleted. Under anaerobic conditions, pyruvate remains in the cytoplasm where it is reduced to lactate. NAD^+ is produced and used to oxidize more glyceraldehyde-3-phosphate in the glycolysis pathway, which produces a small but needed amount of ATP.

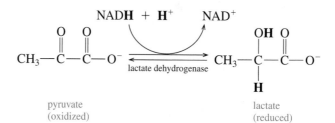

pyruvate
(oxidized)

lactate
(reduced)

The accumulation of lactate causes the muscles to tire rapidly and become sore. After exercise, a person continues to breathe rapidly to repay the oxygen debt incurred during exercise. Most of the lactate is transported to the liver where it is converted back into pyruvate. Under anaerobic conditions, the only ATP production in glycolysis occurs during the steps that phosphorylate ADP directly, giving a net gain of only two ATP molecules:

$$C_6H_{12}O_6 \; + \; 2\,ADP \; + \; 2\,P_i \; \longrightarrow \; 2\,CH_3-\overset{\displaystyle OH}{\underset{|}{C}}H-COO^- \; + \; 2\,ATP$$

glucose   lactate

Bacteria also convert pyruvate to lactate under anaerobic conditions. In the preparation of kimchee and sauerkraut, cabbage is covered with a salt brine. The glucose from the starches is converted to lactate. This acid environment acts as a preservative that prevents the growth of other bacteria. The pickling of olives and cucumbers gives similar products. When cultures of bacteria that produce lactate are added to milk, the acid denatures the milk proteins to give sour cream and yogurt.

Fermentation

MC GOB SELF-STUDY ACTIVITY
Fermentation

Some microorganisms, particularly yeast, convert sugars to ethanol under anaerobic conditions by a process called **fermentation**. After pyruvate is formed in glycolysis, a carbon atom is removed in the form of CO_2 (**decarboxylation**). The NAD^+ for continued glycolysis is regenerated when the acetaldehyde is reduced to ethanol.

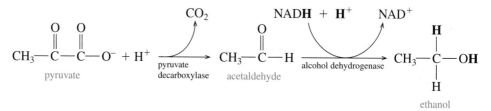

pyruvate   acetaldehyde   ethanol

The process of fermentation by yeast is one of the oldest known chemical reactions. Enzymes in the yeast convert the sugars in a variety of carbohydrate sources to glucose and then to ethanol. The evolution of CO_2 gas produces the bubbles in beer, sparkling wines, and champagne. The type of carbohydrate used determines the taste associated with a particular alcoholic beverage. Beer is made from the fermentation of barley malt, wine from the sugars in grapes, vodka from potatoes, sake from rice, and whiskeys from corn or rye. Fermentation produces solutions up to about 15% alcohol by volume. At this concentration, the alcohol kills the yeast, and fermentation stops. Higher concentrations are obtained by distilling the alcohol.

SAMPLE PROBLEM 22.6

■ **Fates of Pyruvate**

Is each of the following products from pyruvate produced under anaerobic or aerobic conditions?

a. acetyl CoA   **b.** lactate

SOLUTION

a. aerobic conditions   **b.** anaerobic conditions

STUDY CHECK

After strenuous exercise, some lactate is oxidized back to pyruvate by lactate dehydrogenase using NAD^+. Write an equation to show this reaction.

QUESTIONS AND PROBLEMS

■ **Pathways for Pyruvate**

22.39 What condition is needed in the cell to convert pyruvate to acetyl CoA?

22.40 What coenzymes are needed for the oxidation of pyruvate to acetyl CoA?

22.41 Write the overall equation for the conversion of pyruvate to acetyl CoA.

22.42 What are the possible products of pyruvate under anaerobic conditions?

22.43 How does the formation of lactate permit glycolysis to continue under anaerobic conditions?

22.44 After running a marathon, a runner has muscle pain and cramping. What might have occurred in the muscle cells to cause this?

22.45 In fermentation, a carbon atom is removed from pyruvate. What is the compound formed by that carbon atom?

22.46 Some students decided to make some wine by placing yeast and grape juice in a container with a tight lid. A few weeks later, the container explodes. What reaction could account for the explosion?

22.7 Glycogen Metabolism

LEARNING GOAL

Describe the breakdown and synthesis of glycogen.

We have just eaten a large meal that has supplied us with all the glucose we need to produce pyruvate and ATP by glycolysis. Then we use excess glucose to replenish our energy reserves by synthesizing glycogen that is stored in limited amounts in our skeletal muscle and liver. When glycogen stores are full, any remaining glucose is converted to triacylglycerols and stored as body fat as we will see in Chapter 24. When our diet does not supply sufficient glucose, or we have utilized our blood glucose, we degrade the stored glycogen and release glucose.

Glycogenesis

Glycogen is a polymer of glucose with α-1,4 glycosidic bonds and multiple branches attached by α-1,6 glycosidic bonds, as seen in Chapter 15. **Glycogenesis** is the synthesis of glycogen from glucose molecules, which occurs when the digestion of polysaccharides produces high levels of glucose. The synthesis of glycogen starts with the glucose-6-phosphate obtained from the first reaction in glycolysis. (See Figure 22.12.)

 Glycogen Metabolism

FIGURE 22.12

In glycogenesis, glucose is used to synthesize glycogen.

Q **What is the function of UTP in glycogen synthesis?**

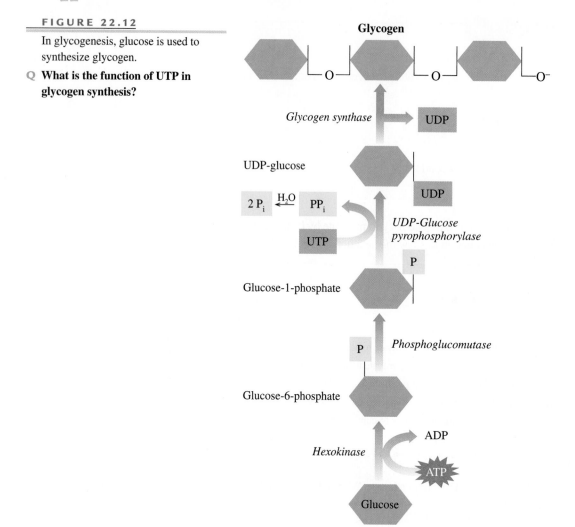

It is converted to an isomer glucose-1-phosphate, which is activated using high-energy UTP (uridine triphosphate) to yield UDP (uridine diphosphate)-glucose. The reaction is driven by the energy released from the hydrolysis of pyrophosphate (PP_i).

$$\text{glucose-6-phosphate} \underset{\text{phosphoglucomutase}}{\rightleftharpoons} \text{glucose-1-phosphate}$$

$$\text{glucose-1-phosphate} + \text{UTP} \xrightarrow{\substack{\text{UDP-glucose} \\ \text{pyrophosphorylase}}} \text{UDP-glucose} + PP_i$$

$$PP_i + H_2O \longrightarrow 2P_i$$

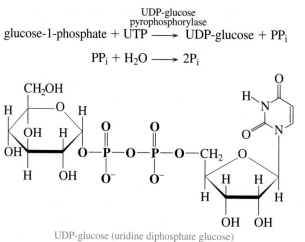

UDP-glucose (uridine diphosphate glucose)

The UDP-glucose attaches to the end of a glycogen chain releasing UDP, which reacts with ATP to regenerate UTP.

$$\text{UDP-glucose} + \text{glycogen} \xrightarrow{\text{glycogen synthase}} \text{glucose—glycogen} + \text{UDP}$$

$$\text{UDP} + \text{ATP} \longrightarrow \text{UTP} + \text{ADP}$$

The overall reaction of glycogenesis starting with glucose is written simply as

$$\text{glucose} \longrightarrow \text{glycogen}$$

Glycogenolysis

Glucose is the primary energy source for muscle contraction and the brain. When blood glucose is depleted, glycogen breaks down to glucose in a process called **glycogenolysis**. Glucose molecules are removed one by one from the end of the glycogen chain and phosphorylated to yield glucose-1-phosphate.

$$\text{glucose—glycogen} + \text{P}_\text{i} \xrightarrow{\text{glycogen phosphorylase}} \text{glucose-1-phosphate} + \text{glycogen}$$

Glucose-1-phosphate is converted to glucose-6-phosphate, which enters the glycolysis pathway to replenish ATP.

$$\text{glucose-1-phosphate} \xrightleftharpoons{\text{phosphoglucomutase}} \text{glucose-6-phosphate}$$

Free glucose is needed for energy by the brain and muscle. While glucose can diffuse across cell membranes, glucose phosphates cannot. Only cells in the liver and kidneys have a glucose-6-phosphatase that hydrolyzes the phosphate to yield free glucose.

$$\text{glucose-6-phosphate} \xrightarrow{\text{glucose-6-phosphatase}} \text{glucose} + \text{P}_\text{i}$$

The overall reaction of glycogenolysis, which converts glycogen to glucose, is written as

$$\text{glycogen} \longrightarrow \text{glucose}$$

Let's summarize the events and compounds involved in the breakdown and synthesis of glycogen as follows.

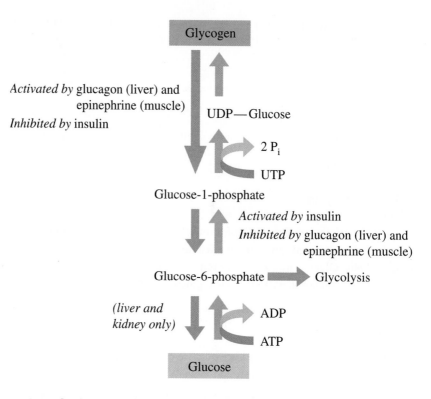

Regulation of Glycogen Metabolism

The brain, skeletal muscles, and red blood cells require large amounts of glucose every day to function properly. To protect the brain, hormones with opposing actions control blood glucose levels. When glucose is low, glucagon, a hormone produced in the pancreas, is secreted into the bloodstream. In the liver, glucagon accelerates the rate of glycogenolysis, which raises blood glucose levels. At the same time, glucagon inhibits the synthesis of glycogen.

Glycogen in skeletal muscle is broken down quickly when the body requires a "burst of energy" often referred to as "fight or flight." Epinephrine released from the adrenal glands converts *glycogen phosphorylase* from an inactive to an active form. The secretion of only a few molecules of epinephrine can break down a huge number of glycogen molecules.

Soon after we have eaten and digested a meal, our blood glucose levels rise. Rising glucose levels stimulate the pancreas to secrete the hormone insulin into our bloodstreams. Insulin promotes the use of glucose in the cells by accelerating glycogen synthesis as well as degradation reactions such as glycolysis. At the same time, insulin inhibits the synthesis of glucose, which we will discuss in the next section.

SAMPLE PROBLEM 22.7

Glycogen Metabolism

Identify each of the following as part of the reaction pathways of (1) glycolysis, (2) glycogenolysis, or (3) glycogenesis.

a. Glucose-1-phosphate is converted to glucose-6-phosphate.
b. Glucose-1-phosphate forms UDP-glucose.
c. An isomerase converts glucose-6-phosphate to fructose-6-phosphate.

SOLUTION

a. (2) glycogenolysis   **b.** (3) glycogenesis   **c.** (1) glycolysis

STUDY CHECK

Why do cells in liver and kidney provide glucose to raise blood glucose levels, but cells in skeletal muscle do not?

QUESTIONS AND PROBLEMS

■ Glycogen Metabolism

22.47 What is meant by the term *glycogenesis*?

22.48 What is meant by the term *glycogenolysis*?

22.49 How do muscle cells use glycogen to provide energy?

22.50 How does the liver raise blood glucose levels?

22.51 What is the function of *glycogen phosphorylase*?

22.52 Why is the enzyme *phosphoglucomutase* used in both glycogenolysis and glycogenesis?

22.8 Gluconeogenesis: Glucose Synthesis

LEARNING GOAL

Describe how glucose is synthesized from noncarbohydrate molecules.

Gluconeogenesis

Glycogen stored in our liver and muscles can supply us with about one day's requirement of glucose. However, glycogen stores are quickly depleted if we fast for more than one day or participate in heavy exercise. Then glucose is synthesized from carbon atoms obtained from noncarbohydrate compounds in a process called **gluconeogenesis**. Most glucose is synthesized in the cytosol of liver cells. (See Figure 22.13.)

Carbon atoms for glucose can be obtained from lactate and food sources, such as amino acids, and glycerol from fats. Each is converted to pyruvate or an intermediate for the synthesis of glucose. Most of the reactions in gluconeogenesis are the reverse of glycolysis and catalyzed by the same enzymes. However, three of the glycolysis reactions are not reversible: the ones catalyzed by hexokinase, phosphofructokinase, and pyruvate kinase, glycolysis reactions 1, 3, and 10, respectively. Different enzymes are used to replace them, but all the other reactions simply reverse glycolysis and use the same enzymes. We will now look at these three reactions in gluconeogenesis that differ from the reactions of glycolysis.

Converting Pyruvate to Phosphoenolpyruvate

To start the synthesis of glucose, two steps are needed. The first step converts pyruvate to oxaloacetate, and the second step converts oxaloacetate to phosphoenolpyruvate. The hydrolysis of ATP and GTP are used to drive the reactions. (See Figure 22.14.)

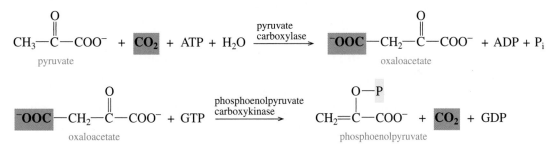

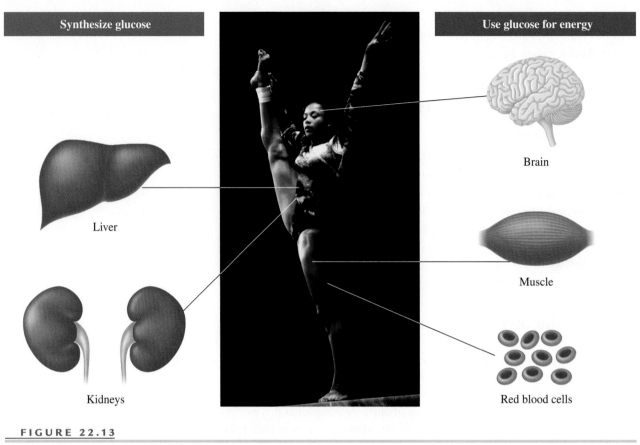

Synthesize glucose	Use glucose for energy
Liver	Brain
Kidneys	Muscle
	Red blood cells

FIGURE 22.13

Glucose is synthesized in the tissues of the liver and kidneys. Tissues that use glucose as their main energy source are the brain, skeletal muscle, and red blood cells.

Q **Why does the body need a pathway for the synthesis of glucose from noncarbohydrate sources?**

Molecules of phosphoenolpyruvate now enter the next five reverse reactions in glycolysis using the same enzymes to form fructose-1,6-bisphosphate.

Converting Fructose-1,6-bisphosphate to Fructose-6-phosphate

The second irreversible reaction in glycolysis is bypassed using *fructose-1,6-bisphosphatase* to cleave a phosphate from fructose-1,6-bisphosphate by hydrolysis with water and releasing the energy that drives the reaction.

$$\text{fructose-1,6-bisphosphate} + H_2O \xrightarrow{\text{fructose-1,6-bisphosphatase}} \text{fructose-6-phosphate} + P_i$$

Then fructose-6-phosphate undergoes a reversible reaction to yield glucose-6-phosphate.

Converting Glucose-6-phosphate to Glucose

In the final reaction, glucose-6-phosphate is converted to glucose by a different enzyme than used in glycolysis. *Glucose-6-phosphatase* catalyzes the hydrolysis of glucose-6-phosphate with water.

$$\text{glucose-6-phosphate} + H_2O \xrightarrow{\text{glucose-6-phosphatase}} \text{glucose} + P_i$$

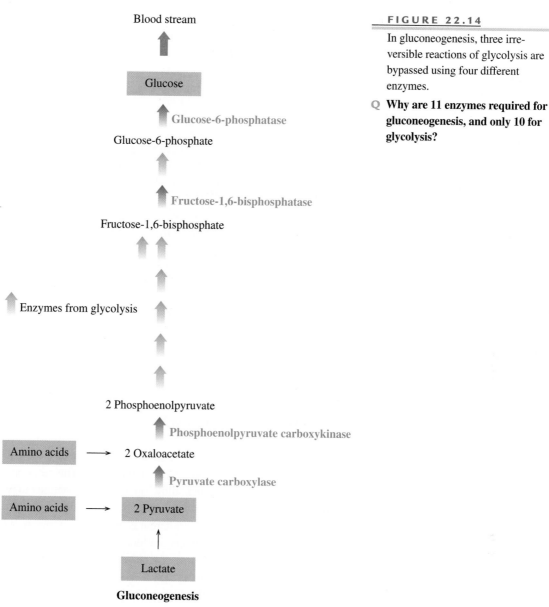

Gluconeogenesis

FIGURE 22.14

In gluconeogenesis, three irreversible reactions of glycolysis are bypassed using four different enzymes.

Q **Why are 11 enzymes required for gluconeogenesis, and only 10 for glycolysis?**

Energy Cost of Gluconeogenesis

The pathway of gluconeogenesis consists of seven reversible reactions of glycolysis and four new reactions that replace the three irreversible reactions. Overall, this synthesis of glucose requires 4 ATPs, 2 GTPs, and 2 NADHs. If all the reactions were simply the reverse of glycolysis, the synthesis of glucose would not be energetically favorable. By using the energy resources and bypassing the three irreversible and energy-requiring reactions, gluconeogenesis becomes favorable in terms of energy. The overall equation for gluconeogenesis is written as follows:

$$\text{2 Pyruvate} + \text{4 ATP} + \text{2 GTP} + \text{2 NADH} + \text{2H}^+ + \text{6H}_2\text{O} \longrightarrow$$
$$\text{glucose} + \text{4 ADP} + \text{2 GDP} + \text{6 P}_i + \text{2 NAD}^+$$

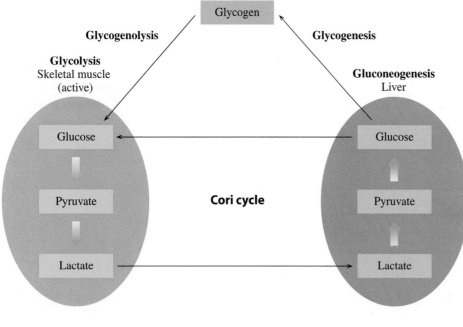

FIGURE 22.15

Different pathways connect the utilization and synthesis of glucose.

Q **Why is lactate formed in the muscle converted to glucose in the liver?**

Lactate and the Cori Cycle

When a person exercises vigorously, the anaerobic conditions cause the reduction of pyruvate to lactate, which accumulates in the muscle. This reaction is necessary to oxidize NADH to NAD^+, which allows glycolysis to continue to produce a small amount of ATP. Lactate is an important source of carbon for gluconeogenesis. Lactate is transported to the liver where it is oxidized to pyruvate, which is used to synthesize glucose. Glucose enters the bloodstream and returns to the muscle to rebuild glycogen stores. This flow of lactate and glucose between the muscle and liver, known as the **Cori cycle**, is very active when a person has just completed a period of vigorous exercise. (See Figure 22.15.)

Regulation of Gluconeogenesis

Gluconeogenesis is a pathway that protects the brain and nervous system from experiencing a loss of glucose, which causes impairment of function. It is also a pathway that is utilized when vigorous activity depletes blood glucose and glycogen stores. Thus, the level of carbohydrate available from the diet controls gluconeogenesis. When a diet is high in carbohydrate, the gluconeogenesis pathway is not utilized. However, when a diet is low in carbohydrate, the pathway is very active.

As long as conditions in a cell favor glycolysis, there is no synthesis of glucose, but when the cell requires synthesis of glucose, glycolysis is turned off. The same three reactions that control glycolysis also control gluconeogenesis, but with different enzymes. Let's look at how high levels of certain compounds activate or inhibit the two processes (See Table 22.2.)

TABLE 22.2

Regulation of Glycolysis and Gluconeogenesis

	Glycolysis	Gluconeogenesis
Enzyme	**Hexokinase**	**Glucose-6-phosphatase**
Activated by	High glucose levels, insulin, epinephrine	Low glucose levels, glucose-6-phosphate
Inhibited by	Glucose-6-phosphate	
Enzyme	**Phosphofructokinase**	**Fructose-1,6-bisphosphatase**
Activated by	AMP	Low glucose levels, glucagon
Inhibited by	ATP	AMP, insulin
Enzyme	**Pyruvate kinase**	**Pyruvate carboxylase**
Activated by	Fructose-1,6-bisphosphate	Low glucose level, glucagon
Inhibited by	ATP, acetyl CoA	Insulin

SAMPLE PROBLEM 22.8

▪ Gluconeogenesis

The conversion of fructose-1,6-bisphosphate to fructose-6-phosphate is an irreversible reaction using the glycolytic enzyme. How does gluconeogenesis make this reaction happen?

SOLUTION

This reverse reaction is catalyzed by a different enzyme *fructose-1,6-bisphosphatase*, which cleaves a phosphate using a hydrolysis reaction, a reaction that is energetically favorable.

STUDY CHECK

Why is hexokinase in glycolysis replaced by glucose-6-phosphatase in gluconeogenesis?

QUESTIONS AND PROBLEMS

▪ Gluconeogenesis: Glucose Synthesis

22.53 What is the function of gluconeogenesis in the body?

22.54 What enzymes in glycolysis are not used in gluconeogenesis?

22.55 What enzymes in glycolysis are used in gluconeogenesis?

22.56 How is the lactate produced in skeletal muscle used for glucose synthesis?

22.57 Indicate whether each of the following activates or inhibits gluconeogenesis:
   a. low glucose levels
   b. glucagon
   c. insulin

22.58 Indicate whether each of the following activates or inhibits glycolysis:
   a. low glucose levels
   b. insulin
   c. glucagon

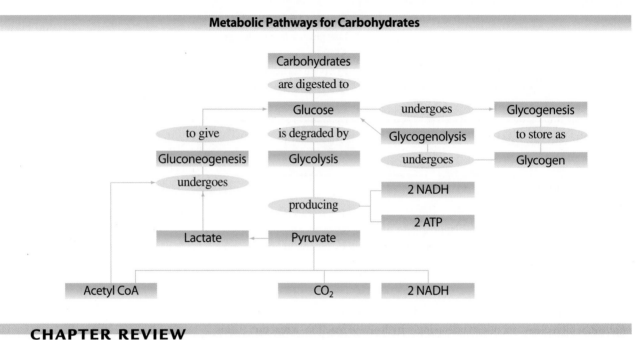

CHAPTER REVIEW

22.1 Metabolism and Cell Structure
Metabolism includes all the catabolic and anabolic reactions that occur in the cells. Catabolic reactions degrade large molecules into smaller ones with an accompanying release of energy. Anabolic reactions require energy to synthesize larger molecules from smaller ones. The 3 stages of metabolism are digestion of food, degradation of monomers such as glucose to pyruvate, and the extraction of energy from the two- and three-carbon compounds from stage 2. Many of the metabolic enzymes are present in the cytosol of the cell where metabolic reactions take place.

22.2 ATP and Energy
Energy obtained from catabolic reactions is stored primarily in adenosine triphosphate (ATP), a high-energy compound that is hydrolyzed when energy is required by anabolic reactions.

22.3 Important Coenzymes in Metabolic Pathways
FAD and NAD^+ are the oxidized forms of coenzymes that participate in oxidation–reduction reactions. When they pick up hydrogen ions and electrons, they are reduced to $FADH_2$ and $NADH + H^+$. Coenzyme A contains a thiol group that bonds with a two-carbon acetyl group (acetyl CoA) or a longer chain acyl group (acyl CoA).

22.4 Digestion of Carbohydrates
The digestion of carbohydrates is a series of reactions that breaks down polysaccharides into hexose monomers such as glucose, galactose, and fructose. These monomers can be absorbed through the intestinal wall into the bloodstream to be carried to cells where they provide energy and carbon atoms for synthesis of new molecules.

22.5 Glycolysis: Oxidation of Glucose
Glycolysis, which occurs in the cytosol, consists of 10 reactions that degrade glucose (six carbons) to two pyruvate molecules (three carbons each). The overall series of reactions yields two molecules of the reduced coenzyme NADH and two ATP.

22.6 Pathways for Pyruvate
Under aerobic conditions, pyruvate is oxidized in the mitochondria to acetyl CoA. In the absence of oxygen, pyruvate is reduced to lactate and NAD^+ is regenerated for the continuation of glycolysis, while microorganisms such as yeast reduce pyruvate to ethanol, a process known as fermentation.

22.7 Glycogen Metabolism
Glycogenolysis breaks down glycogen to glucose when glucose and ATP levels are low. When blood glucose levels are high, glycogenesis converts glucose to glycogen, which is stored in the liver.

22.8 Gluconeogenesis: Glucose Synthesis
When blood glucose levels are low and glycogen stores in the liver are depleted, glucose is synthesized from compounds such as pyruvate and lactate.

SUMMARY OF KEY REACTIONS

HYDROLYSIS OF ATP

$$\text{ATP} + \text{H}_2\text{O} \longrightarrow \text{ADP} + \text{P}_i + 7.3 \text{ kcal (31 kJ)/mole}$$

FORMATION OF ATP

$$\text{ADP} + \text{P}_i + 7.3 \text{ kcal/(31 kJ)/mole} \longrightarrow \text{ATP}$$

REDUCTION OF FAD AND NAD$^+$

$$\text{FAD} + 2\text{H}^+ + 2e^- \longrightarrow \text{FADH}_2$$

$$\text{NAD}^+ + 2\text{H}^+ + 2e^- \longrightarrow \text{NADH} + \text{H}^+$$

HYDROLYSIS OF DISACCHARIDES

$$\text{lactose} + \text{H}_2\text{O} \xrightarrow{\text{lactase}} \text{galactose} + \text{glucose}$$

$$\text{sucrose} + \text{H}_2\text{O} \xrightarrow{\text{sucrase}} \text{fructose} + \text{glucose}$$

$$\text{maltose} + \text{H}_2\text{O} \xrightarrow{\text{maltase}} \text{glucose} + \text{glucose}$$

GLYCOLYSIS

$$\underset{\text{glucose}}{\text{C}_6\text{H}_{12}\text{O}_6} + 2\,\text{ADP} + 2\,\text{P}_i + 2\,\text{NAD}^+ \longrightarrow$$

$$\underset{\text{pyruvate}}{2\text{CH}_3-\overset{\displaystyle O}{\overset{\|}{\text{C}}}-\text{COO}^-} + 2\,\text{ATP} + 2\,\text{NADH} + 4\text{H}^+$$

OXIDATION OF PYRUVATE TO ACETYL CoA

$$\underset{\text{pyruvate}}{\text{CH}_3-\overset{\displaystyle O}{\overset{\|}{\text{C}}}-\text{COO}^-} + \text{NAD}^+ +$$

$$\text{HS}-\text{CoA} \xrightarrow[\text{dehydrogenase}]{\text{pyruvate}} \underset{\text{acetyl CoA}}{\text{CH}_3-\overset{\displaystyle O}{\overset{\|}{\text{C}}}-\text{S}-\text{CoA}} + \text{NADH} + \text{CO}_2$$

REDUCTION OF PYRUVATE TO LACTATE

$$\underset{\text{pyruvate}}{\text{CH}_3-\overset{\displaystyle O}{\overset{\|}{\text{C}}}-\text{COO}^-} + \text{NADH} + \text{H}^+ \longrightarrow$$

$$\underset{\text{lactate}}{\text{CH}_3-\overset{\displaystyle \text{OH}}{\overset{|}{\text{CH}}}-\text{COO}^-} + \text{NAD}^+$$

OXIDATION OF GLUCOSE TO LACTATE

$$\text{glucose} + 2\text{ADP} + 2\,\text{P}_i \longrightarrow 2\,\text{lactate} + 2\,\text{ATP}$$

REDUCTION OF PYRUVATE TO ETHANOL

$$\underset{\text{pyruvate}}{\text{CH}_3-\overset{\displaystyle O}{\overset{\|}{\text{C}}}-\text{COO}^-} + \text{NADH} + 2\text{H}^+ \longrightarrow$$

$$\underset{\text{ethanol}}{\text{CH}_3-\text{CH}_2-\text{OH}} + \text{NAD}^+ + \text{CO}_2$$

GLYCOGENESIS

$$\text{glucose} \longrightarrow \text{glycogen}$$

GLYCOGENOLYSIS

$$\text{glycogen} \longrightarrow \text{glucose}$$

GLUCONEOGENESIS

$$\text{pyruvate (or lactate)} \longrightarrow \text{glucose}$$

$$2 \text{ pyruvate} + 4\,\text{ATP} + 2\,\text{GTP} + 2\,\text{NADH} + 2\text{H}^+ + 6\text{H}_2\text{O} \longrightarrow$$
$$\text{glucose} + 4\,\text{ADP} + 2\,\text{GDP} + 6\,\text{P}_i + 2\,\text{NAD}^+$$

KEY TERMS

acetyl CoA The compound that is formed when a two-carbon acetyl unit from oxidation of pyruvate bonds to coenzyme A.

ADP Adenosine diphosphate, a compound of adenine, a ribose sugar, and two phosphate groups; it is formed by the hydrolysis of ATP.

aerobic An oxygen-containing environment in the cells.

anabolic reaction A metabolic reaction that requires energy.

anaerobic A condition in cells when there is no oxygen.

ATP Adenosine triphosphate, a high-energy compound that stores energy in the cells, consists of adenine, a ribose sugar, and three phosphate groups.

catabolic reaction A metabolic reaction that produces energy for the cell by the degradation and oxidation of glucose and other molecules.

coenzyme A (CoA) A coenzyme that transports acyl and acetyl groups.

Cori cycle A cyclic process in which lactate produced in muscle is transferred to the liver to be synthesized to glucose, which can be used again by muscle.

cytoplasm The material in eukaryotic cells between the nucleus and the plasma membrane.

cytosol The fluid of the cytoplasm, which is an aqueous solution of electrolytes and enzymes.

decarboxylation The loss of a carbon atom in the form of CO_2.

digestion The processes in the gastrointestinal tract that break down large food molecules to smaller ones that pass through the intestinal membrane into the blood stream.

FAD A coenzyme (flavin adenine dinucleotide) for dehydrogenase enzymes that form carbon–carbon double bonds.

fermentation The anaerobic conversion of glucose by enzymes in yeast to yield alcohol and CO_2.

gluconeogenesis The synthesis of glucose from noncarbohydrate compounds.

glycogenesis The synthesis of glycogen from glucose molecules.

glycogenolysis The breakdown of glycogen into glucose molecules.

glycolysis The ten oxidation reactions of glucose that yield two pyruvate molecules.

metabolism All the chemical reactions in living cells that carry out molecular and energy transformations.

mitochondria The organelles of the cells where energy-producing reactions take place.

NAD^+ The hydrogen acceptor used in oxidation reactions that form carbon–oxygen double bonds.

oxidation The loss of hydrogen as hydrogen ions and electrons or the gain of oxygen by a substrate that is degraded to smaller molecules or a coenzyme.

reduction The gain of hydrogen ions and electrons or the loss of oxygen by a substrate or a coenzyme.

UNDERSTANDING THE CONCEPTS

22.59 On a hike, you expend 350 kcal per hour. How many moles of ATP will you use if you hike for 2.5 hr?

22.60 Identify each of the following as a 6-carbon or a 3-carbon compound and arrange them in the order in which they occur in glycolysis.
  a. 3-phosphoglycerate
  b. pyruvate
  c. glucose-6-phosphate
  d. glucose
  e. fructose-1,6-bisphosphate

ADDITIONAL QUESTIONS AND PROBLEMS

For instructor-assigned homework, go to www.masteringgob.com.

22.61 What is meant by the term metabolism?

22.62 How do catabolic reactions differ from anabolic reactions?

22.63 What stage of metabolism involves the digestion of large food polymers?

22.64 What state of metabolism degrades monomers such as glucose into smaller molecules?

22.65 What type of cell has a nucleus?

22.66 What are the organelles in a cell?

22.67 What is the full name of ATP?

22.68 What is the full name of ADP?

22.69 Write an equation for the hydrolysis of ATP to ADP.

22.70 Write the equation for the hydrolysis of ADP to AMP.

22.71 What is the full name of FAD?

22.72 What type of reaction uses FAD as the coenzyme?

22.73 What is the full name of NAD^+?

22.74 What type of reaction uses NAD^+ as the coenzyme?

22.75 Write the abbreviations for the reduced forms of
  a. FAD   **b.** NAD^+

22.76 What is the name of the vitamin in the structure of each of the following?
  a. FAD   **b.** NAD^+ **c.** coenzyme A

22.77 How and where does lactose undergo digestion in the body? What are the products?

22.78 How and where does sucrose undergo digestion in the body? What are the products?

22.79 How do galactose and fructose enter glycolysis?

22.80 What is the general type of reaction that takes place in the digestion of carbohydrates?

22.81 What are the reactant and product of glycolysis?

22.82 What is the coenzyme used in glycolysis?

22.83 In glycolysis, which reactions involve phosphorylation and which reactions involve a direct substrate phosphorylation to generate ATP?

22.84 How do ADP and ATP regulate the glycolysis pathway?

22.85 What reaction and enzyme in glycolysis convert a hexose bisphosphate into two triose phosphates?

22.86 How does the investment and generation of ATP give a net gain of ATP for glycolysis?

22.87 What compound is converted to fructose-6-phosphate by phosphoglucoisomerase?

22.88 What product forms when glyceraldehyde-3-phosphate adds a phosphate group?

22.89 When is pyruvate converted to lactate in the body?

22.90 When pyruvate is used to form acetyl CoA or ethanol in fermentation, the product has only two carbon atoms. What happened to the third carbon?

22.91 How does phosphofructokinase regulate the rate of glycolysis?

22.92 How does pyruvate kinase regulate the rate of glycolysis?

22.93 When does the rate of glycogenolysis increase in the cells?

22.94 If glucose-1-phosphate is the product from glycogen, how does it enter glycolysis?

22.95 What is the end product of glycogenolysis in the liver?

22.96 What is the end product of glycogenolysis in skeletal muscle?

22.97 Indicate whether each of the following conditions would increase or decrease the rate of glycogenolysis in the liver:
a. low blood glucose level
b. secretion of insulin
c. secretion of glucagon
d. high levels of ATP

22.98 Indicate whether each of the following conditions would increase or decrease the rate of glycogenesis in the liver:
a. low blood glucose level
b. secretion of insulin
c. secretion of glucagon
d. high levels of ATP

22.99 Indicate whether each of the following conditions would increase or decrease the rate of gluconeogenesis:
a. high blood-glucose level
b. secretion of insulin
c. secretion of glucagon
d. high levels of ATP

22.100 Indicate whether each of the following conditions would increase or decrease the rate of glycolysis:
a. high blood-glucose level
b. secretion of insulin
c. secretion of glucagon
d. high levels of ATP

CHALLENGE QUESTIONS

22.101 Why is glucose for blood glucose provided by glycogenolysis in the liver, but not in skeletal muscle?

22.102 When does the rate of glycogenesis increase in the cells?

22.103 How do the hormones insulin and glucagon affect the rate of glycogenesis, glycogenolysis, and glycolysis?

22.104 What is the function of gluconeogenesis?

22.105 Where does the Cori cycle operate?

22.106 Identify each of the following as part of glycolysis, glycogenolysis, glycogenesis, or gluconeogenesis:

a. Glycogen is broken down to glucose in the liver.
b. Glucose is synthesized from noncarbohydrate sources.
c. Glucose is degraded to pyruvate.
d. Glycogen is synthesized from glucose.

22.107 One cell at work may break down 2 million (2 000 000) ATP molecules in one second. Researchers estimate that the human body has about 10^{13} cells.
a. How much energy in kcal could be produced by the cells in the body in one day?
b. If ATP has a molar mass of 507 g/mole, how many grams of ATP are hydrolyzed in one day?

ANSWERS

Answers to Study Checks

22.1 The cytoplasm is all the cellular material including cytosol and organelles between the plasma membrane and the nucleus. The cytosol, the aqueous part of the cytoplasm, is a solution of electrolytes and enzymes.

22.2 Adenine, ribose, and three phosphate groups.

22.3 $FADH_2$

22.4 The digestion of amylose begins in the mouth when salivary amylase hydrolyzes some of the glycosidic bonds. In the small intestine, pancreatic amylase hydrolyzes more glycosidic bonds, and finally maltose is hydrolyzed by maltase to yield glucose.

22.5 In the initial reactions of glycolysis, energy in the form of two ATP is invested to activate glucose and convert it to fructose-1,6-bisphosphate.

22.6

$$CH_3-\overset{\overset{\displaystyle OH}{|}}{CH}-\overset{\overset{\displaystyle O}{\|}}{C}-O^- + NAD^+ \xrightarrow{\text{lactate dehydrogenase}}$$

$$CH_3-\overset{\overset{\displaystyle O}{\|}}{C}-\overset{\overset{\displaystyle O}{\|}}{C}-O^- + NADH + H^+$$

22.7 Only liver and kidney cells contain the phosphatase enzyme that converts glucose-6-phosphate to free glucose.

22.8 The reaction catalyzed by hexokinase in glycolysis is irreversible.

Answers to Selected Questions and Problems

22.1 The digestion of polysaccharides takes place in stage 1.

22.3 In metabolism, a catabolic reaction breaks apart large molecules, releasing energy.

22.5 a. (3) smooth endoplasmic reticulum
b. (1) lysosome
c. (2) Golgi complex

22.7 When a phosphate group is cleaved from ATP, sufficient energy is released for energy-requiring processes in the cell.

22.9 a. PEP + H_2O $\longrightarrow$ pyruvate + P_i + 14.8 kcal/mole
b. ADP + P_i + 7.3 kcal/mole $\longrightarrow$ ATP
c. Coupled: PEP + ADP $\longrightarrow$ ATP + pyruvate + 7.5 kcal/mole

22.11 a. coenzyme A   **b.** NAD^+ **c.** FAD

22.13 a. NADH   **b.** FAD

22.15 FAD

22.17 Hydrolysis is the main reaction involved in the digestion of carbohydrates.

22.19 a. lactose
b. glucose and fructose
c. glucose

22.21 glucose

22.23 ATP is required in phosphorylation reactions.

22.25 glyceraldehyde-3-phosphate and dihydroxyacetone phosphate

22.27 ATP is produced in glycolysis by transferring a phosphate from 1,3-bisphosphoglycerate and from phosphoenolpyruvate directly to ADP.

22.29 a. hexokinase   **b.** phosphoglyceratekinase

22.31 a. 1 ATP required
b. 1 NADH is produced for each triose.
c. 2 ATP and 2 NADH

22.33 a. In reaction 1, a hexokinase uses ATP to phosphorylate glucose.
b. In reactions 7 and 10, phosphate groups are transferred from 1,3-bisphosphoglycerate and phosphoenolpyruvate directly to ADP to produce ATP.
c. In reaction 4, the six-carbon molecule fructose-1,6-bisphosphate is split into two three-carbon molecules, glyceraldehyde-3-phosphate and dihydroxyacetone phosphate.

22.35 Galactose reacts with ATP to yield galactose-1-phosphate, which is converted to glucose-6-phosphate, an intermediate in glycolysis. Fructose reacts with ATP to yield fructose-1-phosphate, which is cleaved to give dihydroxyacetone phosphate and glyceraldehyde. Dihydroxyacetone phosphate isomerizes to glyceraldehyde-3-phosphate, and glyceraldehyde is phosphorylated to glyceraldehyde-3-phosphate, which is an intermediate in glycolysis.

22.37 a. activate   **b.** inhibit

22.39 Aerobic (oxygen) conditions are needed.

22.41 The oxidation of pyruvate converts NAD^+ to NADH and produces acetyl CoA and CO_2.

Pyruvate + NAD^+ + CoA $\longrightarrow$

Acetyl CoA + CO_2 + NADH + H^+

22.43 When pyruvate is reduced to lactate, NAD^+ is produced, which can be used for the oxidation of more glyceraldehyde-3-phosphate in glycolysis. This recycles NADH.

22.45 Carbon dioxide, CO_2

22.47 Glycogenesis is the synthesis of glycogen from glucose molecules.

22.49 Muscle cells break down glycogen to glucose-6-phosphate, which enters glycolysis.

22.51 Glycogen phosphorylase cleaves the glycosidic bonds at the ends of glycogen chains to remove glucose as glucose-1-phosphate.

22.53 When there are no glycogen stores remaining in the liver, gluconeogenesis synthesizes glucose from noncarbohydrate compounds such as pyruvate and lactate.

22.55 Phosphoglucoisomerase, fructose-1,6-biphosphate aldolase, triosephosphate isomerase, glyceraldehyde-3-phosphate dehydrogenase, phosphoglyceratekinase, phosphoglyceratemutase, and enolase.

22.57 a. activates   **b.** activates   **c.** inhibits

22.59 120 moles ATP

22.61 Metabolism includes all the reactions in cells provide energy and material for cell growth.

22.63 Stage 1

22.65 Eukaryotic cell

22.67 Adenosine triphosphate

22.69 ATP + H_2O $\longrightarrow$ ADP + P_i + 7.3 kcal (31kJ)/mole

22.71 Flavin adenine dinucleotide

22.73 Nicotinamide adenine dinucleotide

22.75 a. $FADH_2$ **b.** NADH + H^+

22.77 Lactose undergoes digestion in the mucosal cells of the small intestine to yield galactose and glucose.

22.79 Galactose and fructose are converted in the liver to glucose phosphate compounds that can enter the glycolysis pathway.

22.81 Glucose is the reactant and pyruvate is the product of glycolysis.

22.83 Reaction 1 and 3 involve phosphorylation of hexoses with ATP, and reactions 7 and 10 involve direct substrate phosphorylation that generates ATP.

22.85 Reaction 4, which converts fructose-1,6-bisphosphate into two triose phosphates, is catalyzed by aldolase.

22.87 Glucose-6-phosphate

22.89 Pyruvate is converted to lactate when oxygen is not present in the cell (anaerobic) to regenerate NAD^+ for glycolysis.

22.91 Phosphofructokinase is an allosteric enzyme that is activated by high levels of AMP and ADP because the cell needs to produce more ATP. When ATP levels are high due to a decrease in energy needs, ATP inhibits phosphofructokinase, which reduces its catalysis of fructose-6-phosphate.

22.93 The rate of glycogenolysis increases when blood glucose levels are low and glucagon has been secreted, which accelerates the breakdown of glycogen.

22.95 Glucose

22.97 a. increase **b.** decrease
c. increase **d.** decrease

22.99 a. decrease **b.** decrease
c. increase **d.** decrease

22.101 The cells in the liver, but not skeletal muscle, contain a phosphatase enzyme needed to convert glucose-6-phosphate to free glucose that can diffuse through cell membranes into the bloodstream. Glucose-6-phosphate, which is the end product of glycogenolysis in muscle cells, cannot diffuse easily across cell membranes.

22.103 Insulin increases the rate of glycogenolysis and glycolysis and decreases the rate of glycogenesis. Glucagon decreases the rate glycogenolysis and glycolysis and increases the rate of glycogenesis.

22.105 The Cori cycle is a cyclic process that involves the transfer of lactate from muscle to the liver where glucose is synthesized, which can be used again by the muscle.

22.107 a. 20 kcal **b.** 1500 g ATP

23 Metabolism and Energy Production

"I am trained in basic life support. I work with the ER staff to assist in patient care," says Mandy Dornell, emergency medical technician at Seaton Medical Center. "In the ER, I take vital signs, do patient assessment, and perform CPR. If someone has a motor vehicle accident, I may suspect a neck or back injury. Then I may use a backboard or a cervical collar, which prevents the patient from moving and causing further damage. When people have difficulty breathing, I insert an airway—nasal or oral—to assist ventilation. I also set up and monitor IVs, and I am trained in childbirth."

When someone is critically ill or injured, the quick reactions of emergency medical technicians (EMTs) and paramedics provide immediate medical care and transport to an ER or trauma center.

LOOKING AHEAD

Mastering
GOB CHEMISTRY

Visit **www.masteringgob.com**
for self-study materials
and instructor-assigned
homework.

I n Chapter 22 we described the digestion of carbohydrates to glucose and the degradation of glucose to pyruvate. We saw that pyruvate is converted to lactate when no oxygen is available in the cell or to two-carbon acetyl CoA when oxygen is plentiful. Although glycolysis produces a small amount of ATP, most of the ATP in the cells is produced during the conversion of pyruvate, when oxygen is available in the cell. In a process known as *respiration*, oxygen is required to complete the oxidation of glucose to CO_2 and H_2O.

In the *citric acid cycle*, a series of metabolic reactions in the mitochondria oxidize acetyl CoA to carbon dioxide, which releases energy to produce NADH and $FADH_2$. These reduced coenzymes enter the *electron transport chain* or *respiratory chain* where they provide hydrogen ions and electrons that combine with oxygen (O_2) to form H_2O. The energy released during electron transport is used to synthesize ATP from ADP.

23.1 The Citric Acid Cycle

The **citric acid cycle** is a series of reactions that degrades acetyl CoA to yield CO_2 and energy, which is used to produce $NADH + H^+$ and $FADH_2$. (See Figure 23.1.) The citric acid cycle connects the products from stages 1 and 2 with the electron transport chain and the synthesis of ATP in stage 3. As a central pathway in metabolism, the citric acid cycle uses acetyl CoA from the degradation of carbohydrates as well as lipids and proteins.

The citric acid cycle is also called the tricarboxylic acid cycle because citric acid, a tricarboxylic acid, forms in the first reaction. Although citric acid is present as citrate at cellular pH, its acid name is retained. The citric acid cycle is also known as the Krebs cycle for H. A. Krebs, who recognized it as the major pathway for the production of energy.

Overview of the Citric Acid cycle

There are a total of 8 reactions in the citric acid cycle, which we can separate into two parts. In part 1, a two-carbon acetyl group bonds with four-carbon oxaloacetate to yield citrate. Then decarboxylation reactions remove two carbon atoms as CO_2 molecules to give a four-carbon compound. Then the reactions in part 2 convert the four-carbon compound to oxaloacetate, which bonds with another acetyl CoA and goes through the cycle again. (See Figure 23.2.) In one turn of the citric acid cycle, four oxidation reactions provide hydrogen ions and electrons, which are used to reduce FAD and NAD^+ coenzymes.

LEARNING GOAL

Describe the oxidation of acetyl CoA in the citric acid cycle.

 SELF-STUDY ACTIVITY
Krebs Cycle

 The Citric Acid Cycle

Stages of Metabolism

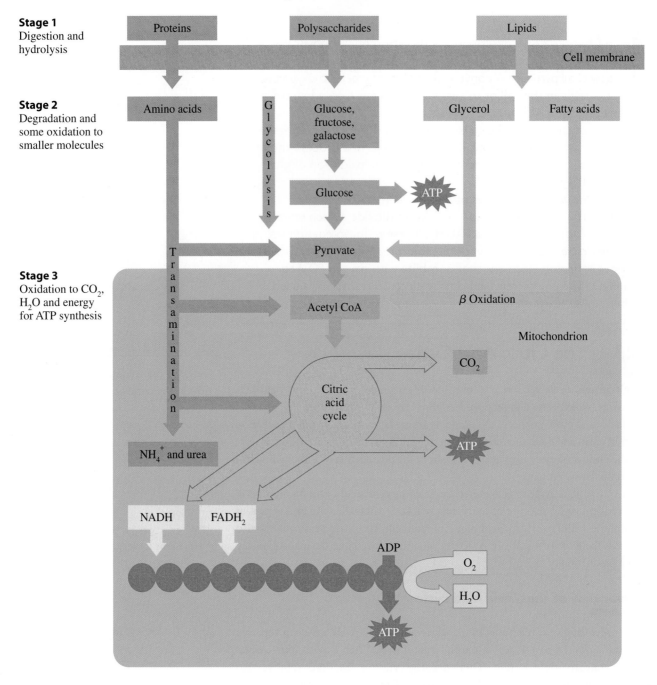

Stage 1
Digestion and hydrolysis

Stage 2
Degradation and some oxidation to smaller molecules

Stage 3
Oxidation to CO_2, H_2O and energy for ATP synthesis

FIGURE 23.1

The citric acid cycle connects the catabolic pathways that begin with the digestion and degradation of foods in stages 1 and 2 with the oxidation of substrates in stage 3 that generates most of the energy for ATP synthesis.

Q **Why is the citric acid cycle called a central metabolic pathway?**

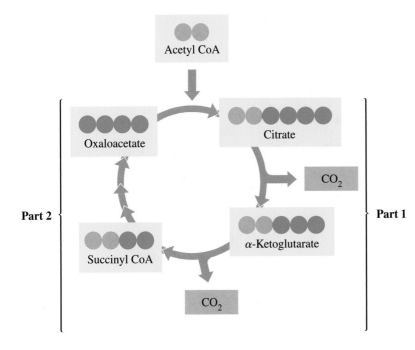

FIGURE 23.2

In part 1 of the citric acid cycle, two carbon atoms are removed as CO_2 from six-carbon citrate to give four-carbon succinyl CoA, which is converted in part 2 to four-carbon oxaloacetate.

Q **What is the difference in part 1 and part 2 of the citric acid cycle?**

Part 1 Decarboxylation Removes Two Carbon Atoms

Reaction 1 Formation of Citrate

In the first reaction, the two-carbon acetyl group in acetyl CoA bonds with the four-carbon oxaloacetate to yield citrate and HS-CoA. (See Figure 23.3.)

The Reactions of the Citric Acid Cycle

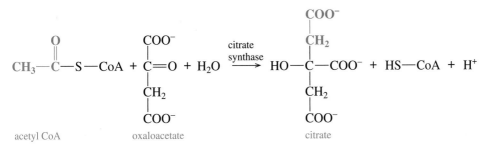

Reaction 2 Isomerization to Isocitrate

In order to continue oxidation, citrate undergoes isomerization in two steps to yield isocitrate. Citrate loses water (dehydration) to yield aconitate, which is rehydrated to form isocitrate. This reaction converts the tertiary hydroxyl (—OH) group in citrate to a secondary hydroxyl group that can be oxidized in the next step.

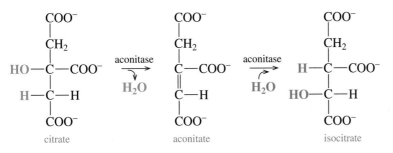

Reaction 3 First Oxidative Decarboxylation (CO_2)

This is the first time in the citric acid cycle that an oxidation and a decarboxylation occur together. The oxidation converts the hydroxyl group to a ketone and the **decarboxylation** removes a CO_2 molecule from a carboxylate group (COO⁻). The

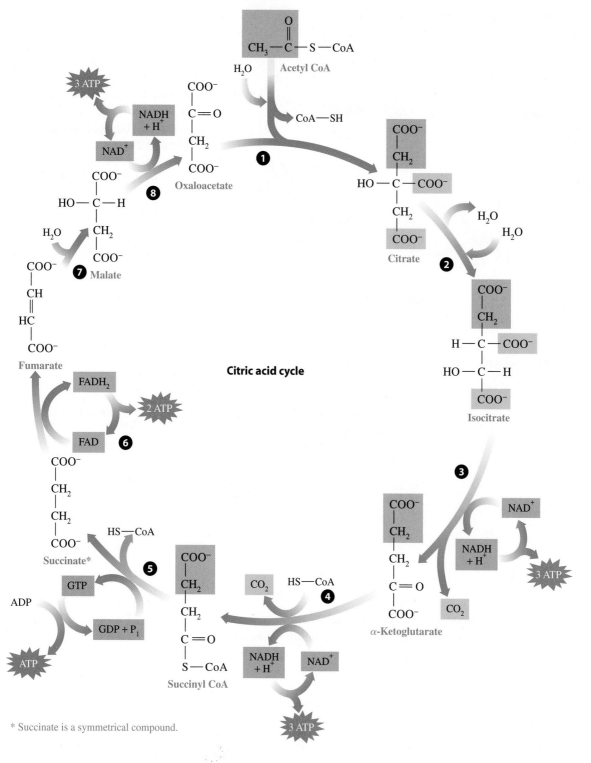

Citric acid cycle

* Succinate is a symmetrical compound.

FIGURE 23.3

In the citric acid cycle, oxidation reactions produce two CO_2, create reduced coenzymes NADH and $FADH_2$, and regenerate oxaloacetate.

Q **How many reactions in the citric acid cycle produce a reduced coenzyme?**

loss of CO_2 yields a five-carbon α-ketoglutarate. The energy from the oxidation is used to transfer hydrogen ions and electrons to NAD^+. We can summarize the reactions as follows:

1. The hydroxyl group (—OH) is oxidized to a ketone (C=O).
2. NAD^+ is reduced to yield NADH.
3. A carboxylate group (COO^-) is removed as CO_2.

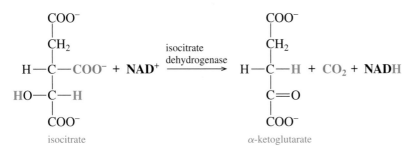

Reaction 4   Second Oxidative Decarboxylation (CO_2)

In this reaction, a second CO_2 is removed as α-ketoglutarate undergoes oxidative decarboxylation. The resulting four-carbon group combines with coenzyme A to form succinyl CoA and hydrogen ions and electrons are transferred to NAD^+. The two reactions that occur are as follows:

1. A second carbon is removed as CO_2.
2. NAD^+ is reduced to yield NADH.

■ Part 2

Reaction 5   Hydrolysis of Succinyl CoA

The energy released by the hydrolysis of the thioester bond in succinyl CoA is used to add a phosphate group (P_i) directly to GDP (guanosine diphosphate). The products are succinate and GTP, which is a high-energy compound similar to ATP.

The hydrolysis of GTP is used to add a phosphate group to ADP, which regenerates GDP for the citric acid cycle. This is the only time in the citric acid cycle that a direct substrate phosphorylation is used to produce ATP.

$$GTP + ADP \longrightarrow GDP + ATP$$

Reaction 6 Dehydrogenation of Succinate

In this oxidation reaction, hydrogen is removed from succinate, to produce fumarate, a compound with a trans double bond. This is the only place in the citric acid cycle where FAD is reduced to $FADH_2$.

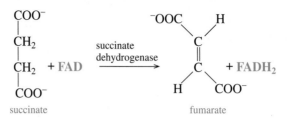

Reaction 7 Hydration

In a hydration reaction, water adds to the double bond of fumarate to yield malate.

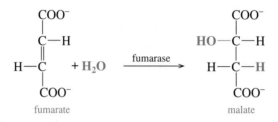

Reaction 8 Dehydrogenation Forms Oxaloacetate

In the last step of the citric acid cycle, the hydroxyl (—OH) group in malate is oxidized to yield oxaloacetate. The coenzyme NAD^+ is reduced to $NADH + H^+$.

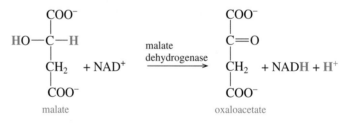

Summary of Products from the Citric Acid Cycle

We have seen that the citric acid cycle begins when a two-carbon acetyl group from acetyl CoA combines with oxaloacetate to form citrate. In part 1 of the cycle, two carbon atoms are removed from citrate to yield two CO_2 and a four-carbon compound that undergoes more reactions in part 2 to regenerate oxaloacetate. In four oxidation reactions, energy is released that reduces three NAD^+ and one FAD. In one reaction, GTP produced by a direct phosphorylation is used to form ATP from ADP. A summary of the products from one turn of the citric acid cycle is as follows:

2 CO_2 molecules
3 NADH molecules
1 $FADH_2$ molecule
1 GTP molecule used to form ATP
1 HS—CoA molecule
$3H^+$

We can write the overall chemical equation for one turn of the citric acid cycle as follows:

$$\text{acetyl-S-CoA} + 3\,\text{NAD}^+ + \text{FAD} + \text{GDP} + \text{P}_i + 2\text{H}_2\text{O} \longrightarrow$$
$$2\,\text{CO}_2 + 3\,\text{NADH} + 3\text{H}^+ + \text{FADH}_2 + \text{HS}-\text{CoA} + \text{GTP}$$

Regulation of Citric Acid Cycle

The primary function of the citric acid cycle is to produce high-energy compounds for ATP synthesis. When the cell needs energy, low levels of ATP stimulate the conversion of pyruvate to acetyl CoA, the fuel for the citric acid cycle. When ATP and NADH levels are high, there is a decrease in the production of acetyl CoA from pyruvate.

In the citric acid cycle, the enzymes that catalyze reactions 3 and 4 respond to allosteric activation and inhibition (Chapter 20). In reaction 3, isocitrate dehydrogenase is activated by high levels of ADP and inhibited by high levels of ATP and NADH. In reaction 4, α-ketoglutarate dehydrogenase is activated by high levels of ADP and inhibited by high levels of NADH and succinyl CoA. (See Figure 23.4.)

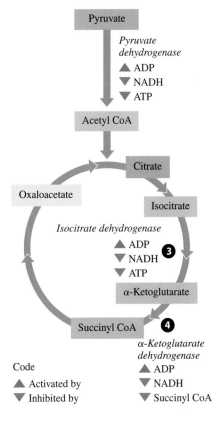

Code
▲ Activated by
▼ Inhibited by

FIGURE 23.4

High levels of ADP activate enzymes for the production of acetyl CoA and the citric acid cycle, whereas high levels of ATP, NADH, and succinyl CoA inhibit enzymes in the citric acid cycle.

Q **How do high levels of ATP affect the rate of the citric acid cycle?**

SAMPLE PROBLEM 23.1

Citric Acid Cycle

When one acetyl CoA completes the citric acid cycle, how many of each of the following are produced?

a. NADH **b.** ketone group **c.** CO_2

SOLUTION

a. One turn of the citric acid cycle produces three molecules of NADH.
b. Two ketone groups form when the secondary alcohol groups in isocitrate and malate are oxidized by NAD^+.
c. Two molecules of CO_2 are produced by the decarboxylation of isocitrate and α-ketoglutarate.

STUDY CHECK

What substance is a substrate in the first reaction of the citric acid cycle and a product in the last reaction?

QUESTIONS AND PROBLEMS

The Citric Acid Cycle

23.1 What other names are used for the citric acid cycle?

23.2 What compounds are needed to start the citric acid cycle?

23.3 What are the products from one turn of the citric acid cycle?

23.4 What compound is regenerated in each turn of the citric acid cycle?

23.5 Which reaction(s) of the citric acid cycle involve oxidative decarboxylation?

23.6 Which reaction(s) of the citric acid cycle involve a dehydration reaction?

23.7 Which reactions of the citric acid cycle reduce NAD^+?

23.8 Which reactions of the citric acid cycle reduce FAD?

23.9 Which reaction in the citric acid cycle involves a direct substrate phosphorylation?

23.10 What is the total NADH and total FADH_2 produced in one turn of the citric acid cycle?

23.11 Refer to the diagram of the citric acid cycle to answer each of the following:
  a. What are the six-carbon compounds?
  b. How is the number of carbon atoms decreased?
  c. What is the five-carbon compound?
  d. Which reactions are oxidation reactions?
  e. In which reactions are secondary alcohols oxidized?

23.12 Refer to the diagram of the citric acid cycle to answer each of the following:
  a. What is the yield of CO_2 molecules?
  b. What are the four-carbon compounds?
  c. What is the yield of GTP molecules?
  d. What are the decarboxylation reactions?
  e. Where does a hydration occur?

23.13 Indicate the name of the enzyme for each of the following reactions in the citric acid cycle:
  a. joins acetyl CoA to oxaloacetate
  b. forms a carbon–carbon double bond
  c. adds water to fumarate

23.14 Indicate the name of the enzyme for each of the following reactions in the citric acid cycle:
  a. isomerizes citrate
  b. oxidizes and decarboxylates α-ketoglutarate
  c. adds P_i to GDP

23.15 State the acceptor for hydrogen or phosphate in each of the following reactions:
  a. isocitrate $\longrightarrow$ α-ketoglutarate
  b. succinyl CoA $\longrightarrow$ succinate

23.16 State the acceptor for hydrogen or phosphate in each of the following reactions:
  a. malate $\longrightarrow$ oxaloacetate
  b. α-ketoglutarate $\longrightarrow$ succinyl CoA

23.17 What enzymes in the citric acid cycle are allosteric enzymes?

23.18 How does NADH affect the rate of the citric acid cycle?

23.19 How do high levels of ADP affect the rate of the citric acid cycle?

23.20 Why does the rate of the oxidation of pyruvate affect the rate of the citric acid cycle?

LEARNING GOAL

■ Describe the electron carriers involved in electron transport.

■ **23.2 Electron Carriers**

At this point, the metabolic cycles of glycolysis, oxidation of two pyruvate, and the citric acid cycle for two acetyl CoA would produce four ATP along with ten NADH and two $FADH_2$ from the degradation of glucose.

From Glucose	ATP	Reduced Coenzymes	
Glycolysis:	2	2 NADH	
Oxidation (2 pyruvate):		2 NADH	
Citric acid cycle (2 acetyl CoA):	2	6 NADH	2 $FADH_2$

Now we will see how the oxidation of these reduced coenzymes provides the energy for the synthesis of considerably more ATP. In the **electron transport chain** or *respiratory chain*, hydrogen ions and electrons from NADH and $FADH_2$ are passed from one electron acceptor to the next until they combine with oxygen to form H_2O. The electron acceptors in this transport system are known as **electron carriers**. The energy released during electron transport is used to synthesize ATP from ADP and P_i, a process called *oxidative phosphorylation*. As long

MC
GOB   Electron Carriers

as oxygen is available for the mitochondria in the cell, electron transport and oxidative phosphorylation function to produce most of the ATP energy manufactured in the cell.

Oxidation and Reduction of Electron Carriers

The electron carriers in electron transport include flavins, iron–sulfur proteins, coenyzme Q, and cytochromes. Each type of electron carrier contains a group or ion that is reduced when electrons are accepted and oxidized when electrons are removed. That is, the transfer of hydrogen ions and electrons from reduced AH_2 to the carrier B forms reduced BH_2 and oxidized carrier A.

oxidized carrier **A** $\qquad$ reduced carrier BH_2

reduced carrier AH_2 $\qquad$ oxidized carrier **B**

There are four types of electron carriers that make up the electron transport system.

1. **FMN (flavin mononucleotide)** is a coenzyme derived from riboflavin (vitamin B_2). FMN contains a flavin ring system that is also found in FAD. (See Figure 23.5.) The reduced product is $FMNH_2$.
2. **Fe–S clusters** is the name given to a group of iron–sulfur proteins that contain iron–sulfur clusters embedded in the proteins of the electron transport chain. The clusters contain iron ions, inorganic sulfides, and several cysteine groups. The iron in the clusters is reduced to Fe^{2+} and oxidized to Fe^{3+} as electrons are accepted and lost. (See Figure 23.6.)
3. **Coenzyme Q (Q or CoQ)** is derived from quinone, which is a six-carbon cyclic compound with two double bonds and two keto groups attached to

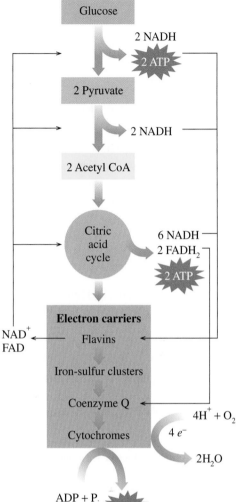

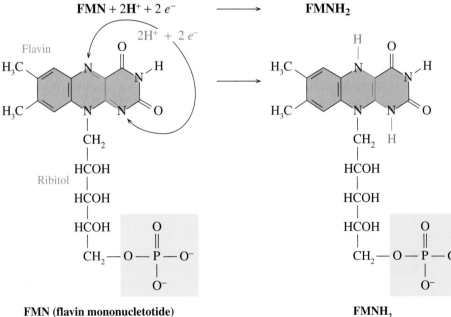

FIGURE 23.5

The electron carrier FMN consists of a flavin ring system containing the reactive center, ribitol, and a phosphate group.

Q **What part of the FMN molecule is reduced when hydrogen ions and electrons are accepted?**

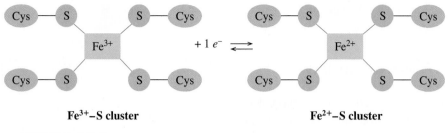

Fe³⁺–S cluster   **Fe²⁺–S cluster**

FIGURE 23.6

In typical iron–sulfur clusters, iron ions bond to sulfur atoms in the thiol (—SH) groups of four cysteine groups in proteins.

Q **In an iron–sulfur cluster, what are the ionic charges of the oxidized and reduced iron ions?**

$$Q + 2H^+ + 2\,e^- \rightleftharpoons QH_2$$

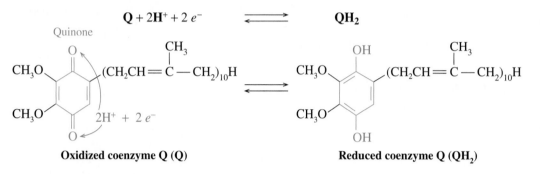

Oxidized coenzyme Q (Q)   **Reduced coenzyme Q (QH₂)**

FIGURE 23.7

The electron carrier coenzyme Q accepts electrons from $FADH_2$ and $FMNH_2$ and passes them to the cytochromes.

Q **How does reduced coenzyme Q compare to the oxidized form?**

a long carbon chain. (See Figure 23.7.) Coenzyme Q is reduced when the keto groups of quinone accept hydrogen ions and electrons.

4. **Cytochromes (cyt)** are proteins that contain an iron ion in a heme group. The different cytochromes are indicated by the letters following the abbreviation for cytochrome (cyt): cyt b, cyt c_1, cyt c, cyt a, and cyt a_3. In each cytochrome, the Fe^{3+} accepts a single electron to form Fe^{2+}, which is oxidized back to Fe^{3+} when the electron is passed to the next cytochrome. (See Figure 23.8.)

$$Fe^{3+} + 1\,e^- \rightleftharpoons Fe^{2+}$$

Oxidation and Reduction of Electron Carriers

SAMPLE PROBLEM 23.2

■ **Electron Carriers**

Give the abbreviation for each of the following carriers:

a. the oxidized form of flavin mononucleotide
b. the reduced form of coenzyme Q

SOLUTION

a. FMN   **b.** QH_2

STUDY CHECK

What is the oxidized form of iron in cyt a_3?

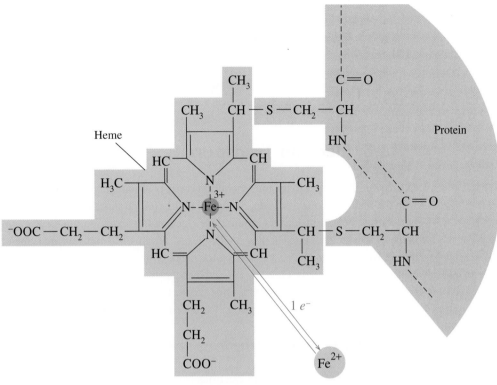

Simplified structure of cytochromes c and c_1

FIGURE 23.8

The iron-containing proteins known as cytochromes are identified as b, c, c_1, a, and a_3.

Q **What are the reduced and oxidized forms of the cytochromes?**

SAMPLE PROBLEM 23.3

■ **Oxidation and Reduction**

Identify the following steps in the electron transport chain as oxidation or reduction:

a. $FMN + 2H^+ + 2\,e^- \longrightarrow FMNH_2$
b. Cyt c $(Fe^{2+}) \longrightarrow$ cyt c $(Fe^{3+}) + e^-$

SOLUTION

a. The gain of hydrogen is reduction.
b. The loss of electrons is oxidation.

STUDY CHECK

Identify each of the following as oxidation or reduction:

a. $QH_2 \longrightarrow Q$
b. cyt b $(Fe^{3+}) \longrightarrow$ cyt b (Fe^{2+})

QUESTIONS AND PROBLEMS

■ **Electron Carriers**

23.21 Is cyt b (Fe^{3+}) the abbreviation for the oxidized or reduced form of cytochrome b?

23.22 Is $FMNH_2$ the abbreviation for the oxidized or reduced form of flavin mononucleotide?

23.23 Identify the following as oxidation or reduction:
  a. $FMNH_2 \longrightarrow FMN + 2H^+ + 2\,e^-$
  b. $Q + 2H^+ + 2\,e^- \longrightarrow QH_2$

23.24 Identify the following as oxidation or reduction:
  a. cyt $c\ (Fe^{3+}) + e^- \longrightarrow$ cyt $c\ (Fe^{2+})$
  b. Fe^{2+}–S cluster $\longrightarrow Fe^{3+}$–S cluster $+\ e^-$

LEARNING GOAL

Describe the role of the electron carriers in electron transport.

MC GOB SELF-STUDY ACTIVITY
Electron Transport

23.3 Electron Transport

In Chapter 22, we saw that a mitochondrion consists of inner and outer membranes with the matrix located between. Along the highly folded inner membrane are the enzymes and electron carriers required for electron transport. Within these membranes, there are four distinct protein complexes. Within each complex are some of the electron carriers needed for electron transport.

Two electron carriers, coenzyme Q and cytochrome c, are not firmly attached to the membrane. They function as mobile carriers shuttling electrons between the protein complexes that are tightly bound to the membrane. (See Figure 23.9.)

Complex I   NADH Dehydrogenase

At complex I, hydrogen ions and electrons from NADH are transferred to FMN. Reduced $FMNH_2$ forms, while NADH is reoxidized to NAD^+, which returns to oxidative pathways such as the citric acid cycle to oxidize more substrates. Within

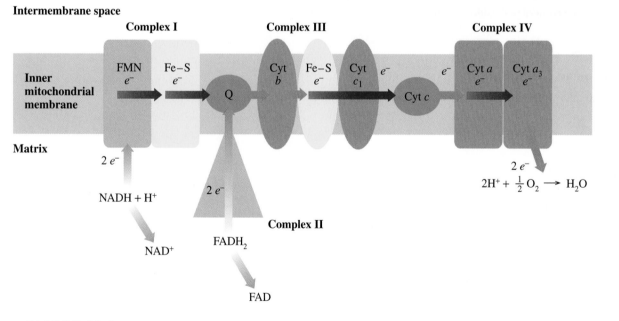

FIGURE 23.9

Most of the electron carriers in the electron transport chain are found in protein complexes bound to the inner membrane of the mitochondria. Two are mobile carriers that carry electrons between the protein complexes.

Q **What is the function of the electron carriers coenzyme Q and cytochrome c?**

complex I, the electrons are transferred from $FMNH_2$ to iron–sulfur (Fe–S) clusters and from Fe–S to coenzyme (Q).

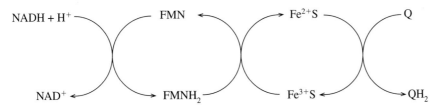

The overall reaction sequence in complex I can be written as follows:

$$NADH + H^+ + Q \longrightarrow QH_2 + NAD^+$$

Complex II Succinate Dehydrogenase

Complex II is used when $FADH_2$ is generated by the conversion of succinate to fumarate in the citric acid cycle. The electrons from $FADH_2$ are transferred to coenzyme Q to yield QH_2. Because complex II is at a lower energy level than complex I, the electrons from $FADH_2$ enter electron transport at a lower energy level than those from NADH.

$$FADH_2 + Q \longrightarrow FAD + QH_2$$

Complex III Co enzyme Q–Cytochrome *c* Reductase

The mobile carrier QH_2 transfers the electrons it has collected from NADH and $FADH_2$ to an iron–sulfur (Fe–S) cluster, and then to cytochrome *b*, the first cytochrome in complex III.

$$QH_2 + 2 \text{ cyt } b \text{ (Fe}^{3+}) \longrightarrow Q + 2 \text{ cyt } b \text{ (Fe}^{2+}) + 2H^+$$

From cyt *b*, the electrons are transferred to an Fe–S cluster, then to cytochrome c_1, and then to cytochrome *c*. Each time an Fe^{3+} ion accepts an electron, it is reduced to Fe^{2+}, and then oxidized back to Fe^{3+} as the electron is passed on along the chain. Cytochrome *c* is another mobile carrier; it moves the electron from complex III to complex IV.

Complex IV Cytochrome *c* Oxidase

At complex IV, electrons are transferred from cytochrome *c* to cytochrome *a*, and then to cytochrome a_3, the last cytochrome. In the final step of electron transport, electrons and hydrogen ions combine with oxygen (O_2) to form water.

$$2H^+ + 2 e^- + \tfrac{1}{2}O_2 \longrightarrow H_2O$$

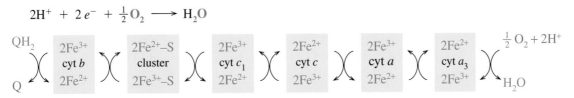

Toxins: Inhibitors of Electron Transport

There are several substances that inhibit the electron carriers in the electron transport chain. Rotenone, a product from a plant root used as an insecticide, blocks electron transport between FMN (complex I) and coenzyme Q. The barbiturates amytal and demerol also inhibit FMN. Another inhibitor is the antibiotic antimycin A, which blocks the flow of electrons between cytochrome *b* and cytochrome *c*₁ (complex III). Another group of compounds including cyanide (CN^-) and carbon monoxide inhibit cytochrome *c* oxidase (complex IV). The toxic nature of these compounds makes it clear that an organism relies heavily on the process of electron transport.

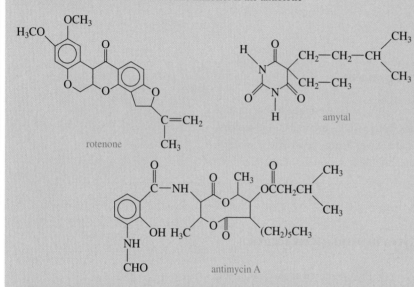

rotenone

amytal

antimycin A

When an inhibitor blocks a step in the electron transport chain, the carriers preceding that step are unable to transfer electrons and remain in their reduced forms. All the carriers after the blocked step remain oxidized without a source of electrons. Thus, any of these inhibitors shut down the flow of electrons through the electron transport chain.

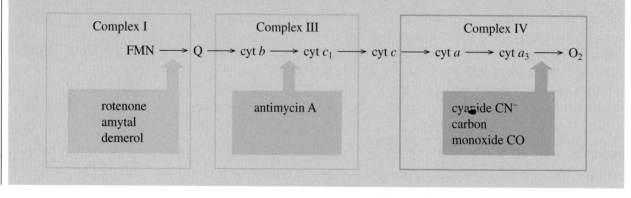

SAMPLE PROBLEM 23.4

MC
GOB   Electron Transport

Electron Transport

Identify the following electron carriers that are mobile carriers.

a. Cyt *c*   **b.** FMN   **c.** Fe–S clusters   **d.** Q

SOLUTION

a. and **d.,** cyt *c* and Q are mobile carriers.

STUDY CHECK

What is the final substance that accepts electrons in the electron transport chain?

QUESTIONS AND PROBLEMS

■ Electron Transport

23.25 What reduced coenzymes provide the electrons for electron transport?

23.26 What happens to the energy level as electrons are passed along the electron transport chain?

23.27 Arrange the following in the order they appear in electron transport: cytochrome c, cytochrome b, FAD, and coenzyme Q.

23.28 Arrange the following in the order they appear in electron transport: O_2, NAD^+, cytochrome a_3, and FMN.

23.29 How are electrons carried from complex I to complex III?

23.30 How are electrons carried from complex III to complex IV?

23.31 How is NADH oxidized in electron transport?

23.32 How is $FADH_2$ oxidized in electron transport?

23.33 Complete the following reactions in electron transport:
    a. $NADH + H^+ +$ _____ $\longrightarrow$ _____ $+ FMNH_2$
    b. $QH_2 + 2$ cyt b $(Fe^{3+}) \longrightarrow$ _____ $+$ _____ $+ 2H^+$

23.34 Complete the following reactions in electron transport:
    a. $Q +$ _____ $\longrightarrow$ _____ $+ FAD$
    b. 2 cyt a $(Fe^{3+}) + 2$ cyt a_3 $(Fe^{2+}) \longrightarrow$ _____ $+$ _____

■ 23.4 Oxidative Phosphorylation and ATP

LEARNING GOAL

Describe the process of oxidative phosphorylation in ATP synthesis.

We have seen that energy is generated when electrons from the oxidation of substrates flow through the electron transport chain. Now we will look at how that energy is coupled with the production of ATP in the process called **oxidative phosphorylation**.

Chemiosmotic Model

SELF-STUDY ACTIVITY
Electron Transport

In 1978, Peter Mitchell received the Nobel Prize for his theory called the **chemiosmotic model**, which links the energy from electron transport to a proton gradient that drives the synthesis of ATP. Three of the complexes (I, III, and IV) extend through the inner membrane with one end of each complex in the matrix and the other end in the intermembrane space. In the chemiosmotic model, each of these complexes act as a **proton pump** by pushing protons (H^+) out of the matrix and into the intermembrane space. This increase in protons in the intermembrane space lowers the pH and creates a proton gradient. Because protons are positively charged, both the lower pH and the electrical charge of the proton gradient produce an electrochemical gradient. (See Figure 23.10.)

To equalize the pH between the intermembrane space and the matrix, there is a tendency by the protons to return the matrix. However, protons cannot diffuse through the inner membrane. The only way protons can return to the matrix is to pass through a protein complex called **ATP synthase**. As the protons flow through ATP synthase, energy generated from the proton gradient is used to drive the ATP synthesis. Thus the process of oxidative phosphorylation couples the energy from electron transport to the synthesis of ATP from ADP and P_i.

The Chemiosmotic Model

$$ADP + P_i + energy \xrightarrow{\text{ATP synthase}} ATP$$

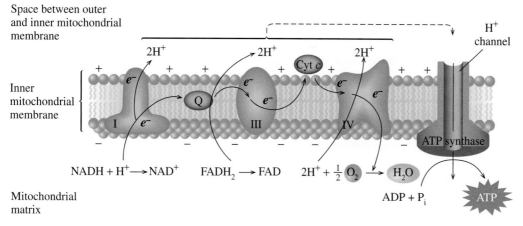

Space between outer and inner mitochondrial membrane

Inner mitochondrial membrane

$$NADH + H^+ \longrightarrow NAD^+ \qquad FADH_2 \longrightarrow FAD \qquad 2H^+ + \tfrac{1}{2}O_2 \longrightarrow H_2O$$

Mitochondrial matrix

FIGURE 23.10

In electron transport, protein complexes oxidize and reduce coenzymes to provide electrons and protons that move into the intermembrane space where they create a proton gradient that drives ATP synthesis.

Q **What is the major source of NADH for the electron transport chain?**

Details of ATP Synthase

MC GOB Power from Protons: ATP Synthase

ATP synthase consists of two enzyme complexes. (See Figure 23.11.) The F_0 complex contains the channel for the return of protons to the matrix. The F_1 section consists of a center subunit (γ) and three surrounding protein subunits, which have three active sites with different shapes or conformations known as loose (L), tight (T), and open (O). As the protons flow through the F_0 channel, the energy released turns the center subunit (γ). We might think of the flow of protons as a stream or river that turns a water wheel. As the center unit supplies energy to the three active sites, their shapes change. ATP synthesis begins when the substrates ADP and P_i enter a loose (L) active site. As the loose (L) site shape converts to a tight (T) shape, ATP is formed. However, the ATP is bound to the active site (T). When energy turns the γ unit, the tight site (T) converts to an open (O) site, which releases the ATP. The open site will convert to an L site and accept new substrates ADP and P_i. According to Paul Boyer, who earned the 1997 Nobel Prize in chemistry for his

FIGURE 23.11

ATP synthase consists of two protein complexes. An F_0 section contains the channel for proton flow, and an F_1 section uses the energy from the proton gradient to drive the synthesis of ATP.

Q **What are the functions of the F_0 and F_1 sections of ATP synthase?**

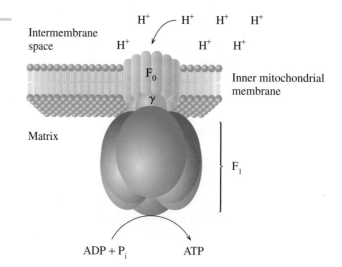

Intermembrane space

Inner mitochondrial membrane

Matrix

$$ADP + P_i \qquad ATP$$

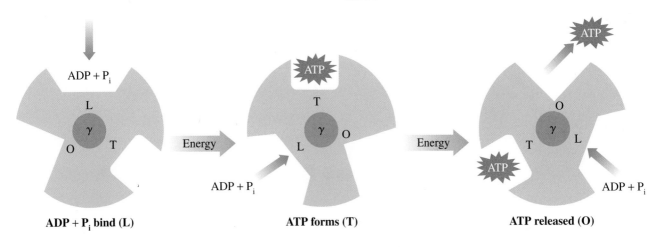

ADP + P$_i$

L

γ

O   T

ADP + P$_i$ bind (L)

Energy

ATP

T

γ O

L

ADP + P$_i$

ATP forms (T)

Energy

ATP

O

γ

T   L

ATP

ADP + P$_i$

ATP released (O)

FIGURE 23.12

In the F$_1$ ATP synthase, ATP is formed when an L active site containing ADP and P$_i$ converts to a T site. When energy from the proton flow in F$_0$ changes the site to an open (O) site, ATP is released.

Q **What shape of an active site in F$_1$ ATP synthase accepts the substrates, and which shape releases the ATP?**

work on ATP synthase, the formation of ATP is spontaneous, whereas its release from the synthase requires the energy supplied by the proton gradient. The following steps summarize the changes in the active site in the synthesis of ATP. (See Figure 23.12.)

■ **STEP 1** The synthesis of ATP begins when ADP and P$_i$ bind to a loose (L) site.

■ **STEP 2** Energy from the γ center now converts all the sites: an L site becomes a tight (T) site, an O site changes to an L site, and a T site changes to an O site. Now the ADP and P$_i$—which are now in a tight (T) site—spontaneously form ATP, which is held tightly in this active site.

■ **STEP 3** Another input of energy from proton flow changes all the active sites again. The ATP site changes to an open (O) site, which has little affinity for ATP, and ATP is released. This site is open once again and ready to accept more ADP and P$_i$.

In summary, the energy from protons flowing through F$_0$ turns the center γ unit in F$_1$, which causes a change in the shapes of the active sites from loose (L), where ADP, and P$_i$ bind, to tight (T) where ATP forms, and to open (O), which releases ATP. This process of oxidative phosphorylation continues as long as energy from the electron transport system is generated, which pumps the protons into the inner membrane space and produces the proton gradient to fuel ATP synthase.

Electron Transport and ATP Synthesis

We have seen that oxidative phosphorylation couples the energy from electron transport with the synthesis of ATP. Because NADH enters the electron transport chain at complex I, energy is released from the oxidation of NADH to synthesize three ATP. However, FADH$_2$, which enters the chain at a lower energy level at complex II, provides energy to drive the synthesis of only two ATP. (See

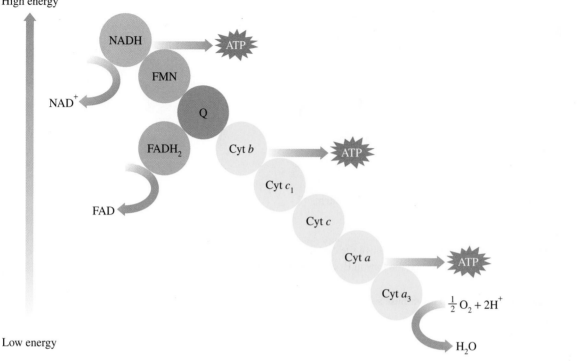

Low energy

FIGURE 23.13

As the energy levels decrease the flow of electrons along the major electron carriers, three of the electron transfers release sufficient energy to drive ATP synthesis.

Q **Why does electron transfer from FADH$_2$ provide less energy than from NADH and H$^+$?**

Figure 23.13.) The overall equation for the oxidation of NADH and FADH$_2$ can be written as follows:

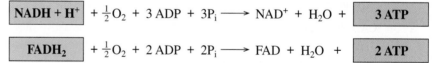

$$\boxed{\textbf{NADH} + \textbf{H}^+} + \tfrac{1}{2}\text{O}_2 + 3\,\text{ADP} + 3\text{P}_\text{i} \longrightarrow \text{NAD}^+ + \text{H}_2\text{O} + \boxed{\textbf{3 ATP}}$$

$$\boxed{\textbf{FADH}_2} + \tfrac{1}{2}\text{O}_2 + 2\,\text{ADP} + 2\text{P}_\text{i} \longrightarrow \text{FAD} + \text{H}_2\text{O} + \boxed{\textbf{2 ATP}}$$

Regulation of the Electron Transport Chain and Oxidative Phosphorylation

The electron transport chain is regulated by the availability of ADP, P$_\text{i}$, oxygen (O$_2$), and NADH. Low levels of any of these compounds will decrease the activity of the electron transport chain and formation of ATP. When a cell is active and ATP is consumed rapidly, the elevated levels of ADP will activate the synthesis of ATP. Therefore, the activity of the electron transport chain is strongly dependent on the levels of ADP for ATP synthesis.

SAMPLE PROBLEM 23.5

■ ATP Synthesis

Why does the oxidation of NADH provide energy for the formation of three ATP whereas FADH$_2$ produces two ATP?

SOLUTION

Electrons from the oxidation of NADH enter the electron chain at a higher energy level than $FADH_2$, providing energy to pump three pairs of protons to the intermembrane. However, $FADH_2$ transfers electrons to Q that go through two sites which pump protons and therefore provide energy for two ATP.

STUDY CHECK

What complexes act as proton pumps?

QUESTIONS AND PROBLEMS

■ Oxidative Phosphorylation and ATP

23.35 What is meant by oxidative phosphorylation?

23.36 How is the proton gradient established?

23.37 According to the chemiosmotic theory, how does the proton gradient provide energy to synthesize ATP?

23.38 How does the phosphorylation of ADP occur?

23.39 How are glycolysis and the citric acid cycle linked to the production of ATP by the electron transport chain?

23.40 Why does $FADH_2$ have a yield of two ATP via the electron chain, but NADH yields three ATP?

HEALTH NOTE

■ ATP Synthase and Heating the Body

Some types of compounds called *uncouplers* separate the electron transport system from the ATP F_0F_1 synthase. They do this by disrupting the proton gradient needed for the synthesis of ATP. The electrons are transported to O_2 in electron transport, but ATP is not formed by ATP synthase.

Some uncouplers transport the protons through the inner membrane, which is normally impermeable to protons; others block the channel in the F_0 portion of ATP synthase. Compounds such as dicumarol and 2,4-dinitrophenol (DNP) are hydrophobic and bind with protons to carry them across the inner membrane. An antibiotic, oligomycin, binds to the F_0 complex, and blocks the channel, which does not allow any protons to return to the matrix. By removing protons or blocking the F_0 channel, there is no proton flow through the F_0 channel to generate energy for ATP synthesis.

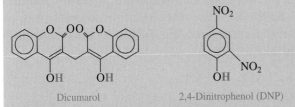

Dicumarol   2,4-Dinitrophenol (DNP)

When there is no mechanism for ATP synthesis, the energy of electron transport is released as heat. Certain animals that are adapted to cold climates have developed their own uncoupling system, which allows them to use electron transport energy for heat production. These ani-mals have large amounts of a tissue called brown fat, which contains a high concentration of mitochondria. This tissue is brown because of the color of iron in the cytochromes of the mitochondria.

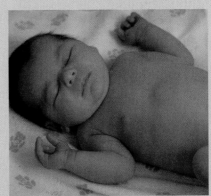

The proton pumps still operate in brown tissue but a protein embedded in the cell wall allows the protons to bypass ATP synthase. The energy that would be used to synthesize ATP is released as heat. In newborn babies brown fat is used to generate heat because newborns have not stored much fat. The brown fat deposits are located near major blood vessels, which carry the warmed blood to the body. Infants have a small mass but a large surface area and need to produce more heat than the adult. Most adults have little or no brown fat, although someone who works outdoors in a cold climate will develop some brown fat deposits.

Plants also use uncouplers. Some plants use uncoupling agents to volatize fragrant compounds that attract insects to pollinate the plants. Skunk cabbage uses this system. In early spring, heat is used to warm early shoots of plants under the snow, which helps them melt the snow around the plants.

23.41 What are the parts of ATP synthase?

23.42 What is the role of each part of ATP synthase in ATP synthesis?

23.43 What type of active site in ATP synthase binds ADP and P_i?

23.44 How is the ATP released from ATP synthase?

LEARNING GOAL

■ Account for the ATP produced by the complete oxidation of glucose.

(MC GOB) SELF-STUDY ACTIVITY
Electron Transport

23.5 ATP Energy from Glucose

Under aerobic conditions, the oxidation of glucose through glycolysis, oxidation to pyruvate, the citric acid cycle, the electron transport chain produces ATP from many NADH molecules and some $FADH_2$. Let's see how much ATP is associated with each of these metabolic cycles.

ATP from Glycolysis

In glycolysis, the oxidation of glucose stores energy in two NADH molecules as well as two ATP molecules from direct substrate phosphorylation. However, glycolysis occurs in the cytoplasm, and the NADH produced cannot pass through the mitochondrial membrane to the electron transport chain.

Therefore the hydrogen ions and electrons from NADH in the cytoplasm are transferred to compounds that can enter the mitochondria. In this shuttle system, dihydroxyacetone phosphate, a compound in glycolysis, is reduced to glycerol-3-phosphate and NAD^+ is regenerated. After glycerol-3-phosphate crosses the mitochondrial membrane, the hydrogen ions and electrons are transferred to FAD. $FADH_2$ is produced along with glycerol-3-phosphate, which returns to the cytoplasm. The overall reaction for the glycerol-3-phosphate shuttle is:

$$NADH + H^+ + FAD \longrightarrow NAD^+ + FADH_2$$

cytoplasm   mitochondria

Therefore the transfer of electrons from NADH in the cytoplasm to $FADH_2$ produces only two ATP, rather than three. In glycolysis, four ATP from two NADH and two ATP from direct phosphorylation add up to six ATP from one glucose molecule.

$$glucose \longrightarrow 2 \text{ pyruvate} + 2\,ATP + 2\,NADH (\longrightarrow 2FADH_2)$$
$$glucose \longrightarrow 2 \text{ pyruvate} + 6\,ATP$$

ATP from the Oxidation of Two Pyruvate

Under aerobic conditions, pyruvate enters the mitochondria, where it is oxidized to give acetyl CoA, CO_2, and NADH. Because each glucose molecule yields two pyruvate, two NADH enter electron transport. The oxidation of two pyruvate molecules leads to the production of six ATP molecules.

$$2 \text{ pyruvate} \longrightarrow 2 \text{ acetyl CoA} + 6\,ATP$$

ATP from the Citric Acid Cycle

One turn of the citric acid cycle produces two CO_2, three NADH, one $FADH_2$, and one ATP by direct substrate phosphorylation. When the NADH and $FADH_2$ enter electron transport, three NADH produce a total of nine ATP molecules, and one

TABLE 23.1

ATP Produced by the Complete Oxidation of Glucose

Reaction Pathway	ATP for One Glucose
ATP from Glycolysis	
Activation of glucose	-2 ATP
Oxidation of glyceraldehyde-3-phosphate (2 NADH)	6 ATP
Transport of 2 NADH across membrane	-2 ATP
Direct ADP phosphorylation (two triose phosphate)	4 ATP
Summary: $C_6H_{12}O_6 \longrightarrow$ 2 pyruvate $+ 2H_2O$ glucose	6 ATP
ATP from Pyruvate	
2 pyruvate $\longrightarrow$ 2 acetyl CoA (2 NADH)	6 ATP
ATP from Citric Acid Cycle	
Oxidation of 2 isocitrate (2 NADH)	6 ATP
Oxidation of 2 α-ketoglutarate (2 NADH)	6 ATP
2 Direct substrate phosphorylations (2 GTP)	2 ATP
Oxidation of 2 succinate (2 $FADH_2$)	4 ATP
Oxidation of 2 malate (2 NADH)	6 ATP
Summary: 2 acetyl CoA $\longrightarrow$ $4CO_2 + 2H_2O$	24 ATP
Overall ATP Production for One Glucose	
$C_6H_{12}O_6 + 6O_2 + 36$ ADP $+ 36$ $P_i \longrightarrow 6CO_2 + 6H_2O + 36$ ATP glucose	

$FADH_2$ produces two more ATP. Thus, one turn of the citric acid cycle generates energy for the synthesis of a total of 12 ATP molecules.

$$
\begin{aligned}
3 \text{ NADH} \times 3 \text{ ATP} &= 9 \text{ ATP} \\
1 \text{ FADH}_2 \times 2 \text{ ATP} &= 2 \text{ ATP} \\
\underline{1 \text{ GTP} \times 1 \text{ ATP}} &= \underline{1 \text{ ATP}} \\
\text{Total (one turn)} &= 12 \text{ ATP}
\end{aligned}
$$

Because two acetyl CoA molecules are produced from each glucose, two turns of the citric acid cycle produces 24 ATP.

$$
\text{Acetyl CoA} \longrightarrow 2CO_2 + 12 \text{ ATP (one turn of citric acid cycle)}
$$
$$
\text{2 Acetyl CoA} \longrightarrow 4CO_2 + 24 \text{ ATP (two turns of citric acid cycle)}
$$

ATP from the Complete Oxidation of Glucose

The details of the reaction pathways given in Table 23.1 show how to determine the total ATP for the complete oxidation of glucose by combining the equations from glycolysis, oxidation of pyruvate, and citric acid cycle. (See Figure 23.14.)

MC GOB ATP Energy from Glucose

SAMPLE PROBLEM 23.6

■ ATP Production

Indicate the amount of ATP produced by each of the following oxidation reactions:

a. pyruvate to acetyl CoA   **b.** glucose to acetyl CoA

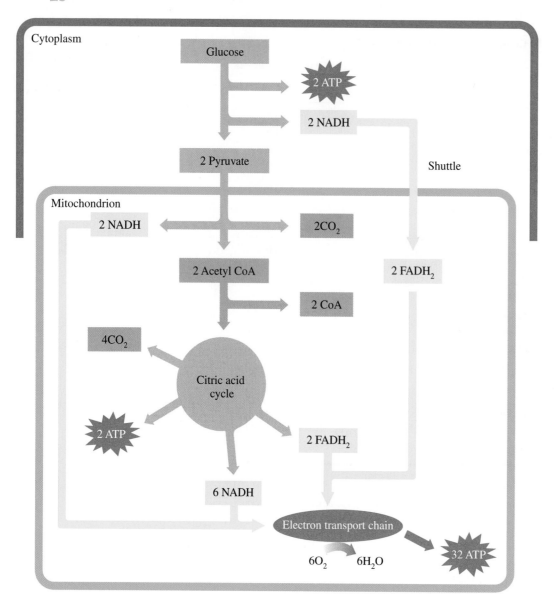

FIGURE 23.14

The complete oxidation of glucose to CO_2 and H_2O yields a total of 36 ATP.

Q **What metabolic pathway produces most of the ATP from the oxidation of glucose?**

SOLUTION

a. The oxidation of pyruvate to acetyl CoA produces one NADH, which yields three ATP.

b. Six ATP are produced from the oxidation of glucose to two pyruvate molecules. Another six ATP result from the oxidation of two pyruvate molecules to two acetyl CoA molecules for a total of 12 ATP.

STUDY CHECK

What are the sources of ATP in the citric acid cycle?

When glucose is not immediately used by the cells for energy, it is stored as glycogen in the liver and muscles. When the levels of glucose in the brain or blood

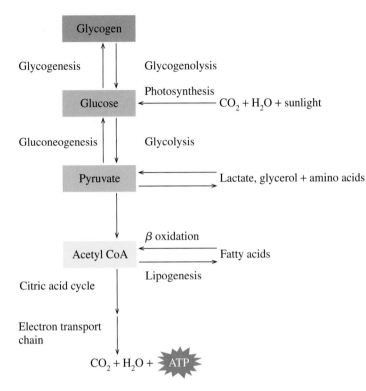

FIGURE 23.15

The ATP level is maintained by a balance of metabolic pathways that increase or decrease the glucose according to the energy requirements in the cell.

Q **What metabolic pathways are stimulated by low ATP levels?**

become low, the glycogen reserves are hydrolyzed and glucose is released into the blood. If glycogen stores are depleted, some glucose can be synthesized from non-carbohydrate sources. It is the balance of all these reactions that maintains the necessary blood glucose level available to our cells and provides the necessary amount of ATP for our energy needs. (See Figure 23.15.)

HEALTH NOTE

■ **Efficiency of ATP Production**

In a laboratory, a calorimeter is used to measure the heat energy from the combustion of glucose. In a calorimeter, one mole of glucose produces a total of 687 kcal (2870 kJ).

$$C_6H_{12}O_6 + 6O_2 \longrightarrow 6CO_2 + 6H_2O + 687 \text{ kcal (2870 kJ)/mole}$$

Let's compare the amount of energy produced from one mole of glucose in a calorimeter with the ATP energy produced in the mitochondria. We can use the energy of the hydrolysis of ATP, which is 7.3 kcal (31 kJ)/mole ATP. Because one mole of glucose generates energy for 36 moles of ATP, the total energy from the oxidation of one mole of glucose in the cells would be 263 kcal (1100 kJ)/per mole.

$$\frac{36 \text{ mole ATP}}{1 \text{ mole glucose}} \times \frac{7.3 \text{ kcal (31 kJ)}}{1 \text{ mole ATP}} = 263 \text{ kcal (1100 kJ) per 1 mole glucose}$$

Compared to the energy produced by burning glucose in a calorimeter, our cells are about 38% (263 kcal/687 kcal) efficient in converting the total available chemical energy in glucose to ATP.

$$\frac{263 \text{ kcal (cells)}}{687 \text{ kcal (calorimeter)}} \times 100 = 38\%$$

The rest of the energy from glucose produced during the oxidation of glucose in our cells is lost as heat.

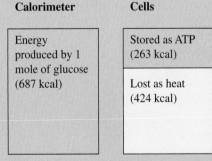

Calorimeter	**Cells**
Energy produced by 1 mole of glucose (687 kcal)	Stored as ATP (263 kcal)
	Lost as heat (424 kcal)

QUESTIONS AND PROBLEMS

◾ ATP Energy from Glucose

23.45 Why does the NADH produced in glycolysis yield only two ATP?

23.46 Under anaerobic conditions, what is the maximum number of ATP molecules that can be produced from one glucose molecule?

23.47 What is the energy yield in ATP molecules associated with each of the following?
a. $NADH \longrightarrow NAD^+$
b. glucose $\longrightarrow$ 2 pyruvate
c. 2 pyruvate $\longrightarrow$ 2 acetyl CoA $+ 2CO_2$

23.48 What is the energy yield in ATP molecules associated with each of the following?
a. $FADH_2 \longrightarrow FAD$
b. glucose $+ 6O_2 \longrightarrow 6CO_2 + 6H_2O$
c. acetyl CoA $\longrightarrow 2CO_2$

CONCEPT MAP

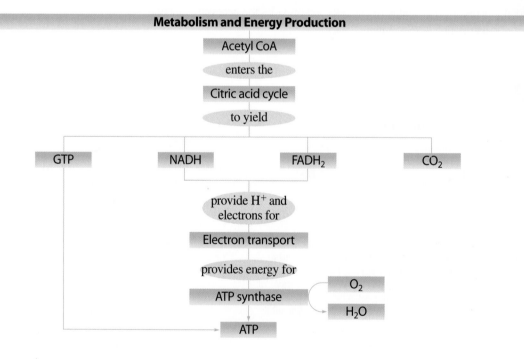

Metabolism and Energy Production

Acetyl CoA

enters the

Citric acid cycle

to yield

GTP   NADH   FADH$_2$ CO$_2$

provide H$^+$ and electrons for

Electron transport

provides energy for

ATP synthase   O$_2$
                      H$_2$O

ATP

CHAPTER REVIEW

23.1 The Citric Acid Cycle
In a sequence of reactions called the citric acid cycle, an acetyl group is combined with oxaloacetate to yield citrate. Citrate undergoes oxidation and decarboxylation to yield two CO_2, GTP, three NADH, and $FADH_2$ with the regeneration of oxaloacetate. The direct phosphorylation of ADP by GTP yields ATP.

23.2 Electron Carriers
Electron carriers that transfer hydrogen ions and electrons include FMN, iron–sulfur clusters, coenzyme Q, and several cytochromes. Both iron–sulfur clusters and cytochromes contain iron ions that are reduced to Fe^{2+} and reoxidized to Fe^{3+} as electrons are accepted and then passed to the next electron carrier.

23.3 Electron Transport
The reduced coenzymes NADH and $FADH_2$ from various metabolic pathways are oxidized to NAD^+ and FAD when their protons and electrons are transferred to the electron transport chain. The final acceptor, O_2, combines with protons and electrons to yield H_2O. At three transfer points in electron transport, the energy decrease provides the necessary energy for ATP synthesis.

23.4 Oxidative Phosphorylation and ATP
The protein complexes in electron transport act as proton pumps to move protons into the intermembrane space, which produces a proton gradient. As the protons return to the matrix by way of ATP synthase, energy is generated. This energy is used to drive the synthesis of ATP in a process known as oxidative phosphorylation. The available ADP and ATP levels in the cells control the activity of the electron transport chain.

23.5 ATP Energy from Glucose
With the exception of the NADH produced from glycolysis, the oxidation of NADH yields three ATP molecules, and $FADH_2$ yields two ATP. The energy from the NADH produced in the cytoplasm is used to form $FADH_2$. Under aerobic conditions, the complete oxidation of glucose yields a total of 36 ATP from the oxidation of the reduced coenzymes NADH and $FADH_2$ by electron transport, oxidative phosphorylation, and from some direct substrate phosphorylation.

SUMMARY OF KEY REACTIONS

CITRIC ACID CYCLE

$$\text{acetyl S-CoA} + 3\,NAD^+ + FAD + GDP + P_i + 2H_2O \longrightarrow 2CO_2 + 3\,NADH + 3H^+ + FADH_2 + HS\text{—}CoA + GTP$$

ELECTRON TRANSPORT CHAIN

$$NADH + H^+ + 3\,ADP + 3\,P_i + \tfrac{1}{2}O_2 \longrightarrow NAD^+ + 3\,ATP + H_2O$$

$$FADH_2 + 2\,ADP + 2\,P_i + \tfrac{1}{2}O_2 \longrightarrow FAD + 2\,ATP + H_2O$$

PHOSPHORYLATION OF ADP

$$ADP + P_i \longrightarrow ATP + H_2O$$

COMPLETE OXIDATION OF GLUCOSE

$$C_6H_{12}O_6 + 6O_2 + 36\,ADP + 36\,P_i \longrightarrow 6CO_2 + 6H_2O + 36\,ATP$$

KEY TERMS

ATP synthase (F_0F_1) An enzyme complex that links the energy released by protons returning to the matrix with the synthesis of ATP from ADP and P_i. The F_0 section contains the channel for proton flow, and the F_1 section uses the energy from the proton flow to drive the synthesis of ATP.

chemiosmotic model The conservation of energy from transfer of electrons in the electron transport chain by pumping protons into the intermembrane space to produce a proton gradient that provides the energy to synthesize ATP.

citric acid cycle A series of oxidation reactions in the mitochondria that convert acetyl CoA to CO_2 and yield NADH and $FADH_2$. It is also called the tricarboxylic acid cycle and the Krebs cycle.

coenzyme Q (CoQ, Q) A mobile carrier that transfers electrons from NADH and $FADH_2$ to cytochrome b in complex III.

cytochromes (cyt) Iron-containing proteins that transfer electrons from QH_2 to oxygen.

decarboxylation A reaction in which a CO_2 molecule is lost.

electron carriers A group of proteins that accept and pass on electrons as they are reduced and oxidized. Most of the carriers are tightly attached to the inner mitochondrial membrane, but two are mobile carriers, which move electrons between the complexes containing the other carriers.

electron transport chain A series of reactions in the mitochondria that transfer electrons from NADH and $FADH_2$ to electron carriers, which are arranged from higher to lower energy levels, and finally to O_2, which produces H_2O. Energy changes during three of these transfers provide energy for ATP synthesis.

Fe–S (iron–sulfur) clusters Proteins containing iron and sulfur in which the iron ions accept electrons from $FMNH_2$ and cytochrome b.

FMN (flavin mononucleotide) An electron carrier derived from riboflavin (vitamin B_2) that transfers hydrogen ions and electrons from NADH entering in the electron transport chain.

oxidative phosphorylation The synthesis of ATP from ADP and P_i using energy generated by the oxidation reactions in the electron transport chain.

proton pumps The enzyme complexes I, III, and IV that move protons from the matrix into the intermembrane space, creating a proton gradient.

UNDERSTANDING THE CONCEPTS

23.49 Identify each of the following as a substance that is part of the citric acid cycle, electron transport, or both.
a. succinate
b. QH_2
c. FAD
d. cyt c (Fe^{2+})
e. cytochrome c oxidase
f. H_2O
g. malate
h. NAD^+

23.50 Complete the names of the missing compounds in the citric acid cycle.

citrate $\longrightarrow$ _____ $\longrightarrow$ _____ $\longrightarrow$ succinyl
CoA $\longrightarrow$ _____ $\longrightarrow$ _____ $\longrightarrow$ malate
$\longrightarrow$ _____

23.51 Arrange the following components of electron transport in order.
a. cyt a_3
b. O_2
c. cyt a
d. cyt c
e. QH_2
f. $FMNH_2$
g. cyt c_1
h. cyt b

23.52 Identify the reactant and product for each of the following enzymes in the citric acid cycle.
a. aconitase
b. succinate dehydrogenase

c. fumarase
d. isocitrate dehydrogenase
e. succinyl CoA synthetase
f. malate dehydrogenase

23.53 For each of the enzymes, indicate which of the following are needed.
1. NAD^+
2. H_2O
3. FAD
4. GDP
a. aconitase
b. succinate dehydrogenase
c. fumarase
d. isocitrate dehydrogenase
e. succinyl CoA synthase
f. malate dehydrogenase

23.54 Identify the type of reaction(s) catalyzed by each of the following enzymes as
1. oxidation
2. decarboxylation
3. hydrolysis
4. hydration
a. aconitase
b. succinate dehydrogenase
c. fumarase
d. isocitrate dehydrogenase
e. α-ketoglutarate dehydrogenase
f. malate dehydrogenase

ADDITIONAL QUESTIONS AND PROBLEMS

For instructor-assigned homework, go to **www.masteringgob.com**.

23.55 What is the main function of the citric acid cycle in energy production?

23.56 Most metabolic pathways are not considered cycles. Why is the citric acid cycle considered to be a metabolic cycle?

23.57 If there are no reactions in the citric acid cycle that used oxygen, O_2, why does the cycle operate only in aerobic conditions?

23.58 What products of the citric acid cycle are needed for the electron transport chain?

23.59 Identify the compounds in the citric acid cycle that have the following:
a. six carbon atoms
b. five carbon atoms
c. a keto group

23.60 Identify the compounds in the citric acid cycle that have the following:
a. four carbon atoms
b. a hydroxyl group
c. a double bond

23.61 In which reaction of the citric acid cycle does each of the following occur?
a. a five-carbon keto acid is decarboxylated
b. a double bond is hydrated

c. NAD^+ is reduced
d. a secondary hydroxyl group is oxidized

23.62 In which reaction of the citric acid cycle does each of the following occur?
a. FAD is reduced
b. a six-carbon keto acid is decarboxylated
c. a carbon–carbon double bond is formed
d. GDP undergoes direct phosphorylation

23.63 Indicate the coenzyme for each of the following reactions:
a. isocitrate $\longrightarrow$ α-ketoglutarate
b. α-ketoglutarate $\longrightarrow$ succinyl CoA

23.64 Indicate the coenzyme for each of the following reactions:
a. succinate $\longrightarrow$ fumarate
b. malate $\longrightarrow$ oxaloacetate

23.65 How do each of the following regulate the citric acid cycle?
a. high levels of NADH
b. high levels of ATP

23.66 How do each of the following regulate the citric acid cycle?
a. high levels of ADP
b. low levels of NADH

23.67 Identify each as part of the structure of one of the following components in the electron transport chain as (1) FMN, (2) Fe–S cluster, (3) CoQ, or (4) cytochrome
a. a heme group
b. contains a ribitol group

23.68 Identify each as part of the structure of one of the following components in the electron transport chain as (1) FMN, (2) Fe–S cluster, (3) CoQ, or (4) cytochrome
  a. contains the three-ring system of flavins
  b. a six-atom ring attached to a long-carbon chain

23.69 Identify each of the following electron carriers as part of a complex or as a mobile carrier. If part of a complex, indicate which one.
  a. CoQ   **b.** Fe–S clusters   **c.** cyt a_3

23.70 Identify each of the following electron carriers as part of a complex or as a mobile carrier. If part of a complex, indicate which one.
  a. cyt b **b.** cyt c **c.** FMN

23.71 Identify the complex where each of the following are oxidized or reduced and complete the equation:
  a. $FADH_2 + Q \longrightarrow$
  b. cyt a (Fe^{2+}) + cyt a_3 (Fe^{3+}) $\longrightarrow$

23.72 Identify the complex where each of the following are oxidized or reduced and complete the equation:
  a. cyt c (Fe^{2+}) + cyt a (Fe^{3+}) $\longrightarrow$
  b. $NADH + H^+ + Q \longrightarrow$

23.73 Complete the following by adding the substances that are missing:

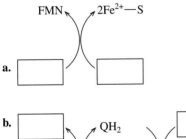

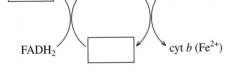

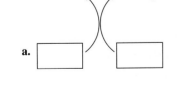

23.74

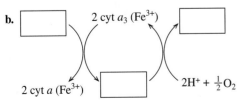

23.75 At which parts of the electron transport system are protons pumped into the intermembrane space?

23.76 What is the effect of proton accumulation in the intermembrane space?

23.77 In the chemiosmotic model, how is energy provided to synthesize ATP?

23.78 In what part of the electron transport chain does the synthesis of ATP take place?

23.79 Why do protons tend to leave the intermembrane space and return to the matrix?

23.80 Why do the enzyme complexes that pump protons extend across the membrane from the matrix to the intermembrane space?

23.81 How many ATP molecules are produced by energy generated when electrons flow from $FADH_2$ to oxygen (O_2)?

23.82 How many ATP molecules are produced by energy generated when electrons flow from NADH to oxygen (O_2)?

23.83 What part of the electron transport chain is inhibited by each of the following?
  a. amytal and rotenone
  b. antimycin A
  c. cyanide and carbon monoxide

23.84 **a.** When an inhibitor blocks the electron transport chain, how are the coenzymes that precede the blocked site affected?
  b. When an inhibitor blocks the electron transport chain, how are the coenzymes that follow the blocked site affected?

23.85 How many ATP are produced when glucose is oxidized to pyruvate compared to when glucose is oxidized to CO_2 and H_2O?

23.86 Why do the two NADH produced in glycolysis provide a net of two ATP and not three?

23.87 Where is ATP synthase for oxidative phosphorylation located in the cell?

23.88 Considering the efficiency of ATP synthesis, how many kcal of energy would be conserved from the complete oxidation of four moles of glucose?

23.89 How is the energy from the proton gradient utilized by ATP synthase?

23.90 How is a proton gradient developed in a cell?

23.91 Why would a bear that is hibernating have more brown fat than one that is active?

23.92 How do the active sites on F_1 ATP synthase change during ATP production?

CHALLENGE QUESTIONS

23.93 Using the value of 7.3 kcal/mole for ATP, how many kcal are conserved as ATP from one mole of glucose in each of the following?
a. glycolysis
b. oxidation of pyruvate to acetyl CoA
c. citric acid cycle
d. complete oxidation of glucose to CO_2 and H_2O

23.94 What percent of ATP energy is conserved from one mole of glucose in problem 23.93a–c?

23.95 What does it mean to say that the cell is 38% efficient in storing the energy from the complete combustion of glucose?

23.96 A student is considering using a 2,4-dinitrophenol, which is an uncoupler, to lose weight. Explain how the uncoupler will affect the body temperature of the student. How would you advise the student if the dosage of DNP needed to lose weight is close to toxic levels?

23.97 If acetyl CoA has a molar mass of 809 g/mole, how many moles ATP are produced when 1.0 μg acetyl CoA completes the citric acid cycle?

ANSWERS

Answers to Study Checks

23.1 oxaloacetate

23.2 Fe^{3+}

23.3 **a.** oxidation   **b.** reduction

23.4 Oxygen (O_2) is the last substance that accepts electrons.

23.5 The protein complexes I, III, and IV pump protons from the matrix to the intermembrane space.

23.6 Three NADH provide nine ATP, one $FADH_2$ provides two ATP, and one direct phosphorylation provides one ATP.

Answers to Selected Questions and Problems

23.1 Krebs cycle and tricarboxylic acid cycle

23.3 $2CO_2$, 3 NADH + $3H^+$, $FADH_2$, GTP (ATP), and HS—CoA.

23.5 Two reactions, 3 and 4, involve oxidative decarboxylation.

23.7 NAD^+ is reduced in reactions 3, 4, and 8 of the citric acid cycle.

23.9 In reaction 5, GDP undergoes a direct substrate phosphorylation.

23.11 a. citrate and isocitrate
b. A carbon atom is lost as CO_2 in decarboxylation.
c. α-ketoglutarate
d. isocitrate $\longrightarrow$ α-ketoglutarate; α-ketoglutarate $\longrightarrow$ succinyl CoA; succinate $\longrightarrow$ fumarate; malate $\longrightarrow$ oxaloacetate
e. steps 3, 8

23.13 a. citrate synthase
b. succinate dehydrogenase and aconitase
c. fumarase

23.15 a. NAD^+ **b.** GDP

23.17 Isocitrate dehydrogenase and α-ketoglutarate dehydrogenase are allosteric enzymes

23.19 High levels of ADP increase the rate of the citric acid cycle.

23.21 oxidized

23.23 a. oxidation   **b.** reduction

23.25 NADH and $FADH_2$

23.27 FAD, coenzyme Q, cytochrome b, cytochrome c

23.29 The mobile carrier Q transfers electrons from complex I to III.

23.31 NADH transfers electrons to FMN in complex I to give NAD^+.

23.33 a. NADH + H^+ + FMN $\longrightarrow$ NAD^+ + $FMNH_2$
b. QH_2 + 2 cyt b (Fe^{3+}) $\longrightarrow$ Q + 2 cyt b (Fe^{2+}) + $2H^+$

23.35 In oxidative phosphorylation, the energy from the oxidation reactions in the electron transport chain is used to drive ATP synthesis.

23.37 As protons return to the lower energy environment in the matrix, they pass through ATP synthase where they release energy to drive the synthesis of ATP.

23.39 The oxidation of the reduced coenzymes NADH and $FADH_2$ by the electron transport chain generates energy to drive the synthesis of ATP.

23.41 ATP synthase consists of two protein complexes, F_0 and F_1.

23.43 The loose (L) site in ATP synthase binds ADP and P_i.

23.45 Glycolysis takes place in the cytoplasm, not in the mitochondria. Because NADH cannot cross the mitochondrial membrane, one ATP is hydrolyzed to transport the electrons from NADH to FAD. The resulting $FADH_2$ produces only 2 ATP for each NADH produced in glycolysis.

23.47 a. 3 ATP   **b.** 2 ATP   **c.** 6 ATP   **d.** 12 ATP

23.49 a. citric acid cycle   **b.** electron transport
c. both   **d.** electron transport
e. electron transport   **f.** both
g. citric acid cycle   **h.** both

23.51 f. $FMNH_2$ **e.** QH_2 **h.** cyt b **g.** cyt c_1
d. cyt c **c.** cyt a **a.** cyt a_3 **b.** O_2

23.53 a. aconitase H_2O
b. succinate dehydrogenase FAD
c. fumarase H_2O
d. isocitrate dehydrogenase NAD^+
e. succinyl CoA synthase GDP
f. malate dehydrogenase NAD^+

23.55 The oxidation reactions of the citric acid cycle produce a source of reduced coenzymes for the electron transport chain and ATP synthesis.

23.57 The oxidized coenzymes NAD^+ and FAD needed for the citric acid cycle are regenerated by the electron transport chain.

23.59 a. citrate, isocitrate
b. α-ketoglutarate
c. α-ketoglutarate, succinyl CoA, oxaloacetate

23.61 a. In reaction 4, α-ketoglutarate, a five-carbon keto acid, is decarboxylated.
b. In reaction 1 and reaction 7, double bonds in aconitate and fumarate are hydrated.
c. NAD^+ is reduced in reactions 3, 4, and 8.
d. In reactions 3 and 8, a secondary hydroxyl group in isocitrate and malate is oxidized.

23.63 a. NAD^+ **b.** NAD^+ and CoA

23.65 a. High levels of NADH inhibit isocitrate dehydrogenase and α-ketoglutarate dehydrogenase to slow the rate of the citric acid cycle.
b. High levels of ATP inhibit the citric acid cycle.

23.67 a. (4) cytochrome   **b.** (1) FMN

23.69 a. CoQ is a mobile carrier
b. Fe–S clusters are found in complex I, III, and IV
c. cyt a_3 is part of complex IV

23.71 a. complex II; $FADH_2 + Q \longrightarrow FAD + QH_2$
b. complex IV; cyt a (Fe^{2+}) + cyt a_3 (Fe^{3+}) $\longrightarrow$ cyt a (Fe^{3+}) + cyt a_3 (Fe^{2+})

23.73

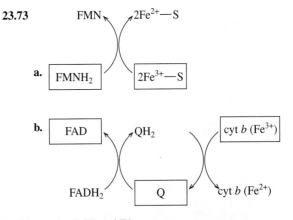

23.75 complex I, III, and IV

23.77 Energy released as protons flow through ATP synthase back to the matrix is utilized for the synthesis ATP.

23.79 Protons flow into the matrix where H^+ concentration is lower.

23.81 Two ATP molecules are produced from $FADH_2$.

23.83 a. NADH dehydrogenase (complex I)
b. electron flow from cyt b to cyt c_1 (complex III)
c. cytochrome c oxidase (complex IV)

23.85 The oxidation of glucose to pyruvate produces 6 ATP whereas the oxidation of glucose to CO_2 and H_2O produces 36 ATP.

23.87 The ATP synthase extends through the inner mitochondrial membrane with the F_0 part in contact with the proton gradient in the intermembrane space, while the F_1 complex is in the matrix.

23.89 As protons from the proton gradient move through the ATP synthase to return to the matrix, energy is released and used to drive ATP synthesis at F_1 ATP synthase.

23.91 A hibernating bear has more brown fat, which can be used during the winter for heat rather than ATP energy.

23.93 a. 6 ~~moles ATP~~ $\times$ 7.3 kcal/~~mole ATP~~ = 44 kcal
b. 6 ~~moles ATP~~ $\times$ 7.3 kcal/~~mole ATP~~ = 44 kcal (2 pyruvate to 2 acetyl CoA)
c. 24 ~~moles ATP~~ $\times$ 7.3 kcal/~~mole ATP~~ = 175 kcal (2 acetyl CoA citric acid cycle)
d. 36 ~~moles ATP~~ $\times$ 7.3 kcal/~~mole ATP~~ = 263 kcal (complete oxidation of glucose to CO_2 and H_2O)

23.95 In a calorimeter, the complete combustion of glucose gives 687 kcal. Using the value for glucose in problem 23.93, 263 ~~kcal~~/687 ~~kcal~~ $\times$ 100 = 38.3%.

23.97 1.5×10^{-8} mole ATP

24

Metabolic Pathways for Lipids and Amino Acids

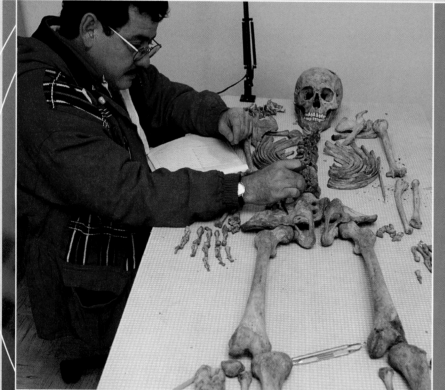

"Chemistry plays an integral part in all aspects of medical legal death investigation," says Charles L. Cecil, forensic anthropologist and medical legal death investigator, San Francisco Medical Examiner's Office. "Crime-scene analysis of blood droplets determines whether they are human or nonhuman, the analyses of toxicological samples of blood and/or other fluids help determine the cause and time of death. Specialists in forensic anthropology can analyze trace-element ratios in bones to identify the number of individuals in mixed human bone situations. These conditions are quite often found during investigations of massive human-rights violations, such as the site of El Mozote in El Salvador."

Forensic anthropologists help police identify skeletal remains by determining the gender, approximate age, height, and cause of death. Analysis of bone can also provide information about illnesses or trauma a person may have experienced.

LOOKING AHEAD

Mastering
GOB CHEMISTRY

Visit **www.masteringgob.com**
for self-study materials
and instructor-assigned
homework.

n previous chapters, we focused on carbohydrates because glucose is the primary fuel for the synthesis of ATP. However, lipids and proteins also play an important role in metabolism and energy production. In this chapter we will look at how the digestion of lipids produces fatty acids and glycerol and digestion of proteins gives amino acids. Almost all our energy is stored in the form of triacylglycerols in fat cells of adipose tissue. Many people go on diets after they discover that adipose tissue can store unlimited quantities of fat. This fact has become quite apparent in the large number of people in the U.S. that are considered obese. When our caloric intake exceeds the nutritional and metabolic needs of the body, excess carbohydrates and fatty acids are converted to triacylglycerols and added to our fat cells.

The digestion and degradation of dietary proteins as well as body proteins provides amino acids, which are needed to synthesize nitrogen-containing compounds in our cells such as new proteins and nucleic acids. Although amino acids are not considered a primary source of fuel, energy can be extracted from amino acids if glycogen and fat reserves have been depleted. However, when a person who is fasting or starving utilizes amino acids as the only source of energy, the breakdown of the body's own proteins eventually destroys essential body tissues, particularly the heart muscle.

24.1 Digestion of Triacylglycerols

LEARNING GOAL

Describe the sites and products obtained from the digestion of triacylglycerols.

Our adipose tissue is made of fat cells called *adipocytes*, which store triacylglycerols. (See Figure 24.1.) Let's compare the amount of energy stored in the fat cells to the energy from glucose, glycogen, and protein. A typical 70 kg (150 lb) person has about 135,000 kcal of energy stored as fat, 24,000 kcal as protein, 720 kcal as glycogen reserves, and 80 kcal as blood glucose. Therefore, the energy available from stored fats is about 85% of the total energy available in the body. Thus body fat is our major source of stored energy.

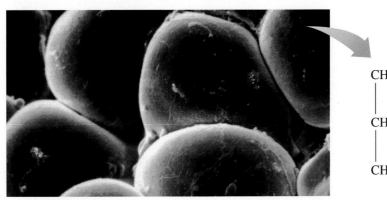
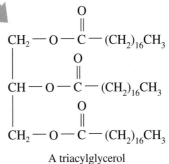

A triacylglycerol

FIGURE 24.1

The fat cells (adipocytes) that make up adipose tissue are capable of storing unlimited quantities of triacylglycerols.

Q **What are some sources of fats in our diet?**

Digestion of Dietary Fats

(MC GOB) Digestion of Triacylglycerols

The digestion of dietary fats begins in the small intestine, when the hydrophobic fat globules mix with bile salts released from the gallbladder. In a process called *emulsification*, the bile salts break the fat globules into smaller droplets called micelles. Then *pancreatic lipases* released from the pancreas hydrolyze the triacylglycerols in the micelles to yield monoacylglycerols and free fatty acids. These digestion products are absorbed into the intestinal lining where they recombine to form triacylglycerols, which are coated with proteins to form lipoproteins called **chylomicrons**. The chylomicrons transport the triacylglycerols through the lymph system and into the bloodstream to be carried to cells of the heart, muscle, and adipose tissues. (See Figure 24.2.) We can write the overall equation for the digestion of triacylglycerols as follows:

$$\text{triacylglycerols} + 2H_2O \xrightarrow{\text{pancreatic lipase}} \text{monoacylglycerols} + 2 \text{ fatty acids}$$

In the cells, enzymes hydrolyze the triacylglycerols to yield glycerol and free fatty acids, which can be used for energy production. The preferred fuel of the heart is fatty acids that are oxidized to acetyl CoA units for ATP synthesis. However, the brain and red blood cells cannot utilize fatty acids. Fatty acids cannot diffuse across the blood-brain barrier, and red blood cells have no mitochondria, which is where fatty acids are oxidized. Therefore, glucose and glycogen are the only source of energy for the brain and red blood cells.

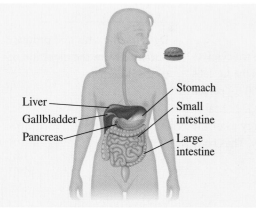

FIGURE 24.2

The triacylglycerols that reform in the intestinal wall from the digestion products monoacylglycerols and fatty acids bind to proteins for transport through the lymphatic system and bloodstream to the cells.

Q **What kinds of enzymes are secreted from the pancreas into the small intestine to hydrolyze triacylglycerols?**

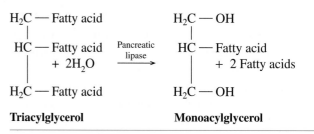

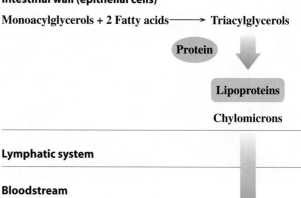

Mobilization of Fat Stores

When blood glucose is depleted and glycogen stores are low, the process of **fat mobilization** breaks down triacylglycerols in the adipose tissue to fatty acids and glycerol. The mobilization occurs when the hormones glucagon or epinephrine are secreted into the bloodstream and bind to receptors on the membrane of adipose cells. This activates enzymes within the fat cells that begin the hydrolysis of a triacylglycerol. A fatty acid is hydrolyzed from carbon 1 or carbon 3 followed by the hydrolysis of the second and third fatty acids.

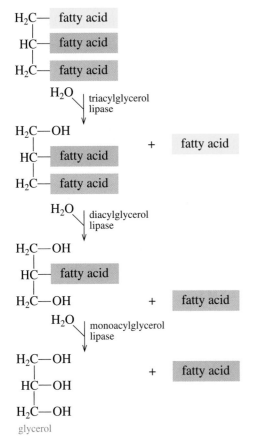

We can write the overall reaction for the mobilization of fats in fat cells as follows:

$$\text{triacylglycerols} + 3H_2O \xrightarrow{\text{lipases}} \text{glycerol} + 3 \text{ fatty acids}$$

The products of fat mobilization, glycerol and fatty acids, diffuse into the bloodstream and bind with plasma proteins (albumin) to be transported to the tissues. Most of the glycerol goes into the liver where it is converted to glucose.

Metabolism of Glycerol

Enzymes in the liver convert glycerol to dihydroxyacetone phosphate in two steps. In the first step, glycerol is phosphorylated using ATP to yield glycerol-3-phosphate. In the second step, the hydroxyl group is oxidized to yield dihydroxyacetone phosphate, which is an intermediate in several metabolic pathways including glycolysis and gluconeogenesis. (See Chapter 22.)

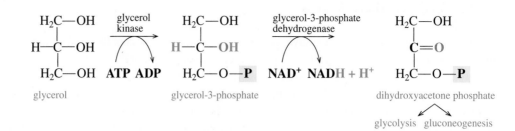

glycerol   glycerol-3-phosphate   dihydroxyacetone phosphate

glycolysis   gluconeogenesis

The overall reaction for the metabolism of glycerol is written as follows:

$$\text{glycerol} + \text{ATP} + \text{NAD}^+ \longrightarrow \text{dihydroxyacetone phosphate} + \text{ADP} + \text{NADH} + \text{H}^+$$

SAMPLE PROBLEM 24.1

■ Fats and Digestion

What are the sites, enzymes, and products for the digestion of triacylglycerols?

SOLUTION

The digestion of triacylglycerols takes place in the small intestine where pancreatic lipase catalyzes their hydrolysis to monoacylglycerols and fatty acids.

STUDY CHECK

What happens to the products from the digestion of triacylglycerols in the membrane of the small intestine?

QUESTIONS AND PROBLEMS

■ Digestion of Triacylglycerols

24.1   What is the role of bile salts in lipid digestion?

24.2   How are insoluble triacylglycerols transported to the tissues?

24.3   When are fats released from fat stores?

24.4   What happens to the glycerol produced from the hydrolysis of triacylglycerols in adipose tissues?

24.5   How is glycerol converted to an intermediate of glycolysis?

24.6   How can glycerol be used to synthesize glucose?

24.2 Oxidation of Fatty Acids

A large amount of energy is obtained when fatty acids undergo oxidation in the mitochondria to yield acetyl CoA. In stage 2 of fat metabolism, fatty acids undergo **beta oxidation (β oxidation)**, which removes two-carbon segments, one at a time, from a fatty acid. The remaining carbon chain is given the symbol R.

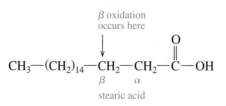

stearic acid

Each cycle in β oxidation produces acetyl CoA and a fatty acid that is shorter by two carbons. The cycle repeats until the original fatty acid is completely degraded to two-carbon acetyl CoA units. Each acetyl CoA can then enter the citric acid cycle in the same way as the acetyl CoA units derived from glucose.

Fatty Acid Activation

Before a fatty acid can enter the mitochondria, it undergoes activation in the cytosol. The activation process combines a fatty acid with coenzyme A to yield fatty acyl CoA. The energy released by the hydrolysis of two phosphate groups from ATP is used to drive the reaction. The products are AMP and pyrophosphate (PP_i), which hydrolyzes to yield two inorganic phosphates ($2\,P_i$).

$$R-CH_2-CH_2-\overset{\overset{\text{O}}{\|}}{C}-OH + ATP + HS-CoA \xrightarrow{\overset{\text{acyl CoA}}{\text{synthetase}}} R-CH_2-CH_2-\overset{\overset{\text{O}}{\|}}{C}-S-CoA + AMP + 2\,P_i + H_2O$$

fatty acid $\qquad\qquad\qquad\qquad\qquad\qquad\qquad$ fatty acyl CoA

Transport of Fatty Acyl CoA

The long hydrocarbon chain (R) prevents the fatty acyl CoA molecule from crossing the inner mitochondrial membrane. However, it is transported into the matrix when it binds with a charged carrier called *carnitine*. (See Figure 24.3.)

$$R-CH_2-CH_2-\overset{\overset{\text{O}}{\|}}{C}-S-CoA + H-\overset{\overset{\displaystyle\overset{+}{N}(CH_3)_3}{\underset{\displaystyle CH_2}{\|}}}{\underset{\displaystyle\underset{COO^-}{\underset{\displaystyle CH_2}{\|}}}{C}}-OH \rightleftharpoons H-\overset{\overset{\displaystyle\overset{+}{N}(CH_3)_3}{\underset{\displaystyle CH_2}{\|}}}{\underset{\displaystyle\underset{COO^-}{\underset{\displaystyle CH_2}{\|}}}{C}}-O-\overset{\overset{\text{O}}{\|}}{C}-CH_2-CH_2-R + HS-CoA$$

fatty acyl CoA $\qquad\qquad\qquad$ carnitine $\qquad\qquad\qquad$ fatty acyl carnitine

After the fatty acyl group is released in the matrix, it recombines with coenzyme A, and the carnitine returns to the inner membrane. While it may seem like a complicated way to move fatty acyl CoA into the matrix, this transport system provides a way to regulate degradation (oxidation) and synthesis of fatty acids. When fatty acids are being synthesized in the cytosol, the transport of fatty acyl CoA into the matrix is blocked, which prevents fatty acid degradation.

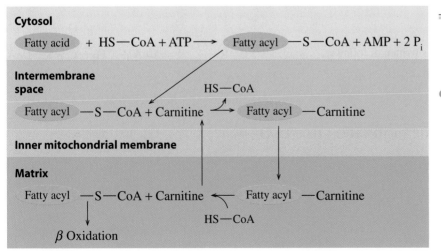

Cytosol

FIGURE 24.3

Fatty acids are activated and transported by carnitine through the inner mitochondrial membrane into the matrix.

Q Why is carnitine used to transport a fatty acid into the matrix?

Reactions of β-Oxidation Cycle

In the matrix, fatty acyl CoA molecules undergo β oxidation, which is a cycle of four reactions that convert the —CH_2— of the β carbon to a β-keto group. Once the β-keto group is formed, a two-carbon acetyl group can be split from the chain, which shortens the fatty acyl group.

β-Oxidation Pathway

Reaction 1   Oxidation (Dehydrogenation)

In the first reaction of β oxidation, the FAD coenzyme removes hydrogen atoms from the α and β carbons of the activated fatty acid to form a trans carbon–carbon double bond, and $FADH_2$.

Reaction 2   Hydration

A water molecule now adds across the trans double bond, which places a hydroxyl group (—OH) on the β carbon.

Reaction 3   Oxidation (Dehydrogenation)

The hydroxy group on the β carbon is oxidized to yield a ketone. The hydrogen atoms removed in the dehydrogenation reduce coenzyme NAD^+ to $NADH + H^+$. At this point, the β carbon has been oxidized to a keto group.

Reaction 4   Cleavage of Acetyl CoA

In the final step of β oxidation, the C_α — C_β bond splits to yield free acetyl CoA and a fatty acyl CoA molecule that is shorter by two carbon atoms. This shorter fatty acyl CoA is ready to go through the β-oxidation cycle again.

The reaction for one cycle of β oxidation is written as follows:

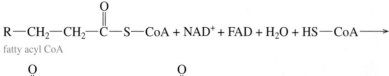

Fatty Acid Length Determines Cycle Repeats

The number of carbon atoms in a fatty acid determines the number of times the cycle repeats and the number of acetyl CoA units it produces. For example, the complete β oxidation of myristic acid (C_{14}) produces seven acetyl CoA groups, which is equal to one-half the number of carbon atoms in the chain. Because the final turn of the cycle produces two acetyl CoA groups, the total number of times the cycle repeats is one less than the total number of acetyl groups it produces. Therefore, the C_{14} fatty acid goes through the cycle six times. (See Figure 24.4.)

Fatty acid	Number of Acetyl CoA	β-Oxidation cycles
myristic acid C_{14}	7 acetyl CoA	6
palmitic acid C_{16}	8 acetyl CoA	7
stearic acid C_{18}	9 acetyl CoA	8

We can write an overall equation for the complete oxidation of myristyl-CoA as follows:

$$\text{Myristyl-CoA} + 6\,\text{HS—CoA} + 6\,\text{FAD} + 6\,\text{NAD}^+ + 6\text{H}_2\text{O} \longrightarrow$$
$$7\,\text{Acetyl CoA} + 6\,\text{FADH}_2 + 6\,\text{NADH} + 6\text{H}^+$$

Oxidation of Unsaturated Fatty Acids

The β-oxidation sequence we have described applies to saturated fatty acids with an even number of carbon atoms. However, the fats in our diets, particularly the oils, contain unsaturated fatty acids, which have one or more cis double bonds. The hydration reaction adds water to trans double bonds, not cis. When the double bond in an unsaturated fatty acid is ready for hydration, an isomerase forms a trans double bond between the α and β carbons, which is the arrangement needed for the hydration reaction.

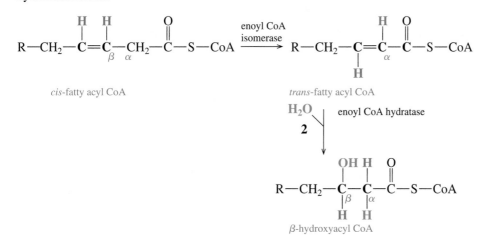

Because the isomerization provides the trans double bond for the hydration in reaction 2, it bypasses the first reaction. Therefore, the energy released by the β oxidation of an unsaturated fatty acid is slightly less because no FADH$_2$ is produced in that cycle.

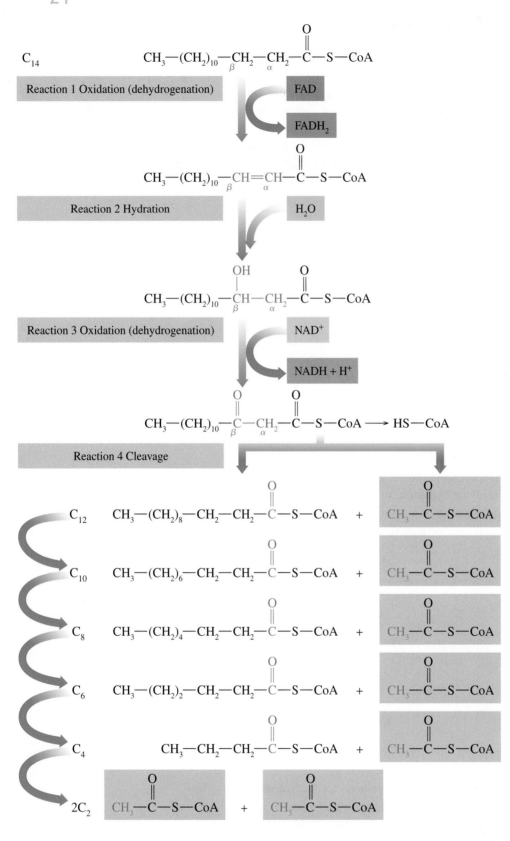

FIGURE 24.4

Myristic acid (C_{14}) undergoes six oxidation cycles that repeat reactions 1–4 to yield 7 acetyl CoA molecules.

Q **How many NADH and FADH$_2$ are produced in one turn of the fatty acid cycle of β oxidation?**

SAMPLE PROBLEM 24.2

▪ β Oxidation

Match each of the following with reactions in the β-oxidation cycle:

(1) first oxidation   (2) hydration   (3) second oxidation   (4) cleavage

a. water is added to a trans double bond

b. an acetyl CoA is removed

c. FAD is reduced to FADH$_2$

d. bypassed during the oxidation of unsaturated fatty acids

SOLUTION

a. (2) hydration   **b.** (4) cleavage

c. (1) first oxidation   **d.** (1) first oxidation

STUDY CHECK

Which coenzyme is needed in reaction 3 when a β-hydroxyl group is converted to a β-keto group?

QUESTIONS AND PROBLEMS

▪ Oxidation of Fatty Acids

24.7 Where in the cell are fatty acids activated?

24.8 What is the function of carnitine in the degradation of fatty acids?

24.9 What coenzymes are required for β oxidation?

24.10 When does an isomerization occur during the β oxidation of a fatty acid?

24.11 In each of the following acyl CoA molecules, identify the β carbon.

$$\textbf{a. } CH_3-CH_2-CH_2-CH_2-CH_2-CH_2-CH_2-\overset{\displaystyle O}{\overset{\|}{C}}-S-CoA$$

$$\textbf{b. } CH_3-(CH_2)_{14}-CH_2-CH_2-\overset{\displaystyle O}{\overset{\|}{C}}-S-CoA$$

$$\textbf{c. } CH_3-CH_2-CH=CH-CH_2-CH_2-CH_2-CH_2-CH_2-\overset{\displaystyle O}{\overset{\|}{C}}-S-CoA$$

24.12 Write the product when each of the following undergoes the indicated reaction in β oxidation:

$$\textbf{a. } CH_3-(CH_2)_{12}-CH=CH-\overset{\displaystyle O}{\overset{\|}{C}}-S-CoA + H_2O \xrightarrow{\text{enoyl CoA hydratase}}$$

$$\textbf{b. } CH_3-(CH_2)_6-CH_2-CH_2-\overset{\displaystyle O}{\overset{\|}{C}}-S-CoA \xrightarrow{\text{acyl CoA dehydrogenase}}$$

$$\textbf{c. } CH_3-(CH_2)_4-\overset{\displaystyle O}{\overset{\|}{C}}-CH_2-\overset{\displaystyle O}{\overset{\|}{C}}-S-CoA + HS-CoA \xrightarrow{\text{thiolase}}$$

24.13 Capric acid, $CH_3(CH_2)_8COOH$, is a C_{10} fatty acid.

  a. Write the formula of the activated form of capric acid.

  b. Indicate the α and β carbon atoms in the fatty acid.

  c. Write the overall equation for the first cycle of β oxidation for capric acid.

  d. Write the overall equation for the complete β oxidation of capric acid.

24.14 Arachidic acid, CH_3—$(CH_2)_{18}$—$COOH$, is a C_{20} fatty acid.
    a. Write the formula of the activated form of arachidic acid.
    b. Indicate the α and β carbon atoms in the fatty acid.
    c. Write the overall equation for the first cycle of β oxidation for arachidic acid.
    d. Write the overall equation for the complete β oxidation of arachidic acid.

LEARNING GOAL

Calculate the total ATP produced by the complete oxidation of a fatty acid.

Fatty Acid Oxidation

24.3 ATP and Fatty Acid Oxidation

We can now determine the total energy yield from the oxidation of a particular fatty acid. In each β-oxidation cycle, one NADH, one $FADH_2$, and one acetyl CoA are produced. From Chapter 23, we know that hydrogen ions and electrons transferred from NADH to coenzyme Q in the electron transport chain generate sufficient energy to synthesize three ATP, whereas $FADH_2$ leads to the synthesis of two ATP. However, the greatest amount of energy produced from a fatty acid is generated by the production of the acetyl CoA units that enter the citric acid cycle. We saw in Chapter 23 that one acetyl CoA leads to the synthesis of a total of 12 ATP.

So far we know that the C_{14} acid produces seven acetyl CoA units and goes through six turns of the cycle. We also need to remember that activation of the myristic acid requires two ATP. We can set up the calculation as follows.

ATP Production for Myristic Acid

Activation	−2 ATP
7 acetyl CoA	
7 acetyl CoA × 12 ATP/acetyl CoA	84 ATP
6 β-oxidation cycles	
6 $FADH_2$ × 2 ATP/$FADH_2$	12 ATP
6 FADH × 3 ATP/NADH	18 ATP
Total	112 ATP

The Energy Yield from Fats

Myristic acid, $C_{14}H_{28}O_2$, has a molar mass of 228 g/mole. We can calculate the ATP produced per gram of the fatty acid as follows:

$$\frac{112 \text{ moles ATP}}{1 \text{ mole myristic acid}} \times \frac{1 \text{ mole myristic acid}}{228 \text{ g myristic acid}} = 0.49 \text{ mole ATP/g myristic acid (fat)}$$

In Chapter 23 we saw that the complete oxidation of glucose generated a total of 36 ATP. Glucose, $C_6H_{12}O_6$, has a molar mass of 180 g. We can calculate the ATP produced per gram of glucose as follows:

$$\frac{36 \text{ moles ATP}}{1 \text{ mole glucose}} \times \frac{1 \text{ mole glucose}}{180 \text{ g glucose}} = 0.20 \text{ mole ATP/g glucose}$$

From these calculations, we see that 1 gram fat produces more than twice the ATP energy as 1 gram glucose. This also means that we obtain more than double the number of nutritional calories from 1 g fat (9 kcal/g fat) than we do from 1 g carbohydrate (4 kcal/g). This is one reason a low-fat diet is recommended when we are trying to lose weight.

HEALTH NOTE

Stored Fat and Obesity

The storage of fat is an important survival feature in the lives of many animals. In hibernating animals, large amounts of stored fat provide the energy for the entire hibernation period, which could be several months. In camels, large amounts of food are stored in the camel's hump, which is actually a huge fat deposit. When food resources are low, the camel can survive months without food or water by utilizing the fat reserves in the hump. Migratory birds preparing to fly long distances also store large amounts of fat. Whales are kept warm by a layer of body fat called "blubber" under their skin, which can be as thick as 2 feet. Blubber also provides energy when whales must survive long periods of starvation. Penguins also have blubber, which protects them from the cold and provides energy when they are incubating their eggs.

Humans also have the capability to store large amounts of fat, although they do not hibernate or usually have to survive for long periods of time without food. When humans survived on sparse diets that were mostly vegetarian, about 20% of the dietary calories were from fat. Today, a typical diet includes more dairy products and foods with high fat levels, and as much as 60% of the calories is/or are from fat. The U.S. Public Health Service now estimates that in the United States, more than one-third of adults are obese. Obesity is defined as a body weight that is more than 20 percent over an ideal weight. Obesity is a major factor in health problems such as diabetes, heart disease, high blood pressure, stroke, and gallstones as well as some cancers and forms of arthritis.

At one time we thought that obesity was simply a problem of eating too much. However, research now indicates that certain pathways in lipid and carbohydrate metabolism may cause excessive weight gain in some people. In 1995, scientists discovered that a hormone called *leptin* is produced in fat cells. When fat cells are full, high levels of leptin signal the brain to limit the intake of food. When fat stores are low, leptin production decreases, which signals the brain to increase food intake. Some obese persons have high levels of leptin, which means that leptin did not cause them to decrease how much they ate.

The causes of obesity have become a major research field. Scientists are studying differences in the rate of leptin production, degrees of resistance to leptin, and possible combinations of these factors. After a person has dieted and lost weight, the leptin level drops. This decrease in leptin may cause an increase in hunger, slow metabolism, and increased food intake, which starts the weight-gain cycle all over again. Currently, studies are being made to assess the safety of leptin therapy following a weight loss.

SAMPLE PROBLEM 24.3

ATP Production from β Oxidation

How much ATP will be produced from the β oxidation of palmitic acid, a C_{16} saturated fatty acid?

SOLUTION

A 16-carbon fatty acid will produce 8 acetyl CoA units and go through 7 β-oxidation cycles, which produces 7 $FADH_2$ and 7 NADH. Each acetyl CoA can produce 12 ATP by way of the citric acid cycle. In the electron transport chain, each $FADH_2$ produces 2 ATP, and each NADH produces 3 ATP.

▪ Fat Storage and Blubber

Obtain two medium-sized plastic baggies and a can of Crisco or other type of shortening used for cooking. You will also need a bucket or large container with water and ice cubes. Place several tablespoons of the shortening in one of the baggies. Place the second baggie inside and tape the outside edges of the bag. With your hand inside the inner baggie, move the shortening around to cover your hand. With one hand inside the double baggie, submerge both your hands in the container of ice water. Measure the time it takes for one hand to feel uncomfortably cold. Experiment with different amounts of shortening.

QUESTIONS

1. How effective is the bag with "blubber" in protecting your hand from the cold?
2. How does "blubber" help an animal survive starvation?
3. How would twice the amount of shortening affect your results?
4. Why would animals in warm climates, such as camels and migratory birds, need to store fat?
5. If you placed 300 g of shortening in the baggie, how many moles of ATP could it provide if used for energy production (assume it produces the same ATP as myristic acid)?

LEARNING GOAL

Describe the pathway of ketogenesis.

(MC GOB) Ketogenesis and Ketone Bodies

ATP Production for Palmitic Acid ($C_{16}H_{32}O_2$)

Activation of palmitic acid to palmitoyl−CoA	−2 ATP
8 acetyl CoA × 12 ATP (citric acid cycle)	96 ATP
7 FADH$_2$ × 2 ATP (electron transport chain)	14 ATP
7 NADH × 3 ATP (electron transport chain)	21 ATP
Total	129 ATP

STUDY CHECK

Compare the total ATP from reduced coenzymes and from acetyl CoA in the β oxidation of palmitic acid.

QUESTIONS AND PROBLEMS

(MC GOB)

▪ ATP and Fatty Acid Oxidation

24.15 Why is the energy of fatty acid activation from ATP to AMP considered the same as the hydrolysis of 2 ATP $\longrightarrow$ 2 ADP?

24.16 What is the number of ATP obtained from one acetyl CoA in the citric acid cycle?

24.17 Consider the complete oxidation of capric acid $CH_3 — (CH_2)_8 — COOH$, a C_{10} fatty acid.
  a. How many acetyl CoA units are produced?
  b. How many cycles of β oxidation are needed?
  c. How many ATPs are generated from the oxidation of capric acid?

24.18 Consider the complete oxidation of arachidic acid, $CH_3 — (CH_2)_{18} — COOH$, a C_{20} fatty acid.
  a. How many acetyl CoA units are produced?
  b. How many cycles of β oxidation are needed?
  c. How many ATPs are generated from the oxidation of arachidic acid?

24.4 Ketogenesis and Ketone Bodies

When carbohydrates are not available to meet energy needs, the body breaks down body fat. However, the oxidation of large amounts of fatty acids can cause acetyl CoA molecules to accumulate in the liver. Then acetyl CoA molecules combine to form compounds called **ketone bodies** in a pathway known as **ketogenesis**. (See Figure 24.5.)

1. In ketogenesis, two molecules of acetyl CoA combine to form acetoacetyl CoA, which reverses the last reaction in β oxidation.
2. The hydrolysis of acetoacetyl CoA forms acetoacetate, a ketone body, which reacts further to produce two other ketone bodies.
3. Acetoacetate can be reduced to yield β hydroxybutyrate, which is considered a ketone body even though it does not contain a keto group.
4. Acetone forms when acetoacetate undergoes decarboxylation and loses CO_2.

Ketone bodies are produced mostly in the liver and transported to cells in the heart, brain, and skeletal muscle, where small amounts of energy can be obtained by converting acetoacetate or β hydroxybutyrate back to acetyl CoA.

β hydroxybutyrate $\longrightarrow$ acetoacetate $\longrightarrow$ acetoacetyl CoA $\longrightarrow$ 2 acetyl CoA

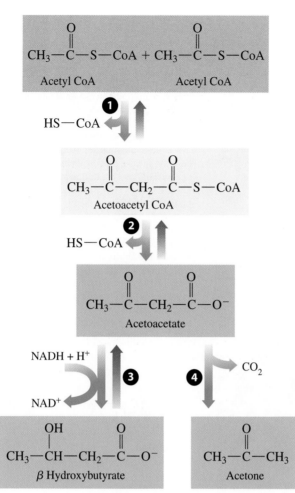

FIGURE 24.5

In ketogenesis, acetyl CoA molecules combine to produce ketone bodies: acetoacetate, β hydroxybutyrate, and acetone.

Q **What condition in the body leads to the formation of ketone bodies?**

Ketosis

When ketone bodies accumulate, they may not be completely metabolized by the body. This may lead to a condition called **ketosis**, which is found in severe diabetes, diets high in fat and low in carbohydrates, and starvation. Because two of the ketone bodies are acids, they can lower the blood pH below 7.4, which is **acidosis**, a condition that often accompanies ketosis. A drop in blood pH can interfere with the ability of the blood to carry oxygen and cause breathing difficulties.

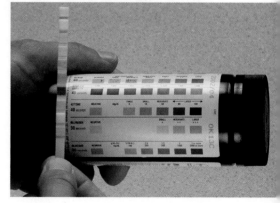

SAMPLE PROBLEM 24.4

■ **Ketogenesis**

When does ketogenesis take place in the liver?

SOLUTION

When excess acetyl CoA cannot be processed by the citric acid cycle, acetyl CoA units enter the ketogenesis pathway where it forms ketone bodies.

STUDY CHECK

What are the names of the three compounds that are called ketone bodies?

Ketone Bodies and Diabetes

The blood glucose is elevated within 30 minutes following a meal containing carbohydrates. The elevated level of glucose stimulates the secretion of the hormone insulin from the pancreas, which increases the flow of glucose into muscle and adipose tissue for the synthesis of glycogen. As blood glucose levels drop, the secretion of insulin decreases. When blood glucose is low, another hormone, glucagon, is secreted by the pancreas, which stimulates the breakdown of glycogen in the liver to yield glucose.

In *diabetes mellitus*, glucose cannot be utilized or stored as glycogen because insulin is not secreted or does not function properly. In Type I, *insulin-dependent diabetes*, which often begins in childhood, the pancreas produces inadequate levels of insulin. This type of diabetes can result from damage to the pancreas by viral infections or from genetic mutations. In Type II, *insulin-resistant diabetes*, which usually occurs in adults, insulin is produced, but insulin receptors are not responsive. Thus a person with Type II diabetes does not respond to insulin therapy. *Gestational diabetes* can occur during pregnancy, but blood glucose levels usually return to normal after the baby is born. Mothers with diabetes tend to gain weight and have large babies.

In all types of diabetes, insufficient amounts of glucose are available in the muscle, liver, and adipose tissue. As a result liver cells synthesize glucose from noncarbohydrate sources (gluconeogenesis) and break down fat, which elevates the acetyl CoA level. Excess acetyl CoA undergoes ketogenesis and ketone bodies accumulate in the blood. The odor of acetone can be detected on the breath of a person with uncontrolled diabetes who is in ketosis.

In uncontrolled diabetes, the concentration of blood glucose exceeds the ability of the kidney to reabsorb glucose, and glucose appears in the urine. High levels of glucose increase the osmotic pressure in the blood, which leads to an increase in urine output. Symptoms of diabetes include frequent urination and excessive thirst. Treatment for diabetes includes a change to a diet to limit carbohydrate intake and may require medication such as a daily injection of insulin.

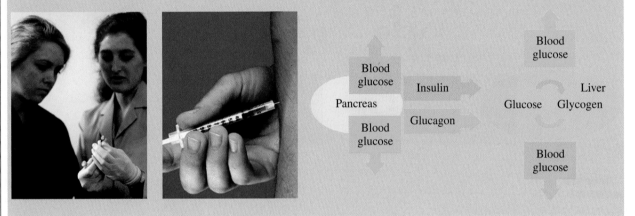

QUESTIONS AND PROBLEMS

Ketogenesis and Ketone Bodies

24.19 What is ketogenesis?

24.20 If a person is fasting, why would they have high levels of acetyl CoA?

24.21 What type of reaction converts acetoacetate to β hydroxybutyrate?

24.22 How is acetone formed from acetoacetate?

24.23 What is ketosis?

24.24 Why do diabetics produce high levels of ketone bodies?

24.5 Fatty Acid Synthesis

When the body has met all its energy needs and the glycogen stores are full, acetyl CoA from the breakdown of carbohydrates and fatty acids is used to form new fatty acids. In the pathway called **lipogenesis**, two-carbon acetyl units are linked together to give a 16-carbon fatty acid, palmitic acid. Although the reactions appear much like the reverse of the reactions we discussed in fatty acid oxidation, the synthesis of fatty acids proceeds in a separate pathway with different enzymes. Fatty acid oxidation occurs in the mitochondria and uses FAD and NAD$^+$, whereas fatty acid synthesis occurs in the cytosol and uses the reduced coenzyme NADPH. NADPH is similar to NADH, except it has a phosphate group.

MC GOB Fatty Acid Synthesis

Acyl Carrier Protein (ACP)

In β oxidation, acetyl and acyl molecules are activated using coenzyme A (CoA). In fatty acid synthesis, acyl compounds also are activated, but by an acyl carrier protein (ACP—SH). In the ACP—SH molecule, the thiol and pantothenic acid found in CoA are attached to a protein.

ACP—SH

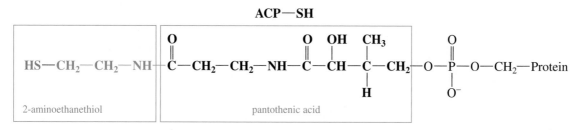

For fatty acid synthesis, the activated forms of malonyl-ACP and acetyl-ACP are produced by transferring an acetyl or acyl group to ACP—SH.

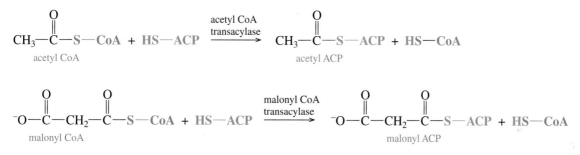

Synthesis of Malonyl CoA

Fatty acid synthesis begins when acetyl CoA combines with bicarbonate to form a three-carbon compound, malonyl CoA. The hydrolysis of ATP provides the energy for the reaction.

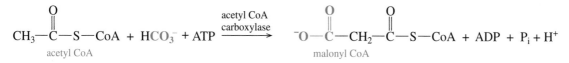

Synthesis of Palmitate

The next four reactions occur in a cycle that adds two-carbon acetyl groups to a carbon chain. (See Figure 24.6.)

Reaction 1   Condensation
Acetyl ACP and malonyl ACP condense to yield acetoacetyl-ACP and CO_2.

Reaction 2   Reduction
The keto group on the β carbon is reduced to a hydroxyl group using hydrogen from the reduced coenzyme NADPH.

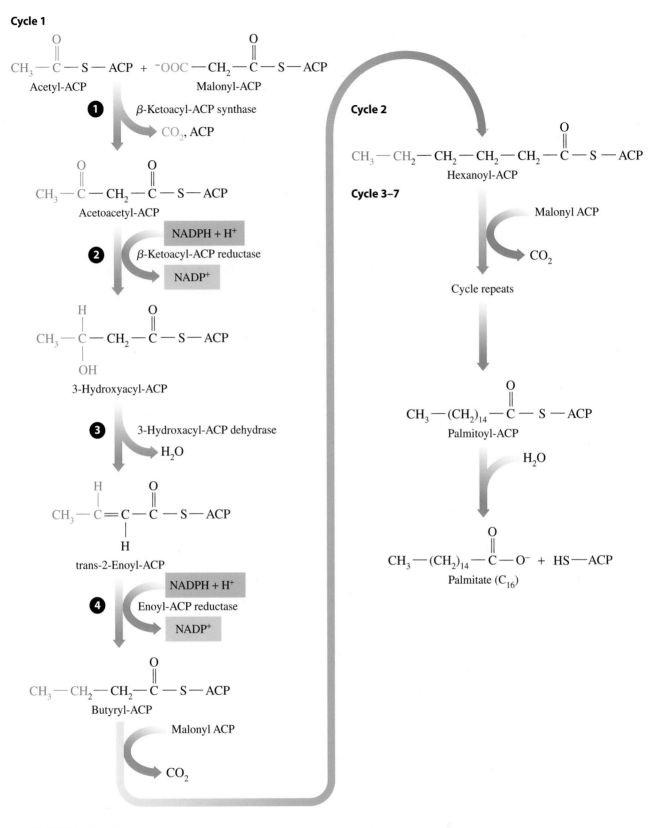

FIGURE 24.6

In fatty acid synthesis (lipogenesis), two-carbon units from acetyl CoA are added together to form palmitate.

Q **Identify the types of reactions in 1–4 as reduction, dehydration, or condensation.**

Reaction 3   Dehydration

The alcohol is dehydrated to form a trans double bond in *trans*-enoyl-ACP.

Reaction 4   Reduction

NADPH reduces the double bond to a single bond, which forms butyryl-ACP, a saturated four-carbon compound.

Cycle Repeats

The cycle repeats as the longer four-carbon butyryl-ACP reacts with another malonyl-ACP to produce hexanoyl-ACP. After seven cycles, the product, C_{16} palmitoyl-ACP, is hydrolyzed to yield palmitate and HSACP.

   The overall reaction starts with acetyl CoA and requires ATP and NADPH. We can write the overall equation for the synthesis of palmitate as follows:

$$8 \text{ acetyl CoA} + 14 \text{ NADPH} + 14\text{H}^+ + 7 \text{ ATP} \longrightarrow \text{palmitate} + 8 \text{ CoA} + 14 \text{ NADP}^+ + 7\text{H}_2\text{O} + 7 \text{ ADP} + 7 \text{ P}_i$$

Longer and Shorter Fatty Acids

Although we have looked at the synthesis of the fatty acid palmitate, shorter and longer fatty acids are also produced in cells. Shorter fatty acids are released before there are 16 carbon atoms in the chain. Longer fatty acids are produced with special enzymes that add two-carbon units to the carboxyl end of the fatty acid chain. An unsaturated cis bond can also be incorporated into a 10-carbon fatty acid followed by the same elongation reactions we have seen.

Regulation of Fatty Acid Synthesis

Fatty acid synthesis takes place primarily in the adipose tissue, where triacyl-glycerols are formed and stored. The hormone insulin stimulates the formation of fatty acids. When blood glucose is high, insulin moves glucose into the cells. In the cell, insulin stimulates glycolysis and the oxidation of pyruvate, thereby producing acetyl CoA for fatty acid synthesis. During lipogenesis, the production of malonyl CoA blocks the transport of fatty acyl groups into the matrix of the mitochondria, which prevents their oxidation.

Comparison of β Oxidation and Fatty Acid Synthesis

We have seen that many of the steps in the synthesis of palmitate are similar to those that occur in the β oxidation of palmitate. Synthesis combines two-carbon units, whereas β oxidation removes two-carbon units. Synthesis involves reduction and dehydration, whereas β oxidation involves oxidation and hydration. We can distinguish between the two pathways by comparing some of their features in Table 24.1.

SAMPLE PROBLEM 24.5

■ Fatty Acid Synthesis

Malonyl ACP is required for the elongation of fatty acid chains. How is malonyl ACP formed from the starting material acetyl CoA?

TABLE 24.1

A Comparison of β Oxidation and Fatty Acid Synthesis

	β Oxidation	Fatty Acid Synthesis (lipogenesis)
Site	mitochondrial matrix	cytosol
Activated by	glucagon	insulin
	low blood glucose	high blood glucose
Activator	coenzyme A (CoA)	acyl carrier protein (ACP)
Initial substrate	fatty acid	acetyl CoA $\longrightarrow$ malonyl CoA
Coenzymes	FAD, NAD^+	NADPH
Types of reaction	oxidation	reduction
	hydration	dehydration
	cleavage	condensation
Function	cleaves two-carbon acyl group	adds two-carbon acyl group
Final product	acetyl CoA units	palmitate (C_{16}) and other fatty acids

SOLUTION

Acetyl CoA combines with bicarbonate to form malonyl CoA, which reacts with ACP to form malonyl ACP.

STUDY CHECK

If malonyl ACP is a three-carbon acyl group, why are only two carbon atoms added each time malonyl ACP is combined with a fatty acid chain?

QUESTIONS AND PROBLEMS

■ **Fatty Acid Synthesis**

24.25 Where does fatty acid synthesis occur in the cell?

24.26 What compound is involved in the activation of acyl compounds in fatty acid synthesis?

24.27 What is the starting material for fatty acid synthesis?

24.28 What is the function of malonyl ACP in fatty acid synthesis?

24.29 Identify the reaction catalyzed by each of the following enzymes:
   (1) acetyl CoA carboxylase   (2) acetyl CoA transacylase
   (3) malonyl CoA transacylase
   a. converts malonyl CoA to malonyl ACP
   b. combines acetyl CoA with bicarbonate to give malonyl CoA
   c. converts acetyl CoA to acetyl ACP

24.30 Identify the reaction catalyzed by each of the following enzymes:
   (1) β-ketoacyl-ACP synthase   (2) β-ketoacyl-ACP reductase
   (3) 3-hydroxyacyl-ACP dehydrase   (4) enoyl-ACP reductase
   a. catalyzes the dehydration of an alcohol
   b. converts a carbon–carbon double bond to a carbon–carbon single bond
   c. combines a two-carbon acyl group with a three-carbon acyl group accompanied by the loss of CO_2
   d. reduces a keto group to a hydroxyl group

24.31 Determine the number of each of the following involved in the synthesis of one molecule of capric acid, a fatty acid with 10 carbon atoms, $C_{10}H_{20}O_2$.
   a. HCO_3^- **b.** ATP   **c.** acetyl CoA
   d. malonyl ACP   **e.** NADPH   **f.** CO_2 removed

24.32 Determine the number of each of the following needed in the synthesis of one molecule of myristic acid, a fatty acid with 14 carbon atoms $C_{14}H_{28}O_2$.

    a. HCO_3^-
    b. ATP
    c. acetyl ACP
    d. malonyl ACP
    e. NADPH
    f. CO_2 removed

24.6 Digestion of Proteins

LEARNING GOAL

Describe the hydrolysis of dietary protein and absorption of amino acids.

The major role of proteins is to provide amino acids for the synthesis of new proteins for the body and nitrogen atoms for the synthesis of compounds such as nucleotides. We have seen that carbohydrates and lipids are the major sources of energy, but when they are not available, amino acids are degraded to substrates that enter energy-producing pathways.

In stage 1, the digestion of proteins begins in the stomach, where hydrochloric acid (HCl) at pH 2 denatures proteins and activates enzymes such as pepsin that begin to hydrolyze peptide bonds. Polypeptides from the stomach move into the small intestine where trypsin and chymotrypsin complete the hydrolysis of the peptides to amino acids. The amino acids are absorbed through the intestinal walls into the blood stream for transport to the cells. (See Figure 24.7.)

SAMPLE PROBLEM 24.6

Digestion of Proteins

What are the sites and end products for the digestion of proteins?

SOLUTION

Proteins begin digestion in the stomach and complete digestion in the small intestine to yield amino acids.

STUDY CHECK

What is the function of HCl in the stomach?

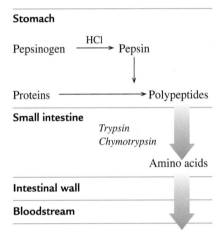

FIGURE 24.7

Proteins are hydrolyzed to polypeptides in the stomach and to amino acids in the small intestine.

Q **What enzyme, secreted into the small intestine, hydrolyzes peptides?**

Protein Turnover

Our bodies are constantly replacing old proteins with new ones. The process of synthesizing proteins and breaking them down is called **protein turnover**. Many types of proteins including enzymes, hormones, and hemoglobin are synthesized in the cells and then degraded. For example, the hormone insulin has a half-life of 10 minutes, whereas the half-lives of lactate dehydrogenase is about 2 days and hemoglobin is 120 days. Damaged and ineffective proteins are also degraded and replaced. While most amino acids are used to build proteins, other compounds also require nitrogen for their synthesis as seen in Table 24.2. (See Figure 24.8.)

Usually we maintain a nitrogen balance in the cells so that the amount of protein we break down is equal to the amount that is reused. A diet that is high in protein has a positive nitrogen balance because it supplies more nitrogen than we need. Because the body cannot store nitrogen, the excess is excreted as urea. A diet that does not provide sufficient nitrogen has a negative nitrogen balance, which is a condition that occurs during starvation and fasting.

TABLE 24.2

Nitrogen-Containing Compounds

Type of Compound	Example
nonessential amino acids	alanine, aspartate, cysteine, glycine
proteins	muscle protein, enzymes
neurotransmitters	acetylcholine, dopamine, serotonin
amino alcohols	choline, ethanolamine
heme	hemoglobin
hormones	thyroxine, epinephrine, insulin
nucleotides (nucleic acids)	purines, pyrimidines

FIGURE 24.8

Proteins are used in the synthesis of nitrogen-containing compounds or degraded to urea and carbon skeletons that enter other metabolic pathways.

Q **What are some compounds that require nitrogen for their synthesis?**

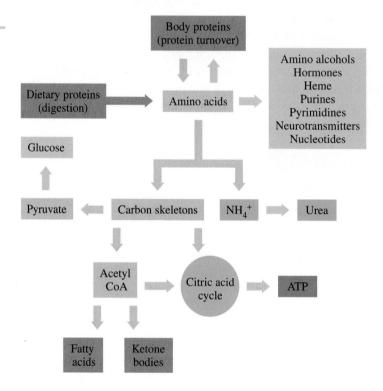

Energy from Amino Acids

Normally, only a small amount (about 10%) of our energy needs is supplied by amino acids. However, more energy is extracted from amino acids in conditions such as fasting or starvation, when carbohydrate and fat stores are exhausted. If amino acids remain the only source of energy for a long period of time, the breakdown of body proteins eventually leads to a destruction of essential body tissues.

SAMPLE PROBLEM 24.7

■ **Nitrogen Balance**

In positive nitrogen balance, why are excess amino acids excreted?

SOLUTION

Because the body cannot store nitrogen, amino acids that are not needed for the synthesis of proteins are excreted.

STUDY CHECK

Under what condition does the body have a negative nitrogen balance?

QUESTIONS AND PROBLEMS

■ Digestion of Proteins

24.33 Where do dietary proteins undergo digestion in the body?

24.34 What is meant by protein turnover?

24.35 What are some nitrogen-containing compounds that need amino acids for their synthesis?

24.36 What is the fate of the amino acids obtained from a high protein diet?

24.7 Degradation of Amino Acids

When dietary protein exceeds the nitrogen needed for protein synthesis, all excess amino acids are degraded in a similar way. The α-amino group is removed to yield a keto acid, which can be converted to an intermediate of other metabolic pathways. The carbon atoms from amino acids are used in the citric acid cycle as well as the synthesis of fatty acids, ketone bodies, and glucose. Most of the amino groups are converted to urea.

LEARNING GOAL

Describe the reactions of transamination and oxidative deamination in the degradation of amino acids.

■ Transamination

The degradation of amino acids occurs primarily in the liver. In a **transamination** reaction, an α-amino group is transferred from an amino acid to an α-keto acid, usually α-ketoglutarate. A new amino acid and a new α-keto acid are produced. The enzymes for the transfer of amino groups are known as transaminases or aminotransferases. We can write an equation to show the transfer of the amino group from alanine to α-ketoglutarate to yield glutamate, the new amino acid, and the α-keto acid pyruvate. The α-keto acid often used in transamination reactions is α-ketoglutarate, which is converted to glutamate.

Transamination and Deamination

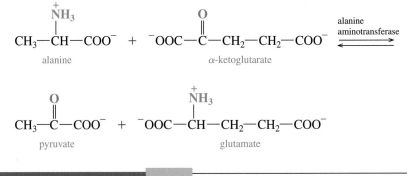

SAMPLE PROBLEM 24.8

■ Transamination

Write the formula of the amino acid and α-keto acid produced from the transamination of oxaloacetate by alanine.

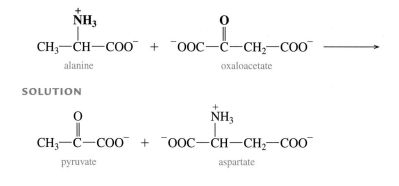

$$\underset{\text{alanine}}{CH_3-\overset{\overset{+}{N}H_3}{\underset{|}{C}H}-COO^-} + \underset{\text{oxaloacetate}}{^-OOC-\overset{O}{\overset{||}{C}}-CH_2-COO^-} \longrightarrow$$

SOLUTION

$$\underset{\text{pyruvate}}{CH_3-\overset{O}{\overset{||}{C}}-COO^-} + \underset{\text{aspartate}}{^-OOC-\overset{\overset{+}{N}H_3}{\underset{|}{C}H}-CH_2-COO^-}$$

STUDY CHECK

What is a possible name for the enzyme that catalyzes this reaction?

Oxidative Deamination

In a process called **oxidative deamination**, the amino group in glutamate is removed as an ammonium ion NH_4^+. This reaction is catalyzed by *glutamate dehydrogenase*, which uses either NAD^+ or $NADP^+$ as a coenzyme.

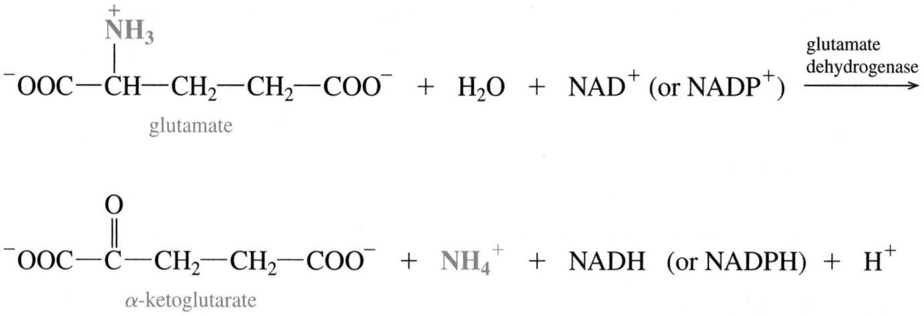

$$\underset{\text{glutamate}}{^-OOC-\overset{\overset{+}{N}H_3}{\underset{|}{C}H}-CH_2-CH_2-COO^-} + H_2O + NAD^+ \text{ (or } NADP^+) \xrightarrow{\text{glutamate dehydrogenase}}$$

$$\underset{\alpha\text{-ketoglutarate}}{^-OOC-\overset{O}{\overset{||}{C}}-CH_2-CH_2-COO^-} + NH_4^+ + NADH \text{ (or NADPH)} + H^+$$

Therefore the amino group from any amino acid can be used to form glutamate, which undergoes oxidative deamination converting the amino group to an ammonium ion. Then the ammonium ion is converted to urea, which we will discuss in the next section.

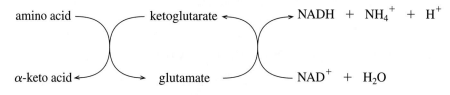

SAMPLE PROBLEM 24.9

Transamination and Oxidative Deamination

Indicate whether each of the following represents a transamination or an oxidative deamination:

a. Glutamate is converted to α-ketoglutarate and NH_4^+.

b. Alanine and α-ketoglutarate react to form pyruvate and glutamate.

c. A reaction is catalyzed by glutamate dehydrogenase, which requires NAD^+.

SOLUTION

a. oxidative deamination   **b.** transamination   **c.** oxidative deamination

STUDY CHECK

How is α-ketoglutarate regenerated to participate in more transamination reactions?

QUESTIONS AND PROBLEMS

■ **Degradation of Amino Acids**

24.37 What are the reactants and products in transamination reactions?

24.38 What types of enzymes catalyze transamination reactions?

24.39 Write the structure of the α-keto acid produced from each of the following in transamination:

a. 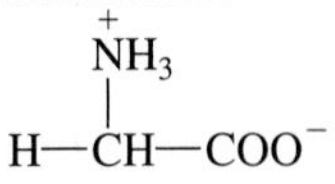 H—CH—COO⁻ glycine   (with $\overset{+}{N}H_3$ group)   **b.** 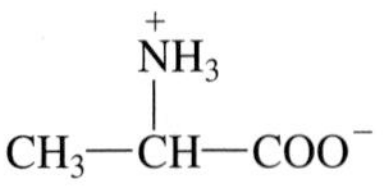 CH₃—CH—COO⁻ alanine   (with $\overset{+}{N}H_3$ group)

c. 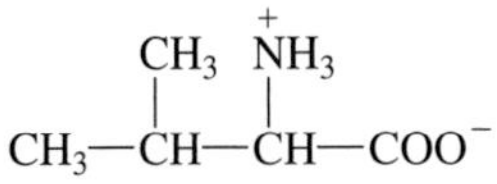 CH₃—CH—CH—COO⁻ valine   (with CH₃ and $\overset{+}{N}H_3$ groups)

24.40 Write the structure of the α–keto acid produced from each of the following in transamination:

a. 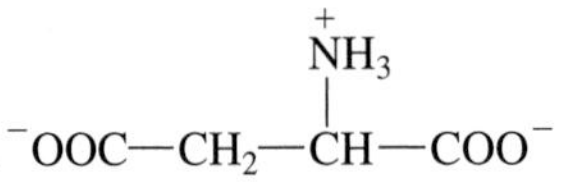 ⁻OOC—CH₂—CH—COO⁻ aspartate   (with $\overset{+}{N}H_3$ group)

b. 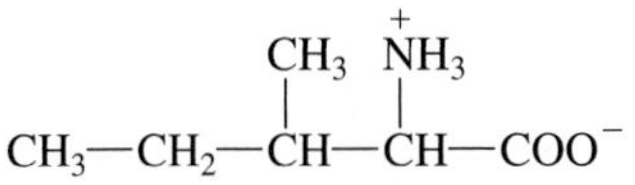 CH₃—CH₂—CH—CH—COO⁻ isoleucine   (with CH₃ and $\overset{+}{N}H_3$ groups)

c. 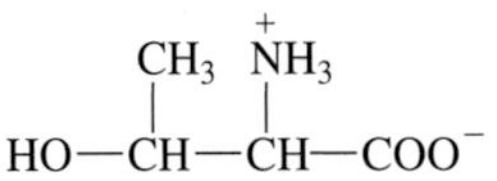 HO—CH—CH—COO⁻ serine   (with CH₃ and $\overset{+}{N}H_3$ groups)

24.41 Write the reaction for the oxidative deamination of glutamate.

24.42 How do the amino groups from all 20 amino acids produce ammonium ions when glutamate undergoes oxidative deamination?

24.8 Urea Cycle

The ammonium ion, which is the end product of amino acid degradation, is toxic if it is allowed to accumulate. Therefore, a pathway called the **urea cycle** detoxifies ammonium ions by converting them to urea, which is transported to the kidneys to form urine.

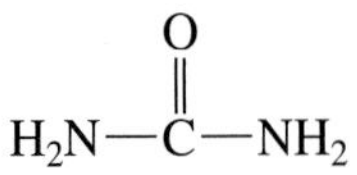

$$H_2N—\overset{\overset{\textstyle O}{\|}}{C}—NH_2$$
urea

In one day, a typical adult may excrete about 25–30 g of urea in the urine. This amount increases when a diet is high in protein. If urea is not properly excreted, it builds up quickly to a toxic level. To detect renal disease, the blood urea nitrogen (BUN) level is measured. If the BUN is high, protein intake must be reduced, and hemodialysis may be needed to remove toxic nitrogen waste from the blood.

LEARNING GOAL

■ Describe the formation of urea from ammonium ion.

Urea Cycle

The urea cycle in the liver cells consists of reactions that take place in both the mitochondria and cytosol. (See Figure 24.9.) In preparation for the urea cycle, the ammonium ions react with carbon dioxide from the citric acid cycle and two ATP to yield carbamoyl phosphate.

$$NH_4^+ + CO_2 + 2\,ATP + H_2O \xrightarrow{\text{carbamoyl phosphate synthetase}} \underset{\text{carbamoyl phosphate}}{H_2N-\overset{\displaystyle O}{\overset{\|}{C}}-O-\overset{\displaystyle O}{\underset{\displaystyle O^-}{\overset{\|}{P}}}-O^-} + 2\,ADP + P_i$$

Reaction 1 Transfer of Carbamoyl Group

In the mitochondria, the carbamoyl group is transferred from carbamoyl phosphate to ornithine to yield citrulline, which is transported across the mitochondrial membrane into the cytosol. The hydrolysis of the phosphate bond provides the energy to drive the reaction.

Reaction 2 Condensation with Aspartate

In the cytosol, citrulline condenses with the amino acid aspartate to form argininosuccinate. The hydrolysis of ATP to AMP and two inorganic phosphates provides the energy for the reaction. The nitrogen atom in aspartate becomes the other nitrogen atom in the urea that is produced in the final reaction.

Reaction 3 Cleavage of Fumarate

The argininosuccinate undergoes a cleavage to yield fumarate, a citric acid cycle intermediate, and arginine.

Reaction 4 Hydrolysis to Form Urea

The hydrolysis of arginine yields urea and ornithine, which returns to the mitochondria to repeat the cycle.

Summary of Urea Formation

The formation of urea starts with two nitrogen atoms, one from NH_4^+ and one from aspartate, and a carbon atom from CO_2. In the urea cycle, a total of four phosphate bonds are hydrolyzed to provide the energy for the reactions. We can write an overall reaction starting with ammonium ion as follows:

$$NH_4^+ + CO_2 + 3\,ATP + \text{aspartate} + 2H_2O \longrightarrow \text{urea} + 2\,ADP + AMP + 4P_i + \text{fumarate}$$

SAMPLE PROBLEM 24.10

■ Urea Cycle

Indicate the reaction in the urea cycle where each of the following compounds is a reactant.

a. aspartate   **b.** ornithine   **c.** arginine

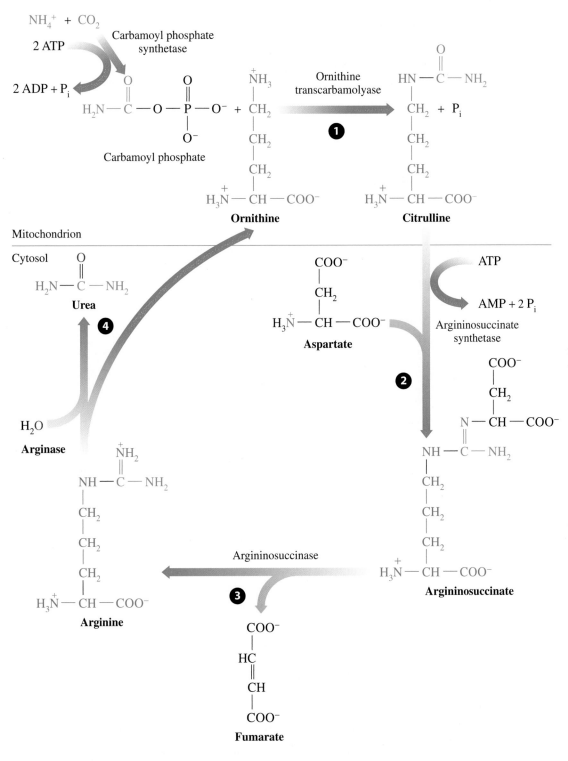

FIGURE 24.9

In the urea cycle, urea is formed from a carbon and nitrogen (blue) from carbamoyl phosphate (initially an ammonium ion from oxidative deamination) and a nitrogen from aspartate (pink).

Q **Where in the cell is urea formed?**

SOLUTION

a. Aspartate condenses with citrulline in reaction 2.
b. Ornithine accepts the carbamoyl group in reaction 1.
c. Arginine is cleaved in reaction 4.

STUDY CHECK

Name the products of each reaction in Sample Problem 24.10.

QUESTIONS AND PROBLEMS

■ **Urea Cycle**

24.43 Why does the body convert NH_4^+ to urea?

24.44 Where is the energy source for the formation of urea?

24.45 What is the structure of urea?

24.46 What is the structure of carbamoyl phosphate?

24.47 What is the source of carbon in urea?

24.48 How much ATP energy is required to drive one turn of the urea cycle?

LEARNING GOAL

Describe where carbon atoms from amino acids enter the citric acid cycle or other pathways.

24.9 Fates of the Carbon Atoms from Amino Acids

The carbon skeletons from the transamination of amino acids are used as intermediates of the citric acid cycle or other metabolic pathways. We can classify the amino acids according to the number of carbon atoms in those intermediates. (See Figure 24.10.) The amino acids with three carbons or those that are converted into carbon skeletons with three carbons are converted to pyruvate. The four-carbon group consists of amino acids that are converted to oxaloacetate, and the five-carbon group provides α-ketoglutarate. Some amino acids are listed twice because they can enter different pathways to form citric-acid cycle intermediates.

A **glucogenic amino acid** generates pyruvate or oxaloacetate, which can be converted to glucose by gluconeogenesis. A **ketogenic amino acid** produces acetoacetyl CoA or acetyl CoA, which can enter the ketogenesis pathway to form ketone bodies or fatty acids.

Carbon Skeletons from the Three-Carbon Amino Acids

 Recycling Amino Acid 'Skeletons'

The carbon atoms from alanine, serine, and cysteine are converted to pyruvate in one step or several steps. Alanine is converted by a simple transamination.

$$\text{alanine} + \alpha\text{-ketoglutarate} \longrightarrow \text{pyruvate} + \text{glutamate}$$

Glycine is converted to serine and then to pyruvate. When tryptophan is degraded, the three carbon atoms that are not part of the ring system form alanine, which goes to pyruvate. Although it has more than three carbon atoms, threonine can be degraded to glycine and acetaldehyde. The oxidation of acetaldehyde gives acetyl CoA.

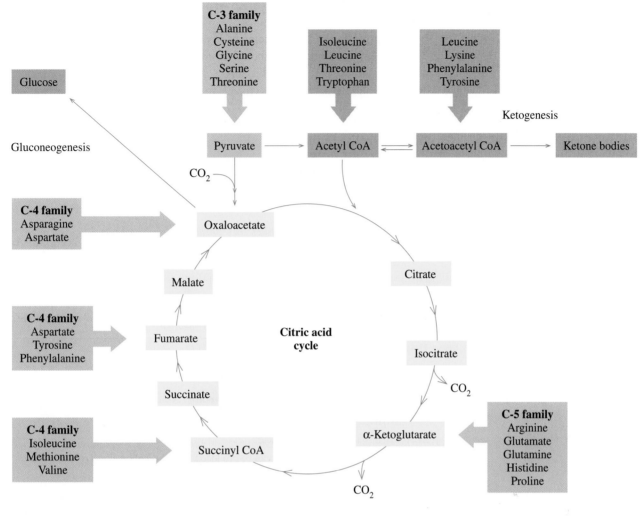

FIGURE 24.10

Carbon atoms from degraded amino acids are converted to the intermediates of the citric acid cycle or other pathways. Glucogenic amino acids (orange boxes) produce carbon skeletons that can form glucose, and ketogenic amino acid (green boxes) can produce ketone bodies.

Q **Why is aspartate glucogenic, but leucine is ketogenic?**

Carbon Skeletons from the Four-Carbon Amino Acids

Oxaloacetate is the α-keto acid that forms when the four-carbon aspartate undergoes transamination.

$$\text{aspartate} + \alpha\text{-ketoglutarate} \longrightarrow \text{oxaloacetate} + \text{glutamate}$$

Asparagine hydrolyzes to give NH_4^+ and aspartate, which goes to oxaloacetate. The degradation pathways of other four-carbon amino acids, isoleucine, methionine, and valine, produce succinyl CoA, another intermediate of the citric acid cycle.

Carbon Skeletons from the Five-Carbon Amino Acids

The five-carbon amino acids glutamine, glutamate, proline, arginine, and histidine are converted to glutamate, which undergoes oxidation–deamination to yield α-ketoglutarate and ammonium ion.

Degradation of Ketogenic Amino Acids

Two ketogenic amino acids, leucine and lysine, require a series of reactions that degrade them to acetoacetate, which is an intermediate in the production of ketone bodies. Lysine also produces acetyl CoA. Some of the carbon atoms of the aromatic amino acids phenylalanine and tyrosine are converted to acetoacetate as well as fumarate.

SAMPLE PROBLEM 24.11

■ **Degradation of Amino Acids**

What degradation products of amino acids are citric acid intermediates?

SOLUTION

Carbon atoms from amino acids enter the citric acid cycle as acetyl CoA, α-ketoglutarate, succinyl CoA, fumarate, or oxaloacetate.

STUDY CHECK

Which amino acids provide carbon atoms that enter the citric acid as α-ketoglutarate?

QUESTIONS AND PROBLEMS

■ **Fates of the Carbon Atoms from Amino Acids**

24.49 What is a glucogenic amino acid?

24.50 What is a ketogenic amino acid?

24.51 What metabolic substrate(s) can be produced from the carbon atoms of each of the following amino acids?
   a. alanine
   b. aspartate
   c. valine
   d. glutamine

24.52 What metabolic substrate(s) can be produced from the carbon atoms of each of the following amino acids?
   a. leucine
   b. asparagine
   c. cysteine
   d. arginine

LEARNING GOAL

Illustrate how some nonessential amino acids are synthesized from intermediates in the citric acid cycle and other metabolic pathways.

24.10 Synthesis of Amino Acids

Plants and bacteria such as *E. coli* produce all of their amino acids using NH_4^+ and NO_3^-. However, humans can synthesize only 10 of the 20 amino acids found in their proteins. The **nonessential amino acids** are synthesized in the body, whereas the **essential amino acids** must be obtained from the diet. (See Table 24.3.) Two amino acids, arginine and histidine, are essential in diets for children due to their rapid growth requirements, but not adults.

Some Pathways for Amino Acid Synthesis

There are a variety of pathways involved in the synthesis of nonessential amino acids. When the body synthesizes nonessential amino acids, the α-keto acid skeletons are obtained from the citric acid cycle or glycolysis and converted to amino acids by transamination. (See Figure 24.11.)

Some of the amino acids are formed from a simple transamination. For example, the transfer of the amino group from glutamate to pyruvate, a three-carbon α-keto acid, produces alanine, an amino acid with three carbons.

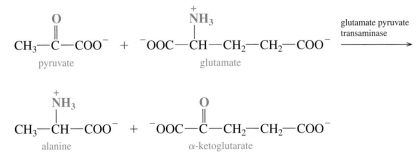

In another transamination using glutamate, the four-carbon oxaloacetate from the citric acid cycle is converted to aspartate.

TABLE 24.3

Essential and Nonessential Amino Acids in Humans

Essential	
arginine*	methionine
histidine*	phenylalanine
isoleucine	threonine
leucine	tryptophan
lysine	valine

Nonessential	
alanine	glutamine
asparagine	glycine
aspartate	proline
cysteine	serine
glutamate	tyrosine

*Essential for children only

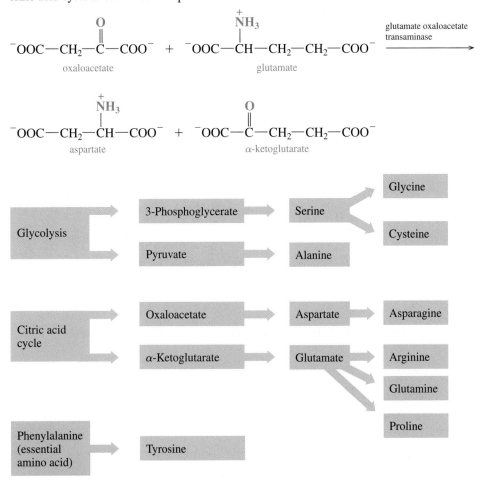

FIGURE 24.11

Nonessential amino acids are synthesized from intermediates of glycolysis and the citric acid cycle.

Q **How is alanine formed from pyruvate?**

These two transaminases are abundant in the cells of the liver and heart, but are present only in low levels in the bloodstream. When an injury or disease occurs, they are released from the damaged cells into the bloodstream. The elevated levels of *serum glutamate-pyruvate transaminase* (SGPT) and *serum glutamate-oxaloacetate transaminase* (SGOT) provide a means to diagnose the extent of damage to the liver or heart.

Constructing the Building Blocks of Proteins

Synthesis of Glutamine

The synthesis of the other nonessential amino acids requires several reactions in addition to transamination. For example, glutamine is synthesized when a second amino group is added to glutamate using the energy from the hydrolysis of ATP.

$$\overset{\overset{+}{N}H_3}{\underset{|}{^-OOC-CH-CH_2-CH_2-COO^-}} \; + \; NH_3 \quad \xrightarrow[\text{ATP} \quad \text{ADP} + \text{P}_i]{\text{glutamine synthetase}} \quad \overset{\overset{+}{N}H_3}{\underset{|}{^-OOC-CH-CH_2-CH_2-\overset{\overset{O}{\|}}{C}-NH_2}}$$

glutamate ⟶ glutamine

Synthesis of Serine and Cysteine

In the synthesis of serine, three steps are required starting with 3-phosphoglycerate from glycolysis. In this pathway, the —OH group of glycerate is oxidized to give an α-keto acid, which undergoes transamination by glutamate accompanied by the loss of the phosphate group.

$$\overset{OH}{\underset{|}{^-OOC-CH-CH_2-O-\overset{\overset{O}{\|}}{\underset{\underset{O^-}{|}}{P}}-O^-}} \; + \; NAD^+ \; + \; \text{glutamate} \; + \; H_2O \; \longrightarrow$$

3-phosphoglycerate

$$\overset{\overset{+}{N}H_3}{\underset{|}{^-OOC-CH-CH_2-OH}} \; + \; NADH \; + \; H^+ \; + \; \alpha\text{-ketoglutarate} \; + \; P_i$$

serine

Once serine is formed, its —OH group is replaced by —SH from a reaction with homocysteine.

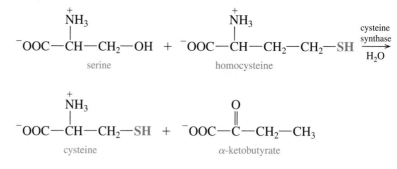

$$\overset{\overset{+}{N}H_3}{\underset{|}{^-OOC-CH-CH_2-OH}} \; + \; \overset{\overset{+}{N}H_3}{\underset{|}{^-OOC-CH-CH_2-CH_2-SH}} \quad \xrightarrow[\text{H}_2\text{O}]{\text{cysteine synthase}}$$

serine   homocysteine

$$\overset{\overset{+}{N}H_3}{\underset{|}{^-OOC-CH-CH_2-SH}} \; + \; \overset{\overset{O}{\|}}{\underset{}{^-OOC-C-CH_2-CH_3}}$$

cysteine   α-ketobutyrate

▪ **Phenylketonuria (PKU)**

In the genetic disease phenylketonuria (PKU), a person cannot convert phenylalanine to tyrosine because the gene for an enzyme in the conversion is defective. As a result, large amounts of phenylalanine accumulate. In a different pathway, phenylalanine undergoes transamination to form phenylpyruvate, which is decarboxylated to phenylacetate. Large amounts of these compounds are excreted in the urine.

In infants, high levels of phenylpyruvate and phenylacetate cause severe mental retardation. However, the defect can be identified at birth, and all newborns are now tested for PKU. By detecting PKU early, retardation is avoided by using a diet with proteins that are low in phenylalanine and high in tyrosine. It is also important to avoid the use of sweeteners and soft drinks containing aspartame, which contains phenylalanine as one of two amino acids in its structure. In adulthood, persons with PKU can eat a nearly normal diet as a long as they are checked for phenylpyruvate periodically.

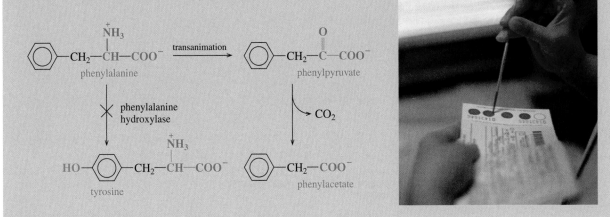

▪ **Synthesis of Tyrosine**

Tyrosine, an aromatic amino acid with a hydroxyl group, is formed from phenylalanine, an essential amino acid.

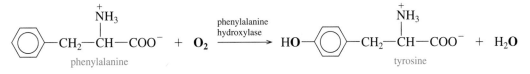

▪ **Synthesis of Amino Acids**

What compound is most often the source of amino groups when transamination is used to synthesize nonessential amino acids?

SOLUTION

Glutamate is the usual source of amino groups in the synthesis of nonessential amino acids.

STUDY CHECK

What are the sources of the substrates used to synthesize nonessential amino acids?

■ Homocysteine and Coronary Heart Disease

In the body, the amino acid methionine is degraded to homocysteine. We obtain most of our methionine, an essential amino acid, from the proteins in meat. In turn, homocysteine can be used to synthesize methionine in a process that requires folic acid and vitamin B_{12} (cobalamin). In a study at Harvard in the 1960s, children suffering from a genetic disorder, *homocystinuria*, were found to have high homocysteine levels. They were also found to have advanced atherosclerosis, which led to strokes and heart attacks early in life. This was one of the first indications of a link between elevated homocysteine levels and heart disease. In the past few years, additional clinical research has indicated that elevated blood levels of homocysteine are associated with increased risk of coronary heart disease, which can also lead to stroke and myocardial infarction (heart attack). It was also found that low levels of folic acid accompanied the elevated levels of homocysteine. This finding suggests that inadequate levels of folic acid limit the synthesis of methionine from homocysteine causing an accumulation of homocysteine.

When the body has adequate amounts of vitamin B_6, B_{12}, and folic acid, the synthesis of methionine maintains proper levels of homocysteine, which we need for healthy tissues. However, a deficiency of any one of these vitamins such as folic acid can lead to increased homocysteine levels, which may be damaging to the heart. Folic acid is recommended as a supplement for pregnant women to avoid folic acid deficiencies during the growth of a fetus.

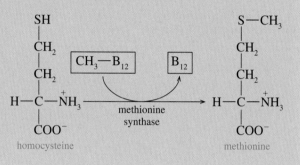

Overview of Metabolism

In these chapters, we have seen that catabolic pathways degrade large molecules to small molecules that are used for energy production via the citric acid cycle and the electron transport chain. We have also looked at the anabolic pathways that lead to the synthesis of larger molecules in the cell. In the overall view of metabolism, there are several branch points from which compounds may be degraded for energy or used to synthesize larger molecules. For example, glucose can be degraded to acetyl CoA for the citric acid cycle to produce energy or converted to glycogen for storage. When glycogen stores are depleted, fatty acids are degraded for energy. Amino acids normally used to synthesize nitrogen-containing compounds in the cells can also be used for energy after they are degraded to intermediates of the citric acid cycle. In the synthesis of nonessential amino acids, α-keto acids of the citric acid cycle enter a variety of reactions that convert them to amino acids through transamination by glutamate. (See Figure 24.12.)

QUESTIONS AND PROBLEMS

■ Synthesis of Amino Acids

24.53 What do we call the amino acids that humans can synthesize?

24.54 How do humans obtain the amino acids that cannot be synthesized in the body?

24.55 How is glutamate converted to glutamine?

24.56 What amino acid is needed for the synthesis of tyrosine?

24.57 What do the letters PKU mean?

24.58 How is PKU treated?

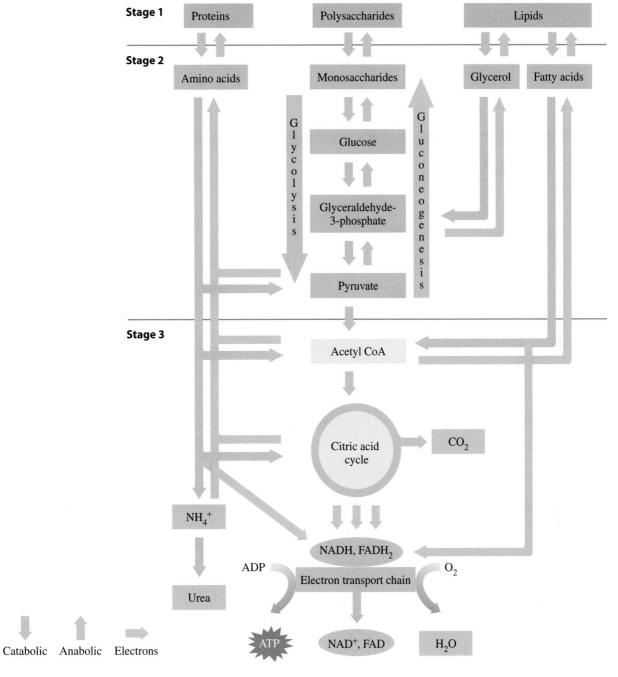

FIGURE 24.12

Catabolic and anabolic pathways in the cells provide the energy and necessary compounds for the cells.

Q **Under what conditions in the cell are amino acids degraded for energy?**

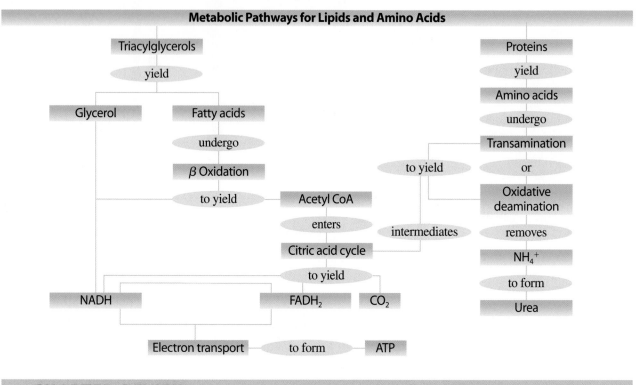

Metabolic Pathways for Lipids and Amino Acids

CHAPTER REVIEW

24.1 Digestion of Triacylglycerols
Triacylglycerols are hydrolyzed in the small intestine to mono-acylglycerol and fatty acids, which enter the intestinal wall and form new triacylglycerols. They bind with proteins to form chylomicrons, which transport them through the lymphatic system and bloodstream to the tissues.

24.2 Oxidation of Fatty Acids
When needed as an energy source, fatty acids are linked to coenzyme A and transported into the mitochondria where they undergo β oxidation. The fatty acyl chain is oxidized to yield a shorter fatty acid, acetyl CoA, and the reduced coenzymes NADH and $FADH_2$.

24.3 ATP and Fatty Acid Oxidation
Although the energy from a particular fatty acid depends on its length, each oxidation cycle yields 5 ATP with another 12 ATP from the acetyl CoA that enter the citric acid cycle.

24.4 Ketogenesis and Ketone Bodies
When high levels of acetyl CoA are present in the cell, they enter the ketogenesis pathway forming ketone bodies such as acetoacetate, which can cause ketosis and acidosis.

24.5 Fatty Acid Synthesis
When there is an excess of acetyl CoA in the cell, the two-carbon acetyl CoA units link together to synthesize palmitate, which is converted to triacylglycerols and stored in the adipose tissue.

24.6 Digestion of Proteins
The digestion of proteins, which begins in the stomach and continues in the small intestine, involves the hydrolysis of peptide bonds by proteases to yield amino acids that are absorbed through the intestinal wall and transported to the cells.

24.7 Degradation of Amino Acids
When the amount of amino acids in the cells exceeds that needed for synthesis of nitrogen compounds, the process of transamination converts them to α-keto acids and glutamate. Oxidative deamination of glutamate produces ammonium ions and α-ketoglutarate.

24.8 Urea Cycle
Ammonium ions from oxidative deamination combine with bicarbonate and ATP to form carbamoyl phosphate, which is converted to urea.

24.9 Fates of the Carbon Atoms from Amino Acids

The carbon atoms from the degradation of glucogenic amino acids enter the citric acid cycle or glucose synthesis, whereas ketogenic amino acids provide acetyl CoA or acetoacetate for ketogenesis.

24.10 Synthesis of Amino Acids

Nonessential amino acids are synthesized when amino groups from glutamate are transferred to an α-keto acid obtained from glycolysis or the citric acid cycle.

SUMMARY OF KEY REACTIONS

DIGESTION OF TRIACYLGLYCEROLS

$$\text{triacylglycerols} + 2H_2O \xrightarrow{\text{pancreatic lipase}} \text{monoacylglycerols} + 2 \text{ fatty acids}$$

MOBILIZATION OF FATS

$$\text{triacylglycerols} + 3H_2O \xrightarrow{\text{lipases}} \text{glycerol} + 3 \text{ fatty acids}$$

METABOLISM OF GLYCEROL

$$\text{glycerol} + ATP + NAD^+ \longrightarrow \text{dihydroxyacetone phosphate} + ADP + NADH + H^+$$

β OXIDATION OF FATTY ACID

$$\text{myristyl-CoA} + 6\,CoA + 6\,FAD + 6\,NAD^+ + 6\,H_2O \longrightarrow 7 \text{ acetyl CoA} + 6\,FADH_2 + 6\,NADH + 6H^+$$

FATTY ACID SYNTHESIS

$$8 \text{ acetyl CoA} + 14\,NADPH + 14H^+ + 7\,ATP \longrightarrow \text{palmitate } (C_{16}) + 8\,CoA + 14\,NADP^+ + 7\,H_2O + 7\,ADP + 7\,P_i$$

TRANSAMINATION

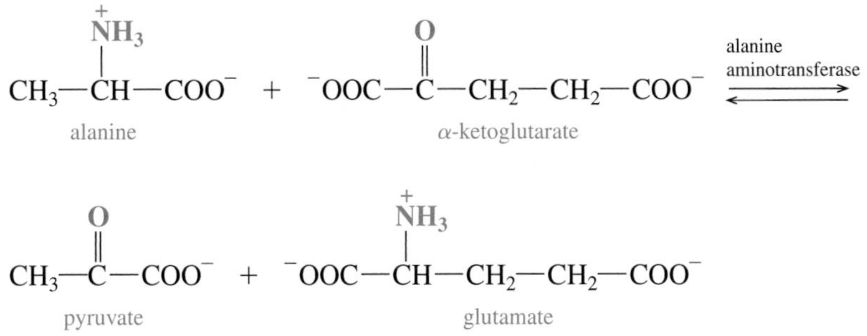

OXIDATIVE DEAMINATION

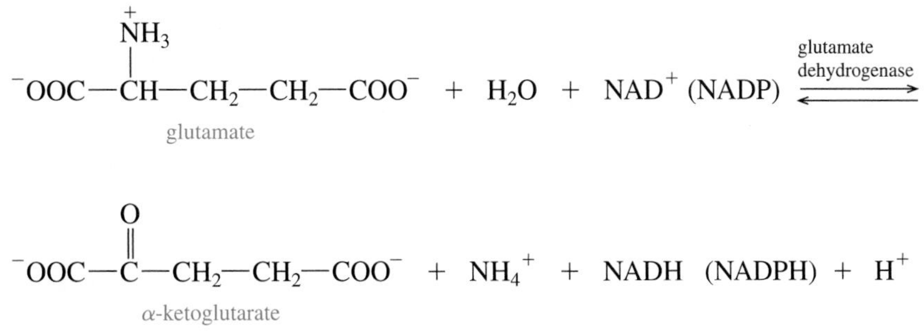

UREA CYCLE

$$NH_4^+ + CO_2 + 3\,ATP + \text{aspartate} + 2H_2O \longrightarrow \text{urea} + 2\,ADP + AMP + 4\,P_i + \text{fumarate}$$

KEY TERMS

acidosis Low blood pH resulting from the formation of acidic ketone bodies.

beta (β) oxidation The degradation of fatty acids that removes two-carbon segments from a fatty acid chain.

chylomicron Lipoproteins formed by coating triacylglycerols with proteins for transport in the lymph and bloodstream.

essential amino acid An amino acid that must be obtained from the diet because it cannot be synthesized in the body.

fat mobilization The hydrolysis of triacylglycerols in the adipose tissue to yield fatty acids and glycerol for energy production.

glucogenic amino acid An amino acid that provides carbon atoms for the synthesis of glucose.

ketogenesis The pathway that converts acetyl CoA to four-carbon acetoacetate and other ketone bodies.

ketogenic amino acid An amino acid that provides carbon atoms for the fatty acid synthesis or ketone bodies.

ketone bodies The products of ketogenesis: acetoacetate, β hydroxybutyrate, and acetone.

ketosis A condition in which high levels of ketone bodies cannot be metabolized, leading to lower blood pH.

lipogenesis The synthesis of fatty acid in which two-carbon acetyl units link together to yield palmitic acid.

nonessential amino acid An amino acid that can be synthesized by reactions including transamination of α-keto acids in the body.

oxidative deamination The loss of ammonium ion when glutamate is degraded to α-ketoglutarate.

protein turnover The amount of protein that we break down from our diet and utilize for synthesis of proteins and nitrogen-containing compounds.

transamination The transfer of an amino group from an amino acid to an α-keto acid.

urea cycle Ammonium ions from the degradation of amino acids and CO_2 form carbamoyl phosphate, which is converted to urea.

UNDERSTANDING THE CONCEPTS

24.59 Lauric acid, CH_3—$(CH_2)_{10}$—COOH, found in coconut oil, is a saturated fatty acid.

    a. Write the formula of the activated form of lauric acid.
    b. Indicate the α and β carbon atoms in the fatty acyl molecule.
    c. Write the overall equation for the complete β oxidation for lauric acid.
    d. How many acetyl CoA units are produced?
    e. How many cycles of β oxidation are needed?
    f. Account for the total ATP yield from β oxidation of lauric acid (C_{12} fatty acid) by completing the following calculation:

activation	$\longrightarrow$	-2 ATP
____acetyl CoA	$\longrightarrow$	___ATP
____$FADH_2$	$\longrightarrow$	___ATP
____NADH	$\longrightarrow$	___ATP
Total		___ATP

24.60 Arachidic acid is a saturated 20-carbon fatty acid found in peanut and fish oils.

    a. Write the formula of the activate form of arachidic acid.
    b. Indicate the α and β carbon atom in the fatty acyl formula.
    c. Write the overall equation for the complete β oxidation of arachidic acid.
    d. How many acetyl CoA units are produced?
    e. How many cycles of β oxidation are needed?
    f. What is the total ATP yield from β oxidation of arachidic acid?

____activation	-2 ATP
____acetyl CoA	___ATP
____$FADH_2$	___ATP
____NADH	___ATP
Total	___ATP

ADDITIONAL QUESTIONS AND PROBLEMS

For instructor-assigned homework, go to
www.masteringgob.com.

24.61 How are dietary triacylglycerols digested?

24.62 What is a chylomicron?

24.63 Why are the fats in the adipose tissues of the body considered as the major form of stored energy?

24.64 How are fatty acids obtained from stored fats?

24.65 Why doesn't the brain utilize fatty acids for energy?

24.66 Why don't red blood cells utilize fatty acids for energy?

24.67 A triacylglycerol is hydrolyzed in fat cells of adipose tissues and the fatty acid is transported to the liver.
a. What happens to the glycerol?
b. Where in the liver cells is the fatty acid activated for β oxidation?
c. What is the energy cost for activation of the fatty acid?
d. What is the purpose of activating fatty acids?

24.68 Consider the β oxidation of a saturated fatty acid.
a. What is the activated form of the fatty acid?
b. Why is the oxidation called β oxidation?
c. What reactions in the fatty acid cycle require coenzymes?
d. What is the yield in ATP for one cycle of β oxidation?

24.69 Identify each of the following as involved in β oxidation or in fatty acid synthesis:
a. NAD^+
b. occurs in the mitochondrial matrix
c. malonyl ACP
d. cleavage of two-carbon acyl group
e. acyl protein carrier
f. acetyl CoA carboxylase

24.70 Calculate the total ATP produced in the complex oxidation of caproic acid, $C_6H_{12}O_2$, with the total ATP produced from the oxidation of glucose, $C_6H_{12}O_6$.

24.71 The metabolism of triacylglycerols and carbohydrates is influenced by the hormones insulin and glucagon. Indicate the results of each of the following as stimulating (1) fatty acid oxidation or (2) the synthesis of fatty acids.
a. low blood glucose
b. glucagon secreted

24.72 Identify each of the following as involved in β oxidation or in fatty acid synthesis:
a. NADPH
b. takes place in the cytosol
c. FAD
d. oxidation of a hydroxyl group
e. coenzyme A
f. hydration of a double bond

24.73 Why is ammonium ion produced in the liver converted immediately to urea?

24.74 The metabolism of triacylglycerols and carbohydrates is influenced by the hormones insulin and glucagon. Indicate the results of each of the following as stimulating (1) fatty acid oxidation or (2) the synthesis of fatty acids.
a. high blood glucose
b. insulin secreted

24.75 Indicate the corresponding reactant in the urea cycle for each of the following compounds:
a. aspartate
b. ornithine

24.76 What compound is regenerated in the urea cycle?

24.77 What metabolic substrate(s) can be produced from the carbon atoms of each of the following amino acids?
a. serine
b. lysine
c. methionine
d. glutamate

24.78 Indicate the products in the urea cycle for each of the following compounds:
a. arginine
b. argininosuccinate

24.79 How much ATP can the degradation product serine provide?

24.80 What metabolic substrate(s) can be produced from the carbon atoms of each of the following amino acids?
a. leucine
b. isoleucine
c. cysteine
d. phenylalanine

CHALLENGE QUESTIONS

24.81 A camel hump contains 14 kg of triacylglycerols.
   a. Using the value of 0.49 mole of ATP per gram of fat, how many moles of ATP could be produced by the fat in the camel's hump?
   b. ATP releases 7.3 kcal/mole. How many kcal are produced by the fat?

24.82 Butter is a fat that contains 80% triacylglycerols and the rest water. Assume the triacylglycerol in butter is glyceryl tripalmitate (C_{16}).
   a. Write an equation for the hydrolysis of the triacylglycerol.
   b. Calculate the ATP yield from the fatty acids in 1 molecule of palmitic acid.
   c. How many kcal are released from the palmitic acid in a 0.50 oz pat of butter?

24.83 Write the structure and name of the amino acid formed when the following α-keto acids undergo transamination with glutamic acid.

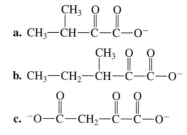

ANSWERS

Answers to Study Checks

24.1 In the membrane of the small intestine, monoacyl glycerols and fatty acids recombine to form triacylglycerols that bind with proteins to form chylomicrons for transport to the lymphatic system and the bloodstream.

24.2 NAD^+

24.3 8 acetyl CoA give 96 ATP
   NADH and $FADH_2$ give 35 ATP

24.4 The ketone bodies are acetoacetate, β hydroxybutyrate, and acetone.

24.5 In each cycle of fatty acid synthesis, a two-carbon atom acyl group from malonyl ACP adds to a fatty acid chain and one carbon atom forms CO_2.

24.6 HCl denatures proteins and activates enzymes such as pepsin.

24.7 In conditions such as fasting or starvation, a diet insufficient in protein leads to a negative nitrogen balance.

24.8 Alanine oxaloacetate transaminase or transaminotransferase.

24.9 The process of oxidative deamination regenerates α-ketoglutarate from glutamate.

24.10 **a.** argininosuccinate
   b. citrulline
   c. urea and ornithine

24.11 Arginine, glutamate, glutamine, histidine, and proline provide carbon atoms for α-ketoglutarate.

24.12 Substrates for the synthesis of nonessential amino acids are obtained from glycolysis and the citric acid cycle.

Answers to Selected Questions and Problems

24.1 The bile salts emulsify fat to give small fat globules for lipase hydrolysis.

24.3 When blood glucose and glycogen stores are depleted.

24.5 Glycerol is converted to glycerol-3-phosphate, then to dihydroxyacetone phosphate, an intermediate of glycolysis.

24.7 In the cytosol at the outer mitochondrial membrane

24.9 FAD, NAD^+, and HS—CoA

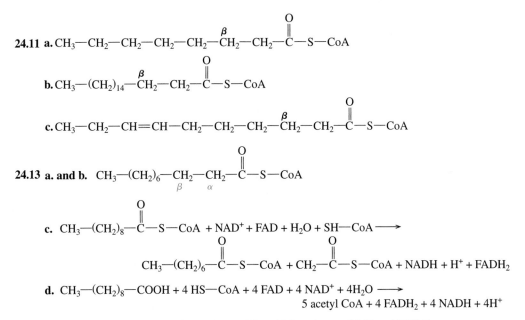

24.11 a. $CH_3-CH_2-CH_2-CH_2-CH_2-\overset{\beta}{CH_2}-CH_2-\overset{O}{\overset{||}{C}}-S-CoA$

b. $CH_3-(CH_2)_{14}-\overset{\beta}{CH_2}-CH_2-\overset{O}{\overset{||}{C}}-S-CoA$

c. $CH_3-CH_2-CH=CH-CH_2-CH_2-CH_2-\overset{\beta}{CH_2}-CH_2-\overset{O}{\overset{||}{C}}-S-CoA$

24.13 a. and b. $CH_3-(CH_2)_6-\underset{\beta}{CH_2}-\underset{\alpha}{CH_2}-\overset{O}{\overset{||}{C}}-S-CoA$

c. $CH_3-(CH_2)_8-\overset{O}{\overset{||}{C}}-S-CoA + NAD^+ + FAD + H_2O + SH-CoA \longrightarrow$

$CH_3-(CH_2)_6-\overset{O}{\overset{||}{C}}-S-CoA + CH_2-\overset{O}{\overset{||}{C}}-S-CoA + NADH + H^+ + FADH_2$

d. $CH_3-(CH_2)_8-COOH + 4 HS-CoA + 4 FAD + 4 NAD^+ + 4H_2O \longrightarrow$
$\qquad\qquad\qquad\qquad\qquad$ 5 acetyl CoA + 4 FADH_2 + 4 NADH + 4H^+

24.15 The hydrolysis of ATP to AMP hydrolyzes ATP to ADP, and ADP to AMP, which provides the same amount of energy as the hydrolysis of 2 ATP to 2 ADP.

24.17 a. 5 acetyl CoA units
b. 4 cycles of β oxidation
c. 60 ATP from 5 acetyl CoA (citric acid cycle) + 12 ATP from 4 NADH + 8 ATP from 4 FADH_2 −2 ATP (activation) = 80 −2 = 78 ATP

24.19 Ketogenesis is the synthesis of ketone bodies from excess acetyl CoA from fatty acid oxidation, which occurs when glucose is not available for energy, particularly in starvation, low-carbohydrate diets, fasting, and diabetes.

24.21 Acetoacetate undergoes reduction using NADH + H^+ to yield β hydroxybutyrate.

24.23 High levels of ketone bodies lead to ketosis, a condition characterized by acidosis (a drop in blood pH values), excessive urination, and strong thirst.

24.25 in the cytosol of cells in liver and adipose tissue

24.27 acetyl CoA, HCO_3^-, and ATP

24.29 a. (3) malonyl CoA transacylase
b. (1) acetyl CoA carboxylase
c. (2) acetyl CoA transacylase

24.31 a. 4 HCO_3^-
b. 4 ATP
c. 5 acetyl CoA
d. 4 malonyl ACP
e. 8 NADPH
f. 4 CO_2 removed

24.33 The digestion of proteins begins in the stomach and is completed in the small intestine.

24.35 Hormones, heme, purines and pyrimidines for nucleotides, proteins, nonessential amino acids, amino alcohols, and neurotransmitters require nitrogen obtained from amino acids.

24.37 The reactants are an amino acid and an α-keto acid, and the products are a new amino acid and a new α-keto acid.

24.39
a. $H-\overset{O}{\overset{||}{C}}-COO^-$ **b.** $CH_3-\overset{O}{\overset{||}{C}}-COO^-$

c. $CH_3-\overset{CH_3}{\overset{|}{CH}}-\overset{O}{\overset{||}{C}}-COO^-$

24.41
$$^-OOC-\overset{\overset{+}{NH_3}}{\overset{|}{CH}}-CH_2-CH_2-COO^- + H_2O + NAD^+ (NADP^+) \xrightarrow{\text{glutamate dehydrogenase}}$$
$\qquad\qquad$ glutamate

$$^-OOC-\overset{O}{\overset{||}{C}}-CH_2-CH_2-COO^- + NH_4^+ + NADH (NADPH) + H^+$$
$\qquad$ α-ketoglutarate

24.43 NH_4^+ is toxic if allowed to accumulate in the liver.

24.45

$$H_2N-\overset{\overset{\displaystyle O}{\|}}{C}-NH_2$$

24.47 CO_2 from the citric acid cycle

24.49 Glucogenic amino acids produce compounds used to synthesis glucose.

24.51 a. pyruvate
b. oxaloacetate, fumarate
c. succinyl CoA
d. α-ketoglutarate

24.53 nonessential amino acid

24.55 Glutamine synthetase catalyzes the addition of an amino group to glutamate using energy from the hydrolysis of ATP.

24.57 phenylketonuria

24.59 a. and b.

$$CH_3-(CH_2)_8-CH_2-\underset{\beta}{CH_2}-\underset{\alpha}{\overset{\overset{\displaystyle O}{\|}}{C}}-CoA$$

c. Lauryl-CoA + 5 CoA + 5 FAD + 5 NAD$^+$ + 5 H$_2$O
$\longrightarrow$ 6 Acetyl CoA + 5 FADH$_2$ + 5 NADH + 5H$^+$
d. Six acetyl CoA units are produced
e. Five cycles of β oxidation are needed
f.

activation	$\longrightarrow$	-2 ATP
6 acetyl CoA × 12	$\longrightarrow$	72 ATP
5 FADH$_2$ × 2	$\longrightarrow$	10 ATP
5 NADH × 3	$\longrightarrow$	15 ATP
Total		95 ATP

24.61 Triacylglycerols are hydrolyzed to monoacylglycerols and fatty acids in the small intestine, which reform triacylglycerols in the intestinal lining for transport as lipoproteins to the tissues.

24.63 Fats can be stored in unlimited amounts in adipose tissue compared to the limited storage of carbohydrates as glycogen.

24.65 The fatty acids cannot diffuse across the blood brain barrier.

24.67 a. Glycerol is converted to glycerol-3-phosphate and to dihydroxyacetone phosphate, which enters glycolysis or gluconeogenesis.
b. Activation of fatty acids occurs on the outer mitochondrial membrane.
c. The energy cost is equal to 2 ATP.
d. Only fatty acyl CoA can move into the intermembrane space for transport by carnitine into the matrix.

24.69 a. β oxidation 　　　　**b.** β oxidation
c. fatty acid synthesis 　**d.** β oxidation
e. fatty acid synthesis 　**f.** fatty acid synthesis

24.71 a. (1) fatty acid oxidation
b. (2) the synthesis of fatty acids

24.73 Ammonium ion is toxic if allowed to accumulate in the liver.

24.75 a. citrulline 　　　**b.** carbamoyl phosphate

24.77 a. pyruvate 　　　**b.** acetoacetyl-CoA
c. succinyl-CoA 　　**d.** α-ketoglutarate

24.79 Serine is degraded to pyruvate, which is oxidized to acetyl CoA. The oxidation produces NADH + H$^+$, which provides 3 ATP. In one turn of the citric acid cycle, the acetyl CoA provides 12 ATP. Thus, serine can provide a total of 15 ATP.

24.81 a. 6900 moles ATP 　　**b.** 50 000 kcal

24.83

a.
$$CH_3-\underset{\underset{\displaystyle CH_3}{|}}{CH}-\underset{\underset{\displaystyle NH_3^+}{|}}{CH}-\overset{\overset{\displaystyle O}{\|}}{C}-O^-\quad \text{valine}$$

b.
$$CH_3-CH_2-\underset{\underset{\displaystyle CH_3}{|}}{CH}-\underset{\underset{\displaystyle NH_3^+}{|}}{CH}-\overset{\overset{\displaystyle O}{\|}}{C}-O^-\quad \text{isoleucine}$$

c.
$$^-O-\overset{\overset{\displaystyle O}{\|}}{C}-CH_2-\underset{\underset{\displaystyle NH_3^+}{|}}{CH}-\overset{\overset{\displaystyle O}{\|}}{C}-O^-\quad \text{aspartic acid}$$

Credits

Chapter 12
p. 445 © 1997 PhotoDisc/Don Tremain.
p. 446 © 1997 PhotoDisc/Tim Hall.
p. 449 Ian O'Leary/Stone.
p. 450 Alastair Shay/Papilio/Corbis.
p. 460 Courtesy of NASA.
p. 465 © Dianora Niccolini/Medical Images, Inc.
p. 470 *left:* DuPont & Company.
top, right: Getty Images/Royalty Free.
bottom, right: PictureQuest/Ingram Publishing.
p. 471 PhotoEdit/Michael Newman.
p. 473 Photo Researchers/Charles D. Winters.

Chapter 13
p. 486 © Tom Pantages.
p. 491 © Al Assid/The Stock Market.
p. 495 Shelley Gazin/Corbis.
p. 496 Owen Franken/Corbis.
p. 501 *top, left:* CORBIS/PictureArts/Robert Fiocca.
top, right: Getty Images/Nancy R. Cohen.
middle, right: Dorling Kindersley/Ian O'Leary.
bottom, left: Peter Arnold, Inc./Ed Reschke.
p. 504 Getty Images/David Lees.

Chapter 14
p. 536 *bottom, middle:* CORBIS.
p. 537 *top, left:* Getty Images.
top, middle: Dorling Kindersley.
top, right: Dorling Kindersley.
bottom, left: Dorling Kindersley.
bottom, right: Dorling Kindersley.

Chapter 15
p. 551 John Wilson White/Addison Wesley Longman.
p. 559 Custom Medical Stocks.
p. 566 © 1997 PhotoDisc.

p. 569 *left:* Richard Hamilton Smith/Corbis.
right: David Toase/ PhotoDisc.

Chapter 16
p. 578 Robert and Linda Mitchell/Robert and Linda Mitchell Photography.
p. 585 *left:* Warren Morgan/Corbis.
p. 599 *left:* Dorling Kindersley.
right: Photo Researchers, Inc./Garo.

Chapter 17
p. 613 Lawson Wood/Corbis.
p. 615 Ralph A. Clevenger/Corbis.
p. 616 George D. Lepp/Corbis.
p. 620 *bottom:* Lori Adamski Peek/Corbis.
p. 630 © C. Raines/Visuals Unlimited.
p. 631 Courtesy of the National Heart, Lung, and Blood Institute/National Institutes of Health.
p. 633 Custom Medical Stock Photo.
p. 637 AFP Photo/Paul Jones/Corbis.
p. 644 *top, left:* Index Stock Imagery, Inc./Lauree Feldman.
top, right: Photo Researchers, Inc./John W. Bova.
bottom, left: Masterfile.
p. 645 *right:* Photolibrary.com/MAGGY OEHLBECK.

Chapter 18
p. 660 Dr. Morley Read/Science Photo Library/Photo Researchers, Inc.
p. 663 *top, left:* © Ed Drews/Photo Researchers, Inc.
p. 664 Michael S. Yamashita/Corbis.
p. 675 *left:* Fundamental Photographs/Richard Megna.
right: Corbis/Ariel Skelley.
p. 676 *bottom:* Dorling Kindersley.
p. 677 Corbis.

Chapter 19
p. 682 Long Horn Cow/Corbis.
p. 694 Jilly Wendell/Stone.
p. 699 *left:* Richard Hamilton Smith/Corbis.
middle: Eric and David Hosking/Corbis.
right: Bruce Wilson/Stone.
p. 701 © Lewin/Royal Free Hospital/Photo Researchers, Inc.
p. 707 Masterfile/Miep van Damm.

Chapter 20
p. 739 *right:* Getty Images/Royalty Free.
p. 740 *left:* Dorling Kindersley/Dave Rudkin.

Chapter 21
p. 752 © 1997 PhotoDisc/ M. Freeman/Photolink.
p. 773 Morton Beebe, S.F./Corbis.

Chapter 22
p. 795 Lori Adamski Peek/Stone.
p. 810 Joel W. Rogers/Corbis.
p. 816 Neal Preston/Corbis.
p. 822 Karen Timberlake.

Chapter 23
p. 845 Philip James Corwin/Corbis.

Chapter 24
p. 856 Thomas Hoepker/Magnum Photos.
p. 857 Hossler/Custom Medical Stock Photo.
p. 867 *left:* Gunter Marx Photography/Corbis.
right: Chase Swift/Corbis.
p. 870 *left:* Custom Medical Stock Photo.
right: SIU/Custom Medical Stock Photo.
p. 887 Custom Medical Stock Photo.
p. 892 *left:* Dorling Kindersley/Philip Dowell.

Glossary/Index

Metric and SI Units and Some Useful Conversion Factors

Length   SI unit meter (m)

1 meter (m) = 100 centimeters (cm)
1 meter (m) = 1000 millimeters (mm)
1 cm = 10 mm
1 kilometer (km) = 0.6214 mile (mi)
1 inch (in.) = 2.54 cm (exact)

Volume   SI unit cubic meter (m^3)

1 liter (L) = 1000 milliliters (mL)
1 mL = 1 cm^3
1 L = 1.06 quart (qt)
1 qt = 946 mL

Mass   SI unit kilogram (kg)

1 kilogram (kg) = 1000 grams (g)
1 g = 1000 milligrams (mg)
1 kg = 2.20 lb
1 lb = 454 g
1 mole = 6.02 × 10^{23} particles
Water
density = 1.00 g/mL

Temperature   SI unit kelvin (K)

°F = 1.8(°C) + 32

$$°C = \frac{(°F - 32)}{1.8}$$

K = °C + 273

Pressure   SI unit pascal (Pa)

1 atm = 760 mm Hg
1 atm = 760 torr
1 mole (STP) = 22.4 L
R = 0.0821 L · atm/mole · K
R = 62.4 L · mm Hg/mole · K

Energy   SI unit joule (J)

1 calorie (cal) = 4.184 J
1 kcal = 1000 cal
Water
Heat of fusion = 80. cal/g
Heat of vaporization = 540 cal/g
SH = 4.184 J/g°C; 1 cal/g°C

Prefixes for Metric (SI) Units

Prefix	Symbol	Power of Ten
Values greater than 1		
giga	G	10^9
mega	M	10^6
kilo	k	10^3
Values less than 1		
deci	d	10^{-1}
centi	c	10^{-2}
milli	m	10^{-3}
micro	μ	10^{-6}
nano	n	10^{-9}
pico	p	10^{-12}